THE HANDBOOK OF HAZARDOUS MATERIALS SPILLS TECHNOLOGY

THE HANDBOOK OF HAZARDOUS MATERIALS SPILLS TECHNOLOGY

Merv Fingas, Ph.D.
Editor-in-Chief

McGRAW-HILL

New York Chicago San Francisco Lisbon London Madrid
Mexico City Milan New Delhi San Juan Seoul
Singapore Sydney Toronto

Library of Congress Cataloging-in-Publication Data

The handbook of hazardous materials spills technology / Merv Fingas, editor-in-chief.
p. cm.
Includes index.
ISBN 0-07-135171-X
1. Hazardous substances—Safety measures—Handbooks, manuals, etc. 2. Hazardous substances—Accidents—Handbooks, manuals, etc. I. Fingas, Mervin F.

T55.3.H3H3515 2001
604.7—dc21 2001034231

McGraw-Hill

A Division of The ***McGraw-Hill*** *Companies*

1 2 3 4 5 6 7 8 9 0 DOC/DOC 0 7 6 5 4 3 2 1

ISBN 0-07-135171-X

The sponsoring editor for this book was Kenneth P. McCombs and the production supervisor was Sherri Souffrance. It was set in Times Roman by Pro-Image Corporation.

Printed and bound by R. R. Donnelley & Sons Company.

This book is printed on recycled, acid-free paper containing a minimum of 50% recycled, de-inked fiber.

CONTENTS

Part 2 NUCLEAR EMERGENCIES

Part 3 RISK ASSESSMENT

Part 4 SPILL COUNTERMEASURES

Part 5 SPILL MODELING

Part 8 PERSPECTIVES ON SPECIFIC CHEMICALS

Chapter 36 Perspectives on Specific Substances: Polychlorinated Biphenyls (PCBs) 36.1

Chapter 37 Sulfur Trioxide and Oleum 37.1

Chapter 38 Perspectives on Specific Substances: Pentachlorophenol 38.1

Chapter 39 Sodium Cyanide: Properties, Toxicity, Uses and Environmental Impacts 39.1

Part 9 CASE HISTORIES

Part 10 CHEMICAL SPILL DATA

CONTRIBUTORS

Jeffrey A. Adams *ENGEO Inc., San Ramon, California* (CHAP. 14)

Ertugrul Alp *Alp & Associates Incorporated, Richmond Hill, Ontario, Canada* (CHAP. 10)

Margaret-Ann Armour *Department of Chemistry, University of Alberta, Edmonton* (CHAP. 29)

J. Arnaldos *Centre d'Estudis del Risc Tecnològic (CERTEC), Universitat Politècnica de Catalunya—Institut d'Estudis Catalans, Barcelona, Catalonia, Spain* (CHAPS. 22, 31, 41)

Andreas F. Bienz *Bienz, Kummer & Partner Ltd, Consulting Engineers—Safety Planning, Risk Management, Zollikerberg, Switzerland* (CHAP. 12)

E. Buglova *Research and Clinical Institute of Radiation Medicine and Endocrinology, Ministry of Health, Minsk, Belarus* (CHAP. 9)

Bertrand Carissimo *Electricité de France, Chatou, France* (CHAP. 19)

J. Casal *Centre d'Estudis del Risc Tecnològic (CERTEC), Universitat Politècnica de Catalunya—Institut d'Estudis Catalans, Barcelona, Catalonia, Spain* (CHAPS. 22, 31, 41)

Philip Chatwin *Department of Applied Mathematics, University of Sheffield, Sheffield, U.K.* (CHAP. 25)

Nijs Jan Duijm *Risø National Laboratory, Roskilde, Denmark* (CHAP. 19)

C. A. Leif Ericson *CHEMTREC, Arlington, VA* (CHAP. 5)

Merv Fingas *Emergencies Science Division, Environment Canada, Environmental Technology Centre, River Road, Ottawa, Ontario* (CHAPS. 16, 30, 32–36, 38, 39, 44)

Ernst Goldschmitt *United Nations Environment Program, Paris* (CHAP. 2)

R. F. Griffiths *Environmental Technology Centre, Department of Chemical Engineering, UMIST, Manchester, United Kingdom* (CHAP. 37)

John S. Gulliver *Department of Civil Engineering, University of Minnesota, Minneapolis, Minnesota* (CHAP. 20)

A. J. (Tony) Haggerty *Hazardous Substance Advisor, Auckland Region, New Zealand Fire Service* (CHAP. 4)

P. F. L. Heilbron *Brazilian Nuclear Energy Commission, Botafogo, Rio de Janeiro, Brazil* (CHAP. 9)

David E. Hibbs *Barr Engineering Company, Ann Arbor, Michigan* (CHAP. 20)

F. Jonckheere *Trade and Environment Services, European Chemical Industry Council/Conseil Européen de l'industrie chimique (CEFIC), Brussels* (CHAP. 8)

T. Kapias *Environmental Technology Centre, Department of Chemical Engineering, UMIST, Manchester, United Kingdom* (CHAP. 37)

Kerry Ketcheson *Emergencies Science Division, Environmental Technology Centre, Environment Canada, Ottawa, Ontario, Canada* (CHAPS. 39, 44)

S. Bruce King *AristaTek, Inc. and Western Research Institute (WRI), Laramie, Wyoming* (CHAP. 23)

Jaakko Kukkonen *Finnish Meteorological Institute, Air Quality Research, Helsinki, Finland* (CHAPS. 27, 40)

Peter O. Kummer *Bienz, Kummer & Partner Ltd., CH-8125 Zollikerberg, Switzerland* (CHAP. 26)

J. F. Lafortune *International Safety Research, Ottawa, Ontario, Canada* (CHAP. 9)

Carol Lau *Environmental Emergencies Branch, Environment Canada, Hull, Quebec* (CHAP. 44)

Richard Lawuyi *Emergencies Science Division, Environment Canada, Ottawa, Ontario, Canada* (CHAPS. 32, 33, 34, 35, 36. 38)

Ken Li *Emergencies Science Division, Environment Canada, Environmental Technology Centre, River Road, Ottawa, Ontario* (CHAP. 16)

Don Mackay *Canadian Environmental Modelling Centre, Trent University, Peterborough, Ontario, Canada* (CHAP. 24)

Michel Marchand *Rue Alain Colas, BP 20 413, 29604 BREST* (*France*) (CHAPS. 3, 7, 43, 45)

R. Martinčič *J. Stefan Institute, Ljubljana, Slovenia* (CHAP. 9)

Gordon McKay *Department of Chemical Engineering, The Hong Kong University of Science and Technology, Clear Water Bay, Kowloon, Hong Kong* (CHAP. 15)

T. McKenna *International Atomic Energy Agency, Vienna* (CHAP. 9)

Stefan Micallef *Regional Marine Pollution Emergency, Response Centre for the Mediterranean Sea (REMPEC), Manoel Island, Gzira, Malta* (CHAP. 1)

H. Montiel *Centre d'Estudis del Risc Tecnològic (CERTEC), Universitat Politècnica de Catalunya—Institut d'Estudis Catalans, Barcelona, Catalonia, Spain* (CHAPS. 22, 31, 41)

Joe C. Y. Ng *Department of Chemical Engineering, The Hong Kong University of Science and Technology, Clear Water Bay, Kowloon, Hong Kong* (CHAP. 15)

M. Nielsen *Risø National Laboratory, Roskilde, Denmark* (CHAP. 18)

Juha Nikmo *Finnish Meteorological Institute, Air Quality Research, Helsinki, Finland* (CHAP. 27)

John S. Nordin *AristaTek, Inc. and Western Research Institute (WRI), Laramie, Wyoming* (CHAPS. 17, 23)

John C. Pine *Professor-Research, Institute for Environmental Studies, Louisiana State University, Baton Rouge, Louisiana* (CHAP. 11)

E. Planas-Cuchi *Centre d'Estudis del Risc Tecnològic (CERTEC), Universitat Politècnica de Catalunya—Institut d'Estudis Catalans, Barcelona, Catalonia, Spain* (CHAPS. 22, 31, 41)

Krishna R. Reddy *University of Illinois at Chicago, Department of Civil and Materials Engineering, Chicago, Illinois* (CHAP. 14)

Thayne Routh *AristaTek, Inc. and Western Research Institute (WRI), Laramie, Wyoming* (CHAP. 23)

K. E. Sellers *AMEC Earth & Environmental, Inc., Westford, Massachusetts* (CHAP. 28)

Rajesh Seth *Canadian Environmental Modelling Centre, Trent University, Peterborough, Ontario, Canada* (CHAP. 24)

David Sheesley *AristaTek, Inc. and Western Research Institute (WRI), Laramie, Wyoming* (CHAP. 23)

Mary-Ann Spicer *Environmental Emergencies Branch, Environment Canada, Hull, Quebec* (CHAP. 44)

Clyde Strong *President, Clyde Strong & Associates, Inc., College Station, Texas* (CHAP. 13)

Paul J. Sullivan *Department of Applied Mathematics, The University of Western Ontario, London, Ontario, Canada* (CHAP. 25)

J. E. S. Venart *Department of Mechanical Engineering, University of New Brunswick, Fredericton, New Brunswick, Canada* (CHAPS. 21, 42)

J. Verlinden *Trade and Environment Services, European Chemical Industry Council/Conseil Européen de l'industrie chimique (CEFIC), Brussels* (CHAP. 6)

J. A. Vílchez *Centre d'Estudis del Risc Tecnològic (CERTEC), Universitat Politècnica de Catalunya—Institut d'Estudis Catalans, Barcelona, Catalonia, Spain* (CHAPS. 22, 31, 41)

Vaughan R. Voller *Department of Civil Engineering, University of Minnesota, Minneapolis, Minnesota* (CHAP. 20)

Thomas K. Warnock *CHEMTREC, Arlington, VA* (CHAP. 5)

PREFACE

The situation with respect to hazardous materials has changed very much in the last 20 years. For example, the sophistication of spill modeling has increased significantly. During the same time, however, the technology for cleaning up chemical spills has changed little. This variance is largely a reflection of technological changes in society as well as funding levels for research into hazardous materials spills.

Hazardous materials have been and will continue to be articles of commerce that, when used correctly, provide many of the materials that are an everyday part of our high standard of living. When hazardous materials are properly handled, they pose no threat or danger. Unfortunately, no matter how careful the planning and how great the care, accidents will occur. It is then critical that we minimize the effects of such accidents on the population and the environment. This can only be done by applying the best available knowledge and technology to the problem.

Systematic work in developing methodologies for containment and mitigation of hazardous materials began in about 1970. Funding for these efforts has been variable since that time. After major, publicized accidents have occurred in an area, funding increases and then wanes as the attention of the public and funders is focused on other problems. It is very important then to capture the knowledge and developments of an era so that they can be transmitted to the next group of spill responders to focus their attention on problems with hazardous materials spills. This Handbook will, we hope, fulfill that role.

This Handbook would not be possible without the many contributors and the efforts of the more than 100 persons who reviewed papers. The contributors represent the world's foremost experts in their fields and bring a great wealth of experience to this Handbook. This expertise is now made available to a broad range of Handbook users and will improve our capability for planning and conducting response to hazardous material incidents.

Merv Fingas, Ph.D.
Editor-in-Chief

P • A • R • T • 1

HAZARDOUS MATERIAL PROGRAMS

CHAPTER 1
PREPAREDNESS FOR AND RESPONSE TO ACCIDENTAL MARINE POLLUTION—A REGIONAL PERSPECTIVE

Stefan Micallef
Regional Marine Pollution Emergency,
Response Centre for the Mediterranean Sea (REMPEC),
Manoel Island, Gzira, Malta

1.1 INTRODUCTION

In March 1967, the tanker *Torrey Canyon* ran aground off the coast of Cornwall in the southwest of England, spilling some 120,000 tons of oil into the sea (IMO, 1997a). It was the biggest marine pollution disaster in history at the time. Perhaps for the first time, the general public was made aware of the threat the marine transport of oil and other products poses to the marine environment.

Since the *Torrey Canyon* spill, the spate of accidents that have occurred shows that the threat from accidental marine pollution remains (*Amoco Cadiz,* 1978, off Brittany, France; *Kark V,* 1989, off the Atlantic coast of Morocco; *Exxon Valdez,* 1989, Prince William Sound, Alaska, U.S.A.; *Haven,* 1991, Genoa, Italy; *Braer,* 1993, Shetland Islands, U.K.; *Sea Empress,* 1996, Milford Haven, Wales, U.K.; and *Erika,* 1999, off Brittany, France).

Although accidents involving the marine transport of oil are recognized as a major threat to the marine environment, concern has also been growing about accidental marine pollution arising from events involving substances other than oil. This concern arises from the fact that many of these substances are far more dangerous to the marine environment on a weight-for-weight basis. Many of the substances carried in bulk form are transported in tankers specially designed for the purpose, and although they are generally smaller than oil tankers, the vessels can carry various substances at the same time, each having different chemical properties.

Chemical substances are also carried in packages such as drums, intermediate bulk containers, and portable tanks. Again, the environmental threat that these substances represent

This chapter reflects the views of the author and not necessarily those of the United Nations or its specialized agencies.

bears no relation to the size of the unit in which they are carried. This is, in fact, one reason why some of these substances are not carried in bulk, but only in package form.

1.2 GLOBAL INITIATIVES FOR DEALING WITH ACCIDENTAL MARINE POLLUTION

The *Torrey Canyon* incident raised certain questions about the measures then in place to prevent oil pollution from ships. It also made the world aware of the inadequacies for providing compensation following accidents at sea. For the next few years, initiatives were taken to find solutions to these questions (IMO, 1989), mainly through the International Maritime Organization (IMO), at the time known as the Inter-Governmental Maritime Consultative Organization (IMCO).

In 1969, the IMO adopted two Conventions. The first, the International Convention Relating to the Intervention on the High Seas, dealt with the rights of nations to take action to prevent, mitigate, or eliminate the danger of oil pollution accidents involving foreign flagged ships outside their territorial waters. It entered into force in 1975. A Protocol extending the Convention to other hazardous substances entered into force in 1983. The second, the International Convention on Civil Liability for Oil Pollution Damage (CLC 69), established a system for providing compensation to victims of oil pollution damage, placing strict liability upon the owner of the ship from which oil was lost.

The 1969 CLC Convention was supplemented by another Convention in 1971, the International Convention on the Establishment of an International Fund for Compensation for Oil Pollution Damage (Fund 71), which was intended to provide further compensation to victims of oil pollution since, although CLC 69 placed the burden of compensation on the ship owner, at the same time it limited the compensation payable. The Fund provided further compensation when the limits laid down by the CLC 69 were exceeded. The Convention entered into force in 1978.

These two Conventions were updated in 1992 by Protocols that increased the compensation limits and broadened the scope of the two instruments. These amended conventions are usually termed 1992 CLC and the 1992 Fund Convention. They entered into force in May 1996.

At the same time, the IMO was also urging governments to establish arrangements for dealing with oil pollution accidents (IMCO, 1969). In November 1968, its Assembly adopted three important interrelated resolutions:

1. Resolution A.148 (ES.IV), *National Arrangements for Dealing with Significant Spillages of Oil*
2. Resolution A.149 (ES.IV), *Regional Co-operation in Dealing with Significant Spillages of Oil*
3. Resolution A.150 (ES.IV), *Research and Exchange of Information on Methods for Disposal of Oil in Cases of Significant Spillages*

In 1969, the first regional agreement for cooperation in combating oil pollution in case of emergency was adopted by the coastal states of the North Sea (Bonn Agreement). This originated following the grounding of the *Torrey Canyon*. The concern that pollution arising from maritime accidents should be mitigated by cooperative action between neighboring states was again reiterated by the IMO in its November 1979 Assembly Resolution A.448 (XI), by which governments were urged first to develop or improve national contingency arrangements to the extent feasible and secondly to develop, as appropriate, joint contingency arrangements at a regional, subregional, or sectoral level or on a bilateral level (IMO, 1979).

The grounding of the tanker *Exxon Valdez* near the Alaskan coast in 1989 revealed once again the need for responding immediately to oil spills. One result of the disaster was the adoption of the IMO's International Convention on Oil Pollution Preparedness Response and Cooperation (OPRC), 1990, which came into force in May 1995 (IMO, 1994). The aim of the Convention was to provide a global framework for international cooperation and mutual assistance in dealing with major oil pollution threats.

The international conference, which adopted the Convention in November 1990, also adopted a number of resolutions, one of which related to the expansion of the scope of the convention to include hazardous and noxious substances. A diplomatic conference in March 2000 adopted a Protocol on Preparedness, Response and Co-operation to Pollution Incidents by Hazardous and Noxious Substances, 2000 (OPRC-HNS 2000 Protocol).

The initiatives by the IMO to tackle pollution were not taken in a vacuum. By the early 1970s, the world had come to realize the problems of the environment and, as a consequence, environmental concern was also increasing. There was also a general feeling that pollution, in particular marine pollution, should be tackled at an international level (IMO, 1992). This global concern culminated in the organization of the United Nations Conference on the Human Environment (UNCHE) in Stockholm in 1972. Of particular relevance to the issue of international cooperation and mutual assistance to abate marine pollution was the recommendation for governments to take early action to adopt "effective national measures for the control of all significant sources of marine pollution and concentrate and coordinate their actions regionally and where appropriate on a wider international basis." The Stockholm conference also led to the establishment of the United Nations Environment Programme (UNEP).

The United Nations Conference on Environment and Development (UNCED), held 20 years later in Rio de Janeiro, Brazil, rekindled public interest in the environment. The conference adopted two conventions: one on biological diversity and the second on climatic change. It also adopted the Rio Declaration on Environment and Development and Agenda 21.

The Rio Declaration contained a set of sixteen principles directed at achieving environmentally sustainable development. The ones often considered of particular significance are the principles of "sustainable development" (Principle 1), the "precautionary approach" (Principle 15) and the "polluter should pay" (Principle 16). Agenda 21, which elaborates more fully on the Rio Declaration, is a comprehensive plan of action for the global community containing numerous chapters. Chapter 17 of Agenda 21 deals with protection of the oceans. Chapter 39 has a specific objective that, through global and multilateral negotiated agreements, international standards for the protection of the environment that take into account the different situations, capabilities, needs, and concerns of all countries, developed and developing, should be promoted (IMO, 1993a).

The United Nations Convention on the Law of the Sea (UNCLOS), which entered into force on November 16, 1994, is considered to provide the umbrella or common framework for the development of laws and policies for managing the oceans (Matthews, 1992). This Convention expressly recognizes regional cooperation on marine environmental protection and management in cases of accidental marine pollution. Article 198 provides that when a state becomes aware of cases in which the marine environment is in danger of being damaged or has been damaged by pollution, it must immediately notify states likely to be affected by such damage and competent international organizations. Article 199 provides that affected States shall cooperate with the competent international organizations, to the extent possible, in eliminating the effects of pollution and preventing or minimizing the damage (IMO, 1997b). States are further required to jointly develop and promote contingency plans for responding to marine pollution incidents. The regional agreements developed for combating accidental marine pollution underline these principles.

1.3 UNEP'S REGIONAL SEAS PROGRAMME

A direct result of the Stockholm conference was the decision in December 1972 by the United Nations General Assembly to establish, in 1973, the United Nations Environment

Programme. The Governing Council of UNEP subsequently identified "Oceans" as one of its priority areas in which efforts would be focused and activities developed (Jeftic, 1992).

The Governing Council of UNEP also endorsed a regional approach to the control of marine pollution and management of marine coastal resources and requested that regional action plans be developed. As a consequence, UNEP's Regional Seas Programme was initiated in 1974. The major objective of the program was to promote regional marine pollution control programs in areas that, for geographical, ecological, or political reasons, were perceived to have common elements and therefore give rise to a regional entity (Hulm, 1983a; Jeftic, 1992).

The program has steadily grown and today covers almost the whole marine global environment, as shown in Fig. 1.1. Among the different sea areas covered by regional programs of action are the Mediterranean, Red Sea and Gulf of Aden, Kuwait Region, Wider Caribbean, West and Central Africa, East Asian Seas, South Pacific, and Black Sea.

The approach taken in developing UNEP's Regional Seas Programme is an action plan tailored to the needs and priorities of a region, but it can also be applied to other regions (Hulm, 1983b; Manos, 1991).

A regional action plan is built on five interdependent components:

1. Environmental assessment where an assessment of the state of marine environment is made through a continuous coordinated pollution research program and exchange of information to identify the problems that need priority attention in the region.
2. Environmental management involving integrated planning activities related to developing and managing coastal areas aimed at controlling and preventing problems.
3. A legal component consisting of an umbrella Convention embodying the general commitments of the parties to the Convention, supported by technical and specific Protocols dealing with individual issues such as: preparedness and response to accidental marine pollution, dumping, seabed activities, all of which are aimed at strengthening cooperation among states in managing the regional pollution problems identified and committing the states to an active program, beyond mere declarations. By ratifying a protocol, a state accepts more specific areas of pollution control.
4. An institutional component consisting of regular meetings of contracting parties to the regional convention, which provides an intergovernmental forum for consultation and

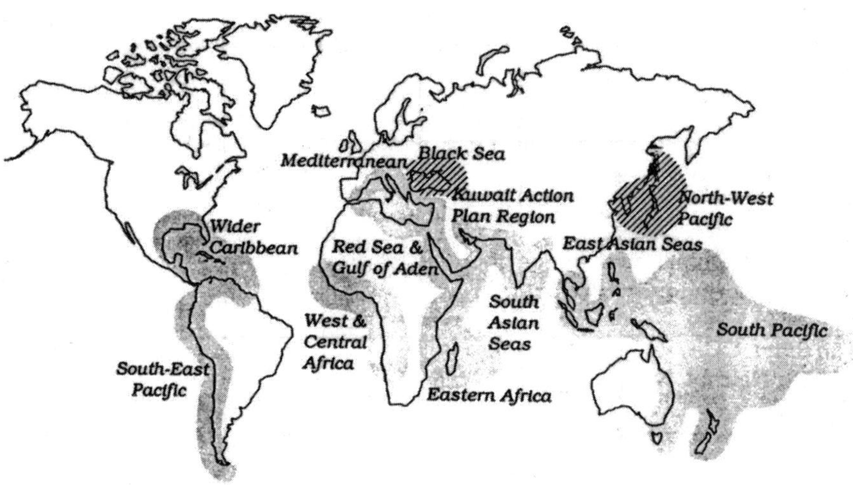

FIGURE 1.1 Regionalization of the oceans. (*Source: The Siren, 1991*).

decision making and a regional coordinating mechanism consisting of a coordinating unit and/or regional activity centers that service specific programs and act as a secretariat for the action plan or parts of it.

5. A financial component that provides the back-up for the other four parts and ensures the continuity of the action plan with funds earmarked for it.

As the United Nations' specialized agency concerned with the prevention of pollution from ships, the IMO plays an important contributory role in UNEP's Regional Seas Programme (IMO, 1997b).

1.4 THE MEDITERRANEAN ACTION PLAN (MAP)

Following a sequence of pleas in the early 1970s for remedial action on an overall Mediterranean scale, including recommendations emerging from various political and technical conferences, one of the first initiatives under UNEP's Regional Seas Programme was assistance to the Mediterranean countries (Saliba, 1989). An intergovernmental meeting held in Barcelona in 1975 approved an action plan for the protection of the Mediterranean Sea within UNEP's Regional Seas Programme and called for a legal framework to be prepared. This action, the Mediterranean Action Plan (MAP), was the first of its kind and has served as a model for establishing regional action plans elsewhere.

The following year in Barcelona, the Convention for the Protection of the Mediterranean Sea against Pollution, known as the Barcelona Convention, was adopted by 16 coastal states as well as the European Economic Community, together with two Protocols—one against dumping from ships or aircraft and the other for cooperation in cases of pollution emergencies.

To date, within the framework of the Barcelona Convention, six Protocols have been adopted, four of which are in force. These cover the following areas of the protection of the marine environment:

1. Limitation of dumping at sea
2. Cooperation in cases of marine pollution emergencies
3. Protection from land-based sources of pollution
4. Protection of special marine areas and biodiversity
5. Control of pollution from offshore activities
6. Prevention of pollution caused by transboundary movements of hazardous wastes

The Secretariat's functions were assigned to UNEP that established a dedicated coordinating unit in July 1982 for the whole of the Mediterranean Action Plan, which was hosted by Greece in Athens (Hulm, 1983b). Apart from its secretariat function, it is also responsible for planning, organization, information, and cooperation with intergovernmental and nongovernmental organizations. In addition, regional activity centers (RACs) dealing with specific components of the Mediterranean Action Plan have been set up. To date, the following have been established:

- Blue Plan Regional Activity Centre (BP/RAC), Sophia Antipolis, France
- Priority Actions Program Regional Activity Centre (PAP/RAC), Split, Croatia
- Specially Protected Areas Regional Activity Centre (SPA/RAC), Tunis, Tunisia
- Regional Marine Pollution Emergency Response Centre for the Mediterranean Sea (REMPEC), Malta.
- Historic Sites Secretariat (RAC/APVM), Marseilles, France

TABLE 1.1 Status of Ratification of the Barcelona Convention and its related Protocols as of 17 February 2000

	Barcelona Convention[a]	Acceptance of amendments	Dumping Protocol[b]	Acceptance of amendments	Emergency Protocol[c]	Land-based sources Protocol[d]	Acceptance of amendments	Specially protected areas Protocol[e]	SPA and biodiversity Protocol[f]	Offshore Protocol[g]	Hazardous wastes Protocol[h]
Contracting Parties											
Albania	X		X		X	X		X			
Algeria	X		X		X	X		X			
Bosnia and Herzegovina	X		X		X	X		X			
Croatia	X	X	X	X	X	X		X			
Cyprus	X		X		X	X		X			
European Community	X	X	X	X	X	X		X	X		
Egypt	X		X		X	X		X			
France	X		X		X	X		X			
Greece	X		X		X	X		X			
Israel	X		X		X	X		X			
Italy	X		X		X	X		X	X		
Lebanon	X		X		X	X		X			
Libya	X		X		X	X		X			
Malta	X	X	X	X	X	X	X	X	X		X
Monaco	X	X	X	X	X	X	X	X	X		
Morocco	X		X	X	X	X	X	X		X	X
Slovenia	X		X		X	X		X			
Spain	X	X	X	X	X	X	X	X	X		
Syria	X		X		X	X		X			
Tunisia	X	X	X	X	X	X	X	X	X	X	X
Turkey	X		X		X	X		X			

Source: UNEP, 2000.

[a]Convention for the Protection of the Mediterranean Sea against Pollution
Adopted (Barcelona): 16 February 1976
Entry into force: 12 February 1978

[b]The Protocol for the Prevention of Pollution of the Mediterranean Sea by Dumping from Ships and Aircraft (Dumping Protocol)
Adopted (Barcelona): 16 February 1976
Entry into force: 12 February 1978

[c]The Protocol concerning Co-operation in Combating Pollution of the Mediterranean Sea by Oil and other Harmful Substances in Cases of Emergency (Emergency Protocol)
Adopted (Barcelona): 16 February 1976
Entry into force: 12 February 1978

[d]The Protocol for the Protection of the Mediterranean Sea against Pollution from Land-based Sources (LBS Protocol)
Adopted (Athens): 17 May 1980
Entry into force: 17 June 1983

[e]The Protocol concerning Mediterranean Specially Protected Areas (SPA Protocol)
Adopted (Geneva): 3 April 1982
Entry into force: 23 March 1986

[f]The Protocol concerning Specially Protected Areas and Biological Diversity in the Mediterranean (SPA & Biodiversity Protocol)
Adopted (Barcelona): 10 June 1995
Entry into force: —

[g]Protocol for the Protection of the Mediterranean Sea against Pollution resulting from Exploration and Exploitation of the Continental Shelf and the Seabed and its Subsoil (Offshore Protocol)
Adopted (Madrid): 14 October 1994
Entry into force: —

[h]Protocol on the Prevention of Pollution of the Mediterranean Sea by Transboundary Movements of Hazardous Wastes and their Disposal (Hazardous Wastes Protocol)
Adopted (Izmir): 1 October 1996
Entry into force: —

TABLE 1.2 Examples of Regional Arrangements/Agreements for Co-operation in Combating Accidental Marine Pollution

Region	Legal instrument	Date of entry into force	States involved/ contracting parties	Organization responsible for secretariat duties	Remarks
North Sea	Agreement for Co-operation in Dealing with Pollution of the North Sea by Oil and by other Harmful Substances (Bonn Agreement).	August 1969	Belgium, Denmark, E.C., France, Germany, Netherlands, Norway, Sweden, U.K.	Secretariat of the OSPAR Commission, London, U.K.	1969 Agreement replaced by the 1983 Agreement which entered into force in 1989.
Nordic Area	Agreement between Denmark, Finland, Norway and Sweden concerning Co-operation in Measures to Deal with Pollution of the Sea by Oil (Copenhagen Agreement).	September 1991	Denmark, Finland, Norway, Sweden		No organization responsible for secretariat duties but direct cooperation between governments.
Baltic Sea Area	Convention on the Protection of the Marine Environment of the Baltic Sea Area (Helsinki Convention), 1974: Annex VI — Co-operation in Combating Marine Pollution.	May 1980	Denmark, Estonia, E.C., Finland, Germany, Latvia, Lithuania, Poland, Russia, Sweden	Baltic Marine Environment Protection Commission, Helsinki, Finland	1974 Convention replaced by the 1992 Helsinki Convention which entered into force in 2000.

Mediterranean Sea	Protocol concerning Co-operation in Combating Pollution of the Mediterranean Sea by Oil and other Harmful Substances in Cases of Emergency.	February 1978	All Mediterranean coastal states and E.C.	UNEP performs secretariat functions through a Co-ordinating Unit for the Mediterranean Action Plan based in Greece	The Regional Marine Pollution Emergency Response Centre for the Mediterranean Sea (REMPEC) was established in 1976 to assist the parties with their commitments under the Protocol.
Kuwait Action Plan Region	Protocol concerning Regional Co-operation in Combating Pollution by Oil and other Harmful Substances in Cases of Emergency.	June 1979	Bahrain, Iran, Iraq, Kuwait, Oman, Qatar, Saudi Arabia, United Arab Emirates	Regional Organization for the Protection of the Marine Environment (ROPME) performs secretariat functions based in Kuwait	
West and Central African Region	Protocol concerning Co-operation in Combating Pollution in Cases of Emergency.	August 1984	Cameroon, Congo, Gabon, Gambia, Ghana, Guinea, Ivory Coast, Nigeria, Senegal, Togo Signatories: Benin, Liberia, Mauritania	UNEP performs secretariat functions through Co-ordinating Unit based in the Ivory Coast	

TABLE 1.2 Examples of Regional Arrangements/Agreements for Co-operation in Combating Accidental Marine Pollution (*Continued*)

Region	Legal instrument	Date of entry into force	States involved/ contracting parties	Organization responsible for secretariat duties	Remarks
Wider Caribbean Region	Protocol concerning Co-operation in Combating Oil Spills in the Wider Caribbean.	October 1986	Antigua and Barbados, Bahamas, British Virgin Islands, Cayman Islands, Costa Rica, Colombia, Cuba, Dominica, Dominican Republic, France, Grenada, Guadeloupe, Guatemala, Guyana, Jamaica, Martinique, Mexico, Netherlands, Netherlands Antilles, Panama, Puerto Rico, St. Croix, St. Lucia, St. Vincent & the Grenadines, Trinidad & Tobago, U.K., U.S.A., Venezuela, E.C.	UNEP performs secretariat functions through a Co-ordinating Unit based in Jamaica	The Caribbean countries drafted a Sub-Regional Oil Spill Contingency Plan for the Islands, States, and Territories of the Wider Caribbean Region (St. Lucia, May 1984) — the "Caribbean Plan," which was updated in November 1992 at a meeting convened by IMO — the Caribbean Island OPRC Contingency Plan. The Regional Marine Pollution Emergency Information and Training Centre for the Wider Caribbean (REMPEITC — Caribbean) was opened in June 1995 under the management of UNEP — CAR/RCU and IMO with support from the Netherlands Antilles, the Netherlands, and the U.S.A.

Northeast Atlantic	Agreement for the Protection of the coasts and waters of the Northeast Atlantic against pollution.	Signed October 1990	Portugal, France, Spain, Morocco, E.C.	See remarks	Centre International de Lutte contre la Pollution de l'Atlantique du N.E. (CILPLAN) performs the secretariat duties as well as the role of an emergency coordinating center.
Denger Plan	Joint Maritime Contingency Plan for Oil Combating revised in February 1993 following its extension to hazardous substances.	September 1982	Germany, Denmark		Its area of application is the German Bight and Western Baltic Sea, and it is aimed at increasing the effectiveness of contingency arrangements already in place.
Manche Plan	Anglo-French Joint Maritime Contingency Plan.	May 1978	France, U.K.		Its area of application is the English Channel or the "Manche." The plan sets out how a joint zone of responsibility should be exercised for incidents where both parties are to be affected. The plan covers both counter-pollution and search-and-rescue operations.

TABLE 1.2 Examples of Regional Arrangements/Agreements for Co-operation in Combating Accidental Marine Pollution (*Continued*)

Region	Legal instrument	Date of entry into force	States Involved/ contracting parties	Organization responsible for secretariat duties	Remarks
Waters of mutual interest to Canada and the U.S.A.	Canada/U.S. Joint Marine Pollution Contingency Plan for Spills of Oil and other Noxious Substances.	June 1974	Canada and U.S.A.		The Canada/U.S. Joint Marine Pollution Contingency Plan is the basis for all contingency planning and emergency response activities along the maritime boundaries of the United States and Canada and includes the following areas: the Atlantic Coast, the Pacific Coast, the Beaufort Sea, the Dixon Entrance of the Pacific Coast, and the Great Lakes.
Bering and Chukchi Seas	U.S./U.S.S.R. Joint Contingency Plan against Pollution in the Bering and Chukchi Seas.	August 1989	U.S.A., U.S.S.R.		The joint contingency plan resulted from the U.S. Coast Guard-headed project entitled "Prevention and Clean-up of Pollution of the Marine Environment from Shipping" under a U.S./U.S.S.R. environmental Agreement.

Source: Beer-Gabel, 2000; IMO, 1985, 1992; OTSOPA, 1995; Skinner and Reilly, 1990; Transport Canada, 1974.

- Environmental Remote Sensing Regional Activity Centre (ERS/RAC), Palermo, Italy
- Regional Activity Centre for Cleaner Production (CP/RAC), Barcelona, Spain

Whereas all RACs are national centers that take on a regional dimension, REMPEC has the status of a UN duty station.

Internal coordination at country level is affected by a MAP national focal point, while national authorities might also be designated as focal points to maintain liaison with individual components of the Plan.

Up to 1978, the MAP was financed mainly from UNEP's Environment Fund with contributions in cash and kind from UN specialized agencies, as well as from other sources. Since 1979, the main cost has been borne by the governments of the region, who contribute in agreed proportions, based on the UN scale, to a Mediterranean Trust Fund (MTF). United Nations specialized agencies also contribute, together with governments hosting regional activity centers that cover part of the cost of hosting these centers in addition to their regular contributions to the MTF.

The introduction of the concept of sustainable development with the approach of the Rio Conference in 1992 led the member states of the Barcelona Convention to draw up an Agenda 21 for the Mediterranean (Vallega, 1995). This concept adapted to the Mediterranean context forms an integral part of the objectives of the MAP. In Barcelona in June 1995, the Mediterranean Action Plan, which was adopted in 1975, was replaced by the Action Plan for the Protection of the Marine Environment and the Sustainable Development of the Coastal Areas of the Mediterranean (renamed MAP Phase II) by the Contracting Parties at a Conference of Plenipotentiaries, which also adopted priority fields of activities for the period 1996–2005 (UNEP, 1997a,b).

As a consequence, the legal instruments have been updated to reflect and ensure coherence with the progress made in international environmental law. The Barcelona Convention was itself thoroughly amended in June 1995 by the same Conference of Plenipotentiaries, while the Dumping Protocol was made more stringent and the Protocol on Specially Protected Areas was extended to include biodiversity. The Protocol dealing with Cooperation in Cases of Emergency is currently under review to include the prevention of pollution by ships. The status of ratification by contracting parties to the Barcelona Convention and its related protocols is shown in Table 1.1.

1.5 REGIONAL AGREEMENTS ON PREPAREDNESS FOR AND RESPONSE TO ACCIDENTAL MARINE POLLUTION

In most sea areas covered by UNEP's action plans, a technical protocol has been adopted providing the legal basis for international cooperation and mutual assistance in combating pollution aimed at mitigating the effects of accidents involving spills (IMO, 1985).

The regional approach of UNEP-sponsored agreements on cooperation in combating accidental marine pollution finds its basis in the 1969 North Sea Agreement, which was also a forerunner of other non-UN regional arrangements or agreements, namely the Helsinki Convention (Hulm, 1983a).

Furthermore, a number of bilateral operational agreements based on the same principles found in regional agreements have been established between neighboring countries. Table 1.2 gives examples of such arrangements or agreements.

By pooling resources and expertise, these types of agreements provide a cost-effective, immediate way of responding to a major spill that cannot be dealt with by any one country on its own.

Furthermore, these agreements provide a legal framework that facilitates:

- Early notification of a pollution incident
- Continuous exchange of information at the preparedness and response phases
- Mutual assistance between neighbouring countries

The common elements to these regional agreements (IMO, 1985) are briefly described below under the following headings:

- Pollution reporting
- Mutual assistance
- National response capability
- Information exchange
- Institutional arrangements

1.5.1 Pollution Reporting

A fundamental feature for the success of a regional agreement is the notification between contracting parties when one party learns of incidents that may affect other contracting parties. All regional agreements include this obligation. They also include the undertaking that a state party to the agreement will instruct masters of ships flying its flag, pilots of aircraft, and, in some cases, offshore operators under its jurisdiction to report to it marine pollution incidents, or to other contracting states likely to be affected by such incidents.

Pollution incident reporting procedures, as well as the standard message formats, have to be established to ensure the rapid transmission of information and requests for assistance.

With a view to harmonizing pollution reporting systems, a standard alert message format (POLREP) has been developed, which is used within a number of regional agreements (IMO, 1995).

1.5.2 Mutual Assistance

Regional agreements contain an undertaking that coastal states party to the agreement use their best endeavors to assist other parties that may request assistance in a marine pollution incident. The term *best endeavors* is generally found to embrace a valuable concept facilitating cooperation, without which there may be a reluctance to negotiate practical arrangements for fear of incurring unacceptable responsibilities and obligations. This is important in regions where the national response capabilities are limited and there is an imbalance in response capability among coastal states. This caveat also implies that an affected state has to assess the situation and determine the kind of assistance that should be requested.

1.5.3 National Response Capability

A national response capability is a prerequisite without which regional agreements and mutual assistance would be ineffective. Many agreements require contracting states to establish and maintain the means to respond to accidental marine pollution. The development of regional agreements has served to encourage the development of national response systems and contingency plans of contracting parties, with regional agreements serving as a supplement to, rather than a substitute for the national response capability of an individual coastal state.

1.5.4 Information Exchange

Regional agreements provide that contracting states exchange information and keep each other informed on:

- The competent national authorities responsible for implementing the regional agreement and in charge of preparedness for and response to accidental marine pollution and for providing mutual assistance, which could be different from the former authority
- Means available for international assistance
- Technological developments in the field of responding to accidental marine pollution
- Operational procedures, guidelines, and regulations put in place relating to preparedness for and response to marine pollution incidents

1.5.5 Institutional Arrangements

All regional agreements are embodied by a treaty that is either an independent legal instrument (e.g., the Bonn Agreement) or part of a comprehensive umbrella Convention, as is the case with UNEP's Regional Seas Programme (e.g., Protocol to the Barcelona Convention dealing with Co-operation in Cases of Emergency). All agreements provide for regular meetings of contracting parties, the purpose of which is to monitor the implementation of the agreement, decide on financial matters related to its implementation, consider revisions or amendments to the agreement, and provide guidance to the secretariat.

Functions of the secretariat are provided primarily by existing organizations, e.g., UNEP in the case of the Barcelona and Cartagena Conventions, or by a regional organization either established or designated by the Convention, e.g., Helsinki Commission; Regional Organization for the Protection of the Marine Environment (ROPME).

In some instances, to support the operational and technical aspects of the regional agreement, a regional center or coordinating unit has also been established (e.g., Marine Emergency Mutual Aid Centre [MEMAC], established in 1983 in Bahrain for the Gulf Region). Centers of this type do not have the resources (equipment and personnel) to carry out operational response at the site of an incident.

A few of the regional agreements have established permanent working groups of experts for the exchange of information on scientific operational and technical matters. These work by correspondence or meet periodically, e.g., the Bonn Agreement Working Group on Operational, Technical and Scientific Questions concerning Counter Pollution Activities (OTSOPA). Conclusions from such meetings are reported to the meetings of contracting parties for approval and any action is incorporated in the workplan of the regional agreement.

1.6 THE MEDITERRANEAN SEA

The Mediterranean Sea consists of a series of interconnecting and adjacent seas (Jeftic et al., 1990), with two major basins, an eastern and a western one. The western basin (approximately 0.85 million km^2) consists of the Alboran Sea, the Algero-Provençal basin, the Ligurian Sea, and the Tyrrhenian Sea, while the eastern basin (about 1.65 million km^2) consists of the Adriatic Sea, the Ionian Sea, the Aegean Sea, and the Levant. These are shown in Fig. 1.2.

The Mediterranean Sea lies 46°N, 30°N, 6°W, and 36°E and is bordered by the African, Asian, and European continents. It is sometimes described as a tideless seawater pool. The surface water current system of the Mediterranean shows a migration of Atlantic water

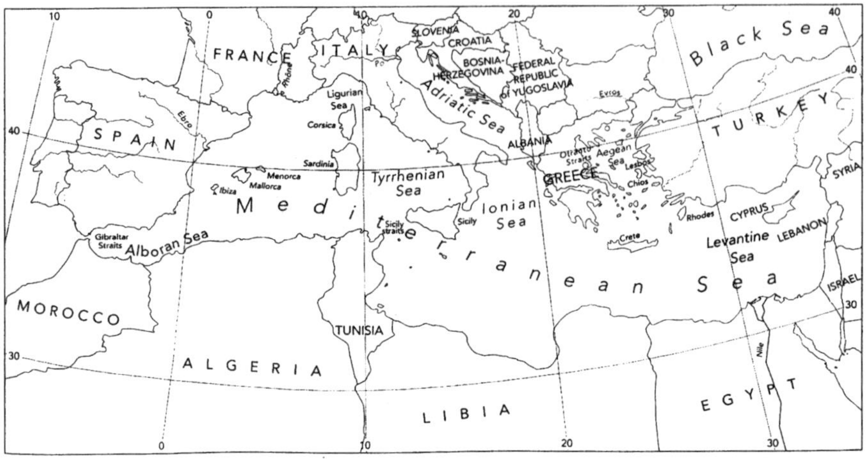

FIGURE 1.2 The Mediterranean Sea. (*Source: European Environmental Agency, 1999*).

entering from the Straits of Gibraltar towards the east with numerous spin-off eddies and the generation of local current systems due to the complexity of the topography and the presence of islands (Fig. 1.3). The return of the Mediterranean water is by way of the Levantine intermediate water for the most part and deep water flowing from east to west and exiting over the Gibraltar sill below the incoming Atlantic surface water jet (European Environment Agency, 1999; Jeftic et al., 1990). Some basic features of the Mediterranean Sea are outlined in Table 1.3.

From the point of view of international relations, the region is enclosed by the three continents of Asia, Africa, and Europe with the coastal states belonging to different political affiliations, such as the Arab League, European Union, and the Organisation of African Unity. The economic status of these states is also quite heterogenous, some being developed while others are developing or in transition.

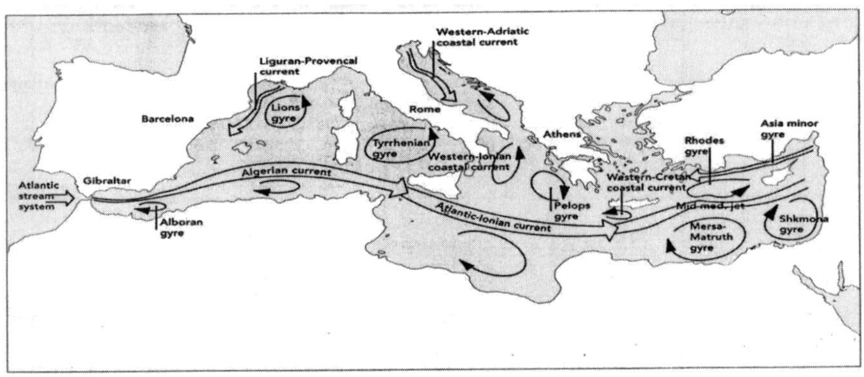

FIGURE 1.3 Major circulation patterns in the Mediterranean Sea. (*Source: Pinardi et al., 1997*).

TABLE 1.3 Characteristics of the Mediterranean Sea Basin

Coastline (including islands)	46,267 km
(excluding islands)	26,993 km
Surface area	2.5×10^6 km^2
	0.7% of world maritime surface
30% of total surface area	contained in 2000–3000 m depth contour
20% of total surface area	contained in less than 200 m depth contour
Volume (total)	3.7×10^6 km^3
50% of total volume	contained 2000—3000 m depth contour
1.5% of total volume	contained in less than 200 m depth contour
Average depth	1500 m
Maximum length (Gibraltar–Syria)	3800 km
Maximum width (France–Algeria)	900 km
Turnover time	80 years
Salinity	Surface water: slightly above 36%
	Deep water: 38.4–39%
Bottlenecks	
Straits of Gibraltar	15 km wide and 290 m deep
Straits of Boniface	
Sea of Marmara (by the Dardanelles)	450 m to 7.4 km wide and 55 m deep
Suez Canal	365–300 m max width and depth dredged to 19.5–20 m
High density of merchant traffic	
Special Area under MARPOL 73/78	Annex I—Regulations for the Prevention of Pollution by Oil

Source: Lloyd's, 1991; Jeftic et al., 1990.

1.7 ESTIMATES OF SEABORNE TRAFFIC

Among the world oceans, the density of maritime traffic in the Mediterranean is particularly significant. Some documented estimates of seaborne traffic are:

- Excluding vessels that are in seaports for any reason, 1,500 cargo vessels of over 150 grt sail the Mediterranean Sea at any one time (IMO, 1990)
- 200 vessels (ferries, passenger ships, and warships) navigate the Mediterranean Sea at the same time (IMO, 1990)
- 2,000,000 commercial ships over 100 grt cross the Mediterranean annually and about 2,000 of these ships, of which 250 to 300 are oil tankers, are at sea at any moment (UNEP, 1989; World Bank/European Investment Bank, 1990)

A recent study has shown that a daily average of 1,900 commercial vessels ply the Mediterranean Sea routes or call at Mediterranean ports (Dobler, 1997). This number is distributed among 490 cargo-carrying vessels: 129 tankers, 232 bulk carriers, and 129 general cargo vessels; and 1,400 passenger ships and car ferries. Based on these figures, it was also estimated that about 30% of the cargo-carrying vessels were in ports, either loading or unloading cargoes or waiting for repairs, and the remaining 70% were at sea, loaded or in ballast.

Specific areas in the Mediterranean that might have a bearing on the accident potential in the region are the so-called bottlenecks:

- Straits of Gibraltar
- Suez Canal
- Dardanelles/Sea of Marmara/Bosphorus complex
- Strait of Boniface

Although traffic transiting the Mediterranean Sea from the Atlantic through the Strait of Gibraltar, from the Black Sea through the Bosphorus, and from the Red Sea through the Suez Canal, might have no direct economic impact on Mediterranean commercial ports, from the point of view of contributing to the risk of an accident, this traffic cannot be ignored. Table 1.4 gives a breakdown of the transits for the Suez Canal in recent years. It has also been estimated that for 1991, 3,549 oil tankers transited the Suez Canal: 1,910 were directed southbound; and 1,639 were directed northbound (INTERTANKO, 1992).

Another significant bottleneck is the Dardanelles/Sea of Marmara/Bosphorus complex, which opens into the Aegean Sea in the northeast Mediterranean. The Bosphorus is a unique waterway in that, while it is bound by the same country, it is in fact an international waterway. Passage through this waterway is regulated by the applicable Conventions of the 1963 Montreux Convention and Article 23 of the 1923 Lausanne Treaty of Peace and by the straits regime of the 1982 UNCLOS Convention in relation to matters falling outside the scope of the two previously mentioned Conventions.

Over the years, there has been a significant increase in traffic. In 1938, an average of 15 ships per day navigated the waterway. This rose to 125 per day in 1995 and 137 in 1996 (Lloyd's, 1997b). Furthermore, it is estimated that at least one large LPG carrier passes through each week (Lloyd's, 1994a).

According to the Bosphorus traffic control center, 38,400 vessels passed through this stretch of water in 1994 (Lloyd's, 1995a), and figures as high as 42,000 vessels per annum have been reported (Lloyd's, 1995b). The urban commuter traffic must be added to these numbers. It is estimated that 1,300 craft daily carry more than 1.5 million people between Istanbul's European and Asian shores (Lloyd's, 1996). It is envisaged that the projected expansion of oil production in the Caspian Sea region will increase dramatically the amount of shipping passing through this unique waterway (Oil and Gas International, 1994).

A bottleneck of recent concern is the Strait of Boniface, where the traffic of laden oil tankers and of ships carrying dangerous chemicals has increased by 40% since 1985 (IMO, 1993b).

1.8 MARITIME PATTERNS IN THE MEDITERRANEAN

Maritime traffic flow in the Mediterranean is characterized by:

- The presence of traffic that transits across the sea, not entering any of the Mediterranean ports
- Four locations for traffic separation schemes

TABLE 1.4 Transits for the Suez Canal

Year	Transits
1988	18,190
1991	17,473
1995	15,051

Source: Lloyd's, 1991, 1997a

- Captive nature of traffic
- Navigation time for this captive traffic between origin and destination is generally not more than two days
- Coexistence of various kinds of vessels
- High density of ports

The current state and expected development of maritime trade in and outside the Mediterranean region indicate that it is and shall remain a major route for maritime transport from the Middle East and North Africa to Europe and North America and vice versa. Figure 1.4 shows the major maritime petrochemical routes in the Mediterranean.

1.9 PORTS' SYSTEM IN THE MEDITERRANEAN

Merchant vessel traffic in the Mediterranean either passes through the Mediterranean or originates from or is directed to the 305 ports scattered along the basin's coastline. There is a variety of port types in the Mediterranean, including:

- Small ports, which, regardless of their size, contribute to the economy of the country
- Single-user complexes whose infrastructure could be owned by a major industrial organization, which could be national, parastatal, or private
- Multipurpose facilities offering a wide range of services to the port user

Other aspects that characterize the Mediterranean ports are:

- The absence of long approach channels
- A high urbanization in the vicinities of the ports

Of the total vessel calls involved in the Mediterranean trade in 1993, 53% of these were made by general cargo vessels, 23% tankers, 13% bulkers, and 7% container vessels (EC, 1995). Of the total vessel movement in that same year, 45% of these were made by general

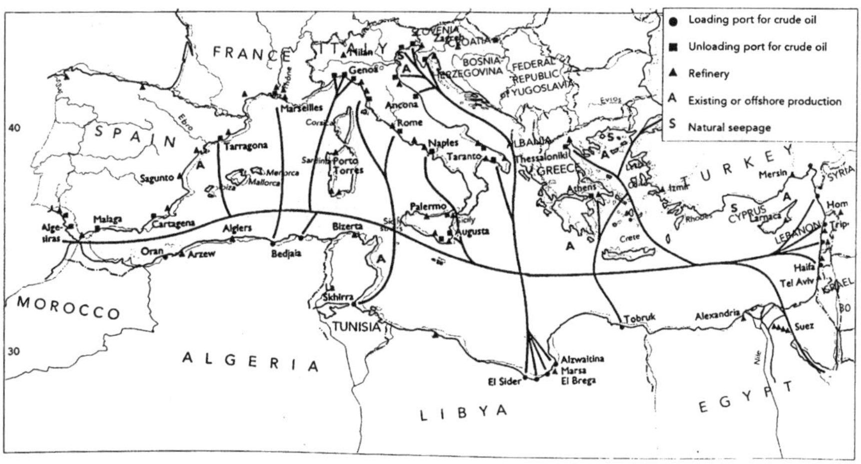

FIGURE 1.4 Activities of the oil industry in the Mediterranean Sea. (*Source: European Environmental Agency, 1999*).

cargo vessels, 25% tankers, 9% bulkers, and 18% container vessels. Although accounting for only 7% of all vessels trading in the Mediterranean region, container vessels account for 18% of all port activity.

1.10 HAZARDOUS CARGO FLOW AND TRADE

It is difficult to find explicit information on the tonnage of hazardous cargo transiting the Mediterranean Sea. However, data collected from official national sources by the United Nations Statistical Division and from estimates based on research done by the United Nations Conference on Trade and Development (UNCTAD) and presented to the IMO show that in 1990, the total amount of goods handled (goods loaded/unloaded) in international maritime trade by Mediterranean countries accounted for 1264×10^6 mt, or approximately 16% of the world's total.

According to more recent data, crude oil is the biggest export commodity of the non-EU Mediterranean coastal states and accounted for 57% of the total exports by maritime means in 1992. Crude oil was in fact three times larger than the most significant export — petroleum products (EC, 1995).

There is no doubt that if plans go ahead for upgrading existing pipelines and developing of new ones in North Africa, the Middle and Near East, crude oil transport strategies in the Mediterranean will alter, in particular the Red and Mediterranean Seas (Lloyd's, 1995c) and those linking the Caspian Sea to the Black and Mediterranean Seas (Oil and Gas International, 1994), as well as LNG-transport strategies originating from the South West Mediterranean (Lloyd's, 1994b).

Very little data exist that help quantify the movement of bulk gases, solids, and liquids at the regional level. From profiles of key ship operators, however, as well as information on marine casualties, gas products such as LPG, LNG, propane and ammonia; solid bulk products such as bauxite, phosphates, and fertilizers; and liquid bulk products such as vegetable oils, alcohols, wines, and inorganic acids are being handled by ports (*Hazardous Cargo Bulletin,* 1998a, b; REMPEC, 1998b). Likewise, data on the movement of packaged dangerous goods are difficult to obtain, although one can presume that products belonging to the nine main classes of the IMO's International Maritime Dangerous Goods Code (IMDG Code) are transported within and through the Mediterranean.

1.11 ACCIDENTAL MARINE POLLUTION IN THE MEDITERRANEAN

Data on accidents causing or likely to cause pollution of the Mediterranean Sea by oil and other harmful substances between 1977 and 1997 show the following (REMPEC, 1998a):

- Accidents involving oil (1977–1997) 242
- Accidents involving other harmful substances (1988–1997) 79
- Accidents involving oil (1996 only) 24
- Accidents involving oil (1997 only) 16
- Accidents involving other harmful substances (1996 only) 8
- Accidents involving other harmful substances (1997 only) 5

Of the 53 accidents registered in 1996 and 1997, 23 (43%) occurred in ports or in their approaches, while 30 (57%) occurred in open sea.

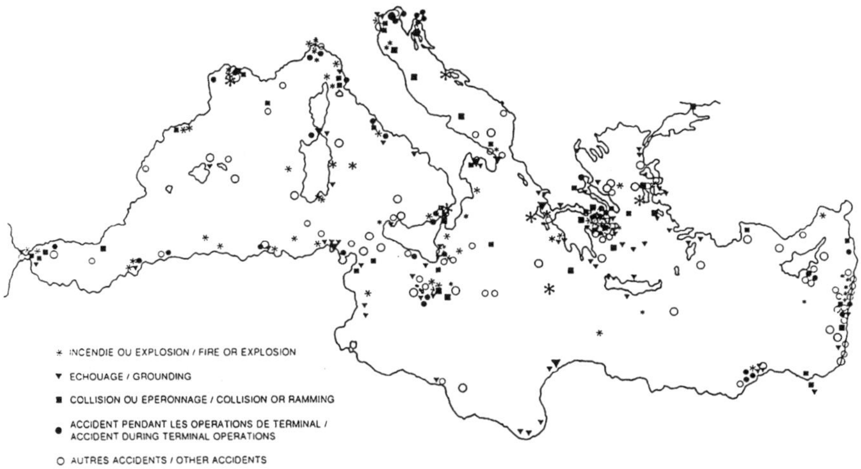

FIGURE 1.5 Location of accidents causing or likely to cause pollution of the Mediterranean Sea, 1977–1997. (*Source: REMPEC, 1998a*).

According to these data, the category "other harmful substances" includes chemical products transported in bulk and as packaged dangerous goods, as well as certain commodities carried in large quantities such as coal and grains, which can also present a certain ecological risk to the marine environment if a cargo load of several thousand tons were to be spilled in shallow areas. Consequently, events involving the spillage or likely spillage of such products are included in the data above. This is justified by the accident involving the M/V *Fenes*, which ran aground in the Strait of Boniface close to a sensitive ecological area, Lavezzi Island, on September 25, 1996, carrying 2,650 tons of wheat (IMO, 1998a). A week later, a storm damaged the ship and the entire cargo of wheat was spilled to depths of 10 to 20 m, over an area of approximately 1 hectare. The wheat immediately smothered the *Posidonia herberium* beds. A further environmental and safety problem was the effect of the organic matter present in the wheat which encouraged the development of sulfate-reducing anaerobic microflora and subsequently promoted the production of hydrogen sulfide from the unlimited supply of sulfate ions in the seawater. The presence of toxic hydrogen sulfide hindered the work of divers involved in the response and affected the marine flora and fauna.

The map in Fig. 1.5 represents the geographical distribution of accidents and illustrates the accident-prone areas in the Mediterranean, namely the approaches of Gibraltar and the Malta–Sicily channel and accesses around the areas of Genoa and Leghorn, Venice and Trieste, Piraeus, Beirut, and Alexandria.

A report presented to IMO identified a number of common issues, mostly relating to the human aspect of ship operations that have contributed to marine casualties worldwide.

Although no in depth study has been undertaken as to the cause of major pollution accidents in the Mediterranean, the findings of this report as summarized in Table 1.5 would probably equally apply to the Mediterranean setting (IMO, 1998b).

1.12 THE EMERGENCY PROTOCOL TO THE BARCELONA CONVENTION

The Mediterranean region was the third region (after the Bonn Agreement for the North Sea and the Helsinki Convention for the Baltic Sea) for which a regional agreement for coop-

TABLE 1.5 Maritime Casualties and Human Contributing Factors

Casualty category	Contributing factors
Collision Contact Stranding/grounding	• Fatigue • Duty system of pilots • Bridge resource management • Management of the ship • Working relationship pilot/bridge team • Hours of darkness/adverse weather
Fire or explosion	• Welding/cutting operations in vicinity of spaces containing flammable gases • Human element aspects
Other	• Tank venting operations

Source: IMO, 1998b

eration in combating pollution in cases of emergency has been adopted, and the first under the UNEP Regional Seas Programme (Sainlos, 1992).

Within the legal framework of the Mediterranean Action Plan, a Conference of Plenipotentiaries of the coastal states of the Mediterranean Region was convened in Barcelona in February 1976. This Conference adopted the following:

- Convention for the Protection of the Mediterranean Sea against Pollution (Barcelona Convention); and
- Two Protocols, one of which concerned Co-operation in Combating Pollution of the Mediterranean by Oil and Other Harmful Substances in Cases of Emergency (known in short as the Emergency Protocol or the Protocol for Co-operation in Cases of Emergency).

For the purpose of this Convention and hence of the Protocols adopted by the Conference and any other subsequent related Protocols, the Mediterranean Sea Area was taken to mean the maritime waters of the Mediterranean Sea proper, including its gulfs and seas, bound to the west by the meridian passing through Cape Spartel lighthouse at the entrance of the Straits of Gibraltar and to the east, by the southern limits of the Straits of the Dardanelles between Mehmetcik and Kumakale lighthouses (UNEP, 1978).

The rationale behind the Emergency Protocol and the elements on which the Protocol are based are those found in other regional agreements for cooperation in combating accidental marine pollution. These consist of building response capabilities, pollution, reporting for early notification of an incident, and mutual assistance. Interestingly, the importance that bilateral and regional agreements have in enhancing the capacity of countries to respond to major pollution accidents is highlighted in article 10 of the OPRC Convention, which stipulates that parties should try to conclude such agreements (IMO, 1991). There are many elements common to both the OPRC and the Emergency Protocol, as shown in Table 1.6.

In order to assist the Mediterannean coastal states in their obligations under the Emergency Protocol, in 1976 the same Conference decided to create the Regional Oil Combating Centre (ROCC), which was established in December 1976 in Malta. The IMO is directly responsible for the technical and administrative management of the Centre. Previously, its financial resources were provided by UNEP's Environment Fund but, as with other components of the Mediterranean Action Plan, since 1979 funding has been provided by the Mediterranean Trust Fund.

In September 1987, the Fifth Ordinary Meeting of the Contracting Parties to the Barcelona Convention held in Athens decided to extend the mandate of the Centre to "other harmful substances" and two years later, the same contracting parties approved new objectives and

TABLE 1.6 Elements Common to OPRC and the Emergency Protocol

Salient features	OPRC convention	Emergency Protocol[a]
International cooperation and mutual assistance	Articles 7, 10	Article 10
Pollution reporting	Article 4	Articles 6b, 7, 8, 9c
Oil pollution emergency plans covering: • oil tankers[b] • fixed or floating offshore facility • seaport/oil handling facility	Article 3	Does not cover ship-based contingency plans; Article 2 defines ports as being an interest of a coastal state that could be threatened, while Article 3 requests states to promote their contingency plans; Article 16 of the Offshore Protocol covers contingency planning to combat pollution from exploration and exploitation of the seabed.
National and regional preparedness and response capability	Article 6	Articles 3, 6a, 9
Technical cooperation and transfer of technology	Articles 8, 9	Article 6c
Institutional arrangements	IMO has been designated	REMPEC has been designated
Application to harmful substances	Through a diplomatic Conference that took place in March 2000	Yes

[a]The Emergency Protocol is currently under review to bring it in line with modern international legal instruments on preparedness and response, as well as to include the prevention of pollution from ships.

[b]Implemented through Regulation 26 of Annex I of MARPOL 73/78.

functions of the Regional Centre and also a change in name to Regional Marine Pollution Emergency Response Centre for the Mediterranean Sea (REMPEC).

In June 1995, MAP Phase II, which had as one of its components the prevention of and response to pollution of the marine environment from sea-based activities, was adopted. This was aimed at proposing strategies and activities that would support and supplement national efforts to promote the prevention of, preparedness for, and response to pollution of the marine environment from sea-based activities (UNEP, 1997). In November 1997, the Contracting Parties to the Barcelona Convention adopted a regional strategy on the prevention of pollution in the marine environment by ships (REMPEC, 1998b). Initiatives are currently underway to revise the Emergency Protocol to introduce the necessary legal provisions as well as to amend the objectives and functions of REMPEC, both of which will give the required legal impetus to implement the strategy.

When the Centre was established, it was considered whether it should be given the means to become an operational institution at the regional level. This seemed an unrealistic option, and it was decided instead to pool resources together in case of a spill of catastrophic consequences. The Centre was therefore assigned the following objectives:

- To provide a framework for the exchange of information on operational, scientific, legal, and financial matters related to preparedness for and response to accidental marine pollution
- To strengthen the preparedness and response capacities of the Mediterranean coastal states
- To facilitate cooperation among the Mediterranean states to respond to accidental marine pollution

The priorities and the program of activities of REMPEC are decided biennially by the contracting parties themselves, taking into consideration the recommendations and proposals made to them by REMPEC's Focal Points. These are the national authorities designated by their respective governments responsible for preparedness, response, and mutual assistance in cases of accidental marine pollution.

1.13 THE ROLE OF REMPEC IN PREPAREDNESS, RESPONSE, AND COOPERATION FOR ACCIDENTAL MARINE POLLUTION

While responsibility of combating a major pollution spill will always rest with the affected coastal state, certain conditions need to be satisfied in order for regional cooperation to be effective in such incidents. The following are imperative:

- The timely dissemination of appropriate information
- The establishment and maintenance of a national organization for preparedness and response
- The development of contacts between national authorities from different coastal states prior to the event
- A basis for cooperation among countries
- The determination of the conditions under which this cooperation is to take place

In conformity with its objectives and taking into consideration the requirements that need to be satisfied, the Centre has developed its activities along four main lines:

1. Dissemination of information
2. Training of personnel

3. Assistance in the preparation of contingency plans
4. Facilitation of cooperation in cases of emergency

1.13.1 Dissemination of Information

In order to respond to the necessity for coastal states to have the information to initiate their own response operations and operations of mutual assistance, the Centre has developed a regional information system. This system is intended to be used by the national authorities of the coastal states, not only in an emergency situation, but also to develop and adapt their national contingency plans taking into consideration their commitments under the Emergency Protocol.

1.13.2 Training of Personnel

The establishment and maintenance of an effective national organization and the subsequent development of subregional and regional systems are dependent on the existence of highly trained personnel. The role of REMPEC in the field of training has always been considered essential. The training program is directed at the following personnel:

- Decision makers/senior officers
- Those involved in on-site operations
- Personnel who provide support in the decision-making process

At the regional level, this comprehensive program consists of general and specialized training courses of a practical nature. The aim of training personnel at the regional level is not only to impart knowledge, give confidence, and instill a sense of intuition, but also to promote contact among those individuals who could be involved in a multilateral response operation. In addition, the Centre also provides assistance in organizing national training courses when requested by individual countries. Since its inception in 1976, as of June 2000, it is estimated that around 1,800 persons from the region have gone through REMPEC's training program.

1.13.3 Assistance in Preparing Contingency Plans

A functional national system is a prerequisite for organizing effective response to accidental marine pollution, and it also forms the basis for efficient regional cooperation. Some of the needs of the coastal states surpass the assistance that can be provided for by the core budget of REMPEC. To satisfy these needs, the Centre has prepared projects based on specific requests of the countries, which are then submitted to external funding sources, in particular those available from the European Union. The projects mainly focus on capacity building at the port, national, and subregional levels.

1.13.4 Facilitation of Cooperation in Cases of Emergency

Regional cooperation at the time of a marine pollution accident can only be effective if it is planned, organized and based on the principle of mutual assistance. For this purpose, a number of recommendations, principles, and guidelines have been prepared by the Centre and subsequently approved by contracting parties that concern finance, administration, customs clearance, and other operational aspects of mutual assistance. To further facilitate cooperation and mutual assistance, REMPEC has developed and keeps up to date a regional communication network to ensure the timely reception and transmission of alerts and requests for assistance. Exercises are periodically organized to test this communications network and information exchange among parties.

Depending on the scale of a pollution accident, the assistance provided to a coastal state can take the form of collecting and providing information including technical advice, assistance in obtaining international aid, and coordinating such aid regardless of whether the resources originate from the governmental or private sector. The state can also activate the Mediterranean Assistance Unit (MAU). The MAU, established in 1993 by a decision of the Contracting Parties to the Barcelona Convention, is made up of a limited number of qualified experts and centers of expertise that can provide advice to national authorities in the event of a pollution accident. The Unit is activated and mobilized at the request of a country under the management of REMPEC, the cost being borne by the Centre's core funding.

In the EU's capacity as a contracting party to the Barcelona Convention, governments of non-Mediterranean EU member states can also provide assistance, with requests for assistance being coordinated between the European Commission and REMPEC.

Recognizing the important contribution that the chemical, oil, and shipping industries can make to preparedness for and response to accidental marine pollution, REMPEC has sought to strengthen mutual cooperation between governments and these sectors of the industries by actively involving certain key organizations representing these sectors in its activities.

1.14 THE STATE OF PREPAREDNESS IN THE MEDITERRANEAN REGION

When the Centre was set up in 1976, only a minority of coastal states were considered to have the means for actual oil spill response. Today, with the assistance of the Centre, the region has reached a certain level of preparedness, in particular for marine oil emergencies. This is shown in Table 1.7.

In addition to the state of preparedness of the individual coastal States to combat major accidental pollution, a number of subregional operational plans have also been established between neighbouring countries for example:

- France, Monaco, and Italy (RAMOGEPOL plan for the Ligurian Sea)
- Cyprus, Egypt, and Israel (for the southeast Mediterranean)

1.15 THE ROLE OF REMPEC IN THE FIELD OF PREVENTION OF MARINE POLLUTION FROM SHIPS

The new amendments envisaged for the Emergency Protocol are expected to expand the Centre's activities, which, in cooperation with the IMO, will be carried out as follows:

- At regional level:
  - To implement the Action Plan concerning the provision of adequate port reception facilities within the Mediterranean region (adopted in Cairo in December 1991)
  - To promote regional cooperation in the field of port state control
  - To promote regional cooperation in monitoring marine pollution from ships, especially illegal discharges, by enforcing the provisions of MARPOL 73/78 more rigorously
  - To develop technical guidelines for the establishment and operation of shore reception facilities for wastes generated by maritime-related activities

TABLE 1.7 The State of Preparedness in the Mediterranean Region

Coastal state	24h contact point[1]	Operational focal point	National contingency plan		Oil clean-up resources[3]		Party to emergency protocol	Sub-regional operational plans	Compensation[4]				Party to OPRC 90[4]
									CLC		Fund		
			Oil	HNS	Tier 1	Tier 2			69	92	71	92	
Albania		YES	DRAFT	DRAFT			YES		YES		YES		
Algeria	[YES]	YES	YES	YES	√		YES		d	YES	d	YES	
Bosnia and Herzegovina		YES					YES						
Croatia	[YES]	YES	YES	YES	√	√	YES		d	YES		YES	YES
Cyprus	YES	YES	YES		√	√	YES	YES	d	YES	d	YES	
Egypt	YES	YES	YES		√	√	YES	YES	YES	YES			YES
France	[YES]	YES	YES	YES	√	√	YES	YES	d	YES	d	YES	YES
Greece	YES	YES	YES	YES	√	√	YES		d	YES	d	YES	YES
Israel	YES	YES	DRAFT	DRAFT	√	√	YES	YES					YES
Italy	YES	YES	YES	YES	√	√	YES	YES	(d)	(YES)	(d)	(YES)	YES
Lebanon	[YES]	YES					YES		YES				
Libya		YES	DRAFT		√		YES						
Malta	YES	YES	DRAFT		√	√	YES		YES	YES	YES	YES	
Monaco	YES	YES	YES[2]	YES[2]	√	√	YES	YES	d	YES	d	YES	YES
Morocco	[YES]	YES	YES	YES	√	√	YES		YES		YES		
Slovenia	[YES]	YES	YES	YES			YES		YES		YES	YES	
Spain	[YES]	YES	YES	YES	√	√	YES		d	YES	d	YES	YES
Syria	YES	YES	DRAFT				YES		YES		YES		
Tunisia	[YES]	YES	YES	YES	√	√	YES		d	YES	d	YES	YES
Turkey	[YES]	YES	DRAFT		√		YES						

Source: ITOFP, 1998; REMPEC

1 [ ] = denotes when 24h contact point responsible for receiving marine pollution accident reports is a different authority to the operational focal point.

2 = included in national contingency plan for France.

3 = a Tier 1 capability indicates that most, if not all, of the oil ports and terminals in a country have resources to combat oil spills associated with routine oil transfer operations usually less than 50 barrels. A tier 2 capability indicates that there are combined resources in a country to respond to a spill (a few hundred barrels) affecting a wider area.

4 = as at July 2000; no Mediterranean coastal state has so far ratified the OPRC/HNS 2000 Protocol

d = denunciation effective.

(d) = Italy denounced membership of the CLC 71 of Fund 71 and became a member of the CLC 92 and Fund 92 in September 2000

- At national level:
  - To assist in the design and implementation of projects related to the establishment and operation of port reception facilities
  - To assist coastal states that face difficulties in ratifying, implementing, and enforcing the relevant international Conventions

1.16 CONCLUSION

Away from the spotlight of the mass media and the commonly adopted gloomy attitude of the Mediterranean being a dead sea, the coastal states and the European Union have, over the years, patiently and steadily woven the threads of their cooperation to implement a regional spill response program with a relatively modest budget. This program has had its successes, but there is still much to be achieved in areas such as maintaining the state of readiness for oil spill response, contingency planning for hazardous substances other than oil, and providing the trained manpower and necessary equipment to implement the plans once a marine chemical emergency occurs.

Future efforts dedicated to preventing marine pollution from ships will lessen the frequency and impact of marine emergencies. Future success in prevention, preparedness, and response is linked to a coordinated approach supported by technical and administrative means together with the continual commitment of states that will implement the measures they themselves adopt. Indeed, this may mean large financial investments for certain coastal states, but if innovative ways are sought to obtain the necessary finances to assist in implementing these measures, this may alleviate part of the burden.

1.17 REFERENCES

Beer-Gabel, J. 2000. "Conventions Régionales relatives à la lutte contre la Pollution des Mers: Panorama des conventions régionales à jour au 12 juillet 1999," *Environnement,* vol. 632, pp. 1–29.

Dobler, J. P. 1997. "Maritime Statistics Required to Monitor the Environmental Impact of Shipping," Paper presented at the International Maritime Statistics Forum, October.

European Commission (EC). 1995. "Conclusions and Recommendations," Regional Conference for the Development of Maritime Transport in the Mediterranean, Barcelona, Spain, April 27–28.

European Environmental Agency. 1998. *State and Pressures of the Marine and Coastal Mediterranean Environment,* Environmental Assessment Series No. 5, Office for Publications of the European Communities, Luxembourg.

Hazardous Cargo Bulletin. 1998a. "Gently Does it," pp. 39–44.

Hazardous Cargo Bulletin. 1998b. "Riders on the Storm," pp. 14–20.

Hulm, P. 1983a. "The Regional Seas Programme: What Fate for UNEP's Crown Jewels?" *Ambio,* vol. 12, pp. 2–13.

Hulm, P. 1983b. *A Strategy for the Seas—The Regional Seas Programme, Past and Future,* UNEP.

Inter-Governmental Maritime Consultative Organization (IMCO). 1969. "6th Session 1969 and 4th Extraordinary Session 1968 (resolutions 146–201)," in *Resolutions and Other Decisions of the Assembly.*

International Maritime Organization (IMO). 1979. "11th Session 1979 (resolutions 410–461)," in *Resolutions and Other Decisions of the Assembly.*

International Maritime Organization (IMO). 1985. *Report of the IMO/UNEP Meeting on Regional Arrangements for Co-operation in Combating Major Accidents of Marine Pollution,* IMO Headquarters, London, April 29–May 3.

International Maritime Organization (IMO). 1989. "Preventing Marine Pollution," *Focus on IMO,* September.

International Maritime Organization (IMO). 1990. "Implementation of Annexes III, IV and V of MARPOL 73/78 including Amendments to the IMDG Code to Cover the Pollution Aspects" (MEPC 30/11).

International Maritime Organization (IMO). 1991. International Convention on Oil Pollution Preparedness Response and Co-operation (OPRC).

International Maritime Organization (IMO). 1992. *Report of the IMO Sub-Regional Meeting of Government-Designated Experts of the Island States and Territories of the Wider Caribbean on Oil Spill Contingency Planning, Curacao,* Netherlands Antilles, November 23–27.

International Maritime Organization (IMO). 1993a. "World Maritime Day 1992; Marine Environment and Development: the IMO role," *IMO News,* no. 3, p. 12.

International Maritime Organization (IMO). 1993b. Resolution A.766 (18)—Navigation in the Strait of Bonifacio; A 18/Res.766, November.

International Maritime Organization (IMO). 1994. "OPRC Convention to Enter into Force," *IMO News,* no. 3, p. 12.

International Maritime Organization (IMO). 1995. *Manual on Oil Pollution: Contingency Planning.*

International Maritime Organization (IMO). 1997a. "30 Years on: What Has Happened Since the *Torrey Canyon, IMO News,* no. 1, p. 28

International Maritime Organization (IMO). 1997b. "Relations with the United Nations and the Specialized Agencies (C/ES.19/19(b)/1)," Note by the Secretary General submitted at the 19th Extraordinary Session of Council of the International Maritime Organization.

International Maritime Organization (IMO). 1998a. "The Shipwreck of the M/V *FENES:* Bacteriological and Chemical Aspects of Wheat Fermentation in a Marine Environment (LC/SG 21/6), submitted by France at the 21st Scientific Group Meeting of the London Convention, April 6–8.

International Maritime Organization (IMO). 1998b. "Flag State Performance Self-assessment Form Approved," *Focus on IMO,* no. 3, p. 2.

International Maritime Organization (IMO). 2000. "Status of Conventions: summary," http://www.imo.org/imo/convent/index.htm.

International Association of International Tanker Owners (INTERTANKO). 1992. "Regional Activities Programme for the Implementation of the Action Plan concerning the Provision of Adequate Port Reception Facilities within the Mediterranean Basin," Discussion Paper.

International Tanker Owners Pollution Federation (ITOPF). 1998. *Country Profiles: A Summary of Oil Spill Response Arrangements and Resources Worldwide; Mediterranean Region.*

Jeftic, L. 1992. "The Role of Science in Marine Environmental Protection of Regional Seas and their Coastal Areas: The Experience of the Mediterranean Action Plan," *Marine Pollution Bulletin,* vol. 25, pp. 66–69.

Jeftic, L., M. Bernhard, A. Demetropoulous, F. Fernex, G. P. Gabrielides, F. Gasparovic, Y. Halim, D. Orthon, and L. J. Saliba. 1990. "State of the Marine Environment in the Mediterranean Region," in *UNEP Regional Seas Report and Studies,* no. 132, UNEP, and *MAP Technical Report Series,* No. 28, UNEP, Athens.

Lloyd's of London. 1991. *Lloyd's Ports of the World,* Lloyd's of London Press.

Lloyd's of London. 1994a. "New Traffic Rules Win Early Backing," *Lloyd's List,* October 10.

Lloyd's of London. 1994b. "Pipeline Signals Breakthrough in Relations," *Lloyd's List,* October 12.

Lloyd's of London. 1995a. "Narrow Margin for Fatal Error," *Lloyd's List,* February 15.

Lloyd's of London. 1995b. "Lonely Are the Brave," *Lloyd's List: Black Sea Ports,* September, pp. 24–25.

Lloyd's of London. 1995c. "Energy Shifts in the Middle Eastern Export Pipeline," *Lloyd's List,* November 6.

Lloyd's of London. 1996. "New Rules Fail to Dispel Safety Fears," *Lloyd's List: Turkish Shipping,* January, p. 15.

Lloyd's of London. 1997a. *Lloyd's Ports of the World,* Lloyd's of London Press.

Lloyd's of London. 1997b. "Strait Talking," *Lloyd's List: BIMCO Istanbul,* June, pp. 4–5.

Manos, A. 1991. "An International Programme for the Protection of a Semi-Enclosed Sea — The Mediterranean Action Plan," *Marine Pollution Bulletin,* vol. 23, pp. 489–496.

Matthews, G. J. 1992. "International Law and Policy on Marine Environmental Protection and Management—Trends and Prospects," *Marine Pollution Bulletin,* vol. 25, pp. 1–4.

Oil and Gas International. 1994. "The Politics of Pipelines," in *Oil and Gas International, Summer 1994,* pp. 76–78.

OTSOPA. 1995. "Information on the U.K. National Strategy, Legislation, Organization, Ship and Aircraft and Equipment (BAWG OTSOPA 95/3/4-E)," Meeting of the Working Group on Operational Technical and Scientific Questions concerning Counter-Pollution Activities, Dunkirk, May 9–12.

Pinardi, N., P. De Mey, G. M. R. Manzella, and A. Ruiz de Elvira. 1997. "The EuroGOOS Mediterranean Test Case: Science and Implementation Plan," in *Operational Oceanography: the Challenge for European Co-operation,* Elsevier Science, Amsterdam.

Regional Marine Pollution Emergency Centre (REMPEC). 1998a. *IMO/UNEP Regional Information System,* Part C, Databanks, Forecasting Models and Decision Support Systems, Section 4, List of Alerts and Accidents in the Mediterranean, REMPEC.

Regional Marine Pollution Emergency Centre (REMPEC). 1998b. *IMO/UNEP Regional Information System,* Part A, Basic Documents, Recommendations, Principles and Guidelines concerning Accidental Marine Pollution Preparedness, Response and Mutual Assistance, REMPEC, January.

Regional Marine Pollution Emergency Centre (REMPEC). 1999. *IMO/UNEP Regional Information System,* Part B, List and Inventories, Section 1, Directory of Competent National Authorities in Charge of Accidental Marine Pollution Preparedness, Response and Mutual Assistance and Other Relevant Information, REMPEC, March.

Sainlos, J. C. 1992. "Regional Co-operation in Case of Emergency and Role of the Regional Marine Pollution Emergency Response Centre for the Mediterranean Sea," *Atti del Convegno Internazionale—Mediterraneo e Caraibi; Due Mari in Pericolo? Sversamenti Accidentali di Idrocarburi ed Emergenze Causate dalle Alghe,* July 4.

Saliba, L. J. 1989. "Regional Measures for Marine Pollution Control in the Mediterranean," *Marine Pollution Bulletin,* vol. 20, pp. 12–17.

The Siren—News from UNEP's Oceans and Coastal Areas Programme. 1991.

Skinner, S. K., and W. K. Reilly, 1990. "Oil Spill Contingency Planning—National Status," Report to the President prepared by the National Response Team, October.

Transport Canada. 1974. *Joint Canada/United States Marine Pollution Contingency Plan for Spills of Oil and Other Noxious Substances,* Transport Canada/Canadian Coast Guard.

United Nations Environment Programme (UNEP). 1978. *Mediterranean Action Plan and the Final Act of the Conference of Plenipotentiaries of the Coastal States of the Mediterranean Region for the Protection of the Mediterranean Sea.*

United Nations Environment Programme (UNEP). 1989. "Transport in the Mediterranean," in M. Grenon and B. Michel, eds., *Futures for the Mediterranean Basin: The Blue Plan,* Oxford University Press, Oxford.

United Nations Environment Programme (UNEP). 1997a. *Mediterranean Action Plan and Convention for the Protection of the Marine Environment and the Coastal Region of the Mediterranean and Its Protocol,* Athens.

United Nations Environment Programme (UNEP). 1997b. *The Mediterranean Action Plan—A Contribution to Sustainable Development in the Mediterranean Basin.*

United Nations Environment Programme (UNEP). 2000. "Status of Signatures and Ratifications of the Convention for the Protection of the Mediterranean Sea against Pollution and its Protocols," http://www.unepmap.org/.

Vallega, A. 1995. "Towards Sustainable Management of the Mediterranean Sea," *Marine Policy,* vol. 19, pp. 47–64.

World Bank/European Investment Bank. 1990. *The Environmental Programme for the Mediterranean—Preserving a Shared Heritage and Managing a Common Resource,* World Bank/European Investment Bank.

CHAPTER 2
AWARENESS AND PREPAREDNESS FOR EMERGENCIES AT LOCAL LEVEL

Ernst Goldschmitt
United Nations Environment Program, Paris

2.1 WHAT IS APELL?

Awareness and Preparedness for Emergencies at Local Level (APELL) is a tool developed by the United Nations Environment Program's Division of Technology, Industry and Economics office (UNEP DTIE) in conjunction with governments and industry. Its purpose is to minimize the occurrence and harmful effects of technological accidents and emergencies, particularly, though not exclusively, in developing countries. APELL provides a well structured, detailed description of how to develop a coordinated, integrated, and well functioning emergency response plan for local communities. The strategy of the APELL approach is to identify and create awareness of risks in an industrialized community, to initiate measures for risk reduction and mitigation, and to develop preparedness for emergencies in industries, the local governments, and the population.

APELL was launched in 1988 following various industrial accidents that had adverse impacts on health and the environment. Well-known examples of such accidents include Bhopal in 1984 and the Sandoz warehouse fire near Basel in 1986, which resulted in extensive contamination of the Rhine.

APELL can be useful in any situation that requires joint planning for disasters by several parties, e.g., government, industry, and local communities. Being aware and prepared means having workable, realistic plans if an accident occurs. It also means creating a better understanding of local hazards, which in turn should lead to action designed to prevent accidents from happening at all.

It is now universally acknowledged that every disaster, whatever the cause, may have an environmental impact. While some major industrial accidents can be contained within the boundaries of the plant, in other cases, there are impacts on the surrounding neighborhood, with adverse short- or long-term consequences affecting life, life-support systems, society, or property. This is even more so for accidents arising from transport of dangerous goods, e.g., by road, rail, or pipeline, through or close to populated areas, since by definition, there is no boundary fence in these cases. The extent of the losses from these accidents depends largely on the actions of the first responders to an emergency, both at the scene of the accident and within the surrounding community.

Clearly, adequate response to such situations calls for cooperation between various institutions and individuals. This can be achieved only if there is awareness within the community of possible risks and of the need for joint preparedness to cope with the consequences of these risks.

2.2 THE OBJECTIVES OF THE APELL HANDBOOK

The *APELL Handbook* describes a process for improving community awareness and emergency preparedness and achieving cooperation between the various parties involved.

The *APELL Handbook* is a detailed guide to a process that trains the reader to:

- Develop and practice a community emergency response plan for any type of accident or disaster.
- Consider all kinds of risks, i.e., hazardous installations, storage facilities and transport of hazardous goods, earthquakes and flooding.
- Identify and initiate measures to reduce risks (see Fig. 2.1 for probability of incidents and seriousness of risks).
- Provide information to the community, thus creating community awareness.
- Combine all technical resources and expertise available in an industrialized community for responding to emergencies.

Any member of the following involved groups may initiate the APELL process: industry managers, local authorities, or community leaders. However, there must then be direct and close interaction between the representatives of the three partners. A bridge is created by means of the APELL Coordinating Group. The Coordinating Group is the critical management team that develops and oversees the APELL process at the local level. Their job is to gather facts and opinions, assess risks, initiate measures for risk reduction, evaluate approaches, and generally organize the personnel and the resources available in the community to produce an emergency response plan.

APELL addresses all emergencies with potential for fire, explosion, spills, or releases of hazardous materials. The possibility of combination accidents should be noted at this point—

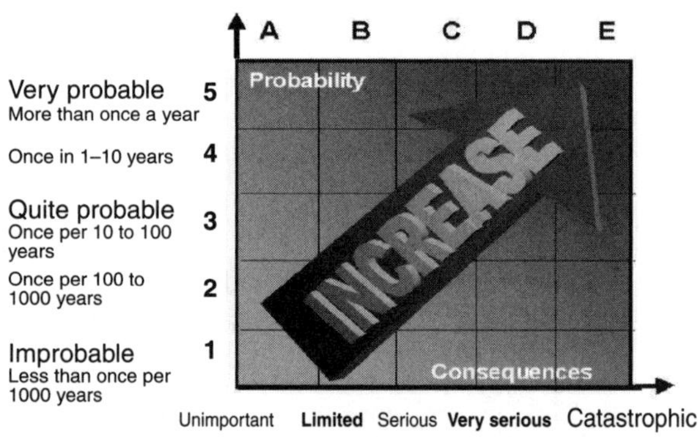

FIGURE 2.1 The key to awareness and preparedness is the identification and evaluation of risks, the probability of an incident, and the potential consequences.

for example, an earthquake that triggers an emergency in a refinery or chemical factory. The determination of which potential hazards should be covered by the APELL process is in principle the result of a risk assessment. In most cases, however, common sense will be sufficient to identify the facilities or areas that present a risk of a major accident. The criteria of accidents (lists of substances and threshold levels) given in international or national regulations or recommendations may also provide guidance.

APELL is flexible. Countries differ in culture, value systems, legal and regulatory requirements, community infrastructure, and response capabilities and resources. Their industries present different potential dangers. However, they have one common need—the need to cope with a major technological accident affecting a local community.

The *APELL Handbook* provides the basic concepts for the development of action plans that can be adapted to local conditions. No legislation or regulations are needed. Since the lessening of health and environmental impacts depends on the speed and scope of the initial local response, local participation is emphasized. It is recognized, however, that national governments and the chief executive officers of industries have a fundamental role in promoting and supporting these local efforts. Industry associations also have an important part to play in encouraging industry participation.

2.3 THE APELL PARTNERS AND THEIR RESPONSIBILITIES

When analyzing past disasters that have caused traumatic loss of lives, serious health problems, environmental damages, or property losses, it becomes obvious that these damages could have been avoided to a large extent. Some of the most common observations are:

- Lack of awareness of actual risks, mainly within the population living adjacent to such risks.
- Inadequate preparedness of emergency responders and the entire organization responsible for response, mitigation, and relief, particularly concerning technological risks (i.e., risk-specific training and suitability of available equipment, coordination between the acting agencies, practising of emergency plans, provision of medical services).
- Inadequate capacity for handling emergencies at the local level. Any kind of emergency requires the most competent, efficient, and immediate response capacity in the community. Emergency response strategies have to focus on strengthening these capabilities at the local level as the main key for effective mitigation.

At the local level, three very important partners must be involved if APELL is to succeed (see Fig. 2.2):

- *Local authorities.* These may include provincial, district, and city or town officials, either elected or appointed, who are responsible for safety, public health, and environmental protection in their area.
- *Industry.* Industrial plant managers from either state-owned or private companies are responsible for safety and accident prevention in their operations. They prepare specific emergency measures within the plant and review their application. But their responsibilities do not stop at the boundary fence. As leaders of industrial growth and development, they are in the best position to interact with leaders of local authorities and community groups in order to create awareness of how the industrial facility operates and how it could affect the environment and to help prepare appropriate community response plans in the event of an emergency. The involvement and active participation of the plant's workforce is also very important.

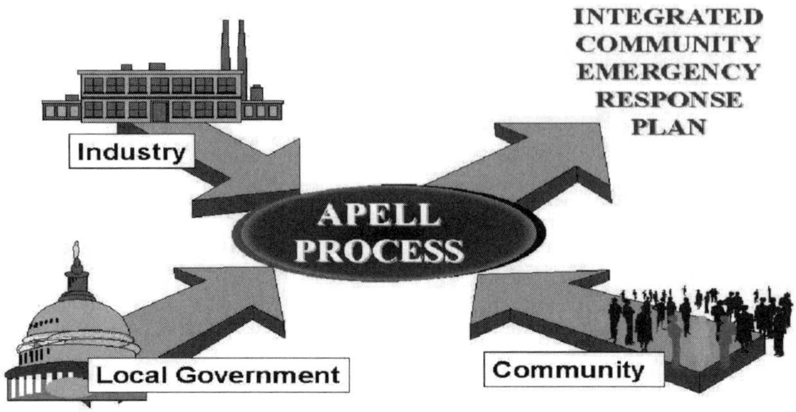

FIGURE 2.2 The basic principles of the APELL process: communication and cooperation.

- *Local community and interest groups.* These include environmental, health, social, and religious organizations and leaders in the educational and business sectors, and the media, who represent the concerns and views of their members or constituents in the community.

There are other partners, such as nongovernmental organizations (NGOs). The APELL process is designed to work with other initiatives to reduce risks and their consequences, *not* to replace them.

2.4 THE APELL PROCESS

The process itself in action is the mainspring of the emergency response plan. The APELL Coordinating Group, in which all stakeholders should be represented, does not itself have any operational role during an emergency but exists to prepare the various partners for their tasks if an accident does occur. Members must be able to command the respect of their various constituencies and be willing to work together in the interests of local safety, well-being, and property security. In particular, local plant managers need to be active participants and local authority and community leaders need to know that they are acting with the blessing and full authority of the most senior managers in their companies. Ideally, the leader of the Coordinating Group should be able to motivate all sections of local society, regardless of cultural, economic, educational, and other differences, and to ensure their cooperation. This needs to be kept in mind when choosing the leader.

The APELL process consists of 10 steps.

1. Identify the emergency response participants and establish their roles, resources, and concerns.
2. Evaluate the hazards and risks that may result in emergency situations in the community.
3. Have participants review their own emergency response plans to ensure a coordinated response.
4. Identify the required response tasks not covered by existing plans.
5. Match these tasks to the resources of the identified participants.
6. Make the changes necessary to improve existing plans, integrate them into an overall community plan, and gain agreement.

7. Commit the integrated community plan to writing and obtain approval from local governments.
8. Educate participating groups about the integrated plan and ensure that all emergency responders are trained.
9. Establish procedures for periodic testing, review, and updating of the plan.
10. Educate the community about the integrated plan.

The APELL process is designed to build on any and all existing emergency plans to create a single coordinated local plan. There may be national government emergency plans in place, but there is always the need for an effective structure at the local level. Industrial facilities should already have on-site emergency plans. Local authorities and rescue services should have plans to deal with the consequences of major emergencies. Local hospitals should certainly have their own major accident plans for dealing with large numbers of seriously injured people. The APELL process ensures that all existing plans contribute to the overall integrated, cooperative plan.

2.5 COMMUNITY AWARENESS

Citizens have the right to know if potentially hazardous materials are being produced, stored, used, or transported in their communities. There is nothing mysterious about a community awareness program. A fenced-in industrial plant can look threatening to the public, but much of the threat disappears when people know what the plant uses and manufactures and that it has a good safety record and an effective emergency plan. People need to be informed about potential risks in order to understand why an emergency plan has been established, how it works, and what action they are expected to take in an emergency.

2.6 APELL WORLDWIDE

Industries all over the world, particularly the chemical and gas industry, have cooperated with UNEP DTIE in supporting the application of APELL.

The APELL concept has been successfully introduced in more than 30 countries and over 80 industrialized communities worldwide: in Latin America (i.e., Brazil, Colombia, Chile, Argentina), in Asia (i.e., China, India, Thailand, Indonesia, and Korea) and in Russia. The guiding principles of the APELL process for emergency planning are also practised in the United States and Canada.

Remarkable changes have occurred in many of the communities that have implemented APELL, such as a general safety consciousness and an increased concern for environmental issues.

APELL is part of a broad cleaner and safer production program that UNEP DTIE has launched with the objective of promoting worldwide sustainable production and consumption patterns.

2.7 THE APELL HANDBOOK AND NEWSLETTER

More than 10,000 copies of the *Apell Handbook* have been distributed throughout the world. The APELL Newsletter is published twice a year as a supplement to UNEP DTIE's quarterly publication *Industry and Environment Review*.

2.8 UNEP DTIE TECHNICAL REPORTS (RELATED TO APELL AND TECHNOLOGICAL ACCIDENTS)

- TR 3, *Storage of Hazardous Materials*
- TR 8, *International Directory of Emergency Response Centres* (in cooperation with OECD, second edition in preparation)
- TR 12, *Hazard Identification and Evaluation in a Local Community* (prepared with help from Sweden)
- TR 19, *Health Aspects of Chemical Accidents*
- TR 21, APELL *Annotated Bibliography* (prepared with help from Canada)
- TR 28, *Safety, Health and Environmental Management Systems* (in preparation)
- TR 35, *TransAPELL*—APELL for accidents arising from dangerous goods transport
- *APELL for Port Areas,* prepared in conjunction with the International Maritime Organization (IMO) and published by IMO
- *APELL World-wide,* national accounts of the development of APELL in 12 countries
- *Management of Industrial Accident Prevention and Preparedness,* a training resource kit for use in universities and colleges
- *LP Gas Safety,* Guidelines for Good Safety Practice in the LP Gas Industry, prepared in conjunction with the World LPG Association.

For copies of these publications or for more information about UNEP, contact: www.uneptie.org.

Forthcoming publications:

- *APELL in the Mining Industry,* with a special focus on tailing dam failures.

CHAPTER 3
RESPONSE OPTIONS FOR ACCIDENTS AND SPILLS OF HAZARDOUS MATERIALS IN FRANCE

Michel Marchand
Rue Alain Colas, BP 20 413, 29604 BREST (France)

3.1 INTRODUCTION

There are many types of accidents and spills involving hazardous materials. In 1992, the Ministry of the Environment listed 710 such accidents, which resulted, separately or simultaneously, in fires and explosions (47%), water pollution (30%), atmospheric pollution (20%), ground pollution (19%) (BARPI, 1992). Fifty-seven percent of these accidents were from industrial establishments. The transport of hazardous materials is often involved (16% of cases). Accidental pollution of unknown origin, which essentially damages natural environments, is caused by a significant number of the cases listed (15%). Agriculture is the source of 6% of these accidents, and 5% are from various trades and activities.

3.2 TRANSPORT OF HAZARDOUS MATERIALS

The transport of hazardous materials represents 45% of the major accidents listed by the Organization for Economic Cooperation and Development (OECD) over the last decade. In France, there were 234 accidents involving the transport of hazardous materials in 1996, of which 40% caused consequences linked to the type of materials being transported (explosion, fire, gas leak, spreading, and related injuries) (MTMD, 1997).

To facilitate the intervention of firefighters at the site of such accidents, special Conventions and Agreements have been established between the public authorities and the industrial manufacturers, users, or carriers of chemical products. As a result, an Aid Convention, TRANSAID, was signed in 1987 by the Interior Minister and the French Chemical Industries Union (UIC), requiring the chemical industries to provide technical aid to the public authorities at the time of an accident involving hazardous materials (Le Guillier, 1997). This includes accidents that happen off industrial sites or outside warehouses, either during transport, by road, rail, or on inland waterways, or in transit, and during handling in stations, ports, airports, or on road platforms. The TRANSAID Convention applies only to the authorities responsible for aid intervention. The substances concerned (identifiable by their UN number or alphabetical order) belong to those classified as explosives, liquid gases, flammable liquids, toxic products, and corrosive products. Radioactive materials, for which there are specific regulatory measures, are not included in the aid Convention. In all, more than 1,000 products are covered by the TRANSAID Agreement.

Firms participating in the aid Convention do so on a voluntary basis (generally adopted within the framework of a commitment to progress). They offer help according to their competence, either by giving information by phone on the product concerned in the accident (role of advisor), or by helping directly through providing personnel and/or material at the site of the disaster (intervening role). Advice by telephone generally involves providing the safety data sheets for the products and specific advice related to the circumstances of the accident.

The second level of aid, on-site intervention, necessitates an application from the public authorities for the requisitioning of both personnel and material means from the voluntary firms. A total of 173 firms are members of the TRANSAID Convention. Moreover, specialized aid agreements have already been signed by the Interior Minister and the professional unions of manufacturers of the following chemicals: chlorine, cyanhydric acid, ethylene, alkyl lead compounds, and liquefied gases.

These agreements, dating from the 1970s, are integrated into the TRANSAID Convention, which is the French Chemical Industries Union's (UIC) commitment to progress ("responsible care"), which applies not only to French authorities in the case of an accident during the transport of hazardous materials in France, but also to authorities in European countries within the framework of the ICE network (International Chemical Environment), started by the CEFIC (Conseil Européen des Industries Chimiques). In the latter, the TRANSAID file is used by the Cedre (Centre de documentation, de recherche et d'expérimentations sur les pollutions accidentelles des eaux), which is, in the name of the UIC, the French national point of contact in the ICE network.

In addition, the CEFIC has provided the public authorities of the member states of the European Union with a system of intervention cards used during emergencies for accidents involving the transport of hazardous materials. This system, called ERIC (Emergency Response Intervention Cards), provides instructions for fire and emergency services, either on paper or on computer software (http://www.ericards.net). The distribution of the software in France is the result of cooperation between the public authorities (community financing), CEFIC (conception), the UIC (reproduction, distribution), and the national federation of firefighters (advice). The ERIC system is complementary to the TRANSAID aid system, offering to evaluate the situation and risks involved or to provide operational assistance in the case of an accidental situation. On the basis of the name of the substance or its UN number, it is possible to obtain the following information on more than 1,800 substances:

- Characteristics of the substance
- Potential risks
- Personal protection
- General intervention
- Control of spreading
- Firefighting
- First-aid treatment in cases of intoxication
- Recuperation
- Cleanup precautions after an intervention
- Cleaning of equipment used

3.3 TRANSPORT BY RAIL

Dangerous goods are not the cause of accidents, but risks are increased when the rail system breaks down. For this reason, the transport of hazardous materials by rail is the object of specific actions and measures taken by the French Rail Company (SNCF) together with the public authorities, professions involved, and other European rail networks (International Union of Railways, 1995). In 1993, the SNCF transported 17.6 million tons (Mt) of hazardous materials in approximately 350,000 shipments, most of them in tankers (95%). The main products transported are liquid oil products (7.4 Mt), chemical products (6.4 Mt), and liquid oil gases (1.6 Mt) (SNCF, 1994). The level of safety for transporting these goods by rail cannot be judged by the number of accidents involving hazardous materials, estimated at 66 accidents over a period of 10 years (1982–1991). Over the same period, there were 2,080 accidents involving road transport of hazardous materials, 15 times more than the amount using rail transport. However, the study of risks of hazardous materials transported by rail, especially on the Paris–Lyon–Marseille line, shows that the probability of a train accident on this line is not to be ignored (0.1 accident/year) (Chapron et al., 1994).

The demand for safety led the SNCF to create an operational service, the "Présence Frêt," in 1993, whose two aims were to constantly follow the movement of rail cars loaded with hazardous merchandise and to provide useful information to the response team at the time of accidents or incidents involving hazardous materials. The service is located in Dijon, a central point of national hazardous material traffic, and consists of about 20 people linked to the central SNCF computers and computerized software dealing with hazardous products. Events are classified into three levels, corresponding to the seriousness of the events: *incident* (small leak, seepage, drop by drop), *accident* (smoke, odor, shock, derailing), and *serious accident* (accident made worse by fire, explosion, serious derailing, pollution, or victims).

3.4 WATER POLLUTION

When water pollution is caused by an accidental spill of chemical products or an accident involving hazardous materials, the public authorities can call the Cedre for assistance. An operational service, available 24 hours a day, provides permanent and free assistance in the form of information on spilled products (gasoline, chemical products) and on actions to take in cases of accidental water pollution (Chemical Industries Union, 1998). The Cedre has the computerized means of answering such demands. The Cedre is also the National French contact within the framework of the European aid network ICE, established by the CEFIC for accidents involving hazardous materials. The Cedre, which depends on the Ministry of the Environment, has direct links with operators in charge of steps taken against accidental pollution of sea and mainland water. The Cedre was set up more than 20 years ago following the major oil pollution of the *Amoco Cadiz* on the French coast. Its present interventions cover accidental oil and chemical pollution in sea and mainland waters.

3.5 STATISTICAL ANALYSIS OF THE ACCIDENTS

In 1992, the French Ministry of the Environment created the Bureau d'analyse des risques et pollutions industrielles (BARPI). This Department centralizes and analyzes data related to accidents, serious pollution, and fires linked to what may be industrial, commercial, agricultural, or other types of activities, particularly the transport of hazardous materials (BARPI, 1992). Information is listed on the ARIA database (Analyse, Recherche et Information sur les Accidents) started in 1993. The ARIA database allows for a multi-criteria selection and statistical analyses on one or several of the 2,200 descriptors used. In 1992, 710 accidents involving hazardous materials were listed in France, the origins and effects of which are stated in the introduction. This statistical analysis of accidents reinforces a program of risk prevention using an assessment of repeated events.

In addition to its national function, the OECD, the Economic Commission for Europe, and the United Nations (EC-UN) have also made BARPI responsible for providing and promoting information given to these organizations by member countries. It groups together a subset of 40 countries in Western Europe, Eastern Europe, the United States, and Canada. The BARPI directly cooperates with the European Community system, which, according to European Directive SEVESO, obliges the member states to submit an analysis report in cases of major accidents. For that purpose, the database MARS (Major Accident Reporting System), which is run by the community research center ISPRA (Italy), allows information concerning reported accidents to be centralized (Kirchsteiger, 1997).

3.6 REFERENCES

Bureau d'analyse des risques et pollutions industrielles (BARPI). 1992. "Analyse des Risques et Pollutions Industrielles," Ministère de l'Environnement, Service de l'environnement industriel, BARPI, Paris, France.

Chapron, M., F. Dischert-Rancillac, and M. Fargeaudou. 1994. Trafic de matières dangereuses sur l'axe ferroviaire PLM. Etude de risques, *Sécurité,* no. 13, pp. 27–31.

Chemical Industries Union. 1998. "Le rôle du Cedre en cas de pollution accidentelle des eaux," Union des Industries Chimiques (UIC), La lettre du Département Technique, Paris, France, no. 52, pp. 3–4.

International Union of Railways. 1995. "Le transport ferroviaire des marchandises dangereuses en Europe," Union Internationale des chemins de fer, Paris, France.

Kirchsteiger, C. 1997. "The Functioning of the Major Accident Reporting System in the European Union Seminar on Lessons Learned from Accidents," Linz, Austria, October.

Le Guiller. 1997. "TRANSAID: une assistance de l'industrie chimique dans les accidents de TMD," *Le Sapeur-Pompier,* no. 887, pp. 772–775.

Mission des Transports des Matières Dangereuses (MTMD). 1997. "Les accidents de transport de marchandises dangereuses par voie routière et ferroviaire en France," Bilan 1996, Ministère de l'Equipement, des Transports et du Logement, Paris, France.

SNCF. 1994. "Les exemples de progrès du transport des marchandises dangereuses (TMD) par la SNCF," Doc. SNCF, Paris, France.

CHAPTER 4
HAZARDOUS MATERIALS EMERGENCIES IN NEW ZEALAND

A. J. (Tony) Haggerty
Hazardous Substance Advisor, Auckland Region, New Zealand Fire Service

4.1 NEW ZEALAND—A BACKGROUND

New Zealand is a collection of South Pacific islands about the size of Great Britain, 2,000 km east of Australia. The two main islands are simply known as the North Island (pop. 2.8 million) and the South Island (pop. 0.9 million), and to the south of the South Island is Stewart Island (pop. 400) a haven for fishermen and tourists.

For many years New Zealand was perceived as a traditionally agricultural economy that supplied the United Kingdom with lamb, beef, and dairy products. There were few major industries, and those that did exist consisted of dairy factories (milk, butter, and cheese production) and freezing works (abattoirs). The only large-scale chemicals used were nitric acid, sodium hydroxide, and sodium hypochlorite for cleaning the dairy factories and ammonia in the refrigeration systems of freezing works. Although the economy was predominantly agricultural, chemicals were being used in other areas, but they were seen as part of the agricultural scene. Sulphur was imported to make sulphuric acid, which was then converted to ammonium sulphate. Ammonium nitrate was used to fertilize the soil, and selenium was added to soils in the volcanic central plateau of the North Island to bring them from their barren state to a level of productivity. This allowed the planting of pine trees, which led to the development of a timber industry, which in turn led to a paper-making industry using tons of chemicals, including chlorine.

Despite the predominance of agriculture, there are pockets of heavy industry scattered through the country. There is an aluminum smelter at Bluff in the deep south and an oil refinery at Marsden Point in the north. Just south of Auckland there is a steel mill extracting iron directly from the iron sands of the west coast. In 1964 Dow formed an alliance with a local manufacturer, Ivon Watkins (Ivon Watkins Dow [IWD]), to manufacture pesticide chemicals in the heart of the New Zealand dairy industry in the province of Taranaki in the western North Island. Here, among other things, they produced the infamous chloro-phenoxies 2,4 D and 2,4,5 T. 2,4,5 T was particularly essential to New Zealand farmers because it was the only economical means of killing woody pests such as gorse. Gorse was imported from England as a hedging material that because of its thorns does not need to be particularly big to keep stock under control. In the relatively mild climate of New Zealand, with very few severe frosts to kill it off, it rapidly covered hillsides and paddocks and threatened to drive out the stock. New Zealand farmers clung to the use of this very high-purity 2,4,5 T for as long as possible. IWD was one of the last manufacturers in the world to succumb to international pressure to have it banned.

The oil crisis of the mid-1970s prompted the development of the gas and oil fields that had been known to exist in the Taranaki area for over 100 years. Suddenly this most rural area of New Zealand, which had been known for its butter and cheese, suddenly became the energy capital of the country. Two major gas and oil wells, Maui and Kapuni, were opened up, and natural gas, liquid petroleum gas, and some light crude were pumped to processing plants. The light crude is sent off to supplement the imported crude at the country's only oil refinery, in Whangarei. The natural gas is sent to gas utilities in the North Island, where in the early days much of it was reformed to be fed into the low-pressure gas streams from the old gas works. Some of the methane is sidetracked and used in the manufacture of methanol, which has become a major export item. More methane is converted in a unique process to synthetic gasoline. This is probably the only plant of this type in the world.

The liquid petroleum gas (LPG) was initially distributed by road tanker, primarily for use as a heating fuel where natural or coal gas are not reticulated. When the price of gasoline rises, LPG is used as an automotive fuel at a savings of about a third. Compressed natural gas (CNG) is also used when the savings can be as much as 50%. This booming industry led to the establishment of gas storage, both LPG and CNG, at service stations throughout the country, much to the concern of surrounding residents.

In the early 1980s, there was a strong anti-LPG lobby from those who were aware of incidents such as those in Kingman, Arizona, Waverley, Tennessee, and the Spanish campsite disaster at Los Alfraques. The New Zealand Fire Service came under particular pressure because the public looked to it as the expert in handling emergencies. In response, the Fire Service produced operational orders and guidelines that recommended evacuation distances of up to a mile based on the effects of these incidents. This did little to allay public fears because no one wanted to live within a mile of a proposed two-ton (1,200 U.S. gallon) tank. Nobody seemed to appreciate that the incidents they read about overseas usually involved much larger quantities—i.e., 20 tons at Los Alfraques and 90 tons at Kingman.

Even firefighters, to whom this was all quite new, were concerned and took every precaution at even the smallest of incidents.

Another byproduct of the gas industry, carbon dioxide, was separated from the hydrocarbons and was used to make urea fertilizer. The hydrogen byproduct of this reaction was combined with atmospheric nitrogen to produce ammonia.

It was hoped at the time that this sudden surge in chemical would turn around New Zealand's economy. Unfortunately, by the time it was completed, the oil price had dropped and the great bonanza never materialized. Nevertheless, it did provide New Zealand with a degree of buffering from world prices, but not the complete independence that was hoped for.

4.2 THE NEW ZEALAND FIRE SERVICE

Fire departments were formed in many small towns in New Zealand in the early 1800s. The first municipal fire department appeared in Auckland, New Zealand's largest city, around 1850 and vied with the old insurance company brigades until 1874, when the two groups formed one brigade with some of their funding coming from insurance companies and the balance from local councils.

In the late 1940s a serious department store fire in Christchurch, with a severe loss of life, led to the formation of a national body, the New Zealand Fire Service Council. Its role was to coordinate the brigades and introduce standardized equipment. In 1976, following another series of fires where brigades failed to liaise and request support, the brigades were amalgamated into a national Fire Service.

The Fire Service Act of that time gave chief officers the powers to protect life and property from fire and to act where life and property were in danger but with no direct reference to hazardous materials. At this time chemical incidents started to be recognized and the Fire Service started to attend motor vehicle accidents and perform extrications as a matter of course, although they had attended and dealt with spills for many years before.

In 1990 the Act was revised to give the Fire Service full power to act at hazardous materials emergencies.

The Police are still legally the prime agency in dealing with hazardous materials emergencies, although in practice the New Zealand Fire Service actually handles the spills while the Police handle any breaches of legislation and control the surrounding area, giving the Fire Service a secure area to work in.

Hazardous materials incidents, as we would call them now, were apparently virtually unknown in New Zealand before 1973. However, a stroll through local history references and fire brigade records show that there were several "chemical" fires, at least in the Auckland province. One of the earliest occurred in 1894.

4.3 EARLY CHEMICAL FIRES

The first recorded hazardous materials emergency in New Zealand occurred in April 1894, when a fire broke out in Kempthorne and Prossers drug and pharmaceutical warehouse. The New Zealand Herald reported: "A lead of hose was passed in, but though the branchmen wore respirators, the fumes of the acid well nigh suffocated them, and they had to be relieved by others every few minutes and led out into the open air."

The following day the paper reported that several of the firemen were "temporarily disabled from their normal duties." The paper went on to suggest that "there should be a bylaw or other provision to prevent the storage of combustible acids in the heart of the city."

In 1899 firefighters suffered injury to their eyes from acid used inadvertently to fight a fire. They drafted from a 400-gallon tank on the premises, presuming it to be water, but in fact it was an acid solution.

In 1928 a major fire occurred at an oil refinery less than half a mile from downtown Auckland. Storage tanks ignited, and the fire was so severe that one firefighter lost his life.

The next 35 years seem to have been quiet, but then, from 1963 until 1973, a series of fires occurred involving hazardous materials. Even so, they caused little comment for their chemical content.

In 1963 a fire was reported on the MV *Suffolk* as it approached Auckland harbor. The ship was carrying general cargo, chemicals, and explosives. The Auckland brigade did not have a fire boat, so firemen went to sea on harbor tugs and dealt with it in the same way as any other fire at that time, by pouring on water. Fortunately, the fire was confined to a single hold, which contained only jute sacks, cork, and timber. The chemicals on board were ammonium chloride and sodium carbonate, which other than making access difficult were unaffected. Fortunately, the explosives were not affected.

In 1966 a 4.5-million liter capacity tank exploded while being filled with kerosene from the SS *Korenia*. There were 1.2 million liters of fuel in the tank at the time. Although the tank had two fixed foam pourers, one was damaged in the blast and so firefighters supplemented the foam attack by using hand-held deliveries fired through holes in the roof. The alarm had been raised at 6:15 on a Sunday evening and was under control by 7:38, with damage limited to the one tank.

In 1968 a shed full of black powder exploded at an ammunition store quite close to the city and beside the city jail at Mount Eden. Fortunately, the store had been placed in a natural extinct volcanic crater and the force of the blast went largely upwards. Damage to

FIGURE 4.1 Winstone's oil fire, Freemans Bay, 1928. Fireman R. Beuth killed while fighting fire.

surrounding buildings including the jail was minimal, although debris was scattered for some distance.

In 1970 a chemical warehouse and plastic fabricator caught fire, completely destroying the building. The site was owned by a subsidiary of New Zealand's only agrichemical manufacturing company at that time. Besides the plastic manufacturing, the warehouse contained a range of pesticides, about 2,000 aerosols and drums of general chemicals, including some acids and alkalis.

Fumes and thick smoke filled the air around the plant, but even so this provoked very little comment about toxic smoke after the event, unlike today's fires.

Thirteen firefighters, one ambulance officer, and three bystanders were hospitalized after exposure to the fumes but released the next day. After discussion with the National Poisons Centre, the hospital medical superintendent issued a statement declaring that the fumes were an irritant rather than toxic. This indicates the low level of concern for the environment and personal well-being 30 years ago.

What was also interesting in conversations with firefighters and their wives years after the fire were the health effects, such as lethargy, listlessness, and irritability, suffered by some of them for a long time afterwards. Firefighters noted gaps in their memory of events after the fire, and yet no mention was ever made of chemical poisoning. Much of this was put down to the stresses of attending several large fires in extremely hot weather. The brigade had just been restructured following a long fight for improved working hours and pay.

In 1971 the lid blew off a 90,000-liter tank containing vinyl acetate monomer. It was later thought that the vapor above the 22,000 liters of liquid in the tank started to polymerize after moisture got into the tank through a defective drier on a tank breather. Although the tank was small diameter, the lid came right off and drifted like a frisbee across the tank farm, fortunately without hitting any other tanks.

In this case there were no fixed foam lines and other chemical tanks were situated quite close by. Because aerial application of foam from hydraulic platforms was only moderately successful, two firefighters advanced a foam line up an outside staircase and dropped it over the rim of the tank. It was secured and the crew withdrew. It proved very effective and the fire was extinguished in under 15 minutes. The firefighters involved received recognition awards for their bravery.

4.3.1 Parnell Fumes Incident—1973

Parnell is an inner city residential suburb close to the wharf. In 1973 it was still reasonably populated, unlike the commercial area it has since become.

This is the first solely chemical incident in which the brigade was involved and is still viewed as the landmark incident that led to the revision of hazardous substances legislation in New Zealand.

In February 1973 the cargo ship the SS *Good Navigator* had left San Francisco with a mixed cargo including drums of a cotton defoliant that were stored on the forward deck. Two days out, she struck a typhoon-strength storm that dislodged some deck cargo and damaged the drums of defoliant. The damage to the vessel was negligible and she steamed on to Auckland, her first destination.

On arrival, a strange odor was noticed and various authorities were called in for advice. The "wharfies" (dock workers) refused to work the ship because of the offensive odor, although after much discussion they decided that for another $2 an hour they could work. The ship's agents were reluctant to have the offending cargo unloaded because it was destined for Australia. Health and Dangerous Goods officials attended, but it was the first time they had been called upon to deal with an incident on such a scale. The black drums were identified with just the word "merphos" stenciled in white on the black drums. There were no toxic labels and no manifest declaring the cargo as dangerous. Attempts to identify the chemical proved fruitless because there were no New Zealand agents and the Australian consignees were refusing to disclose the technical name on the grounds of commercial sensitivity. All they would say was that it was not particularly harmful. Information gleaned locally suggested that it might in fact be highly toxic. After a whole day of discussion, it was agreed that the drums should be unloaded and the undamaged drums sent to the agent's warehouse and the damaged ones sent to a nearby warehouse in Parnell. When the truck arrived at the warehouse, it was parked in the open. It seems that the delivery was not supervised by the authorities, and indeed at the later inquiry it came out that there was confusion as to what should happen to the drums.

When the truck driver arrived the following morning, he was instructed to take the truck from outside the warehouse to a piece of nearby vacant land. Another truckload of drums that were thought to be sound was found to be leaking. The driver took them to a dangerous goods store to unload them but was told to take them to the same piece of vacant land that the others were on.

In the hot summer sun, the drums soon became heated, increasing the odor in the surrounding area. The warehouse manager called the fire brigade, which duly arrived and donned the gas-tight suits that they used to deal with ammonia leaks. They reidentified the trade name on the drum and confirmed the leaks but were not really geared up to do much more.

Meanwhile, the citizens of Parnell were starting to be affected by the smell and were turning up at local hospitals in great numbers. As the chemical was unknown, doctors could only treat the patients symptomatically. Much of the first day was spent evacuating residents while the authorities attempted to get a positive lead on the nature of the material. After much questioning of local chemical experts, it was concluded that the substance was a thio organo phosphate, i.e., an organo phosphate with some of the oxygen replaced by sulphur. We now know that these materials can have extremely strong odors detectable at parts per billion. Generally, the odor comes not from the chemical but from mercaptans produced as contaminants in the manufacture. Mercaptans are used as stenching agents in gas plants. It was eventually identified through analysis as tributyl phosphorothioite, although some sources identified it as phosphorothioate.

Despite the brigade's best efforts to stem the leaks and reseal the damaged drums, the odor continued to spread and more and more people reported to the hospitals. There was so much congestion that many were treated on stretchers in the car park.

The other problem starting to emerge was who was responsible for dealing with the incident and more importantly who was going to pay for it. The Fire Service did not feel that it was their responsibility because legislation in New Zealand at that time only gave them authority to deal with fires or the threat of fire and this was neither. The Police felt it was not under their jurisdiction, because, although they were charged with protecting life and property from danger, they were not sure that dealing with chemical spills came within that definition. The Civil Defence Act, however, gave the controller of Civil Defence ultimate powers once a declaration was made, which allowed him to direct anyone to do almost anything. The controller was usually designated as the local mayor, who knew little or nothing about controlling emergencies, but the declaration gave him the power to spend whatever money was necessary on equipment and resources and subsequently to reclaim it from central government. He duly made the declaration and then immediately gave the power of control to the Police, who passed on the responsibility of dealing with the chemical to the Fire Service.

Now they could go out and requisition trucks and diggers to move the material and the chemicals necessary to neutralize the spills. It took nearly five days to resolve this incident. It became clear that because the material had been moved, there were several sites that needed decontaminating. The truck originally transporting the drums was contaminated and needed to be treated. Both the ship and the wharf were contaminated. The hospital, which had been totally overwhelmed, was also collecting the odor from contaminated firefighters and members of the public arriving for treatment. Over the five days, around 4,000 were treated. Some people who were close to the incident undoubtedly were affected, if only by the terrible stench. Although others came from far away and in an upwind direction, if they said they felt ill, they had to be seen. It was important to identify the genuine cases because in the worst cases they needed to be treated with the antidote atropine to restore their cholinesterase.

The "neutralizer" for organo phosphates is fortunately quite simple. Because sodium hydroxide (caustic soda) solution breaks it down quite readily, every contaminated item had to be treated with caustic solution. It was at this point that some of the purely academic scientists recommended that the truck be cut into small pieces to be soaked in decontaminant. The Chief Fire Officer simply had his men pour caustic over every surface of the truck, which ran onto the contaminated soil underneath, helping to decontaminate it. The soil was then trucked to a secure landfill and buried along with more caustic. The ship was sand blasted and steam cleaned from stem to stern and the residue was washed into the harbor. Environmental awareness was not too high in those days.

A Royal Commission of Enquiry was set up to investigate the circumstances of the incident and the subsequent handling. The Commission generally did not criticize any actions taken by emergency services, but it did make some strong recommendations on how such incidents might be better handled in the future. As a result, emergency services coordinating groups were set up with joint exercises focused on nonfire emergencies. There was also a hazardous materials advisory group formed, which could give advice to emergency services when such events occurred in the future. These groups have proved extremely beneficial.

It was from that point on that the Fire Service was called more and more often to chemical spills and their expertise increased. A new position was established to act as a liaison between the scientists and the firefighters. The Technical Liaison Officer (TLO) position was filled initially by a Fire Service officer with chemical expertise and subsequently by civilians. This post was the equivalent of what we now call a Hazmat Responder. Many of the problems during the incident at Parnell were caused by poor communications between scientists and the emergency services, and the TLO position has eliminated most of these problems.

Although the brigade was now attending hazardous materials incidents as a matter of course, the decade from 1973 to 1984 was a reasonably quiet time for chemical incidents but a bad time for fires generally. There were two major fires at one printing company. A furniture factory was completely gutted and a warehouse full of paraffin wax and jute sacks burnt like a large candle.

4.3.2 Revertex Fire—1984

One of the biggest chemical fires to date in New Zealand occurred in a resin and emulsion blending plant in 1984. The plant was not large by world standards but was a major New Zealand supplier of blended resins and ran a 24-hour operation. The fire occurred in the reactor tower, which fed raw material and additives in at the top, dropped them into the reactor kettle, and at the end of the reaction time dropped the product into a solvent-blending tank at the base. At the time of the fire, the resin cook was running through a steady heating phase and the solvent, a hydrocarbon blend, had been loaded into the blending tank ready for the next phase.

The night staff was just a skeleton crew that kept the resin kettles in operation. It consisted of a supervisor and three operators. At about 4.30 a.m., they broke for refreshments and retired to the restroom, which was only about 10 meters from the reactor but in a separate building. The reaction was one that they had done hundreds of times before and required very little supervision, so the reactor tower was empty of people.

At 4:50 a.m., there was an explosion that appears to have occurred in the space between the reactor tower and the walls of the building. The wall nearest the restroom fell over as a slab and came to a stop leaning on the restroom walls. The three with their backs to the door heard the bang and ran through the door and ducked under the fallen wall. They managed to escape successfully. The worker facing the door had seen a flash and threw a chair through a window to escape to the street.

The ensuing fire was successfully contained but burned for about 4 to 5 hours. The damage to surrounding buildings was confined to windows and doors being sucked out by the rush of air into the explosion. The heat of the fire was sufficient to cause semiburied tanks to rupture.

Although this was a major fire involving petrochemicals, the chemical effect seemed negligible. Odors from acrylates in the plant were detected up to 8 km away. Fish were killed in a nearby harbor, but this was attributed to the large volumes of fresh water going into saltwater.

An even more significant chemical fire in New Zealand's history occurred later that year at an ICI warehouse on the banks of the Tamaki River.

4.3.3 ICI Warehouse Fire 1984

On December 21, 1984, at about 5 p.m., a fire started alongside a pallet of calcium hypochlorite in a corner of the store. The cause was thought to have been a small spillage from a container, which then reacted with some other organic material and ignited. Two employees attempted to extinguish the fire with portable extinguishers. The building was not sprinkler protected and was not required to be under the codes of the time. Unfortunately, one of the workers tried to escape across an open mezzanine landing above the fire. The fire flashed over and he suffered 80% burns. He died four months later.

The section of warehouse involved in the fire contained a wide variety of materials ranging from pesticides such as paraquat and 245 T to fiberglassing resins and cylinders of sterethox, a mixture of ethylene oxide and CFCs. The calcium hypochlorite, which was not supposed to be in that area, helped to fuel the fire and made it difficult to extinguish. Thirty-one-ton chlorine cylinders stored close to the building caused great consternation but because of wind direction were not affected.

The fire burned for about four hours, during which time the light steel truss roof collapsed into the fire, making access very difficult. At the height of the fire, 5,000 liters per second of water were being poured into the fire, and a large proportion of this was flowing out of the debris into the nearby Tamaki River. Fortunately, the river is still tidal at this point, with an approximately three-meter tide fall. As the tide was high for the greater duration of the fire, the runoff was held in the estuary. As the tide fell, it took the pollution out to sea in a moderately concentrated slug, which then dispersed. Apart from two trees that died on the riverbank at the point where the runoff flowed into the river, there was very little environmental damage. The company commissioned an independent ecologist to survey the area over the next 12 months. Little damage was found to the immediate environment, and no signs were found of long-term effects on the next generation of plant and marine life.

The same could not be said for firefighters who were involved. Immediate effects were irritation, particularly of the legs, from slipping into deep puddles of contaminated water. All firefighters wore self contained breathing apparatus (SCBA) most of the time but not necessarily all of the time. There does not seem to have been any respiratory damage.

Over the ensuing two to three years there was an ongoing battle as to whether firefighters had been affected by the fire. An inquiry was eventually held into the health of firefighters who had attended the fire, which concluded that most of the problems after the initial physical effects were brought about by worry that they may have been affected.

The fire generated a huge amount of interest throughout the country and dozens of medical insurance claims were made. Those who were involved are still being monitored, but there have been no signs of anything unusual to date.

Another incident occurred in 1985 when 400 liters of chlorosulphonic acid spilled from a partially overturned container. Chlorosulphonic acid in contact with moisture gives off clouds of sulphur dioxide and hydrogen chloride. Seventy people were evacuated for about eight hours while the spill was dealt with.

4.4 STORAGE AND HANDLING OF HAZARDOUS MATERIALS IN NEW ZEALAND

Because of New Zealand's relatively small size, chemicals of different types are often warehoused and stored on the same site. In the good places, segregation is well maintained, usually by having separate storage buildings or areas for each class of chemical. Some other places are not so good. Difficulties in control sometimes come about because of the fragmented legislation that has grown up in New Zealand over the years.

Up until early 2001, flammable materials, including gases, were controlled by the Dangerous Goods Act and Regulations administered by the Department of Labour. This legislation evolved, as in most countries, from the need to control the storage and sale of gasoline with the advent of the motor car. Over the years this has expanded to include gases, flammable solids, and oxidizers. The Department of Labour already controlled explosives under its own Act and Regulations and so it had control of all materials with a flammable nature or an ability to cause fire.

The Dangerous Goods Division of the Department of Labour controlled all of these materials by specifying the construction of buildings and segregation from other activities. They also controlled the containers and tanks used for storing flammables or explosives. The compliance was usually in line with international practice such as UN codes.

Toxic, corrosive, and radioactive materials were under the control of the Ministry of Health. Toxics and corrosives were controlled by the Toxic Substances Act and Regulations, while radioactive materials came under the Radiation Protection Act and Regulations.

The Toxic Substances Act evolved in the late 1970s from an old Poisons Act, which was designed to control the packaging, labeling, and selling of poisons, which was done primarily through pharmacies. The Poisons Act placed very little control on storage or bulk handling other than to require that poisons be "securely" stored and kept out of the reach of children. Even when control was introduced through the Toxic Substances Act, the wording still seemed geared to relatively small quantities.

Around these two core Acts (the Dangerous Goods Act and Toxic Substances Act) were acts such as the Pesticides Act, which imposed additional controls on toxic pesticides, and the Health Act, which required the reporting of poisonings. The building regulations of the time recognized flammable liquids only as a hazard, and no controls were placed on buildings storing other substances.

To set up and run a hazardous materials warehouse in New Zealand is a bureaucratic nightmare with a myriad of Acts covering different aspects. And this was before there was any serious environmental protection legislation. It was in this environment that the ICI warehouse fire occurred.

4.5 LEGISLATION REVIEW

The ICI fire, like the incident in Parnell 12 years before, served to focus concern on the legislation for controlling hazardous materials. The event brought a realization that there were some shortcomings in the legislation.

In 1985 the wheels were set in motion to review all of New Zealand's legislation related to hazardous materials. The intention was to produce a single piece of legislation for all classes of hazardous material. "One Act, one authority" was the catch phrase of the time. Over the next 15 years most of New Zealand's safety and environmental legislation was reviewed and rewritten.

Most of the recent reviews of legislation around the world have tended towards a non-prescriptive approach, and New Zealand has followed this trend. So where the old legislation might prescribe that a building should be constructed of a four-hour fire-rated material and should be at least a fixed distance from other buildings or public places, the new legislation would require that persons or property not be injured in trying to exit from a site and that there be minimal effects on those not on the site. Such nonprescriptive legislation allows innovation in techniques and materials. The difficult side is that it requires those giving approvals to have some real knowledge of building safety, chemicals, and combustion chemistry and physics and then to be able to trust their own judgment rather than relying on figures in a book.

One of the first new pieces of legislation to be introduced was an amendment to the Transport Act in 1989 imposing controls on the Transport of Hazardous Substances on Land. There had always been some legislation, but it was dispersed through several other Acts and agencies. Now the police had powers in their own right.

Next came the Building Act of 1991, which replaced New Zealand Standard NZS1900-1964. In essence, this Act requires a building to be safe and to have a safe means of egress based on the activity in the building and the construction. Specific sections deal with buildings storing hazardous substances. How this is achieved is not prescribed but must be justifiable. This usually means these buildings must be built according to another recognized standard or that the building proposal has been thoroughly tested. "Acceptable solutions" are published with the Act, which are intended as a guide but more often than not become the only solution. Human nature doesn't like making decisions.

The Resource Management Act also came into effect in 1991. This is an all-encompassing piece of legislation aimed at protecting the environment not only from accidental damage such as spills but also from deliberate acts such as building unsightly buildings or additions to existing buildings. The environmental protection sections are working well.

In 1992 the Health and Safety in Employment Act clarified the requirements for a safe workplace and among other things gave employees the right to know about hazards that may affect them. This included chemical hazards and made it obligatory to have material safety data sheets available in the workplace.

All of this new legislation tidied up the building, workplace, and transport controls, but the base Dangerous Goods and Toxic Substances Acts still remained and distorted some of the authority in the new legislation. Since 1987 committees and groups have been working towards producing an up-to-date and comprehensive Hazardous Substances law that would be all-encompassing. The Hazardous Substances and New Organisms Act was introduced in 1996.

4.6 HAZARDOUS SUBSTANCES AND NEW ORGANISMS ACT

The Hazardous Substances and New Organisms Act set up an Environmental Resource Management Authority (ERMA) that controls and monitors any new Regulations to be made under the Act.

The New Organisms sections of the Regulations came into force in 1998, no doubt encouraged by the looming on the horizon of genetically modified plants and animals, although its real purpose of monitoring the introduction of any new plant or species into New Zealand is much more mundane. Any new species of tree, plant, or animal goes through an assessment process. For example, this law will ensure that we never have snakes in New Zealand, or for that matter any other species that may endanger our indigenous species of plant or animal.

The Hazardous Substance Regulation sections, which at the time of writing are still in the process of being enacted, will formalize classification and labeling according to best international practice. They will require an appropriate standard for storage, packaging, and identification and also require emergency plans to be in place and tested as appropriate. For instance, the classification and labeling will follow the Global Harmonization criteria, which were agreed upon at the Rio Conference in 1992.

Early drafts of the Regulations had a requirement for ecotoxic data to be supplied with every application for the importation or manufacture of a new chemical. When the cost of providing such information, which did not appear to be required anywhere else in the world in such detail, was assessed, it was agreed that generic or best practice information would be acceptable in most cases. If a completely new type of chemical were to be proposed, then the ecotoxic data would probably be required.

We believe that the Hazardous Substances Regulations will be among the best and most complete available anywhere and will give manufacturers and importers very clear direction on what can and cannot be introduced into New Zealand and under what controls and conditions. The Regulations will provide clear guidance on the controls that will be required for those substances that are permitted. There is no intention to be overrestrictive, and for low-level substances, the Regulations will only require the sorts of controls that are often already required in other legislation but not clearly spelled out. At the time of writing, there is trepidation about the effects of this legislation, but overall simplification should overcome any concerns about tightened requirements.

4.7 CONCLUSIONS

New Zealand is entering the 21st century looking very different from the rural economy of 160 years ago. It underwent its industrial revolution quite late in life, during the 1960s and

1970s. Fortunately, the environmental protection movement was gathering strength at that time, which means that New Zealand has avoided many of the polluting practices that have affected older nations. Emergency services are also evolving and developing strategies that protect the environment. The days of the big washdown are long gone. Containment and controlled disposal are all part of a modern firefighter's armory.

CHAPTER 5
CHEMTREC (THE CHEMICAL TRANSPORTATION EMERGENCY CENTER)

C. A. Leif Ericson and Thomas K. Warnock
CHEMTREC, Arlington, VA

5.1 INTRODUCTION

The Chemical Transportation Emergency Center (CHEMTREC) was established as a public service by the American Chemistry Council (ACC) in 1971. It was organized in response to several accidental chemical releases for which responders could not obtain adequate information to mitigate the hazards. ACC's approximately 190 member companies represent more than 90% of the productive capacity for basic industrial chemicals in the United States.

CHEMTREC is a 24-hour resource center that provides technical and emergency response information, shipper contact, and medical assistance for incidents involving chemicals and other hazardous materials to first responders, the transportation industry, medical professionals, and others. The Center also provides access to industry mutual aid programs and is involved in prevention by providing training assistance and materials and educational programs.

CHEMTREC also assists shippers of hazardous materials to satisfy regulatory requirements related to shipping chemicals and assists in chemical incidents and medical emergencies not related to transportation. Anyone with a chemical emergency may call, free of charge, to receive appropriate emergency response advice.

CHEMTREC provides immediate technical assistance from chemical product safety specialists, emergency response coordinators, toxicologists, physicians, and other industry experts. The Center serves as a communications link for the responder, facilitating the rapid transfer of information and allocation of resources. Calls are received from all over the world, including from ships at sea. Translation services are available for callers who do not speak English. ACC employees who are fluent in foreign languages also provide translation assistance.

The emergency response information provided by CHEMTREC comes from a number of sources. Some is from Material Safety Data Sheets (MSDSs), which are product-specific documents containing health and safety information provided by the manufacturer. Shippers provide MSDSs on their products to CHEMTREC for distribution to responders during an emergency. Other sources of information include the North American Emergency Response Guidebook (NAERG), standard industry text references such as the Emergency Action

FIGURE 5.1 CHEMTREC Emergency Call Center, Arlington VA.

Guides published by the Association of American Railroads (AAR), and programs on CD-ROM such as Tomes.

CHEMTREC's Emergency Services specialists are hazardous materials professionals, many of whom are experienced in the emergency response field. They attend a 40-hour, technician-level course on hazmat response as well as additional specialized training courses throughout their career. They have access to an extensive database of contacts with manufacturers, shippers, carriers, and contractors if further chemical or technical assistance is required.

As shown in Table 5.1, CHEMTREC receives more than 100,000 calls each year and handles nearly 20,000 incidents annually. The number of employees at the Center has increased by 50% in the last three years as their services become better known and the number of calls increases. In addition, shippers and carriers are now required to report 100% of incidents through CHEMTREC and Responsible Care outreach to carriers and the industry.

CHEMTREC can be reached 24 hours a day, every day of the year, from anywhere in the United States, Canada, Puerto Rico, and the Virgin Islands through a toll-free emergency number (1-800-424-9300). Callers outside the United States and ships at sea can contact the Center using CHEMTREC's international and maritime number (1-703-527-3887). Emergency collect calls are accepted.

5.2 CHEMTREC'S RESPONSE TO AN EMERGENCY CALL

When CHEMTREC receives an emergency call, an Emergency Services specialist obtains the caller's name and organization, telephone and fax number, the location of the incident, the shipper, consignee, and carrier, the product name, and the nature of the incident. CHEMTREC then provides immediate emergency response information about the chemical(s) involved, often obtained from the manufacturer's product-specific MSDS in the Center's library of nearly 2.5 million such documents. These are accessed in seconds and faxed to responders on-scene at a hazardous materials incident. Other sources of information are other technical text references and computer databases.

The details of the incident are quickly relayed to the shipper or manufacturer of the product involved, and the carrier can be notified as necessary. The shipper is put in contact with the incident scene to provide follow-up technical advice and assistance, including dispatching personnel to the scene if required.

TABLE 5.1 Total Calls to CHEMTREC (1995 to 1998)

1995: 44,179 total calls (25,020 emergency-related, 19,159 non-emergency)

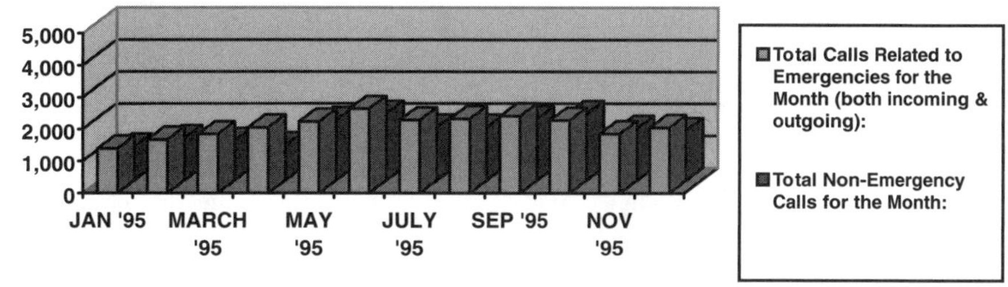

1998: 110,973 total calls (84,685 emergency-related, 26,288 non-emergency)

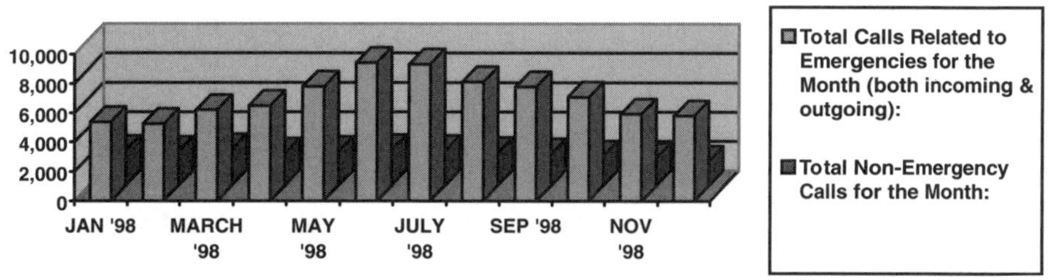

FIGURE 5.2 CHEMTREC's original reference library: Tub files with documents on 1,700 different products.

While not a government agency, CHEMTREC maintains contacts with several U.S. government response organizations that provide response information and assistance. Through a Memorandum of Understanding, the U.S. Department of Transportation recognizes that CHEMTREC will assist any caller with a chemical emergency. The Center can contact those government agencies that ship hazardous materials if their shipments are involved in accidents while in transit. CHEMTREC also liaises with emergency call centers in countries outside the United States.

It is important to identify the shipper and the specific product name to minimize the time needed to provide information and assistance in an emergency. Shipping papers, which normally accompany hazardous materials in transportation, are the primary source for this information. If shipping papers are unavailable, rail car numbers, truck numbers, container markings, and carrier names can help to identify cargoes and shippers. Shippers want to know about emergencies quickly so they can provide needed emergency response information and solve customer service problems.

Even when only limited information is available about the product or the nature of the problem, personnel at the Center can usually provide valuable information and guidance by using the network of company contacts, extensive reference library, product-specific mutual-aid networks, and other resources. The more detailed the information provided, however, the faster and better service can be provided.

The CHEMTREC Emergency Assistance Guidelines form is included in this chapter. This form can be reproduced and used as a checklist for gathering and organizing information during a hazardous materials incident.

5.3 MEDICAL EMERGENCIES

CHEMTREC can assist those providing emergency medical treatment for chemical exposure. Assistance is provided by medical specialists from the manufacturer, the shipper, or those under contract to CHEMTREC. This assistance is best used by the physician at the receiving hospital. While the Center can provide assistance to emergency medical service personnel on the scene if necessary, CHEMTREC normally advises on-scene personnel to prepare and transport the patient in accordance with local protocols. The physician at the receiving medical facility should then contact CHEMTREC for further medical assistance.

FIGURE 5.3 CHEMTREC's data center includes optical disk storage equipment for 2.5 million MSDSs.

5.4 NEW HAZARDS: RESOURCES FOR INCIDENTS INVOLVING WEAPONS OF MASS DESTRUCTION (WMD)

With the growing threat posed by terrorism and weapons of mass destruction (WMD), CHEMTREC has developed resources to meet the response community's need for information about these new hazards. In addition to assembling product information on chemical and biological agents that can be immediately relayed to responders, CHEMTREC has an up-to-date network of contacts with military, health services, and law enforcement personnel. The Center can quickly put emergency responders in touch with experts in this field.

The U.S. Coast Guard National Response Center (NRC) must also be notified of all known or suspected incidents involving acts of terrorism and weapons of mass destruction occurring in the United States. NRC then notifies the appropriate federal authorities. While CHEMTREC does not make that notification, it can arrange a conference call between responders and the NRC. Responders can contact the NRC directly by calling 1-800-424-8802.

5.5 PRODUCT-SPECIFIC, MUTUAL-AID NETWORKS

Mutual aid networks are organized by chemical industry trade associations or by agreement between producers to ensure that resources are available for mutual assistance in case of an emergency involving specific chemical products. CHEMTREC serves as a central reporting and coordination point for several of these networks. When advised of an incident, shippers can instruct CHEMTREC to activate a mutual-aid program to dispatch specialized teams to the scene for incidents involving chlorine, compressed gases, vinyl chloride, pool chemicals, and many other products.

For more information on CHEMTREC mutual-aid programs, contact CHEMTREC's Director of Information Services and Marketing at 1-703-741-5503, by fax at 1-703-741-6503, or by e-mail at CHEMTREC@americanchemistry.com.

5.6 EMERGENCY SERVICES SPECIALISTS

CHEMTREC's Emergency Services specialists are people with extensive experience in civilian and military emergency response. The current staff of 14 Emergency Services specialists includes active and retired firefighters, former U.S. Coast Guard personnel, retired military explosive ordnance disposal personnel, and a chemist.

Emergency Services specialists receive ongoing training throughout their career. In addition to a monthly refresher/drill program, they attend advanced training courses in telecommunications skills, crisis management, and hazardous materials response. Each year, several personnel receive specialized training on railroad and intermodal equipment at the Association of American Railroads Transportation Test Center training facility.

While CHEMTREC personnel do not actually go to the scene of hazardous materials incidents, this training coupled with prior experience allows them to understand conditions faced by responders and to provide guidance on technical issues, such as selecting air monitoring or personal protective equipment.

Most of CHEMTREC's Emergency Services specialists work a rotating shift of two 12-hour days, two 12-hour nights, and four days off. In addition to the 12 personnel assigned to shifts, two people work on weekdays during peak periods. The staffing schedule ensures that there are always enough personnel to handle the number of calls, which can be as many as 250 calls per day.

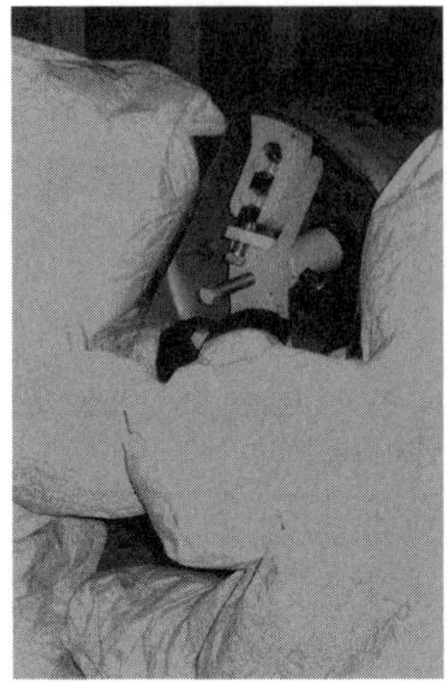

FIGURE 5.4 A CHEMTREC Emergency Services specialist (with an industry response team member) receives training in chlorine leak control.

5.7 TRAINING RESOURCES

The following emergency response training resources are available.

- *Handling Hazardous Materials Incidents: A Lending Library of Audiovisual Training Programs for Emergency Response Personnel.* Several of the programs in this catalog were developed by emergency response experts from ACC member companies. Others are produced by other chemical industry trade associations, carriers, and carrier associations, or hazardous materials training consultants. These materials provide responders with technical information and instruction on safely mitigating incidents involving a variety of hazardous materials. Library materials are available on loan at no cost. Many of the programs can

FIGURE 5.5 Handling Hazardous Materials Incidents.

also be purchased and some can be reproduced at no charge. For a copy of the catalog, please contact CHEMTREC at 1-703-741-5726 or by fax at 1-703-741-6082.

- *CHEMTREC Workshops.* These workshops are hands-on training courses for emergency responders. Two classes are offered—a basic course and a 36-hour course designed for intermediate and advanced responders. In both classes, students participate in realistic scenarios and learn how to mitigate incidents involving various trucks, rail cars, and other

CHEMTREC® CMA

EMERGENCY ASSISTANCE GUIDELINES

IN THE EVENT OF A CHEMICAL EMERGENCY, CHEMTREC CAN EXECUTE A THOROUGH EMERGENCY RESPONSE EVEN WITH VERY LITTLE INFORMATION. **IT IS NOT NECESSARY TO HAVE ALL OF THE FOLLOWING INFORMATION FOR ASSISTANCE TO BE GIVEN**, BUT MORE DETAIL WHEN AVAILABLE CAN BE VERY HELPFUL.

- ❑ CALLER'S NAME AND TITLE ____________
- ❑ CALLER'S COMPANY OR ORGANIZATION ____________
- ❑ CALLER'S LOCATION ____________
- ❑ AT LEAST ONE CALLBACK NUMBER, WITH AREA CODE. (______) ____________
  (IF YOU USE A CELLULAR PHONE, POST THE NUMBER IN CASE THE CALLER IS NOT FAMILIAR WITH IT.)
- ❑ DISPATCH CENTER NUMBER, IF AVAILABLE/APPROPRIATE (______) ____________
- ❑ FAX NUMBER (______) ____________
- ❑ LOCATION OF THE INCIDENT/WEATHER CONDITIONS ____________
- ❑ TIME INCIDENT OCCURRED (OR ESTIMATE) ____________
- ❑ SHIPPING PAPERS ____________
  - ❑ UN, NA (PLACARD) OR STCC NUMBER OF THE PRODUCT(S) ____________
  - ❑ NAME OF THE PRODUCT(S), PREFERABLY A TRADE NAME ____________
  - ❑ CARRIER ____________
  - ❑ SHIPPER AND POINT OF ORIGIN ____________
  - ❑ CONSIGNEE AND DESTINATION ____________
- ❑ TYPE OR DESCRIPTION OF CONTAINER/PACKAGE ____________
- ❑ CONTAINER NUMBERS AND/OR MARKINGS ____________
- ❑ BRIEF DESCRIPTION OF INCIDENT AND ACTIONS TAKEN ____________
- ❑ NUMBER AND TYPE OF INJURIES/EXPOSURES ____________
- ❑ AMOUNT OF PRODUCT(S) INVOLVED AND RELEASED ______ / ______
- ❑ IS THERE SPECIFIC INFORMATION NEEDED AS A PRIORITY? ____________
  (MSDS, PROTECTIVE CLOTHING INFO, MEDICAL ASSISTANCE)
- ❑ ARE ANY INDUSTRY REPRESENTATIVES ON SCENE OR HAVE ANY BEEN CONTACTED? (DRIVER, PLANT MANAGER, ETC.) ____________

FIGURE 5.6 CHEMTREC Emergency Assistance Guidelines.

FIGURE 5.7 Training workshop at the Association of American Railroads Transportation Test Center in Pueblo, Colorado.

containers. The classes also cover working in the Incident Command System, container features, containment and confinement, and personal protective equipment. For more information on training opportunities, contact CHEMTREC at 1-703-741-5259 or by fax 1-703-741-6259.

- *The North American Emergency Response Guidebook.* In 1996, the U.S., Canadian, and Mexican governments developed a single Emergency Response Guidebook for North America. The new book replaces previous versions of both the U.S. Department of Transport (DOT) Emergency Response Guidebook and Canada's Dangerous Goods Initial Emergency Response Guide. This book is available free of charge to emergency response organizations through DOT-designated coordinators in each state. CHEMTREC personnel review proposed revisions of this publication prior to release. CHEMTREC is listed in the document as an organization that can be called for assistance in a hazardous materials emergency.

5.8 DRILLS FOR EMERGENCIES

In many cases, CHEMTREC is called for the first time during an actual emergency. If CHEMTREC is involved in a drill for a hazardous materials response, an additional dimension of realism can be added to the exercise. CHEMTREC handles drills in exactly the same manner as an actual emergency. By working together in advance, emergency responders will better understand what resources and services are available. This facilitates smoother operations if an actual incident occurs.

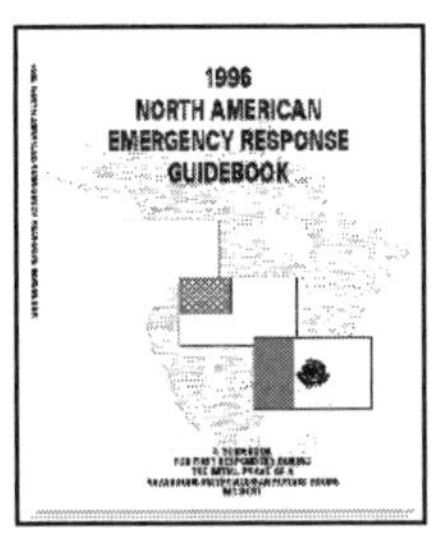

FIGURE 5.8 The North American Emergency Response Guidebook.

CHEMTREC's Hazardous Materials Exercise Scenario form, which is included at the end of this chapter, asks for the basic details about the exercise (who, what, where, and when) and a local contact person. The expected time of the drill must be included, as well as the time that participants are expected to call CHEMTREC. Because an emergency services specialist is assigned to each exercise, it is *very important* that an accurate time be given for an incoming call. If a call does not come in to the center at the indicated time, the staff person will attempt to reach the drill coordinator. If CHEMTREC is unable to participate for any reason, every effort will be made to contact the coordinator in advance.

The completed form must be faxed or e-mailed back to CHEMTREC at least 48 hours before the drill or mailed at least 10 days before the drill. The fax number is 1-703-741-6090, the e-mail address is CHEMTREC@americanchemistry.com, and the mailing address is 1300 Wilson Boulevard, Arlington, VA 22209. Call the Center at 1-703-741-5525 to confirm that the form was received. For a drill involving specific products or chemicals, it can also be arranged for a product shipper or manufacturer to participate in the drill.

CHEMTREC must be advised about drills as early as possible to allow for scheduling. CHEMTREC must also be advised if the drill is canceled or postponed for any reason, so that any special scheduling can be changed. The Center sends out a brief survey form after a drill to evaluate their performance and to obtain feedback on ways to improve their service. Further information is available by contacting the Senior Staff Assistant at 1-703-741-5525, or by fax or e-mail.

5.9 TYPICAL CHEMTREC INCIDENTS

CHEMTREC is committed to providing emergency response personnel, law enforcement agencies, and other interested persons with information and advice for dealing with emergencies associated with hazardous materials. The following incidents are representative of the emergencies handled at the Center every day.

- A driver for a trucking company called CHEMTREC from a highway in New Mexico to report that tetramethyl ammonium hydroxide was leaking from a tote bin inside his trailer. He could not identify the shipper or consignee for his load, but did provide his company's safety department contact and his trailer and tractor number. The Emergency Service specialist contacted the carrier's safety department and identified the shipper and the consignee. The shipper was contacted and asked to call the carrier's safety department. The safety manager for the carrier called back stating that the shipper had contacted him to provide product information and advising that the local fire department had arrived on the scene. He also advised that the consignee was sending a cleanup crew. CHEMTREC was also contacted by the New Mexico State Police, who were put in touch with the shipper to answer their concerns about the product.
- A fire brigade in South Africa called the CHEMTREC Emergency Center reporting that a truck carrying purified terephthalic acid had been involved in an accident. An undetermined amount of product had spilled onto the highway and the shipper was not known. The caller wanted information on the product and advice on mitigation. While the caller was on the line, the Emergency Services specialist faxed out an MSDS from a basic producer in the United States. CHEMTREC offered to contact the basic producer and the caller accepted the offer. The specialist then passed all known details to the U.S. producer of the chemical, who agreed to assist. The producer later called back to advise that a chemist from the U.S. producer's plant had contacted the fire brigade to provide needed information.
- A trucking company dispatcher called CHEMTREC to report that an unknown product was leaking from a trailer parked at a truck stop in Arizona. After the local fire department was called, they called CHEMTREC for assistance in identifying the products in the trailer. As there were multiple shippers' products in the truck, CHEMTREC asked that the bill of

CHEMTREC Hazardous Materials Exercise Request

Thank you for requesting **CHEMTREC** assistance for your hazardous materials exercise. To ensure the exercise proceeds smoothly, please provide us with the following information. We suggest that you contact us at (703) 741-5524 the business day before the drill to confirm the drill or let us know if the drill has been cancelled or rescheduled.
IMPORTANT: **WHEN CALLING, USE THE CHEMTREC EMERGENCY NUMBER (800) 424-9300 AND CLEARLY STATE TO THE COMMUNICATOR ANSWERING THAT THE CALL IS FOR A DRILL.**

DATE OF EXERCISE:____________________

TIME OF EXERCISE:__________ ☐EASTERN ☐CENTRAL ☐MOUNTAIN ☐PACIFIC ☐OTHER (specify)

DRILL COORDINATOR NAME/TITLE:____________________

ORGANIZATION:____________________

ADDRESS:____________________ CITY/STATE/ZIP:____________________

TELEPHONE:____________________ FAX:____________________

PRODUCT(S) INVOLVED:____________________
(Chemical/product name and/or UN number)
CONTAINER TYPE/SIZE(S):____________________

AMT. OF PRODUCT INVOLVED:____________________ RELEASED:____________________

LOCATION OF INCIDENT:____________________

INCIDENT SITE TELEPHONE:____________________ FAX:____________________

BRIEF DESCRIPTION OF INCIDENT:____________________

____

____

EXPECTED TIME THAT DRILL PARTICIPANTS WILL CALL CHEMTREC:__________
☐EASTERN ☐CENTRAL ☐MOUNTAIN ☐PACIFIC ☐OTHER (specify)

ARE ANY OF THE FOLLOWING PARTICIPATING?

☐MANUFACTURER: ____________________

☐SHIPPER: ____________________

☐CARRIER: ____________________

PLEASE PROVIDE THE NAME(S) AND NUMBER(S) FOR ANY SHIPPER, CARRIER OR MANUFACTURER REPRESENTATIVES PARTICIPATING, AS WELL AS ANY SPECIAL INSTRUCTIONS FOR THIS EXERCISE.

____

____

____

PLEASE RETURN THIS FORM BY FAX AT LEAST 48 HOURS IN ADVANCE, OR BY MAIL 10 DAYS IN ADVANCE TO:
CHEMTREC
1300 WILSON BOULEVARD, ARLINGTON, VIRGINIA 22209
voice:(703)741-5525 ***fax:(703)741-6090*** ***e-mail: chemtrec @ americanchemistry.com***

FIGURE 5.9 CHEMTREC Hazardous Materials Exercise Request.

lading be faxed to the Emergency Center and determined that there were two shippers involved, shipping hydrofluoric acid solutions and ethylene glycol. The Emergency Services specialist contacted each shipper, requesting that each company contact the reporting carrier's dispatcher. In turn, the shippers called back to advise that their respective emergency coordinators had made contact and that responders had been given information on the product.

- CHEMTREC received a call from a railroad reporting a train derailment. The report stated that derailed tank cars contained sulfuric acid, a naphtha/xylene mixture, and residue of silicon tetrachloride. It was reported that an unknown amount of product was leaking from the rail cars. Adverse weather conditions in the area were hampering response and evaluation of the incident. CHEMTREC faxed MSDSs to the local fire department and to the nearest poison control center and arranged a conference call between the shippers, the fire department, and railroad representatives.

CHAPTER 6
SQAS: SAFETY AND QUALITY ASSESSMENT SYSTEMS FOR THE TRANSPORT/STORAGE/HANDLING OF CHEMICALS

J. Verlinden
Trade and Environment Services, European Chemical Industry Council/Conseil Européen de l'industrie chimique (CEFIC), Brussels

6.1 BACKGROUND

The chemical industry in Europe often uses the logistics services offered by third parties to store, handle, and transport feedstock, intermediates, and finished chemical products. Chemical companies have always needed to be assured that these operations were carried out in a safe and high-quality manner with due regard for the protection of employees, the public, and the environment. In the past, this assurance has often been obtained by individual chemical companies undertaking periodic audits of their logistics service providers, leading to a fragmented approach and a multiplicity of auditing programs, which has been costly and inefficient for both industries.

In the early 1990s, chemical companies recognized the need to take a fresh look at the safety, quality, and environmental aspects related to the provision of logistics services. Within the framework of the Responsible Care Programme, initiatives were started that have since evolved into a number of safety and quality assessment systems (SQAS), each related to a particular transport mode or logistics operation.

6.2 OBJECTIVE

Safety and quality assessment systems enable chemical companies to have the quality and safety management systems of their logistics service providers assessed in a uniform manner, thereby avoiding multiple assessments by individual chemical companies. The results of an SQAS assessment carried out by an independent body do not lead to a certificate but are used by an individual chemical company to evaluate the performance of service providers according to its own requirements. Among other tools and criteria, SQAS help in the process of selecting suppliers and in defining areas of improvement for each of them.

Although this assessment system does not guarantee the safety, quality, and value of the service provided, it does offer a mechanism for promoting and monitoring continuous improvement. The system not only offers impartial information on strengths and weaknesses directly to the service provider but also provides the opportunity for customer-specific feedback by the chemical company that evaluated the assessment results. This creates or strengthens a foundation for true partnership that is mutually beneficial.

Safety and quality assessment systems are the "prevention" part of the European Chemical Industry Council's International Chemical Environment (ICE) initiative, complementing the ICE Emergency Response Programme in the field of chemical transport safety.

6.3 ASSESSMENT DOCUMENTS

The assessment documents are developed by CEFIC working groups made up of representatives from both chemical companies and service providers. This guarantees that the concerns, knowledge, and expertise of the overall industry are fully taken into account.

An SQAS package consists of a questionnaire and corresponding guidance notes for the assessor. Each question must be answered with "1" (yes), "0" (no), or "na" (not applicable), based on verbal replies by the auditee and objective evidence seen by the assessor. There is no mechanism within SQAS for differentiated scoring of individual questions relative to their importance. However, chemical companies can apply their own company-specific weighting factors when evaluating the SQAS results.

For certain SQAS packages, there are questions for which the aspects of Safety & Health (SH), care for the Environment (E), and Quality (Q) are considered separately. One question can therefore require up to three replies. Most SQAS questionnaires contain the following three categories of questions.

1. "M" questions represent basic safety requirements that should be met by all logistics service providers. They cover items that are a statutory requirement in most countries in Europe, but not necessarily in all countries.
2. "I" questions refer to good industry practices according to which the chemical industry requires its logistics service providers to act.
3. "D" questions cover items for further improvements.

An SQAS questionnaire typically covers areas such as management policy, training and recruitment, safety, health and environmental procedures, safety equipment, emergency response, customer focus, equipment maintenance and inspection, operational instructions, communications, security, and site inspection (where appropriate).

6.4 ASSESSMENT PROCESS

The following are the steps in a typical SQAS process.

1. In responding to a request from chemical companies, the logistics service provider usually contracts an independent qualified assessor to carry out an SQAS assessment and become familiar with the appropriate SQAS package.
2. The assessor carries out the assessment, producing a factual, signed and dated report, which is supplemented with eventual comments from the logistics service provider. The report does not give recommendations for improvement nor does it constitute a certificate.

3. Depending on the SQAS package involved, the logistics service provider supplies, at his or her discretion, a copy of the report to any requesting chemical company or agrees that the assessment data can be accessed from a central database.
4. The individual client evaluates the factual assessment results against his or her own requirements and agrees with the logistics service provider on priorities for continuous improvement.
5. The logistics service provider arranges for periodic reassessment.

Unless mentioned otherwise, e.g., marine bulk, tank storage terminals, the chemical industry considers that all aspects of the service provided by a logistics service provider should undergo a full SQAS assessment every three years. It is left to the discretion of the logistics service provider to arrange for intermediate follow-up assessments, e.g., in accordance with ISO 9000 procedures.

6.5 BENEFITS OF SQAS

An SQAS assessment provides the following benefits:

- One concerted industry approach, which encourages mutual understanding
- Objective evaluation adapted to individual needs
- Systematic focus on issues requiring attention
- Cost-effectiveness (in money and time)
- An essential tool in the process of continual improvement

6.6 OVERVIEW OF SQAS PACKAGES

The following is a general overview of the key aspects of each SQAS package.

6.6.1 Marine Bulk

Two assessment documents have been developed addressing chemical tankers and liquefied gas carriers.

The questionnaires address the status of the management and operational standards of a vessel in areas such as certification, manning, navigation, mooring, cargo operations, firefighting, pollution prevention, and hull and superstructure.

In order to operate the scheme, the Chemical Distribution Institute (CDI) was set up as a foundation and incorporated under Dutch law in 1994 by 13 founding members. By early 1999, this number had risen to 29 chemical companies and 313 ship owners had made an operational agreement with CDI.

The role of CDI is to establish an inspection system that provides inspection data on chemical and gas ships. This information is accessible (electronically) only to chemical companies participating in CDI. The CDI also qualifies, trains, and accredits inspectors. By early 1999, more than 2,200 inspections had been carried out worldwide (continuing at a rate of 75 inspections per month) and 50 inspectors had been trained and accredited. The results of a ship inspection remain valid for 1 year.

Further information is available from:

Chemical Distribution Institute
St. Martins House Business Centre
Ockham Road South
East Horsley
Surrey, U.K. KT24 6RX

Tel 44-(0)1483-281268
Fax 44-(0)1483-285474
E-mail: chemdist@aol.com
http://www.enter.co.uk/cdi

6.6.2 Tank Storage Terminals

The following two assessment documents have been developed:

1. A management questionnaire (630 questions) addressing management responsibility, training, personnel, procedures, maintenance, etc.
2. A technical questionnaire (935 questions) addressing site firefighting, tank cleaning, waste treatment, pumping stations and transfer lines, jetty, etc.

In order to operate the scheme, the activities of CDI were extended with a CDI-T part in 1997 with identical objectives as those of SQAS Marine Bulk. These objectives are:

- To establish an inspection system that provides inspection data on bulk storage terminals to participating chemical companies. Unlike the CDI scheme for marine bulk vessels, this information, accessible to chemical companies participating in CDI-T only, is not presently accessible electronically in view of the limited number of chemical bulk storage terminals in Europe.
- To qualify, train, and accredit inspectors

By mid-1999, 13 chemical companies and 14 terminal operators were participating in CDI-T. By the end of 1999, 16 terminals (out of a total of 60 that could be subject to an assessment in Europe) had been assessed and 10 inspectors had been trained and accredited. The results of a terminal assessment remain valid for 27 months.

6.6.3 Road Haulage

The SQAS package for road haulage operations primarily deals with the managerial aspects of a transport company. However, a depot inspection is part of the assessment in order to verify that the management system has been effectively and consistently implemented and that the company is providing a high level of operational quality and safety.

SQAS documents for road haulage (questionnaire and corresponding guidance document) were first issued in September 1994 and revised in the following manner in 1997:

- Taking into account comments from users (chemical and haulage companies and certification bodies) collected during three years of operating practice
- Introducing a substantial number of questions related to the protection of the environment following the growing importance of tools such as the ISO 14001, Environmental Management System Standard
- Introducing the modular approach as described here

Modular Approach. More and more transport companies now offer a range of logistics services such as tanker cleaning and intermodal transport. It became clear that a number of these companies would be unnecessarily duplicating assessments as the various SQAS packages had a substantial number of common questions.

These common or core questions address the same aspects of the safety, health, environmental, and quality management system of any company providing logistics services. They have now been put into a separate questionnaire and guidelines, i.e., one questionnaire (with corresponding guidelines) with all core questions. Questions (and corresponding guidelines) that relate to a specific logistics service have been put in separate documents.

This prevents a road haulage company that also operates a cleaning station from being subject to two full assessments: SQAS Road and SQAS Cleaning Stations. With this modular approach, one single core assessment can be followed by two separate but specific assessments of the road transport and the cleaning station activities.

This modular approach now applies to:

- Road haulage
- Cleaning stations
- Rail carriers (not yet issued)
- Intermodal service providers
- Intermodal railway operators

Both the core and the road-specific documents (questionnaire and guidelines) are available in several languages. More guidance on the questionnaire and the assessment process, specific to SQAS Road, is given in the guidelines. This means that a complete SQAS assessment document for road haulage consists of a completed core document (226 questions) and a completed road-specific document (284 questions). The 510 questions related to road haulage have been mapped against ISO 9002 (1994) and ISO 14001 (1996).

An SQAS assessment of a road hauler is carried out by an assessor of an ISO 9000/14000 certifying body. The list of qualified assessors from several certification bodies is regularly updated. Although this list should not be considered as a formal accreditation, it is highly recommended that haulers select an assessor from this list and verify that they effectively comply with the qualification criteria mentioned in this list.

Between 1995 and 1998, approximately 400 assessments were carried out covering most European countries. The list of assessed haulers is regularly updated and provides the name of the haulage company, the location(s) assessed, the year of the assessment, the name of the assessing certification body, and the name of the assessor. The results of the assessment are not included on this list, nor are they stored in a central database, but they can be obtained from the individual road haulage company.

6.6.4 Cleaning Stations

In cooperation with EFTCO (European Federation of Tank Cleaning Organisations), the modular approach, as described for SQAS Road Haulage, has also been adopted for SQAS Cleaning Stations. This means that a complete SQAS assessment document for cleaning stations consists of a completed core document (226 questions) and a completed cleaning stations-specific document (223 questions). The list of qualified assessors with corresponding recommendations, applicable for SQAS Road, can also be used for SQAS Cleaning Stations. The list of cleaning stations assessed since 1998 is updated regularly.

6.6.5 Intermodal Transport

The modular approach described for SQAS Road Haulage has also been adopted for SQAS Intermodal Transport. Two documents have been developed covering the different aspects of intermodal transport: the intermodal service provider and the intermodal railway operator. This means that a complete SQAS assessment document for intermodal transport consists of a completed core document (226 questions) and a completed intermodal transport-specific document (181 or 131 questions).

An intermodal transfer terminal can be assessed by using the SQAS Transfer Terminal questionnaire of SQAS Marine Packed Cargo (see Section 6.6.7). This questionnaire does not make use of the modular concept. The list of qualified assessors with corresponding recommendations, applicable for SQAS Road, can also be used for SQAS Intermodal Transport.

6.6.6 Rail Carriage

In cooperation with UIC (Union Internationale de Chemins de Fer/International Association of Railway Companies a SQAS package is being developed for rail carriers. It will include all 226 core questions (see Section 6.6.3), supplemented by approximately 400 questions specific to rail operations. A draft version is being tested in a number of countries, and the package is expected to be launched in early 2000.

6.6.7 Marine Packed Cargo Operations

The CEFIC Marine Packed Cargo (MPC) Working Group initially looked at ferry operations, but it was soon evident that a number of other sectors also needed to be covered. As a result, a specific SQAS package was developed including assessment questionnaires for the following six elements within the marine packed cargo operations chain:

1. Freight forwarder
2. Port agency
3. Container stuffing and stripping contractor
4. Transfer terminal
5. Shipping line
6. Ships (ferries, container ships, general cargo)

Because road and rail operations had already been addressed in separate SQAS documents, the Working Group did not get involved with these aspects although they form part of the whole supply chain. The package was launched in March 1999.

A consolidated questionnaire including the questions (not the Guidelines) of all packages (except Ship) has also been developed. A substantial number of questions are common among the questionnaires and therefore the use of this working document will help to minimize duplication for larger companies that carry out several MPC activities. The Transfer Terminal questionnaire can also be used to assess land-based intermodal transfer terminals.

The required qualifications, training, and experience of the assessors are described in the respective documents. With the exception of Ships, an assessor of an ISO 9000/14000 certifying body carries out the SQAS assessment. In the case of Port Agencies or Forwarders, however, self-assessment of the subsidiary may be considered, provided the subsidiary operates under the same safety and quality management system as the head office. In the case of Ships, the chemical company carries out the SQAS assessment on a spot basis.

6.6.8 Barges

This package has been developed by a group of representatives from both oil and chemical companies with the objective of setting up a scheme for the inspection of oil, chemical, and gas barges operating on inland waterways in Europe, particularly on the Rhine.

The questionnaire, originally containing 92 questions, is currently being revised to address more chemical-specific issues and will include approximately 200 questions. Questions relate to barge certificates, manning, health/safety/environmental aspects, firefighting, navigation, cargo operations, etc.

The inspection scheme started in mid-1998 and is operated by EBIS (European Barge Inspection Scheme), which is currently made up of 19 oil and chemical companies. Inspectors, trained and accredited by EBIS, carry out the barge inspections. The scheme operates a database of inspection reports, which are electronically accessible only to members participating in EBIS.

More information (in 4 languages) about the questionnaire, the list of accredited inspectors, the list of members, and the EBIS procedures, is available at http://www.ebis.nl.

6.6.9 Chemical Distributors

This package was developed jointly by representatives from the chemical industry (CEFIC) and the distributing industry (FECC—European Federation of Chemical Distributors—http://www.fecc/org/). In contrast with other SQAS documents, this document is structured around the 8 Guiding Principles of the Responsible Care/Distribution Programme of the International Council of Chemical Trade Associations (ICCTA). These Principles and the corresponding 52 self-assessment questions have been made more explicit by integrating a number of more detailed questions. This makes it possible:

- To measure a distributor's compliance with these Principles as part of the Responsible Care/Distribution Programme
- To provide a common tool for suppliers of chemicals to evaluate, against their individual requirements, the health, safety, and environmental performance of their distributors in one single assessment

The document (216 questions), called ESAD (European Single Assessment Document), was launched in early 1999. The questionnaire concentrates on managerial aspects but also contains two appendices:

1. A site-inspection questionnaire (69 questions) addressing a number of hardware aspects in order to get an impression of how well policies and procedures are actually being complied with
2. An additional questionnaire (98 questions) for completion by distributors of Food, Cosmetic, and/or Pharma Grade products only. In view of the specific requirement, this assessment needs to be carried out by a pharmacist/chemist with at least two years of practical experience with current Good Manufacturing Practices (GMPs).

Although a list of qualified assessors is not yet available, it can be assumed that until sufficient experience has been gained, the assessors qualified for SQAS Road assessments are also qualified for an ESAD assessment.

6.7 HOW TO USE SQAS

The safety and quality aspects of the logistics operations of a chemical company are normally addressed by its management system. Such a system is expected to identify, evaluate, and continually reduce the risks posed to humans, the environment, or property by the transport and storage of the company's chemical products. Such a process typically consists of the following sequential steps, in which the application of SQAS is described in steps 5 and 6 in this section.

6.7.1 Step 1—Listing the Chemicals

At least all transported chemicals that are classified as dangerous according to the United Nations Recommendations on the Transport of Dangerous Goods—Model Regulations ("Orange Book") should be considered. At a further stage, companies may decide to also take into account the chemicals that are not classified by these Recommendations.

6.7.2 Step 2—Ranking the Chemicals by Hazard

Experts in international transport have classified dangerous goods into three degrees of acute danger—high, medium, and low—according to their inherent potential to affect adversely humans, the environment, or property. This potential is based on a number of factors, such as vapor pressure, flash point, toxicity, and corrosivity. This is reflected in the packing groups (I, II, III) of the UN Recommendations or in the corresponding letters (a, b, c) of international transport agreements such as ADR or RID. This categorization makes it easy to rank the chemicals being distributed into three groups of hazards.

6.7.3 Step 3—Ranking the Chemicals by Exposure/Impact Potential

If not properly confined during transport or in the event of an accident, there is a potential for chemicals to adversely impact the public, the environment, or property. In order to rank the chemicals according to their exposure/impact potential, a major indicator is (estimated) ton-km per year. This may be adjusted, however, by company-specific weighting factors, for example related to storage or transport in densely populated or environmentally sensitive areas.

This relative ranking results in grouping the transported chemicals into three broad categories of potential exposure/impact: high, medium, and low.

6.7.4 Step 4—Prioritizing Risk Reviews

The two rankings of each chemical can be put into a hazard/exposure matrix as shown in Fig. 6.1. This offers a simple and practical way for prioritizing and grouping the review of the risks posed by the transport of chemicals.

Initially, attention should be focused on the group of chemicals situated in the black area. These should first be subject to a risk assessment and, in the future, should undergo more frequent reviews than the other groups. At a later stage, the group of chemicals in the gray area should be addressed. Individual companies may introduce their own or additional criteria for prioritization.

	High			
Hazard	Medium			
	Low			
		Low	Medium	High
		Exposure/Impact Potential		

FIGURE 6.1 Hazard/exposure matrix.

6.7.5 Step 5—Conducting Risk Reviews

This step involves a more detailed consideration of the exposure/impact assessment combined with the assessment of the likelihood of accidents and resulting consequences. This helps to evaluate the acceptability of current practices and procedures.

The following are some typical tools and processes for conducting such reviews.

- Consult historical accident data and reports from the company's own database or from publicly available databases such as the European Community Documentation Centre on Industrial Risk[1] or the U.S. DOT accident database.[2] These provide information, such as relative accident rates per transport mode or likelihood of leaks resulting from an accident, which can then be used as input to risk-calculating computer models.
- Exchange information within sectorial associations (e.g., CEFIC Sector Groups) about lessons learned from accidents and effects of training programs.
- Apply "what if" scenarios.
- Request the results of SQAS assessments, offering information about the procedures and practices operated by the companies that provide logistics services, e.g., information related to emergency preparedness, accident reporting, training of personnel, maintenance of equipment, and raising safety and quality awareness.

In order to extract the relevant information from these assessment reports, each chemical company will need to develop a template. This template is a list of those questions from the SQAS questionnaire for which a positive reply is requested, and is normally the result of an in-depth discussion among company logistics experts. It is suggested that chemical companies not yet familiar with this process and getting involved with SQAS consider the set of "M" (statutory requirements) and "I" (good industry practices) questions marked in most SQAS packages as a first step in the development of their template.

Figure 6.2 illustrates how the results could be presented in order to facilitate the evaluation of the assessment data. The *x*-axis lists the references of the chapters in the SQAS questionnaire, e.g., training and security. The *y*-axis is the percentage of negative replies in each chapter for each category of question, i.e., "M," "I," and optionally "D." High scores indicate that steps for improvement will need to be discussed with the assessed company.

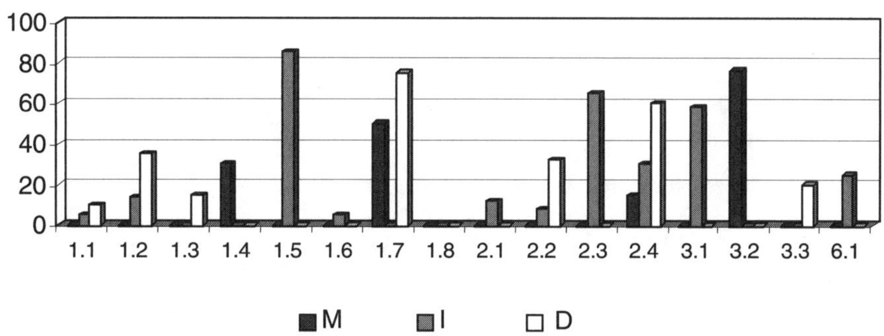

FIGURE 6.2 Evaluation of assessment data.

[1] CDCIR, JRC, Ispra (I)—Tel +39-332-789244, Fax +39-332-789007, e-mail claudio.carnevali@jrc.it.

[2] U.S. Department of Transportation, Office of Hazardous Materials Safety, hazmat.dot.gov/spills.htm.

It should be noted that a company might have various templates depending on the hazard and exposure ranking of the products involved and may therefore consider a stepwise implementation of SQAS, e.g., starting with bulk transport by road of its most hazardous chemicals.

6.7.6 Step 6—Developing and Implementing Risk-Reduction Alternatives

Depending on the need to reduce the risks identified, a number of alternatives will need to be formulated and implemented. The following options will result from consulting databases or reports and the exchange of experience and information within the industry sector:

- Container selection and design (e.g., construction material, fittings, insulation)
- Selection of transport mode (road, rail, inland waterway, sea, pipe)
- Route selection (e.g., highways only, avoiding tunnels and big agglomerations)
- Alteration of chemical and physical properties (e.g., diluting, inhibiting)
- Selection of package type (e.g., portable tank, IBC, drum)

In addition, the evaluation of the SQAS assessment results will lead to the following additional set of risk-reducing proposals:

- Revision of operating procedures, particularly those related to loading and unloading
- Availability of adequate and regularly tested emergency response plans (with the logistics service provider as well as with the chemical company)
- Establishment of proper accident reporting procedures and analysis of root causes
- Training of personnel, especially when conditions or procedures have changed (involving own as well as subcontracted personnel)
- An element in the selection of logistics service providers, as SQAS assessment results allow a chemical company to benchmark them according to specific safety and quality requirements

It is absolutely essential that feedback be given to the SQAS-assessed logistics service provider in such a way that both parties agree upon priorities and targets for any action to rectify deficiencies or to implement improvements. This fundamental objective of continual improvement is a shared issue and requires input and commitment from both parties. Consequently, this joint effort will be mutually beneficial and lead to a productive partnership. Even before being requested, responsible logistics service providers may send a copy of the assessment results to their major chemical customers.

6.7.7 Follow-up

The implementation of the actions defined in step 6 needs to be monitored and followed by a new cycle of risk review whereby a number of products may have moved in the hazard/exposure matrix from the black area into the gray or white area. This repetitive process will make information that is more pertinent available, focusing attention on specific areas and therefore helping to reduce further the risk related to the transport of chemicals. It is expected that this process will result in a decrease in the number of logistics accidents, which could be used as one objective yardstick to measure the improvements achieved.

6.8 REFERENCES

Recommendations on the transport of dangerous foods. Model Regulators. Eleventh revised edition. ISBN 98-1-139067-8. United Nations, New York and Geneva. 1999.

CHAPTER 7
REGULATIONS FOR TRANSPORTING HAZARDOUS MATERIALS IN FRANCE

Michel Marchand
Rue Alain Colas, BP 20 413, 29604 BREST (France)

7.1 INTRODUCTION

Each year about 115 million tons of hazardous materials are transported in France, 78 million tons of gasoline and 37 million tons of chemical products (Sari, 1990). In 1997, the total port traffic in most metropolitan ports amounted to an annual sum of 324 million tons, with 150 million tons of petroleum products and 12 million tons of chemical products (Babillot and Marchand, 1999).

When the industrial age began in the second half of the 19th century, public authorities began to control the transport of hazardous materials by water and road (Law of June 18, 1870) and by rail (Regulation of November 12, 1897). French legislation has changed since then. These regulations were abolished on April 15, 1945, when a Regulation for the Transport of Hazardous Materials came into force, still called the TMD or RTMD Regulation. The common regulations for ground transport come from the Law of February 5, 1942, which regulates loading, unloading, packing, storing, and handling conditions. The regulations and ensuing changes have been developed by the Interministerial Commission for the Transport of Hazardous Materials (CIMITD), whose mission is to adapt regulations to technical changes in the context of international regulations. The Commission is based on work done by the Mission of Transport of Hazardous Materials, which depends directly on the

TABLE 7.1 Annual Transport of Hazardous Materials in France (million tons/year)

	Petroleum products	Chemical products
■ Ports	150	12
■ Road	62	24
■ Rail	12	11
■ Waterways	4	2

Source: Sari, 1990 and Babillot and Marchand, 1999.

Ministry of Equipment and Transport. French regulations concerning the transport of hazardous materials are RTMDR for road transport and RTMDF for rail transport. Regulations for the transport of hazardous materials (RTMD) do not apply to rigid transport (gas pipelines, oil pipelines), which are treated according to specific laws concerning the type of materials (petroleum products, gas, or chemical products). Nuclear material falls under a special category with special measures; all major transport of radioactive materials is directed by the Ministry of the Interior's Direction of Defence and Civil Safety.

French regulations on transporting hazardous materials are classified into two large groups, depending on whether transport is domestic or international.

7.2 DOMESTIC TRANSPORT

When the view of the interministerial commission for transporting hazardous materials (CITMD) was adopted in December 1998, it was decided to adopt the same rules for transporting hazardous materials in France, whether national or international (J.O., 1999). These new conditions apply to road, rail, and domestic transport. French regulations, therefore, comply with international road (ADR), rail (RID), and waterway (ADNR) regulations.

7.2.1 International Road Transport

This is subject to an international Convention, the European Agreement concerning International Road Transport of Hazardous Materials (ADR), developed under the auspices of the United Nations' European Economic Commission (EEC/UNO) and signed in Geneva in 1957. The ADR, which came into force in France on January 29, 1968, contains the text of the agreement and two annexes.

The text of the agreement outlines the measures that apply to the transport of hazardous goods: materials and packing, transport materials, loading and unloading, information concerning transport, and transport and parking. Annex A, concerning hazardous materials and objects, gives the categories and particular instructions according to the nine transport classifications: packing conditions, registration and hazard labeling, notification in the transport documents, and instructions for empty packaging. Annex B concerns the material transported and the transport itself, including general and particular conditions according to transport classification, as well as a number of technical appendices concerning transport (tanks, tank containers, electric equipment, approved vehicle certificate, and driver training).

7.2.2 Rail Transport

This is subject to the Regulation concerning International Rail Transport (RID), dating from 1883. Thirty-six countries have joined the RID (UIC, 1995). The RID is made up of three distinct sections relating to both general and particular conditions according to transport classifications and the technical appendix. General instructions of the RID provide a certain number of essential guidelines concerning various operators and carriers, materials and packing, loading and unloading, and information about transport and parking.

The second part of the RID is devoted to the study of various hazard classifications, according to the same classification of materials and transport conditions applicable to packs, inscriptions and labeling, material and means of transport, and the illegality of collective loading and empty packages.

Finally, the RID contains 11 technical appendices covering a variety of aspects. The RID edition, published by the French Rail Company (SNCF), contains an alphabetical list of the materials, giving the hazard identification number for each product, its material identification number (UN number), its classification, hazard label, and the NHM code (Nomenclature Harmonisée Marchandises).

7.2.3 Waterway Transport

Rules for domestic transport are subject to regulations concerning the transport of hazardous materials on the Rhine, the ADNR Law, which was adopted by the Central Commission for Navigation on the Rhine (CNNR) on December 1, 1993.

7.3 INTERNATIONAL TRANSPORT

7.3.1 Port Transit

Port installations must conform to a certain number of instructions and safety measures during entrance and exit of goods in and out of ports, loading and unloading of ships and traffic circulation, and handling and storing of goods in port zones. These instructions are in the Regulation for the Transport and Handling of Hazardous Materials in Sea Ports, still called the RPM regulation. The RPM is completed by local regulations in the form of bylaws so as to respect the differences of each port. The RPM includes general instructions applicable to all hazardous goods and specific instructions applicable to groups of materials for which similar or identical safety measures are to be taken. The regulation is generally applied to operations carried out off the ships, meaning that it concerns measures to be taken on land or in the port itself. The Port Authority is responsible for controlling the application of port regulation.

Safety measures to be taken on board ship must comply with international conventions adopted and put into force in the French merchant navy. They concern the 1974 SOLAS Convention for the Safety of Human Life at Sea (the establishment of an acceptable level of safety onboard ship), the MARPOL 73/78 Convention for the Prevention of Sea Pollution by Ships, and the 1972 Container Safety Convention (CSC).

The IMDG Code (International Maritime Dangerous Goods Code) establishes international rules for maritime transport of dangerous goods. It is aimed at facilitating the application of the SOLAS Convention for the safety of human life at sea concerning the maritime transport of dangerous goods in packs or in solid form in bulk. It is also designed to provide a basis for national regulations for countries that practice such transport. Annex III of the MARPOL 73/78 Convention addresses the prevention of pollution by hazardous substances carried at sea in packages and draws on the recommendations of the IMDG Code. In the framework of the CSC Convention, a certain number of organizations are authorized to carry out controls and to issue CSC approval of international containers transported by sea and land.

To avoid discrepancies between national regulations and international measures for maritime transport of dangerous goods, French shipping regulations (RMF) have adopted measures of the IMDG Code, certain points being completed by instructions specific to France. This is the Bureau des Affaires Maritimes (Centre de Sécurité des Navires), responsible for verifying that the shipping of dangerous goods conforms to the instructions of the IMDG Code.

7.3.2 Air Traffic

Until 1983, air transport of hazardous materials was controlled by a law dating from August 22, 1957, stating that air companies could not transport hazardous materials without special authorization from the Director General of Civil Aviation. Authorized permission was given when companies transported the materials in accordance with the International Air Transport Association (IATA), applicable to almost all air companies. In accordance with work carried out by the IATA in 1981, the International Civil Aviation Organization (ICAO), an institution linked to the United Nations, adopted a document entitled Technical Instructions for Safe Air Transport of Hazardous Materials. These technical instructions, which must be respected by professional personnel involved in the air transport of hazardous materials, are applicable in France according to the law of January 14, 1983. The ICAO regulations include nine sections particularly concerned with the classification of dangerous goods, packing instructions, the responsibility of the sender and the user, and required training. The 34th edition of the IATA regulations of 1993, concerning the transport of dangerous goods, covers all the ICAO instructions. Operators are advised to use the IATA text, which is easier to handle.

7.4 REFERENCES

Babillot, P., and M. Marchand, Institut Français de l'Environnement (IFEN). 1999. "Traffic maritime et pollution du milieu marin," *Les Données de l'Environnement,* no. 44, 4 pp, 1999.

Journal Officiel (J.O.). 1999. "Arrêté du 17 décembre 1998." no. 1, January 1, pp. 53–75 (road), 75-82 (rail), 82–85 (waterway).

Sari, J. C. 1990. "Risques technologiques majeurs. Scénarios d'accidents de transport de matières dangereuses," in *Le Transport des Matières Dangereuses,*" Journée d'Etude et de Perfectionnement, CIC Marseille, pp. 49–54, June 6.

Union Internationale des chemins de fer (UIC). 1995. "Le transport ferroviaire des marchandises dangereuses en Europe," UIC, Paris.

CHAPTER 8
INTERNATIONAL CHEMICAL ENVIRONMENT (ICE)—EMERGENCY RESPONSE

F. Jonckheere
Trade and Environment Services, European Chemical Industry Council/ Conseil Européen de l'industrie chimique (CEFIC), Brussels

8.1 GENERAL CONCEPT

8.1.1 Objective

Within the framework of responsible care, the chemical industry makes every effort to transport goods to and from its manufacturing sites and storage locations safely and in full accordance with relevant regulations and codes of practice. If, nevertheless, a chemical distribution incident happens, the chemical industry will provide information, practical help, and, if necessary and possible, intervention equipment to the competent emergency authorities in order to minimize adverse effects.

International chemical environment (ICE) is the cooperative program of chemical companies established to achieve this goal. It aims to set up a framework for effectively providing this competent assistance in each European country by:

- Using the existing emergency response schemes of individual chemical companies
- Building on existing local, regional, and product-related emergency response schemes (e.g., for chlorine, isocyanates, and ethylene oxide)
- Cooperating with national authorities through the National Chemical Federation
- Communicating and exchanging information with other national ICE schemes
- Promoting mutual assistance within the chemical industry

This framework or national ICE scheme is intended only for distribution incidents, such as those outside manufacturing sites, and is formalized in a Protocol between the National Chemical Industry Federation and the national competent authorities, such as police and fire brigades.

8.1.2 Level of Chemical Industry Support

Participation in a national emergency scheme is voluntary but requires the commitment of a company to provide help in (land-based) distribution incidents at the request of the competent authorities. Taking into account the capabilities and resources of a company, there are three levels of intervention:

Level 1: remote product information and general advice by telephone or fax

Level 2: advice from an expert at the scene of an incident

Level 3: assistance with personnel/equipment at the scene of an incident

Such a commitment is first applied to products manufactured by the company itself. This is normally incorporated into the company's own distribution emergency response scheme. The emergency services may also require advice or help if the product supplier is not known or cannot be contacted. In such cases, companies may offer assistance on the basis of a prior arrangement with the national ICE scheme. However, safeguarding their own site facilities takes priority over level 2 and level 3 intervention.

The ultimate responsibility for any intervention on the site remains with the competent emergency authorities.

The key document in the Protocol between the competent national authorities and the chemical industry (represented by the National Chemical Federation) is the list of participating companies. Details for each participant are provided, such as address, telephone and fax numbers, range of products (hazardous products are identified by the four-digit UN number), time availability, intervention equipment, and area of intervention (if limited). This normally includes a map showing the location of the participating companies so that the competent authorities can contact the site nearest to the transport accident.

8.1.3 National ICE Center

The focal point of a national ICE scheme is the national ICE center, which the emergency authorities can call when:

- The supplier cannot be contacted.
- An incident happens involving international movements, necessitating contacts with national ICE centers in other countries.
- Mutual assistance within the national ICE scheme needs to be mobilized.
- The product or the producing company cannot be readily identified because of missing information.

When called by the authorities, the national ICE center will provide initial telephone advice for the immediate control of the incident in the local language. It will promptly alert the producing company, obtain further information (possibly via other national ICE centers), or mobilize mutual assistance. To do this, the center has at its disposal appropriate communications equipment, a library of reference books or databases, and up-to-date telephone and fax numbers of contacts within the chemical industry. The center is manned 24 hours a day by at least one person who can communicate in English as well as the local language(s) in order to facilitate communication between the national ICE centers.

8.1.4 Information Sources

Safety data sheets (SDSs) are the main source of information. Participating companies must therefore ensure that these sheets are accessible for their own products at all times at the locations indicated as company contacts in the national ICE scheme. A series of reference books or databases are available at national ICE centers to provide initial advice, or the centers may have access to a set of SDSs.

In 1993, the European Chemical Industry Council (ECIC) published a guidance document on the use of SDS entitled *Distribution Emergency Response—Guide to Use Safety Data Sheets.*

8.1.5 Typical Flow of Response

The typical flow of response in transport emergencies is illustrated in Fig. 8.1. However, each country can adapt the operation of a national ICE scheme to its own specific needs and practices already in place.

8.1.6 Liability

All chemical companies that may provide assistance or advice in case of distribution incidents should confirm with their insurance companies that their policies do cover potential claims arising from such service for their own products. Those companies that may be involved in giving assistance or advice with respect to products other than their own should notify their insurance companies specifically of this fact and obtain confirmation that their general public liability policy will provide coverage for these activities. This should not lead to increases in insurance premiums.

In 1994, CEFIC published a brochure entitled *Distribution Emergency Response—Legal and Financial Aspects of Mutual Assistance between Companies.*

8.1.7 Responder Profile

In order to provide competent assistance, responders in the national ICE center or in a company must meet a number of minimum requirements, either through training or experience. These requirements have been agreed upon within the chemical industry and are outlined in a CEFIC publication entitled *Distribution Emergency Response—Guidelines for Use by the Chemical Industry,* published in 1993.

8.1.8 Membership

A national ICE scheme is a voluntary initiative normally open to all manufacturers and distributors of chemical products. It is administered by the national Chemical Industry Federation, which keeps a register of participating members and arranges for financial contributions. Up-to-date information about available assistance is communicated promptly to the national competent authorities.

8.2 CURRENTLY OPERATIONAL NATIONAL ICE SCHEMES AND CENTERS

Because all national ICE schemes fit into the concepts already described, the following concentrates on their specific aspects. "Companies" should be understood as "sites" or

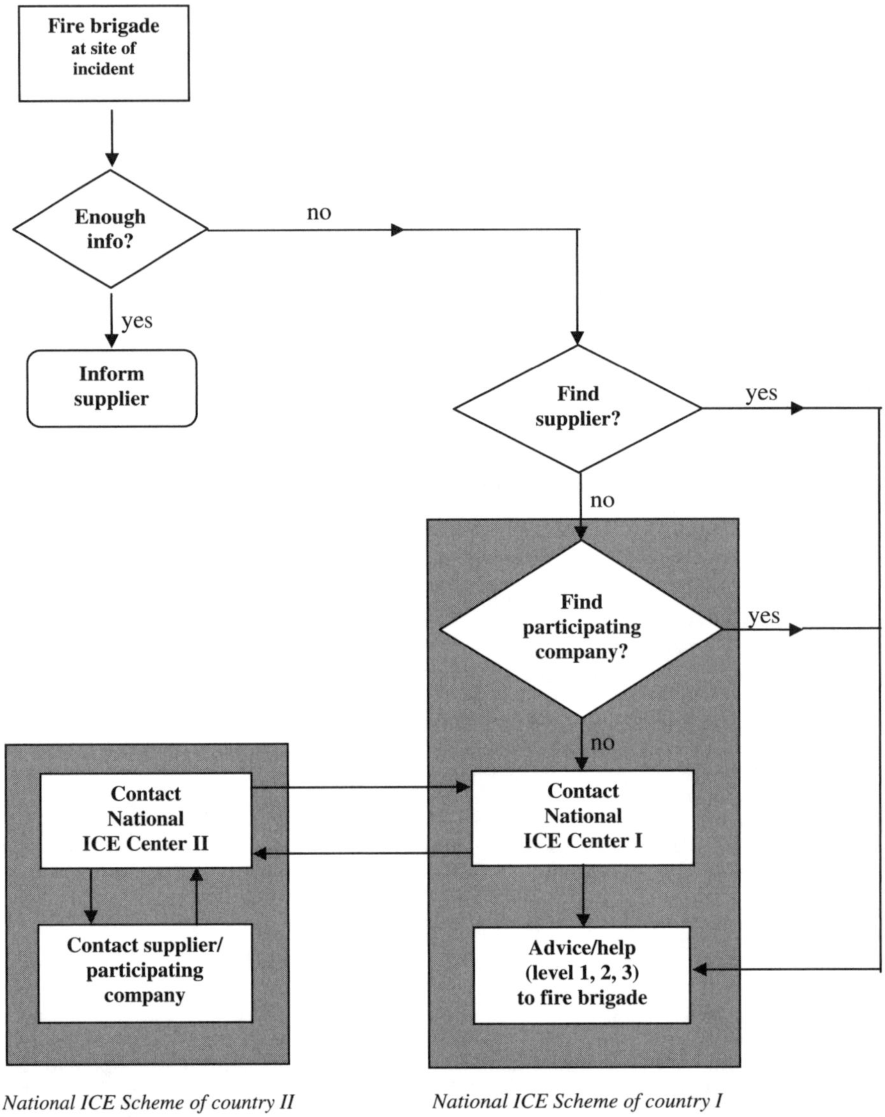

FIGURE 8.1 Flow of response in transport emergencies.

"contact points." In fact, one company might have various manufacturing sites that participate in the national ICE scheme of the same country.

8.2.1 Germany

In Germany, the national ICE scheme is called TUIS (Transport Unfall Information und Hilfeleistungs system). There are 10 centers:

- BASF—Ludwigshafen (acting as the national ICE center for international contacts)
- Bayer—Leverkusen
- Hoechst—Frankfurt
- Henkel—Düsseldorf
- Hüls—Marl
- Infra Leuna—Leuna
- Merck—Darmstadt
- Schering—Berlin
- Wacker—Chemie—Burghausen
- BASF—Schwarzheide

An official Protocol has been signed between the German Chemical Industry Federation (VCI) and the competent authorities of the various Bundesländer. TUIS is recognized as a valuable source of assistance in chemical transport accidents and is to be called upon by the emergency authorities when required.

Some 130 companies are listed in the TUIS Manual, which specifies such information as addresses, contact numbers, level of assistance, and products (by name and UN number). Competent authorities may directly contact the nearest company listed in the TUIS Manual for the product(s) involved or one of the 10 centers that can act for all products. All three levels of assistance are provided.

Further information can be obtained from VCI (H. Koehler) at telephone 49-(0) 69-25560; fax 49-(0) 69-225561607; or e-mail koehler@vci.de.

8.2.2 Austria

In Austria, the national ICE scheme is called TUIS (Transport Unfall Information und Hilfeleistungs system), with 34 companies participating. These companies are listed in a manual and may be contacted directly by the national competent authorities (such as police and fire brigades) in case of transport accidents. All three levels of assistance are provided.

Based on an agreement between the Austrian Chemical Industry Federation (FCIO) and the German Chemical Industry Federation, BASF, Ludwigshafen (Germany), also acts as national ICE center for Austria.

Further information can be obtained from FCIO (E. Tomschik) at telephone 43-(0) 1-50105; fax 43-(0) 1-50206280; or e-mail TomschikE@fcio.wk.or.at.

8.2.3 Belgium

In Belgium, the national ICE scheme is called BELINTRA (Belgian Intervention System for Transport Accidents). Under a formal agreement between the Belgian Chemical Industry Federation (Fedichem) and the competent authorities, the authorities can contact either BASF, Antwerp (Dutch-speaking), or Solvay, Jemeppe (French-speaking). They will provide assistance themselves or call on one of the approximately 30 participating companies for support. All three levels of assistance are provided.

BASF, Antwerp, acts as national ICE center for international contacts. Further information can be obtained from Fedichem (R. Dillen) at telephone 32-(0) 2-2389711; fax 32-(0) 2-2311301; or e-mail rdillen@fedichem.be.

8.2.4 Czech Republic

In the Czech Republic, the national ICE scheme is called TRINS and 10 companies participate. Under a formal agreement between the Czech Chemical Industry Federation (SHCP) and the competent authorities, the emergency authorities may either directly contact the companies listed in a Manual or contact the Czech national ICE center in case of an accident. All three levels of assistance are provided. The emergency department of the chemical company Chemopetrol, Litvinov, acts as the national ICE center. Further information can be obtained from SCHP (I. Zika) at telephone 420-(0) 2-67154131; fax 420-(0) 2-67154130; or e-mail mail@schp.cz.

8.2.5 Denmark

In Denmark, the national ICE scheme is called RVK and 24 companies participate. There is a formal agreement between the Danish Chemical Industry Federation (FDKI) and the Emergency Management Agency (Copenhagen) of the Danish Ministry of Interior on emergency response by the chemical industry. The Agency maintains a 24-hour service to fire brigades by providing relevant information on chemicals. Under the agreement, the Agency can request advice or assistance (at all three levels) from any of the 24 companies, according to the product(s) involved in the accident. The Agency also acts as the national ICE center.

Further information can be obtained from FDKI (J. Jessen) at telephone 45-33151748; fax 45-33151722; or e-mail jj@fdki.dk.

8.2.6 Finland

In Finland, the national ICE scheme is called FINTERC. Some 30 companies participate, including the major Finnish chemical producers. Under a formal agreement between the Finnish Chemical Industry Federation (KT ry) and the public Helsinki Area Emergency center of the Finnish Ministry of Interior, the Center deals with all requests for information from emergency authorities in case of chemical transport accidents. The Center is linked to the Finnish Register of Chemical Products, containing the safety data sheets of chemicals that may endanger health or the environment. KT ry has provided the Center with the list of names and contacts (per product) of the experts within the chemical industry who can be called upon in case of accidents. All three levels of assistance are provided. The Center also acts as national ICE center.

Further information can be obtained from KT ry (S. Loikkanen) at telephone 358-(0) 9-172841; fax 358-(0) 9-630225; or e-mail seppo.loikkanen@kemia.ttliitot.fi.

8.2.7 France

In France, the national ICE scheme is called Transaid and some 160 companies participate. Under a formal agreement between the French Chemical Industry Federation (UIC) and the Ministry of the Interior, the competent authorities can call participating companies for assistance. Whereas level 1 assistance will be provided at all times without any formalities, assistance at level 3 involves the requisitioning of intervening persons by the authorities according to an official procedure.

All operational data related to Transaid, such as company addresses and contact numbers, level of assistance, product names, and corresponding UN numbers, are accessible by the competent authorities via the French electronic Minitel system. Based on an agreement between the French Chemical Industry Federation and Cedre (Marine Pollution Expert Centre in Brest, Brittany), Cedre assumes the role of national ICE center for contacts with national ICE centers in other countries.

Further information can be obtained from UIC (D. Rain) at telephone 33-(0) 1-46531100; fax 33-(0) 1-46531105; or e-mail drain@uic.fr.

8.2.8 Hungary

In Hungary, the national ICE scheme is called VERIK and seven companies participate. Under a formal agreement between the Hungarian Chemical Industry Federation (MAVESZ) and the competent authorities, the emergency authorities may either directly contact the companies listed in a manual or contact the Hungarian national ICE center. All three levels of assistance are provided. The Fire Department of the Hungarian Oil and Gas Corporation (Szazhalombatta) acts as the national ICE center.

Further information can be obtained from MAVESZ (L. Bondar) at telephone 36-(0) 1-3438920; fax 36-(0) 1-3430980; or e-mail l.bondar.mavesz@mail.datanet.hu.

8.2.9 Italy

In Italy, the national ICE scheme is called SET (Servizio Emergenze Trasporti) and participation is restricted to companies adhering to the responsible care program of the Italian Chemical Industry Federation (Federchimica). This program calls for provision of 24-hour, level-1 assistance as a mandatory requirement. Seventy companies are now participating. Under a formal agreement between Federchimica and the competent authorities, the Emergency Authorities may either directly contact a participating company or contact the Italian national ICE center. This centre will make sure that a company participating in SET and competent for the product(s) involved in the transport accident, provides the requested level of intervention to the emergency authorities. All three levels of assistance are provided and a list of standard intervention equipment for the 17 companies that provide level-3 support is included in the SET Manual. The emergency Department of the chemical company Enichem (Porto Marghera) acts as the national ICE center.

Further information can be obtained from Federchimica (R. Mari) at telephone 39-02-268101; fax 39-02-26810209; or e-mail tes1@federchimica.it.

8.2.10 Netherlands

In the Netherlands, the national ICE scheme is called TRC (Transport Responsible Care) and some 15 companies participate. There is a formal agreement between the Dutch Chemical Industry Federation (VNCI) and DCMR (Dienst Centraal Milieubeheer Rijnmond, Rotterdam). On behalf of the Dutch Ministry of the Interior, DCMR maintains a 24-hour service to fire brigades on emergency response by providing relevant information on chemicals. Under this agreement, DCMR can request advice from any of the participating companies, according to the product(s) involved in the accident. DCMR also acts as national ICE center.

Further information can be obtained from VNCI (D. Schuddebeurs) at telephone 31-(0) 70-3378787; fax 31-(0) 70-3203903; or e-mail schuddebeurs@vnci.nl.

8.2.11 Spain

The Spanish Chemical Industry Federation (FEIQUE), in cooperation with the Spanish Ministry of the Interior, has set up a national ICE scheme called CERET. A national ICE center has been established within the Ministry of the Interior in Madrid that became fully operational in 1999. The center maintains a 24-hour service to competent authorities by providing relevant information on chemicals. Under the CERET scheme, the participating companies (currently 90) provide emergency information about all their transported products to the database of the national ICE center. In case of an accident, the national ICE center can request further assistance from a participating company, depending on the product(s) involved in the accident. All three levels of assistance are provided.

Since early 1997, a regional scheme, called ICECAT and covering the Catalan region, has been operational. It is the result of a formal agreement between the Catalan Chemical Industry Federations (FEDEQUIM and AEQT) and the Catalan Civil Protection center in Barcelona according to a concept that is identical to the future CERET. Some 20 companies based in the Catalan region are participating in ICECAT. In addition to its regional role, the center in Barcelona acted as national ICE center for Spain until CERET became fully operational. It now operates as a subsidiary center.

Further information can be obtained from FEIQUE (F. Perez Garcia) at telephone 34-(0) 91-4317964; fax 34-(0) 91-5763381; or e-mail feique@interbook.net.

8.2.12 Sweden

In Sweden, the national ICE scheme is called ERC (Emergency Response Center). Under a formal agreement between the Swedish Chemical Industry Federation (Kemikontoret) and the Swedish Poison Information Center (GIC) in Stockholm, the center can be contacted in case of a chemical accident. The Poison Center provides only level-1 assistance. Information and advice are based on detailed knowledge of the health hazards of chemical products and appropriate first aid measures. The Poison Information Center has more than 30 year's experience in providing such advice, and chemical companies regularly supply them with updated information on their products.

Because advice on fire, explosions, and other areas is the responsibility of the emergency authorities, the Poison Information Center refers these matters immediately to the officer on duty of the Swedish National Rescue Services Agency. Questions concerning environmental impact may also have to be directed to industry experts. Kemikontoret provides the center with the names of experts in approximately 100 chemical companies who can be contacted if necessary. The Poison Information Center also acts as national ICE center.

Further information can be obtained from Kemikontoret (U. Nyman) at telephone 46-(0) 8-7838000; fax 46-(0) 8-6636323; or e-mail unn@chemind.se.

8.2.13 Switzerland

In Switzerland, the national ICE scheme is called Chemiefachberatung. The Swiss Chemical Industry Federation (SGCI/SSCI) provides the fire brigades with a list of names of experts in 35 companies who can be contacted for advice on health and environmental aspects in case of a chemical transport accident. All three levels of assistance are provided. The emergency department of the chemical company Novartis in Basel assumes the role of national ICE center for international contacts with national ICE centers in other countries.

Further information can be obtained from SGCI/SSCI (A. Grieder) at telephone 41-(0) 1-3681711; fax 41-(0) 1-3681770; or e-mail alfons.grieder@sgci.ch.

8.2.14 United Kingdom

In the United Kingdom, the national ICE scheme is called CHEMSAFE. British domestic transport legislation requires a 24-hour emergency response telephone number to be displayed on vehicles carrying dangerous goods in bulk. The provision of level 1 response (either through in-house resources or by using a third-party agency) is a mandatory requirement for all members of the U.K. Chemical Industries Association (CIA). Companies are also required to provide SDS information on all their products to the National Chemical

Emergency Centre (NCEC) at Culham, Oxfordshire, which assumes the role of national ICE center.

Level 2 response is provided either by the product owner or by some 30 industry operational sites/centers across the country. These constitute a 24-hour mutual assistance network for other CIA members. Arrangements for level-3 response are normally provided through the contract haulier or, should a spillage occur, by companies specialized in the disposal of chemical waste. The NCEC in Culham acts as national ICE center.

Further information can be obtained from CIA (A. Robinson) at telephone 44-(0) 171-8343399; fax 44-(0) 171-8344469; or e-mail robinsona@cia.org.uk.

P • A • R • T • 2

NUCLEAR EMERGENCIES

CHAPTER 9
RESPONSE TO NUCLEAR OR RADIOLOGICAL EMERGENCIES

T. McKenna
International Atomic Energy Agency, Vienna

J. F. Lafortune
International Safety Research, Ottawa, Ontario, Canada

R. Martinčič
J. Stefan Institute, Ljubljana, Slovenia

E. Buglova
Research and Clinical Institute of Radiation Medicine and Endocrinology, Ministry of Health, Minsk, Belarus

P. F. L. Heilbron
Brazilian Nuclear Energy Commission, Botafogo, Rio de Janeiro, Brazil

9.1 INTRODUCTION

9.1.1 The Nature of Radiation Hazard

Thousands of devices containing potentially dangerous amounts of radioactive material are used in the world. However, the safety record for the use of radioactive materials is incredibly good. As shown in Table 9.1 there are only about two deaths or serious injuries worldwide each year from accidental radiation exposure. This is because most countries have rules requiring that dangerous amounts of radioactive material (and chemicals) be carefully controlled at all times.

Emergencies involving radioactive material are very similar to those involving hazardous materials. In both cases, serious health effects can result. In a radiation and chemical emergency, that hazard comes from living tissue being damaged when it is exposed to either hazardous chemicals or radiation, and the danger increases with increasing amounts of material and time of exposure. There are both chemicals and radioactive materials that can be very hazardous in very small amounts, and there are facilities containing large amounts of chemicals and radioactive material that could result in hazardous exposures at 100 to thousands of meters from the source.

TABLE 9.1 Early Deaths from Radiation Accidents (1945–2000)[a]

Year	Location	Radiation source	Fatalities	
			Worker	Public
1945	U.S.A.	Critical assembly	2	
1958	Yugoslavia	Experimental reactor	1	
1958	U.S.A.	Critical assembly	1	
1960	U.S.S.R.	Suicide Cs-137	1	
1961	U.S.S.R.	Submarine reactor	8	
1961	U.S.A.	Small military power reactor	3	
1961	Switzerland	Tritiated paint	1	
1962	Mexico	Lost radiography source		4
1963	China	Seed irradiator		2
1964	West Germany	Tritiated paint	1	
1964	U.S.A.	Uranium recovery plant	1	
1972	Bulgaria	Suicide	1	
1975	Italy	Food irradiator	1	
1978	Algeria	Lost radiography source		1
1981	U.S.A.	Industrial radiography	1	
1982	Norway	Instrument sterilizer	1	
1983	Argentina	Research reactor	1	
1984	Morocco	Lost radiography source		8
1985	U.S.A.	Accelerator	2 (?)	
1986	Chernobyl, U.S.S.R.	Nuclear power plant	31[b]	
1987	Goiânia, Brazil	Stolen teletherapy source		4
1989	El Salvador	Industrial sterilizer	1	
1990	Israel	Industrial sterilizer	1	
1991	Belarus	Industrial sterilizer	1	
1992	China	Lost sealed source		3
1994	Estonia	Lost sealed source		1
1996	Georgia	Lost radiography source		2 (?)
1999	Japan	Criticality	1	
2000	Thailand	Lost sealed source	1	
Total:			61	25

[a] Does not include the patients who died because of misadministration.
[b] Includes members of the fire brigade who responded from off-site; includes 2 who died due to explosion and 1 due to thermal burns.

The response to radiation and chemical emergencies is also very similar. In both cases, our senses (e.g., smell or sight) may not be able to detect hazardous levels of the material. Therefore, the initial response is often carried out based on secondary indications of the hazards such as signs or placards indicating the presence of a hazardous material, appearance of medical symptoms in exposed individuals, or readings from specialized instruments.

In both a radiation and chemical emergency, the response attempts to:

1. Control the source of the hazard (release or exposure)
2. Protect emergency personnel responding to the emergency
3. Protect the public

The actions taken to protect the public are also very similar for both chemical and radiation emergencies. Actions are taken to prevent exposure, remove any contamination, and treat exposure. The actions used to prevent exposure are the same: evacuation, sheltering, decontamination, and restrictions on contaminated food. The actions to protect workers are also the same: use of protective clothing and respiratory protections and specialized training and instruments. Local officials and emergency service personnel will typically carry out the initial response to many chemical and radiological emergencies.

What are the differences between emergencies involving hazardous amounts of chemicals and radioactive materials? The basic differences are:

1. Responders generally have no experience with radiation emergencies, as these emergencies are very rare.
2. Very small amounts of radioactive material and radiation (unlike many chemicals) can be immediately detected with simple, commonly available instruments.
3. The health effects resulting from radiation exposures will not appear for days, weeks, or even years.
4. The public, media, and responders often have an unrealistic fear of radiation due to lack of information.

9.1.2 Uses of Radioactive Materials

This section will describe the uses of radioactive material. It will not describe other sources of radiation exposure such as X-ray machines used in hospitals, as these machines can be turned off, whereas radioactive sources cannot.

Radioactive material is everywhere. Our bodies, houses, air, and food are radioactive. Devices containing small amounts of radioactive material can be found in homes, workplaces, or schools, and radioactive materials are transported continuously and safely throughout the world. We are bombarded by radiation from space constantly. Radioactive materials are widely used every day in our society. Nuclear power plants are used throughout the world to generate electricity. Radioactive material is used by the military and industry and in research. In addition, radioactive materials are widely used in medicine to diagnose and treat diseases.

Common Uses. One of the most common uses of radioactive materials is in medicine, where they have been used since about 1900. Radioactive materials are used in most hospitals and in many small clinics for two different types of medical applications. The first is diagnosis, during which a small amount of radioactive material is injected into a person. As the amount of radioactive material used does not cause harm to the person, these uses have not resulted in any serious emergencies. The second type of medical use is for treatment. During treatment, radioactive material is used, in most cases, to destroy cancer cells or tissue. If the sources can kill or destroy cancers or other organs, they can also cause injury or death if improperly controlled.

Brachytheraphy uses radioactive needles or catheters to place a small amount of very highly radioactive material very close to the tissue to be destroyed. There have been emergencies resulting from brachytheraphy sources being left in the person or being lost. These emergencies have not caused serious health effects among the general public, medical personnel, or responders. However, failing to remove a brachytheraphy radioactive source has resulted in fatal exposures to the person being treated.

Teletherapy uses a radioactive source outside the body to destroy cancerous tumors, and the source must therefore be much larger than that used in brachytheraphy. The radioactive source used in a teletherapy unit is very dangerous and can result in lethal doses in less than an hour if not shielded. A typical teletherapy source is a cylinder about 5 cm in diameter and 5 cm long. The source is housed in a large shielded device called the teletherapy head. A serious emergency happened in Goiânia, Brazil, in 1987 when an abandoned teletherapy head was stolen, taken to a scrap yard, and disassembled. The thieves thought that the metal components were valuable. The radioactive source was removed and some of the 100 g of radioactive material were dispersed. Brazilian officials became aware of this situation when a doctor realized he was treating effects of radiation exposure in new patients. Once alerted, officials acted quickly to identify and isolate the important sources of radiation exposure and treat those exposed. The response continued for months until the cleanup was completed and public confidence was reestablished. This event resulted in 4 deaths, 20 people being hospitalized, contamination of several square blocks of a major city, monitoring of 100,000 people, and 370 trucks full of contaminated waste.

Industrial radiography is another common use of potentially very hazardous amounts of radioactive material. A portable camera that contains a radioactive source is used to test welds, pipes, and so on at construction sites. These cameras weigh 25 to 250 kg, but the radioactive sources they contain can be less than 2 cm long and 1 cm in diameter. These sources are sometimes attached to cables that carry them remotely to where they will be used. There have been cases when the radioactive source became disconnected from the cable and was lost. Serious injuries and deaths have resulted from people finding these lost sources and placing them in their pockets or taking them home.

Other portable devices containing potentially hazardous amounts of radioactive material are also used in oil exploration, mining, and construction. These portable devices are often stolen because they appear to be valuable construction equipment or are in or attached to a truck being stolen. In some instances, thieves have removed and discarded the sources from the device (shielding) in public places. In some cases, prompt action by public officials to alert the public (and thus the thieves) of the hazard has resulted in the thieves telling officials where to find the sources. In at least two cases, the highly publicized arrival of national-level monitoring teams and aircraft to look for the sources convinced the thieves to return the sources and seek medical treatment.

There have been events in Mexico, Turkey, and Thailand resulting in fatalities from stolen, lost, or abandoned sources. These types of events can occur in virtually any country.

There are also potentially hazardous radioactive devices (gauges) permanently installed in facilities to measure the thickness of steel, the levels in tanks, or flows through pipes. In some cases, these devices have been left in place when the facility was abandoned or demolished and sold for scrap and subsequently melted down at scrap yards, resulting in the production of radioactive steel and contamination of the facility.

While several million packages containing radioactive material are safely transported every year, accidents do occur. There are approximately two emergency responses in the United States each day involving some type of radioactive material. Typically these involve a package carrying small quantities being punctured or crushed. The typical response is isolation and cleanup of a small area. Due to careful labeling and packaging, there have not been any transportation accidents resulting in serious health effects due to radiation exposure.

There are also many uses of radioactive materials that pose no radiological threat under any circumstances. These are called consumer products and include items such as gun sights, smoke detectors, and exit signs. While consumer products are not hazardous, they have resulted in emergencies. A sign containing tritium (a radioactive form of water) was stolen by children and brought to their home. In this case, the sign was opened, resulting in tritium contamination of the house. While this posed no health risk, it caused a major emergency because of the public's fear of any amount of radiation, and local officials were forced to respond as if there were a true radiological hazard. This demonstrates one of the unique characteristics of events involving radioactive materials. Emergency responses may be needed not because of the radiological risk but because of the perceived risk on the part of the public, media, or officials. In these cases, the response is intended to address the public concern.

The hazard levels of various common uses of radioactive material if the source becomes unshielded or ruptured are shown in Table 9.2.

TABLE 9.2 The Radiological Hazard from Unshielded or Ruptured Radioactive Devices

Device	Hazard
Brachytheraphy source >400 GBq[a] (10 Ci) Radiographic source >400 GBq (10 Ci) Self-shielded irradiator Teletherapy source Well logging source >40 GBq (1 Ci)	*High* Lethal doses in less than an hour of handling an unshielded source
Brachytheraphy source >40 GBq (1 Ci) Fixed gauge (level, density, thickness, etc.) X-ray fluorescence analyzer	*Moderate* Lethal doses would require hours or days of casual possession but severe tissue damage would occur within minutes of holding an unshielded source
Radiographic source <40 GBq (1 Ci) Well logging source <40 GBq (1 Ci)	*Low* Lethal doses very unlikely but severe tissue damage would occur within hours of holding an unshielded source
Depleted uranium (shielding, ballast, etc.) Moisture density gauge Static eliminator Tritium exit sign Pu pacemaker	*Minimal* Significant health effects impossible from radiation exposure

[a] Section 2.1

Large Facilities. Research indicates (NRC, 1988, 1990) that nuclear power plants or facilities storing large amounts of nuclear waste from reprocessed nuclear fuel pose the only risk of early deaths off-site resulting from a radioactive release. Other facilities containing large amounts of radioactive material, such as industrial irradiators, can result in serious injuries or deaths on-site.

There are over 400 commercial nuclear power plants (NPPs) operating in the world. Unlike a coal- or gas-fueled plant, the NPP energy does not come from the chemical combustion of fuel but from a nuclear reaction that results in the fission (or splitting) of fuel atoms. Within a few days of operation, the nuclear reaction in the core of the reactor will produce sufficient radioactive materials (fission products) to cause serious effects off-site if not contained. Consequently, nuclear power plants are built with numerous systems and barriers in place to prevent a serious release of radioactive materials. Analyses (NRC, 1990a) of NPP accidents show that, for the most severe possible releases, protective actions (e.g., evacuation of the people within several kilometers of the plant) should begin within one or two hours of the start of the emergency to be effective. Extensive emergency preparations have therefore been made both at the nuclear power plant site and in the surrounding areas. This typically includes:

1. The designation of an emergency-planning zone around each plant in which provisions have been made to notify the public promptly, e.g., with sirens, and implement evacuation and sheltering of the public
2. Specialized training and radiation protection for those who will provide emergency services (e.g., fire and police protection)
3. Emergency centers at which the on- and off-site response will be coordinated
4. A center for coordination of the information provided to the media; and

5. Provisions to promptly measure the amounts of radioactive material released and implement additional protective measures for the public if needed.

Most countries have established a system to classify significant nuclear emergencies at large facilities. Under this system, the declaration of a certain class of emergency will immediately initiate a coordinated response, with each response agency immediately taking preplanned actions. Upon detection, the plant will classify the event and on- and off-site authorities will immediately respond to correct the problem and protect the public and emergency workers. The system used in the United States, which is similar to that recommended by the International Atomic Energy Agency (IAEA), is discussed in Section 9.8.

The Chernobyl nuclear power plant accident released immense amounts of radioactive material into the environment. The release was carried high into the atmosphere by the heat generated by the accident and carried by the wind away from populated areas. This prevented hundreds of early deaths or injuries off-site. As the release moved away from the reactor site, radioactive particles were deposited (on the ground, trees, people, etc.), exposing people at greater distances to significant amounts of contamination on the ground. Drinking of contaminated milk and eating of contaminated food from family gardens resulted in a significant increase in thyroid cancer rates among children more than 350 km from the site. Contamination levels in agricultural products more than a 1,000 km away exceeded national restriction standards. Consequently, in many countries, emergency preparations have been made hundreds of kilometers from NPPs to deal with potential food contamination resulting from a release.

Accidents at NPPs can also result in very high doses of radiation on-site. The only early deaths resulting from such accidents have occurred among plant personnel or off-site fire fighters responding on-site. In the Chernobyl nuclear power plant accident, 28 people responding on-site died from radiation exposure.

9.2 BACKGROUND

9.2.1 Quantities and Units

All matter is composed of elements, which consist of characteristic atoms, also called nuclides. Atoms contain a positively charged nucleus and electrons, which carry negative electric charges. The nucleus is composed of positively charged protons and electrically neutral neutrons.

Nuclides of an element that have the same number of protons but different number of neutrons are called isotopes of that element. Some nuclides are stable but many are not; these are called radionuclides. The process of transformation is called decay. Radionuclides may decay by emitting an electron, i.e., a beta particle, or photons (gamma or X rays); or an alpha particle consisting of two protons and two neutrons. These processes are termed radioactivity. There can be several different isotopes for an element. Different isotopes are indicted by the symbol for the element with the atomic mass number for the isotope. For example, the important radioactive isotopes of iodine (I) are ^{129}I, ^{131}I, ^{132}I, ^{134}I, and ^{135}I. Each of these isotopes will act the same chemically but can be considerably different radiologically. ^{131}I and I-131 are the two common ways of indicating a particular isotope.

All radionuclides are uniquely identified by three characteristics: the type of radiation they emit, the energy of radiation, and the rate at which the spontaneous transformation occurs—the activity. The activity is expressed in a unit called the becquerel (Bq).[1] One becquerel equals one transformation per second. Different radionuclides are transformed at

[1] Activity was formerly expressed in a unit called the curie, 1 Ci = 3.7×10^{10} Bq.

different rates, and each radionuclide has its own characteristic transformation rate. The time required for the activity of a radionuclide to decrease to one-half of its value by decay is called the half-life. The half-life of a radionuclide is a unique, unalterable property of the radionuclide. Values for various radionuclides range from fractions of a second to millions of years.

When radiation passes through matter, it deposits some of its energy in the material by ionization or excitation of the absorber atoms. It is this ionization of tissue, accompanied by chemical changes, that causes the harmful biological effects of radiation. Radiation damage depends on the absorption of energy from the radiation or on the dose of radiation received. The basic unit of radiation dose is expressed in terms of absorbed energy per unit mass of matter, such as tissue. Absorbed dose is expressed in a unit called the gray (Gy).[2]

The damage-producing potential of a given absorbed dose depends on the type of radiation. An absorbed dose from alpha particles, for example, is more harmful than the same dose from beta radiation. To put all ionizing radiation on the same ground with regard to potential of causing harm, a radiation-weighted dose called equivalent dose is introduced. The unit is the sievert (Sv).[3] The equivalent dose is equal to the absorbed dose multiplied by a radiation-weighting factor. For gamma rays, X rays, and beta particles, the factor is set to 1. For alpha particles, for example, that factor is 20. Equivalent dose is the basic quantity used to assess exposure and determine the level of protection required.

In addition, the risk of harm is not the same for various tissues in the body. For example, it is lower for the bone surfaces than for the breast. This can be dealt with by taking the equivalent dose in each of the major organs and tissues of the body and weighting it by a factor related to the risk associated with that organ or tissue (tissue-weighting factor). The sum of the weighted equivalent doses is called the effective dose. Thus, the effective dose broadly represents the risk to health from any exposure to ionizing radiation.

Exposure of a person may be external or internal and may be incurred by various exposure pathways (see Section 9.3). External exposure may be due to direct irradiation from the source, airborne radionuclides in the air (immersion or exposure to an overhead plume), or radionuclides deposited onto the ground and onto a person's clothing and skin. Internal exposure follows from the inhalation of radioactive material either directly from a plume or resuspended from contaminated surfaces, from the ingestion of contaminated food and water, or through contaminated wounds. Total effective dose can be calculated by taking into account all dominant exposure pathways by which persons were exposed.

9.2.2 Health Effects

As previously stated, the process of ionization changes atoms and molecules. If cellular damage does occur and is not adequately repaired, it may prevent the cell from surviving or reproducing, or it may result in a viable, but modified cell. Radiation-induced effects of concern in emergency response fall into two general categories: deterministic and stochastic effects.

Deterministic Effects. The function of most organs and tissues of the body is unaffected by the loss of a small or sometimes even a substantial number of cells. If enough cells are lost, however, and the cells are important, there will be observable harm, reflected in a loss of tissue function. The probability of causing such harm is zero at small doses of radiation, but above some level of dose (the threshold) it increases to unity (100%). Above the threshold, the severity of the harm increases with dose. This type of effect is called deterministic because it is sure to occur if the dose is large enough and is higher than threshold. If the

[2] Absorbed dose was formerly expressed in a unit called the rad, 1 rad = 0.01 Gy.

[3] Equivalent dose was formerly expressed in a unit called the rem, 1 rem = 0.01 Sv.

loss of cells can be compensated for by the repopulation of cells, the effect will be relatively short lived. Deterministic effects, also called early or acute health effects, usually occur soon, i.e., hours to weeks after exposure. Thresholds differ depending on the organ, dose rate, medical treatment, and other factors.

If people of varying susceptibility are exposed to radiation, the threshold in a given tissue for deterministic effects of sufficient severity to be observable will be reached with smaller doses in more sensitive individuals.

Examples of deterministic effects are temporary or permanent sterility in the testes and ovaries; depression of the effectiveness of the blood-forming system, leading to a decrease in the number of a blood cells; and cataracts. A special case of deterministic effect is the acute radiation sickness resulting from acute whole body irradiation.

The organs of the body have different sensitivities to radiation and may receive considerably different doses depending on the chemical form and exposure pathway of the radioactive material. For example, if inhaled, radioactive iodine may dissolve and travel through the blood to concentrate in the thyroid, thus giving a very high dose to this sensitive organ while delivering a much smaller dose to other organs. Critical organs can be identified for many radionuclides. If the dose is kept below the threshold for deterministic health effects in this critical organ, then deterministic health effects will be prevented in all the organs.

Deterministic health effects occur when an organ receives a very high dose over a short period. As an example, the whole body must receive in one hour a dose over 1,000,000 times the radiation received normally (from natural sources) in one hour before deterministic effects will appear.

The first objective of the radiological response to an emergency is to take actions to prevent doses that could result in deterministic health effects. For the purpose of emergency response it is also necessary to note that if the doses are caused by an identified event, it will usually be possible to identify the affected individuals. This will allow them to receive specialized treatment. Also, some deterministic effects have characteristics that distinguish them from similar effects due to other causes, which may help to identify the affected individuals. The presence of an uncontrolled dangerous source in the public domain has, in many cases, been first indicated by the appearance of deterministic effects.

Stochastic Effects. Except as a result of serious accidents and the unwanted but inevitable irradiation of healthy tissues in radiotherapy, the doses incurred by humans are not large enough to produce deterministic effects. The main practical interest in the risks of radiation lies in the region of lower doses and dose rates that are experienced in radiation work or in other situations of everyday life.

Sometimes irradiation will not kill the affected cells, but may only alter them. A viable but modified somatic cell may still retain its reproductive capacity and may give rise to a clone. If the clone is not eliminated by the body's defence mechanisms, after a prolonged and variable period of delay termed the latent period, it may result in the development of malignant conditions, usually termed cancers, which are the principal late somatic effects of exposure to radiation. In contrast to deterministic effects, it is assumed that there is no threshold of dose below which stochastic effects (e.g., cancer) cannot occur. These effects do not occur in every exposed individual; the probability that an individual or one of his or her descendants may develop one of these effects increases with the dose received. Thus, even if the dose is very small, the person still has a chance, albeit a very small one, of incurring such an effect.

There is a considerable latent period between radiation exposure and the appearance of cancer. For most cancers in adults, the latent period is at least 10 years, or even longer. The shortest latent period is for leukemia and thyroid cancer (3 to 5 years). The appearance of radiation-induced cancers follows additive or multiplicative models of prediction with absolute or relative risks as main parameters. Assessment of the risk coefficients is based on the follow-up of exposed persons through epidemiological studies.

If the damage occurs in a cell whose function is to transmit genetic information to later generations, the effects, which may be of different kinds and severity, will be expressed in the progeny of the exposed person. This type of stochastic effect is called a hereditary effect.

The cancers caused by radiation, with or without the contributions of other agents, are not distinguishable from those that occur owing to other causes, or "spontaneously." By undertaking an epidemiological study on a large population group, however, it is possible to determine additional cancers resulting from the exposure. It follows, therefore, that the numbers of additional cancers resulting from exposure can be predicted, but not the individuals within the group that would be affected.

The second objective of radiological response is to take reasonable steps to reduce the risk of stochastic health effects. "Reasonable" means taking actions that do more good than harm. Any protective action has its own cost and risk. There will be a point at which taking a protective action will cause more harm than the radiation that the action is intended to prevent. For example, people were relocated from areas with dose rates due to Chernobyl accident similar to dose rates found elsewhere in Europe. This resulted in considerable psychological harm. Clearly this is an example of an unreasonable protective action. After considerable study, international organizations have established guidance for taking protective actions that will do more good than harm. This guidance will be discussed in Section 9.5.1.

9.2.3 Types of Emergencies

As seen in Section 9.1.2, the range of potential emergencies involving ionizing radiations is enormous, ranging from a major reactor accident to accidents involving small amounts of radioactive material. In general, emergencies may be classified into two broad categories, nuclear and radiological having in common the radiological nature of the threat. The arbitrary distinction between a radiological and a nuclear emergency is explained in this section.

Nuclear Emergencies. A nuclear emergency is a reactor accident or an accident at reprocessing plants or other large nuclear facilities. It is one that involves the nuclear fuel cycle (e.g., uranium, plutonium, thorium) and the potential for criticality. Examples of nuclear emergencies include the Three Mile Island accident in 1979 and the Chernobyl accident in 1986. The nuclear fuel damage that occurred during both of these events released fission products consisting of noble gases, iodine, and particulates and, in the case of Chernobyl, actinides to the environment.

An accident involving the detonation with partial nuclear yield of a nuclear weapon is also considered a nuclear accident.

The potential for health hazards is greatest for nuclear emergencies because the affected area can extend over hundreds of square kilometers and thousands of people can be affected.

Radiological Emergencies. A radiological emergency is one that involves sources other than nuclear fuel. The most common type of radiological emergency is the dispersion of and contamination from a single source, e.g., caesium, or the mishandling of a sealed source, e.g., iridium used in industrial gamma radiography. The dispersion of material from a nuclear weapon without a nuclear yield is also considered a radiological accident.

Accidents with radioactive sources or material include found radioactive material or contaminated areas or items, a lost or missing radioactive source, unshielded source, accidents in a laboratory, transport accidents involving radioactive sources or material, accidents with X-ray machines and particle accelerators, and an accidental reentry of a nuclear-powered satellite, which may lead to impact on the earth's surface and the spread of contamination. Radiological emergencies that could result from deliberate acts, such as terrorist activities or illicit trafficking, also fall within this category, although the security aspects of the response to such events are not within the scope of this chapter.

A nuclear accident that results in very serious consequences off-site at facilities located 100 to 1,000 km outside a given country is unlikely to have consequences significant enough to warrant urgent protective actions such as evacuation or sheltering in that country. However, it can still significantly affect the food chain, in some cases requiring the control of national food and water supplies. It can also have an indirect impact through, for example, food and supplies imported from affected countries, nationals living in affected countries or wanting to visit affected countries, and possibly from contaminated transport vehicles entering the country. Transboundary effects also may result from an accident at a facility located on or near major bodies of water. Radioactive material released in such an accident may be transported some distance from the accident site by water currents. Transboundary effects can also occur from such accidents as fire involving radioactive material.

9.3 EXPOSURE PATHWAYS

Exposure pathway is the term used to describe how the tissues in a person's body are exposed to the radiation. The exposure pathway is important because it determines effective protective actions for the public and emergency workers and the methodology for assessing the potential radiological consequences of the material. The importance of the various exposure pathways is determined by:

1. The radiological properties of the material (e.g., gamma, beta, alpha emitter, half-life)
2. The physical (e.g., gaseous, liquid, solid) and chemical properties of the material
3. The dispersal mechanism (e.g., airborne, carried by people)

A nuclear or radiological accident will involve the potential for exposure to radiation arising primarily from alpha or beta particles and gamma rays. In many cases, it would be a combination of those sources.

Gamma radiation is penetrating and represents the most common external radiation hazard, often referred to as "external exposure." Gamma emitters are commonly used in medicine and industry where the penetrating radiation is needed (see Section 9.1.2). Because it can penetrate matter, gamma radiation can be dangerous at considerable distances. It is thus the primary source of early deaths and injuries during radiation accidents. Commonly available instruments can easily detect dangerous levels of gamma radiation.

The external dose from a gamma emitter can be reduced in two basic ways: shielding (sheltering) and moving away (evacuation). Shielding can be placed between a person and the gamma radiation to absorb the radiation. As a general rule, the denser the shielding material the more effective it will be in reducing the external dose. As an example, the gamma dose rate from ^{60}Co (a strong gamma emitter) will be reduced by half by about 1 cm of lead, 2 cm of iron, 5 cm of concrete, or 10 cm of water. While shielding (or sheltering) could be used in an emergency, it must be done with care since the effectiveness of shielding can only be fully assessed by use of radiation monitoring equipment. The dose rate from a gamma emitter also decreases with distance. Doubling the distance from the gamma source will reduce the dose rate to ¼. This is called the inverse square law. Table 9.3 shows the relative dose rates at various distances from gamma emitter and demonstrates two very important points. First it shows that very high doses can be received by holding or carrying a gamma source. The dose rate at 1 cm (distance to tissue in a hand holding a gamma source) is 10,000 times the dose rate at 1 m. Thus the dose to the skin, hand, or other tissue in contact or within a few centimetres of even a low hazard source (see Table 9.2) can result in serious injuries within hours. Such events continue to result in horrific injuries. This demonstrates the danger of a lost or stolen high activity gamma source. The second point

TABLE 9.3 Gamma Dose Rate with Distance from Gamma Emitters

Distance (cm)	Relative dose-rate
1	10,000
2	2,500
5	400
10	100
100 (1 m)	1
1000	0.01

demonstrated by Table 9.3 is that distance can be used as a very effective protective measure. If it is suspected that an item is a gamma emitter, the best immediate action is to move away. Shelter and evacuation are also effective in reducing the external exposure in the event of an atmospheric release.

Beta particles can only penetrate through small amounts of matter, and thus the external dose is limited to the skin. In extreme cases, beta emitters have caused severe skin burns (e.g., among the firefighters at Chernobyl). Skin contamination could result from being in a cloud (or plume) of radioactive material and keeping the skin covered would be an effective protective action from the beta emitters in a plume. The more likely cause of serious skin contamination is unknowingly handling a ruptured radioactive source containing a beta emitter. Washing with soap and water is very effective in removing dangerous levels of skin contamination. However, early (deterministic) injuries or deaths from beta or alpha contamination are very rare.

Alpha particles are not an external radiation hazard, but they can result in significant internal dose if inhaled or ingested. In addition, beta emitters can be major sources of dose to internal organs if inhaled or ingested. Once the material enters the body, different radionuclides will concentrate in different organs thus greatly increasing the dose to these organs. If the material is airborne it can be inhaled. Airborne radioactive materials can result from an accident at a facility (e.g., nuclear power plant), a fire containing radioactive material, or resuspension of material deposited on the ground. However, resuspension in most cases is not an important source of dose.

Ingestion is often a very important source of dose, occurring when contaminated food is eaten or from contamination on the hands. Ingesting milk contaminated by the Chernobyl accident has caused many cases of thyroid cancer in Belarus and Ukraine. Very high, possibly fatal doses have resulted from ingestion of contamination from a ruptured source. If contamination is suspected, wash your hands, do not eat or smoke, keep your hands away from your mouth, and advise the public to do the same. However, early (deterministic) injuries and deaths from ingestion of radioactive material are very rare.

There are two basically different types of uncontrolled releases of radioactive material resulting in serious exposure from accidents: airborne release and nonairborne exposure. These are discussed in Sections 9.3.1 and 9.3.2. The historically important exposure pathways for these types of releases are shown in Table 9.4 along with some examples of accidents.

9.3.1 Non-airborne Exposure

Most radiological emergencies do not involve an airborne release of radioactive material. Typically, a radioactive source used in industry or medicine is lost or stolen. In this case humans are the most important method of spread and movement of the material in the

TABLE 9.4 Historically Important Exposure Pathways

Type of release	Pathways of concern	Examples
Nonairborne exposure	External exposure from: • Handling or carrying a radioactive source • Being in the vicinity of radioactive source • Ground contamination (e.g., from ruptured source) • Material on the skin (e.g., from ruptured source) Internal exposure from: • Ingestion of contamination on hands	Goiânia accident–spread of ^{137}Cs due to rupture of a teletherapy sealed source
Airborne release	External dose from: • The plume (cloud) • Radioactive material deposited on the ground • Radioactive material deposited on the skin	Chernobyl, Three Mile Island, or Windscale reactor accidents
	Internal exposure from: • Inhalation of radioactive material in the plume • Ingestion of food directly contaminated by plume and milk or meat from grazing animals	Tyumensk reactor fuel reprocessing waste tank explosion

environment. The Goiânia accident involved a radioactive source being carried, opened, then intentionally spread on the bodies and eaten. The source of exposure could be a point source (e.g., small highly radioactive object) or, as was seen at Goiânia, could ultimately involve a large and complex area of contamination if the source is ruptured. These types of radiological accidents have been the primary source of deaths among the public, as shown by Table 9.1.

The most important sources of dose in these cases have been direct exposure from being near the source (e.g., left in a bedroom) and contact dose from the sources being carried (e.g., by hand or in pockets). Lethal exposures are possible within minutes from handling some sources (see Table 9.2). This shows the importance of warning the public when a high- or moderate-risk source is lost or stolen. In most cases, serious exposures can be stopped by simple actions taken by the public, such as moving away from the source and washing their hands, once they realize the danger.

9.3.2 Airborne Release

An emergency at a nuclear power plant or some other facilities (e.g., large spent fuel storage or reprocessing facilities) can result in a dangerous amount of radioactive material being released into the atmosphere, forming a plume. As the plume travels, radioactive material is deposited on the ground, structures, etc. The deposition of radioactive material can be increased by rain forming hot spots, or areas with higher radiation or contamination levels. The plume and deposited material are the primary sources of dose to the public in the event of a large atmospheric release.

Figure 9.1 shows the most important pathways of exposure following a large atmospheric release. The dose comes mostly from five pathways:

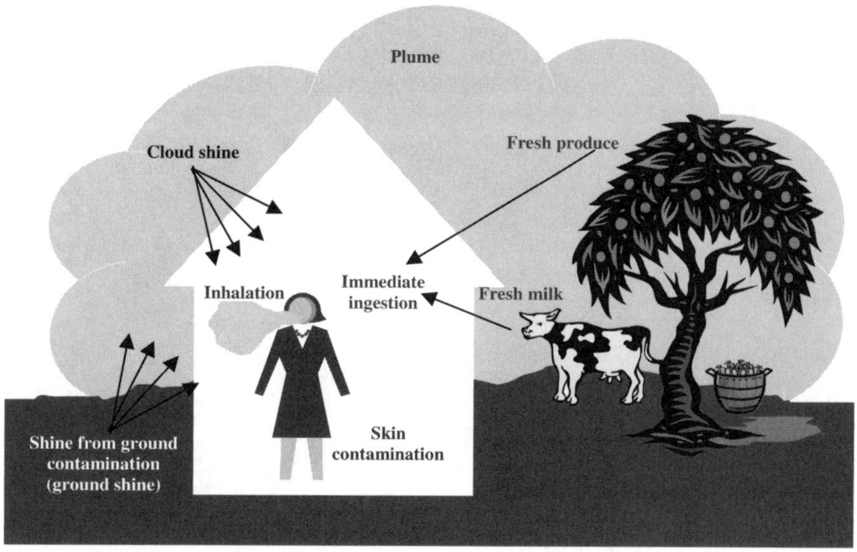

FIGURE 9.1 Major pathways of exposure following a large atmospheric release.

1. External gamma radiation from the plume, called cloud shine
2. External gamma radiation from radioactive material deposited on the ground, called ground shine
3. Inhalation of radioactive material in the plume
4. External beta and gamma radiation from radioactive material deposited on the skin
5. Ingestion of contaminated food and milk

During a release, the dose from cloud shine, ground shine, skin contamination, and inhalation is predominant. After the plume has passed, the dose from ground shine and ingestion of contaminated food and milk becomes predominant.

The plume will travel downwind and the concentration of radioactive materials will tend to decrease as it travels further from the plant. As the concentration of radioactive materials in the plume decreases, the dose rate to the affected population will also decrease. Thus, those who are further away from the plant will generally be at less risk of deterministic (early) health effects. While the exposures further from the plant are small, they all add to the chance of getting cancer (stochastic effects). Since the total amount of human exposure is larger further from the plant (large number of people exposed to small amounts of radiation), this is where most cancers will occur. Following the Chernobyl release the vast majority of the excess thyroid cancers caused by the accident occurred between 50 and 350 km from the plant.

The off-site doses and deposition from an airborne release can be very complex, as demonstrated by the Chernobyl accident (Fig. 9.2). People were relocated from hot spots (i.e., areas with high levels of contamination) that were more than 300 km from the plant. Clearly, extensive environmental monitoring will be needed to characterize the contamination resulting from a release.

Exposure from external gamma, skin contamination, and inhalation can be prevented or reduced by what are referred to as *urgent protective actions* (see below, Public Protective Actions).

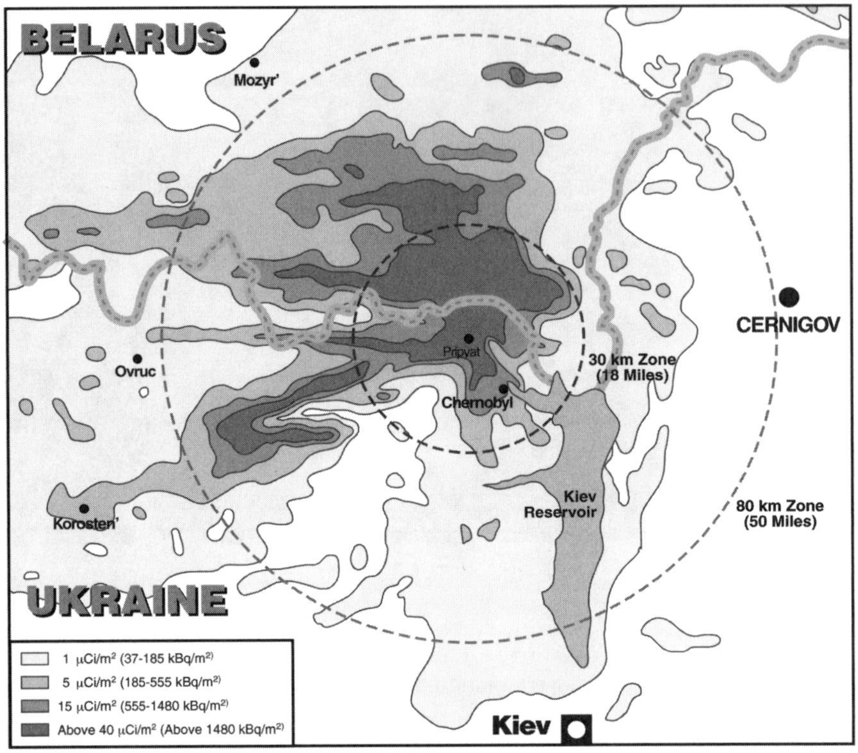

FIGURE 9.2 Map of ^{137}Cs deposition from the Chernobyl accident.

9.4 PREPAREDNESS REQUIREMENTS

This section discusses the preparedness and response requirements that are set to reduce the impact of radiological and nuclear emergencies. Preparedness requirements describe what must be in place prior to an emergency in terms of arrangements, plans, procedures, personnel, training, and equipment. Response requirements describe what must be achieved during an emergency and follow a performance-based approach reflecting the recent international guidance on the subject (IAEA, 1996b, 1997b).

9.4.1 Radiological and Nuclear vs. Conventional Emergencies

There are very few differences between radiological/nuclear emergencies and other conventional emergencies as discussed in Section 9.1.1. Indeed, international guidance on the subject actively promotes the integration of radiological/nuclear emergency plans with existing national, regional, local, and facility arrangements for dealing with all other types of emergencies (IAEA, 1997a).

This does not preclude the fact that there are some notable differences between radiological/nuclear and conventional emergencies. The first of these is perception, which plays a large role in the response to a radiological/nuclear emergency. People's perception of a radiological/nuclear emergency amplifies the impact, biases decisions, worsens the trauma associated with the emergency, and can result in the mobilization of disproportionate levels of resources and capabilities. Another significant difference is the delayed impact that is produced by radiological/nuclear emergencies. The deterministic effects do not appear for days or even weeks, and the stochastic effects are "real" but are indiscernible until several years later. This complicates decision making and the implementation of protective actions, partially because people have difficulty visualizing the risk areas and feeling that they are truly being protected. Contamination also represents a major potential factor in nuclear/radiological emergencies. Very small amounts of contamination can be measured, and while in most cases it poses very little risk, it can result in extensive response actions and complicate post-emergency efforts. Finally, nuclear/radiological accidents can have measured impacts over extremely wide areas extending to several tens or hundreds of kilometers. As a result, there is an extensive need for resources and resource coordination, often across national boundaries.

The planning for radiological/nuclear emergencies should be done in concert with the planning for conventional emergencies. It should rely on the same basic organization infrastructure and planning strategy while recognizing the technological differences that exist when dealing with radiological/nuclear types of emergencies.

9.4.2 Risk Evaluation and Planning Zones

Emergency plans should be based on a sound understanding of the risk, event frequencies, and consequences. Plans for events with high to moderate frequency should be detailed and should aim at reducing stochastic effects as much as reasonably possible. Plans for severe accidents with lower frequencies of occurrence should focus on the prevention of early deterministic effects (ICRP, 1991; IAEA, 1994).

Therefore, a risk assessment should be carried out to identify the uses (practices) of radioactive or nuclear materials requiring emergency planning. The risk assessment does not necessarily need to be fully quantitative, but it should at a minimum include:

1. Identification of all the hazards associated with the practice
2. Process review to determine how the hazards can lead to consequences
3. Estimation of the likelihood of accidents
4. Estimation of the consequences of accidents

In general, more complex processes (e.g., nuclear reactor, isotope production facility) require a more in-depth and quantitative risk assessment. The results of the risk assessment should be documented in a technical planning basis document, which typically includes:

1. A description of the site
2. A description of the accident sequences for which emergency planning is required
3. Event frequencies or likelihood
4. Typical timing of accidents
5. Typical consequences of accidents in terms of dose to unprotected individuals
6. An appreciation of the importance of various exposure pathways, e.g., external exposure, inhalation, or ingestion
7. A discussion on the effectiveness of various protective actions, such as sheltering, stable iodine intake, and evacuation
8. Emergency planning zones, for which detailed plans will be developed for given emergency response strategies
9. The optimal emergency response strategy, which describes the expected timing of required emergency response actions and their sequence

The accidents considered should include design basis accidents and severe accidents. Design basis accidents are event sequences used to determine the requirements for safety systems. Safety systems are designed to prevent significant fuel damage (in the nuclear industry) or significant dispersion of radioactive material (in other activities). Hence, by definition, design basis accidents do not cause significant off-site consequences. The U.S. siting guide stipulates that design basis accidents should not lead to doses higher than 250 mSv (25 rem) at the site boundary (NRC, 1984a). These accidents may still require the implementation of emergency response actions in order to reduce the risk of stochastic effects. Severe accidents are those at which safety systems fail to prevent significant fuel damage or dispersion of radioactive material. For this type of accident, deterministic health effects are possible, and plans should aim at preventing such impacts. However, in general, severe accidents are extremely unlikely.

The timing of accidents can be determined through a severe accident progression analysis. This can be done using a probabilistic safety analysis level 1, through the use of failure mode and effects analysis (FMEA) techniques, or hazard operability studies (HAZOPs). The important parameters are the time between the recognition of an accident by the operator (i.e., the declaration of an emergency) and the onset of the release, the release rate, and the duration of the release.

The consequences of accidents are expressed in terms of dose, which can be calculated using standard atmospheric dispersion and dose calculation software, such as those based on the Canadian standard (CSA, 1991). Stochastic doses are usually expressed in terms of the effective and thyroid doses for emergencies that involve the release of radioactive iodine. For emergencies that do not involve iodine, the critical organ dose would be based on the main radionuclide most likely to be released. For the calculation of deterministic health effects, it is important to use organ-specific equivalent doses because the effective dose concept is not applicable to doses this high. It is also important to consider the rate at which the dose is received because the thresholds for deterministic effects vary with the exposure rate.

The exposure pathway is important because it affects the way the dose is received, the rate at which it is received, and hence the effectiveness of any protective action. For example, if most of the dose comes from external exposure to gamma radiation from the passing plume, evacuation after the plume has passed will not be effective. On the other hand, if the dose comes mainly from exposure to ground deposition, sheltering during plume passage followed by an evacuation will be a very effective protective action.

The effectiveness of urgent protective actions, namely sheltering, stable iodine intake, and evacuation, varies as a function of the parameters already discussed. By reducing the appropriate component of the dose received, it becomes possible to determine the effectiveness of a single protective action, or a combination thereof. For example, numerous publications provide dose-reduction factors of sheltering in a typical North American dwelling for the external exposure pathway of between 0.8 and 0.05, depending on the dwelling type and exposure pathway. An evacuation after the plume's passage would virtually eliminate the ground exposure dose but not the plume exposure dose.

The above analysis leads to the determination of planning zones, which define where efforts should be directed during planning. These do not represent "response" zones, which depend on the specific circumstances at the time of the accident. Depending on the type of practice, up to three conceptual planning zones can be considered. Although definitions vary, the International Atomic Energy Agency (IAEA) has adopted the following nomenclature: a precautionary action zone (PAZ), an urgent protective action planning zone (UPZ), and a longer-term protective action planning zone (LPZ). These zones are described in this section.

Precautionary Action Zone (PAZ). A precautionary action zone is an area where there is the potential for severe deterministic health effects if protective actions are not taken before or shortly after the start of a severe release. The precautionary action zone is therefore also a response zone. In this zone, automatic protective actions are taken to reduce the risk of exposure to levels above the deterministic dose thresholds, i.e., to prevent deaths and injuries. The precautionary action zone is determined on the basis of the releases expected for the most severe accidents. The zone is delimited based on the distance at which prompt protective actions are needed to prevent exposure above the deterministic dose thresholds. Within this zone, evacuation or substantial sheltering may be the only effective protective actions and should be implemented as much as possible before a release takes place.

Planning based on conservative assumptions, those leading to the worst consequences, can lead to taking unnecessary protective actions, e.g., evacuation, with its associated cost and risks. They can also result in plans that divert resources from the population at greatest risk (those closest to the plant). Consequently, PAZ size should be determined based on reasonable assumptions to include:

1. Average weather conditions
2. Average individual (e.g., adults) being exposed
3. People acting normally (e.g., inside most of the time)

By definition, the risk of deterministic health effects occurring outside the PAZ is very low.

Urgent Protective Action Planning Zone (UPZ). This zone is the area for which detailed urgent protective action plans, such as sheltering, stable iodine intake, and evacuation, should be developed to minimize the risk of stochastic effects. The UPZ is difficult to define because there is no cut-off value below which stochastic effects do not occur. Therefore, the determination of the UPZ requires an assessment of the cost/benefit value of planning beyond certain distances. By definition, planning does not provide a significant risk-reduction benefit beyond the UPZ. In addition, it is generally believed that detailed planning for the UPZ would provide an adequate basis for the expansion of response actions outside that zone should the need arise.

Several methods are available for determining the size of the UPZ. The UPZ size can be based, for example, on any of the following:

1. The distance at which intervention levels are not exceeded for most accidents and most weather scenarios, e.g., 90% of all events and 90% of all weather scenarios.
2. The distance at which the intervention levels are not exceeded for most of the risk, e.g., by considering all accidents and weather scenarios that constitute 90% of the total risk (where risk is frequency times consequence).
3. The distance at which the probability of exceeding a certain dose (e.g., the intervention level for evacuation) for all event types drops significantly. This means that beyond that distance, the benefit/cost ratio of emergency planning also drops significantly; and
4. The distance at which the conditional dose drops below the intervention levels. The conditional dose is a probabilistic quantity that takes into account all accident types and all possible weather scenarios. It corresponds to the dose that would be received by an individual at a given location given that an accident has happened.

Generic intervention levels are described in Section 9.5.1.

Longer-Term Protective Action Planning Zone (LPZ). This zone is the area for which detailed longer-term protective action plans, such as relocation, resettlement, and agricultural countermeasures, should be developed to minimize the risk of stochastic effects. As for the UPZ, the LPZ is difficult to define due to the absence of a stochastic effect threshold. In addition, decisions on longer-term protective actions are typically made over the weeks following the accident and there is time to adjust the sampling and assessment strategies associated with the LPZ. Therefore, the size of the LPZ is not as critical as that of the UPZ and adequate preparations within a reasonable distance should allow longer-term protective actions to be smoothly extended outside that zone during an emergency.

9.4.3 Threat Categories

In general, emergency planning could be different for each practice. However, this can be simplified by grouping practices into five threat categories, each presenting common features in terms of the magnitude and timing of the hazard. Table 9.5 defines the five emergency threat categories (IAEA, 1997a). Categories I through IV represent decreasing levels of threat and therefore decreasing emergency preparedness and response requirements. Category IV is the minimum level of threat assumed to exist everywhere and thus *always applies* possibly along with other categories. Category V is a special category and *may apply* along with other categories. Note that these threat categories are only used as a convenient way to provide guidance on planning and are not used during an accident. These categories apply to both facilities or uses and governmental jurisdiction for which various levels of planning are warranted.

This categorization does not replace the need for conducting a risk assessment for each type of activity, but it does allow a prompt and approximate determination of the preparedness and response needs for each practice.

TABLE 9.5 Categories of Nuclear and Radiological Threat

Threat category	Description
I	Nuclear installations for which events that could give rise to severe deterministic health effects[a] off-site are postulated or have occurred in similar installations, including very low-probability events.
II	Installations for which events that can give rise to off-site doses warranting urgent protective actions consistent with international standards[b] are postulated or have occurred in similar installations. This category (as opposed to category I threats) has no credible events postulated that could give rise to off-site doses resulting in deterministic health effects.
III	Installations for which events that could give rise to on-site doses resulting in deterministic health effects are postulated or have occurred within similar installations. This category (as opposed to category II threats) has no credible events postulated for which urgent off-site protective actions are warranted.
IV	Minimum level of threat assumed for all countries and jurisdictions. This category also includes (1) facilities[c] for which events could give rise to doses or contamination warranting urgent protective actions consistent with international standards on-site but for which no credible events are postulated that could result in severe deterministic effects and (2) mobile practices using dangerous sources.
V	Areas that could be contaminated to levels necessitating food restrictions consistent with international standards as a result of events at installations in threat categories I or II, including installations in nearby countries.

Source: IAEA, 1997a.

[a] Doses in excess of those for which intervention is expected to be undertaken under any circumstances, Schedule IV of IAEA (1996b).

[b] Schedule V of IAEA, 1996b.

[c] This includes medical, industrial, and research uses of sources or radioactive material for which events warranting emergency intervention to include medical misadministration are possible.

Threat category I includes facilities with the potential for very large releases resulting in serious deterministic health effects off-site. This applies to nuclear power plants, research reactors with power levels greater than approximately 100 MW(th), large spent fuel storage facilities, and large radioactive reprocessing waste storage facilities. In all these facilities, the size and composition (source term) of a severe release is probably unpredictable and could result in very complex contamination, and doses off-site. On-site there may be very high dose rates (e.g., >10 Gy/h), beta emitter contamination, and other hazardous conditions (e.g., live steam) in areas requiring actions by the staff to mitigate the accident.

Threat category II includes facilities with the potential for releases resulting in off-site doses above intervention levels but with little or no threat of doses resulting in deterministic health effects off-site. This applies to reactors with power levels 2 to 100 MW(th), cooling pools containing spent reactor fuel equivalent to a 10 to 3,000 MW(th) reactor core, facilities storing significant amounts of unsealed radioactive material, and facilities with potential for accidental criticality or large gamma emitters located near the site boundary. On-site there may be very high dose rates, beta emitter contamination, and other hazardous conditions (e.g., live steam) in areas requiring actions by the staff to mitigate the accident.

Threat category III includes facilities without significant off-site risk but with the potential for accidents resulting in deterministic health effects on-site. This applies to research reactors of less than 2 MW(th), critical assemblies, and facilities with a potential for unshielded dose rates of more than 10 Sv/h at 30 cm or with moderate inventory of unsealed radioactive material. Jurisdictions that provide fire, police, or medical support to these facilities also require this level of planning. On-site there may be high dose rates, beta emitter contamination, or other hazardous conditions in areas requiring actions by the staff to mitigate the accident.

Threat category IV is the minimum level of planning and applies to all countries. In general, this applies to uncontrolled source(s), transport accident, and nuclear-powered satellite reentry. Thus the threat category IV addresses the planning for response to accidents involving dangerous sources, such as those used for radiography that have been lost, abandoned, stolen, or illegally brought into the country (uncontrolled sources); transport accidents involving radioactive material; and reentry of nuclear powered satellites. It also includes weapons accident (plutonium dispersal) and other unanticipated events.

Threat category V is for areas that are far enough away from category I or II facilities not to require implementation of urgent protective actions such as evacuation, relocation, and sheltering but where there is a potential for food and foodstuff contamination that calls for the implementation of agricultural countermeasures and foodstuff monitoring and control.

Table 9.6 shows how the planning zones apply to each threat category.

9.4.4 General Emergency Preparedness Requirements for All Threat Categories

Plans and capabilities for all aspects of response must be developed for all threat categories of radiological and nuclear emergency preparedness, regardless of the category. Because these aspects are conceptually the same for any type of emergency, radiological and nuclear emergency response plans should be integrated with the national and local emergency response infrastructure as much as practicable. These plans should cover the following aspects:

1. Authority, command, and control
2. Organizational responsibilities
3. Preparedness and response coordination
4. Plans and procedures
5. Logistical support and facilities

TABLE 9.6 Planning Zone Size vs. Threat Categories

Threat category	Precautionary action planning zone size (PAZ)	Urgent protective action planning zone size (UPZ)	Longer-term protective action planning zone size (LPZ)
I	3 to 5 km	10 to 25 km	50 to 100 km
II	Limited to the facility itself.	0.5 to 2 km	5 to 20 km
III	Limited to the facility itself.	Not required.	Not required.
IV	For transportation events 50 to 500 m around the accident site and up to 1000 m downwind.	Not required.	Not required.
V	Not required.	Not required.	Part of the country that is within approximately 300 km of the reactor.

Source: IAEA, 1997a.

6. Communications
7. Training and exercises
8. Public education

These arrangements should enable response organizations to perform all required emergency functions, including:

1. Initial accident assessment and classification
2. Notification and activation
3. Accident mitigation
4. Public instructions
5. Public protective actions
6. Protection of emergency workers
7. Psychosocial impact mitigation

The degree of preparedness and the type of response required depend on the potential magnitude of emergencies, or in other words, on the threat category. In the following discussion, the needs for each threat category are examined.

9.4.5 Emergency Preparedness Requirements for Threat Categories I and II

Threat categories I and II are the most demanding in terms of planning and response capabilities because of the severity and geographical extent of the potential accidents. This requires planning and coordination efforts at all jurisdictional levels, i.e., national, regional (e.g., provincial or state level), local (e.g., municipality, regional municipality, or county), and at the facility itself. Unless all plans and procedures at all levels are properly coordinated, the response to category I and II accidents is unlikely to be effective.

Planning Elements

Authority, Command and Control. Emergencies involving threat category I or II practices, e.g., a NPP, can have a significant impact over a wide territory, thereby directly affecting several jurisdictions and large populations. One of the most difficult challenges with this type of situation is ensuring that there is a clear authority for determining what actions should be carried out and, in particular, what public protective actions should be implemented.

Typically, the facility operator has full authority for response actions within the facility itself, including other actions that can mitigate the accident at the source, e.g., operating procedures aimed at reducing potential radioactive releases to the environment. Because the facility organization usually has the highest degree of readiness, it is often also the authority for recommending what immediate public protective actions should be implemented in the area closest to the facility (e.g., within the PAZ). Local or regional off-site authorities, however, usually have command and control over the implementation of public protective actions. Once activated, they may also have decisional authority over those public protective actions. Therefore, it is important that the decision-making process be properly documented and understood by all response organizations.

Depending on the scale and dynamics of the emergency, the authority in charge of command and control, and decision making may vary. For example, in the case of emergencies causing a limited geographic impact, the local emergency management organization may have full authority over the entire response. If the severity of the consequences were to increase, however, and several local areas were affected, the regional or national emergency management organization might need to take over the decisional authority and/or the command and control of operations. Such cases should be anticipated, and the process for the transfer of authority should be well documented in the plans. The fact that this transfer may occur during the emergency should also be considered.

Organizational Responsibilities. As for any type of emergency, it is important that preparedness and response tasks be clearly assigned. There are two major challenges related to organizational responsibilities in preparation for radiological/nuclear emergencies.

The first challenge relates to the very nature of radiological/nuclear emergencies. Because they are *perceived* to be quite different from other types of emergencies, there is a tendency to create an organizational emergency structure that is completely unlike the one already established to deal with other environmental or conventional emergencies. In general, this is counterproductive and inefficient. As much as possible, the organizational emergency preparedness and response structures for radiological and nuclear emergencies should be integrated with those in place for other types of emergencies.

The second challenge relates to the complexity of decisions that must be made during radiological/nuclear emergencies and the large number of organizations that may have jurisdiction over such decisions. For example, decisions regarding public health may involve the national and provincial/state health ministry, the national and provincial/state environment ministries, the national and provincial/state ministries for agriculture and food products, the nuclear regulatory organization, and the local governments. Acts and statutes that deal with the authorities and responsibilities of these various organizations are not always clear, or clearly interpreted. Quite often, decisions need to involve more than one government department, ministry, or agency. When dealing with large-scale operations, resources from more than one organization or jurisdiction may be required. Therefore, it is important that the roles and responsibilities of all organizations be clearly defined and understood, and that appropriate functional coordination mechanisms be established.

Preparedness and Response Coordination. Effective and practical preparedness and response coordination mechanisms must be established for any type of emergency. In the case of threat categories I and II, coordination arrangements must be formalized within each major emergency response organization, at each major jurisdictional level involved in the response, and between organizations and jurisdictional levels.

Within the facility itself, particularly those with large staff and compartmentalized departments, coordination mechanisms should be developed between all individuals and departments with a major role during the response.

Coordination agreements between the facility and the off-site authorities should also be developed. These will usually address the levels at which notification must occur, the type and format of the information that needs to be provided by the facility, the communication mechanisms, and the assistance that will be provided by the facility staff and organizations to support the off-site planning and response efforts. These agreements should also address the need for the off-site authorities to handle the exit of facility staff in case of a site evacuation of nonessential personnel.

At the operational level, coordination should address the need for using common units, equipment, and communication networks. For instance, when several organizations are involved in the measurement of radioactivity (e.g., airborne, deposited on the ground, or contained in water and foodstuff), there should be preestablished and coordinated methods for monitoring and sampling. These methods should consider measurement techniques, equipment type, sensitivity and detection effectiveness, the interpretation of data, and response thresholds or levels (i.e., operational intervention levels). Communications present yet another challenge. When several organizations must coordinate their efforts during major operations, communicating effectively becomes paramount. This means establishing common communication frequencies or central dispatch systems.

Coordination arrangements should also include common maps and reference grid points. This is particularly important when geographic-based data must be communicated between organizations. If several teams are involved in survey and sampling, maps could also include preestablished survey and sampling points. This can help coordinate and optimize the overall survey strategy.

When local plans may have an impact on other municipalities or jurisdictions (e.g., for hosting a potentially large number of evacuees), mutual aid agreements should be established as part of the planning effort. When the consequences of an accident may affect another jurisdiction, province/state, or country, agreements should be in place to ensure the prompt and effective coordination of communications and operations, including notification. These agreements must take into account possible differences in intervention levels adopted by the different jurisdictions. When a planning zone includes another country, the arrangements in place for notifying the population in the planning zone that is outside of the accident country should afford the same level of readiness as for those in place within the accident country.

Plans and Procedures. All response organizations should have emergency response plans and procedures in place. Several plan models offer suggestions as to plan content and procedure format, but these will not be covered here. The particular challenge for threat categories I and II relates once again to the large number of organizations that are often involved in response activities. This introduces the need for coordination plans and a concept of operations. The difficulty lies in determining what periodic changes need to be made to this multitude of plans. Complications also arise when carrying out and tracking these changes and ensuring that all organizations are working from the same set of plans.

Establishing common coordination plans and a common concept of operations generally necessitates the formation of emergency preparedness committees with representation from all major organizations and jurisdictional levels involved in the response. Generally, there would be a single overall coordination plan for all organizations and jurisdictions, which could vary, however, depending on the existing political structure. For example, in countries where the province/state has overall authority over the plans and response, the overall coordination plan would likely be developed and promulgated at that level. In this particular case, the national plan would have to take into account—and comply with—the regional disparities. Ideally though, all provinces/states should work on a common set of intervention levels, health standards, and an interjurisdictional concept of operations.

Ensuring that all organizations work from the same version of the plans requires a well-structured and maintained configuration management system. Such a system should not delay the introduction of critical changes and modifications to the plans and procedures.

Logistics Support and Facilities. All response organizations should have suitable equipment and facilities for supporting their functional role. This should include the operational rudiments for dealing with any conventional emergency, e.g., command posts, transportation, traffic control resources, emergency shelters, and communications. For threat categories I and II, the radiological hazard and potential involvement of large populations call for special requirements.

Within the facilities considered in the plans, arrangements should be made to provide emergency personnel with a room where they are adequately protected from external irradiation and airborne contamination. This means that the rooms and/or building provided for emergency personnel and ongoing operating staff must be shielded and equipped with isolated air intake, filtered air intake, or autonomous air supply systems. For operating staff, this could be achieved by providing an alternative control room from which it is possible to control all essential plant safety functions. There should also be alternate off-site facilities for the nonoperating emergency management personnel. Nonessential personnel should have access to emergency shelters or provisions should be made to evacuate them promptly in case of an accident.

During an accident at a threat category I or II facility, laboratory analyses of radioactive samples must be done. Because the site itself may be contaminated, laboratory facilities for such analyses must be established outside the potentially contaminated zone, i.e., outside the urgent protective action zone.

Command posts and/or staging areas for off-site operational organizations should also be established outside the urgent protective action zone. Suitable equipment for dose control, contamination monitoring, and decontamination should be available at all staging areas for managing of emergency personnel entering or leaving the urgent protective action zone.

People who are evacuated from a potentially contaminated zone must be monitored for contamination and assessed for potential overexposure to radiation. This also applies to nonessential personnel leaving the plant. Because of the potentially large number of people involved, this generally requires that prepositioned reception centers be established outside the UPZ. These reception centers should be designed to process the percentage of the population within that zone that is expected to use them and should provide the following basic services:

1. Contamination control of people and vehicles
2. Decontamination of people and vehicles
3. First aid
4. Exposure screening
5. Registration and inquiry
6. Counseling and social services
7. Child and infant care
8. Evacuee assembly area
9. Adequate power, water, sewer, sanitation, waste disposal (including hazardous waste), and heating, ventilation, and air conditioning services
10. Adequate access and exit routes
11. Adequate off-loading and parking areas
12. Health and safety arrangements and systems
13. Directions for evacuee centers (emergency shelters)
14. Transportation alternatives for confiscated, contaminated vehicles
15. Access to telephones
16. Information on how to stay informed and when to return to the affected areas

Stable iodine prophylactic has proven to provide effective protection of the thyroid against the intake of radioactive iodine. For threat categories I and II, where the release of radioactive iodine to the environment is possible, the population within the urgent protective action zone should be given access to stable iodine. In order to be effective, stable iodine must be administered at the time of exposure or within a few hours from exposure. Approximately 12 hours after the initial exposure, stable iodine is no longer beneficial.

Communications. Good communication systems contribute a great deal to achieving an effective response. In a large-scale radiological or nuclear emergency, it is generally anticipated that common communication systems will be vulnerable to failures due to overloads. This applies to both landlines and cellular systems. Hence, provisions should be made to provide key players with robust and reliable communication channels. This may include protected or direct lines between the facility and off-site authorities and between off-site response organizations. While satellite telephones are now being used for such communications, initial trials have revealed limitations with their directional sensitivity. Microwave links are an effective alternative where applicable.

All main communication systems should be tested periodically and contact numbers updated on a quarterly basis.

Training and Exercises. Training and exercises are an essential part of an effective emergency preparedness programme. In general, it is recommended that training and exercises for facility personnel be held several times a year. For off-site organizations, this constitutes a substantial investment in finances and human resources. The training and exercise program for off-site organizations should take into account the role of each organization, its critical needs for responding to radiological or nuclear emergencies, and the value of the training and exercises. For example, police forces may not need to obtain special training in traffic control related to nuclear emergencies if their normal duties already involve this type of activity. They should, however, undergo specific radiological hazard awareness training.

In general, each facility within threat categories I and II should hold one annual large-scale coordination exercise involving off-site authorities. The training and exercises should take into account the needs of the public living close to the facility.

In some countries, annual exercises involve the full participation of designated segments of the population. In several Eastern European countries, sheltering and evacuation exercises are held annually. In most Western countries, this is not practical for cultural and economic reasons. When large-scale exercises are conducted, however, they are often widely publicized and the facility's surrounding population is kept informed of the results.

Public Education. Ensuring that the public reacts appropriately and in accordance with the plan and instructions of decision makers is important in successfully managing emergencies affecting the population. A spontaneous and uncontrolled evacuation, for example, may cause more harm than benefit. Although it is difficult, if not impossible, to control this, educating the public on the risks of radiological or nuclear accidents can help ensure an appropriate reaction during an emergency. Therefore, public education is an important factor in influencing the public's reaction to such emergencies. Nowhere is this truer than during a radiological or nuclear emergency.

The public should receive effective information on the plans and provisions in place in case of an emergency. They should be informed of the protective actions and their effectiveness. They should also be sensitized to the importance of obeying instructions provided by the authorities during an emergency. The public will need to know who these authorities are and how instructions are to be relayed. Addressing the following questions with the public will benefit response operations. How will they be informed? Where should the public go when instructed to evacuate? What should they do when told to shelter? What instructions related to stable iodine distribution could they be asked to follow? How will other family members be protected (e.g., schoolchildren, patients)?

An effective public education program can be achieved, e.g., appropriate material can be inserted into free calendars, telephone directories, etc. It is also effective to convey the educational material through community (e.g., doctors, teachers) or religious leaders.

Response Elements

Initial Accident Assessment and Classification. Because accidents in threat categories I and II are typically complex, it may take time before the initiating event is clearly understood. Depending on the progression of the accident, operators can have difficulty assessing its severity. There is also a tendency to wait until the accident is understood before declaring an emergency. In some cases, facility emergency plans wait for a release to occur before declaring an off-site emergency. This can delay on-site response and, more importantly, off-site readiness to implement effective protective actions. During the Three Mile Island accident, it took the plant operators several days to realize that a significant portion of the core had melted. During the Chernobyl accident, operators did not initially believe that a large release was underway.

Assessing the severity of an accident can be a complicated task. Procedures should be in place to allow an assessment based on simple and available measurements of processes, such as reactor core temperature, radiation inside the plant, and availability of major process and safety systems. During severe accidents, difficulties usually arise when some instrumentation becomes unavailable or out of range. This should be taken into account in the accident assessment procedures.

Another challenge lies in relating the accident assessment to an adequate level of response. It can be difficult to communicate accident severity to other emergency groups. Prompt initiation of a coordinate response is more difficult still, unless a simple way to categorize or classify the event is established. Facilities in threat categories I and II should therefore have a clearly defined emergency classification system.

The IAEA recommended classification system for threat category I and II facilities is described in Table 9.7.

There may be additional classes of emergencies, provided that they are clearly understood by both on-site and off-site authorities and allow operators to make prompt determinations. Classification levels should be identical for both on-site and off-site emergency response

TABLE 9.7 Classification System for Threat Categories I and II

Emergency class or condition	Definition
General Emergency	Events resulting in the actual or substantial risk of a release requiring implementation of urgent protective actions off-site. This includes: • Actual or projected damage to the core or large amounts of spent fuel; and • Detection of radiation levels off-site warranting implementation of urgent protective actions.
Site Area Emergency	Events resulting in a major decrease in the level of protection available to the public or on-site personnel. This includes: • A major decrease in the level of protection provided to the core or large amounts of spent fuel; • Conditions in which any additional failures could result in damage to the core or spent fuel; and • High doses on- or off-site approaching the urgent protective actions interventions levels.
Facility Emergency	Events resulting in a major decrease in the level of protection for people on-site. This class of emergency does not represent an off-site threat.
Alert	Decreased level of safety or unknown events that warrant increased readiness or further assessments.

Source: IAEA, 1997a, b.

organizations. Upon declaration of an emergency class, each response organization should immediately take its assigned actions for that class.

The emergency classification should not be confused with the International Nuclear Events Scale (INES) (IAEA, 1997c). INES is designed to indicate how serious an event was *after it is understood* and is not the basis for the response. Determining the INES rating is impossible early in an event and does not form part of the initial response, and determining the rating should not delay any response actions.

Notification and Activation. If there is a potential for off-site impacts, notification and activation remain the same regardless of the type of facility involved, whether nuclear or conventional. There should be a clear and effective way for key response organizations to be alerted of an emergency and to mobilize required personnel, 24 hours a day, seven days a week. The requirement for notification and activation should be linked to the emergency classification. For example, in many cases, facility operators are required to inform off-site authorities of any event leading to an alert, and to notify them of a site area emergency or a general emergency. Emergency response organizations are normally required to be partially activated during a site area emergency and fully activated during a general emergency.

The timing requirement for notification varies depending on the type of facility and the emergency class. Notification must be prompt enough to enable response organizations to perform their required functions effectively and should be based on the emergency classification. This emphasizes the need for an easy-to-use classification procedure.

Accident Mitigation. For threat category I and II facilities, accident mitigation refers to the need for emergency operating procedures dealing with upset events and accidental conditions. Such procedures constitute a standard licensing requirement and are normally tested on simulators as part of the operator certification and recertification program.

Emergency operating procedures are often based on design basis accidents, i.e., accidents that have been analyzed and are the basis for the design of safety systems. However, as demonstrated by the accidents at the Three Mile Island NPP (1979), the Chernobyl NPP (1986), and the Tokaimura processing facility in Japan (1999), accidents can evolve to conditions that are beyond those which were the basis for the design. This is the reason for the recent emphasis on developing procedures for the full spectrum of accidents to include those with a very low probability of occurring. These procedures are called severe accident management guidelines and are required by international requirements (IAEA, 2000c) and are being implemented in nuclear facilities around the world.

Public Instructions. When an accident with the potential for off-site impacts occurs, the affected public must be alerted promptly and be provided with instructions on appropriate protective actions. If protective actions are not required, the public will still require instructions in order to reduce their concern and prevent them from taking inappropriate actions when an emergency situation is announced or reported by the media or when friends tell them.

The method used for promptly informing the public should depend on:

1. The population characteristics (density, transients, language) around the facility
2. The size of the emergency planning zones
3. The expected dynamics of possible accidents
4. The amount of time required for implementing protective actions after an emergency is declared

Those parameters should be defined in the technical planning stage. Methods for alerting and instructing the public include (but are not limited to):

1. Sirens
2. Sirens with a public address system
3. Automatic activating radios

4. Automatic telephone dial systems
5. Other types of telephone-based community alerting systems, e.g., those using digital lines and connected to dedicated displays in the homes
6. House-to-house notification (in remote areas)

In designing the public alerting system, care must be taken to ensure that the system can reach the entire target public, including those with special needs, such as the hearing impaired. The system must also address the needs of industries within the planning zones, where noise levels may reduce the effectiveness of an exterior siren system.

The system should also take into account the need to provide instructions to members of the public who are outside the planning zone to prevent them from reacting in a way that will hinder the effective implementation of protective actions within the zone. Indeed, the unexpected evacuation of a large number of people outside the zone may prevent the orderly evacuation of people coming from within the zone.

When designated media are called upon to alert the public, clear agreements and preset messages should be established.

Experience has shown that siren systems with modulating signals are not always effective in conveying the right information. Apparently, members of the public have trouble understanding or remembering what the modulations signify. Therefore, a system that combines *attention-grabbing* and verbal or text-based information is the most effective for alerting the public and giving instructions.

The accident at Three Mile Island demonstrated that the public will not follow official recommendations unless national and local officials:

1. Provide consistent assessment and recommendations (speak with one voice)
2. Promptly and consistently explain the basis for revisions
3. Monitor and discuss different opinions and assessments being presented in the media

Consequently, immediately after the pubic is alerted they must be provided with coordinated information on emergency, the risks and protective actions. Provisions must also be in place to promptly address concerns raised by other sources through the media. It is crucial that all official assessments and recommendation be coordinated. This is best accomplished by having only one spokesperson.

Public Protective Actions. Public protective actions can be divided into two types: *urgent protective actions* and *longer-term protective actions.* Urgent protective actions must be implemented promptly to be effective and thus require quick decision-making and are applicable over a short period of time (days/weeks). Longer-term protective actions are intended to reduce the dose from long-term exposure; they will be applied over months to years and implemented after radiological and psychological impact is assessed.

Urgent Protective Actions. Urgent protective actions include sheltering, evacuation, and, where radioactive iodine may be released, stable iodine administration. They also include immediate bans on potentially contaminated food and water.

Sheltering involves staying indoors away from exterior walls, with windows closed and sealed and the ventilation turned off. It is an effective measure against external radiation from a radioactive plume, as well as inhalation. It is also an effective measure against irradiation from radioactive ground contamination. Reduction factors for inhalation vary with the building type and can be as low as 0.005, but the effectiveness of sheltering against inhalation decreases with time as air ingress slowly contaminates the inside of the structure. Table 9.8 shows examples of estimated protection factors. Normally, sheltering should not be implemented for more than one or two days.

The administration of stable iodine (nonradioactive) can be very effective against radioactive iodine intake, which tends to concentrate in the thyroid gland. Figure 9.3 shows the variation of the protection factor as a function of the time at which the stable iodine was

TABLE 9.8 Reduction Factors for Various Types of Buildings

Protective action	Cloud shine	Ground shine
Normal activities	1.0	0.7
Sheltered in wood-frame house	0.9	0.4
Sheltered in block-brick house	0.8	0.2
Sheltered in basement of two-story house	0.4	0.03
Sheltered in multistory building	0.2	0.005

Source: NRC, 1984b, 1996.

administered after exposure to radioactive iodine. Obviously to be effective stable iodine must be taken before or shortly after intake of radioactive iodine. Since stable iodine only protects the thyroid, it is not a substitute for evacuation or shelter.

Evacuation is the most common public protective action for emergencies involving possible release of hazardous materials. Evacuations are common and people do not panic and travel during an evacuation is safer than normal travel. Evacuation is the most effective protective action, provided that it can be implemented before or soon after a release begins. Evacuation after a release is an effective countermeasure against irradiation from contaminated ground, which can be an important source of exposure. Evacuation can be effective even if carried out in a plume for a long-duration release and for areas close to the release. Normally, evacuation should not be considered for more than seven days.

If evacuation is initiated during or after a release, special precautions must be in place to monitor evacuees for contamination and, if required, to decontaminate them prior to directing them to evacuation centers or other emergency shelters. This unveils one of the many challenges associated with an evacuation during a radiological or nuclear emergency: a potentially large number of people and vehicles may need to be monitored and decontaminated. In addition to effective traffic control plans, measures and equipment capabilities should be

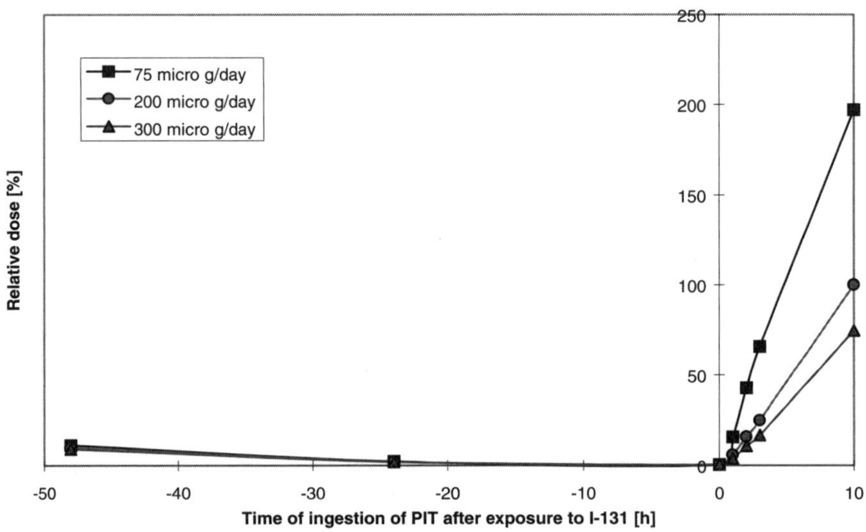

FIGURE 9.3 Relative dose (%) following exposure to ^{131}I relative to time of ingestion of stable iodine (PIT—Potassium iodide tablets). (*Source: Turai and Kanyar, 1986*)

established to handle the large number of people who may need to be processed and to prevent bottlenecks during the evacuation process.

As large areas may need to be evacuated, diversion routes should be included in evacuation plans.

Other measures that may need to be considered include banning food and water, sheltering livestock, and protecting animal feed and providing uncontaminated feed, especially to milk-producing cows, goats, and sheep.

A major problem with many emergency plans is that urgent protective action cannot be implemented promptly because the decision-making and planning process begins when the emergency occurs. During the Chernobyl accident urgent protective actions were late or not taken at all. This could have resulted in hundreds of deaths if the wind had been blowing in a different direction and did result in over a thousand unnecessary thyroid cancers. Urgent protective actions must be integrated into protective action strategy that will be implemented promptly. The strategy must be based on an analysis of the possible emergencies and effectiveness of the protective actions when applied in the local area. Factors such as availability of transport and effectiveness of shelters and other local characteristics must be considered. Provisions must be in place to implement the strategy promptly. This is best accomplished by preplanning to implement the strategy immediately upon declaration of an emergency that is a serious and immediate off-site threat (e.g., general emergency). IAEA (IAEA, 1997b) developed a basic strategy for severe nuclear power plant accidents. The strategy calls for prompt implementation of the following upon declaration of a general emergency at an NPP:

1. Evacuation or substantial shelter for the population within the PAZ
2. Sheltering, in their homes, of the population within the UPZ
3. Restrictions on eating possible contaminated food out to 300 km
4. Prompt monitoring of the UPZ and beyond to identify and evacuate hot spots
5. Taking thyroid blocking

This strategy is based on analysis of the effectiveness of urgent protective actions and the Chernobyl experience. Figure 9.4 shows the results of such an analysis (NRC, 1990a) for various urgent protective actions that could be taken to protect the public in response to the most severe type of NPP release. The numbers are the probability of a person receiving a dose to the whole body (bone marrow) in excess of the threshold for early deaths (>2 Sv) at various distances. Case 1 shows the risk is small beyond eight km even if no urgent protective actions are taken. In addition this shows, for areas within five km, that the risk of deaths can be reduced to almost zero by starting evacuation at walking speed one hour before the release (case 4) and substantially reduced by substantial sheltering in a large building (case 3). Even walking out in the plume (case 5) is better than basement shelter in a normal home (case 2). This analysis assumes that the *evacuation is conducted at walking speed* and all people in areas with significant levels of contamination are evacuated within six hours.

This analysis and resulting protective actions strategy are valid only for NPP accidents; analysis of other types of facilities could result in significantly different conclusions.

Longer-Term Protective Actions. Longer-term protective actions include relocation, resettlement, agricultural countermeasures, and medical follow-up of the affected population. These measures are implemented in the recovery phase of the accident, after the release has stopped. There will be tremendous pressure from the public, media, and political officials after the initial response to take actions to return things to normal. During the Chernobyl accident there were no criteria in place for implementation of longer-term protective actions. Consequently, the criteria were developed in the charged political atmosphere immediately after the accident. The result was a process that continues until today. The longer-term program of relocation and compensation was implemented in the hope of reducing public

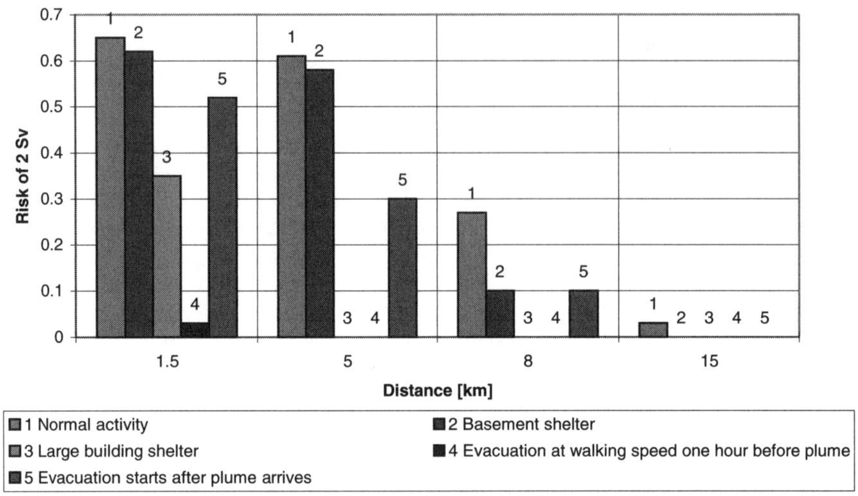

FIGURE 9.4 Probability of exceeding the early death threshold for the worst reactor accident and different urgent protective actions. (*Source: NRC, 1990a*)

concern. These programs were based on perceived impact and not on actual or projected consequences. The result were programs that many feel have done more harm than good. They resulted in relocation from areas with radiation levels lower than natural radiation levels in other parts of Europe and in compensation programs that could continue for generations. The end result was significant psychological and financial harm that the affected governments are still struggling with. The Chernobyl accident response showed that longer-term actions taken based on perceived threats and based on criteria and processes developed during the event can result in actions that do more harm than good.

It is not easy to plan in detail for longer-term protective actions, but some basic planning is essential. The basic criteria and process to be followed must be established before an emergency. IAEA and other international organizations have developed criteria for taking longer-term protective actions that will do more good than harm. These are discussed in Section 5.

Key aspects that should be part of the emergency plan for any government near a threat category I facility are:

1. Defining a transition organization to cover key recovery roles such as authority for decisions (note that this is not necessarily the authority charged with implementing urgent protective actions)
2. Criteria for taking longer-term protective action consistent with international guidance
3. A process for calculation of clear operational intervention levels based on the criteria (see Section 5 for details on intervention levels)
4. Surveying and sampling plans
5. A process to determine where monitoring indicates the operational criteria are exceeded
6. A process for revising the operational intervention levels
7. A process for implementation of long-term actions (e.g., relocations)
8. Criteria for decontamination and disposal of contaminated material

9. A plan for providing long-term financial assistance and compensation (short and long term) to address actual impacts (e.g., loss of wages, compensation of property) of the event that are developed considering the long term psychological impact
10. A process for medical follow-up of affected population and emergency workers
11. A process to identify and address adverse psychological impacts

Protection of Emergency Workers. By definition, emergency workers are those people charged with attempting to reduce the impact of an accident on personnel, the population, and/or the environment. Some of these workers risk exposing themselves to significant doses, while most will only be expected to receive relatively low doses.

During the first few days of Chernobyl accident, 28 emergency workers and plant staff received lethal doses and then died. This included members of the off-site fire brigade. This tragedy was a result of not monitoring their dose and their not being properly trained. To prevent such tragedies in the future, the protection of emergency workers must be part of any emergency planning. All workers with the potential for receiving very high doses must continuously monitor their dose and be provided with dose guidance and training. This must include off-site personnel who may respond on-site, such as fire brigades, law enforcement, or medical teams.

According to IAEA (IAEA, 1996b), emergency workers include anyone needed to implement urgent protective action (e.g., bus drivers) or to take action to mitigate the accident (e.g., on-site damage control). IAEA states that emergency workers should be aware of the risks and be trained. IAEA has also established guidance concerning appropriate dose limits for emergency workers depending on their functions. The dose to anyone in general should not exceed their annual occupational dose limit. The only exception is for undertaking life-saving actions or actions that can avert a large collective dose or prevent the development of catastrophic conditions. In these cases, the dose received should be kept below 10 times the annual dose limit and the worker should understand and accept the risk incurred. Operational levels that relate to this guidance and that are measurable in the field by the workers must also be developed. The workers must be trained to monitor their dose and on the action to take if the operational guidance is exceeded. When the emergency is over and recovery operations such as decontamination and waste disposal have been initiated, the normal occupational exposure limits should apply.

In addition, as appropriate for their functions, protective clothing, continuous communications, respiratory protection, and stable iodine tablets, if radioactive iodine may be present, should be provided to emergency workers. If the entry into contaminated areas is anticipated, an access control point should be established; protective clothing, personal decontamination equipment (e.g., emergency showers), and spare clothing should be available. All the hazards possible while performing their response duties (e.g., toxic atmospheres) should be addressed.

IAEA has developed guidance on operational levels and equipment for emergency workers (IAEA, 1997a, 1997b).

Mitigation of Psychosocial Effects. The Chernobyl and Goiânia accidents demonstrated that psychosocial impacts often outweigh the radiological health effects. Psychosocial effects are difficult to mitigate because they are primarily related to perceived risk.

Emergency plans should explicitly recognize the significance of the psychosocial dimension and make provisions to reduce the detrimental effects it can create. Measures that should be considered include:

1. Informing the affected population accurately and promptly of the accident's progression, the risks involved, and the protective actions being taken
2. Keeping sheltered, evacuated, or relocated people informed of the expected time at which they may return to normal activities and/or return to their homes
3. Ensuring that the public trusts the response organizations and is kept aware that measures are indeed being taken to protect them; and

4. Providing means to reassure people who believe that they themselves, their dwellings, and/or their properties may have been exposed or contaminated

9.4.6 Emergency Preparedness Requirements for Threat Category III

Threat category III presents a much lower risk to the public than categories I and II. Here, the main threat is to facility workers. While the discussion on categories I and II applies in concept, emergency preparedness arrangements are quite different. The following discussion deals with aspects that are specific to threat category III. For general concepts, the reader should refer to Section 9.4.5.

Planning Elements

Authority, Command, and Control. This aspect of the plans will be limited to the facility's internal organization and must be clearly defined because the emergency response authority may differ from the normal operations authority. The emergency director/controller who is authorized to make key operational and strategic decisions during an emergency should be a person who has the expertise, knowledge, and capabilities to manage crisis situations. One of the greatest challenges is to ensure that those responsible for emergency response and those responsible for normal operation do not interfere with each other.

Logistical Support and Facilities. Logistics and equipment requirements for threat category III facilities are limited compared to categories I and II. All equipment and emergency management facilities will normally be located within the facility itself. Based on experience, two areas require particular attention. The first has to do with equipment supply and storage. Equipment and backups should be stored in areas deemed accessible during any emergency situation. The second involves the location of emergency management centers. These centers must be located in an area of the facility that is expected to be accessible following any type of accident. They should be protected with special shielding and/or air filtration equipment.

Public Education. Public education is not normally required for a threat category III facility. However, as we have discussed, the public reaction to a perceived threat can be dramatic. To limit the psychological effects and public over reaction during an emergency limited public education is advisable. Members of the public living near the facility, the facility staff who are not involved in the radiological operation, and those who occasionally visit the facility should be informed of the risks involved, the emergency response arrangements in place, and the actions to be taken in response to alerts.

Response Elements

Initial Accident Assessment and Classification. Initial accident assessment and classification is as important for this category as it is for categories I and II. The classification scheme used for category III is often based on the following:

1. *Alert:* events that have or may have reduced the level of safety of people on-site or involving requests to off-site emergency services
2. *Facility emergency:* events warranting implementation of protective actions on-site

Protection of Emergency Workers. Emergency workers on-site could be exposed to very high doses. The same basic requirements apply for these workers as for threat category I and II workers. This includes those responding from off-site into the facility (e.g., firefighters).

Public Instructions, Public Protective Actions, and Mitigation of Psychosocial Impacts. There is generally no need for public instruction or protective action plans for threat category III. Mitigation of psychosocial impacts is also not a consideration beyond providing timely information to counter rumors and misinformation.

9.4.7 Emergency Preparedness Requirements for Threat Category IV

Transportation Accidents. Transportation accidents involving radioactive sources may call for emergency service personnel such as firefighters, police officers, and paramedics from various areas and jurisdictions. The main challenge in planning for a transportation accident involving radioactive material comes if the event cannot be addressed by local officials responsible for hazardous material events or if there is considerable pubic interest or concern. In these cases a large number of staff may be called upon to respond, with their general lack of training in radiation incidents.

The most effective way of planning for such accidents is to rely on the infrastructure that is already in place for handling transportation accidents involving conventional hazardous materials and to provide a national or regional capability to provide additional support personnel trained and equipped for dealing with radiation emergencies.

Authority, Command and Control. Authority, command, and control for immediate actions in a transportation accident will normally rest with the organization responsible for responding to conventional hazardous material accidents. Once the situation has been stabilized and the spread of contamination has been confined, these organizations will often turn the response over to qualified experts from regional or national agencies. If public protective actions are required, however, their implementation will require the effective coordination of expert agencies and local authorities. Plans should include clear mechanisms for this coordination to take place, recognizing the delays that may be involved in mobilizing regional or national response organizations.

Organizational Responsibilities. These should be as close as possible to the organizational responsibilities in effect for response to accidents involving conventional hazardous materials. The responsibilities of the owner of the radioactive material (the consignor), the carrier, and the support agencies should be clearly defined and understood by all.

Preparedness and Response Coordination. In most cases, regional or national organizations will respond in support to local emergency services. Given the large number of localities and emergency services in place in most countries, coordination becomes difficult. Coordination can be improved at the planning stage by developing national, state, provincial, or regional guidelines/procedures and to ensure that they are distributed to all local emergency services. Central organizations such as CANUTEC in Canada, that have a 1-800 telephone number, act as the national coordinator for assistance to local authorities in case of transportation accidents anywhere in the country.

In Illinois, for example, an authorized officer from the Illinois Department of Nuclear Safety must accompany all significant shipments of radioactive material. This officer provides the initial radiological expertise at the scene of an accident and also provides the interface between the local emergency services and the central specialist organizations.

Often, the owner and the carrier are not prepared to deal with the radiological aspects of the accident. In these cases, there should be preplanned arrangements in place to promptly acquire and dispatch the required resources. These may include a national team with the capability to conduct radiological monitoring, assess the health and safety of workers and public, and conduct decontamination and recovery.

Plans and Procedures. Plans and procedures for transportation accidents involving radioactive material should be integrated into standard plans for conventional hazardous material accidents.

The owner and carrier should also have plans and procedures, up-to-date versions of which should be submitted to the agency that regulates the transport of radioactive material and to the central organization that coordinates the provision of assistance in case of a transportation accident (if applicable).

Radioactive materials being shipped are clearly labeled in accordance with international standards (IAEA, 1996a). Shipping papers, which describe the type of material, the quantity involved, and the UN number, accompany any significant quantity of radioactive material. Canada, Mexico, the United States, and many other countries have published guidance documents, referenced to the UN numbers, which describe the response to accidents involving different hazardous materials including radioactive materials. Millions of copies of these documents have been distributed to local fire departments, hazardous material response teams, and police departments throughout the world. The local agency responsible for responding to hazardous materials events must be provided with guidance on how to respond to transportation events tired to the UN number. IAEA is also working on guidance on the protective actions for the public and responders for each UN number associated with radioactive material.

Logistical Support and Facilities. The only specific requirements for the threat category IV are for survey and cleanup equipment such as dosimeters, survey and contamination meters, protective gear, decontamination supplies, and recovery tools at the national level. In several areas, particularly those in which the risk of a transportation accident is high, local response services are equipped with dosimeters and survey meters. If so, training (and periodic refresher training) must be provided, equipment must be periodically verified and calibrated, and the emergency services staff must be provided with valid procedures.

Training and Exercises. Training specific to radiation should be integrated into existing training program for conventional hazardous material incident response. This training must be frequent and accompanied by several drills, otherwise emergency services staff who are unfamiliar with and afraid of radiation may be reluctant to respond to an accident involving radioactive material.

The effectiveness of such training can be improved by:

1. Keeping the training as simple as possible
2. Making the procedures similar to those for conventional hazardous material accidents
3. Emphasizing the effectiveness of conventional procedures designed to deal with hazardous material emergencies
4. Emphasizing the effectiveness of universal precautions
5. Providing prompt and effective access to expert support in case of accidents

Public Education. There is no specific public education requirement for category IV.

Initial Accident Assessment and Classification. First responders will base their initial assessment and actions on the placards, UN numbers and other observables (fire or signs of a spill). Radiological monitoring will be needed if a ruptured package is suspected. Operational intervention levels (see Section 9.5.1) for transportation accidents must be developed in advance in order to be able to interpret monitoring results.

In November 1999, about 20 delegates from various countries, including the United States, France, and the United Kingdom, attended an international technical committee meeting at the International Atomic Energy Agency. These delegates suggested that there was no need to develop a special classification system for transportation accidents involving radioactive materials. Therefore, if it is required the regular accident evaluation and classification system should be used.

Public Protective Actions. Because the location of a potential transportation accident is not known in advance, planning for public protective actions is limited to establishing distances within which the public should be evacuated. These distances, provided in IAEA (2000a) for the UN numbers discussed above under Plans and Procedures, are summarized in Table 9.9.

TABLE 9.9 Evacuation Distances in Case of a Transportation Accident with Radioactive Material

Situation	Evacuation distance
Intact package with a class I (white), class II (yellow and white), or class III (all-yellow) label	Immediate area surrounding the package
Damaged package with a class I (white), class II (yellow and white), or class III (all-yellow) label	30-m radius or at readings of 100 μSv/h
Undamaged common source (consumer item) such as smoke detector	Immediate area surrounding the source
Other unshielded or unknown source (damaged or undamaged)	30-m radius or at readings of 100 μSv/h
Spill	Spill area plus 30 m of surrounding area
	Spill area plus 100 m of surrounding area
Major spill	
Fire, explosion or fumes	300-m radius
Spent fuel	100-m radius
Plutonium spill	100-m radius
Explosion/fire involving nuclear weapons (no nuclear yield)	1000-m radius

Plans should include guidance for implementation of evacuation of the public, to be conducted by first responders, that is keyed to observables (e.g., fire) and the UN numbers. This could be based on the guidance being developed by IAEA discussed above under Initial Accident Assessment and Classification. In addition, first responders should be informed of the relative risks of radiation exposure and that they should not delay life saving in the event of an immediate threat to life (such as a fire) because of a potential for radiation exposure or contamination. The procedures should also include provisions for gathering the public in a safe area and recording the names of members of the public who may have been within the distance limits at the time of the accident. If contamination is suspected, the public should be monitored for contamination and, if required, decontaminated. The plans should include provisions for medical follow-up and counseling of exposed and contaminated members of the public and emergency workers.

Protection of Emergency Workers. Protective equipment for responding to accidents involving conventional hazardous materials is quite effective against radioactive contamination. If airborne contamination is suspected, respiratory protection and dosimeters should be worn, keeping in mind that this gear will not protect workers against external gamma radiation.

An access control point with provisions for contamination monitoring should be arranged. No one should leave the area until he or she has been controlled for contamination and, if required, decontaminated. If the prompt arrival of radiological expertise is not expected at the site of an accident, plans should require a precautionary decontamination of personnel.

Descriptions of the hazards (radiological and other) to responders and appropriate protective equipment and actions to be taken by first responders should be developed for each UN number and set of observable conditions (e.g., fire). This could be based on the guidance being developed by IAEA discussed above under Initial Accident Assessment and Classification. This basis guidance should be provided to the first responders who may respond to an emergency involving hazardous material. Radiological teams should be provided with the same basic protection as for threat category I and II emergency workers.

Mitigation Psychosocial Effects. Apart from one possible exception, there is no need for mitigation of psychosocial effects for threat category IV. In highly publicized cases, such as the transport of spent fuel or the transport of plutonium fuel from nuclear weapons, emergency plans become necessary and are generally highly publicized by the media. In these cases, emergency plans should include provisions to monitor for possible psychosocial effects caused by heightened sensitivity and fear. Increased outreach, education, and other methods can be used to counter these effects.

Contamination from an Unknown Source. The most famous accident included in threat category IV occurred in Goiânia, Brazil, in 1987. A medical source of caesium was abandoned, taken, scraped, and disassembled. The breached source contaminated a large area of the city. Several people died, more than 100 people were treated for contamination, and more than 100,000 people were monitored for contamination. The emergency was discovered by a doctor when one of the victims brought him part of the radioactive source and told him, "I think this is making us sick."

This type of event can occur anywhere, especially if a system for control of radioactive sources is not in place or not effective.

Authority, Command and Control, Organizational Responsibilities, and Coordination of Preparedness and Response. In general, a sound plan at the national or regional level is the only emergency preparedness arrangement that can be effective against this type of accident. Because the authorities faced with the immediate problem typically do not have the required radiological expertise to deal with the accident, this situation presents the same complications as those for transportation accidents involving radioactive material.

During Goiânia, problems occurred because early in the event national officials directed the response from a location hundreds of kilometers away. These problems were resolved when an intergraded national/local command system was developed in the vicinity of the accident. National plans should recognize the role of local authorities in dealing with the immediate impact and provide them the prompt assistance from qualified organizations. Although ideally central government agencies should rarely take control over the situation, in reality, the lack of knowledge and expertise available at the local levels often forces these agencies to assume a predominant role in the management of the response. The plans should provide procedures for the transfer of responsibilities from local to higher-level authorities or for a joint command located in the vicinity if the event has national significance or local resources are overwhelmed.

Accident Mitigation. Threat category IV accidents should be treated like any other spill of conventional hazardous material. The risks involved are similar and include the potential contamination of people, resuspension of ground contamination, and ingestion of the contamination.

The priorities in this case are the prompt isolation and confinement of source of exposure (find the source) and significant contamination, the identification and control of potentially exposed individuals, and the decontamination of people and affected areas.

Public Protective Actions. These events may require monitoring large areas and screening a potentially large number of people. An effective way to be prepared for these rare events is to develop a network of existing qualified personnel (e.g., personnel from NPPs, research institutions, and universities) to be mobilized and coordinated on short notice.

Evacuation and relocation decisions will be based on environmental monitoring and operational intervention levels (see Section 9.5.1). People should be monitored for both external and internal contamination. When an individual is externally contaminated, his or her home and workplace must also be monitored. Monitoring for external contamination is done with hand-held contamination monitors or portal monitors. Portal monitors have the advantage of processing a large throughput.

Internal contamination can be monitored with:

1. Whole-body counters, which measure the radiation emitted from the body
2. Bioassays, which measure the quantities of radioactive material present in the blood, urine or other human excretas
3. Special measuring devices that measure the activity present in certain organs with a detector that is placed close to the organ (e.g., thyroid counters)

Limits for external contamination are generally taken as background, which means that any detectable activity should be removed. In practice, the limit is set at 50% above background. Limits for internal contamination are set by the country's health or radiological regulatory agencies.

If external or internal contamination is found, the individual may have received a significant dose. Therefore, plans should provide specialized medical resources to assess the dose through such techniques as chromosome aberration counts. If these resources are not available in the country, plans should include provisions for obtaining assistance from a neighboring country (through bilateral agreements) or from international agencies such as the World Health Organization (WHO) or the International Atomic Energy Agency's emergency response network (ERNET) (IAEA, 2000b).

Protection of Emergency Workers. Emergency workers would be subject to the same provisions as those established for threat categories I and II (see above, Protection of Emergency Workers).

Mitigation of Psychosocial Effects. Psychosocial effects are one of the most significant impacts caused by an accident involving contamination from an unknown source. Plans should address this by ensuring the availability of:

1. Timely and accurate information pertaining to the risks involved and the actions being taken by authorities
2. Access to qualified health specialists
3. Access to social workers
4. In some cases, access to religious leaders who are kept well informed of the risks and issues related to the emergency (this is particularly applicable in countries where religious leaders are considered the best—and sometimes the only—trustworthy source of information)

Lost or Stolen Source. Planning for lost or stolen source incidents is similar to planning for contamination from an unknown source (see discussion above).

Authority, Command and Control, Organizational Responsibilities, and Coordination of Preparedness and Response. Until the source is found, the authority for operations will normally rest with the police or internal security services. Once the source is found, the situation is similar to the case discussed previously for contamination from an unknown source.

Public Instructions and Public Protective Actions. As discussed, prompt action by public officials to alert the public (and thus the thieves) of the hazard has resulted in the thieves telling officials where to find the sources. In at least two cases, the highly publicized arrival of national-level monitoring teams and aircraft to look for the sources convinced the thieves to return the sources and to seek medical treatment.

If a source is stolen, inadvertently taken, or lost, the person responsible or others may develop symptoms of overexposure (e.g., skin burns). In several incidents of this type, the source was retrieved because someone reported to a clinic with these symptoms. Doctors often do not have the knowledge to recognize these symptoms as being caused by radiation.

Plans should therefore call for promptly making public announcements if a moderate- or high-hazard device (see Table 9.2) is lost or stolen. In addition, these plans should call for informing hospitals of the symptoms of radiation exposure so that affected people can be identified and the source can be found.

Satellite Reentry. Several satellites carrying high-hazard sources have reentered, resulting in debris landing on an area, called the footprint, larger than the area of Austria. None of these has resulted in a serious exposure. The main difference between this and other category IV events is that there is usually some warning time to activate the plan and refine it prior to the satellite reentry. This warning also will heighten public concern. In addition, the operator, through the IAEA, may provide information on the hazard and an estimate of the footprint location. However, experience shows the estimates of the location of the reentry footprint are very imprecise or wrong.

Experience has also shown that reasonable protective actions can only be taken if the area of impact can be accurately identified. People within the footprint then are told to report and not to handle suspicious objects. Other protective action, such as restrictions on food, is not warranted until monitoring is conducted. Once a suspicious object is identified, the event is very similar to the case discussed above involving contamination from an unknown source.

Plans for satellite reentry are normally developed for implementation by national authorities and should include provision to address heightened public concerns.

9.4.8 Emergency Preparedness Requirements for Threat Category V

A threat category V accident would result in only very small levels of airborne and ground contamination, with the main concern being on the food chain. Therefore, planning for a category V emergency is similar to planning for categories I and II emergencies, except that urgent protective actions are not required and there is more time to implement protective actions. For threat category V areas, a capability should be in place for effectively taking agricultural countermeasures to include restriction of the consumption, distribution, and sale of locally produced foods and agricultural products following a release. This capability should include OILs for deposition exposure rates, deposition densities, and food concentrations, means to revise the OILs, timely monitoring for ground contamination, careful and timely sampling and analysis of food and water, and means to enforce agricultural countermeasures locally and at international borders.

Between the time the notification of an accident is made and the time the impact on the country is felt, plans should focus on:

1. Mobilization of survey and sampling teams
2. Early detection of plume arrival
3. Provisions to provide prompt instructions to agriculture industry on action to take
4. Precautionary implementation of food control measures in areas where it is anticipated that OILs for food consumption will be exceeded
5. Implementation of methods for enforcement of restriction of agricultural products that exceed the OILs
6. Implementation of methods for compensation for contaminated products

9.4.9 Medical Preparedness

Effective medical response is a necessary component of an overall response to nuclear or radiological emergencies. Medical response to radiation accidents generally represents a difficult challenge to the authorities due to the complexity of the situation, often requiring highly qualified specialists and organizational and material resources. Therefore, an adequate medical preparedness is needed.

Medical preparedness begins with an awareness of what type of ionizing radiation and radioactive materials are used in a country and where they are used. This is accomplished by identifying the threat categories for facilities and areas in the country.

Minimum Level of Preparedness—Threat Category IV. There have been several events during which physicians failed to diagnose radiation-induced injuries caused by lost or stolen sources. If these injuries had been promptly diagnosed additional deaths and injuries could have been prevented. Consequently, general and emergency medical practitioners should be aware of the medical symptoms indicative of a radiological emergency, of notification arrangements and of the appropriate immediate actions. A simple poster with photos of injuries along with a description of radiation-induced symptoms has been developed by IAEA for this purpose.

Threat category IV events can occur anywhere but are very rare; consequently, the medical community is inexperienced in dealing with them. There have been events during which persons exposed to lost sources have received inappropriate treatment resulting in unnecessary suffering. During the event in Goiânia, some medical professionals refused to treat exposed people due to fear. It is therefore recommended that a limited national capability be in place to provide the appropriate treatment for those members of the public and emergency workers who may have become exposed or contaminated. This shall also include provisions to obtain (possibly through IAEA or WHO) consultation on treatment for any exposure that can result in severe tissue damage or life-threatening doses (severe deterministic health effects) from physicians with experience in dealing with such injuries. The national response plan needs to identify organizations, plans, and procedures for providing such assistance.

In the planning stage, the following should be prepared: a list of medical facilities at the local, regional, and national levels, a list of specialized medical facilities in other countries, a list of medical and support staff with telephone numbers and addresses in each respective location, a list of specialized medical centers for treating patients with radiation-induced skin lesions or immunosuppression, a list of equipment and supplies needed for emergency response, and agreements with ambulance transport services.

In the case of threat category IV accidents, a local practitioner should be ready to institute life-saving measures, assist in decontamination procedures, initiate minimal supportive treatment, collect pathological specimens (e.g., blood, excreta), arrange the transfer of patient to a special hospital, and convey the necessary information for continuous treatment. All major hospitals should make some provision to accommodate exposed or contaminated individuals so that first aid can be given.

It is recommended that the following minimum staffing, equipment, and supplies are prepared in advance in at least one facility in the country.

Staffing. At least two physicians and three support staff should be trained annually in the risks, precautions, and treatment of exposed and contaminated patients. In addition, a radiation specialist experienced in dealing with radiation and contamination (e.g., from a research facility) should be assigned to the facility at the time of the event. Provision should also be in place to train additional medical staff at the time of the event if necessary.

Equipment and Supplies

1. Means to control contamination and handle contaminated samples and waste
2. Radiation survey and contamination instruments
3. Personal protective equipment
4. Medical equipment and drugs necessary for treatment of contaminated person
5. Plastic covers for preventing spread of contamination
6. Personal decontamination equipment
7. Radiation warning labels and signs
8. Provision to obtain consultations (e.g., through IAEA)

The reader is referred to IAEA documents (IAEA, 1988, 1997b, 1998a, b, 2000b) for further guidance.

Threat Categories I to III. Threat categories I to III facilities can have very high dose rate on-site. Therefore for each of these facilities, a nearby medical facility should be prepared to treat a limited number of contaminated and exposed people that meets the recommendations for the national facility listed above under Minimum Level of Preparedness—Threat category IV. In addition, the treatment for a high exposed patient is determined based on the exposure received. Consequently, the facility must have procedures, immediately after an event, to gather information helpful in estimating the dose received.

Threat category I facilities can also have releases resulting in severe exposure and contamination off-site. Consequently plans should be in place to screen exposed people and assign them to hospitals and other facilities for treatment. Due to the very low probability of such an event, reliance should be placed on use of existing facilities that have made a minimum level of preparations to include some staff training.

9.5 RESPONSE REQUIREMENTS

9.5.1 Intervention Criteria and Decision Making

The following primary objectives of emergency response are:

1. All possible effort should be made to prevent deterministic health effects
2. Reasonable effort should be made to reduce stochastic effects such as cancers

Intervention or action levels are values, in terms of radiation dose or concentrations in food, that are used to determine when protective actions should be taken to meet these objectives. Different countries and organizations use different names for these levels. For example, in the United States intervention levels are referred to as protective action guides (PAG) and action levels for food as derived intervention levels (DIL).

Obviously, any immediate protective actions should be directed toward meeting the first objective by keeping the dose below the thresholds for deterministic health effects.

For ease of implementation, a single set of intervention levels should be established that meets both these objectives. Establishing intervention levels that meet the first objective is easy. The doses or food concentrations at which deterministic health effects would occur are very high. These values would be about a million times those seen normally in an hour. But since any dose is assumed to increase the risk of cancers, the generic intervention levels are determined by the second objective. Consequently, the intervention levels will be established at levels well below those at which deterministic, early health effects would be expected, even for such sensitive populations as pregnant women and children.

Establishing intervention levels that meet the second objective is difficult because it is assumed that any dose, no matter how small, may result in an increased risk of cancer. Since it will be impossible to prevent all exposures if there is release, the issue is when it is "reasonable" to take a protective action. Any protective action, such as evacuation, has its own risks and costs. At some point the negative impact of the protective action will be greater than that of the dose that the protective action is attempting to prevent. The guiding principle in meeting the second objective is therefore to establish intervention levels that do *more good than harm.* In response to the Chernobyl accident, some governments took actions to relocate people from areas where the radiation levels were no higher than those found naturally in other parts of the world. This relocation caused people to lose their jobs and social ties. The ultimate result was a reduction in life expectancy due to the psychological and sociological impact of the protective actions. And tragically, these protective actions did not reduce the risks of cancers among those relocated, because the majority of the dose was already received before the relocation. This is a clear example of taking protective actions that did more harm than good.

To determine the appropriate intervention and action levels to meet the second objective requires a careful examination of the cost in terms of money and health impact of protective actions compared to the risks of radiation exposure. IAEA (IAEA, 1994) provides a detailed

TABLE 9.10 Generic Intervention Levels Recommended by IAEA

Protective action	Generic intervention level[a]	Period
Shelter	10 mSv	2 days
Evacuation	50 mSv	1 week
Iodine prophylaxis	100 mSv	
Temporary relocation	30 mSv	First month
	10 mSv	Any subsequent month
Permanent relocation	1 Sv	Life time (50 to 70 years)

Source: IAEA, 1996b.
[a] Dose averted (prevented) by action during the period specified

example of such an analysis. Through careful analysis, numerous international and national organizations have determined intervention levels that meet both the response objectives while doing more good than harm. The dose-related intervention levels are in terms of the dose averted by the action. The cancer risk from exposure obviously cannot be reduced by any protective actions taken after the dose received.

The generic intervention levels recommended by IAEA, WHO, and other international organizations for various protective actions are given in Table 9.10. These generic intervention levels (GIL) were also endorsed by numerous international organizations. Table 9.11 shows the internationally recommended generic action levels (GAL) for determining when food should be restricted. These generic levels are consistent with those adopted by many countries, including the United States. These intervention and action levels are referred to as generic because they may require adjustment. For example, if restricting foods will result in starvation, then the food action levels will need to be adjusted. It is important to realize that these levels are very conservative. In other words, it is unlikely that even much higher doses or higher food concentrations than those specified in the intervention levels would result in any detectable increase in cancers following a major release. Therefore, it is both inappropriate and possibly hazardous to use intervention levels that are considerably lower than those established by IAEA. If, for example, levels 1/10 of those recommended by IAEA were used, the resulting protective action would clearly do more harm than good.

Generic intervention levels or action levels cannot be used directly in an emergency, but they provide a basis for calculating values or developing tools such as computer programs

TABLE 9.11 Generic Action Levels Recommended by IAEA for Restriction of Food

	Recommended values (kBq/kg)	
Radionuclide	Food destined for general consumption	Milk, infant's food, and drinking water
^{134}Cs, ^{137}Cs, ^{103}Ru, ^{106}Ru, ^{89}Sr	1	1
^{131}I	1	0.1
^{90}Sr	0.1	0.1
^{241}Am, ^{238}Pu, ^{239}Pu, ^{240}Pu, ^{242}Pu	0.01	0.001

Source: IAEA, 1996b.

that would be used during the actual emergency. Consequently, the IAEA and most countries have calculated quantities that would be used during emergencies that are surrogates for the generic intervention levels or action levels. These are values that can be measured by environmental monitoring instruments. These quantities go by many different names. IAEA refers to them as operational intervention levels (OILs). They are also called derived response levels (DRLs) and derived intervention levels (DILs). In every case, they are values that, if measured in the environment, indicate that protective actions should be taken at that location.

Default operational intervention levels should be calculated in advance for the various types of releases that could occur. These default values would be used initially to assess environmental data. Some of the IAEA default OILs for a nuclear power plant release are shown in Table 9.12. As soon as possible, operational intervention levels should be recalculated based on the actual composition of the release and these values should replace the default operational intervention levels if appropriate. Examples of the detailed procedures used to calculate operational intervention levels are provided in (IAEA, 1997b). This reference also provides generic operational intervention levels for a nuclear power plant release and the IAEA generic intervention levels.

Operational intervention levels are not restricted to power-reactor emergencies. They can be calculated for all other types of activities based on the radionuclides present, the exposure pathways and the types of measurement instruments available.

In summary, each country that could be impacted by major atmospheric release or contaminating event should have established intervention and action levels to be used as a basis for implementation of protective actions. Default operational intervention levels should be established for the mixture of radionuclides that may be released during an emergency. These levels can then be used promptly during an emergency to determine whether environmental measurements indicate that protective actions should be taken.

In addition, there should be procedures that allow these values to be revised to account for the actual composition of the release and other important conditions at the time of an emergency. It is very important that default values and procedures for their revision be established before an emergency. The Chernobyl accident showed that if you wait until an emergency occurs to determine such criteria, political considerations could influence the

TABLE 9.12 Example of Default Operational Intervention Levels for a Reactor Accident

Measurement	Default operational intervention levels		Protective action
Gamma dose rate from deposition	1 mSv/h		Evacuate or provide substantial shelter
	0.2 mSv/h		Consider relocation
	1 μSv/h		Restrict immediate consumption of potentially contaminated food produced in area until sampled
	General food	Milk	
Deposition densities	10 kBq/m^2 ^{131}I 2 kBq/m^2 ^{137}Cs	2 kBq/m^2 ^{131}I 10 kBq/m^2 ^{137}Cs	Restrict immediate consumption unless it results in food shortages
	General food	Milk	
Food, water, or milk concentrations	1 kBq/kg ^{131}I 0.2 kBq/kg ^{137}Cs	0.1 kBq/kg ^{131}I 0.3 kBq/kg ^{137}Cs	

Source: IAEA, 1997b.

development of the criteria. This can result in criteria that do more harm than good when implemented.

9.5.2 Assessment of Facility or Source Conditions

Experience and research have shown that the most likely events that can result in deterministic health effects (deaths and injuries) are:

1. Reactor accidents or accidents at fuel reprocessing facilities storing very large quantities of radioactive materials
2. Accidents involving sealed sources used in radiography or external treatment, such as teletherapy units

In the second case, only gamma dose-rate instruments and operational intervention levels are needed to determine where actions are needed. The first category of accidents, however, is much more difficult to assess. For such events, protective action must be taken before or shortly after a very severe release to be effective in reducing the risk of severe deterministic health effects near the facility. The strategy for taking protective actions for severe reactor accidents is discussed above under Public Protective Actions.

Before the accident at Three Mile Island Unit 2 (TMI-2) in 1979, protective action decisions would often be based on real-time environmental measurement of dose rates following a release. Once this dose rate was measured, a projected effective dose could be calculated. This effective dose could then be compared to the intervention levels and the appropriate protective action could be selected. There is a serious problem with this approach: by definition, environmental measurements are obtained *after* a release. Thus, they cannot be used to initiate protective actions *before* the release. Moreover, even if field measurements are taken shortly after release initiation, much time can be consumed in the process of selecting and implementing appropriate protective responses. After gamma dose-rates are assessed, it is necessary to select an action, obtain the concurrence of off-site authorities, and transmit warnings to the population at risk—who must prepare to evacuate and then drive out of the risk area. The result is that for severe releases the protective action may be taken too late to be effective.

Nuclear power plants are constructed very safely with numerous barriers and systems that must fail in order for a severe release to occur. For most severe accidents, it is estimated that it will take two or more hours from the start of the event until a major release commences. There is considerable instrumentation in the plants that will allow the plant operators to detect dangerous conditions before a release. Consequently, it was recognized that a system could be designed to detect dangerous plant conditions before a significant release and initiate the appropriate protective actions before that release.

The IAEA (IAEA, 1997b), the United States, and others have developed guidance or requirements for nuclear power plants and other threat category I and II facilities to establish emergency classification systems for which various levels of off-site response are preplanned.[4] The U.S. system is discussed in Section 9.8 and is very similar to that recommended by IAEA. Under these systems, each emergency class is defined by emergency action levels (EALs) that are based on control room instrumentation that would indicate the class of emergency and these EALs are incorporated into each plant's emergency operating procedures. The most serious emergency class in the IAEA system is a general emergency, which would be declared when plant conditions indicate that severe core damage is imminent

[4] The emergency response classification system should not be confused with the IAEA/NEA International Nuclear Event Scale (INES). The INES is designed for communicating the severity or estimated severity of an event to the public and cannot be the basis for emergency response actions.

or in progress and, thus, a very real potential exists for severe off-site health effects. A general emergency would warrant immediate implementation of protective action by off-site authorities. While some events have been postulated that could cause very rapid releases, most severe accidents studied by the U.S. Nuclear Regulatory Commission (NRC) could be classified as general emergencies well before a major release occurs.

Considerable attention has also been given to the use of computer dose projections as the basis for initiating off-site protective actions. However, for very severe accidents, dose projections would be available too late and would not be adequate to initiate effective off-site protective response. To be useful, dose projections require an accurate estimate of the amount of material to be released and must project, with confidence, where the release will travel. The problem is that it is virtually impossible to predict the time or measure the magnitude of a severe release. Once the release does occur, its movement through the atmosphere will be very complex. In addition, a severe release may also last several days, ultimately impacting areas in every direction around the plant. The complexity of the resulting contamination (and dose) can be seen in Fig. 5.2, which shows the extent of contamination from the Chernobyl accident. Consequently, dose projections based on models will probably be of little value in the early phase of a severe accident.

Once radioactive material is in the environment, monitoring will be one of the principal methods of determining which protective actions are needed.

9.5.3 Emergency Monitoring

Ionizing radiation cannot be directly detected by the human senses, but it can be detected and measured by a variety of other means. To obtain the data or information required in an emergency, different types of measurements can be performed and different physical quantities can be measured. Nowadays, many measuring methods and techniques are available and, in general, several techniques can be used to measure a given physical quantity. In the development of an emergency monitoring strategy, national and international aspects must be considered. The reasons for performing emergency monitoring must be the basis of the strategy, and the intended uses of the results should guide the choice of monitoring priorities and the technical details of what monitoring is performed.

There are several objectives of emergency monitoring. The first is to provide information for emergency classification. As soon as they are available, monitoring data will guide decision makers on the need to take protective actions and interventions on the basis of operational intervention levels (OILs). Environmental data will also provide the required information to prevent the spread of contamination and protect emergency workers. After the initial prompt response phase of an emergency, environmental data will enable technical and decision-making staff to assess more accurately the level and degree of hazards as well as the nature, extent, and duration of the hazard. Finally, environmental data will allow an assessment of the efficiency of remedial measures such as decontamination procedures.

The actual measurements to be made in an emergency include ambient dose rate and dose measurements, airborne radionuclide concentration measurements, environmental deposition measurements, food, water, and environmental contamination measurements, individual dose measurements, and object-surface contamination measurements.

Ambient dose-rate and integrated ambient dose measurements may be performed using several different types of dose-rate instruments from fixed stations, from monitoring teams with portable instruments, or from aerial measurements. The physical quantity being measured can vary and depends on the instrument used. However, it is desirable to measure ambient equivalent dose-rate or dose. The results are expressed in Sv/h or the equivalent (μSv/h, mSv/h, etc.) for external gamma dose-rates and in Sv or the equivalent for integrated doses. These measurements are generally used to support decisions concerning the implementation of various protective actions.

Measurements of airborne aerosols and gases are generally made to provide early warning of releases from unmonitored or as-yet unreported sources. The physical quantity measured is activity concentration. The units used are usually Bq/m^3 or the equivalent (μBq/m^3, mBq/m^3, etc.). These measurements are generally radionuclide-specific, but may also be of gross alpha/beta activity. They are more sensitive than dose-rate measurements.

Environmental deposition measurements refer to the amount of radioactive material that has been deposited on ground or surfaces. These measurements are performed to characterize deposition, to determine the physical extent and profile of deposition, to support dose calculations and intervention decisions, and for public information purposes. The physical quantity that is measured is the radionuclide-specific surface activity. The results are usually expressed in Bq/m^2. These measurements may also refer to the specific activity of soil, water, or plants. In this case, sampling of rain, run-off water, soil, grass, crops, etc., is required. The results are then expressed in Bq/kg or Bq/l. These measurements may give total surface activity or radionuclide-specific activity. Such results can be used to guide the initial food sampling efforts, for example, by concentrating in areas where the ground contamination exceeds preset limits.

In measuring food, water, and environmental contamination, a sampling and in many cases sample preparation is required. This is performed in fixed laboratories, although some measurements could be performed in mobile laboratories. The quantity measured is activity concentration, expressed generally in Bq/kg or Bq/l. This can be gross activity (alpha, beta, or gamma) or can be radionuclide-specific.

Individual doses may arise from external irradiation or from intakes of radionuclides, primarily by ingestion or inhalation. In general, doses to individuals cannot be measured directly. Some combination of measurement and assessment is required. For example, personnel involved in an emergency usually wear personal dosimeters, and their external individual doses are measured and recorded. Dosimeters record the external dose, not the internal dose. Doses from intakes of radionuclides have to be assessed by other means. External individual doses to members of the exposed public, however, are not measured directly and individual doses must be assessed completely. If the dose is relatively low, the relevant quantity is effective dose or equivalent dose to an individual tissue expressed in Sv or equivalent. Organ equivalent doses (thyroid, skin) are measured using specific counting techniques. At higher doses and dose rates at which deterministic effects may be of concern, the relevant quantity is absorbed dose (measured in Gy). External dose or dose-rate measurements—made, for example, by thermoluminescent dosimeter or a dose rate monitor—will provide a good indication of the doses absorbed by the whole body from penetrating gamma radiation. For intakes, some other means of assessing body or organ radionuclide content is required. This may be an in-vivo technique, such as whole body gamma spectrometry, or excreta analysis. Other indirect assessment of individual dose can be made utilizing air concentration, ground deposition, and food or environmental concentration measurements.

During any release of radioactive materials, surfaces of objects or equipment may become contaminated. The objective of these measurements is to identify those objects that should be decontaminated, disposed of as waste, or treated in some other controlled fashion. The physical quantity that is generally measured is surface activity. Measurements may be radionuclide-specific or can also be assessed based on dose-rate measurements. The results are generally expressed in Bq/m^2 or Sv/h or equivalent, respectively.

In general, several techniques can be used to measure a given physical quantity. For example, ground contamination levels can be measured in situ or by sampling as gross beta activity or as radionuclide-specific activity, depending on the counting equipment used, and several different types of counting equipment can be used for the same measurement (IAEA, 1999). The most commonly used techniques are listed in Table 9.13 and generic guidance on monitoring in a nuclear and radiological emergency is given in Table 9.14.

As previously discussed, operational intervention levels should be calculated that would indicate when different types of protective actions are needed. These operational intervention levels will be actual readings that can be obtained by environmental monitoring instruments. The values and the instruments used are determined by the nature of the radioactive release. For example, operational intervention levels based on gamma dose rates can be used to determine whether evacuations and relocations are warranted following a major release from

TABLE 9.13 Type of Measurements and Basic Features

Physical quantity (unit)	Measuring technique	Instruments	Advantages	Disadvantages	Remarks
Type of measurements: Gamma dose-rate and dose measurements					
• Ambient equivalent dose rate • Air kerma rate (Sv/h) or equivalent	• Stationary, automatic gamma monitoring system	• Energy-compensated GM counters • Proportional counters • Ionization chambers • NaI(Tl) detectors	• Automatic alarm possible • Rapid overview over wide areas • Time trends can be followed	• Could be complex and costly • Representative siting is critical	• Height above background is an important parameter • Differences in calibration may lead to non-comparable results
	• Portable or mobile measurements		• Allows locally detailed surveys	• Doses to personnel in case of high dose rates	
	• Aerial measurements	• NaI(Tl) detectors • Proportional counters	• Covers large areas	• Complex calibration procedure • Costly	
• Ambient equivalent dose (Sv) or equivalent	• Integrated dose measurements	TLDs	• Cheap • Easy to use and transport • Flexible use	• No alarm • No dose-rate profile • Require processing	• Representative siting is critical
Type of measurements: Airborne radionuclide concentration measurements					
• Activity concentration (Bq/m^3) or equivalent	• Stationary filter stations equipped for on-line measurement	• On-line gamma spectrometer	• Alarm function • Radionuclide-specific • Time trends can be followed	• Expensive	• Aerosol filters sample only the particulate portion of iodine, missing the gaseous portion
		• Proportional counter for on-line gross beta measurement	• Cheap • Alarm function • Time trends can be followed	• Not radionuclide-specific	
	• Stationary filter stations requiring filter collection for measurement	• Gamma spectrometer in laboratory	• Radionuclide-specific • Provides accurate results	• Sample collection, transportation, preparation, and measurement is time-consuming • Experienced staff required	• Iodine sampling (elementary and organic) with impregnated charcoal and aerosol filters
	• Stationary filter stations equipped with advanced sampling device and monitors for iodine	• On-line iodine monitor	• Time trends can be followed	• Costly	
	• Mobile air-sampling stations and analysis of a filter sample	• Proportional counter for gross beta measurements • Gamma spectrometer	• Spatially flexible	• Do not run continuously	
	• Aerial sampling at high altitudes and analysis of a filter sample	• Gamma spectrometer in laboratory	• Activity concentrations at various elevations could be measured	• Contamination of aircraft • Very expensive	

Type of measurements: Environmental deposition measurements

• Surface activity concentrations (Bq/m^2) or equivalent	• In-situ measurement of surface activity on the ground	• In-situ gamma spectrometer (HPGe or NaI(Tl) detector)	• Can give reliable data on radionuclide deposition • Can give fast estimation of radionuclide composition	• Demanding calibration procedure • Require experienced staff • Risk of contamination of equipment • Limited to gamma emitting radionuclides • Normally liquid N_2 required	• Calibration must match the depth distribution of radionuclides in the soil • Not useful in ambient equivalent dose-rates above 30 to 50 $\mu Sv/h$ without detector shielding
	• Aerial measurement of surface activity on the ground	• Gamma spectrometer system with NaI(Tl) detector and GPS	• Detailed information on spatial inhomogeneities in deposition • Rapid • Radionuclide-specific	• Costly • Limited radionuclide identification • Ground reference calibration needed • Limited to gamma emitting radionuclides	• Experienced staff needed
		Gamma spectrometer with HPGe detector and GPS		• Costly • Ground reference calibration needed • Limited to gamma emitting radionuclides • Normally liquid N_2 required	
• Activity concentration (Bq/kg) or (Bq/l) or equivalent	• Environmental sampling (soil, vegetation, water) and laboratory measurements	• Gamma spectrometer with HPGe detector	• Can give fast estimates of radionuclide composition	• Rain samplers must be pre-installed • Requires sample preparation • Limited to gamma emitting radionuclides	• Need to specify dry or wet sample weight • Normally liquid N_2 required

Type of measurements: Food, water and environmental contamination measurements

• Activity concentration (Bq/kg) or (Bq/l) or equivalent	• Gamma spectrometry	• Gamma spectrometer with HPGe detector	• Excellent nuclide identification	• Need experienced personnel	• Normally liquid N_2 required
		• Gamma spectrometer with NaI(Tl) detector	• Simple operation, high sensitivity	• Limited radionuclide identification for complex spectra	

TABLE 9.13 Type of Measurements and Basic Features (*Continued*)

Physical Quantity (Unit)	Measuring technique	Instruments	Advantages	Disadvantages	Remarks
	• Beta spectrometry	• Liquid scintillation counter	• Large-scale automatic measurements • Method for low energy beta emitters	• Limited radionuclide identification capability • Relatively complex sample preparation	
	• Alpha spectrometry	• Semi-conductor counter (Si diode)	• Spectral information	• Labor intensive • Significant sample preparation necessary • Long counting times necessary	• Experienced personnel needed
	• Gross beta measurements without radiochemical separation	• Proportional counter	• High intensity	• Not radionuclide-specific	
	• Gross beta measurements with radiochemical separation		• Radionuclide-specific	• Labor intensive	
	• Gross alpha measurements	• Proportional counter	• Short counting time • Screening possible	• Not radionuclide-specific	
Type of measurements: Individual dose measurements					
• External gamma dose (Sv or Gy) or equivalent	• Wearing personal dosimeters	• TLD	• Cheap		
		• Electronic dosimeter	• Direct display • Alarm function	• Costly	
• Surface activity concentration (external contamination) (Bq/m^2) or equivalent	• Alpha monitoring	• Scintillation or proportional counter	• Inexpensive • Rapid • Immediate results	• Not radionuclide-specific • Very sensitive to distance from surface to monitor • Monitors are fragile • Not very accurate	• Requires trained personnel
	• Beta/gamma monitoring	• GM, proportional or scintillation counters		• Not radionuclide-specific	
• Activity (in the body) (Bq) or equivalent	Internal contamination screening	• Contamination monitors or dose-rate instruments • Thyroid monitor	• Quick and very portable • Equipment inexpensive and can be used for other purposes • Results available immediately • Large throughput possible • Quite sensitive	• Not radionuclide-specific • No automatic storage of information	

	Whole body, thyroid, or chest counting	• Whole-body counter (spectrometer) with NaI(Tl) or/and HPGe detectors	• Radionuclide-specific • Can be whole body or specific organs • Very sensitive • Short measurement time (5 to 10 mins in emergency) • Equipment is quite robust	• Possible shielding required to ensure sensitivity • Fairly expensive • Demanding calibration procedure • Limited to gamma emitting radionuclides • Interpretation of mixed radionuclide spectra quite difficult if using NaI(Tl) detectors	• Requires experienced personnel • Equipment requires maintenance • Liquid N_2 required for HPGe detectors
• Activity (in excretion samples) (Bq) or equivalent	Laboratory analyses of excretion (nose blow, urine, feces)	• Gamma spectrometers • Liquid scintillation counters	• Radionuclide-specific • Samples can be transported to distant laboratories	• Samples require special handling • Long delay (possibly days/weeks) for results • Problems with sample contamination • Samples may be biological hazard • Analyses are often expensive	• Considerable expertise required for some analyses • Transport of samples requires careful planning
• Individual accumulated dose (Sv) or equivalent	• Biological dosimetry (Cytogenic analysis)		• Applicable in connection with evaluation of accidental exposure	• Limited sensitivity	• Doses above 100 mSv
Type of measurements: Object-surface contamination measurements					
• Surface activity concentration (Bq/m²) or equivalent	• Alpha monitoring	• Scintillation or proportional counter	• Inexpensive • Rapid • Immediate results	• Not radionuclide-specific • Monitors are fragile • Not very accurate	• Requires trained personnel • Very sensitive to distance from surface to monitor
	• Beta/gamma monitoring	• GM, proportional, or scintillation counters		• Not radionuclide-specific	

TABLE 9.14 Generic Guidance on Emergency Monitoring

Emergency type	Type of radiation measurements	Objectives
Nuclear accident		
After declaration of alert	1. Gamma/beta dose-rate measurements	1. To detect major release from the facility and to locate plume direction
During a release	1. Gamma/beta dose-rate measurements (from plume and ground deposition) 2. Airborne radionuclide concentration measurements	1. To identify where gamma dose rates indicate urgent protective actions are warranted 2. To determine radionuclide mix and measure radionuclide concentrations in air
After release has ended or after plume passage	1. Environmental deposition measurements 2. Gamma dose measurements in the environment 3. Food, water, and environmental contamination measurements 4. Individual dose measurements	1. To implement protective actions 2. To determine deposition maps for ^{131}I and ^{137}Cs and other important radionuclides 3. To identify radionuclide mix in deposition 4. To control personal exposure and contamination 5. To determine food and drinking water contamination 6. To assess doses to the public 7. To plan follow-up countermeasures and longer term protective actions
Radiological accident		
Misplaced, lost or stolen source	1. Gamma/beta dose-rate measurements (by foot, vehicle, or aerial survey)	1. To locate the source
Found source or contamination	1. Gamma/beta dose-rate measurements 2. Ground and object contamination measurements 3. Food, water, and environmental contamination measurements 4. Individual dose measurements	1. To set security and safety perimeter 2. To implement immediate protective actions 3. To identify the source or contamination 4. To determine contaminated areas and/or objects 5. To control personal exposure and contamination 6. To plan recovery and cleanup operations
Unshielded sealed source	1. Gamma/beta dose-rate measurements 2. Object contamination measurements 3. Individual dose measurements	1. To set security and safety perimeter 2. To implement immediate protective actions 3. To check for possible contaminated surfaces and/or objects 4. To control personal exposure 5. To plan source recovery

TABLE 9.14 Generic Guidance on Emergency Monitoring (*Continued*)

Emergency type	Type of radiation measurements	Objectives
Damaged sealed source	1. Gamma/beta dose-rate measurements 2. Object contamination measurements 3. Individual dose measurements	1. To set security and safety perimeter 2. To implement immediate protective actions 3. To determine contaminated areas and/or objects 4. To control personal exposure and contamination 5. To plan recovery and cleanup operations
Unsealed source accident	1. Airborne radionuclide concentration measurements 2. Gamma/beta dose-rate measurements 3. Ground and object contamination measurements 4. Food, water, and environmental contamination measurements 5. Individual dose measurements	1. To set security and safety perimeter 2. To implement immediate protective actions 3. To determine air contamination 4. To determine contaminated areas and/or objects 5. To control personal exposure and contamination 6. To plan recovery and cleanup operations
Dispersion of alpha emitters	1. Ground and object contamination measurements 2. Food, water, and environmental contamination measurements 3. Individual dose measurements	1. To implement immediate protective actions 2. To determine air contamination 3. To determine contaminated areas and/or objects 4. To control personal contamination 5. To plan recovery and cleanup operations 6. To plan post-accident activities (follow-up) and longer term protective actions
Nuclear powered satellite re-entry	1. Gamma/beta dose-rate measurements (by aerial survey, by foot) 2. Ground and object contamination measurements 3. Food, water, and environmental contamination measurements 4. Individual dose measurements	1. To locate debris 2. To implement immediate protective actions 3. To determine contaminated areas and/or objects 4. To control personal contamination 5. To plan recovery and cleanup operations 6. To plan post-accident activities (follow-up) and longer term protective actions
Trans-boundary impact	1. Gamma/beta dose-rate measurements 2. Airborne radionuclide concentration measurements 3. Environmental deposition measurements 4. Gamma dose measurements in the environment 5. Food, water, and environmental contamination measurements 6. Individual dose measurements	1. To implement protective actions 2. To determine ground contamination 3. To identify radionuclide mix 4. To determine food and drinking water contamination 5. To assess doses to the public 6. To plan follow-up measures and longer term protective actions

a nuclear power plant. Alpha emitter contamination (e.g., plutonium) cannot be assessed using gamma dose rates. Operational intervention levels for this type of accident might involve alpha exposure rates.

Examples of the types of radiation measurements that can be used as operational intervention levels for different types of emergencies are shown in Table 9.15. The table also shows the distance at which these measurements could be made. There are basically four different types. The first three are gross alpha, beta, and gamma. "Gross" refers to the fact that the radiation measured does not discriminate between radiation energy (i.e., is not radionuclide-specific). As seen from the last column in Table 9.15 it is difficult to detect alpha and, to a certain extent, beta levels that might indicate the need for protective actions because these particles can only be detected at relatively short distances. Gross gamma measurements, however, are by far the easiest to perform and, as can be seen by the table, can be easily used to detect significant contamination levels at considerable distances. The last type of measurement, in situ gamma spectrometry, is used to determine the actual radionuclide mixture of the contamination. This type of measurement is needed when food and milk should be restricted. This instrumentation is expensive, and highly technical and these measurements require considerably more time than measurements of gamma dose-rates.

The types of measurements needed for fixed facilities can be determined in advance based on the likely composition of any release. For an event involving a radioactive substance of unknown composition, truly dangerous levels can typically be detected by gross gamma measurements. These should then be followed up by beta and alpha measurements.

9.5.4 Public Protective Actions

In this section we examine the most common protective actions and the practical issues associated with their implementation in the context of a nuclear or a radiological emergency. Because not every protective action is applicable to every type of accident, the discussion will also highlight the types of situations for which a particular protective action should be considered.

In many respects, the protective actions applicable to radiological/nuclear emergencies are the same as those used for conventional hazardous material emergencies. Therefore, the focus here will be exclusively on the key practical aspects that are directly related to the nuclear or radiological nature of an emergency.

A given protective action is seldom used by itself. A protective action strategy, combining two or more of the protective actions, is usually the most effective way to protect individuals. As discussed in Section 9.3, a protective action depends on the exposure pathway.

Time, Distance and Shielding—Radiation Protection Basics for Sealed Sources. A sealed source in most cases is a small radioactive object that is sealed to prevent dispersal of the radioactive material. Consequently, the only important exposure pathway is external shine, much like the shine from a light bulb. A radioactive source emits alpha, beta, or gamma radiation. Some specialized sources such as an americium-beryllium neutron generator also emit neutrons.

As discussed in Section 9.3, basically, individuals can effectively protect themselves by:

1. Limiting the time they spend near the source—the dose received is proportional to the time of exposure.
2. Increasing their distance from the source—the rate of exposure is proportional to the inverse square of the distance from the source for a point source emitting gamma radiation (see Table 9.3); for alpha and beta radiation, which only travels a few centimeters (for alpha) or a few meters (for beta) in air, the dose rate falls much more rapidly with

TABLE 9.15 Instruments Needed to Detect Levels of Dose or Contamination (OILs) in the Field

Measuring techniques	Needed to detect levels of concern: Used for OIL levels for determining need for protective actions	Instruments	Distance
Alpha monitoring	Evacuation to prevent inhalation dose resulting from resuspenstion of ground contamination resulting from an accident involving Plutonium dispersal (e.g., nuclear weapon accident).	Alpha scintillation or proportional counters	5 to 10 cm
Beta monitoring	Decontamination due to deposition on the skin resulting from a release from reactor or spent reactor fuel.	GM or proportional counters	5 to 100 cm
Gamma dose-rate survey	Evacuation or resettlement due to external dose from the plume or ground deposition from release from reactor.	GM, proportional counters, or scintillation counters	meters to kilometers
	Evacuation due to external dose from a radiography or teletherapy source.		
In situ gamma spectrometry	Restriction of food due to deposition following a release from a reactor spent fuel or fuel reprocessing.	Portable gamma spectrometer with HPGe or NaI(Tl) detector	1 m

distance; for large-scale surface contamination, the rate of exposure decreases approximately linearly with distance.

3. Placing some shielding between themselves and the source—the relationship between shielding and protection depends on the type of radiation emitted by the source and the nature of the shielding material used.

Because alpha radiation is stopped by a few centimeters of air, shielding is unnecessary. Beta radiation travels only up to a few meters in air, depending on its energy, and is effectively stopped by a thin layer of plastic or metal. On the other hand, high-energy beta radiation can produce low-energy gamma radiation (X rays) when it strikes another material. Appropriate protective clothing can significantly reduce the exposure from beta radiation. The fraction of beta radiation that is absorbed by typical protective clothing is shown in Table 9.16.

Shielding material does not stop gamma radiation; it merely attenuates it. Indeed, gamma radiation is able to penetrate most materials, with lead being one of the exceptions. The effectiveness of the shielding depends on the thickness of material and on the energy of the incident gamma radiation. For example, for ^{60}Co, with an energy of just over 1 MeV, reducing the dose rate by one half would require approximately 1 cm of lead, 1.5 cm of iron, 5 cm of concrete, or 10 cm of water.

Neutrons can travel several meters in most materials. As they do, they undergo successive collisions, which have the effect of slowing them down and reducing both their energy and potential harmful effects. As they slow down, neutrons can also be absorbed. The best shields against neutrons are made of material that contains light atoms (e.g., hydrogen). Water and polyethylene are good examples of effective shields.

TABLE 9.16 Absorption of Beta Radiation from ^{90}Sr

Material	Percent absorption
Safety glasses lens	95
Full-face respirator lens	80
Plastic suit	10
Rubber gloves	10
Cotton gloves	0
Disposable hood (for a suit)	0
Coveralls	0

Source: Burnham, 1992.

Sheltering. As discussed above under Public Protective Actions, sheltering can be effective protection from exposure from the cloud shine, ground shine, and inhalation dose resulting from a plume. The degree of effectiveness depends on the building type.

Sheltering, i.e., confining people indoors with windows closed and ventilation turned off and closed, should be implemented downwind of a radioactive plume, or where contamination outside buildings is found on the ground, as long as the measure is justified in terms of intervention levels or operational intervention levels. In the precautionary action zone, substantial sheltering or evacuation should be considered in the event of a severe emergency (e.g., general emergency).

As shown in Table 9.8, a substantial shelter (e.g., the basement of a two-story masonry house) can be up to 60% effective in reducing the dose received from the cloud shine. However, a typical wood frame house provides very little protection. The protection of sheltering from receiving an internal dose from inhalation decreases rapidly with time as the inside of the house becomes contaminated through natural air ingress. For example, most North American houses have an air exchange rate of two air changes per hour. Sealing doors and windows can reduce this. The air exchange rate is much lower for modern, highly insulated dwellings and large complexes. Figure 9.5 illustrates typical reduction factors[5] for various rates of air change and release duration (equivalent to the duration of sheltering), which are calculated assuming that the airborne contaminants are not filtered out as they leak into the house. As can be seen from the figure, the effectiveness of sheltering against inhalation for a dwelling with one air change per hour is almost negligible after 24 hours. Nevertheless, sheltering remains very effective for short releases. Once the plume has passed, ventilation of the dwelling can help get rid of the contamination that may have seeped in.

Evacuation. Evacuation can be the most effective protective action from all the exposure pathways resulting from a plume. In general, evacuation is the preferred protective action for areas close to any source of hazardous material release, provided that it can be implemented safely and promptly. For example, in the precautionary action zone, evacuation or substantial shelter is appropriate for a severe accident. However, the effectiveness of an evacuation depends on the nature of the release, time to initiate and complete, distance from the source, duration of the release, and other factors. The effectiveness of evacuation relative to shelter must be determined by careful study for each threat category I facility. These studies have come to some counterintuitive conclusions. For example, as shown above under Public Protective Actions, walking (e.g., evacuation by foot) in a plume from close to a

[5] A reduction factor of 1 corresponds to no protection at all.

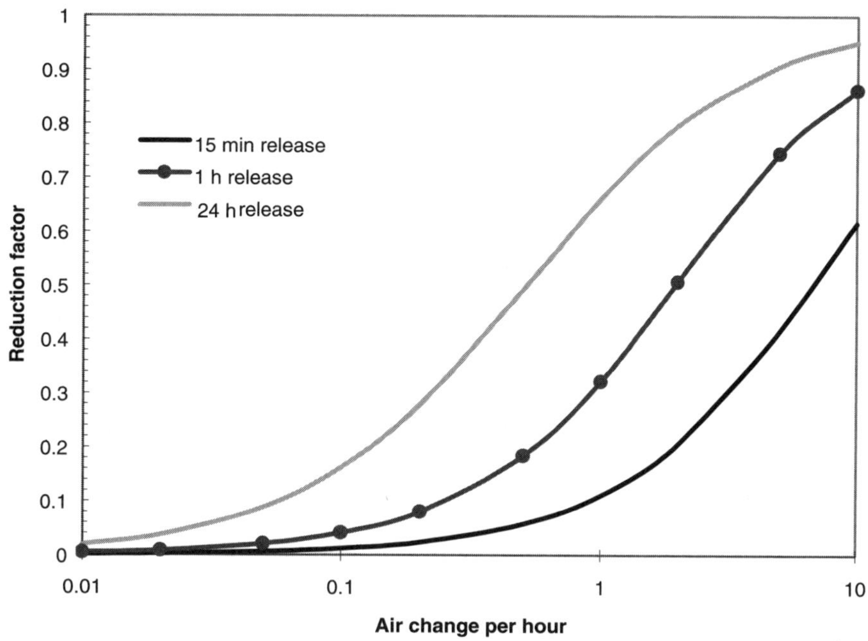

FIGURE 9.5 Typical reduction factors for rates of air exchange and duration of release.

reactor release may be more effective than shelter in a frame house. However, studies have also shown that shelter is more effective for a plutonium (Pu) release if the evacuation cannot be completed before exposure. This difference arises because for a reactor release shine dose is most important and for a Pu release inhalation is most important.

If an evacuation takes place before a release, there is no risk of personal contamination. In this case, the evacuation is operationally identical to any evacuation for conventional hazards.

If an evacuation takes place after the plume has passed, there is a risk that the evacuees may already be contaminated or may get contaminated during the evacuation. In this case, it will be necessary to establish monitoring points along the evacuation route(s). The monitoring points should be outside the affected area, but not too far, so as to minimize the spread of contamination. Both vehicles and people should be monitored and then decontaminated if contamination is detected. Alternative transportation arrangements should be provided if the immediate decontamination of vehicles is not a practical solution. Alternate clothing may also need to be provided.

As discussed above under Public Protective Actions, evacuation during a release should normally be avoided. A possible exception is when the release is expected to last a long time and the dose received from the plume and ground contamination during the evacuation is expected to be less than the dose that would be received by staying sheltered for the duration of the release.

Stable Iodine. Stable iodine is only effective in protecting the thyroid from inhalation or ingestion of radioactive iodine. Therefore, it is normally considered for reactor accidents. Because it is only effective when administered before exposure or shortly after exposure (see Fig. 9.3), the distribution strategy can have a great impact on the success of this protective action. In addition, because it only protects the thyroid, administration of stable iodine must be accomplished such that it will not delay implantation of substantial shelter or evacuations within PAZ intended to prevent early deaths.

Dosage depends on the type of tablets used and on the age group of the people to whom it is administered. The World Health Organization has published guidelines addressing this topic (WHO, 1999). Instructions from the manufacturer should also be consulted.

In practice, for a nuclear reactor emergency, stable iodine should be stocked and distributed at the time of an accident to all staff and emergency workers who are expected to work within the potentially affected area. Stocks should be maintained and replaced after the shelf time recommended by the manufacturer has been exceeded.

Distribution of stable iodine to the general population is much more complex. Basically, three types of distribution strategies exist:

1. Predistribution
2. Wide-scale distribution at the time of an accident
3. Distribution at the time of an accident from local stockpiles

Predistribution to all members of the population living in the planning zone means that there will be a minimal delay between the order to take the iodine and its administration. However, there are several potential problems associated with this strategy:

1. The perception of risk associated with the predistribution
2. Possible misplacing of supplies
3. The need to replace the supplies periodically
4. The need to ensure that all new arrivals in the area receive their supply
5. The difficulty in dealing with transient populations

Wide-scale distribution at the time of an accident can be a formidable task involving a great deal of logistics and time, thereby delaying and reducing the effectiveness of the action. Distribution from local stockpiles at the time of an accident is an interesting compromise between the first two options. The management of stockpiles is easier than dealing with a wide-scale predistribution and is less time consuming.

Relocation and Resettlement. Relocation and resettlement are protective actions intended to reduce the dose from long-term exposure (months to years) from ground contamination resulting from a plume or other contaminating event. Consequently, decisions on implementation do not need to be made promptly.

As discussed above under Public Protective Measures, during the Chernobyl response, relocations and resettlement may have been performed that did more good than harm. These are drastic measures that should only be considered in extreme cases where contamination is such that people would have to be prevented from returning to the affected area for several weeks or more. Before a decision to relocate or resettle people can be reached, international criteria and the long-term social and psychological impact must be considered. In addition, attempts at decontaminating the affected area(s) would have to be seriously considered. Disruptions in industry and key installations as well as the social cost of relocation or resettlement need to be examined by the government. Depending on the area, relocating residential housing while keeping industries and key installations operational can also be a valid option.

Food Control. Food restrictions are used to protect from ingestion of locally produced or imported contaminated food. The implementation of food control depends on many factors, including (but not limited to):

1. The international standards and agreements
2. The season
3. The type of soil and food grown
4. The availability of replacement food
5. The availability of alternative feed (for animals)
6. The nonradiological impact (e.g., economic, health effects)
7. The practicality and effectiveness of decontamination measures such as decontamination of the ground as well as cleaning the food before consumption
8. Alternative (nontraditional) use of the land
9. The possible dilution of the contamination by mixing the contaminated food with uncontaminated supplies

It is prudent to impose an immediate ban on food grown in affected areas until the radionuclide concentrations can be carefully assessed. As a guide, it is suggested that priority for food control and sampling be established when gamma readings near the ground exceed 1 μSv/h. However, this does not apply to pure alpha or beta emitters (for which the readings at ground level would be negligible), nor does it imply that the other areas are safe.

For example, the iodine operational intervention level for immediate restrictions on food consumption (e.g., milk from grazing cows) is 10 kBq/m^2. The dose-rate conversion factor for external exposure from a uniformly contaminated surface is 1.33 10^{-6} mSv/h/(kBq/m^2). Therefore, for a ground contamination equal to the operational intervention level, the dose rate at 1 m from the ground would be 13.3 10^{-6} mSv/h. This is much lower than the typical background of 10^{-4} mSv/h and would not be detectable using hand-held gamma dose-rate meters.

Example of Protective Action Strategies. The various protective action strategies used for different types of emergencies are listed in Table 9.17. These are only examples and should not be viewed as quick recipes for dealing with accidents. The actual strategy depends on the risk assessment, the technical planning basis, and the situation at the time of the accident. Figure 9.4 provides an example of the assessment of various protective actions for a reactor accident.

9.5.5 Protection of Emergency Workers

During the first few days of the Chernobyl accident, 28 emergency workers and plant staff received lethal doses and then died from radiation exposure. This was a result of workers not monitoring their exposure and not being properly trained and equipped. Inhalation and external exposure were important sources of doses, and burns to the skin resulting from beta contamination were a major contributor to many of the fatalities.

To prevent such tragedies, the protection of emergency workers must be part of any emergency planning. All workers with the potential for receiving very high doses must continuously monitor their doses, and be provided with turn-back guidance, training and protective equipment. This must also include off-site personnel such as fire brigades who may respond on-site.

Determining the turn-back guidance is not easy if inhalation exposure is a possible hazard since self-reading dosimeters monitor only external exposure. For such cases, emergency worker turn-back guidance has been calculated. IAEA turn-back guidance for emergency workers is provided in Table 9.18. Emergency workers should make all reasonable efforts not to exceed these values.

TABLE 9.17 Examples of Protective Action Strategies

Nuclear accident	Upon declaration of a general emergency: 1. Evacuate or substantial shelter of the precautionary action zone (PAZ) 2. If not practical, shelter the PAZ, until evacuation is possible 3. Shelter in the urgent protective action zone (UPZ) 4. Monitor the UPZ starting with downwind 5. If OILs are exceeded, evacuate or adjust protective actions 6. When the release is over, monitor ground contamination and take ground samples 7. Evacuate if required 8. Carefully analyze ground sample measurements 9. Determine if relocation is required 10. Decontaminate if justified
Found shielded source	1. Evacuate to safe distance around the source 2. Check for contamination 3. If contamination is found, go to "contamination accident" 4. Establish access control point 5. Implement dose control measures 6. Recover the source 7. Medical follow-up of potentially exposed persons
Contamination accident	1. Evacuate to safe distance around the source 2. Confine people who were within the evacuation distance until they can be monitored 3. Measure contamination 4. Adjust safe distance if contamination is found beyond it 5. Establish access control point 6. Implement dose control measures 7. Implement contamination control measures 8. Provide medical follow-up to potentially exposed persons 9. Clean up
Fire involving radioactive sources	1. Evacuate to safe distance around the source and downwind from the fire 2. Ensure no one is under the visible smoke 3. Confine people who were within the evacuation distance until they can be monitored 4. Measure contamination on the ground and in the air 5. Adjust safe distance if contamination is found beyond it 6. Establish access control point 7. Implement dose control measures 8. Implement contamination control measures 9. Provide medical follow-up to potentially exposed persons 10. Clean up if contamination is present

Emergency worker turn-back guidance is given as an integrated external dose on a self-reading dosimeter. Values in Table 9.18 have been calculated to account for the inhalation doses from a reactor core melt accident, assuming that thyroid blocking (stable iodine) has been taken. Note that skin contamination can also be a major source of dose and can lead to deterministic health effects for workers in highly contaminated areas if they are not provided with adequate protective clothing. Also note that if adequate respiratory protection is worn, the inhalation dose can be reduced significantly, in accordance with the filtration effectiveness of the equipment used.

Emergency worker turn-back doses are to serve as guidance and are not limits. Judgment must be used in their application. If analysis of air samples or other conditions results in emergency worker turn-back dose guidance that are significantly different from those in Table 9.18, then revised guidance should be used. Note that Table 9.18 is applicable only for a reactor accident. Different turn-back guidance may be needed for different accidents, depending on the contribution of the inhalation dose.

Once the early phase of the accident is over, the total dose incurred (during the early phase) must be confirmed before an emergency worker is allowed to perform activities that may result in additional dose.

The following are some basic rules for emergency workers.

1. ALWAYS be aware of the hazards that you may encounter in the field and take the necessary precautions.
2. NEVER attempt any field activities without the appropriate safety equipment, and always know how to use this equipment.
3. Conduct ALL activities so that exposures are maintained as low as reasonably achievable.
4. BE AWARE of turn-back levels as shown in Table 9.18. Emergency worker turn-back doses are to serve as guidance and not limits. Judgment must be used in their application.
5. DO NOT linger in areas where the dose rate is 1 mSv/h or greater.
6. BE CAUTIOUS proceeding to areas where the dose rate is greater than 10 mSv/h.
7. DO NOT proceed to areas in which the dose rates exceed 100 mSv/h unless otherwise directed by qualified health physicists.
8. USE time, distance, and shielding to protect yourself.
9. PREPLAN entry into high dose-rate areas in conjunction with your supervisor.

TABLE 9.18 IAEA Reactor Accident Dose Guidance for Emergency Workers

Tasks	Total effective dose guidance (mSv)	Turn-back guidance[a] [mSv]
Type 1:		250[b]
Life-saving actions	<500[b]	
Type 2:		
Prevent serious injury	<100	<50
Avert a large collective dose		
Prevent the development of catastrophic conditions		
Type 3:		
Short-term recovery operations	<50	<25
Implement urgent protective actions		
Monitoring and sampling		
Type 4:		
Longer-term recovery operations	Occupational exposure guidance	
Work not directly connected with an accident		

[a] It is supposed that thyroid blocking was taken before exposure. If no thyroid blocking is provided divide the turn-back guidance by 5; if respiratory protection is provided, multiply the turn-back guidance by 2.

[b] This dose can be exceeded if justified *but* every effort shall be made to keep dose below this level and certainly below the thresholds for deterministic effects. The workers should be trained in radiation protection and understand the risk they face. They must be volunteers and be instructed on the potential consequences of exposure.

10. DO NOT take unnecessary risks. DO NOT eat, drink, or smoke in any contaminated areas.
11. WHEN in doubt, seek advice from your team leader or coordinator.

9.5.6 Medical Management

The medical preparedness needed for different threat categories is discussed in Section 9.4.9. This section will discuss general concepts. The basic principles of the medical handling of exposed persons are based, to a large degree, on the methods used for handling other types of accidents, taking into account the specificity of the possible health effects of radiation and problems with contamination.

Accidents resulting in deterministic health effects will be very rare, and usually this will occur among employees or other professionals. However, in the case of a lost or stolen source, limited number of the general public may receive doses that can lead to deterministic health effects. Such a situation requires special medical care and supportive treatment for the early effects of acute radiation. In the event of internal exposure, especially by long-lived radionuclides, decorporation might be considered, even if the dose is below the threshold for deterministic health effects. The decision about decorporation levels should be based on committed equivalent dose to the organs and the effective committed dose.

Medical handling in an emergency situation is normally divided into medical care on-site (more often for workers) and off-site (for workers and affected population). To organize the off-site medical response, it is recommended that a system of off-site medical assistance for radiation emergencies be established under the supervision of national health authorities, i.e., Ministry for Public Health. The general structure of such a system is shown in Fig. 9.6.

The Ministry for Public Health is usually responsible for providing advice to other governmental departments on the health implications of any exposure to radiation. It is also responsible for ensuring that plans exist to provide treatment, monitoring, and health advice

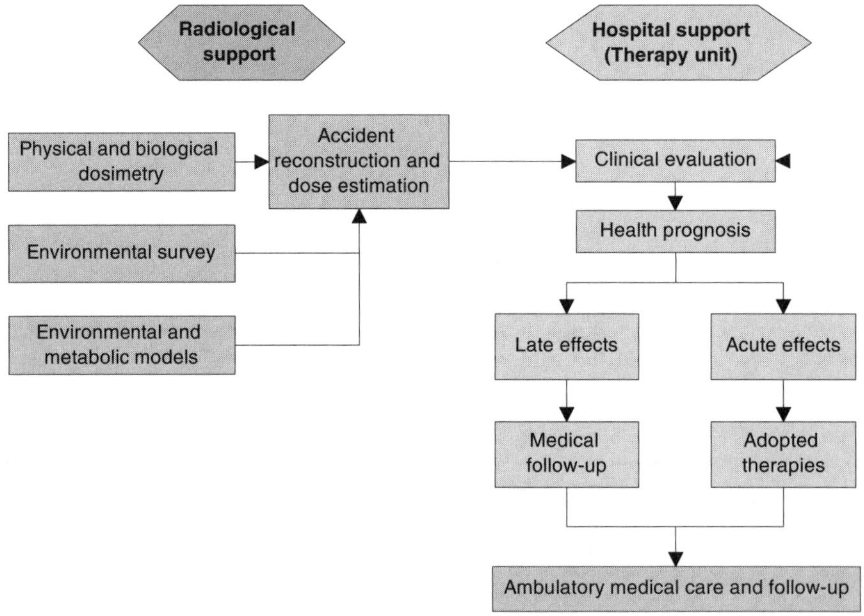

FIGURE 9.6 System for medical assistance in radiological emergencies.

to the public and to persons who have been contaminated or exposed to radiation or those who fear they may have.

In general, there are three levels of response:

1. First aid provided at the scene of the accident
2. Initial medical examination, detailed investigation, and medical treatment in a general hospital
3. Complete examination and treatment in a specialized medical center for treatment of radiation injuries

Management of Victims at the Scene of the Accident. At facilities with radioactive sources, trained personnel on every shift should normally provide any first aid required. In case of serious injury, medical personnel from suitable off-site medical centers should be available. The purposes of medical response on-site are to treat traumatic injuries, to assess contamination and perform limited decontamination. If anyone receives high doses exceeding threshold for deterministic effects, it is usually recommended that he or she be transported directly to a highly specialized medical hospital for complete medical examination, treatments, and assessment of the dose.

At the scene of many radiation accidents, e.g., transport accident, first aid is provided by emergency services, such as police, civil defence, or other responders. There is no risk from exposure to those treating a contaminated or exposed individuals. However, there have been cases of emergency personnel refusing to treat contaminated individuals due to unrealistic fears of exposure or social pressures. Medical and emergency services personnel should therefore be trained to deal with a radiological emergency.

All persons involved in a radiological accident should be carefully interviewed to provide a detailed description of the emergency situation, positions of persons at the scene of an accident, and time spent there. This is necessary for the purpose of dose reconstruction.

For situations involving a large number of exposed persons, triage may be necessary. These are actions to sort the patients into classes on the basis of their injury and/or disease. This is done to expedite clinical care, and maximize the use of the available clinical services and facilities. For example, after the Goiânia accident, 112,800 persons were triaged. The existing medical facilities can be used effectively to perform triage if provided with criteria.

Any person who is externally contaminated or who is suspected of being contaminated should be confined in a special area to prevent the spread of contamination. He or she should be decontaminated as soon as possible. Priority should go to persons who are heavily contaminated and to those who have open wounds or contamination near the mouth and face, in order to reduce the risk of internal contamination. Usually it is done in the hospital.

Management of Victims at the First Off-Site Stage. The task of medical staff at the first off-site stage should be to identify the type, origin, severity, and urgency of the cases. The basic principle is that treatment of serious or life-threatening injuries must take priority over other actions. The following is a simple classification system.

Persons with symptoms of radiation exposure: Patients should be transported urgently to a specialized hospital after appropriate medical care. Experience has shown that localized external exposure, often without radioactive contamination, is the most common consequence of radiological accident. In most cases, the treatment can be offered in hospital units specifically identified for this purpose as part of a medical emergency plan.

Persons with combined injuries (*radiation plus conventional trauma*): Treatment of such patients has to be individualized in accordance with the nature and grade of the combined injury. Usually, a combination of radiation exposure with mechanical, thermal, or chemical injuries may worsen prognosis.

Persons with external and/or internal contamination: These individuals need to be monitored to assess the degree of contamination, if any. Decontamination facilities will be required. Contamination alone, without physical injury or a significant dose from external radiation, may be sufficient to cause an acute effect to the patient but not to attendants. Decontamination is required to prevent or reduce further exposure, reduce the risk of inhalation or ingestion of contaminating material, and reduce spreading of contamination.

Persons with potential radiation symptoms: These individuals do not require immediate medical treatment but require urgent evaluation of the levels of dose. Because of this, the medical staff should have sufficient knowledge, developed procedures, equipment, and supplies to perform the first biological, medical examinations and analysis, which are necessary immediately after the accident.

Unexposed persons with conventional trauma: Patients should be taken to the specialized hospital where the medical treatment can be adapted to the symptoms.

Persons believed to be uninjured and unexposed: Patients are normally sent home. Sometimes medical follow-up should be provided to ensure that the first assessment was correct and to evaluate the dose more accurately.

At all stages of medical care, the treatment of highly contaminated individuals will require special facilities or isolated facilities with the special procedures that limit the spread of contamination and disposal of contaminated waste. For the detection of radioactive contamination, radiation equipment should be available, such as specialized radiation monitoring instruments, whole body counter, and iodine thyroid counter. Usually a radiation protection officer or health physicist performs the measurements. For the purpose of dose reconstruction, different instruments and methods can be used, such as electronic paramagnetic resonance (EPR) spectrometry and cytogenetic dosimetry. Because of this, collection of various tissues (blood, hair, and teeth) and clothes of exposed persons should be organized. Provisions (plastic bags, labels, etc.) should be made in advance.

Medical staff dealing with contaminated persons should wear protective clothing (overalls, masks, plastic gloves, and overshoes, as required) and personal dosimeters and should be monitored for possible contamination. Provisions for changing clothes, necessary stocks of clothes, and places for washing for staff should be made in advance. Contaminated clothing should be carefully removed and discarded in well-marked plastic bags. Dry decontamination using a towel may be a practical way to decontaminate a person if no showers are available. Contaminated individuals should shower, using mild soap as required to wash off the contamination. Harsh scrubbing is not recommended, as it may injure the skin and lead to internal contamination.

General Management of Human Contamination—Basic Principles

1. Before any action for decontamination is taken, careful and detailed monitoring is the first priority.
2. Protect yourself with gloves and apron or complete surgical clothing and mask, depending on size of the areas to be decontaminated.
3. If clothing of the victim is contaminated, remove it carefully and slowly so that deposited material does not become airborne.
4. If the hair is heavily contaminated, cutting it off may be the simplest and most effective solution.
5. Place all contaminated materials, clothing, linen, swabs, etc. into large, impervious plastic bags and seal carefully. Collecting contaminated cleaning fluids is also desirable but is often not practical.

Experience shows that the face and hands are the most likely areas to be affected. During the survey, any wounds or abrasions should be carefully noted because these provide possible direct transportation of the radionuclides to other parts of the body. If any damage of the skin surface is found, it should be immediately covered with waterproof adhesive plaster. The wound and its surrounding area should be decontaminated first.

Rules of Human Decontamination. Simply showering and changing clothes can typically remove dangerous amounts of skin contamination. For an event resulting in very large numbers of people being contaminated, provision should be made to implement these simple measures. However if there are only a limited number of contaminated people, it is desirable, for medical and psychological reasons, to remove as much contamination as reasonable. Most radioactive contamination does not readily penetrate the skin, as it tends to stay on the thin film of oil that covers the outer layer of the skin and the openings of follicles. One of the exceptions is tritium, which can penetrate the oily layer and then be absorbed through the intact skin if the contact time is long enough. Decontamination methods are based on removing this oily film with soap and detergents. The following steps should be used in decontamination of an individual:

1. Remove contamination around the various orifices of the body first.
2. Start the decontamination procedures from the periphery of the contaminated area and work gently towards the center.
3. Wash gently with warm water and mild soap. Scrubbing the area with anything other than a soft brush could irritate the skin and add the risk of absorption.
4. If soap and water fail, a weak solution of detergent may be used in the same way.
5. If contamination persists, use a saturated solution of potassium permanganate or ordinary household bleaches, which removes some of the outer layer of the skin. These substances should not be used near the eyes or on the hair. The potassium permanganate solution should be left on for a few minutes only, until the skin is deeply discolored. It is then washed off and allowed to dry; the resulting pigmented area is then treated with a 10% solution of sodium metabisulphite to remove the coloration. If the contamination still persists, these procedures may be repeated.
6. If the skin becomes tender or red, the procedure must be stopped. In such circumstances, it is good practice to cover the area with a lanolin-containing cream followed by an impervious dressing. The decontamination should continue when the skin will allow further attempts at decontamination (usually the following day).
7. Another method for decontamination of small areas with residual decontamination is to cover the area with adhesive plaster and leave it for a day or two. The residual contamination will often come off with the removed plaster.
8. The described method will remove practically all types of contamination, and it is rarely necessary to try another method. If, after several attempts at decontamination, radioactive material is still present, it will probably be necessary to scrub the skin, using abrasive powders. However, integrity of the skin should be preserved.
9. Decontamination should generally be repeated until measurements indicate background levels.

Management of Victims at the National Level. At the national level, specialized assistance must be provided to victims with acute radiation syndrome or serious radiological injuries of the skin. For this purpose, highly specialized hospitals with various departments (hemotology, hemotherapy, intensive care, and plastic surgery) must be identified and agreement developed to treat highly exposed persons at such hospitals. If the capability to treat high exposures is not available in the country, this can be obtained at WHO Collaborating Centers (Argentina, Australia, Brazil, France, Germany, Japan, Russia, and the United States), which can be requested through IAEA.

Medical staff and support personnel should be trained in the purposes and principles of radiation protection, health consequences of the exposure, and methods for dealing with exposed and/or contaminated persons. The training should include drills and exercises in medical response and in performing contamination monitoring, decontamination, interviews, etc.

Managers at every designated medical facility are responsible for the following aspects:

1. Designation and, if necessary, additional training of appropriate staff
2. Development of detailed emergency plan and procedures
3. Indication of space where reception and treatment can take place
4. Provision for special equipment and its maintenance and all necessary materials

9.6 EMERGENCY WASTE MANAGEMENT

Nuclear or radiological accidents can give rise to large amounts of radioactive wastes. These wastes must be managed and dealt with in a safe manner.

Radioactive wastes resulting from cleanup and recovery operations can be organic in origin (contaminated animals, food, soil, etc.) or inorganic material, such as equipment, building material, and houses. A large amount of compressible wastes can be also generated, such as gloves, clothes, and material contaminated during the response itself.

The removal methodology, required recovery equipment, and transport mechanisms for dealing with such waste must be determined in relationship to the quantity of radioactive material involved, its activity levels, and the availability of either interim or long-term storage facilities within the jurisdiction in which the waste is generated. In some instances, such as low-level spills, it may be possible to collect all or most contaminated waste relatively easily. In others, however, it may be impossible to remove any material quickly due to the volume of contaminated waste involved or high activity levels. There also may not be an available repository for the material.

One of the first steps in managing radioactive wastes is to estimate potential quantities. This information is required in order to:

1. Select appropriate storage facilities and locations.
2. Carry out the long-term safety assessment of the repository that will be used for the final disposal.
3. Estimate the radioactive contamination that may remain in the environment after cleanup operations are completed (e.g., to check for unknown contaminated areas).
4. Select an adequate type of packaging for transportation; this is necessary because the activity that is planned to be transported may exceed the regulatory limits for the transportation of radioactive material (in this case, it would be necessary to adopt special safety precautions or to reevaluate the transport strategy; sometimes authorities may be willing to waive normal shipment requirements to expedite removal of the wastes from the emergency scene; packaging may sometimes have to be improvised).

To manage the radioactive wastes adequately and ensure that packaging follows regulatory requirements, practical guidelines and appropriate forms must be developed and distributed to team members responsible for cleanup and recovery operations. Segregation and treatment guidelines must also be developed before the cleanup and recovery work begins. For example:

1. Compressible waste should not be mixed with uncompressible waste.
2. Organic and inorganic waste should not be mixed.
3. Organic waste (such as animals) should be treated (e.g., with calcium-activated charcoal for dehydration and gas retention).

Information about each packaged quantity of waste material must be thoroughly documented before shipment, including activity levels of the material involved as well as survey readings to be affixed to the outside of the shipping container. A copy of the documentation must accompany each shipment from the scene to its disposal or storage destination. Four levels of information should be included in the documentation:

1. Place and date of recovery
2. Type of packaging, number of packages, and type of waste
3. Surface and 1-m exposure rates of the package; instrument used for the measurements
4. Accountability for completing the forms (e.g., date, person name, position, signature)

The date is important for low- and intermediate-level waste so that decay corrections can be applied. The location allows a comparison of where the wastes were collected with the history of the accident. The type of packaging (dimension, shielding, etc.) and the type of waste (material, density) can be used, together with the exposure rates, to estimate the activity of the radioactive waste in the package. The surface and 1-m exposure rates are needed for labeling the package for transport (transport category).

The 1-m exposure rate should be used instead of the surface exposure rate when:

1. The measurement is done in a location that is far from the contaminated site and is therefore not affected by the high background.
2. The waste is not homogeneously distributed.

If the measurements are made in situ (i.e., in a high background area), using the surface exposure rate is a better option since it is less affected by the surrounding contamination.

Ideally, the type of package should be selected from those already developed for the nuclear industry (e.g., 200-liter drums used in a nuclear power plant—usually a type A package). Containers can also be used for large contaminated pieces, house debris, contaminated furniture, etc. If a new packaging is required (e.g., a metal box), its dimensions should be standardized as much as possible with respect to existing standard packages. For example, a box of $1.2 \times 1.2 \times 1.2$ m, or 1.7 m^3, is equivalent in volume of four 200-liter drums. This could be important for the structural stability and volume control (minimizing the void fraction) of the final repository.

There are several computer models for package activity estimation based on exposure rates. Models can be simple if build-up factors are not taken into account, or very sophisticated if the Compton effect is included.

To answer the question "How clean is clean enough?," clearance limits should also be established by the competent authority. The values should be derived based on conservative scenarios taking into account the main exposure pathways. Diluting the wastes for the sole purpose of meeting regulatory requirements should be avoided.

In general, existing waste storage facilities should be used. In some extreme cases, however, it may be necessary to build a provisional storage site. This was the case after the Goiânia accident. As a result of the accident, approximately 3,500 m^3 of wastes had to be removed. The wastes were temporarily stored in open air on concrete platforms occupying an area of about 8.5×10 m^2 at a site near the village of Abadia de Goias, 23 km away from the center of Goiânia, a city of 1 million inhabitants.

Deciding on a provisional wastes storage site can be a very complex process. The following aspects should be taken into consideration:

1. Location—far from highly populated areas, yet not too far to reduce transport requirements
2. Sources of power
3. Drainage systems

4. Site security and access control
5. Protection against natural disasters
6. Configuration of the waste containers to minimize exposure rates at the site perimeter

Once all contaminated waste materials are removed from the scene to the extent practical, the area should be resurveyed.

When wastes cannot be removed from the scene and may pose a continuing exposure risk, temporary shielding and security should be provided at the scene.

9.7 INTERNATIONAL CONVENTIONS

The International Atomic Energy Agency is the depository of several international Conventions, covering safety, safeguards, and liability issues. Four Conventions are relevant to nuclear and radiological emergencies:[6]

1. Convention on Early Notification of a Nuclear Accident
2. Convention on Assistance in the Case of Nuclear Accident or Radiological Emergency
3. Vienna Convention on Civil Liability for Nuclear Damage
4. Convention on Nuclear Safety

The reader is referred to the following Internet address for the full text of Conventions: http://www.iaea.org/worldatom/Documents/Legal/.

9.7.1 Convention on Early Notification of a Nuclear Accident (Notification Convention)

The Convention on Early Notification of a Nuclear Accident entered into force in September 1986 (IAEA, 1987) following the Chernobyl nuclear plant accident. This Convention establishes a notification system for nuclear accidents, which have the potential for transboundary release that could be of radiological safety significance for another state. It requires states to report the accident's time, location, radiation releases, and other data essential for assessing the situation.

The preamble of the Notification Convention mentions that the states parties to the Convention are convinced of the need for States to provide relevant information about nuclear accidents as early as possible in order that transboundary radiological consequences can be minimized. Furthermore, the Notification Convention applies in the event of any accident involving facilities or activities of a state party from which a release of radioactive material occurs or is likely to occur and which has resulted or may result in an international transboundary release that could be of radiological safety significance for another state. The facilities and activities referred to are the following:

1. Any nuclear reactor wherever located
2. Any nuclear fuel cycle facility
3. Any radioactive waste management facility

[6] When a state signs a Convention, it indicates an intention to be bound by it and agrees not to introduce legislation that will conflict with the Convention nor prevent accession at a later date. When a state ratifies or accedes to a Convention, it becomes bound by it and must have introduced national legislation to ensure compliance with the Convention.

4. The transport and storage of nuclear fuels or radioactive wastes
5. The manufacture, use, storage, disposal, and transport of radioisotopes for agricultural, industrial, medical, and related scientific and research purposes
6. The use of radioisotopes for power generation in space objects

The provision and transmission of information places obligations on a state party as well as on the International Atomic Energy Agency Secretariat.

According to Notification Convention, a state party shall:

1. Forthwith notify, directly or through the IAEA, those states that are or may be physically affected and the IAEA of the nuclear accident, its nature, the time of occurrence, and its exact location where appropriate; and
2. Promptly provide the states, directly or through the IAEA, with such available information relevant to minimizing the radiological consequences in those states.

The IAEA, according to Notification Convention, has to inform states parties, member states, and other states that may be physically affected of a notification received and promptly provide any state party, member state, or relevant international organizations, upon request, with the information. The information to be provided will include the following data as then available to the notifying state party:

1. The time, exact location where appropriate, and nature of the nuclear accident
2. The facility or activity involved
3. The assumed or established cause and the foreseeable development of the nuclear accident relevant to the transboundary release of the radioactive materials
4. The general characteristics of the radioactive release, including, as far as is practicable and appropriate, the nature, probable physical form, chemical form, and quantity, composition, and effective height of the radioactive release
5. Information on current and forecast meteorological and hydrological conditions necessary for forecasting the transboundary release of the radioactive materials
6. The results of environmental monitoring relevant to the transboundary release of the radioactive materials
7. The off-site protective actions taken or planned
8. The predicted behavior over time of the radioactive release

Such information will be supplemented at appropriate intervals by further relevant information on the development of the emergency situation, including its foreseeable or actual termination. Information received may be used without restriction, except when such information is provided in confidence by the notifying state party.

The Notification Convention also assigns obligations to the IAEA and the states parties that are not for response but for preparedness issues such as the following.

International Atomic Energy Agency. The IAEA is responsible for the collection and dissemination of information concerning experts, equipment, and materials that could be made available in the event of nuclear accidents or radiological emergencies, and methodologies, techniques and available results of research relating to response to nuclear accidents or radiological emergencies.

State Parties. These are responsible for making known the point of contact that is responsible for issuing and receiving the notification. Information must be available continuously, i.e., it must be manned 24 hours a day and seven days a week. The staff receiving a notification should have both the authority and means to promptly activate their own emergency response system as well as to make and receive requests for assistance in the case of a nuclear accident or radiological emergency. The national contact point[7] should be embedded into the national system for emergency response (not only nuclear or radiological emergencies).

As of October 9, 2000, there were 86 states parties to the Notification Convention. The Food and Agriculture Organization of the United Nations (FAO), the World Health Organization (WHO), and the World Meteorological Organization (WMO) are also Parties to the Notification Convention.

9.7.2 Convention on Assistance in Case of a Nuclear or Radiological Emergency (Assistance Convention)

Adopted in 1986 following the Chernobyl nuclear plant accident, this Convention (IAEA, 1987) sets out an international framework for cooperation among states parties and with the IAEA to facilitate prompt assistance and support in the event of nuclear accidents or radiological emergencies. It requires states to notify the IAEA of their available experts, equipment, and other materials for providing assistance. In case of a request, each state party decides whether it can render the requested assistance as well as its scope and terms. Assistance may be offered without costs taking into account, inter alia, the needs of developing countries and the particular needs of countries without nuclear facilities. The IAEA serves as the focal point for such cooperation by channeling information, supporting efforts, and providing its available services.

The Assistance Convention defines the exact responsibilities of the state party asking for assistance, and also of the states parties that are asked to provide the assistance. The IAEA will make available appropriate resources, transmitting the request to other states parties and international organizations and coordinating the assistance on an international level, if so requested.

The IAEA is obliged to:

1. Keep a list of contact points and competent authorities
2. Collect and disseminate information concerning
3. Assist, on request, in preparing both emergency plans and appropriate legislation; developing appropriate training programs; transmitting requests for assistance and relevant information in the event of an accident; developing appropriate radiation monitoring programs, procedures, and standards; and conducting investigations into the feasibility of establishing radiation monitoring systems
4. Make available resources, allocated to that purpose, to conduct an initial assessment of the accident or emergency
5. Maintain liaison with relevant international organizations

The states parties can request the IAEA to:

1. Assist a state party or a member state when requested in any of the following or other appropriate matters: (a) preparing emergency plans in the case of nuclear accidents and radiological emergencies and also preparing the appropriate legislation; (b) developing appropriate training programs for personnel to deal with nuclear accidents and radiological

[7] The IAEA maintains an up-to-date list of national contact points and Competent Authorities and makes it available to all member states and states and international organizations parties to the Conventions. The national competent authority is the authority or body within a state that can verify or arrange for verification of any data provided from that state concerning nuclear accidents or radiological emergencies. Furthermore, the authority or body concerned has to be in the appropriate position within the country for sending or providing information during a nuclear accident or radiological emergency.

emergencies; (c) transmitting requests for assistance and relevant information in the event of a nuclear accident or radiological emergency; (d) developing appropriate radiation monitoring programs, procedures, and standards; and (e) conducting investigations into the feasibility of establishing appropriate radiation monitoring systems; make available to a state party or a member state requesting assistance in the event of a nuclear accident or radiological emergency appropriate resources allocated for the purpose of conducting an initial assessment of the accident or emergency

2. Offer its good offices to the states parties and member states in the event of a nuclear accident or radiological emergency
3. Establish and maintain liaison with relevant international organizations for the purposes of obtaining and exchanging relevant information and data, and make a list of such organizations available to states parties, member states, and the aforementioned organizations.

As of October 9, 2000, there were 82 states parties to the Assistance Convention. The Food and Agriculture Organization of the United Nations (FAO), the World Health Organization (WHO), and the World Meteorological Organization (WMO) are also parties to the Assistance Convention.

9.7.3 Vienna Convention on Civil Liability for Nuclear Damage

Following the Chernobyl accident, the IAEA initiated work on all aspects of nuclear liability with a view to improving the basic Conventions and establishing a comprehensive liability regime. In 1988, as a result of joint efforts by the IAEA and OECD/NEA, the Joint Protocol Relating to the Application of the Vienna Convention and the Paris Convention was adopted. The Joint Protocol established a link between the Conventions, combining them into one expanded liability regime. Parties to the Joint Protocol are treated as though they were parties to both Conventions, and a choice of law rule is provided to determine which of the two Conventions should apply to the exclusion of the other in respect of the same incident.

In September 1997, a significant step forward was taken in improving the liability regime for nuclear damage. At a Diplomatic Conference at IAEA Headquarters in Vienna, September 1997, delegates from over 80 states adopted the Protocol to Amend the 1963 Vienna Convention on Civil Liability for Nuclear Damage and also adopted the Convention on Supplementary Compensation for Nuclear Damage. The Protocol sets the possible limit of the operator's liability at not less than 300 million special drawing rights (SDRs) (roughly equivalent to 400 million U.S. dollars). The Convention on Supplementary Compensation defines additional amounts to be provided through contributions by states parties on the basis of installed nuclear capacity and UN rate of assessment. The Convention is an instrument to which all states may adhere regardless of whether they are parties to any existing nuclear liability conventions or have nuclear installations on their territories. The Protocol contains, inter alia, a better definition of nuclear damage (addressing also the concept of environmental damage and preventive measures), extends the geographical scope of the Vienna Convention, and extends the period during which claims may be brought for loss of life and personal injury. It also provides for jurisdiction of coastal states over actions incurring nuclear damage during transport. Taken together, the two instruments substantially enhance the global framework for compensation.

9.7.4 Convention on Nuclear Safety

The Convention on Nuclear Safety was adopted in Vienna in June 1994. The Convention was drawn up during a series of expert level meetings from 1992 to 1994 and was the result of considerable work by governments, national nuclear safety authorities, and the IAEA's

Secretariat. Its aim is legally to commit participating states operating land-based nuclear power plants to maintaining a high level of safety by setting international benchmarks to which states would subscribe.

The obligations of the parties cover, for instance, sitting, design, construction, operation, the availability of adequate financial and human resources, the assessment and verification of safety, quality assurance, and emergency preparedness.

The Convention is an incentive instrument. It is not designed to ensure fulfilment of obligations by parties through control and sanction but is based on their common interest to achieve higher levels of safety, which will be developed and promoted through regular meetings of the parties.

The Convention obliges parties to submit reports on the implementation of their obligations for peer review at meetings of the parties to be held at the IAEA. This mechanism is the main innovative and dynamic element of the Convention.

9.7.5 The Role of IAEA in a Nuclear or Radiological Emergency

Throughout the world, there are a large number of nuclear power reactors and fuel processing facilities that have requirements to develop and maintain site-specific emergency preparedness and response plans. While the responsibility for the emergency response and public protection under these plans remains at the national or regional level, the International Atomic Energy Agency and its staff are expected to respond in some capacity in the event of a serious problem.

For many years, the IAEA has been providing assistance to member states related to nuclear accidents and radiological emergencies. The assistance provided has included:

1. Technical advice on emergency planning, preparedness, and response
2. Radiological surveys
3. Source recovery
4. In situ verification of the radiological conditions and related technical advice
5. Facilitation of the provision of medical advice and assistance with cases of suspected radiation exposure

Through liaison officers, the IAEA will keep in contact with the accident state, the affected states, with all other states, and the relevant international organizations (World Meteorological Organization, etc.). By frequent releases of information, the IAEA will inform the states parties and member states about the progression of a nuclear accident or radiological emergency. Information will always be analyzed and double checked before release, but urgent information will not be delayed unreasonably.

Requests for information will be answered either directly through the liaison officers or, if they might be of interest to a larger group, by a fax distribution to all the contact points.

The Agency can either conduct an initial assessment of the event itself or can contact other states parties to the Assistance Convention to determine whether they are in a position to offer the assistance requested. In the latter case, the agency will coordinate the assistance.

The responsibilities entrusted to the Agency for the purpose of implementing the Convention on Early Notification of a Nuclear Accident and the Convention on Assistance in the Case of a Nuclear Accident or Radiological Emergency have necessitated the establishment of a focal point within the Secretariat to which the Agency's member states, parties to the two Conventions, and relevant international organizations can promptly and effectively direct its notification (in the case of an accident) or event reports, requests for emergency assistance, requests for information, etc.

For this purpose and to facilitate the coordination of actions within the Secretariat, in 1986, the Agency's Emergency Response Center (ERC) was established and designated by the Director General to serve as a center for management and control of the Agency's response during nuclear accident or radiological emergencies anywhere in the world. This Center is located at the Agency Headquarters in Vienna, Austria. During normal operation, this Center is under the supervision of the Emergency Preparedness and Response Unit, Radiation Safety Section, Division of Radiation and Waste Safety of the Department of Nuclear Safety.

The Emergency Preparedness and Response Unit (EPRU) is responsible for maintaining and updating the Agency's Emergency Preparedness and Response System. It is the focal point in the IAEA for all emergency preparedness and response activities. The EPRU has a program of work dedicated to improve member states' current deficiencies in emergency preparedness and ensure the IAEA's quick response to a notification or request for assistance from a member state in case of an emergency.

The ERC's telecommunications systems allow the IAEA to fulfill its communications-related obligations, responsibilities, and functions under the provisions of the Emergency Conventions. This system operates a 24-hour contact point for notification or request for assistance. Using communication carriers, the ERC can rapidly communicate with more than 300 contact points worldwide.

The request for IAEA's assistance under the terms of the Assistance Convention has to be in a form of written communication.

Under the terms of the Assistance Convention, the ERC expects to receive a request for assistance from the Agency's member states. If the situation requires, however, the ERC will receive in the same format or by any other means of communication, request for emergency assistance from a non-member state. In the request, the following information should be provided:

1. *Radiological emergency:* nature of the event, location, time of its occurrence, name and full address of the organization in charge of the response actions, and name and contact numbers of the person assigned as counterpart to the Agency's requested emergency assistance
2. *Type(s) of assistance required:* aerial survey, radiation monitoring, radionuclide identification, source recovery, radiation safety assessment and advisory, medical support and/or advisory, bioassay support and/or advisory, radiopathology support and/or advisory, biodosimety support and/or advisory, waste safety support and/or advisory, and other(s), which should be specified.

It is essential for every message to contain the name of the sender and the contact telephone and/or facsimile numbers. Messages arriving at the ERC in languages other than English may be delayed until a proper translation is done. If the language of the message is in any other language than one of the Agency's official languages, an additional delay between the receipt of the message and any subsequent action may occur. Therefore, as far as practicable, the use of English is strongly recommended in order to avoid delays in dealing properly and promptly with any notification or request for assistance.

IAEA has provided detailed instructions for reporting emergencies to the IAEA and for requesting assistance through IAEA (IAEA, 2000d).

9.8 NATIONAL REQUIREMENTS AND GUIDANCE—UNITED STATES

The United States Environmental Protection Agency (EPA) and Department of Health and Human Services (HHS) have established guidance concerning the need to take protective actions in the event of the release of radioactive material into the environment. The United

TABLE 9.19 U.S. EPA PAGs for the Early Phase of a Nuclear Incident

Protective action	PAG projected dose	Comments
Evacuation (or sheltering)[a]	1 to 5 rem [b,d]	Evacuation (or, for some situations, sheltering[a]) should normally be initiated at 1 rem
Administration of stable iodine	25 rem[c]	Requires approval of state medical officials

Source: EPA, 1992.

[a] Sheltering may be the preferred protective action when it will provide protection equal to or greater than evacuation, based on consideration of factors such as source term characteristics, and temporal or other site-specific conditions.

[b] The sum of the effective dose resulting from exposure to external sources and the committed effective dose incurred from all significant inhalation pathways during the early phase. Committed equivalent dose to the thyroid and to this can may be five and 50 times larger, respectively.

[c] Committed equivalent dose to the thyroid from inhalation.

[d] 100 rem = 1 Sv.

States Nuclear Regulatory Commission (NRC) has established detailed requirements concerning emergency preparedness for facilities and uses of radioactive material that it licenses. The NRC licenses the vast majority of facilities and materials that represent a radiological risk in the United States. The United States Department of Transportation (DOT) has published guidance for response to transportation accidents involving radiological materials.

The EPA has established levels called protective action guides (PAGs) at which various protective actions should be taken following an atmospheric release (EPA, 1991). The protective action guides are generic intervention levels and thus are the dose that can be averted or prevented by the action. Doses received before taking the actions are not considered because any action after the exposure will not be effective in reducing any risk of cancers. PAGs have been established for early phase actions, which are intended to reduce the dose during or shortly after the release. The early-phase PAGs are summarized in Table 9.19. PAGs have also been established for the intermediate phase, which is the first year following implementation of the early phase protective actions. The intermediate phase PAGs are summarized in Table 9.20.

The U.S. Department of Health and Human Services has established concentrations of radionuclides in food at which the food should be restricted. These concentrations are generic

TABLE 9.20 U.S. EPA PAGs for Exposure to Deposited Radioactivity during the Intermediate Phase of a Nuclear Incident

Protective action	PAG (projected dose)[a]	Comments
Relocate the general population[b]	>2 rem	Beta dose to skin may be up to 50 times higher
Apply simple dose-reduction techniques[c]	<2 rem	These protective actions should be taken to reduce doses to as low as practicable

Source: EPA, 1992.

[a] The projected sum of effective dose from external gamma radiation and committed effected dose from inhalation of resuspended materials, from exposure to intake during the first year. Projected doses refer to the dose that would be received in the absence of shielding from structures or the application of dose reduction techniques.

[b] Persons previously evacuated from areas outside relocation zone defined by the PAG may return to occupy their residences. Cases involving relocation of persons at higher risk from such actions (e.g., patients under intensive care) should be evaluated individually.

[c] Simple dose-reduction techniques include scrubbing and or flushing hard surfaces, soaking or plowing soil, conducting minor removal of soil from spots where radioactive materials have concentrated, and spending more time than usual indoors or in other low exposure rate areas.

TABLE 9.21 U.S. HHS Derived Intervention Levels (DIL) (Bq/kg) (radionuclide groups, most limiting of all diets)

Radionuclide group	Derived intervention levels (Bq/kg)
^{90}Sr	160
^{131}I	170
Cs group	1200
$^{103}Ru^{a}$	6800
$^{106}Ru^{a}$	450
Pu + Am group	2

[a] Due to the large differences in DILs for ^{103}Ru and ^{106}Ru, the individual concentrations of ^{103}Ru and ^{106}Ru are divided by their respective DILs and then added. The sum must be less than one.

action levels called derived intervention levels (DILs) and are shown in Table 9.21. DILs are established for all radionuclides expected to be released in any major nuclear or radiological emergency. This includes releases from nuclear power plants, nuclear fuel cycle facilities, and nuclear satellites.

It is important to note that the EPA's PAGs cannot be used directly to interpret environmental measurements. For this purpose, derived response levels (DRLs, another name for operational intervention levels) have been established for environmental survey instruments indicating when protective actions should be taken to meet the PAGs. The DRLs will be different for different types of releases and different times during the emergency. Default DRLs for nuclear power plant release are shown in Table 9.22.

The DRLs established by HHS are also of only limited value early in an emergency because they require a time-consuming laboratory analysis. Consequently, the DRLs in terms of quantities to be directly measured in the environment have been established for use early in an emergency for determining where a locally produced food might need to be restricted. The DRLs for restriction on food are typically for deposition densities of a radionuclide. Exposure rates can only be used to do a preliminary assessment because deposition-warranting restriction of food may not be detectable on gamma dose-rate measurement instruments.

The EPA has also established dose limits for emergency workers. The limits depend on the type of activity being performed by the worker and are summarized in Table 9.23.

TABLE 9.22 Reactor Accident Exposure Rate DRL

Protective action	Exposure rate [mR/h] DRL (1 to 7 days after a release)
Evacuate to prevent early health effects	500
Evacuation to meet PAG	10
Relocation to meet PAG	5
Ingestion restriction	>background

Source: DOE, 1995.

TABLE 9.23 U.S. EPA Guidance on Dose Limits for Workers Performing Emergency Services

Dose limit[a] (rem)	Activity	Condition
5	All	
10	Protecting valuable property	Lower dose not practicable
25	Life saving or protection of large populations	Lower dose not practicable
>25	Life saving or protection of large populations	Only on a voluntary basis to persons fully aware of the risks involved

Source: DOE, 1995.

[a] Sum of external effective dose and committed effective dose to nonpregnant adults from exposure and intake during an emergency situation. Workers performing services during emergencies should limit dose to the eyes to three times the listed value and to any other organ (including skin and body extremities), 5 to 10 times the listed value. These limits apply to all doses from an incident, except to those received in unrestricted areas to members of the public during the intermediate phase of the incident.

The NRC has established detailed emergency preparedness requirements for their licensees where large releases of radioactive materials can occur no matter how small the probability of such an event (NRC, 1980, 1996). This includes nuclear power plants and fuel cycle facilities. The United States Department of Energy (DOE) for their major facilities has implemented similar preparations. The basic objectives of these preparations are to:

1. Prevent early deaths or injuries (deterministic health effects)
2. To keep the doses and food concentrations below the EPA PAGs and HHS DILs

To meet these goals, the NRC philosophy is that protective action should be taken as soon as dangerous conditions are detected in a facility. These protective actions can then be revised based on environmental measurements if there is a major release.

The preparations include establishment of emergency planning zones around these facilities. There are two zones. The *plume emergency planning zone* (plume EPZ) typically includes the area within 10 miles of a nuclear power plant. Within this zone, preparations are made to promptly implement sheltering and evacuation. The second zone, called the *ingestion zone,* extends out to 50 miles. Within this zone, detailed preparations are made to protect the food supply.

The NRC has established systems for classifying emergencies at major facilities. The NRC classification system for nuclear power plant emergencies is outlined in Table 9.24. The system has criteria both for plant conditions and radiological conditions for implementation of various response actions. The most severe emergency is the general emergency. This level is declared if there is core damage in the reactor because this could result in a very serious and unpredictable release. It is also declared if radiation levels are detected in the environment warranting the implementation of early protective actions.

The basic protective action strategy in the event of a general emergency is shown in Fig. 9.7.

Over the past 25 years there have been only two emergencies of the highest classification involving commercial nuclear power plants, at Three Mile Island and Chernobyl. In the United States, with about 100 nuclear power plants, there is about one site area emergency every few years. The United States has about two to five alerts, the lowest level, each year. In most cases these emergencies involve severe weather conditions such as hurricanes, which warrant an increased level of preparedness.

The U.S. Department of Transportation (DOT), in cooperation with the Canadian and Mexican governments, has established guidance for the actions to be taken in the event of a transportation accident involving any hazardous material to include radioactive material.

TABLE 9.24 The NRC Classification System for Nuclear Power Plant Emergencies

	Event Conditions		Response actions	
	Plant	Radiological	Plant (on-site)	Off-site
Unusual event[a]	Unusual event but considered significant	No significant release expected	Notify off-site officials	None
Alert	Potential significant decrease in safety	Off-site doses a small fraction of the PAGs	Partially activate to correct problem	Increased readiness
Site area emergency	Major decrease in level of safety—one more failure or remaining at this state for extended time could result in core damage (General Emergency)	Off-site doses a fraction of the PAGs	• Fully activate response • Evacuate non-essential personnel • Implement radiation protection for staff • Conduct monitoring off-site	• Fully activate response • Alert the population to the problem
General emergency	Actual or projected severe core damage	Off-site doses exceeding the PAGs	• Recommend to off-site that they implement pre-determined protective actions off-site	• Activate public alerting system • Evacuate the area near the plant and shelter remainder of plume EPZ • Place grazing animals on stored feed

Source: NRC, 1980.
[a] Not considered an emergency.

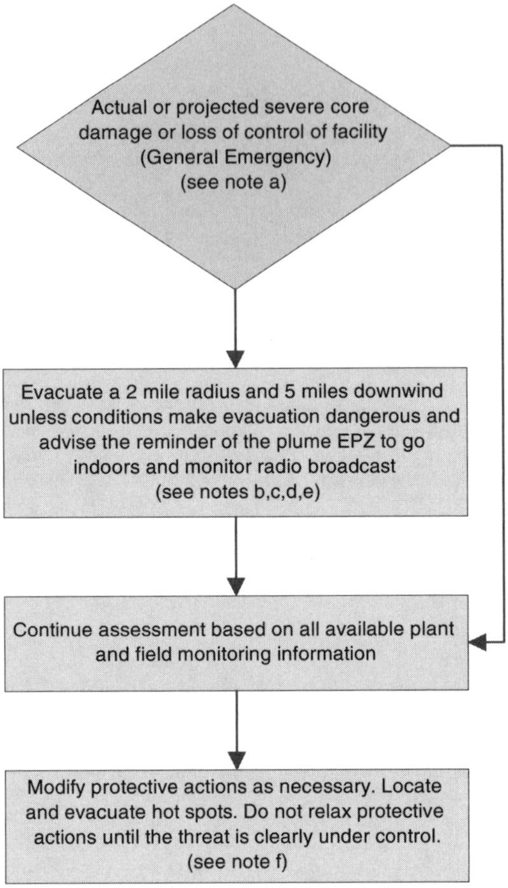

a Severe core damage is indicated by instrumentation readings in the control room.
b Distances are approximate; actual distances will be determined by the size of the evacuation zones established by local governments that are based on geopolitical boundaries.
c If travel conditions are dangerous, shelter rather than evacuate the population.
d Transportation-dependent persons should be advised to remain indoors until transportation arrives.
e Shelter may be the appropriate protective action for controlled releases of radioactive materials if there is assurance that the release is the short-term (puff release) and any area near the plant cannot be evacuated before the plume arrives.
f Based on derived intervention levels (DRLs), based on EPA PAGs.

FIGURE 9.7 U.S. NRC protective actions for severe reactor accidents. (*Source: NRC, 1996*)

Actions taken to fight fires, contain spills, and protect the public and emergency workers are keyed to the shipping name or UN number for the hazardous material. This guidance has been widely distributed to first responders.

Following a major release or contamination event, an environmental monitoring will be used to determine the need for response actions. In the United States, monitoring would first be conducted by the operator of the facility, next by state and local governments, and finally by the federal government. Monitoring efforts at the national (federal) level are directed by Department of Energy (DOE) as part of their Federal Radiological Monitoring and Assessment Center (FRMAC) program. This includes the ability to field numerous monitoring teams and conduct airborne surveys. In addition, the program has established detailed guidance used to assess environmental data.

9.9 SOURCES OF ADDITIONAL GUIDANCE

The guidance presented here has been based, to a large extent, on guidance developed by the International Atomic Energy Agency. IAEA has established a program of developing an integrated set of guidance based on technical analysis and operational experiences responding to real emergencies. This includes general guidance on development of an adequate response program (IAEA, 1997a) and example procedures for assessing and responding to nuclear and radiological emergencies (IAEA, 1997b, 1999, 2000a). In addition, guidance has been developed on the medical aspects (IAEA, 1998a, b) of the response to such emergencies. These documents are continuously being revised and updated to reflect the latest technical findings and operational experience. The reader is referred to these documents for further guidance.

9.10 REFERENCES

Burnham, J. U. 1992. *Radiation Protection,* Rev. 3. New Brunswick Power Corp., Point Lepreau.

Canadian Standards Association (CSA). 1991. *Guidelines for Calculating Radiation Doses to the Public from a Release of Airborne Radioactive Material under Hypothetical Accident Conditions in Nuclear Reactors,* CAN/CSA-N288.2-M01, Toronto.

International Atomic Energy Agency (IAEA). 1987. *Convention on Early Notification of a Nuclear Accident and Convention on Assistance in the Case of a Nuclear Accident or Radiological Emergency,* Legal Series No. 14, Vienna.

International Atomic Energy Agency (IAEA). 1988. *Medical Handling of Accidentally Exposed Individuals,* Safety Series No. 88, Vienna.

International Atomic Energy Agency (IAEA). 1994. *Intervention Criteria in a Nuclear or Radiation Emergency,* Safety Series No. 109, Vienna.

International Atomic Energy Agency (IAEA). 1996a. *Regulations for the Safe Transport of Radioactive Material,* Safety Standards Series No. ST-1, Vienna.

International Atomic Energy Agency (IAEA). 1996b. *International Basic Safety Standards for Protection against Ionizing Radiation and for the Safety of Radiation Sources,* Safety Series No. 115, IAEA, Vienna.

International Atomic Energy Agency (IAEA). 1997a. *Method for the Development of Emergency Response Preparedness for Nuclear or Radiological Accidents,* IAEA-TECDOC-953, Vienna.

International Atomic Energy Agency (IAEA). 1997b. *Generic Assessment Procedures for Determining Protective Actions during a Reactor Accident,* IAEA-TECDOC-955, Vienna.

International Atomic Energy Agency (IAEA). 1997c. *Subject Categories and Scope Description,* INIS Reference Series No. 3, IAEA-INIS-3 (Rev. 8), Vienna.

International Atomic Energy Agency (IAEA). 1998a. *Diagnosis and Treatment of Radiation Injuries,* Safety Report Series No 2, Vienna.

International Atomic Energy Agency (IAEA). 1998b. *Planning the Medical Response to Radiological Accident,* Safety Report Series No 4, Vienna.

International Atomic Energy Agency (IAEA). 1999. *Generic Procedures for Monitoring in a Nuclear or Radiological Emergency,* IAEA-TECDOC-1092, Vienna.

International Atomic Energy Agency (IAEA). 2000a. *Generic Procedures for Assessment and Response during a Radiological Emergency,* IAEA-TECDOC-1162, Vienna.

International Atomic Energy Agency (IAEA). 2000b. *Emergency Response Network, Emergency Preparedness and Response Series,* EPR-ERNET 2000, Vienna.

International Atomic Energy Agency (IAEA). 2000c. *Safety of Nuclear Power Plants: Operation,* IAEA Safety Standards Series, Requirements, NS-R-2, Vienna.

International Atomic Energy Agency (IAEA). 2000d. *Emergency Notifications and Assistance Technical Operations Manual, Emergency Preparedness and Response Series,* EPRENATOM, Vienna.

U.S. Department of Energy (DOE). 1995. *FRMAC Assessment Manual,* DOE/NV 11718-061, Washington, DC.

U.S. Environmental Protection Agency (EPA). 1992. *Manual of Protective Action Guides and Protective Actions for Nuclear Accidents,* EPA 400-R-92-001, Rev. 1, Washington, DC.

U.S. Nuclear Regulator Commission (NRC). 1980. *Criteria for Preparedness and Evaluation of Radiological Emergency Plans and Preparedness in Support of Nuclear Power Plants,* NUREG-0654, Rev 1, Washington DC.

U.S. Nuclear Regulator Commission (NRC). 1984a. *Reactor Site Criteria,* Title 10, CFR, Part 100, Washington DC.

U.S. Nuclear Regulator Commission (NRC). 1984b. *Dose Calculations for Severe LWR Accidents,* NUREG-1062, Washington DC.

U.S. Nuclear Regulator Commission (NRC). 1988. *A Regulatory Analysis on Emergency Preparedness for Fuel Cycle and Other Radioactive Materials Licensees,* NUREG-1140, Washington, DC.

U.S. Nuclear Regulator Commission (NRC). 1990a. *Severe Accident Risks: An Assessment for Five U.S. Nuclear Power Plants,* NUREG-1150, Washington, DC.

U.S. Nuclear Regulator Commission (NRC). 1996. *RTM-96, Response Technical Manual,* NUREG/Br-0150, Vol. 1, Rev. 4, Washington DC.

U.S. Nuclear Regulator Commission (NRC). 1996. *Criteria for Preparedness and Evaluation of Radiological Emergency Plans and Preparedness in Support of Nuclear Power Plants,* NUREG-0654, Supp 3, Washington, DC.

Turai, I., and B. Kanyar, 1986. "Compartment Modelling Study of Stable Iodine Prophylaxis in Relation to the Daily Iodine Supply," *Acta Physica Hungarica,* vol. 59, nos. 1–2, pp. 43–46.

World Health Organization (WHO). 1999. *Guidelines for Iodine Prophylaxis following Nuclear Accident,* Geneva.

P • A • R • T • 3

RISK ASSESSMENT

CHAPTER 10
RISK ASSESSMENT AND MANAGEMENT WITH A FOCUS ON RELEASES OF HAZARDOUS MATERIALS

Ertugrul Alp
Alp & Associates Incorporated, Richmond Hill, Ontario, Canada

10.1 RISK ASSESSMENT AND MANAGEMENT

10.1.1 Why Do We Live with Risk?

Industrial facilities, such as chemical plants, mines, and transportation systems, serve a useful purpose in society by providing products that the society needs or wants. They also form the backbone of today's economies by providing jobs and generating wealth and help finance the governments by paying taxes.

The operation of these facilities, however, also poses risks to the environment, the public, the employees, and indirectly to the owners, if something goes wrong. This is because these facilities often require the use of hazardous materials and/or disruption of the environment by the very nature of the processes they employ for making the products that the society uses. Hazardous by-products are also formed, which need to be discharged into the environment or stored for long periods of time before discharge. Hence, there are certain hazards that are inherent in the operation of these facilities. The broader society accepts having to deal with these hazards in order to gain the benefits from the products made by these facilities.

10.1.2 How to Deal with Risk?

Risk management is the process whereby the risks of operating these facilities are balanced against the benefits gained. There are many established methods of analyzing, evaluating, and controlling risks. We will discuss a risk management process model and some of these methods employed by industries. However, we will first clarify some of the terminology and concepts.

- *Hazard:* A chemical, physical, social, or political condition that has the potential to cause damage or any kind of harm to people, property/production, or the environment
- *Risk:* A measure of potential damage due to a hazard, taking into account both the likelihood and magnitude of damage
- *Safety:* A judgment of the acceptability of risk

A thing is "safe" if its risks are judged to be acceptable. There are degrees of risk, and consequently there are degrees of safety.

The concept of risk includes five components:

- *Hazard* inherent in an activity that is otherwise deemed to be beneficial
- A potential *undesirable event,* which has the capability to bring out the hazard
- Adverse *consequence* of that undesirable event
- Uncertainty as to whether or not the undesirable event will happen (*likelihood*)
- *Perception* about the combination of the above

We base our decisions on perception. Accurate understanding of the inherent hazards, the potential undesirable events, and consequences and the likelihood of these events will lead to more balanced perceptions and hence to better decisions in managing that activity. It is important to recognize that, although it is necessary to do a good job in understanding these components of risk for achieving high performance in risk management, this is by no means sufficient. Other factors that influence perception, such as outrage, must also be taken into account.

We will now discuss these components and factors in some detail.

10.1.3 Risk Analysis and Assessment

The process of understanding the hazards, events, consequences, and likelihood is called *risk analysis.* Its primary components are hazard identification, consequence analysis, frequency analysis, and risk estimation. These are shown in Fig. 10.1 in flowchart form. Once the risks are understood, decision making involves evaluating acceptability of these risks (*risk evaluation*) and examining different *risk control* options for reducing the risks that are deemed to be too high, taking into account the cost of implementing these measures (cost/benefit analysis). Often, the term *risk assessment* refers to the risk *analysis* and the risk *evaluation* steps taken together.

Let us now examine the steps within risk analysis in some detail.

Hazard identification refers to the identification of hazards and potential hazardous events that could occur due to the operation of the facility. This step requires a good knowledge of the manufacturing processes used in the facility, the hazardous materials used, and their toxicological properties.

Consequence analysis refers to understanding the damage that the hazardous events could inflict on the receptors which might be *exposed* to the events. Consequence analysis can be based on historical experience or may involve the use of sophisticated mathematical models. Identification of the receptors that are valued by the society is an important component of this step.

Frequency analysis refers to understanding the likelihood of the hazardous events.

Risk estimation refers to putting the frequency and consequence components together, either through a formula such as $Risk = frequency \times consequence$, or by placing the hazardous event on a scatter graph or matrix of frequency versus consequence.

For the remainder of this discussion, we shall limit ourselves to people as the valued receptors, although many of the concepts will be equally applicable to the ecosystem or property.

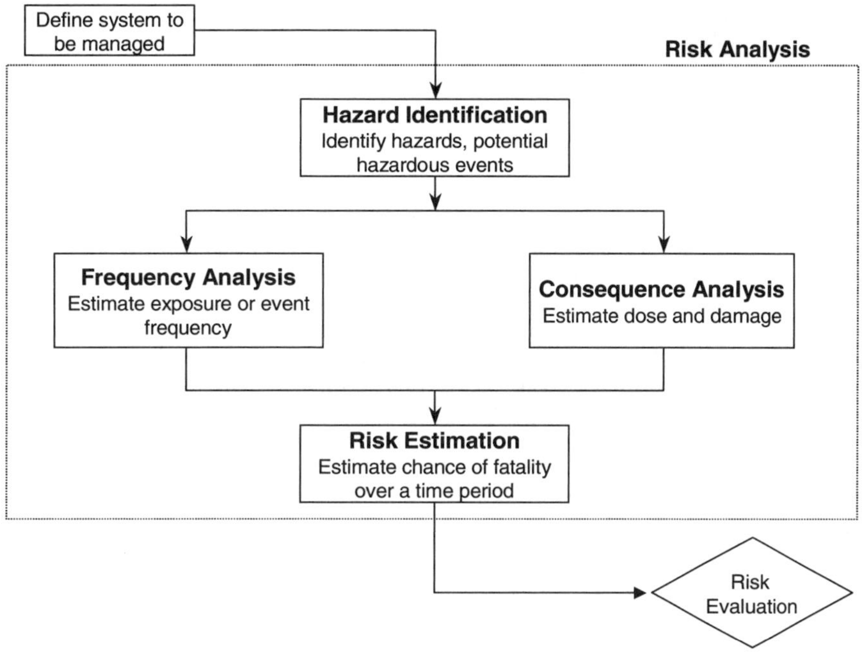

FIGURE 10.1 Framework for risk analysis.

Health Risk versus Safety Risk. At this stage, it is also important to distinguish between assessments of *health risk* and *safety risk*. The terminology used in these two different types of assessments tends to be similar, with the same word sometimes having different meanings. This tends to create much confusion, even among experienced practitioners.

Health risk, also called *chronic risk,* relates to fatalities or illnesses due to long-term exposure of individuals to hazardous materials discharged into the environment more or less continuously or routinely.

Safety risk, also called *acute risk,* relates to "immediate" fatalities or injuries due to accidental events, such as fire, explosion, falling debris, or a concentrated toxic gas cloud. It should be noted that some accidental undesirable events might also lead to health risks due to environmental contamination during that single event and long-term exposure of the receptors to this contamination afterwards.

In *health risk analysis,* the significant parameter is the total dose of the hazardous material received by an individual over a lifetime through repeated exposures to a contaminant that is in the environment more or less all the time (the contaminant may be coming from a single facility or from different sources). Total dose is estimated from frequency of exposure, exposure time per exposure, and level of concentration that the individual experiences during each exposure. Total dose is then used to calculate a consequence, usually expressed in terms of "the chance of fatality of an individual" due to exposure to that contaminant. This parameter is also called the individual risk. The total annual number of fatalities that might result due to exposure to that same contaminant is called the societal risk. The process of calculating frequency of exposure is called frequency analysis. The process for calculating the risk parameters is sometimes also called risk estimation or risk characterization. This process involves examining the pathways whereby a given contaminant from a given facility can enter the environment, estimating the amount of material entering the environment, modeling its transport and dispersion in the environment, modeling how this contaminant might get into the food chain, the water we drink, and the air we breathe, and modeling the damage that the contaminant might cause on exposed individuals of different susceptibilities. This would also take into account where these receptors live relative to the facility and their lifestyle.

In *safety risk analysis,* the important parameters are the frequency of occurrence of an accidental event and the single dose received by an individual during that single event. That single dose is then used to calculate an event consequence, which is usually expressed in terms of the chance of fatality of the exposed individual from that event. The consequence analysis process involves estimating the pathways whereby the material can be released from the facility, the amount released, how fast and how far it can travel, and the damage it can inflict on exposed individuals in its path. For fires, we are interested in thermal radiation. For explosions, we are interested in overpressures. For toxic gas clouds, we are interested in concentration in air, etc. Frequency of the event is estimated using frequency analysis techniques, and is generally expressed in terms of events per year. The consequence parameter is then multiplied by the frequency parameter, to arrive at the event individual risk, expressed in units such as chance of fatality of an individual per year as a result of a given type of potential event at a given facility. This parameter varies as a function of location relative to the facility. At a given location, the sum of the event individual risks for all possible events at that facility is called the facility individual risk. The total annual number of fatalities that might result from one type of event is called the event societal risk. The total number of fatalities that might result from all types of events is called the facility societal risk. The process of calculating the risk parameters is called risk estimation.

No wonder there is so much confusion about risk. What would *you* call the sum of facility societal *safety* risk and societal *health* risk from the facility?

10.1.4 Risk Evaluation

Once we estimate the risk, the next step in the risk management process is to evaluate its acceptability or tolerability. The basic principles, depicted in Fig. 10.2, are relatively straightforward.

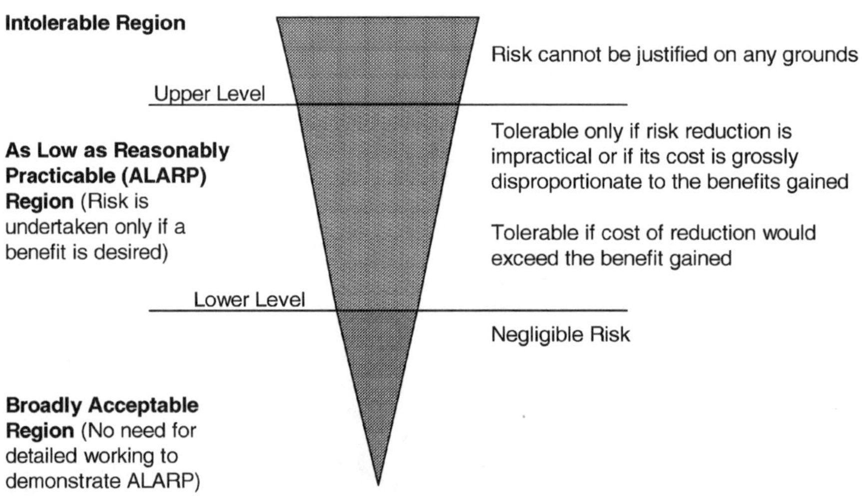

FIGURE 10.2 Framework for risk acceptability (tolerability) criteria.

In general, there is a level of risk that people will not tolerate going above. Similarly, there is a lower level below which people will not be concerned. People will tolerate higher levels of risk if they are exposed to it voluntarily. Conceptually, these principles are more or less universal. The problem comes when we try to define these levels, because they differ from one individual to the other and from society to society. Nevertheless, there are some established guidelines, which are different for health and safety risks.

For health risks, a commonly used guideline for individual health risk is 10^{-6} (involuntary exposure), expressed in terms of chance of fatality of an exposed individual over a *lifetime,* due to a given industrial activity or chemical. This value is for both the upper and lower level of risk (i.e., no ALARP region).

For safety risks, the criteria for individual safety risk varies among jurisdictions, ranging from 10^{-6} to 10^{-8} for the lower level and from 10^{-4} to 10^{-6} for the upper level (again for involuntary exposure), expressed in terms of chance of fatality of an exposed individual over a *year*. The Canadian guidelines developed by the Major Industrial Accidents Council of Canada are shown in Fig. 10.3 as an example of how these can be used for land use planning (MIACC, 1994).

Guidelines are also available for societal risk, commonly expressed in terms of complementary cumulative frequency distributions. An example is shown in Fig. 10.4.

10.1.5 The Risk Management Process

Risk analysis and risk evaluation are part of a broader process of decision making and implementing risk control measures for managing risks. A model for this process that has been used successfully by companies and government in Canada is shown in Fig. 10.5. Other models are also available (e.g., CAN/CSA, 1997).

This process starts with identifying the stakeholders of a given activity, such as building an industrial facility. The next step is to define the limits of the activity to be managed and to identify the stakeholders' needs. The importance of this step cannot be overemphasized if acceptance by all stakeholders, such as the affected public, businesses, and government agencies, is important to the success of the activity. A significant component of stakeholder participation is risk communication, which will be discussed in the next section.

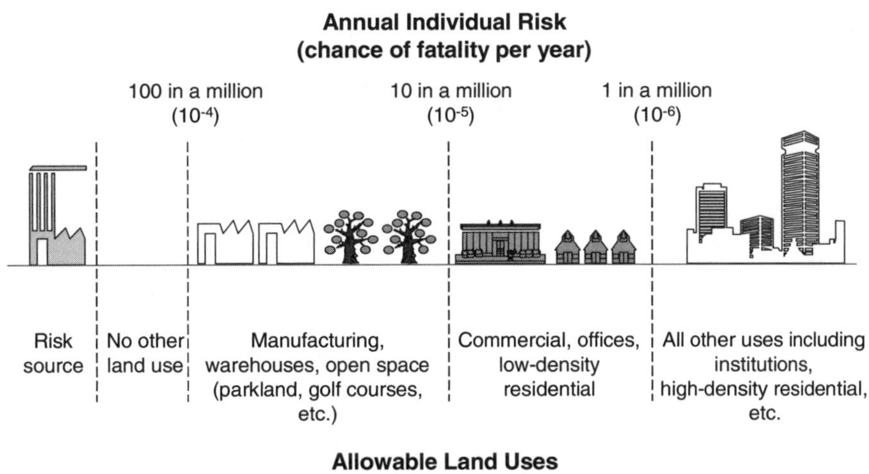

FIGURE 10.3 MIACC's risk acceptability guidelines for land-use planning (1994).

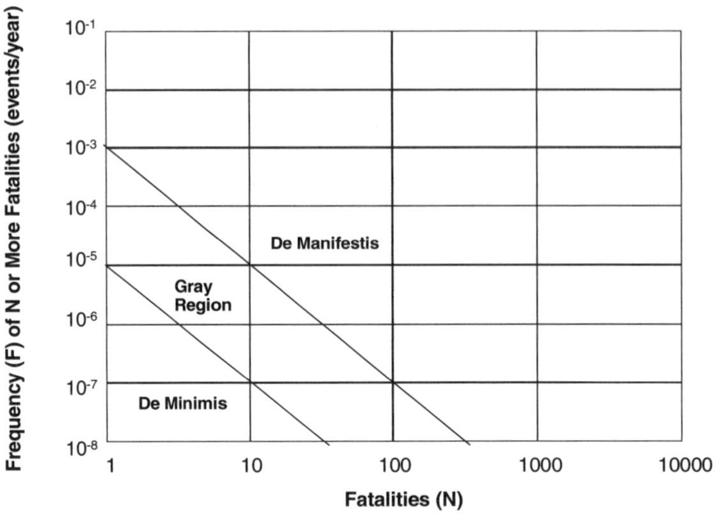

FIGURE 10.4 Example off-site societal risk guidelines.

Once the stakeholder participation process is established and the activity is defined, the next step is analyzing the risks of that activity. The steps involved in risk analysis have already been discussed in some detail. The risk evaluation step follows the risk analysis step. Participation of the stakeholders is again critical in evaluating acceptability or tolerability of these risks. Guidelines established by others should only be used for demonstration purposes,

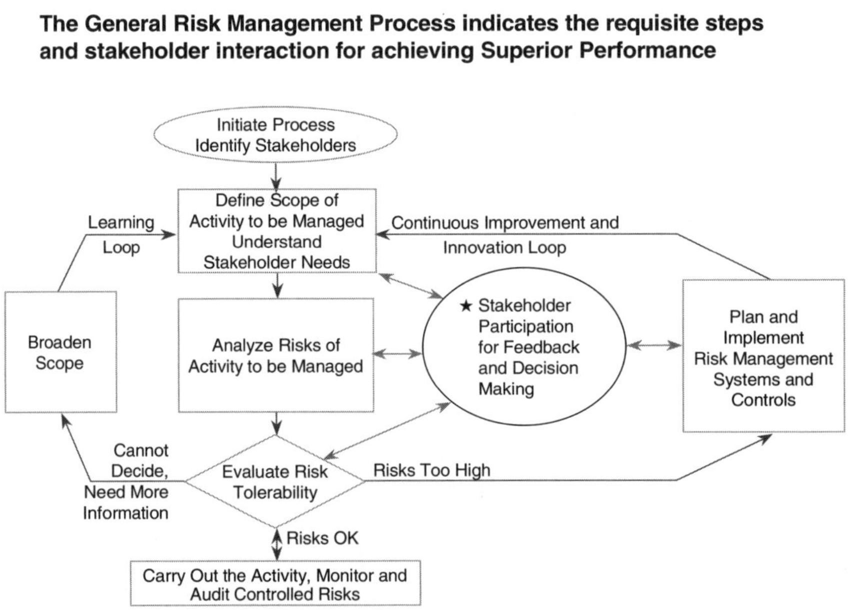

FIGURE 10.5 The general risk management process.

and not for pressing for decisions or acceptance by all stakeholders. The reasons for and importance of this statement will become clearer in the risk communication section.

At the risk evaluation step, if the risks are found to be unacceptable, then further risk control measures will be necessary. These include measures for preventing incidents and impacts (such as implementing process safety and loss management systems for minimizing chances and consequences of hazardous material releases), and measures for responding to incidents should they occur. Cost/benefit considerations will also play a role in selecting the risk control measures for implementation. Each additional risk control measure changes the risk level, and the process of improvement continues until the risks become acceptable to the stakeholders. The involvement of the stakeholders in each step in the process is indicated in Fig. 10.5 by double-headed arrows.

Once the risks are deemed to be "OK," then the activity is carried out. The risks are continuously monitored through safety, environmental, insurance, and financial audits to ensure that the activity is carried out as intended. If risks are found to increase to an unacceptable level, this can be recognized through the audit process and corrected through additional risk control measures, if necessary.

If a decision to proceed one way or the other cannot be made at the risk evaluation step due to lack of sufficient information, then the process calls for further study and analysis.

Let us now focus on the stakeholder participation component and risk communication, which play a significant role in the overall process.

10.1.6 Stakeholder Participation and Risk Communication

If you make a list of risks in order of how many people they kill each year, and then list them again in order of how alarming they are to the general public, the two lists will be very different. Risk managers in industry and government often deduce from this that public perception of risk is uninformed or irrational. But a better way to conceptualize the problem is that the public defines risk more broadly than the risk assessment profession.

It has already been mentioned that although it is necessary to do a good job in understanding the technical components of risk to achieve high performance in risk management, this is by no means sufficient. Other factors that influence perception (therefore influencing the decision-making process) and that are commonly summarized as "outrage," must also be taken into account.

People perceive certain activities to be high risk (dangerous) even though the technical risk may be low. This occurs when people *feel* outrage. The factors that cause outrage include whether the exposure is voluntary versus coerced, familiar versus unfamiliar, controlled by individual versus controlled by others, fair versus unfair, and imposed by institutions that are trustworthy versus untrustworthy.

Efforts to explain the risk are unlikely to succeed as long as the outrage is high. To decrease public concern about low-risk activities, risk managers must therefore work to diminish outrage.

The following are the four stages of risk communication that progressively improve stakeholder understanding and help achieve rational decision making.

1. *Stonewall stage:* No communication—Ignore the public.
2. *Missionary stage:* One-way communication—Show the public why you are right and they are wrong.
3. *Dialogue stage:* Two-way communication—Learn from the public the ways in which they are right and you are wrong.
4. *Organizational stage:* Ongoing communication—Become the sort of organization that finds dialogues with the public natural.

Here are the seven cardinal rules of risk communication:

1. Accept and involve each stakeholder as a legitimate partner.
2. Plan carefully and evaluate your efforts.
3. Listen to the stakeholders' specific concerns.
4. Be honest, frank, and open.
5. Coordinate and collaborate with other credible sources.
6. Meet the needs of the media.
7. Speak clearly and with compassion.

And the papal rule:

8. Know what you are talking about.

Application of these principles throughout the risk management process will help ease the tensions and result in high performance in risk management for companies and government organizations (much of the above material is based on Covello et al., 1988).

10.1.7 Risk Management Systems and Controls

Risk can be reduced by decreasing the likelihood and/or consequences of hazardous events through implementation of risk management systems and controls (see box in Fig 10.5). Risk control measures can be broadly classified into:

- Safety management of the hazardous facility. This includes process safety management practices, such as technological measures (e.g., design changes and inventory reduction), risk elimination (avoidance), risk transfer (insurance), and management measures (e.g., auditing, inspection, maintenance, training and work practices).
- Incident management, such as emergency response, emergency response plans and exercises.
- Land-use restrictions.

The facility safety management type of risk control measures can generally only be taken by the organization operating the hazardous facility. These include process safety, occupational health and safety, and environmental management systems and controls. Incident management issues are generally addressed by municipal fire departments or hazardous materials teams in cooperation with the operators of the facility. Municipal planners, often in consultation with all stakeholders, including companies operating the hazardous facilities, address the third type of risk control measures.

Another set of commonly used terminology for risk control is prevention, preparedness and response. Facility safety management and land-use restrictions can generally be considered as prevention measures, whereas incident management methods can be considered as preparedness and response.

Certain costs are associated with risk control measures. By estimating the risk reduction possible for each option, the costs and benefits for each option can be assessed and informed decisions can be made on which option should be selected. A good understanding of the factors contributing to risk (and of the frequency and consequences of potential hazardous events) through risk analysis is essential for this purpose.

10.2 RISK ANALYSIS TECHNIQUES

Let us now review the tools that are available to us in understanding the risks. We will be focusing on the tools used in safety risk assessments related to spills of hazardous materials with the potential to cause fatalities or injuries, damage the environment or property, or result in loss of reputation of the organization operating a hazardous facility.

10.2.1 Qualitative versus Quantitative Techniques for Safety Risk Assessments

Risk analysis and evaluation can take place at different levels of detail, depending on the requirements at hand. In general, the available techniques can be grouped into three categories:

1. Qualitative techniques, consisting of techniques for primarily hazard identification, such as SLRA (screening level risk analysis), checklist, what-if, what-if/checklist, HAZOP (guide-word hazard and operability study), and FMEA (failure modes and effects analysis).
2. Semi-quantitative techniques, consisting of index and matrix methods. Examples are Dow Fire and Explosion Index, Mond Index, and Pipeline Index for risk ranking of different facilities of the same type, and the matrix method for risk ranking of potential hazardous events that a given facility or organization can face.
3. Quantitative techniques, consisting of frequency and consequence modeling and risk estimation techniques.

The first group of techniques is primarily for identifying hazards and potential hazardous events that can occur at a given facility. Often, if the person or persons undertaking the qualitative analysis can quickly come up with recommendations for reducing the risk or mitigating the hazard, these recommendations will also be noted and action will be taken without further (qualitative or quantitative) analysis. Again, often the recommendations for action do not consider importance of a hazard relative to others and the cost of the recommended measures other than in a superficial manner based on the experience of the team undertaking the study. In the last 10 years or so, in order to get around this difficulty and bring some degree of objectivity into the analysis, these qualitative techniques have been coupled with the matrix method for risk ranking. This combination has become a very powerful and practical tool for routine application in the design and upkeep of hazardous facilities. We will describe these methods in some detail starting with the next section.

The index methods are also very powerful methods for understanding the risks of a facility and comparing one facility to others of the same type (e.g., a pipeline segment against another pipeline segment using the pipeline index, or one refinery against another using the fire and explosion index, but not a pipeline segment against a refinery). This is because, as the name of the method implies, the end result of applying the index methods is an index, often without any units of measurement associated with it. Hence, each indexing system is consistent within itself, but not when comparing different types of facilities, and not when trying to make cost-benefit comparisons on a global basis.

Quantitative methods of risk analysis, on the other hand, do provide results in readily understandable terms, such as fatalities per year, injuries per year, lost production per year, cleanup cost per year. As such, they are indispensable for undertaking cost/benefit analyses of risk reduction or risk control measures for organizations operating hazardous facilities and also serve as a reminder of potential costs to society as a whole of releases of major hazardous materials. Another area where quantitative tools have been used very effectively is in land-use planning regarding location of hazardous facilities relative to other forms of development. The quantitative estimation of the consequence component of risk again provides invaluable information to emergency response planners in deciding what kind of events they should prepare for and what kind and size of resources they would need should the emergency event occur.

10.2.2 Using a Hierarchy of Risk Analysis Techniques for Decision Making

Risk analysis is a tool for decision making. As evidenced by the previous section, a large number of techniques is available. However, not all techniques are appropriate or useful for all cases. A hierarchy of techniques from simple to more complex should be utilized for decision making, depending on whether the hazards are well understood, the magnitude of consequences, and the cost of the risk control or mitigation measures that are being considered for a given situation.

The following hierarchical protocol has been found to be very effective in risk management decisions at hazardous facilities.

Referring back to the risk management process model discussed earlier (Fig 10.5), the starting point for analyzing the risks of an existing or proposed facility is to conduct a screening level risk analysis (SLRA), which is discussed in the next section. This is especially true for an organization that is at an introductory stage in its application of risk analysis tools.

The result of an SLRA is a list of potential hazardous events that could occur at that facility, all ranked using a risk matrix (made up of frequency and consequence components), based on the experience and knowledge of the analysts undertaking the assessment. The ranking process involves assigning each hazardous event to an appropriate frequency and consequence class.

The SLRA should be followed with more detailed (qualitative and/or quantitative) analyses for hazards that warrant further examination for decision making and for development of risk control or mitigation measures.

As facilities become familiar with the screening level risk analysis method, they may start using other methods as well, with each method having its own advantages and disadvantages.

One of the above methods—the guide word "HAZOP"—is recommended for detailed qualitative analysis if a detailed hazard identification is warranted at the end of an SLRA.

Qualitative analysis generally:

- Is conducted by walk-through inspections and in sit-down team sessions
- Is relatively quick
- Does not require high-level technical and scientific skills
- Is based on the personal knowledge and experience of the team.

As such, there may be large uncertainties associated with the assignment of undesirable events into frequency/consequence classes. For some events of interest (e.g., high risk, high consequence, high mitigation cost), it may be desirable to understand their frequencies and/or consequences with a higher level of accuracy than would be possible with a qualitative analysis based on personal knowledge.

For events where higher accuracy is desirable, well-established quantitative techniques are available for improving the accuracy of the analysis. For such events (especially those with off-site consequences), the next steps are:

- Estimating their frequency using established techniques (e.g., fault trees, event trees; see below)
- Estimating hazard levels as a function of distance from event location using mathematical models (e.g., dispersion models for toxic gases or vapors; also see below)

  This quantitative (frequency, consequence) information is used to confirm or correct the earlier qualitative assignments and ranking, and to feed into emergency response plans. Once event risks are quantified in appropriate units, this information will also lend itself to:

- Making risk/cost/benefit analyses
- Estimating total facility risk for:

- Comparing with acceptability guidelines (such as those in Figs. 10.3 and 10.4); and/or
- Understanding the total financial risk exposure of projects or facilities

10.2.3 Screening Level Risk Analysis

Screening level risk analysis (SLRA) is a systematic screening process to identify hazards and prioritize risks for developing risk control measures.

The SLRA protocol follows the manufacturing process (business process) at the facility (start with inputs and track their movement in the facility) to ensure systematic coverage. It also allows participation by employees through interviews.

The following are the steps in the SLRA protocol:

- Collect and review background material on the facilities to be examined.
- Divide the facility into appropriate process sections, considering all process areas, operational steps or procedures, and ancillary services.
- Review the hazardous properties and compatability of materials handled in each section through a checklist of the hazards and potential consequences of materials in each section, considering the quantities involved.
- Complete a compatibility matrix for the materials in each section.
- Examine the major pieces of equipment in each section through a checklist of the hazards and significant losses associated with this equipment.
- Examine the potential for external events, such as earthquake or floods, through a checklist of such events.
- Review the effectiveness of the site's process safety by becoming familiar with the physical safeguards and management systems in place.
- List potential hazardous events for each section (following the process, answer the question "what can go wrong?") Events may include equipment failure, fire/explosion, toxic gas clouds, toxic liquid/dust spills, and other accidents, such as falls or falling of heavy objects.

Equipment failures that may not in themselves lead to safety or environmental implications can also be included if the scope of the analysis includes business interruption and production losses, thus helping to identify priorities for effective maintenance programs.

The generation of the hazardous event list is done through:

- Inspection of each section or operational step
- Reviews of previous hazard studies
- Interviews and brainstorming with knowledgeable individuals, such as operators, design and process engineers, instrumentation and maintenance personnel

This last step allows employees to participate in the risk assessment process and helps in their buy-in of the results of the analysis.

Once the hazardous events are identified and listed, then the consequences of each hazardous event are delineated in terms of impacts on different risk receptors (see below for an example scheme).

The following are the remaining steps in the SLRA protocol:

- Classify each hazardous event into (see below for an example methodology):
  - Frequency categories, according to the likelihood of the event
  - Consequence categories, according to the expected severity of the event
- Prioritize the significance of the event using a risk-ranking matrix (see below).

TABLE 10.1 Example Frequency Categories

Category	Description
1	Not expected to occur during the facility lifetime (<0.02/year)
2	Expected to occur no more than once during the facility lifetime (0.02 to 0.05/year)
3	Expected to occur several times during the facility lifetime (0.05 to 1/year)
4	Expected to occur more than once in a year (>1/year)

- Develop scope for more detailed hazard identification (e.g., guide word "HAZOP"), and quantitative analysis, such as consequence modeling for establishing hazard zones for emergency response planning purposes.
- Identify opportunities for improvement (e.g., physical plant design/maintenance/technology, employee training, management systems/integration, emergency response).
- Assign priorities to improvement opportunities based on the priority of the risk they are designed to address.
- Follow up.

TABLE 10.2 Example Consequence Categories

Category	Public Consequences
1	No injury or health effects
2	Minor injury or health effects
3	Injury or moderate health effects
4	Death or severe health effects
Category	Consequences in terms of Employee Safety
1	No injury or occupational safety impact
2	Minor injury or minor occupational illness
3	Injury or moderate occupational illness
4	Death or severe occupational illness
Category	Environmental Consequences (cleanup cost/compensation cost/regulatory fines)
1	Less than $1,000
2	Between $1,000 and $10,000
3	Between $10,000 and $100,000
4	Above $100,000
Category	Consequences in terms of Production Loss
1	Less than 8 hours
2	Between eight hours and 24 hours
3	Between 24 hours and one week
4	More than one week
Category	Consequences in terms of Capital Loss, Facility/Equipment Damage
1	Less than $10,000
2	Between $10,000 and $100,000
3	Between $ 100,000 and $500,000
4	Above $500,000
Category	Loss of Market Share (Dollars)
1	Less than 1% of annual revenue
2	Between 1% and 10% of annual revenue
3	Between 10% and 25% of annual revenue
4	More than 25% of annual revenue

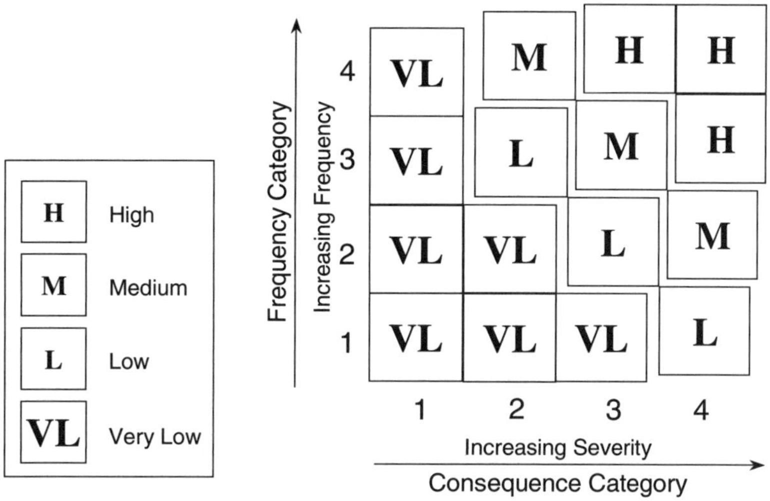

FIGURE 10.6 Example risk-ranking matrix.

10.2.4 Frequency/Consequence Categorization and Risk Ranking

Each hazardous event identified as part of the hazard identification step can be categorized using broad categories of frequency and consequence. Consequences are treated in six different categories: consequences to the public, the environment, the employees, production loss, capital loss, and finally loss of market share as a result of loss of goodwill in the marketplace. Example definitions of these categories are presented in Tables 10.1 and 10.2. It should be noted that these definitions, especially those for production loss, capital loss, and environmental damage, should be reconfirmed by senior management based on the risk tolerance of the organization. For example, if there is a zero-tolerance policy regarding environmental consequences, then the dollar value should be established at very low levels for each category. Environmental consequences may also be expressed in other than financial terms if desired.

Each hazardous event, once categorized, can then be represented on a risk matrix shown in Fig. 10.6, and prioritized with respect to the urgency of risk control measures that should be implemented to reduce the risk from that particular type of event. A commonly used set of definitions for each risk category on this matrix is given in Table 10.3.

10.2.5 Other Popular Hazard Identification Methods: Checklist, What-if, HAZOP, FMEA

A number of qualitative hazard identification methods are available to the risk analyst. Some of the more popular ones are discussed here. Further detailed information can be found in CCPS (1992). These methods become more powerful tools if they are coupled with the matrix-based risk-ranking scheme previously described.

Checklist Analysis. A checklist analysis uses a predefined documented list of items or questions to assess the integrity of systems, processes, procedures, or facilities. The checklist questions are typically answered yes or no. Using checklists is a very prescriptive technique that is frequently applied to test for compliance with standards or regulations.

Checklists provide an opportunity to take advantage effectively of previous experience. Questions may be structured based on previous incident data to seek out the potential for

TABLE 10.3 Example risk-ranking Categories (CCPS, 1992).

Code	Category	Description
H	High (unacceptable)	Should be mitigated with engineering and/or administrative controls to a risk ranking of L or less within a specified period such as 6 months.
M	Medium (not desirable)	Should be mitigated with engineering and/or administrative controls to risk ranking of L or less within a specified time period such as 12 months.
L	Low (conditionally acceptable with controls)	Should be verified that procedures or controls are in place.
VL	Very low (acceptable as is)	No mitigation required.

specific events that can lead to a release of hazardous substances. The corollary to this is that the questions are only formulated based upon previous experience. Checklists are therefore limited by their authors' experience and knowledge. The technique is not particularly effective in identifying new or previously unrecognized events that can lead to a release. Also, checklists may not apply to new or unique facilities or processes.

The rigor and thoroughness of checklists are directly proportional to their length. Unfortunately the longer the checklist, the more tedious it is to apply.

Checklists by their very nature tend toward being facility or process specific. They do, however, have the advantage that they can be prepared by experienced staff but implemented by those with less experience. They can be customized to specific industries, processes, or companies. A specific example of this would be the development of a checklist relating to the use of chemicals, particularly chlorine, in the treatment of water for swimming pools.

Within the limitations described above, particularly the potential to miss new hazards, checklists can be a powerful, cost-effective hazard identification technique.

Checklist analysis		
Advantages	Limitations	Applicability to hazard identification
• Easy to use (requires less experience) • Can be quick • May be customized • Cost effective	• Still somewhat limited by team experience • Repetition may lead to errors • Minimum level of hazard identification • May not identify new hazards	• Custom checklists for specific industries, processes, and companies • Best applied to processes where hazards are well understood

What-if Analysis. What-if analysis is appropriate for application to new and unusual circumstances as well as existing operations. It is essentially a brainstorming approach in which the participants review the process or facility and repeatedly ask the question "what if?" Through their imagination and experience and through asking questions of themselves and voicing concerns to their peers, the participants identify events that may lead to undesirable consequences. It is preferable that the participants be both experienced with and knowledgeable of processes or facilities at least similar to those under review.

As a hazard evaluation technique, the strength of a what-if analysis is its ability to identify hazards, particularly those in new or unusual circumstances. It may be used to target specifically the consequences defined in the scope of this chapter, that is, releases of hazardous substances that can lead to fatalities or injuries among people as our valued risk receptors.

What-if analysis can therefore be a powerful procedure for identifying specific events that can lead to hazardous substance releases. It is, however, an analysis technique that is constrained by the knowledge, capability, and experience of the participants. If the participants do not have the appropriate knowledge or experience, they may not have the ability to identify events that have not yet occurred.

A what-if analysis is less structured than most of the other techniques. The lead analyst may open the session with a few generic questions. However, other questions will be developed and new scenarios explored as the brainstorming progresses. The what-if approach works well for evaluating procedures.

What-if analysis		
Advantages	Limitations	Applicability to hazard identification
• Easy to use • Adaptable to specific scenarios • Works well for new and unusual scenarios	• Limited by participant experience • Unstructured, challenging to retain focus	• Works well for evaluating procedures • Works well for new and unusual circumstances

Failure Modes and Effects Analysis (FMEA). A failure modes and effects analysis (FMEA) is more structured than a what-if analysis. The technique provides for a rigorous analysis of equipment to identify single failure modes that can lead to an effect or incident. The failure mode provides a description of how the equipment failed (e.g., open, closed, stopped, running, on, off). The effects provide a description of the undesired consequence or incident. As a hazard evaluation technique, it may also be used to prioritize the criticality of each effect.

This technique is adaptable to identify the undesirable consequences within the scope of this guide. That is, an analysis can be undertaken on equipment pertinent to hazardous substances. The effects analyzed can be limited to releases that will lead to potential for fires fueled by flammable liquids and flammable liquefied gases, the release of toxic liquefied gases, and the evaporation of volatile toxic liquids.

The technique focuses on single failures of equipment. Its weakness is that it does not recognize multiple failures or multiple causes of incidents. It is also inadequate for identifying failures resulting from human error or procedural weaknesses.

Failure modes and effects analysis (FMEA)		
Advantages	Limitations	Applicability to hazard identification
• Structured and rigorous	• Limited to identification of single failures • Does not recognize multiple causes • Does not examine human factor or procedural causes	• Works well for evaluating specific items of equipment for physical integrity

Hazard and Operability Study (HAZOP). The hazard and operability study (HAZOP) uses rigorous analytical methods to identify deviations from the design intent through a detailed analysis of the facility design. A multidisciplined team, lead by a risk specialist, rigorously analyzes the design, normally using piping & instrumentation diagrams (P&IDs). The analysis is structured around a systematic review of the design using guide words applied to preselected system nodes. The guide words (no, more, less, as well as, part of, reverse, other than) are applied to parameters such as flow, level, and pressure at the specific design nodes. The guide words provide structure to ensure a systematic approach. The HAZOP technique can be viewed as a structured brainstorming session.

Since the original guide word HAZOP technique was developed, some companies have developed the knowledge-based HAZOP technique. It relies on the knowledge of team members to identify hazards with the design or operation of the facility, and as such can be considered as more of a safety review/checklist technique. We shall focus on the guide word technique in the remainder of this section.

The outcome of a HAZOP study includes the identification of hazards and operability problems. A study may lead to recommendations to change the design or operating procedures or identify the need for further information and study. It is important to note that the prime purpose of the study is to identify hazards, not to redesign the process and develop solutions.

Although originally conceived as a process for identifying hazards and operability problems in new, untested designs, HAZOP has been shown to be an effective method for reviewing existing operations. It should be recognized that a HAZOP study is a rigorous review of detailed design and, as such, requires extensive resources in both manpower and time. A typical study may occupy five to seven multidisciplined specialists for a number of weeks.

Hazard and operability analysis (HAZOP)

Advantages	Limitations	Applicability to hazard identification
• Structured brainstorming • Applicable to existing and especially novel or untested designs • Addresses design and procedures in multi-disciplined manner	• Time consuming and has large manpower needs • Normally assumes initial design intent was correct (the team could also identify that the design intent was incorrect) • Potential to be tedious, may lead to errors	• Works well for multidiscipline evaluation of detail design and procedures

10.2.6 Index Methods

A number of index methods have already been listed. Of these, the pipeline index will be described here because it is possible to demonstrate the essence of index methods in a very brief manner.

The pipeline index consists of a dimensionless parameter that represents the reciprocal of failure frequency and another dimensionless parameter that represents the inherent hazard associated with the material being transported.

The inverse frequency factor is then divided by the consequence parameter to arrive at a dimensionless inverse risk parameter. Thus, a higher value of the resultant parameter indicates higher safety of a pipeline segment.

The inverse frequency factor can have a value between two arbitrary integers selected by the developer of the index, and it is made up of factors such as susceptibility of the pipeline to third party damage or corrosion, as well as including factors such as the strength of the design and operating systems. A high number implies a smaller potential for failure.

The index is a very powerful tool for comparing different pipeline segments. However, because the index is dimensionless, it is not possible to undertake cost/benefit analyses or make comparisons, for example, with other modes of transport using this technique.

10.2.7 Quantitative Methods of Safety Risk Analysis

Quantitative methods aim at expressing risk in understandable units. As such, the frequency and consequence of a hazardous event are first quantified and then multiplied together to arrive at the commonly used risk parameter

Recalling the definitions of individual and societal risk that we used in the previous section:

Event individual risk at a receptor point	= event frequency ×	event individual consequence at that receptor point
event societal risk	= event frequency ×	event societal consequences

Note the location-dependent nature of individual risk versus the aggregate nature of societal risk.

The total facility individual risk is then the sum of all the event risks at a receptor point. Repeating the process at different receptor points will generate a risk curve, where risk generally decreases with increasing distance from the risk source. The units of this risk measure can be expressed as the annual chance that a person living near the hazardous facility might die due to potential accidents in that facility. (The MIACC risk acceptability guidelines are for a specific receptor location and not for a specific receptor that may spend some of his or her time away from that receptor location. Hence, the risk calculation should also assume continuous exposure of the receptor.)

In the calculation of the total facility risk, it is important that all significant representative accident scenarios be identified. Due to the large number of potential scenarios in complex installations, events with similar consequences are normally grouped together to reduce the amount of effort required to quantify their consequences. Then a representative scenario is selected for each event category and is assigned the *total* frequency of *all* events falling into that category of events.

The implications of this effort-saving step must be understood and evaluated very carefully by all stakeholders.

In public risk assessments, the representative scenario selected for each event category is generally the worst case in that category of events. This is done to ensure that the risk estimates are conservative (i.e., risks are overestimated) so that public safety is not compromised. If the number of event categories used in the assessment is too small, this standard practice of risk analysis may lead to unrealistically high risk estimates, thus losing their usefulness in decision making. A balance must therefore be sought between the amount of effort spent (which is proportional to the number of event categories used) and the degree of overprediction tolerable in the risk estimates.

For events with little or no dependence on meteorology and wind direction (such as explosions) in facilities that can be considered as point sources (such as chemical plants and storage facilities), the mathematical expression for the total individual risk is relatively straightforward:

$$I(R;R') = \Sigma\, F_h\, P_{e,h}(R;R') \tag{10.1}$$

Here $P_{e,h}(R;R')$ denotes the probability of hazardous effect (e.g., fatality) at receptor location R due to the risk source at R' and hazardous event h, F_h denotes the annual frequency of the hazardous event h, the multiplication of the two gives the event individual risk at receptor point R, and the sum is over all the event categories. In this discussion, $P_{e,h}(R;R')$ is assumed to be known from consequence modeling (see also further below), and the process of estimating total individual risk is described in graphical form in Fig. 10.7, where risk from two events, each with its own risk distribution, is added together at each distance from the event location to arrive at the total risk at that distance.

For meteorology- and wind direction-dependent events (such as gas clouds), the treatment is more complex, requiring consideration of joint frequency of occurrence of different weather conditions with wind direction.

For a hazard that depends on meteorology and wind direction (such as a toxic cloud), the individual risk can be estimated by (using the coordinate system shown in Fig. 10.8):

$$I_{e,h,m}\,(R;R') = \int_0^{2\pi} F_h\, P_{e,h,m}\,(R;R')\; \varphi_{i,m}(\theta)\; d\theta \tag{10.2}$$

where $P_{e,h,m}(R,R')$ is the probability of effect e (e.g., death) at R due to event h taking place at R'

$I_{e,h,m}(R;R')$ is individual risk at R due to risk source at R'

F_h is the frequency of event h

$\varphi_{i,m}(\theta)\, d\theta$ is the proportion of time the wind spends between θ and $\theta + d\theta$ within wind sector i during meteorological condition m (m defined by atmospheric stability and wind speed combination).

The total risk for that event h over all (M) meteorological conditions is then:

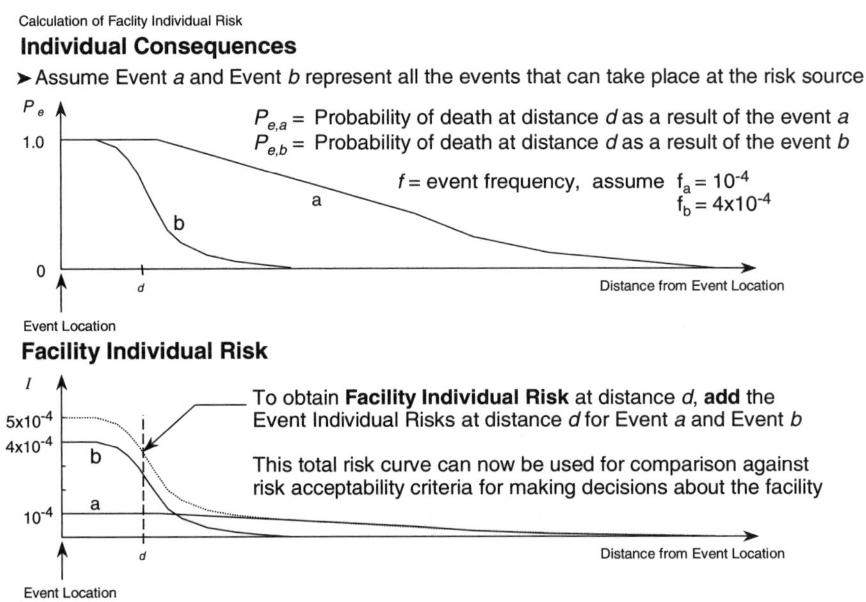

FIGURE 10.7 Estimation of total facility individual risk (for events with little or no dependence on meteorology and wind direction–explosions, pool fires, simple jet fires).

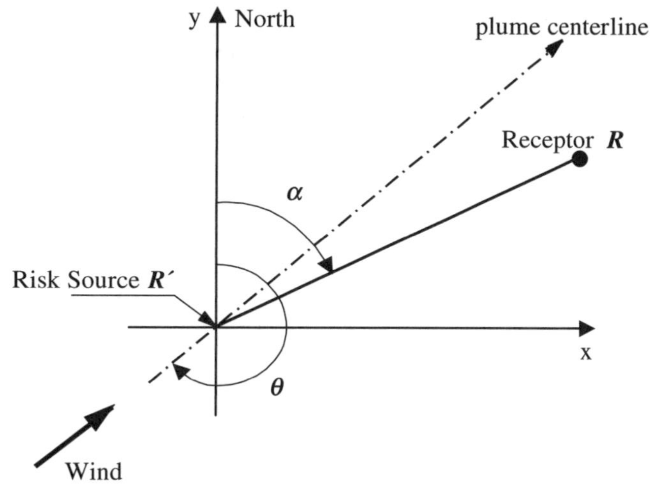

FIGURE 10.8 Coordinate system.

$$I_{e,h}(R;R') = \sum_{m=i}^{M} I_{e,h,m}(R;R') \tag{10.3}$$

For hazardous installations of a linear nature, such as pipelines and rail, truck, and marine tanker transportation corridors, the estimation of individual risk also involves an integration along the corridor. For a linear risk source of length L', the individual risk at R can be calculated by integrating the individual risk due to an infinitesimal risk source of length ds' at location R' (see Fig. 10.9).

$$I_{e,h,m}(R;L') = \int_{l_0}^{l_0+L'} F'_h(R') \left[\int_0^{2\pi} P_{e,h,m}(R;R')\ \varphi_{i,m}(\theta)\ d\theta \right] ds' \tag{10.4}$$

where $F'_h(R')$ is the per-unit-length event frequency.

Equation (10.3) again applies for obtaining the total risk for the event h over all meteorological conditions.

These formulations can be readily extended to estimation of societal consequences and societal risk as shown in Alp and Zelensky (1996). The basic formula for calculating societal consequences of a hazardous event is demonstrated in Fig. 10.10. The societal risk can then

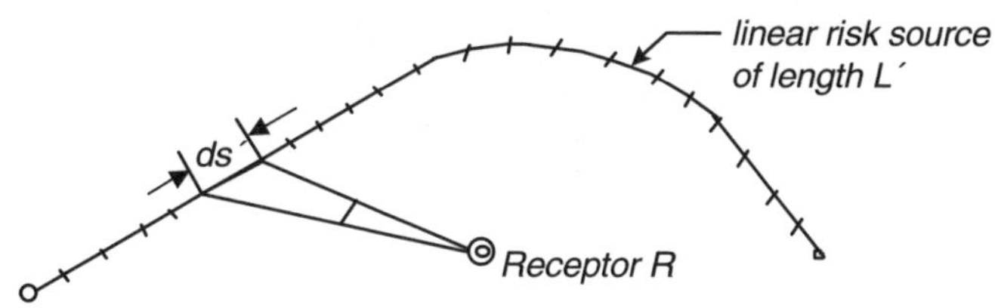

FIGURE 10.9 Estimation of individual risk for a risk source of a linear nature.

$$C_{e,h,m,\theta}(R') = \iiint P_{e,h,m}(R;R')\rho(R)\,dx\,dy\,dz$$

$$\rho(R) = \text{population density at receptor } R$$

➤ The integral is over all receptors

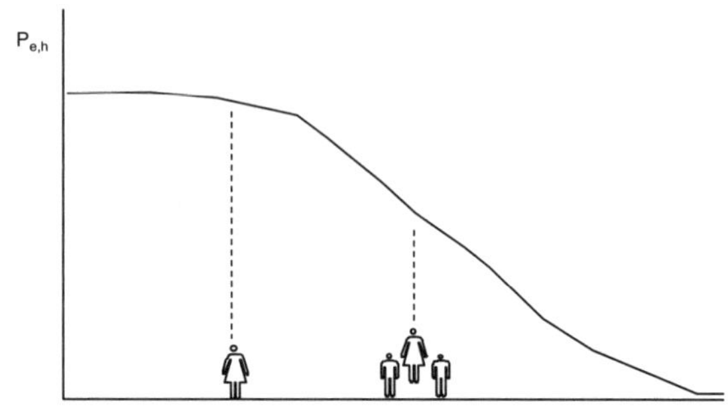

FIGURE 10.10 Calculation of societal consequences of a hazardous event.

be calculated simply by multiplying the societal consequence by the corresponding event frequency.

For linear sources, it should be noted that the societal risk is directly proportional to the length of the source similar to that shown by Eq. (10.4).

Practical tips for using these formulations and their limitations are also presented in Alp and Zelensky (1996). Other simplified formulations can be found in CCPS (1989a, 1995b).

The above discussion assumed that we know the quantities:

- $P_{e,h}(R;R')$—the probability of hazardous *e*ffect (e.g., fatality) at receptor location R due to the risk source at R' and hazardous event h
- F_h—the annual frequency of the hazardous event h.

Let us now discuss the methods available to us for estimating these two quantities.

10.2.8 Frequency Analysis Techniques

The risk presented by a facility is dependent on the frequency at which an undesired event can be expected to occur and the adverse consequences that could result from the event. For this purpose, the undesired events that were identified using the techniques described as qualitative must be analyzed to determine their expected frequency.

Many techniques are available to estimate the frequency of a given type of major hazardous event occurring at a specific facility, on a transportation corridor, or along a pipeline. (For detailed descriptions see TNO, 1988, and CCPS, 1989a, 1995b.) These vary in complexity and type of information they yield. Their selection depends on the desired outcome and the amount of time and effort available. The techniques we will consider are:

- Historical data analysis
- Fault tree analysis
- Event tree analysis
- Human reliability analysis
- External events analysis

All of these techniques rely on past experience to a certain extent. Fault and event trees are the most common frequency modeling techniques for complex situations that require tracking of chains of events. Human reliability analysis and external events analysis can be considered essentially as components of fault and event tree analysis, with the information generated from their application to be fed into the fault and event trees.

An important factor to be considered in deciding the amount of effort to spend in improving the accuracy of event frequencies is whether the numbers will be used in an absolute sense or for a comparative exercise. For complex land-use decisions regarding pipelines, dangerous goods transportation corridors, or industrial plants, accuracy of the frequencies is important because the end result will be compared to the established risk-acceptability guidelines for land use. This may not be the case for the addition of controls or mitigation for the specific purpose of reducing overall risk. In such a study, it may be more important to be consistent in approach than to seek high levels of accuracy in the frequency estimates.

Historical Data Analysis. Use of historical data in the estimation of hazardous event frequencies is a suitable approach if the operating experience of the equipment is sufficient to produce a statistically meaningful database.

Historical data can be used in two different ways:

- To estimate directly the frequency of the hazardous event of interest (top event) identified in the hazard identification step of the risk analysis; and
- To estimate frequencies of events or causes that contribute to the occurrence of the top event

The latter is generally used in conjunction with fault trees.

There are generic types of equipment that are used in the various industries (e.g., pumps, valves, pipelines). Industry-average failure frequency rates are available for these pieces of equipment. However, not all facilities experience failures at the same rate. These rates can vary considerably depending on site or company conditions, such as:

- Management practices
- Operating practices
- Appropriateness of design and construction materials
- Level of testing, inspection, and maintenance
- Equipment age
- Severity of operating conditions
- Nature of the materials handled

Therefore, it is best to use site- or company-specific release data if they are available. However, any given site or company will not generally experience a significant number of major events to form a statistically significant database. In this case, it will be necessary to use general industry data for overall failure rates as a first approximation.

When using general industry data, it is common practice to adjust the data up or down by up to an order of magnitude based on engineering judgment, depending on the specific site or company conditions.

A common problem in generating frequency information for hazardous events is lack of divisor information. In a risk analysis, the unit of frequency needed for estimating annual risk is events/year. For example, for a hydrocarbon storage tank fire in a facility containing three tanks, we would need base frequency information in terms of tank fires/tank-operating-year and multiply this quantity by three to obtain the frequency of the event we are looking for in terms of tank fires/year in that three-tank facility. Historical data from government or industry sources may be available in terms of tank fires/calendar year in Canada or in the world but not in terms of tank fires/tank-operating-year. In order to estimate the base frequency information tank fires/tank-operating-year from tank fires/calendar year in Canada or in the world, one would need to divide it by the number of operating tanks in that calendar year in Canada or in the world (hence the term *divisor*), but this parameter is generally not easily available. The same situation generally exists for all industries, including rail transportation, pipeline transportation, chemical industries, and others. Some industries and government agencies have commissioned special studies to resolve this issue and provide base frequency information that can be directly used in risk analyses. Examples of such data are presented further below.

In any case, use of fault and event trees is strongly recommended instead of such approaches in all quantitative risk assessments.

Fault Tree Analysis. When failure rate data are not available for the undesired event or the top event, or its accuracy is not judged to be sufficient, the event frequency can be estimated using analytical methods, specifically fault tree analysis (FTA). FTA uses a "backward logic," which begins with an undesired event (e.g., a release of a hazardous material from containment), analyzes the system to determine the basic cause(s) of the undesired event, and enables the user to quantify the likelihood of the top event. This is done through a "top down" tree, with the branches identifying the main causes and influencing factors contributing to the top event. The tree-like or branching investigation of each scenario gives rise to the name "fault trees." Since the method is deductive, it focuses on the particular event in question, thereby eliminating time spent following trains of thought that do not lead to hazardous situations.

FTA is a tool employed in the analysis of complex systems to estimate the likelihood of a hazardous event. It has been applied, for example, in safety evaluations of nuclear power plants, space missions, air, rail, highway, marine and pipeline transport, liquefied natural gas, chemical manufacturing, and other hazardous material facilities. With this method, all material, personnel, and environmental factors of a complex system can be systematically presented. A well-constructed fault tree enables us to discover failure combinations that would not normally be discovered and provides for both qualitative and quantitative evaluation.

FTA starts with a particular undesired top event, such as a flammable material release and fire or explosion from a particular system. It then breaks down the causes of that event into all the identifiable contributing sequences, and each sequence is separated into all necessary components or events. The presentation of all this information is facilitated by the use of a logic diagram, or fault tree. The fault trees are generally developed only as far as necessary down to a level at which failure or event frequencies are known with a reasonable degree of accuracy from past experience or historical data. The elemental parts of a fault tree at the bottom level are known as basic events.

The symbols commonly used in developing fault trees are shown in Fig. 10.11.

To quantify a fault tree, failure rates are assigned to the basic events at the bottom levels of the tree. The occurrence rates for human error and equipment failure used in the fault trees are based either on information reported in the literature, specific facility or company history, or analyst estimates that combine information supplied by the company (operating procedures, personnel organization and experience, and design information) with information from other sources in the literature. If available, it is best to use site-specific failure data when quantifying the tree. This data is often available from preventive maintenance records or from a review of incident reports.

The sequence of events forms pathways, along which are found AND or OR gates. These gates connect the basic initiating event and contributing events to the higher-order events. When all of a set of lower-order events must occur in order for the next-higher-order event to occur, these events are joined by an AND gate. Multiplying the probabilities of each event in the set (sometimes coupled with one frequency) yields the likelihood of the next higher event. When the occurrence of any one of the set of lower-order events is sufficient for the next-higher-order event to take place, the events in the set are joined by an OR gate, and their probabilities or frequencies are added. Likelihood of the top events is generally ex-

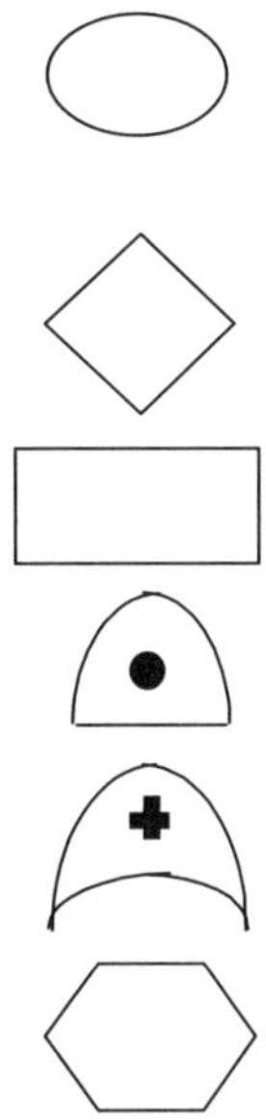

Basic Event: Circles or ovals represent basic events. These are basic initiating faults requiring no further development. They include initiating events with yearly rates of occurrence, and response (demand) events with failure rates of occurrence per demand, i.e., conditional on the prior initiating and contributing events having occurred.

Undeveloped Event: Diamonds represent events that are not developed further either because the event contributes insignificantly to the top event or information relevant to further development is unavailable.

Rectangles represent states that are the product of several initiating and/or contributing events through an "AND" or "OR" gate and may therefore have rates of occurrence either yearly or per demand.

"AND" gate – the rates of occurrence on the incoming branches are multiplied. All input events must occur for the output event to occur.

"OR" gate – the rates of occurrence on the incoming branches are added. Occurrence of any one input event can result in the output event.

Hexagons represent numbers of components and serve as multipliers.

FIGURE 10.11 Symbols used in fault trees.

pressed as a yearly rate, e.g., 10^{-4} chance of occurrence per year (once in every 10,000 operating years on average).

Since the likelihood of each top event (accident scenario) is to be expressed as a yearly rate, no more than one event leading into an AND gate can have a probability expressed as a frequency. Otherwise, the overall rates will be in terms of something similar to "occurrence rate per year squared"—a meaningless concept. Thus, at most one event leading into an AND gate can be expressed as a frequency; the remaining events are expressed as conditional probabilities, or probabilities per demand.

At OR gates, it is essential that all the events entering the gate be quantified in the same units, i.e., as either frequencies or probabilities, since they are to be added. The next higher-order event will be in the same units as the events preceding it.

One of the most common mistakes is to multiply two or more frequencies together, yielding meaningless results. Hence, we sometimes use appropriate symbols to clarify which events are given as frequencies and which are probabilities. We also try to ensure independence between events, particularly during quantification.

Figure 10.12 gives an example of a simple fault tree. The resulting fault tree provides a visual understanding of the basic causes of the undesired event and a structure that can be quantified to arrive at the expected frequency of the undesired event.

The literature contains several examples of fault trees that have been developed for specific situations. For example, the Health and Safety Executive of the United Kingdom has developed a fault tree for the BLEVE of a butane storage bullet (Blything and Reeves, 1988). They consider the ways in which a leak may occur and progress to a BLEVE and assign failure data to arrive at a top event frequency in the range of 10^{-8} to 10^{-6}/vessel-year. Other examples can be found in CCPS (1989a).

Event Tree Analysis. Event tree analysis is a forward-looking method that takes an initiating event, identifies post-initiating event influencing factors, and combines the information into a logic tree in which the occurrence of each influencing factor is either "true" or "false."

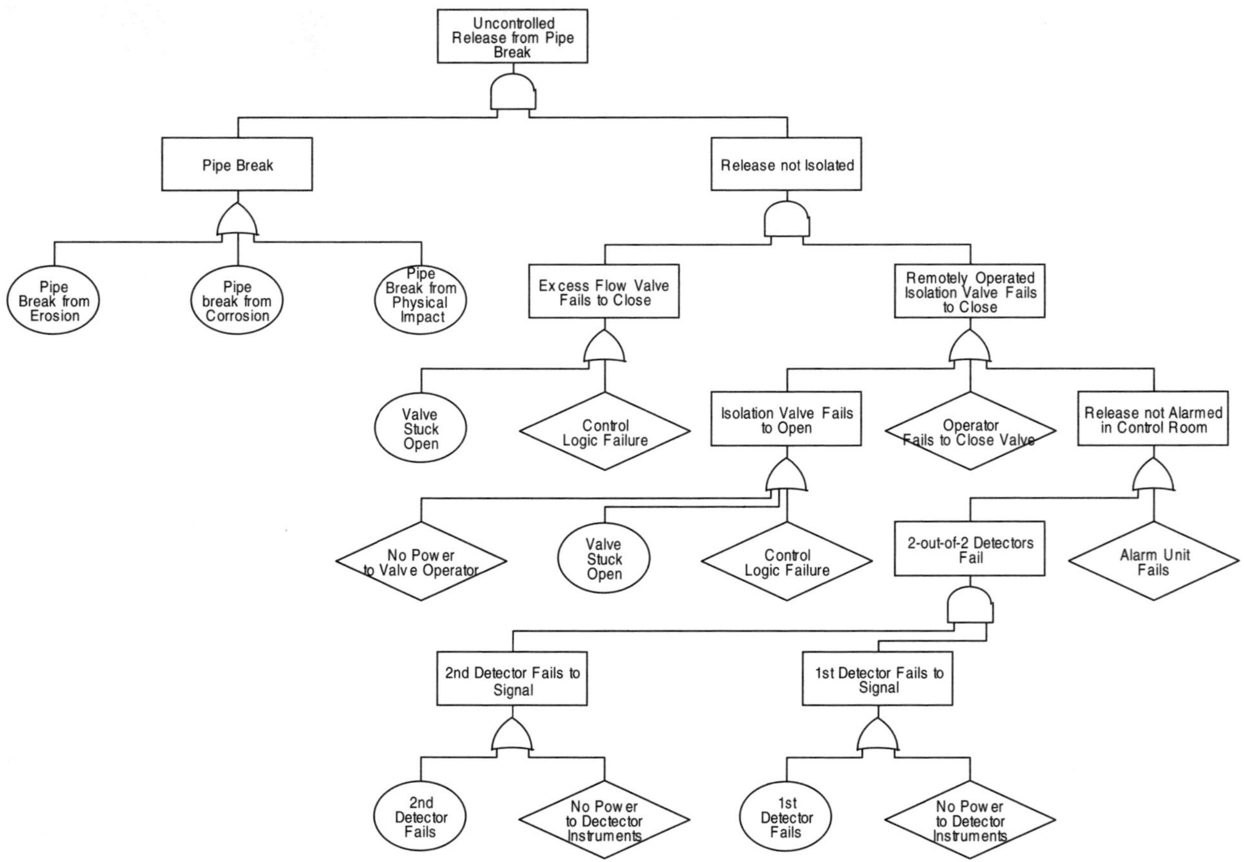

FIGURE 10.12 Example of a fault tree.

Two types of event trees are commonly used in risk assessments: pre- and postincident event trees.

Preincident event trees are generally used to develop and track the responses of a control system after the system has failed, this failure being the initiating event. A typical example is shown in Fig. 10.13. Each possible outcome following the initiating event is tracked with a series of positive or negative branches, examining what would happen if the next line of defense functions as designed or fails to function, each with its associated probability of failure. In this way, likelihood of undesirable outcomes, such as releases of hazardous materials, can be estimated.

Postincident event trees are used to track possible outcomes following a hazardous material release or other top events examined by a fault tree, and to estimate the frequencies of these outcomes. A typical example is given in Fig. 10.14, which shows the possible outcomes of a propane release from pressure-liquefied storage.

Using event trees to develop scenarios during the hazard identification step in a risk analysis is attractive in that it helps organize the thinking process and helps the analyst see how all the parts fit together. The conditional probabilities used in quantifying the branches of the event trees are commonly based on historical data and engineering judgment, also taking into account the types of ignition sources in the surrounding area for flammable releases.

Human Reliability Analysis. This component of frequency analysis refers to quantitative examination of human responses to given routine and emergency situations that require human intervention. The information is generally in the form of conditional probability of failure to respond in the appropriate manner to a given signal or failure to perform a certain task correctly. Failure rates will be higher for high-stress situations and will also depend on environmental conditions, timing of events, experience, availability of written procedures, and training levels. In risk analyses, this information is used in fault and event trees.

Human reliability analysis is an important component of risk analysis. Reviews of past accidents show that human error accounts for the vast majority of these events. The technique most widely used for estimating human error probabilities is called THERP (Swain and Guttman, 1983). The method uses event trees drawn in a different format to arrive at a human error probability. See Fig. 10.15 for an example. In these event trees, failure paths branch right and success paths branch left.

External Events Analysis. This component of frequency analysis considers the impact of external events (such as earthquakes, tornadoes, floods, aircraft crashes, terrorism, and vandalism) as initiating events to undesirable event scenarios. Quantitative frequency information is then used in fault and event trees.

Quantification Based on Experience. The availability of concrete failure data is rare for most operations, although there are standard databases available. The best data to use for plant fault trees are those which refer directly to the site in question. Equipment failures can often be determined from maintenance records; component failures can often be determined from warehouse records that show the use of a given product, e.g., solenoid valves, over a period of time.

If such information is not readily available, it is still possible to quantify a given cause by asking other staff members how often they would expect a failure to occur using a Delphi technique. Such an estimate is often accurate enough, especially for an initial evaluation.

The following is one possible procedure for obtaining this data:

1. Select a group of experts (usually three or more).
2. Solicit, in isolation, their independent estimates on the value of a particular parameter and reasons for their choice.

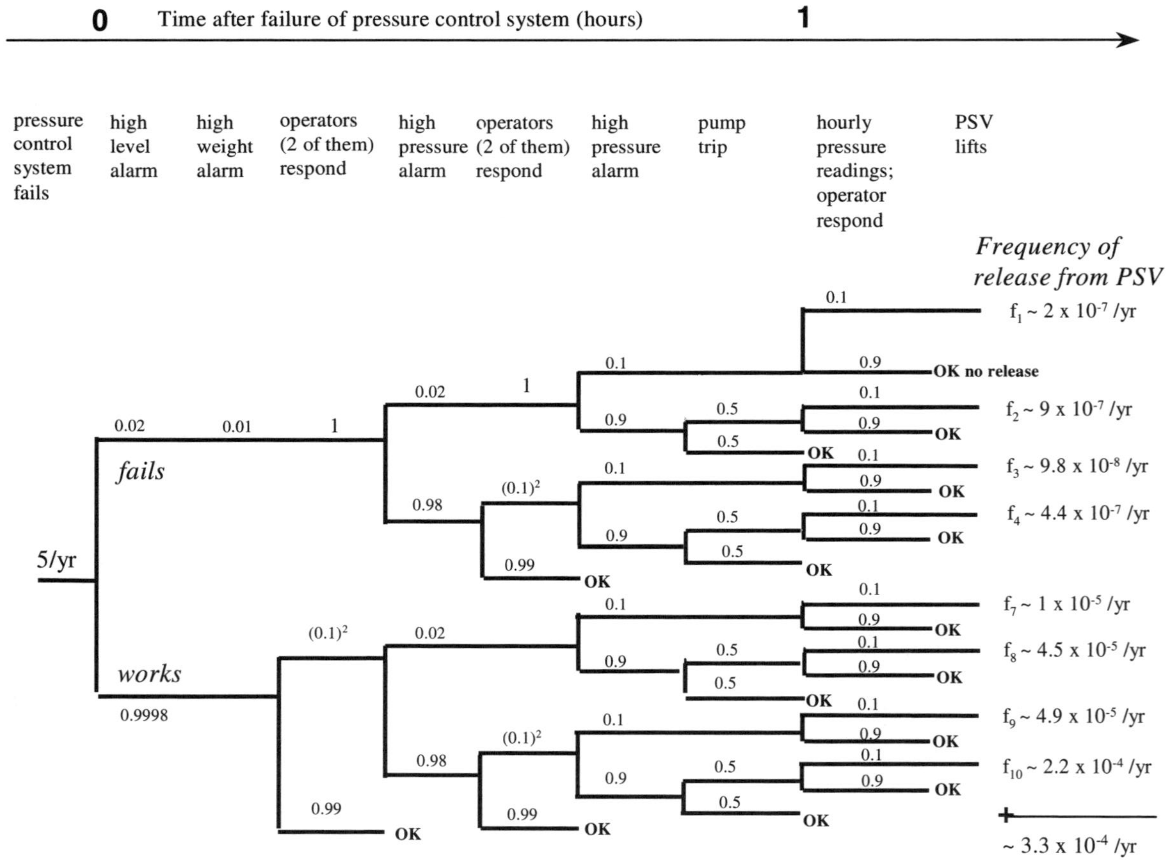

FIGURE 10.13 Example of a preincident event tree (for failure of a pressurized liquid storage vessel pressure control system resulting in release from a pressure safety valve).

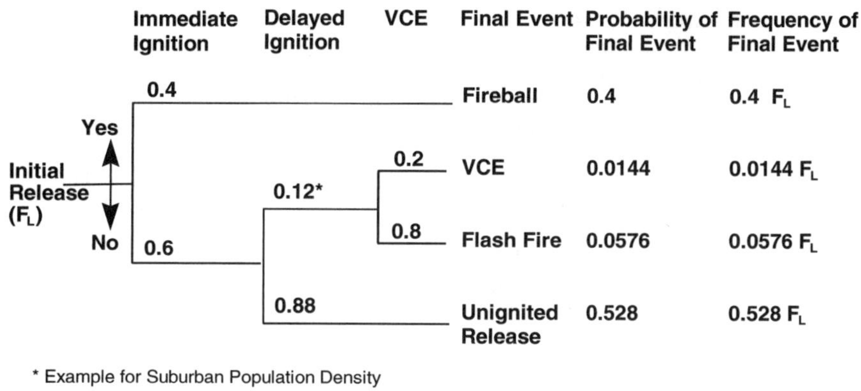

FIGURE 10.14 Example of a postincident event tree following a large release of liquefied petroleum gas.

3. Provide initial results to all experts and allow revisions to the initial estimates. In some cases, the experts are brought together and allowed to discuss the basis for their estimates in order to try to reach a consensus.
4. Use the average of the final estimates as the best estimate of the parameter. Use the standard deviation of the estimates as a measure of the uncertainty.

If it is determined that the estimated event plays a critical part in the undesired event frequency, then time can be spent to obtain more detailed data from records, from supplier data, or by testing the components.

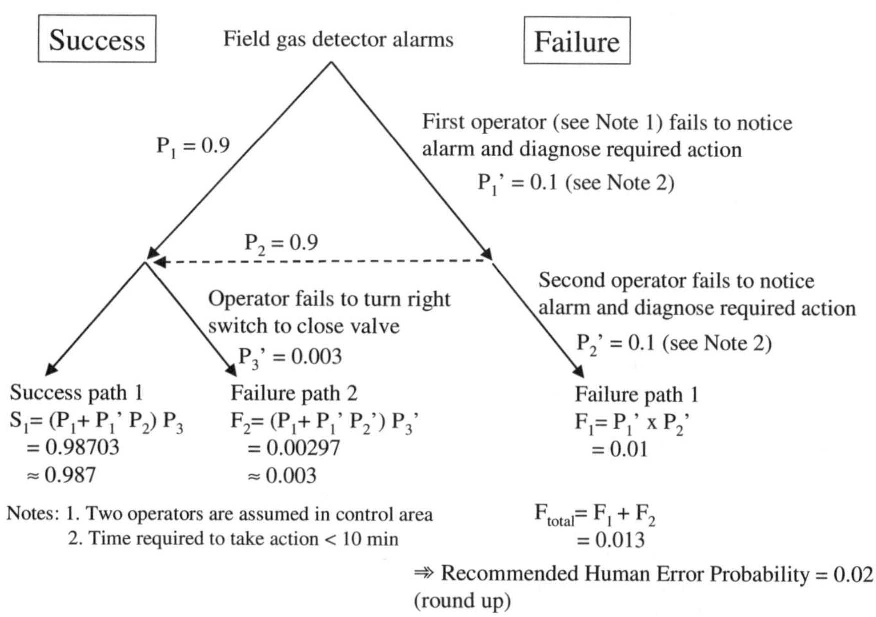

FIGURE 10.15 Example of a human reliability event tree (for failing to close a remotely operated isolation valve given a field gas detector alarm).

Available Data on Event Frequencies. A number of sources are available where historical data on undesirable event frequencies can be found. These data sources can be grouped into the following three categories:

Component Failure Rate and Human Error Data. These sources provide generic data on failure rates of components such as valves, flanges, and pipes on a per unit time or per demand basis as appropriate. Commonly used references include CCPS (1989b), Lees (1986, 1996), Rijnmond (1982), EPRI (1981), CONCAWE (1982), HSE (1996a, 1996b, 1997), IEEE (1983), Johnson and Welker (1981), OREDA (1997), U.S. NRC (1975), Swain and Guttman (1983).

Industry Top Event Frequency Data. These data sources provide information on industry-average frequencies of top events that could be used directly in risk analyses. They are published in the following manner.

- As part of industry guidelines or recommended practices (e.g., API, 1995)
- By government agencies, e.g., Alberta Energy Utilities Board annual sour gas incident data reports, Transportation Safety Board annual reports containing air, marine, rail accident statistics, U.K. Health and Safety Executive annual reports containing offshore accident and release statistics, and others, such as:
  - The National Transportation Safety Board (U.S.)
  - The Federal Railroad Administration (FRA, U.S.)
  - The Federal Transit Administration (FTA, U.S.)
  - The Federal Highway Administration (FHWA, U.S.)
  - The Research and Special Programs Administration (RSPA, U.S.)
  - The Office of Pipeline Safety (U.S., part of the RSPA)
  - The Bureau of Transportation Statistics (U.S.)
  - The Office of Hazardous Materials Safety (U.S.)
  - The Transportation Research Record (National Research Council, Transportation Research Board, U.S.)
  - The International Transportation Safety Association (ITSA)
  - The Railroad Commission of Texas
- By consulting companies under contract to government agencies (e.g., Worldwide Offshore Accident Databank—WOAD), or
- As occasional research papers or reports (e.g., papers on rail and truck accident rates, reports on pipeline accident rates from groups such as the Institute for Risk Research, University of Waterloo)

Selected industry-average information from various sources can be directly used in risk analyses. Due to the large variability in actual experience from facility to facility and from company to company, extreme caution should be exercised in use of such industry-average data. Use of facility-specific data in combination with fault and event trees is recommended wherever possible.

Government of Canada Accident Data. Various government agencies in Canada collect and report information on accidents. These provide country-wide incident information on a yearly basis but are not directly useful in risk analyses because they do not include appropriate divisor data (see above, Historical Data Analysis). These include the following:

- National Analysis of Trends in Emergencies System (NATES) Database

  The NATES database was established in 1973 by Environment Canada to record information from voluntary reporting of pollution incidents involving hazardous substances. The database contains spill information entered under a number of data fields, including location, material spilled, quantity, cause, source, and sector.

  NATES captures the most significant of the spill events reported each year. For the sake of clarity, the name "NATES" is used to encompass all of the data sources for various analyses of trends undertaken by Environment Canada. However, NATES is only one of the data sets used; data are also obtained through the Department's cooperative agreements with the provincial and territorial reporting agencies and other government departments.

- National Pollutant Release Inventory (NPRI)

  In addition to NATES, Environment Canada also maintains a national database called the National Pollutant Release Inventory (NPRI). It is designed to collect and make available to the public, on a yearly basis, comprehensive national data on releases to air, water, and land, transfers in waste, and ongoing emissions of specified substances. Under the authority of the Canadian Environmental Protection Act (CEPA), owners or operators of facilities that manufacture, process, or otherwise use one or more of the 176 specified substances under prescribed conditions are required to report to the NPRI. The NPRI reports for the years 1994 and 1995 can be found on the Environment Canada website (http://www.ec.gc.ca/pdb/npri/).

  One of the main differences between NATES and NPRI is that reporting to NATES is voluntary, while reporting to the NPRI is mandatory. Also, NPRI covers all emissions including spills, whereas NATES covers only spills. In addition, the thresholds and reporting criteria exempt many fixed facilities from reporting to NPRI, whereas all spills may be reported to NATES.

- Dangerous Goods Accident Information System (DGAIS)

  Transport Canada maintains the Dangerous Goods Accident Information System (DGAIS). All transportation incidents resulting in spills must be reported to the Transport of Dangerous Goods Directorate by the person responsible for the dangerous goods consignment at the time of the incident. Since July 1985, dangerous goods incident information has been submitted under the reporting requirements of Section IX of the Transportation of Dangerous Goods Regulations.

- National Environmental Emergencies System (NEES)

  Recognizing the incompatibilities among the various Environment Canada regional spill databases, the Environmental Emergencies Program began developing the National Environmental Emergencies System (NEES) in the fall of 1993. This system incorporates historical data tables from the regional systems, as well as the NATES database, and data from various contributing agencies.

Uncertainties in Estimating Frequency. The greatest influence on uncertainty in risk results can be attributed to uncertainties in frequency estimates arising from:

- Uncertainties in modeling
- Errors in modeling
- Omissions in modeling of safety features
- Uncertainties in failure data

Each of these can cause the estimated frequency to deviate from the true mean frequency. Uncertainties in modeling occur for a variety of reasons. The analyst may not have sufficient design, layout, or operating information to develop accurate logic tree models. For example, the total length of pipeline in a plant may not be known with confidence. Another type of uncertainty may arise from taking short cuts in the modeling in order to simplify the effort required. Usually conservative assumptions can be made for the above factors.

Errors in modeling may arise if due care is not taken in developing fault/event tree models or identifying appropriate failure mechanisms. For an example, let us consider an area that has two toxic gas detectors, either of which can close an emergency isolation valve. In the fault tree, both detectors would have to fail for leak detection failure to occur and the events would be "ANDed" together. The fault tree should be described down to the power supply level. This would ensure that the Boolean solution to the fault tree would capture the case in which power is supplied to the detection equipment from the same power bus. Another example is the case of sour gas pipeline failure rates. Because of the corrosive nature of sour gas, these failure rates are significantly higher than for sweet gas pipelines. Using a sweet gas pipe failure rate for sour gas service will underpredict the frequency of releases.

Failing to acknowledge safety features can cause a hazardous event frequency to be overestimated significantly (by up to two orders of magnitude or more). The magnitude of this uncertainty alone may be greater than the cumulative uncertainties in all other estimates. If the results are acceptable, then there is no need for a second iteration and the analyst would be certain that frequency and risk have not been underestimated.

In the above factors, the analyst has control over the uncertainties. However, when it comes to failure data based on historical observations, the analyst has little control over the uncertainties. These data tend to be generic (i.e., "average") and limited. The unique conditions at a specific plant (e.g., component service, age, or environmental conditions) may not be captured in the data. In addition, there may not be data available for all components or component failure modes. Inevitably, approximations are made; these should be made conservatively. Failure rates that are available will also have significant uncertainties—divisors (i.e., component years of service) may not be well known or the number of component failures in the database may be underreported. This is particularly important if using generic hazardous event frequencies (i.e., BLEVEs per tank-year) in that they are unlikely to capture the design, layout, operational, and mitigation features of a particular plant. Here, the uncertainty in the frequency estimates may be so significant as to render the risk results meaningless.

In summary:

- There are a number of different factors that may result in uncertainties in frequency estimates.
- In most situations, uncertainties in frequency estimates are the greatest contributor to uncertainties in risk estimates.
- Analysts experienced in frequency estimation can minimize these uncertainties and ensure, where possible, that they err on the safe side (i.e., overestimating frequency without going overboard).
- In choosing the method for delivering frequency estimates, while the use of fault/event logic trees is superior to deriving hazardous event frequency from generic historical data, it is also more costly.

10.2.9 Consequence Analysis Methods

Once the hazardous events are identified, the next step in the risk analysis is to analyze their consequences, i.e., estimate the magnitude of damage to the receptors of interest should those hazardous events occur.

Consequence estimation can be accomplished by:

- Comparison to past incidents
- Expert judgment
- Using mathematical models (consequence modeling), which can be at various levels of detail and sophistication

Consequence modeling is an analytical approach used to determine the possible physical effects resulting from the release of a hazardous substance. The inputs to this analysis include the physical, chemical, and toxicological characteristics of the hazardous substance and the characteristics of the system in which it is contained (e.g., pressurized vessel, pipe, reactor, bulk carrier container).

The decision on the level of sophistication of the consequence analysis depends on the desired objective and accuracy of the results. If the results and the insight gained from the modeling will be used for emergency response planning or siting of critical plant units within an industrial facility or in complex land use decisions, then detailed modeling should be preferred. If a quick assessment is all that is required, such as deciding whether to install an additional risk control measure such as an isolation valve, then sophisticated models may not be necessary (especially if the cost of the risk control measure is not large compared to the cost of the analysis). The rule of thumb is to undertake detailed modeling if the cost of the risk control measure under consideration by the decision maker, or the potential consequence cost of not implementing the risk control measure, is much higher than the cost of the modeling.

The credibility of any given modeling result depends on the credibility of the release scenario (hazardous event) chosen, the supporting assumptions made in the analysis, and the technical merits of the model itself. Numerous studies have attempted to develop comprehensive consequence models for the hazards of interest; however, due to the wide range of variables that may affect the behavior of hazardous releases, there is no single model that will satisfy all situations. There is a wide range of available models that may be based on simple or complex equations, state-of-the-art research, and actual field test results. When the intent of the modeling exercise is to use the results to support decisions, it is important that the decision-makers or at least their advisors understand the key considerations that have gone into the development of these models. This knowledge will help with the model selection and establish confidence in the final results. Considerable judgment is required to assess which models are appropriate and relevant to a particular situation. A good understanding of the underlying physics of the scenario is essential to the success of model selection.

Since modeling results are highly sensitive to supporting assumptions, consistency is best achieved by having the same person(s) carry out the modeling calculations. The assumptions are generally left to the discretion of the modeler; they must be consistent with the laws of physics and should reflect experience for the system or location under study. They should also recognize the broader experience gained from the history of previous events.

This section describes the important underlying physical mechanisms for some of the more common types of hazardous material releases and gives guidance on the type of models that should be used to provide an acceptable level of accuracy in estimates of event consequences (and hence individual risk, which is the desired end point for comparison against the MIACC risk acceptability guidelines).

The focus is on estimating concentrations of toxic or flammable gases in the atmosphere, thermal radiation (heat intensity) levels from fires, and explosion overpressures. Each of these effects is capable of causing serious injuries or fatalities. Results are normally expressed at selected receptor locations and, for time-varying hazards, as a function of time.

Consequence modeling generally involves three distinct steps:

1. Estimate the source term (source term modeling), i.e., how much material in what form (gas/liquid/two-phase) is being released from containment as a function of time, and develop the release scenarios or possible hazard outcomes (toxic cloud, fire, explosion, etc.) following the release. (A powerful tool to develop and keep track of possible outcomes following a release is an event tree; event trees are commonly used in quantifying the frequency of these various outcomes and were therefore described in Section 10.2.8.)
2. Estimate the hazard level (hazard modeling) as a function of time and at selected receptor locations, i.e., estimate:
   - Ambient concentrations for a toxic or flammable gas release (for modeling the effects of a toxic cloud or flash fire)
   - Thermal radiation flux for fires (for a jet fire, pool fire, or fireball)
   - Overpressure for explosions (for a confined explosion, boiling liquid expanding vapor explosion [BLEVE], or vapor cloud explosion [VCE])

3. Estimate the damage level on the selected receptor, based on the hazard level at the receptor location (vulnerability modeling).

In scenario development, usually only minimum allowance is made for active hazard mitigating factors such as emergency shutdown or isolation devices, alarms or emergency response plans, especially in initial stages of analysis to reduce the complexity of the analysis, thereby saving costs. This is the approach suggested in the 1996 US EPA risk management program legislation and will tend to build a degree of conservatism in the results (i.e., overestimation of risk). When communicated appropriately, this approach may give additional comfort to the stakeholders in making any decisions if even these overestimates are within acceptable limits. If, however, the risks estimated using this assumption turn out to be unrealistically high, further detailed analysis, which would take into account active mitigation systems and their failure frequencies, is then recommended.

Physical Mechanisms and Parameters Important for Determining Source Terms and Outcomes of Hazardous Material Releases. The total mass of the release and its rate of release are probably the most important parameters that influence the hazard zone associated with a release.

A release rate will normally vary with time and as a function of hole size and location, containment conditions, system inventory, and external conditions. If a mixture is involved, the composition of the release may also vary with time.

It is important to establish the initial release rate (over the first 10 seconds or so) for releases of high-pressure gases or liquefied gases from pipelines and vessels because these initial release rates often control the maximum extent of the hazard zone. One of the major difficulties is in dealing with two-phase flow since there is a high degree of variability in how various models predict these release rates.

In the case of an instantaneous release, the source strength is specified in terms of the total mass released. For a continuous release, the source strength is a function of outflow expressed in terms of mass per unit of time. In order to determine the strength of the source, the physical state of a contained substance must be defined and described. The physical properties of the substance, together with containment pressure and ambient temperature, determine the physical state.

In the case of a continuous release, it is necessary to determine whether a gas, a liquefied gas, or a liquid is being released. The release rate from a breach in the containment wall will generally be proportional to the square root of the pressure difference between the containment and outside pressures and the area of the opening. If there is no liquefied gas in the system, and if no new material is being supplied from within the system, the containment pressure will start decreasing as soon as the breach takes place. As a result, the strength of the source will decrease as a function of time.

If the release point is located above the liquid level in the vessel, vapor outflow will occur. In the case of a pressure-liquefied gas, the liquid in the vessel will start boiling as a result of the drop in pressure as the liquid-vapor system tries to reach equilibrium at saturated vapor pressure. The necessary heat of evaporation will be drawn from the liquid in the vessel, the liquid thus cooling down to its boiling point at the (dropping) vessel pressure. The source strength of the releasing vapor, being a function of the vessel pressure (unless it is choked flow at the hole, in which case it is a function of temperature), is controlled by a balance between the amount of material escaping from the vessel and heat transfer from the surroundings. For relatively large release rates (large hole sizes), the temperature of the liquid will quickly reach its boiling point at near-atmospheric pressure; after this occurs, the source strength will be controlled primarily by heat transfer from the surroundings and will remain relatively constant until all liquid is depleted. Also, for large hole sizes, the boiling inside the vessel will be rapid, possibly resulting in frothing of the material; this may lead to release of some liquid along with the vapor, even when the hole is located above liquid level.

If the release point is located below the liquid level, liquid outflow will occur. In the case of a pressure-liquefied gas, the escaping liquid will rapidly flash. As the pressure of the escaping material drops to atmospheric as it is going through the hole, some of the material will become vapor, absorbing heat from the liquid and cooling it to its boiling point at atmospheric pressure (or even below its boiling point as the droplets further evaporate), resulting in a cold two-phase jet. Depending on the liquid content of the two-phase jet, some of the liquid may fall to the ground and form a liquid pool. However, in most situations, as the rapidly expanding vapor forms, it will break up any adjacent liquid particles into very small droplets (flash atomization), commonly referred to as aerosol. Thus, assumption of no rain-out is recommended in modeling such conditions.

The aerosol (droplets that do not fall to the ground), together with the vapor, will form a cloud that is colder and denser than the air around it. This heavy cloud will spread laterally under the influence of gravity and takes much longer to disperse than a buoyant gas cloud. The cold liquid from the liquid pool and from the droplets within the cloud will continue to evaporate and contribute to the cloud. Quasiadiabatic evaporation of the aerosols will tend to reduce the cloud temperature to below the normal boiling point. Any humidity in the air that gets entrained into the cloud may condense, forming small water or ice droplets, and then reevaporate farther downwind. This humidity may also react with the hazardous material, thus influencing the behavior of the cloud through, for example, heat of reaction effects.

If the cloud is flammable and ignites, it could result in a flash fire or vapor cloud explosion, depending on the degree of confinement, the degree of turbulence and mixing, and the total flammable mass within the cloud.

When a refrigeration-liquefied gas is quickly released, it boils off and generally forms a cold, heavy gas cloud. A liquid pool may also form. The boil-off is primarily controlled by heat transfer from the surfaces that the liquid contacts. The gas cloud will contain fewer aerosols than a release from pressure-liquefied state.

A flammable gas cloud, if ignited at a distance from its release location, may burn back to its source and result in a jet fire and/or a pool fire, depending on the conditions at the source.

The source modeling for the above situations should consider the thermodynamics and dynamics of what happens inside the vessel and to the released material once it is outside containment, the heat transfer between the vessel and its surroundings, and the heat transfer between the released material and its surroundings, along with the appropriate chemical/physical reactions and mass balances.

A fire-induced BLEVE is a physical explosion that can occur when flame impingement locally overheats the vapor space of a storage vessel containing a liquefied material under pressure. As a result of the increased temperature, the vessel pressure will increase due to the higher vapor pressure. For vessels that lack adequate pressure relief, rupture can occur due to local overheating because the metal may be sufficiently weakened that it is unable to withstand even the normal vessel design pressures. When contents are noncombustible (e.g., water), a mechanical explosion (liquid expanding rapidly into vapor) will occur. When flammable, as with hydrocarbons, a fireball will also follow.

Missiles (projectiles) may also cause injuries or fatalities at considerable distances from source, depending on the energy of an explosion and the mechanical integrity of the system in which it occurs. Missiles are more likely to occur as a result of a BLEVE. The risk of direct impact at any specified location is primarily a function of the frequency distribution of ranges of missiles.

It is important to recognize that some fire hazards also produce toxic combustion products, which may require modeling. This will involve estimation of the source strength followed by dispersion modeling, plume rise being an important factor. Estimating the source strength for such incidents presents particular challenges due to uncontrolled conditions of such fires.

For releases into water bodies, the modeling of spread and trajectory of liquids should take into account not only evaporation, but also dissolving and sinking of the hazardous materials. For liquid releases, the modeling of evaporation should take into account evaporative cooling, since this will affect the evaporation rate.

TABLE 10.4 Models Required for Different Hazards

Hazard category	Potential outcome	Model required
Flammable liquids, including those liquefied by refrigeration	Pool fire	Liquid discharge, liquid spread, pool fire
	Flash fire	Liquid discharge, liquid spread, evaporation/boil-off, (passive or heavy), gas dispersion
Flammable gases, liquefied by compression	Boiling liquid expanding vapor explosion (BLEVE)	BLEVE
	Fireball	Fireball
	Jet fire	Two-phase discharge, jet fire
	Vapor cloud explosion (VCE)	Two-phase discharge, heavy gas dispersion and/or VCE
	Flash fire	Two-phase discharge, rain-out, evaporation/boil-off, heavy gas dispersion
	Pool fire	Two-phase discharge, rain-out, liquid spread, pool fire
Flammable gases, gas under pressure	Fireball	Gas discharge, fireball
	Flash fire	Gas discharge (passive or heavy), gas dispersion
	Jet fire	Gas discharge, jet fire
Toxic liquids, including those liquefied by refrigeration	Toxic vapor cloud from liquid pool	Liquid discharge, liquid spread, evaporation/boil-off (passive or heavy), gas dispersion
Toxic gases liquefied by compression	Toxic gas cloud	Two-phase discharge, rain-out, liquid spread, evaporation/boil-off, heavy gas dispersion
Toxic gases, gas under pressure	Toxic gas cloud	Gas discharge (passive or heavy), gas dispersion
Toxic combustion products	Toxic gas cloud	Model for the combustion process, gas dispersion

Table 10.4 summarizes the outcomes of events resulting from the release of hazardous substances and the types of models required for their consequence analysis.

Gas Dispersion. This group of models describes the atmospheric dispersion of clouds of gases and gas/aerosol mixtures. The objective of these models is to estimate the variation of concentration in air of the released material as a function of time and distance from release location. Further information can be found in CCPS (1995a).

Dispersion calculations should take into consideration the time-varying rates of gas release (source strength), as well as atmospheric turbulence conditions.

Atmospheric turbulence is primarily a function of:

1. Solar heating/radiative cooling (determined by cloud cover, surface cover, latitude from the equator, time of day, and time of year)
2. Wind speed

3. Surface roughness
4. Terrain
5. Height from ground

In general, less information is required about site source details to calculate consequences of a given release scenario at larger distances from the release location.

For toxic materials, hazard level end points are generally on the order of a fraction of a percent of material by volume in air. Thus, the assessment of toxic hazards involves modeling large dilutions in the atmosphere, typically on the order of 1000:1 or larger. Such high dilutions require cloud travel over relatively large distances from the source, and it is atmospheric dispersion (together with total quantity of emitted gas and its release rate) that primarily controls the extent of the hazard zone; source characteristics such as height, geometry, and release velocity are generally not significant.

For flammable materials, hazard level end points are generally on the order of several percent. The assessment of flammable hazards typically involves modeling dilutions on the order of 10:1 or 100:1. For the same amount of material released, the hazard zones are therefore much smaller than for toxic clouds and source characteristics will play a more important role, along with atmospheric dispersion. A knowledge of momentum jet and buoyant versus dense gas plume mixing is important to these dispersion calculations. Key variables include site source details such as:

1. Source diameter
2. Initial jet density, velocity, and orientation
3. Proximity and shape of impeding obstacles and confining structures
4. Initial chemical reactions, droplets, aerosols, and initial fallout

In modeling the behavior of gas clouds, it is very important to select between passive/buoyant and dense gas dispersion, as appropriate for the situation.

The passive gas dispersion models are usually based on the Gaussian plume model. In Gaussian models, atmospheric dispersion is taken into account through empirical dispersion coefficients that vary by atmospheric turbulence class (stability class) and distance from source. Dilution by the wind is taken into account through division by wind speed. No consideration, however, is given to the difference of the density between the ambient air and the gas, other than to calculate an initial plume rise if the release is hot (buoyant plumes rise according to relatively well-established approximations and then behave as a plume characterized by Gaussian concentration profiles). Because of this, these models must only be used for gas mixtures with a density approximately the same as that of air.

The Gaussian model is based on analytical solutions of the general transport equation for a point source. Steady continuous and instantaneous releases (plumes and puffs) can be modeled. For time-varying continuous releases, time-integrated puff models should be used. For area sources, such as a liquid pool, two approaches are possible. In the first approach, an imaginary point source is assumed upwind of the actual source, so that the width of the cloud matches the source dimensions at the site of the actual source. The second approach is based on an area integration of the point source equations over the source area.

Heavy gas dispersion models are used if the gas is expected to exhibit heavy gas behavior upon release. Not all gases with a density greater than that of the surrounding air will exhibit heavy gas behavior. In order to establish whether a release will exhibit heavy gas behavior, empirical formulations taking into account the density difference, release rate, source diameter, and wind speed are available.

Heavy gas clouds tend to slump, flow down sloping ground, and spread in a radial direction because of gravity, even on flat ground. In contrast to a passive gas, the gas released may spread against the direction of the wind. Downwind from the source area, a dense gas will lead to a wide low-lying cloud, which is more difficult to disperse than a passive gas cloud. Eventually, the dispersion of the cloud becomes passive due to dilution. The cloud may also lift off and rise depending on the material and atmospheric conditions.

Most releases are influenced by buildings or structures either at the source or during the dispersion of the plume. Most available models are unable to handle these complexities well; they are suitable only for dispersion over flat, homogeneous terrain.

Since most model formulations contain a division by wind speed to account for dilution by the wind, they become increasingly conservative and unreliable for calm situations with wind speeds less than 1 to 1.5 m/s.

Recommended Atmospheric Conditions for Use in Risk Analysis. The main input parameters used in dispersion models for estimating the downwind extents of hazard zones are atmospheric stability, wind speed, and wind direction. For any location in Canada, joint frequency distribution for these variables can be obtained from the Atmospheric Environment Service of Environment Canada, in terms of:

- The six Pasquill–Gifford stability classes (A–F)
- A selected number of wind speed classes, including calms (typically five)
- 36 or 16 wind direction classes

Full treatment of the range of these inputs requires a large number of modeling runs, and this is often unnecessary for arriving at risk estimates with reasonable accuracy. Thus, the following approach is recommended for a simplified analysis:

- Use only two meteorological conditions for modeling purposes:
  - Stability D and wind speed 3 m/s (D3), with an associated probability of 0.7
  - Stability F and wind speed 1.5 m/s (F1.5), with an associated probability of 0.3
- Use a uniform wind rose (equal probability in all directions), unless certain wind directions are significantly different, such as in a deep valley. In this case, use of the local wind rose appropriate for each grouping of stability classes is recommended.

Toxic Effects. Toxic substances can have immediate and severe physiological effects on people and may ultimately cause death. Toxic effects are derived from a wide range of combinations of concentration and exposure times. A relatively short-term, high-level exposure to a highly hazardous substance such as methyl isocyanate is very different from a long-term exposure to low concentrations of benzene. In consequence modeling exercises for accidents, the usual focus is on highly toxic substances that can kill or cause serious injury over a relatively short time frame of minutes or hours. Such substances are normally referred to as acutely toxic. Experience has shown that the types of acute toxicity most likely to result in human fatalities in an industrial emergency are highly irritating or corrosive substances like chlorine, ammonia, or hydrogen fluoride or fast-acting nervous system toxins like hydrogen sulphide.

The relationship between concentration and effects on humans is highly nonlinear, i.e., a doubling of concentration will generally result in more than a doubling of the damage. This nonlinearity requires the estimation of peak and time-mean concentrations through modeling of concentration fluctuations for improved accuracy in predicted toxic effects.

For rapid releases of toxic materials into the atmosphere, such as from a sour gas pipeline burst or rupture of a chlorine vessel, the passage time of the toxic cloud over a receptor point may be relatively short. This affords an opportunity for sheltering within buildings until the cloud passes over the receptor. During the passage of the cloud, the infiltration rate into a building will be a function of the type of building, whether the windows are open or closed, and whether the air exchange fans are operational. The buildup of concentration inside the building will be slower than outside. Once the cloud passes, however, the concentration within the building will remain higher than outside for a period of time. Modeling of these situations will provide critical input to emergency response planning near areas where large releases of hazardous materials are possible.

Thermal Radiation Effects. Thermal radiation effects arise from flash fires, pool fires, jet fires, or fireballs. These involve the combustion of flammable mixtures. The intensity of thermal radiation (measured in terms of thermal radiation flux or energy per unit area and time) at a receptor outside a fire depends on its distance from the fire, the flame height, flame emissive power, and atmospheric transmissivity.

Flash Fires. For flash fires, the controlling factor for the amount of damage that a receptor will suffer is whether the receptor is physically within the burning cloud or not (and, if within the cloud, whether the clothing ignites or not). This is because most flash fires do not burn very hot and the thermal radiation generated outside of the burning cloud will generally not cause significant damage due to the short duration. Thus, modeling of flash fire consequences consists primarily of an exercise in dispersion modeling, the hazard zone being essentially the extent of the flammable zone of the cloud.

Other Types of Fires. For the other types of fires, available models are broadly classified as either point-source models (simple or with multiple sources) or view-factor models based on either an equivalent radiator or a solid flame approach. They differ in their required input parameters according to the type of fire and the level of detail and complexity inherent in the inputs and submodels needed to describe the physical event.

Point-source models are generally less complex than the view-factor models. They are appropriate when the receptor is sufficiently separated from the fire that the specific shape and size of the fire is no longer important. In contrast, view-factor models allow the geometry of the flame, as well as the receptor configuration, to be taken into account in the estimation of thermal flux. These are therefore more applicable to cases where the receptor is close to the fire and/or when the geometric details of the fire are important (e.g., wind effects, receptor orientation).

Explosion Effects. Explosion overpressure effects that are of interest here result either from the rapid combustion of a fuel–air mixture (confined explosion or VCE), or a sudden release of pressure energy (BLEVE).

BLEVE. For BLEVEs, the available models are based on the similarity of the blast waves in the far field to those generated by high-explosive detonation. The compressed gas's stored energy is first calculated based on pressure at the time of burst. The energy of explosion is obtained as the difference between the initial and final states, assuming isentropic expansion. This energy contributes primarily to the production of a blast wave and missiles. The fraction of pressure energy that contributes to the blast wave can be taken to be about 40%. Overpressure and impulse are then read from charts that relate detonation-blast parameters to charges of high explosive with the same energy. In the near field, this similarity to high explosives is not valid, and correction factors based on numerical simulations should be used.

Missile damage from BLEVEs is more difficult to model and of relatively little importance in risk assessments. A statistical account of the extent of missile damage from actual BLEVEs involving primarily LPG is described in Lees (1996).

Confined Explosions. Confined explosions occur when a flammable mixture in a confined space is ignited. A typical example is the case of a flammable liquids tank. When these tanks are emptied, residual liquid may evaporate and form a flammable mixture in the tank. If ignited, a confined explosion would be produced. The modeling of confined explosion effects is analogous to the modeling of BLEVEs. Here the explosion energy released is obtained from the enthalpy of combustion.

VCE. For a fuel/air mixture outside containment, conditions favoring a VCE as opposed to a flash fire include:

1. The mass of the cloud (e.g., five metric tons appears to be a lower limit for propane vapor cloud explosions outside containment)

2. Flame speed
3. Degree of confinement
4. Degree of turbulence in the cloud

A rapid violent release, if not ignited immediately, may result in sufficient mixing through self-generated turbulence for explosive conditions to occur. The portion of the vapor cloud within the explosive range at the time of ignition will contribute directly to the explosion. The resulting overpressure at a given point is a function of:

1. Distance from source
2. Fuel properties
3. Mass of the cloud
4. Degree of confinement (affected by the presence of obstacles)

Two different types of models are generally used in practice for estimating VCE overpressures at a distance from a source:

1. The TNT equivalency method relates the explosive potential of a release to the total quantity of fuel in the vapor cloud, whether or not it is within flammable limits. The explosive power of the vapor cloud is expressed as an energy equivalent amount of TNT located at the center of the cloud. The value of the proportionality factor is determined from damage patterns observed in a large number of similar incidents involving vapor cloud explosions. Calculated blast overpressures tend to be high near the cloud center (regardless of physical surroundings), and a gradual decay is observed as distance from the cloud center increases. This translates into a localized high-damage zone with low to moderate damage in outlying areas.

   It is important to apply conservative values to the proportionality constants used for the TNT method. An explosion efficiency of 0.06 to 0.10 should be used even in areas that are not tightly confined. Scaling factors should be averaged among several literature sources and used to calculate overpressure profiles. These data are often material-specific and, if not averaged, could introduce additional errors.
2. The multienergy method reflects current consensus that one of the controlling factors of severe explosions is turbulence. One source of such turbulence is the high-velocity flow of fuel being ejected from a pressurized system. Explosive combustion rates may develop in such a turbulent fuel–air mixture. Another source of turbulence is combustion within a partially confined/physically obstructed environment. The expansion of combustion gases against a confining structure can cause exponential increases in the combustion rate and an overall increase in overpressure. The explosive power of a vapor cloud is determined primarily by the energy of fuel present in the confined areas of a vapor cloud.

It should be noted that in cases where VCEs may be possible, the footprint of the flash fire zone (the zone within the lower flammability limit [LFL] of the material) should also be estimated and used in the overall risk estimation with its corresponding frequency.

Recommended Source Term and Hazard Models. It is difficult to specify the use of any model for a particular application without assessing all the features of each model and establishing the end purpose for which the modeling will be used. Table 10.5 provides a list of model references from the public domain that are suitable for modeling hazardous releases.

Other excellent references on consequence modeling techniques are CCPS (1989a, 1994, 1995c), Lees (1996), and TNO (1992a, b).

The following are some of the available computer models that attempt to combine a number of the above (and other) models to provide complete modeling packages for users.

TABLE 10.5 Recommended Source Term and Hazard Models

Discharge models, liquid	Bernoulli equation
Discharge models, two-phase	Fauske and Epstein (1988)
Discharge models, gas	Sonic discharge equations (for choked flow)
Heavy gas dispersion models	Britter and McQuaid (1988)
	SLAB (Ermak, 1990)
	HGSYSTEM (Post, 1994)
Passive gas dispersion models	Gaussian plume, puff or integrated-puff equations
Concentration fluctuations in atmosphere	Wilson (1986), CCPS (1995c)
Building infiltration models	Basic mass balance with air exchange
Liquid pool spread models	Cavanaugh et al. (1994)
Liquid pool evaporation models	Cavanaugh et al. (1994)
Rainout models	SUPERCHEMS (1998)
Jet fire models	CCPS (1994), Lees (1996), Baker et al. (1983)
Pool fire models	CCPS (1994), Crocker and Napier (1986)
BLEVE models	CCPS (1994)
VCE models	TNT (Baker et al., 1983)
	Multienergy (Van den Berg, 1985)
Toxic combustion products models	TNO (1992a)

- *Public Domain Models*:
  - ALOHA — NOAA
  - HGSYSTEM — Shell Research (Post, 1994)
- *Proprietary Models*:
  - CHARM — Radian Corp.
  - PHAST — DNV Technica Ltd.
  - SUPERCHEMS — Arthur D. Little
  - TRACE/SAFER — SAFER

Vulnerability Modeling. Once the hazard level (i.e., concentration, thermal radiation flux, or overpressure) is estimated at a receptor point following a hazardous event, the next and final step in consequence modelling is estimating the level of damage to the receptor. Although we will focus on fatality as the effect of interest, injury or other effects such as property damage can also readily be estimated using the methods described here.

For all hazards except flash fires, there are two commonly used methods for estimating level of damage on the receptor:

1. The fixed-limit method
2. The PROBIT method

The fixed-limit method consists of comparing the estimated average (or maximum) hazard level to which a receptor is exposed against fixed limits which are available from the literature. For example, for toxic clouds, estimated concentration levels can be compared to IDLH (immediately dangerous to life and health) levels to establish whether fatality or serious injury might occur at a receptor point. Table 10.6 contains the commonly used fixed-limit values for some toxic materials and for fires and explosions.

The advantage of the fixed-limit method is its simplicity. Its disadvantage is that it can be very misleading for time-varying hazards, which is generally the case under major accident conditions.

For example, a short-duration toxic release may be over in a matter of minutes, with exposure duration of receptors not much longer than that. The time variation of concentration

TABLE 10.6 Table of Inhalation Exposure Limits for Selected Hazardous Materials

MIACC LIST 1 Substances (MIACC, 1994)	IDLH (ppm)	ERPG (ppm)		
		1	2	3
Acetaldehyde	CA (2,000)	10	200	1000
Acetylene	N.D.			
Ammonia	300			
Arsine	CA (3)			
Benzene	CA (500)	50	150	1000
Bromine and bromine solutions	3	0.2	1	5
Butane and butane mixtures	N.D.			
Chlorine	10	1	3	20
Cyclohexane	1,300 (LEL)			
Ethylbenzene	800 (LEL)			
Ethylene	Not listed			
Ethylene dichloride	CA (50)			
Ethylene oxide	CA (800)	N/A	50	500
Fluorine	25	0.5	5	20
Gasoline	CA (N.D.)			
Hydrogen chloride/acid	50	3	20	150
Hydrogen fluoride/acid	30	2	20	50
Hydrogen sulfide	100	0.1	30	100
Liquefied petroleum gases	2,000 [LEL]			
Mercury	10 (mg/m^3) (as Hg)			
Methane	Not listed			
Naphtha, petroleum naptha or naptha solvent	1,000 [LEL]			
Nitric acid, fuming or red fuming	25			
Propane, and propane mixtures	21,000 [LEL]			
Propylene oxide	CA (400)	50	250	750
Sodium chlorate	Not listed			
Sulfur dioxide	100	0.3	3	15
Sulfuric acid, fuming	15 (mg/m^3)	2 (mg/m^3)	10 (mg/m^3)	30 (mg/m^3)
Tetraethyl lead	40 (mg/m^3) (as PB)			
Toluene	500	50	300	1000
Vinyl chloride	CA (N.D.)			
Xylene	900			

IDLH: Immediately dangerous to life and health. Data taken from NIOSH (1994). According to this Guide, IDLH exposure is a condition "that poses a threat of exposure to airborne contaminants when that exposure is likely to cause death or immediate or delayed permanent adverse health effects or prevent escape from such an environment." IDLH values are based on the effects that might occur as a consequence of a 30-minute exposure.

ERPG: Emergency response planning guidelines taken from AIHA (1998).

ERPG-1: The maximum airborne concentration below which it is believed nearly all individuals could be exposed for up to one hour without experiencing other than mild transient adverse health effects or perceiving a clearly defined objectionable odor.

ERPG-2: The maximum airborne concentration below which it is believed nearly all individuals could be exposed for up to one hour without experiencing or developing irreversible or other serious health effects or symptoms that could impair their abilities to take protective action.

ERPG-3: The maximum airborne concentration below which it is believed nearly all individuals could be exposed for up to one hour without experiencing or developing life-threatening health effects.

CA: Refers to substance NIOSH considers to be potential human carcinogen. Any IDLH values accompanying a "CA" notation were determined before NIOSH began considering carcinogenic effects.

LEL: Indicates that the IDLH was based on 10% of the lower explosive limit for safety considerations even though recent toxicological data indicated that irreversible health effects or impairment of escape existed only at higher concentrations.

N.D.: IDLH has not yet been determined.

at a given receptor point may look like that shown in Fig. 10.16. Typical dispersion models will output what is shown as the time average in this figure.

If one compares the maximum in this curve to a fixed-limit value, such as the IDLH (which is a 30-minute exposure limit for exposed persons to be able to leave the area unaided, but could nevertheless lead to severe injuries), this could lead to severe overestimates of fatality risk since the actual exposure to these levels could be only a few minutes as opposed to 30 minutes.

Any use of the ERPG (emergency response planning guidelines) values will be even more conservative and will not be appropriate for estimating fatalities. This information is provided in Table 10.6 only for reference purposes and completeness.

The recommended method and a more appropriate one is to use the PROBIT method, which can readily handle time-varying situations, including concentration fluctuations (Alp et al., 1990).

To apply this method, a hazard load L is estimated at each receptor point:

$L = \int C^n\, dt$ for toxic clouds (C is the time-varying concentration at the receptor point, estimated by the dispersion model)

$L = \int I^n\, dt$ for thermal radiation hazards (I is the time-varying thermal radiation flux resulting from the fire)

$L = P_o$ for explosion hazards (P_o is the overpressure resulting from the explosion)

Here, the integration essentially represents the total amount of contaminant or thermal energy received by the receptor (weighted by the power n), and n is an empirical PROBIT parameter appropriate for the chemical and type of hazard. The integration is performed over the time of exposure during the hazardous event. (Effect of evacuation or sheltering in a building can thus be incorporated into the results if desired.)

We then estimate the PROBIT (probability unit) Y:

$$Y = k_1 + k_2 \ln L \tag{10.5}$$

where k_1 and k_2 are additional empirical PROBIT parameters.

The values of the PROBIT parameters for some common toxic chemicals are given in Table 10.7. Information for some other toxic chemicals and description of an approximate

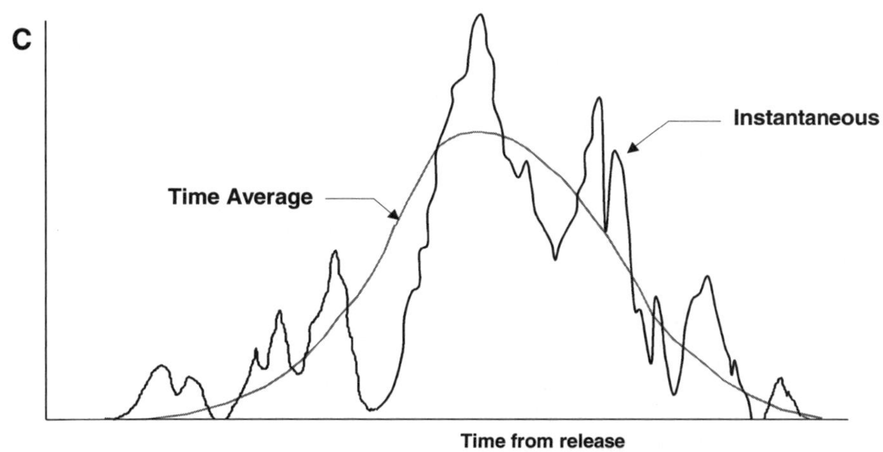

FIGURE 10.16 Typical time variation of concentration at a down-wind receptor following a short-duration release.

TABLE 10.7 PROBIT Constants for Lethal Toxicity (CCPS, 1989a)

Substance	k_1 (ppm)	k_2 (ppm)	n (min)
Acrolein	−9.931	2.049	1
Acrylonitrile	−29.42	3.008	1.43
Ammonia	−35.9	1.85	2
Benzene	−109.78	5.3	2
Bromine	−9.04	0.92	2
Carbon monoxide	−37.98	3.7	1
Carbon tetrachloride	−6.29	0.408	2.50
Chlorine	−8.29	0.92	2
Formaldehyde	−12.24	1.3	2
Hydrogen chloride	−16.85	2.00	1.00
Hydrogen cyanide	−29.42	3.008	1.43
Hydrogen fluoride	−35.87	3.354	1.00
Hydrogen sulfide	−31.42	3.008	1.43
Methyl bromide	−56.81	5.27	1.00
Methyl isocyanate	−5.642	1.637	2
Nitrogen dioxide	−13.79	1.4	1
Phosgene	−19.27	3.686	1
Propylene oxide	−7.415	0.509	2.00
Sulfur dioxide	−15.67	2.10	1.00
Toluene	−6.794	0.408	2.50

method for estimating these parameters from LC50 data can be found in TNO (1992b). PROBIT constants for fire and explosion effects are given in Table 10.8.

Y can then be related to the level and type of hazardous effect in terms of percent probability of death, injury, or damage occurring ($P_{e,h}$) using Table 10.9.

Flash Fires. For flash fires, the maximum extent of the hazard zone is generally based on the lower flammable limit (LFL) of the material. Sometimes LFL/2 is also used to take into account the possibility of having high-concentration pockets of gas that might result from concentration fluctuations in the atmosphere. However, this is not the whole story. Ignition of a gas cloud can occur as the leading edge of the cloud reaches an ignition source and the cloud will burn towards the source. Hence, the flash fire will only affect the area between the ignition point and the release location. By estimating the probability of ignition as the cloud reaches each ignition source, one can estimate the probability of affecting any receptor as a function of distance from the release point.

A common assumption for probability of fatality for people caught in a flash fire is 10% for those having protective (fire-retardant) clothing (such as NOMEX suits) and 90% for those without such protection. Both are somewhat on the conservative side and include major injury, which will lead to overestimates of risk of fatality.

Uncertainties in Consequence Modeling. Uncertainties in consequence estimation arise due to uncertainties in modeling the source term, the migration of a hazard away form the hazard source (hazard modeling), the effects of a level of hazard on receptors (vulnerability modeling), and assumptions made about the degree of protection afforded to receptors. As discussed here, these uncertainties are for the most part treated conservatively.

In the estimation of consequences, a major source of uncertainty is the modeling of the source term. The source term describes the rate of release of material from containment and into the carrying medium (e.g., atmosphere). In effect, the source term determines the amount of the material released.

TABLE 10.8 PROBIT Constants for Fire and Explosion Damage

Type of injury or damage	Hazard load L	k_1	k_2
Fire			
Burn deaths	$\frac{1}{10^4} \int I^{4/3}\, dt$	−14.9	2.56
Explosions			
Death (lung hemorrhage)	P_0	−77.1	1.91
Eardrum rupture	P_0	−15.6	1.93
Deaths from impact	J	−46.1	4.82
Injuries from impact	J	−39.1	4.45
Injuries from fragments	J	−27.1	4.26
Structural damage	P_0	−23.8	2.92
Glass breakage	P_0	−18.1	2.79

Source: Lees, 1986.
t = exposure time (s)
I = radiation intensity (W/m^2)
P_0 = peak overpressure (N/m^2)
J = impulse (N s/m^2)

There are a number of uncertainties related to source term, including:

- Hole characteristics: size, location, shape
- Orientation of the release: vertical, horizontal
- Degree of pooling of flashing two-phase discharges
- Degree and size of confinement: release outdoors/indoors, into a dyked area
- Amount of material involved

With the above, conservative assumptions can usually be made to avoid underestimation of consequences. For example, one can assume that a tank/system is full when the accident happens and that the entire inventory is released. Also, pooling of liquid from a two-phase release can be neglected. These conservative assumptions, however, need to be made with

TABLE 10.9 Transformation of PROBITS to Percentages

%	0	2	4	6	8
0	—	2.95	3.25	3.45	3.59
10	3.72	3.82	3.92	4.01	4.08
20	4.16	4.23	4.29	4.36	4.42
30	4.48	4.53	4.59	4.64	4.69
40	4.75	4.80	4.85	4.90	4.95
50	5.00	5.05	5.10	5.15	5.20
60	5.25	5.31	5.36	5.41	5.47
70	5.52	5.58	5.64	5.71	5.77
80	5.84	5.92	5.99	6.08	6.18
90	6.28	6.41	6.55	6.75	7.05
99	7.33	7.41	7.46	7.65	7.88

Source: CCPS, 1989a.

care. For example, a toxic gas released indoors will likely result in a substantial hold-up of material initially within the indoor volume and will subsequently be vented relatively slowly to the outdoors. If it is assumed that a release indoors occurs in the same way as one outdoors, the consequences may be substantially overestimated.

In consequence modeling, it is also important that appropriate hazard migration models be used. An error often made is the use of a Gaussian dispersion model for gas clouds that are initially heavier than air. This would lead to underestimating hazard distances. State-of-the-art models have undergone various degrees of validation. However, due to the vast number of atmospheric, terrain, and material factors that complicate the prediction of hazard levels, these models are not capable of accurately predicting experimental observations over a large range of conditions. In fact, models that predict experimental observations within a factor of two are considered reasonably good.

Also, due care must be taken in specifying model input parameters. Factors such as atmospheric transmissivity in fire models, or ambient temperature in modeling cold heavier-than-air gas clouds affect model results. They are typically specified conservatively to avoid a large number of model runs.

Modeling health effects from various hazard levels is a difficult task. Risk assessments are typically based on the risk of death or serious injury. Obviously there are no experimental data available on the dose-response relationship of material concentration and exposure duration, thermal radiation intensity or blast overpressures on humans. What little there is has been inferred from actual accidents. Models that predict the impact of exposure to hazardous materials are heavily influenced by animal experiments. Typically, they have large safety factors built in. It is believed that models based primarily on exposure of experimental animals are conservative when applied to humans, especially when, on a body weight difference, the animals are much smaller than humans. In fact, many will argue that they are too conservative. These estimates are difficult to make, and unfortunately little can be done to improve the degree of uncertainty.

Finally, in modeling the impact of a hazard level on a human, it is typically assumed that the exposed individual is outdoors and stationary (does not retreat or take shelter). In reality, in Canada, individuals are indoors most of the time. This affords considerable protection from exposure to most types of hazards. Taking credit for sheltering in a risk assessment may improve the accuracy of the results by reducing the conservatism, but it increases the cost of the assessment because additional consequence modeling runs are required.

Summary of Consequence Modeling. Depending on the type of hazard, and at each receptor of interest, the following steps must be taken.

- Estimate the hazard level, using mathematical models (selection of the appropriate model is critical for accuracy of risk assessment):
  - Concentration, for gas clouds
  - Thermal radiation flux, for fires
  - Overpressure and impulse, for explosions
- Calculate load, PROBIT, and probability of "effect" (although we have focused on "fatality" as the "effect" of interest, injury or other effects such as property damage can also readily be estimated using the methods described here) at each receptor point (also called event individual consequence).
- Alternatively, compare the hazard level to an appropriate fixed limit, and thus estimate the probability of effect.
- For flash fires, estimate ignition probability as a function of ignition sources in the downwind direction, and thus estimate the probability of effect.
- The number of receptors affected can then be calculated by integrating the product of individual consequence and population density over the exposed area, if an estimate of

event societal consequences is within the scope of the analysis (Alp and Zelensky, 1996). Local municipal planning, land use, and population maps should be considered as data sources when identifying receptors. In addition, local off-site surveys and reviews may be considered to assess the density of permanent habitation and the frequency of habitation by transient community members.

It is often desirable to estimate the impacts of hazardous events in terms of their financial cost for the purposes of cost–benefit analysis and making decisions for risk mitigation and control, such as in terms of emergency resources planning, physical design changes, or making management system changes. The extension of the analysis to financial cost is relatively straightforward in principle. When issues such as human life and environmental quality enter into the picture, however, such practices generate controversy among practitioners, decision makers, and other stakeholders.

10.2.10 Risk Presentation and Evaluation of Acceptability

The methods described here can be used for estimating frequency and consequence of undesirable events. The formulations described in Section 10.2.7 can then be used to estimate risk.

A number of methods of presenting risk results are available for decision makers.

For individual risk, the usual ones are (see Fig. 10.17):

- Individual risk profiles (sometimes also called transects), which depict the variation of individual risk (or its logarithm) as a function of distance from the risk source as an x–y graph
- Individual risk contours, which show the variation of risk around a risk source generally as contours superimposed on maps

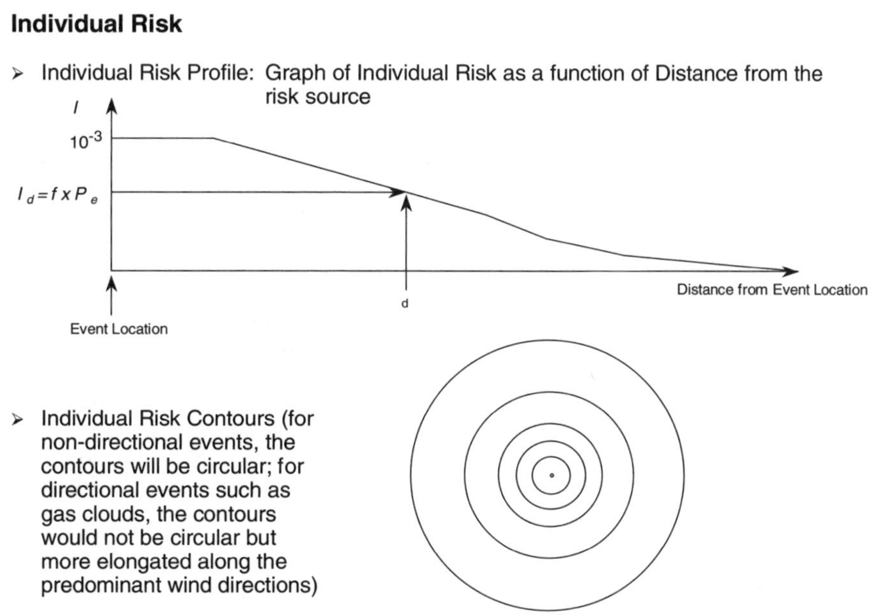

FIGURE 10.17 Representation of individual risk.

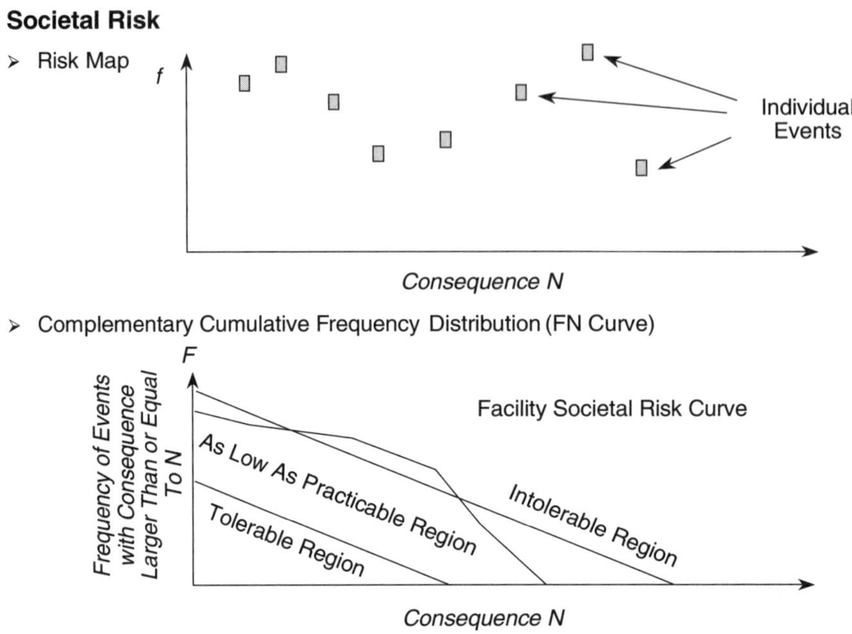

FIGURE 10.18 Representation of societal risk.

For societal risk, the common methods of presentation are the (see Fig. 10.18):

- *f–N* graphs (also called risk maps), which are simply *x–y* scatter diagrams showing each (frequency, consequence) pair for each hazardous event (*N* being expressed as total fatalities, injuries, or $ cost of a given type of event)
- *F–N* graphs, which are complementary cumulative distributions of the events in the *f–N* graphs, *F* having units of annual frequency of events with consequence greater than or equal to *N*.

In Section 10.1, we discussed the principles of risk evaluation. With reference to that section, land use planning often involves use of individual risk contours and comparing them to guidelines such as those shown in Fig. 10.3. In some jurisdictions, the FN form of societal risk is used together with guidelines such as in Fig. 10.4.

To assist emergency response planning, representations on maps of consequence contours are very useful, as they provide information about hazard zones. Another useful presentation method for this purpose is the risk map shown in Fig. 10.17, which puts the various hazardous events in perspective relative to others.

The FN method of presenting risk is very powerful, especially when financial consequences of hazardous events are being demonstrated to upper management of companies or organizations. Such graphs showing the risk reductions that would be realized by implementing certain risk reduction measures under consideration, can convey powerful messages to decision makers and justify approving investments for such measures.

10.3 ACKNOWLEDGMENTS

Some of the material in this section was developed while the author was the Chair of the Risk Assessment Expert Committee of the Major Industrial Accidents Council of Canada

(MIACC), with contributions and comments from the Committee members. The material was to be published by the now-defunct MIACC as the next generation of Canadian risk analysis guidelines. The contributors to the original material of the guidelines include: Derek Atkinson, Robert Beatty, David DeGagne, Vince Gagner, Ed Gulbinas, Matthew Hilbert, William Hume, Nigel Hyatt, Brian Kelly, Jean-Paul Lacoursiere, Marcello Oliverio, Gerry Philips, Cathy Preston, James Pirie, James Reid, John Shrives, and Michael Zelensky. The original material has been edited and new material has been added by the author for the purposes of this Handbook.

10.4 REFERENCES

American Industrial Hygiene Association (AIHA). 1998. *Emergency Response Planning Guidelines and Workplace Environmental Exposure Level Guides Handbook,* AIHA, Fairfax, VA.

Alp, E. 1996. "Risk Assessment Methodologies for Pipelines," Paper presented at the ASME International Pipeline Conference, June 10–14, Calgary, AB.

Alp, E., and M. J. Zelensky. 1996. "Risk Quantification for Meteorology- and Direction-Dependent Hazards due to Point and Linear Risk Sources," *J. Loss Prev. Process Ind.,* vol. 9, pp. 135–145.

Alp, E., M. J. E. Davies, R. G. Huget, L. H. Lam, and M. J. Zelensky. 1990. *GASCON2—A model to Estimate Ground-Level H_2S and SO_2 Concentrations and Consequences from Uncontrolled Sour Gas Releases.* Concord Environmental Report CEC.J1692, prepared for the Alberta Energy Resources Conservation Board, October.

American Petroleum Institute (API). 1995. *Management of Hazards Associated with Location of Process Plant Buildings.* 1st ed. API Recommended Practice 752. Washington, DC.

Baker, W. E., P. A. Cox, P. S. Westine, J. J. Kulesz, and R. A. Strehlow. 1983. *Explosion Hazards and Evaluation,* Elsevier, New York.

Blything, K. W., and A. B. Reeves. 1988. *An Initial Prediction of the BLEVE Frequency of a 100 Te Butane Storage Vessel.* Her Majesty's Stationery Office, London.

Britter, R. E., and J. McQuaid. 1988. *Workbook on the Dispersion of Dense Gases,* Health and Safety Executive, U.K.

Canadian Standards Association (CSA). 1997. *Risk Management: Guideline for Decision-Makers,* CAN/CSA Q850.

Cavanaugh, T. A., II, J. H., Siegell, and K. W. Steinberg. 1994. "Simulation of Vapour Emissions from Liquid Spills." *J. Hazardous Materials*, vol. 38, pp. 41–63.

Center for Chemical Process Safety (CCPS). 1989a. *Guidelines for Chemical Process Quantitative Risk Analysis.* A.I.Ch.E., Center for Chemical Process Safety, New York.

Center for Chemical Process Safety (CCPS). 1989b. *Guidelines for Process Equipment Reliability.* A.I.Ch.E., Centre for Chemical Process Safety, New York.

Center for Chemical Process Safety (CCPS). 1992. *Guidelines for Hazard Evaluation Procedures,* A.I.Ch.E., Center for Chemical Process Safety, New York.

Center for Chemical Process Safety (CCPS). 1994. *Guidelines for Evaluating the Characteristics of Vapor Cloud Explosions, Flash Fires, and BLEVEs.* A.I.Ch.E., Center for Chemical Process Safety, New York.

Center for Chemical Process Safety (CCPS). 1995a. *Understanding Atmospheric Dispersion of Accidental Releases.* A.I.Ch.E., Center for Chemical Process Safety, New York.

Center for Chemical Process Safety (CCPS). 1995b. *Chemical Transport Risk Analysis.* A.I.Ch.E., Center for Chemical Process Safety, New York.

Center for Chemical Process Safety (CCPS). 1995c. *Concentration Fluctuations and Averaging Time in Vapor Clouds.* A.I.Ch.E., Center for Chemical Process Safety, New York.

CONCAWE. 1982. *Methodologies for Hazard Analysis and Risk Assessment in the Petroleum Refining and Storage Industry*, prepared by CONCAWE's Risk Assessment Ad-hoc Group, Report No. 10/82, The Hague, December.

Covello, V. T., P. M. Sandman, and P. Slovic. 1988. *Risk Communication, Risk Statistics, and Risk Comparisons: A Manual for Plant Managers.* Chemical Manufacturers Association, Washington, DC.

Crocker, W. P., and D. H. Napier. 1986. "Thermal Radiation Hazards of Pool Fires and Tank Fires," *Hazards in the Process Industries: Hazards IX*. I. Chem. E. Symp. Series No. 97, UMIST, Manchester, pp. 159–184.

Electric Power Research Institute (EPRI). 1981. *Component Failure and Repair Data for Coal-Fired Power Units*, prepared by Fluor Power Services, Inc., EPRI AP-2071, RP239-2, Topical Report, EPRI, October.

Ermak, D. L. 1990. *User's Manual for SLAB: An Atmospheric Dispersion Model for Denser-than-Air Releases*, Lawrence Livermore National Laboratory.

Fauske, H. K., and M. Epstein. 1988. "Source Term Considerations in Connection with Chemical Accidents and Vapour Cloud Modeling," *J. Loss Prev. Process Ind.*, vol. 1, pp. 75–83.

HSE. 1996a. *Offshore Accident and Incident Statistics Report 1996,* Ref. OTO 96 955. UK Health and Safety Executive Information Centre, Sheffield.

HSE. 1996b. *Offshore Hydrocarbon Releases Statistics 1996,* Ref. OTO 96 954, UK Health and Safety Executive Information Centre, Sheffield.

HSE. 1997. *Offshore Hydrocarbon Releases Statistics 1997,* Ref. OTO 97 950. UK Health and Safety Executive Information Centre, Aberdeen.

Institute of Electrical and Electronic Engineers (IEEE). 1983. *Guide to the Collection and Presentation of Electrical, Electronic, Sensing Component, and Mechanical Equipment Reliability Data for Nuclear Power Generating Stations,* Std. 500.

Johnson, D. W., and J. R. Welker. 1981. *Development of an Improved LNG Plant Failure Rate Data Base,* GRI-80/0093. Final Report, Gas Research Institute, September.

Lees, F. P. 1986. *Loss Prevention in the Process Industries*. Butterworth, London.

Lees, F. P. 1996. *Loss Prevention in the Process Industries,* 2d ed. Butterworth-Heinemann, London.

Major Industrial Accidents Council of Canada (MIACC). 1994. *Hazardous Substances Risk Assessment: A MiniGuide for Municipalities and Industry.*

National Institute of Occupational Safety and Health (NIOSH). 1994. *Pocket Guide to Chemical Hazards.* NIOSH, Washington, DC.

OREDA. 1997. *Offshore Reliability Data*. 3rd Edition. Hovik, Norway.

Post, L. 1994. *HGSYSTEM3.0: User's Manual.* Shell Research Limited, Chester, U.K.

Rijnmond. 1982. *Risk Analysis of Six Potentially Hazardous Industrial Objects in the Rijmond Area, A Pilot Study,* Report to the Rijmond Public Authority, presented by COVO Steering Committee. D. Reidel, Dordrecht.

Swain, A. D., and H. E. Guttman. 1983. *Handbook of Human Reliability Analysis with Emphasis on Nuclear Power Plant Applications,* NUREG/CR-1278, U.S. Nuclear Regulatory Commission, Washington, DC, August.

TNO. 1988. *Methods for Determining and Processing Probabilities.* Committee for the Prevention of Disasters Caused by Dangerous Substances, CPR 12E, The Labour Inspectorate, The Netherlands.

TNO. 1992a. *Methods for the Calculation of Physical Effects,* Committee for the Prevention of Disasters, CPR 14E, The Director-General of Labour, The Netherlands.

TNO. 1992b. *Methods for the Determination of Possible Damage to People and Objects Resulting from Releases of Hazardous Materials.* Committee for the Prevention of Disasters, CPR 16E, The Director-General of Labour, The Netherlands.

Van den Berg, A. C. 1985. "The Multi-Energy Method—A Framework for Vapor Cloud Explosion Blast Prediction," *J. Haz. Mat.*, vol. 12, pp. 1–10.

U.S. Nuclear Regulatory Commission. 1975. *Reactor Safety Study, An Assessment of Accident Risks in U.S. Commercial Nuclear Power Plants, Appendix III—Failure Data*, WASH-1400 (NUREG-75/014), October.

Wilson, D. J. 1986. *Plume Dynamics and Concentration Fluctuations in Gas Emissions.* Alberta Environment Report, Contract No. 840702, Edmonton, April.

Worldwide Offshore Accident Database (WOAD). Developed and operated by Det Norska Veritas (DNV), website: www.dnvda.com/offshore.

CHAPTER 11
LIABILITY IN EMERGENCY MANAGEMENT

John C. Pine
Professor-Research, Institute for Environmental Studies, Louisiana State University, Baton Rouge, Louisiana

11.1 INTRODUCTION

For many years, states and their local governmental units enjoyed extensive freedom from civil claims or suits seeking financial compensation for the acts of governmental agencies, officials, employees, or volunteers. This protection or immunity was based on a belief that neither the King nor his representatives could do any wrong. The King as sovereign was thus immune (sovereign immunity). This concept of protecting the state and its representatives from civil claims was adopted in the United States. Under this doctrine, individual citizens or businesses harmed by the actions of state or local agencies and their representatives are prohibited from recovering damages from the government. Under sovereign immunity, those harmed by a governmental unit could only recover if the public entity voluntarily recognized the claim and agreed to pay. The effect of sovereign immunity was that public jurisdictions were immune from suit (Reynolds, 1982).

From the early 1960s through the mid-1980s, the law in each state changed, making state and local governments more accountable for their actions. The law was changed either by the state legislatures, by decisions of the state courts, or by changes to the state constitution. Today, sovereign immunity no longer provides total protection for state and local public officials, employees, and volunteers. State law, usually in the form of a state tort claims statute and state emergency management statute, provides the basis for civil actions for damages involving emergency management activities against state and local jurisdictions. The law in each state, including statutes, court decisions, and the state constitution, defines the extent to which state and local governments may be subjected to civil suits for money damages.

The goal of this section is to provide those involved in emergency management with a basic understanding of liability issues involved in emergency planning, response, recovery, and mitigation. It is to provide an introduction to the nature and scope of governmental liability under civil law in emergency management programs and activities. Because of the potential exposure of public agencies to civil claims for money damages, public employees and officials need to understand the law as it affects their jurisdiction. This section is also intended to clarify steps that may be taken to reduce individual and organizational vulnerability to claims of liability.

Whether liability exists depends on the specific circumstances in a situation and the law in the state. The legal principles presented here are applicable in any state, but they must be viewed with an understanding that the law may be different in a specific state. Questions to be addressed include:

- Can a public organization or official be held liable for money damages under state law?
- How do the courts determine if liability exists?
- What type of acts could lead to liability under state law?
- What types of immunity exist for governmental agencies or their representatives?
- What defenses exists for governmental units, officers, employees, or representative?
- What types of damages or relief may be recovered?
- Who pays if the courts determine liability?
- Can local governments be held accountable for emergency planning and response activities established by federal law (Occupational Safety and Health Act or the Emergency Planning Community Right to Know Act)? How do federal hazardous materials requirements affect state and local emergency management organizations?

One of the goals of any public official should be to understand the circumstances from which claims for money damages arise and what proactive steps can be taken to prevent individual and business losses due to the actions of public agencies in a disaster or emergency. The strategy should be to plan, organize, implement, and control public programs and activities in such a manner as to avoid losses.

11.2 THE THREAT

Can an agency, government employee, or volunteer be sued? The answer in most cases is yes. There are few rules and restrictions that will prevent a person or organization from being named in a legal action seeking damages and reimbursement to remedy a supposed wrong. However, whether the suit will be successful is another issue.

While most people are familiar with the liability that commonly revolves around traffic accidents, claims may also be brought based on the use of equipment, chemicals, and other products and services blamed for injuries.

11.3 LIABILITY UNDER STATE LAW

A tort is an action that harms another. It occurs when a person acts or fails to act, without right, and as a result another is harmed. Torts involve civil actions for personal injuries or property damage rather than a criminal action or a contractual claim. Torts encompass a very broad area of the law, including intentional acts, negligence, and strict liability (Oleck, 1982).

Tort law is defined at the state level by statutes, court decisions, and constitutional provisions; it applies to government entities, individual citizens, and businesses. The law of torts protects individual and business interests from harm and provides a means for those harmed by another to seek compensation for their loss. Tort liability claims also provide a basis for distributing losses to those who are responsible for the harm. Tort law thus provides a systematic means for analyzing and resolving liability claims, while protecting both the interests of the person injured and the governmental jurisdiction. Torts encompass a very broad area of the law, including (Oleck, 1982):

- Intentional acts
- Negligence
- Strict liability

Tort liability includes *intentional acts* that harm others. Intentional torts include trespass, assault and battery, intentional infliction of emotional distress, defamation, and invasion of privacy.

A second type of liability includes *strict liability*. It is known as liability without fault and relates to situations where one is held responsible for the consequences of his or her actions or omissions, regardless of fault or exercise of due care. Strict liability was first applied in cases involving abnormally dangerous activities such as blasting, but it has achieved significantly broader application in the law of products' liability and workers' compensation.

The third form of liability is the major focus of this section. This includes claims of negligence, which are unintentional acts or omissions that cause harm to another. Negligence involves an unintentional but wrongful action or inaction by one person that harms another.

11.4 NEGLIGENCE: DETERMINE IF LIABILITY EXISTS

The legal standard in negligence cases has often been stated as simply that a person has a duty to exercise that degree of care, skill, and diligence that a reasonable or prudent person would exercise under similar circumstances. This rule, as applied to governmental entities, must be understood in terms of the essential elements of negligence. These elements include:

- *Duty:* the existence of a duty to conform to a defined standard of care that is owed to a particular party, either established by statute, defined by common law (based upon judicial decisions), or established by policy by the governmental entity
- *Breach:* a failure to conform to that standard of care, or a failure to carry out the duty
- *Damage:* actual loss or damage to the injured party(ies)
- *Causation:* a causal connection between the act of the governmental employee, official, or agency body and injury to a third party(ies)

All negligence cases have these elements in common, and absence of proof of any one element will defeat a finding of liability (American Law Institute, 1965).

11.4.1 Duty

A duty of care may be imposed by common law or by statute or may be created by the voluntary assumption of a responsibility. The common law generally imposes a duty of care where a person is required to recognize that his or her conduct involves a risk of harm to another. The common law duty requires a person to use a reasonable degree of attention, perception, memory, knowledge, intelligence, and judgment in his or her actions. This is often referred to as due care.

The common law duty of care arises in almost every aspect of a public official's duties and responsibilities. Employees have a common law duty to repair or place off limits premises and equipment that are in a dangerous condition. Public safety employees have a duty to protect persons in their care and custody, including protective custody. Similarly, a public works department, when informed of a road hazard, has a duty under common law to repair defects.

This duty or standard of conduct may also be imposed and defined by statute rather than just by common law. Statutory duties include traffic codes, motor vehicle maintenance codes, workplace safety requirements, park construction and maintenance standards, environmental regulations, and inspection requirements. Local emergency management agencies may have a statutory requirement (duty) to develop a local emergency response plan. These legal requirements may or may not specifically mention liability but simply establish a legal duty. The Wyoming Emergency Management Act provides an example of a state statutory duty to plan.[1] This act requires the preparation of a local disaster and emergency management plan. Other states also require local jurisdictions to develop a local emergency response plan.

11.4.2 Breach of Duty

Liability based on negligence means that there has been a failure to act according to a standard of care. It is the failure to conform to prevailing professional practice or to act as a reasonable or prudent person under the same circumstances. The standard of care is critical to the determination of a breach of duty and a finding of negligence. Although the standard may be established by the governmental actor's profession, the courts allow the injured party an opportunity to show that the actions of the governmental official or organization were unreasonable. Since a determination of breach of duty may be subjective, the courts may ultimately have to determine if the actions of the government actor were reasonable.

In order to determine if a duty exists, the courts may review previous court decisions in a state or from other states to clarify the scope of the law. However, emergency managers should make comparisons to any given case with caution. First, decisions in negligence cases are almost always dependent on findings of fact in the case, the particular conduct of the alleged wrongdoer, and whether the acts of the wrongdoers are the cause in fact of the injury suffered by the complaining party. Factual differences between cases, no matter how slight, may lead to different conclusions regarding liability. Second, comparisons to cases in other states may be based on statutory language or common law that differs from state to state.

A municipality has a duty to repair a known defect in roads, bridges, or walkways in a reasonable period of time. Examples of emergency management duties could unclude any of the following:

1. Warning residents of a foreseeable flooding threat
2. Supervising volunteer emergency workers serving a public agency[2]
3. Training employees or volunteers in shelter operations and procedures
4. Responding to citizen notification of downed trees or utility lines
5. Investigating complaints by citizens
6. Responding to calls for assistance when public agencies state they will provide assistance

Liability may hinge on whether employees of the organization were informed that a hazard existed and that the hazard posed a significant threat to others.

---

[1] Wyo. Stat. Section 19-13-10.

[2] For an example of a state statutory immunity provision for response to a hazardous materials incident see Wis. Stat. Ann. Section 895.48(2) (1993), which creates a "good Samaritan" rule granting immunity from civil liability to persons who provide assistance or advice in handling problems in emergency or potential emergency situations relating to the threatened or actual discharge of hazardous substances. The immunity extends to mitigation of effects, prevention, cleanup, and disposition. The Act excludes persons who cause and are liable for the discharge or whose actions are grossly negligent, reckless, or intentional as well as those who are compensated for their efforts from the grants of immunity.

11.4.3 Loss

To recover in a tort claim, the injured party must demonstrate that he or she has suffered some loss as well as the amount and severity of that loss. This loss could be bodily harm such as a broken bone or strained muscle, property damage requiring automobile or home repairs, or a demonstrated loss of business. The claimant can show any single type of loss or a combination of any of the three. The key is that the claimant's loss be not hypothetical, but real.

11.4.4 Causal Connection

The final element of negligence requires proof that the claimant's injury, damage, or loss is related to the acts or the inaction of the government agency or agent (defendant). For example, a citizen injured when he or she slipped on a wet floor in a public building would attempt to show that personal injuries resulted when staff who knew of a wet floor in a public area failed to act. This means that an employee failed to warn of a wet floor or to repair the cause of the wet floor. The claimant must demonstrate that the injuries resulting from the slip and fall on a wet floor were caused by the actions of the government agency and not the careless actions of the claimant.

This causal connection is critical since a fall could result even on a dry floor from a person's lack of attention. A citizen could have been carrying a large package, lost her balance, and slipped. For a finding of negligence against a public agency, the claimant must show that there was connection between the actions or inaction of the staff of a public entity and the injury. The key is the failure of an employee or volunteer to carry out his or her duty to warn or clean up a wet spot on the floor, leading to the claimant's fall.

11.4.5 Case Example

An auto accident involving a public employee in a government vehicle and a citizen in a nonemergency situation provides a clear illustration of the principles of negligence.

- *Presence of a duty:* A public employee, like any driver, has a duty to drive in a safe manner even in an emergency. When entering an intersection, the public safety vehicle has an obligation to ensure that other vehicles are yielding to the emergency vehicle.
- *Failure to carry out duty:* If public employees or volunteers drive a vehicle in a reckless manner or exceed the posted speed limit, they have failed to carry out their duty.
- *Injury:* If a collision occurs between the public employee and a citizen resulting in personal injuries to the claimant and their car, then loss has occurred.
- *Link or causal relationship between the injury and acts of the public employee:* The claimant must show that their loss resulted from the careless driving of the public employee rather than his or her own lack of attention or skill.

For a finding of negligence to occur, each of these four elements must be present. If any of the four elements is not present, negligence cannot be found present.

In *Hamilton v. City of Shreveport*,[3] The city was liable for flooding homes since the city lowered a floodgate without notice in order to raise the water level of the lake. The city of Shreveport was not acting in an emergency and had adequate time to consider alternative

[3] Louisiana Court of Appeals, 180 50.2d 30

strategies to reduce potential flooding that might not lead to flooding of homes near the river. The Louisiana Court of Appeals determined that the city had a duty to warn citizens of the action it was to take.

11.5 HOW COULD LIABILITY ARISE?

11.5.1 Planning

In many states, local agencies are required by law to prepare an emergency response plan. A failure to prepare a plan could leave a community open to claims. Emergency plans that are out of date, have not been based on realistic assumptions, inflate capabilities, do not involve all operations areas and functions of governmental entities, or are not fully developed may form the basis for a claim. It may be more difficult to recover from a local jurisdiction in a suit based on a failure to prepare, maintain, or exercise a good plan. The court would have to determine if the local community fulfilled its duty when the plan required by law was determined inadequate in a disaster.

Plans are not static documents. They require change. A regular review process should be established and updates prepared, tested, and communicated to other local units. There should be an updating procedure to ensure that the new or revised parts of the plan are disseminated to all involved in emergency management. Once a plan is developed, it should be followed. Freelancing or performing procedures that may be opposite or contrary to the planning document without just cause could result in harm to the community. Having a realistic, effective, and current plan is one of the best ways for a community to reduce its liability exposure.

11.5.2 Training

Employees and volunteers must be trained to perform their jobs and fulfill their responsibilities in a safe manner. Training requirements may be required by state law or imposed by local regulation. These regulations may cover anyone who is expected to participate in emergency response activities. In addition, steps should be taken to ensure that the training participants understand and can function with the training that they have received.

11.6 TYPES OF IMMUNITY FROM LIABILITY

Three types of immunity exist for governmental agencies and officials:

1. Discretionary immunity
2. Governmental or proprietary immunity for specific functions of a jurisdiction
3. Statutory immunity as provided by state law

11.6.1 Discretionary Immunity

Public officials in each state enjoy immunity while exercising discretionary judgments including policy-making decisions. Discretionary immunity evolves from the judgmental decision-making process of public officials and employees. The intent of discretionary immunity is to free the public official from fear of tort liability if that judgment results in harm to another. The discretionary immunity provisions in each state evolved from a concern that decision makers would not make effective decisions for fear of liability. This form of liability is not intended to exempt public officials from liability but to protect them when making public policies.

Public policy decisions could involve the determination to evacuate a community or put shelter in place. These actions establish a public policy and are discretionary in nature. They are not operational in nature and do not state how citizens should evacuate or the specific steps one should take to shelter in place. Examples of discretionary immunity include:

- The formation of basic governmental policy, programs, or objectives
- The act of public officials in setting priorities rather than actions that implement a decision made by others
- The act requires the exercise of basic policy evaluation, judgment, and expertise on the part of a government official
- The governmental agency involved possesses the requisite constitutional, statutory, or other lawful authority to carry out the challenged act[4]

The above actions are general concepts to provide a basis for identifying what is a discretionary action and thus immune from liability. The emergency manager should review the specific definition of discretionary acts under his or her state law to ensure that the action is covered. Discretionary immunity protects the public official, employee, or volunteer, but also the organization. Where discretionary immunity blocks a claim against the governmental actor, it also prohibits the action against the agency. Officials who understand the scope and extent of their authority and exercise rational basis for decisions are protected from claims for exercising public policy decisions.

11.6.2 Governmental Immunity

The law in 14 states recognizes a special form of immunity for some activities of public agencies. It distinguishes between governmental functions, which are traditionally performed by the government, and those functions that are proprietary in nature or performed traditionally by the private sector. Under the governmental function theory, core governmental functions, such as public safety, firefighting, police activities, health and building inspections, and the collection of taxes, are mandated responsibilities that can be performed only by governmental units. Because of the unique role that these essential functions have in the community, public agencies and employees enjoy immunity from claims of negligence under state law. In those states that recognize this form of immunity, governmental units and employees enjoy immunity or limited liability from tort liability even though the governmental actor may have been negligent (Reynolds, 1982). For states that do not recognize a distinction between governmental and proprietary immunity, this form of immunity is not available.

Proprietary functions, however, have no special immunity attached to the activity. Proprietary activities may be performed by either a public or private organization. Examples of proprietary activities include transit systems, parking garages, medical care, recreation services, and sewer and garbage collection. For these proprietary activities, public agencies and employees have no special immunity and are as accountable for their actions as a private employer.

[4] *Evangelical United Brethren Church v. State,* 407 P.2d 440 (Wash. 1965).

11.6.3 Statutory Immunity

Each state legislature has adopted statutory provisions recognizing immunity in specific public activities even when the actor has been negligent. Immunity from claims of negligence is recognized in selected states for child care social service workers providing protection services, teachers who report child abuse, emergency response activities in hazardous materials accidents, firefighters, and inspections or licensing activities. Statutes granting immunity from personal liability for acts associated with these designated activities are intended to create qualified immunity for those they protect. These immunity provisions extend protection to negligent acts but not to actions for gross negligence or intentional actions intended to harm another.

Many states have adopted immunity provisions in emergency management activities. These provisions are often found in the state emergency management act. A critical element of these provisions involves the defining of an emergency. If the emergency activity is not included in the definition of emergency, the immunity provision does not apply. The state immunity provisions in state emergency management acts may be broadened to include any emergency management activity or restricted to actions only during a declared emergency (Pine, 1995). In addition, state statutes may provide immunity for specific activities including providing assistance in a hazardous materials spill.[5]

Following are sample state emergency provisions:

> *"Emergency Management"* means the preparation for and carrying out of all emergency functions essential to the recovery, and restoration of the economy by supply and resupply of resources for the urgent survival and military needs caused by enemy attack, sabotage, civil disorder, or other hostile action, or by fire, flood, earthquake, or other natural causes and other technological, industrial, civil, and political events. These functions include without limitation the coordination of fire-fighting services, police services and health services, rescue, engineering, attack warning, communication, evacuation, welfare services, transportation, public utilities, and other services necessary for the civilian protection.[6]
>
> *"Immunity from Liability"* a) All activities relating to emergency management are governmental functions. The state, any political subdivision, state agencies, and except in willful misconduct, gross negligence or bad faith, any emergency management worker complying with or reasonably attempting to comply with this statute is not liable for the death of or injury to persons or for damage to property as a result of the activity or training.
>
> b) Any requirement for a license does not apply to any authorized emergency management worker in the course of performing practices as a professional, teaching, training, mechanized or during an emergency management emergency in training for an emergency or during an emergency management exercise.
>
> c) "Emergency management worker" includes any full- or part-time paid or volunteer or auxiliary employee of any state or political subdivision performing emergency management services in this state subject to the order of the state or any political subdivision.[7]

The above immunity provision appears to provide extensive protection for officials, employees, and volunteers. Actually, the immunity is limited by the *Wyoming Governmental Claims Act*, which waives immunity from suit for the operation of vehicles, health care, and law enforcement.[8] Other states provide statutory immunity for the activities of a firefighter.[9]

---

[5] See note 2 above for an example of a state statutory immunity provision for response to a hazardous materials incident.

[6] Wyoming Emergency Management Act, Wyo. Stat. Section 19-13-102 (1998).

[7] Wyoming Emergency Management.

[8] Wyo. Stat. Section 1-39-101 (1977).

[9] For an example of immunity for firefighters see La. Rev. Stat. Ann. Section 37:1735 (1992): Grants immunity from civil liability to volunteer firefighters rendering emergency or rescue services in the course of performing their duties. First aid care given to persons is excluded from immunity, as is grossly negligent or intentional conduct.

La. Rev. Stat. Ann. Section 37:1735 (1992): Any volunteer fireman who renders emergency or rescue services while in the performance of his duties at the scene of an emergency shall not be individually liable for civil damages as a result of acts or omissions. The provision includes volunteer firemen.

TABLE 11.1 State Emergency Management Acts

Alabama Emergency Management Act. Ala. Code Section 31-9-1 to 24 (1955).
Alaska Disaster Act. Alaska Stat. Section 26.23.10 to 230 (1977).
Alaska Civil Defense Act. Alaska Stat. Section 26.20.010 to 200 (1977).
Arizona Disaster Act. Ariz. Rev. Stat. Ann. Section 26-301 to 317.
Arkansas Emergency Management Act. Ark. Stat. Ann. Section 12-75-101 to 130 (1985).
California Emergency Services Act. Cal. Gov't. Code 8550 to 8668 (1970).
Colorado Civil Defense Act. Col. Rev. Stat. Section 24-33.5-901.
Connecticut Emergency Services Act. Conn. Gen. Stat. Ann. Section 28-1 et seq.
Delaware Civil Defense Act. Del. Code Ann. Tit. 20 Section 3101 to 3133.
Florida State Emergency Management Act. Fl. Stat. 252.32 to 252.61 (1983).
Georgia Emergency Management Act. Ga. Code Ann. Section 38-3-20 et seq. (1981)
Hawaii Civil Defense and Emergency Act. Hawaii Rev. Stat. Section 127-1 et seq. and 171–85.
Idaho State Disaster Preparedness Act. Idaho Code Section 46-1001 to 1018 (1975).
Indiana Civil Defense and Disaster Law. Ind. Code Section 10-4-1-1 to 27 (1975).
Iowa Disaster Services Act. Iowa Code Ann. Section 29C.1 to 16 (1976).
Kansas Emergency Preparedness Act. Kan. Stat. Ann. Section 48-904.
Kentucky Disaster and Emergency Response Act. Ky. Rev. Stat. Section 39.400 (1974).
Louisiana Civil Defense Act. La. Rev. Stat. Ann. Section 29:601 et seq.—Chapter 4 State Civil Defense Agency (1950). And The Louisiana Disaster Act of 1974. LA. Rev. Stat. Ann. 29:701–715. Hazardous Material Act. La. Rev. Stat. Ann. 32:1520.
Maine Civil Emergency and Preparedness Act. Me. Rev. Stat. Ann. Tit. 37-B Section 701 to 824 (1987).
Maryland Emergency Management and Civil Defense Act. Md. Code Ann. Article 16A Section 1 to 36.
Massachusetts Emergency Management Act. Mass. Gen. Laws Ann. Ch. 31 Section 12.
Michigan Emergency Preparedness Act. Mich. Comp. Laws Section 30.401 (1960).
Minnesota Emergency Services. Minn. Stat. Ann. Section 12.01 to 46 (1979).
Mississippi Emergency Management Act. Miss. Code Ann. Section 33-15-3.
Missouri State Emergency Management. Mo. Rev. Stat. Section 44.010
Montana Disaster and Emergency Services Act. Mont. Code Ann. 10-3-101 to 609.
Nebraska Disaster and Civil Defense Act. Neb. Rev. Stat. Section 81-829.37 to 68 (1973).
Nevada Emergency Management Act. Nev. Rev. Stat. Section 414.020 to 160 (1975).
New Hampshire Civil Defense Act. N. H. Rev. Stat. Ann. 107:1 - 25 (1973).
New Jersey Emergency Services Act. NJ Stat. Ann. Section 52:14E-1 (1972).
New Mexico Civil Emergency Preparedness Act. NM Stat. Ann. Section 12-10-1 to 10.
New York State and Local Natural and Man-Made Disaster Preparedness Act. NY Executive Law Art. 2-B Section 20. N. Y. Stat Stat. Section 9101 to 9190 (1951).
North Carolina Emergency Management Act. N.C. Gen. Stat. Section 166A-1. et seq. (1977).
North Dakota Disaster Management. N.D. 37-17.1-01 to 22 (1985).
Ohio Civil Defense Act. Ohio Rev. Code Ann. Section 5915.01 to .99.
Oklahoma Civil Defense and Emergency Resources Management Act. Okla. Stat. Ann. Tit. 63 Section 683.2 (1967).
Emergency Management and Services Act. Or. Rev. Stat. Section 401.015 to .580 (1983).
Pennsylvania Emergency Management Services Code. 35 Pa.C.S.A. Section 7101
Rhode Island Civil Defense Preparedness Act. R.I. Gen. Laws Section 30-15-1.
South Carolina Emergency Preparedness Act. S.C. Code Ann. 25-4-420 to 460.
South Dakota Emergency Disaster Services Act. S.D. Codified Laws 33-15-1 to 28.
Tennessee Emergency Management Act. Tenn. Code Ann. Section 58-2-103.
Texas Disaster Act. Tx Rev. Civ. Stat. Ann. Chapter 418.001 to 173 (1975).
Utah Disaster Response and Recovery Act. Utah Code Ann. Section 63-5a-1 to 11.
Vermont Civil Defense Act. Vt. Stat. Ann. Tit. 20 Section 1 to 29.
Commonwealth of Virginia Emergency Services and Disaster Law. Va. Code Section 44-146.13 to 29 (1973).

For examples of state statutes recognizing immunity for volunteers, see the following:

> (a) A qualified volunteer shall not be held vicariously liable for the negligence of another in connection with or as a consequence of his or her volunteer activities. (b) Nothing in this chapter shall be construed to limit the liability of a person acting outside the scope of the volunteer program. (c) Nothing in this chapter shall be construed to limit the liability of any volunteer agency. [Ark. Stat. Ann. Section 16-6-104 (1987)]
>
> A qualified volunteer shall not be liable in damages for personal injury or property damage sustained by one who is a participant in, or a recipient, consumer, or user of, the services or benefits of a volunteer by reason of any act or omission of a qualified volunteer in connection with the volunteer except as follows:
>
> **(1)** Where the qualified volunteer is covered by a policy of insurance, in which case liability for ordinary negligence is limited to the amount of the coverage provided
>
> **(2)** Where the qualified volunteer acts in bad faith or is guilty of gross negligence
>
> **(3)** Where the qualified volunteer negligently operates a motor vehicle, aircraft, boat or other powered mode of conveyance. If the actionable conduct of the qualified volunteer is covered by a policy of liability insurance, whose liability for ordinary negligence shall be limited to the amount of the coverage provided
>
> **(4)** Where the qualified volunteer negligently performs professional services extended to an individual, which the qualified volunteer is licensed under state law to perform, including but not limited to legal, medical, engineering, and accounting services. [Ark. Stat. Ann Section 16-6-105]

11.6.4 Exceptions to Immunity

The immunity provisions of almost all state statutes include a statement that provides that the acts of employees or representatives of the governmental jurisdiction will be protected from civil suit except in the case of willful misconduct, gross negligence, or intentional harm to others. Willful or wanton misconduct involves highly unreasonable conduct or actions that are an extreme departure from ordinary care. Gross negligence involves unreasonable conduct but actions that are not taken with the intent to cause injury to another. Gross negligence is more than mere thoughtlessness or inattention or a mere mistake resulting from inexperience or confusion; these actions would be viewed by the courts as negligence and covered by any statutory immunity provisions under state law (Pine, 1997).

While there may be laws in your state to protect or limit your liability from suit, you may still be required to go to court and prove that you were covered by the immunity provisions at the time that the alleged wrong occurred. The immunity provisions that may be present in your state are best viewed as a defense to be used against, rather than a prohibition of, tort suits.

11.7 DEFENSES IN LIABILITY CLAIMS

Governmental jurisdictions may avoid liability by using two major types of defenses: denial defenses and affirmative defenses. A denial directly disputes the allegation by the plaintiff that the defendant has behaved negligently. In effect, the defendant is claiming that he or she (or the governmental unit) has acted with reasonable care. When the defendant denies plaintiff's allegation of negligence, the burden of proof remains on the plaintiff to show negligence, including duty, breach of duty, legal causation, and injury. Although not obligated to do so, the defendant usually offers proof of his or her actual conduct, showing that it conformed to the required standard of care. Unless plaintiff's proof outweighs the defendant's on this issue, there is no finding of liability.

Affirmative defenses may allow the defendant to avoid liability, even where his or her conduct is negligent. These defenses include, but are not limited to:

- Absolute or statutory immunity (as noted in the above section)
- The settlement of the claim
- Filing or bringing the suit after the statute of limitations (which bars the action)
- Assumption of the risk giving rise to the injury
- The person bringing suit contributed to the injury (strict contributory negligence)

The *settlement of the claim* at any time before the dispute is formally resolved by the court terminates the action. A settlement agreement may be reached before formal hearings are begun or at any time during the court proceedings. The agreement resolves the dispute and includes conditions such as the payment of a portion of the damages requested by the plaintiff. The settlement may be reviewed by the court but bars future action on the dispute.

Statutes of limitations are legislative enactments that impose time restrictions on filing a notice of a claim or suit. Statutory periods of limitations on the bringing of actions do not ordinarily begin to run until the injured party has been harmed and has a basis for the claim. But once the time restriction concludes, the suit is barred. Since the running of the statutory period could bar plaintiff's claim, even though he was unaware of his injury, some states have held the running of the time requirement does not begin until plaintiff discovers, or with reasonable diligence, should have discovered, the injury. The statute of limitations in many states for claims against a governmental entity may vary from six months to one or two years.

Assumption of risk means that the injured party has voluntarily expressed in advance or implied that he or she was aware of the risks involved and agreed to take his or her chances of being injured. The assumption of risk may be by express agreement, implied agreement, or simple awareness (knowledge) of an obvious risk. Assumption of risk is often included in the statutory or comparative negligence provision in many states. An example of this concept could include a citizen refusing to evacuate as a hurricane approaches the community, even when local officials mandate an evacuation from the area.

Contributory negligence states that a plaintiff may not recover damages for injuries suffered as a result of the defendant's actions if the plaintiff's conduct was also negligent and contributed to his or her injuries. Under a strict standard of contributory negligence, which is applicable in some states, the plaintiff's own negligent conduct may bar recovery entirely, even though the defendant was negligent and that negligence contributed to the plaintiff's injury. An alternative to this strict standard of contributory negligence approach may be used apportioning damages in direct proportion to the relative fault of the parties. This doctrine of comparative negligence which applies in many states does not avoid liability or bar the plaintiff from recovering damages for injuries incurred if those injuries were caused, in part, by the defendant's conduct. Rather, the plaintiff's recovery is reduced in some proportion, based upon his or her fault, if that fault contributed to the injuries.

In some states a modified doctrine is used by the courts that applies a "50%" rule, or an "equal to or greater than" system. Under this modified approach, a plaintiff recovers unless his negligent conduct is equal to or greater than that of defendant, in which case the plaintiff recovers nothing.

11.8 DETERMINING LIABILITY

To clarify the scope of liability under state law, local officials, volunteers, employees, and members of local emergency planning committees should focus on the following points:

1. *Is sovereign immunity abolished, retained, or partially waived under state law?* If sovereign immunity is retained, abolished, or partially waived, state statutes and court decisions will clarify what immunity exists in specific programs or actions. Does the state have a comprehensive tort liability statute or a few provisions of the state code that describe the liability of government units? (see Table 11.2 for a list of state tort law references.)
2. *Does state law recognize discretionary immunity?* Almost all states recognize immunity for public officials exercising discretionary actions and making decisions.
3. *Does the state recognize the distinction between governmental and proprietary activities, and offer immunity for governmental functions?* States that recognize immunity for governmental functions may specifically include state emergency management programs and activities as governmental functions.
4. *What is considered a disaster under your state emergency management act?* Many state emergency management acts provide immunity for state and local jurisdictions involved in emergency management activities. The immunity provisions may apply to both natural and technological disasters if the term *disaster* is defined in the statute. Most state emergency management statutes define the term *disaster* in the introduction to the statute. Review the introductory section of your state emergency management act to see if disaster includes both technological and natural events.

11.9 TYPES OF DAMAGES AND RELIEF

The injured party seeks relief in the form of a court order (equitable relief), financial compensation, or both forms of relief. Equitable relief in a tort case is usually an injunction ordering the defendant to cease and activity or to act in a specific manner. Money damages are sought in most tort cases involving injury to a person or an individual's property and are commonly the principal form of relief. Money damages are awarded first as compensation, or restitution for the harm suffered by the injured party. Money damages are classified into two major categories: compensatory (including nominal, general, and special damages) and punitive.

- *Compensatory damages* are for the recovering party's actual out-of-pocket medical expenses; temporary or permanent disability; lost earnings (present and future); loss of consortium; and, in wrongful death cases, damages, including funeral expenses, and pain and suffering of survivors.
- *Nominal damages* are awarded where no real or substantial loss has resulted, but where the defendant has technically harmed the plaintiff. In this case the plaintiff may be awarded $1.00.
- *General damages* are those that compensate the injured party for actual losses which are the normal result of the actual injuries suffered (including pain and suffering).
- *Special damages* are those that compensate for a loss but do not normally flow from the injuries sustained in all typical cases. Such damages may include the aggravation of a preexisting condition, loss of business revenue, and other damages, which flow from special conditions of the case or the special situation of the plaintiff.
- *Punitive damages* are awarded where the defendant's conduct was considered by the court to be intentional, willful, reckless or to constitute gross negligence. Punitive damages are intended to deter future conduct of the type at issue. This form of damages is not generally available against a public entity.

TABLE 11.2 References to State and Local Governmental Tort Claims Statutes

Alabama	Local Governmental Claims Provisions. Ala. Code Section 11-47-190 (1994), Ala. Code Section 11-93-3, Ala. Code Section 26-14-9 (1965), and Ala. Code Section 41-4-301 (1990).
	Alabama State Board of Adjustment Act. Ala. Code Section 41-9-60 (1994).
Alaska	Alaska Local Government Tort Claims Act. Alaska Stat. 09.65.070 (1994).
	Alaska State Tort Claims Act. Alaska Stat. 09.50.250 (1992).
Arizona	Arizona State and Local Tort Claims Act. Ariz. Revs. Stat. Ann. Section 12.820 et seq. (1992).
Arkansas	Arkansas Local Governmental Tort Claims Act. Ark. Stat. Ann. Section 21-9-301 (1993).
	Arkansas State Claims Commission. Ark. Stat. Ann. Section 21-9-201 (1991).
California	Tort Claims Act Calif. Govt. §810 et. seq. (West 1963).
Connecticut	Local Governmental Tort Claims Act. Conn. Gen. Law 7-465. (1989).
	State Tort Claims Act. Conn. Gen. Stat. Ann. 4-141 et seq. (1995).
Delaware	The State Tort Claims Act. Del. Code Ann. Title 10 Section 4001 et seq. (1979).
District of Columbia	District of Columbia Tort Claims Act. D.C.C.E. 1-1201 et seq. (1974).
Florida	Florida Tort Claims Act. Fla. Stat. 768.28 et seq. (1996).
Georgia	Georgia local government liability provisions. Ga. Code Section 36-33-1.
	Georgia State Claims Advisory Board. Ga. Code Ann. Section 28-5-60 (1987).
Hawaii	Local government liability provisions. Hawaii Rev. Stat. Section 657 and 663 (1986).
	Hawaii State Tort Liability Act. Hawaii Rev. Stat. Section 662-1 et seq. (1991).
Idaho	Idaho Tort Claims Act. Idaho Code Section 6-901 (1971).
Illinois	Local Government and Governmental Employees Tort Immunity Act. Ill. Revs. Stat. Ch. 745 Section 10/ 1-101 et. seq. (1965).
	Illinois Court of Claims Act. Ill. Rev. Stat. Ch. 705 Section 505/ 1 etc. (1996).
Indiana	Indiana Tort Claims Act. Ind. Code Section 34-4-16.5-1 et seq. (1974).
Iowa	Iowa Tort Liability of Governmental Subdivisions Act. Iowa Code Ann. Section 670.1 to 670.13 (1989).
Kansas	Kansas Tort Claims Act. Kansas Stat. Ann. Section 75-6101 et seq. (1981).
Kentucky	Local government liability provisions. Ky. Rev. Stat. Section 65.2002, 411.100, 411.188, and 413.125.
	Kentucky Board of Claims Act. Ky. Revs. Stat. Ch. 44.110(2) (1985).
Louisiana	Louisiana Governmental Claims Act. La. Rev. Stat. Ann. 13: 5101 (1996).
Maine	Maine Tort Claims Act. Maine Rev. Stat. Ann. Title 14:8102 (1995).
Maryland	Maryland local government liability provisions. Md. Cts. & Jud. Proc. Code Article 23A Section 1A (1990), Md. Cts. & Jud. Proc. Code Ann. Section 5-300 to 401 (1996), Md. Ann. Code Article 23A Section 1B (1979).
	Maryland Tort Claims Act. Md. Code Ann. State Government Section 12-101 to 12-406 (1990).
Massachusetts	Massachusetts tort claims provisions. Mass. Gen. Laws Ann. Chapter 258 Section 2 (1993).
Michigan	Michigan tort liability claims provisions. Mich. Comp. Stat. Section 691.1400 et seq. (1991).
Minnesota	Minnesota Local Government Tort Claims Act. Minn. Stat. 466.01 et seq. (1996).
	Minnesota State Government Tort Claims Act. Minn. Stat. Ann. Section 3.732 et seq. (1996).
Mississippi	Mississippi tort claims act Miss. Code Ann. Section 11-46-1 et seq. (1993).
Missouri	Missouri tort claims provisions. Mo. Revs. Stat. 537.600 (1989), 71.207, 292.602 3(1) (1996), 71.185, 537.610 (1989).

TABLE 11.2 References to State and Local Governmental Tort Claims Statutes (*Continued*)

Montana	Montana Governmental Tort Liability Act. Mont. Code Ann. 2-9-101 to 2-9-318 (1991).
Nebraska	Nebraska Political Subdivision Tort Claims Act. Neb. Revs. Stat. Section 13-902 et seq. (1992).
	Nebraska State Tort Claims Act. Neb. Revs. Stat. 81-8,209 et seq. (1991)
Nevada	Nevada Tort Claims Act. Nev. Rev. Stat. Section 41.031 et seq. (1995).
New Hampshire	Local government tort claims provisions. N.H. Rev. Stat. Ann. Section 507-B:2 (1991), 231.92, 228.37, 97:3, 31:104, 412.3, and 507-B:4.
	State Claims Board. N.H. Revs. Stat. Ann. Section 541-B:2 (1994).
New Jersey	The New Jersey Tort Liability Act. N.J. Stat. Ann. Section 59:1-2 et. seq (1995).
New Mexico	New Mexico Tort Claims Act. N.M. Stat. Ann. Section 41-4-1 to 29.
New York	Tort claims provisions. N.Y. Court Claims Article 2 Act Section 8 (1984).
	N.Y. General Municipal Law Section 50 (1909). N.Y. General Municipal Law Section 50 et seq. (McKinney Consolidated Laws of New York Annotated)
North Carolina	Local government tort claims provisions. N.C. Gen. Stat. Section 115c-43 (1984) and N.C. Gen. Stat. Section 115C-259 (1984).
	The North Carolina Industrial Commission. N.C. Gen. Stat. Section 143-291 et seq. (1988).
North Dakota	North Dakota Local Governmental Tort Claims Act. N.D. Cent. Code Section 32-12.1-01 et seq. (1995)
	North Dakota State Tort Provisions. N.D. Cent. Code Section 32-12.2-02 et seq. (1995), N.D. Cent. Code Section 54-14-03.
Ohio	Ohio Local Government Tort Claims Act. Ohio Rev. Code Ann. Section 2744.01 et seq. (1990).
	State Claims Act.
	Ohio Rev. Code Ann. Section 2743.02 et seq. (1989).
Oklahoma	Oklahoma State Tort Claims Act. Okla. Stat. Ann. Title 51 Section 152 et seq. (1994).
Pennsylvania	Political Subdivision Tort Claims Act. Pa. Stat. Ann. 42 Section 8541 et seq. (1995).
	Pennsylvania State Tort Claims Act. Pa. Stat. Ann. Title 42 Section 8521 et seq. (Purdon 1985).
Rhode Island	Rhode Island Tort Claims Act. R.I. Gen. Laws Section 9-31-1 et seq. (1984).
South Carolina	South Carolina Tort Claims Act. S.C. Code Ann. Section 15-78-10 to 190. (1986)
South Dakota	South Dakota Tort Claims Act. S.D. Codified Laws Section 3-21-1 et seq. (1985).
Tennessee	Tennessee Local Governmental Tort Liability Act. Tenn. Code Ann. Section 29-20-101 et seq. (1984).
	Tennessee State Claims Board. Tenn. Code Ann. Section 9-8-101 (1985).
Texas	Tort Claims Texas Tort Claims Act. Title 5. Civil Practices and Remedies—Governmental Liability Chapter 101 Civil Practices and Remedies—Section 101.001 (1995).
Utah	Utah Governmental Immunity Act. Utah Code Ann. Section 63-30-1 et seq. (1966).
Vermont	Tort claims provisions. Vt. Stat. Ann. Title 19 Section 1371 (1984), 29 Section 1401 (1984) and 19 Section 1371 (1984)
	Vermont State Tort Claims Act. Vt. Stat. Ann. Title 12 Section 5601. et seq.,

TABLE 11.2 References to State and Local Governmental Tort Claims Statutes (*Continued*)

Virginia	Local governmental tort claims provisions. Va. Code Section 15.1-506 (1962), 15.1-7.01 (1987), 10-62, 44-146.36 (1989), and 8.01-224 (1984). Tort Claims against the Commonwealth of Virginia Act. Va. Code Section 8.01-195.1 et seq. (1991)
Washington	Local government tort claims provisions. Wash. Rev. Code Ann. 4.96.010 (1993). Tort claims provisions. Wash. Revs. Code Ann. Section 4.92.110.
West Virginia	Governmental Tort Claims and Insurance Reform Act. W.Va. Code 29-12A-1 to 17 (1986). West Virginia Court of Claims Act. W.Va. Code Section 14-2-4 et seq. (1991)
Wisconsin	Wisconsin Tort Claims Act (Local). Wis. Stat. Ann. Section 893.80 (3) (1995) Wisconsin State Claims Act. Wis. Rev. Stat. 775.01
Wyoming	Wyoming Governmental Claims Act. Wyo. Stat. Chapter 39 Section 1-39-101 (1977). (State and local application)

11.9.1 Limits on Damages

Many states have imposed statutory limits upon the amount of compensatory damages that may be recovered in tort cases against governmental units. These statutory caps may provide a limit on awards for individual claims and limits for each incident or occurrence. These limits may range in amount from $25,000–$50,000 per individual claim, to $100,000, or up to $1,000,000 per occurrence (Pine, 1998).

The limitation imposed by these statutory provisions may be waived in some states by the purchase of insurance. This exception to the statutory cap, which is recognized in some states, is seen as a voluntary waiver of immunity to the extent of the coverage provided by the insurance policy in many states.

11.10 INDEMNIFICATION OF EMPLOYEES AND VOLUNTEERS—WHO PAYS?

Official representatives of a governmental unit who are named individually in a tort action are generally entitled to protection against personal financial loss or indemnification, with regard to both attorney's fees and judgments that might be awarded against them. In some states, the suit must be filed against the public entity itself, and not the employee or the government agency representative. Almost all states recognize that the governmental unit is liable for the negligent acts or omissions of its agents or employees who are acting within the scope of their duties as public employees. The employee in this context includes not only paid staff, but also volunteers. Elected officials who receive no pay and volunteers would thus be included in this definition of employee. The liability for the employee's actions is passed on to the governmental unit as employer, under a theory generally known as "vicarious liability." In many states the vicarious liability of the governmental employer is specified by statute.

The governmental entity may not be liable for an employee's actions if the employee acted outside the scope of his or her duties or with an intent (malice) to harm another, or if the actions were made with reckless disregard for the rights of others. Thus, damages must be paid by the individual. Most state indemnification statutes provide that where the employee acted with malice or the employee's actions were outside the scope of the job, the employer will not defend the employee and will not pay a judgment rendered against the employee. Under these circumstances, the individual government employee is liable.

11.11 FEDERAL PLANNING REQUIREMENTS

Threats posed by hazardous materials have been increasingly visible since the 1970s. Hazardous materials incidents can be deadly or cause extensive injury and property damage and can occur in any community. Federal legislation adopted in 1986 requires state and local jurisdictions to enhance hazardous materials planning. The Emergency Planning Community Right to Know Act (EPCRA) mandates the creation of local emergency planning committees. The Act also requires the local emergency planning committees to prepare an emergency plan for possible releases of hazardous substances, and make information on toxic chemicals in the community available to the public.[10] EPCRA is the primary vehicle for facilitating emergency planning, preventing accidents, and preparing facilities and the surrounding community to respond to a release of hazardous material.

EPCRA requires states to create a state emergency response commission (SERC). The SERC designates local emergency planning districts and committees to be in charge of developing an emergency response plan for their communities. The SERC supervises the activities of the local communities and reviews local plans. As a part of the planning process, local districts and communities identify available resources that can be called on to respond to hazardous materials emergencies. Local communities maintain information on hazardous materials transported through, stored in, or produced in the community. Local planning committees are required to review the plan annually, conduct exercises of the plan, and make recommendations to the state commission, local government agencies, and facilities with respect to resources that may be required to implement the plan effectively. Facilities that produce, store, or use extremely hazardous substances are required to cooperate in the planning and information-gathering process.

EPCRA provisions require the local committee to make specified information available to the general public during normal working hours. The information includes the local emergency response plan, documents indicating what hazardous materials exist in the community, and forms describing the nature of toxic chemical releases. Civil penalties exist for any person or business other than a government entity that violates the reporting requirement of this act. A fine up to $25,000 may be assessed by a federal district court for a failure to report toxic and hazardous chemicals. An additional fine of up to $10,000 may be assessed against any person other than a governmental entity who fails to submit materials safety data sheets or provide information to health professionals (sections 11021 and 11-43[b]).

A citizen may file a civil action in federal district court under EPCRA against the administrator of the U.S. Environmental Protection Agency (EPA), a state governor, or the state emergency response commission for failure to respond to a request for information within 120 days after the date of receipt of the request (section 11046). A citizen may also file a claim in federal district court and possibly under state law against a state emergency response commission for failure to fulfill its obligations under this statute. The statute, however, authorizes only injunctive relief against a state commission; it does not authorize money damages or penalties. The court may thus issue an injunction, or court order, requiring the commission to comply with the statute. EPCRA does not create a federal cause of action for citizens who wish money damages from the commission, individual members of commissions or communities, or the governor.

[10] 42 U.S.C. Ann. Section 1103.

EPCRA does not create a federal cause of action, or legal basis for a suit, for citizens who wish money damages from the state commission, individual members of commissions or committees, or the governor. Unless there is a cause of action or statutory basis for a suit under state law, the claim against the commission or local government will be dismissed (section 11046[f]). The EPCRA therefore does not impose liability on local governmental units or officials for a failure to plan effectively for hazardous materials emergencies. Although federal law imposes emergency planning and recordkeeping requirements on state and local government officials, the basis of liability remains an issue of state law.

In addition to EPCRA, the Occupational Safety and Health Administration (OSHA) and EPA require each employer, whether governmental or private, to develop specific hazardous materials response plans. OSHA Regulations 29 CFR 1010.120(q)(1) require emergency response agencies to develop an emergency response plan (ERP). These plans must include hazardous materials response standard operating procedures (SOP). An SOP is a series of guidelines that describe how a hazardous materials incident will be managed by the agencies involved. To be effective, a hazardous materials SOP must be integrated into the operational system of the local emergency response committee (LEPC) plan. The LEPC plan must be available to the public, emergency response agencies, the medical community, and the media. This availability is provided in EPRCA. If an employer or emergency response organization fails to comply with the administrative requirements of these federal laws and regulations, OSHA and EPA may levy fines or penalties against the governmental unit (Pine, 1989). These federal requirements impose obligations on the employer and do not provide for civil remedies against first responders or public officials.

11.12 ROLE OF LEGAL COUNSEL

Local governments retain the services of a person who has a clear understanding of the law of torts under state lay. Whether the person is a full-time employee or paid under a contract, the general counsel is a key person to know. This individual knows the state and local laws that apply to local government and its employees. In order to provide the best possible defense, employees and volunteers of a governmental body need to explain what they do, how they operate, and the scope of their intended actions. The legal counsel thus needs to be informed of operational procedures and policies. A communication link must be established between the emergency management staff and legal counsel. This link is one of the best resources for dealing with liability issues.

11.13 CONCLUSIONS

Statutes granting immunity from personal liability are intended to create qualified immunity for those they protect. The law extends protection for ordinary negligent acts but not personal liability for gross negligence or intentional torts. Since these immunity provisions are state statutes, they apply only to state tort actions and have no effect on federal actions against governmental units or officials when a federal claim is made in a federal court or filed in state district court. Government officials should understand that there are limits to the immunity provided for public agencies and officials under state law.

The legal counsel for the local government may be one of the most important persons to consult to understand what liability issues face local officials and what steps can be taken to reduce exposure to lawsuits. Regular contact with legal counsel may help to minimize exposure to suits. Legal counsel can gain a better perspective on the complex nature of the emergency management process of the local jurisdiction.

The governing body is generally liable for providing legal counsel to its employees and representatives who are serving in their official capacity and representing the government agency as an elected official or a volunteer. These provisions may not apply where there were willful or intentional actions that went beyond the scope of their job descriptions and duties. Employees, officials, or representatives of public organizations may be held liable for conduct that is reckless, unjustifiable, or intentional or actions that are well beyond what the reasonable and prudent person would have done under the circumstances.

11.14 KEY TERMS

Case law: The collected decisions of courts

Common law: The body of law that is formulated from case law, as opposed to statutes generated by legislatures.

Comparative negligence: Negligence is measured in terms of percentage. Any damages allowed shall be diminished in proportion to the amount of negligence attributable to the person for whose injury, damage, or death recovery is sought. Many states are replacing statutes of contributory negligence with comparative negligence.

Contributory negligence: A legal doctrine accepted in some states which provides that if a plaintiff's own negligence contributed, even in just a small way, to the plaintiff's damages, the plaintiff will not be able to recover against a negligent defendant.

Defendant: The party sued by the plaintiff.

Immunity: Freedom from liability.

Intentional tort: A tort committed on purpose, such as an assault.

Joint and several liability: A type of liability in which more than one person is liable for damages, but in the event one party cannot pay, the other party is liable for the entire judgment.

Judgment: The court's final decision.

Liability: The obligation of one person to another under the law.

Plaintiff: The person who commences the lawsuit.

Malfeasance: Improperly performing an act that in itself is lawful and proper.

Sovereign immunity: A doctrine of law, existing in various forms in various jurisdictions, which holds that governments are immune from suits unless they agree to be sued.

Statute: A law passed by the legislature.

11.15 REFERENCES

American Law Institute. 1965. *Restatement of the Law: Torts,* 2d ed., St. Paul, MN.

Pine, J. C., 1989. *Tort Liability in Emergency Planning,* Technical Assistance Bulletin 7, Chemical Emergency Preparedness and Prevention, U.S. Environmental Protection Agency, Washington, DC.

Pine, J. C. 1995. "Liability Issues in Emergency Management," *Safety Science,* March.

Pine, J. C. 1995. "Liability Issues in Natural Disasters," *International Journal of Mass Emergencies,* Vol. 15, No. 3.

Pine, J. C., 1998. *Tort Liability Today: A Guide for State and Local Governments,* 5th ed., Public Risk Management Association, Washington, DC.

Occupational Safety and Health Administration (OSHA). 1970. Emergency Responder Planning and Training Regulations, 29 CFR 1010.120(q)(1).

Oleck, H. L. 1982. *Tort Law Practice Manual,* Prentice-Hall, Englewood Cliffs, NJ.

Reynolds, O. M. 1982. *Handbook on Local Government Law,* 2d ed., West, St. Paul, MN.

Wyoming Emergency Management Act, Local Planning Requirement. Wyo. Stat. Section 19-13-102 (1998).

CHAPTER 12
RISK-BASED CONCEPT FOR THE SAFETY ASSESSMENT OF HANDLING OF AMMUNITION AND EXPLOSIVES BY SWISS FORCES AND ADMINISTRATION

Andreas F. Bienz
Bienz, Kummer & Partner Ltd,
Consulting Engineers—Safety Planning, Risk Management,
Zollikerberg, Switzerland

12.1 THE REASONS

In the late 1940s, four catastrophic accidents occured in ammunition storage magazines in Switzerland, killing several people and causing more than CHF 100 million (Swiss francs) of damage in the surroundings and to the installations. In addition, roughly 10,000 tons of ammunition were lost. As a consequence, the government established an Ammunition Storage Board uniting military and civilian logistic leaders and safety experts. After the investigation of the accidents, one of the first tasks of this board was to review and tighten the safety regulations. In essence, these regulations were deterministic and relied mainly on measures: they followed the traditional and widely accepted principles of safety distances to inhabited buildings and of various hazard categories for different types of ammunition. As another consequence, the construction and manufacture of the ammunition were improved, less dangerous materials were used, and the quality of the stored ammunition was continually controlled.

However, after only a few years, these regulations proved to be too inflexible to respond properly to the new problems which evolved. The amount of ammunition to be stored and its explosive content steadily increased. Military readiness requirements called for additional storage space closer to populated areas. At the same time, a great number of residential, public, and industrial buildings, leisure installations, and roads were built closer and closer to the existing storage installations. And finally, the financial funds were limited as always and everywhere.

The Ammunition Storage Board did not want to bury its head in the sand and increase the number of waivers. It therefore decided to investigate the problem thoroughly and in all its dimensions in order to find the correct answers to the actual questions and, based on

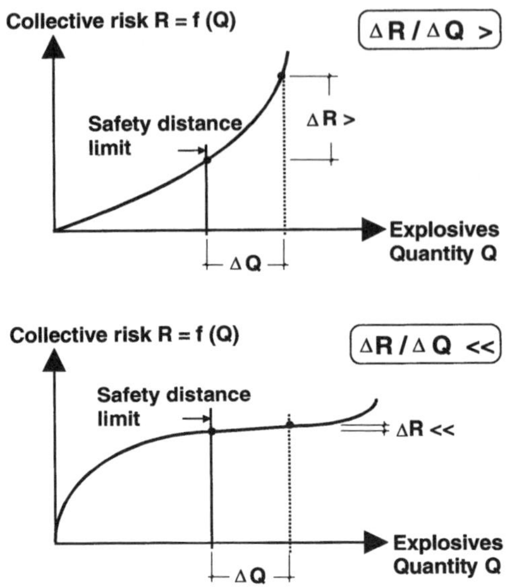

FIGURE 12.1 Favorable and unfavorable risk/filling relationships of two different magazines.

them, work out new regulations that would really help. The general direction was shown by the first experimental risk analyses, performed in 1970 for a few existing underground magazines, the capacity of which was actually limited by the former safety regulations and not for reasons of space. With the risk/filling relationship calculated individually for every magazine, taking into account the actual human activities in the surroundings, it was shown distinctly which magazines could be filled beyond the limit given by the regulation without a bad conscience and which could not (Fig. 12.1).

12.2 THE COURSE OF ACTION

For the program in mind, the Ammunition Storage Board looked for external expertise. It founded a subcommittee for conceptual decisions, study groups with military and civilian officials as well as experts, and obtained the necessary funds. In essence, the resulting organization still exists today.

The problem was tackled along the whole front, the questions arising being numerous and contrary. There were the military requirements such as readiness and protection against enemy weapon effects, the cost, and, of course, the safety aspects. Obviously all these aspects should have been met at the same time, and solutions could be found not by setting absolute requirements but instead by weighing up the three aspects and uniting them in an optimum solution.

Concerning safety, there was no better way than to introduce the quantitative risk concept and risk/cost criteria. This was a completely new method that was possible because the civilian laws and guidelines did not focus on the safety of ammunition and explosives in the military area at that time. The Swiss Explosives Law [1] released (and still releases) the military from its substantive contents but commits the government to issuing its own regulations.

Many methodological, technical, and political questions arose, such as risk calculation procedures or data on spreading of hazardous effects, effects on humans, probabilities of explosions, representative explosive quantities, etc., and safety criteria. Much technical data came out of literature studies. Other data were gained through model or even full-scale tests. The safety criteria were discussed thoroughly and issued by the responsible authorities.

Of course, not all questions were anticipated in the very beginning and, due to the limited funds, capacity and capability, they had to be answered one after the other and little by little. The priorities were given by the practical needs: new magazines had to be built, i.e. they had to be designed according to the requirements and sited safely.

The basic work on methodology and data and the conceptual decisions were applied immediately to actual examples. Thus, the storage regulations [2], which were written and established rather quickly (1975), had already been successfully tested in practice when they came into force.

Initiated by a project of a new assembly plant in one of the ammunition factories of the Department of Defence, risk analysis was also applied to ammunition and explosives handling in factories. This activity also lacked adequate regulations and criteria that would have allowed safe and economical installations and operations. A study group was founded in the mid-1970s, and a concept study for safety assessment in ammunition factories was worked out [3]. A risk analysis of all the working places of one of the factories underlined the feasibility of the quantitative approach in this field of activities [4]. The findings of this study were applied to the assembly plant project. The missing technical data for constructions and risk analysis were acquired through model tests.

A couple of years later, the concept study was adapted to technical regulations [5], and the safety criteria were harmonized with those of the storage activity. On the upper level the risk based concept (RSCAE—Risk based Safety assessment Concept for Ammunition and Explosives) was enacted in a directive of the Chief of General Staff that covered the whole range of handling of ammunition and explosives in the Department of Defence [6].

Today, the RSCAE has been carried out in storage, manufacture/demilitarization, and testing/passing as well as in special problems. Because of financial priorities, it has not been introduced in transport and in the armed forces, where safety is dependent on a very dense net of detailed deterministic regulations. At the moment, a study of the transport is ongoing and should show how to proceed in carrying out the RSCAE.

In addition, models, data, and criteria in the existing technical regulations have to be revised and adjusted periodically according to the advancing state of knowledge.

12.3 THE METHODOLOGICAL CONCEPT

12.3.1 The Three Main Points of View

As many discussions show, the term of safety is not always understood definitely and clearly. Current deterministic safety concepts leave open which safety they are good for. But three important points of view can be distinguished and have to be considered according to the Swiss RSCAE (Fig. 12.2):

- The first viewpoint is that of the endangered **individual.** His primary interest is on his own hazard, which he judges based upon his own standard, regardless of how many other people are endangered and how.
- The second viewpoint is that of the anonymous **society** at large, which is first of all interested in the total extent of a hazard, e.g., as it appears in accident statistics.
- Apart from this, the **people responsible** for the dangerous activity are interested in limiting hazards in such a way that public opinion does not question the specific dangerous activity.

FIGURE 12.2 Main viewpoints on safety.

The Department of Defence feels generally more exposed to the public response and is much more afraid of catastrophic accidents that cause a great deal of discussion than of smaller but more frequent accidents creating less stir.

12.3.2 Dichotomy of the Safety Assessment

The RSCAE distinguishes distinctly between the objective, factual part of the safety question and the subjective part, which is determined by social values (Fig. 12.3). Very often, conventional deterministic concepts mix these aspects and cause numerous misunderstandings. Corresponding to the structure of the safety question, the risk analysis and the risk appraisal are separate tasks.

12.3.3 Quantification of the Hazard

A very important element is the quantitative description of the hazard using the terms of risk. Only if hazards are expressed quantitatively can they be compared with other hazards of the same or other activities, and only in this way can the benefit of safety measures be shown reasonably.

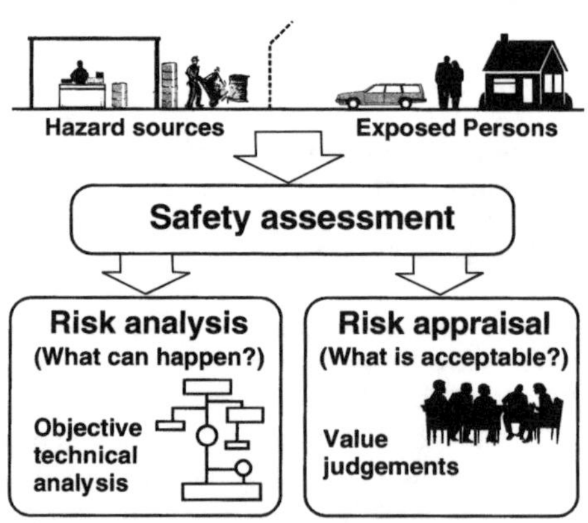

FIGURE 12.3 Dichotomy of safety assessment.

Simply said, the risk is given by the product of the probability or frequency and the expected damages or consequences of the event concerned (Fig. 12.4). The hazards of explosions would usually be described with sufficient accuracy as a fatal risk.

Corresponding to the above-mentioned three aspects of the safety problem, individual risk, real collective risk, and perceived collective risk are identified. The real collective risk of an event—a statistical expected value—is given by the sum of all individual risks caused by the event. The perceived collective risk is the real collective risk increased by an aversion function that takes into account that the reaction of the public is much more violent to rare events with large consequences than to more frequent events with fewer consequences per accident.

12.3.4 Methodical Procedure of the Risk Analysis

The aim of a risk analysis is to calculate these risks. The risk analysis is a systematic procedure of four steps (Fig. 12.5):

1. *Event analysis:* Possible events are identified and described concerning location, type of reaction, probability, and size (quantity of explosives).
2. *Effect analysis:* The dangerous effects of the possible events to persons in the surroundings such as fragments, debris, airblast, etc. are determined.
3. *Exposure analysis:* Places, protection, and history of possibly exposed persons in the hazardous areas are investigated.
4. *Risk calculation:* The parameters deduced from the previous steps are connected mathematically.

12.3.5 Levels of Risk Appraisal

While the risk analysis aims at identifying the characteristics of a particular case in question, risk appraisal has to be seen from a wider perspective (Fig. 12.6). Methodologically, it is

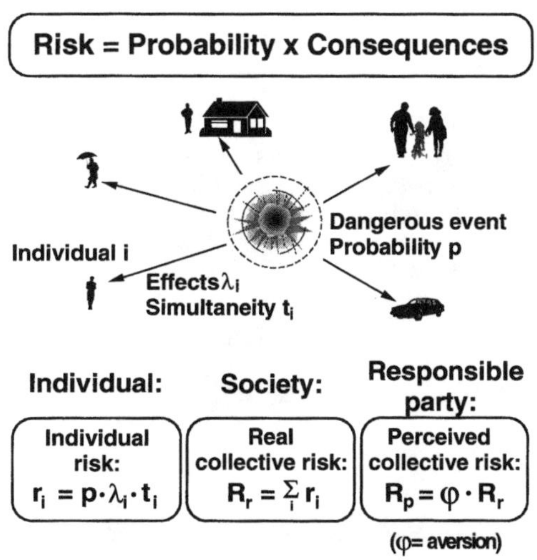

FIGURE 12.4 Levels of risk appraisal.

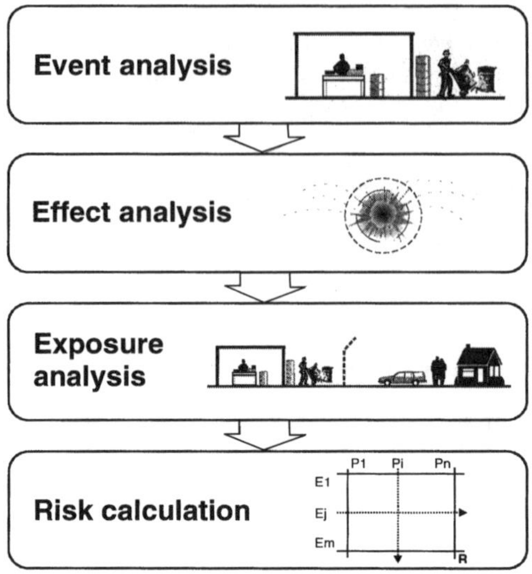

FIGURE 12.5 Steps of risk analysis procedure.

necessary to distinguish between the establishment of long-term safety criteria by the responsible persons and the proof of safety in practical cases when the safety analyst has to prove that the remaining risks do not exceed the limits.

12.3.6 Differentiation of the Risk Bearers

The RSCAE considers that the acceptance of risks depends on the relationship of the exposed person to the hazardous activity and to what extent the person is able to influence his risk (Fig. 12.7).

For example, risks voluntarily taken are considered acceptable on a much higher level than those risks unintentionally run. The difference may be about a factor of 1,000.

The Swiss safety criteria are based on a simplified model that distinguishes four categories of risks: (1) voluntary, (2) high self-determination, (3) low self-determination, and (4) involuntary. Two of them (3 and 4) are relevant in the field of safety of explosives and ammunition. The risk of third persons in the surroundings of a storage or a factory belongs to category 4. The risks of people earning their money by working with dangerous goods with a low ability to influence and a low degree of self-determination, but a certain perceived benefit (directly and indirectly involved persons), are assigned to 3.

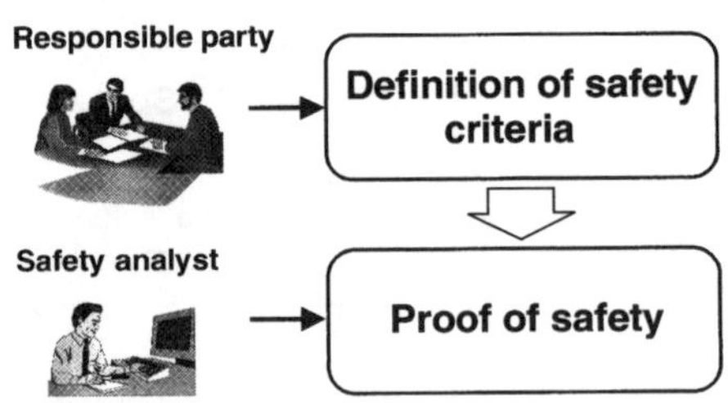

FIGURE 12.6 Levels of risk appraisal.

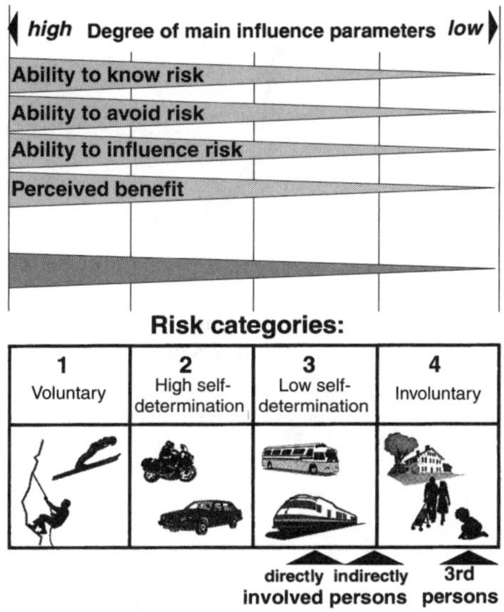

FIGURE 12.7 Differentiation of risk bearers.

12.3.7 Types of Safety Criteria

The main aspect under which individual risk has to be appraised is equity: nobody should bear a higher risk than any other person in the same situation. So the adequate safety criteria for individual risks are upper limiting values (Fig. 12.8).

In the case of the collective risk, however, upper limits are no longer reasonably applicable. It can be shown that upper limits would prevent from producing the minimum risk for a given investment for safety measures. In addition, there is no common basis for risks from different sources of hazard. Being aware that at the very end safety is a function of financial means, it seems plausible to rely on how much society is (and should be) willing to pay for the safety of its members. There is no question that financial means for safety measures are definitely limited, and everybody agrees that the resources available should be spent so as to achieve the maximum safety (or the minimum risk) overall. Thus, limiting collective risks is a typical optimization problem. When appraising the collective risk of an activity, the investment for safety has to be related to the achieved risk reduction, as it is shown in the risk/cost diagram in Fig. 12.9. One has to go on with safety measures until a

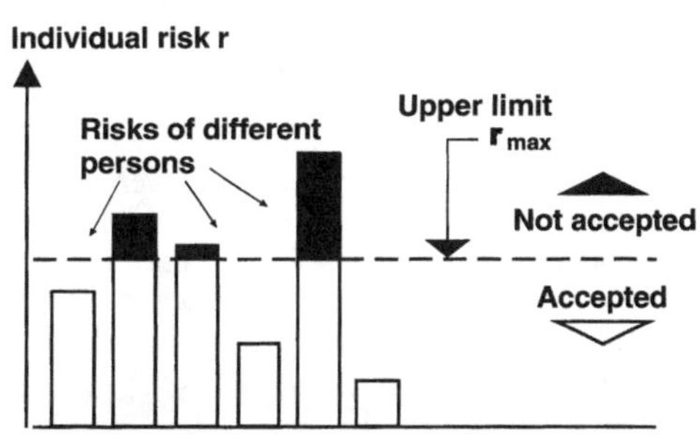

FIGURE 12.8 Limiting individual risks by upper limiting values.

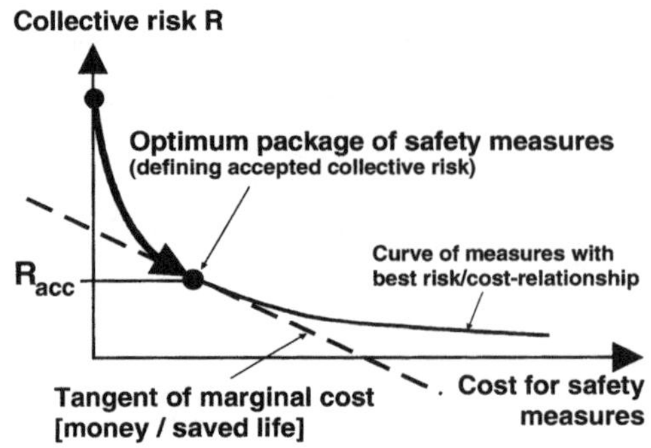

FIGURE 12.9 Limiting collective risks in the specific case by marginal cost for safety measures.

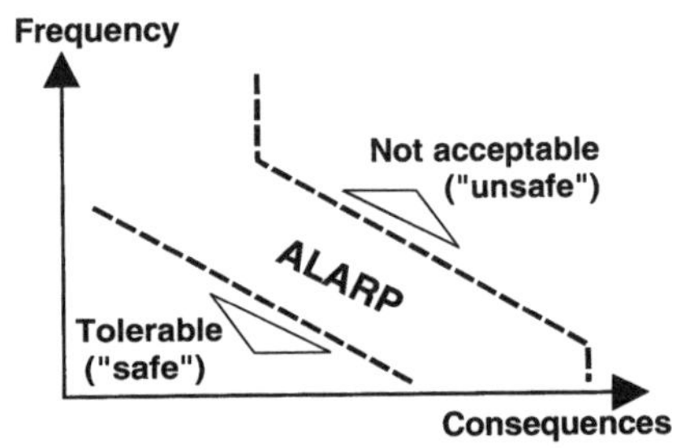

FIGURE 12.10 Frequency/consequence safety criteria and ALARP principle for the collective risk.

certain risk/cost ratio is attained that has to be respected by everyone. So the basic principle in limiting the collective risk is the willingness-to-pay-approach, and the quantitative safety criterion limiting the collective risk is actually a marginal-cost value.

It has to be emphasized that the willingness-to-pay principle differs completely from attempts to use the monetary value of a human life or even life years as a safety criterion. There will never be a reasonable and ethically indisputable answer to that question. Ten different people who would be willing to quantify the values of the lives of 10 different persons might give 100 different answers. But even people who would refuse to answer this question for ethical reasons will find it necessary to spend money to prevent people from becoming victims. But, with regard to the value and object of the dangerous activity as well as the limited funds, they would not spend an infinite amount and would want to realize the minimum risk with the money spent.

The marginal-cost approach is also totally different from the attempts to limit the collective risk (expressed in a frequency/consequence value couple) in a frequency/consequence diagram that gives the quantitative safety criteria by two lines enclosing an ALARP (As Low As Reasonably Practicable) area, separating "safe" from "unsafe" (Fig. 12.10). These lines marking off what is obviously considered tolerable and obviously not acceptable usually lie rather far apart. However, the actual cases usually lie in between (i.e., in the ALARP area), still leaving the safety analyst or the responsible authority without clear instructions for deciding on "safe" or "unsafe." On the other hand, the marginal-cost principle really helps, independent of the level of the system (i.e., factory or pressing machine), quantifying the notions "approriate" and "reasonable," which the judge would apply after an accident,

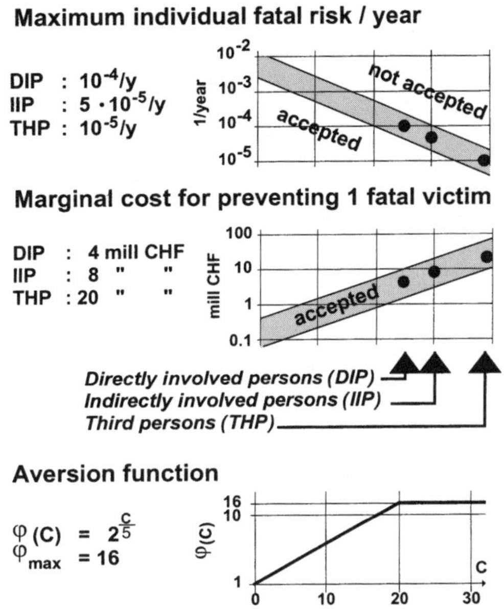

FIGURE 12.11 Quantitative safety criteria for the handling of ammunition and explosives in the military field in Switzerland.

clearly and in advance. It has to be added that the frequency/consequence-method can be outwitted by suitably modeling the hazardous system in the risk analysis [7].

12.3.8 Quantitative Safety Criteria

The quantitative safety criteria for the handling of ammunition and explosives in the military field in Switzerland that were laid down by the responsible authority are shown in Fig. 12.11:

- The maximum acceptable individual fatal risks are:
  10^{-4}/year for directly involved persons[1]
  $5 \cdot 10^{-5}$/year for indirectly involved persons[2]
  10^{-5}/year for not involved third persons[3]
- The marginal costs for preventing one fatality are:
  CHF 4 million (= € 2.7 million or US$ 2.4 million in 2001) for directly involved persons
  CHF 8 million (= € 5.3 million or US$ 4.7 million in 2001) for indirectly involved persons
  CHF 20 million (= € 13.3 million or US$ 11.8 million in 2001) for third persons
- The aversion function weighing up the actual consequences to the perceived consequences is φ (C) = $2^{C/5}$, where φ_{max} = 16.

[1] Such as the worker at the high-explosive pressing machine.

[2] Such as the secretary in an ammunition factory.

[3] Such as the car driver driving along an ammunition magazine.

12.4 THE REGULATIONS AND ORGANIZATION

The development of the regulations did not take the ordinary course from top to bottom. The urgent needs came from the practical side. Real needs could not wait for the judicial and administrative background to be set up.

On the highest level there is the Swiss Explosives Law [1], including rather traditional deterministic decrees concerned with limited explosive quantities, safety distances, and qualitative catalogues of safety measures. This law releases the military forces and the military administration from the substantive contents of the law and decrees but commits the government to issuing its own concepts and regulations for this area. The government assigned this duty to the Department of Defence, which designated the Chief of General Staff as the person responsible for the safety of the handling of ammunition and explosives by the forces and the administration.

The Chief of General Staff enacted the *Directives Concerning the Safety of the Handling of Ammunition and Explosives by the Military Forces and the Military Administration* (WSUME) [6]. These directives lay down:

- The general safety goal (e.g., the protection of human life and prevention of injury)
- The safety assessment concept and planning mode based on quantitative risk analysis
- The quantitative safety criteria
- Duties and responsibilities of the subordinate spheres
- The policy on information to the public

The *Technical Regulations for the Storage of Ammunition* (TLM 75) are the regulations for ammunition and explosives storage in peacetime, an activity with which several agencies of the Department of Defence are concerned. TLM 75 consists of five parts:

Part 1: General Principles

Part 2: Safety Assessment

Part 3: Planning and Construction of Magazines

Part 4: Storing of Ammunition

Part 5: Storing of Ammunition by the Troops

Part 2 [8] contains the detailed methodology, data, and criteria for the risk analysis and risk appraisal due at different stages of planning and operating magazines.

These parts are renewed periodically with regard to the improvements of methodology, models, data, and criteria as well as with respect to new needs, developments in the civilian branch of safety and public opinion.

For the safe handling of ammunition and explosives in the ammunition and propellant factories, the chief of the agency concerned enacted the *Directives for Safety in Federal Armament Factories with Explosives Hazards* (WAE) [5]. It consists of two parts: Part I: General Principles and Responsibilities, and Part II: Guidelines to Perform Safety Assessments. With respect to methodology and criteria, the WAE is also applicable in the field of testing/passing and other special problems.

For transport, the third large-scale activity with ammunition and explosives, the concepts and the regulations have not yet been worked out, mainly for reasons of financial support and capacity. Transport was given second priority because there are many civilian regulations that apparently must and can be followed without severe problems. A preliminary step will soon be taken to investigate whether transport is safe regarding the quantitative safety concept.

Because many different military and civilian government agencies are involved with ammunition and explosives handling, coordination is necessary for economical management with maximum safety benefit. For this purpose, several permanent bodies have been founded over the years. The top level committee is the Committee for the Safety of Handling Ammunition and Explosives by Forces and Administration (the former Ammunition Storage Board). This committee advises the Chief of General Staff concerning storage and ammunition and explosives safety. It leads and coordinates the management of storage, development and revisions of regulations, and basic research in this field. There is also a Safety Deputy and a Project Committee with advisory functions.

The decisions are supported by a number of study groups. The main technical (permanent) body is the Working Group for Basics, which is concerned with methodology, models, data, and criteria for safety assessment and risk management on several levels of ammunition and explosives handling. It is the oldest study group, having performed the first risk analyses nearly 30 years ago. In addition, there are nonpermanent working groups who are responsible for the regulations, as well as a couple of management teams.

12.5 ILLUSTRATING APPLICATIONS

12.5.1 Transport of Dangerous Goods

This is a classic example from the early 1970s. A state-owned Swiss ammunition factory used (and still uses) an outside storage installation for its explosives products (Fig. 12.12). They were transported on a public road by trucks and by railway on an industrial track along the same road. This road and the track crossed a main road that was, at that time, one of the most important transit routes from southern to northern Europe! Some 20 trucks and about 25 railway wagons met several thousand vehicles every day at this crossing. Thus, it seemed only matter of time before a crash between a vehicle and an explosives transport occurred that could lead to a catastrophic explosion.

In the Department of Defence, people were aware of the problem, saw the solution in building an underground passage for the main road, and made a corresponding agreement with the state government. However, in the process of planning the underpass, they became aware that it would cost at least CHF 7 million (equivalent to about U.S. $5 million). Therefore it was decided to perform a risk analysis of the situation and study cheaper alternatives [9].

The investigation of the situation without any safety measurements showed that all individual risks were far smaller than the accepted limits. As for the collective risk (Fig. 12.13),

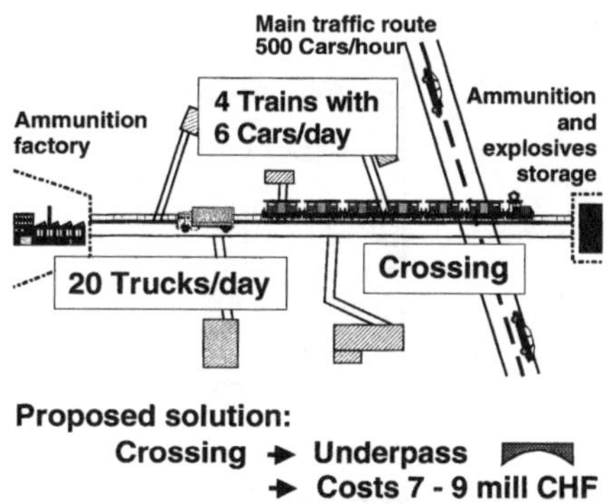

FIGURE 12.12 Case study: explosives transport safety problem and proposed solution.

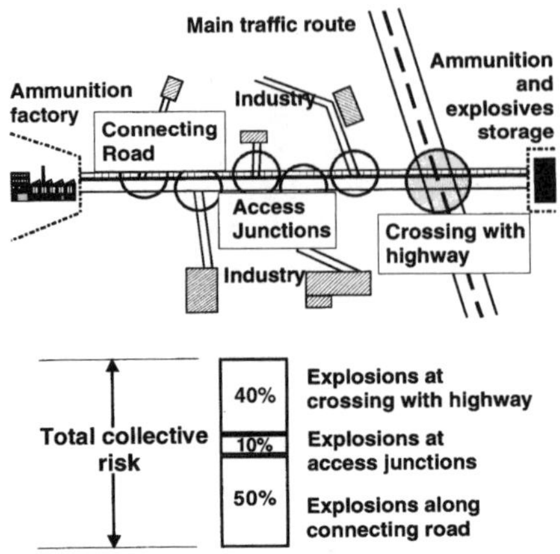

FIGURE 12.13 Case study explosives transport: sources of collective risk.

the crossing with the main road caused only about 40% of the total risk to transport. About 50% of the risk was due to the traffic along the road used by the transport vehicles, and the remaining 10% was caused by the access road intersections leading to the neighboring factories. When the underpass had been considered the best solution, nobody would have imagined that the crossing represented less than half of the real problem.

Subsequently, a great number of safety measures were studied, their risk reduction calculated, and their cost estimated. Figure 12.14 shows the principal measures represented in a risk/cost diagram. The most interesting measures are those with the best risk/cost ratio. It is readily apparent that the underpass is ill advised due to the high cost and the limited risk-reduction effect.

This risk/cost diagram convinced not only the responsible members of the Department of Defence but also the Minister of the state government, who was persuaded that the constructing of the underground passage would represent a considerable waste of money. Con-

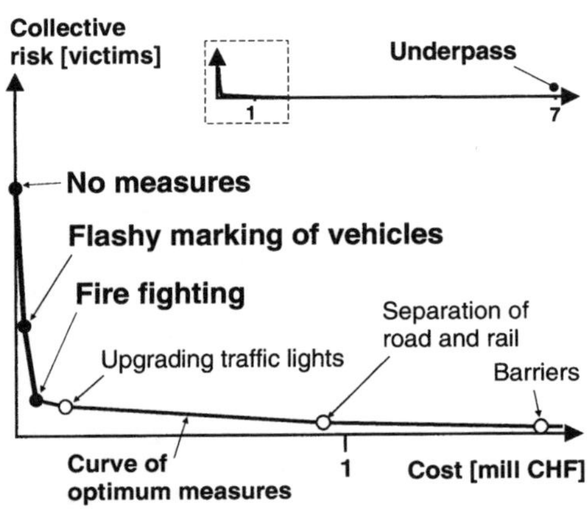

FIGURE 12.14 Case study explosives transport: risk/cost diagram for alternative safety measures.

sequently, this plan was dropped and the proposed alternative measures were taken instead. They have proved their effectiveness over the last 25 years. Thus, the investigation finally resulted in an increase of safety by a factor of about 7.5 and savings of several millions of CHF.

With their decision, the representatives of the Department of Defence and the state government implicitly determined the level of accepted risk of this activity. This decision was not a very difficult one, as a package of safety measures was chosen that was better and at the same time less expensive. Experience shows that politicians are well able to read and interpret a risk/cost diagram in a sensible way.

12.5.2 Prevention of Artillery Shell Duds

This example also shows that the RSCAE can prevent the squandering of money on ineffective safety measures. In Switzerland, which is a small country, the artillery training of the troops cannot be conducted in absolutely isolated shooting ranges. It has to take place in normally utilized and populated areas, not without a number of precautions, of course. Consequently, there are a few hundred duds a year, which normally are blown up immediately by the troops themselves or by special dud-disposal teams of the defense administration. As the duds sometimes cannot be found immediately, especially those in snow, from time to time accidents have occurred resulting in personal injuries or even fatalities.

The Department of Defence was offered a device that, once fitted into artillery fuses, would make the duds ineffective a few minutes after impact. The appropriate agency assessed the device positively on the technical level but was in doubt as to the safety gain. A risk analysis was performed [10] and the risk reduction was compared to the cost. It showed that the risk reduction would not justify the cost (Fig. 12.15); it would need to be seven times higher or the cost seven times lower. As a result, the dud-safety device was not procured.

12.5.3 Propellant Supply System

As a consequence of the disastrous accident in Lapua, Finland, in 1976 [11], where an explosion in a cartridge loading facility killed 40 and injured more than 70 persons, the safety of similar installations in Swiss ammunition factories was examined. After some immediate safety measures had been taken—derived from a purely qualitative analysis—the question remained whether the present manual supply should be replaced by a pneumatic system similar to the new one in Lapua.

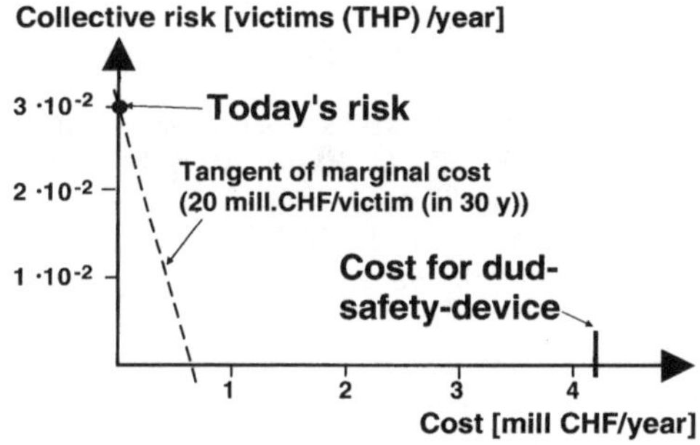

FIGURE 12.15 Risk/cost diagram for dud-safety device in artillery fuses (THP = third persons).

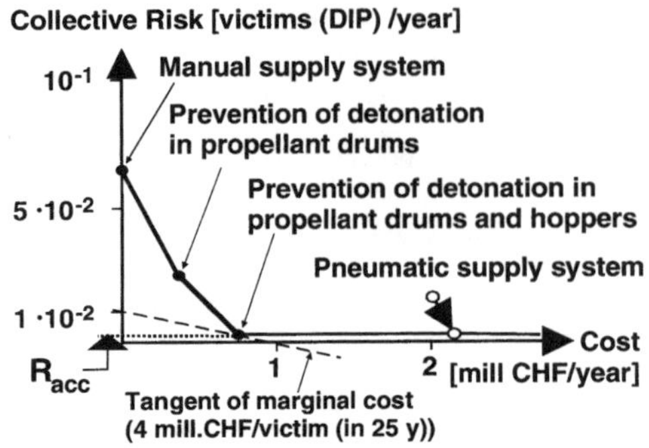

FIGURE 12.16 Risk/cost diagram for propellant supply system (DIP = directly involved persons).

The related cost was estimated to be almost CHF 2 million. Since this investment could not be justified from an operational point of view, and since the production of this particular item was planned to be given up in the foreseeable future, the problem boiled down to the question of whether the gain in safety would justify this investment.

The results of the quantitative risk analysis [12] are shown in a risk/cost diagram in Fig. 12.16: The pneumatic system with an estimated cost of about CHF 2 million decreased the risk by about 80%. Assuming a lifetime of 25 years for this new system, a mean expected value of about 1.3 human lives could have been saved with this safety measure in this period.

Applying the above-mentioned willingness-to-pay criterion, the risk reduction would have justified this investment. But the question remained whether the safety could be improved more effectively by other safety measures. In order to answer this question, the contributing factors to the collective risk of the manual supply system were analyzed. It was demonstrably the case that:

- Safety measures are required to reduce the risks caused by the manual supply system.
- The costs for the originally planned pneumatic supply system are not justifiable.
- Safety can be enhanced more effectively and at lower cost if the powder drums and the hoppers above the loading machines are modified so that fire hazards cannot turn into detonations.

Both safety measures were put into practice.

12.5.4 Explosives Magazine for an Ammunition Factory

According to the RSCAE, explosives magazines are not standardized as, for example, NATO principles [13] or U.K. ESTC (Explosive Safety Transport Committee) [14] call for. Instead, construction regulation TLM 75 Part 3 defines a range of solutions for a few typical sizes and construction types, out of which the magazine in question can be optimized taking into account the actual readiness and capacity requirements as well as the actual conditions of the location.

Thus, at the end of the planning of a shallowly buried magazine for explosives of an ammunition factory, the problem came down to the question of whether one or two (propagation safe) chambers with the same total capacity should be built and whether they required automatic fire extinguishing installations. The risk/cost diagram (Fig. 12.17) clearly showed that the optimum solution was one chamber with a fire extinguishing installation [15].

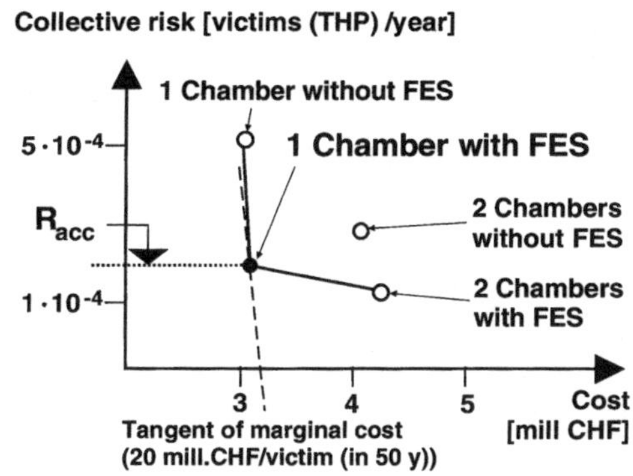

FIGURE 12.17 Risk/cost diagram for different magazine concepts (THP = third persons; FES = fire extinguishing system).

12.5.5 Fire Extinguishing Systems in Artillery Bunkers

In Switzerland, a number of artillery bunkers are used for training with live ammunition. The question arose whether ultra-high-speed fire extinguishing systems should be installed to increase the safety of the troops. A fire hazard from the considerable amount of propellant in the closed area cannot be excluded completely but would have catastrophic consequences for the shooting crew (up to more than 20 victims) as the accident in a Swedish coast artillery bunker in 1984 [16] or the catastrophe in the guntower of the U.S.S. *Iowa* in 1987 [17] demonstrated. The investigation clearly showed (Fig. 12.18) that such installations have to be planned to meet the safety criteria.

12.5.6 Ammunition Storage

Over the last two decades, all under- and above-ground magazines in Switzerland have been investigated according to the RSCAE. The hazards are evident. For example, the difference between the highest and the lowest risk turned out to be about a factor of 1,000! Based on these investigations, safety improvement programs were developed and put into practice.

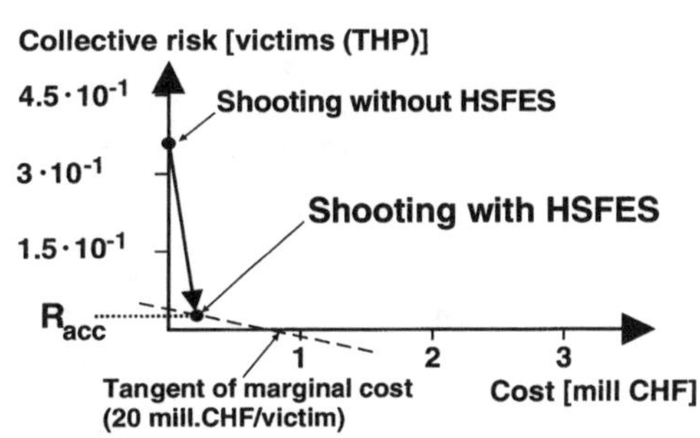

FIGURE 12.18 Risk/cost diagram for high-speed fire extinguishing system (HSFES) in artillery bunkers (THP = third persons.

The quantitative approach for safety assessment allowed purposeful investments to achieve an optimum solution reflecting safety gain, investment, and a problem-oriented realization procedure. A comparison of an actual situation covering 20 existing above-ground magazines assessed according to the RSCAE and according to NATO Safety Principles [13] (or roughly, a modern version of the former Swiss regulation) made the savings of the Swiss approach clear (Fig. 12.19). According to the RSCAE, 19 of the 20 magazines were approved and covered the required capacity completely. By comparison, NATO principles approved only 7 of the existing magazines. For the remaining 75% of the required capacity, 23 new magazines would have been needed, resulting in costs of about CHF 10 million.

Transferred to the whole storage system, the savings would be many times higher. In this context it is interesting to note that the first underground storage built according to the new concept in the early 1970s cost about CHF 10 million less than the version according to the former regulation.

12.5.7 Planning of New Manufacturing Facilities

The RSCAE has also proven flexible and useful for the planning of new facilities for ammunition manufacturing. The main safety problems in ammunition and explosives manufacturing are connected with the personnel working with explosive items and the range of potential hazardous effects of numerous concentrations of such goods. In addition, persons in the the surroundings of factories can be endangered. The flexibility of the RSCAE allows an optimum solution to be found, reflecting operational requirements, technical installations, and building structures as well as safety and cost.

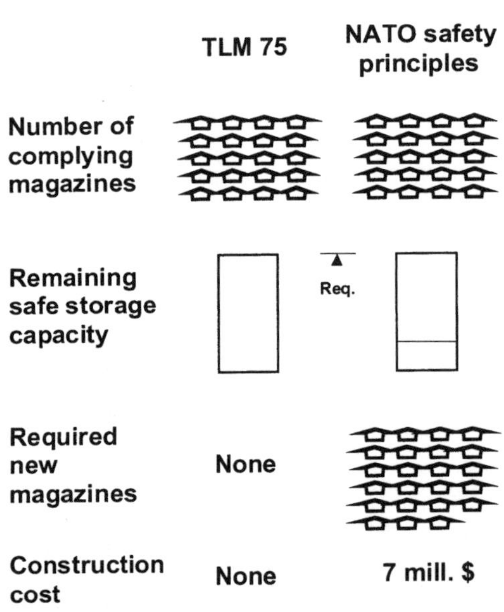

FIGURE 12.19 Comparison of the Swiss concept versus NATO principles.

12.6 BENEFIT

Three main advantages of the concept, regulation, and organization presented, which have been achieved over the years:

1. The methodological concept of quantitative safety assessment makes the hazards of ammunition and explosives handling comprehensibly and intelligibly visible and comparable. The responsible people know the responsibilities they actually take.
2. There are regulations that can actually be applied. There are distinctly fewer waivers compared to previous experience.
3. The necessary safety level can be achieved economically, e.g., so far the flexible assessment concept has allowed solutions resulting in savings of millions and millions of CHF.

12.7 REFERENCES

1. **Bundesgesetz über explosionsgefährliche Stoffe (Sprengstoffgesetz)**
Die Bundesversammlung der Schweizerischen Eidgenossenschaft
Bundesgesetz 941.41, 25.03.77.
2. **Technische Vorschriften für die Lagerung von Munition (TLM 75)**
Teil 1 (Rev 86): Allgemeine Grundsätze
Teil 2 (Rev 90): Sicherheitsbeurteilung von Munitionslagern
Teil 3 (Rev 86): Planung und Projektierung von Munitionslagern
Teil 4 (Rev 93): Friedensmässige Belegung von Munitionslagern
Teil 5: Sicherheitstechnische Vorschriften fürden Umgang mit Munition bei der Truppe
Eidgenössisches Militärdepartement
01.01.87 KLASSIFIZIERT
3. **Konzept für die Sicherheitsbeurteilung in explosivstoffverarbeitenden Betrieben**
Gruppe für Rüstungsdienste, Direktion der Abteilung der Militärwerksätten
Basler & Hofmann AG, Zürich
B 670-3, 01.03.76
4. **Grobrisikoanalyse Beispiel M+FA**
(im Rahmen des Konzeptes für die Sicherheitsbeurteilung in explosivstoffverarbeitenden Betrieben)
Gruppe für Rüstungsdienste, Arbeitsgruppe Sicherheitskonzept/CH
Basler & Hofmann AG, Zürich
Leimer, P.; Cruchaud, M.; Moser, F.; Senn, J-P.; Pulfer, Ch.; Schneider, Th.; Bienz, A.
B 670-4, 01.03.76
5. **Weisungen betreffend die Arbeitssicherheit in R+B mit Explosionsgefahr**
Teil I (WAE) Weisung Nr. 5/10: Allgemeine Grundsätze und Verantwortlichkeiten
Teil II (RSB) Weisung Nr. 5/11: Richtlinien für die Durchführung von Sicherheitsbeurteilungen
Gruppe für Rüstungsdienste/Bundesamt für Rüstungsbetriebe
15.12.86 KLASSIFIZIERT
6. **Weisungen über das Sicherheitskonzept für den Umgang mit Munition und Explosivstoffen in Armee und Militärverwaltung (WSUME)**
Generalstabschef Schweizer Armee
25.03.91
7. **What is Wrong with Criterion FN-Lines for Judging the Tolerability of Risk**
Society for Risk Analysis
Evans, Andrew W.; Verlander, Neville Q.
Risk Analysis, Vol. 17, No. 2, 1997, 21.02.97

8. **Technische Vorschriften für die Lagerung von Munition (TLM 75)**
Teil 2 (Rev 90): Sicherheitsbeurteilung von Munitionslagern
Schweizerische Armee
01.07.92 KLASSIFIZIERT

9. **Sicherheitsmässige Beurteilung des Explosivstofftransportes zwischen der M+FA und der Anlage Rynächt**
Gruppe für Rüstungsdienste, Abteilung für Militärwerkstätten
Basler & Hofmann AG, Zürich
Bohnenblust, Hans; Bienz, Andreas
I 986-1, 06.02.79

10. **Blindgängersichere Zünder für Artilleriegeschosse**
Beurteilung der sicherheitsmässigen Notwendigkeit
Gruppe für Rüstungsdienste - Rüstungsamt 2/CH
Bienz, Kummer & Partner AG
Bienz, A.
I 121-1, 26.02.92

11. **Report on the Explosion Accident at the Lapua Cartridge Factory in Finland, 13th April 1976**
Ministry of Defence, Finland
Karhulahti, Jorma
01.01.79

12. **Sicherheitsmässige Beurteilung der geplanten pneumatischen Pulverzuführung bei der GP 11**
Eidg. Munitionsfabrik Thun/CH
Ernst Basler & Partner AG
Janser, Paul; Bienz, Andreas; Willi, Walter
I 1131.70-1, 01.02.82

13. **Allied Ammunition Storage and Transport Publication 1 (AASTP-1)**
Manual of NATO Safety Principles for the Storage of Military Ammunition and Explosives
NATO Headquarters, Brussels/Belgium
AASTP-1 01.05.1992 NATO unclassified

14. **Quantity-Distances for Military Explosives**
Explosive Safety Transport Committee (ESTC)/UK
ESTC/220/Leaflet No. 5-Part 2 // File No. ESTC/226/16, 06.1979 KLASSIFIZIERT

15. **Eidgenössische Munitionsfabrik Thun**
Unterfluhranlage Glütschbachtal
Generelles Sicherheitskonzept
Amt für Bundesbauten, Abteilung Ingenieurwesen
Ernst Basler & Partner AG
AB/KP/ms
TM 3163-1, 01.07.1982

16. **Explosion in a Swedish Coast Artillery Bunker**
Personal Communication
Newspaper Reports, 01.01.84

17. **U.S.S. IOWA Explosion**
Sandia National Laboratories' Final Technical Report
US Department of Energy
Sandia National Laboratories, Albuquerque/NM, Livermore,/CA
Diverse
GAO/NSIAD-91-4S, 01.08.1991

P • A • R • T • 4

SPILL COUNTERMEASURES

CHAPTER 13
CONTAINMENT AND CONFINEMENT OF HAZARDOUS CHEMICAL SPILLS

Clyde Strong
President, Clyde Strong & Associates, Inc., College Station, Texas

13.1 CONTAINMENT VERSUS CONFINEMENT

Containment refers to actions taken to keep a material in its container, such as plugging or patching a leaking 55-gallon drum. *Confinement* refers to the procedures taken to keep a material in a defined or localized area once it has been released. Confinement is used when the chemical has breached the container and it is desirable to keep the material within defined boundaries. Confinement may be viewed as a defensive activity, and containment as an offensive activity. Through the effective use of containment and confinement, the spread of the spill or release is minimized; people, property, and the environment are protected; and cleanup and recovery operations become more efficient.

13.1.1 Major Factors for Consideration

Two major points should be considered when developing containment/confinement strategies. Safety of response personnel is the primary consideration for any response action. While responders are taught to enter from an upwind, uphill, upstream position to deal with emergency situations, they may have to respond downwind, downhill, and downstream to contain/confine a spill. Responders may be required to work in, or at least in close proximity to the material that is spilling or being released. Proper monitoring for situations such as high concentrations of flammable vapors should be done to ensure that the safety of responders is not compromised. The level and type of personal protection chosen for entry must also be suitably matched to the chemical properties, concentration, and duration of exposure. Decontamination for all contaminated responders and equipment should be anticipated and planned before entry. The second major consideration for containment/confinement strategies should be the effect these actions will have on other response and cleanup operations and the environment. Confining a spill in the wrong location can lead to a greater environmental impact due to limited access for recovery equipment and delays in removal and cleanup.

Containing a pressure release may not be wise if pressure continues to build in the container and the pressure cannot be controlled in other ways.

13.1.2 Other Factors

Other factors should be kept in mind with regard to containment/confinement actions. Prior to proceeding with containment/confinement, every effort should be made to minimize the rate of leakage from the container (e.g., rolling the drum so that the leak is pointing up) or to reduce the pressure (e.g., blocking in lines and shutting down pumps). Compatibility between the containment/confinement device or material and the chemical is also a concern because incompatibility could cause harmful reactions or even ignition in some cases. The length of time containment/confinement methods continue to be effective may be greatly reduced when incompatibilities exist. Proper bonding and grounding procedures should be used when handling flammables, and spark-resistant tools should be used. What will ultimately be done with the material or container once containment/confinement has been achieved? How long will it be before transfer operations commence or the vacuum trucks arrive? These issues and many others should be considered as a part of the overall containment/confinement strategy.

13.2 CONFINEMENT OF CHEMICAL SPILLS ON LAND

Generally speaking, chemical spills that occur on land do not have as much potential for spreading and traveling away from the source as spills into waterways. Spills on land are usually visible to the responder, which is helpful in predicting spill behavior and maintaining the safety of response personnel and the public. This is not to say that spills on land do not present difficult confinement problems. On porous soil, liquids can exhibit rapid vertical movement, contaminating large volumes of soil and possibly affecting groundwater supplies. In urban areas, spills may find their way into storm sewer systems or utility excavations, spreading contamination below the surface and making confinement and recovery difficult. Spills into storm drains or sewers could also lead to additional hazards when vapors from the spilled chemical are concentrated in the confined space.

13.2.1 Spill Location and Characteristics

A spill's potential to exhibit rapid horizontal or vertical movement must be well understood in order for the spill to be confined in a timely and appropriate way. As one might expect, spills on hard surfaces such as concrete, asphalt, and clay have a greater tendency to move horizontally, whereas spills on more porous substrates such as sand and gravel are more likely to move vertically. Whether or not the soil is wet, existing vegetation, soil cracking, and soil cultivation all influence horizontal and vertical spill movement. Characteristics of the spill material such as water solubility, temperature, and viscosity also influence horizontal and vertical flow.

13.2.2 Spill Barrier Techniques

Confinement of spills on land is accomplished primarily by capturing the spill in some kind of excavation or depression or by setting up some type of physical barrier. Barriers may be used to stop the spill movement or to divert the spill depending on what is most critical to the particular situation. In the "stop" mode, a barrier is placed in the path of the spill to block the spill movement completely and hold it for recovery. In the diversionary mode, a barrier is placed to prevent spilled material from moving into areas where it might cause greater environmental harm or greater risks to life and property. Many factors such as spill volume, properties of the spilled chemical, time since the spill occurred, available equipment and manpower, weather conditions, and spill location will determine how barriers can be used most effectively for spill confinement.

13.2.3 Types of Barriers

There are many types and forms of physical barriers for spill confinement. Natural depressions, ditches, or basins, as well as specially constructed holding ponds at the scene, can serve as barriers for spills. Dikes and dams can be constructed out of readily available materials often found at the scene, including sand, soil, and gravel. Barriers can also be constructed with commercially available products in the form of mats, plugs, sorbents, gels, dikes, and foams.

Use of Natural Barriers and Catch Basins. Spilled liquids will naturally move down gradient and can be made to collect in already existing depressions, ditches, or pits. In some cases, it may be necessary to dig channels to encourage a spill to move into these areas. When time permits, holding pits should be lined with material impervious to the chemical. For liquids that float on water and are not greatly soluble or reactive with water, water may be added to the holding basin to reduce migration of the spill material into the soil. Careful thought should be given to the overall effect of the addition of water because waste disposal problems could be increased. It may be good to cover the collected material. If noxious or flammable vapors are being produced, many types of firefighting foams and even special foams for strongly acidic or basic spills are available.

Factors to Consider. Although the use of excavations and catch basins as a confinement method is not complicated, some important factors must still be considered to determine if this is a viable method to confine the spill. Not all surfaces are suitable for excavation, and natural depressions are not always present in the spill area. Large amounts of material may have to be removed in a short period of time for effective spill confinement, meaning that heavy equipment and skilled operators may be required on short notice. This is not always possible under emergency conditions.

13.2.4 Dikes and Dams

Along with collection basins, various types of dikes and dams may be used to confine a spill. In some cases, simple earthen dikes can be quickly constructed at the scene of the spill. In other cases, sandbags may be used to stop the spill's movement. These methods often lead to a large volume of contaminated material for disposal as the spilled liquid is sorbed. Dirt or sand dikes are also easily washed away if the flow increases. Being readily available, however, these materials can and usually should form the first line of defense until other materials are brought to the scene.

13.2.5 Commercial Devices and Materials

A number of commercially available devices and materials may also be used to confine or divert a spill. Several manufacturers produce small portable sprayer systems that supply a quick-setting polyurethane foam. The foam expands and hardens in about one minute, forming a liquid-tight barrier on concrete, asphalt, and other hard surfaces. It could be useful for sealing off drains or to erect a confining dike ahead of the spill. A major limitation is the inability of the foam material to stick to a wet surface or to be applied through a liquid. Therefore, it cannot be used to block a flowing stream of liquid already going down a drain. Cold temperatures can also affect performance.

Various types of chemical-resistant putties and mastics can be used to seal around drain openings and to form small dikes. Some manufacturers supply pre-formed mats of these materials in sizes designed to completely cover drain openings. These mats are set over the opening and pressed down to form a liquid-tight seal. Other commercially available mats are constructed of special elastomers (rubber-like materials) with a chemical-resistant cover. Their heavy weight and ability to spread around and over what they are laid upon make them excellent deflection barriers. Sheets of polyethylene or similar materials may also be used to cover drains but are difficult to keep in place in high winds, and often soil or other materials must be spread over and around them to achieve a liquid-tight seal. Inflatable plugs are also available, primarily for sealing circular openings such as culverts, pipes, and manholes. These low-pressure plugs are similar to an automobile tire in that they are steel-belted for toughness and puncture resistance. Ribs on the outer surface of these plugs help ensure a solid, nonslip fit in the opening. They come in many sizes, from 4 to 96 in. in diameter, and can be fitted with drain valves.

Gelling agents to solidify or immobilize chemical spills have been around for some time and comprise another class of physical barriers. Gelling agents are applied to the spill or set out in the path of the spill as a dike. When they are applied to the spill, physical absorption occurs, binding the chemical liquid in the matrix of the gelling agent and rendering the liquid immobile. Gelling agents can be found for water-based chemicals, as well as hydrocarbons. Ideally, a gelling agent will gel many times its weight in liquid and thereby minimize the amount that needs to be applied. They may be applied with pressurized spraying systems such as dry chemical fire extinguishers or by simply pouring from a container.

Commercial sorbents have some use as physical barriers, although they are used primarily for recovering spills. Application of sorbent mats or pads may prevent or at least reduce the amount of material that enters drains. Sock-like tubes filed with sorbent material may be used to seal around drains and to stop the flow of liquid across a hard surface such as a warehouse floor. Tubes filled with heavy sorbent materials such as powdered clays provide better barriers than light sorbent materials, although the lighter weight sorbents usually have much greater absorption efficiency. In addition to sorbent tubes, there are specially designed rubber diking devices for use on concrete floors and other hard, smooth surfaces. These spill-blocking devices are formulated from a gummy, elastomeric material that sticks and seals when pressed down. Connectors are available so that a number of lengths of this diking material can be joined together. This diking device stands about 2¼ in. high and could reasonably be expected to hold about 2 in. of liquid without difficulty.

13.3 CONFINEMENT OF CHEMICAL SPILLS ON WATER

Confinement techniques for chemical spills that enter waterways must be developed around three types of chemical behavior: spills that float, spills that sink, and spills that mix. Once the spill enters a waterway, the spill material is subject to the same movement and dispersal forces as the body of water. Currents and wind can cause spills to move and spread over wide areas relatively quickly and make rapid containment all the more important. Floating spills are usually visible and can be monitored with relative ease as they move on the water. Since the fastest flow rates occur at the surface of rivers and streams, floating spills travel at speeds equal to the greatest surface current speed.

Confinement of sinking spills may be complicated by the inability to visually observe and monitor their movement in the water. Irregularities in the bottom, such as debris and rock piles, deep holes, and ledges, can slow the spill movement for heavier-than-water chemicals. Drag due to the spill moving across the bottom substrate may also reduce the speed at which the spill can travel. Water-soluble spills quickly disperse throughout the water column, and their movement will generally be indistinguishable from the water flow. Discoloration, cloudiness, or other visible effects may help track the spill movement. Monitoring for pH may help determine spill movement of acids and bases, and other water quality tests might have application for tracking spills of certain chemicals.

13.3.1 Methods for Confining Floating Spills

Methods used to confine floating hazardous chemical spills are more readily available and specialized than for other types of spills. Most of the techniques and procedures have been developed during the proper management and confinement of oil spills, and are directly applicable to confining floating chemical spills.

Mechanical Booms. In large waterways, mechanical booms are the most efficient tool presently available for confining floating spills. Figure 13.1 shows the general features of a mechanical boom. The upper freeboard component provides flotation and prevents splash-over. The skirt provides a barrier against subsurface loss of the spill. Booms come in sections typically 50 to 100 ft in length with end connectors for linking multiple sections together. Booms are available in a wide range of sizes and designs, so that it is possible to match the boom to various water conditions including fast currents and high waves. The outer fabric covering of the boom can be constructed from a variety of materials such as polyvinyl chloride, polyurethane, and neoprene. When extended contact time between the boom and chemicals is expected, compatibility between the chemical and the boom fabric should be considered. Mechanical booms do have their limitations. Manpower requirements can be substantial, as well as the need for boats with sufficient power to tow and anchor the boom. Training is needed in the techniques for proper deployment. Once deployed, booms may impede navigation. Booms collect everything that floats, not just the spill material, so large amounts of contaminated debris may also have to be handled.

Pneumatic Barriers. Pneumatic or "bubble" barriers create a curtain of bubbles and surface turbulence that has the ability to contain floating spills at low current speeds (0.5 mph or less). Figure 13.2 illustrates the pneumatic barrier. A perforated manifold is set across the waterway and compressed air is forced through the piping. A curtain of air bubbles rises to the surface, creating turbulence, a countercurrent effect, and an elevation in the water level along the length of the manifold. Large compressors are needed, and this equipment may have to be operated for an extended period of time. The bubble barrier is not a hindrance

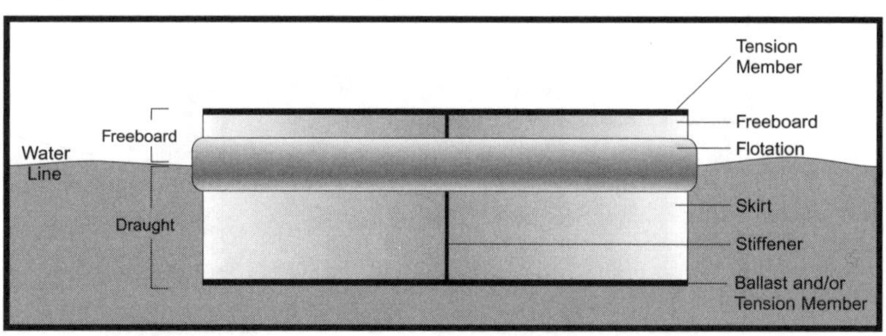

FIGURE 13.1 General features of a mechanical boom.

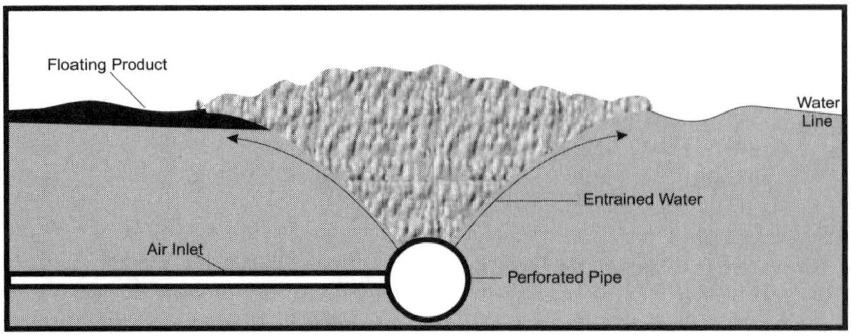

FIGURE 13.2 Pneumatic or bubble barrier.

to navigation, but neither does it prevent the passage of large pieces of contaminated debris through the barrier. This method has not been widely used in spill response because of its limited effectiveness in fast currents and the equipment required for deployment.

High-Pressure Water Jet Systems. High-pressure water jet systems employ spray nozzles set above the water surface on floating rafts or other structures. The nozzles are positioned so that they spray in an overlapping, fan-shaped pattern across the surface of the water. They are directed downward at an angle to create a countercurrent to the water movement. Water pressure to the nozzles must be maintained for the duration of the confinement operation or until other methods are employed. With sufficient water pressure, confinement can be accomplished at current speeds greater than 1.0 mph.

Underflow Dams. While mechanical boom, pneumatic barriers, and water jet systems may be adapted to almost any size of waterway, a number of simple confinement methods, such as the underflow dam, are best suited to smaller waterways. As shown in Figure 13.3, an underflow dam uses a series of pipes extending through an earthen barrier to confine a floating spill. The pipes allow water to pass through the barrier, and when the pipes are positioned with the upstream end lower than the downstream end, the water level is raised above the intake of the pipes and the floating spill is retained. In order to control the flow of water, a sufficient number of pipes must be installed in the dam. The width, depth, and speed of the water can be used to calculate the flow rate and thus determine the number of pipes that need to be used. The use of elbows on the upstream end of the pipes will help prevent the formation of a vortex that can sometimes occur, pulling surface material down

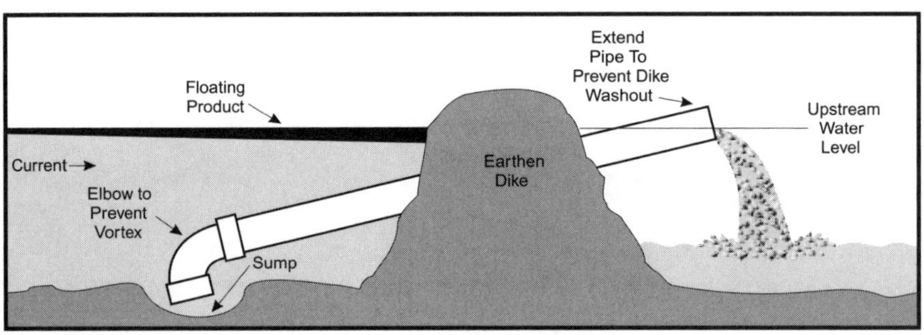

FIGURE 13.3 Underflow dam.

and through the pipes. Soil should be tightly compacted around each pipe, because water tends to try to find its way through the dam along the pipes. The ends of the pipes on the downstream side should extend far enough out from the dam to prevent backwashing and weakening of the barrier. Once in place, the underflow dam should be periodically monitored and any rise in the water level noted. The ends of the pipes can become partially blocked due to silting or clogging with debris. When this happens, the dam is subject to overflow and washout.

Overflow Dams. Instead of the water flow being directed through the dam, water may be brought over or around the dam through siphon tubes or via subsurface pumping while the floating spill material is confined. As with the underflow dam, major consideration has to be given to the volume of water that must be moved past the barrier. With a pumping system, plans must be made for maintaining the pump for the duration of the spill recovery. Backup systems would be advisable in most cases.

Other Methods. Other methods for confinement of floating spills in small waterways include containment weirs and sorbent filter fences. Weirs are simply barriers to divert the flow of water. For floating spills, water is typically diverted under the weir and the chemical is retained on the surface. Figure 13.4 shows a weir constructed from plywood or other suitable piece of lumber. Handles may be attached to assist in raising or lowering the weir. To be effective, the weir should extend no further below the surface of the water than necessary to confine the spill. The farther the weir extends below the surface, the more the water flow is restricted and the more likely it becomes that the spill will be entrained under the weir. Weirs are very useful for deflecting a spill out of the main flow of the creek or stream and into a collection sump set into the bank.

Sorbent filter fences utilize fencing material or netting to place sorbent material in the path of the spill. Loose-floating sorbents are typically employed on the upstream side of the netting or between two pieces of netting. The "sandwich" method is useful for creating a very thick bed of sorbent material. The sorbent material must frequently be removed and replaced as it becomes saturated. The use of straw bales is not recommended for waterways where the contaminated bales might sink and become difficult to remove.

13.3.2 Methods for Confining Sinking Spills

Confining spills that sink in the water and do not appreciably mix is similar to confining floating spills, in that barriers can be placed in the path of the spill to restrict or prevent its movement. It is much more difficult to confine sinking spills, however, due to the complexity of placing barriers along the bottom of a waterway, especially when there is substantial water

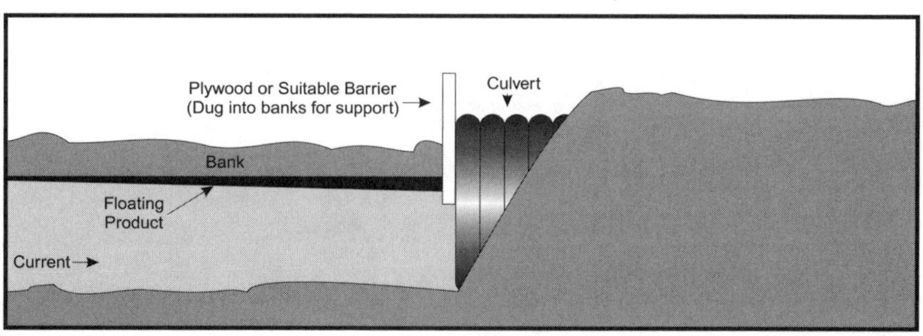

FIGURE 13.4 Culvert weir.

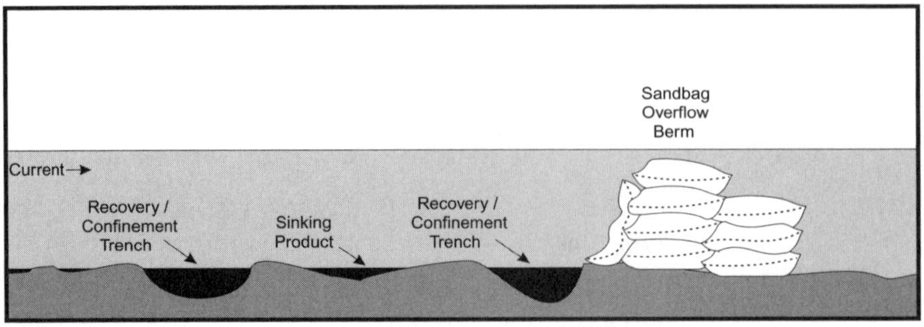

FIGURE 13.5 Subsurface trenches and berm.

flow. In large, fast river systems, there are no suitable methods to ensure spill confinement of chemicals that are heavier than water. In such cases, the material must be transferred from the damaged vessel or container promptly to reduce the amount of spillage as much as possible.

Existing Water Controls. In some waterways, controlling the flow of water can confine heavier-than-water spills. Tidal streams can simply be diked. Waterways used for irrigation may have a system of diversion gates that can be closed or a pumping system that can be shut down. Typically, streams, creeks, and rivers are not controlled and require a different approach.

Subsurface Trenches and Berms. When the waterway is small and slow moving, construction of trenches or depressions across the bottom ahead of the spill may help confine it. Berms set on the downstream edge of these depressions will further enhance confinement. Sandbags are preferred over loose soil or mud for constructing the berm, because they are less susceptible to being washed away (see Fig. 13.5). Time limitations for construction of these barriers, suitable excavation equipment, deep water, wide streams, and rocky bottom substrate all present serious logistical problems to the use of this method.

Overflow Dams. Overflow dams have also been used with some success for heavier-than-water spills. Figure 13.6 shows a typical design. Water is moved past the dam using a pump or siphon tubes to draw water from close to the surface. As with floating spills, the pump or siphon system must be capable of moving a sufficient volume of water past the dam.

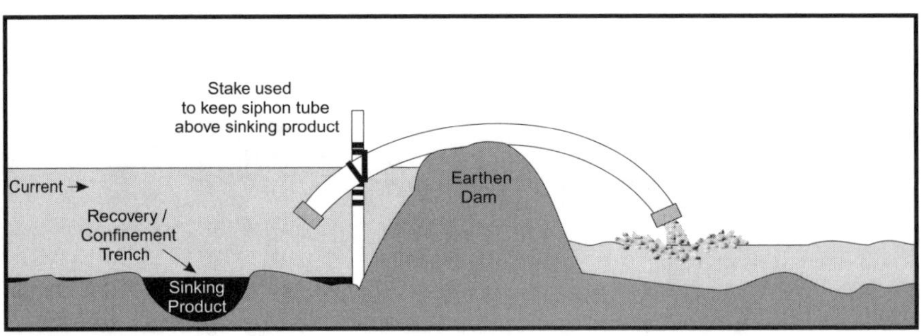

FIGURE 13.6 Overflow dam using siphon tube.

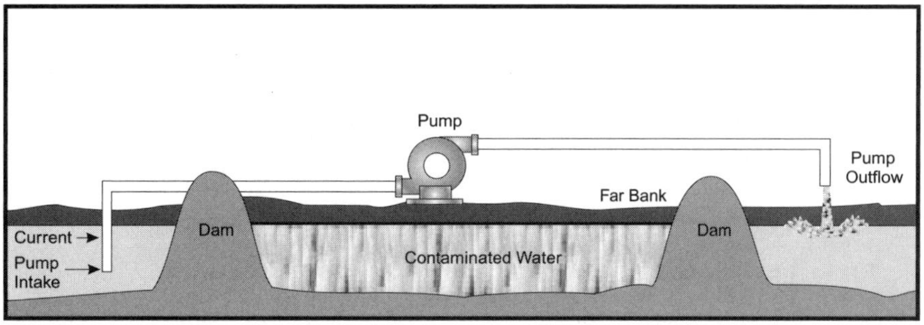

FIGURE 13.7 Isolate and divert.

Sealed Booms. The sealed boom is a modified form of the containment boom that extends the skirt of the boom as much as 25 ft into the water. Heavy chain extends along the entire length of the bottom of the boom skirt, and carbon dioxide-activated anchor points are used to drive the chain into soft mud or sand. Divers are used to discharge the carbon dioxide cartridges and ensure that the chain provides a good seal on the bottom of the waterway. Expense, deployment time, subsurface debris, hard bottom substrates, moving water, and excessive water depths are major drawbacks to this method.

13.3.3 Methods for Confining Water-Soluble Spills

Water-soluble spills create confinement problems quite different from those created by floating or sinking spills. The entire volume of contaminated water must be isolated or confined to be totally effective. This can be accomplished in two ways.

Isolate and Divert. In order to isolate and divert a spill, it is necessary to dam up the waterway both below the spill and above it. Pumps and a series of pipes are then used to divert the clean water around the contaminated area that has been isolated (see Fig. 13.7). Pumps must be of sufficient capacity to handle the volume of water coming downstream and must remain operational for the duration of the recovery and cleanup.

Diversion Channels. Construction of diversionary channels is a second method for isolating water-soluble spills. As with the pumping method, dams are set above and below the spill. A channel is constructed, allowing clean water to move around the contaminated area while the spill is being treated. Both of these procedures are limited to fairly small, low-volume waterways. The practicality of these methods is also highly dependent on the spill location, the availability of equipment, and the speed with which they can be implemented.

13.4 CONTAINMENT OF PRESSURE LEAKS

Before any containment technique is attempted, other options should be considered, such as closing valves, tightening fittings, moving the container, and transferring the product to control the leak. If containment is necessary, every effort should be made to reduce the pressure in the vessel before the containment operation is performed. It is usually necessary to determine the source of the leak prior to containment so that the patch or cap can be applied. This may not always be easy, especially with small leaks of compressed gasses.

Looking for areas where escaping vapors have blown away dirt or debris, listening for the sound of escaping gas, and using soap-bubble solutions may assist in determining the source of the leak. Visibility could also be a problem under some circumstances when a heavy cloud of vapors is present. Devices to blow or suck away the vapors should be considered to improve visibility during the containment operation. As with confinement, the compatibility of the containment materials and equipment with the leaking product is a major consideration.

13.4.1 Enclosure Devices

Enclosure devices, sometimes referred to as cylinder coffins or cylinder recovery vessels, are cylinder-shaped metal shells with hinged, gasket-equipped hatches. A leaking pressure cylinder can be totally encapsulated by inserting it into the enclosure device and bolting down the hatch. These devices are designed to withstand the pressures exerted by the contents of the leaking cylinder and are equipped with valves that allow the product contained in the enclosure device to be recovered. Figure 13.8 shows a typical enclosure device.

Enclosure devices are commercially available for the chlorine and sulfur dioxide industries that utilize 100-lb and 150-lb DOT 3A480 cylinders. Other facilities have constructed their own enclosure devices to meet the demands posed by their particular pressure vessels. It is important that the shell, hatch, gaskets, and valves on these devices be product-compatible and able to withstand the anticipated pressures.

13.4.2 Capping Kits

Capping kits are available for only a small number of valve systems, most notably chlorine, sulfur dioxide, and vinyl chloride. Commercial capping kits are available for 150-lb cylinders, ton containers, tank trucks, and tank cars of chlorine. Due to the weight of the equipment and the procedures necessary to apply a capping kit, it is always recommended that at least two responders carry out the operation. Ropes and pulleys may be helpful for moving the capping kit to the top of tank cars and trucks. Because caps have a limited ability to hold pressure, their effectiveness in containing leaking materials is reduced and the amount of time responders have to transfer the contents of the container or effect other control measures is limited. On tank cars, tank trucks, and barges, when a capping kit is used, the hatch cover over the valves and fittings must remain open. This can be a problem if the vessel needs to be moved and will encounter low underpasses and bridges enroute. Transportation regulations may prevent movement of a capped vessel. When the safety relief system is capped, other safety relief systems for controlling pressure buildup should be

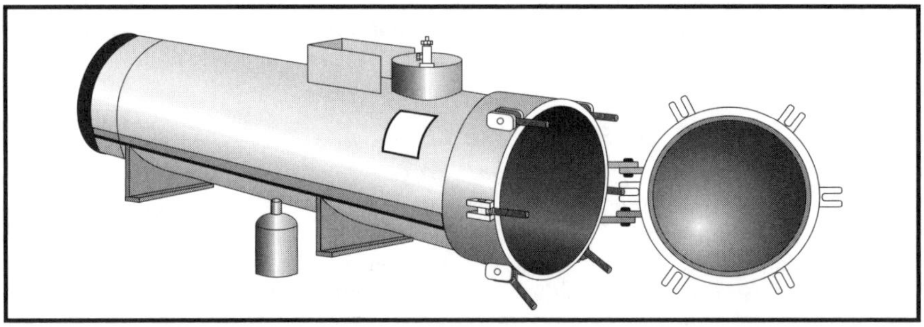

FIGURE 13.8 Enclosure device.

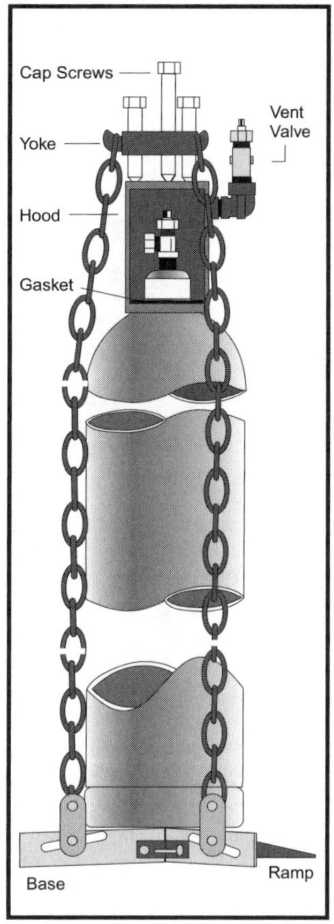

FIGURE 13.9 Typical capping kit for 150-lb cylinders.

established or methods undertaken to monitor the pressure and prevent a rise in pressure. Caps must be vented when being installed in order to get a good seal around the gaskets and are equipped for such an operation. There may be considerable corrosion or pitting in the area onto which the cap gasket must be placed. Scraping and sanding may be necessary to achieve a gas-tight seal. Figure 13.9 depicts a typical capping kit for 150-lb cylinders, and Fig. 13.10 shows a typical capping kit for tank cars or tank trucks.

13.5 CONTAINMENT OF NONPRESSURE LEAKS

Containment of nonpressure leaks is typically referred to as patching and plugging the leak. Any size of nonpressure container, from a bulk storage tank to a pail, can be patched or plugged using these methods. The procedures and materials used are relatively simple and straightforward, but some precautions and guidelines can make the operation safer and more effective. With small containers such as drums, pails, and possibly even tote tanks, try to position the container by tipping or rotating so that the leak is above the liquid level. Then make sure the container is securely supported in place before beginning to contain the leak.

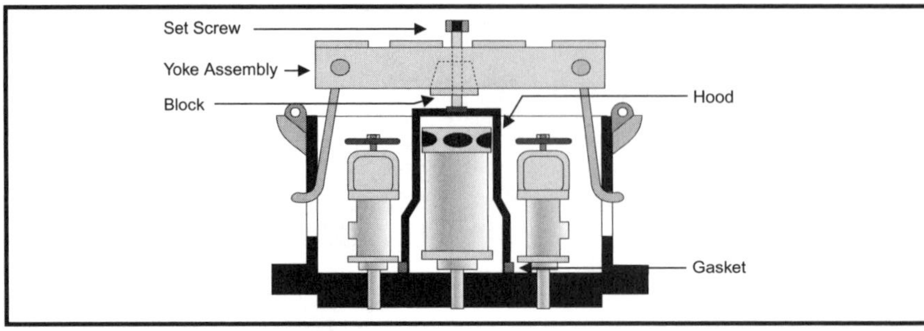

FIGURE 13.10 Typical capping kit for tank cars or tank trucks.

Try to clean up, sorb, cover, divert, or neutralize any product that responders will potentially be working in around the container. Not only is there a possible hazard from the chemical properties, but the chemical may also present other problems, such as poor footing for the responders. Questions need to be answered, such as what is going to be done with the contents of the container after it is patched, how long the patch or plug will be expected to hold, and whether the patch or plug is compatible with the chemical being contained. Often a patch or plug can only be expected to slow down the leak. Small catch basins such as cans or buckets can often be positioned to catch drips escaping past the patch or plug.

The best advice when it comes to patching and plugging is to make it as quick and simple as possible. If a handful of putty will stop the leak temporarily so that a transfer can be made, why spend the time and effort to apply all the straps and hoses and take all the time to position and inflate a rubber patch? However, when the hole is large and the transfer operation will be delayed for some time, using an inflatable patch may be worth the time and effort necessary to apply it. Consideration must always be given to the limited dexterity and visibility of the responders who will be applying the containment device or material and to minimizing their time in the hazard area.

13.5.1 Devices and Materials for Containment of Nonpressure Leaks

The choice of materials and gadgets for containing a nonpressure leak is almost endless. Wooden objects such as pencils and sticks, handkerchiefs and rags, chewing gum, soap, metal screws, rubber balls, and various putties and epoxies have all been used at one time or another to contain a leak. Commercial patch kits are available from simple (a bag of wooden plugs and wedges) to elaborate (inflatable patches and plugs with straps, hoses, and regulators). Homemade patch kits can also be easily put together and tailored for the chemicals and containers likely to be encountered. Figure 13.11 illustrates some of most common patching and plugging devices used by responders.

Wooden Cones and Wedges. Plugs made of softwood provide very simple and effective containment devices. Softwood is preferred over hard wood due to its greater ability to conform to the opening and hold in place. Redwood is a very good material for such purposes. It is readily available and has none of the rosin of pine or cedar. Wooden plugs are porous and will soak up the product and allow it to leak through after a while, so product should be removed as soon as possible. When more than one wooden plug is applied, such as with a large opening, putty or other conforming material should be placed between the wooden plugs to prevent the product from leaking between the plugs. It is probable that the wooden plug will not completely stop the leak due to splits and irregularities around the leak opening. Putties, mastics, and epoxies can be used to fill in these small gaps around the perimeter of the wooden plug and make the containment more effective.

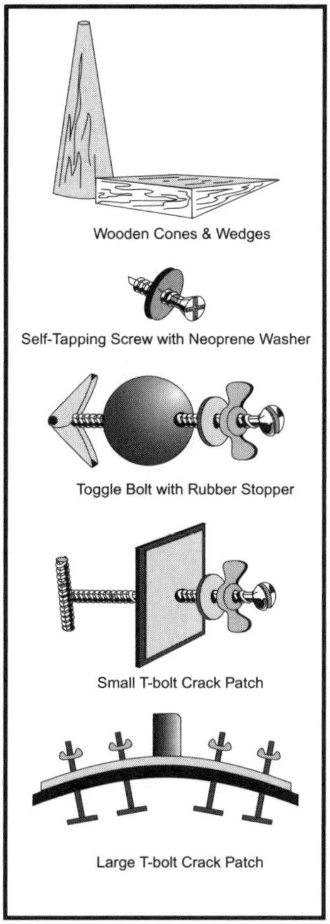

FIGURE 13.11 Common patching and plugging devices.

Foam Plugs and Patches. Cone- and wedge-shaped plugs can be made from closed-cell foam materials such as neoprene and polyethylene. These materials offer the advantage of being able to be compressed to fit into the opening and then being able to expand when released to fit tightly in the opening and provide a good seal. These plugs do not absorb the liquid and can be cut down easily with a knife when the length interferes with other operations, such as overpacking. Putties and mastics may still be needed around the edges of these foam plugs to complete the seal. Foam plugs are available commercially in patch kits, and large uncut sheets of closed-cell neoprene can be obtained from shoe repair shops.

Screws, T-bolts, and Toggle Bolts. Various types of screws, T-bolts, and toggle bolts can be used to fabricate plugs in combination with neoprene lab stoppers and pieces of neoprene or other rubber sheeting. Sheet metal screws with a neoprene washer are very useful for plugging small punctures. Self-tapping screws used for attaching corrugated metal siding and roofs are also useful and come with a rubber washer already on the screw. Many find a nutdriver or socket more convenient for driving the screws than a typical screwdriver. T-bolt and toggle bolt combinations with rubber stoppers are found in most patch kits and are effective for plugging leaks. T-bolts have a perpendicular crosspiece welded to the bolt. Toggle bolts have a "toggle" on the bolt that is spring-loaded and allows the wings to be squeezed together to fit into smaller openings. Both are cumbersome to apply and are subject to leaking through the stopper along the threads of the bolt. Putty can be applied to minimize this leakage. Wing nuts are preferable to screwdrivers for tightening.

Bandages and Wraps. In some cases, it may be preferable to apply a patch in the form of a bandage or wrap for leak containment. A number of commercial devices are available for this purpose that utilize a series of straps to hold the patch in place. Once the patch is in position over the leak, the straps are tightened, applying pressure to the patch and stopping the leak. Some patches are inflatable, which allows a more uniform pressure to be applied over the entire area of the patch and creates a better chance for successful leak stoppage. These patches come in a variety of sizes depending on the size of the leak to be stopped and the type of container to be patched. Protective covers are sometimes used over the primary patch to protect it for a longer time against the damaging action of the product.

Putties, Mastics, and Epoxies. No patch kit is complete without some sort of putty material for patching and filling in around other primary plugs and patches. One of the most useful materials is a putty made with a bentonite clay base, cornstarch, and polymeric ingredients. It adheres to most surfaces, whether dirty, rusty, oily, or contaminated with chemicals. It is also resistant to most chemicals. Putties are easy to apply with gloved hands and are quick and simple for responders. Even on some large diameter openings, putties can be very effective in containing the leak. They are designed so that they must be kept moist. Over a period of hours on a hot day, putties may begin to dry out and crack. Wetting the surface of the putty with a gentle water mist from time to time may prevent this. The putty surface can also be covered with something like plastic wrap, petroleum jelly, or grease to prevent drying. Epoxy putties are designed so that when the two parts are kneaded together, a reaction occurs that causes the material to harden, sometimes within just a few minutes. Epoxies can create problems when the two parts are kneaded together while gloves are worn. The epoxy sometimes sticks to gloves, making it difficult to remove and apply to the leak. The inability of epoxies to adhere to painted, contaminated, or dirty/rusty surfaces may also be limiting.

Other Devices and Materials. Anything that can be applied to the leak to stop or slow down the release without making things worse would be considered acceptable as patching /plugging material. Tubeless tire patches, chemical-resistant tapes, lead wool, quick-setting urethane foam, plumber's plugs, foam insulation material, and a whole host of other materials and devices may be considered.

13.6 CONCLUSIONS

Containment/confinement of hazardous chemical spills are a very critical part of the overall response strategy. The methods, materials, and devices used in the containment/confinement part of the operation must be chosen with the objectives for the overall operation in mind. Many of the strategies for containment and confinement are not complicated, but they are limited by the properties of the spill or leak and the conditions and location in which the incident occurs. The safety of response personnel, compatibility of materials and equipment, logistics for acquiring and applying materials and devices, and ultimate disposition of the contained or recovered material are all important considerations. Response personnel should be knowledgeable about the materials, containers, equipment, and techniques they may be called upon to use, as well as the limitations they may face.

CHAPTER 14
CLEANUP OF CHEMICAL SPILLS USING AIR SPARGING

Krishna R. Reddy
University of Illinois at Chicago, Department of Civil and Materials Engineering, Chicago, Illinois

Jeffrey A. Adams
ENGEO Inc., San Ramon, California

14.1 INTRODUCTION

Air sparging is an in situ technique to remediate sites that have been contaminated with petroleum products and chlorinated solvents as a result of accidental spillage and leaking underground storage tanks. This technique was developed in the mid-1980s in Germany for the treatment of saturated soils and groundwater contaminated with volatile organic compounds (VOCs). This chapter provides an overview of the air sparging technology, including (1) air sparging technology, including applicability and advantages and disadvantages; (2) fundamental processes involved during air sparging process; (3) air sparging system design, operation, and monitoring; (4) mathematical models to describe air sparging process; (5) performance of air sparging systems in the field; (6) use of air sparging with certain modifications and/or complementing technologies; and (7) other issues, such as cost and regulatory issues.

14.2 TECHNOLOGY DESCRIPTION

14.2.1 How It Works

In a typical field system, as depicted in Fig. 14.1, compressed air is delivered to a manifold system, which in turn delivers the air to an array of air injection wells. The wells inject the air into the subsurface below the lowest known point of contamination. Due to buoyancy, the air will begin to rise toward the surface and, through a variety of mechanisms, will either remove or assist in degradation of the contamination. As the contaminant-laden air rises towards the surface, it will eventually reach the vadose zone. At this point, either the contaminated air may be collected with a soil vapor extraction (SVE) system or, if vadose zone

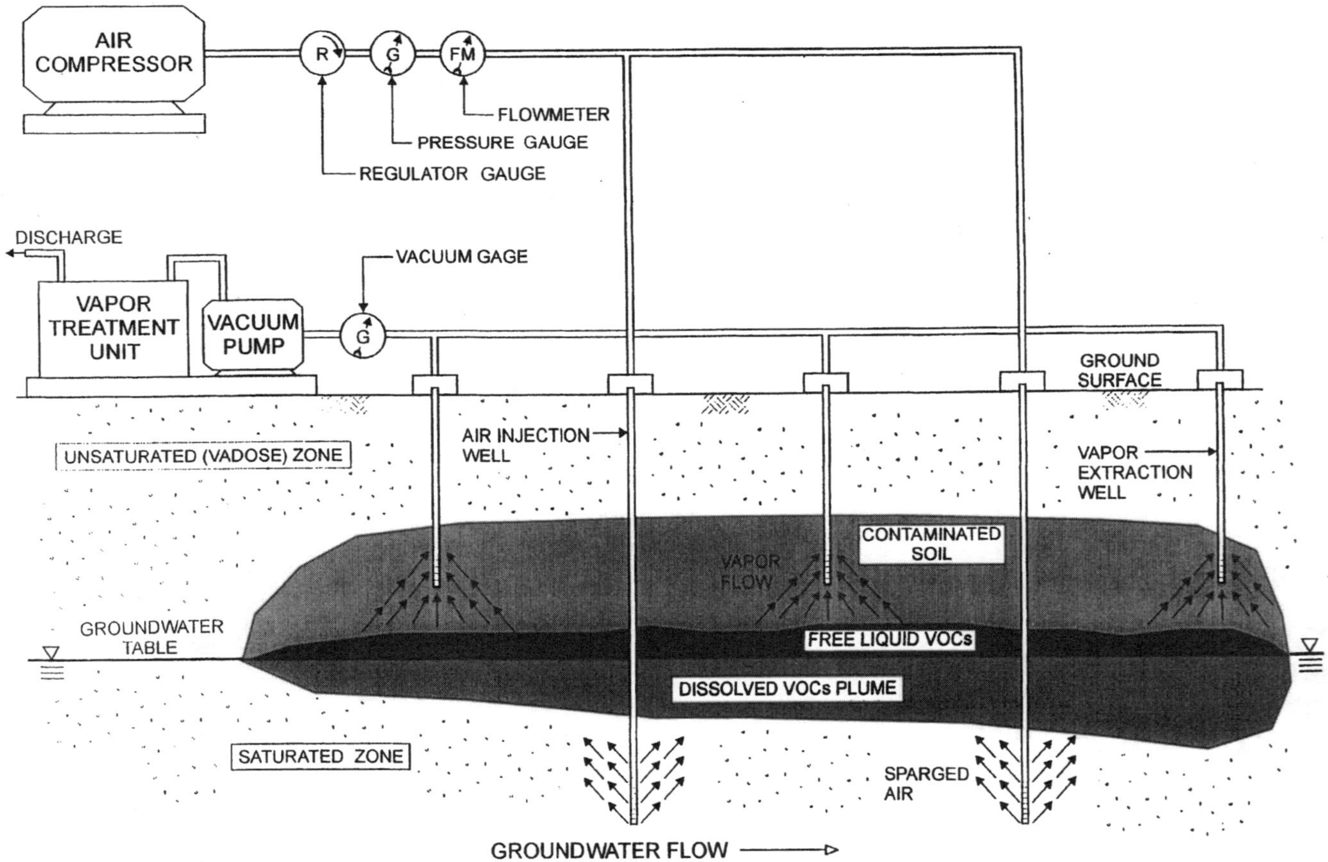

FIGURE 14.1 Schematic of air sparging field system.

conditions permit, the native subsurface microbial population may degrade the vapors into non-toxic products.

14.2.2 Applicability

Geology/Hydrogeology. Prior to the implementation of an air sparging system, a detailed geologic and hydrogeologic characterization of the contaminated site must be performed. Because the patterns of air flow resulting from air injection are controlled by the permeability of a soil, the stratigraphy of the soil profile must be assessed. The permeability of soil within each layer must be assessed to ensure that effective air flow will result; soil permeability should be at least 10^{-3} cm/s for adequate air flow to occur (Bausmith et al., 1996; Loden, 1992). Because air flow may be impacted by the presence of adjacent layers of differing permeability (Ji et al., 1993), the thickness and properties of each layer need to be determined. Additionally, lenses or inclusions of soil with different properties than the surrounding soil may impact air flow and overall remedial performance and must be delineated.

The hydrogeologic conditions of the site are also an important consideration. When applying air sparging, the depth to the water table should be at least 1.5 m (Bausmith et al., 1996), but the most successful applications are made when the depth to the water table is at least 10 ft (Miller, 1996). If depth to groundwater is less than 10 ft, the surface may be capped to prevent uncontrolled vapor migration through the surface (Miller, 1996). The properties of the aquifer material need to be assessed to determine groundwater circulation and other transport phenomena that may occur during air injection, including hydraulic conductivity and groundwater gradients that may induce groundwater flow. Additionally, the groundwater flow velocity and direction must be assessed to ensure that the air sparging system is designed to prevent down gradient contaminant migration away from the treatment zone (Reddy and Adams, 2000).

Contaminants. Air sparging is applicable to a class of contaminants known as volatile organic compounds (VOCs). Common VOCs that are the target of air sparging include petroleum products, including gasoline and its BTEX (benzene, toluene, ethylbenzene, and the xylenes) constituents, and chlorinated solvents, including trichloroethylene (TCE) and tetrachloroethylene (PCE). A compound is capable of being remediated through air sparging if it is deemed strippable; that is, if it has a Henry's constant greater than 10^{-5} atm-m^3/mol and vapor pressure greater than 1 mm Hg (Bausmith et al., 1996). Bass and Brown (1997) have reported that, after a review of case studies, it appears that air sparging is more successful at chlorinated solvent sites than at petroleum sites.

Once contamination is released into the soil, it may exist in a variety of phases. It can exist as a nonaqueous or free (pure) phase liquid, dissolved in the groundwater, in vapor phase within the soil gas, bound to the organic matter of the soil, trapped within the mineral phase of the soil matrix, or within the microbial phase (Yang et al., 1995; Davis, 1997). Before it comes into contact with the water table, it will travel or reside in the vadose zone in the form of a nonaqueous phase liquid (NAPL). NAPLs fall into two categories: light (LNAPLs) and dense (DNAPLs). The distinction is simple: LNAPLs are less dense than water while DNAPLs are more dense than water. LNAPLs will spread laterally on top of the water table, while DNAPLs will continue migrating downward through the groundwater (Panday et al., 1994). During their migration through the vadose zone, NAPLs may become trapped in soil pores in the form of blobs. This transport and deposition will continue until all the NAPL becomes trapped or a low-permeability stratum is reached, preventing further migration (Powers et al., 1991). Once into the groundwater, LNAPL will be transported by water table fluctuation and ambient groundwater flow, while DNAPL will migrate to the bottom of the aquifer, where it is able to mound and spread by the resulting gradients (Panday et al., 1994; Hunt et al., 1988a). Air sparging is shown to be applicable to remove both

LNAPLs and DNAPLs existing as nonaqueous or dissolved phase (Adams and Reddy, 1999, 2000).

Advantages and Disadvantages. There are many advantages of using air sparging, including:

1. As an in situ process, it can be implemented with minimal disturbance to site operations.
2. It has been proven to be efficient—generally treatment time is less than one to three years under optimal conditions.
3. It is simple—readily available equipment can be used and installation is easy.
4. It is cost-effective—treatment cost ranges from $20–50/ton of saturated soil, which is significantly less than other treatment technologies.
5. No removal, treatment, storage or discharge of groundwater is required.
6. It isflexible enough to address contaminants in nonaqueous and dissolved phase.
7. It can be complemented with other technologies, such as soil vapor extraction and bioremediation to increase remedial efficiency.

Although the advantages of air sparging are numerous, the following disadvantages must be kept in mind:

1. The spreading of the contaminants to the clean areas or zones due to excessive pressurization or the presence of low-permeable formations or covers above the contaminated areas
2. The uncontrolled migration of harmful vapor to undesirable receptors (e.g., residential basements, subways, utility tunnels)
3. Lack of rational design accounting for the fundamental contaminant removal processes
4. Difficulty in using conventional monitoring technologies due to transient effects of sparged air on pressures and water levels at monitoring locations

14.3 FUNDAMENTAL PROCESSES/PHENOMENA

When injected into the subsurface, air will enter into pores with lower capillary pressure (larger pore sizes) and produce a fingering pattern of preferential pathways due to low viscosity air invading the higher viscosity groundwater (Ahlfeld et al., 1994a). The mode of air migration and the zone of influence are the primary concerns during air sparging. Air sparging owes its effectiveness to a variety of mechanisms that occur during this subsurface air movement. The mechanisms can be classified into three categories: mass transfer mechanisms, mass transformation mechanisms, and mass transport mechanisms.

Mass transfer mechanisms include volatilization, dissolution, and adsorption/desorption. Biodegradation is the mass transformation that occurs during air sparging. The three mass transport mechanisms consist of advection, dispersion, and diffusion. Air sparging is a dynamic process; during different stages of a remedial program, different mechanisms will control remedial rates and efficiency. The contributions of each of the mechanisms will also vary from site to site and will ultimately determine the effectiveness of an air sparging remedial program.

The migration of air in the subsurface as well as the different mass transfer, mass transformation, and mass transport mechanisms responsible for the contaminant removal during air sparging are described in this section.

14.3.1 Air Migration

Air sparging is most effective in coarser soils because better air saturation and density will result (Baker and Benson, 1996). Unfortunately, permeability cannot be too high or else the injected air will move upwards to the surface without spreading laterally and thus will impact only a small volume of soil. Additionally, the mode of air travel within the subsurface is determined by soil permeability. In coarse soils such as gravels, the air will migrate in bubble form. Bubble form is desired because it maximizes the surface area through which mass transfer and removal is to occur. In less permeable soils, such as sands and silts, the air will migrate through a channel network formed during injection. The extent of the channel network is dependent on the permeability and the air flow being used; coarser sands and higher flow rate will lead to the development of higher channel densities.

A soil must be not only permeable but also relatively homogenous for proper air flow to occur. The air injected into the subsurface will seek the path of least resistance as it migrates away from the well. If a soil is too heterogeneous, large regions of less permeable soil will be left devoid of any air contact (Baker and Benson, 1996).

To help investigate the effect of soil permeability and heterogeneity on subsurface air flow, Ji et al. (1993) performed a series of laboratory tests. Using a two-dimensional Plexiglas tank and different sizes of glass beads to represent a saturated soil matrix, they were able to simulate a variety of subsurface conditions. When 4-mm diameter glass beads were used (representing a gravel), the injected air was seen to travel in bubble form with very little lateral spreading from the injection well. The use of 0.75-mm diameter beads yielded air flow in the form of channels, and a higher density of channels was observed with increased air flow. When 2-mm diameter glass beads were used, a transition pattern was observed. When lenses, layers, and other heterogeneous structures in the soil matrix were included, it was observed that air flow was very sensitive to such inclusions and would follow paths of lesser resistance, leaving regions in the vicinity of such inclusions untouched by the injected air. Similar conclusions were drawn based on the laboratory testing of different types of soils and sediments by Semer et al. (1996, 1998), Reddy and Adams (1999), and Peterson et al. (1999).

14.3.2 Mass Transfer Mechanisms

Volatilization. Volatilization has been shown to be the dominant removal mechanism occurring during the use of air sparging (Unger et al., 1995; Johnson, 1998). It is defined as the partitioning of a contaminant from the aqueous or nonaqueous phase into the vapor phase. During air sparging, the injected air provides both a medium and a transport network to allow for such partitioning to occur. Additionally, the injected air disturbs the equilibrium that exists between the liquid and vapor phases. When the vapor phase is pulled away from its liquid source, the resulting nonequilibrium will force additional partitioning into the vapor phase. Generally speaking, the higher the vapor pressure of a contaminant, the easier it is to remove through volatilization. For a given contaminant, however, volatilization is two orders of magnitude lower in soil than when the contaminant resides in a liquid (Davis, 1997).

Henry's Law. For a given contaminant, the vapor phase and liquid phase fractions under equilibrium conditions are determined by Henry's law, which can be written as follows:

$$P_a = C_{\text{water}} H_c \tag{14.1}$$

where P_a = partial pressure of the gas (atm)
C_{water} = aqueous-phase molar concentration of VOC (mol/m^3)
H_c = Henry's law constant ($\text{atm-m}^3\text{/mol}$)

The units of Henry's law may be altered; a dimensionless Henry's law constant may be used to relate aqueous-phase concentration (mg/L) to vapor-phase concentration (mg/L). Unfortunately, due to the difficulties of determining the solubility and vapor pressure of a variety of compounds, the Henry's coefficient has proven to be difficult to measure, with reported values often differing by two orders of magnitude or more (Davis, 1997).

For a given compound, a higher Henry's coefficient will indicate that more of the compound will exist in the vapor phase at equilibrium. Similarly, higher Henry's coefficients for a given contaminant result in more of the contaminant being accepted by rising air (Semer and Reddy, 1997). For a given compound, the Henry's law constant will increase with increasing temperature. Even though subsurface temperatures remain nearly constant year round in a majority of cases, methods such as steam injection or soil heating may offer advantages. Increasing temperatures, however, may be of little significance because most volatilization takes place during the early stages of air sparging (Sellers and Schreiber, 1992; Unger et al., 1995) and temperature increases may not help to remove contamination at later stages.

In order for Henry's law to be valid in determining the rate of volatilization for a given contaminant, a condition of equilibrium must exist. For this to occur, there must be an adequate amount of residence time for air within the groundwater. Unfortunately, short residence times, discrete channels, high velocities, and the lengths of travel paths prevent equilibrium from occurring within the subsurface (Ahlfeld et al., 1994b; Sellers and Schreiber, 1992). Therefore, Henry's law does not give a good approximation for volatilization rates, and models based on Henry's law may not provide good system approximations (Sellers and Schreiber, 1992).

Raoult's Law. When subsurface NAPL-phase contamination exists, it is rarely in the form of a single compound. Distilled petroleum products, for instance, contain a great number of constituent compounds with widely ranging properties. In order to help determine the composition of vapor from a multicompound NAPL-phase contaminant, Raoult's law is used, which may be written as follows:

$$P_i = V_{pi} X_i \tag{14.2}$$

where P_i = partial pressure of compound i in the gas phase
V_{pi} = vapor pressure of pure compound i
X_i = mole fraction of compound i in the liquid mixture

During volatilization, the compounds with higher volatility and higher solubility will be removed first, leaving behind compounds with lower volatility and solubility (Hayden et al., 1994). Because less volatile and soluble compounds remain, a tailing effect will be observed in removal as more effort is required to remove less volatile contaminants.

Volatilization is the dominant removal mechanism in the early stages of air sparging, but its importance is diminished with increasing remedial time. When air injection begins, contaminant near a channel will be easily volatilized. If a high-density channel network exists within the subsurface, this will account for a high percentage of removal and a rapid reduction in initial contaminant levels. Eventually, as the contaminant in the vicinity of channels is removed, removal rates decrease as contamination is forced to diffuse towards a channel. Thus, volatilization and overall removal rates drop when the concentration gradient drops at the air–water interface (Ahlfeld et al., 1994a).

As a demonstration, Braida and Ong (1997) performed a laboratory study of the spatial effects of volatilization. Using a small apparatus and an Ottawa sand with D_{50} of 0.2 mm, removal rates of BTEX compounds at a variety of concentrations were analyzed. Flows ranging from 460 to 2,200 m/day were used. During the first three hours of operation, volatilization was much faster than diffusion, leading to a sharp concentration gradient at the air-water interface. After three hours, however, volatilization rates dropped as diffusion became the controlling process. Contaminants with higher aqueous diffusivity were found to be removed easier.

Dissolution. Dissolution plays an important removal role during air sparging because it assists in the volatilization and ultimate removal of the contaminant. While volatilization controls the process during the early stages of an air sparging program, contaminant dissolution from the NAPL phase to the aqueous phase is a controlling factor in later phases (Unger et al., 1995). Solubility limits vary greatly for different contaminants, but even in the case of compounds with low solubility, maximum contaminant levels for drinking water are often at least two orders of magnitude lower than solubility limits (Miller et al., 1990). Yet the solubility limit of a contaminant can often indicate whether a specific compound may be removed in a timely fashion. Burchfield and Wilson (1993) used a model and reported that an increase in the Henry's coefficient of a given contaminant did not greatly affect its removal rate using air sparging, but an increase in the solubility of a contaminant greatly increased its rate of removal.

When a contaminant enters into a soil, it will often become trapped as a blob within a single-pore or multiple-pore spaces. Once these nonaqueous phase blobs are created, they need to dissolve out of the immiscible phase into the aqueous phase, at which time they may be transported in the aqueous phase and eventually volatilized (Marley, 1992). Until this occurs, however, an immobile NAPL blob can act as a sink/source for organic solutes, enhancing retardation and delaying removal (Brusseau, 1992). When considering trapped NAPL blobs, Malone et al. (1993) reported that they may be placed into two categories: some will exist in a fast condition during which the NAPL readily exchanges mass with the aqueous phase. This occurs when there is a high NAPL surface area-to-volume ratio. The remaining NAPL will exist in a slow condition, during which there will be a slow, reversible mass transfer between the two phases. This situation occurs when the NAPL blob features a low surface area-to-volume ratio. It is obvious that surface area is an important factor because it is across this interface that dissolution will take place.

Powers et al. (1991) presented a formula to determine the interfacial area across which mass transfer may occur. The specific interfacial area between the NAPL and aqueous phases was found to be a function of soil porosity, NAPL saturation, the surface-to-volume ratio of the NAPL blob, and the fraction of blob surface area exposed to mobile water, written as follows:

$$a_o = n\, S_o\, (A/V)\, f \tag{14.3}$$

where a_o = specific area between the two phases
n = soil porosity
S_o = degree of NAPL saturation
(A/V) = surface area to volume ratio for the blob
f = fraction of blob surface area exposed to mobile water

NAPL size has been shown to be an important parameter (Gomez-Lahoz et al., 1994). Hunt et al. (1988a, b) found that the maximum NAPL blob length was a function of the surface tension between the NAPL and the water, the permeability of the soil, the viscosity of water, the Darcy velocity, and the throat radius of the pore. It was found that blob size and NAPL residual saturation could only be reduced by increasing the Darcy velocity or reducing the surface tension between the two phases.

When multiple contaminants are present in the subsurface, there may be a synergistic effect allowing for increased or decreased solubility compared to when the contaminants exist alone. If gasoline contains methyl tertiary butyl ether (MTBE), solubility of the BTEX contaminants is greatly increased. Additionally, Geller and Hunt (1993) presented a dissolution relationship relating to multiple contaminants. The aqueous equilibrium concentration of a single compound was found to be a function of its mole fraction in the contaminant mixture. As the more soluble compounds dissolve, the less soluble compounds remain, but their mole fraction within the mixture is increased, forcing dissolution. This relationship does not hold if a contaminant's aqueous solubility is less than 0.001 mole fraction.

There is a growing body of evidence that dissolution is a nonequilibrium process. Voudrias and Yeh (1994) reported that the concentrations of aqueous-phase toluene in flowing water below floating NAPL-phase toluene pools were much lower than the solubility limit. Brusseau (1992), Szatkowski et al. (1995), and Hunt et al. (1988a, b) have reported that NAPL-aqueous phase partitioning is a rate-limiting, nonequilibrium process. This is supported by discrepancies between theoretical and field-measured solubilities. Some reasons that discrepancies may occur include the rate-limited mass transfer between the NAPL and aqueous phase, physical bypassing of the aqueous phase around contaminated low-permeability regions, non-uniform flow due to soil heterogeneity, and synergistic effects between compounds (Powers et al., 1991). It is suggested that nonequilibrium descriptions are needed for high pore water velocities, hydrophobic solutes, high dispersivities, as well as small spills, large blobs, low residual saturations, and heterogeneous aquifers (Powers et al., 1991).

After air sparging has ended and the subsurface is allowed to reach equilibrium once again, additional dissolution of residual contaminants may take place, serving to increase groundwater concentrations. This effect is known as rebound. Rebound is largest in areas that were subjected to low air flux during injection (Gomez-Lahoz et al., 1994). Bass and Brown (1997) reported the following equation for rebound:

$$\text{Rebound} = \frac{\log\left(\dfrac{C_r}{C_f}\right)}{\log\left(\dfrac{C_o}{C_f}\right)} \tag{14.4}$$

where C_r = dissolved concentration during postmonitoring
C_o = initial dissolved concentration
C_f = final dissolved concentration

If the rebound is less than 0.2, permanent reductions in contaminant concentrations have been achieved, but a rebound value greater than 0.2 is regarded as substantial rebound (Bass and Brown, 1997).

Adsorption/Desorption. When contaminants are released into the subsurface, they may sorb onto the particles of the soil matrix. Adsorption may become substantial when organic soil content or clay is present. Organic contaminants will be subject to increased adsorption with increased organic content. Adsorption may be a reversible process; if driven, contaminants are able to desorb from the soil matrix and into the aqueous phase. If the ratio of adsorption to desorption is less than unity, the soil has no retention capability and adsorption indeed becomes a reversible process.

Adsorption can be a difficult process to predict accurately. Brusseau (1992) has reported that equilibrium cannot be assumed for adsorption/desorption, and differences between predicted and actual rates may differ between one and three orders of magnitude. The differences may be due to the effects of competitive sorption or to the enhancement caused by cosolutes (Brusseau, 1992). Additionally, organics may bind to organic soil matter more tightly with time (Davis, 1997).

In order to help predict adsorption, the octonol/water partition coefficient, K_{ow}, is used. It is defined as the ratio of the equilibrium concentrations of a dissolved solute in a system of octonol and water, two immiscible solvents. The partition coefficient is helpful in predicting soil adsorption; a higher K_{ow} indicates a more hydrophobic compound that will adsorb more readily to organic soil matter (Davis, 1997). This in turn has been used to estimate an adsorption coefficient by different formulas (Fetter, 1999). One such example is (Karickhoff et al., 1979):

$$K_d = 0.6\, f_{oc}\, K_{ow} \tag{14.5}$$

where K_d = adsorption coefficient
f_{oc} = organic fraction in soil

For any sizable amount of adsorption to occur, the soil must contain at least 0.1% organic matter (Davis, 1997). The amount of organic matter in deep soils is low, often less than 0.1% (Perlinger and Eisenreich, 1991). Clayey soils, however, may be prevalent in deep subsurface. The amount of adsorption depends on the type of clay minerals present in the soils. Adsorption to montmorillonite clay is higher than illite clay, which in turn causes higher adsorption than kaolinite clay (Semer and Reddy, 1997). Additionally, the amount of adsorption is also dependent on the water content of the soil because the contaminant must compete with water for adsorption sites (Semer and Reddy, 1997).

Different isotherms are used to represent adsorption under different conditions (Fetter, 1999). One that is commonly used to represent non-polar organic contaminant adsorption to soil is the Freundlich isotherm (Semer and Reddy, 1997). The isotherm may be represented as:

$$C_s = K\, C_e^N \tag{14.6}$$

where C_s = mass of contaminant adsorbed per mass of sorbent
K = sorption equilibrium constant
C_e = solution concentration at equilibrium after sorption
N = constant describing adsorption intensity

The higher the K and the N values, the greater the sorption capacity of the medium. The equation may be linearized by expressing it in logarithmic form. When this is done, N is an indicator of whether adsorption remains constant at different equilibrium concentrations (Semer and Reddy, 1997). If N is equal to one, a linear relationship exists between equilibrium concentrations and the amount of mass adsorbed; if N is less than one, adsorption decreases with increased equilibrium concentration levels; and if N is greater than one, adsorption increases with increased equilibrium concentration levels (Semer and Reddy, 1997).

Adsorption may also be modeled as a nonequilibrium process using nonequilibrium kinetic equations. In a kinetic model, the solute transport equation is linked to an appropriate equation to describe the rate that the solute is sorbed onto the solid surface and desorbed from the surface (Fetter, 1999). Depending on the nonequilibrium condition, the rate of sorption may be modeled using an irreversible first-order kinetic sorption model, a reversible linear kinetic sorption model, a reversible nonlinear kinetic sorption model, or a bilinear adsorption model (Fetter, 1999).

In order for air sparging to be successful, desorption must occur if organic or clayey soils are present at a given site. Although organic matter is usually confined to the upper layers of a soil stratum, clay may be present at all depths. If desorption takes place at an acceptable rate, air sparging will be successful at remediating contaminated saturated soils. If desorption does not occur, saturated soils will remain contaminated and will act as a source to further contaminate the surrounding groundwater.

14.3.3 Mass Transfer Mechanism: Biodegradation

Biodegradation is different from the other removal mechanisms that act on subsurface contamination during the air sparging process. While other mechanisms act to transfer the contaminant out of the subsurface, biodegradation acts to transform the contaminant into non-toxic products. Biodegradation can often account for a substantial portion of the remedial process by acting on the less volatile, more strongly adsorbed contaminants during the later stages of remediation and when dissolved concentrations have decreased (Miller, 1996; Johnson, 1998). Biodegradation becomes a significant removal mechanism when dissolved contaminant concentrations are less than 1 mg/L (Johnson, 1998). Additionally, middle distillate fuels, including diesel fuel and kerosene, are not easily volatilized and rely upon biodegradation for treatment (Hoeppel et al., 1991). Biodegradation occurs when native microbial communities within the subsurface interact with and metabolize organic contaminants. An electron acceptor (oxidizing agent) receives electrons from the electron donor (Middleton et al., 1995). The contaminant is mineralized into end products that are based both on the type of contaminant and the oxidizing agent. Biodegradation rates may be monitored in a number of ways, including the degree of consumption of electron donors, the degree of consumption of electron acceptors, and the production of metabolic byproducts (Chapelle et al., 1996). The process may involve direct mineralization or the transformation of the contaminant into intermediate compounds before mineralization is completed. Malone et al. (1993) reported that benzene, toluene, and the xylenes are degraded into intermediate compounds before they are mineralized into carbon dioxide and water.

Optimal subsurface conditions are necessary in order for biodegradation to occur successfully. While suitable microbial populations or nutrient levels may be a limiting factor to the rate of biodegradation (Widdowson and Aelion, 1991), the type of electron acceptor is often the controlling factor. The most efficient electron acceptor for biodegradation is oxygen. Aerobic biodegradation utilizes oxygen as an electron acceptor. Additionally, nutrients such as phosphorus and nitrogen necessary for cell growth and reproduction must be present if bioremediation is to be successfully implemented. During aerobic degradation of benzene, the following reaction occurs (Chiang et al., 1989):

$$C_6H_6 + 2.5\ O_2 + NH_3 \rightarrow C_5H_7O_2N\ \text{(cells)} + CO_2 + H_2O$$

When sufficient oxygen is lacking and anoxic conditions exist within the subsurface, biodegradation becomes less efficient. During anoxic conditions, other subsurface electron acceptors are utilized. Following oxygen as the preferred subsurface electron acceptor, in descending order of efficiency, possible electron acceptors include nitrate, ferric iron, sulfate, and carbon dioxide (Middleton et al., 1995). Despite preferences in anoxic conditions, field tests have shown that sulfate reduction is the primary electron accepting process (Chapelle et al., 1996; Thierrin et al., 1995). By adding oxygen to the subsurface through air sparging to create oxic conditions, more efficient, aerobic degradation is allowed to occur (Crocetti et al., 1992).

In some cases, oxygen is needed in order for any sufficient biodegradation to occur. While toluene has been shown to degrade using a variety of electron acceptors, benzene will not degrade to any measurable extent without the presence of oxygen (Middleton et al., 1995; Thierrin et al., 1995). Chapelle et al. (1996) reported that in field tests, benzene degradation rates were low under nitrate-amended conditions. Patterson et al. (1993) reported in another field study that benzene persisted in anoxic conditions but was easily degraded under oxygenated conditions. Thus, the presence of oxygen may be a rate-limiting factor for efficient biodegradation (Crocetti et al., 1992).

In order for aerobic biodegradation to occur, there must be an adequate level of oxygen within the soil. Miller (1996) reported that 2 to 4% oxygen by volume is needed within the subsurface for sustaining biodegradation to occur. Unfortunately, this level of subsurface oxygen may be difficult to achieve. It is estimated that between 3 to 3.5 g of oxygen are needed to completely mineralize hydrocarbons (Johnson et al., 1993). Oxygen can be supplied to the subsurface through the infiltration of precipitation, but this oxygen is quickly utilized (Chapelle et al., 1996). Groundwater may be a poor medium through which to transport oxygen; Hoeppel et al. (1991) reported that 75,000 kg of groundwater are needed to break down 1 kg of hydrocarbons. Therefore, the delivery of oxygen to the subsurface cannot be achieved by injecting saturated groundwater, but through the use of air sparging, oxygen levels can be increased to increase the rate of biodegradation and ultimately improve remedial efficiency. When injecting air into the subsurface, oxygen levels can reach 8 to 10 mg/L, while the injection of pure oxygen can increase levels to 40 mg/L (Johnson et al., 1993). The injection of hydrogen peroxide can increase subsurface oxygen levels to 500 mg/L (Johnson et al., 1993).

Field studies have indicated successful increases of biodegradation during the use of air sparging. Covell and Thomas (1997) reported on the use of air sparging at an underground coal gasification site. Groundwater was used to help control air flow as well as deliver nutrient (ammonium phosphate) to the subsurface. After six months, over 80% of the benzene initially present at the site was removed. It was determined that volatilization was the primary initial removal mechanism; while biodegradation served as an effective polishing step.

Yaniga and Smith (1984) reported on the restoration of an aquifer that had been contaminated with gasoline-range hydrocarbons. The soils consisted of a heavy silt loam over fractured shale. A number of methods were employed at the site, including excavation of the upper soils and the inclusion of an infiltration gallery with air sparging and an air stripping tower. Even though the sparger had to be frequently cleaned due to biological fouling, a 50–85% reduction was achieved within 11 months. Because of clogging, the sparge point had trouble delivering oxygen to the subsurface. Therefore, a solution of 100 mg/L hydrogen peroxide was also injected. This served to increase oxygen delivery while reducing clogging of biological material.

While BTEX compounds can be rapidly degraded under aerobic conditions (Malone et al., 1993; Patterson et al., 1993), aerobic biodegradation of chlorinated solvents is more difficult, often requiring the presence of cosubstrates for degradation. They may, however, be degraded under anaerobic conditions, especially by methanogenic bacteria. Lombard et al. (1994) reported on the use of air sparging with the use of methane as the injected gas at the Savannah River Site in Aiken, South Carolina. Approximately 3,500,000 lb. of trichloroethylene and tetrachloroethylene were detected in the subsurface. A methane–air mixture was injected in order to stimulate the native methantropic bacteria. Extraction was performed at a rate 20% higher than injection to control site behavior as well as prevent an explosive condition (greater than 5% methane).

Longer exposure of contaminants to the subsurface can actually sometimes help to increase the rate of biodegradation. Crocetti et al. (1992) reported that soils contaminated with heating oil had heterotropic bacteria populations one to two orders of magnitude higher than background soils. Patterson et al. (1993) found that during a field study, degradation rates increased with time, especially in the case of toluene, possibly due to increased degradation efficiency of the resident bacteria. Jin et al. (1994) demonstrated that in three soils contaminated with toluene (sterile soil, soil preexposed to toluene, and soil not previously exposed to toluene), degradation rates increased with increased exposure time due to the buildup of an enriched microbial population.

14.3.4 Transport Mechanisms

In addition to the mass transfer and transformation mechanisms that occur during air sparging, VOC transport mechanisms play a role in ensuring remediation. The important transport mechanisms that occur during air sparging include advection, dispersion, and diffusion.

Advection and Dispersion. Advection is defined as the transport or movement of vapors or liquids in air or water due to the response to a pressure gradient (Bear, 1972; Fetter, 1999). The migration due to advection is dependent on the pressure gradient and the soil permeability. The soil permeability is dependent on soil grain size and grain size distribution, the type and structure of the soil, the soil porosity, and the water content of the soil. While both groundwater and soil gas flow are governed by Darcy's law, mass flux due to advection in one dimension can be written as (Fetter, 1999):

$$F_v = v\, n_e\, C \tag{14.7}$$

where F_v = mass flux
v = flow velocity
n_e = effective porosity
C = contaminant mass per unit volume of solution

For one-dimensional transient conditions and uniform velocity, the advective transport equation may be written as:

$$\frac{\partial C}{\partial t} = -v \frac{\partial C}{\partial x} \tag{14.8}$$

Dispersion is defined as the mixing or spreading of the contaminant in either the groundwater or soil gas due to differences in micro-scale flow velocities. Once again, soil permeability affects dispersion as pore size, path length, and pore friction create these differences in flow velocity. Dispersion that occurs along the flow path is known as longitudinal dispersion, while flow occurring perpendicular to the flow path is known as transverse dispersion. Both advection and dispersion are important mass transport mechanisms. Hein et al. (1994) reported that model simulations indicated that mass transfer/transport is dominated in part by gaseous advection and dispersion. Pepper and Stephenson (1995) reported that groundwater transport is advection-dominated, while Wilson et al. (1994a) reported that injection wells should be operated in a manner that will generate the maximum amount of dispersive mixing, a situation that can be accomplished using pulsed air injection.

Advection and dispersion act to disrupt the saturated zone by creating turbulence. The turbulence created by air injection serves to disrupt equilibrium. This action can help introduce water into dead-end pores previously occupied by NAPL ganglia as well as create groundwater flow or circulation. Such groundwater movement will aid dissolution by transporting dissolved contaminant away from NAPL sources, creating concentration gradients that will force additional dissolution. This type of behavior will also aid in desorption as concentration gradients will be established near the surface of soil particles with sorbed contaminants.

Caution should be used, however, when inducing subsurface turbulence. While this turbulence allows for desirable effects, too much movement may force unwanted migration into areas previously free of contamination. Additionally, some mechanisms, especially adsorption/desorption, are reversible, and thus advection/dispersion can act to trap contaminant in dead-end pores as well as force additional contaminant adsorption. Therefore, subsurface air flow should be carefully monitored to help minimize any negative effects of advection/dispersion.

Diffusion. Diffusion is defined as the migration of contaminant from regions of high concentration to regions of low concentration. The diffusive flux is governed by Fick's law and also depends on tortuosity, or the length of the transport path, within the soil (Bear, 1972; Fetter, 1999). Diffusion is the primary mechanism for the removal of NAPL trapped in dead-end pores. Diffusion must occur in order for movement to occur from such trapped sites into regions where other processes may occur for removal. It is also a means of transporting contaminant from source locations to air channels in order for volatilization or biodegradation to take place. Diffusion is an extremely slow process; Ahlfeld et al. (1994a) reported that in a water column with a 20-in. radius, 78 years would be required to reach a concentration of 5 parts per billion if flow took place in a 1-in.-radius channel, 1.7 years if the channel radius were increased to 2 in., and 27 days if the channel radius was increased to 5 in. Therefore, diffusion often becomes a rate-limiting process (Ahlfeld et al., 1994a).

While diffusion rates in gas can be calculated to a reasonable degree of accuracy, it is more difficult to do so in liquids (Davis, 1997). Diffusion is between three and four orders of magnitude slower in liquids than in gas (Petersen et al., 1994). Additionally, diffusion was determined to be affected by porous media (Davis, 1997), and diffusion rates in porous media will decrease with increased saturation (Petersen et al., 1994). Davis (1997) reported that soil gas diffusion can be approximated as follows:

$$\frac{D_s}{D_a} = \frac{a^2}{n^{2/3}} \tag{14.9}$$

where D_s = soil diffusion
D_a = air diffusion
a = volumetric air content for soil
n = soil porosity

Diffusion of gas within soil is important as it will help determine any type of air or vapor movement. Additionally, trapped gas may also be removed through diffusion (Fry et al., 1995).

14.4 SYSTEM DESIGN AND IMPLEMENTATION

When designing an air sparging system, a number of site-specific and operator-controlled parameters must be addressed. A flowchart summarizing air sparging design process is shown in Fig. 14.2. The soil profile needs to be explored in order to determine what soils exist. The hydrologic conditions at the site also need to be analyzed. Questions need to be answered regarding site contamination, including contaminant type and properties, concentrations, and locations. If the contaminants present at the site are conducive to removal using air sparging, then system parameters such as the number, spacing, and depth of air injection wells need to be determined. Flow rates and pressures need to be determined, as well as the details of the soil vapor-extraction system. Once the system is designed, the performance of the system needs to be monitored to ensure adequate performance. If the system is not performing well, changes need to be made and performance reevaluated. This iterative design process continues until the air sparging system is performing to expectations.

14.4.1 Equipment

A major advantage of air sparging is that it requires only simple materials to construct a field system. No special equipment needs to be designed; and all equipment used is easily obtained. Equipment required for the air injection system and soil vapor extraction system is summarized in this section.

Injection System. An air compressor is used to force air into the air injection wells. The compressor must be capable of injecting air at a proper flow rate under suitable air pressure. The number of wells that are fed by a given compressor also dictates the size of the air compressor. Typically, reciprocating or rotary screw compressors are used during field application (Marley et al., 1992a). Regardless of what type is used, the compressor must be oil-free in order to prevent the delivery of additional contamination to the subsurface (Johnson et al., 1993).

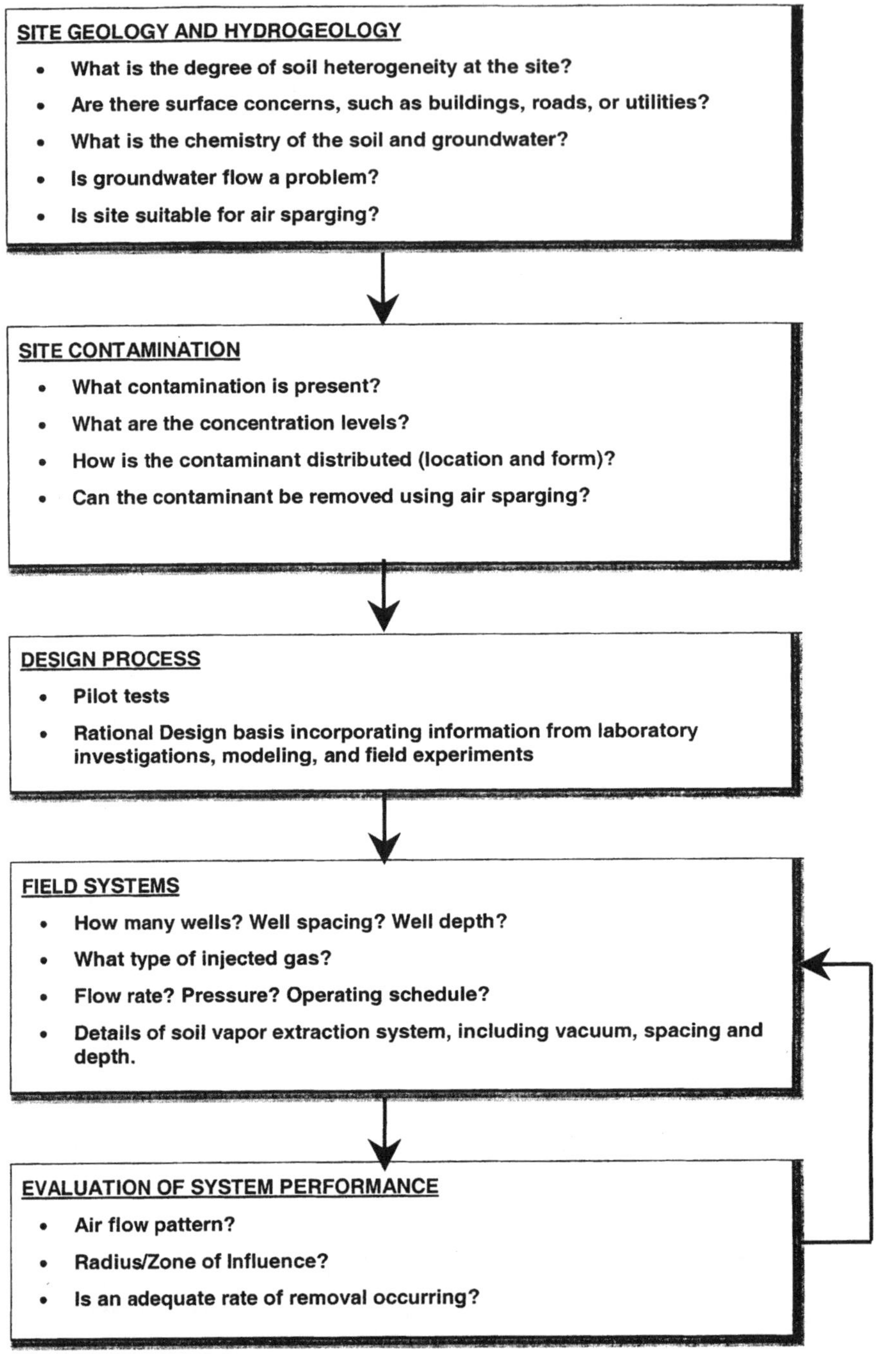

FIGURE 14.2 Design process for air sparging systems.

The air leaves the compressor at a flow rate and pressure that is controlled through the use of regulators. It passes into a manifold system used for delivery to the injection wells. Because the air leaving the compressor may be high in temperature, rubber hose or metal pipes are often used in the manifold system (Marley et al., 1992a). Valves are included in the manifold system to ensure that the desired amount of air is delivered to each well. The injection well usually consists of 1- to 4-in.-diameter PVC pipe, although stainless steel may be used if steam or heated air is to be injected (Johnson et al., 1993). One- to 2-in. in diameter PVC pipe is often used because it is cheaper to install (Johnson et al., 1993). A borehole for well placement is often drilled, but the use of direct push methods for well placement may provide significant cost savings (Joiner and O'Keefe, 1996). The well screen is placed below the lowest known point of contamination to ensure that the entire plume is addressed.

Nyer and Suthersan (1993) stated that injection should be at least 1 to 2 ft below the lowest point of contamination, but typical injection points during field applications are 10 ft below the lowest point of contamination (Bruell et al. 1997). Loden (1992) reports that wells are effective in delivering air to a depth of 150 ft, but if depths are below 40 ft, a series of nested wells should be used (Nyer and Suthersan, 1993). The well screen is usually 1 to 3 ft long (Marley et al., 1992a). The screens help minimize any effects of clogging as well as provide for even air distribution. The well is backfilled with 6 in. to 2 ft of sand or gravel to act as a filter and is then sealed with a bentonite seal to prevent air from short-circuiting to the surface. The remaining annulus is grouted to the surface to assist in providing a seal (Loden, 1992).

Soil Vapor Extraction System. Once the air is injected into the saturated subsurface, it will migrate toward the surface due to the effect of buoyancy. Eventually the contaminated injected air will enter the vadose zone. At this point, a soil vapor extraction (SVE) system may be implemented. Soil vapor extraction applies a vacuum to the vadose zone to assist in collecting vapors. The vacuum will help vapor collection as well as assist in preventing unwanted off-site migration. Equipment used in a soil vapor extraction system is similar to that used in an air sparging system, except that a vacuum pump is used instead of an air compressor. Additionally, if the vapors being collected by the soil vapor extraction system are moist, a dewatering system may be employed (Johnson et al., 1993).

The extraction flow rate is very important in order to control air and contaminant vapor migration. Simply stated, the extraction rates must be higher than the injection rates in order to prevent soil pressure buildup. Loden (1992) reported that the ratio of extraction to injection commonly used during field application are between 4:1 and 5:1. Johnson et al. (1990) reported that typical extraction rates are between 100 and 1,000 standard cubic feet per minute (scfm).

14.4.2 Operational Controls

In addition to proper system design and layout, proper choices regarding the operating parameters of an air sparging system are essential to the success of the remediation program.

Injection Media. The first choice that must be made is the type of gas to be injected into the subsurface. The choice of gas is especially important in ensuring the delivery of oxygen to the subsurface in order to stimulate biodegradation. Air is the gas most commonly used in remedial programs. Air is low in cost and not explosive or flammable. When air is injected into the subsurface, dissolved oxygen levels may be raised from typical concentrations of about 2 mg/L to between 9 and 10 mg/L, thereby providing suitable oxygen levels for biodegradation (Johnson et al., 1993).

If higher levels of dissolved oxygen are desired, pure oxygen or air enriched with oxygen may be used. When pure oxygen is injected into the subsurface, dissolved oxygen levels may reach 40 mg/L (Johnson et al., 1993). Even if the reported oxygen levels can be delivered to the subsurface, Bausmith et al. (1996) reported that high levels of oxygen could stimulate such rapid biological growth that injection wells may become clogged with biomass. Additionally, the elevated levels of oxygen can force soluble manganese or iron present in the groundwater to precipitate into insoluble manganese dioxide (MnO_2) or ferric hydroxide ($Fe(OH)_3$). Calcium may precipitate as insoluble calcium carbonate ($CaCO_3$) in the presence of carbon dioxide (Bausmith et al., 1996). If chlorinated solvents are present at a site, aerobic degradation becomes difficult. These solvents, however, may be anaerobically degraded, especially by methantropic bacteria. In such situations, methane can be injected into the subsurface to stimulate methanogenic degradation (Leonard et al., 1994).

Another alternative is the use of ozone for injection, a method that is being increasingly adopted for treating both contaminated soils and groundwater. The injection of ozone allows subsurface ozonation to occur. Ozonation is an advanced chemical oxidation process that may be used to break complex organic contaminants into less persistent or less toxic molecules (Nelson and Brown, 1994). The ozone accomplishes this by assisting in the degradation of double-bonded organics (Leonard et al., 1994).

Steam may be used in place of air or oxygen for injection into the subsurface. The injection of steam increases the temperature of the subsurface, helping to improve the removal efficiency of semivolatile organic compounds (SVOCs) as well as stimulating microbial activity in colder soils and climates (Leonard et al., 1994). Steam has been used for years by the oil industry to recover crude oil from the subsurface. By 1979, 20% of the oil produced in California had been extracted from the subsurface using steam injection (Falta et al., 1992). During the use of steam injection, contaminant mass transport occurs by advection in all three phases and by multicomponent diffusion in the gas phase (Falta et al., 1992). Heat transfer to the subsurface occurs due to conduction, multiphase convection, and gaseous diffusion (Falta et al., 1992).

To study the performance of steam injection, Hunt et al. (1988b) performed laboratory tests in a 91-cm Pyrex horizontal glass column. Ottawa sand was contaminated with trichloroethylene, benzene, and toluene and then subjected to both water (pump-and-treat) injection and steam injection. The water injection proved to be ineffective, even when flow rates of up 15 m^3/day were used. When steam was injected, the contaminant was displaced as a slug just ahead of the steam front at flow velocities of up to 1.5 m/day. It was demonstrated that only one pore volume of fluid was needed to achieve cleanup using steam. Additionally, the use of steam allowed for the possibility of actual NAPL recovery.

Flow Rate/Mode of Injection. Once the type of gas to be injected is determined, the flow rate and mode of injection must be determined. Loden (1992) reported that flow rates of 2 to 16.5 scfm are typical for field application. Nyer and Suthersan (1993) reported that when soil vapor extraction systems are used at a site, injection flow rates between 4 and 10 scfm are used. It has been reported that any flow rates beyond contaminant diffusion kinetics are a waste of effort (Roberts and Wilson, 1993; Reddy and Adams, 1999; Adams and Reddy, 1999). Additionally, Rutherford and Johnson (1996) found that oxygen transfer into groundwater may actually be impeded by an injection flow rate that is high because the air will act to push the groundwater away from the point of injection, decreasing interfacial transfer area and oxygen transfer.

When injecting air, either continuous or pulsed air flow may be used. Pulsed flow involves using a cycle that consists of injection followed by a time of cessation. After the period of cessation has elapsed, air is once again injected and the cycle is repeated. Wilson et al. (1994a, b) reported that pulsing helps improve removal efficiency by inducing mixing. McKay and Acomb (1996) stated that frequent pulsing is needed to optimize oxygen distribution within the subsurface. Reddy and Adams (1998) and Adams and Reddy (1999) found that pulsed air injection improves removal efficiency in finer soils, whereas it does not affect the removal in coarser soils.

Injection Pressure. When air sparging is performed, air must be injected under proper pressure. If air is injected under pressure that is too low, it will be unable to enter into the subsurface. If air is injected at pressure that is too high, the contaminant plume may be forced to migrate into previously uncontaminated areas. Additionally, soil heaving or fracturing may result within the subsurface if injection pressure is too high. The necessary pressure needed for air to enter into the subsurface is the sum of hydrostatic pressure at the injection point, friction due to exiting from the well, and the capillary pressure due to the interface of two fluids within the porous media (Johnson et al., 1993; Ahlfeld et al., 1994a; Bausmith et al., 1996). One pound per square inch of pressure is needed to overcome every 2.3 ft of static water above the injection point (Bausmith et al., 1996). The capillary or air entry pressures that must be overcome may range between a few centimeters of pressure head in coarse, sandy soils to several meters in low-permeability clay (Johnson et al., 1993). Wehrle (1990) reported that additional air pressure will be needed to overcome capillary pressure when the D_{50} of a soil is smaller than 0.8 mm.

Vapor Treatment. When soil vapor extraction is used, the contaminated vapors that have been collected must be addressed. The simplest solution is to release the contaminated vapors directly to the atmosphere. Using this option, the effluent gases are passed through a diffuser stack and released into the air. This is applicable only when local regulations allow and compliance with air emission standards can be ensured. When vapor treatment is needed, a popular option is the use of carbon adsorption. The vapors are passed through an activated carbon matrix, effectively removing contaminated vapors from the effluent gas. Either virgin or recharged carbon may be used during application (Loden, 1992). Thermal incineration is another option. Temperatures between 1,000 and 1,400°F are used to destroy the contaminants, providing contaminant destruction rates of 95–100% (Loden, 1992). Care must be taken not to exceed the explosive limit; this can be avoided by diluting and mixing the vapors with fresh air (Loden, 1992). Lower temperatures may be used with the use of catalytic oxidation. Because a catalyst is employed, temperatures may be lowered to values of 600–800°F. Such a process allows for contaminant destruction efficiencies of up to 85% (Loden, 1992). Additionally, if off-vapors provide proper conditions, internal combustion engines may be used to provide power while treating the vapors.

14.5 PREDICTIVE MODELING

Mathematical models for predicting air sparging performance will eliminate a great deal of the guesswork involved in system component design and layout. Several researchers have developed mathematical models to simulate air sparging performance (Marley et al., 1992b; Sellers and Schreiber, 1992; Wilson, 1992; Burchfield and Wilson, 1993; Roberts and Wilson, 1993; Gomez-Lahoz et al., 1994; Reddy et al., 1995a, b; Reddy and Zhou, 1996; Lundegard and Andersen, 1996; McCray and Falta, 1997; Wilson et al., 1994a, b; Wilson and Norris, 1997; Rabideau and Blayden, 1998; and van Dijke and van der Zee, 1998).

Simple models may be used to study specific mass transfer, transformation, or transport processes, or more rigorous models may be used to study macroscale performance. Unfortunately, because air sparging creates dynamic, nonequilibrium processes, models based on simplifying assumptions may provide misleading results. Additionally, many models have yet to be validated with controlled laboratory test data and field data.

14.6 PERFORMANCE ASSESSMENT

Monitoring of air sparging systems installed at different sites provided valuable information on the air sparging process, particularly the radius or zone of influence and contaminant removal behavior, which are briefly explained in this section (Bruell et al., 1997; Bass and Brown, 1997).

14.6.1 Radius of Influence / Zone of Influence

Arguably the most important factor related to air sparging is the radius of influence (ROI), defined as the lateral distance from the center of the sparge well to the edge of the region that is impacted by the injected air. The region of soil that is delineated by the ROI is known as the zone of influence (ZOI); it is the actual volume of soil that is impacted by the injected air. Both the radius and the zone of influence are of the utmost importance for the successful design and implementation of an air sparging system. Since the radius and zone indicate the footprint of treatment that a given well creates, it assists in the proper spacing and layout of wells. If the design radius of influence is underestimated, the wells will be spaced closer together than necessary, leading to an overdesigned and overpriced remedial system. If the radius of influence used for design is overestimated, the wells will be spaced too far apart, leaving regions of the contaminated soil profile untreated.

The radius and zone of influence are heavily dependent on soil permeability. The injected air rises towards the surface due to buoyancy, but the conical zone is established because of lateral air migration. The lateral air migration takes place because of differences between the horizontal and vertical permeability. Because horizontal permeability may be 10–1,000 times greater than vertical permeability (Nyer and Suthersan, 1993), the vertical resistance is high enough to overcome buoyancy and force some lateral movement. Overall soil permeability also dictates the size of the radius/zone of influence; the less permeable a soil, the larger the radius/zone of influence. Nyer and Suthersan (1993) reported that the angle of the zone tangent with the vertical ranges from 15° for coarse gravels to 60° for silty sands. Zumwalt et al. (1997) performed laboratory tests with homogenous, isotropic sands and measured an angle of 26.5°.

Since the radius of influence is so important in the design of a field system, its accurate measurement is also crucial to ensure proper performance during operation. Several different methods are available to facilitate field measurement of the radius/zone of influence. Water table mounding is defined as the increase in the height of the water table due to the additional pressure caused by the injection of air. It can be measured rather easily through the use of monitoring wells. Unfortunately, the increase in water table height is a transient situation; the increase in table height creates a hydraulic gradient within the table that will eventually level out the table, creating a steady-state condition. At a remedial site in Florence, Oregon, Lundegard and LaBrecque (1995) measured a radius of influence of 63 ft 30 minutes after injection began, but the increase in water table elevation had disappeared within 3 to 4 hours.

A similar method that may be employed is the observation of air bubbles within groundwater monitoring wells. The bubbles are the result of the intersection of an air channel or flowing bubbles with the wall of the well. The bubbles provide direct evidence that air flow is occurring within the subsurface of the well. In a related application, Kraus et al. (1997) reported on the progress of remediation at a field site. The remedial site was at a lower elevation than the surrounding area and often became submerged with surface runoff following summer rainstorms. While the site was submerged, the effluent air was observed exiting the soil profile because it bubbled through the water lying above. Radii of influence measuring between 4.5 and 12 m were observed during injection.

The mounding of groundwater is essentially the measure of water table encroachment into the vadose zone. Other measures of the radius/zone of influence are possible, including the measurement of VOC vapors in the vadose zone, the increase of dissolved oxygen, the extent of tracer gases that may have been included in the air injection process, and the increase of overall soil gas pressure within the vadose zone. Caution should be exercised, however, when using vadose detection methods because differences exist between the permeabilities of the saturated zone and the vadose zone. Any type of measurement that requires measurement of gas within the vadose zone may be skewed by the migration patterns that occur and may not be representative of conditions within the saturated zone (Javanmardian and Kremesec, 1995). McCray and Falta (1997) used a model known as T2VOC to analyze a subsurface flow pattern occurring during air injection. They found that positive pressure measurements made below the water table closely corresponded to subsurface gas distribution and the zone of influence, showing that soil gas pressure measurement is an easily implemented technique that is a good monitor of the radius/zone of influence.

Other gas measurements may be made within the saturated zone. Using monitoring devices placed directly in the saturated zone, increases in dissolved oxygen or carbon dioxide caused by the sparging process may be measured. The injected air will cause volatilization of the subsurface contaminants to occur; therefore, measurements of contaminant vapors within the subsurface may help determine the extent of the zone of influence. During injection, tracer gases may be sparged into the groundwater. One such gas that has been widely used is sulfur hexafluoride (SF_6). Sulfur hexafluoride is virtually nonreactive, has very similar dissolution characteristics to those of oxygen, and is detectable at very low concentrations (Javanmardian and Kremesec, 1995). The changes in soil gas composition may also be indirectly determined through the measurement of major consumers/producers of soil gas, the native microbial population.

Some of the above techniques, including measurements of water table mounding, soil gas pressure, and tracer gases are indirect indicators of subsurface air flow and the radius/zone of influence; caution should be exercised when using such methods (Lundegard and LaBrecque, 1995). Newer, more innovative techniques for measuring subsurface air flow are being increasingly used. Goldflam et al. (1997) implemented radar at a remedial site. They used a RAMAC 60 MHz center borehole radar to determine both the extent of air flow and the zone of influence. This system provided definition on the scale of 10 cm due to the spacing of the rays. Another innovative technique that has seen widely used recently is electrical resistivity tomography (ERT). During implementation, an electrode pair is placed within the subsurface, and a current is passed between the two electrodes. Air insulates electricity, but groundwater is a conductor of electricity (Schima et al., 1996). Subsurface electrical resistance is a function of saturation, porosity, and clay content. The resistance to current flow is measured, with regions of high resistance indicating regions of air flow. During measurement, a resistivity distribution is created, and a finite element mesh is generated for analysis.

Lundegard and LaBrecque (1995) performed an extensive field test in order to compare several methods for radius of influence measurement with electrical resistivity tomography. The field site (Florence, Oregon) consisted of a sandy, isotropic, homogenous aquifer that had been contaminated with hydrocarbons, especially gasoline. The water table was located at a depth of 17 ft below the ground surface, and a groundwater flow gradient of 0.006 ft/ft was measured. The permeability of the aquifer soils was approximately 2×10^{-2} cm/s. Air was injected into the subsurface using a 2-in. diameter PVC well that was screened for 2 ft beginning at a depth of 14 ft below the water table. In addition to the use of the measurement of groundwater mounding previously mentioned, dissolved oxygen levels and increases in soil gas pressure (at least 0.01 inches of water) were measured, as well as the use of ERT. The ERT system, which had a definition on the scale of 1 to 2 ft, revealed a zone of influence parabolic in shape with a radius of influence of approximately 8 ft. Because rather large changes in resistance were measured, it was determined that a high-density air flow channel network had been established. It was also observed that air was entering the soil at the bentonite seal. When comparing all of the methods used during the test, it was determined that the radius of influence determined using ERT was two to eight times smaller than with the other methods, indicating that the other methods may seriously overestimate the radius of influence of a given well.

Johnson et al. (1995) also performed a comparative study to determine how different measurement techniques may offer different values for the radius of influence. A sparge well was screened at 6 ft below ground surface. Direct push methods were used for the monitoring of the radius/zone of influence. A neutron probe revealed a radius of influence of 3 m with highly scattered air flow. Electrical resistivity tomography was applied, revealing the same result. The use of sulfur hexafluoride as a tracer gas also gave a radius of influence of 3 m, but the zone of influence appeared to be skewed in shape. Dissolved oxygen levels within the groundwater were monitored, and a radius of influence of between 6 and 9 m was determined. Because of induced groundwater circulation, groundwater that had come into contact with the injected air may have flowed away from the well, being detected at further distances and providing a larger radius of influence than that determined with the other methods.

14.6.2 Contaminant Removal

A number of methods can be employed to monitor remedial progress, including the use of monitoring wells, groundwater analysis, and effluent vapor analysis. Selected case studies are summarized in Table 14.1 and in this section to provide information on expected contaminant removal using air sparging systems.

Ardito and Billings (1990) utilized a subsurface volatilization and ventilation system (SVVS), which featured a nested well pair consisting of a sparge well and an extraction well. Three site applications were reported. At Isleta, New Mexico, a 90% cost reduction per percent benzene reduction was realized. At Bernanillo, New Mexico, between 12,000 and 13,000 gallons were remediated to nondetectable levels. At Arneal, New Mexico, concentrations of 23,000 to 43,000 ppb benzene dropped to 730 to 6,300 ppb after application of SVVS and free-product skimming.

Brown et al. (1991) reported on the remediation of a former dry cleaning facility that was contaminated by a combination of PCE, TCE, DCE, and heating oil. After pilot tests were run, a design radius of influence of 50 ft was used. Injection was performed at 16 standard cubic feet per minute under a pressure of 10 in water column. Within the first six weeks, 900 lb of PCE and TCE were removed, and a 98% reduction was achieved in 125 days.

Marley (1992) reported on the use of air sparging at a BTEX-contaminated site. Coarse sand extended to 20 ft below ground surface, which in turn was underlain by denser sand. The water table was between 15.5 and 16 ft below ground surface. A pilot test determined no appreciable benefit from pulsed air injection, but a full-scale system was installed featuring pulsed air injection due to operating concerns. Shallow wells were run with three hours injection followed by nine hours of down time, while the deep wells were operated with a regime of six hours on, six hours off. In 60 days, 5 to 10 lb of hydrocarbon were removed.

Leonard and Brown (1992) reported on the use of air sparging applied to a site contaminated with toluene and chlorinated solvents. It was found that 75% of the contaminant resided in the saturated zone. A sparge well was placed at 14.5 ft below ground surface (8.5 ft below the water table). Within three months, 7% of the contaminant was removed.

Deaver and Tworkowski (1994) reported on a remedial program at an adhesives manufacturing facility. The site consisted of 56,000 μg/L of 1,1,1-TCA and 4,000 μg/L of 1,1,1-DCE. SVE was shown to clean the vadose zone, but it did not address the saturated zone. A single well was placed in the spill area, leading to a drop from 2,000 μg/L to 200 μg/L.

Damera et al. (1997) used air sparging at a site contaminated by a multiple fuel spill. The site consisted of 3 to 10 ft of silty clay over fine to medium sand, with the water table at 29 ft. Within a year and a half, 57,000 lb of hydrocarbons were removed from the site due to biodegradation and volatilization. An analysis was performed, determining that 61% was removed due to volatilization.

TABLE 14.1 Selected Air Sparging Case Studies

Author	Contaminant	Soil	Number of wells/ depth	Operation	Results
Marley et al., 1992	TCE	Fine/very fine sand, silts	7 wells 15 ft B.W.T.	30 days 3 to 10 scfm	4 lb removed
Marley et al., 1992	BTEX 20,000 to 30,000 μg/L	Sand/gravel over fine sand	7 wells 6 ft B.W.T. 6 deep wells	60 days 2 to 6.5 scfm	5 to 10 lb removed 600 μg/L in groundwater
Basinet and Wollenberg, 1997	Gasoline	Fine sand, silty clays	1 A.S. well 18 ft B.G.S.	1 year	50% reduction
Covell and Thomas, 1997	Benzene	0–10 ft clay/silt 10–20 ft silty sandstone 28–31 ft coal seam	5 A.S. wells 4 SVE wells	3.5 months on 1 month off 2.5 months on	80% benzene reduction
Damera et al., 1997	BTEX 1 to 2 mg/L MTBE 7010 μg/L	Silty clay 3–10 ft sand, clay, gravel water table at 29 ft	3 SVE, A.S. clusters	1.5 years	BTEX non-detect. at 9 mos. MTBE 2450 μg/L
Klemm et al., 1997	Gasoline	12–15 ft sand/gravel over slate water at 1 ft	63 wells at 15 ft B.G.S.	2 years	Benzene 98% reduction BTEX 99.3% reduction
Gordon, 1998	Chlorinated solvents	Medium to coarse sand, sandy clay	134 A.S. wells, 58 SVE wells	450 days	1100 lb removed, most solvents at non-detectable levels

14.7 MODIFIED/RELATED TECHNOLOGIES

Even though traditional air sparging (vertical well, injection of air) has proven to be successful in a wide variety of field applications, alternative strategies exist that can make air sparging applicable to other conditions. Sometimes other remedial technologies such as pump and treat and bioremediation are used in conjunction with the air sparging technology to address site-specific contaminant conditions.

14.7.1 Horizontal Wells

Horizontal wells may be implemented under a wide range of field applications, but they are especially useful in treating long, shallow spills, such as those that may occur under leaking above-ground or in-ground pipelines (Bausmith et al., 1996). There are four proven methods for horizontal well placement: directional drilling, trenching, boring, and backhoe excavation. These placements and subsequent improvements have made the use of horizontal wells an attractive option with many benefits over the use of vertical wells (Plummer et al. 1997). One horizontal well can replace several vertical wells, leading to a reduction in operating expense that can help offset the additional placement costs. Because of the continuous nature of a horizontal well, the injected air is able to interact with the subsurface contaminant over a larger surface area, accelerating removal. The continuous nature of a horizontal well also allows for easier interception of a migrating contaminant plume (Mast and Koerner, 1996). Additionally, the flexibility of directional drilling allows for the placement of injection wells in locations that may be difficult or impossible with the use of vertical wells, especially in cases of surface or subsurface obstructions caused by roads, buildings, or utilities.

Despite the possible advantages that the use of horizontal wells offers, horizontal wells do have drawbacks that must be dealt with during design and operation (Mast and Koerner, 1996). The placement of horizontal wells becomes impractical with increasing depth, especially when techniques other than directional drillings are used. Their use may become cost-prohibitive when placement depths greater than 15 ft are required. Therefore, when contamination is detected at great depths, vertical wells should be used. In a horizontal well, it is often hard to provide for even air injection and flow over the length of a well, leading to uneven air coverage within the subsurface. Even though horizontal wells may provide a large area of coverage, the use of a vertical well in lower permeability soils provides a sizeable zone of treatment, making the use of a horizontal well unnecessary. Wilson et al. (1994b) analyzed the performance of horizontal versus vertical wells using a computer model and found that vertical wells actually performed better because the groundwater circulation path that a horizontal well created spread the contaminant out too far to be remediated efficiently.

To analyze the performance difference between horizontal and vertical wells, Plummer et al. (1997) performed laboratory tests in both saturated beach sand and glass beads to analyze the air flow created by both types of wells. Their apparatus consisted of a two-dimensional Lexan tank topped by a manifold/flowmeter system designed to measure regional effluent gas flow rates. They found that under the same air injection pressure, a horizontal well allowed for one and a half to two times higher air flow than a vertical well. Additionally, more groundwater was displaced during the use of a horizontal well, indicating the existence of more extensive air flow.

Past work in the field using horizontal wells has demonstrated successful implementation. Ghandehari et al. (1994) reported on the use of a horizontal well at a remedial site in Charlotte, North Carolina. The site, consisting of clayey silts with a shallow water table (7 to 15 ft below ground surface), was contaminated with BTEX constituents and methyl tertiary butyl ether (MTBE). A 2-in. diameter PVC vapor extraction measuring 230 ft in length was placed 4 to 5 ft below ground surface. A 2-in. diameter PVC injection well measuring 310 ft in length was placed 40 ft below grade. The injection well was screened over 130 ft of its length. The radius of influence was measured in a variety of ways, including soil air pressure, dissolved oxygen, groundwater mounding, and native bacterial populations, and was found to measure approximately 40 ft. Within one year, BTEX concentrations dropped from 65 to 100 ppb to non-detectable levels. Reductions in MTBE concentrations were also seen.

14.7.2 Sparge Trenches

Sparge trenches or curtains may be used instead of horizontal and/or vertical wells. Trenches and curtains are especially useful when the contaminant plume is migrating in a region of high groundwater flow or where the prevention of off-site migration is of the utmost concern. The typical volume flux into a trench is on the order of 0.015 to 0.5 m^3/m^2 per day (Pankow et al., 1993). After digging, the trench is shored up with sheet piling and may be left open or backfilled with soil. The backfill must be of equal or higher permeability in order to prevent the flowing groundwater from circumventing the treatment zone. Additionally, cutoff wells are often used to ensure the groundwater is flowing in the proper direction. Along the wall, 5 to 20% needs to consist of gates to prevent groundwater from building up, which could possibly cause flow in undesired directions (Pankow et al., 1993). Larger dimensions actually do not lead to greater efficiency, only to increased volumes of water to be treated. The parameters that have the greatest effect on efficiency include the Henry's constant(s) of the contaminant(s) and the gas sparge rate (Pankow et al., 1993).

Marley et al. (1994) reported on the typical field application of a trench. The site consisted of clayey silts with an unconfined, perched aquifer with the water table residing 2 ft below ground surface. The trench used was built in 40-ft segments for a total of 360 ft. The trench measured 1 meter in width, and air was injected at a rate of 0.005 m^3/s per meter of trench. Volatile VOCs were emitted directly to the atmosphere. Bittner and Hoffman (1996) reported on the use of an aeration curtain at a site in Utah. The subsurface was contaminated with trichloroethylene with concentrations as high as 1,000 $\mu g/L$. Air was injected into the trench 21.7 L/s. Approximately 90% of the initial concentrations were removed within 48 hours of operations, and 99% was removed within 8 days.

14.7.3 Biosparging

Biosparging is an alternative to the use of an air sparging/soil vapor extraction system. During the implementation of biosparging, lower injection flow rates are used to prevent the buildup of excess soil gas pressure. The injected air increases the concentration of dissolved oxygen within the groundwater, stimulating aerobic biodegradation of subsurface by the native microbial population. To prevent the buildup of pressure, flow rates of approximately 1 scfm are used (Nyer and Suthersan, 1993). The viability of biosparging is dependent on subsurface conditions; in addition to the introduction of dissolved oxygen to serve as an electron acceptor, a suitable microbial population as well as essential microbial nutrients must be present in sufficient quantity. When the subsurface conditions are favorable, the contamination may be degraded and mineralized into harmless by-products such as carbon dioxide and water, depending on the specific metabolic reaction occurring.

14.7.4 Related/Complementary Technologies

Air sparging may be combined with other technologies such as pump and treat and bioremediation to address certain site-specific conditions, including the type and distribution of the contaminants as well as site hydrogeologic conditions.

14.8 SUMMARY

Air sparging has proven to be an effective technique for the remediation of VOC-contaminated saturated soils and groundwater. Due to its use of readily available equipment and its overall remedial efficiency, air sparging can offer significant cost savings compared to other remedial technologies. The operator-controlled parameters allow for flexible implementation, and an air sparging system may be combined with a soil vapor extraction system or used for biosparging. To ensure efficient performance, a detailed geologic and hydrogeologic characterization must be performed before system implementation. The site characterization will also provide important information regarding the relative contribution of critical mass transfer, transport, and transformation parameters that occur during the use of air sparging.

Several air sparging systems have been designed, operated, and monitored in the field under different contaminant and hydrogeologic conditions. The performance data from these sites have helped in understanding the advantages and limitations of the technique as well as provided a database to use as guidance for future air sparging systems. Air sparging is being used increasingly because of its low cost and significantly reduced treatment time compared to the very high cost and long treatment time associated with conventional remediation technologies. In general, acceptance of the air sparging technology by the regulatory agencies has been high.

Air sparging is still a relatively new technology, and much more needs to be learned about how it works and how it can best be applied. Most of the laboratory work performed has included simplifications that may restrict the applicability of the results to field situations. Often, field studies have not been monitored and documented enough to understand the processes of air sparging. Many models developed to date include unrealistic simplifications or have not been validated with laboratory or field investigations. Therefore, in order to advance the state of the art regarding air sparging, controlled laboratory tests studying a wide range of operator-controlled and site-specific parameters and a careful assessment of field systems need to be performed to develop realistic mathematical models that will adequately describe and predict behavior during the implementation of air sparging.

14.9 REFERENCES

Adams, J. A., and K. R. Reddy. 1999. "Laboratory Study of Air Sparging of TCE-Contaminated Saturated Soils and Groundwater," *Ground Water Monitoring and Remediation,* vol. 19, no. 3, pp. 182–190.

Adams, J. A., and K. R. Reddy. 2000. "Removal of Dissolved and Free-Phase Benzene Pools from Groundwater Using In Situ Air Sparging," *Journal of Environmental Engineering,* vol. 125, no. 8, pp. 697–707.

Ahlfeld, D. P., A. Dahmani, and W. Ji. 1994a. "A Conceptual Model of Field Behavior of Air Sparging and its Implications for Application," *Ground Water Monitoring Review,* vol. 14, no. 4, pp. 132–139.

Ahlfeld, D., A. Dahmani, G. Hoag, M. Farrell, and W. Ji. 1994b. "Field Measurements of Air Sparging in a Connecticut Site: Results and Comments." in *Proceedings of the Conference on Petroleum Hydrocarbons and Organic Chemicals in Ground Water: Prevention, Detection, and Restoration,* Houston, pp. 175–190.

Ardito, C. P., and J. F. Billings. 1990. "Alternative Remediation Strategies: The Subsurface Volatilization and Ventilation System," in *Proceedings of Petroleum Hydrocarbons and Organic Chemicals in Ground Water: Prevention, Detection, and Restoration,* pp. 281–296.

Baker, D. M., and C. H. Benson. 1996. "Review of Factors Affecting In-Situ Air Sparging," in *Proceedings of 1996 ASCE Specialty Conference,* Washington, DC, pp. 292–310.

Basinet, R., and J. Wollenberg. 1997. "Petroleum Hydrocarbon Clean-up Using Soil Vapor Extraction and Air Sparging Via Horizontal Wells at a Commercial Navy Site in San Diego, CA," in *Proceedings of the Fourth International Symposium on In-Situ and On-Site Bioremediation,* Batelle Press, Columbus, OH, vol. 1, pp. 205–208.

Bass, D. H., and R. A. Brown. 1997. "Performance of Air Sparging Systems—A Review of Case Studies," in *Proceedings of the Fourth International Symposium on In-Situ and On-Site Bioremediation,* Batelle Press, Columbus, OH, vol. 1, pp. 117–122.

Bausmith, D. S., D. J. Campbell, and R. D. Vidic. 1996. "In Situ Air Stripping: Using Air Sparging and Other In Situ Methods Calls for Critical Judgments," *Water Environment and Technology,* February, pp. 45–51.

Bear, J. 1972. *Dynamics of Fluids in Porous Media,* American Elsevier, New York.

Bittner, P. R., and D. A. Hoffman. 1996. "Non-aqueous Phase Liquids (NAPLs) in Subsurface Environment: Assessment and Remediation," in *Proceedings of 1996 ASCE Specialty Conference,* Washington, DC, pp. 776–787.

Braida, W. J., and S. K. Ong. 1997. "Factors Affecting Air Sparging Remediation Technologies," in *Proceedings of the Fourth International Symposium on In-Situ and On-Site Bioremediation,* Batelle Press, Columbus, OH, vol. 1, pp. 187–192.

Brown, R., C. Herman, and E. Henry. 1991. "The Use of Aeration in Environmental Cleanups," in *Proceedings of the Haztech International Pittsburgh Waste Conference,* Pittsburgh, pp. 2A.1–2A.42.

Bruell, C. J., M. C. Marley, and H. H. Hopkins. 1997. "American Petroleum Institute Air Sparging Database," *Journal of Soil Contamination,* vol. 6, no. 2, pp. 169–185.

Brusseau, M. L. 1992. "Rate-Limited Mass Transfer and Transport of Organic Solutes in Porous Media that Contain Immobile Immiscible Organic Liquid," *Water Resources Research,* vol. 28, no. 1, pp. 33–45.

Burchfield, S. D., and D. J. Wilson. 1993. "Groundwater Cleanup by In-situ Air Sparging. IV. Removal of Dense Non aqueous Phase Liquid by Sparging Pipes," *Separation Science and Technology,* vol. 28, nos. 17/18, pp. 2529–2551.

Chapelle, F., P. Bradley, D. Lovley, and D. Vroblesky. 1996. "Measuring Rates of Biodegradation in a Contaminated Aquifer Using Field and Laboratory Methods," *Ground Water,* vol. 34, no. 4, pp. 691–698.

Chiang, C. Y., J. P. Salanitro, E. Y. Chai, J. D. Colthart, and C. L. Klein. 1989. "Aerobic Biodegradation of Benzene, Toluene, and Xylene in a Sandy Aquifer Data Analysis and Computer Modeling," *Ground Water,* vol. 27, no. 6, pp. 823–834.

Covell, J. R., and M. H. Thomas. 1997. "Air Sparge and Bioremediation Demonstration at an Underground Coal Gasification Site," in *Proceedings of the Fourth International Symposium on In-Situ and On-Site Bioremediation,* Batelle Press, Columbus, OH, vol. 1, pp. 173–178.

Crocetti, C. A., C. L. Head, and A. J. Ricciardelli. 1992."Aeration-Enhanced Bioremediation of Oil-Contaminated Soils: A Laboratory Treatment Study," in *Proceedings of Petroleum Hydrocarbons and Organic Chemicals in Ground Water: Prevention, Detection, and Restoration,* NGWA, Dublin, OH, pp. 427–440.

Damera, R., M. R. Hill, D. Murali, and R. McDermott. 1997. "Rapid Clean-up of a Multiple Fuel Spill," in *Proceedings of the Fourth International Symposium on In-Situ and On-Site Bioremediation,* Batelle Press, Columbus, OH, vol. 1, pp. 109–204.

Davis, E. L. 1997. *How Heat Can Enhance In-Situ Soil and Aquifer Remediation: Important Chemical Properties and Guidance on Choosing the Appropriate Technique,* EPA/540/S-97/502. Washington, DC.

Deaver, G. H., and R. J. Tworkowski. 1994. "Application of Air Sparging to 1,1,1-trichloroethane Contaminated Soil—A Success Story," in *Proceedings of 1st International Congress on Environmental Geotechnics,* BiTech, Richmond, VA, pp. 201–208.

Falta, R. W., K. Pruess, I. Javandel, and P. A. Witherspoon. 1992. "Numerical Modeling of Steam Injection for the Removal of Non aqueous Phase Liquids from the Subsurface 1. Numerical Formulation," *Water Resources Research,* vol. 28, no. 2, pp. 433–449.

Fetter, C. W. 1999. *Contaminant Hydrology,* Prentice-Hall, Upper Saddle River, NJ.

Fry, V. A., J. D. Istok, L. Semprini, K. T. O'Reilly, and T. E. Buscheck. 1995. "Retardation of Dissolved Oxygen due to a Trapped Gas Phase in Porous Media," *Ground Water,* vol. 33, no. 3, pp. 391–398.

Geller, J. T., and J. R. Hunt. 1993. "Mass Transfer from Non aqueous Phase Organic Liquids in Water-saturated Porous Media," *Water Resources Research,* vol. 29, no. 4, pp. 833–845.

Ghandehari, M., B. Kelly, A. Holt, S. Hines, and J. Doesburg. 1994. "In Situ Remediation of Groundwater using Horizontal Well Air Injection," in *Proceedings of Petroleum Hydrocarbons and Organic Chemicals in Ground Water: Prevention, Detection, and Restoration,* NGWA, Dublin, OH, pp. 221–235.

Goldflam, R., M. Van Benthem, and A. G. Finci. 1997. "Use of Borehole Radar for Air Sparging Operations Monitoring," in *Proceedings of the Fourth International Symposium on In-Situ and On-Site Bioremediation,* Batelle Press, Columbus, OH, vol. 1, pp. 259–264.

Gomez-Lahoz, C., J. M. Rodriguez-Maroto, and D. J. Wilson. 1994. Groundwater Cleanup by In-Situ Air Sparging. VII. Volatile Organic Compounds Concentration Rebound Caused by Diffusion after Shutdown," *Separation Science and Technology,* vol. 29, no. 12, pp. 1509–1528.

Gordon, M. J. 1998. "Case History of a Large-Scale Air Sparging Soil Vapor Extraction System for Remediation of Chlorinated Volatile Organic Compounds in Ground Water," *Groundwater Monitoring and Remediation,* vol. 18, no. 2, pp. 137–149.

Hayden, N. J., T. C. Voice, M. D. Annable, and R. B. Wallace. 1994. "Change in Gasoline Constituent Mass Transfer during Soil Venting," *Journal of Environmental Engineering,* vol. 120, no. 6, pp. 1598–1615.

Hein, G. L., N. J. Hutzler, and J. S. Gierke. 1994. "Quantification of the Mechanisms Controlling the Removal Rate of Volatile Contaminants by Air Sparging," in *Proceedings of the National Conference on Environmental Engineering,* Boulder, CO, pp. 556–563.

Hoeppel, R. E., R. E. Hinchee, and M. F. Arthur. 1991. "Bioventing Soils Contaminated with Petroleum Hydrocarbons," *Journal of Industrial Microbiology,* vol. 8, pp. 141–146.

Hunt, J. R., N. Sitar, and K. S. Udell. 1988a. "Nonaqueous Phase Liquid Transport and Cleanup 1. Analysis of Mechanisms," *Water Resources Research,* vol. 24, no. 8, pp. 1247–1258.

Hunt, J. R., N. Sitar, and K. S. Udell. 1988b "Nonaqueous Phase Liquid Transport and Cleanup 2. Experimental Studies," *Water Resources Research,* vol. 24, no. 8, pp. 1259–1269.

Javanmardian, M., and V. J. Kremesec. 1995. *Air Sparging/Biosparging Pilot Test and Design Guidance,* Amoco Corporation, July.

Ji, W., A. Dahmani, D. P. Ahlfeld, J. D. Lin, and E. Hill, III. 1993. "Laboratory Study of Air Sparging: Air Flow Visualization," *Ground Water Monitoring Review,* vol. 13, no. 4, pp. 115–126.

Jin, Y., T. Streck, and W. A. Jury. 1994. "Transport and Biodegradation of Toluene in Unsaturated Soil," *Journal of Contaminant Hydrology,* vol. 17, pp. 111–127.

Johnson, P. C. 1998. "Assessment of the Contributions of Volatilization and Biodegradation to In-situ Air Sparging Performance," *Environmental Science Technology,* vol. 32, no. 2, pp. 276–281.

Johnson, P. C., C. C. Stanley, M. W. Kemblowski, D. L. Byers, and J. D. Colthart, 1990. "A Practical Approach to the Design, Operation, and Monitoring of In situ Soil-venting Systems," *Ground Water Monitoring Review,* vol. 10, no. 2, pp. 159–178.

Johnson, R. L., P. C. Johnson, D. B. McWhorter, R. E. Hinchee, and I. Goodman. 1993. "An Overview of In Situ Air Sparging," *Ground Water Monitoring Review,* vol. 13, no. 4.

Johnson, P. C., R. L. Johnson, C. Neaville, E. E. Hansen, S. M. Stearns, and I. J. Dortch. 1995. "An Assessment of Conventional In Situ Air Sparging Tests," *Ground Water,* vol. 35, no. 5, pp. 765–774.

Joiner, D. P., and T. L. O'Keefe. 1996. "Direct Push Technology in Air Sparging Pilot Studies for Groundwater Remediation," *Iron and Steel Engineer,* vol. 73, no. 12, pp. 53–58.

Karickhoff, S. W., D. S. Brown, and T. A. Scott. 1979. "Sorption of Hydrophobic Pollutants on Natural Sediments," *Water Resources,* vol. 13, pp. 241–248.

Klemm, D. E., S. Lummus, and S. Eaton. 1997. "Air Sparging in Various Lithologies: 3 Case Studies," in *Proceedings of the Fourth International Symposium on In-Situ and On-Site Bioremediation,* Batelle Press, Columbus, OH, vol. 1, pp. 193–198.

Kraus, J., S. Nelson, P. Boersma, and A. Maciey. 1997. "Comparison of Pre/post-sparging VOC Concentrations in Soil and Groundwater," in *Proceedings of the Fourth International Symposium on In-Situ and On-Site Bioremediation,* Batelle Press, Columbus, OH, vol. 1, pp. 123–128.

Leonard, W. C., and R. A. Brown. 1992. "Air Sparging: An Optimal Solution," in *Proceedings of the Conference on Petroleum Hydrocarbons and Organic Chemicals in Ground Water: Prevention, Detection, and Restoration,* Houston.

Leonard, W. C., S. R. Thompson, and R. A. Brown. 1994. "Air Sparging Reduces Solvent Contamination," *Environmental Solutions,* June, pp. 69–71.

Loden, M. E. 1992. *A Technology Assessment of Soil Vapor Extraction and Air Sparging,* EPA/600/R-92/173, Cincinnati, OH.

Lombard, K. H., J. W. Borthen, and T. C. Hazen. 1994. "The Design and Management of System Components for In situ Methanotrophic Bioremediation of Chlorinated Hydrocarbons at the Savannah River Site," in *Air Sparging for Site Remediation,* ed. R. H. Hinchee, Lewis, Ann Arbor, MI, pp. 81–96.

Lundegard, P. D., and G. Andersen. 1996. "Multiphase Numerical Simulation of Air Sparging Performance," *Ground Water,* vol. 34, no. 3, pp. 451–460.

Lundegard, P. D., and D. LaBrecque. 1995. "Air Sparging in a Sandy Aquifer (Florence, Oregon, U.S.A.): Actual and Apparent Radius of Influence," *Journal of Contaminant Hydrology,* vol. 19, pp. 1–27.

Malone, D. R., C. M. Kao, and R. C. Borden. 1993. "Dissolution and Biorestoration of Nonaqueous Phase Hydrocarbons: Model Development and Laboratory Evaluation," *Water Resources Research,* vol. 29, no. 7, pp. 2203–2213.

Marley, M. C. 1992. "Air Sparging in Conjunction with Soil Vapor Extraction for Source Removal at VOC Spill Sites," in *Proceedings of 5th Annual Conference on Hydrocarbon-Contaminated Soils,* Lewis, Chelsea, MI, pp. 579–590

Marley, M. C., D. J. Hazebrouck, and M. T. Walsh. 1992a. "The Application of In situ Air Sparging as an Innovative Soils and Ground Water Remediation Technology," *Ground Water Monitoring Review,* vol. 12, no. 2, pp. 137–145.

Marley, M. C., F. Li, and S. Magee. 1992b. "The Application of a 3-D Model in the Design of Air Sparging Systems," in *Proceedings of Focus on Eastern Ground Water Issues,* pp. 377–392.

Marley, M., F. Li, E. Droste, and R. Cody. 1994. "The Design of an In-situ Sparging Trench," in *Proceedings of Petroleum Hydrocarbons and Organic Chemicals in Ground Water: Prevention, Detection, and Restoration,* NGWA, Dublin, OH, pp. 579–590.

Mast, V., and K. Koerner, 1996. "Horizontal vs. Vertical Wells: Which to Use and When," *Soil and Groundwater Cleanup,* April, pp. 10–16.

McCray, J. E., and R. W. Falta. 1997. "Numerical Simulation of Air Sparging for Remediation of NAPL Contamination," *Ground Water,* vol. 35, no. 1, pp. 99–100.

McKay, D. J., and L. J. Acomb. 1996. "Neutron Moisture Probe Measurements of Fluid Displacement during In-situ Air Sparging," *Ground Water Monitoring Review,* vol. 16, no. 4, pp. 86–94.

Middleton, A. C., A. W. Lawrence, D. J. Morgan, M. G. Less, and T. D. Hayes. 1995. "Biosparging Strategies for Contaminant and Remediation of Organic Contaminant Groundwater Plumes at E & P Sites Using Either Vertical or Horizontal Sparge Wells," Institue of Gas Technology, Des Plaines, IL.

Miller, C. T., M. M. Poirier-McNeill, and A. S. Mayer. 1990. "Dissolution of Trapped Nonaqueous Phase Liquids: Mass Transfer Characteristics," *Water Resources Research,* vol. 26, no. 11, pp. 2783–2796.

Miller, R. R. 1996. *Air Sparging,* Groundwater Remediation Technologies Analysis Center, Pittsburgh.

Nelson, C. H., and R. A. Brown. 1994. "Adapting Ozonation for Soil and Groundwater Cleanup," *Environmental Engineering,* November.

Nyer, E. K., and S. S. Suthersan. 1993. "Air Sparging: Savior of Ground Water Remediation or Just Blowing Bubbles in the Bathtub?" *Ground Water Monitoring Review,* vol. 13, no. 4, pp. 87–91.

Panday, S., Y. S. Wu, P. S. Huyakorn, and E. P. Springer. 1994. "A Three-dimensional Multiphase Flow Model for Assessing NAPL Contamination in Porous and Fractured Media, 2. Porous Medium Simulation Examples," *Journal of Contaminant Hydrology,* vol. 16, pp. 131–156.

Pankow, J. F., R. L. Johnson, and J. A. Cherry. 1993. "Air Sparging in Gate Wells in Cutoff Wells and Trenches for Control of Plumes of Volatile Organic Compounds (VOCs)," *Ground Water,* vol. 31, no. 4, pp. 654–663.

Patterson, B. M., F. Pribac, C. Barber, G. B. Davis, and R. Gibbs. 1993. "Biodegradation and Retardation of PCE and BTEX Compounds in Aquifer Material from Western Australia Using Large-Scale Columns," *Journal of Contaminant Hydrology,* vol. 14, pp. 261–278.

Pepper, D. W., and D. E. Stephenson. 1995. "An Adaptive Finite-Element Model for Calculating Subsurface Transport of Contaminant," *Ground Water,* vol. 33, no. 3, pp. 486–496.

Perlinger, J. A., and S. J. Eisenreich. 1991. "Sorption of Alkyl Benzene to Mineral Oxides," in *Organic Substances and Sediments in Waters,* ed. R. A. Baker, Lewis, Chelsea, MI, pp. 49–78.

Petersen, L. W., D. E. Rolston, P. Moldrup, and T. Yamaguchi. 1994. "Volatile Organic Vapor Diffusion and Adsorption in Soils," *Journal of Environmental Quality,* vol. 23, pp. 799–805.

Peterson, J. W., P. A. Lepczyk, and K. L. Lake. 1999. "Effect of Sediment Size on Area of Influence During Groundwater Remediation by Air Sparging: A Laboratory Approach," *Environmental Geology,* vol. 38, pp. 1–6.

Plummer, C. R., J. D. Nelson, and G. S. Zumwalt. 1997. "Horizontal and Vertical Well Comparison for In-situ Air Sparging," *Ground Water Monitoring Review,* vol. 17, no. 1, pp. 91–96.

Powers, S. E., C. O. Loureiro, L. M. Abriola, and W. J. Weber, Jr. 1991. "Theoretical Study of the Significance of Nonequilibrium Dissolution of Nonaqueous Phase Liquids in Subsurface Systems," *Water Resources Research,* vol. 27, no. 4, pp. 463–477.

Rabideau, A. J., and J. M. Blayden. 1998. "Analytical Model for Contaminant Mass Removal by Air Sparging," *Ground Water Monitoring and Remediation,* vol. 18, no. 4, pp. 120–130.

Reddy, K. R., and J. A. Adams. 1999. "System Effects on Benzene Removal from Saturated Soils and Groundwater Using Air Sparging," *Journal of Environmental Engineering,* vol. 124, no. 3, pp. 288–299.

Reddy, K. R., and J. A. Adams. 2000. "Effects of Groundwater Flow on Remediation of Dissolved Phase VOC Contamination Using Air Sparging," *Journal of Hazardous Materials,* vol. 72, nos. 2/3, pp. 147–165.

Reddy, K. R., and J. Zhou. 1996. "Finite Element Modeling of In-situ Air Sparging for Groundwater Remediation," in *Proceedings of the 2nd International Congress on Environmental Geotechnics,* Balkema, Rotterdam, pp. 299–304.

Reddy, K. R., S. Kosgi, and J. Zhou. 1995a. "A Review of In-situ Air Sparging for the Remediation of VOC-Contaminated Saturated Soils and Groundwater," *Hazardous Waste and Hazardous Materials,* vol. 12, no. 2, pp. 97–118.

Reddy, K. R., J. Zhou, and S. Kosgi. 1995b. "New Model to Simulate Air Sparging for Groundwater Remediation," in *Proceedings of 5th WERC Development Conference,* Las Cruces, NM, pp. 299–308.

Reddy, K. R., R. Semer, and J. A. Adams. 1999. "Air Flow Optimization and Surfactant Enhancement to Remediate Toluene-Contaminated Saturated Soils Using Air Sparging," *Environmental Management & Health,* vol. 10, no. 1, pp. 52–63.

Roberts, L. A., and D. J. Wilson. 1993. "Groundwater Cleanup by In-situ Air Sparging. III. Modeling of Dense Nonaqueous Phase Liquid Droplet Removal," *Separation Science and Technology,* vol. 28, no. 5, pp. 1127–1143.

Rutherford, K. W., and P. C. Johnson. 1996. "Effects of Process Control Changes on Aquifer Oxygenation Rates during In-situ Air Sparging in Homogenous Aquifers," *Ground Water Monitoring Review,* vol. 16, no. 4, pp. 132–141.

Schima, S., D. J. LaBrecque, and P. D. Lundegard. 1996. "Monitoring Air Sparging Using Resistivity Tomography," *Ground Water Monitoring Review,* vol. 16, no. 2, pp. 131–138.

Sellers, K. L., and R. P. Schreiber. 1992. "Air Sparging Model for Predicting Groundwater Cleanup Rate," in *Proceedings of Petroleum Hydrocarbons and Organic Chemicals in Ground Water: Prevention, Detection, and Restoration,* Houston.

Semer, R., and K. R. Reddy. 1997. "Mechanisms Controlling Toluene Removal from Saturated Soils during In Situ Air Sparging," *Journal of Hazardous Materials,* vol. 57, nos. 1–3, pp. 209–230.

Semer, R., J. A. Adams, and K. R. Reddy. 1996, "Surfactant Enhanced Air Sparging for Groundwater Remediation: Preliminary Results," in *Proceedings of the Fourth Great Lakes Geotechnical/Geoenvironmental Conference on In-Situ Remediation of Contaminated Sites,* Chicago, pp. 197–213.

Semer, R., J. A. Adams, and K. R. Reddy. 1998. "An Experimental Investigation of Air Flow Patterns in Saturated Soils during Air Sparging," *Geotechnical and Geological Engineering Journal,* vol. 16, no. 1, pp. 59–75.

Szatkowski, A., P. T. Imhoff, and C. T. Miller. 1995. "Development of a Correlation for Aqueous-Vapor Phase Mass Transfer in Porous Media," *Journal of Contaminant Hydrology,* vol. 18, no. 85–106.

Thierrin, J., G. B. Davis, and C. Barber. 1995. "A Ground-water Tracer Test with Deuterated Compounds for Monitoring In Situ Biodegradation and Retardation of Aromatic Hydrocarbons," *Ground Water,* vol. 33, no. 3, pp. 469–475.

Unger, A. J. A., E. A. Sudicky, and P. A. Forsyth. 1995. "Mechanisms Controlling Vacuum Extraction Coupled with Air Sparging for Remediation of Heterogeneous Formations Contaminated by Dense Nonaqueous Phase Liquids," *Water Resources Research,* vol. 31, no. 8, pp. 1913–1925.

Van Dijke, M. I. J., and S. E. A. T. M. van der Zee. 1998. "Modeling of Air Sparging in a Layered Soil: Numerical and Analytical Approximations," *Journal of Geophysical Research,* vol. 34, no. 3, pp. 341–353.

Voudrias, E. A., and M. F. Yeh. 1994. "Dissolution of a Toluene Pool under Constant and Variable Hydraulic Gradients with Implications for Aquifer Remediation," *Ground Water,* vol. 32, no. 2, pp. 305–311.

Wehrle, K. 1990. "In-situ Cleaning of CHC Contaminated Sites: Model-Scale Experiments Using the Air Injection (In-Situ Air Stripping) Method in Granular Soils," in *Proceedings of Contaminated Soil,* ed. F. Arendt, M. Hinsenveld, and W. J. van den Brink, Kluwer, Dordrecht, pp. 1061–1062.

Widdowson, M. A., and C. M. Aelion. 1991. "Application of a Numerical Model to the Performance and Analysis of an In Situ Bioremediation Project," *In Situ Bioremediation,* ed. R. E. Hinchee and R. F. Olfenbuttel, Butterworth-Heinemann, Boston, pp. 227–244.

Wilson, D. J. 1992. "Groundwater Cleanup by In-situ Air Sparging. II. Modeling of Dissolved Volatile Organic Compound Removal," *Separation Science and Technology,* vol. 27, no. 13, pp. 1675–1690.

Wilson, D. J., and R. D. Norris. 1997. "Sparging and Biosparging: Insights through Mathematical Modeling," in *Proceedings of the Fourth International Symposium on In-Situ and On-Site Bioremediation,* Batelle Press, Columbus, OH, vol. 1, pp. 147–152.

Wilson, D. J., C. Gomez-Lahoz, and J. M. Rodriguez-Maroto. 1994a. "Groundwater Cleanup by In-situ Air Sparging. VIII. Effect of Air Channeling on Dissolved Volatile Organic Compounds Removal Efficiency," *Separation Science and Technology,* vol. 29, no. 18, pp. 2387–2418.

Wilson, D. J., J. M. Rodriguez-Maroto, and C. Gomez-Lahoz. 1994b. "Groundwater Cleanup by In-situ Air Sparging. VI. A Solution/Distributed Diffusion Model for Nonaqueous Phase Liquid Removal," *Separation Science and Technology,* vol. 29, no. 11, pp. 1401–1432.

Yang, X., L. E. Erickson, and L. T. Fan. 1995. "A Study of the Dissolution Rate-Limited Bioremediation of Soils Contaminated by Residual Hydrocarbons," *Journal of Hazardous Materials,* vol. 41, pp. 299–313.

Yaniga, P. M., and W. Smith. 1984. *Aquifer Restoration via Accelerated In-Situ Biodegradation of Organic Contaminants,* Ground Water Technology, Chadds Ford, PA.

Zumwalt, G. S., A. P. Krishna, and J. D. Nelson. 1997. "Air Distribution within a Sparging Cone of Influence," in *Proceedings of the Fourth International Symposium on In-Situ and On-Site Bioremediation,* Batelle Press, Columbus, OH, vol. 1, pp. 141–146.

CHAPTER 15
SORBENTS FOR CHEMICAL SPILL TREATMENT

Gordon McKay and Joe C. Y. Ng
Department of Chemical Engineering,
The Hong Kong University of Science and Technology,
Clear Water Bay, Kowloon, Hong Kong

15.1 SORBENTS AND SORPTION MECHANISM

15.1.1 Sorption Mechanism

The application of adsorption to chemical spills can be highly effective, but the nature of the spill must first be carefully analyzed. Adsorption is the ability of a solid material, usually in the form of a powder or granular particles, to attract and attach solute/solvent molecules and ions to its surface. This effectively concentrates the components (or certain components) in a chemical spill onto the solid, thus transferring the pollution from the liquid phase to a significantly smaller volume of the solid phase. The disadvantage of adsorption is that it produces a contaminated solid material that must be collected and either processed for purification or landfilled. The advantage of the method is that a carefully selected porous solid can adsorb large quantities of pollutants in a chemical spill. Figure 15.1 shows that, provided that the pollutant molecules are small enough to penetrate the fine porous structure of a microporous adsorbent, the adsorbent will potentially have a large capacity.

In addition to a large available adsorption surface area, there must be an affinity between the spilled pollutant molecules and the adsorbent surface sites. This affinity can be chemical (chemisorption), physical van der Waals forces (physisorption), or the exchange of ions (ion exchange).

15.1.2 Adsorbents and Applications

The application of adsorbents can be classified according to the adsorbents themselves, the chemicals to be cleaned up, the type of spill (i.e. large or small scale) and whether the spill liquid can be pumped to containers for conventional treatment or the type of chemical/water interaction (Robinson, 1979). Chemicals that mix with water can be treated in conventional adsorption systems, or adsorbents can be spread onto the spill and mixing achieved by agitation. For chemicals that float, booms, blankets, fibers, or powders are available. Applying sorbents to chemicals that sink is the most difficult application. Weighted sorbent bags and sacks have been developed. For this purpose, in addition, the use of dense adsorbents

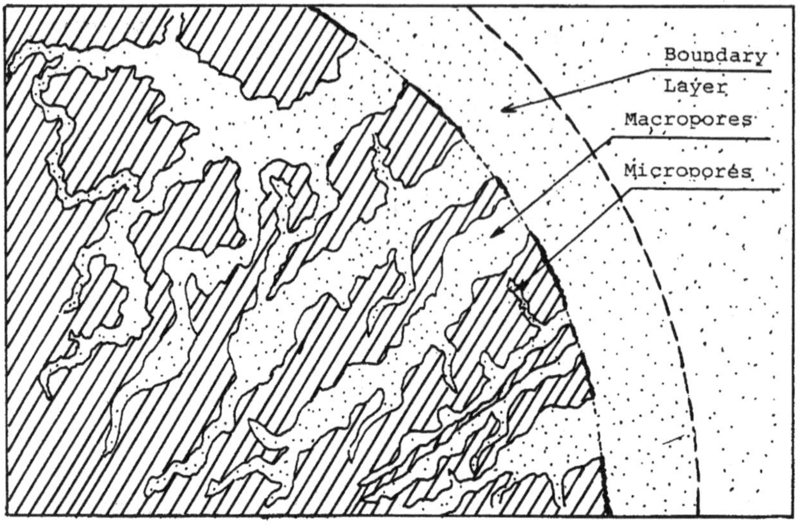

FIGURE 15.1 Pollutant molecules adsorbing on internal pore sites.

and the technique of injecting adsorbent slurries below the surface of the water have been studied (Suggs, 1972).

The types of adsorbents available for chemical spill cleanup will be reviewed in this section, and some of the characteristics and applications of specific adsorbents will be discussed.

The main adsorbents in commercial use and potential new ones will be discussed in the following categories:

- Activated carbon
- Polymer fibers
- Molecular sieves
- Lignite
- Ion exchangers
- Chitin and chitosan
- Sorbent clays
- Woodmeal
- Polyurethane foam
- Bone char
- Macroreticular resins
- Bagasse pith
- Activated alumina
- Polyolefins–ethylene, isobutylene, methacrylates and styrenes
- Silica gel
- Others

Activated Carbons. As there are so many varieties of activated carbons available on the market, there is probably a carbon that can sorb any chemical involved in a spill. However, sorption capacity and economics dictate the selection. Active carbons are highly porous carbonaceous sorbents with large surface areas. The carbon surface can be basic or acidic, hydrophilic or hydrophobic, or oleophilic or lipophobic. The exact properties depend on the source of the carbon, the activation temperature, and the activation chemicals used.

In a recent paper (Allen et al., 1997) over 30 sources of origin for active carbon production were described. The main sources of origin for active carbon production for commercial-scale production are coal, coconut shell, wood, and peat. Physical activation is achieved using steam and/or carbon dioxide, and chemical activation has been performed using inorganic acids (HCl, H_2SO_4, and H_3PO_4), alkalis (NaOH, KOH, and K_2CO_3), and several salts such as $ZnCl_2$, $FeCl_3$, $FeSO_4$, $AlCl_3$, and $Fe_2(SO_4)_3$. In addition to their abundance, other benefits of activated carbons are their relative stability, range of bulk density from 90 to 2,000 kg/m^3 (float or sink), and availability in powders, granules, spheres, rods, sheets, and fibers. The range of organic chemicals that can be readily adsorbed by active carbons includes aromatic solvents, polynuclear aromatics, chlorinated aromatics, phenolics, aromatic amines, high MW aliphatic amines, surfactants, fuels, most dyestuffs, chlorinated solvents, aliphatic acids, aromatic acids, and pesticides. The range of inorganic sorbates includes metal ions in metal salt solutions, radioactive materials and inorganic cyanides (this process involves sorption and chemical reaction so great care is required).

The properties of a series of well-established water-treatment carbons are summarized in Table 15.1.

Limitations to active carbon usage include compounds that have low MWs and high polarity (EPA, 1991). In addition, active carbons should not be used in streams with suspended solids $\geq$50 mg/dm^3 and oil and grease content $>$10 mg/dm^3, which may cause fouling. High levels of organics such as 1,000 mg/dm^3 may result in rapid exhaustion of the bed, but this is not determined for chemical spill cleanup applications. Compounds not well adsorbed require substantial amounts of activated carbon, and this increases the treatment costs significantly. Some examples of activated carbon performance data are presented in Table 15.2 (O'Brien and Fisher, 1983).

Peat

Formation. Peat is 80–90% water in its natural state. The solid component is derived from the partially decomposed residue of dead plants and the remains of decay microorganisms. Only a short depth below the surface of a peat bog, there is virtually no oxygen because the wetness limits air access. This oxygen deficiency gives rise to predominantly anaerobic

TABLE 15.1 Properties of Some Water-Treatment Carbons

Carbon type	F100	F200	F300	F400
Iodine number, min^m	850	850	950	1050
Methylene blue number, min^m	200	200	230	260
Abrasion number, % max^m	75	75	75	75
B.E.T. surface area, m^2 g^{-1}	900	900	1,000	1,100
Particle density, g m^{-1}	1.25	1.25	1.20	1.20
Mean particle diameter, mm	1.60	1.00	1.20	1.20
Phenol loading at 1 mg L^{-1}		4.30	4.33	
C.T.C. activity, %	75		68	

Source: Chemviron Carbon Technical Manual.

TABLE 15.2 Performance Data at Selected Sites (O'Brien, 1983)

Source of contaminants	Typical influent conc. (mg/L)	Typical effluent conc. (μg/L)	Carbon usage rate (lb/1000 gal)	Total contact time (min)
Truck spill				
Methylene chloride	21	<1.0	3.9	534
1,1,1-trichloroethane	25	<1.0	3.9	534
Rail car spills				
Phenol	63	<1.0	5.8	201
Orthochlorophenol	100	<1.0	5.8	201
Vinylidine chloride	2.4	<10.0	2.1	60
Ethyl acrylate	200	<1.0	13.3	52
Chemical spills				
Carbon tetrachloride	130–135	<1.0	11.6	262
Dichloroethyl ether	1.1	<1.0	0.45	16
Dichloroisopropyl ether	0.8	<1.0	0.45	16
DBCP	2.5	<1.0	0.7–3.0	21
On-site storage tanks				
Cis-1,2-dichloroethylene	0.5	<1.0	0.8	64
Tetrachloroethylene	7.0	<1.0	0.8	64
Isopropyl alcohol	0.2	<10.0	1.54	36
Acetone	0.1	<10.0	1.54	36
1,2-dichloroethylene	0.5	<1.0	1.0	52
Landfill site				
TOC	20	<5000	1.15	41
Chloroform	1.4	<1.0	1.15	41
Carbon tetrachloride	1.0	<1	1.15	41
Gasoline spills, tank leakage				
Benzene	9–11	<100 total	<1.01	214
Methyl t-butyl ether	0.030–0.035	<5.0	0.62	12
Trichlioloethylene	0.050–0.060	<1.0	0.62	12
Chemical by-products				
Di-isopropyl methyl Phosphonate	1.25	<50	0.7	30
Dichloropentadiene	0.45	<10	0.7	30
Chemical landfill				
1,1,1-trichloroethane	0.060–0.080	<1.0	<0.45	30
1,1-dichloroethylene	0.005–0.015	0.005	<0.45	30

Source: O'Brien and Fisher, 1983.

decay processes in the bog. The position of peat in the era of time is shown in Table 15.3, relative to coal and lignite.

Chemical Composition and Other Properties. The composition of peat varies considerably. Chemically, peats are largely organic material. When burned, they leave ash residue (2–10% by weight) and have calorific values of 7,000 $\pm$100 kJ kg^{-1}. The main varieties of peat can be distinguished by their acidity and ash content.

Peat is derived from living plants, which principally contain proteins, carbohydrates, lipids, and polyphenols such as lignin. Small amounts of nucleic acids, pigments, alkaloids,

TABLE 15.3 Classification of Peat by Age

Classification	Material	Approximate age
Fossil	Bituminous coal	<300 million years old
Aged	Lignite, brown coal peat	6–40 million years old
Recent	Straw, wood coconut shells, pith, etc.	10,000 years old

vitamins, and other organic substances are also present, along with inorganic materials. During decomposition, triglycerides hydrolyze, yielding fatty acids and glycerol. In anaerobic conditions of decay, the glycerol is readily consumed as a carbon and oxygen source for microorganisms. The residual fatty acids, unaltered waxes, and steroids persist as relatively stable components of the peat. Humic acids are thought to originate directly from lignin or as microbial product.

In the past 20 years, the chemical and physical characteristics of peat, and in particular sphagnum peat moss, have generated significant interest in the environmental field.

Oil Spillages. The problems of significant oil spills in the ocean and their subsequent migration onto beaches and coastal areas are well documented. It has been shown that peat moss has a much stronger absorption affinity for oil than straw, which is often used for oil spill treatment and can absorb up to eight times its own weight of oil (D'Hennezel and Coupal, 1972). Microscopic studies of peat moss show a highly porous and fibrous material, and this property, associated with the phenomemon of selective absorption, has been used to absorb oil.

Only a limited number of field tests have been undertaken, but it has been found that beach cleaning is relatively easy with peat moss. Raking the peat moss and the oil provides mechanical energy which facilitates absorption. The mixture does not stick to the tools and the technique is valid where raking is possible.

For treating oil slicks at sea, future proposals include the design of absorption booms of peat and finding a use for the peat–oil mixture. Burning is a possibility, but this can contribute to air pollution.

Phosphate Removal. The use of soils for wastewater treatment is an old practice that is receiving increasing attention as a solution to nutrient removal problems. A paper from Finland by Surakka and Kamppi (1971) describes the use of a ditched peatland area used for wastewater treatment at the village of Kesalahti since 1957. The wastewater is infiltrated through peat soil by pumping raw sewage to a large storage ditch in the peat bog and the wastewater then percolates through the peat to intercept ditches 20 m away. After 14 years of operations, results have been good in the removal of phosphorus (82%), nitrogen (90%), BOD (95%), and pathogenic bacteria (99+%).

A peat and a peat–sand filter has been used to remove phosphorus and organic matter from wastewaters on a pilot plant scale. A full-scale field plant has been proposed by Farnham and Brown (1972). More recently, Viraraghavan and Kikkeri (1988) have used peat filters to purify wastewaters from the food industry.

Treatment of Slaughterhouse Wastewaters. Wastewaters from slaughterhouses contain proteins, iron, fats, and phosphorus. Different slaughterhouse wastes will also differ in quality due to many factors, such as the quality and quantity of animals slaughtered, various procedures used, working time, and the quantity of water used.

Experiments were made on slaughterhouse wastes with *Sphagnum* peat, and it was found that a 0.15 m deep layer of peat removes disturbing impurities up to 90%, and after filtration the wastewater resembles household wastewater (Silvo, 1972).

Textile Effluents. Textile effluents contain a wide range of chemicals from processes such as sizing, scouring, dyeing, bleaching, and finishing. A process using peat has been developed for the treatment of such effluent: the Hussong-Couplan water treatment system, developed

TABLE 15.4 Test Results for a Typical Textile Effluent

Characteristic	Before	After
COD (ppm)	1200	85
BOD (ppm)	150	8
TOD (ppm)	1200	146
Phosphates (total) (ppm)	33.6	0.76
Suspended solid (ppm)	216	4

by Leslie (1974). The purification process is based on the scrubbing action of a moving mat of peat as the effluent is passed through.

In tests performed on actual textile effluents with an operational water treatment system, color was reduced by as much as 99.6% and turbidity by 100%. Other data are listed in Table 15.4. Further tests on textile effluents indicate the success of the system in treating heavy metal pollutants. The results of these tests are shown in Table 15.5.

Color Removal from Aqueous Spills. Several papers have been published demonstrating the effectiveness of peat as an adsorbent for removing dyestuffs in aqueous solutions. A thorough study involves determining equilibrium isotherms, batch contact time studies, and fixed-bed column breakthrough curves. Figure 15.2 shows the adsorption isotherms for Basic Blue 3. Basic Red 22, and Basic Yellow 21 onto *Sphagnum* peat (Allen, 1987; Allen et al., 1989a, b). These plots represent the amount of dyestuff adsorbed per unit mass of peat in equilibrium with the dye solution.

Removal of Metal Ions. The natural capacity of peat for heavy metal retention has been recognized in many studies which have investigated the removal of Hg, Cd, Zn, Cu, Fe, Ni, Cr(VI), Cr(III), Ag, Pb, and Sb. These studies found sorption to be quite high due to the polar character of peat and its weak acid ion exchange properties (Zhipei et al., 1984; Sharma and Forster, 1993; Ho et al., 1994, 1995; Gosset et al., 1986; Chen et al., 1990).

Figure 15.3 (Brown et al., 1992) shows the equilibrium isotherms for the sorption of copper and cadmium ions onto peat at 20°C. The figure shows the amount of metal ion sorbed (q_e μmol/g) at the liquid phase equilibrium metal ion concentration (C_e mmol/dm^3). The maximum saturation capacities for copper and cadmium ions on peat are 270 and 180 μmol/g respectively. These results indicate the considerable potential for peat in metal ion removal.

TABLE 15.5 Test Results for Metallic Pollutants in Textile Effluents

Metal	EPA effluent limitations schedule A	Before	After
Cadmium	0.1	25	0.1
Chromium (+6)	0.05	300	0.04
Chromium (+3)	0.25	300	0.25
Copper	0.2	250	0.2
Iron	0.5	31.5	0.25
Lead	0.05	8.4	0.025
Nickel	1.00	67.5	0.05

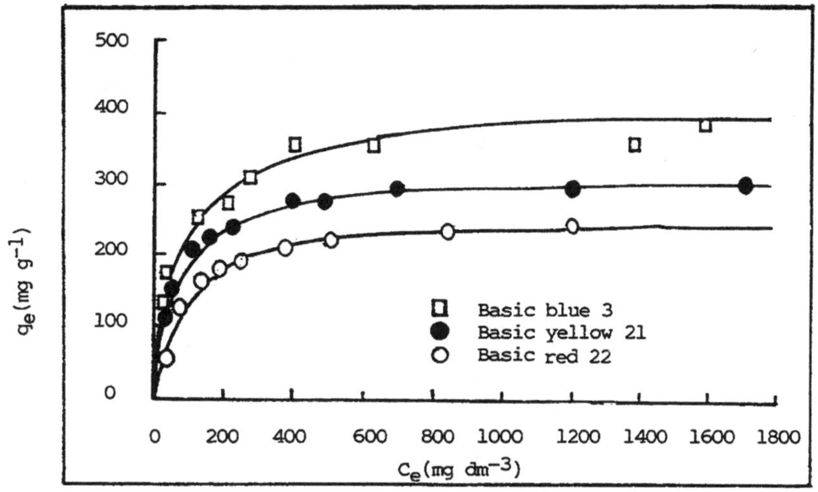

FIGURE 15.2 Equilibrium isotherms for various dyes on peat.

Molecular Sieves. One of the most widely used of the synthetic adsorbents is the range of zeolite metal aluminosilicates known as molecular sieves. These are synthesized adsorbents that have exact pore sizes caused by the driving off of water molecules by heating. The zeolites consist of porous aluminosilicate frameworks of SiO_4^- and AlO_4^- forming polyhedra. The gaps in the framework lattice permit the molecules to pass through to the body of the framework. The sieves can separate according to molecular size and by molecular polarity. Molecular sieves can therefore be tailor-made for particular applications. They are used for

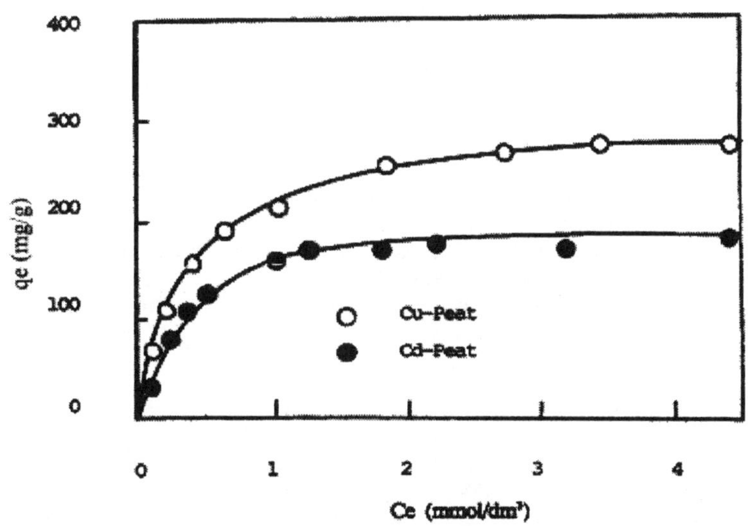

FIGURE 15.3 Sorption of cadmium and copper onto peat.

dehydration of gases and liquids and the separation of gaseous and liquid hydrocarbon mixtures. Molecular sieves can be regenerated by heating or by elution.

Adsorption from liquids into molecular sieve zeolites is increasingly being used in separation and purification processes. Broughton et al. (1970) and Broughton and Gembichi (1984) reported the separation by adsorption of liquid hydrocarbon mixtures and fructose–glucose mixtures by simulated moving bed applications (Luis, 1999; Rios et al., 1998).

Silicalite is a microporous crystalline silica that has received attention in both gaseous and liquid applications. The hydrophobic nature of silicalite has led to its use in the adsorption of ethanol from dilute aqueous solutions (Oulman and Chriswell, 1982; Milestone and Bibby, 1981; Bui et al., 1985). The adsorption of methanol, ethanol, acetone, toluene, and cyclohexane in silicalite crystals has also been investigated.

Ion Exchangers. Ion exchangers, resin or zeolite, have been judged most advantageous for use against ammonia, concentrated sulphuric acid, sodium hydroxide, potassium cyanide, cadmium chloride, oxalic acid, ethylene diamine, sodium alkylbenzenesulfonates, and phenol. They are sorbents for the ionic, most polar, and soluble solutes and complement polymeric sorbents for covalent nonmixers.

Several different ion-exchange resins have been used to remove carbolic oil (phenol) from wastewater and drinking water (Bauer et al., 1975). There are ion-exchange materials available for ionic, most polar, and soluble chemicals. Hazardous chemicals that sink, with their inherent low solubilities, could preclude the use of ion exchange in many cases. Pilie et al. (1975) investigated the use of ion-exchange resins as a substitute for activated carbon in the treatment of a heavy metal spill in water.

Polyurethane and polypropylene sorbents, macroeticular resins, zeolites, and ion-exchange resins cover the sorption spectrum of activated carbon, but in each of the sorption ranges, a particular one or another of these can be more advantageous or more feasible, for use in amelioration, than carbon.

Ion-exchange resins are uniformly available on a nationwide basis, due in large part to their wide range of use in water purification and treatment. On a cost and flexibility basis, ion-exchange resins must be considered inferior to activated carbon.

Sorbent Clays. Clays sorb like zeolites except that the sorption channels are interstitial between polysilicate sheets that are forced apart by the sorbate in a swelling action. Sorbent clays have been used for sorption in a variety of large-scale separations and for removing phenol from effluent. Some specific clays are listed and some of these materials do exhibit ion exchange properties.

Fuller's Earths. These are natural clays that contain magnesium aluminium silicates. The clay is given an open, porous structure by heating and drying processes. The adsorbent can be regenerated by washing and burning the organic contaminants that adhere to the clay surface. Fuller's earths are used in decolorizing and drying of oils.

Bentonite. Bentonite is a clay that requires activation by acid washing before it exhibits adsorptive properties. It is normally used as a fine powder for decolorizing and clarifying liquids and is discarded after use.

Bauxite. Bauxite is a naturally occurring alumina that requires thermal activation to facilitate its use as an adsorbent. It can be reactivated by heating. Other sorbent clays include kaolin and china clay.

Polyurethane Foams. These foams can be manufactured to any range of pore forms, from highly porous to nonporous. Polyurethane does not have the same range of sorption potential as activated carbon, but it is an effective sorbent for benzene, kerosene, naphtha, hexane, *n*-butyraldehyde, dimethylsuphoxide, epichlorohydrin, and phenol (Bowen, 1970). The sorption action is a combination of absorption and adsorption, but is mostly absorption depending on the pore structure. The polyurethane foam belts laden with recovered pollutant chemical can be squeezed and the removed sorbate pumped to storage (Kinoshita and Katsumata, 1999;

Chouno, 1998). The sorption capacity is a function of open-volume porosity, foam densities ranging from 16–640 kg/m^3, and the pore size distribution. Polyurethane foams can pick up from 0.1–80 times their own weight of spilled pollutant chemical depending on the affinity, pore size, and distribution.

For hazardous chemicals that float on water and vaporize, and to which existing devices for oil recovery apply, polyurethane foam can be applied in the float and mix. Devices for recovery of sorbate-loaded floating polyurethane can be readily applied, if fire hazards are controlled.

Macroreticular Resins. Polymerization of linear polymers with bifunctional monomers added during the reaction produces a reticulated structure that is similar to a macroporous sponge. Typical macroreticular polymers in this class are polymethyl methacrylate-co-ethylene dimethoacrylate and polystyrene-co-divinylbenzene. Substitution of polar groups in the monomers enables ion-exchange resins to be produced, such as polystyrenesulphonate.

The polymer resins sorb liquids or solutes like ethyl acetate, hexane, and ethylene diamine and dissolved solids like aldrin, sodium alkylbenzenesulfonates, and phenol. If sulponated, they exchange ions like cadmium more readily than the unsulphonated resin. Thus the macroreticular resins are more versatile than polyurethane foam, but not as universally so as activated carbons.

Application, recovery, and regeneration equipment and techniques would resemble those described for carbon and for ion-exchange sorbents. Limitations occur because of desorption sensitivity to oxidation. Thermal desorption is too hazardous for use with flammable sorbates. Effluent desorption seems to have some potential.

Activated Alumina. This adsorbent is manufactured by heating a precipitated mixture of alumina mono-and trihydrates up to 400°C. The heating process drives off the moisture, leaving an open porous structure with a high internal surface area. Alumina is widely used as a desiccant and can be regenerated by evaporation of the adsorbed species. Alcoa S-100, a commercial transition alumina, is treated with NaOH solution and dried overnight at 110°C. This activated alumina can adsorb hydrogen chloride gas from a nitrogen carrier. Isotherms for adsorption using aluminas treated with the different concentrations of NaOH have been determined (Fleming, 1987). Monolayer surface coverage for all materials is 2.0 w/w% at low HCl partial pressure. Considerable difference in HCl uptake among the materials is observed in the range of HCl partial pressure from 10^{-4} to 10^{-2} bar. Adsorption capacities increase for the low sodium materials linearly over this range and then approach an asymptotic limit at around 7 w/w% HCl saturation.

Ethylene dichloride is widely used in the petrochemical industry as a refrigerant in cooling systems and as a feed material for the production of polyvinyl chloride. During typical processing of the ethylene dichloride, it can become mechanically and thermally degraded. Low levels of hydrochloric acid, water, and iron chloride (ferric) can contaminate the ethylene dichloride. An alumina can be modified chemically to adsorb these contaminants. One such alumina is Selexorb-HCl. The adsorption isotherms for HCl, H_2O, and iron are presented in Fig. 15.4 (Sood and Fleming, 1987). The isotherms were measured as multicomponent mixtures. At 25°C, the Selexsorb-HCl adsorbent had a capacity of 0.8% Fe, 34% HCl, and 8% H_2O by weight. An increase in adsorption temperature to 40°C was found to increase the iron and HCl capacity and to cause a smaller increase in water capacity. Removal efficiencies for the three contaminants were 98.2% for Fe, 99.4% for HCl, and 17.9% for H_2O. The alumina can be regenerated by methanol rinse or by thermal regeneration.

Silica Gel. Silica gel is a common adsorbent, used mainly in granular form. It is made by heating an acidified sodium silicate gel to about 350°C. Silica gel is a hard, white, glassy, highly porous, granular solid. Silica is used chiefly for dehydration of air and other gases but is also used in some liquid operations. Silica can adsorb organics such as toluene and xylene from heptane in single-component and binary systems (Matz and Knaebel, 1991).

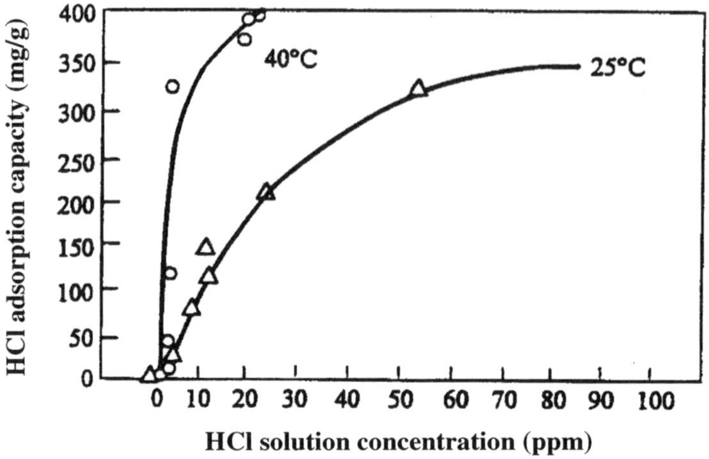

FIGURE 15.4 HCl adsorption isotherms.

Sorbsil silica gel is reported to adsorb basic dyes as shown in Table 15.6 (Alexander et al., 1978). Some characteristics of silica sorbents are shown in Table 15.7.

Polymer Fibers. A wide range of polymeric sorbents has been developed in conjunction with oil spill recovery work, including polyethylene and polypropylene fibres. These have been manufactured in a number of forms, including belts, pillows, pads, and roll blankets. Due to the oleophilic properties of these materials, they will likely find at least some applicability to spills of liquid organics as well as to spills of selected organic solids.

Many of these polymeric materials are available from oil spill contractors in ready-to-use form, and their availability through this network makes the transfer of material and technology to the control of hazardous materials a simpler task than it otherwise would be.

As is the case with other polymers, byproduct polypropylene and recycled polyproplyene are commonly cheaper than new polymer of the same type and grade. Correspondingly, polypropylene tailored to specifications or delivered in unusual form is often much more expensive. Waste polypropylene is sometimes usable with little loss of control performance

TABLE 15.6 Langmuir and Freundlich Constants for the Adsorption of Astrazone Blue on Silica

	Freundlich Constants		Langmuir Constants		
Particle size, μm	n, g L^{-1}	K, mg g^{-1}	B, L mg^{-1}	Q_0, mg g^{-1}	Dye adsorbed, mg m^{-3} [a]
150–180	2.90	2.5	0.0038	34.5	67.6
180–250	3.33	2.9	0.0038	30.3	75.0
250–355	4.81	2.5	0.0063	16.1	60.0
355–500	4.50	2.7	0.0079	13.5	67.5
500–710	4.17	2.1	0.0085	11.8	81.0
710–850	6.80	2.0	0.0117	6.3	67.0

[a] Initial dye concentration C_0 = 200 mg L^{-1}.

TABLE 15.7 Silica-Based Sorbents

Silica coating	Surface area, $m^2\ g^{-1}$	Average pore diameter, Å
C_1	27.5	180
C_8	27.5	200
C_{18}	27.5	190
PhP	10.0	390

capability. Overall, especially for application in existing devices, polypropylene is more feasible for present day use than macroreticular resins of higher versatility. It is a tenable alternative for polyurethane or activated carbon for the sorption of chlorine, kerosene, naphtha: solvent, aldrin, dimethyl sulfoxide, and epichlorohydrin.

Polyolefins, Polyethylenes, Polyisobutylenes, Poly(Methyl Methacrylates), and Poly(Styrene Sulfonates). In addition to polyurethane foams, polyethylene and polypropylene fibers can be manufactured as alternative polymeric sorbents. However, based on sorbent capacity, cost, and availability, these materials would be considered inferior to polyurethane.

Thus, use of polyolefins against chlorine, polyethylenes and polyisobutylenes against aldrin, poly(methyl methacrylates) against sodium alklbenzenesulfonates, and poly(styrene sulfonates) against cadmium chloride, though possible, is not advantageous for amelioration except where local availability may compensate.

Lignite. Lignite, a member of the solid fuel family, is often referred to as brown coal. Table 15.8 shows that lignite fits between peat and bituminous coals in terms of its properties. It is drier than peat and has a higher carbon content.

Lignite and coal are formed from decaying trees and plants. The decaying vegetation falls to the ground and, if it is preserved by water, is slowly oxidized by microorganisms to form peat. Peat becomes lignite and coal when it becomes buried below the earth's surface. Heat gradually dries out the peat and causes the chemical changes that result first in lignite and then ultimately in coal. Coal formation requires the vegetation to reach temperatures of about 150°C, so the peat must become buried to a depth of around 5 km. Lignite formation needs a lower temperature, and thus lignite is usually found fairly close to the surface and can be recovered by opencast methods, whereas coal normally is buried deeply and has to be mined underground.

During the transformation into lignite, the vegetation loses hydrogen and oxygen atoms and the lignite now contains 65–70% carbon, about 5% hydrogen, and 25–30% oxygen (by weight). Much of the oxygen is present in acidic groups. Lignites contain a little paraffinic material–and generally some waxes and resins–but are mainly composed of aromatic struc-

TABLE 15.8 Technical Properties of Solid Fuels

Fuel	% Moisture	% Carbon (daf)	Calorific value ($kJ\ kg^{-1}$)
Peat	> 75	50–60	7,000
Lignite	30–70	60–70	17,000
Bituminous coal	10	80	36,000
Anthracite	2	90	37,000

tures based on benzene and naphthalene rings. Many different structures are present in lignite. Lignites have a rich chemistry; they appear to consist of a rigid, complicated lattice in which molecules are trapped.

Lignite possesses adsorptive properties. The surface area, porosity, functional groups, and calcium and magnesium ions all contribute to these properties. Lignite has a strong affinity for basic dye (Allen et al., 1989). The lignite-cationic dye sorption process is facilitated by the presence of the humic acid groups in the lignite. Different ions will experience different physical and electrical attractive and repulsive forces according to their structure, molecular size, and functional groups. The process is a combination of ionic attraction/repulsion, hydrogen bonding, ion-dipole forces, covalent bonds, and Van der Waals forces.

The adsorptive powers of lignite have been shown to reduce the COD of an effluent. Lignite treated with 50% sulphuric acid at 100°C is partly activated. It is claimed that activated carbon reduces the COD of slaughterhouse wastes by up to 60%. Lignite and lignite-derived activated carbons also exhibit adsorption of organics such as benzoic acid and chlorobenzoic acids. The retention of metallic cations by lignites is well established (Allen et al., 1992; Ibarra et al., 1979). The adsorption of copper, cadmium, lead, aluminum, and zinc is shown in Fig. 15.5.

Chitin and Chitosan. Chitin is the second most widely occurring natural carbohydrate polymer next to cellulose. Chitin is a long, unbranched polysaccharide and can be regarded as a naturally occurring derivative of cellulose, where the C2 hydroxy group has been replaced by the acetyl amino group —$NHCOCH_3$. The primary unit in the chain polymer is 2-deoxy-2-(acetyl-amino) glucose. These units are linked by $\beta(1 \rightarrow 4)$ glycosidic bonds forming a long-chain linear polymer having degrees of polymerization around 2,000 to 4,000. It is insoluble in almost all solvents except strong mineral acids, notably nitric and hydrochloric and certain other solvents, due to the presence of the carbonyl functionality in the acetyl group. The carbonyl group is responsible for hydrogen bonds, resulting in a rigid structure, less permeable to water and other reagents, as in the case with cellulose.

Chitin is widely distributed throughout nature, being the main structural polysaccharide that forms the characteristic exoskeleton of most of the invertebrates. High concentrations of up to 85% are found in Arthropoda (especially the edible crab), which are particularly able to synthesize chitin (Muzzarelli, 1976).

Chitosan is an aminopolysaccharide and consists of [(1 → 4)-2-amino-2-deoxy-β-D-glucan] monomer units. Like chitin, it is nontoxic and biodegradable and has a molecular

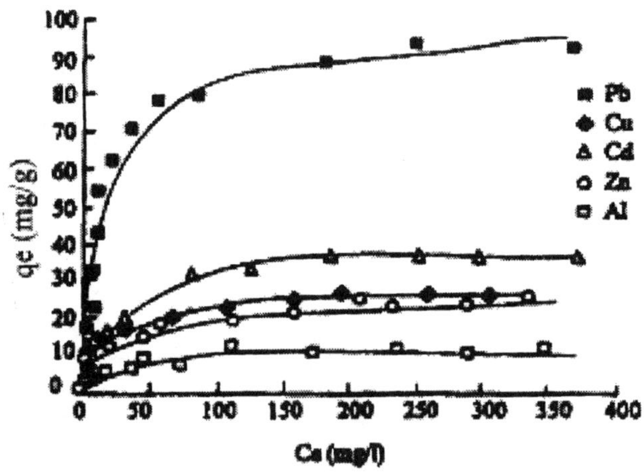

FIGURE 15.5 Capacity isotherms for single-component metals on lignite.

weight (about 120,000) depending on the source of chitin and the severity of *N*-deacetylation reaction. It is commercially produced by the chemical modification of chitin by alkaline hydrolysis. This cleaves the acetyl group, leaving the free amine, and depending on controllable process parameters, varying degrees of deacetylation can be achieved. There are numerous methods for deacetylating chitin, and a number have been outlined by Muzzarelli (1976). The systematic repetition of the alkali deacetylation produces chitosan with degrees of a deacetylation up to 90%, but also leads to hydrolysis of the glycosidic linkages and a much reduced molecular weight.

Adsorption of metal cations is one of the most beneficial functions of chitin and chitosan. An extensive review has been made by Muzzarelli (1973, 1976). Marucca et al. (1982) examined adsorption of chromium under various conditions. Koyama and Taniguchi (1986) studied the effect of deacetylation of chitin on adsorption and the importance of high hydrophilicity. Eiden et al. (1980) reviewed the uptake of lead (II) and chromium (III) by chitin and chitosan as a function of concentration. Hexavalent chromium interaction with chitosan has also been studied by Onsoyen and Skavgrud (1990). Other recent studies on adsorption include iodine, nickel, copper, zinc, and cadmium.

The mechanism of complex formation of metals with chitosan is manifold and is probably dominated by different processes such as adsorption, ion exchange, and chelation under different conditions. In Fig. 15.6, it can be seen that the sorption capacity of chromium and cadmium is similar, while zinc has approximately double the affinity of these two metal ions for chitosan. Chitosan has almost three times the removal capacity for copper sorption than that for cadmium or chromium. These effects have proved complex to interpret but are a function of a number of parameters: ionic radii; ionic charge; electron structure and possibly some hydration capacity of the metal ions: solution pH and nature; and availability of sites for chitosan.

Muzzarelli (1976) gave the collection rates of metal ions onto chitosan as $Cu^{2+} > Zn^{2+} > Cd^{2+}$ as the same as the Irving and Williams series, and there was a relation between the order and the second ionization potentials.

Woodmeal. Woodmeal has been used as an adsorbent for basic and acid dyes. The uptake of acid dye Telon Blue 25 is shown for Sitka spruce wood in Fig. 15.7 (McKay and McConvey, 1981, 1985). The results show that spruce wood does have an affinity for basic dyes, although this is much less than for active carbons, peats, and lignite. Figure 15.7 also

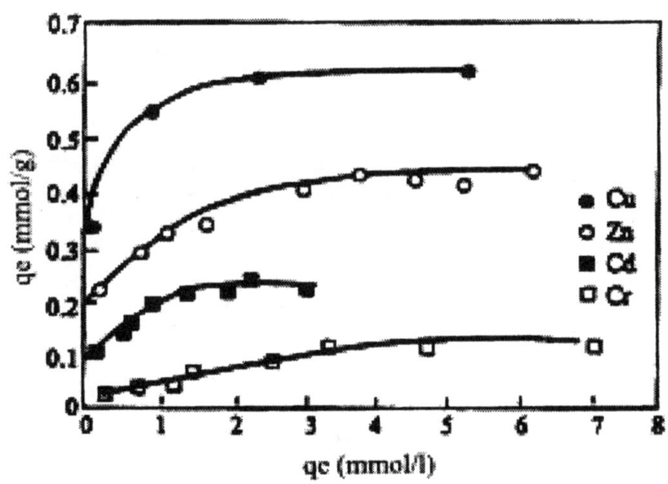

FIGURE 15.6 Metal ion uptake onto chitosan.

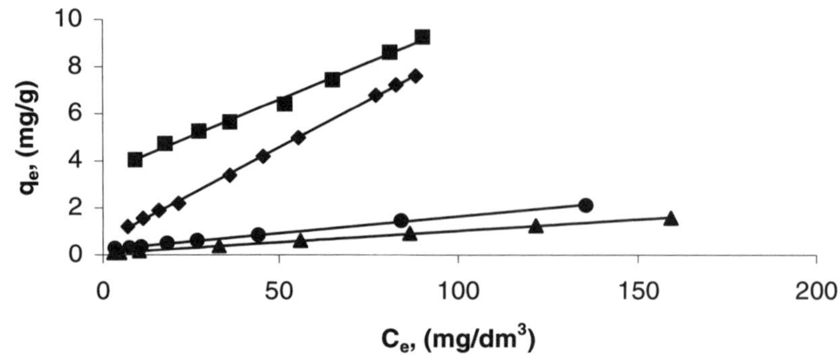

◆ AB25/Peat, Experimental Data — AB25/Peat, Langmuir Equation; ▲ BB69/Peat, Experimental Data — BB69/Peat, Langmuir Equation; ■ AB25/Wood, Experimental Data — AB25/Wood, Langmuir Equation; • BB69/Wood, Experimental Data — BB69/Wood, Langmuir Equation

FIGURE 15.7 Langmuir isotherms for THE sorption of BB69 and AB25 onto peat and wood.

shows linearized plots of the Langmuir equation for Basic Blue 69 dye and Acid Blue 25 dye on both peat and wood.

Bone Char. Bone charcoal is a carbonaceous substance derived from the carbonization of selected grades of animal bones by heating dry bones in an airtight iron retort at 500–700°C for about four to six hours. Comparing the capacity of metal ions removal with activated carbon, bone charcoal provides not only a porous carbon surface for physical adsorption, but also a hydroxyapatite lattice—$Ca_{10}(PO_4)_6(OH)_2$ for ion exchange of metal ions. Based on these properties, this sorbent should have excellent adsorption capacities for metal ions. The characteristics of typical bone char are shown in Table 15.9.

Traditionally, bone char has been used for decolourizing sugar solutions in the sugar refining industry. But recently it has been shown to have a significant sorption capacity for metal ions (Raouf and Daifullah, 1977; Cheung, 1999). The isotherms for the sorption of copper, cadmium, and zinc ions onto bone char are shown in Fig. 15.8.

Bagasse Pith. Bagasse pith is a waste product from the sugar refining industry. It is the name given to the residual pulp remaining after the sugar has been extracted. Bagasse pith is composed largely of celluose, pentosan, and lignin as shown in Table 15.10. Pith does

TABLE 15.9 Physical and Chemical Properties of Bone Char

Chemical composites		Physical properties	
Parameter	Limits	Parameter	Limits
Acid insoluble ash	3 wt % max	Bulk density (dry)	640 kg/m^3
Calcium carbonate	7–9 wt %	Carbon surface area	50 m^2/g
Calcium sulfate	0.1–0.2 wt %	Moisture	5 wt % max
Carbon content	9–11 wt %	Pore size distribution	7.5–60,000 nm
Hydroxyapatite (Tricalcium phosphate)	70–76 wt %	Pore volume	0.225 cm^3/g
Iron—as Fe_2O_3	<0.3 wt %	Total surface area	100 m^2/g

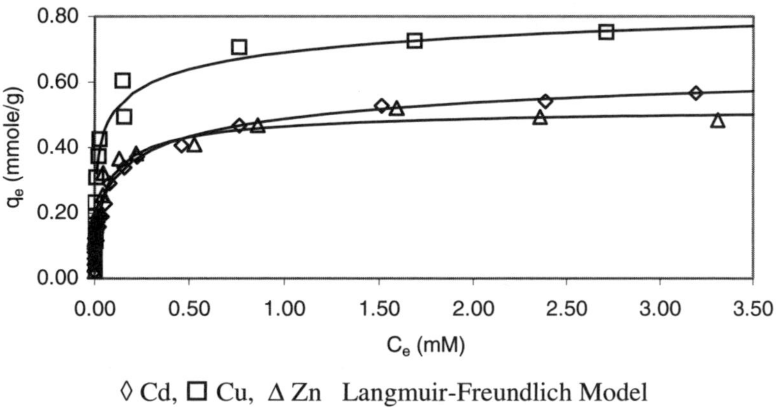

FIGURE 15.8 Sorption of copper, cadmium, and zinc ions onto bone char.

exhibit adsorptive properties. McKay et al. (1997) reported the adsorption of three dyes onto pith. The adsorption of basic dyes is shown in Fig. 15.9.

Other Potential Adsorbents. While activated carbon is the most widely used adsorbent, in the past 10 years considerable attention has been directed towards low-cost biosorbents. Activated carbon is expensive, and an alternative inexpensive adsorbent could drastically reduce the cost of an adsorption system. Many waste or naturally occurring materials have been investigated to assess their suitability. For water pollution control, the use of low-cost natural materials for the removal of copper has been studied for several materials. Other potential sorbents include peat, anaerobically digested sludge, kaolin and montmorillonite clay, treated bagasse, treated acacia bark, treated laurel bark and treated techtona bark, fly ash, *Penicillium spinulosum,* dyestuff-treated (Red) rice hulls and dyestuff-treated (Yellow) rice hulls, resins moss *Calymperes delessertii Besch,* water hyacinth (*Eichornia crassipes*), *Rhizopus arrhizus, Cladosporium resinae* and *Penicillium italicum,* tea leaves, amorphous iron hydroxide, and activated carbon.

In McKay et al. (1999), another 40 potential sorbents are listed. This single adsorbate case emphasizes the future potential of finding new, cheaper sorbents for chemical spills in the future. The main problem is the time required for screening all these sorbents for the vast array of chemicals currently in use.

Selecting Sorption for Spill Treatment. Spills can occur during transportation, storage, and chemical processing. They may result from transport accidents, such as collisions, from storage vessels and containers, due to poor maintenance/operator error, or due to fire or

TABLE 15.10 Chemical Analysis of the Bagasse Pith

Determination	%
α-cellulose	53.7
Pentosan	27.9
Lignin	20.2
Alcohol/benzene solubility	7.5
Ash	6.6

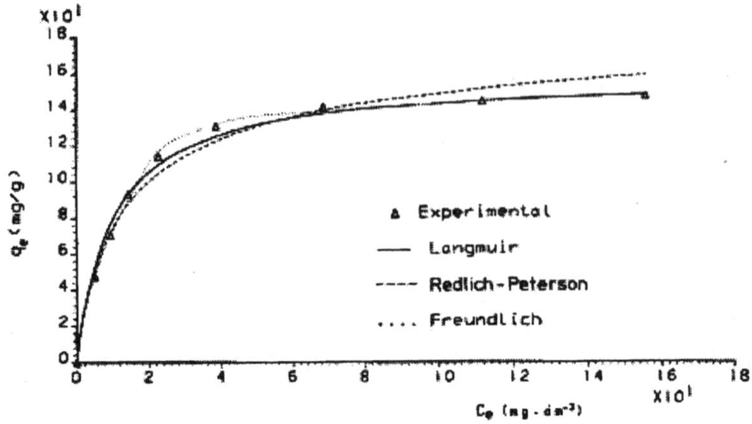

FIGURE 15.9 Comparison of theoretical isotherm curves with experimental results for basic Blue 69 dye on pith.

explosion during processing. The spillage can occur on land/road or at sea, depending on the transport method; it can occur at the manufacturing site and it could be a large or small spill. The decision to use sorption for spill treatment is made based on the type of spill and the conditions under which it has occurred.

If the spillage is large and at sea or off-site, the logistics of transporting large amounts of adsorbent materials may prove too difficult and too complex. In addition, harvesting the used ("spent") adsorbent is difficult. Adsorbent booms may be useful to limit the migration of these types of spillage.

Small-scale spillages from leaking or broken containers on the road or in the air can readily be treated by having a specially selected adsorbent spill kit on the transport vehicle. This same approach would also apply to small spillages on-site. On a multipurpose chemical site, these "small spills kits" would contain a selection of adsorbent materials as outlined in Section 15.2.1.

In the case of chemical spills due to process upsets, particularly fire and explosion, the chemical spillage on-site may be extensive and the water used on the fire can become contaminated with one or more process chemicals. Containment for fire water runoff should be planned for all manufacturing plants at the design stage. Containment volume should also be provided on-site to ensure that no contaminated water leaves the site after a small or a major incident. Such water should not be flushed into a ditch, sewer, or drain or off the road since this would spread the chemical farther. The key feature of a well-designed chemical spill or contaminated water containment system is the ability to pump the spilled liquid into recovery containers for treatment. These large containers of contaminated liquid can then be treated by the conventional adsorption contacting systems as described in Section 15.3.

15.1.3 Sorbent Characterization Tests

In order to assess the suitability of a sorbent for a particular chemical spill application, it is important to carry out specific tests for the specific application. This type of information is available only for a limited number of commercially available chemical spill sorbents. There are several specific properties that also indicate sorbent's usefulness in certain applications. Because detailed analysis has already been carried out (Brady and McKay, 1995), only major tests are listed here.

Total Surface Area

- The Brunauer Emmett Teller (BET) method based on nitrogen adsorption (Brunnauer et al., 1940)
- Isotherm measurements (Kienle and Bader, 1980)
- Adsorption measurements using other solutes (Paryjczak, 1985)

Porosity

- Particle porosity (DIN 4188)
- Bed voidage (Brunnauer et al., 1940)
- Pore size distribution (Brunnauer et al., 1940)

Attrition Resistance

- Mechanical strength (Schneider, 1979; ASTM D 3802)

Density

- Apparent density (Czaplinski, 1965)
- Packed density (ASTM D 2854-70)
- Absolute density (Czaplinski and Lason, 1965)

Electron Microscopy

- High-resolution electron microscopy (HRTEM) (Evan and Marsh, 1979)

Physiochemical Characterization

- Carbon tetrachloride activity test (Patrick, 1995)
- Dye adsorption-methylene blue test (DAB 7)
- Iodine sorption value (AWWA B 600-78)
- Iodine number (AWWA B 600-78)
- Baylis's phenol test (DIN 19603)

Moisture

- Humidity/water (ASTM D 2867-70)

Chemical Characterization

- Adsorbent pH (Jankowska et al., 1991)
- Elemental analysis (Vogel, 1989)
- X-ray elemental analysis (Mering and Schoubar, 1968)
- Carbon, hydrogen, nitrogen elemental analysis (Vogel, 1989)
- Surface group IR analysis (Vogel, 1989)
- End group titration (cation exchange capacity) (Vogel, 1989)

Carbonaceous Adsorbents

- Ash (ASTM D 2866-70)
- Volatile matter (BS 1016 Part 4)

- Molasses test (Vogel, 1989)
- Caramel test (Vogel, 1989)
- Determination of gold adsorption capacity (*K*-value) of activated carbon (Vogel, 1989)

15.2 PLANNING FOR SPILLS AND CLEANING UP

Because other sections of this Handbook deal more specifically with cleanup methods, policies, and detailed spill procedures, this section only emphasizes some key points related to adsorption cleanup. The safety, health, and environment department must prepare formal written procedures for all types of chemical spillage. The specification of sorbent type must be based on an expert analysis of the MSDS (material safety data sheet) for every chemical onsite. The manufacturers' technical data or MSD sheets provide information for the site-specific safety plan, such as: flash point, ignition temperature, solubility, toxicology, density, reactivity, and chemical compatibility. The written procedures must cover all chemicals on site and all possible spill scenarios, particularly "What is the worst incident that could happen?" Prevention techniques must obviously be in place and emergency plan procedures carefully prepared and practiced.

15.2.1 Sorbent-Based Spill Kit

The following is a list of materials that could be included in a sorption kit for chemical spills. The list is by no means comprehensive or specific; this can only be achieved by a thorough analysis of the MSD sheets for all the chemicals on site. Many other items are needed in a sorbent-based spill kit, but they are not specific to sorption cleanup and are covered elsewhere in this Handbook.

- Neutralizer sorbent mixes should be included for acids and bases.
- Sorbent for solvents, particularly those that minimize flammable vapors.
- Vermiculite, sand, clays, etc. are useful for less hazardous spills
- Calcium bentonite is often incorporated into chemical spill kits because it is one of the most cost-effective sorbents and can be used on most chemical spills, with the exception of hydrofluoric acid.
- Dry angular grain clay-type sorbents are useful for traction in slippery areas and as an adsorbent.
- Sorbent spill pillows or booms can adsorb, neutralize, and reduce flammability, corrosivity, reactivity, and toxicity of liquid acids, caustics, and solvents.
- Diking booms are long sorbent synthetic pulp-filled tubes.
- Vapor barrier sorbent blankets and rolls are used to sorb and suppress acidic vapors (including HF), caustics, and solvents
- Mercury sorbent and collector, which is a dedicated special vacuum cleaner suitable for mercury use, should be included.
- Chemical-resistant, disposable sponge mop, spill squeeze mop, large sponges, dustpan and brushes, all made out of nonsparking materials, enable residual liquors to be sorbed and used/spent sorbent particles to be swept up.
- Heavy-duty plastic bags and labels are used to collect the contaminated sorbent particles.

- A nonmetallic scoop is used to sprinkle sorbent particles onto the spills and mix sorbents with liquid.
- A shovel, which should be nonsparking and chemically resistant, is used to collect larger quantities of used sorbent.

15.2.2 Sorbent Cleanup Procedures

The following steps are based primarily on the use of sorbent; other steps impacting on the use of sorbents are briefly reviewed.

Notification. Initially a request for help is placed by any available means–this could be to a police or other emergency hotline service on-site or off-site. The emergency response team (ERT) will be activated and attend the scene of the spill. An immediate assessment of the chemicals involved and extent of the spill are necessary in order to select the most effective chemical spill treatment and, in particular, the most suitable sorbent.

First Aid and Personal Protective Equipment. The ERT members must attire themselves in the appropriate personal protective equipment (PPE), move any injured personnel from the affected area, and administer the correct first aid treatment as the top priority.

Spill Containment. The chemical spill should be contained as much as possible on the site where the spill occurs. The spilled chemical must be prevented from entering storm drains, wells, ditches, streams, rivers, or other water systems. This may involve repositioning the leaking container or applying a seal to the leak from the repair/patch kit, or separating the leaking container from the other containers.

In order to confine the chemical spill, the spill area should be encircled with a dike of sand or other sorbent material. If the spill involves a small watercourse, a dam of activated carbon or other sorbent can be used to filter the water. For larger waterways, a log boom or baled straw may be used to contain the spill. If possible, the flow of clean water should be diverted around the spill. Sorbent materials are usually used to cover liquid spills. On most major chemical sites, there is usually sufficient containment volume to contain all chemical spills and handle contaminated water runoff from firefighting/sprinkler systems.

Site Cleanup. In order to apply sorption successfully for large-scale spills or large-scale volumes of contaminated water, it is necessary to pump as much of the spilled liquid into recovery containers so that it can be treated by conventional adsorption processes described in Section 15.3.

For small-scale spills, the procedure is to use the appropriate sorbent to soak up the spill. Only the minimum amount of sorbent should be used; the method of determining this is covered in Section 15.3. The sorbent should be spread around the perimeter of the spill and swept toward the center. The adsorbent should be shoveled into leak-proof container(s) for subsequent disposal.

Decontamination. Sorbents are useful for decontaminating small amounts of residual chemicals after the cleanup process from soil, roadways, inside truck/aircraft containers, concreted surfaces (or other materials) on-site, tools, and other non-porous areas. Soil should be removed to a level below the contaminated area and placed in drums for subsequent treatment or disposal. In the case of roadways, floors, and other nonporous surfaces, the appropriate adsorbent should be spread onto the spill and worked into the surface using a coarse broom. The sorbent should remain in situ for at least two hours. The sorbent should be then collected by first spreading fresh sorbent material around the periphery of the spill area, sweeping inwards toward the centre, and shoveling into labeled plastic bags or drums. Dampening of the sorbent is often useful to minimize the dust levels and prevent it from becoming airborne. This procedure is usually repeated. The area is finally rinsed with water, taking adequate precautions to avoid runoff.

Evaluation of Sorbent Capacity. The quantity of pollutant that a sorbent can adsorb is defined by the equilibrium isotherm. This is a mathematical relationship between the liquid phase concentration of the spill chemical, C_e, and the solid phase concentration of the spill chemical on the sorbent, q_e, when the contacting system has reached equilibrium. Figure 15.10 shows this relationship for cadmium salt in solution that has been spilled.

This analysis is best explained by considering that most of the spilled solution, V dm^3, can be pumped into a storage tank and a mass of peat, M g, must be added to remove the Cd^{++} ions to achieve a certain concentration level $C_{e,1}$, from the original concentration in the spill tank, C_0. The schematic for this process is equivalent to adsorption in an agitated batch tank.

The schematic diagram for a single-stage adsorption is shown in Fig. 15.11. The solution to be treated contains V dm^3 water, and the pollutant concentration is reduced from C_0 to $C_{e,1}$ mg solute per gram of solution. The amount of adsorbent added is M g of adsorbate free solid, and the solute concentration on the adsorbent increases from q_0 to $q_{e,1}$ mg solute per gram adsorbent. If fresh adsorbent is used, $q_0 = 0$. The mass balance equates the solute removed from the liquid to that picked up by the solid.

$$V(C_0 - C_{e,1}) = M(q_{e,1} - q_0) = Mq_{e,1} \tag{15.1}$$

Rearranging Eq. (15.1):

$$-\frac{M}{V} = \frac{(C_0 - C_{e,1}}{q_{e,1}} \tag{15.2}$$

In addition to the graphical solution, the final effluent concentration $C_{e,1}$ can be calculated analytically. In most practical cleanup situations, equilibrium will not be reached, but in a well-designed process with effective adsorbent selection, 75–90% of the equilibrium adsorption capacity may be achieved. For the purpose of simplification, this mathematical analysis assumes that equilibrium is reached.

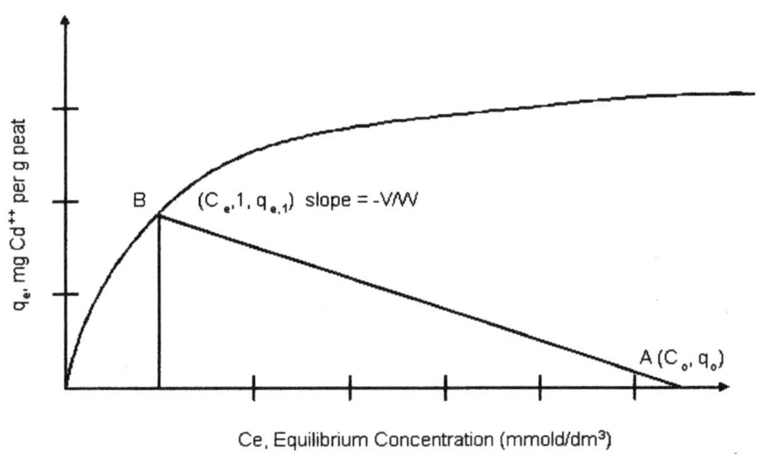

FIGURE 15.10 Equilibrium isotherm and batch sorption operating line.

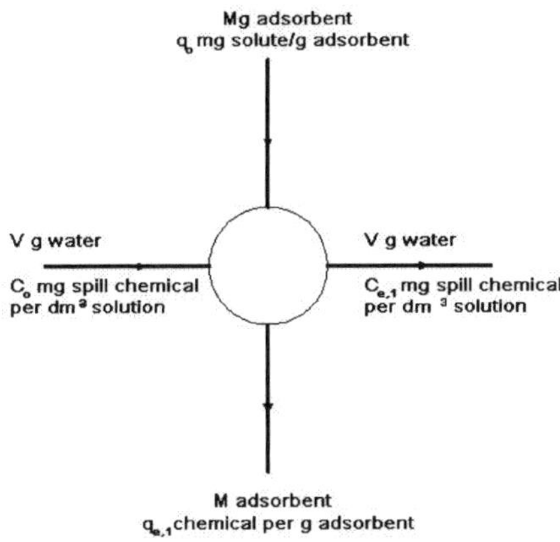

FIGURE 15.11 Single-stage batch adsorption for the removal of cadmium ions in solution.

At this stage in the analysis, in order to solve Eq. (15.2) for M, it is necessary to use an isotherm equation to determine $q_{e,1}$. Three of the most common isotherm equations are Langmuir, Freundlich, and Redlich-Peterson, as shown in Eqs. (15.3), (15.4), and (15.5), respectively.

$$q_e = \frac{K_L C_e}{1 + a_L C_e} \tag{15.3}$$

$$q_e = K_F C_e^{1/n_F} \tag{15.4}$$

$$q_e = \frac{K_R C_e}{1 + a_R C_e^{n_R}} \tag{15.5}$$

K_L and a_L are conventional Langmuir constants; K_F and n_F are conventional Freundlich constants. Their numerical values are normally found by linearizing Eqs. (15.3) and (15.4). K_R, a_R, and n_R are constants in the Redlich-Peterson isotherm and are determined by optimizing the fit between experimental equilibrium data points and theoretical data points generated by Eq. (15.5).

In the present example, using the cadmium chemical spill problem, the Langmuir equation provides a good correlation of the experimental equilibrium isotherm data. Equation (15.3) can be substituted into Eq. (15.2) and the mass of sorbent, M g peat, to bring the liquid phase concentration down to $C_{e,1}$, can be determined from Eq. (15.6).

$$-\frac{M}{V} = \frac{(C_0 - C_{e,1})(1 + a_L C_{e,1})}{K_L C_{e,1}} \tag{15.6}$$

This equation represents the unique solution for this example of a cadmium solution spill treatment. It is presented schematically in Fig. 15.10 by the line AB, which is an operating line of slope $-V/M$ and coordinates $A(C_0, q_0)$ and $B(C_{e,1}, q_{e,1})$.

With this isotherm approach coupled with the mass balance equation, the capacity of the sorbent for a specified concentration of chemical spill can be obtained and the amount of sorbent required to remove the hazardous solute can be determined.

15.3 DESIGN OF SORPTION CONTACTING SYSTEMS

15.3.1 Sorption Contacting Systems

The main sorption contacting methods are:

- Single-stage agitated batch adsorbers
- Multistage agitated batch adsorbers
- Pulsed beds
- Fixed bed adsorbers
- Steady state moving beds
- Fluidized bed adsorbers
- Filtration-type adsorbers

Single-stage Agitated Batch Adsorbers. A typical batch process is shown in Fig. 15.12. The effluent to be treated and the sorbent are intimately mixed in the treating tank (agitated contacting tank) for a fixed contact time to enable the system to approach equilibrium, after which the thin slurry is filtered to separate the solid sorbent and sorbate from the liquid. The equipment may be made multistage by providing additional tanks and filters. If the operation is to be made continuous or semicontinuous, centrifuges or a continuous rotary filter may be substituted for the filter press. The adsorbent is usually applied in the form of a finely ground powder, and the time required for the adsorbent and liquid to come to substantial equilibrium depends primarily on the concentration and particle size of the solid, the viscosity of the liquid, solute concentration, and the extent of agitation. Agitation should be vigorous to ensure rapid contact of the sorbent particles with the liquid.

Multistage Crosscurrent Batch Sorption. The efficiency of the contaminated spill solution is realized by treating with separate small batches of sorbent rather than in a single-stage batch process. When using expensive sorbents, it is important to operate as economically as possible with this process, the greater the saving the larger the number of batches, but

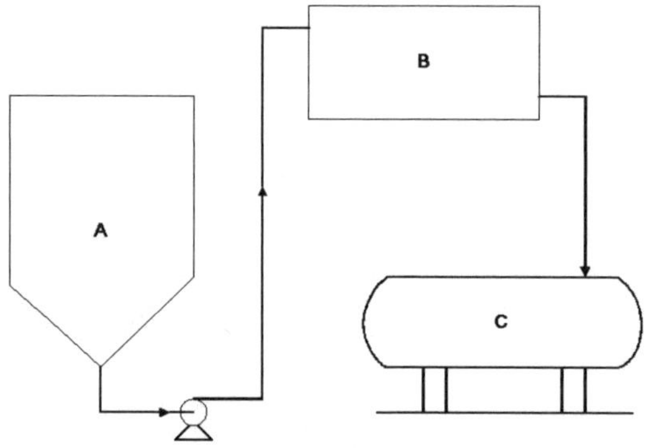

FIGURE 15.12 Batch contact filtration. Schematic arrangement for single-stage batch sorption. (*a*) Agitated batch contacting tank; (*b*) filter press; (*c*) filtrate storage tank.

filtration and handling costs increase rapidly. As a result, usually only two stages are used. The schematic for this process mode is shown in Fig. 15.11.

Stages 1 and 2 treat the same amount of solution, V. The operating process diagram is constructed as shown in Fig. 15.13 for a two-stage process. The approach is the same as for the single-stage operating diagram, and the operating lines will be parallel when the same quantity of adsorbent is used. The operating diagram assumes fresh sorbent is added at each stage and therefore $q_0 = 0$. The mass balance equation for stage 1 is

$$V(C_0 - C_{e,1}) = M_1(q_{e,1} - q_0) \tag{15.7}$$

and for stage 2,

$$V(C_{e,1} - C_{e,w}) = M_2(q_{e,2} - q_{e,1}) \tag{15.8}$$

Two-Stage Crosscurrent Optimization Based on Minimizing the Quantity of Sorbent. An important economic criterion in multistage adsorption is to determine the minimum quantity of an adsorbent to treat a fixed volume of chemical spill at a specified contaminant concentration.

Two-Stage Crosscurrent Optimization. When fresh adsorbent is used at each stage, $q_0 = 0$ and using an isotherm expression, to describe equilibrium, the least total amount of adsorbent can be calculated analytically. In this example, the Freundlich expression, Eq. (15.5), is used. For stage 1,

$$\frac{M_1}{V} = \frac{(C_0 - C_{e,2})}{q_{e,1}} = \frac{(C_0 - C_{e,1})}{K_F C_{e,1}^{1/n_F}} \tag{15.9}$$

K_F and n_F are the Freundlich isotherm constants. For stage 2,

$$\frac{M_2}{V} = \frac{(C_{e,1} - C_{e,2})}{q_{e,2}} = \frac{(C_{e,1} - C_{e,2})}{K_F C_{e,2}^{1/n_F}} \tag{15.10}$$

The total amount of sorbent used.

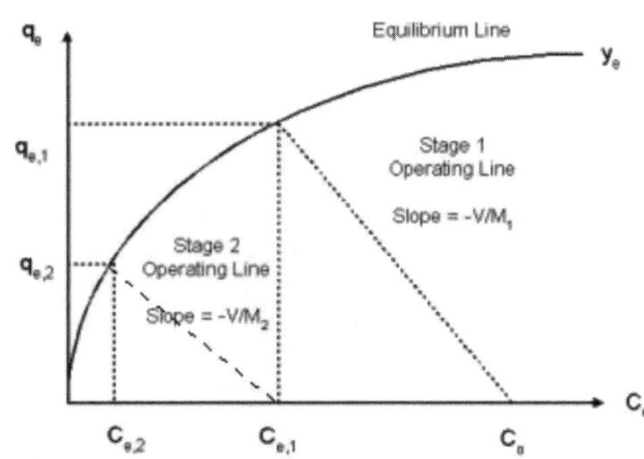

FIGURE 15.13 Two-stage crosscurrent adsorption process.

$$\frac{M_1 + M_2}{V} = K_F \left(\frac{C_0 - C_{e,1}}{C_{e,1}^{1/nF}} + \frac{C_{e,1} - C_{e,2}}{C_{e,2}^{1/n_F}} \right) \tag{15.11}$$

In order to minimize the total amount of sorbent, it is necessary to differentiate the following Eq. (15.11):

$$\frac{d\{(M_1 + M_2)/V\}}{dq_{e,1}} = 0 \tag{15.12}$$

and since for a given system K_F, n_F, C_0, and $C_{e,2}$ are constant,

$$\frac{(C_{e,1}^{1/n_F})}{C_{e,2}} - \frac{1}{n_F}\frac{C_0}{C_{e,1}} = 1 - \frac{1}{n_F} \tag{15.13}$$

The liquid phase contaminant concentration, $C_{e,1}$, can now be obtained from Eq. (15.13) using the standard trial-and-error solution method. Furthermore, a graphical solution may also be used to solve Eq. (15.13), and this is shown in Fig. 15.14. The two axes represent the fraction of contaminant not adsorbed after stage 1, $C_{e,1}/C_0$, and stage 2, $C_{e,2}/C_0$. Since the final condition, $C_{e,2}/C_0$, is known, and the Freundlich constant, n_F, is known (obtained from a Freundlich linearized plot of the equilibrium data), then $C_{e,1}/C_0$ can be read directly from Fig. 15.14 and $C_{e,1}$ is obtained. It has been proposed (Lerch and Rathowsky, 1967) that countercurrent batch adsorber operation is usually more advantageous than the application of crosscurrent split sorbent operation or crosscurrent split solution flow.

Two-Stage Countercurrent Sorption. A two-stage countercurrent batch sorption system is shown in Fig. 15.15. In stage 2, M g fresh sorbent is added to the chemical spill solution, V dm^3, and the concentration change of the contaminant is from $C_{e,1}$ to $C_{e,2}$.

The contaminant-laden, $q_{e,1}$, sorbent is then used to treat V dm^3 fresh solution from C_0 to $C_{e,1}$ solute per dm^3 spill solution, the spent adsorbent is discharged at $q_{e,2}$ mg solute per gram sorbent. The overall solute mass balance is given in Eq. 15.14

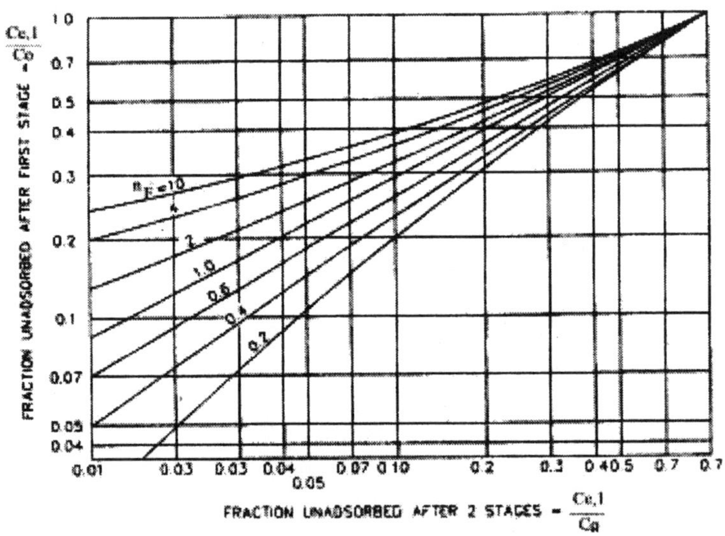

FIGURE 15.14 Graphical solution to determine minimum sorbent required in two-stage crosscurrent operation.

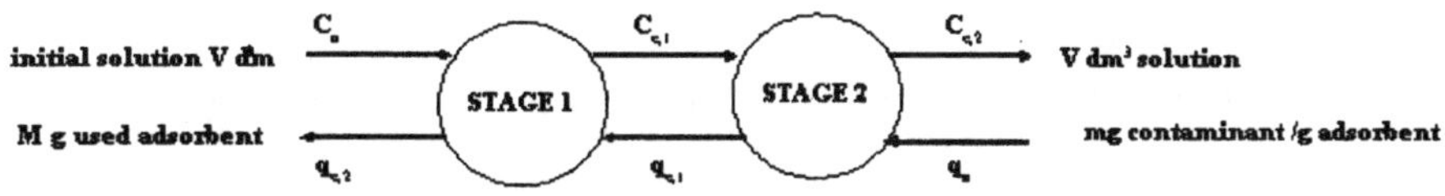

FIGURE 15.15 Schematic for two-stage countercurrent sorption.

$$M(q_{e,1} - 0) = V(C_{e,1} - C_{e,2}) \tag{15.14}$$

The operating line diagram is similar to Fig. (15.13). The slope of the operating line is for the overall cascade system, M/V, and for the two individual stages the slopes are both also equal to $-M/V$ (for equal masses of adsorbent and equal masses of solution).

Two-Stage Countercurrent Batch Sorption. The sorbent/solution ratio for a fixed set of process conditions can be determined by trial and error using the graphical technique. This is carried out by graphically fitting two steps between the equilibrium curve and the initial and final solute-liquid concentration boundary concentrations. The operating line can then be drawn, and this provides the coordinates, $C_{e,2}$, $q_{e,1}$, of the intermediate concentration, the sorbent/solution ratio is determined from the slope of the operating line. Figure 15.16 shows the operating lines for the two stages on the equilibrium curve.

In the present case study, it is assumed that the equilibrium curve follows the Freundlich equation (although any best-fit isotherm equation should be selected). The intermediate concentration, $q_{e,1}$, can be calculated from Eq. (15.15) for a Freundlich isotherm:

$$q_{e,1} = K_F C_{e,1}^{1/n_F} \tag{15.15}$$

This value of the solid phase contaminant loading is coupled with the mass balance equation as follows:

$$\frac{M}{V} = \frac{C_0 - C_{e,2}}{q_{e,2}} = \frac{C_0 - C_{e,2}}{K_F C_{e,1}^{1/n_F}} \tag{15.16}$$

The mass balance for the operating line in the second stage is:

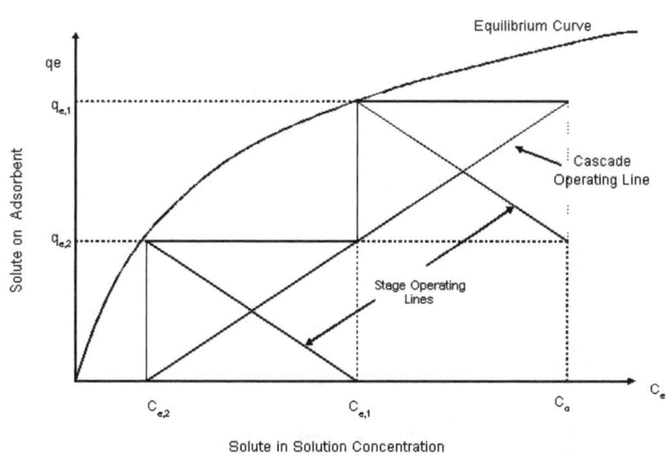

FIGURE 15.16 Operating diagram for two-stage cocurrent sorber.

$$V(C_{e,1} - C_{e,2}) = Mq_{e,2} = MK_F C_{e,2}^{1/n_F} \tag{15.17}$$

Combining Eqs. (15.16) and (15.17), the term V/M can be eliminated to give:

$$\frac{C_0}{C_{e,2}} - 1 = \frac{C_{e,1}^{1/n_F} \cdot C_{e,1}}{C_{e,2} \cdot C_{e,2}} - 1 \tag{15.18}$$

In process design, the boundary conditions are usually specified and $C_{e,1}$ can be determined. Therefore, the ratio V/M can be determined. A graphical solution is available for obtaining the intermediate concentration $C_{e,1}$; this is shown in Fig. 15.17.

Although this procedure can be used in the design of the multistage batch sorption process, economics usually (but not always) limit the number of contact stages to two.

Multistage Batch Adsorber Optimization Based on Minimizing Contacting Time. Batch sorber design has mainly concentrated on reducing adsorbent costs, which is particularly relevant when expensive sorbent materials such as active carbon, silica, zeolites, and resins are used. But for cheaper adsorbents, minimizing the contact time for a fixed percentage of pollution removal using a fixed mass of sorbent will result in being able to process more batches of polluted wastewater per day, thus enabling the required treatment plant items to be reduced in size, with a decrease in the plant capital cost.

The cost and performance of product/equipment/system or the mode of application are always of concern in controlling process efficiency. Therefore, the sorption capacity and required contact time are two of the most important parameters to understand in a sorption process. Equilibrium analysis is fundamental to evaluate the affinity or capacity of a sorbent. However, thermodynamic data can only predict the final state of a system from an initial

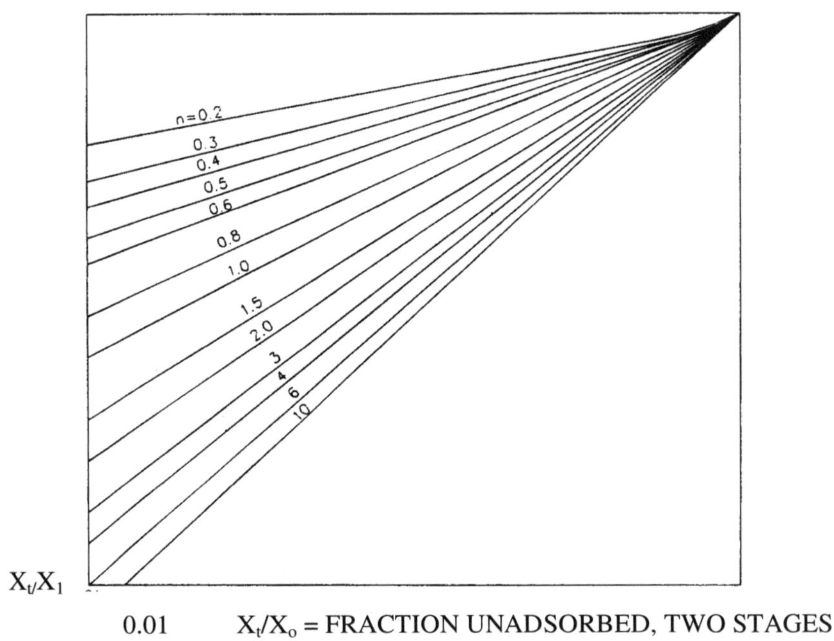

FIGURE 15.17 Graphical solution to determine the intermediate solute concentration in countercurrent adsorption. (*Source: Treybal, 1987*)

nonequilibrium mode. It is therefore important to determine how sorption rates depend on the concentrations of sorbate in solution and how rates are affected by sorption capacity or by the character of sorbent in terms of kinetics. From the kinetics analysis, the solute uptake rate, which determines the residence time required for completion of the sorption reaction, may be analyzed and established.

In order to minimize contact time, it is necessary to use a contact time model, which may be a kinetic model (first order, second order, Elovich kinetics, etc.) or a mass transport diffusion model (film, pore, surface, pore–surface). The present case study is based on a pseudo-second order kinetic model (Ho and McKay, 1998), and the kinetic equation is:

$$\frac{t}{q_t} = \frac{1}{k_2 q_e^2} + \frac{t}{q_e} \tag{15.19}$$

This relationship enables the amount of chemical contaminant, q_t, adsorbed to be predicted as a function of time, t. The equations are developed generally for an n-stage batch sorption system, and the case study is for the treatment of dye solution spillage using peat as a cleanup sorbent.

The solution to be treated contains V dm^3 solution, and the dye concentration is reduced for each stage from C_{n-1} to C_n mg dm^{-3}. The amount of sorbent added is M g with solid phase dye concentration on the peat q_0 mg g^{-1}. The dye concentration on the sorbent increases from q_0 to q_n mg g^{-1} sorbent. The mass balance equation gives:

$$V(C_{n-1} - C_n) = M(q_n - q_0) \tag{15.20}$$

When fresh sorbents are used at each stage and the pseudo-second order rate expression is used to describe equilibrium in the two-stage sorption system, then the mass balance equation can be obtained by combining Eqs. (15.19) and (15.20):

For fresh adsorbent:

$$q_0 = 0 \tag{15.21}$$

Then:

$$C_n = C_{n-1} - \frac{Mkq_n^2 t}{V(1 + kq_n t)} \tag{15.22}$$

The total amount of dye removal can be calculated analytically as follows:

$$\sum_{n=1}^{n} (C_{n-1} - C_n) = \sum_{n=1}^{n} \frac{Mkq_n^2 t}{V(1 + kq_n t)} \tag{15.23}$$

The dye removal, R_n, in each stage can be evaluated from the equation as follows:

$$\sum_{n=1}^{n} R_n = \frac{100\, Mt}{VC_0} \sum_{n=1}^{n} \frac{kq_n^2}{1 + kq_n t} \tag{15.24}$$

Therefore, it is useful for process design purposes, if q_e and k can be expressed as a function of C_0 for sorption of dye onto peat and wood as follows:

$$q_e = A_q C_0^{B_q} \tag{15.25}$$

$$k = A_k C_0^{B_k} \tag{15.26}$$

Substituting the values of q_e and k from Eqs. (15.25) and (15.26) into Eqs. (15.23) and (15.24), equations (15.23) and (15.24) can be represented as follows:

$$R_n = \frac{100\ S(A_kC_{n-t}^{B_k})(A_q(C_{n-t}^{B_q})^2t}{LC_0[1 + (A_kC_{n-1}^{B_k})(A_qC_{n-1}^{B_q})t]} \tag{15.27}$$

$$\sum_{n=1}^{n} R_n = \frac{100\ St}{LC_0} \sum_{n=1}^{n} \frac{(A_kC_{n-t}^{B_k})(A_qC_{n-1}^{B_q})^2}{[1 + (A_kC_{n-1}^{B_k})(A_aC_{n-t}^{B_q})t]} \tag{15.28}$$

In the case of a two-stage countercurrent sorption system, the design procedure is now outlined. For example, 5 m^3 of spill solution are to be treated. The amount of sorbent added is 10 kg in each of the two stages and the BB69 initial concentration is 200 mg dm^{-3} in the first stage. A series of contact times from 10 minutes up to 220 minutes in 10-minute increments has been considered in stage one of a two-stage sorption system. Therefore, in Figs. 15.19 and 15.20, each system number is based on a 10-minute contact time interval. In the first adsorber, for example, a system number 6 implies the first adsorber contact time; t_1, is $6 \times 10 = 60$ minutes. Therefore, the contact time in the second adsorber, t_2, is the time required, T minutes, to achieve a fixed total % dye removal minus the contact time in the first adsorber stage t_1.

$$T = t_1 + t_2 \tag{15.29}$$

For N systems, t_1 becomes 10 N minutes and

$$T = 10\ N + t_2 \text{ (for a fixed \% removal)} \tag{15.30}$$

The total contact time is calculated for each system number for a fixed percentage of dye removal. The T values are plotted against the system number, as shown in Figs. 15.18 and Figure 15.19, and the minimum contact time may be determined.

Figure 15.18 shows the time for 90% BB69 removal using peat for each stage and the total reaction time of the two-stage countercurrent batch sorption process. Figure 15.19 shows the minimum contact time of the two stages in series for the two-stage sorption of BB69/ peat.

Fixed-Bed Sorption. Normally in practical cases, the concept of fixed-bed adsorbers is usually expressed in graphical terms by the breakthrough curve concept. The dynamic adsorption system is represented in Fig. 15.20. Effluent enters the column at solute concentration C_0, and a concentration gradient or profile is established within a finite zone (break-

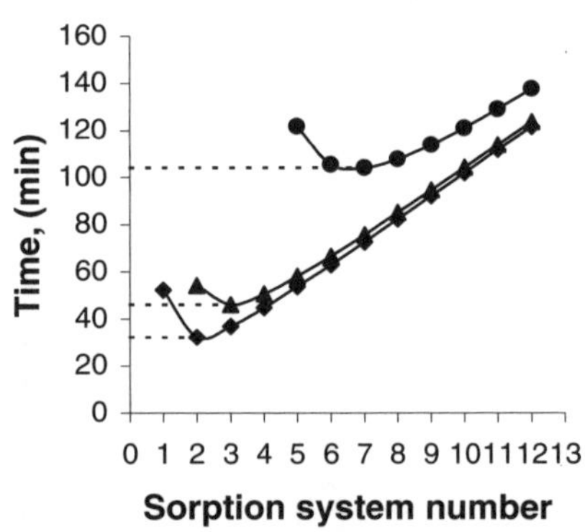

FIGURE 15.18 Comparison of 90% BB69 removal time of each stage in two-stage BB69/ peat process. ◆ Stage 1, BB69/peat; ● stage 2, BB69/peat; ▲ 90% removal, BB69/peat.

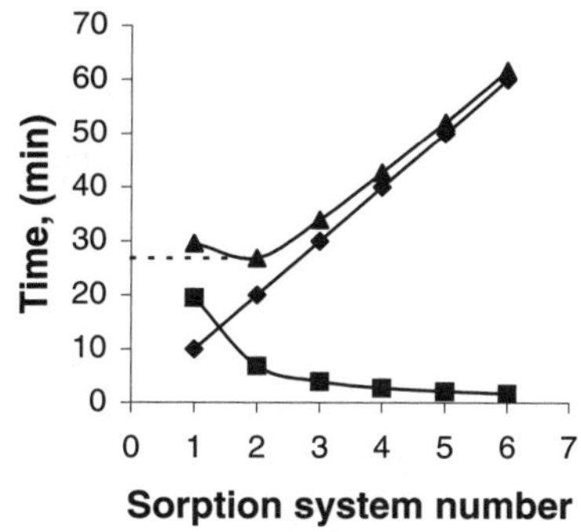

FIGURE 15.19 Minimum contact time for various percentages BB69 removal in a two-stage BB69/peat process. ◆ 93% removal, BB69/peat; ▲ 96% removal. BB69/peat; ■ 99% removal, BB69/peat.

through zone or mass-transfer zone [MTZ]). In this zone, the concentration of effluent in solution changes from C_0 to C_e, where C_e should be close to zero. The loading of solute on the solid also changes within the mass-transfer zone from q_e to 0. Since the system is a dynamic one, the mass-transfer zone moves steadily from the influent end of the adsorber bed to the outlet end. On the mass transfer zone curve shown in Fig. 15.20, point *a* represents an effluent concentration C_e and point *d* represents a solute concentration C_0, that is, the effluent entering the column.

In Fig. 15.20, the wave front or breakthrough curve can be expressed in terms of time, *t*. The area below the wave (*a, g, d, e, f, a*) reflects unused sorbent capacity, and the (*a, g, d, e, f, a*)/(*a, b, d, e, f, a*) ratio is the fraction of unused sorbent bed in the breakthrough curve. A vertical straight line drawn through *g*, the 50% breakthrough point, and that area (*a, f, c, b, a*) = (*a, g, d, b, a*) and (*f, e, d, c, f*) = (*a, g, d, e, a*) yields an equivalent

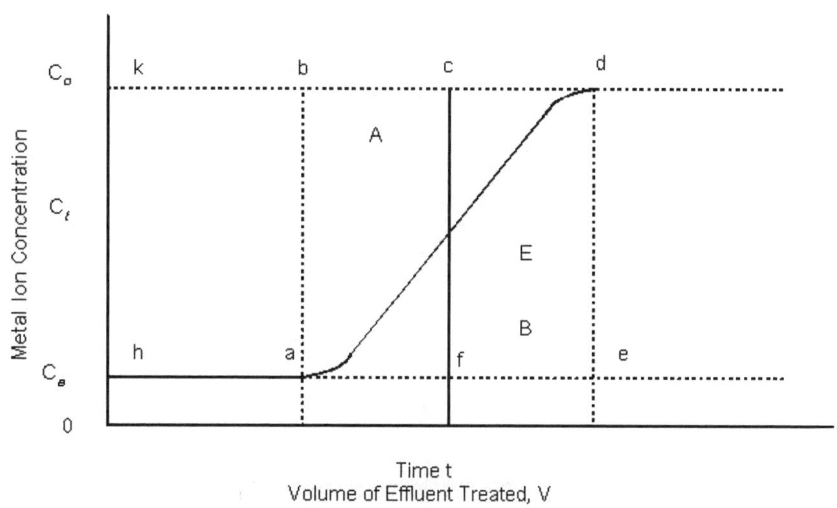

FIGURE 15.20 The breakthrough curve mass-transfer zone.

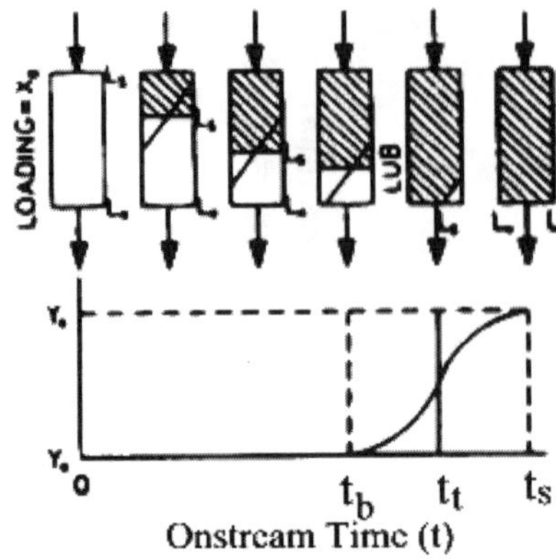

FIGURE 5.21 Stable breakthrough curve in sorbent bed and position of the stoichiometric front (horizontal lines) relative to the stable mass-transfer front during dynamic adsorption.

stoichiometric front for the system at time, t_s. The rectangle (h, f, c, k, h) corresponds to sorbent at its equilibrium loading, q_e; it is defined as the equivalent equilibrium section and is specified in terms of the length of this section, Z_e. The area (f, e, d, c, f) corresponds to adsorbent at its initial loading, q_0; it is specified in terms of the length of unused bed, Z_0.

The adsorbent bed at breakthrough actually comprises an equilibrium zone and a mass-transfer zone. The system is shown schematically in Fig. 15.21 in which a stable breakthrough curve is depicted moving at uniform velocity through an adsorber and the stoichiometric transfer front superimposed on the actual transfer front. The results for the adsorption of phenol onto activated carbon are shown in Fig. 15.22. The typical S-shaped breakthrough curves are shown at four bed heights. $C_0 = 400 \times 10^{-3}$ kg phenol/kg solution and $F = 1.67 \times 10^{-6}$ m^3/s.

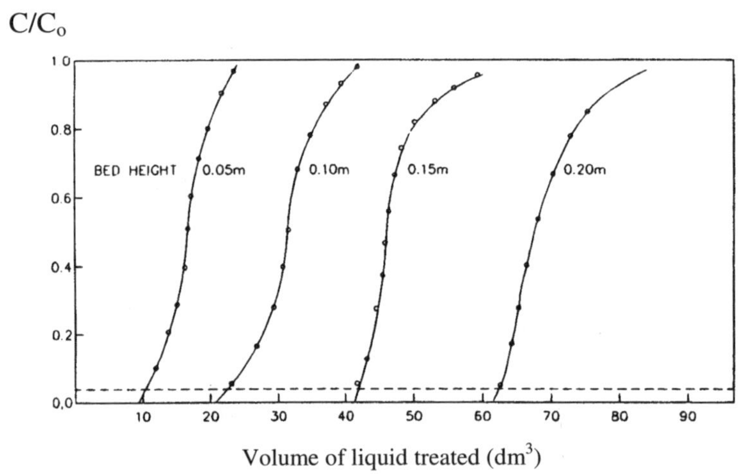

FIGURE 15.22 Fixed-bed sorption of phenol using activated carbon. Breakthrough curves at four bed heights are shown.

Fixed-bed columns for liquid-phase sorption are cylindrical vertical columns; they may be operated in upflow or downflow mode. Single beds or multiple-bed arrangements may be used; the range includes:

- Fixed beds in series–upflow; downflow
- Fixed beds in parallel–upflow; downflow

Fixed-bed Column Operation. A further major advantage of fixed-bed is that sorption/exchange columns can be skid-mounted so that they can be readily transported to a chemical spill anywhere onsite.

Fixed-bed sorption columns can be operated as single units or multiple units. The columns can operate by upflow or downflow, and multiple-column systems can operate in series or parallel. Figure 15.23 shows columns in series operating downflow. Each bed can be replaced as a complete, separate unit. When breakthrough occurs in the last column, the first column will be in equilibrium with the influent, thus achieving maximum adsorbent sorption capacity and utilization. This may be as a result of economic necessity. This "first column" will then become the "last column" in the system.

Figure 15.24 shows expanded upflow columns operating in parallel with a single downflow column. These are staggered when put into operation. Downflow operation is used on large-volume plants, and the breakthrough curve should be steep to maximize adsorbent utilization.

A single-column system could be considered for the following reasons: if the breakthrough curve is steep; if the sorbent will last so long at the desired processing conditions that the cost of replacing or regenerating it represents only a minor part of the operating costs; or if the capital cost of additional columns cannot be justified because not enough adsorbent cost can be saved to pay for additional equipment.

Comparing upflow and downflow columns provides advantages and disadvantages for each mode of operation, including the choice of gravity feed and pumped feed. With downflow operation, filtration is more effective and suspended solids will be trapped at the top of the bed. Pumps would be required to handle a significant pressure drop build-up, and backwashing facilities and pipe work are needed. Piping is simpler in upflow systems because the direction of flow is the same for adsorption and washing cycles. Pressure drops are lower because of bed expansion, and less downtime and wash water are required. Upflow systems would more commonly have gravity feed than downflow systems, but they are not as effective in the filtration of suspended solids.

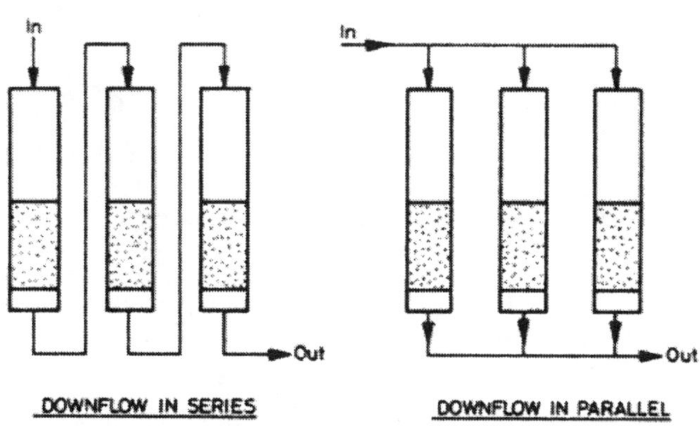

FIGURE 15.23 Fixed-bed sorption columns in series and parallel operating downflow.

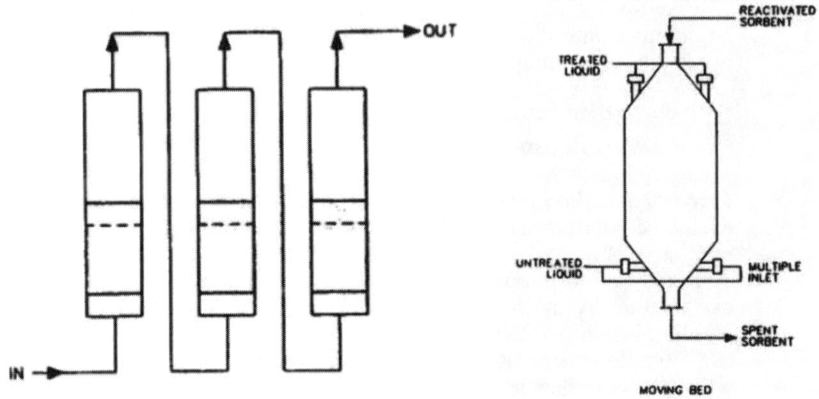

FIGURE 15.24 Expanded upflow bed and downflow column.

Breakthrough Curve–Bed Depth Service Time (BDST) Model. In the operation of a fixed-bed adsorption column, the service time, t, of the bed can be related to the bed depth, Z, for a given set of conditions by a model and equation called the bed depth service time model (BDST). The BDST offers a rapid method of designing fixed-bed columns. The influent solute concentration, C_0, is fed to the column, and it is desired to reduce the solute concentration in the effluent to a value not exceeding C_b. At the beginning of the operation, when the adsorbent is still fresh, the effluent concentration is actually lower than the allowable concentration, C_b, but, as the operation proceeds and the sorbent reaches saturation, the effluent concentration reaches C_b. This condition is called the break point.

Hutchins (1993) proposed a linear relation between the bed depth and service time, which can be written as:

$$t = \frac{N_0 \Delta Z}{C_0 F} - \frac{1}{K_a C_0} \ln\left(\frac{C_0}{C_b} - 1\right) \tag{15.31}$$

N_0 is the adsorption capacity, K_a is the adsorption rate parameter and F is the flow rate. The critical bed depth, Z_0, is the theoretical depth of carbon sufficient to prevent the solute concentration from exceeding the C_b value at $t = 0$. By letting $t = 0$, Z_0 is obtained by solving Eq. (15.31) for Z. The final result is:

$$Z_0 = \frac{F}{(K_a N_0)} \ln\left(\frac{C_0}{C_b} - 1\right) \tag{15.32}$$

Equation (15.32) enables the service time, t, of an adsorption bed to be determined for a specific bed depth, Z, of adsorbent. The service time and bed depth are correlated with the process parameters; the initial pollutant concentration, solution flow rate, and adsorption capacity are the factors in the first term on the right-hand side of Eq. (15.31). The second term on the right-hand side of Eq. (15.31) represents the time required for the pollutant to establish its breakthrough curve; that is, it represents that part of the bed that is not saturated when the pollutant concentration in the solution leaving the bed is above the breakthrough concentration, C_b.

$$t = m_x Z - X_x \tag{15.33}$$

and the intercept of this equation represents:

$$X_x = F\left[\ln\left(\frac{C_0}{C_b} - 1\right)\right] \tag{15.34}$$

The gradient of Eq. (15.33) may be used to predict the performance of the bed if there is a change in the initial solute concentration, C_0, to a new value, C'_0. Hutchins (1993) proposed that the new gradient, M'_x, can be written as:

$$m'_x = m_x \frac{C_0}{C_b} \tag{15.35}$$

If the equation is developed for a standard chemical spill concentration C_0, it can easily be modified for any spill concentration C'_0, and the new intercept is:

$$X'_x = X_x \frac{C_0 \ln}{C_b \ln}\left(\frac{\dfrac{C'_0}{C_b} - 1}{\dfrac{C_0}{C_b} - 1}\right) \tag{15.36}$$

The application of the model to predict design performance is shown in Fig. 15.25 for a parachlorophenol spill. The figure shows BDST results obtained by the theoretical lines, and these are compared with experimental points represented by the various symbols. It can be seen that the theoretical equations, derived from the BDST analysis, accurately predict experimental results for various possible spill concentrations of parachlorophenol. Agreement between theoretical lines and experimental points, represented by the symbols, is very good.

Fixed-Bed Optimization Using the Empty Bed Residence Time Procedure. Many variables can influence the design of fixed-bed sorbers (FBD_s), and it is advisable to use the most economical conditions. The empty bed residence time (EBRT), provides a simplified procedure for minimizing sorbent usage.

Two types of variables are recognized as important in optimizing the sorption system. First, there are primary process variables that are mainly determined by sorbent exhaustion

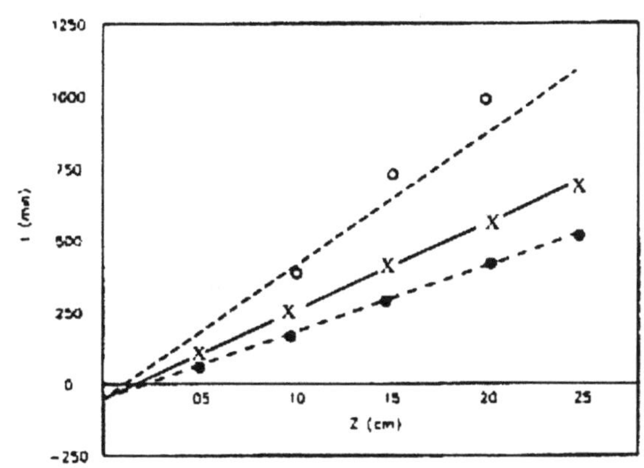

—x Co = 200 x 10^{-3} kgm^{-3}; —— ° Co = 300 x 10^{-3} kgm^{-3}; —— • Co = 500 x 10^{-3} kgm^{-3}

FIGURE 15.25 Theoretically predicted lines for different *p*-chlorophenol concentrations on carbon–predicted from solid line of concentration 400 × 10^{-3} kgm^{-3}.

rates. Second, there are those process variables that affect the operating costs as a result of changes in the required level of effluent purity, influent liquid purity, temperature and viscosity, as well as sorbent particle size.

For a given system, in which the liquid flow rate, impurity concentrations, and sorbent characteristics are fixed, the capital and operating costs are almost entirely dependent on the following two primary variables: sorbent exhaustion rate, which is usually expressed as kilograms of adsorbent used per volume of liquid treated, and superficial liquid retention time, empty bed contact time, EBCT, sometimes referred to as empty bed residence time, which is the time that the liquid would take to fill the volume of the sorbent beds sand and is a direct function of liquid flow rate and adsorbent volume.

For a given system, the total capital cost is established primarily by the volume of the adsorbent beds and, usually to a lesser extent, by the size of an adsorbent reactivation furnace, if one is to be used. The operating costs are determined by the sorbent exhaustion rate since the large variable is usually the cost of the makeup sorbent.

The relation between these variables can be established by tests and plotted as shown in Fig. 15.26. For a given system to achieve a given performance, there is a single line relating these two variables, called the operating line. The operating line concept can be used to optimize the basic design to achieve the lowest cost. The operating line approaches a minimum on both axes in Fig. 15.26. The minimum exhaustion rate for a given sorption duty is that which is achieved when the exhausted adsorbent is in equilibrium with the influent liquid, and the minimum retention time represents the minimum time for solution to pass through the minimum volume of carbon necessary to achieve the desired level of purity at infinitely high sorbent exhaustion rates. The arrangement of the sorber vessel influences the operating line, as shown in Fig. 15.27.

Data for use in the EBRT process are obtained from the BDST analysis. A specified percentage breakthrough concentration is selected and plotted as a BDST equation using experimental/pilot plant results. From the plots at various sorbent bed heights, the EBRT and the adsorbent exhaustion rate are obtained. The points are then plotted on a graph and the "best" asymptotes are drawn to the curves, as shown in Fig. 15.27. The optimum minimum values for EBRT and sorbent exhaustion rate are obtained from the intercept of these curves with axes.

Figures 15.28 and 15.29 show how the fixed-bed breakthrough curves were employed to obtain data for plotting the operating line. This involves fixing the defined breakthrough

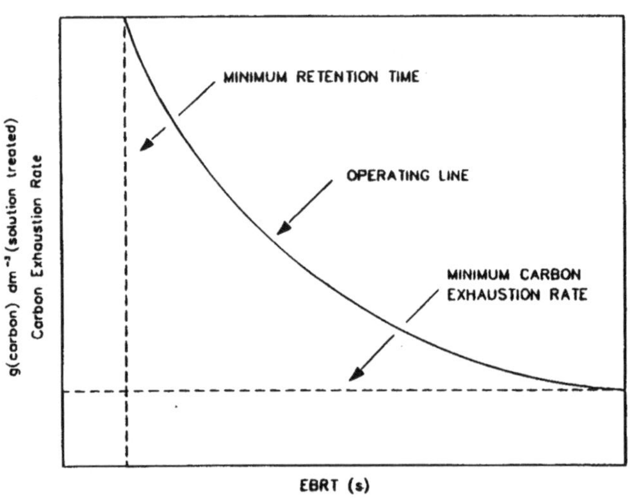

FIGURE 15.26 Example of EBRT operating line plot.

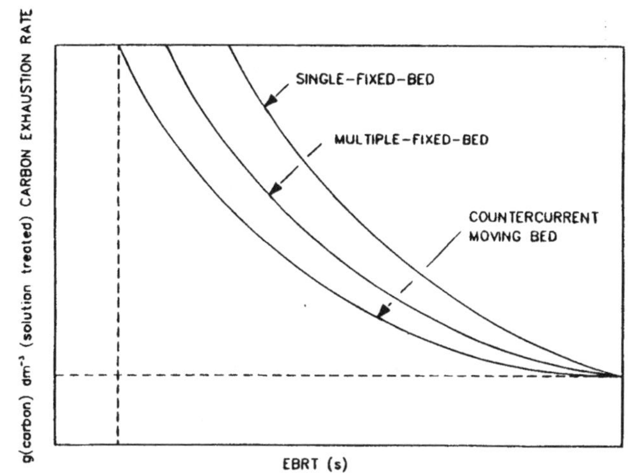

FIGURE 15.27 Effect of sorber vessel on the EBRT operating line.

value, e.g., 10% for parachlorophenol and 50% for cadmium ions, by drawing a horizontal line to cut the curves at a fixed percentage breakthrough. The volume of effluent treated at each bed height then yields the sorbent exhaustion rate.

The operating line always approaches a constant value of sorbent exhaustion rate as the EBRT increases, so it is possible to predict the minimum sorbent exhaustion rate.

Typical operating line plots are shown in Fig. 15.30 and 15.31. The optimum conditions are obtained from the figures and used to obtain the lowest cost design. Since both capital and operating costs are involved, these must be reduced to a common annual cost (Allen et al., 1967). The annual capital value varies from 5 to 40% of total cost depending on the application and the relation to EBRT as shown in Fig. 15.32. Other cost data based on operating fixed-bed GAC adsorption systems are available (Adams and Clark, 1991).

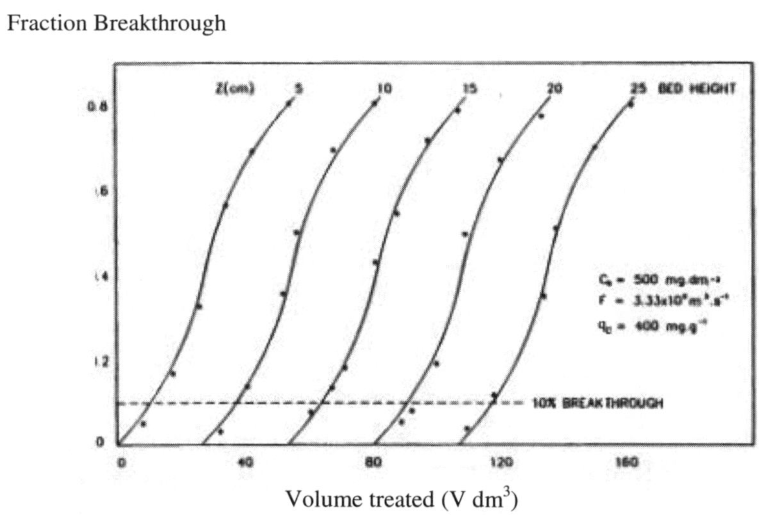

FIGURE 15.28 Adsorption of *p*-chlorophenol onto activated carbon.

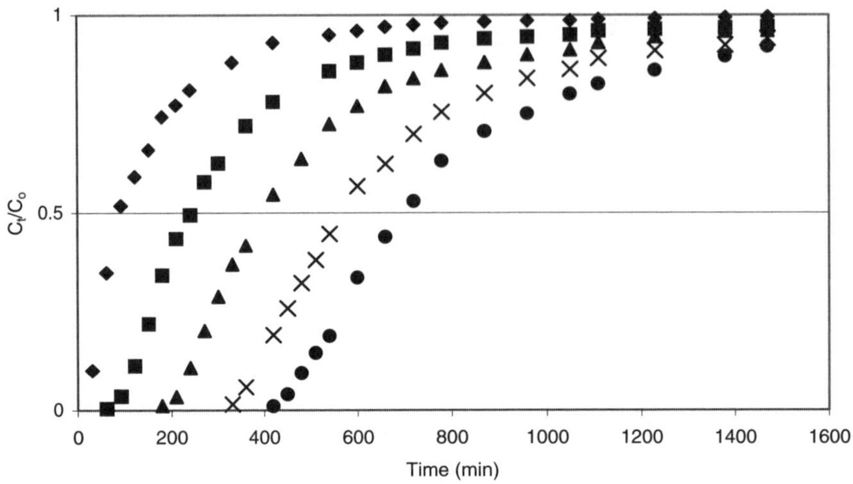

♦ 5 cm, ■10 cm, ▲ 15 cm, X 20 cm, ● 25 cm

FIGURE 15.29 Breakthrough curves for bone char column (500–710 μm) with 4 mmol/L cadmium ions in solution flowing at 50 mL/min.

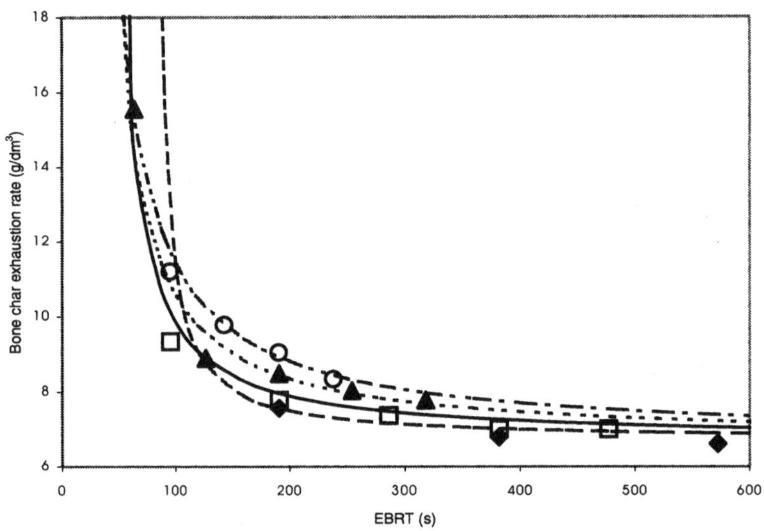

♦ 25 mL/min, □50 mL/min, ▲ 75 mL/min, o 100 mL/min, – – – Operating line for 25 mL/min, —— Operating line for 50 mL/min, ------ Operating line for 75 mL/min, – - - – Operating line for 100 mL/min

FIGURE 15.30 Effect of flowrate on bone char exhaustion rate for cadmium ion removal. C_0 = 3 mmol/L; average D_p = 605 μm; at 50% breakthrough.

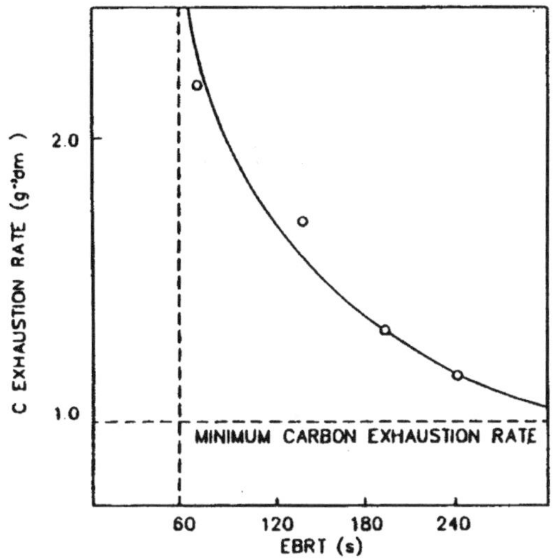

FIGURE 15.31 Effect of percentage breakthrough on carbon exhaustion rate for *p*-chlorophenol removal. $C_0 = 200 \times 10^{-3}$ kg m^{-3}, $d_p = 605 \times 10^{-6}$ m.

Pulsed Beds. A moving or pulsed-bed system may be used in which some carbon is removed at intervals from the bottom of the column and replaced at the top by fresh adsorbent. The rate at which the adsorbent is replenished should be balanced by the rate at which the adsorbent is used; in practice, the mass transfer zone should be held at a constant position within this bed.

Pulsed beds are normally operated with the columns completely full of adsorbent so there is no free board to allow bed expansion during operation. This prevents mixing, which would disturb the sorption zone and reduce the adsorption efficiency. Usually, the adsorbent is maintained in "plug" flow so that a sharp adsorption zone will be retained. The pulsed bed

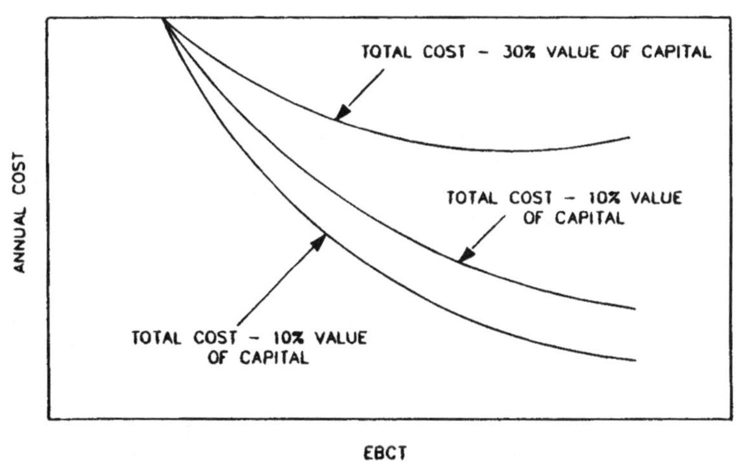

FIGURE 15.32 Annual cost to remove ABS from water.

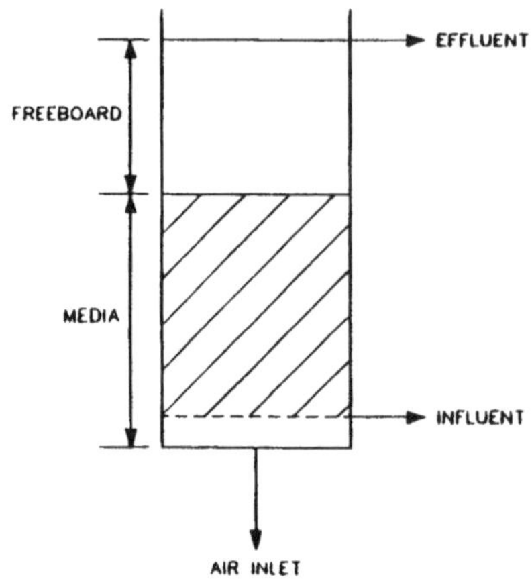

FIGURE 15.33 Schematic diagram of pulsed-bed sorption unit.

can be operated and designed to come closest to completely exhausting the bed with a minimal capital investment. A schematic diagram of a pulsed bed sorption unit is shown is Fig. 15.33.

Fluidized Beds. In wastewater treatment systems, it is advantageous to keep the particle size as small as possible so that high rates of sorption may be achieved. Columnar operation, in which the solution to be treated flows upward through an expanded bed of the particular adsorbent, is one method of taking advantage of small particle size. The main problems with the use of small particles in fixed beds are excessive head loss, air binding, entrainment of fine particles in the effluent carryover, and fouling with particulate matter. The design of a continuously operating countercurrent fluidized bed is relatively simple (Kunii and Levenspiel, 1969), although the solution of the design equations can become quite complex (McKay and Alexander, 1977).

15.3.2 Adsorbent Cost Considerations

The full economic analysis of using sorption for chemical spill treatment is highly complex for large spills that can be pumped to storage and treatment systems. The costs of the optimized treatment plant (considering adsorbent capacity optimization and minimizing contact time–to minimize equipment size), the cost of adsorbent regeneration or disposal, and the cost of purchasing the adsorbent must all be analyzed and correlated. In order to demonstrate the adsorbent cost-capacity relationship, a simple model was developed for the removal of dye from a dye solution spill (Allen, 1995).

A series of experiments were undertaken using similar conditions and using Basic Blue 3 (Astrazone Blue) to compare adsorbent costs using several materials. A uniform range of adsorbent particle size, between 150 μm and 250 μm, was selected for all materials. Various masses of adsorbent were agitated with a constant volume and constant concentration of dye solution. The isotherms for the various dyes on different adsorbents were determined and plotted as q against C, where q is milligrams of dye adsorbed per gram adsorbent and C is the concentration of dye at equilibrium in solution. Operating conditions were selected so that the isotherms reached a plateau in order that the saturation value of the adsorption capacity, q, could be determined. The values were used to assess the quantity of adsorbent required to remove 1 kg of dye. These quantities have been used as a basis for costing the sorption process.

The relative costs of the sorbents for the two basic dye systems are shown in Table 15.11 together with the sorption costs of removing 1 kg of dye. Carbon was taken as the standard, having a comparative unit cost per kilogram, and the relative costs of the other adsorbents are shown in the third column of Table 15.11. The mass (kg) of adsorbent required to remove 1 kg of dye is obtained from the dye isotherms and consequently, using columns 3 and 5, a comparative cost of different adsorbents to remove 1 kg of dye may be obtained; this is shown in column 4. The benefits of using peat, based on this simple analysis, are apparent.

15.3.3 Adsorbent Regeneration and Disposal

The fate of the spent adsorbent after being used for treating a chemical is an important economic issue. For large-scale spill cleanup, substantial quantities of adsorbent are required, and this is costly if expensive adsorbents are used and the material goes directly to disposal (landfill or incineration) without the possibility of adsorbent regeneration and/or chemical recovery of the spill contaminant material. For chemical plants in which adsorption processes are a routine part of daily process operations, economic usage rates are well established. For example, in the case of expensive adsorbents like active carbons and zeolites, if the usage rates are over 100 to 200 kg per day, then it is uneconomical to use for landfill. For cheaper adsorbents, such as wood, peat, and straw, usage rates can be at least five times greater. Since chemical spills are not regular process operations, these operating costs are not as critical for one-off chemical spills, although the concept of using the most economic adsorption process and cheap adsorbent is always applicable. Spent adsorbent treatments fall into two main categories: disposal and regeneration systems.

Disposal Methods

Landfill. Spent adsorbents from chemical spills are frequently landfilled, although legislation in recent years has made this option more difficult and more controlled. The type of landfill (lined or unlined) depends on the type of adsorbent, types of contaminants ad-

TABLE 15.11 Adsorption Costs of C.I. Basic Blue 3

Absorbent material	Units (q)	Comparative cost (relative to activated carbon): Per kg of adsorbent	Comparative cost (relative to activated carbon): To remove 1 kg of BB3	Mass of adsorbent required to remove 1 kg of dye
Activated carbon	448	1.00	1.000	2.232
Peat	375	0.04	0.048	2.667
Silica	87.5	1.50	7.682	11.43
Pith	62.0	0.04	0.286	16.00
Lignite activated carbon	560	0.85	0.674	1.770
Char	125	0.25	0.896	8.000
Lignite	392	0.15	0.171	2.551
Fuller's earth	500	0.66	0.600	2.000

sorbed, and particularly the leachate characteristics of the sorbent-sorbate system. There are severe penalties if metals or organic compounds can leach into natural watercourses, and if leaching is a problem, it may be necessary to consider encapsulation of the spent adsorbent.

Incineration. For carbonaceous or cellulosic-based sorbents loaded with certain classes of contaminants, incineration provides an opportunity to destroy the contaminants/adsorbent and generate waste heat. For certain organics, such as chlorinated solvents or amines, special incineration and treatment systems should be employed due to the potential formation of NO_X, hydrogen chlorine, dioxins, and furans.

Regeneration Systems

Thermal Regeneration. Various types of furnace are used for thermal regeneration, including rotary kilns and multiple hearth furnaces. Hot gases pass through the furnace and may burn off or simply vaporize the sorbed contaminants. Burn-off processes usually result in a 10 to 15% weight loss of active carbons but this still gives a six- to tenfold usage capacity.

If the contaminant sorbates are relatively, i.e., with boiling points less than 200°C, quite effective regeneration can be achieved using steam. This process can be carried out either in situ, using the fixed bed column system, or by removing the spent adsorbent (for example, after batch adsorber filtration) to a furnace-type regeneration system. The contaminant-laden steam can then be condensed at a suitable temperature to condense and recover the volatiles from the chemical spill. Further separation can be applied if necessary, such as distillation.

Regeneration by pH Change. Gosset et al. (1986) observed that, except for nickel, which seems so strongly complexed on peat that at pH 1.2–2.0 half of the maximum sorption capacity is attained, metals may be easily removed from peat by acid treatment. Aho and Tummavuori (1984) studied the effect of the repeated use of peat columns on the structure of the peat, its capacity to remove metals, acid elution profiles, and the maximum column flow rate. After several cycles of sorption, elution and washing with distilled water it was found that the ion-exchange capacity of the peat remained within narrow limits during repeated Cu^{2+} sorptions, thus the cations sorbed can be released with a small volume of acid and the peat sorber repeatedly used. However, the release of bound ions is strongly dependent on the chemical composition of the stripping solution.

Infrared Regeneration. A number of infrared furnaces have been developed for adsorbent regeneration and which are similar in principle to thermal regeneration. The spent adsorbent is fed into the furnace and falls onto a woven-wire mesh conveyor belt and is formed into a layer 1.5 to 2.5 cm thick. The layer of adsorbent moves through the furnace under the infrared heat sources. During this process, the adsorbent undergoes drying, pyrolysis, and reactivation. In the final stage, water quenching is applied prior to transport to storage or directly to the adsorption system.

Biological Regeneration. The use of biosorption and biological filters for the destruction of many organic pollutants is well established. Biological solutions can be used to digest and regenerate adsorbed organics from chemical spills by pumping these solutions through the fixed beds. The advantage is that this method can also be used in situ (Cheremisinoff and Cheremisinoff, 1993; Clark and Lykins, 1989).

Supercritical Fluid Extraction. The properties of cryogenic fluids under supercritical conditions give them considerable potential for regenerating spent adsorbents. The solution characteristics of the fluid should be compatible with the adsorbed components and, when the fluid is pumped through the adsorbent bed, it will dissolve the adsorbed spill components. The supercritical fluid can be evaporated easily due to its volatility, and in some cases both the solvent and the contaminant from the spill can be recovered. The high solvation character of the supercritical fluids is due to low intermolecular distances between the solvent molecules. This novel process is expensive at present and therefore the spilled material should be high value added and recoverable, such as pharmaceutical products.

15.4 ACKNOWLEDGMENT

Joe Ng wishes to acknowledge the support of The Croucher Foundation, Hong Kong, for this research.

15.5 USEFUL WEBSITES

What is Spill-Sorb?
http://spillsorb.com/peat/what_spill-sorb.htm

Hazardous Materials Spill Kit Instruction
http://www.astate.edu/docs/admin/es/spill.htm

Basic Steps for Emergency and Spill Response
http://www.orcbs.msu.edu/chemical/chp/section4.html

Advanced Sorbent Technology
http://www.thomasregister.com/olc/sorbent/intro1.thm

Sphag Sorb
http://www.chemspill.com/sphagsorb.htm

C-180 Inorganic Microporoous Adsorbent Material
http://www.buscom.com/membrane/c180.html

C-132A Sorbent Materials for Non-Spill Applications: Market Study
http://www.buscom.com/archive/c132A.html

Safety Catalog
http://www.vwrsp.com/catalog/safety/class_203276.html

United States Environmental Protection Agency, Site Superfund Innovative Technology Evaluation, Emerging Technology Summary, Demonstration of Ambersorb 563 Adsorbent Technology, EPA/540/SR-95-516, 1-5, 1995
http://www.epa.gov/clariton/clhtml/pubtitle.html

15.6 REFERENCES

Adams, I. Q., and R. M. Clark. 1991. "Evaluating the Costs of Packed Tower Aeration and GAC for Controlling Selected Organics," *Journal American Water Works Association,* vol. 83, no. 1, pp. 49–57.

Aho, M., and J. Tummavuori. 1984. "On the Ion-Exchange Properties of Peat–IV. The Effects of Experimental Conditions on Ion-Exchange Properties of Sphagnum Peat," *Suo,* vol. 35, pp. 47–53.

Alexander, F., V. J. P. Poots, and G. McKay. 1978. "Adsorption Kinetics and Diffusional Mass Transfer Processes during Colour Removal from Effluent Using Silica," *Industrial Engineering Chemistry (Fundamental),* vol. 47, no. 4, pp. 406–410.

Allen, J. B., R. S. Joyce, and R. H. Kasch, 1967. "Dyes in Textile Effluents," *Journal of Water Pollution Control Federation,* vol. 39, pp. 217–226.

Allen, S. J. 1987. "Equilibrium Adsorption Isotherms for Peat," *Fuel,* vol. 66, pp. 1171–1176.

Allen, S. J. 1995. "Types of Adsorbent Material," *in Use of Adsorbents for the Removal of Pollutants from Wastewaters,* ed. G. McKay, CRC Press, Boca Raton, FL.

Allen, S. J., G. McKay, and K. Y. H. Khader. 1989a. "Multi-component Sorption Isotherms of Basic Dyes on Peat," *Environment Pollution,* vol. 52, no. 1, pp. 39–53.

Allen, S. J., G. McKay, and K. Y. H. Khader. 1989b. "Equilibrium Adsorption Isotherms for Basic Dyes onto Lignite," *Journal of Chemical Technology and Biotechnology,* vol. 45, pp. 291–302.

Allen, S. J., P. Brown, G. McKay, and O. J. Flynn. 1992. "An Evaluation of Single Resistance Transfer Models in the Sorption of Metal Ions by Peat," *Journal of Chemical Technology and Biotechnology,* vol. 54, pp. 271–276.

Allen, S. J., L. J. Whitten, M. Murray, O. Duggan, and P. Brown. 1997. "The Adsorption of Pollutants by Peat," *Journal of Chemical Technology and Biotechnology,* vol. 68, no. 4, pp. 442–452.

ASTM D 2854-70, Standard Test Method for Apparent Density of Activated Carbon.

ASTM D 3802, Standard Test Method for Ball-Pen Hardness of Activated Carbon.

AWWA B 600-78, Powdered Activated Carbon.

Bauer, W. H., D. N. Borton, and J. J. Bulloff. 1975. *Methods and Devices for Amelioration of Discharges of Hazardous Chemicals on Water,* CG-D-38-76.

Bowen, H. J. M. 1970. "Absorption by Polyurethane Foams: New Method of Separation," *Journal of Chemical Society,* vol. A, no. 7, pp. 1082–1085.

Brady, M., and G. McKay. 1995. "Characterization of Adsorbents," in *Use of Adsorbents for the Removal of Pollutants form Wastewater,* ed. G. McKay, CRC Press, Boca Raton, FL, p. 39.

Broughton, D. B. and S. A. Gembichi. 1984. "Fundamentals of Adsorption," in *Engineering Fundamental Conference,* ed. A. L. Myers and G. Belfort, pp. 115–124.

Broughton, D. B., R. W. Neuzil, J. M. Pharis, and C. S. Brearly. 1970. "Parex Processes for Recovering *p*-xylene., *Chemical Engineering Processing,* vol. 66, pp. 70–75.

Brown, P., O. Flynn, G. McKay, and S. J. Allen. 1992. "The Evaluation of Various Sorbents for the Removal of Heavy Metals from Wastewaters," in *International Chemical Engineering Research Conference,* Dublin, pp. 152–154.

Brunnauer, S. H., P. H. Emmett, and, E. Teller. 1940. "Classification of Isotherm Shapes," *Journal of the Chemical Society,* vol. 62, pp. 1723–1735.

Bui, S., X. Verykios, and R. Mutharasan. 1985. "In-situ Removal of Ethanol from Fermentation Broths. I. Selective Adsorption Characteristics," *Industrial and Engineering Chemistry Process Design and Development,* vol. 24, no. 4, pp. 1209–1213.

Cheremisinoff, N. P., and Cheremisinoff, P. N. 1993. *Carbon Adsorption for Pollution Control,* PTR Prentice Hall, Englewood Cliffs, NJ.

Chouno, T. 1998. "Polyurethane Foams Containing Ceramic-Carbon Granules and Showing Good Deodorizing Property and Their Manufacture," Jpn. Kokai Tokkyo-Koho, Japanese Patent 10292104, November.

Clark, R. M., and B. W. Lykins, 1989. *Granular Activated Carbon: Design, Operation, and Cost,* Lewis, Chelsea, MI.

Chen, X. H., T. Gosset, and D. R. Theevenot. 1990. "Batch Copper Ion Binding and Exchange Properties of Peat," *Water Research,* vol. 24, pp. 1463–1471.

Cheung, C. W. 1999. "Sorption Kinetic of Metal Ions on Bone Char," M. Phil. thesis, Department of Chemical Engineering, The Hong Kong University of Science and Technology, Kowloon, Hong Kong, SAR, China.

Czaplinski, A. 1965. In *Archiwum Gornictwa,* vol. 10, p. 239.

Czaplinski, A. and M. Lason. 1965. In *Archiwum Gornictwa,* vol. 10, p. 53.

DAB 7, *Deutsches Arzneibuch,* 7th ed.

DIN 4188, *Drahtsiebboden für Analysensiebe*

DIN 19603, *Aktivkohle zur Wasseraufbereitung Technische Lieferbedingungen.*

d'Hennezel, F., and B. Coupal. 1972. "Peat Moss: A Natural Adsorbent for Oil Spills," *Journal of Canadian Petroleum Technology,* vol. 1, pp. 51–53.

Eiden, C. A., C. A. Jewell, and J. P. Wightman. 1980. "Interaction of Lead and Chromium with Chitin and Chitosan," *Journal of Applied Polymer Science,* vol. 25, pp. 1587–1599.

Environmental Protection Agency (EPA) 1991. *Granular Activated Carbon Treatment,* Engineering Bulletin, EPA No. 540291024, National Service Center for Environmental Publications, Cincinnati.

Evans, M., and H. Marsh. 1979. Chapter 4: Pore Size Distribution by HREM. In *Characterization of Porous Solids,* ed. K. K. Rouquérol, K. S. W. Sing, and H. Kral, Society of Chemical Industry, London.

Farnham, R. S., and J. L. Brown. 1972. "Use of Peat and Peat-Sand Filtration Media," In *Proceedings 4th International Peat Congress,* vol. 4, pp. 271–286.

Fleming, H. L. 1987. "Surface Energetics in the Adsorption of Hydrogen Chloride onto Transition Aluminas," *American Institute of Chemical Engineers Symposium Series,* vol. 259, no. 83, pp. 11–27.

Gosset, T., J. L. Trancart, and D. R. Thevenot. 1986. "Batch Metal Removal by Peat: Kinetics and Thermodynamics," *Water Research,* vol. 20, pp. 21–26.

Ho, Y. S., and G. McKay. 1998. "The Kinetics of Sorption of Basic Dyes from Aqueous Solution by Sphagnum Moss Peat," *Canadian Journal of Chemical Engineering,* vol. 76, pp. 822–827.

Ho, Y. S., D. A. J. Wase, and C. F. Forster. 1994. "The Adsorption of Divalent Copper Ions from Aqueous Solution by Sphagnum Peat Moss," *Transactions of Institution of Chemical Engineers—Part B: Process Safety and Environmental Protection,* vol. 72, no. B, pp. 185–194.

Ho, Y. S., D. A. J. Wase, and C. F. Forster. 1995. "Batch Nickel Removal from Aqueous Solution by Sphagnum Moss Peat," *Water Research,* vol. 29, pp. 1327–1332.

Hutchins, R. A. 1993. "New Method Simplifies Design of Activated-Carbon Systems," *Chemical Engineering,* vol. 80, pp. 133–138.

Ibarron, J. V., K. Oscar, and J. M. Gavilan. 1979. "Retention of Metallic Cations by Lignites and Humic Acids," *Fuel,* vol. 58, pp. 827–830.

Jankowska, H., A. Swiatkowski, and J. Choma. 1991. *Active Carbon,* Ellis Horwood, New York.

Kienle, H. von, and E. Bader. 1980. *Aktivkohle und ihre industrielle Anwendung,* Ferdinand Enke, Stuttgart.

Kinoshita, H., and S. Katsumata. 1988. *"Flexible Polyurethane, Foams for Printing Ink Absorbents," Jpn. Kokai Tokkyo Koho,* Japanese Patent 10110053, April.

Koyama, Y., and A. Taniguchi. 1986. "Studies of Chitin. Homogeneous Cross-linking of Chitosan for Enhanced Cupric Ion Adsorption," *Journal of Applied Polymer Science*, vol. 31, pp. 1951–1954.

Kunii, D., and O. Levenspiel. 1969. *Fluidization Engineering,* John Wiley & Sons, New York.

Leslie, M. E. 1974. "Peat Filtration Plant for the Treatment of Textile Effluents," *American Dyestuffs Report,* vol. 63, pp. 15–21.

Luis, C. M. and M. G. Dolores. 1999. "Treatment of Water by Adsorption (II). Processes in Treatment Plants," *Ingenieria Quimica,* vol. 30, pp. 81–187.

Marucca, R., J. P. Whiteman, and B. J. Suder. 1982. "Interaction of Heavy Metals with Chitin and Chitosan," *Journal of Applied Polymer Science,* vol. 27, pp. 4827–4837.

Matz, M. J. and K. S. Knaebel. 1991. "Recycled Thermal Swing Adsorption–Separation of Binary and Ternary Mixtures," *Industrial and Engineering Chemistry Research,* vol. 30, pp. 1046–1054.

McKay, G., M. El-Geundi, and M. M. Nassar. 1997. "Equilibrium Studies for the Adsorption of Dyes on Bagasse Pith," *Adsorption Science and Technology,* vol. 15, pp. 251–270.

McKay, G. and F. Alexander. 1977. "Kinetics of the Removal of Basic Dye from Effluent Using Silica. II Fluidised Bed Studies," *Chemical Engineering,* vol. 319, pp. 244–247.

McKay, G., Y. S. Ho, and J. C. Y. Ng. 1999. "Biosorption of Copper from Wastewater: A Review," *Separation and Purification Methods,* vol. 28, no. 1, pp. 87–125.

McKay, G., and I. F. McConvey. 1981. "The External Mass Transfer of Basic and Acidic Dyes on Wood," *Journal of Chemical Technology and Biotechnology,* vol. 31, pp. 401–408.

McKay, G. and I. F. McConvey. 1985. "Adsorption of Acid Dye onto Woodmeal by Solid Diffusional Mass Transfer," *Chemical Engineering Process,* vol. 19, pp. 285–295.

Mering, J., and D. Schoubar. 1968. "Chemical Composition of Adsorbents Using X-Ray Elemental Analysis," *Journal of Applied Crystallography,* vol. 1, pp. 153–162.

Milestone, N. B., and D. M. Bibby. 1981. "Concentration of Alcohols by Adsorption on Silicalite," *Journal of Chemical Technology and Biotechnology,* vol. 31, pp. 732–736.

Muzzarelli, R. A. A. 1973. "Natural Chelating Polymers," in *Analytical Chemistry,* ed. R. Belche and H. Freiser, Pergamon Press, Oxford.

Muzzarelli, R. A. A., ed 1976. *Chitin,* Pergamon Press, Oxford.

O'Brien, R. P., and J. L. Fisher, 1983. "There Is an Answer to Groundwater Contamination," *Water/Engineering and Management,* May pp. 30–70.

Onsoyen, E., and O. Skaugrud. 1990. "Metal Recovery Using Chitosan," *Journal of Chemical Technology and Biotechnology,* vol. 49, pp. 395–404.

Oulman, C. S., and C. D. Christwell. 1987. "Analytical Investigation of the Properties and Uses of a New Hydrophobic Molecular-Sieve," *Atlanta,* vol. 29, no. 6, pp. 447–452.

Paryjczk, T. 1985. *Gas Chromatograpy in the Study of Adsorption and Catalysis,* Ellis Horwood, Chichester, PQN, Warsaw.

Pilie, R. J., R. E. Baier, R. C. Ziegler, R. P. Leonard, J. G. Michalovic, S. L. Peck, and D. H. Bock. 1975. *Methods to Treat, Control and Monitor Spilled Hazardous Materials,* EPA-670/2-75-042.

Raouf, M. W. A., and A. A. M. Daifullah. 1997. ''Potential Uses of Bone Charcoal in the Removal of Antimony and Europium Radioisotopes from Radioactive Wastes,'' *Adsorption Science Technology,* vol. 5, no. 8, pp. 559–569.

Rios, J. A., D. A. Coyle, C. A. Durr, and B. M. Frankie. 1998. ''Removal of Trace Mercury Contaminants from Gas and Liquid in the LNG and Gas Processing Industry,'' in *Proceedings of 77th Annual Convention, Gas Producers Association,* March 16–18, pp. 191–196.

Robinson, J. S., *Hazardous Chemical Spill Cleanup,* Noyes Data Corp., Park Ridge, NJ.

Schneider, F. 1979. *Sugar Analysis,* ICUMSA, Peterborough, pp. 221–222.

Sharma, D. C., and C. F. Forster. 1993. ''Removal of Hexavalent Chromium Using Sphagnum Peat Moss,'' *Water Resarch,* vol. 27, pp. 1201–1208.

Silvo, O. E. J. 1972. ''Some Experiments on Purifications of Waste Waters from Slaughterhouses with Sphagnum Peat–Peat Chemistry and Physics,'' *Proceedings Fourth International Peat Congress,* Otaniemi, Finland, vol. 4, pp. 311–318.

Sood, A., and H. L. Fleming. 1987. ''Purification of Liquid Ethylene Dichloride by Adsorption,'' *American Institute of Chemical Engineering Symposium Series,* vol. 259, no. 83, pp. 40–51.

Suggs, J. D. 1972. *Mercury Pollution Control in Stream and Lake Sediments,* Water Pollution Control Research Series, 16080 HTD-03/72, Office of Research and Monitoring, U.S. Environmental Protection Agency.

Surakka, S., and A. Kamppi. 1971. ''Infiltration of Wastewater into Peat Soil,'' *Suo,* vol. 22, pp. 51–57 (English summary).

Treybal, R. E. 1987. *Mass-Transfer Operations,* 3d ed. (reprint), McGraw-Hill, New York.

Viraraghaven, T., and S. R. Kikkeri. 1988. ''Peat Filtration of Food-Processing Wastewaters,'' *Biological Wastes,* vol. 26, pp. 151–155.

Vogel, A. S. 1989. *Vogel's Textbook of Quantitative Chemical Analysis,* 5th ed., Longman Scientific & Technical, Harlow, Essex, and John Wiley & Sons, New York.

Voice, T. C. 1989. ''Activated-Carbon Adsorption,'' in *Standard Handbook of Hazardous Waste Treatment and Disposal,* ed. H. M. Freeman, McGraw-Hill, New York.

Zhipei, Z., U. Junlu, W. Zengui, and C. Piya. 1984. ''A Preliminary Study of the Removal of Pb^{2+}, Cd^{2+}, Zn^{2+}, Ni^{2+} and Cr^{6+} from Wastewaters with Several Chinese Peats,'' in *Proceedings of Seventh International Peat Congress,* vol. 3, pp. 147–152.

CHAPTER 16
ANALYSIS OF HAZARDOUS MATERIALS USING EQUIPMENT IN VEHICLE-PORTABLE (LEVEL 3) SPILL RESPONSE VEHICLES

Ken Li and Merv Fingas
Emergencies Science Division, Environment Canada,
Environmental Technology Centre, River Road,
Ottawa, Ontario

16.1 INTRODUCTION

Since 1980, Environment Canada's Emergencies Science Division (ESD) has been conducting projects to develop and systematically evaluate various analytical equipment for on-site spill response. For convenience, four levels of emergency response are defined, each with a severity level requiring a different degree of analytical sophistication: level 1, pocket-portable; level 2, personal-portable; level 3, vehicle-portable, and level 4, fixed-laboratory facilities, usually off-site (Fingas et al., 1991). This section will focus on vehicle-portable (level 3) type of response, which requires equipment generally too heavy to be carried by personnel. Although less mobile and convenient than portable instruments, vehicle-borne equipment is much more powerful and versatile. In ESD's mobile installation, they consist of essentially laboratory-grade instruments with perhaps simplified functionality for ease of operation and maintenance. These instruments have greater analytical capability and can provide automation for higher sample throughput. A much greater choice of instrumentation is also available at a reasonable cost, without the premium usually associated with portable instruments. Moreover, consumables and parts are readily available, thus keeping maintenance costs low.

The vehicle and support equipment have been described in detail in Li et al. (1995a). Briefly, the laboratory is built on a 1986 Ford E-350 cube van that was custom installed with a cabin with interior dimensions 3.7 m (L) by 2 m (W) by 2 m (H). The cabin houses two laboratory benches, a gas cylinder storage rack, a ductless fumehood, and under-counter drawers/cabinet storage facilities. The vehicle is cooled with a rooftop air conditioning unit and heated by a propane space heater. Instrumentation is powered by a 6.5-kW onboard generator, the output from which is stabilized by two Sola power conditioners. There is also a sophisticated air handling system designed to pressurize the interior of the van and minimize air infiltration, in case the vehicle has to be located in a high hazard location. Outside air is filtered through HEPA and charcoal filter for personnel protection. The entire laboratory is designed to operate in a completely self-reliant mode. Land operation is possible using a standard four-prong, 60-amp 240 VAC power cord.

The main analytical instrumentation in the vehicle is a benchtop GC/MS by Hewlett Packard, HP 5890 GC/5970 MSD (mass selective detector). A basic configuration for analyzing liquid samples includes a heated injection port and a capillary fused-silica column interfaced directly to the MS via a heated transfer line. The MS is a quadrupole design operated under vacuum provided by a diffusion pump and backed by a mechanical rotary pump. System operation and data analysis are performed by a Pentium-level personal computer loaded with proprietary software and a NBS spectral library to aid with identification of unknowns. Depending on the mission, a simpler installation may consist of a GC equipped with an appropriate detector such as the electron capture detector (ECD).

Although mostly used for liquid injection, the system is versatile enough to allow introduction of gaseous samples. Hence, by combining with a few relatively simple extraction/concentration techniques, multimedia samples collected on-site (soil, water, or air) can be extracted and analyzed with a high degree of precision and with more than adequate detection limit. The ability to provide high-quality data plus rapid sample turnaround is essential for proper assessment of the risks to the population or the environment.

We will examine the rationale for maintaining such a capability and the different analytical methodology and instrumentation that can be employed to equip such a laboratory. We will concentrate on instrumentation that is relevant to field analytical work and present our experience of operating them. Some specific examples of deployment will be used to illustrate its application.

16.2 METHODOLOGY

16.2.1 Extraction Techniques

Soil. Ultrasonic extraction is carried out by a Bransonic ultrasonic bath (model 8200, 650 W). An appropriate weight of soil or solid sample (usually 1 gm) is weighed into a 40-mL screw cap vial. Acetone is added in the ratio 10:1 and, if desired, a surrogate standard is added. The surrogate standard is an organic compound chemically similar to the target analyte, which can be detected accurately in the analysis. From its recovery, the efficiency of sample extraction or loss from additional workup can be assessed. The sample is sonicated for 10 to 15 minutes. After allowing 5 minutes for the suspension to settle, a 1-μL aliquot is withdrawn with a syringe and injected into an analytical instrument such as a GC/MSD. If the supernatant is not clear, the raw extract can be filtered through a disposable syringe-type filter. For multiple injections, an automated liquid sampler (ALS) can be used, in which case a 1-mL aliquot is transferred to a 1.5-mL ALS vial. An internal standard (IS) can also be added prior to analysis. Similar to the surrogate standard, the IS is an organic compound not present in the sample extract and from its response, the stability of the instrument and variation in the analytical condition such as any evaporation loss can be acertained. Together with the surrogate recovery, a high degree of certainty is obtained and forms the basis of an essential data quality protocol.

Mechanical shaking extraction can also be carried out using the same solvent and procedure. In the case of heavy clay, clean glass beads can be added to the soil and shaken to break up the aggregate. It should be pointed out that the quick extraction techniques outlined here are generally adequate for surface contamination, as in spill-related samples, which are invariably surface wipes, vegetation, or soils. The main objective for screening is qualitative or semiquantitative analysis. Exhaustive extractions are usually not required and can actually be undesirable because they might coextract excessive background material, which requires extensive sample cleanup, adding to the sample turnaround time.

GC/MSD instrument detection limit is typically 1 to 5 μg/mL (ppm) operated in scan mode. For a 1-g sample extracted into 10 mL, the method detection limit is about 10 μg/g (ppm) which is sufficient for screening purposes. Table 16.1 summarizes typical recovery figures using this simplified extraction procedure. For soils fortified with 100 ppm of diesel with matrix ranging from sand to peat moss, recovery varies from 68 to 83%. To test the rapid extraction further, an actual sample contaminated with well-aged diesel was analyzed and showed a value of 54% of that obtained by more vigorous extraction (2,254 ppm versus 4,200 ppm).

Water. Simple liquid partition is carried out by shaking an appropriate volume of water with a 1⁄50 volume of hexane (e.g., 2 mL hexane for 100 mL water in a disposable crimp-top vial). Similar to soil extractions, a surrogate standard can be added to the sample. After shaking for one to two minutes, a 1-mL aliquot of the hexane layer is withdrawn and transferred to a 1.5-mL ALS vial. An IS is added prior to injecting on the GC/MSD. Because of the space limitation and time constraints, all extractions of soil/water samples are carried out using disposable vials since washing of regular lab glassware is not practical. Without the benefit of concentration, method detection limit for water is typically 0.2 ppm. Water fortified with a few representative contaminants was used to evaluate the quick extraction method. For pesticides- and gasoline-contaminated water, recoveries range from 67 to 81% (Table 16.1).

Air. A grab sample can be collected by a preevacuated steel Summa canister or by drawing air through a glass bulb sampler equipped with Teflon stopcocks at both ends, using a personal sampler pump. If desired, an IS (dissolved in methanol) can be added to the bulb

TABLE 16.1 Recovery of TPH from Soil/Water using Rapid Extraction Methods

(A) Spike recovery

Matrix	Spiked with	% Recovery of diesel
Soil		
Silica deactivated with 20% water	Diesel, final concentration = 100 ppm	76 (n = 3)
Sea sand	Diesel, final concentration = 100 ppm	83 (n = 3)
Peat moss	Diesel, final concentration = 100 ppm	68 (n = 3)
Water	Gasoline, final concentration = 18 ppm	67 (n = 5)
Water	Gasoline, final concentration = 100 ppm	81 (n = 4)
Water	Pesticide mixture, final concentration = 5 ppm	72 (n = 4)

(B) Actual sample

TPH-contaminated soil from Nordegg, Alberta		TPH as Diesel, ppm
Trial 1		2,562
Trial 2		2,194
Trial 3		2,250
Trial 4		2,256
Trial 5		2,008
	Mean	2,254
	RSD	8.8

by introducing through a septum injection port. A 1-mL aliquot is withdrawn using a gas-tight syringe and injected into the GC/MSD. To improve peak shape, the gaseous sample can be compressed using a syringe with a stopcock before injection. Because no concentration is involved, the direct analysis is only practical for very high ambient concentration.

16.2.2 Instrumental Extraction and Concentration Techniques

An important advantage of a mobile laboratory is the flexibility of using laboratory-grade automated extraction instrumentation to achieve more precise and powerful extraction, on a level rivalling that of fixed labs. Together with a greater degree of automation, rapid on-site analyses can be carried out that significantly increase the sample throughput. Among the many instrumental extractors, the following is a report of those that have been evaluated by ESD in the past. The discussion is limited to what is practical, bearing in mind the limitation of the confined work space and general lack of support facilities in the mobile van.

Supercritical fluid extraction (SFE) is a solventless extraction technique that utilizes carbon dioxide under supercritical conditions to solvate organic analytes from a solid matrix. Carbon dioxide is a popular SFE fluid because of its relatively low critical temperature (31.3°C) and pressure (72.9 atm), rendering instrument design simple and inexpensive. It is also nontoxic and generally available at high purity. In an off-line extraction, the SFE fluid simply depressurizes via a restrictor into a small volume of organic solvent in which the target analytes accumulate. Some SFE instruments employ solid sorbent on which analytes are cold-trapped or chemically trapped onto a coating on the sorbent particles such as octydecylsilane. The trap is subsequently rinsed with a solvent that is then injected and analyzed.

An obvious advantage of SFE is the fact that little or no solvent is required: the fluid reverts back to a gas, thus leaving the extracted analyte behind as a condensed phase or dissolved in a small volume of solvent that is ready to inject. Minimizing use of solvent produces significant savings in cost and at the same time greatly reduces solvent storage and disposal problems. Since extraction is carried out in a closed system under controlled temperature and pressure, the analyst has much more control over the extraction process. By adjusting the density of extraction fluid, the fluid can be manipulated to resemble solvents of various solvent strengths (e.g., low-density fluid resembles hexane, whereas at high density, it resembles dichloromethane). A selective extraction can easily be carried out by first extracting with a low-density fluid to remove the easily extractable from the solid matrix. This is followed by stepping up the density, which then extracts the next class of analytes, and so on. Since the bulk of extractable is removed at an early stage, the analytes of interest can be isolated in a relatively cleaner medium, which should be beneficial in the subsequent analysis.

The HP7673 SFE has been evaluated (Li et al., 1995b, c). By itself, the HP SFE is a very powerful integrated extraction system that is capable of extracting eight samples unattended. In the course of an extraction cycle, extracts are stored in injection-ready vials, which are subsequently removed and analyzed. The usefulness of this instrument is further enhanced when it is configured with a proprietary HP bridge system, which links the SFE to the HP 5890/5971 MSD, thus making it a totally automated integrated system. The SFE is set up via the bench supervisor software, which allocates resources and controls the timing of both instruments. The vials containing the extracts are passed onto the GC/MSD by the robotic arm of the GC ALS in real-time so that an entire set of eight solid samples can be extracted and analyzed by multiple analytical methods and the data reported without any human intervention.

Matrices extracted in the work cited include spiked soil material, a wood-preservative-contaminated soil, NBS-certified urban dust, and sediments. Good recoveries were reported for spiked samples and generally acceptable agreement was found between the SFE results and reference values. Table 16.2 shows results of a creosote-contaminated soil reference material available from Fisher Scientific. Low recovery for the light compounds indicated volatility loss during trapping, whereas low recovery for high boiling compounds was indicative of incomplete extraction from the soil matrix. For relatively clean soot-laden filter cuttings collected from oil burn experiments, there was good agreement with Soxhlet extraction values.

Headspace (HS) analysis of VOC is conveniently carried out by the static headspace method, in which the water sample is incubated at a known temperature for a specific period of time. An aliquot of HS developed over the liquid phase is injected into a GC/MSD using a gas-tight syringe. An automated HS sampler such as the HP19395A headspace analyzer can also be used. The instrument couples to the GC simply by inserting the heated transfer line into the injector and can hold 21 samples. An internal standard of d8-toluene is added to all samples and standards alike. Samples are loaded onto the constant heating time (CHT) magazine, with which each sample is equilibrated at 85°C for a nominal time of 40 minutes in the sample carousel heated by a silicone oil bath. Soil samples can also be analyzed by first extracting in water.

Since only vapor sample is analyzed, HS analysis is considered a "clean" technique. The bulk of the water and soil do not contribute to any significant background, leading to much longer column life. Disposable HS vials are also used, which minimizes any cross-

TABLE 16.2 PAH Results in Environmental Samples by SFE
Standard reference material: SRS 103-100 (creosote-contaminated soil)
Four determinations

	Reference value (μg/g)	Found (μg/g)	Recovery, %	RSD, %
Naphthalene	35	3	9	42
1-Methylnaphthalene	60	6	10	25
Acenaphthene	627	234	37	1
Fluorene	443	315	71	2
Anthracene	431	382	89	1
Phenanthrene	1,925	1,670	87	4
Fluoranthene[a]	1,426	1,395	98	1
Pyrene	1,075	1,116	104	3
Benzo(a)anthracene	264	296	112	16
Chrysene	316	239	76	3
Benzo(a)pyrene	97	44	45	22
Indeno(1,2,3-cd)pyrene	32	7	22	40

Soot on filters from oil burn experiments

	Soxhlet value (μg)	SFE value (μg)
Filter #1	197	197
Filter #2	166.0	162
Filter #3	107.0	109

[a] Compounds with noncertified value.

contamination. Although sensitivity is not as high as with the dynamic HS (purge and trap) method, the sensitivity may be adequate for screening spill samples. The method detection limit is typically 0.05 to 0.1 ppm for BTEX.

Microwave-assisted process (MAP) is a rapid solid extraction technique by which the sample suspended in a polar solvent is subjected to microwave irradiation (Paré et al., 1991). Extraction is carried out in a specially designed, closed Teflon vessel. At elevated temperature and pressure, rapid solubilization of analytes occurs in a much shorter time than with conventional extractions (typically 10 to 15 minutes). Two types of instrumentation exist: one using an enclosed system and the other based on focused microwave energy in an open vessel. With open-vessel MAP, some unique microwave-assisted reaction can be used to advantage in certain extractions. The lower extraction temperature compared to closed-vessel MAP also means there is less artifact formation and breakdown of thermally labile components, as demonstrated with a diesel-contaminated soil sample in an internal study. In both cases, semivolatile environmental contaminants such as PAH, PCB, pesticides, pentachlorophenol (PCP), and hydrocarbon fuel can be extracted readily from a variety of environmental matrices including sand, glass-fiber filters, soil, and peat moss.

Validation of the MAP extraction was performed using standard reference materials from lake and harbor sediments, creosote-contaminated soil, and air particulate. Although only one set of extraction parameters was used, extraction recoveries were largely satisfactory, despite the low levels of some contaminants. Repeatability and accuracy seem to be comparable to conventional extraction procedures. A clear advantage over conventional extraction lies in the economy of solvent usage and fast extraction time (Li et al., 1996).

Accelerated solvent extraction (ASE) is a relatively recent advance in sample preparation for trace environmental analysis. This technique uses conventional solvents at elevated pressures and temperatures to extract solid samples quickly. The process takes advantage of the increased analyte solubilities at temperatures well above the boiling points of common solvents. Under these conditions, the kinetic processes for the desorption of analytes from the matrix are accelerated. Currently a commercial unit is available in which automated extractions can be carried out on 24 samples sequentially (Richter et al., 1995, 1996). This technique offers some significant advantages over SFE and MAP. SFE uses supercritical CO_2, which is a nonpolar fluid, whereas MAP requires the presence of a polar solvent that couples with microwave to promote heating. By comparison, ASE uses the same solvent as traditional Soxhlet extractions, which means a direct transfer of methodology is feasible without any of the restrictions or limitations of the two other methods. Method development time is therefore shortened.

The applicability of ASE for the quantitative extraction of different classes of semivolatiles, including PAHs, phenols, PCBs, and total petroleum hydrocarbons, has been evaluated (Li et al., 1998). For that study, two conventional SFE systems, Suprex Prep Master and SFE/50, were modified to function as ASE systems. The rationale was based on the SFE's ability to withstand high pressure as well as elevated temperature. Solid samples were weighed into the SFE thimble and extracted using conventional extraction solvents instead of supercritical carbon dioxide. Parameters such as extraction temperature and effect of modifiers were investigated. Although limited by the 150°C maximum oven temperature, it was found that effective extraction could still be carried out in less than 25 minutes for all the compounds studied. The technique was applied to different real matrices contaminated with hydrocarbons, PAHs, and phenols. Validations of the technique were performed using standard reference materials. Recoveries for these matrices were good (>75%), and precision (RSD) was generally less than 10%. Extensive comparisons with sonication and MAP extractions were made, and recoveries were found to be comparable to MAP and superior to sonication. Only sequential extraction can be carried out on the present ASE system, but given the rapid nature of the process, at least 10 samples can be completed in a working day.

Solid-phase microextraction (SPME) is a simple, time-saving extraction technique developed by Janusz Pawliszyn and associates at the University of Waterloo, Ontario, Canada

(Pawliszyn, 1995). In SPME, a phase-coated fiber housed within a syringe is exposed to the sample medium (water or air), allowing the analytes to be adsorbed on the fiber coating. For organic compounds having high partition coefficients, adsorption equilibrium can be attained within minutes. After sample adsorption, the fiber is withdrawn into the needle, which is then inserted into the heated injection port of the GC. The adsorbed analytes are thermally desorbed and analysis proceeds in the same manner as in normal GC/MS analysis.

SPME has many advantages in spill response work. Most analytes readily partition into the fiber coating, which functions as both a collector and concentrator. For example, in water analysis, it is far quicker than conventional methodology, which requires taking a discrete sample, performing a liquid/liquid extraction, cleanup, and concentration. For SPME, once the fiber has been properly conditioned by inserting it in a heated injection port of a GC, it can be sealed and transported to the field and is ready to use. Multiple samples can easily be taken due to its shirt-pocketable size and low unit cost. It can be analyzed in any GC with a heated inlet and, since no solvent is involved, it is compatible with all detectors. This versatility is significant from a field analysis standpoint because the analytical instrument can remain in the same configuration without any modification, thus suffering no loss in productivity. Solvent usage is also entirely eliminated, thus reducing storage and disposal problems in a field laboratory.

SPME fiber assembly is commercially available with different chemistries (ranging from polar to nonpolar) and coating thickness to handle analytes with diversified equilibrium characteristics. Our experience with the SPME has been favorable, with diverse applications, such as measuring solvent vapor (Li et al., 1999a), analyzing ultra-trace amounts of PCB in lake water (Llompart et al., 1998) and rapidly determining VOC from water and soil (Llompart et al., 1997a, b, 1999).

A recent application involved a quick phenol analysis from an industrial accident at a pulp and paper mill near Temiscaming, Quebec. Ottawa river samples were collected and analyzed on-site by quick test kits (Hach DR2000, method #470). Additional water samples were brought back and confirmational analyses were performed by SPME, using a polar acrylate fiber and GC/MSD. The water samples were modified with salt addition and PH adjustment. A target list of 18 phenol compounds can be detected at 10 ppb level. None of the water samples had any detectable amounts of phenol as found by the quick screening tests. There were, however, traces of trichlorophenol (1 ppb or less) and pentachlorophenol (30 ppb or less), both well below the Quebec provincial criteria of 3 ppm (Lambert, 1999).

16.2.3 Hyphenated Techniques

From a GC equipped with a suitable sample inlet and detector, more sophisticated, dedicated analytical systems can be constructed to tackle a variety of tasks. In essence, an extraction/concentration device can be interfaced (hyphenated) to the analytical system so that once the required method or sequence is set up, a solid sample is loaded into an extraction cell and from then on analysis can proceed unattended and the end product will be an analytical report ready to be used by the spill responder. Since the extracted analytes are transferred to the GC without any dilution, improved sensitivity is an immediate benefit. Significant savings in time and reduced contamination can result since extraction and analysis are undertaken in one system without any further intervention. Direct analysis can also mean lessening the use of disposable glassware. Some of the following are simple hyphenated techniques requiring no major modification of the GC, while others require modification to the gas lines of the inlet system and careful consideration of the flow/pressure of the two systems.

Thermal extraction (TD) is a solventless extraction technique by which the soil sample is ballistically heated to vaporize organic contaminants. The desorbed chemicals are swept into and retained by a primary trap. This is in turn flash-heated to desorb the chemicals into a GC. To improve the chromatography, a secondary trap (usually cryogenically cooled) can be used. The condensed chemicals are desorbed rapidly into the GC to reduce peak broadening. A Tekmar 3000 Thermal Extractor has been successfully interfaced to a HP 5890 GC with FID.

In addition to using the TD for direct analysis of soils, the trap is packed with tenax for VOC analysis in air. In that application, air is drawn through the tenax trap at a known rate. The tenax tube is then inserted into the desorption chamber and subjected to heating cycles, which transfers the adsorbed VOC from the tenax tube to the primary/secondary traps. Demonstration runs made to sample air in an office outside a cool room storage for solvents showed fugitive solvent vapor in the office area (Li et al., 1995b). In the limited confines of the mobile lab, the large cylinder of liquid nitrogen used to cool the secondary trap may not be practical. Without cryogenic cooling, the analytes are then limited to the relatively high boiling compounds, such as PAH or PCB.

On-line SFE-GC/MSD: The solventless SFE has been demonstrated to be a useful on-site extraction technique. An even more useful technique is the direct coupling of the SFE to a GC/MS. By means of a suitably constructed fused silica restrictor/interface that links the two instruments together, the extracted target analytes are carried from the SFE and deposited directly onto the analytical GC column via the injection port of the GC. Since SFE and GC/MS operate under drastically different pressure conditions, different interfacing techniques were investigated. A fused silica restrictor inserted directly into an on-column injector seemed to give acceptable results. Because of the concentration factor, direct SFE-GC/MS can offer a potential 1000-fold increase in sensitivity over off-line collection with a significant savings in turnaround time: a complete analysis can be performed within one hour.

We have attempted to link up the two systems but due to difficulties in constructing the optimal interface, which caused frequent overpressure of the MSD, quantitative analysis was only marginal. On the other hand, qualitative analysis was demonstrated to be feasible from sample amounts as little as 1 mg of soil for PCP, thus making such an installation very useful for quick identification of spill chemicals in solid samples (Li and Fingas, 1999).

16.3 RECENT FIELD DEPLOYMENT OF THE MOBILE LABORATORY

16.3.1 Total Petroleum Hydrocarbon (TPH) Analysis, Camp Fortune Remediation Site

In the summer of 1994, the mobile lab was used to assess a hydrocarbon contamination problem at a National Capital Commission remediation site in Quebec. The site was used to store suspected TPH-contaminated soil. The objective was to provide training for some staff members and evaluate the rapid extraction procedures for soils.

Soil samples were extracted by three methods in acetone/hexane: method 1 mixed the soil using a vortex homogenizer for 2 minutes; method 2 in an 60°C oven for 5 minutes; method 3 by microwaving for 30 seconds at 50% power level in an appliance-type microwave oven. Blank soils (silica deactivated with 20% water and sea sand) were first spiked with diesel (the suspected TPH contamination) to evaluate validity of the methods, together with an actual sample (Nordegg, Alberta) known to be contaminated with heavily weathered diesel. Results are shown in Table 16.3. Spiked recovery for all three methods ranged from 109 to 139%. As with the Nordegg soil, there appeared to be little difference between the methods. The method using a simple test tube vortex mixer has the advantage of being the easiest to carry out and was used for the Camp Fortune soil testing. None of the soil samples had measurable amounts of diesel at a method detection limit of 10 ppm.

TABLE 16.3 Comparison of Rapid Extraction Methods for TPH in Soil

(A) Spike recovery

	% Recovery of diesel		
	Method 1[a]	Method 2[b]	Method 3[c]
Spiked silica deactivated with 20% water	120	113	113
Spiked loamy soil	139	109	115

(B) Actual sample

TPH-contaminated soil from Nordegg, Alberta	TPH as Diesel				
		ppm		Mean	RSD, %
Trial 1	2,562	2,360	2,660	2,527	3
Trial 2	2,194	2,420	2,470	2,361	8
Trial 3	2,250	2,165	2,600	2,338	11

[a] Method 1 Homogenized by vortex, 2 min in acetone/hexane.
[b] Method 2 60°C oven, 5 min in same solvent.
[c] Method 3 Microwave, 30 sec at 50% power in same solvent.

16.3.2 Sault Ste. Marie Transformer Rebuilding Project

From December 1997 to January 1998, ESD was contracted by Natural Resources Canada to perform on-site PCB analyses in supporting a transformer rebuilding project at Sault Ste. Marie, Ontario. Samples generated on site consisted of the original transformer oil, surface wipes, varsol rinsing of the transformer tank and cooling fins during decontamination, and the backfill R-Temp fluid. To validate the proposed rebuilding method, extensive analysis of the rinse samples was carried out. After washdown, surface wipe samples showed surface concentration of several hundred μg Aroclor 1254/100 cm^2, well below the target value of 1,000 μg/100 cm^2. Because of inaccessibility of the interior of the fin banks, exhaustive rinsings were performed since they comprised a significant percentage of the internal surface area. Final rinses of each of the seven fin banks of transformer 1 still showed the presence of PCB, ranging from 80 to 590 ppm (μg/mL) with a mean value of 280 ppm.

Upon completion of rebuilding, analysis of the R-Temp retrofill fluid showed the concentration essentially remained unchanged from 5 ppm at the initial power-up to 7 ppm after five months of continuous operation (far below the target level of 50 ppm). By comparison, during decontamination of the second transformer, the final fin rinse had a mean concentration of 54 ppm. However, the R-Temp showed an initial concentration of 38 ppm and remained essentially unchanged after approximately four months of operation. The high voltage compartment showed an initial concentration of 37 ppm and increased to 126 ppm, indicating that the compartment might have been missed during washing.

On site, extensive comparison of gas chromatograph (GC) analysis of the transformer fluid and the quick test Clor-N-Oil kit was also carried out and showed generally good agreement. The use of an on-site GC was crucial in providing rapid and accurate analysis of various samples, thus enabling quick modifications to the rinsing strategies in order to meet the target PCB level. For projects of this nature, a GC equipped with electron capture detector (ECD) was far superior to quick test kits because it provided the selectivity and sensitivity necessitated by the diverse nature of the sample media (Li et al., 1999b).

16.4 CONCLUSIONS

While commercially available portable analytical instruments are becoming smaller and more powerful, at present the choice is still relatively limited. There is also the price premium usually associated with portability. The examples given here show that there are still significant advantages to operating a properly designed and operated mobile laboratory. By combining simple extraction techniques and the use of GC/MS, a variety of chemicals in different types of sample media can readily be identified and quantitated. With laboratory-grade instruments as the basic building block, very powerful automated extraction/analysis systems can be set up in accordance with the requirements of the mission. Because the same type of laboratory instruments are used in a fixed lab, no additional learning or operator interface is required. The possibility of using common extraction and analytical methods saves time when developing methods for field application. As parts and consumables are shared, it further reduces operating inventory cost and simplifies operation of a spill response laboratory.

16.5 REFERENCES

Fingas, M. F., K. Li, M. Goldthorp, and J. R. J. Paré. 1991. "Progress on the Development of Spill Analytical Techniques," in *Proceedings of the Eighth Technical Seminar on Chemical Spills*, Environment Canada, Ottawa, ON, pp. 183–187.

Lambert, P. 1999. "Phenol and Formaldehyde Incident, Tembec Pulp and Paper Facility, Temiscaming, Quebec. May 1999," Report to Environment Canada, Quebec Region, June.

Li, K., and M. F. Fingas. 1999. "Coupled Supercritical Fluid Extractor-GC/MS for Quick Extraction of Organic Analytes from Solids," in *Proceedings of the Sixteenth Technical Seminar on Chemical Spills*, Environment Canada, Ottawa, ON, pp. 57–67.

Li, K., M. F. Fingas, J. M. R. Bélanger, and J. R. J. Paré. 1995a. "The Vehicle-Portable Analytical System of the Emergencies Science Division of Environment Canada," *Journal of Hazardous Materials*, vol. 43, pp. 141–154.

Li, K., M. F. Fingas, J. M. R. Bélanger, and J. R. J. Paré. 1995b. "The Use of Thermal Extraction in Spill Response," in *Proceedings of the Twelfth Technical Seminar on Chemical Spills*, Environment Canada, Ottawa, ON, pp. 191–211.

Li, K., M. F. Fingas, J. M. R. Bélanger, and J. R. J. Paré. 1995c. "Evaluation of the Hewlett Packard HP7680T Automated Supercritical Fluid Extractor System," in *Proceedings of the Twelfth Technical Seminar on Chemical Spills*, Environment Canada, Ottawa, ON, pp. 173–189.

Li, K., J. M. R. Bélanger, M. Llompart, R. Turpin, R. Singhvi, and J. R. J. Paré. 1996. "Evaluation of Rapid Solid Sample Extraction Using the Microwave-Assisted Process (MAP) under Closed-Vessel Conditions," in *Spectroscopy*, vol. 13, pp. 1–13.

Li, K., M. Landriault, M. F. Fingas, and M. Llompart. 1998. "Accelerated Solvent Extraction (ASE) of Environmental Organic Compounds in Soils Using a Supercritical Fluid Extractor," in *Proceedings of the Fifteenth Technical Seminar on Chemical Spills*, Environment Canada, Ottawa, ON, pp. 115–128.

Li, K., M. Goldthrop, S. Whiticar, P. Lambert, and M. Fingas. 1999a. "Solvent Vapour Monitoring in Work Space by Solid Phase Micro Extraction," in *Proceedings of the Sixteenth Technical Seminar on Chemical Spills,* Environment Canada, Ottawa, ON, pp. 1–12.

Li, K., M. Landriault, M. F. Fingas, and A. Potvin. 1999b. "On Site PCB Analysis in Support of a Transformer Rebuilding Project," in *Proceedings of the Sixteenth Technical Seminar on Chemical Spills*, Environment Canada, Ottawa, ON, pp. 129–145.

Llompart, M., K. Li, and M. F. Fingas. 1997a. "The Application of Solid Phase Micro Extraction (SPME) for Spill Emergency Work, Part 1: HS-SPME Analysis of Volatiles and Semi-volatiles in Soil," in *Proceedings of the Fourteenth Technical Seminar on Chemical Spills*, Environment Canada, Ottawa, ON, pp. 83–91.

Llompart, M., K. Li, and M. F. Fingas. 1997b. "The Application of Solid Phase Micro Extraction (SPME) for Spill Emergency Work, Part 2: HS-SPME Analysis of Volatiles and Semi-volatiles in Water and Air," in *Proceedings of the Fourteenth Technical Seminar on Chemical Spills,* Environment Canada, Ottawa, ON, pp. 93–104.

Llompart, M., F. Fingas, and K. Li. 1998. "Determination of PCB in Water Samples by Headspace Solid Phase Micro Extraction (HSSPME) under Non-Equilibrium Conditions," in *Proceedings of the Fifteenth Technical Seminar on Chemical Spills*, Environment Canada, Ottawa, ON, pp. 83–92.

Llompart, M., K. Li, and M. F. Fingas. 1999. "Headspace Solid Phase Micro Extraction (HSSPME) for the Determination of Volatile and Semi-volatile Pollutants in Soils," *Talanta*, vol. 48, pp. 451–459.

Paré, J. R. J., M. Sagouin, and J. Lapointe. 1991. U.S. Patent No. 5,002,784, March 26 (and selected international counterparts).

Pawliszyn, J. 1995. "New Directions in Sample Preparation for Analysis of Organic Compounds," *Trends in Analytical Chemistry*, vol. 14, no. 3, pp. 113–122.

Richter, B. E., J. L. Ezzell, D. Felix, K. A. Roberts, and D. W. Later. 1995. "An Accelerated Solvent Extraction System for the Rapid Preparation of Environmental Organic Compounds in Soil," *American Laboratory,* vol. 7, pp. 24–28.

Richter, B. E., B. A. Jones, J. L. Ezzell, N. L. Porter, N. Avdalovic, and C. Pohl. 1996. "Accelerated Solvent Extraction: A Technique for Sample Preparation,"*Analytical Chemistry*, vol. 68, no. 6, pp. 1033–1039.

P • A • R • T • 5

SPILL MODELING

CHAPTER 17
PRACTICAL USES OF AIR PLUME MODELING IN CHEMICAL EMERGENCIES

John S. Nordin
AristaTek, Inc. and Western Research Institute (WRI), Laramie, Wyoming

17.1 METHODS

During a chemical emergency, a toxic or hazardous cloud of gas, vapor, or particulates may potentially or actually be released to the atmosphere. The release may be a direct release of gas, evaporation from a spilled pool, an aerosol or droplets from a tank or pipe under pressure, or as a result of fire or explosion. Sometimes no chemical has actually been released but there is potential for a release because of an accident or a nearby fire or terrorist threat. In the United States, users of large quantities of hazardous chemicals are required by the 1990 Clean Air Act Amendments (CAAA) to model potential releases of these chemicals, including catastrophic releases under worst-case conditions. The modeling results are made available to emergency responders as part of their risk management plans, even though no release has actually taken place.

When the chemical is released to the air, a cloud or plume forms that tends to disperse as the cloud travels downwind. The two dispersion mechanisms accounted in the popular gas dispersion models are (1) turbulence from air instability resulting from solar heating and (2) turbulence resulting from wind blowing across the terrain. Many of the popular models do not account for (1) buoyancy resulting from a fire or fires, (2) rainout of droplets or larger particulates, (3) effect of releases during a precipitation event, (4) variable terrain, (5) highly reactive chemicals with air moisture or oxygen, (6) a zero-wind condition, or (7) multiple sources. If these situations occur, worst-case situations could be examined; for example, the user might model both the chemical and the reaction product(s) with water for the case of a highly reactive chemical with water. Eventually, far downwind from the source the concentration drops below the level of concern.

The model user is faced with the tasks of (1) identifying the chemical, (2) determining how much is released or could be potentially released to the atmosphere, (3) assessing meteorological conditions, (4) assessing terrain conditions, (5) assessing other factors such as fire or a potential explosion, (6) selecting an appropriate model, and (7) establishing a level of concern. First responders make decisions on establishing an initial isolation zone and protective action distances and whether to evacuate the public or have some people remain on site inside a building with windows closed until the toxic cloud passes. These tasks are not trivial. Even basic information such as the name of the chemical(s), its release rate, or the total amount present may not be known.

During a chemical emergency, the first responder also does not have the time to select an appropriate model and learn how to use it. If the accident occurs at a fixed U.S. facility, the modeling may already be incorporated as part of the facility risk management plan. Also, the facility should have made available lists of large volumes of chemicals and where they are stored (a requirement of the Community Right to Know regulations) to fire departments and emergency responders. The *1996 North American Emergency Response Guidebook* (NAERG96) and the updated version *2000 Emergency Response Guidebook* (2000ERG) gives information on initial isolation distances and protective action distances for spills from large and small containers in the case of a transportation accident. Computer software programs, such as ALOHA, designed to run on a desktop or laptop computer are available. The *2000 Emergency Response Guidebook* and the U.S. *National Institute for Occupational Safety and Health (NIOSH) Pocket Guide to Chemical Hazards* are available on CD-ROM. Western Research Institute has developed software for a palmtop (hand-held) computer containing emergency response information including modeling capabilities. These are all useful tools in emergency situations.

17.2 CONCENTRATION LEVELS OF CONCERN FOR TOXIC CHEMICALS

When a toxic gas or vapor is released to the atmosphere, the chemical will eventually disperse and reach some concentration at which the risk to the general population is minimal. A major and controversial part of the emergency planning process is establishing guidelines as to what these concentrations or levels of concern should be and duration. These information is used for determining when to evacuate or in some instances when to advise people to stay indoors with windows closed until the danger has passed.

The American Industrial Hygiene Association (AIHA) of Akron, Ohio, has developed a series of emergency response planning guidelines (ERPG) for three different levels of concern, all based on a one-hour airborne exposure to the chemical. As of 2000, ERPG values have been developed for almost 100 chemicals, with about 10 added each year. The levels are defined as follows:

ERPG-1: The maximum airborne concentration below which most individuals could be exposed for up to one hour without experiencing anything other than mild transient adverse health effects or perceiving a clearly defined objectionable odor

ERPG-2: The maximum airborne concentration below which most individuals could be exposed for up to one hour without experiencing or developing irreversible or other serious health effects, or symptoms that could impair their ability to take protective action

ERPG-3: The maximum airborne concentration below which most individuals could be exposed for up to one hour without experiencing or developing life-threatening health effects

The U.S. Department of Energy Subcommittee on Consequence Assessment and Protective Actions (SCAPA) has developed Temporary Emergency Exposure Limits (TEEL) for approximately 1,200 chemicals as approximations to ERPG-1, ERPG-2, and ERPG-3 values (Craig et al., 2000) to be used as temporary guidance by DOE and its contractors until the peer reviewed ERPG values developed by AIHA are published. The TEEL and ERPG values can be obtained from the SCAPA website (http://www.scapa.bnl.gov/teels.htm).

The following are concentration limits representing levels of concern:

The U.S. Occupational Safety and Health (OSHA) regulations[1] specify time-weighted average concentrations (OSHA-TWA) in the workplace for an eight hour per day, 40 hours per week exposure. The National Institute for Occupational Safety and Health (NIOSH) also specifies time-weighted average concentrations for an 8 hours per day, 40 hours per week exposure which are sometimes different from the OSHA regulations. The values established by OSHA are used to determine whether workers, including response personnel, should wear personnel protective equipment.

The immediate danger to life and health (IDLH) concentrations represent the maximum concentration from which a worker could escape in 30 minutes without experiencing any escape-impairing (e.g., severe eye or respiratory irritation) or irreversible health effects. The IDHL concentrations do not consider long-term adverse effects such as cancer. The definition of IDHL was originally based on the U.S. Mine Safety and Health Administration stipulation (30 CFR Part 11.3(t)) to ensure the ability of a worker to escape in 30 minutes in case respiratory protective equipment fails. NIOSH reviewed and revised the IDHL values, adding to it the criterion that the IDHL value must not exceed 10% of the lower explosive limit (LEL) even though relevant toxicological data indicated irreversible health effects or impairment of escape existed at higher concentrations.

The ERPG-2 level was used in preparing the NAERG96 and 2000ERG for defining protective action distances. If ERPG-2 level numbers were not available, NAERG96 and 2000ERG used 0.01 times the LC_{50} value for one-hour exposure of the chemical to rats. If the exposure time was different than one hour, the LC_{50} was divided by the exposure time (in hours) if less than one hour, or divided by the square root of the time if greater than one hour. The LC_{50} is the median (50%) lethal concentration. For the initial isolation distances, NAERG96 and 2000ERG uses the ERPG-3 level where available.

The toxic endpoint used in U.S. EPA Risk Management Plans under 40 CFR Part 68 is either based on ERPG-2 values or is 0.1 times the IDHL value or (for a few chemicals) is estimated at some fraction of LC_{50} values for some animal.

The ALOHA version 5.2.1 dispersion modeling incorporated as a part of CAMEO uses IDHL values as a default condition. The user may override these values and model to some other toxic endpoint.

Toxic inhalation endpoints for selected chemicals are compared in Table 17.1. These numbers, current as of 1996, are subject to revision. The numbers apply to most healthy adults. Some people will be more sensitive than others, including the elderly, infants, chemical-sensitive individuals, and individuals whose immune system is impaired.

17.3 SOURCE TERM

Air plume modeling of a chemical release accident requires knowledge of the release rate and the release situation. During an actual emergency, even if the chemical involved is identified, the release rate in all likelihood cannot be determined because safety considerations do not allow approach near the source. Instead, the responder might identify the chemical and size of the container or containers and assume that all of the chemical is released at once if there is danger of explosion. If there is no danger of explosion, the modeling may be done, assuming all of the gas is released over a 10-minute period. If the spill is a liquid that evaporates on the ground, the release rate is the evaporation rate. Mathematical expressions are available that estimate an evaporation rate based on pool area; these may be embodied in popular dispersion models such as ALOHA or in hand-held computers with appropriate software such as PEAC. If the pool area is not known, sometimes it is assumed

[1] Under 29 CFR Part 1910; CFR = Code of Federal Regulations.

TABLE 17.1 Toxic Inhalation Endpoints for Selected Chemicals

Chemical	ERPG-1	ERPG-2	ERPG-3	OSHA TWA	NIOSH TWA	IDLH	EPA 112(r)
Acetaldehyde	10	200	1000	200	Ca	2000	76 mg/m³
Acrolein	0.1	0.5	3	0.1	0.1	2	1.1 mg/m³
Ammonia	25	150	750	ST 50	25	300	140 mg/m³
Benzene	25	150	1000	0.1	0.1 Ca	500	
Bromine	0.2	1	5	0.1	0.1	3	6.5 mg/m³
1,3-Butadiene	10	200	5000	1000	Ca	2000	
Carbon disulfide	1	50	500	20	1	500	160 mg/m³
Carbon monoxide	200	350	500	50	35	1200	
Carbon tetrachloride	20	100	750	ST 10	2 Ca	200	
Chlorine	1	3	20	1	0.5	10	8.7 mg/m³
Chlorotrifluoroethylene	20	100	300	no data	no data		
Dimethyldichlorosilane	0.8	5	25	5 (HCl)	5 (HCl)		36 mg/m³
Dimethylamine	1	100	500	10	10	500	
Ethylene oxide	na	50	500	1	Ca	800	90 mg/m³
Formaldehyde	1	10	25	0.75	Ca	30	12 mg/m³
Hydrogen chloride	3	20	100	5	5	50	30 mg/m³
Hydrogen cyanide	na	10	25	10	ST 4.7	50	
Hydrogen fluoride	2	30	50	3	3	30	16 mg/m³
Hydrogen sulfide	0.1	30	100	20	10	100	42 mg/m³
Isobutyronitrile	10	50	200	n.l.	8	30	140 mg/m³
Methanol	200	1000	5000	200	200	6000	
Methyl isocyanate	0.025	0.5	5	0.02	0.02	3	1.2 mg/m³
Methylene chloride	200	750	4000	500	Ca		
Methyltrichlorosilane	0.5	3	15	see HCl	see HCl		18 mg/m³
Monomethylamine	10	100	500	10	10	100	
Phenol	10	50	200	5	5	250	
Phosgene	na	0.2	1	0.1	0.1	2	0.81 mg/m³
Phosphine	na	0.5	5	0.3	0.3	50	3.5 mg/m³
Propylene oxide	50	250	750	100	Ca	400	590 mg/m³
Styrene	50	250	1000	100	50	700	
Sulfur dioxide	0.3	3	15	5	2	100	7.8 mg/m³
Sulfuric acid	2 mg/m³	10 mg/m³	30 mg/m³	1 mg/m³	1 mg/m³		10 mg/m³
Toluene	50	300	1000	200	100	500	
Trimethylamine	0.1	100	500	n.l. (5)	10		
Vinyl acetate	5	75	500	n.l. (10)	4		260 mg/m³

na = not appropriate

Concentrations in ppm unless otherwise noted. See below to convert ppm to mg/m³.

n.l. = not listed (number in brackets is PEL)

ST = short-term or ceiling value during a normal work day

Ca = NIOSH potential occupational carcinogen (if no value listed, NIOSH recommends occupational exposure to the lowest feasible concentration)

HCl means that for the chemicals listed react with air moisture producing HCl and the numbers listed are for HCl

The listed ERPG-2 and ERPG-3 values for ammonia and phosphine reflect the 1999 revision.

Concentrations in ppm may be converted to mg/m³ by multiplying by MW/24.45 where MW is the molecular weight, e.g.,

$$\text{mg/m}^3 = \text{ppm (MW)}/24.45$$

Conversion of mg/m³ to ppm is not appropriate for particulates, metal fumes, or chemicals whose molecular weight is uncertain.

that all of the container liquid spills onto the ground to a depth of one centimeter. Mathematical expressions are also available to calculate a discharge from a pipe or a hole in a tank under pressure. The dense gas model SLAB allows the user four choices: (1) a puff (instantaneous) release, (2) an evaporating pool, (3) a horizontal jet, and (4) a vertical jet.

The responder should also be aware of special circumstances. For example, some chemicals react with water (from air moisture or if the spill is over water), in which case the modeling should be done with the reaction products with water. In addition, release of some cryogenic gases may result in ground-hugging aerosols that may increase concentrations downwind more than that predicted by models. The toxic cloud plume may also be confined by a valley or by buildings and travel further downwind.

17.4 EVACUATE OR REMAIN IN PLACE

Sometimes in the case of a chemical spill a decision must be made whether to evacuate the public or to have people remain inside buildings with windows closed until the toxic cloud has passed. If there is time to evacuate, this is usually the choice that is made. Evacuation may be ordered if no spill has actually occurred but there is potential for a release, as in a transportation accident or fire. But sometimes the toxic cloud forms quickly or there may be a wind shift so that there is no time to evacuate. Responders may also recommend that people remain indoors with windows closed in fringe areas, such as (1) in areas not directly downwind or (2) when the toxic cloud has become diluted to the point where outdoor concentrations are below ERPG-2 levels, or (3) when the toxic cloud passage is very brief. If it is practical to evacuate, this is the preferred choice.

Air plume modeling can help in the process of deciding whether to evacuate or remain in place, but the modeling should not replace common sense. An appropriate model is used to predict the maximum outdoor concentration and cloud duration at the building location. The assumption is usually made that the air recirculation rate within the building is ample but the building air intake is closed (windows closed). The simplest calculation predicts a concentration C_t within the building after time t (t = outside toxic plume cloud duration in hours):

$$C_t = C_p\,(1 - e^{-Dt})$$

where C_p = maximum plume concentration outside predicted by the model and D = number of building volume air exchanges per hour. For many public buildings, $D = 1$ may be used as a default condition (e.g., one building volume exchange of air per hour). For example, if the cloud duration is predicted by the model to be 30 minutes and the maximum outdoor concentration is 100 ppm, at the end of 30 minutes the concentration inside the building is predicted to reach 39 ppm. If the cloud duration is predicted to be 5 minutes, the indoor concentration is only 8 ppm at the end of 5 minutes. The calculations show that if the cloud is of short duration, it may be better for people to remain in place inside the building. If the cloud duration is one hour or less, the concentration might be matched up with ERPG-2 levels as the level of concern. But if the cloud duration is expected to be several hours, the OSHA-TWA or other more conservative number might be more appropriate to use as the level of concern.

The ALOHA model incorporates indoor building concentrations as part of its calculation methodology.

17.5 RISK MANAGEMENT PLANS

In the United States, the Clean Air Act Amendments of 1990 (section 112 r) require that facilities that store or use large quantities of hazardous chemicals complete risk management plans (RMPs) and make them available to firefighters and emergency respondors before any incident. An offsite consequence analysis in case of a potential spill or chemical accident is a part of the RMP. Details of how the offsite consequence analysis is to be completed depend on quantities that may be stored; these details may also be revised and updated by the U.S. Environmental Protection Agency (U.S. EPA) as defined in their 40 CFR Part 68 regulations. It is the responsibility of the facility to make sure it is in compliance with the latest regulations.

The offsite consequence analysis requires that the facility determine (1) the worst-case consequence distance, where all of the chemical is released under an F atmospheric stability condition and a wind speed of 1.5 m/s (meters per second); and (2) alternative release scenarios. This is done for each chemical. The worst-case and alternative release scenarios require that the toxic plume from the hypothetical spill be modeled to a toxic endpoint (in mg/m^3) and the downwind distance to that endpoint be determined. The facility must also identify offsite receptors within the circle defined by the worst case and alternative release scenarios. Offsite receptors include public receptors (list population within circle, identify schools, commercial or industrial areas, etc.) and environmental receptors (wildlife sanctuaries, preserves, national forests, state parks, etc.). The facility does not need to consider the consequence of a toxic plume on the offsite receptors, only identify them.

The alternative release scenario could include a worst-case release amount but with passive (e.g., dikes or sumps) or active responses (e.g., a deluge system, emergency shutdown system) in place, or a release based on past history. The modeling is done under either typical meteorological conditions or a D atmospheric stability and a wind speed of 3 m/s.

For the worst-case scenario for a flammable substance, the modeling assumes either a vapor cloud explosion or BLEVE. The endpoint distance for a vapor cloud explosion is 1 psi over pressure. For a BLEVE, the distance is at 5 kw/m^2 for 40 seconds, or a thermal equivalent to receive second degree burns. For a worst-case release scenario involving toxic gas, the modeling is based on all of the gas released in 10 minutes. For a worst-case release scenario involving a liquid that evaporates, the modeling may be for a 10-minute or 60-minute release duration.

The U.S. EPA provides look-up tables to give distances to the toxic endpoint. To use, the facility must determine a release rate and the toxic endpoint for the chemical under question. For a flammable material, distances are computed to the lower flammable limit. Additional tables and equations are provided for calculating distances for vapor cloud explosions and BLEVEs. Separate tables are provided for dense gases and neutrally buoyant gases, urban or rural conditions, and a D stability at 3 m/s wind speed or F stability at 1.5 m/s wind speed. The U.S. EPA also provides tables listing toxic endpoints for the chemicals regulated by 40 CFR Part 68.

The EPA lookup tables for dense gases are based on the SLAB dense gas model, developed by Lawrence Livermore National Laboratories (as of 1998). The EPA lookup tables for neutrally buoyant gases are based on a Gaussian model using Beals dispersion coefficients (documentation in U.S. Air Force, 1971). The facility may use other appropriate models such as ALOHA instead of the EPA lookup tables.

17.6 NORTH AMERICAN EMERGENCY RESPONSE GUIDEBOOK

The *North American Emergency Response Guidebook* is a pocket-sized book developed jointly by Transport Canada, the U.S. Department of Transportation (DOT), and the Secretariat of Communications and Transportation of Mexico. The guidebook is intended for use by first responders who arrive at the scene of a transportation accident involving dangerous materials. The intent is to update this guidebook every three years and place one copy at no charge in each emergency service vehicle throughout North America, through voluntary cooperation of key government agencies. This guidebook aids first responders in (1) quickly identifying the specific or generic hazards of the material(s) involved in the incident and (2) protecting themselves and the general public during the initial response phase of the incident. The "initial response phase" means that period, following arrival at the scene of an incident, during which the presence and/or identification of the hazardous material is confirmed, protective actions and area securement are initiated, and assistance of qualified personnel is requested.

In the United States, Occupational Safety and Health Administration regulations (under 29 CFR Part 1910.120) require that first responders be trained in the use of the *North American Emergency Response Guidebook.*

The 1996 version (NAERG96) and 2000 version (2000ERG) are divided into several sections. The 2000 version is also available on CD-ROM. The first section gives general information on the hazard classification system for shipping dangerous goods, including a table of placards referenced to an initial response guide number. The second section lists hazardous materials according to the placard ID number (a four-digit number that appears on the shipping placard) and cross-references this number with the material name and guide number. For example, the placard ID number 1090 refers to acetone and guide number 127. In the next section, the hazardous material is listed alphabetically by name and cross-referenced to the placard ID number and guide number. The following section gives information on (1) potential hazards, (2) public safety, and (3) emergency response for each guide number. For example, guide number 127 is for the general class of materials under "flammable liquids, polar/water-miscible." Potential hazards include both fire or explosion hazards and health hazards. Public safety includes (1) emergency response telephone numbers, (2) protective clothing, and (3) evacuation distance for a fire or large spill. Emergency response includes information for fighting fires, containment for a spill or leak, and first aid. A list of water-reactive materials is also presented for spills on water; the list also gives the toxic vapor produced.

The major changes in the 2000 version are (1) chemical warfare agents are included, (2) a few additional placard numbers for additional categories are included, (3) the initial isolation zone and protective action distances for some chemicals have been changed, and (4) the guide numbers for a very small number of chemicals have been changed. The 2000 version is available at no charge from the Internet (http://hazmat.dot.gov/gydebook.htm).

Some of the chemicals and placard ID numbers listed in the guidebook are highlighted in color. For the highlighted chemicals, the user is directed to another section listing initial isolation and protective action distances in miles and kilometers (or feet and meters). The initial isolation distance is defined by a circle, the center of which is at the spill site. The protective action distance is defined by a square that has the center of one end located at the spill site and the other side downwind. The square assumes that random changes in wind direction confine the vapor plume to an area within 30° on either side of the predominant wind direction, resulting in a crosswind protective action distance equal to the downwind protective action distance. This is shown in Fig. 17.1.

For the initial isolation and protective action distances, the user must choose between a large spill (>55-gallon drum equivalent) or a small spill. For the protective action distance, the user must also choose between day and night conditions.

Technical documentation for the model used in NAERG96 is described in Dunn et al. (1996). The initial isolation and protective action distances were obtained using a Gaussian dispersion model assuming that the gas is neutrally buoyant. The argument is given that dense gas effects are relatively unimportant at the distances of interest compared with other uncertainties in the analysis. The ERPG-2 value is used as the basis of concentration for the protective action distance. If no ERPG-2 value is available and no surrogate is available, then the concentration is based on 0.01 times the LC_{50} value for a one-hour exposure. The ERPG-3 value is used as the basis of concentration for the initial isolation zone. Again, if no ERPG-3 value is available and in the absence of other information, the ERPG-3 is set equal to five times the ERPG-2 value.

The NAERG96 also provides a section on dangerous water-reactive materials for spills over water giving toxic vapor(s) produced. For example, a spill of methyltrichlorosilane over water produces the toxic vapor HCl. Technical documentation for this section is in Carhart

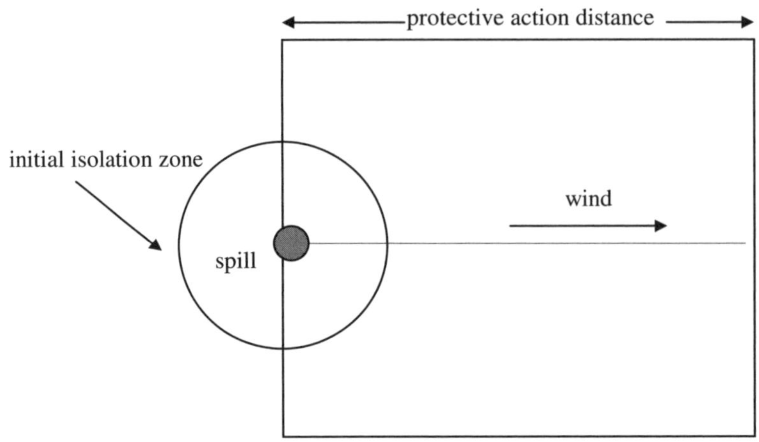

FIGURE 17.1 Defining initial isolation zone and protective action distances.

et al. (1996). The 2000ERG provides initial isolation zone and protective action distances for water-reactive materials in separate categories of spills in water or spills on land.

The Gaussian dispersion models used in the guidebook may be a puff or continuous release For the puff mode, a 15-minute concentration averaging time is used. The corresponding ERPG-2 concentration value for 15 minutes is assumed to be twice the one-hour concentration. The 15-minute ERPG-3 concentration is also assumed to be twice the one-hour concentration.

In order to use the Gaussian dispersion model equations, knowledge of parameters such as wind speed, boundary layer height (for stable atmospheric conditions), inversion height (unstable or neutral conditions), Monin-Obukhov length, friction velocity, and convective velocity (unstable conditions) are required. While a first responder might ballpark estimate a wind speed, the other parameters are not obvious. Dunn et al. (1996) describe what they call a meteorological processor, which estimates these parameters from routine weather observations. The meteorological processor includes a surface energy budget model to calculate friction velocity and Monin-Obukhov length and an integral model for unstable conditions to calculate the inversion height and convective velocity. For the stable atmospheric conditions (defined by a positive value of Monin-Obukhov length), the boundary layer height is calculated from the wind speed and Monin-Obukhov length.

The NAERG96 methodology examined weather data from 61 stations (Dunn et al., 1996, references 64 stations) to come up with a statistical average for daytime (unstable and neutral atmospheric conditions) and nighttime (stable) conditions. Each station represented approximately 40,000 to 44,000 measurements over a five-year period. Then a maximum concentration probability function was computed for each station at the 90% level, meaning, for 90% of the time, the meteorology was such that the protective action distance was equal or less than the normalized concentration (concentration divided by release rate). This was done for both day and night conditions for each station. Then the 61 (or 64) stations were weighted to come up with a single representation of normalized concentration as a function of distance for puff and continuous releases, daytime and nighttime conditions, representing four conditions. The final result Fig. 4.7 in Dunn et al. (1996) were four plots of distance X vs C_m/q for X between 0.1 and 10 miles. In their plots, C_m = maximum centerline ground level concentration (ERPG-2 or ERPG-3 value for protective action distance or initial isolation zone) and q = release rate. For a puff release, q was the total release divided by 900 seconds. The plots are reproduced in Fig. 17.2. In these plots, Cm/q has the units of seconds per cubic meter (s/m^3) and the protective action distance is in miles. It is noted that the continuous release and the 15-minute instantaneous release give almost the same plot.

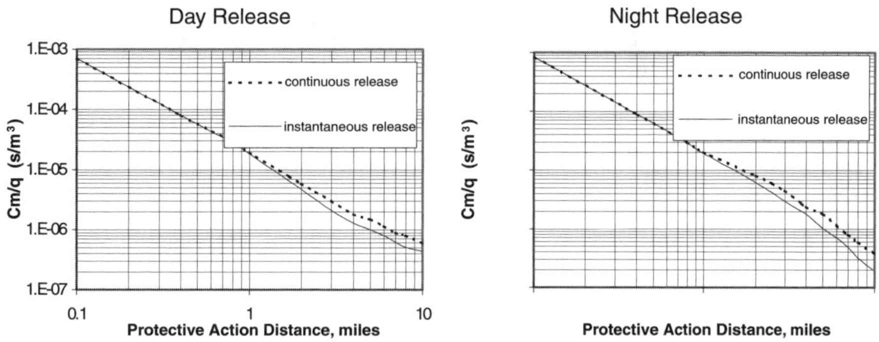

FIGURE 17.2 Downwind concentrations vs protective action distances.

The NAERG96 and 2000ERG simplify the results further by providing first responders with only two categories for the initial isolation distance and four categories for the protective action distance. The categories are small or large spills and (for the protective action distance) daytime or nighttime conditions. Small means 55 gallons as from a drum or a release from a single standard cylinder, and large means a representative large container (e.g., a tank truck) for that chemical. The NAERG96 and 2000ERG do not state the release rate used for each chemical for the small and large spills. No distinction is made between 15-minute instantaneous releases and continuous releases. The result is tables that are quick and simple for first responders to use.

The four choices allowed for protective action distance for a given spilled chemical are conservative for most circumstances. But the tables do not provide answers for different situations such as a catastrophic tank car spill or very small leak, an evaporating pool, spills under very stable atmospheric conditions, or distinctions between a rural spill over flat land or in an urban area. The toxic cloud dispersion is very different for a cloudy, windy night than for a calm clear night. Having many categories to cover these cases, however, would result in a document too cumbersome to use in emergency situations.

17.7 HAND-HELD COMPUTERS WITH DISPERSION MODELING CAPABILITY

Under a U.S. Department of Energy Contract, Western Research Institute (WRI) of Laramie, Wyoming, surveyed over 100 spill accidents where public evacuations occurred and determined a need for responders to have a hand-held computer tool that incorporates a chemical database, information on respirators and protective clothing, and dispersion modeling capabilities. WRI noted that even though communities sometimes had modeling capabilities such as CAMEO or ALOHA available, the models were not used at the time of the spill. The people familiar with the computer software were not available at the time of the incident, and sometimes responders were unable to determine the chemicals and amounts released.

WRI determined that the requirements of the computer tool were that it should (1) be small and rugged for field use, including operation under adverse weather conditions, (2) provide a database for hazardous chemicals including information on respirators and protective clothing, (3) have dense gas and neutrally buoyant dispersion modeling capabilities, (4) provide fast answers, including the ability of the responder to model several "what-if" situations, and (5) could be used by someone under the stressful condition of an emergency situation. The user should have the flexibility to model any spill size, under rural or urban conditions and under a variety of meteorological conditions and not be limited to four choices as in NAERG96 or 2000ERG. The dispersion modeling should be as powerful as ALOHA but designed for field use by a responder with minimal computer skills.

As of the year 2000, hand-held computer versions were available using CE Windows and the Apple Newton Message Pad 130, which featured the Newton 2.0 operating system. Additional versions became available in 2001. The system is designated the PEAC (palmtop emergency action for chemicals) management tool. Information programmed into the PEAC tool included the following:

- Database of over 10,000 chemicals and synonyms, including DOT Guide information, and physical properties such as flash point, lower explosion limit, and boiling point, including the information contained in the *1997 NIOSH Pocket Guide to Chemical Hazards,* the 2000ERG, and other sources.
- NFPA hazard rating designations
- Specific chemical protective clothing information
- Respirator recommendations
- Built-in computations for fire-foam applications
- Database of levels of concern based on ERPG-1, ERPG-2, ERPG-3, IDLH, eight-hour worker exposure limits, or short-term exposure limits (STEL) for different chemicals
- Gas dispersion modeling capabilities that compute protective action distances based on either levels of concern stored in the system or a user-supplied level of concern
- Internal computations for chemical release rate for discharges from a pipe, hole in a tank, or evaporating pool
- Built-in calculator
- Metric/English conversions
- Storage of national and user-specified emergency phone numbers

The user inputs the chemical (by name or DOT number), geographic location, time of day, weather conditions (estimation of wind speed, temperature, cloud cover), ground surface features, type of release (e.g., hole or pipe discharge, a major rupture as in a BLEVE), and container size and type. If a liquid pool is formed that evaporates, the user may input an estimate of the pool area. The user also specifies whether the protective action distance is to be based on ERPG, IDHL, STEL, eight-hour worker exposure TWA, or a specified level of concern. For ground cover features, the user is allowed three choices: flat, cropland/light residential, and urban/forest. Default conditions are shown on the computer screen if the user does not know what information to input. The process of the user inputting conditions and modeling takes less than a minute, with the actual computational time about one second.

A patented feature is how gas dispersion modeling, including dense gas modeling, is programmed into the unit to enable rapid calculations. The field responder can compute protective action distances based on different levels of concern, different release rates, or all of the chemical released at once, or different meteorological conditions, and have all of this information quickly. The ability to go through several different scenarios in a couple of minutes or less in the field helps facilitate the decision-making process for public evacuation. This is especially useful if the responder does not know the chemicals present or is lacking other information; the responder can use the tool to go through a number of what-if situations quickly.

Units have been sold to fire departments and other emergency response agencies through AristaTek, Inc., Laramie, Wyoming, which sells the PEAC tool. They have been used in training exercises and in the field.

17.8 EXAMPLE MODELING OF ACTUAL RELEASES

17.8.1 Example 1: Kit Fox Field Demonstration Tests at the DOE HazMat Spill Center

From August through September 1995, approximately 90 releases of carbon dioxide were completed at the Department of Energy HazMat Spill Center near Mercury, Nevada. The purposes of the tests were to gather chemical cloud dispersion data under a variety of surface roughness and meteorological conditions and compare with available gas dispersion models. The carbon dioxide, which simulated a toxic gas spill, was released at ground level for periods of time ranging from a 20-second puff to several minutes. The carbon dioxide in the resulting cloud was measured by arrays of sensors located 25, 50, 100, and 225 m (meters) downwind. The test results (as of 2000) are available to the general public on the Internet (www.westernresearch.org) with analyses or summaries presented in several technical papers (King et al., 1999; Spicer and Havens, 1999; Hanna and Chang, 1999).

Sample carbon dioxide sensor plots for one of the releases (test day 8, release number 11) are given in Fig. 17.3. For this release, 1.49 kg/s of carbon dioxide were released at ground level for 240 seconds using a specially designed release system that permitted a sharp on/off cutoff of the gas. The surface roughness was 0.02 m; the roughness was created by using arrays of rectangular sheets orientated perpendicular to the wind. The wind speed measured at the 2-m height at a location within the roughness elements was 2.3 m/s. The calculated Monin-Obukhov length was 27 m (indication of a D transitional to E atmospheric

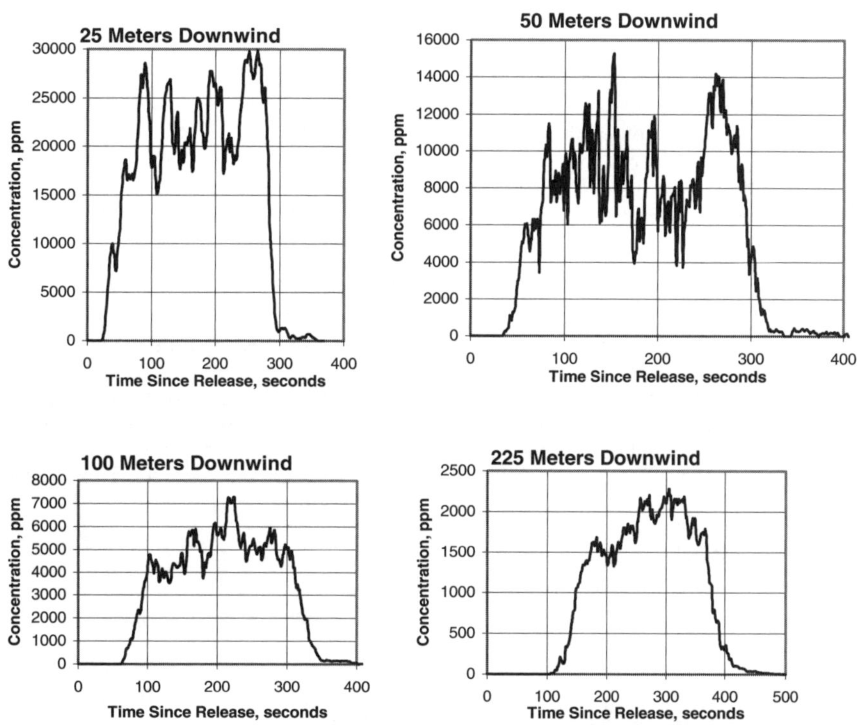

FIGURE 17.3 Carbon dioxide sensor plots for sample Kit Fox tests.

TABLE 17.2 Time-Averaged Concentrations (ppm) for Kit Fox Release 11, Test Day 8

Distance	One-second peak	One-minute average	Ten-minute average	One-hour average
25 m	29,769	24,989	10,991	1,528
20 m	14,171	10,513	4,564	617
100 m	7,280	5,865	2,629	333
225 m	2,283	2,076	1,085	220

stability) based on temperature and wind speed sensors within the roughness array. Other sensors, placed outside the roughness array, indicated a Monin-Obukhov length of about 12 m (E atmospheric stability). The plots (sensor readings) are zeroed out for background carbon dioxide. The sensor time plots presented are for sensors located closest to the centerline of the carbon dioxide cloud. The time scale is seconds from the release start. The concentration is in parts per million (ppm) by volume.

The sensor time plots presented are typical of all releases during the Kit Fox tests, in that (1) concentrations decreased farther downwind from the source; (2) the cloud size increased laterally, vertically, and duration farther from the source; and (3) the peak (one-second) concentration to one-minute average concentration decreased farther from the source. All of these observations were due to mixing of the chemical plume with the surrounding air. For this test, the release duration was 240 seconds but the plume duration at 225 m downwind was almost 400 seconds. Other releases showed that the plume duration increased as the atmospheric stability increased from D to F. Also, the plume arrival time was greater than what would be predicted from wind speed measurements (e.g., for the plots presented, the plume should have arrived at the 225 m distance 98 seconds after the start of the release, but the plume arrive at about 120 seconds). This was in part due to lower wind speeds closer to the ground and possibly also due to dense gas effects near the source.

Of interest are the average concentrations for the distances downwind (Table 17.2).

Obviously, the 10-minute and one-hour concentration averages are much less than the peak one-minute average because the spill duration was only 4 minutes and the plume duration at the sensors was 7 minutes or less. But the peak one-second concentration was significantly greater than the peak one-minute average concentration. This has important implications because ERPG-2 and ERPG-3 values represent one-hour exposure but actual instantaneous concentrations, especially near the source, are usually much higher. This is of special concern with toxic chemical compounds, especially nerve gases and other chemical agents, where the harm may come with a single breath of air.

The concentrations predicted by several popular gas dispersion models at the same distances downwind are presented in Table 17.3. The popular gas dispersion models selected were (1) neutrally buoyant Gaussian dispersion, (2) dense gas model SLAB, and (3) ALOHA version 5.2. When the model called for a concentration-averaging time, a one-minute average

TABLE 17.3 Carbon Dioxide Concentrations (ppm) Predicted by Popular Gas Dispersion Models

Distance downwind	Gaussian E stability	SLAB	ALOHA version 5.2
25 m	50,600	27,000	29,000
50 m	12,600	10,500	11,000
100 m	3,100	3,850	3,300
225 m	600	1,040	800

was used. The Gaussian dispersion model used Briggs' dispersion coefficients for rural conditions and an E atmospheric stability.

In this example, ALOHA version 5.2 selected the dense gas model DEGADIS, developed at the University of Arkansas (Spicer and Havens, 1989). ALOHA also gives the option of using a Gaussian model, which gave the same answer as the Gaussian E stability tabulated above. In the example presented, both dense gas models SLAB and DEGADIS accurately predicted the one-minute average concentrations at 25 and 50 m downwind. The measured one-minute average concentrations were higher than predicted at 100 m and 225 m downwind. The comment is made that micrometeorology has an important role on the local plume cloud shape and concentrations. The example is chosen for illustrative purposes and no general conclusion should be drawn on the ability of the models to predict accurately a toxic cloud resulting from a chemical release on the basis of the information presented.

17.8.2 Example 2: Oleum Spill at Richmond, California

This example was chosen because chemical spills and resulting dispersion are sometimes more complex than can be accounted for in popular gas dispersion models.

On July 26, 1993, at the General Chemical Corporation just outside Richmond, California, workers were starting to unload oleum from a 100-ton tankcar. Oleum is a mixture of sulfuric acid and sulfur trioxide. Because oleum is a very viscous liquid at room temperature and does not flow readily, the workers followed a standard operating procedure of heating the oleum by running steam through heating coils on the tankcar. Shortly after 7 a.m., a safety release valve unexpectedly blew out even though the tankcar's pressure gauge read only 55 psi. The result was a steady stream of vapor that began to escape through the three-inch valve. A thick white cloud formed around the tankcar and began drifting downwind. The cloud hugged the ground initially but then rose to a height of about 1,000 ft (300 m). The workers were able to escape safely. By the time the chemical cloud towered high above the chemical plant, people began to arrive at the local hospitals complaining of stinging eyes and lungs, nausea, and vomiting. The weather conditions at the time of the incident were 70°F, partly cloudy skies, and winds from the southwest at five to seven knots.

Evans (1999) discusses the use of the ALOHA model to model this incident. The author points out that oleum is a mixture of two chemicals (sulfur trioxide and sulfuric acid), but ALOHA is (as of 1999) designed to model release and dispersion of pure chemicals only. Therefore, oleum was not in the ALOHA database. The user may still model the release as sulfur trioxide, but ALOHA displayed a warning message saying that sulfur trioxide reacts with any water to produce sulfuric acid and water, and therefore ALOHA cannot model chemically reactive substances and cannot accurately predict the air hazard. Fortunately, in this incident the plume cloud escaping from the oleum tankcar was clearly visible. Visual observation allowed responders to track its movements more accurately than any air dispersion model could.

For all practical purposes, sulfur trioxide was the constituent released from the heated tank car. The behavior of the resulting plume suggested that initially sulfur trioxide behaved as a dense gas, but the chemical reacted with air humidity, producing sulfuric acid and heat. The heat caused the plume to rise. Eventually the plume cooled, resulting in sulfuric acid decending towards people on the ground. This kind of behavior is not accounted for in the popular gas dispersion models.

Yet there may be a need to model the situation. In this Richmond, California, case the plume was visible, but another such accident might occur at night or there may be a transportation accident or threat of fire and the public should be evacuated as a precaution even though little or no chemical has yet been released. The NAERG96, under oleum, cross-references to guide number 137 and ID number 1831; oleum is also highlighted, indicating that the chemical is listed in the table for initial isolation and protective action distances. Under the "large spill" category, day conditions, for ID number 1831, the initial isolation distance is 185 m (600 ft); the protective action distance is 0.6 km (0.4 miles). These distances may be too low. One problem in using the NAERG96 tables is that we don't know what a large spill is or how much chemical is released or could potentially be released. Another problem is that NAERG96 is applicable to transportation accidents and does not apply to situations where the tank car is heated as during a chemical transfer operation or a fire. The configuration is a 100-ton tank car containing an unknown amount of oleum that contains an unknown percentage of sulfur trioxide. The responder might assume a worst case, where the tank car is full and all of the chemical is released as sulfur trioxide. Assuming that 30 tons of sulfur trioxide (30% of the oleum is sulfur trioxide) is released over 10 minutes and the ERPG-2 endpoint of 10 mg/m^3 for sulfuric acid is used (based on sulfur trioxide converted to sulfuric acid), the Dunn et al. (1996) graph (Fig. 17.2) predicts a protective action distance of 10 miles.

Either the PEAC tool can pull up the same NAERG96 or 2000ERG information for oleum for large day spills or the user can model as a release of sulfur trioxide to a 10 mg/m^3 equivalent sulfuric acid level of concern. The user may choose a different level of concern. A warning message is displayed that sulfur trioxide reacts with water, producing heat and sulfuric acid. The model stored in the PEAC tool assumes that the sulfur trioxide or sulfuric acid stays at ground level and behaves like a gas and does not consider buoyancy effects. If all of the sulfur trioxide (30 tons) is released within a short period of time, a message is displayed showing the protective action distance in excess of 10 km (7 miles), with a caution message in using the results.

17.8.3 Example 3: Chlorine Release at Morristown, Tennessee Water Filtration Plant on September 2, 1987

This example was chosen because the approximate amount released was known and the resulting five-mile-long plume was described and photographed in a magazine (Ryan, 1988). The description of the situation was suggestive of an F atmospheric stability condition.

The chlorine leak occurred from one of two one-ton capacity chlorine tanks connected with a manifold at a water treatment plant. A room chlorine monitor and alarm signaled employees that a leak occurred at 4:50 a.m. on September 2, 1988. However, chlorine concentrations in the room were too great for employees to enter. The Morristown fire department was notified at 5:07 a.m.; the fire department entered the room using self-contained breathing apparatus but was unable to cap the leak. The leak was described as a chlorine liquid jet escaping from the tank. The liquid dropped to the floor and vaporized. The chlorine corroded the electrical components, eventually starting a fire in the transformer room. Power to the plant was cut off at 7 a.m. Chorine hydrate slush began to build up in the area. At noon a team from the chlorine supplier arrived on site and capped the leak. Before the leak was capped, an estimated 2,400 to 3,000 pounds of chlorine had escaped.

TABLE 17.4 Chlorine Concentrations (ppm) Predicted by Popular Gas Dispersion Models

Distance downwind	Gaussian F stability	SLAB	ALOHA version 5.2
100 m	1000	400	240
500 m	37	50	19
1000 m	8.3	19	6.3
2000 m	5.1	7.1	2.4
4000 m	1.0	2.7	0.8
8000 m (5 miles)	0.18	1.0	—

The published aerial photograph of the chlorine cloud showed that the cloud was 5 miles long and 1 mile wide and followed the terrain, seeking valley areas. The initial cloud that escaped from the building that housed the chlorine tank hugged the ground and was only 2 feet high. Within 30 minutes after the spill, the cloud advanced 0.25 miles from the building. After 3.5 hours, the cloud was 5 miles long and 10 feet high. The winds were described as calm, and the humidity was high. As the morning progressed and the sun began to heat the ground, the cloud height increased to 30 feet, and then the cloud dissipated.

The cloud forced the evacuation of 4,000 people, including 131 patients from a nursing home, and closed three schools.

The initial 3.5 hours when the cloud was formed is of interest in modeling this release. The nighttime conditions and shallow depth of the cloud were indicative of very stable atmospheric conditions. After 3.5 hours, the cloud had progressed 5 miles, which is equivalent to 0.6 m/s average travel time. It is not clear how much the cloud traveled due to following the valley seeking a lower elevation and how much is due to air movement, but a low wind speed is indicated. For the purpose of modeling, an F atmospheric stability and a wind speed equal to the average cloud movement of 0.6 m/s was chosen. For the purpose of modeling, 2,000 lb were assumed to be released in 3.5 hours, which is equivalent to an average release rate of 0.072 kg/s.

The Briggs dispersion coefficients (Briggs 1973, cited in Reynolds 1992) for "F Stability" and open country conditions were used for the Gaussian model. The default surface roughness of 3 cm recommended by the ALOHA version 5.2 model was used for ALOHA; the same surface roughness was used for SLAB. ALOHA version 5.2 did not accept a wind speed of 0.6 m/s but accepted a wind speed of 0.62 m/s as a minimum condition; also, the model did not calculate a concentration at 8,000 m downwind.

In Table 17.1, an ERPG-2 value of 3 ppm is listed for chlorine; the most conservative model (SLAB) indicated a distance of 2.5 miles (4,000 m) for evacuation purposes. Observations indicated that the plume cloud extended 5 miles. At 5 miles SLAB predicted the concentration dropped to 1 ppm chlorine, which is the OSHA-recommended worker eight-hour exposure limit. SLAB predicted the cloud width should be roughly 0.25 miles and the maximum height should be roughly 50 ft, when the concentration dropped off to 1 ppm. In modeling a real-world spill, allowances should be made for uncertainties in typography and surface roughness and local meteorology. Chlorine also reacts with air moisture and particulates in the air so concentrations cannot be related directly to physical observations of the plume shape.

17.9 REFERENCES

Briggs, G. A. 1973. "Diffusion Estimation for Small Emissions," ATDL Contribution File 79, Atmospheric Turbulence and Diffusion Laboratory.

Carhart, R. A., W. A. Freeman, and A. J. Policastro. 1996. *Technical Report Documentation to Support the 1996 North American Emergency Response Guidebook. Part II. Toxic Gases Produced into Natural Waters in Table "List of Dangerous Water-reactive Materials,"* Office of Hazardous Materials Technology, U.S. Department of Transportation, Washington DC, September.

Craig, D. K., J. S. Davis, D. J. Hansen, A. J. Petrocchi, T. J. Powell, and T. E. Tuccinardi, Jr. 2000. "Derivation of Temporary Emergency Exposure Limits (TEELs)," *Journal of Applied Toxicology,* vol. 20, pp. 11–20.

Dunn, W. E., D. F. Brown, and A. J. Policastro. 1996. *Technical Documentation in Support of the 1996 North American Emergency Response Guidebook,* Office of Hazardous Materials Technology, U.S. Department of Transportation, Washington DC, May.

Evans, M. 1999. "Ask Dr. ALOHA," *Cameo Today,* vol. 9, no. 2, p. 1.

Hanna, S. R., and J. C. Chang. 1999. "Testing of the HEGADAS Model Using the Kit Fox Field Data," in *Proceedings of International Conference and Workshop on Modeling the Consequences of Accidental Releases of Hazardous Materials,* American Institute of Chemical Engineers, New York.

King, S. B., D. Sheesley, T. Routh, and J. Nordin. 1999. "The Kit Fox Field Demonstration Project and Data Set," in *Proceedings of International Conference and Workshop on Modeling the Consequences of Accidental Releases of Hazardous Materials,* American Institute of Chemical Engineers, New York.

National Institute for Occupational Safety and Health (NIOSH). 1997. *NIOSH Pocket Guide to Chemical Hazards,* U.S. Department of Health and Human Services, Washington, DC (CD-ROM version 1999).

Ryan, G. 1988. "Out from Under a Cloud," *Fire Engineering,* February, pp. 22–29.

Reynolds, R. M. 1992. "ALOHA (Areal Locations of Hazardous Atmospheres) 5.0. Theoretical Description," NOAA Technical Memorandum NOS ORCA-65, National Oceanic and Atmospheric Administration, Seattle, WA.

Spicer, T., and J. Havens. 1989. *User's Guide for DEGADIS 2.1 Dense Gas Dispersion Model,* EPA Report EPA 450/4-89-019, U.S. Environmental Protection Agency, Cincinnati, OH.

Spicer, T., and J. Havens. 1999. "Description and Analysis of Atmospheric Dispersion Tests Conducted by EPA at the DOE Hazmat Spills Center," in *Proceedings of International Conference and Workshop on Modeling the Consequences of Accidental Releases of Hazardous Materials,* American Institute of Chemical Engineers, New York.

U.S. Air Force. 1971. *Guide to Local Diffusion of Air Pollutants,* Technical Report 214, U.S. Air Force, Air Weather Service, Scott Air Force Base, IL.

CHAPTER 18
SPREADING OF COLD DENSE CLOUDS

M. Nielsen
Risø National Laboratory, Roskilde, Denmark

18.1 INTRODUCTION

Emissions from a pressure-liquefied or refrigerated gas storage often produce a cloud which is both heavier and colder than the ambient air. Horizontal spreading is enhanced by gravity, whereas vertical mixing is reduced by density stratification similar to dispersion of an ordinary dense gas cloud. The objectives of this section are to identify the differences induced by the temperature deficit. First, the cloud density depends on a heat balance, which if the entrained air is moist, must include heat of condensation. Another significant difference is the heat convection from the warm ground to the cold gas mixture that gradually modifies the cloud temperature and thereby the density. Finally, the heat convection is an additional source of turbulence which tends to enhance the mixing. From a risk-analysis point of view, the thermally-induced processes are mitigating factors.

18.2 DENSITY CALCULATIONS

18.2.1 Homogeneous Equilibrium

Numerical models of cold dense gas dispersion predict cloud dynamics, temperature, and phase composition in parallel to its dispersion. The documentation often contains a reference to the homogeneous equilibrium assumption (HEA). This means that the aerosol and gas phases are supposed to have the same temperature and that the contaminant partial pressure in the gas phase is equal to the saturation vapor pressure. HEA is a computationally convenient assumption since it allows calculation of the thermodynamic state by a submodel independent of cloud dynamics. The appropriate thermodynamic relations depend on the contaminant properties, e.g., its water solubility, and it is practical to implement these as exchangeable modules in the dispersion code. HEA is not strictly valid during aerosol evaporation, since the exchange of heat and matter between the two phases requires spatial gradients in the boundary layer surrounding each aerosol (Vesala and Kukkonen, 1992). It does, however, predict the mixture density with sufficient accuracy except when aerosols are very large, e.g., greater than 100 μ for ammonia (Kukkonen et al., 1994).

TABLE 18.1 Cloud Composition before and after Aerosol Formation

		Before condensation	At equilibrium
Vapor	Gas	c	$(1-\alpha)\cdot c$
	Dry air	$(1-q)\cdot(1-c)$	$(1-q)\cdot(1-c)$
	H_2O	$q\cdot(1-c)$	$(1-\alpha_w)\cdot q\cdot(1-c)$
	Total	1	$1-\alpha\cdot c-\alpha_w\cdot q\cdot(1-c)$
Aerosol	Gas	0	$\alpha\cdot c$
	H_2O	0	$\alpha_w\cdot q\cdot(1-c)$
	Total	0	$\alpha\cdot c+\alpha_w\cdot q\cdot(1-c)$

18.2.2 The Mole Budget

The rapid expansion associated with a flash-boiling jet from a pressure-liquefied storage fragmentizes the remaining liquid phase into tiny airborne aerosols, which follow the gas flow. In the following, we assume perfect mixing of contaminant and humid air, although it should be said that part of the liquid might rain out under certain conditions, such as when a two-phase jet hits an obstacle.

Two-phase density calculations require some bookkeeping, as shown in Table 18.1. This table describes mixture composition by its overall contaminant concentration, c, the water concentration of the diluting air, q, and the degrees of condensation (α, α_w) for contaminant and water vapor, respectively. The mole budget provides useful ratios, e.g., the overall degree of condensation ($\alpha_{tot} = \alpha \cdot c + \alpha_w \cdot q \cdot (1 - c)$. When the terms in the right-hand column are normalized by the contents of each phase, we obtain gas-phase concentrations (χ, χ_a, χ_w) [mole mole^{-1}] of contaminant, dry air, and humidity, respectively, and liquid-phase concentrations (X, X_w) [mole mole^{-1}]. These concentrations are related by $\chi + \chi_a + \chi_w = 1$ and $X + X_w = 1$.

The gas-phase mixture is considered an ideal gas, and in this case Dalton's law states that concentrations are equal to partial pressures divided by the overall pressure p (N m^{-2}). According to HEA, these partial pressures are equal to the saturation pressures of the liquid aerosols. The appropriate description of such saturation pressures depends on the circumstances (see Table 18.2). A hydrocarbon gas does not readily dissolve in water, and therefore two sets of immiscible aerosols will exist in independent equilibrium with the gas phase. Raoult's law describes equilibrium over dilute mixtures, whereas equilibrium over nonideal binary solution requires contaminant-specific empirical models. An example of the latter is Wheatley's model, which states that:

$$p_{sat}(X, T) = p_0 \exp\left\{-\frac{A(X)}{T} + B(X)\right\} \tag{18.1}$$

This is almost the same as the often-used exponential approximation, except that A and B here are functions of the liquid composition rather than fixed constants. The model was calibrated by experimental data on aqueous solutions of hydrogen fluoride (Wheatley, 1986) and ammonia (Wheatley, 1987).

TABLE 18.2 Homogeneous Equilibrium Models

Immiscible aerosols	$\chi = p_{sat}(T)/p$
Raoult's law	$\chi = X \cdot p_{sat}(T)/p$
Wheatley's binary aerosol model	$\chi = X \cdot p_{sat}(T, X)/p$

18.2.3 The Enthalpy Budget

Phase composition depends on mixture temperature T (K), which is found by the enthalpy budget:

$$(T - T_a) \cdot [cMc_p + (1 - c) \cdot ((1 - q) \cdot M_a c_p^a + q \cdot M_w c_p^w)] + \Delta H_{\text{con}} = \Delta H \tag{18.2}$$

Here ΔH is the mixture enthalpy [J kg^{-1}] defined with the air temperature, T_a, and pressure, p, as the reference state, and emission from a liquefied gas storage therefore results in negative mixture enthalpy $\Delta H < 0$. The symbols (M, M_a, M_w) and (c_p, c_p^a, c_p^w) refer to molar weights [mole kg^{-1}] and heat capacities [J (K kg)$^{-1}$], and ΔH_{con} is the enthalpy change by aerosol formation.

$$\Delta H_{\text{con}} = c\,\alpha ML + (1 - c)q\alpha_w\,M_w L_w + \alpha_{\text{tot}}\Delta H_{\text{mix}} \tag{18.3}$$

Here (L, L_w) are latent heats [J kg^{-1}] of contaminant and water, respectively, and ΔH_{mix} is the heat of reaction in nonideal liquid mixtures, e.g., estimated by Wheatley's model. We neglect the kinetic energy term in Eq. (18.2) since this will be insignificant compared to the heat of evaporation, even in a flash-boiling jet (Nielsen et al., 1997).

The specific volume of the two-phase mixture is the sum of the specific volumes of the two phases:

$$\rho^{-1} = (1 - \alpha_{\text{tot}}) \cdot \rho_{\text{gas}}^{-1} + \alpha_{\text{tot}} \cdot \rho_{\text{liq}}^{-1} \tag{18.4}$$

where (ρ, ρ_{gas}, ρ_{liq}) are densities [kg m^{-3}] of the mixture and its two phases. The liquid-phase density may be regarded as constant, whereas the ideal gas law determines the gas-phase density:

$$\rho_{\text{gas}} = \frac{p}{RT}(\chi M + \chi_a M_a + \chi_w M_w) \tag{18.5}$$

where $R = 8.314$ J (K mole)$^{-1}$ is the universal gas constant. Numerical solution of mixture composition generally involves an iteration, since equilibrium saturation pressure depends on temperature which in turn depends on the degree of condensation.

Density calculation of chemically active mixtures relies on mole budgets similar yet slightly more complex than those shown in Table 18.1. The composition will depend on the degree of reaction, which in the ideal case, is determined by the law of mass action. The heat of reaction will contribute to the enthalpy budget in Eq. (18.2).

18.2.4 Dry Adiabatic Mixing

As the cloud is mixed with air, its temperature will rise to the dewpoint where the aerosols evaporate. The enthalpy of condensation will cease to contribute to the heat balance $\Delta H_{\text{mix}} = 0$ when the mixture becomes a homogeneous gas phase and density calculation becomes much simpler. Consider further the idealized case of adiabatic mixing in which the cloud receives heat from no other sources than the diluting air. The mixture enthalpy will then relate to the source enthalpy by $\Delta H = c \cdot \Delta H_0$, and inserting these simplifying conditions in the above equations, we obtain the following formula:

$$\frac{\Delta\rho}{\rho_a} = \frac{M_a + c \cdot \Delta M}{M_a + \dfrac{c \cdot M_a \Delta H_0}{[(1 - c) \cdot M_a c_p^a + c \cdot Mc_p]}} - 1 \approx c \cdot \left\{\frac{\Delta M}{M_a} - \frac{\Delta H_0}{M_a c_p^a T_a}\right\} \tag{18.6}$$

Here $\Delta\rho$ is the density difference between mixture and air and ΔM is the molar weight difference between contaminant and air now including humidity. The linearization is justified for a dilute mixture $c << 1$. The density effect of a source enthalpy deficit is seen to be equivalent to excess molar weight and consequently we define an 'effective' molar weight by:

$$M_* = M - \frac{\Delta H_0}{c_p^a T_a} \tag{18.7}$$

This may be used for scaling isothermal wind tunnel simulations of two-phase releases, still assuming dry adiabatic mixing.

18.2.5 Wet Adiabatic Mixing

Figure 18.1 is a comparison of the simple M_* approximation and more refined phase-transition models. The mixing is assumed to be adiabatic and the release conditions correspond to the most humid case of the FLADIS ammonia field experiments (Nielsen et al., 1997; Nielsen and Ott, 1996). Wheatley's model (solid line) is the most accurate one since this includes the hygroscopic effect of ammonia. This solution may be divided into three domains: dry mixing, nearly pure-water aerosols, and nearly pure ammonia aerosols. Experimentation with the model input reveals that atmospheric moisture affects the aerosol formation in two ways: the *relative* humidity determines the limit of transition between the

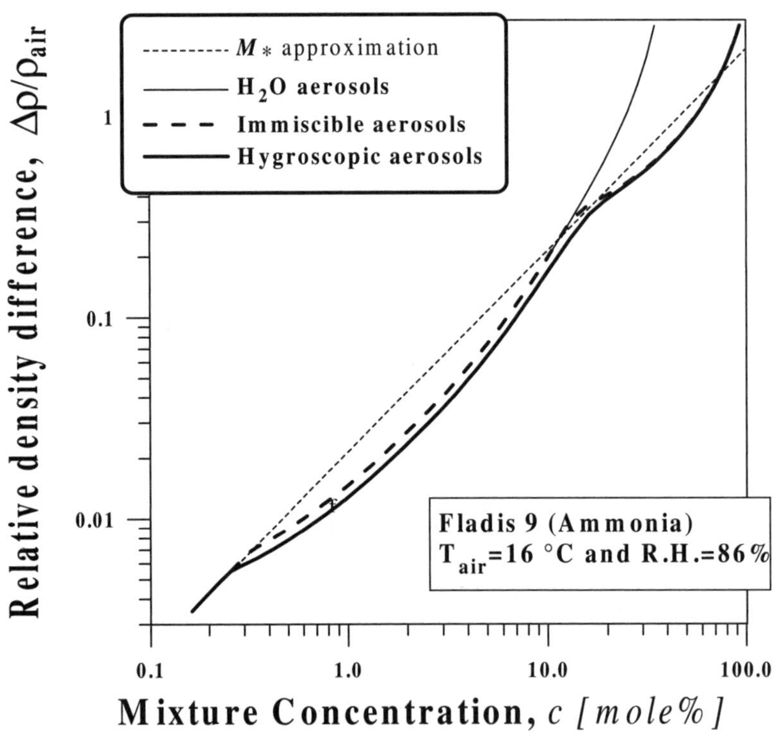

FIGURE 18.1 Four models of the density difference of a two-phase mixture of ammonia and humid air as a function of concentration.

dry and wet mixing, while the *absolute* humidity, depending on air temperature, determines the magnitude of deviation from dry mixing. The immiscible aerosol model (dashed line) is doing surprisingly well, with just a slight overprediction of the density in the domain of almost pure-water aerosols. Most dense gas sources provide substantial initial dilution, which makes the initial domain with almost pure contaminant aerosols irrelevant for the dispersion process. Therefore, the relatively simple pure-water condensation model (thin line) is usually adequate for practical dispersion calculations. It is in fact applied by the publicly available dense-gas dispersion models HEGADAS (Witlox, 1994), DEGADIS (Spicer and Havens, 1986), and SLAM (Ermak, 1990), in contrast to DRIFT (Webber et al., 1992) which offers a wide range of exchangable phase-transition modules. The M_* approximation (dotted line) describes the domain of dry mixing quite well but has a deviation of up to 78% in the domain of almost pure-water aerosols. The test scenario is, however, a demanding one, partly because of the high relative humidity and partly because of the large heat of evaporation and low molar weight of ammonia. The M_* approximation will be more successful for high-molar-weight compounds released in less humid air.

18.3 BUOYANCY CHANGE BY SURFACE HEAT FLUX

In the nonisothermal laboratory experiments of Meroney and Neff (1986), heat transfer from the surface was found to reduce cloud density. These authors even observed a cold and initially heavy methane plume lift off the ground at the downwind distance where heat transfer from the wind tunnel floor had made the plume buoyant. Ruff et al. (1988) measured the heat balance of a cold nitrogen gravity current and identified heat transfer from the wind tunnel floor as the main cause of the volume-integrated buoyancy change. The dynamic effect of this was clearly demonstrated by the experiments of Gröbelbauer (1995), who studied a similar gravity current advancing over a surface with controllable temperature, and the front velocity was observed to slow down when the heat transfer was turned on.

Britter (1987) reviewed the scaling laws for simultaneous dense-gas dispersion and surface heat transfer and concluded that additional constraints had to be imposed on the thermal diffusivity and heat capacity of the simulant gas in the limit of forced convection. He found that free convection was impossible to model correctly. These scaling-law difficulties were also recognized by Meroney and Neff (1986), who estimated that heat-transfer effects should be weaker though probably still significant in full-scale releases in the atmosphere. Kunsch and Fanneløp (1995) found and experimentally verified an analytical solution for simultaneous heat transfer and gravitational spreading in a calm environment. The significant transfer mechanism in an emission of cryogenic gas (LNG) was identified as free convection. With a well-insulated surface the entrainment was found most intense in the region just behind the advancing front and the ground surface temperature under the cloud decreased substantially.

18.3.1 Diabatic Mixing

The surface heat flux depends on cloud temperature, and therefore it is a function of cloud dilution and time history of the heat flux upstream of the observation point. The effect of surface heat flux has to be considered in the context of dispersion and to this end we shall apply a simplistic box model containing the essential cloud dynamics. The heat flux φ [W m^{-2}] is parameterized by:

$$\varphi = \rho[(1 - c) \cdot c_p^a + c \cdot c_p] \cdot c_h u(T - T_s) \tag{18.8}$$

where c_h is an exchange coefficient, u is the plume velocity, and $T - T_s$ is the temperature difference between the cloud and surface. A momentum balance involving the source and

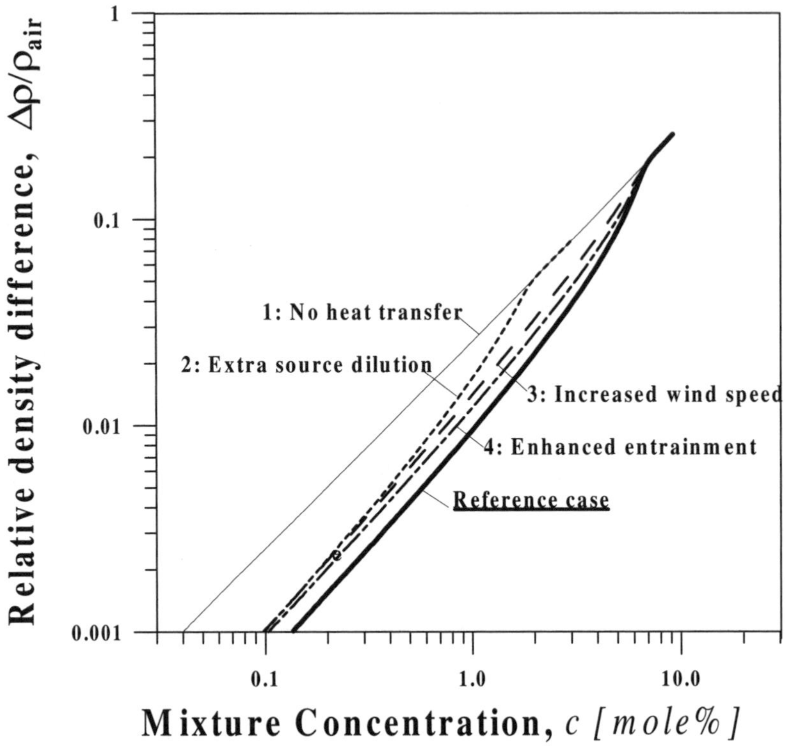

FIGURE 18.2 Predicted heat-flux effect on the density difference of a propane plume as a function of concentration. Release conditions are given in Table 18.3.

entrained air and surface friction determines the plume velocity. Cloud dilution is determined by the entrainment function of Britter (1988)

$$\frac{u_e}{u_*} = \frac{2.85}{6.95 + Ri_{u_*}} \tag{18.9}$$

where u_e is the entrainment rate (m s^{-1}) into the cloud and Ri_{u_*} is the cloud Richardson number, defined by the height of the cloud h (m) and the in-plume friction velocity u_* [m s^{-1}].

Figure 18.2 presents the predicted plume density as a function of concentration using the release conditions listed in Table 18.3. For simplicity, we restrict ourselves to dry mixtures and chose the initial condition as the thermodynamic state in which all liquid material has evaporated and the temperature is the boiling point T_b. The corresponding initial concentration c_0 is given by:

TABLE 18.3 Release Conditions for Calculations Presented in Fig. 18.2

Reference case	$\dot{m} = 3$ kg s^{-1} C_3H_8(l) $u_{10} = 2$ m s^{-1}	$z_0 = 0.01$ m $T_a = 288$ K	$p_a = 100$ kNm^{-2} R.H. = 0%
Comparison cases	1: $\varphi = 0$	2: 3 × dilution 3: u_{10} doubled	4: u_e doubled

$$c_0 ML = (T_b - T_a) \cdot [(1 - c_0) \cdot M_a c_p^a + c_0 \cdot Mc_p] \tag{18.10}$$

The model is initiated at this condition and approaches lower concentrations during the dispersion process, i.e., the calculation progess toward the lower-left corner of Fig. 18.2. The density of the reference case (thick solid line) is significantly different from the case of no heat transfer (case 1, thin solid line), and the effect is seen to accumulate. This development is most significant at the beginning of the dispersion process, when the temperature difference is large. In the later stages of the dispersion process, the mixture has lost about half of its initial buoyancy difference. The heat-transfer effect becomes less significant with enhanced source dilution (case 2). The plume from an elevated release will warm up before touching the ground and this moderates the heat-transfer effect. Other moderating factors are increased wind speed (case 3) and enhanced entrainment rate (case 4). The model did not suggest a significant dependence on release rate or initial plume width.

18.3.2 Enthalpy Budgets in the Field

Laboratory experiments indicate a significant effect of the surface heat flux on cloud density. However, in light of the known scaling-law difficulties (Britter, 1987), it is necessary to seek additional evidence from large-scale field experiments.

Variable atmospheric wind conditions makes the integral heat balances quite difficult, and as an alternative, Nielsen and Ott (1999) preferred to test whether *local* cloud enthalpies were in accordance with the assumption of adiabatic mixing. Figure 18.3 illustrates how this local enthalpy was found. The gas concentration c and cloud temperature T were measured at close positions, whereas the ambient temperature T_a was represented by the measurements of an unexposed reference thermometer at the top of the mast. In order to adjust the pretrial enthalpy deficit to zero, the temperature signal was corrected for its pretrial offset relative

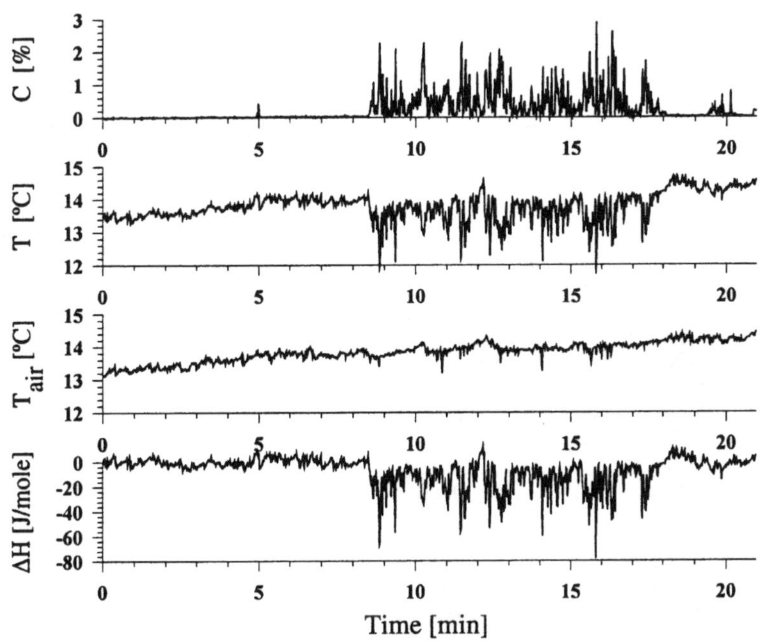

FIGURE 18.3 EEC57, 38-m distance, 1 m above terrain: time series of concentration and temperature, ambient temperature, and the derived enthalpy time series.

to the reference thermometer. The water content of the air q was calculated by upwind measurements of the relative humidity. The reference signal was low-pass filtered, leaving only the general trend of the ambient temperature. The time responses of the other signals were matched by moving average filters. The left-hand side of Eq. (18.2) then calculated the enthalpy time series shown at the bottom of the figure, using Eq. (18.3), the homogeneous equilibrium assumption, and Wheatley's (1987) model for aerosol formation.

The data for this analysis were obtained from the liquefied propane experiments made within Project MTH-BA of the Major Technological Hazards program of the European Commission (Heinrich and Scherwinski, 1990; Nielsen, 1991) and from the Desert Tortoise liquefied ammonia experiments (Goldwire et al., 1985; Koopman et al., 1986). The release conditions are summarized in Table 18.4. The sources in EEC55 and DT3 were downwind horizontal pointing nozzles producing flash-boiling jets with little rainout. In EEC57, liquefied propane expanded inside a cyclone and escaped with no net momentum and some initial rainout. Some of the release parameters in the table are estimates (Nielsen and Ott, 1999).

Figure 18.4 shows a scatter plot of 10-second block-average values of enthalpy and concentration. According to a "null-statement" of adiabatic mixing, we would expect ΔH equal to $c \cdot \Delta H_0$, so that the points should lie on a straight line. Indeed, they seem to do so, but the slopes of this lines are inconsistent with adiabatic mixing. Estimates of local ratios are obtained by linear regression of the type $c = \beta \cdot \Delta H$, forced through the point $(c, \Delta H) = (0, 0)$. The results of this and similar analyses for the other signal pairs are plotted in Fig. 18.5. Uncertainties are evaluated by the residual variance between observations and regression lines. Dashed horizontal lines indicate the enthalpy-to-concentration ratios corresponding to adiabatic mixing, i.e., the source enthalpy ΔH_0. The plot for the ammonia experiment also includes the enthalpy to concentration ratio where, according to Eq. (18.7), the effective molar weight M_* becomes equal to the molar weight of the ambient air. This is the limit at which an initially dense plume would be changed into a buoyant one.

The observations of $\Delta H/c$ lie significantly above the source enthalpy ΔH_0 and demonstrate that mixing was not adiabatic. The observation height or the presence of obstacles in the

TABLE 18.4 Release Conditions

Trial	EEC55	EEC57	DT3
Gas	Propane	Propane	Ammonia
Release type	Jet	Cyclone	Jet
Nozzle diameter (m)	0.0155	–	0.095
Jet momentum F_{jet} (kN)	≈0.25	–	≈11.4
Liquid fraction α	100%	100%	100%
Rain-out fraction f	0%	≈33%	≈5%
Source temperature T_0 (K)	287	287	295
Exit pressure P_0 (Bar)	10.0	9.3	11.2
Source enthalpy ΔH_0 (kJ mole^{-1})	≈−16.3	≈−13.3	≈−21.4
Release rate $\dot{m}$ (kg s^{-1})	3.0	3.9	133
Wind speed u (m s^{-1})	3.2 at 6 m	2.4 at 6 m	7.4 at 2 m
Friction velocity u_* (m s^{-1})	≈0.19	≈0.16	0.45
Surface roughness (mm)	6	6	3
Monin-Obukhov length (m)	≈−90	≈−20	570
Atmospheric stability	≈D	≈C	D
Cloud cover	100%	75%	70%
Air temperature (K)	283	287	307
Atmospheric pressure (hPa)	1025	1025	907
Relative humidity	99%	93%	13%

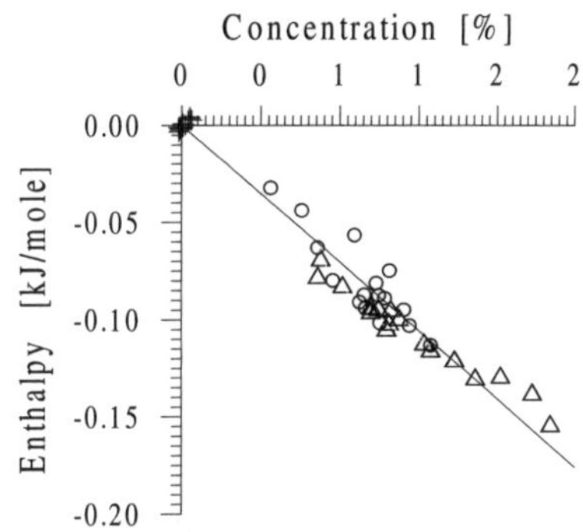

FIGURE 18.4 Local correlation between 10-second block averages of enthalpy and concentration. The regression line is forced through $(c, \Delta H) = (0, 0)$.

MTH BA experiments did not significantly influence the enthalpy-to-concentration ratio. In the two jet releases, the enthalpy increased with downward distance, probably as a result of accumulated heat transfer to the plume from the ground. The release rates and meteorological conditions were comparable in the two MTH BA experiments, and the cause of the different enthalpy-to-concentration ratios must be the different sources. The near-source entrainment was most efficient with a jet release, and presumably the jet was warmer than the plume when it first touched the ground. In addition, the ground contact area upwind of the measuring positions was smaller for the jet than for the plume. In other words, the heating from the ground depends on the initial mixing and therefore on source parameters such as momentum.

Table 18.5 shows the influence on the "effective" molar weight M_* as defined by Eq. (18.7) with the release enthalpy substituted by the observed enthalpy-to-concentration ratios. The effect on the density difference between cloud and ambient air $\Delta\rho$ after aerosol evap-

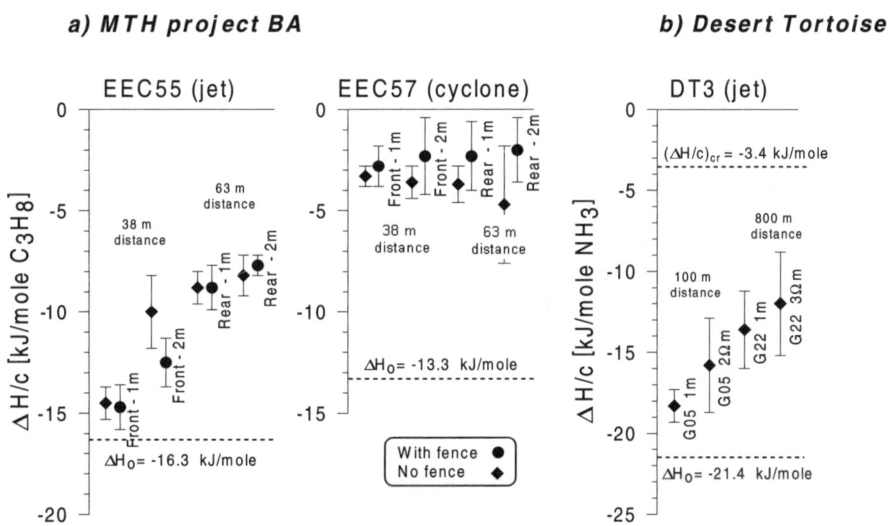

FIGURE 18.5 Observed enthalpy-to-concentration ratios.

TABLE 18.5 The Effective Molar Weight M_* at the Source and in the Field and the Corresponding Reduction in Excess Density

	Source	In-field		$\Delta\rho$ reduction
EEC55	110 g mole^{-1}	78 g mole^{-1}(38 m)	68 g mole^{-1}(63 m)	40% → 52%
EEC57	98 g mole^{-1}	57 g mole^{-1}(38 m)	58 g mole^{-1}(63 m)	≈59%
DT3	87 g mole^{-1}	77 g mole^{-1}(100 m)	61 g mole^{-1}(800 m)	17% → 38%

oration is evaluated by Eq. (18.6). The density difference is the key parameter in dense-gas dispersion and the magnitude of the thermally induced buoyancy reductions is significant.

18.3.3 Surface Temperature Change

A natural question is how long the soil can sustain the heat flux to the gas cloud. Nielsen and Ott (1999) studied the problem of linear heat diffusion in a homogeneous semisolid coupled with forced convection from the surface setting the surface flux proportional to the temperature difference between surface and gas layer. The temperature difference and surface flux after sudden exposure to a gas cloud of constant temperature was found[1] to be

$$\Delta\tilde{T}(\tilde{t}) = \tilde{\varphi}(\tilde{t}) = \exp\tilde{t}\cdot\operatorname{erfc}\tilde{t}^{1/2} \qquad \text{with } \operatorname{erfc}(x) = \frac{2}{\sqrt{\pi}}\int_x^\infty \exp(-\xi^2)\,d\xi \tag{18.11}$$

using the dimensionless heat flux $\tilde{\varphi}(\tilde{t}) = \varphi(t)/\varphi_0$, dimensionless temperature difference $\Delta\tilde{T}(\tilde{t}) = \Delta T(t)/\Delta T_0$, and dimensionless time $\tilde{t} = t\varphi_0^2/(\rho_s c_s \lambda_s \Delta T_0^2)$. The physical parameters are the initial heat flux φ_0 (Wm^{-2}), initial temperature difference between cloud and surface ΔT_0 (K), plus the density ρ_s (kg m^{-3}), heat capacity c_s [J (kgK)$^{-1}$], and thermal conductivity λ_s [W (mK)$^{-1}$] of the soil. The solution in Eq. (18.11) is shown in Fig. 18.6.

The Desert Tortoise included heat-flux sensors buried just beneath the surface, and Nielsen and Ott (1999) fitted a solution of the sub-surface flux measured at the centerline on 100 m downstream of the source in trial DT3. The best fit indicated that the dimensionless exposure time was $\tilde{t} \approx 0.24$, implying that the local surface flux decreased to 62% of its initial value φ_0.

18.4 MIXING OF COLD DENSE GAS CLOUDS

Vertical mixing of a stratified gas cloud is associated with work against gravity, and this effect damps the turbulent kinetic energy level. The local mixing rate depends on the vertical gradients of the concentration gas, which in turn depend on the diffusion. Three-dimensional numerical models usually describe these phenomena by $k - \varepsilon$ turbulence closure. Here we restrict ourselves to much simpler models. The prime objective is to illustrate that heat convection from the ground enhances the turbulent mixing.

[1] The sign of the exponent in the argument of the complementary error function in Eq. (18.11) was incorrect in the original paper.

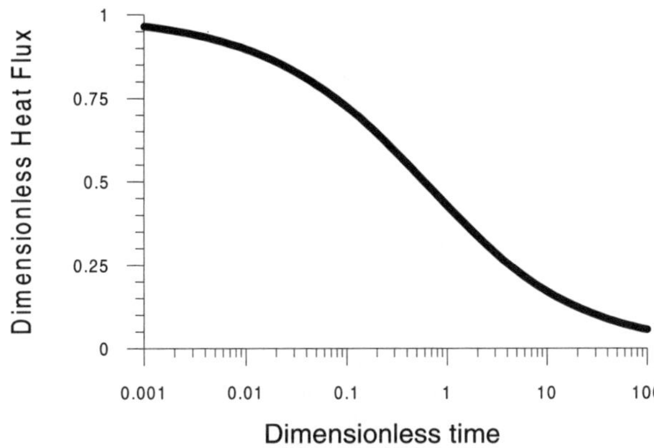

FIGURE 18.6 The normalized surface heat flux $\tilde{\varphi}_0(\tilde{t})$ according to Eq. (18.11).

18.4.1 Entrainment

In dense gas dispersion box models, the mixing process is simplified to a flux of diluting air across a virtual interface surrounding a well-mixed gas cloud. The mixing rate is referred to as the entrainment velocity u_e (m s^{-1}). The interface is just a model concept, and the concentration distributions presented to the model user are usually based on similarity profiles superimposed on the box model.

Equation (18.9) gave a typical example of an entrainment function, where the Richardson number in the denominator is a measure of cloud stability.

$$Ri_{u_*} \equiv \frac{\Delta\rho g h}{u_*^2} \tag{18.12}$$

where $\Delta\rho$ is a characteristic excess density, g is the gravity, and h is a characteristic layer height. Most model developers match the product of the box height and concentration to the depth-integrated gas flux. Britter (1988) set the box model concentration equal to the ground concentration, giving a relatively low layer height, whereas van Ulden (1983) matched the center of gravity, giving a low box model concentration. The friction velocity defined by the turbulent momentum flux $u_*^2 = -\overline{u'w'}$ is either that of the atmospheric boundary layer or a local friction velocity typical for the gas layer. It is worth noting that the calibration coefficients of the individual entrainment function depend on such model definitions.

Heat convection from the ground to a cold dense gas cloud is an additional source of turbulent kinetic energy. The velocity scale for this process is defined by

$$w_*^3 = \frac{\varphi \cdot gh}{\rho c_p \cdot T} \tag{18.13}$$

where c_p [J (kgK)$^{-1}$] is the heat capacity and T (K) is the absolute cloud temperature.

Eidsvik (1980) proposed an entrainment model where the combined effect of mechanical and convective turbulence production was expressed by the turbulence kinetic energy parameterized by $e = 1.7u_*^2 + 0.5w_*^2$

$$\frac{w_e}{\sqrt{e}} = \frac{2.5}{8.7 + Ri_e} \qquad \text{with } Ri_e = \frac{\Delta\rho h}{e} \tag{18.14}$$

where u_* is the in-plume friction velocity. Without heat convection $w_* = 0$ this equation takes the same form as Eq. (18.9). The estimate of the in-plume turbulence in the SLAB model (Ermak, 1990) includes the front velocity of the spreading plume $u_f \cong \sqrt{\Delta\rho gh}$ and the velocity slip between the plume and the ambient air and δu, and the limit of passive dispersion is modeled in a way that accounts for ambient stability.

18.4.2 Calibration of Entrainment Functions

Most entrainment functions contain many adjustable parameters, and unfortunately there is not much turbulence data from large-scale field experiments. Model developers have therefore been forced to look for alternative reference cases. One strategy is to fit the entrainment function to certain limits, e.g., the four scenarios listed in Table 18.6.

The first case in the table is the neutrally buoyant surface plume, the growth rate of which is proportional to the turbulent friction velocity u_*. The proportionality factor depends on the exact box model height h definition. The value in Table 18.6, $\lambda_1 \approx 0.75$, is calculated from Sutton's (1953) analytical solution using van Ulden's (1983) interface definition, which fixes the height h to twice the center of gravity. The advantage of this interface definition is that it gives the box model the right potential energy. The more common choice of h equal to the center of gravity leads to $\lambda_1 \approx 0.35$.

The entrainment rate in the limit of strongly stratified shear flow was determined in the laboratory experiment by Kato and Phillips (1969). The setup was an annular tank with stratified water and constant shear stress induced by a moving screen at the surface. This produced a turbulent well-mixed upper layer, which gradually entrained the quiescent stratified fluid below.

The entrainment rate in the limit of weak free convection is Bo Pedersen's (1980) interpretation of Farmer's (1975) measurements of the development of a thermal profile in an ice-covered lake in the spring season. The solar heating near the surface produced a well-mixed convective layer, and the ice sheet prevented additional turbulence production by wind shear. The mixing rate was deduced from the vertical variation of the phase of the diurnal component of the temperature signals.

The entrainment rate in the limit of strong free penetrative convection was measured in the laboratory experiment of Deardorff et al. (1980), in which initially stratified water was heated from the bottom. The turbulence developed a well-mixed lower layer, which gradually entrained the quiescent fluid above.

Jensen (1981) simplified the turbulent kinetic energy equation to:

TABLE 18.6 Empirical Mixing Rates in Special Reference Cases

Situation	Rate	Conditions	Parameters
Passive dispersion of surface plume	$\frac{u_e}{u_*} \approx \lambda_1$	$Ri_{u_*} \rightarrow 0$ and $w_* = 0$	$\lambda_1 = 0.75$ (Nielsen, 1998)
Stratified shear flow	$\frac{u_e}{u_*} \approx \frac{\lambda_2}{Ri_{u_*}}$	$Ri_{u_*} >> 0$ and $w_* = 0$	$\lambda_2 = 2.5$ (Kato and Phillips, 1969)
Weak free convection	$\frac{u_e}{w_*} \approx \lambda_3$	$Ri_{w_*} \rightarrow 0$ and $u_* = 0$	$\lambda_3 = 0.37$ (Bo Pedersen, 1980)
Strong free convection	$\frac{u_e}{w_*} \approx \frac{\lambda_4}{Ri_{w_*}}$	$Ri_{w_*} >> 0$ and $u_* = 0$	$\lambda_4 = 0.25$ (Deardorff et al., 1980)

$$\overbrace{\frac{\partial \bar{e}}{\partial t}}^{t_1} + \overbrace{\frac{\partial \bar{u}_j \bar{e}}{\partial x_j}}^{t_2} = \overbrace{u_*^2 \frac{\partial \bar{u}_i}{\partial x_j}}^{t_3} + \overbrace{\frac{\overline{\rho' u_j' g_j}}{\bar{\rho}}}^{t_4} - \overbrace{\frac{\partial \overline{u_j' u_i' u_i'}}{\partial x_j}}^{t_5} - \overbrace{\frac{\overline{\rho' u_j'}}{\bar{\rho}}}^{t_6} - \overbrace{\bar{\varepsilon}}^{t_7} \tag{18.15}$$

where the indices refer to three orthogonal directions. The terms in the equation are the temporal change t_1, advection t_2, work by friction t_3, work by gravity t_4, turbulent diffusion t_5, work by pressure perturbations t_6, and the energy dissipation rate ε (m^2s^{-3}). A crude scale analysis of this equation leads to:

$$\overbrace{c_1 \frac{e - e_0}{h} u_e}^{t_1 + t_2} \approx \overbrace{c_2 \frac{u_*^3}{h}}^{t_3} + \overbrace{c_6 \frac{w_*^3}{h} - \frac{\Delta\rho g u_e}{\rho}}^{t_4} + \overbrace{c_3 \frac{e^{3/2}}{h}}^{t_5 + t_6} - \overbrace{c_4 \frac{e^{3/2}}{h}}^{t_7} \quad \text{with } e = u_*^2 + c_5^2 w_*^2 \tag{18.16}$$

where the vertical gradients are parameterized by the layer height, h, and the entrainment velocity is used to find the temporal change. The buoyancy term t_4 is split into two parts: energy production by heat convection and energy consumption by entrainment. The last three terms are all proportional to the cube of a velocity scale divided by a length scale. Rearrangement of Eq. (18.16) leads to the following entrainment function:

$$\frac{u_e}{\sqrt{e}} = \frac{c_2 \left(\dfrac{u_*}{\sqrt{e}}\right)^3 + c_6 \left(\dfrac{w_*}{\sqrt{e}}\right)^3 + c_3 - c_4}{c_1 \left(1 - \dfrac{e_0}{e}\right) + Ri_e} \quad \text{with } Ri_e \text{ based on } e = u_*^2 + c_5^2\, w_*^2 \tag{18.17}$$

Jensen (1981) originally set $c_5 = 1$ and neglected the buoyancy production by convection ($c_6 = 0$), and in order to avoid a singularity for Jensen and Mikkelsen (1984) found it necessary to set $e_0 = 0$. At first sight, this may seem a bold assumption of an ever-quiescent ambient fluid, but the motivation was simply to match the function to the passive limit shown in Table 18.6. Such pragmatism is permitted since in the limit of passive dispersion the energy budget degrades to a balance between production and dissipation with insignificant energy feedback by entrainment.

In search of a solution including turbulence production by heat convection, we need additional boundary conditions. The first assumption is that energy diffusion and pressure transport cancel each other ($c_3 = 0$). The second assumption is that the ratio between energy dissipation t_7 and turbulence production $t_3 + t_{4a}$ is fixed:

$$\frac{c_4(u_*^3 + c_5^3 w_*^3)}{c_2 u_*^3 + c_6 w_*^3} = 1 - R_f^T \tag{18.18}$$

This approach is based on the bulk flux Richardson number R_f^T, which is defined as the ratio of energy recovery due to entrainment and the energy production. It serves as an efficiency factor empirically known to be $R_f^T \approx 0.045$ for subcritical flows and $R_f^T \approx 0.18$ for supercritical ones (Bo Pedersen, 1980). Dense gas dispersion usually falls in the latter category. A solution is possible only with a slightly modified velocity scale $e = (u_*^3 + c_5^3 w_*^3)^{2/3}$ and when the empirical constants in Table 18.6 obey the relation:

$$\lambda_2 \lambda_3^3 = \lambda_4 \lambda_1^3 \tag{18.19}$$

It happens that they almost do, e.g., if the weak convection limit is altered from $\lambda_3 \approx 0.37$ to $\lambda_3 \approx 0.34$. The general solution (Nielsen, 1998) is:

$$\frac{u_e}{\sqrt{e}} = \frac{\lambda_2}{\lambda_2/\lambda_1 + Ri_e} \quad \text{with } Ri_e \text{ based on } e = \left(u_*^3 + \frac{\lambda_4}{\lambda_2} w_*^3\right)^{2/3} \tag{18.20}$$

After insertion of the values in Table 18.6, this becomes

$$\frac{u_e}{\sqrt{e}} = \frac{2.5}{3.3 + Ri_e} \quad \text{with } Ri_e \text{ based on } e = (u_*^3 + 0.1w_*^3)^{2/3} \tag{18.21}$$

18.4.3 In-plume Turbulence

The entrainment functions in the previous section require estimates of the in-plume velocity scales u_* and w_*. Jensen (1981) made an analogy with the atmospheric surface layer in which the velocity and temperature profiles are:

$$u(z) = \frac{u_*}{\kappa}\left\{\ln\left(\frac{z}{z_0}\right) - \psi_m\left(\frac{z}{L}\right)\right\} \quad \text{and} \quad T(z) - T_s = \frac{T_*}{\kappa}\left\{\ln\left(\frac{z}{z_0}\right) - \psi_h\left(\frac{z}{L}\right)\right\} \tag{18.22}$$

where T_* is a temperature linked to the surface heat flux by $\varphi = \rho c_p u_* T_*$, z_0 is the surface roughness, ψ_m and ψ_h are empirical diabatic correction functions (e.g. Paulson, 1970), and L (m) is the Monin-Obukhov length. The Monin-Obukhov length is defined by the turbulent fluxes and this leads to the equation:

$$\frac{z}{L} = Ri_{\Delta T}\frac{\left[\ln\left(\frac{z}{z_0}\right) - \psi_m\left(\frac{z}{L}\right)\right]^2}{\ln\left(\frac{z}{z_0}\right) - \psi_h\left(\frac{z}{L}\right)} \quad \text{with} \quad Ri_{\Delta T} = \frac{T(h) - T_0}{T}\frac{gz}{u(z)^2} \tag{18.23}$$

where $Ri_{\Delta T}$ is a bulk Richardson number for the convection. The equation was solved for the in-plume Monin-Obukhov length, setting the reference height equal to the box-model height $z = h$. Two theoretical objections to Eq. (18.23) are that the surface layer profiles in Eq. (18.22) are based on an unfulfilled assumption of constant turbulent fluxes and that the diabatic correction functions, density ψ_m and ψ_h, only account for the thermal buoyancy effect.

As an example, consider a propane cloud from a liquefied gas storage container with $\Delta H_0 = -16$ kJ/kg released in an atmosphere with temperature $T_a = 290$ K and wind speed $u = 2$ m/s over a smooth surface with $z_0 = 0.01$ m. If mixing is adiabatic and the gas layer height and concentration are $h = 1.5$ m and $c = 2\%$, we find $\Delta T = 11$ K and $Ri_{\Delta T} = 0.145$. The in-plume stability that satisfies Eq. (18.23) is $h/L = -0.64$, and inserting this value in Eq. (18.22), we obtain $(u_*, w_*) = (0.195, 0.225)$ m/s. Repeating the calculation with negligible temperature difference $\Delta T = 0$ K would lead to $(u_*, w_*) = (0.16, 0.0)$ m/s. The density parameter for the gas layer is $\Delta\rho/\rho_a = 0.05$ and Eq. (18.21) predicts the entrainment rates $u_e = 0.145$ m/s and 0.115 m/s with and without heat convection, i.e., a difference of 28%.

The scarcity of field data makes it difficult to validate Jensen's (1981) method. The only field experiments with in-plume turbulence measurements are the Thorney Island isothermal Freon/NO_2 experiments (Mercer and Davies, 1987) and the MTH BA liquefied-propane experiments (Nielsen, 1991). Thorney Island trials 45 and 47 were made with low ambient wind speed. The in-plume[2] turbulence nearly died out, and much of the mixing in these experiments was probably driven by turbulence produced at the fronts of the plume. The observed turbulence level was higher in the propane experiments, as might have been expected for lower release rate, higher wind speed, and heat convection. The lack of heat-flux measurements makes it difficult to quantify the relative significance of these causes.

18.5 CONCLUSIONS

The spreading of a cold dense gas cloud is not much different from the spreading of an isothermal dense gas cloud. The dispersion process is modified indirectly, however, since a cold cloud may lose part of its buoyancy by aerosol condensation or heating from the ground. The condensation effect is temporary, whereas the heat input from the ground is irreversible. The loss of buoyancy has been observed in the laboratory and large-scale field experiments. As the ground temperature decreases, the heat flux to the gas cloud will gradually lose its significance.

Heat convection seems to enhance turbulent mixing, although the available field measurements lack important details. Model developers have therefore been forced to calibrate their box model entrainment functions by comparing predicted and observed concentration fields, i.e., taking a global attitude to model calibration. This is not quite satisfactory since errors in one part of the dispersion model could be compensated by totally independent errors.

18.6 REFERENCES

Bo Pedersen, F. 1980. *A Monograph on Turbulent Entrainment and Friction in Two-Layer Flow*, Series paper 25, Technical University of Denmark, Institute of Hydrodynamics and Hydraulic Engineering.

Britter, R. E. 1987. *Assessment of the Use of Cold Gas in a Windtunnel to Investigate the Influence of Thermal Effects on the Dispersion of LNG Vapour Clouds,* CUED/A-AERO/TR-14-1987, Cambridge University, Engineering Department.

Britter, R. E. 1988. ''A Review of Some Mixing Experiments Relevant to Dense Gas Dispersion,'' in *Stably Stratified Flow and Dense Gas Dispersion*, ed. J. S. Puttock, Clarendon Press, pp. 1–38.

Deardorff, J. W., G. E. Willis, and B. H. Stockton. 1980. ''Laboratory Studies of Entrainment Zone of a Convectively Mixed Layer,'' *Journal of Fluid Mechanics,* vol. 100, pp. 41–62.

Eidsvik, K. J. 1980. ''A Model of Heavy Gas Dispersion in the Atmosphere,'' *Atmospheric Environment,* vol. 14, pp. 769–777.

Ermak, D. L. 1990. *User's Manuals for SLAB: An Atmospheric Dispersion Model for Denser-Than-Air Releases,* UCRL-MA-105607, Lawrence Livermore National Laboratory, Livermore, CA.

Farmer, D. M. 1975. ''Penetrative Convection in the Absence of Mean Shear,'' *Quarterly Journal of the Royal Meteorological Society,* vol. 101, pp. 869–891.

Goldwire, H. C., T. G. McRae, G. W. Johnson, D. L. Hipple, R. P. Koopman, J. W. McClure, L. K. Morris, and R. T. Cederwall. 1985. *Desert Tortoise Series Data Report—1983 Pressurized Ammonia Spills,* UCID-20562, Lawrence Livermore National Laboratory, Livermore, CA.

Gröbelbauer, H. P. 1995. *Experimental Study on the Dispersion of Instantaneously Released Dense Gas Clouds,* Diss. ETH No. 10973, Swiss Federal Institute of Technology, Zurich.

Heinrich, M., and R. Scherwinski. 1990. *Propane Releases under Realistic Conditions—Determination of Gas Concentrations Considering Obstacles,* Report 123UI00780, TÜV Norddeutchland, Germany.

[2] This project is most renowned for instantaneous releases. The final campaign applied a continuous gas source.

Jensen, N. O. 1981. "Entrainment through the Top of a Heavy Gas cloud," in *Air Pollution Modeling and Its Application I,* ed. C. de Wispelaere, vol. 1, Plenum Press, New York, pp. 477–487.

Jensen, N. O., and T. Mikkelsen. 1984. "Entrainment through the Top of a Heavy Gas Cloud, Numerical Treatment," in *Air Pollution Modeling and its Application III,* ed. C. de Wispelaere, vol. 5, Plenum Press, New York, pp. 343–350.

Kato, H., and O. M. Phillips. 1969. "On the Penetration of the Turbulent Layer into a Stratified Fluid," *Journal of Fluid Mechanics,* vol. 37, pp. 643–665.

Koopman, R. P., T. G. McRae, H. C. Goldwire, D. L. Ermak, and E. J. Kansa. 1986. "Results of Recent Large-Scale NH_3 and N_2O_3 Dispersion Experiments," in *Heavy Gas and Risk Assessment III,* ed. S. Hartwig, Battelle Institute, Frankfurt am Main, Germany, pp. 137–156.

Kukkonen, J., M. Kulmala, J. Nikmo, T. Vesala, D. Webber, and T. Wren. 1994. "The Homogeneous Equilibrium Approximation in Models of Aerosol Cloud Dispersion," *Atmospheric Environment,* vol. 28, pp. 2763–2776.

Kunsch, J. P., and T. K. Fanneløp. 1995. "Unsteady Heat-Transfer Effects on the Spreading and Dilution of Dense Cold Clouds," *Journal of Hazardous Materials,* vol. 43, pp. 169–193.

Mercer, A., and J. K. W. Davies. 1987. "An Analysis of the Turbulence Records from the Thorney Island Continuous Release Trials," *Journal of Hazardous Materials,* vol. 16, pp. 21–42.

Meroney, R. N., and D. E. Neff. 1986. "Heat Transfer Effects during Cold Dense Gas Dispersion: Wind-Tunnel Simulation of Cold Gas Spills," *Journal of Heat Transfer*, vol. 108, pp. 9–15.

Nielsen, M. 1991. "Dense Gas Field Experiments with Obstacles," *Journal of Loss Prevention in the Process Industries,* vol. 4, pp. 29–34.

Nielsen, M. 1998. *Dense Gas Dispersion in the Atmosphere,* Risø-R-1030(EN), Risø National Laboratory, Denmark.

Nielsen, M., and S. Ott. 1996. *Fladis Field Experiments—Final Report,* Risø-R-898(EN), Risø National Laboratory, Denmark.

Nielsen, M., and S. Ott. 1999. "Heat Transfer in Large-Scale Heavy-Gas Dispersion," *Journal of Hazardous Materials,* vol. 67, pp. 41–58.

Nielsen, M., S. Ott, H. E. Jørgensen, R. Bengtsson, K. Nyrèn, S. Winter, D. Ride, and C. Jones. 1997. "Field Experiments with Dispersion of Pressure Liquefied Ammonia," *Journal of Hazardous Materials,* vol. 56, pp. 59–105.

Paulson, C. A. 1970. "The Mathematical Representation of Wind Speed and Temperature Profiles in the Unstable Atmospheric Surface Layer," *Journal of Applied Meteorology,* vol. 9, pp. 857–861.

Ruff, M., F. Zumsteg, and T. K. Fanneløp. 1988. "Water Content and Energy Balance for Gas Cloud Emanating from a Cryogenic Spill," *Journal of Hazardous Materials,* vol. 19, pp. 51–68.

Spicer, T. O., and J. A. Havens. 1986. "Development of a Heavier-than-Air Dispersion Model for the US Coast Guard Hazard Assessment Computer System," in *Heavy Gas and Risk Assessment III,* ed. S. Hartwig, Battelle Institute, Frankfurt am Main, Germany, pp. 73–121.

Sutton, O. G. 1953. *Micrometeorology,* McGraw-Hill, New York.

van Ulden, A. P. 1983. "A New Bulk Model for Dense Gas Dispersion Spread in Still Air," in *Atmospheric Dispersion of Heavy Gas and Small Particles,* ed. G. Ooms and H. Tennekes, Springer Verlag, Berlin, pp. 419–440.

Vesala, T., and J. Kukkonen. 1992. "A Model for Binary Droplet Evaporation and Condensation, and Its Application for Ammonia Droplets in Humid Air," *Atmospheric Environment,* vol. 26A, no. 9, pp. 1573–1581.

Webber, D. M., S. J. Jones, G. A. Tickle, and T. Wren. 1992. *A Model of a Dispersion Dense Gas Cloud, and the Computer Implementation II. Steady Continuous Releases,* SRD-R587, UK Atomic Energy Authority, Safety and Reliability Directorate.

Wheatley, C. J. 1986. *A Theory of Heterogeneous Equilibrium between Vapour and Liquid Phases of Binary Systems and Formulae for the Partial Pressures of HF and H_2O Vapour*, SRD-R357, UK Atomic Energy Authority, Safety and Reliability Directorate.

Wheatley, C. J. 1987. *Discharge of Liquid Ammonia to Moist Atmospheres—Survey of Experimental Data and Model for Estimating Inital Conditions for Dispersion Calculations,* SRD-R410, UK Atomic Energy Authority, Safety and Reliability Directorate.

Witlox, H. W. M. 1994. "The HEGADAS Model for Ground-Level Heavy-Gas Dispersion—I. Steady-State Model," *Atmospheruc Environment,* vol. 28, no. 18, pp. 2917–2932.

CHAPTER 19
EVALUATION METHODOLOGIES FOR DENSE GAS DISPERSION MODELS

Nijs Jan Duijm
Risø National Laboratory, Roskilde, Denmark

Bertrand Carissimo
Electricité de France, Chatou, France

19.1 INTRODUCTION

Since about 1975, much research has been devoted to consequence modeling. This has involved the modeling, mathematically as well as physically, of the chemical and physical phenomena associated with major industrial hazards (MIH). Such models are used primarily in risk assessments for safety reports and by safety officers. These models may therefore influence very important decisions, such as the design or authorization of chemical plants. Proper attention should therefore be paid to the quality of these models.

The quality of consequence models, especially dense gas dispersion models, started to be assessed around 1980. From that time, experimental data, both at laboratory scale and at field scale, have been gathered. Among the more famous large-scale experiments are those performed by Lawrence Livermore National Laboratory in the United States (with names like Desert Tortoise, Coyote, and Burro) and those by Shell and UK Health and Safety Executive in Europe (Maplin Sands, Thorney Island). These experiments resulted in a considerable improvement of the available models, reducing the range of variation between the predictions of the different models (McQuaid, 1983).

As modeling capabilities increase and more complex problems are addressed, however, there remains a serious concern related to model quality and the possible limits for areas of application. Since the Thorney Island trials, the European Commission (Directorate General XII, EC DGXII) alone has supported two major series of field tests in parallel with wind tunnel and modeling work. In addition, there are many heavy gas dispersion (HGD) models of varying quality and applicability, ranging from simple box models through more sophisticated shallow layer-type models to fully three-dimensional, computational fluid dynamics (CFD) models. Each type of modeling has its merits and disadvantages, and the abundance of models presents a baffling situation for those having to choose and use models. A structured and unified approach to evaluation and quality enhancement is important in this field.

An initiative was taken by the EC DGXII, which led to the study carried out by Britter (1992). This study provided a framework for managing the quality and evaluation of technical models. As a result, a Model Evaluation Group (MEG) was set up by the European Commission. The group produced a generic evaluation protocol for consequence models (EC, 1994a)

In 1993, the Model Evaluation Group set up an expert group on heavy gas dispersion, one of the aims of which was to produce an evaluation protocol specific to HGD models. The protocol produced by the group was tested in a small evaluation exercise. This protocol was taken further by the project Scientific Model Evaluation Techniques Applied to Dense Gas Dispersion Models in Complex Situations (SMEDIS) (Daish et al., 1999). About 30 different institutes contributed to this project, with their models supported by the European Commission.

19.2 EVALUATION

Several evaluation studies and model intercomparison studies have been done since 1980 (e.g., Mercer, 1988; Hanna et al., 1991; Brighton et al., 1994). Hanna et al. (1993) carried out a seminal validation study on a number of commonly used dense gas dispersion models. This study focused on flat, unobstructed terrain and included only some elements of scientific assessment. These evaluation exercises focused on comparison of calculated data and measured data, using some statistical measure to express the degree of agreement between the different sets of data.

Like Hanna (1991), we want to stress beforehand that a statistical evaluation is only one part of the evaluation of dispersion models. Evaluation of the scientific basis, solution techniques, and the model description is at least equally important in assessing the validity or, more generally, the usefulness of a model.

The purpose of model evaluation is twofold. For use in an operational environment, it can assist interested parties in choosing a correct model and interpreting output results. More precisely, model evaluation should do the following:

- It should produce measures of model quality that may be communicated to users and other interested parties. The evaluator (which is the editor of the evaluation document) ensures that the results of the model evaluation are accessible to non-expert users. The document provides the interested party with the information on the bounds of applicability.
- It should allow a referee to consider the applicability of a model to specific situations and to assess the weight that should be given to the model results.

The second purpose of model evaluation is to encourage appropriate model improvements in a cost-effective way. An evaluation can:

- Encourage the management of model quality
- Reduce distortions in output results by the use of different models
- Identify possible areas of improvement in a model
- Help to identify shortcomings in data sets requiring further experimental effort

Model quality needs to be evaluated and communicated to interested parties in a structured way in order to ensure acceptance and usefulness. The MEG advises a structure consisting of three elements: assessment, verification, and validation (EC, 1994b). In this chapter, we adopt the terminology as suggested by the MEG protocol, i.e., model evaluation includes the whole review of any model with respect to proper scientific formulation (assessment), correct coding (verification), and comparison with experimental data (validation). These stages are defined in Table 19.1. We are aware that other terms are sometimes used and that there is some philosophical criticism of the use of the term *validation*.

TABLE 19.1 Definition of Terms

Model evaluation	This includes examination of a model according to a set of well-defined rules embodied in the evaluation protocol. Although the scientific assessment of the model is the central activity, "evaluation" is used to mean the entire range of activities before, during, and after the scientific assessment.
Validation	This is the process of comparing the predictions of a model, which has been run to simulate a given event, with the observations made in connection with the same event.
Verification	This is the process of comparing the implementation of a model with its mathematical basis. Most commonly, this refers to checking that a computer implementation of a model (computer software) is an accurate representation of the algorithms in the model.

The assessment is itself composed of two parts. The scientific assessment identifies the mathematical and physical algorithms of the model. The user-oriented assessment concerns operational aspects such as costs and ease of use.

Verification provides evidence that the model has been checked on a correct coding of algorithms, databases, and interfaces.

Validation is described in the MEG documents as "getting the model to predict experimental results." The validation supplies the evidence that the output results of the model compare well with experimental data. The usefulness and accessibility of the experimental data are quality aspects of the comparison.

Typical stakeholders in model evaluation are the model developers, safety officers in industry and consultants applying models during preparation of safety reports, and authorities facing approval of these safety reports. The preferable approach is that either a model developer or a user is able to undertake the evaluation. An independent agent with appropriate expertise can also undertake the evaluation. Because the question of impartiality may be raised, the relation between the evaluator and the model should be documented. In each case, however, the evaluation report should be fully verifiable (e.g., by means of an audit).

Prior to any evaluation exercise, the context, scope, and aim of the exercise should be made explicit. It is important to define the area of application for the evaluation in relation to the originally intended use of the model. It is also important to consider a priori the use of the evaluation results.

These aspects and the other procedures to be followed during the exercise will be laid down beforehand in an evaluation protocol. The next two sections will provide more details on the scientific assessment and the validation stages. An outline of a general evaluation protocol for dense gas dispersion models will be presented and some previous evaluation studies will be discussed.

19.3 SCIENTIFIC ASSESSMENT

Until now, most evaluation studies and model intercomparisons relied on comparison of the outputs of models with observations or other model results. However, there is a limited amount of observations of sufficient quality compared to the complexity of the phenomena, not to mention the inherent variability of atmospheric dispersion, which makes it difficult to measure. Therefore, such validation studies alone do not provide sufficient proof of a model's quality and its capabilities to address problems outside the range of the validation data sets. A scientific assessment of the model is an important means to obtain information about the capabilities, reliability, and quality of a model. The scientific assessment seeks:

1. To clearly identify the scientific basis of the model and how this scientific basis is implemented
2. To provide an overview of the capabilities and limitations of the model, based on the chemical/physical aspects ("features") that are explicitly addressed in the model
3. To provide a judgment on the adequacy of the scientific basis and its implementation in relation to the present state of the art both for the model as a whole and for the individual features

Items 1 and 2 can be achieved by writing a comprehensive description of the physical, chemical, and mathematical basis of the model in a unified format or protocol that systematically addresses the different features relevant to heavy gas dispersion. A unified format, in the form of a template to be filled for each model, will also enable an easy comparison between different models. The effort required to produce such a description depends on the quality and completeness of the available documentation of the model. Hopefully, future model developers will follow the format of the protocols that have been developed by now. In any case, it is to the advantage of a model if the original documentation is as detailed and comprehensive as possible.

Performing the review as intended under item 3 is a more problematic task. Competence to judge scientific quality requires a sound expert knowledge of dense gas dispersion as a specific scientific domain linked to fluid dynamics, thermodynamics, and meteorology. Generally accepted scientific principles are based on verification and falsification of theories and assumptions. However, nowadays dense gas dispersion models are complex systems that cannot be simply verified or falsified. Scientific assessment requires that the following aspects be considered carefully:

- The traditional method of scientific assessment is the peer review. It is recommended that the evaluator set up a peer review panel (where the evaluator can be a member, but not necessarily) to review the documentation produced during the assessment. It should be recognized, however, that peer reviews are subjective and tend to be conservative, i.e., innovative solutions may be unjustly rejected.
- Detailed descriptions and validations of single parts of a model in its original documentation, e.g., using analytical solutions or specific well-controlled experiments, may be in favor of the model, showing the thoroughness of the developer in developing the model.
- For the same reason, the amount and quality of the literature to which the model developer refers and the developer's argumentation for certain approaches as compared to alternative methods may be in favor of the model.
- The number and quality of publications based on the model and citations to these publications may give an indication of the model's quality. With respect to the quality of a publication, peer-reviewed articles in international journals are normally ranked highest. It provides a means to include the referees' judgment in the review. Whether or not a model is published in a refereed article therefore makes a differences, but the correlation between scientific quality and the *number* of international articles is low in general. One should also consider that a citation does not necessarily mean an approval of the cited material (Christiansen and Hansen, 1993).

19.3.1 Review of Capabilities and Physical–Chemical Aspects

The structured description of the model provided by the scientific assessment is especially useful for addressing the capabilities of the model. General capabilities relate to the source properties, the topography, and atmospheric conditions:

- Duration, size, and momentum of the source
- Contents of the cloud, thermodynamic properties and chemical composition of the released material, aerosol behavior, chemical reactions, deposition, and heat transfer
- Domain geometry (slopes, buildings, and similar obstacles)
- Atmospheric conditions (stability, humidity, wind speed, and turbulence)
- Ground conditions (surface roughness and heat fluxes)
- Minimum and maximum distances from the source

Aspects to be included in the scientific assessment depend on the type of model. Models can be distinguished on the basis of the spatial dimensions of the independent variables:

- 1D (downwind distance or travel distance/time is the independent variable, lateral similarity profiles of concentration are often included)
- 2D (either two horizontal dimensions—typical for shallow layer models—or downwind distance or travel time and plume height—typical for plume trajectory models—are independent variables)
- 3D (explicit calculation in three directions)

Models can also be distinguished according to the solution technique:

- Phenomenological models and screening tools, in which the dispersion behavior is described by a series of nomograms
- Integral models (e.g., box models for instantaneous releases and steady plume models for continuous releases)
- Shallow-layer models
- Computational fluid dynamics CFD models

The following are the categories of physical and chemical aspects to be reviewed for all classes.

- *General:*
  - Parameterization of processes and possible discontinuities introduced when different expressions are used
  - Limiting or asymptotic behavior, e.g., dense to passive; short finite release to instantaneous release
- *Fluid dynamics and diffusion processes:*
  - Representation or calculation of atmospheric flow, including its turbulence and diffusion characteristics
  - Turbulence closure used, if any
  - Formulation of interaction of atmospheric flow with terrain and/or obstacles, including friction and modification of the flow and turbulence
  - Coupling with other dispersion regimes, especially neutral and positively buoyant ones
  - Representing source geometries by approximating them with available source-term simplifications in the model
  - Modeling approach: Lagrangian/Eulerian treatment, deterministic/stochastic treatment
  - Concentration fluctuations, plume meander
  - Averaging considerations—averaging assumed in model, averaging time for output

- *Thermodynamic processes:*
  - Formulation of heat transfer (sensible and latent) from various types of surface to the cloud
  - Formulation of treatment of contaminant aerosol, including possible interaction with ambient water (vapor and liquid)
  - Correlations for thermodynamic properties, e.g., density, vapor pressure
- *Chemical and other processes:*
  - Formulation of chemical reactions, whether spontaneous or with the environment (e.g., water vapor in atmosphere reacting with HF)
  - Correlations for chemical properties
  - Radioactive decay
- *Initial/boundary conditions:*
  - Comment on characterization of source, including any link between dispersion model source and different types of primary source (jet, leakage, etc.)
  - Coupling with source term model

Tables 19.2 to 19.7 provide checklists of the physical aspects that need to be addressed for the various models.

19.3.2 Outcome of the Scientific Assessment

The scientific assessment leads to a structured presentation of the scientific basis of the model. This structured presentation allows a systematic comparison of the capabilities of different models based on the features and aspects that are explicitly taken care of in the model.

The outcome of the scientific assessment preferably also includes an expert judgment of the scientific basis of the model and its implementation. Such an assessment will always be subjective because objective indicators for scientific quality do not exist. Therefore, the model developer/provider should always be given the opportunity to comment on the assessment as part of the evaluation report.

19.4 VALIDATION

There is much more experience with validation than with scientific assessment. This validation can either be qualitative, for example by means of scatterplots of observed versus modeled concentration, or more quantitative, usually by means of statistical measures of the difference between the two.

TABLE 19.2 Phenomenological Models and Screening Tools

In general, the models provide an empirical relationship between the concentration decay of the gas cloud and the downwind distance from the source, in terms of dimensionless quantities. The model type is characterized by the choice of included parameters. The most important parameters are:

- The choice of scaling for dimensionless quantities
- Gravity constant
- Gas density relative to ambient air
- The released volume (instantaneous) or release rate (continuous)
- Ambient wind velocity

TABLE 19.3 Box Models

Box models are used to describe instantaneous heavy gas releases (puffs). Physical aspects of interest for a box model are:

- Entrainment of air into the cloud.
- The expressions for entrainment into the edges and/or the top of the cloud are often different in different models and should be reviewed.
- Heat transfer between the substrate and the cloud. The transfer can be a sensible heat flux or a latent (evaporative) heat flux. In the latter case, it can be the evaporation of aerosols in the gas or the condensation of ambient water vapor.
- Chemical reactions.
- The vertical wind profile and atmospheric stability.
- The frontal spreading velocity.
- The form and the length scale used for both the horizontal and vertical concentration distributions.

In some cases the box model approach has been extended for the simulation of a continuous or time-varying release. The total release is then divided into a number of puffs, each of which is considered as a separate release. The model evaluator should then discuss the modeling of gravitational slumping and mixing in the along-wind direction.

Although readily available procedures for validation studies do not yet exist, when comparing model results with experimental data, three major problems occur:

1. The level of spatial and temporal detail one wants to retain in the comparison
2. The selection of experimental data
3. The statistical measures used to quantify the result of the comparison

19.4.1 Spatial and Temporal Detail

The highest level of detail occurs if one tries to compare model results and experimental data paired in time and space, i.e., each point value from a series $C_{\text{observed}}(t, x, y, z)$ is compared with the corresponding point value from the series $C_{\text{model}}(t, x, y, z)$. Due to the

TABLE 19.4 Steady Plume Models

Steady plume models are developed in the same way as box models, with the variables depending on the downwind distance instead of time. Physical aspects of interest are the same items as in Table 19.3 for box models. The following are some other aspects for steady plume models:

- Time-varying releases are simulated by a superposition of steady plumes, which are formed at subsequent time intervals. The model evaluator should then discuss the modeling of gravitational slumping and mixing in the along-wind direction.
- The steady plume models cannot describe instantaneous releases because the initial slumping phase is not incorporated. This can be remedied by adding a front-end box model for the gravity-dominated stage.
- The transition to passive dispersion.
- The similarity profiles of the heavy gas concentration in the plume and relation to the algorithm of air entrainment.

TABLE 19.5 Integral Plume Models

The integral plume theory is intended for application on lofted plumes. Physical aspects of interest are:

- Entrainment of air into the cloud. Points to check are (1) the presence or absence of an entrainment term involving atmospheric turbulence and (2) the presence or absence of an entrainment term due to plume turbulence far away from the source.
- The inclusion of a drag force in the conservation equations of momentum.
- The presence of a zone of flow establishment.
- The shape of the plume cross section. The models employ either a circular or elliptical cross section.
- The profile of the crosswind variables around the centerline of the plume
- Chemical reactions
- The vertical wind profile and atmospheric stability.

inherent variability of dispersion in the atmosphere and therefore the fundamental impossibility of specifying initial and boundary conditions well enough to be able to reproduce $C_{\text{observed}}(t, x, y, z)$ completely, ensemble averages have to be constructed in the form of some processing of the data, both model data and experimental data.

Due to limited resolution in experiments, some averaging over time and volume is involved. Normally, some time averaging is required also to average out turbulence and wind-direction fluctuations that cannot be resolved experimentally and in the boundary conditions for the model.

Another simplification that is often applied is to compare some derived characteristics as a function of distance from the source, mainly the maximum concentration (often called maximum arc-wise concentrations) and the crosswind width of the plume or cloud. For instantaneous releases, the dose (integral of concentration over time) can be used instead of concentration. This procedure can be very effective because the major concern related to heavy gas dispersion is the assessment of hazard distances and the area affected by the released material. Duijm et al. (1996) conclude that for a proper conclusion of the overall

TABLE 19.6 Shallow-Layer Models

The shallow-water method is based on vertical integration of the cloud properties, reducing the problem to a two-dimensional set of conservation equations (mass, species, downwind and cross-wind momentum, and energy). Some models consider only downwind distance as the independent variable. Physical aspects of interest are (see also Table 19.3 for box models):

- Entrainment of air into the cloud.
- The frontal spreading velocity.
- The transition from plume to puff mode. Close to the source the release can resemble a plume, but the model switches to transient puff mode when the steady-state period is over.
- Heat transfer between the substrate and the cloud.
- Chemical reactions.
- The vertical wind profile and atmospheric stability.
- Shear stress. The conservation equations contain downwind and crosswind friction terms.
- Vertical concentration profiles and the definition of cloud height.

TABLE 19.7 CFD Models

Physical aspects of interest for CFD models are:

- The turbulence closure model. Different levels of turbulence closure are used to obtain the eddy diffusion coefficients. The turbulence models differ in the assumptions on isotropy of turbulence.
- The simplifications in the conservation equations. Examples of different approximations are: hydrostatic, boussinesq, anelastic.
- The type of domain mesh: structured or unstructured grid.
- The discretization method of the equations. Finite-difference methods, finite-element, and finite-volume methods are all used.
- Terrain/obstacle effects refer to obstructions in the flow field and to variations in the terrain height.
- Meteorological fields used as initial and boundary conditions.
- Momentum, heat, and mass (e.g., deposition) transfer between the cloud and its surroundings.

behavior of the model, it is necessary that comparison of maximum arc-wise concentration or dose be combined with comparison of the plume width or cloud width and length, respectively.

Using arc-wise concentrations and plume width also allows the random effect of wind-direction variations to be circumvented by using the concept of moving frame averaging of the experimental data (if sufficient crosswind sensors are available) (see Nielsen and Ott, 1996).

For "simple" situations where plume direction is dominated by wind direction only, using arc-wise maximum concentrations together with plume width is the recommended procedure. However, for other situations, such as dispersion close to buildings or in complex terrain, part of the model performance depends on the correct prediction of the plume path, or there may not even be a unique plume path. In those situations, comparisons of concentration or dose paired in space should be included. In those cases, the experimental data points (sensor positions) included in the comparison should be restricted to points that give significant information about the location and behavior of the plume or cloud, i.e., the number of zero-level measurements should be restricted, though zero-level measurements contain valuable information.

19.4.2 Data Set Selection

The selection of the data set depends primarily on the type of model and the aspect of the model that one wants to test, e.g., dispersion over flat terrain (a considerable volume of data exists for this situation), dispersion close to buildings, or dispersion in the presence of heat transfer.

The data set should preferably cover a reasonably wide spatial range in a structured way. To allow comparison with arc-wise maximum concentrations, the sensor arrangement should include at least four crosswind or arc-wise sensors with less than one standard deviation of plume width between them, although more than six sensors per arc are recommended. Preferably there is at least one position at each arc where vertical concentration distributions are available. For field data, frequency responses of sensors between 0.1 and 1 Hz are fine for general use (some averaging will often be performed afterwards), unless concentration fluctuations are part of the validation. The relevant time scale for sensor response is the time of flight u/x at a downwind distance, x.

Attention should be paid to the origin and quality of the data and, if secondary sources are used, one should be aware of any data processing that has been applied, such as time averaging and rejection of certain data series. The REDIPHEM database (Nielsen and Ott,

1995) contains one-second block-averaged data, but otherwise it keeps very close to the original data sets. In the modelers data archive by Hanna et al. (1991) the information is much simplified and therefore convenient to handle. It consists of tables of the highest average concentrations observed at various distances and plume widths based on average concentrations. For a discussion of quality aspects of data sets and documentation, see Nielsen and Ott (1995). Other information on available databases can be obtained from Mercer et al. (1998) and Carissimo et al. (2001).

An essential quality factor is the reliability of the information on boundary conditions, such as wind speed, stability, and local surface roughness. The reliability of the source term and release rate information is extremely important. Doubts on the real value of release rates have in fact reduced the potential of some large-scale experiments.

Another issue related to data selection is whether to include laboratory data, especially results from wind-tunnel experiments. The advantages of wind-tunnel results are that boundary conditions are well controlled, large volumes of good quality data are available, and specific aspects, especially effects of obstacles such as buildings, can be studied in detail. On the other hand, meteorological conditions, such as wind meandering, are not fully "real" (this may also be an advantage), there may be scaling effects related to small values of the Reynolds number, and heat transfer, droplet evaporation, and chemical processes cannot normally be reproduced.

19.4.3 Statistical Parameters

Analyzing the performance of a dispersion model by comparing the model output with observed values should result in the establishment of some performance measure. Often such a performance measure consists of one or more statistical parameters, but it can also be a written statement or a graphical presentation. Here we restrict ourselves to statistical measures, keeping in mind that they should be applicable to whole data sets as well as to specific subsets of data in order to examine trends of the performance with some external parameters (e.g., atmospheric stability or downwind distance).

Specifying "performance" of a model in an objective way is not a straightforward process. It can have several aspects. We specify the following requirements;

- A performance measure should indicate whether the model in general under- or overpredicts.
- A performance measure should indicate the level of scatter, i.e., the random deviation from the average under- or overprediction.

If models are evaluated over the full downwind range, i.e., over a range of concentrations varying by some orders of magnitude, the performance measure should weight all pairs of observation and prediction equally, independent of the absolute value of the concentrations. This is the preferred method if a validation is based on arc-wise concentrations and plume width. If the validation is based on spatially paired data points, the disadvantage of this method may be that mismatch between model and observations at the edge of the plume may dominate over relatively good agreement at the center of the plume. Performance measures that consider the absolute difference between model calculation and observation can be used if the range of maximum concentrations in the database does not change too much, e.g., when one is interested in the concentrations close to a single building. A discussion of different performance measures can be found in Duijm et al. (1996)

A pair of performance measures that considers the absolute difference between model and observations are the fractional bias and the Normalized mean square error (see Hanna et al., 1991), which are only "normalized" by the data set-averaged concentration. Averaging over the data set is indicated by the operator $< >$:

Fractional bias: $\mathrm{FB} = 2\,\dfrac{\langle C_p\rangle - \langle C_o\rangle}{\langle C_p\rangle + \langle C_o\rangle}$

Normalized mean square error: $\mathrm{NMSE} = \dfrac{\langle (C_p - C_o)^2\rangle}{\langle C_p\rangle\langle C_o\rangle}$

In view of the requirement of equal weight irrespective of the absolute value of the concentration, Duijm et al. (1996) introduced the mean relative bias, together with the mean relative square error:

Mean relative bias: $\mathrm{MRB} = \left\langle 2\,\dfrac{C_p - C_o}{C_p + C_o}\right\rangle$

Mean relative square error: $\mathrm{MRSE} = \left\langle 4\left(\dfrac{C_p - C_o}{C_p + C_o}\right)^2\right\rangle$

The MRB ranges (similarly to the FB) from -2 to 2, with an optimum value of 0. A negative MRB indicates underprediction and a positive MRB indicates overprediction.

The above measures are based on the difference between observation and prediction. Another approach is to consider the ratio of the observation and the prediction C_o/C_p. This puts inherently equal weight on the pairs irrespective of the absolute values.

The simplest set of performance measures using ratios is based on sign tests. To those belong the fraction of overprediction (the fraction of pairs for which $C_p > C_o$) and the factor of n statistics (the fraction of pairs for which the ratio is between n and $1/n$):

Fraction of overprediction: $\mathrm{FOEX} = \dfrac{N_{C_p>C_o}}{N} - 0.5$

Factor of n: $\mathrm{FA}n = \dfrac{N_{1/n<C_o/C_p<n}}{N}$

The FOEX ranges from -50% to $+50\%$, with an optimum value of 0. The FAn ranges from an optimum value of 100% to 0. The factor of 2 (FA2) is most often referred to in dense gas evaluation studies (Hanna et al., 1991). The sign tests have the advantage of being distribution-free.

One of the evaluation systems developed by Hanna et al. (1991) is based on the geometric mean bias (MG) and the geometric mean variance (VG). By first taking the logarithm, extreme ratios have a less than proportional weight in the final result:

Geometric mean bias: $\mathrm{MG} = \exp(\langle \ln C_o/C_p\rangle)$

Geometric mean variance: $\mathrm{VG} = \exp(\langle (\ln C_o/C_p)^2\rangle)$

The MG varies between 0 and ∞ and the optimum value is 1. Values smaller than 1 indicate overprediction by the model and values larger than 1 indicate underprediction. In order to obtain symmetry in the final results, we prefer ln(MG), which ranges from $-\infty$ to ∞ with an optimum value of 0. In this number, a same amount of under- or overprediction leads to the same absolute values (but with opposite sign). VG is the variance counterpart of MG.

The advantage of the MRB over the MG is that the MRB accepts if one of C_o or C_p has a zero value. Data pairs where either C_p or C_o is zero contain valuable information, although instrument properties should be considered when interpreting low-level measured concentrations. The MRB is discriminative for C_p/C_o ratios between 0.1 and 10 but little outside this range.

Duijm et al. (1996) compared different performance measures. Based on this study, the MRB, MRSE, and FA2 parameters are recommended, though the MG and VG can replace the MRB and MRSE. These parameters lead to the most consistent ranking of model performance. The FA2 is an easy-to-understand measure and takes into account the effects of bias and variance in a straightforward manner.

The total volume of experimental data is still small compared to the variation in conditions, and it appears that the numeric results of the performance measures depend on the data sets selected. Therefore, models and model performance numbers should always be compared using exactly the same data sets. The value of the performance measures have little absolute meaning, i.e., one should never rank models on performance measures obtained from different evaluation exercises.

In some cases it can be useful to group data sets to improve the statistical significance of the results. For example, in SMEDIS, data sets have been grouped by category of complex effects (aerosols, obstacle, terrain) (Carissimo et al., 2001).

A statistical model evaluation exercise cannot be based on assessing performance numbers alone. One should try to explain why certain models respond in a certain way. Graphical displays can be helpful to explore the characteristics of models and their behavior for certain data sets and provide additional information to the performance numbers.

19.5 EVALUATION PROTOCOL

The protocol described in this section is based on (though not identical to) the heavy gas dispersion model evaluation protocol developed by the Heavy Gas Dispersion Expert Group (Mercer et al., 1998). This protocol provides guidance on how a dispersion model should be evaluated and the steps that must be taken. It follows the guidelines of the Model Evaluation Group (EC, 1994b) for structuring an evaluation protocol and extends these to heavy gas dispersion modeling. A protocol can be shaped to contain more specific considerations and actions. An example of such a protocol was developed during the SMEDIS project (Daish et al., 1999).

The protocol divides the evaluation procedure into six steps:

1. Model description
2. Database description
3. Scientific assessment
4. User oriented assessment
5. Verification
6. Validation

These steps are described in each section of the protocol together with guidance and practical information on the steps. The evaluation report should preferably follow this protocol. In the introduction to the evaluation report, the relation between the evaluator and the model should be explained, as well as the aim, context, and framework of the evaluation.

19.5.1 Model Description

The model description provides a brief identification of the model to the interested party. The interested party is either a (potential) user or a referee. The identification text contains the following information.

The name, version number, and release date: It is essential that the dispersion model be clearly identified by name and version number.

The area of application: The applicability of a dispersion model for a particular risk analysis problem depends on the actual conditions in the near and far field.

The name, address, telephone and fax number of the originating organization.

The source and costs of the model.

The model type: Heavy gas dispersion models can be divided into different classes according to the number of dimensions of the independent variables (1D to 3D) and/or the scientific basis and solution technique (e.g., integral model).

Hardware requirements: Information must be provided on the following hardware:

- The processor (personal computer, workstation, minicomputer or mainframe)
- Memory requirements (RAM)
- Storage device requirements (disc space)
- Other devices (visual display units, graphic cards, printers, plotters, and mice)

Software requirements: The software can impose requirements on:

- The operating system, identified with name and version number
- Graphic packages
- Drivers

Quality assurance: Reference should be made to the guidelines or standards used during development of the model and software, such as the EC guidelines for model development or international quality standards.

A list of references to relevant publications: The list of references should refer to all relevant model documentation: user manuals, tutorials, and technical reference manuals. The description should also provide the interested party with relevant references of publications in the open literature. Heredity, i.e., the relation between the model and other and previous models, is of particular importance in the area of heavy gas dispersion modeling. The lists of references should contain at least the scientific source of the algorithms in the model.

19.5.2 Database Description

The database description identifies the data to be included in the *validation* and describes the most important aspects. The database description is composed of two forms. The identification form contains the following information for each data set.

Identification Form

Data Type. The text should indicate the name or identifier of each data set and the date of data collection. The data sets may originate from several sources. For each source, the type should be indicated. It can be:

- Analytic results
- Results from existing (more sophisticated) models
- Laboratory experiments
- Large-scale field experiments
- Accident reports

Data Ownership and Accessibility. The intention is a text that clearly states where the data resides and in what format and who owns the data.

Origin of the Data. Because much useful data already exists in databases, the database description should contain, wherever possible, the following two statements for each data set:

1. A statement that allows the data to be traced back to a primary source. The source of the original data should be clear
2. A statement that specifies the possible loss of information between the primary source and the presented data set. The statement should mention any processing of the data that has been carried out: deletion of data points, truncation of time series, etc.

Quality Assurance. The intention is to describe what quality assurance (if any) has been performed on the data and any uncertainties about the data. Examples of topics that are addressed here are:

- The calibration of the detectors
- The correction of data for detector errors
- Error bars or uncertainty values for the measured quantities
- An estimate of the inherent variability in the measured variables

Second Form

A second form is added to the identification form in order to state clearly any processing of the data that has been carried out and the reasons for using the data set in the validation. This text contains the following information.

The Appropriateness of the Data. A statement must be given that explains why the data set is acceptable for use in the model evaluation. Any properties of the data that limit their use in the evaluation process must be stated. Properties of the data that make it particularly useful in the evaluation process should also be stated.

The Features and Parameters Covered by the Data Set. A statement is required to specify the ranges of input parameters of the model that are covered by the database. This information could be in a raw variable form or in a dimensionless form. A stepwise approach is suggested. The first step is to identify the component features in the model. A feature defines a scientific area, such as:

- Physics of the atmosphere
- Heavy gas dynamics
- Thermodynamics
- Chemical reactions

The second step is to determine the important parameters for each feature.

19.5.3 Scientific Assessment

The scope of the scientific assessment must be clearly stated. This statement provides the user with information on any of the model's aspects that have been assessed. A motivation for the assessment should be included. In some cases it may be desirable to evaluate isolated aspects of a large model, in which case the method used to isolate the aspects should also be discussed.

The scientific assessment is made up of the following:

A Comprehensive Description of the Model. The scientific assessment by the model evaluator supplements the information in the model description, which is usually based on the documentation from the model developer. The assessment should:

- Be limited to documenting the dominant underlying scientific basis. Peripheral parts of the model are split off prior to this activity.
- Review the algorithms for the dominant process. This can be the complete dispersion algorithm or a particular aspect.
- State explicitly the values of any empirical constants in the model. The constants may have been tuned at the development stage to give the best possible agreement with one or more data sets. If so, the model evaluator should identify the data sets.

An Assessment of the Scientific Content. The scientific content of the model should be assessed. Based on the description of the model, the evaluator is required to provide a judgment of the adequacy of the underlying scientific basis and its use in the model, for example compared to the current state of the art. The evaluator should have knowledge of and access to additional information in scientific review papers and other publications in the literature.

Limits of Applicability. The limits of applicability of the model should be discussed. This is based on an assessment of the features that are explicitly included in the model and the adequacy of the modeling of these features. These features can also include integrated source term models.

Limitations and Advantages of the Model. In discussing the limitations and advantages of the model, it is sometimes necessary to document particular positive and negative aspects of the model. For example, models may range from a screening tool to a research tool, or a model can be designed to be always on the conservative side. If this is the case, then the statements have to be preceded by clear and sensible arguments.

Any Special Features. The formulation should be described for any special features, e.g., the capability to deal with obstacles and/or complex terrain.

Possible Improvements. Suggestions for possible improvements in the model are best included in the scientific assessment.

19.5.4 User-Oriented Assessment

A "user" could be a person in industry (or a consultant) responsible for carrying out a safety analysis and/or preparing a safety report, or a person in a regulatory authority responsible for assessing such safety reports and/or carrying out risk assessments. It is important to state the aims and the scope of the user-oriented assessment and discuss any limitations of the model. The statement provides the user with information on the model's aspects that are assessed. A motivation for the assessment should be included. Again, the evaluator should state the type of application that is being made—e.g., as a screening tool or, at the other extreme, as a research tool.

The user-oriented assessment describes how easy the model is to use on aspects such as:

The Associated Documentation. The MEG documents suggest that the user-oriented assessment should cover the aspects "fitness-for-purpose" and "ease-of-use" (EC, 1994a, b). Fitness-for-purpose refers to the degree to which the model is able to provide the necessary results for solving a specific problem. The documentation associated with the model should discuss fitness-for-purpose in a way that is easily accessible to the user, i.e., the educational level of the documentation has to be assessed by the model evaluator. The documentation should supply example calculations to illustrate the model use and list addresses of the developers, expert groups, or user groups for additional user support.

Installation Procedures. The documentation on installation requirements and procedures should be complete and clear and preferably give step-by-step guidance.

A Description of the User Interface. Ease-of-use is also a requirement of the user interface. The interface can be interactive (real time), either text based or with the help of a GUI, or batch oriented.

Guidance in Selecting Model Options. An important aspect for evaluation is the degree to which the model's user manual or the help facility (in case of a computer implementation) gives guidance in selecting model options. Other means to assist the user are computer-oriented training facilities and short courses.

Guidance in Preparing Input Data. The user manual of the model or the help facility (in case of a computer implementation) should give guidance in the preparation of input data. Aspects that can be addressed are:

- The degree to which the input processor checks on typing errors and allows the user to make corrections.
- Initiation of variables with default values.
- The presence of a log file allows the user to check the input data. For batch-oriented as well as for interactive calculations, it is advised that the log file can also be used as an input file.
- A facility for archiving of input/output data helps the user to store results in a structured way.

The possibility of coupling the model to source term models or the integration of source term models (such as jet models and pool evaporation) should also be addressed here.

Checks to Verify Whether the Model Is Used beyond Its Scope. The input processor and the core (dispersion) program should perform checks at two levels: (1) on the validity and range of the inputs and (2) on the obtained values of variable parameters in a program run and should give warnings to the user and eventually stop the calculation.

Clarity and Flexibility of Output Results. Complete output results are preferably stored in a file with only a summary displayed on screen. In heavy gas dispersion modeling, the applicability is commonly the description of hazard effects (toxicity, flammability) for use in risk analysis.

Dispersion model output results relevant for continuous releases and clouds of simple plume-like shape are:

- $C(x)$: the profile of concentration versus distance on the cloud centerline axis
- $Y(x, c)$: the cross-wind distance to a given concentration level (c) as a function of along-wind distance (x)
- M: the mass of gas between the lower and upper flammability limits. The quantity is needed for estimating the explosive contents of the cloud.

For more general cloud shapes, the model can supply $C(x, y, z)$, the concentration distribution in three dimensions. This form of output of the results retains the entire information in the calculation but require the user to perform some postprocessing to obtain $C(x)$, $Y(x, c)$, and M.

In time-varying or instantaneous releases, the quantities are all functions of the time t. Two more relevant quantities for time-dependent releases and simple cloud shapes are:

- $D(x)$: the (toxic) dose versus distance on the cloud centerline axis. The (toxic) dose depends of course on the exposure time, t_{exp}, being the time between cloud arrival and evacuation. The dose is defined as the integral of C over the time interval between cloud arrival and evacuation, and the toxic dose is the integral of C^n over the time interval between cloud arrival and evacuation. The real number n depends on the toxic chemical substance.
- $Y(x, d)$: the cross-wind distance to a given dose level (g) as a function of along-wind distance (x).

For more general cloud shape, for example if the cloud is split by buildings, the model can supply $D(x, y, z)$, the toxic dose in three dimensions.

Relevant concentration levels for toxic levels can be very low, down to 1 ppm, so the output results of the dispersion model should cover the far field.

The transient output can be reduced by taking, for the time period of interest, the maximum occurring values of $C(x, t)$, $Y(x, c, t)$, and $M(t)$—next to the (toxic) dose $D(x)$.

Facilities are needed for the postprocessing of output results in order to produce lists, graphs, contour maps of concentration distributions, etc. The facility can either be a dedicated postprocessor or commercially available programs (spreadsheets, databases, etc.).

Unambiguous and Understandable Error Messages. Error messages should be self-explanatory or refer to the user manual.

Computational Costs. The aspects of computational costs (in terms of running times) can be illustrated in the documentation with benchmark tests for various scenarios. If the program operates in interactive mode, then the response times have to be short.

Possible Improvements. Suggestions for possible improvements in the ease-of-use category are best located in the user-oriented assessment.

19.5.5 Verification

An assessor has to ensure that the computer code is producing output in accordance with the model specifications. It is emphasized that the undertaking of model verification is an extremely tedious task. In the case of a computer implementation, the task is to prove in a step-by-step notation that the code does what it is supposed to do. The common practice is that developers take a less rigorous approach and perform verification by individually testing sections of the computer code (e.g., subroutines). In addition, developers should illustrate the model implementation by putting comment lines within the code. In general, the model evaluator will appeal to the developer to provide information on the extent the code has been verified.

- It is often possible to carry out checks for internal consistency, e.g., mass and/or mass flux balances.
- It may also be possible to run the code for "simple" scenarios for which an analytic solution may be available.

- Also, for such simple scenarios, the results of the code could be compared with the results of any workbook procedures, such as the Britter-McQuaid Workbook (Britter and McQuaid, 1988).
- The behavior of the model in certain limiting conditions could also be considered. For example, for a passive release, does the model give results that are consistent with a "simple" passive dispersion model?
- Automatic tools can be used in some cases to check the correct types of variables and the correct branching of conditional tests.

19.5.6 Validation

The aims and intent of the model validation should be stated. The text should list the model parameters that are to be tested. The validation is of particular importance because it can provide conclusive evidence for model selection to interested parties. A validation starts with selecting a set of various data sets of known quality and suitable for that purpose (e.g., sufficient number of measurements). At the start one also selects a set of statistical measures defined for the comparison of the model output (such as the calculated peak concentration C_p of heavy gas at a location) and the observed values (such as the measured peak concentration C_o).

In the following text, the tasks needed to ensure a satisfactory validation are explained.

The Database Selection. An important item to be addressed in the model evaluation is the components from the database that will be selected for the validation procedure. The item includes statements on:

- The adequacy and appropriateness of the database scenarios for the intended validation
- Treatment of source terms and of the meteorology
- The inclusion of laboratory and field experiments within a single validation
- The scaling aspects of laboratory and field experiments
- The weight attached to data components
- The assumptions in the input data for a model calculation. The statement on input preparation is needed in cases where the database variables contain insufficient information to perform the calculation. The variables taken from the database in order to generate input data are called the independent variables. If the independent variables available from the database are not in the form required by the model, then some modification of the model or some preprocessing of the input may be required.

The Model Characteristics. A selection should be made of the variables to be compared with the output data of the model. These variables are called dependent variables. Some consideration and explicit documentation is required concerning the selection. In heavy gas dispersion modeling, the application is commonly the description of hazard effects (toxicity, flammability) for use in risk analysis. The validation attaches weights to specific variables in a manner that depends on the model characteristics. Variables that are directly relevant to the intended (possibly restricted) use of the model should be weighted more heavily than those of peripheral interest.

An Estimation of the Model Uncertainty. With regard to the model, a specific quantitative assessment is required as to the uncertainty in the input data and the output data. The following are some sources of uncertainty in the output results.

- In numerical models, the grid size and time step have a direct effect on the convergence and accuracy of the output data.
- The modeling of stochastic processes (e.g., turbulence) may also cause deviations in the output data.
- The model's scientific basis may contain erroneous assumptions.
- The effect of the uncertainty in the input data on the output data should be estimated.

An Estimation of the Uncertainty in the Data. With regard to the data sets in the database, a specific quantitative assessment is required as to the uncertainty in the dependent and independent variables. Whenever possible, this information should be contained in the database description.

Use of Code Comparison Exercises. Instead of validating a single model, a series of models can be compared with the same data sets. This provides the evaluator with information about the performance of various models. If the models are carefully chosen this can also establish a state-of-the-art reference to test other models.

The Results of the Validation. The validation leads to a series of calculated values (e.g., concentrations) and a corresponding series of observed values. These values can be compared using different statistical techniques. The statistical analysis can lead to a single set of validation measures, such as a geometric mean bias and mean variance. Alternatively, and more informatively, it can be pursued to look for trends in the comparison as a function of relevant parameters such as distance, wind speed, and density of the released material.

As well as these quantitative measures, it is sometimes informative to produce scatter plots of predicted versus observed values or to plot the agreement between model and observations as a function of relevant parameters.

Conclusions That May Be Drawn. Evaluators can draw their own conclusions from the statistical analysis results of the validation procedure. Alternatively, they can just present the results of the validation procedure and leave it to the interested party to select the best model for a specific application.

In view of the relative scarcity of experimental data in relation to the large number of parameters that play a role in dense gas dispersion, statistical measures do not have an absolute meaning in the sense that by changing the validation data set, the statistical measures can change significantly. Therefore, statistical performance measures of different models can only be compared if exactly the same data sets are used for all models.

Recommendations. The evaluator can close the validation with recommendations on model improvements or experiments for extending the database.

19.6 EXAMPLES OF EVALUATION EXERCISES

19.6.1 Study by Hanna et al.

The extensive work of Hanna et al. (1991, 1993) is the first model evaluation project in which standard objective means of validating heavy-gas dispersion models were applied. The performance evaluations were based on a modeler's data archive, which contains data sets from eight field experiments. This allowed Hanna et al. to obtain conclusive results on the performance of 14 dispersion models. Six models were ranked as "better" models. These models have their concentration predictions within a factor two of observations in 70–80%

of all cases. There appeared to be no correlation between the quality and complexity of the models.

19.6.2 REDIPHEM

The REDIPHEM project was aimed at two aspects of model evaluation. First, a database system was designed and implemented to make experimental data sets available that are useful for evaluation dense gas dispersion models (Nielsen & Ott, 1995). This database has proven to be an excellent platform for dissemination of experimental data related to dispersion. In parallel, the general MEG evaluation protocol (EC, 1994a) was modified to specifically address dense gas dispersion models. A limited, by no means comprehensive, attempt was made to perform a scientific assessment of a number of models developed or improved under support from the European Commission's research programs (Bakkum et al., 1996).

19.6.3 Heavy Gas Dispersion Expert Group

The Heavy Gas Dispersion Expert Group (HGDEG), which was set up by the European Commission's Model Evaluation Group, developed an evaluation methodology based on the REDIPHEM activities and tested it through a small evaluation exercise. Cole and Wicks (1995) give details and experience of this limited exercise. The objective was to develop a protocol for evaluation of the models and make an overview of the experimental data available for such an evaluation. It was concluded that a more extensive evaluation exercise is needed and that the scientific evaluation requires more attention (Mercer et al., 1998; Duijm et al., 1997).

19.6.4 SMEDIS

Continuing the activities from the HGDEG, SMEDIS is an ongoing research project funded under the European Commission's Environment and Climate Research and Technical Development Program. Its main objective is to develop a methodology for the scientific evaluation of dense gas dispersion models and to test this methodology by actually carrying out the scientific evaluation of a large number of models currently available in Europe. The project is focusing on situations in which complex effects such as aerosols, topography, and obstacles are important as well as "simple" situations. The Health and Safety Executive (HSE, UK) coordinate the project. Thirteen partners cooperate in the project.

SMEDIS is the first project to combine scientific assessment with validation against observed data, including these complex effects, and to apply its procedure to a large number (almost 30) of dense gas dispersion models—the majority in use across Europe. Furthermore, the goal of SMEDIS is to encourage continual model improvement, rather than to rank a set of models at one instant in time, by leaving in place a protocol and archived database of test cases, which can be used by all DGD developers and users in the future.

Daish et al. (1999) have given a general description of the entire project. The scientific assessment is based on a further specification of the protocol and the issues described in this chapter. Carissimo et al. (2001) present results of the first phase of the validation part of the exercise. In this publication models are grouped together according to their type (work book, integral models, shallow-layer models, or CFD), so the performance of individual models is hidden. Results are presented for different release scenarios, covering dispersion over flat terrain without complex effects, dispersion close to obstacles, dispersion of aerosol clouds, and dispersion over complex terrain (see Table 19.8). Contrary to the study by Hanna et al. (see Section 19.6.1), improved performance is observed with increasing model complexity

TABLE 19.8 First Phase of Results from the SMEDIS Study. FA2 Statistics for Arc-wise Maximum Concentrations for Models Grouped According to Type

	No complex effects	Dispersion close to obstacles	Aerosol effects	Dispersion over complex terrain
Work book (Phenomenological Models and Screening Tools)	40%	42%	43%	Not applicable
Integral models (box models, steady plume models, and integral plume models)	74%	79%	69%	33%
Shallow-layer models	65%	53%	32%	50%
Computational fluid dynamics models	No validation runs performed	89%	75%	71%

Source: Carissimo et al., 2001.

on average, although this table does not exclude that some "simple" models perform as well for some scenarios as more complex models.

19.6.5 Accident Phenomenological and Consequences Assessment

The U.S. Department of Energy (DOE) established the Accident Phenomenology and Consequence (APAC) Methodology Evaluation Program to identify and evaluate methodologies and computer codes to support accident phenomenological and consequence calculations for both radiological and nonradiological materials at DOE facilities and to identify development needs. The program did not include validation, but some models were selected to run test problems. One hundred and thirty-five models were considered and 24 models were finally included in the evaluation (this program did not specifically aim at dense gas dispersion). The scientific assessment was based on the following 10 model attributes or features:

1. Source term algorithm
2. Input parameters required to run the transport and dispersion model
3. Dispersion submodel type
4. Model capabilities/physics
5. Transport submodel
6. Meteorological input
7. Health consequences submodel
8. Source/receptor mitigation measures
9. Output capabilities
10. Uncertainty analysis

It is recommended that the study be completed with a validation exercise (Lazaro et al., 1997).

19.7 REFERENCES

Bakkum, E. A., N. J. Duijm, W. F. J. M. Engelhard, and C. J. H. van den Bosch. 1996. *REDIPHEM Dense Gas Dispersion Models—Summary Descriptions and Recommendations for Use,* TNO Report TNO-MEP—R96/203, Apeldoorn, The Netherlands.

Brighton, P. W. M., A. J. Byrne, S. J. Jones, D. M. Webber, R. P. Cleaver, P. Courtiade, A. Girard, D. Nedelka, C. Proux, B. Crabol, V. Lhomme, R. D. Fitzpatrick, and A. Mercer. 1994. "Comparison of Heavy Gas Dispersion Models for Instantaneous Releases," *Journal of Hazardous Materials,* vol. 36, pp. 193–208.

Britter, R. E., and J. McQuaid. 1988. *Workbook on the Dispersion of Dense Gases,* HSE Contract Research Report No 17/88, Health and Safety Directorate, Sheffield, U.K.

Britter, R. E. 1992. *The Evaluation of Technical Models Used for Major-Accident on Hazard Installations,* Office for Official Publications of the European Communities, L-2985 Luxembourg, EUR 14774.

Carissimo, B., S. F. Jagger, N. C. Daish, A. Halford, S. Selmer-Olsen, K. Riikonen, J. M. Perroux, J. Würtz, J. Bartzis, N. J. Duijm, K. Ham, M. Schatzmann, and R. Hall. 2001. "The SMEDIS Database and Validation Exercise," *International Journal of Environment and Pollution,* vol. 16, pp. 1–6.

Christiansen, J. K., and H. Foss Hansen. 1993. *Forskingsevaluering i Teori og Praksis,* Samfundslitteratur, Frederiksberg (Copenhagen), Denmark.

Cole, S. T., and P. J. Wicks. 1995. *Model Evaluation Group Seminar—The Evaluation of Models of Heavy Gas Dispersion,* Office for Official Publications of the European Communities, L-2985 Luxembourg, EUR 16146.

Daish, N., R. E. Britter, P. F. Linden, S. F. Jagger, and B. Carissimo. 1999. "SMEDIS: Scientific Model Evaluation Techniques Applied to Dense Gas Dispersion Models in Complex Situations," in *Proceedings of the CCPS/AIChE Conference on Modeling Hazardous Releases,* San Francisco, September.

Duijm, N. J., S. Ott, and M. Nielsen. 1996. "An Evaluation of Validation Procedures and Test Parameters for Dense Gas Dispersion Models," *Journal of Loss Prevention in the Process Industries,* vol. 9, no. 5, pp. 323–338.

Duijm, N. J., B. Carissimo, A. Mercer, C. Bartholome, and H. Giesbrecht. 1997. "Development and Test of an Evaluation Protocol for Heavy Gas Dispersion Models," *Journal of Hazardous Materials,* vol. 56 no. 3, pp. 273–285.

European Commission (EC). 1994a. *Model Evaluation Protocol,* Model Evaluation Group, European Communities, Directorate General XII, Science Research and Development, Version 5, May, Brussels.

European Commission (EC). 1994b. *Guidelines for Model Developers,* Model Evaluation Group, European Communities, Directorate General XII, Science Research and Development, Version 5, May, Brussels.

Hanna, S. R., D. G. Strimaitis, and J. C. Chang. 1991. *Evaluation of Commonly Used Hazardous Gas Dispersion Models,* Final report, contract F08635-89-C-0136, Tyndall AFB, FL.

Hanna, S. R., J. C. Chang, and D. G. Strimaitis. 1993. "Hazardous Gas Model Evaluation with Field Observations," *Atmospheric Environment,* vol. 27a, pp. 2265–2285.

Lazaro, M. A., K. Woodard, S. R. Hanna, D. J. Hesse, J.-C. Huang, J. Lewis, and C. A. Mazzola. 1997. *Model Review and Evaluation for Application in DOE Safety Basis Documentation of Chemical Accidents—Modeling Guidance for Atmospheric Dispersion and Consequence Assessment. Accident Phenomenology and Consequence Assessment (APAC) Methodology Evaluation,* Working Group 6 Report. Argonne National Laboratory, Environmental Assessment Division, Report ANL/EAD/TM-75, NTIS, U.S. Dept. of Commerce, Springfield, VA.

McQuaid, J. 1983. "Observations on the Current Status of Field Experimentation on Heavy Gas Dispersion," in *Atmospheric Dispersion of Heavy Gas and Small Particles,* ed. G. Ooms and H. Tennekes, Springer Verlag, Berlin.

Mercer, A. 1988. "Methods of Validating Models of Dense Gas Dispersion: A Review," in *Stably Stratified Flow and Dense Gas Dispersion, Proceedings of the IMA Conference,* Chester, 1986, ed. J. S. Puttock, pp. 169–203.

Mercer, A., C. Bartholome, B. Carissimo, N. J. Duijm, and H. Giesbrecht. 1998. *CEC Model Evaluation Group, Heavy Gas Dispersion Expert Group, Final Report,* Office for Official Publications of the European Communities, L-2985 Luxembourg, EUR 17778 EN.

Nielsen, M., and S. Ott. 1995. *A Collection of Data from Dense Gas Experiments,* Risø National Laboratory, Report Risø-R-845(EN), Roskilde, Denmark.

CHAPTER 20

MODELING OIL SPILLS ON RIVER SYSTEMS: EVALUATION OF AQUEOUS CONCENTRATIONS

John S. Gulliver and Vaughan R. Voller
Department of Civil Engineering, University of Minnesota, Minneapolis, Minnesota

David E. Hibbs
Barr Engineering Company, Ann Arbor, Michigan

20.1 INTRODUCTION

Accidental releases of buoyant oils, fuels, and chemicals into water bodies can be catastrophic events for the aquatic biota. Riverine spills can be particularly hazardous. Unlike spills onto oceans, in which the water column is subjected to almost an infinite dilution, rivers are shallow and confined by the river bed and banks. Consequently, aqueous concentrations resulting from riverine spills can be several orders of magnitude higher than similar spills onto oceans.

Spills of any sparingly soluble buoyant compound initially form a slick on the water surface. The slick spreads across the water surface and drifts downstream as the various compounds in the slick evaporate and dissolve into the water column. To predict the impact of spills on the river biota, the concentration of compounds dissolved in the water column must be assessed. There are many models that look at the hydrodynamics of the oil slick (Yapa and Shen, 1994). These models focus on the tracking, spreading, and containment of the oil slick and are applicable to wide rivers where the two-dimensionality of the slick is important. The focus of this chapter is not on the spread of the oil slick per se, but on the effect that spread may have on the biota of the river. This requires a detailed tracking of how the aqueous concentrations develop as the slick moved downstream. Modeling the evolution of the aqueous concentrations requires an adequate accounting of the various transport coefficients that control the process. The contribution of this chapter is the identification of these parameters, citations of appropriate literature sources, and, most importantly, a sensitivity analysis that identifies the coefficients that have the greatest impact on simulation results. The model described was developed primarily for application to smaller, nonnavigable rivers where the assumptions of a cross-sectional, well-mixed water body and slick are usually valid. When the cross-sectional uniform water body and oil slick assumptions are not valid, a combination of a two-dimensional flow slick (Yapa et al., 1994; Shen et al.,

1995) and the aqueous transport and transfer processes described in this chapter would be required.

A spill of JP-4 jet fuel is used to illustrate the operation of the model. The rate coefficients used in this example spill are most applicable for smaller rivers, which exhibit a fair degree of meandering and are often sheltered from wind by the riverbanks and vegetation. In keeping with all environmental fate and transport modeling, the predicted aqueous concentrations have an inherent level of uncertainty due to bias in the model input parameters. Since the user-specified rate constants are seldom measured but are themselves often estimated from predictive relationships, these input parameters can often be in significant error. Depending on the particular parameter, the uncertainty associated with each input parameter can have an impact on the predicted aqueous concentrations ranging from drastic to insignificant. Within the context of a JP-4 jet fuel spill into a small river system, sensitivity analysis to identify the model parameters and/or processes that are the most important in determining aqueous concentrations of contaminants is a key component in the testing of the model.

20.2 THE DUAL-PHASE MODEL

The model is established on a mixture of Eulerian and Lagrangian coordinate systems. The river is approximated as a series of completely mixed cells (typically 10–1,000 m in length) fixed in position, as shown in Fig. 20.1. The slick is approximated as a series of completely mixed cells that move across the water surface in a Lagrangian coordinate system. This treatment of the slick as a series of moving cells allows for spatial variation in the concentration of the slick. The application of the model, per se, is to situations where both flow and slick can be described as one-dimensional. This occurs when the slick is spread completely across the river, as with relatively narrow streams. The length of the river from the spill site before the one-dimensional assumption can be applied is approximately:

$$L \cong \frac{L_w U_{\text{slick}}}{dl/dt} \tag{20.1}$$

where U_{slick} is the streamwise velocity of the centroid of the slick, L_w is the largest transverse length across the river from the location of the spill, and dl/dt is the spreading velocity of the slick. If the spill occurred on one bank of the river, L_w would be the river width.

All of the described processes for the evolution of aqueous concentration, however, would apply to both a one-dimensional and a two-dimensional spill model.

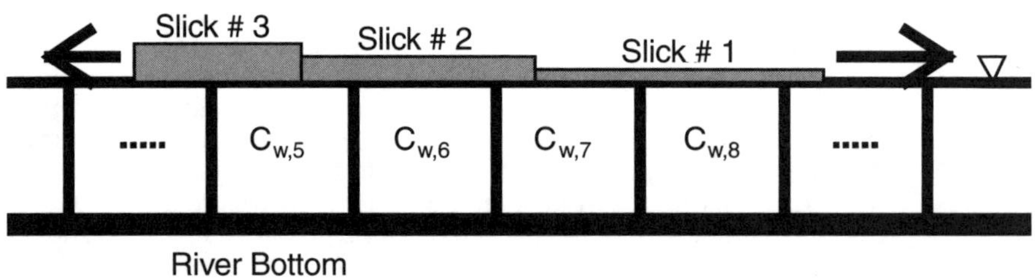

FIGURE 20.1 Numerical approximation of river cross-section with multiple surface slicks.

20.2.1 Movement of the Surface Slick

The movement of the slick on the water is largely a function of the velocity of the water surface and the direction and magnitude of the wind:

$$U_{\text{slick}} = \alpha_{\text{vel}} U + a_{\text{drift}} U_{\text{wind}} \tag{20.2}$$

where U = the cross-sectional mean river velocity
U_{wind} = the streamwise component of the wind speed vector
α_{vel} = the velocity profile correction factor ($\alpha_{\text{vel}} \cong 1.1$ [Addison, 1941])
α_{drift} = the wind drift coefficient ($\alpha_{\text{drift}} \cong 0.03$–$0.04$ [Wu, 1983])

For smaller rivers, which are often tortuous and sheltered, the wind drift term in Eq. (20.2) is dropped. The velocities of the leading and trailing edges of the slick are:

$$U_{\text{leading}} = 1.1\ U + \alpha_{\text{drift}} U_{\text{wind}} + \frac{dl}{dt} \tag{20.3}$$

and

$$U_{\text{trailing}} = 1.1\ U + \alpha_{\text{drift}} U_{\text{wind}} + \frac{dl}{dt} \tag{20.4}$$

where dl/dt is the spreading rate of the slick. While the oil is spilling onto the water surface, the leading edge of the slick is allowed to drift and spread downstream at a velocity determined from Eq. (20.3) and the trailing edge of the slick is assumed to be fixed in position. The oil is added to the trailing end of the overall slick, i.e., into the upstream-most individual slick. This treatment of the spilling process reasonably simulates the behavior of a slick forming on a river. During the early stages of a riverine spill, dl/dt is on the order 0.1 m/s (Fay, 1971). Thus, the spreading velocity is less than typical water surface velocities, and the slick is not likely to spread upstream from the point of spilling. Once the spilling stops, the trailing edge of the slick is released to drift and spread at a velocity determined from Eq. (20.4). The position of the leading and trailing edges of the overall slick are determined from Eqs. (20.3) and (20.4) and the boundaries of the individual slicks are interpolated from the spreading rate of the leading edges based on their distance from the midpoint of the overall slick.

20.2.2 Mass Balance for the Slick Phase

The mass of the slick decreases over time as the compounds dissolve into the water phase and evaporate into the atmosphere. A mass balance for each compound in the slick is written as:

$$\frac{dm}{dt} = \int_{\substack{\text{length}\\\text{of slick}}} \left[-k_{\text{dis}} W(X_o C_w^{\text{sat}} - C_w) - k_{\text{evap}} W \left(\frac{X_o P^o M_i}{RT} \right) \right] dx + r \tag{20.5}$$

where C_w = the aqueous concentration
C_w^{sat} = the aqueous saturation concentration
k_{dis} = the dissolution rate coefficient
k_{evap} = the evaporation rate coefficient
m = the mass of the compound in the slick
M = the compound's molecular weight

r = the release rate (i.e., rate at which the compound is spilled)
P^o = the vapor pressure of the pure compound
R = the universal gas constant
t = time
T = the absolute temperature of the interface
x = distance downstream
X_o = the mole fraction in the slick
W = the width of the slick

This one-dimensional model formulation assumes the slick covers the water surface from bank to bank, therefore W is also the width of the river. The product X_oP^o represents the vapor pressure of a compound at the slick–air interface based on Raoult's law. The mass flux rate is integrated over the length of the slick since W and C_w can vary as a function of distance. The concentration of each compound in the slick can then be expressed as a mole fraction:

$$X_o = \frac{m(M)^{-1}}{\sum_{\text{all compounds}} m(M)^{-1}} \tag{20.6}$$

20.2.3 Transport Equation for Aqueous Phase

The concentrations of compounds dissolved in the water change over time and distance as the compounds dissolve from the slick into the water, volatilize from the water to the atmosphere, and disperse in the river. The one-dimensional advection-diffusion equation for each compound is written as:

$$\frac{\partial(AC_w)}{\partial t} + \frac{\partial(QC_w)}{\partial x} = \frac{\partial}{\partial x}\left(AD_L\frac{\partial C_w}{\partial x}\right) + \phi k_{\text{dis}}W(X_oC_w^{\text{sat}} - C_w) - (1 - \phi)\, k_{\text{vol}}WC_w \tag{20.7}$$

where A = the cross-sectional area of river
D_L = the longitudinal dispersion coefficient
Q = discharge
k_{vol} = the volatilization rate coefficient
ϕ = a phase marker ($\phi = 1$ if there is a slick at location x; $\phi = 0$ otherwise)

When the surface slick is present at location x, there is an exchange between the slick and the water. When the surface slick is not present at location x, there is an exchange between the water and the atmosphere.

The approximation of the river as a series of discrete well-mixed cells introduces additional dispersion into the model. Even if a value of $D_L = 0$ is input, some dispersion will still be predicted by the model. Banks (1974) developed a mixed cell model which may be used to quantify this numerical dispersion:

$$D_{L-\text{num}} = 0.5\ U\ \Delta x \tag{20.8}$$

where Δx is the length of the river control volume. Equation (20.8) is based on an infinite number of cells and is approximately valid as long as roughly 10 or more cells are used in the model (Levenspiel, 1962). Since Eq. (20.7) is often approximated as a linear partial differential equation where superposition applies, the numerical dispersion and the user-specified dispersion are nearly additive. Thus, the value of longitudinal dispersion that should be specified in the model to obtain the true dispersion, D_L, can be estimated from the expression:

$$D_{L-\text{model}} = D_L - D_{L-\text{num}} = D_L - 0.5\ U\ \Delta x \tag{20.9}$$

where $D_{L-\text{model}}$ is the longitudinal dispersion coefficient that is supplied to the model in Eq. (20.8).

20.2.4 Solution Algorithm

Equations (20.3) through (20.7) form the basis of the model. A formulation of Eq. (20.7) is needed for each compound that is to be modeled. A pair of Eqs. (20.5) and (20.6) are needed for each compound and for each individual slick used in the simulation. No direct solution for the equation set is available, so the equations are solved numerically. At each time step, the extent and location of the oil slick is determined from Eqs. (20.3) and (20.4). The mass of each compound in the oil phase is then determined by an explicit time integration of Eq. (20.5), i.e., the values of C_w and X_o are taken from the previous time step. The mole fraction of each compound in the slick is then determined from Eq. (20.6). Next, the aqueous concentration of each compound at each location in the river is determined by a fully implicit time integration of Eq. (20.7), i.e., the current values of C_w and X_o are used in the calculation.

20.3 A SAMPLE APPLICATION

As an example of the capabilities of the model, consider a spill of 10,000 kg of JP-4 jet fuel released over a period of 10 minutes onto a river of the scale typical of a tributary to a navigable river. The approximate composition of the fuel, the physical properties of the compounds, the river flow data, and the model parameters for the simulation are listed in Table 20.1. Most of the jet fuel is composed of several relatively insoluble long-chain aliphatics (Burris and MacIntyre, 1987). For simplicity, these aliphatic compounds are grouped and modeled as a single compound with properties averaged from the individual aliphatic compounds.

The results of the jet fuel spill simulation are shown in Figs. 20.2 through 20.5. Figure 20.2 shows the aqueous concentrations of two compounds, ethylbenzene and 1-methylnaphthalene, plotted versus distance downstream at times of 10, 20, 40, and 60 hours after the spill occurred. The concentration profiles are nearly Gaussian in shape, typical of pulse loadings in rivers. The profiles tend to broaden over time due to the streamwise mixing in the river, approximated by the longitudinal dispersion term in Eq. (20.7). Notice that the concentration profiles are slightly skewed to the downstream direction. Since the slick is drifting downstream at a rate slightly higher than the mean river velocity, the compounds in the slick are constantly dissolving into the river slightly downstream of the aqueous concentration peak, creating a skewed profile.

The peak aqueous concentrations of all eight compounds in the jet fuel are plotted versus time in Fig. 20.3. The peak aqueous concentrations of the more volatile compounds (e.g., toluene and methylcyclohexane) reach maximums within the first 10 hours, then decrease as these volatile compounds rapidly evaporate from the slick and are no longer available to dissolve into the water. The aqueous concentration of each compound continues to rise until that compound has evaporated from the slick. Once a compound has been removed from the slick, the peak aqueous concentration of that compound decreases as its concentration is diluted by longitudinal dispersion and as it volatilizes from the water to the atmosphere. In this example, the least soluble compounds also are least volatile and therefore tend to remain in the slick for a longer time and continue to dissolve into the water column, resulting in

TABLE 20.1 Input Parameters Used in Sample Application

Properties	Toluene	Ethylbenzene	n-Butylbenzene	Tetralin
X (−)[a]	0.04331	0.04785	0.03879	0.02922
MW (g/mol)	92.1	106.2	134.2	132.2
ρ (kg/m^3)[b]	870	870	860	970
C_w^{sat} (mol/L)[a,b]	$10^{-2.50}$	$10^{-2.80}$	$10^{-3.97}$	$10^{-3.47}$
P^o (atm)[b,d]	$10^{-1.42}$	$10^{-1.90}$	$10^{-2.86}$	$10^{-3.27}$
D_{water} (m^s/s)[e]	9.1×10^{-10}	8.2×10^{-10}	7.0×10^{-10}	7.6×10^{-10}
D_{oil} (m^2/s)[f]	1.5×10^{-9}	1.4×10^{-9}	1.1×10^{-9}	1.2×10^{-9}
D_{air} (m^2/s)[g]	8.1×10^{-6}	7.4×10^{-6}	6.4×10^{-6}	6.8×10^{-6}

Properties	1-Methylnaphthalene	1,4-Dimethylnaphthalene	Methylcyclohexane	Aliphatics[h]
X (−)[a]	0.02666	0.02613	.2123	.5759
MW (g/mol)	142.2	156.2	98.2	145.7
ρ (kg/m^3)[b]	1,002	1,000	770	731
C_w^{sat} (mol/L)[b,c]	$10^{-3.67}$	$10^{-4.22}$	$10^{-3.77}$	$10^{-6.52}$
P^o (atm)[b,d]	$10^{-4.07}$	$10^{-4.60}$	$10^{-1.23}$	$10^{-2.77}$
D_{water} (m^s/s)[e]	7.4×10^{-10}	6.8×10^{-10}	8.2×10^{-10}	6.2×10^{-10}
D_{oil} (m^2/s)[f]	1.2×10^{-9}	1.1×10^{-9}	1.4×10^{-9}	1.0×10^{-9}
D_{air} (m^2/s)[g]	6.6×10^{-6}	6.2×10^{-6}	7.7×10^{-6}	5.8×10^{-6}

River properties	Rate constants[i]
$Q = 3$ (m^3/s)	$k_{evap} = 5.27 \times 10^{-4}$ (m/s)
$A = 10$ (m^2)	$k_{dis} = 1.50 \times 10^{-6}$ (m/s)
$W = 10$ (m)	$k_{vol} = 1.20 \times 10^{-5}$ (m/s)
$D_L = 25$ (m^2/s)	**Model parameters**
$U_{wind} = 2$ (m/s)	$\Delta t = 60$ (sec)
$k_{L-oxygen} = 2.47 \times 10^{-6}$ (m/s)	$\Delta x = 100$ (m)
$T = 20°C$	

[a] Burris and MacIntyre, 1987.
[b] Lide, 1993.
[c] Schwarzenbach et al., 1993.
[d] Lyman et al., 1990.
[e] Approximated using Hayduk and Laudie, 1974.
[f] Approximated using Wilke and Chang, 1995.
[g] Approximated using Fuller et al., 1966.
[h] Average properties of the 5 aliphatic compounds in JP-4 fuel, Burris and MacIntyre, 1987.
[i] Approximate values. Actual values determined for each compound based on wind speed, reaeration rate, and physical properties of individual compounds.

higher aqueous concentrations. Consequently, the highest aqueous concentrations resulting from this spill are not of the most soluble compounds, but rather of some of the least soluble and least volatile compounds (e.g., 1-methylnaphthalene and 1,4-dimethylnaphthalene).

Figure 20.4 shows the mass of each compound in the slick as a function of time. Figure 20.5 shows the mole fraction of each compound in the slick plotted versus time. For simplicity, this simulation was run using only one slick. The composition of the slick changes drastically with time. Notice that the evaporation and dissolution flux rates of compounds from the slick, as inferred from the slopes of the lines in Figs. 20.4 and 20.5, increase as a compound's mole fraction in the slick increases. This behavior exemplifies the fact that the dissolution and evaporation rates are governed not only by environmental parameters but also by slick composition. Since the slick composition is constantly changing over time, the concentration of compounds in the slick must be modeled in conjunction with the aqueous concentrations.

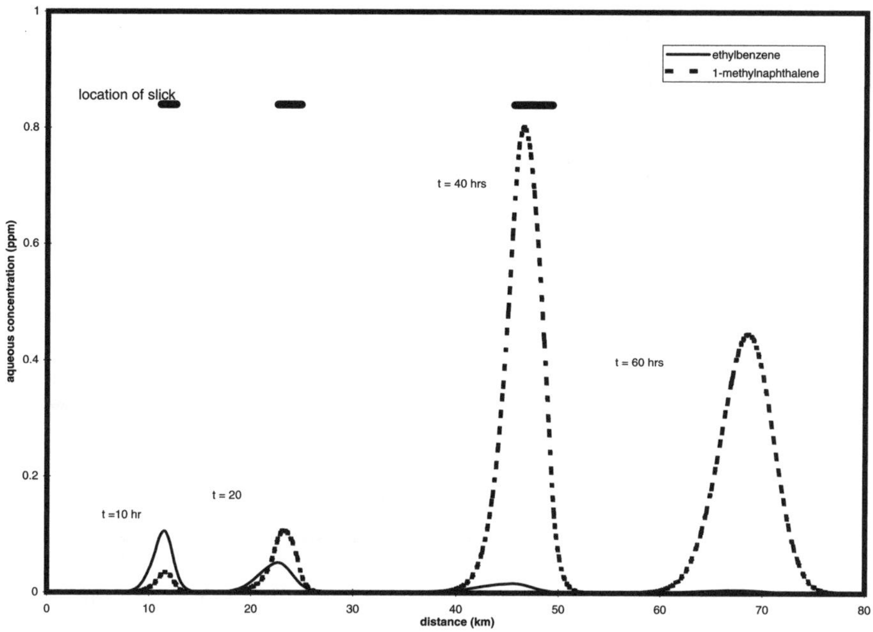

FIGURE 20.2 Results of jet fuel spill simulation: aqueous concentration profiles for ethylbenzene and 1-methylnaphthalene.

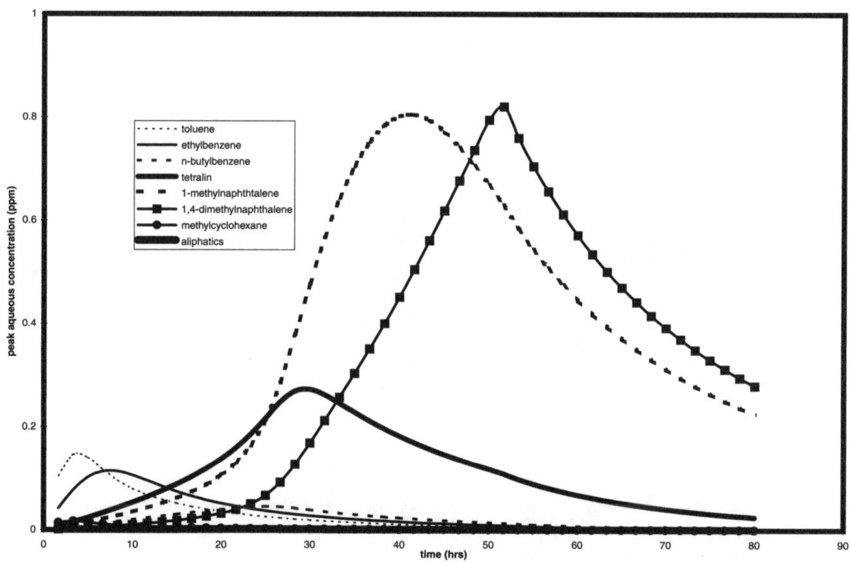

FIGURE 20.3 Results of jet fuel spill simulation: peak aqueous concentrations of compounds.

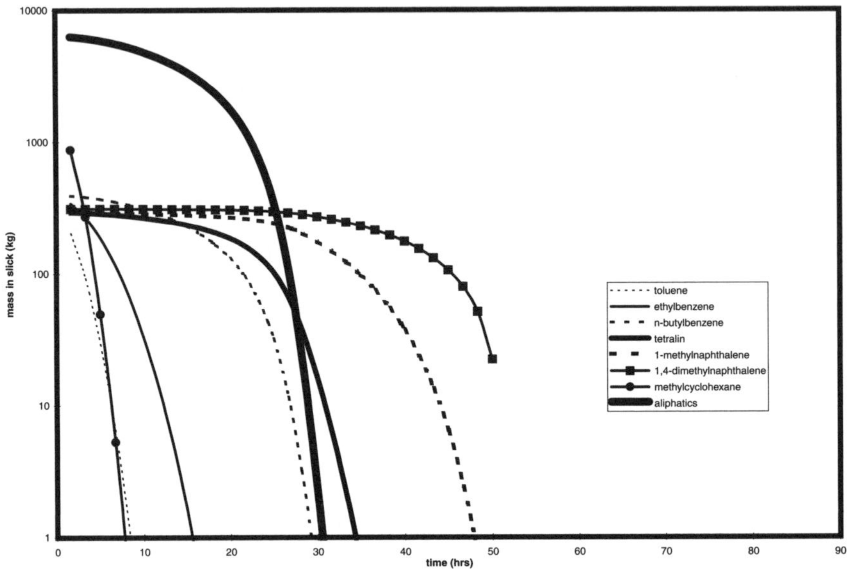

FIGURE 20.4 Results of jet fuel spill simulation: mass of compounds in slick.

As shown in Fig. 20.5, the composition of the slick, and therefore the nature of the slick, can change drastically over the duration of the spill as the more volatile, more soluable, and lower molecular weight compounds evaporate and dissolve from the slick. Most notably, the specific gravity of the slick increases over time. At 30 hours, the specific gravity of the slick reaches 1.0 and the slick should break up into globules dispersed in the water column or

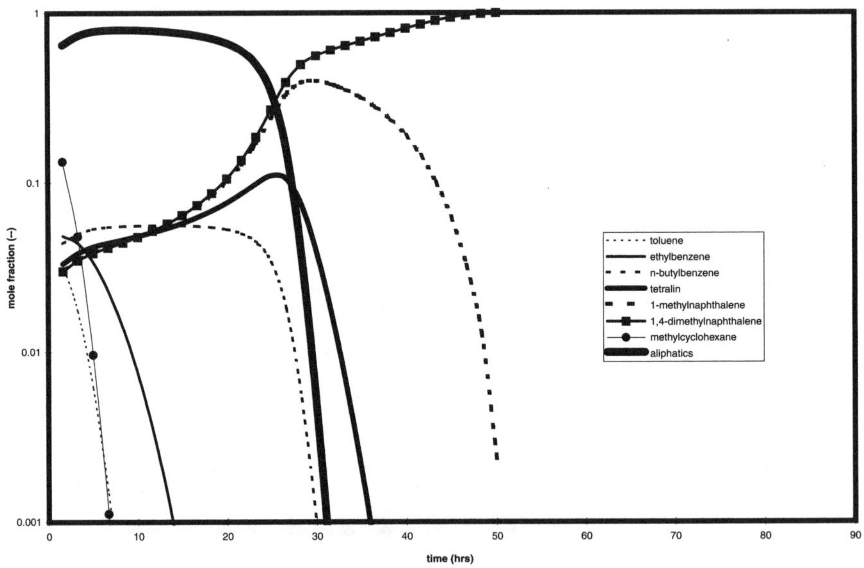

FIGURE 20.5 Results of jet fuel spill simulation: mole fraction of compounds in slick.

sink to the bottom. Even if the slick breaks up, however, the aqueous concentration of the compounds remaining in the oil phase will continue to rise, though perhaps not precisely as shown in Fig. 20.3. If the slick degrades to small globules suspended in the water column, the aqueous concentrations of the compounds in the oil phase will increase more rapidly than predicted in Fig. 20.3 due to the increased surface area of the globules compared to the surface slick. If the globules sink, they will not be drifting downstream suspended in the water column, and therefore the peak concentrations will likely be lower than those predicted by Fig. 20.3 which assumes that the slick is drifting at roughly the same velocity as the river surface. The viscosity of the slick also increases over time, leading to an increase in resistance to transfer in the oil phase. Additionally, the surface tension and the activity coefficients of the compounds in the slick will change over time, altering the spreading rate and the slick–water and slick–air equilibrium partitioning of the compounds. Thus, the slick composition can affect the physical characteristics of the slick and, consequently, the aqueous concentrations resulting from the spill.

20.4 SENSITIVITY ANALYSIS

Using the hypothetical spill of 10,000 kg of a JP-4 jet fuel, the uncertainty associated with each model parameter is estimated and the sensitivity of the predicted aqueous concentrations to variations in the input parameters over the estimated range of uncertainty is presented. As a summary of the main findings, estimates of the uncertainties in the user-supplied parameters and the resulting variations in the highest aqueous concentrations are given in Table 20.2.

20.4.1 Longitudinal Dispersion

Longitudinal dispersion accounts for the dilution of the cross-sectional average concentration of compounds dissolved in the water due to mixing in the streamwise direction. The longitudinal dispersion coefficient, D_L, can be estimated as (Fischer et al., 1979):

$$D_L = \frac{0.011\ U^2 W^2}{h\sqrt{g\,h\,S}} \tag{20.10}$$

where h = the average depth of river
g = acceleration due to gravity
S = the slope of the water surface

TABLE 20.2 Summary of Sensitivity Analysis. Confidence Interval of Input Parameters and Resulting Range of Maximum Aqueous Concentrations are Relative to the Original Values

User-supplied parameter	Approximate uncertainty of input parameter	Resulting range of maximum aqueous concentration
D_L	± Factor of 4	±30%
k_{evap}	± Factor of 4	±50%
k_{vol}	± Factor of 3	±5%
k_{dis}	± Factor of 20	±2000%
C_w^{sat}	± Factor of 2	±200%
dl/dt	± Factor of 2	±10%

Fischer et al. (1979) list numerous predictive equations for longitudinal dispersion, and perhaps not inconsequentially, estimates of D_L from stream parameters are notoriously inaccurate. Fischer et al. state that the recommended relationship, i.e., Eq. (20.10), will generally predict D_L only within a factor of 4.

Figure 20.6 shows the highest aqueous concentrations of four compounds that occur during the duration of the JP-4 jet fuel spill simulation, using the same parameters listed in Table 20.1. The highest aqueous concentrations are plotted as a function of D_L over the range of uncertainty of the predicted value (i.e., from approximately 25 to 400% of the original value of $D_L = 25\ \mathrm{m^2/s}$). Higher values of D_L tend to dilute the dissolved compounds in the streamwise direction, yielding lower maximum aqueous concentrations. Lower values of D_L correspond to less dilution, and therefore higher aqueous concentrations. The maximum aqueous concentration of both the soluble/volatile compounds (toluene and ethylbenzene) and the relatively insoluble/nonvolatile compounds (1-methylnaphthalene and 1,4-dimethylnaphthalene) vary by approximately 30% from the high end value of D_L to the low end. Therefore, longitudinal dispersion is a fairly significant process, and whenever possible, the value of D_L should be measured by performing a dye study on a river reach (Fischer et al., 1979). Dye study measurements can also be used to precisely determine the spatial mean value of U (Kilpatrick and Wilson, 1989).

20.4.2 Evaporation Rate

Evaporation is generally described by a resistance in series model, commonly used to describe mass-transfer between phases. The overall transfer coefficient depends on the rate transfer coefficients on either side of the interface:

$$\frac{1}{k_{\mathrm{evap}}} = \frac{P^o M_i}{k_{oa} R T \rho_o} + \frac{1}{k_G} \tag{20.11}$$

where k_{oa} is the oil-film transfer coefficient at the oil-air interface, k_G is the gas-film transfer

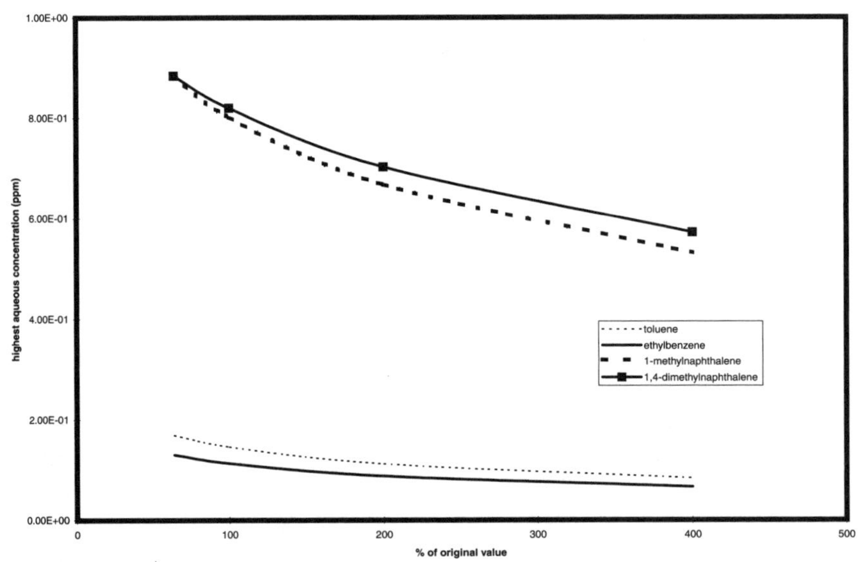

FIGURE 20.6 Sensitivity analysis: effect of longitudinal dispersion on maximum aqueous concentration.

coefficient, and ρ_o is the density of the oil. For thick oil layers, the oil-film transport coefficient controls the evaporation of hydrocarbons due to the relatively high vapor pressures (Thibodeaux and Carver, 1997). However, in most riverine spills, the oil layer is so thin (<1 mm) that the resistance to transfer in the oil film is small, even if transport is assumed to be only by molecular diffusion, and k_{evap} can be approximated by k_G (Mackay and Yeun, 1983).

The gas-side mass transfer coefficient for compound i can be estimated from the wind function coefficient by Ryan et al. (1974), which relates evaporative heat flux to a vapor pressure difference:

$$k_o = f(w_z)\left(\frac{Sc_i}{Sc_{\text{water}}}\right)^{-0.67} \tag{20.12}$$

where k_G is the gas-side mass transfer coefficient for compound i (m/s), $f(w_z)$ is the wind function (m/s). The coefficient, $f(w_z)$, is a weak function of temperature but is nearly constant for the range of temperatures typically seen in rivers. Sc_{water} is the Schmidt number of water in air ($Sc_{\text{water}} = \nu/\text{D}$, where ν is kinematic viscosity of air and D is diffusivity of water in air), and Sc_i is the Schmidt number of compound i in air. The ratio of Schmidt numbers in Eq. (20.12) converts the gas-side mass transfer coefficient for water vapor to an equivalent coefficient for compound i.

The -0.67 power on the ratio of diffusivities has been used by several investigators, including Mackay and Yeun (1983), Mackay and Matsugu (1973), and Goodwin et al. (1976), and is generally associated with a "dirty" or fixed surface. The function $f(w_z)$ is used to describe the influence of natural convection, wind, and waves on evaporation. The two known sets of field experiments to characterize the wind function in a sheltered stream resulted in the relationships of Gulliver and Stefan (1986) and Jobson and Keefer (1979). Gulliver and Stefan's relation is used herein:

$$f(w_z) = [8.57\ U_9 + 14.7\ (\Delta\theta)^{1/3}] \times 10^{-4} \tag{20.13}$$

where U_9 is the wind speed at a height of 9 m (m/s), and $\Delta\theta$ is the virtual temperature difference (Ryan et al., 1974) between the water surface and the air at a height of 2 m (°C).

Many existing spill models use a relationship for k_G developed by Mackay and Matsugu (1973) from pan evaporation and wind tunnel experiments. Values of k_G used in this analysis are based on measurements taken in sheltered streams and are typically two to five times lower than those for spills in open areas determined by Mackay and Matsugu. Regardless of the relationship used to predict the value of k_G, there is likely a large uncertainty due to its dependence on the local air turbulence level. Under similar wind velocities at a height of 10 m, the wind velocity profile over open water may be significantly different than the velocity profile over a river due to the topography and the vegetation on the river banks. Even on open lakes, the measured evaporation transfer coefficient can vary as much as $\pm 100\%$ for a given wind speed (Adams et al., 1990). Thus, evaporation is a complicated and site-specific process. The confidence interval of the predicted evaporation-rate constant for rivers is estimated to be plus or minus a factor of 4. The confidence interval for the value of k_G is likely smaller for wide rivers in areas without significant vegetation or changes in topography.

Figure 20.7 shows the highest aqueous concentrations of four compounds resulting from a series of JP-4 jet fuel spill simulations over the anticipated range of uncertainty in k_G. The maximum aqueous concentrations range from 130 to 60% of the original value as the value of k_G is varied from 50 to 400% of the original value. The value of k_G indirectly affects the aqueous concentrations by forcing changes in the composition of the slick. At low values of k_G, the compounds evaporate more slowly and remain in the slick for a longer time. In the slick, compounds continue to dissolve, increasing the aqueous concentration. At high values of k_G, the compounds evaporate more quickly and are available to dissolve for less time, creating lower aqueous concentrations.

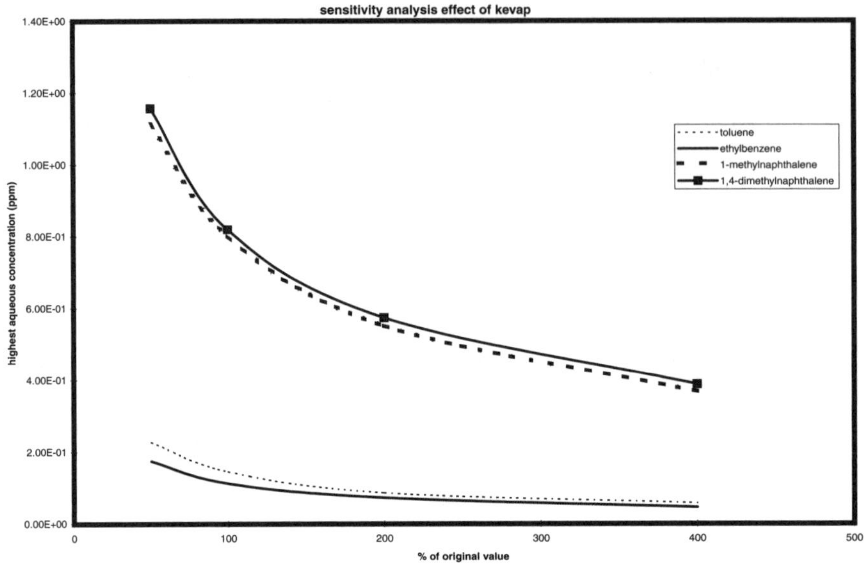

FIGURE 20.7 Sensitivity analysis: effect of variations in evaporation rate on maximum aqueous concentration.

20.4.3 Volatilization Rate

Volatilization from the water column (rather than from the slick) is also generally described by a two-resistance model:

$$\frac{1}{k_{\text{vol}}} = \frac{1}{k_L} + \frac{R\,T}{H\,k_G} \tag{20.14}$$

where k_L is the liquid-side mass transfer coefficient for compound i and k_G is determined from Eq. (20.12). Values of k_L used in the model are adapted from the stream reaeration coefficient of Cadwallader and McDonnell (1969):

$$k_{L\,\text{oxygen}} = 2.15 \times 10^{-3}\,(SU)^{0.5} \tag{20.15}$$

where $k_{L\,\text{oxygen}}$ = the liquid mass transfer coefficient for oxygen at 20°C (m/s)
S = the average slope of river
U = the average streamwise velocity (m/s)

Several empirical relationships for reaeration coefficients were recently reviewed and tested by Moog and Jirka (1995), who found that the form of the relationship given in Eq. (20.15) best characterized stream reaeration rates. A liquid mass transfer coefficient for each compound, $k_{L\,i}$, can then be determined from a ratio of the Schmidt numbers (Mackay and Yven, 1983):

$$k_{Li} = k_{L\,\text{oxygen}} \left(\frac{Sc_i}{Sc_{\text{oxygen}}}\right)^{-0.5} \tag{20.16}$$

where $Sc_{\text{oxygen}} = \nu_{20}/D_{\text{oxygen}}$, $Sc_i = \nu/D_i$, ν is the kinematic viscosity of water at the modeled temperature, ν_{20} is the kinematic viscosity of water at 20°C, D_{oxygen} is the diffusivity of oxygen in water, and D_i is the diffusivity of component i in water.

Estimates of volatilization rates from stream parameters are also notoriously inaccurate. Moog and Jirka (1995) found that even the best empirical relationships have a 95% confidence interval of greater than a factor of 3. Values of k_{vol} for a given river reach can be measured by conducting a gas tracer study when greater accuracy is desired (Kilpatrick et al., 1989; Hibbs et al., 1998).

Figure 20.8 shows the highest aqueous concentrations resulting from a series of JP-4 jet fuel spill simulations over the anticipated range of uncertainty in the value of k_{vol}. The highest aqueous concentrations of all the compounds are essentially unaffected by variations in the value of k_{vol} from 50 to 300% of its original value. The insensitivity of the aqueous concentrations to variations in k_{vol} can be explained by examining the mass balance of the compounds dissolved in the water. From Eq. (20.7), the aqueous concentration of a compound is a function of both the dissolution flux rate and the volatilization flux rate. Even though the value of k_{vol} is typically of the same order as k_{dis} or larger, the concentration difference driving the mass transfer is usually much greater for dissolution than it is for volatilization. Due to the extremely low concentrations of most compounds dissolved in the water after a spill, the volatilization flux is typically very small, and the aqueous concentrations are fairly insensitive to variations in the volatilization rate.

20.4.4 Dissolution Rate

Like evaporation and volatilization, dissolution is also commonly described by a two-resistance model:

$$\frac{1}{k_{\text{dis}}} = \frac{1}{k_w} + \frac{C_w^{\text{sat}} M_o}{k_o \rho_o} \tag{20.17}$$

where k_w is the water film transfer coefficient, k_o is the oil (slick) film transfer coefficient,

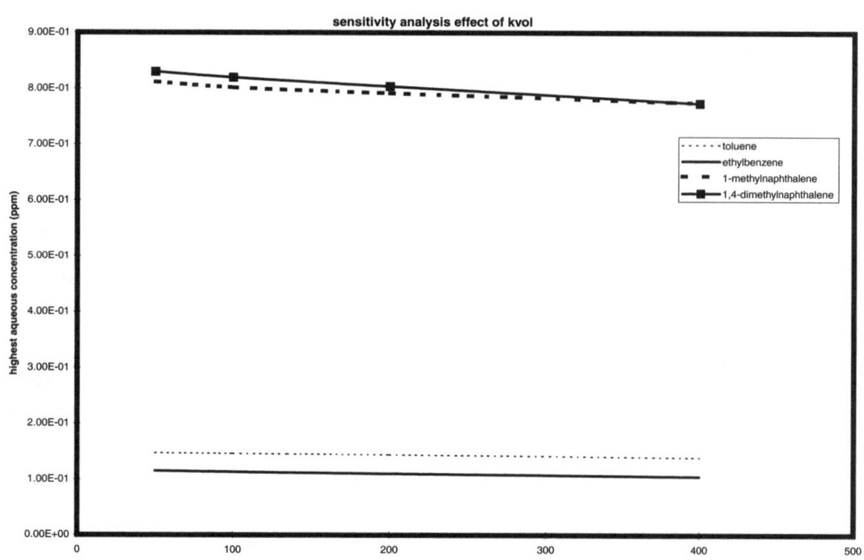

FIGURE 20.8 Sensitivity analysis: effect of variations in volatilization rate on maximum aqueous concentration.

ρ_o is the density of the slick, and M_o is the average molecular weight of the slick. In most instances, the resistance to transfer in the oil film can be ignored due to the small value of C_w^{sat} for most hydrocarbons, such that $k_{dis} \approx k_w$. Values of k_o and k_w can be estimated from reaeration rates. Cohen et al. (1980, 1978) measured dissolution rates of phenol from a surface slick and volatilization rates of benzene from the water to the atmosphere in a wind tunnel at wind speeds of 0 to 9 m/s. The measured water-film volatilization rate coefficients of benzene, $k_{L\ \text{benzene}}$, were found to be roughly 8 times the water-film dissolution rate coefficients for phenol, $k_{w\ \text{phenol}}$, at all but the highest wind speeds when it was thought that the oil slick had a significant dampening effect on the turbulence at the oil-water interface:

$$k_{L\ \text{benzene}} = 8\ k_{w\ \text{phenol}} \tag{20.18}$$

Since the molecular diffusivities of phenol and benzene in water are similar, the rate coefficients for phenol and benzene are nearly interchangeable:

$$k_{L\ \text{benzene}} = 8\ k_{w\ \text{benzene}} \tag{20.19}$$

Values of k_w can then be estimated from values of $k_{L\ \text{oxygen}}$ by accounting for the differences in molecular diffusivities between benzene and oxygen using Eq. (20.16):

$$k_{w\ \text{benzene}} = \frac{k_{L\ \text{oxygen}}}{8} \left(\frac{Sc_{\text{benzene}}}{Sc_{\text{oxygen}}} \right)^{-0.5} \tag{20.20}$$

More generally, the water-film dissolution rate coefficient for any compound, i, can be estimated from the reaeration rate coefficient, again by correcting for the differences in molecular diffusivities:

$$k_{wi} = \frac{k_{L\ \text{oxygen}}}{8} \left(\frac{Sc_i}{Sc_{\text{oxygen}}} \right)^{-0.5} \tag{20.21}$$

Cohen et al. (1980, 1978) also found the measured water-film volatilization rate coefficients of benzene to be roughly 20 times the measured water-film dissolution rate coefficients for phenol. Similarly, the oil-film dissolution rate coefficient for any compound, i, can be estimated as:

$$k_{oi} = \frac{k_{L\ \text{oxygen}}}{20} \left(\frac{Sc_i}{Sc_{\text{oxygen}}} \right)^{-0.5} \tag{20.22}$$

Values of $k_{L\ \text{oxygen}}$ can be obtained indirectly from stream parameters using Eq. (20.13), or more directly from volatilization measurements by conducting a gas tracer study (Kilpatrick et al., 1989; Hibbs et al., 1998).

Hibbs and Gulliver (1999) recently found that when the turbulence responsible for mass transfer at the water surface is generated from the bottom, such as in a stream or in a stirred reactor, the near-surface turbulence that dominates the mass transfer process on the water side of the interface is essentially unaffected by the presence and/or the properties of an oil slick. Thus, the water-film dissolution rate coefficient can be estimated directly from the air–water liquid–film coefficient, corrected by a ratio of Schmidt number to the $-\frac{1}{2}$ power.

$$K_{wi} = K_{L\ \text{oxygen}} \left(\frac{Sc_i}{Sc_{\text{oxygen}}} \right)^{-0.5} \tag{20.23}$$

Equation (20.23) was also verified with the stirred beaker data of Southworth et al. (1983).

The laboratory results of Southworth et al. (1983) and Hibbs and Gulliver (1999), without wind, are much different than the wind-influenced results of Cohen et al. (1978), and thus

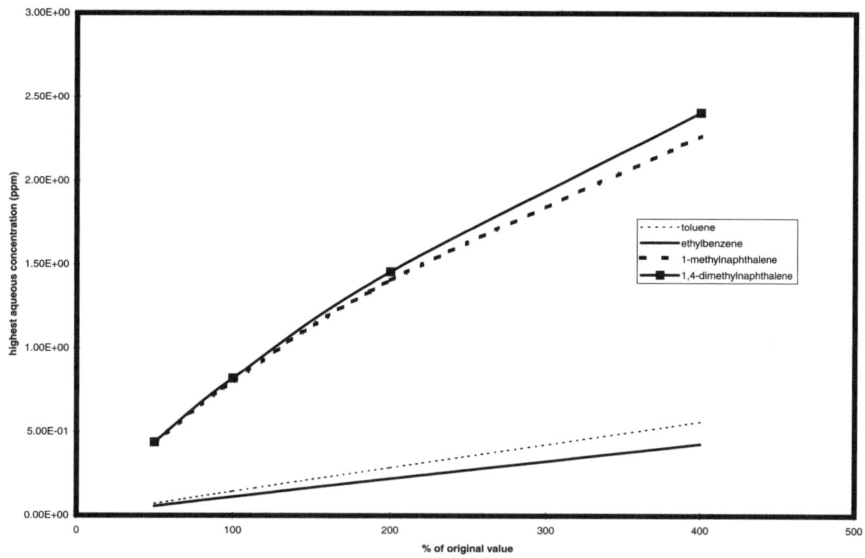

FIGURE 20.9 Sensitivity analysis: effect of variations in dissolution rate on maximum aqueous concentration.

there is no definitive relationship to use in the estimation of k_{dis}. Additionally, there are no known field studies from which to estimate the uncertainty in the value of k_{dis} on rivers. Many existing oil spill models assume constant values for the film coefficients, even though Lamont and Scott (1970) have shown that the values of the individual film coefficients depend on the level of turbulence on either side of the oil–water interface. Shen and Yapa (1988) used $k_{dis} = 1.0$ cm/hr for all applications in their riverine oil spill model. Herbes and Yeh (1985) used values of $k_o = 0.5$ cm/hr and $k_w = 2.0$ cm/hr for navigable rivers, which for most compounds gives $k_{\text{dis}} \cong 2.0$ cm/hr. The relationships used in this analysis are based on laboratory experiments relating k_{dis} to k_{vol}, but the dependence of k_{dis} on k_{vol} varies by more than a factor of 8, depending on the importance of the wind. As previously mentioned, predictive relationships for reaeration coefficients are usually only accurate to within a factor of 3. Therefore, the overall uncertainty in the estimation of k_{dis} is a factor of 3 times a factor of 8, or roughly a whopping factor of 20 to 30. When wind is not a factor in the reaeration coefficient, the overall uncertainty of plus or minus a factor of 3 is more appropriate.

Figure 20.9 shows the highest aqueous concentration resulting from a series of JP-4 jet fuel spill simulations with the value of k_{dis} varied from 50 to 300% of its original value. The slopes of the lines in Fig. 20.9 are nearly 1:1. Thus, over the estimated range of uncertainty in the value of k_{dis}, the maximum aqueous concentrations can be expected to range from roughly 5 to 2000% of their original values, or over several orders of magnitude. This presents a serious problem for spill modelers.

20.4.5 Saturation Concentration

Measured values of the aqueous solubility of pure compounds are available in the literature for most hydrocarbons of interest in the environment (Schwarzenbach et al., 1993). However, solubilities of compounds not available in the literature must be estimated. Depending on the compound, empirical relationships based on molecular structure can predict the aqueous

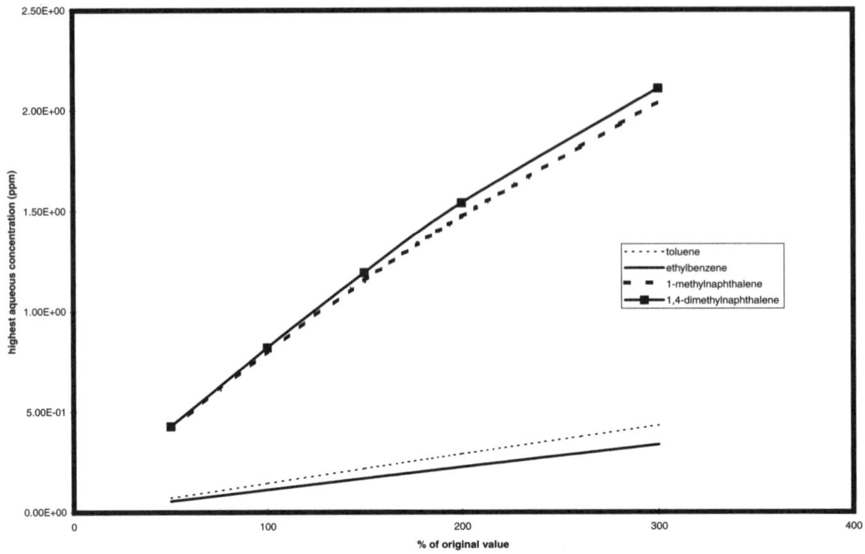

FIGURE 20.10 Sensitivity analysis: effect of variations in saturation concentration on maximum aqueous concentration.

solubility within ±25% to ±300% (Lymann et al., 1990). Additionally, for mixtures of structurally dissimilar compounds, Burris and MacIntyre (1987) have shown that the aqueous solubilities can be 1.5 to 2.5 times higher due to increased activities in the hydrocarbon phase. In this analysis, the uncertainty in C_w^{sat} is assumed to be plus or minus a factor of 2.

Figure 20.10 shows the highest aqueous concentrations resulting from a series of JP-4 jet fuel spill simulations over the anticipated range of uncertainty in the values of C_w^{sat}. The slopes of the lines in Fig. 20.10 are nearly 1:1. As the values of C_w^{sat} are varied from 50 to 200% of the original values, the resulting maximum aqueous concentrations range from roughly 50 to 200% of their original values. Thus, the aqueous concentrations of all compounds are extremely sensitive to variations in the saturation concentration.

20.4.6 Slick Spreading Velocity

Oils spread across a water surface due to differences in density and surface tension between the slick and the water. Spreading is resisted by inertial and viscous forces. Except during the initial minutes following an extremely large spill, the extent of the slick due to spreading can be determined by the larger of an expression equating the surface tension and viscous forces (Fay, 1971):

$$l = k_{1_t} \left(\frac{\sigma^2 t^3}{\rho^2 \nu} \right)^{0.25} \tag{20.24}$$

or an expression equating the gravitational and viscous forces (Fay, 1971):

$$l = k_{1_\nu} \left(\frac{\Delta\, g\, V^2\, t^{1.5}}{\nu^{0.5}} \right)^{0.25} \tag{20.25}$$

where l = the length of slick resulting from spreading in one direction
k_{1t} = the spreading law coefficient for surface tension spreading (k_{1t} = 1.33 [Garrett and Barger, 1970])
$k_{1\nu}$ = the spreading law coefficient for viscous spreading ($k_{1\nu}$ = 1.5 [Fay, 1971])
σ = the net surface tension ($\sigma = \sigma_{\text{air-oil}} + \sigma_{\text{oil-water}} - \sigma_{\text{air-water}}$)
ρ = the density of water, ν is the kinematic viscosity of water
Δ = the ratio of density difference between water and oil to density of water
g = gravity, t is time
V = the volume of slick per unit length normal to the direction of spreading (per unit width of the river)

Equations (20.24) and (20.25) were developed for spills of constant volume, constant surface tension, and low viscosity on calm water. The effects of wind and currents on spreading rates are not well studied and are difficult to estimate. Therefore, the quantifiable uncertainty in the spreading rate lies in the estimation of the parameters used in Eqs. (20.24) and (20.25). The transition from a viscous spread, i.e., Eq. (20.25) to a surface tension spread, i.e., Eq. (20.23) occurs rapidly for most spills, and the spreading rate is described by Eq. (20.24). Since the density and viscosity of water can be estimated fairly confidently, most of the uncertainty in the spreading rate lies in the estimation of the net surface tension, specifically in the estimation of the air–oil surface tension and the oil–water surface tension. There is also an uncertainty in the applications of the slick-spreading model to a cross-sectional nonuniform velocity profile, where the nonuniformities would add to the spreading. In this case, the slick would experience a longitudinal dispersion in addition to the water. This phenomenon is not a component of the sensitivity analysis.

There are two sources of uncertainty in estimating an interfacial surface tension for an oil slick. First, there can be a substantial error in estimating the initial value of the surface tension. For most organic compounds, the air–oil surface tension can be predicted within $\pm 20\%$ and the oil–water surface tension can be predicted within $\pm 50\%$ (Lyman et al., 1990). Second, the surface tension of a slick changes over time as the volatile and more soluble compounds selectively evaporate and dissolve from the slick. However, the range of surface tensions for most organic compounds is fairly narrow. Values for the air–oil surface tension for most organics range from approximately 0.02 to 0.04 N/m, and for oil–water surface tension from approximately 0.020 to 0.035 N/m (Lyman et al., 1990). This indicates that the uncertainty in the interfacial surface tension, and also in the overall spreading rate, is roughly a factor of 2.

Figure 20.11 shows the highest aqueous concentrations resulting from a series of JP-4 jet fuel spill simulations over the anticipated range of uncertainty in the values of the spreading rate. Over the range of spreading rates simulated, the highest aqueous concentrations of all compounds are fairly insensitive to the spreading rates. Variations in the spreading rate ranging from 50 to 200% of its original value produce changes in the maximum aqueous concentration ranging from 90 to 110% of their original values. Higher spreading rates create a larger slick and more surface area for both dissolution and evaporation. A large rate of evaporation would tend to produce lower aqueous concentrations since the slick would be on the water surface for a shorter time. However, a larger dissolution rate would tend to produce higher aqueous concentrations. It is believed that effect of slick size on dissolution and evaporation counteract each other, producing little net change in the highest aqueous concentration due to variations in the spreading rate.

20.5 SPATIAL VARIATIONS IN SLICK COMPOSITION

The preceding application was modeled using only one slick, and thus it was assumed that the concentration of compounds within the slick was uniform over the entire length of the

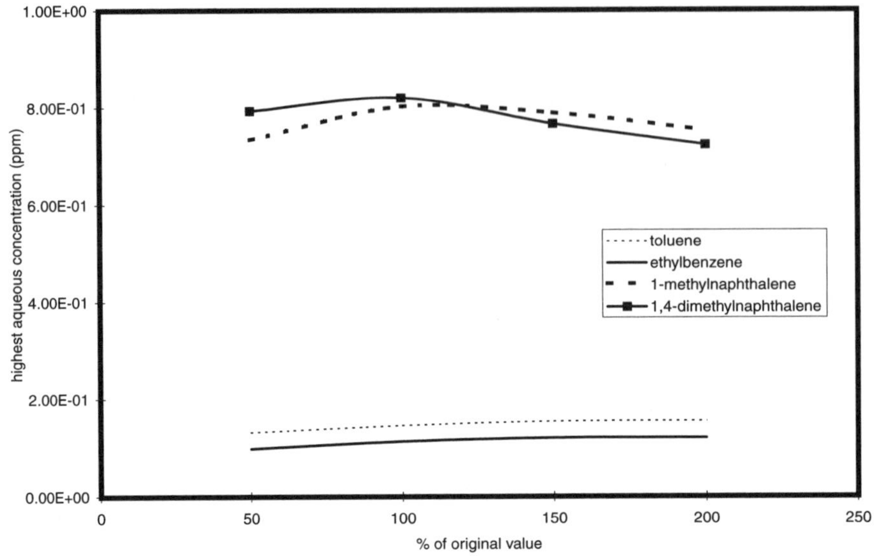

FIGURE 20.11 Sensitivity analysis: effect of variations in spreading rate on maximum aqueous concentration.

slick. However, for slicks that are several kilometers in length or that are being fed slowly by a point source such as a leaky pipe, the concentration of compounds within the slick may in fact vary with distance downstream. Since the aqueous concentration is controlled largely by the concentration within the overlying slick, the assumption of a uniform concentration over the entire length of the surface slick may introduce significant errors in the prediction of the aqueous concentration.

Assuming that the evaporation rate coefficient, k_{evap}, is constant over the entire length of the slick, a streamwise variation in the concentrations of compounds within the slick can be caused in only two ways. First, a variation in slick composition could be caused by spilling a mixture of compounds into one end of a slick while compounds rapidly and selectively evaporate and dissolve from the slick. In such a case, the composition of the slick near the spilling point would resemble the composition of the spilled product, while the portions of the slick farther downstream from the spilling point could contain less of the more volatile compounds. Second, a variation in slick composition may be caused by variations in the aqueous concentration of a compound beneath the slick. If the aqueous concentration beneath the slick varies significantly from one end of the slick to the other, the dissolution rate, which is driven by the difference between the actual water concentration and the equilibrium concentration in the water, would also vary from one end of the slick to the other. Over time, the variation in dissolution rate could create a variation in slick composition.

To investigate the conditions under which the composition could vary from one end of the slick to the other due to rapid evaporation, consider the following analysis of a binary spill. Let compound 1 be volatile and semisoluble, and let compound 2 be perfectly nonvolatile and insoluble. The mass balance for compound 1 in the farthest downstream end of the slick can be simplified from Eqs. (20.5) and (20.6) by assuming that the evaporative flux is much greater than the dissolution flux, and by setting $M_1 = M_2$:

$$\frac{dm_1}{dt} = -k_{evap} \frac{m_1}{m_1 + m_2} \frac{p_1^o M_1}{R\,T} \tag{20.26}$$

where t is the time since the product was spilled onto the water surface. Integrating Eq.

(20.26) from $t = 0$ to t_s and from $m_1 = m_1^o$ to m_1, the time to achieve a specified reduction in the initial, or the spilled, concentration of compound 1 in the slick can be estimated as:

$$t_s = \frac{R\,T}{P_1^o\,M_1\,k_{\text{evap}}}\left[(m_1^o - m_1) - m_2 \ln \frac{m_1}{m_1^o}\right] \tag{20.27}$$

Thus, larger variations in the concentration of a compound within a slick (as indicated by m_1/m_1^o) will occur when the duration of the spilling period is long, i.e., t_s is large, when one of the compounds evaporates quickly, i.e., $P_1^o \times k_G$ is large, or when the slick is thin, i.e., m_1 and m_2 are small. Inserting values of $m_1^o = 100$ g/m^2 and $m_2 = 900$ g/m^2, i.e., slick thickness approx. = 1 mm, $k_{\text{evap}} = 5 \times 10^{-4}$ m/s, $P_1^o = 10^{-1}$ atm, and $M_1 = 100$ g/mol into Eq. (20.27), the time to achieve a 10% reduction in the concentration of compound 1 in the downstream end of the slick is on the order of $t_s = 10$ min.

The second possible cause of streamwise variations in concentrations of compounds in the slick could be the variation in dissolution rates due to streamwise gradients in the aqueous concentration beneath the slick. Hibbs et al. (1999a), however, have shown that significant streamwise concentration gradients within the slick are not likely to be caused by concentration-driven variations in dissolution fluxes. Thus, spatial variations in slick composition are likely due to the rapid evaporation of volatile compounds from the slick.

20.5.1 Simulation of Short-Duration Spill

The results of a series of simulations of the jet fuel spill using the multiple slicks are consistent with the above analysis. The concentrations of the more volatile compounds in the downstream end of the slick decrease only slightly during the 10 minutes that the jet fuel is being spilled onto the water. Figure 20.12 shows the maximum aqueous concentration of four compounds resulting from a 10-minute spill plotted as a function of the number of

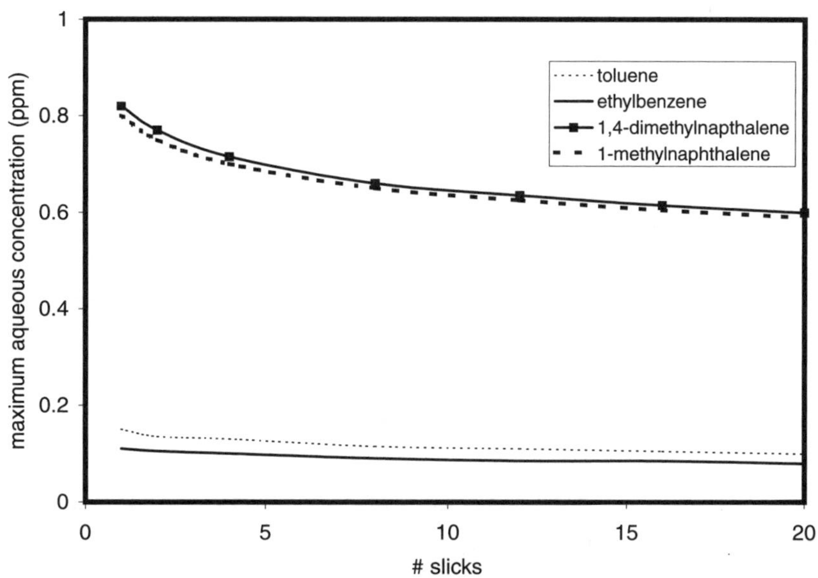

FIGURE 20.12 Multiple-slick model: effect of the number of slicks on the maximum aqueous concentration for a release of 10,000 kg of jet fuel over 10 min.

slicks used in the model. As more slicks are used in the model, larger concentration gradients develop in the slick and the maximum aqueous concentration decreases.

The variations in the slick concentration that occur while the slick is being formed affect not only the maximum highest aqueous concentrations resulting from the spill, but also the location and time at which the maximum aqueous concentration occurs after a spill. Figure 20.13 shows the model simulation results for a spill of 10,000 kg of jet fuel over a period of 10 minutes. The peak aqueous concentrations of ethylbenzene and 1-methylnaphthalene are plotted against time for simulations using the single-slick model and a 10-slick model. Similar to what was shown in Fig. 20.12, the peak concentrations of both compounds are roughly 25% lower when 10 slicks were used than when only 1 slick was used in the simulation. Also, the highest aqueous concentration occurs roughly 25% earlier when 10 slicks are used in the simulation than when only 1 slick is used. The decrease in the maximum aqueous concentration and the shift in the time of occurrence of the maximum aqueous concentration can be attributed to the selective evaporation of the more volatile compounds over the length of the slick, during the period that the fuel is being spilled onto the water surface. We were surprised that the results of the single-slick model and the multi-slick model differed for a spill of such short duration. We thus investigated the mechanisms in some detail.

Using the 10-slick model for this 10-minute spill, each individual slick has fuel added to it for 60 seconds. The slick then drifts downstream while the more volatile compounds rapidly evaporate, creating a concentration gradient in the overall slick. The effective length of the overall slick for dissolution of the volatile compounds becomes shorter as the more volatile compounds evaporate from the downstream end of the slick. At roughly six hours, the effective length of the slick has become so small that the dispersion in the river dilutes any further dissolution from the slick, and the peak aqueous concentration of ethylbenzene begins to decline. In the single-slick model, as the concentration of ethylbenzene is uniform throughout the slick, it has a long effective length for dissolution. The peak aqueous concentration is able to increase until roughly eight hours, after which the dispersion in the river overpowers the declining dissolution flux.

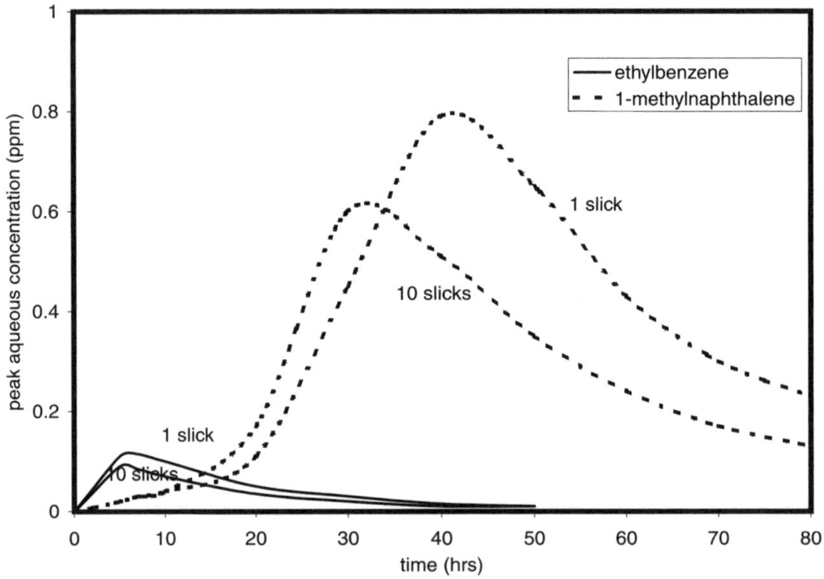

FIGURE 20.13 Multiple-slick model: peak aqueous concentrations resulting from a release of 10,000 kg of jet fuel over 10 min.

While the effective length of the slick for the volatile compounds is decreasing, the effective length for dissolution of the nonvolatile compounds is increasing. The peak aqueous concentration of 1-methylnaphthalene in the 10-slick model shown in Fig. 20.13 rises slightly faster than the peak aqueous concentration of the single-slick model. The 1-methylnaphthalene dissolves and volatilizes faster from the effectively longer 10-slick model than from the single-slick model. The 1-methylnaphthalene is thus completely gone from the slick faster in the 10-slick model than in the single slick model, resulting in a lower maximum aqueous concentration.

20.5.2 Simulation of Long-duration Spill

The same selective evaporation phenomena illustrated in Figs. 20.12 and 20.13 can be seen in a similar plot of peak aqueous concentrations resulting from a 10-hour release of 10,000 kg of jet fuel, shown in Figs. 20.14 and 20.15. Figure 20.14 shows the maximum aqueous concentration of four compounds plotted as a function of the number of slicks used in the model. As in the 10-minute spill simulation, the highest aqueous concentration of all compounds decreases as the number of slicks in the model increases.

Figure 20.15 shows peak aqueous concentrations of ethylbenzene and 1-methylnaphthalene plotted against time for simulations using the single-slick model and a 10-slick model. In the 10-minute release shown in Fig. 20.13, each slick in the 10-slick model was filled with jet fuel for 1 minute before being released. In the 10-hour spill simulation shown in Fig. 20.15, each slick is also filled with jet fuel for 1 hour before being released. During that 1-hour time period, a significant portion of the more volatile compounds evaporates from the slick. By the time the slick is released to drift downstream, most of the volatile compounds have already evaporated from the slick. Any remaining dissolution of the more volatile compounds is diluted by dispersion. Consequently, the maximum aqueous concentration of ethylbenzene occurs very early after the spill. With the single-slick model, new jet fuel is uniformly distributed over the entire slick during the 10-hour spilling period.

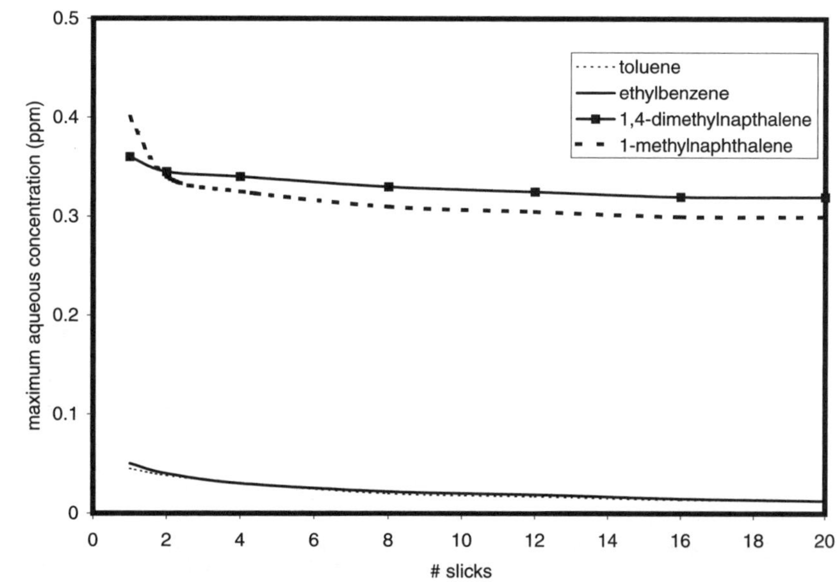

FIGURE 20.14 Multiple-slick model: effect of the number of slicks on the maximum aqueous concentration for a release of 10,000 kg of jet fuel over 10 h.

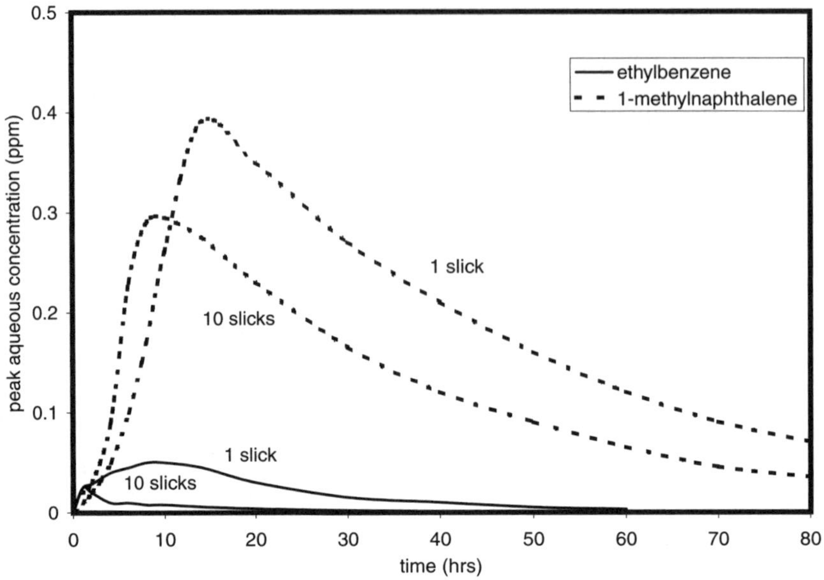

FIGURE 20.15 Multiple-slick model: peak aqueous concentrations resulting from a release of 10,000 kg of jet fuel over 10 h.

Consequently, the peak aqueous concentration of ethylbenzene continues to rise for 10 hours, after which the peak concentration is diluted by dispersion in the river.

The maximum aqueous concentrations shown in Figs. 20.12 and 20.14 appear to decrease logarithmically as a function of the number of slicks used in the model, suggesting that there is a diminishing rate of return on the quality of the model prediction with the increasing number of slicks used in the model. The decrease of the maximum aqueous concentration for each compound can be described by the empirical equation:

$$\frac{dC_{\max}}{dn} = \beta(C_{\max} - C_{\infty}) \tag{20.28}$$

where $C_{\max}$ = the maximum aqueous for a given number of slicks
n = the number of slicks used in the simulation, β is a fitted constant
C_{∞} = a fitted constant that represents the maximum aqueous concentration for an infinite number of slicks

Table 20.3 shows values of β and C_{∞} for four compounds determined from regressions of Eq. (20.28) against the data shown in Figs. 20.12 and 20.14. Table 20.3 also shows the maximum aqueous concentrations plotted in Figs. 20.12 and 20.14 as a percentage of C_{∞}. For the 10-minute release simulation, the concentrations of all four compounds were within 50% of C_{∞} using only 1 slick, within 10% of C_{∞} using 10 slicks, and within 1% of C_{∞} when using 20 slicks. For the 10-hour release simulation, the maximum aqueous concentrations of the volatile compounds varied from nearly four times the value of C_{∞} using only 1 slick to within 5% using 20 slicks, while the range of variation of the maximum aqueous concentrations of the nonvolatile compounds was similar to that of the 10-minute release simulation. Thus, the number of slicks used in the simulation can have a significant impact on the predicted aqueous concentrations, especially in models of slow, continuous spills containing highly volatile compounds.

TABLE 20.3 Extrapolation of C_{max} to an Infinite Number of Slicks. C_{max} is Determined from Spill Simulations. C_{∞} and β are Determined from a Curve Fit of C_{max} Versus the Number of Slicks Using Eq. (20.28).

	C_{∞} (ppm)	β	C_{max} as a percentage of C_{∞} (i.e., $C_{max}/C_{\infty} \times 100$) 1 slick	2 slicks	5 slicks	10 slicks	20 slicks
10-minute release							
Toluene	0.103	−0.193	142	133	120	108	101
Ethylbenzene	0.081	−0.195	140	130	119	107	101
1-Methylnaphthalene	0.585	−0.194	138	128	118	108	100
1,4-Dimethylnaphthalene	0.603	−0.213	136	127	116	107	100
10-hour release							
Toluene	0.012	−0.214	386	325	224	141	105
Ethylbenzene	0.015	−0.253	337	283	181	135	96
1-Methylnaphthalene	0.297	−0.889	134	113	105	100	97
1,4-Dimethylnaphthalene	0.314	−0.312	114	108	105	101	99

20.6 CONCLUSIONS

Oil spill modeling needs to be performed *before* a spill occurs as part of a planning process for emergency response. This avoids the "chicken-with-its-head-cut-off" look that is often apparent in spill response teams. There will not be sufficient time following a spill to collect appropriate data and bring a model up to speed. Oil spill models are also valuable in a forensic investigation following an oil spill event.

In rivers and streams, the largest impact of an oil spill is often the aqueous concentrations that result from dissolution of the slick. The water cannot leave the slick through mixing with a surrounding water body, and aqueous concentrations of toxic compounds in the slick continue to rise. The modeling team must understand the importance of accurate interfacial chemical transfer rates between the slick and the water.

This chapter discussed the best physical characterizations of chemical transfer rates and performed a sensitivity analysis on the more significant parameters. Table 20.2 indicates that aqueous concentrations resulting from spills of hydrocarbons into rivers were most sensitive to the following, listed in descending order of significance:

Dissolution rate coefficients

Aqueous concentration in equilibrium with the oil slick

Evaporation rate coefficient of the oil slick to the atmosphere

Longitudinal dispersion coefficient

Spreading rate of the slick

Volatilization rate between the water and the atmosphere

For spills that are longer in duration and cannot be simulated by a pulse, streamwise gradients caused by the rapid evaporation of the more volatile components can have a significant impact on aqueous concentrations. A multiple-slick model can account for these streamwise gradients in the slick.

For any planning of an oil spill into rivers or streams, a field investigation needs to be undertaken. This investigation involves determining the travel time of the reach, the longitudinal dispersion coefficient of the reach, and the gas transfer coefficient of the reach. An aqueous concentration model with acceptable results can then be formulated from the relationships and references provided in this chapter.

20.7 ACKNOWLEDGMENTS

This model was developed with support from a Section 104 grant distributed by the University of Minnesota Water Resources Center, a doctoral dissertation fellowship from the University of Minnesota Graduate School, and the National Science Foundation under grant No. BES-9522171.

20.8 REFERENCES

Adams, E. E., D. J. Cosler, and K. R. Helfich. 1990. "Evaporation from Heated Water Bodies: Predicting Combined Forced plus Free Convection," *Water Resources Research,* vol. 26, pp. 425–435.

Addison, H. 1941. *Hydraulic Measurements*, John Wiley & Sons, New York, pp. 155–156.

Banks, R. B. 1974. "A Mixing Cell Model for Longitudinal Dispersion in Open Channels," *Water Resources Research,* vol. 10, pp. 357–358.

Burris, D. R., and W. G. MacIntyre. 1987. "Water Solubility Behavior of Hydrocarbon Mixtures Implications for Petroleum Dissolution," in *Oil in Freshwater,* ed. J. H. Vandermeulen and S. E. Hrudey, Pergamon Press, New York, pp. 85–94.

Cadwallader, T. E., and A. J. McDonnell. 1969. "A Multivariate Analysis of Reaeration Data," *Water Research,* vol. 3, pp. 731–742.

Cohen, Y., W. Cocchio, and D. Mackay. 1978. "Laboratory Study of Liquid-Phase Controlled Volatilization Rates in the Presence of Wind Waves," *Environmental Science and Technology,* vol. 12, 553–558.

Cohen, Y., D. Mackay, and W. Y. Shiu. 1980. "Mass Transfer Rates between Oil Slicks and Water," *Canadian Journal of Chemical Engineering,* vol. 58, pp. 569–575.

Fay, J. A. 1971. "Physical Processes in the Spread of Oil on a Water Surface," in *Prevention and Control of Oil Spills,* American Petroleum Institute, Washington, DC, pp. 463–467.

Fischer, H. B., E. J. List, R. C. Y. Koh, J. Imberger, and N. H. Brooks. 1979. *Mixing in Inland and Coastal Waters,* Academic Press, New York, pp. 136–138.

Fuller, E. N., P. D. Schettler, and J. C. Giddings. 1966. "A New Method for Prediction of Binary Gas-phase Diffusion Coefficients," *Industrial and Engineering Chemistry,* vol. 58, 19–27.

Garrett, W. D., and W. R. Barger. 1970. "Factors Affecting the Use of Monomolecular Surface Films to Control Oil Pollution on Water," *Environmental Science and Technology,* vol. 4, pp. 123–127.

Goodwin, S. R., D. Mackay, and W. Y. Shiu. 1976. "Characterization of the Evaporation Rates of Complex Hydrocarbon Mixtures under Environmental Conditions," *Canadian Journal of Chemical Engineering,* vol. 54, pp. 290–294.

Gulliver, J. S., and H. G. Stefan. 1986. "Wind Function for a Sheltered Stream," *Journal of Environmental Engineering,* vol. 112, pp. 387–399.

Hayduk, W., and H. Laudie. 1974. "Prediction of Diffusion Coefficients for Non-electrolysis in Dilute Aqueous Solution," *AiChE Journal,* vol. 20, pp. 611–615.

Herbes, S. E., and G. Yeh. 1985. "A Transport Model for Water-soluble Constituents of Synthetic Oil Spills in Rivers," *Environmental Toxicology and Chemistry,* vol. 4, pp. 241–254.

Hibbs, D. E., and J. S. Gulliver. 1999. "Dissolved Rate Coefficients for Surface Slicks on Rivers," *Water Research,* vol. 33, pp. 1811–1816.

Hibbs, D. E., K. L Parkhill, and J. S. Gulliver. 1998. "Sulfur Hexafluoride Gas Tracer Studies in Streams," *Journal of Environmental Engineering,* vol. 124, pp. 752–760.

Hibbs, D. E., J. S. Gulliver, V. R. Voller, and Y.-F. Chen. 1999a. "An Aqueous Concentration Model for Riverine Spills," *Journal of Hazardous Materials,* vol. 64, pp. 37–53.

Hibbs, D. E., J. S. Gulliver, V. R. Voller, and Y.-F. Chen. 1999b. "Processes controlling aqueous concentrations for riverine spills," *Journal of Hazardous Materials,* vol. 64, pp. 57–73.

Jobson, H. E., and T. N. Keefer. 1979. "Modeling Highly Transient Flow, Mass, and Heat Transport in the Chattahoochee River Near Atlanta, Georgia," U.S. Geological Survey, Professional Paper No. 1136.

Kilpatrick, F. A., and J. F. Wilson, Jr. 1989. "Measurement of Time of Travel in Streams by Dye Tracing," in *TWRI Book 3,* U.S. Geological Survey, Denver, CO, chap. A9.

Kilpatrick, F. A., R. E. Rathbun, N. Yotsukura, G. W. Parker, and L. L. DeLong. 1989. "Determination of Stream Reaeration Coefficients by Use of Tracer," in *TWRI Book 3,* U.S. Geological Survey, Denver, CO, chap. A18.

Lamont, J. C., and D. S. Scott. 1970. "An Eddy Cell Model of Mass Transfer into the Surface of a Turbulent Liquid," *AIChE Journal,* vol. 16, pp. 513–519.

Levenspiel, O. 1962. "Comparison of the Tanks-in-Series and the Dispersion Models for Non-ideal Flow of Fluid," *Chemical Engineering Science,* vol. 17, pp. 576–577.

Lide, D. R., ed. 1993. *CRC Handbook of Chemistry and Physics,* CRC Press, Ann Arbor, MI, pp. 3-1–3-521.

Lyman, W. J., W. F. Reehl, and D. H. Rosenblatt. 1990. *Handbook of Chemical Property Estimation Methods,* American Chemical Society, Washington, DC.

Mackay, D., and R. S. Matsugu. 1973. "Evaporation Rates of Liquid Hydrocarbon Spills on Land and Water," *Canadian Journal of Chemical Engineering,* vol. 51, pp. 434–439.

Mackay, D., and A. T. K. Yeun. 1983. "Mass Transfer Coefficient Correlations for Volatilization of Organic Solutes from Water," *Environmental Science and Technology,* vol. 17, pp. 211–217.

Moog, D. B., and G. H. Jirka. 1995. "Analysis of Reaeration Equations Using Mean Multiplicative Error," in *Air-Water Gas Transfer,* ed. B. Jähne and E. C. Monahan, ASCE, New York, pp. 101–111.

Ryan, P. J., D. R. F. Harleman, and K. D. Stolzenbach. 1974. "Surface Heat Loss from Cooling Ponds," *Water Resources Research,* vol. 10, pp. 930–938.

Southworth, G. R., S. E. Herbes, and C. A. Allen. 1983. "Evaluating a Mass Transfer Model for the Dissolution of Orgaincs from Oil Films into Water," *Water Research,* vol. 17, pp. 1647–1651.

Schwarzenbach, R. P., P. M. Gschwend, and D. M. Imboden. 1993. *Environmental Organic Chemistry,* John Wiley & Sons, New York, pp. 107–108, 618–624.

Shen, H. T., and P. D. Yapa. 1988. "Oil Slick Transport in Rivers," *Journal of Hydraulic Engineering,* vol. 114, pp. 529–543.

Shen, H. T., P. D. Yapa, and B. Z. Zhang. 1995. "A Simulation Model for Chemical Spills in the Upper St. Lawrence River," *Journal of Great Lakes Research,* vol. 21, no. 4, pp. 652–664.

Thibodeaux, L. J., and J. C. Carver. 1997. "Hindcasting Volatile Chemical Emissions to Air from Ponded Recycle Oil," *Environmental Progress,* vol. 16, no. 2, pp. 106–115.

Wilke, C. R., and P. Chang. 1995. "Correlation of Diffusion Coefficients in Dilute Solutions," *AIChE Journal,* vol. 1, pp. 264–270.

Wu, J. 1983. "Sea-Surface Drift Currents Induced by Wind and Waves," *Journal of Physical Oceanography,* vol. 13, pp. 1441–1451.

Yapa, P. D., and H. T. Shen. 1994. "Modeling River Oil Spills: A Review," *Journal of Hydraulic Research,* vol. 32, pp. 765–782.

Yapa, P. D., H. T. Shen, and K. Angammana. 1994. "Modeling Oil Spills in a River-Lake System," *Journal of Marine Systems,* vol. 5, no. 3.

CHAPTER 21

BOILING LIQUID EXPANDING VAPOR EXPLOSIONS (BLEVEs): THE CAUSES AND CONSEQUENCES—RECENT EVIDENCE

J. E. S. Venart
Department of Mechanical Engineering, University of New Brunswick, Fredericton, New Brunswick, Canada

21.1 PREFACE

A BLEVE has been defined as "an explosion resulting from the failure of a vessel containing a liquid at a temperature significantly above its boiling point at normal atmospheric pressure" (CCPS, 1994). This section advances possible explanations for the very complex fluid structure interactions (FSI) observed in the BLEVE event and supports the hypotheses with detailed reexaminations of recent experimental data (Roberts et al., 1995a–d), and new physical interpretation and metallurgical appraisals of these same trials. The detailed reanalyses of the catastrophic failures of these four 4.5-ton water capacity LPG vessels with various fills subjected to jet fire attack indicates that the severity of the event and the intensity of the fireballs formed is *not* necessarily a function of the superheat of its contents but may have more to do with the initiating mode of vessel failure and the thermohydraulic state of the contents at final failure.

The mechanism of vessel failure appears to be a two-step process: The formation of an initiating overpressure crack in the high-temperature, vapor-wetted walls of the vessel, followed by the final catastrophic "unzipping" of the containment and a nearly instantaneous release of its contents. The distribution and flashing of the lading causes a fireball if the contents are flammable. The failure of the vessel and the surface emissive power of the BLEVE fireball do not appear to be directly related to the superheat of the contents at failure and indeed may be most severe for conditions when the vessel fails while undergoing a pressure reduction at low superheat.

Possible reasons for the final rapid failure of the vessel may be structural instability of the vessel, rapid overpressurization due to a dynamic head space impact of the two-phase swell initiated upon a depressurization (initiated by the formation of a thermal crack or tear which arrests), or the rapid quenching of its crack tip, due to the two-phase discharge, that results in large local thermal stresses that cause the uncontrolled vessel failure.

21.2 INTRODUCTION

A major objective of current research programs in pressure liquefied gas (PLG) safety has been to develop, verify, and validate models for the loss of containment (LOC) failures in partially filled, accidentally heated PLG storage and process vessels, whether for transport or for in on- or offshore situations, e.g., separators, blow-down tanks.

Some current work has seen the carrying out of extensive and carefully designed and performed large-scale field trials (e.g., Johnson et al., 1990; Roberts and Beckett, 1996), as well as other, less well characterized, smaller experiments (e.g., Venart et al., 1993; Birk and Cunningham, 1994). In addition, hypotheses for the cause(s) have been developed and tested (Sumathipala et al., 1992; Birk, 1995; Birk and Cunningham, 1996; Kielec and Birk, 1997; Miller and Birk, 1997). Unfortunately most of this work has not been able to address the fundamental question as to *how* the vessels fail and *why* the BLEVE event exhibits such diverse fluid-vessel reactions. Failures range from total catastrophic loss of containment to the venting of the contents through an arrested crack and only a partial loss of containment. A further question is why the fireballs formed differ so markedly in form and intensity.

21.3 BACKGROUND

In the fire attack of a PLG vessel, flames first heat up the vapor space walls, which increase in temperature more rapidly than the liquid-wetted sides since heat transfer to them is resisted by the low thermal conductivity of the vapor and its lack of motion due to its thermal stability. On the liquid-wetted walls, heat transfer initially occurs by free convection and later by subcooled and then saturated pool boiling; only rarely is the critical heat flux exceeded for most fluids and fire exposures. As heating progresses, both the liquid and vapor portions of the lading stratify with the temperature at the liquid surface, setting the vapor pressure inside the tank. With continued thermal exposure, the subcooled liquid core can become homogeneous as boiling proceeds from subcooled to saturated boiling. This usually occurs within the stratified liquid zone after the pressure relief valve (PRV) opens and later within the bulk liquid when pressure falls.

With time, the pressure rises to the set pressure of the PRV, which opens (the valve is theoretically sized to maintain the vessel at a safe pressure irrespective of its thermal exposure: (API 520, 1990, API 521, 1993, ASME, 1992 and NFPA 58, 1998)). Then, depending on fill and heating, the valve may cycle or remain open in its attempt to maintain the pressure of the contents at its design setting. If fill is low—the liquid-wetted area of the tank is small—the evaporation rate may not exceed the capacity of the valve and the pressure can remain constant with only partial valve lift, or the valve may cycle. As fill increases, wetted surface area increases, and the evaporation rate now can exceed the valve's design relief capacity, especially since the exiting vapor may be severely superheated due to vapor stratification. In this case, there will be an increase in pressure with time until the level falls and along with it its evaporation rate. In all instances, the opening of the valve will first depressurize the vapor, which will then be followed by the formation of a two-phase swell within the now superheated liquid. If fill is sufficient, the valve intake can continue to be vapor or, with greater swell, two-phase; conditions that will vary with valve size and fire exposure (Sumathipala et al., 1992).

Two-phase valve flow can be caused either by entrainment or because the vessel becomes two-phase full. In the entrained case, mist/droplet flow usually results. If the vessel becomes two-phase full, churn-turbulent, bubbly two-phase flow through the valve may occur. In any case, the choke pressure for such flows is greater, and the choke velocity substantially less, than those for any prior superheated vapor flow. The pressure-relieving capacity of the valve can now become compromised as the two-phase fluid exits with a much lower enthalpy, though greater density, than to the earlier superheated vapor discharge (Sumathipala et al., 1990).

With continuing fire attack, the vapor metal walls weaken and commence plastic deformation at the hottest locations. This eventually leads to the formation of a crack that will cause further depressurization of the contents and an even greater two-phase swell of the contents. The size of initial fissure formed should be a function of the metal temperature, the fill level, and the available energy in the vapor because only the vapor is immediately available to perform the necessary plastic work on the metal. Crack development during this process should be relatively slow with choked, nearly isothermal vapor flow conditions being established as the crack lengthens (γ, the isentropic index, is just slightly greater than one for superheated Propane vapor: Huber et al., 1996). Any crack formed should arrest locally, in the stronger, thicker, and lower temperature, but still very ductile surrounding metal, upon the initial vapor pressure unloading (Venart, 1998).

Once formed, the structural stability of this opening now becomes a matter of vessel loading, the dynamics of the subsequent thermohydraulics, exiting flow, and/or the local cooling of the metal surrounding the crack. With a liquid overpressure, a crack type of failure would most probably occur, whereas with a gas or vapor, the failure is likely to be catastrophic due to the differences in the speed of sounds in the different media: liquid, vapor, or two-phase versus the speed of crack propagation. In subcooled liquid LPG, the speed of sound is about 650 m/s and that of the vapor approximately 200 m/s; in the liquid case, the pressure vessel is able to unload rapidly, whereas in the vapor case this unloading is less rapid since the ductile crack propagation or tearing speed is the same order as the velocity with which the pressure wave is moving in the media. With the contents homogeneous two-phase, there is no unloading of the vessel because the speed of sound in the mixture is much less than that for the vapor and very much less than for the liquid and so the vessel walls cannot unload at all and any cracks will propagate the entire extent of the vessel and may indeed exhibit brittle characteristics.

If, in the initial stages of crack development, mist/droplet flow issues through the opening, cooling times of the adjacent hot vapor walls can be long since there will be little liquid contact with the heated metal. With sufficient fill, impact of a low void fraction swell on the superheated head of the vessel and quenching of the hot metal by direct liquid contact and /or its water hammer-like pressure impulse can occur, perhaps catastrophically restarting the crack (Venart and Ramier, 1998). These latter effects may be amplified by the interaction of both the thermal/hydraulic effects and the geometry of the head space (in both horizontal cylindrical and spherical vessels the vapor regions usually comprise convergent sections that lead to any crack) and the recompression and shock of the contents (Venart and Ramier, 1998; Campbell and Pitcher, 1954).

Whether the cooled crack is now stable, in a fracture mechanics sense, due to its size (McClintock and Argon, 1966) or becomes unstable due to pressure or fluid impact loading, and/or its imposed thermal stress (Goodier, 1957) leads to differing fluid-vessel interactions. On one hand, we may have a long-duration two-phase discharge through a stable crack with the vessel left intact. Alternatively, an apparent instantaneous catastrophic vessel failure can occur. A relatively long-time, two-step LOC failure is also possible. Both of these latter cases will yield a BLEVE. There have been many examples of all types of these adaptive fluid-structure behaviors in the process safety BLEVE literature (e.g., Pietersen, 1988).

It is our view that such a two-step process may be the *cause* of *all* BLEVEs. In other words a "leak before break" (LBB) crack initiator followed by a total loss of containment (LOC) of the PLG vessel is the normal sequence in the development of a BLEVE. The consequence of such a progression of events is that the time delays involved can influence both the pressure at failure and the boiling process within the remaining liquid. If the contents are experiencing a continued increase in pressure, despite the additional relief provided by the crack, the bulk contents will still be subcooled and there will be few bubble nuclei within the fluid because boiling will only be occurring on the liquid-wetted walls of the vessel. If the pressure is constant or falling, however, the contents will be homogeneously boiling and have many uniformly distributed bubbles within the bulk fluid. Catastrophic LOC under these conditions and with the liquid's abrupt depressurization and superheating allows the internal bubbles to play a significant role in the loading of the vessel and in the consequent development of the fireball. First, since the contents are homogeneous, vessel unloading is not possible due to the fact that the speed of crack propagation is greater than the speed of sound in the two-phase fluid and thus the vessel rapidly fails catastrophically. And then, with the unzipping of the vessel, the now unconstrained high-pressure bubbles expand and burst, shattering their surrounding superheated liquid host completely. The result of this will be a fine, mechanically distributed, high-velocity, evaporating aerosol. Any fireball formed from such a rapidly developing droplet cloud could thus involve the total contents of the vessel and not just the adiabatic flash fraction usually presumed for fireball development (Roberts, 1982). The nearly instantaneous nature of its deflagration now perhaps could also develop significant overpressures such as noted in Pietersen, 1988.

Fireball characteristics—size, duration and surface emissive power (SEP)—should therefore be functions not only of the mass of the liquid involved (Roberts, 1982) but also the time delay from the LBB initiator to the final LOC and whether the LOC occurs with the contents still increasing in pressure and prior to the liquid contents becoming homogeneous. If vessel LOC occurs with a stratified liquid layer and a subcooled core under increasing pressure, the fireball should be less buoyant and have an appreciable flash fraction and/or rainout and thus a lower SEP than in a case with dropping pressure and therefore homogeneous boiling.

Whether the vessel fails completely as a result of severe quenching of the superheated vapor-space metal, the imposed thermal stresses or the dynamics, i.e., the water-hammer like impact of the swell upon the already damaged shell and its fracture mechanic criticality, the time scales for the two-step processes envisaged could range from the near zero to tens of milliseconds (for an immediate quenching case) to the ten of seconds (for the mist-flow cooling case). Indeed, two-step failures have been noted previously but not explained (e.g., Birk and Cunningham, 1996).

Clear examples of all of the above types of BLEVEs have been found in the reexamination of the U.K. Health and Safety Laboratory's (HSL) test records of recent propane jet fires (Venart, 1998). This information has been supplemented by additional analyses involving physical as well as macro- and metallurgical examinations of the tank remnants, frame-by-frame analysis of the video records, and a comprehensive thermohydraulic and metallurgical interpretation of the data and vessels (Venart, 1996; Sherrard, 1997).

21.4 RECENT HSL LARGE-SCALE EXPERIMENTS

As part of the Commission of the European Community (CEC) Science and Technology for Environmental Protection (STEP) program and with the sponsorship of the U.K. Health and Safety Executive's (HSE) Technology and Health Sciences Division, a joint project (STEP-CT90-098) was set up to investigate the hazard consequences of jet-fire interactions with vessels containing pressurized liquids (JIVE). The HSE's Health and Safety Laboratory (HSL) was contracted to investigate the thermal response of propane tanks when subjected to jet-fire attack and to assess the effectiveness of mitigation techniques.

The HSL's Process Safety Section undertook four field experiments on the thermal response of partially filled 4.5-ton water capacity horizontal propane tanks to a jet fire. The jet fire consisted of an ignited, horizontal flashing liquid propane jet at a flow rate of about 1.5 kg/s from a nozzle equivalent to a 12.7 mm diameter hole. The nozzle was placed 4.5 m from the front surface and 1 m below the axial center of the tanks at about the still-air lift-off position of the flame. Vessel exposures were about 200 kW/m^2, more than twice that for a fully engulfing hydrocarbon pool fire.

The target tanks were standard 1.2-m diameter 4546-L water capacity (3864-L propane—85% fill) LPG vessels just over 4 m long with semiellipsoidal end caps. The center barrel was 3.276 m long and was constructed of two rolled and longitudinally butt-welded plates with a band-reinforced circumferential butt weld. The walls were almost all 7.1-mm-thick, low-carbon, low-alloy steel (Bodycote, 1997). The tanks were fitted with standard-sized (1.5-in. NTP ASME/BS 500 3090) pressure relief valves.

All vessels were fitted with external wall thermocouples (3 mm, stainless steel, sheathed type K). Interior fluid temperatures, vapor and liquid, were measured on several levels at a single, near-central, vertical position corresponding to the fill-level gauge connection with similar, though 1.5-mm diameter, thermocouples. Both vapor and liquid pressures were determined from remote calibrated pressure transducers. The target tank was mounted on a frame supported on four fire-protected load cells; these were used to determine the variation in propane mass with PRV action.

All transducers were monitored using a remote computer data acquisition system. There were separate video recordings (two to five cameras) made (Nind, 1996) as well as infrared thermal image records of the fireballs formed (Hawksworth and Brearley, 1996). British Gas additionally made wide- and narrow-angle thermal radiation measurements (Gosse and Pritchard, 1996).

Commercial grade propane was utilized for both the jet fire fuel and tank contents. Vessel fills of 20, 41, 60, and 85 volume percent were examined under jet fire attack of the ignited flashing discharge of subcooled liquid propane at about 0.9 MPa.

Separate HSL documents (Roberts and Beckett 1996; Roberts et al., 1995a–d) as well as the video records (Nind, 1996; Hawksworth and Brearley, 1996) provide greater detail and archive the data obtained. Further physical, macro- and micrometallurgical examinations of the tank remnants were also undertaken, in addition to extensive video and thermohydraulic analyses of the data sets (Venart, 1996; Sherrard, 1997) and are summarized in Venart (1998).

21.4.1 Vessel Failures

The vessels all failed catastrophically in less than five minutes and resulted in boiling liquid expanding vapor explosions (BLEVEs) and fireballs.

21.4.2 Results

Detailed discussion of the results, derivations (Tables 21.1 to 21.3), and interpretations for only two of the trials, the 20 and 85% volume fills, follow. Greater detail may be found in the original references and Venart (1998). Short summaries for the other trials (41 and 60% volume fills) are also given.

Examination of the thermohydraulic data (Table 21.1) indicates that, as fill level decreases, time to first vent increases (multiple vents were observed for the 85% fill), the rate of pressure

TABLE 21.1 Thermohydraulic Response of JIVE Vessels

Tank (fill %)	C (20)	A (41)	B (60)	D (85)			
1. Valve operation							
(a) cycles	1	1	1	1	2	3?	open
(b) $(P/T)_{initial}$ (barg/°C)	7.9/19	8.4/20	7.7/20	8.7/18			
(c) P_{open} (barg)	18.6	18.8	18.1	18.3	16.0	14.8	
(d) P_{close}	17.2	16.5	14.1	13.9	–	–	
2. Pressurization (dP/dt)							
(a) before valve open	0.095	0.092	0.106	0.15	–	–	
(b) after valve open	–ve	0.035	0.047				0.1

TABLE 21.2 JIVE Vessel Thermal Response at Failure

Tank (fill %)			C (20)	A (41)	B (60)	D (85)
Tank-wall thickness (mm)			7.1	7.5	7.1	7.1
Liquid space outside metal temperatures (°C)			257 ± 47	141 ± 68	116 ± 24	159 ± 27
Vapor space metal temperatures (°C)	Front	T_{mean}	711 ± 93	645 ± 46	559 ± 49	734 ± 89
		T_{min}	516	573	485	644
		T_{max}	870	704	641	848
	Back	T_{mean}	564 ± 86	424 ± 136	646 ± 94	663 ± 112
		T_{min}	479	304	531	508
		T_{max}	697	694	798	848
Failure pressure (barg)			16.5	21.3	18.6	24.4

increase prior to vent decreases, the rate of pressure increase after continuous venting decreases, and the pressure at failure diminishes. Most of this behavior can be explained in terms of the variation in liquid-wetted wall with fill level (Sumathipala et al., 1992) though some observations are in conflict with other workers, most notably Birk (1995) and his coworkers.

The times to vessel failure are similar (251 ± 28 seconds) all with comparable maximum and mean vapour wall temperatures (810 ± 74° and 611 ± 90°C, Table 21.2).

Table 21.3 records the size of the initial ruptures formed in these trials. The size of the openings were estimated from the tank remnants and taken to be the extent of the feather-edged ruptures up to the point of fast fracture initiation as 45° shear lips with little plastic contraction. A derived equivalent von Mises stress (0.866 σ_{hoop}), for a vessel with end shells under pressure alone and the rupture crack stress intensity factors at the commencement of fast fracture failure are also indicated. At best these indices are only simple indicators of the actual stress and stress intensity levels due to the extreme temperature gradients present in the shell at the time of failure. Large radial, axial, and circumferential temperature differences exist that impose significant local thermal strain upon the vessel (e.g., Tan et al., 1999; Tan and Venart, 1999). So though approximate, and unlike the work of Kielec and Birk (1997), this indicator should provide a better approximation for a failure criteria of LPG vessels under such extreme fire exposures since the mean equivalent stress (142 MPa: Table 21.3) does correspond to the ultimate stress at the maximum mean temperature (~140 MPa at 684°C: Table 21.2) for all tests. The importance of the role of the significant temperature

TABLE 21.3 Initial Crack Size, Failure Stress, and Stress Intensity Factor; JIVE Tanks

Tank (fill %)	P_{f_i} (barg)	t (mm)	σ_{eq} (MPa)	a_c (mm)	K_1 (MPa $\sqrt{m}$)
C (20)	17.2	7.1	126	278	166
A (41)	20	7.5	138	178	146
B (60)	17.5	7.1	128	290	174
D (85)	24	7.1	176	145	168
				mean	164 ± 12

P_{f_i} ≡ initial failure pressure; t ≡ wall thickness; σ_{eq} ≡ 0.866(σ_{hoop}); K_1 ≡ equivalent elastic stress intensity factor; 2_{a_c} ≡ measured size of initial crack.

variations in the span-wise directions of the tank's vapor walls (Table 21.2) should not, however, be overlooked as the source of the initiating rupture failures.

20% Full Trial (Tank C, Figure 21.1, Roberts et al., 1995c). In this trial, a 1.52-kg/s flashing liquid propane jet fire brought the 20% full tank, initially containing 453 kg of commercial propane, to failure in just over 250 seconds. The PRV opened at a pressure of 18.6 barg in 112 seconds and then remained partially open, venting its contents at the rate of 1.26 kg/s. Just before failure, the pressure started to fall gradually and a maximum shell temperature of 870°C was recorded. Vessel failure was initiated by a longitudinal rupture just to the back side of the top of the tank. This crack was 556 mm long (Table 21.3) and commenced forming about 635 mm to the right of the left endcap weld some 200 mm circumferentially behind the top. Visual and video observations indicated that, when the PRV operated, this gave a vertical jet of flame approximately 10 m high with a lift-off distance of about 2 m. After 250 seconds, the tank failed catastrophically at a pressure of 16.5 barg, as pressure was slowly falling.

The composite figures of pressure, mass, internal fluid, and exterior shell temperatures indicate that, at approximately 197 seconds, the recorded pressures (Fig. 21.1*a*) gradually commence falling from 17.2 barg, where they had been constant for over 50 s, until at failure they were 16.5 barg. About the same instant, however, all internal temperatures (Fig. 21.1*b*) rise at least 10°C in about 8 s and then fall to regain their previous values or trends. These factors, in addition to the video record, suggest strongly that an initiating fracture developed and the additional relief provided by its opening allowed the valve to close; hence the momentary (approximately 16 seconds) excursion in internal temperatures. These observations suggest that the rupture crack formed around this time and then remained stable for about 40 seconds prior to catastrophic failure of the vessel by a fast shear fracture (~50 ms, i.e., about one to two video frames).

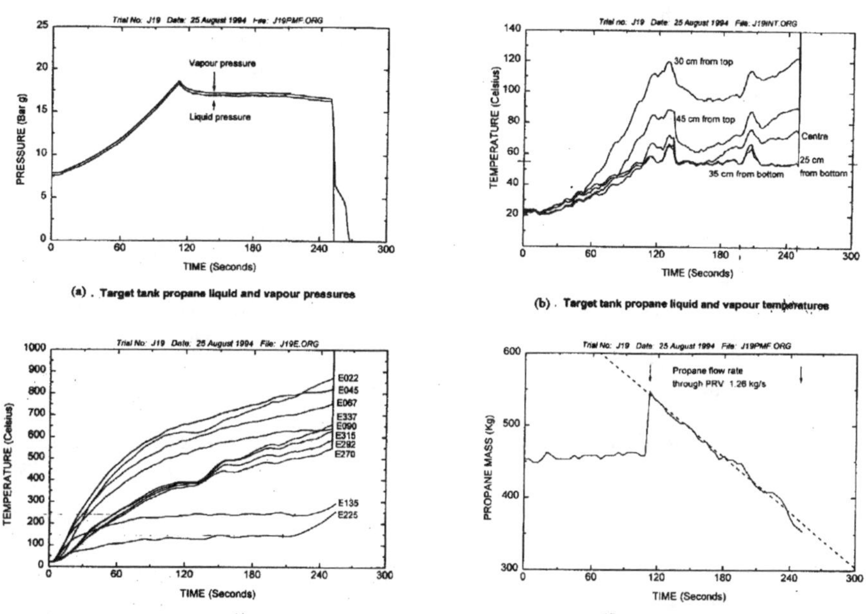

FIGURE 21.1 Selected composite results 20% fill trial: (*a*) pressure; (*b*) interior temperatures; (*c*) exterior wall temperatures; (*d*) mass (manually logged). (*Source: Roberts et al., 1995c*)

The examination of the video records (Nind, 1996) by time averaging (150 and 250 frames; 5 and 8.5 s respectively) at instances between 50 and 30 s prior to catastrophic rupture, then setting a threshold level between 150 and 160 out of 255, supports the above possibility since the films, for all camera angles, show at least a 6 to 10-second period where there is a reduction in apparent PRV flare size and intensity for this interval; such a period very much exceeds the prior and later turbulent flare fluctuations, due to wind, and even suggests a jet with greater vertical momentum than with the original PRV discharge after crack formation. Figure 21.2*a* is an example of the time-averaged raw video extract for camera 2, and Fig. 21.2*b* shows a similar image set with a threshold of 156 for camera 3. In both figures, the reduction in flare size within the specified time frame is obvious.

The additional depressurization caused by the initial crack also helps explain the increase in liquid-wetted wall temperatures of some 50°C commencing some 40 seconds prior to failure (Fig. 21.1*c*); the progressive reduction in pressure, due to the additional area for discharge, causes more vigorous boiling and thus an increase in the fluid-wall temperature difference. There was a failure of the automatic recording system for mass, and so this record (Fig. 21.1*d*) was logged manually and therefore little detail can be inferred.

85% Full Trial (Tank D, Fig. 21.3, Roberts et al., 1995d). In this the last experiment of the test series, a 1.68 kg/s flashing liquid propane jet fire brought the 85% full propane tank, initially containing 1932 kg of commercial propane, to failure in 254 seconds. The PRV opened at a pressure of 18.6 barg after 68 seconds and then cycled open and partially shut at least twice before remaining open (Fig. 21.3*a*), venting its contents at the rate of approximately 1.28 kg/s. At failure, the pressure was 24.4 barg with a maximum recorded shell temperature of just under 850°C (Fig. 21.3*c*).

Vessel failure was initiated by a small longitudinal rupture near the top of the tank. This crack, which was 290 mm long, commenced forming about 880 mm to the left of the center weld and about 50 mm circumferentially forward of the top. Visual and video observations indicated that when the PRV operated, this gave a vertical jet of flame approximately 10 m high with a lift-off distance of about 2 m. After 254 seconds, the tank failed catastrophically.

The composite figures of pressure, mass, internal fluid, and external shell temperatures indicate no unusual characteristics in the recorded parameters such as were noted previously in the 20% fill trial. However, a detailed frame-by-frame analysis of the video record clearly shows that the fast fracture phase of vessel failure was preceded, at about 1 to 2 seconds (approximately 25 to 50 video frames), with the formation of a crack that permitted a vapor and then two-phase discharge to occur through the rupture.

Figure 21.4 illustrates the development of the crack discharge and the resulting development of its supplemental flare taken from the frame-by-frame analysis of the video. At about 01:04 seconds, just to the left of the PRV flare, an ignited gas discharge is apparent, which then proceeds to develop and enlarge (01:14 to 01:24 seconds), becomes two-phase (02:09 seconds), and somewhat diminishes (perhaps due to two-phase choking), until finally the vessel bursts just before 03:04 seconds. These video frame sequences seem to confirm that a two-stage process of rupture crack formation and arrest occurs prior to the catastrophic failure of the vessel by a fast plane stress shear fracture.

41% and 60% Full Trials (Tanks A and B, Roberts et al., 1995a, b). In these two trials the flashing liquid propane jet fire brought the 40.9% and 60% full vessels to failure in just less than 5 minutes.

For the 40% full case, the PRV opened at a pressure of 18.8 barg at 130 seconds and then remained open, venting its contents at the rate of about 1.41 kg/s. At failure, the pressure was 21.3 barg with a maximum shell temperature of just over 700°C located near the center top of the shell. Vessel failure was initiated by a longitudinal rupture near the top of the tank. This opening, which was originally 356 mm long, commenced forming about 420 mm to the right of the center weld and some 250 mm circumferentially forward of the top. Visual and video observations indicated that when the PRV operated, this gave a vertical

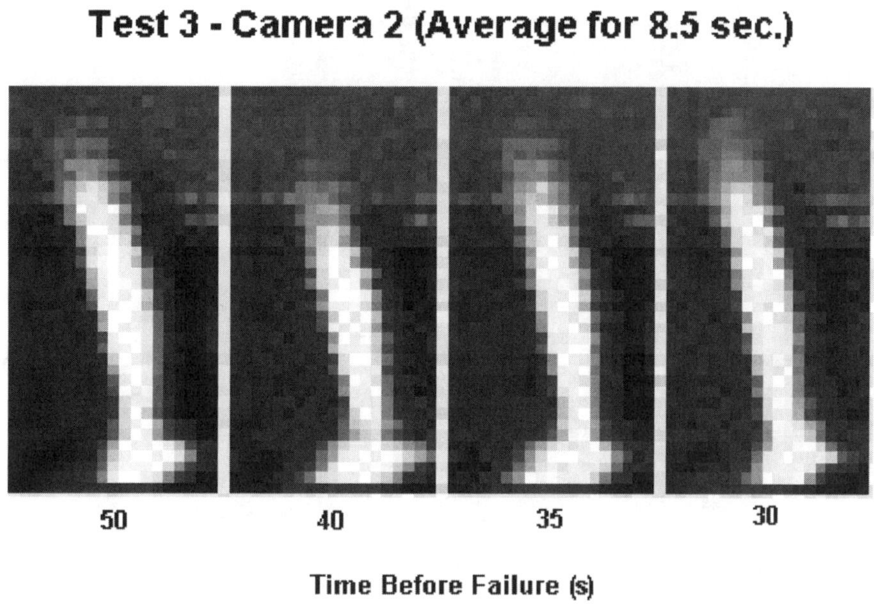

(a)

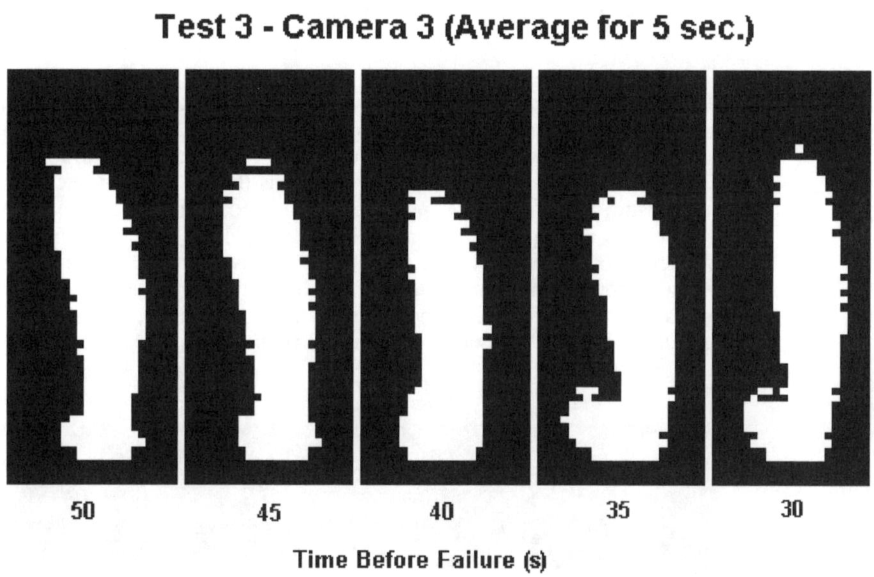

(b)

FIGURE 21.2 (*a*) Time-averaged (255 frame) raw video of PRV/crack jet fire, camera 2, 20% fill, 50, 40, 35, and 30 seconds before catastrophic BLEVE failure; (*b*) Threshold (156) of time-averaged video, camera 3, 20% fill.

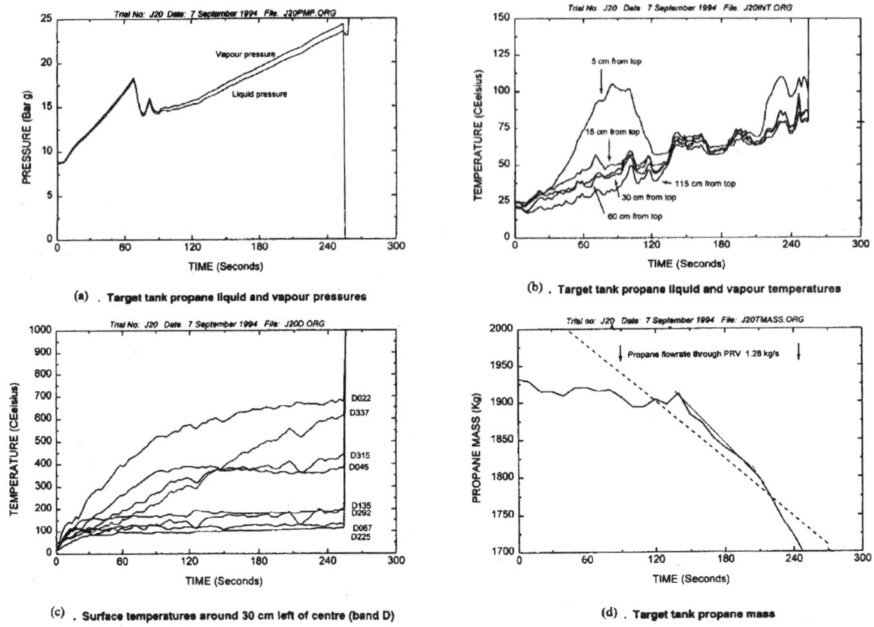

FIGURE 21.3 Selected composite results 85% fill: (*a*) pressure; (*b*) interior temperatures; (*c*) exterior wall temperatures; (*d*) mass (manually logged). (*Source: Roberts et al., 1995d*)

jet of flame approximately 10 m high with a lift-off distance of about 2 m. The flare increased in size and intensity after 250 seconds and the tank failed catastrophically at 285.5 seconds. The composite figures of pressure, temperatures, both interior fluid and exterior shell for this test indicated that at approximately 250 seconds there was an inflection point in the recorded liquid pressure and its rate of increase diminished. Further, commencing at about the same time there was an apparent progressive increase in the mass flow rate recorded through the PRV. These observations, taken with those determined from a frame-by-frame analysis of the visual record for two cameras, similar to that used earlier in the 20% fill trial, indicated an increase in PRV flare size and intensity at 250 seconds, supporting the suggestion that a rupture crack formed around this time and then remained stable for some 15 seconds prior

FIGURE 21.4 Selected video frame sequences; 85% fill PRV and crack discharge flares, time in seconds to commencement of BLEVE.

to catastrophic failure of the vessel by a fast plain stress shear fracture. Additional, but perhaps less convincing, evidence comes from the audio track from the video for this trial; here the intensity and pitch of the PRV discharge dropped in level at about the same instance as the flare increased in intensity. This observation provides some confirmation to firefighters' anecdotal adage about dealing with fires involving PLGs: "When you hear a reduction in pitch of the PRV discharge, run like hell!"

In the 60% full experiment, a 1.59 kg/s flashing liquid propane jet fire brought the 60% full tank, originally containing 1364 kg of commercial propane, to failure in 217 seconds. The PRV opened at a pressure of 18.1 barg at 109 s and then remained open, venting its contents at the rate of about 0.85 kg/s. At failure, the pressure was 18.6 barg with a maximum recorded shell temperature of just under 800°C. Vessel failure appeared to have been initiated by a complex series of longitudinal rupture cracks near the top of the tank. These cracks had a combined feather-edged length of some 580 mm and commenced forming about 1,200 mm to the right of the center weld and on the top center of the vessel. Visual and video observations indicated that when the PRV operated, this gave a vertical jet of flame approximately 10 m high with a lift-off distance of about 2 m. At 217 seconds the tank failed catastrophically at a pressure of 18.6 barg. On failure, the tank split into three sections with about the left-hand two-thirds of the vessel rocketing 447 m, an end cap being thrown 309 m, and approximately one third of the barrel opening out flat in the target area. The composite figures of pressure, mass, internal fluid, and external surface temperatures indicated that at approximately 190 seconds, the recorded liquid and vapor temperatures all dropped some 20°C in about 15 seconds and then slightly recovered some 10°C just at failure. The time-averaged video record for this trial, with only one camera available for analysis, however, is less clear than in the previous instances. The observations suggest that an opening developed about 10 seconds before failure and the crack so formed remained stable prior to catastrophic failure of the vessel by a fast shear fracture that bifurcated and caused two pieces of the tank to rocket off significant distances.

Other Observations. All the vessels were fabricated from a low-alloy low carbon steel (0.21 to 0.25 % total C, 0.65 to 0.82 % Mn, 0.01 to 0.03 % Si, 0.015 to 0.021 % S, 0.014 to 0.035 % P, 0.02 % Cr, 0.01 to 0.02 % Ni, < % 0.01 Mo (Bodycote, 1997); determined A_1 (austenite start) and M_s (martensite start) temperatures from these compositions were 715.8 (±1.6) and 408.8 (±9.3)°C, respectively.

Vapor wall metal temperatures (Table 21.2) were extremely variable due to the impact of the jet fire. It was not unusual for temperatures to drop locally by up to 150°C away (<500 mm) from the initial rupture site. Metal strength would thus be affected and any overpressure crack formed could possibly arrest in the cooler, thicker, stronger, and tougher wall as the vessel unloads due to the initial local vapor depressurization prior to two-phase repressurization (Venart and Ramier, 1998). The plastic deformation of selected initial crack cross-sections exhibit progressive area reductions, with greater amounts occurring on the heated side (Venart, 1996; Sherrard, 1997). This can be attributed to wall thinning and bulging while under the initial complex triaxial stress state imposed by the internal pressure, support loads, and thermal conditions. Microhardness examinations of the metal samples taken near the originating rupture crack tips for all four tanks showed statistically significant increases in hardness towards the crack tips. Metallographic macro and micrographs did not show any direct signs of quenching (i.e., an acicular structure of transformation products), although in earlier studies of a similar failure in a full-size rail tank car, transformation products were found at the subcritical crack initiation region along with similar increases in hardness (Anderson and Norris, 1974).

In the B (60% fill) tank samples, a circumferential rolling direction for the plate forming the section of the barrel where the cracks originated was indicated (Venart, 1996). The originating ruptures formed in this section may thus possibly be linked to the early crack bifurcation that was unique to this trial. This process caused a portion of the tank to circumferentially separate and rocket nearly 450 m. It is well known that fracture toughness is a

TABLE 21.4 Fireball Surface Emissive Powers (SEP), Duration and Size

Tank (fill %)	Mass (kg)	SEP (kW/m^2) mean	SEP (kW/m^2) max	Duration (s)	Diameter (m)	Height (m)
C (20)	279	410	640	3.5	45	45
A (41)	710	278[a]	484[a]	5.5	45	70
B (60)	1272	365	550	6.5	55	75
D (85)	1708	350	580	7	45	45

[a] British Gas Measurements

function of plate rolling direction; the low toughness direction is parallel to the rolling direction. In previous work on gas and liquid pressurized lines, significant variations in strength and toughness were observed due to rolling direction sufficient to cause the bifurcation of longitudinal cracks to form guillotine breaks similar to the case of the 60% full vessel (Wilkowski, 1991).

Summary. The thermohydraulic, metallurgical, and video film evidence for these experiments indicate that crack development and propagation during vessel failure occurred in the following distinct steps:

1. A rupture stage, where a crack opens up and becomes stable, forming an opening for a vapor or two-phase jet discharge additional to that of the PRV.
2. A final fast fracture stage, where the initial crack rapidly propagates into the metal as a stress shear failure (usually) along the entire length of the tank and then circumferentially at the end caps. Additionally:
3. Rolling direction of the shell appears to be implicated in the "rocketing" for one vessel.

The consequence of this type of failure is called a BLEVE. The process causes the remaining contents of the vessel to be rapidly released as a superheated liquid aerosol, which may then ignite and form a fireball.

Fireball Characteristics. The fireball characteristics, shown in Table 21.4, indicate that, as the mass of material involved in the BLEVE increases, the duration and size of the fireball becomes larger, although, its shape becomes less spherical and more vertically elongated. From the video analyses, it is also apparent that there is a greater ground flash fraction for large fills and these require significant times before ignition. The mean surface emissive powers (SEP) of the fireballs, at their maximum size, also decreases despite the significant increases in saturation pressure and thus liquid superheat of the contents. The 85% fill fireball develops much more slowly and exhibits both an average and a top 10% SEP (max) nearly 20% less than that exhibited in the 20% fill experiment. This is contrary to the expectation of the models of Roberts (1982) and Shield (1993) as well as recent evidence from both deflagrating and detonating fuel–air clouds (Dorofeev et al., 1995; Makhviladze et al., 1999).

21.5 DISCUSSION AND CONCLUSION

The analyses and interpretations of the HSL JIVE tank failure data present a coherent and plausible case that can best be explained by the formation of a fracture that arrests and then is reinitiated for some reason and then becomes critical. For the transitions to fast fracture that were observed in all these experiments, there must be some physical reason for crack reinitiation. Once cracks are arrested in a ductile material, reinitiation requires significant further energy. It is our view that the thermohydraulic, video, and physical data support the possibility of a crack tip quenching mechanism as the most probable cause; the metallurgical data for this are less conclusive, however, despite some earlier supporting evidence.

Taken together, the situation described is not unlike that envisaged in the highly transient "rewet" situation in high heat flux forced convective boiling; it now occurs, however, within a stressed and damaged pressure vessel. The dependence of the time delay on liquid fill would appear to be a function of the heat transfer and the distance to the boiling LPG surface. Due to entrainment, flow through the initial rupture crack could be "mist/droplet" two-phase flow in the case of low fills or for fills greater than about 50% "churn-turbulent bubbly" flow. In this latter case, impact of the two-phase swell on the hot vapor space walls, as a result of the additional pressure relief from crack formation, may be possible and very rapid cooling of the metal surface could therefore take place due to direct liquid-metal contact.

The influence of such a two-step vessel failure on fireball formation appears to be significant. Some of the observations made here contradict earlier works based on liquid superheat and the results from single-step failures and thus will be important in designing further experiments and developing realistic fireball and BLEVE models.

21.6 ACKNOWLEDGMENTS

The experimental field work described and interpreted here was performed as part of the Commission of the European Community (CEC) Science and Technology for Environmental Protection (STEP) program and with the sponsorship of the United Kingdom Health and Safety Executive (HSE) Technology Sciences Division. The program (STEP-CT90-098) investigated the hazard consequences of jet-fire interactions with vessels containing pressurized liquids (JIVE). Dr. T. A. Roberts, HSL, Buxton, provided access to the data sets and gave permission for the removal of samples and the metallurgical examinations. Marc LeBlanc and Jennifer Sherrard helped in the preparation and analyses of the metallographic and hardness specimens. Dr. C. Thornley, New Brunswick Research and Productivity Council, assisted in part of the examination and analyses of the metallographic samples. The views and interpretations put forward in this section are those of the author alone. The original work in this section was reported first in a report to the HSL entitled "HSL JIVE Tank Failures; Physical and Metallurgical Examinations," and significant portions have been presented at the IChemE Hazards XV and the ASTM STP 1336, Very Large-Scale Fires conferences.

21.7 REFERENCES

American Petroleum Institute (API). 1990. *Guide for Pressure-Relieving and Depressuring Systems,* 3d ed., API Recommended Practice 520, Washington, DC, November.

American Petroleum Institute (API). 1993. *Sizing, Selection, and Installation of Pressure-Relieving Devices in Refineries,* Part 1—Sizing and Selection," 6th ed., API Recommended Practice 521, Washington, DC, July.

American Society of Mechanical Engineers (ASME). 1992. "Rules for Construction of Pressure Vessels, Division 1," ASME Boiler and Pressure Vessel Code, Section VIII, ASME, New York, July.

Anderson, C., and E. B. Norris. 1974. *Fragmentation and Metallurgical Analysis of Tank Car RAX 20,* Report No. FRA-OR&D 75-30, U.S. Department of Transporation, FRA, April.

Birk, A. M. 1995. "Scale Effects with Fire Exposure of Pressure-Liquefied Gas Tanks," *Journal of Loss Prevention in the Process Industries,* vol. 8, no. 5, pp. 275–290.

Birk, A. M., and M. H. Cunningham. 1994. "The Boiling Liquid Expanding Vapor Explosion," *Journal of Loss Prevention in the Process Industries,* vol. 7, no. 6, pp. 474–480.

Birk, A. M., and M. H. Cunningham. 1996. "Liquid Temperature Stratification and Its Effect on BLEVEs and Their Hazards," *Journal of Hazardous Materials,* vol. 48, pp. 219–237.

Bodycote Materials Testing. 1997. *Chemical Test Reports,* Galt Laboratory, August 26.

Campbell, I. J., and A. S. Pitcher. 1954. "Shock Waves in a Liquid Containing Gas Bubbles," *Proceedings of Royal Society,* vol. A 243, pp. 534–545.

Center for Chemical Process Safety (CCPS). 1994. *Guidelines for Evaluating the Characteristics of Vapor Cloud Explosions, Flash Fires and BLEVEs,* CCPS, AIChE, New York, p. 6.

Dorofeev, S. B., V. P. Sidorov, A. A. Efimenko, A. S. Kochurko, M. S. Kuzenetsov, B. B. Chaivanov, D. I. Matsukov, A. K. Pereverzev, and V. A. Avenyan. 1995. "Fireballs from Deflagration and Detonation of Hetrogeneous Fuel-Rich Clouds," *Fire Safety Journal,* vol. 25, pp. 323–336.

Goodier, J. N. 1957. "Thermal Stress and Deformation," *Journal of Applied Mechanics, Trans. ASME,* September, pp. 467–474.

Gosse, A. J., and M. J. Pritchard. 1996. *A Study of the Radiative Characteristics of the Fireball Produced following the BLEVE Failure of a Pressurised LPG Vessel; a Record of Measurements Made by British Gas as Part of a CEC Co-funded Research Project,* British Gas Report GRC R 1197, January.

Hawksworth, S. J., and D. Brearley. 1996. *Thermal Imaging of Propane Fireballs as Part of the Hazardous Consequences of Jet-fire Interactions with Vessels Containing Pressurised Liquids Project,* HSL Report R04.029, IC/06/01, February.

Huber, M., J. Gallagher, M. McLinden, and G. Morrison. 1996. "NIST Thermodynamic Properties of Refrigerants and Refrigerant Mixtures Database (REFPROP)," Version 5.0, February, U.S. Department of Commerce, NIST, SRDP, Gaithersburg, MD.

Johnson, D. M., M. J. Pritchard, and M. J. Wickens. 1990. *Large Scale Catastrophic Releases of Flammable Liquids,* British Gas Report to the CEC, Contract EV4T.0014.UK.

Kielec, D. J., and A. M. Birk. 1997. "Analysis of Fire Induced Ruptures of 400-L Propane Tanks," *ASME Journal of Pressure Vessel Technology,* vol. 119, pp. 365–373.

Makhviladze, G. N., J. P. Roberts, and S. E. Yakush. 1999. "Combustion of Two-Phase Hydrocarbon Fuel Clouds Released into the Atmosphere," *Combustion and Flame,* vol. 118, pp. 585–605.

McClintock, F. A., and A. S. Argon (eds.) 1966. *Mechanical Behavior of Materials,* Addison-Wesley, Reading, MA.

Miller, T., and A. M. Birk. 1997. "A Re-examination of Propane Tub Rockets Including Field Trial Results," *ASME Journal of Pressure Vessel Technology,* vol. 119, pp. 356–364.

National Fire Protection Association (NFPA). 1998. NFPA 58, Liquefied Petroleum Gas Code, NFPA, Quincy, MA.

Nind, M. J. 1996. "Video Record BLEVE Tests (Great Longstone); All Camera Views from Ignition to Explosion," HSL Visual Presentation Services, Sheffield, UK, January. (video July 1996).

Pietersen, C. M. 1988. "Analysis of the LPG-Disaster in Mexico City," *Journal of Hazardous Materials,* vol. 20, pp. 85–107, 1988.

Roberts, T., H. Beckett, G. Cooke, and D. Brown. 1995a. *Jet Fire Impingement Trial on a 41% Full, Unprotected, 2 Tonne Propane Tank,* HSL Report R04.029, IR/L/PH/95/11, Buxton, UK, July.

Roberts, T., H. Beckett, G. Cooke, and D. Brown. 1996b. *Jet Fire Impingement Trial on a 60% Full, Unprotected, 2 Tonne Propane Tank,* HSL Report R04.029, IR/L/PH/95/12, Buxton, UK, August.

Roberts, T., H. Beckett, G. Cooke, and D. Brown. 1995c. *Jet Fire Impingement Trial on a 20% Full, Unprotected, 2 Tonne Propane Tank,* HSL Report R04.029, IR/L/PH/95/13, Buxton, UK, October.

Roberts, T., H. Beckett, G. Cooke, and D. Brown. 1995d. *Jet Fire Impingement Trial on a 85% Full, Unprotected, 2 Tonne Propane Tank,* HSL Report R04.029, IR/L/PH/95/14, Buxton, UK, October.

Roberts, T., and H. Beckett. 1996. *Hazard Consequences of Jet-Fire Interactions with Vessels Containing Pressurised Liquids: Project Final Report,* HSL Report R04.029, PS/96/03, Buxton, UK, February.

Roberts, A. F. 1982. "Thermal Radiation Hazard from Releases of LPG from Pressurised Storage," *Fire Safety Journal,* vol. 4, pp. 197–212.

Sherrard, J. 1997. "Metallurgical and Physical Measurements of HSL JIVE Samples," UNB FSC, Fredericton, NB, September.

Shield, S. R. 1993. "A Model to Predict the Radiative and Blast Hazards from LPG BLEVE's," *AIChE Symposium Series,* vol. 89, pp. 139–149.

Sumathipala, K., J. E. S. Venart, and F. R. Steward. 1990. "Two-Phase Swelling and Entrainment during Pressure Relief Valve Discharge," *Journal of Hazardous Materials,* vol. 25, pp. 219–236.

Sumathipala, K., G. Hadjisophocleous, N. Aydemir, C. Yu, A. Sousa, F. Steward, and J. E. S. Venart. 1992. "Fire Engulfment of Pressure-Liquefied Gas Tanks: Experiments and Modeling," in *Fire Hazard and Fire Risk Assessment,* ASTM STP 1150, pp. 106–121.

Tan, D. M., and J. E. S. Venart. 1999. "LPG Tank Failure: Ste. Elizabeth de Warwick, Quebec, June 27, 1993; the Lessons to be Learnt," in ASME PVP 396 ed. W. L. Cheng, pp. 129–136.

Tan, D. M., K. F. Sollows, and J. E. S. Venart. 1999. "Dragon: LPG Rail Tank Failure Analysis," 8392 21 2101 ASME PVP 396, ed. W. L. Cheng, pp. 121–128.

Venart, J. E. S. 1996. *HSL JIVE Tank Failures; Physical and Metallurgical Examinations,* UNB Fire Science Centre, UNB FSC: R-96-09-23 (revised), Fredericton, NB, November.

Venart, J. E. S. 1998. "Boiling Liquid Expanding Vapor Explosions (BLEVE): Possible Failure Mechanisms", in *Very Large-Scale Fires,* ASTM STP 1336, ed. N. R. Keltner, N. J. Alvares, and S. J. Grayson, pp. 112–132.

Venart, J. E. S., and S. A. Ramier. 1998. "Boiling Liquid Expanding Vapor Explosions: The Influence of Dynamic Re-pressurization and Two-phase Discharge," in ASME PVP 377-2 ed. C. R. Kleijn and S. Kawano, pp. 249–254.

Venart, J. E. S., G. A. Rutledge, K. Sumathipala, and K. Sollows. 1993. "To BLEVE or Not to BLEVE: Anatomy of a Boiling Liquid Expanding Vapor Explosion," *AIChE Process Safety Progress,* vol. 12, no. 2, pp. 67–70.

Wilkowski, G. M. 1991. "Anisotropic Fracture Toughness Effects on Failure Modes of Piping," *ASME Journal of Pressure Vessel Technology,* vol. 113, pp. 154–158.

CHAPTER 22
MODELING AND UNDERSTANDING BLEVEs

J. Casal, J. Arnaldos, H. Montiel, E. Planas-Cuchi, and J. A. Vílchez
Centre d'Estudis del Risc Tecnològic (CERTEC),
Universitat Politècnica de Catalunya—Institut d'Estudis Catalans,
Barcelona, Catalonia, Spain

22.1 INTRODUCTION

Boiling liquid expanding vapor explosions (BLEVEs) are one of the most severe accidents that can occur in the process industry or in the transportation of hazardous materials. Strictly speaking, these explosions do not necessarily imply thermal effects. However, in most cases the substance involved is a fuel that causes a severe fireball after the explosion. Usually BLEVE refers to the combination of these two phenomena, BLEVE and fireball, i.e., to an accident simultaneously involving mechanical and thermal effects.

BLEVEs occur with a certain frequency: the substances that can lead to them (butane, propane, vinyl chloride, chlorine, etc.) are relatively common in the industry, as well as the installations in which they can happen (tanks and tank cars). They can have diverse origins, such as runaway reactions and collisions, but the most frequent one is the action of fire on a container. Table 22.1 (Prugh, 1991) shows a list of the most significant BLEVEs that have occurred between 1926 and 1986. As can be seen, most of these involved fatalities. Another source (Londiche and Guillemet, 1991) mentions 900 fatalities and over 9,000 injured in 77 BLEVEs occurring between 1941 and 1990.

In this section, the main features of BLEVEs and fireballs are discussed and a practical methodology for estimating their effects is described.

The information provided in Table 22.1 allows a simple statistical analysis to determine the most frequent external causes of BLEVEs. The results obtained can be seen in Table 22.2.

Some of these causes can occur simultaneously (for example, among the 18 cases of fire, 5 happened immediately after a derailing). Nevertheless, this information shows the importance of fire, overfilling, and runaway reactions both in fixed installations and during transport.

22.2 DESCRIPTION OF THE PHENOMENON

If a tank containing a pressurized liquid is heated—for example, due to the thermal radiation from a fire—the pressure inside it will increase. At a certain moment, its walls will not be

TABLE 22.1 The Most Significant BLEVE Accidents from 1926 to 1986

Date	Place	Cause	Material	Tons	Fatalities
12/13/1926	St. Auban, France	Overfilling	Chlorine	25	19
5/28/1928	Hamburg, Germany	Runaway reaction	Phosgene	10	10
5/10/1929	Syracuse, NY, U.S.A.	Explosion (H_2)	Chlorine	25	1
12/24/1939	Zarnesti, Rumania	Overfilling	Chlorine	10	60
7/29/1943	Ludwigshafen, Germany	Overfilling	Butadiene	16	57
11/5/1947	Roemo, Finland	Overfilling	Chlorine	30	19
7/28/1948	Ludwigshafen, Germany	Overfilling	Ethyl ester	33	209
7/7/1951	Port Newark, NJ, U.S.A.	Fire	Propane	2,600	0
4/4/1952	Walsum, Germany	Overfilling	Chlorine	15	7
1/4/1954	Institute, WV, U.S.A.	Runaway reaction	Acrolein	20	0
1/8/1957	Montreal, Quebec, Canada	Fire	Butane	?	1
1958	Alma, MI, U.S.A.	Overfilling	Butane	55	1
6/28/1959	Meldria, U.S.A.	Derailing	Propane	55	23
8/18/1959	Kansas City, U.S.A.	Fire	Gasoline	20	5
4/17/1962	Doe Run, MO, U.S.A.	Runaway reaction	Ethylene oxide	25	1
1/4/1966	Feyzin, France	Fire	Propane	1,000	18
1/1/1968	Donreith, U.S.A.	Derailing (fire)	Ethylene oxide	2	0
8/21/1968	Lieven, France	Mechanical	Ammonia	20	5
1/2/1969	Repcelak, Hungary	Overfilling	Carbon dioxide	35	9
1/25/1969	Laurel, MS, U.S.A.	Derailing (fire)	Propane	65	2
2/18/1969	Crete, NE, U.S.A.	Derailing	Ammonia	65	8
4/29/1969	Cumming, IA, U.S.A.	Derailing	Ammonia	?	?
9/11/1969	Glendora, CA, U.S.A.	Fire	Vinyl chloride	55	0
6/21/1970	Crescent City, IL, U.S.A.	Derailing (fire)	Propane	275	0
1/19/1970	Baton Rouge, LA, U.S.A.	Overpressure	Ethylene	4	0
10/19/1971	Houston, TX, U.S.A.	Derailing (fire)	Vinyl chloride	50	1
2/9/1972	Tewksbury, MA, U.S.A.	Collision	Propane	28	2
3/30/1972	Rio de Janeiro, Brazil	Fire	Propane	1,000	37
9/21/1972	Turnpike, NJ, U.S.A.	Collision	Propylene	18	2
11/2/1972	San Antonio, TX, U.S.A.	Corrosion	Carbon dioxide	0.01	0
7/5/1973	Kingman, AZ, U.S.A.	Fire	Propane	100	13
1/11/1974	West St. Paul, MN, U.S.A.	Fire	Propane	27	4
2/12/1974	Oneonta, NV, U.S.A.	Derailing (fire)	Propane	288	0
7/29/1974	Pueblo, CO, U.S.A.	Fire	Propane	80	0
4/29/1975	Eagle Pass, TX, U.S.A.	Collision	Propane	18	16
12/14/1975	Niagara Falls, NY, U.S.A.	Runaway reaction	Chlorine	20	4
5/11/1976	Houston, TX, U.S.A.	Collision	Ammonia	20	6
8/31/1976	Gadsden, AL, U.S.A.	Fire	Gasoline	4	3
1977	Cartagena, Colombia	Overpressure	Ammonia	7	30
2/22/1978	Waverly, TN, U.S.A.	Derailing	Propane	45	12
7/11/1978	Els Alfacs, Spain	Dilatation/overpressure	Propylene	23.6	216
5/30/1978	Texas City, TX, U.S.A.	Fire	Butane	1,500	7
8/30/1979	Good Hope, LA, U.S.A.	Ship collision	Butane	120	12
8/1/1981	Montonas, Mexico	Derailing	Chlorine	110	29
1/19/1982	Spencer, USA	Overheating	Water	0.3	7
12/11/1982	Taft, LA, U.S.A.	Runaway reaction	Acrolein	250	0
7/12/1983	Reserve, LA, U.S.A.	Runaway reaction	Chlorobutadiene	1	3
10/4/1983	Houston, TX, U.S.A.	Overfilling	Methyl bromide	28	2
11/19/1984	Mexico City, Mexico	Fire	Propane	3,000	500
1/28/1986	Kennedy Space Center, FL, U.S.A.	Fire	Hydrogen	115	7

Source: Prugh, 1991.

TABLE 22.2 The Most Frequent External Causes of BLEVEs

Cause	%
Fire	26
Derailing	20
Overfilling	18
Runaway reaction	12
Collision	10
Overpressure	6
Other	8

able to withstand the high stress and they will collapse (the steel typically used for the construction of LPG vessels may fail at pressures of about 15 atm, when the temperature of the walls reaches approximately 650°C). This is most likely to occur in the top section of the container, where the walls are not in contact with the liquid and therefore not cooled by it; the temperature of the walls will increase and their mechanical resistance will decrease (Birk, 1995). Instead, the wall in contact with the liquid will transfer heat to the liquid, thus maintaining a much lower temperature. If a safety valve opens, the boiling liquid will have a stronger cooling action due to the heat of evaporation.

Upon failure, due to the instantaneous depressurization, the temperature of the liquid will be greater than that corresponding to the new pressure according to the saturation curve in the P-T diagram. In this unstable condition, it is called superheated liquid. Liquids normally can withstand a small amount of superheating, which in certain experimental conditions can be extended far above the atmospheric-pressure boiling point (Prugh, 1991). However, there is a limit to superheat, called the superheat temperature limit (different for each substance). Then, if the temperature of the liquid at the moment of depressurization is higher than the superheat temperature limit, a violent and instantaneous flash of a fraction of the liquid and a superheated liquid vapor explosion will take place; a biphasic liquid/vapor mixture will then be released. This phenomenon occurs in a very short time (1 ms). The significant increase in the liquid's volume when it vaporizes—1,750 times in the case of water and 250 times in the case of propane—plus the expansion of the previously existing vapor, will give rise to a strong pressure wave (explosion, bursting of the container) as well as to the breaking of the container into several pieces, which will be propelled considerable distances. Experimental work performed with small 1-l vessels (Mcdevitt et al., 1990) has shown that when there is a break in the vessel, the pressure drops slightly and then rises up to a maximum; the initial depressurization brings the fluid near the break to a superheated state, thus causing a local explosion.

If the substance involved is not combustible, the pressure wave and the missiles will be the only effects of the explosion. This could happen if a steam boiler (water steam) exploded. If the substance is a fuel, however, as often happens in the process industry (for example, liquefied petroleum gas, such as ethylene or propane), the mixture of liquid/gas released by the explosion will probably ignite, giving rise to a fireball of approximately hemispherical shape, initially at ground level. The effect of the thermal radiation in this first stage, which is usually only a couple of seconds, is very important. The whole mass of fuel can burn only at its periphery because there is no air inside the mass (the mixture is outside the flammability limits).

In fact, not all the fuel initially contained in the tank is involved in this fire. Some of the fuel is entrained in the wake formed by the flying fragments. In one case (Mexico City, 1984), it has been suggested that a portion of the liquid was thrown significant distances without being ignited, which caused local fires (this effect has not been mentioned in any other case). This decreases the amount of fuel contained in the fireball and also affects its dimensions and the duration of the fire.

Later on, the turbulence of the fire entrains air into the fireball. Simultaneously, the thermal radiation vaporizes the liquid droplets and heats the mixture. As a result of these processes, the whole mass turbulently increases in volume, evolving towards an approximately spherical shape that rises, leaving a wake of variable diameter. Such fireballs can be very large, causing a very strong thermal radiation. The combined action of BLEVE and fireball can be summarized therefore in the following effects:

- Thermal radiation
- Pressure wave
- Flying fragments

The mode in which these effects actuate varies: punctual or directional in the case of projectiles, and zonal—covering a given surface—in the case of thermal radiation and blast.

It is worth noting that it is practically impossible to establish the exact instant at which the explosion will take place. Twenty years ago it was believed that once the emergency started—for example, from the instant in which fire started to impinge on a container—there was a certain time available before the explosion. It therefore seemed that diverse measures could be taken to prevent the explosion (for example, firefighters could refrigerate the tank with hoses). However, as more information was gathered on actual accidents, it became evident that this time could be extraordinarily short; in the San Juan Ixhuatepec accident in Mexico City (Pietersen and Cendejas, 1985), the time elapsed between the first explosion (which caused the fire) and the first BLEVE was only 69 seconds. The various stages of a BLEVE/fireball are shown in Fig. 22.1.

The instant at which a BLEVE can occur in a tank exposed to fire depends on the following factors:

1. Thermal flux from the fire, which will be a function of the distance from the flame to the tank and will depend on whether there is flame impingement and the type of flame (pool fire, torching, etc.)
2. Diameter of the tank
3. Tank fill level
4. Release capacity of safety valves
5. Existence of a layer (with a certain thickness) of isolating material (passive protection)

Theoretically, an insulated container should resist the effect of the flames from a pool-fire (thermal flux of approximately 100 $kW \cdot m^{-2}$) for two hours. In the case of a jet fire, the thermal flux increases significantly (up to 350 $kW \cdot m^{-2}$). In these conditions, some BLEVEs have occurred in the first minutes. For the development of this type of accident, the following times have been suggested (Nazario, 1988): flame impingement from a jet fire, 5 minutes; flame impingement with turbulent flames, 30 minutes (this value agrees with that proposed by ASTM (ASTM, 1983), 20 to 30 minutes). Although this time can vary with the features of the installation (insulating layer, cooling devices), it is evident that other factors can decrease it significantly (partial destruction due to impacts or pressure wave, for example). The most cautious practice, therefore, is to take into account that the explosion can occur at any moment from the beginning of the emergency. The exclusion area should therefore be rapidly evacuated.

FIGURE 22.1 BLEVE-fireball: loss of containment, fire, tank heating, bursting and evolution of fireball.

22.3 CONDITIONS REQUIRED FOR A BLEVE TO OCCUR

22.3.1 Superheating and Depressurization

While the explosion of a tank containing a pressurized flammable liquid will almost always lead to a fireball, the explosion cannot always be considered strictly a BLEVE. To qualify as this type of explosion, the following conditions must be met (Reid, 1979; Mañas, 1984; Bestratén and Turmo, 1991a; Birk, 1995):

- Significant *superheating* of the liquid. Most liquefied gases under fire attack (LPG, ammonia, chlorine) fulfil this condition; it can also be fulfilled by other liquids contained in closed containers that undergo anomalous heating, for example due to a fire; and, as stated before, water can also be at this condition upon instantaneous depressurization.
- Instantaneous *depressurization*. This phenomenon is usually related to the type of failure of the vessel. The sudden pressure drop in the container upon failure causes the liquid superheat. If the liquid superheat is significant, the flashing may be explosive.

When these two conditions are met, a practically instantaneous evaporation of the contents takes place, with the formation of a large number of boiling nuclei in all the liquid mass (homogeneous nucleation). In these conditions, the velocity at which the volume increases is extraordinary and the explosion is therefore very violent. Strictly speaking, this is the phenomenon associated with the BLEVE explosion.

22.3.2 Temperature and Superheating Limit Locus

Diverse authors have suggested procedures to establish the superheat temperature limit and the superheating limit locus that determine, for each substance, the conditions under which a BLEVE can occur. Reid (1976, 1979) made a significant contribution to this field.

The theoretical superheating limiting conditions at which spontaneous homogeneous nucleation will exist in all the liquid mass can be established from the tangent line to the vapor pressure-temperature curve at the critical point. This represents the limit to which the liquid may be heated before spontaneous nucleation occurs with a vapor explosion (Reid, 1979). This is shown in Fig. 22.2.

The relationship between the vapor pressure and the temperature is established by the Antoine equation:

$$\ln P = -\frac{A}{T} + B \tag{22.1}$$

The tangent to the saturation curve at the critical point is obtained by calculating the derivative of pressure with respect to temperature:

$$\frac{dP}{dT} = A\frac{P}{T^2} \tag{22.2}$$

By applying this expression to the critical point,

$$\frac{dP_c}{dT_c} = \frac{P_c A}{T_c^2} = \text{tg}\ \alpha \tag{22.3}$$

This expression gives the slope of the line tangent to the saturation curve at the critical point. The equation of this straight line is:

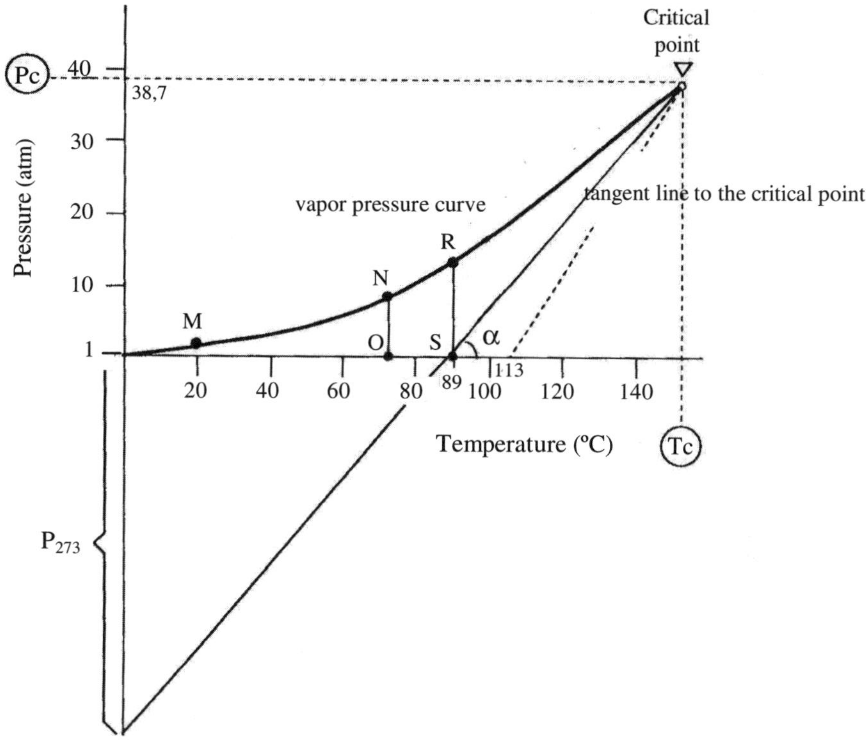

FIGURE 22.2 Saturation curve for butane and limiting conditions for BLEVE.

$$P = \text{tg } \alpha \cdot T + b \tag{22.4}$$

An example will show how this expression can be used. The value of the superheating limiting value will be calculated for butane. The equilibrium data corresponding to the critical point and to atmospheric pressure are:

$$P_c = 38.7 \text{ atm} \qquad T_c = 425.8 \text{ K}$$

$$P = 1 \text{ atm} \qquad T = 272.5 \text{ K}$$

By introducing these data in the Antoine equation, the values of constants A and B for butane are found (the pressure expressed in atm and the temperature in K):

$$A = 2769 \qquad B = 10.16$$

The slope of the tangent to the saturation curve at the critical point is therefore:

$$\text{tg } \alpha = 38.7 \cdot \left(\frac{2769}{425.8^2}\right) = 0.591$$

And the value of the ordinate at the origin, b, can be found by again introducing the values corresponding to the critical point, thus obtaining $b = -213$. In this way, the equation of the tangent line is obtained; its intersection with the horizontal line at $P = 1$ gives a temperature of 89°C (Fig. 22.2). Therefore, for a vessel containing liquid butane, the minimum temperature required to reach (upon vessel failure) a superheating degree causing spontaneous nucleation (and therefore BLEVE) is 89°C; at this temperature, the spontaneous nucleation would occur at 1 atm.

For the coordinate system used, with the origin at 0°C, the ordinate at the origin will be the pressure corresponding to 273 K,

$$P_{173} = 0.591 \cdot 273 - 213 = -52 \text{ atm}$$

Take for example a vessel containing butane at room temperature (20°C), in which liquid and vapor are at equilibrium at an absolute pressure of 2 atm (point M in Fig. 22.2). If, due to the thermal radiation from a fire, the temperature increases to 70°C, the pressure inside the vessel will be 8 atm (point N). If, at these conditions, the vessel bursts (due to the failure of the material or an impact, for example), there will be an instantaneous depressurization from 8 atm to the atmospheric pressure. At the atmospheric pressure, the temperature of the liquid-vapor mixture will be −0.5°C (point O in Fig. 22.2) and the depressurization process corresponds to the vertical line between N and O. As this line does not reach the tangent to the saturation curve at the critical point, the conventional theory states that there will be no BLEVE strictly speaking: although there will be a strong instantaneous vaporization and even an explosion, nucleation in all the liquid mass will not occur.

Instead, if during the heating process the liquid temperature reaches, for example, 89°C (point R in Fig. 22.2), during the depressurization the tangent line will be reached (point S in Fig. 22.2). In this case the conditions required (superheating) by the aforementioned spontaneous homogeneous nucleation would exist and a BLEVE explosion would occur.

However, this conventional theory, although accepted by many authors, fails to explain some of the BLEVEs that have occurred.

In fact, the use of the tangent line to the saturation curve at the critical point as the limiting value for the occurrence of a BLEVE implies a margin of safety. The experimental data seem to indicate that the difference between the real superheating limit required to originate a BLEVE and the value thus obtained is in the range of 15 to 25°C (Table 22.3). This is because when the substance is depressurized to reach the superheating limit line, there is also a slight decrease in its temperature. Thus, to reach the intersection point between the tangent at the critical point and the straight line at $P = 1$ atm (superheat temperature limit), it is necessary to start from a higher temperature. If, from this new temperature, a vertical line is drawn, a new limit temperature will be obtained for the line $P = 1$ atm. Connecting this new point to the critical point, a new superheating limit line will be obtained which is nearer to the real situation than the tangent line. The following expression has been suggested (Sigalés and Trujillo, 1990) for the calculation of this superheating limit:

$$T_g - T_s = 0.82206\ T_{cr} - 0.89485\ T_s \tag{22.5a}$$

where T_g is the superheat temperature limit and T_s is the saturation temperature corresponding to atmospheric pressure. For many organic and inorganic liquids the superheat temperature limit values are within 88 to 92% of the critical temperature. Reid (1976) used the equation-of-state of Redlich-Kwong to obtain the following expression:

$$T'_g = 0.895 \cdot T_{cr} \tag{22.5b}$$

The difference between the temperatures calculated from these two expressions ranges from −1°C to +8°C for the substances included in the table (Armet, 1997).

However, this is a theoretical treatment and some authors (e.g., Casal et al., 1999) doubt the applicability of these predictions to a real case. In the heating of a large vessel by a fire, there are probably some aspects that can make it difficult to predict the BLEVE occurrence. Amongst these we can cite the liquid temperature stratification, which can significantly affect the thermodynamic conditions inside the vessel. The content of a tank is not heated uniformly

TABLE 22.3 Calculation of the Superheat Temperature Limit for Diverse Substances

Substance	Formula	T_s, K	T_c, K	P_{cr}, atm	ln P_{cr}	A	B	tg α	b, atm	T_R, K	T_g, K	T'_g, K	$T'_g - T_g$, K
Water	H_2O	373	647	217.7	5.38312	4741	12.71	2.4656	−1378	559	571	579	8
Carbon dioxide	CO_2	195	304	73	4.29046	2333	11.96	1.8429	−487	265	270	272	2
Ammonia	NH_3	240	406	112.3	4.72117	2771	11.55	1.8878	−654	347	359	363	4
Phosgene	$COCl_2$	281	455	56	4.02535	2958	10.53	0.8001	−308	386	404	407	3
Methane	CH_4	111.5	191	45.8	3.82428	1024	9.18	1.2856	−200	156	169	171	2
Ethane	C_2H_6	184.4	305	48.8	3.88773	1812	9.83	0.9499	−241	255	270	273	3
Ethylene	C_2H_4	169.1	282.7	50.9	3.92986	1654	9.78	1.0534	−247	235	250	253	3
Propane	C_3H_8	231.1	369.8	43	3.76120	2317	10.03	0.7286	−226	312	328	331	3
Propylene	C_3H_6	225	365.3	45	3.80666	2230	9.91	0.7520	−230	307	324	327	3
n-Butane	C_4H_{10}	272.5	425.8	38.7	3.65584	2767	10.15	0.5906	−213	362	379	381	2
n-Pentane	C_5H_{12}	309.3	470.2	33	3.49651	3160	10.22	0.4717	−189	403	419	421	2
n-Hexane	C_6H_{14}	342	507.8	29.5	3.38439	3545	10.37	0.4056	−176	436	453	454	1
n-Heptane	C_7H_{16}	371.3	539.8	26.8	3.28840	3911	10.53	0.3597	−167	467	483	483	0
n-Octane	C_8H_{18}	398.8	569.2	24.7	3.20680	4272	10.71	0.3257	−161	497	510	509	−1
Ethyl eter	$(C_2H_5)_2O$	307.6	467	35.5	3.56953	3217	10.46	0.5236	−209	401	416	418	2
Chlorine	Cl_2	238.4	419	93.5	4.53796	2510	10.53	1.3368	−467	350	370	375	5

during a fire. In diverse fire tests conducted on LPG tanks (Towsend et al., 1974; Appleyard, 1980; Birk and Cunningham, 1996), the temperature of the liquid varies from the tank bottom (where the liquid is cooler) to the top (where the liquid is warmer) (Birk, 1995). This temperature stratification is due to buoyancy effects. Therefore, the temperature stratification plays an important role in case of an accident. This is a field in which there is still interesting research to be done.

22.4 ESTIMATION OF BLEVE EFFECTS

22.4.1 Thermal Radiation

When a BLEVE explosion involves a flammable substance, it is usually followed by a fireball and intense thermal radiation will be released. The thermal energy is released in a short time, usually less than 40 seconds (although this time is a function of the mass in the tank). The phenomenon is characterized from the first moments by strong radiation; this eliminates the possibility of escaping for the persons nearby (who also will have suffered the effects of the blast).

The parameters that must be evaluated for predicting the effects of a fireball are the diameter, duration, and thermal radiation at any given distance. In this section, a methodology is described to estimate these values.

Diverse authors have proposed correlations for the prediction of the diameter and duration of a fireball originated by a given mass of fuel (CCPS, 1994). Most of them have the following general expression:

$$D = a \cdot M^b \tag{22.6}$$

$$t = c \cdot M^e \tag{22.7}$$

where a, b, c, and e are empirical or semiempirical constants. A comparative study of 16 of these expressions has been made (Satyanarayana et al., 1991) (Table 22.4).

These authors compared the predictions from the diverse equations to real data from several explosions. The statistical analysis showed that the "best" correlation for the estimation of fireball diameter was the one proposed by Gayle (2), followed by those proposed by Marshall, Roberts, TNO, and High. Concerning the duration of the phenomenon, no comparison was possible, as there were no real data in the literature. Therefore, the expression proposed by Gayle could also be used. In another analysis carried out with 23 equations, Capdevila (1994) showed that the best ones were those proposed by Gayle (2), Marshall, Roberts, Moorhouse, TNO, High and Clay et al. Other authors (e.g., Lefin et al., 1993) found significant differences between the correlations published.

Therefore, it is rather difficult to establish which is really the best equation. Actually, and according to the previously comparative analysis, the diameter and duration of a fireball can be predicted using the following correlations:

$$D = 6.14 \cdot M^{0.325} \tag{22.8}$$

$$t = 0.41 \cdot M^{0.340} \tag{22.9}$$

where the units are m (D), kg (M), and s (t).

It is worth noting, however, that there are very few experimental data available to back up this type of comparative analysis. Furthermore, these data—obtained from real accidents in the case of large fireballs—are not always accurate, as often the films are incomplete or

TABLE 22.4 Equations for Estimating Fireball Diameter and Duration

Author	Reference	a	b	c	e
Gayle (1)	(Bagster and Pitblado, 1989)	3.68	0.326	0.245	0.356
Gayle (2)	(Bagster and Pitblado, 1989)	6.14	0.325	0.410	0.340
Brasie	(Bagster and Pitblado, 1989)	3.80	0.333	0.300	0.333
Marshall	(Bagster and Pitblado, 1989)	5.50	0.333	0.380	0.333
Roberts	(Lees, 1980)	5.80	0.333	0.450	0.333
SRD[a]	(Bagster and Pitblado, 1989)	6.00	0.333	0.005	—[c]
Fay-Lewis	(Fay and Lewis, 1977)	6.36	0.333	2.570	0.167
Hardee	(Bagster and Pitblado, 1989)	6.24	0.333	1.110	0.167
Hasegawa	(Bagster and Pitblado, 1989)	5.28	0.277	1.099	0.097
Hasegawa-Sato	(Hasegawa and Kato, 1978)	5.25	0.314	1.070	0.181
Moorhouse	(Bagster and Pitblado, 1989)	5.33	0.327	0.923	0.303
TNO	(Bagster and Pitblado, 1989)	6.48	0.325	0.852	0.260
Maurer	(Lihou and Maunde, 1982)	3.51	0.333	0.320	0.333
High	(Lihou and Maunde, 1982)	6.20	0.320	0.490	0.320
HSCC[b]	(Lihou and Maunde, 1982)	6.45	0.333	5.530	0.333
API	(Kayes, 1985)	5.33	0.327	1.089	0.327

[a] Safety and Reliability Directorate.
[b] Hot shell cold core model.
[c] M is used as $\log_{10} M$

bad. The significant difficulty involved in experiments on a large scale complicates the study of fireball accidents that happen from time to time in the industry or in the transportation of certain materials.

In fact, the lack of accuracy is not due only to the differences in the predictions from the diverse correlations. Another factor influencing it is the estimation of the fraction of the overall mass of fuel that really is involved in the fireball. As happens in many cases of risk analysis, the inaccuracy arises from the definition of the problem itself. It should be taken into account that some fuel has been leaving the vessel through the safety valves from the moment in which they opened; the amount released will depend on the time elapsed between this moment and that of the explosion. Furthermore, more fuel is entrained in the wake of the propelled fragments. The final result is that it is impossible to accurately establish the mass of fuel that will actually contribute to the fireball. This difficulty is found in the criteria recommended by different authors. Nazario (1988) suggests that the mass corresponding to the maximum capacity of the vessel should be used, and Pietersen and Cendejas (1985) recommend 90% of this value; other authors consider that only two-thirds or three-quarters of the initial fuel mass is finally involved in the fireball.

However, in calculating the fireball's diameter, most correlations include the mass of fuel affected by an exponent equal to one-third; this reduces considerably its influence on the value of D. This influence has been calculated (Calpe and Casal, 1989) for a maximum mass of 15,000 kg (Table 22.5).

Finally, the lack of accuracy is also due to the fire wake left by the fireball, which can reach a significant size; this modifies the flame surface and consequently the radiation that will reach a given point.

In any case, the correlations mentioned in the previous paragraphs allow an approximate estimation of the size of the fireball. It should be taken into account, furthermore, that its size and position change continuously; therefore, the thermal radiation is not constant. The available films of BLEVE accidents show that the fireball grows quickly up to its maximum diameter, remaining at this diameter for a short time and then dissipating. Usually the cal-

TABLE 22.5 Influence of the Initial Mass of Fuel on the Fireball Diameter (values corresponding to a maximum mass of 15,000 kg of LPG)

%	*D, m*
100	140
90	135
75	127
67	123

culation of the radiation received by a given target is performed supposing that the fireball reaches its maximum size immediately after reaching a certain height.

To estimate the radiation received by a surface located at a given distance, the solid body model can be applied:

$$I = \tau \cdot F \cdot E_p \tag{22.10}$$

It is necessary, therefore, to know the value of the emissive power (E_p), the view factor (F), the atmospheric transmissivity (τ), and the distance between the flame and the target. To know this distance, it is necessary to estimate the height at which the fireball is located. In fact, this height is a function of the specific volume and the latent heat of vaporization of the fuel; therefore, strictly speaking, it varies with the substance. This is not usually taken into account. Diverse correlations have been proposed to estimate this height; one of the most simple ones is the following:

$$H = 0.75\ D \tag{22.11}$$

where H is the height at which the center of the fireball is located (in m) and D is its diameter, calculated with Eq. (22.8).

Fay and Lewis (1977) proposed another correlation:

$$h = 12.7\ V_i^{1/3} \tag{22.12}$$

where h is the height of the top of the fireball (in m) and V_i is the initial volume of vapor in the fireball (in m^3).

Both expressions have been compared to the values corresponding to three real cases; the results can be seen in Table 22.6 (the heights correspond to the top of the fireball; note that if Eq. (22.11) is used, $(D/2)$ must be added to obtain the value of h).

TABLE 22.6 Predicting the Height (Top of the Flame) of the Fireball

Accident	Fuel	Mass, kg	*h, m* (observed)	*h, m* (Eq. 22.11)	*h, m* (Eq. 22.12)
Crescent City[a]	Propane	35,000	230	230	315
Priolo	Ethylene	80,000	225	301	435
Priolo	Propylene	50,000	250	258	357

[a] Lewis, 1991.

The results from Eq. (22.11) are much closer to the observed values than those predicted by the other correlations; although only three cases imply a low statistical significance, this equation—which, furthermore, is the simplest one—can be recommended.

Once more, we do not know for certain what fraction of the energy released is emitted as thermal radiation. In fact, this is the most important uncertainty in the calculation of the thermal radiation from a fireball. The following correlation has been proposed (Roberts, 1982) to estimate this value:

$$\eta = 0.27 \cdot P_o^{0.32} \qquad \text{(maximum value of } \eta \text{ limited to 0.4)} \tag{22.13}$$

where P_o is the pressure (relative) in the vessel just before the explosion, in MPa. This expression was obtained from experimental data at laboratory scale, with amounts of fuel of a few kilograms; its validity at real scale has not been tested. The value of η ranges from 0.13 to 0.35, according to diverse authors, and from 0.24 to 0.40 according to others. In any case, its maximum value is 0.4. From this radiation coefficient and the overall combustion energy, the energy radiated can be deduced. Then, the emissive power can be calculated from the following equation:

$$E_p = \frac{\eta\, M\, H_c}{\pi\, D^2\, t} \tag{22.14}$$

where H_c is the combustion heat (kJ · kg^{-1}) and t the time of duration of the fireball (s). Because the value of η can be inaccurate, another possibility is to use directly an arbitrary value of E_p in the range of 200 to 350 kW · m^{-2} according to some authors (Bagster and Pitblado, 1989) and 200 kW · m^{-2} according to others (Pietersen and Cendejas, 1985).

The maximum view factor is that corresponding to a sphere and a plane surface perpendicular to its radius. Due to the geometrical simplicity of this system, this factor can be calculated with a very simple equation:

$$F = \frac{D^2}{4r^2} \tag{22.15}$$

where r is the distance between the surface receiving the radiation and the center of the fireball [$(R + x)$ in Fig. 22.6]. For other positions of the surface of the target, the value of F must be corrected by using the angle formed by the surface and the surface perpendicular to the radius (see the example at the end of this section).

Finally, the atmospheric transmissivity can be estimated from the following equation:

$$\tau = 2.02 \cdot (P_w \cdot x)^{-0.09} \tag{22.16}$$

22.4.2 Mechanical Energy Released in the Explosion

When a vessel bursts in a BLEVE explosion, the mechanical energy contained inside is released (note that the units of pressure are energy per unit volume). The substance contained in the vessel instantaneously increases in volume due to the expansion of the vapor already existing in the vessel at the moment of the explosion and the superheated liquid, which undergoes a partial vaporization practically instantaneously (flash).

The energy released in a BLEVE explosion is distributed among the following:

- The energy of the pressure wave
- The kinetic energy of the projectiles

- The potential energy of the fragments (deformation plastic energy absorbed by the fragments)
- The heating of the environment

The relative distribution of the energy will change in relation to the particular conditions of the explosion. It is very difficult to establish accurately which amount of energy will contribute to the pressure wave (Capdevila, 1994). An important aspect is the type of failure (fragile or ductile). When the tank wall is heated and stressed, it may begin to thin due to plastic creep, and fissures may form. If this effect is very localized (for example, due to a jet fire) the fissure may stop growing as it enters thicker and stronger material (Birk, 1995). However, this fissure may result in sudden depressurization and a strong flashing effect in the liquid. As a result, pressure recovery can take place and this can restart the crack and finally lead to tank bursting and BLEVE.

If the thermal effects are more widespread (as for example, in the case of a tank engulfed in a pool fire (Planas-Cuchi et al., 1996)), then the fissure may continue to grow and the tank may burst, resulting in a BLEVE. The propagation speed will depend on the mode of failure. If the wall thickness has decreased by plastic creep, the crack speed will be different than if the crack must propagate through thicker material by shear failure (Birk, 1995). There is an upper limit to the crack velocity, related to the material yield strength and density (Baum, 1982). The actual propagating velocity is usually less than the limiting velocity; it can reach velocities up to 200 $m \cdot s^{-1}$.

It has been suggested that in a fragile failure of a vessel, 80% of the energy released contributes to the creation of the pressure wave. In the case of a ductile breaking—in which large fragments of the vessel are propelled—the energy in the pressure wave is only 40%. In both cases, the rest of the energy becomes kinetic energy of the fragments, as the fourth contribution (heating of the environment) is negligible.

In fact, most vessels or tanks are constructed with materials that are ductile at the operating conditions. A fragile failure is found only in very special conditions, when the stress reached by the material is much higher than its plastic limit. This only happens with tempered steel and glass. Therefore, BLEVE explosions usually consist of ductile breaking.

Concerning the vapor initially in the vessel, the energy released in its expansion (from the breaking pressure in the vessel up to the atmospheric pressure) is:

$$E_v = m(u_1 - u_2) \qquad (22.17)$$

where E_v = the energy released in the expansion of the vapor (kJ)
m = the mass of vapor already existing in the vessel at the moment of the failure (kg)
u_1 = the internal energy of the vapor under the conditions at which the vessel bursts ($kJ \cdot kg^{-1}$)
u_2 = the internal energy of the vapor after the expansion up to atmospheric pressure ($kJ \cdot kg^{-1}$).

Supposing that the expansion is isoentropic—due to the velocity at which it takes place—and that the vapor behaves as an ideal gas, this energy is (Prugh, 1991):

$$E_v = 10^2 \cdot \left(\frac{P \cdot V}{\gamma - 1}\right) \cdot \left(1 - \left(\frac{P_a}{P}\right)^{(\gamma-1)/\gamma}\right) \qquad (22.18)$$

where P_a is the atmospheric pressure (bar), V is the initial volume of vapor (m^3), γ is the ratio of specific heats, and P is the pressure (bar) in the vessel just before the explosion.

This energy can be expressed as TNT equivalent mass by using the adequate energy conversion factor (approximately 1,120 cal per gram of TNT),

$$W_{\mathrm{TNT}} = \left(\frac{0.021 \cdot P \cdot V}{\gamma - 1}\right) \cdot \left(1 - \left(\frac{P_a}{P}\right)^{(\gamma-1)/\gamma}\right) \tag{22.19}$$

where W_{TNT} is the equivalent mass of TNT (kg).

Furthermore, if the vessel contained superheated liquid—as in the case of a BLEVE explosion—the released energy can be estimated approximately by using the same method. In this case, it must be taken into account that the mass of liquid will partly vaporize suddenly when reaching atmospheric pressure. The volume of this vapor at the pressure in the vessel just before the explosion must then be calculated; adding this fictitious volume to the real one, the equivalent mass of TNT will be:

$$W_{\mathrm{TNT}} = \left(\frac{0.021 \cdot P \cdot V^*}{\gamma - 1}\right) \cdot \left(1 - \left(\frac{P_a}{P}\right)^{(\gamma-1)/\gamma}\right) \tag{22.20}$$

where V^* is the volume of vapor in the vessel plus the volume (at the pressure inside the vessel) of the vapor generated in the explosion, in m^3:

$$V^* = V + V_1\, f\left(\frac{\rho_1}{\rho_v}\right) \tag{22.21}$$

V is the volume of vapor inside the vessel before the explosion, V_1 is the volume of liquid in the vessel before the explosion (m^3), and f is the vaporization fraction (flash), i.e., the fraction of liquid which vaporizes in the depressurization; its value can be calculated with the following expression:

$$f = 1 - \exp\left(-2.63(C_p/H_v)(T_c - T_b) \cdot (1 - ((T_c - T_o)/(T_c - T_b))^{0.38})\right) \tag{22.22}$$

where T_c is the critical temperature of the substance (K), T_b is the boiling temperature of the substance at atmospheric pressure (K), T_o is the temperature of the substance in the moment of the explosion (K), and H_v is the enthalpy of vaporization of the substance (kJ · kg^{-1}).

In fact, the blast contribution from the liquid will be affected by the lack of homogeneity in liquid temperature, decreasing as liquid temperature stratification increases (since liquid stratification reduces the average liquid temperature, the vaporization fraction will decrease). However, the actual knowledge of the liquid stratification phenomenon does not allow the suitable correction to be introduced into the blast calculation.

22.4.3 Pressure Wave

The pressure wave generated by the explosion can be estimated from the equivalent TNT mass. This method implies a certain inaccuracy because in the BLEVE explosion of a vessel the energy is released at a lower velocity than in a TNT explosion and also because the volume of the vessel is much larger than that which would have the equivalent amount of a conventional explosive. Nevertheless, the method is simple and allows useful estimations.

Due to the fact that the volume initially occupied by the energy released in the explosion is much larger than that which would occupy the equivalent mass of TNT, a correction must be made on the distance from the explosion center to the place in which the pressure wave must be estimated.

This correction is carried out by using the scaled distance, d_n, based on the similitude principle proposed by Hopkinson in 1915, according to which when two explosive charges of similar geometry and of the same explosive but different sizes, detonate in the same atmosphere, similar pressure waves are generated at the same scaled distance. This principle can also be applied to two different explosives, taking into account the fact that two types of explosion with the same overpressure give rise to the same effects. Because overpressure is a function of the distance and two different explosions do not cause the same overpressure at the same distance from the center of the explosion, the scaled distance is defined as that at which the overpressure has the same value for both explosions. The scaled distance is related to the real distance and to the equivalent TNT mass by the cubic root law,

$$d_n = \frac{d}{(\beta \cdot W_{\mathrm{TNT}})^{1/3}} \tag{22.23}$$

where d_n = the scaled distance ($\mathrm{m \cdot kg^{-1/3}}$)
β = the fraction of the energy released converted in pressure wave
d = the real distance (from the center of the explosion) at which the overpressure must be estimated (m).

From the value of d_n it is possible to estimate the overpressure by using a graph as shown in Fig. 22.3.

22.4.4 Missiles

Because of their random behavior, projectiles from BLEVEs are one of the most difficult hazards to quantify (Birk, 1996). The fragments thrown by the explosion have a restricted and directional action, but with a larger radius of destructive effects than the pressure wave and the thermal effects of the fireball. These fragments can cause a domino effect if they destroy other tanks or equipment. The velocity required by a fragment to penetrate another similar tank ranges from 4 to 12 $\mathrm{m \cdot s^{-1}}$, and the maximum velocity that can be reached by the fragments in a BLEVE explosion—a function of the conditions at which the explosion occurs, the volume of vapor initially contained in the vessel, and the shape of the vessel—ranges from 150 to 200 $\mathrm{m \cdot s^{-1}}$.

There are basically two kinds of projectiles from BLEVEs, as in the case of conventional explosions of containers:

1. Primary projectiles, which are major pieces of the container
2. Secondary projectiles, which are generated by the acceleration of nearby objects (pipes, bars, bricks, etc.)

The number of primary projectiles (i.e., major pieces of the tank) will depend on the type of failure, the shape of the vessel, and the severity of the explosion. Typically, a BLEVE will involve a ductile failure; the cracks will propagate at lower velocity and without branching. The number of fragments will be less than if it were a fragile failure. The number of projectiles will be in the range of 2 to 15 (Baum, 1988, has reported that typically there will be less than five projectiles).

In the case of cylindrical tanks, the initial crack usually follows an axial direction and then changes and follows a circumference (following, for example, a welding); thus, the vessel is usually broken into two pieces: the bottom of the tank and the rest of the vessel.

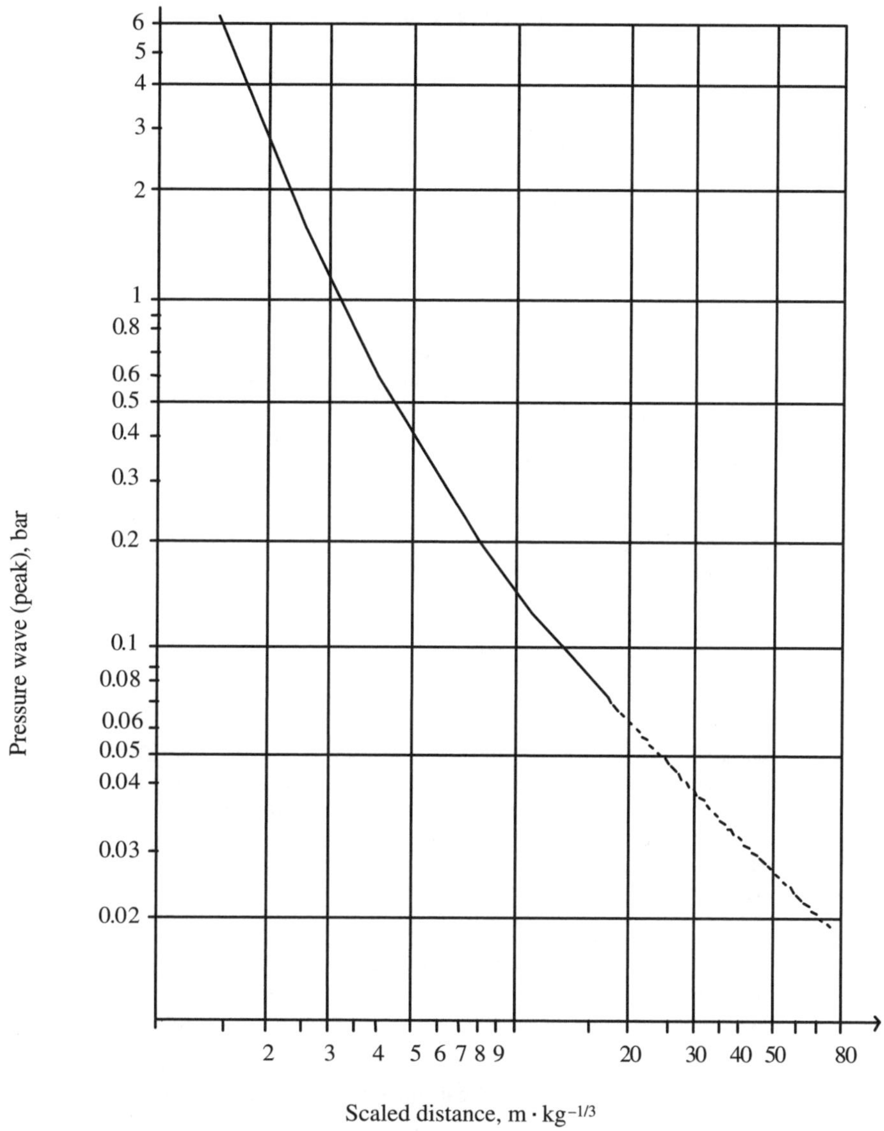

FIGURE 22.3 Pressure wave as a function of the scaled distance. (*Source: Van den Berg and Lannoy, 1993*)

The results of an analysis on 130 BLEVE cases, most of which occurred in cylindrical tanks, are shown in Table 22.7.

In cylindrical tanks there can also be three projectiles. If there are three fragments, there can be two types of failure. The vessel can be divided into two bottoms and the central body, or it can be first divided into two fragments, one bottom and the rest, and then this second fragment can be divided through the imaginary line that would separate the liquid and the vapor as shown in Fig. 22.4. The bottom usually breaks by the welding; if there is no

TABLE 22.7 Number of Projectiles in BLEVE Accidents

	With projectiles	Without projectiles
Due to fire	89	24
Without any fire	17	—

Source: Holden and Reeves, 1985.

welding, we can suppose that it will break at a distance from the end equal to 10% of the total length of the vessel.

Concerning the direction, the projectiles will probably follow the direction of the cylinder axis. Data from 15 accidents (Holden and Reeves, 1985) provided the data in Table 22.8, taking into account the 45° sectors at each side of the cylinder. Data obtained experimentally from 13 BLEVEs of 400-l vessels (Birk, 1996) gave somewhat different results; in some cases, this author observed that the tank remained flattened on the ground with both ends attached.

The distance reached by projectiles from cylindrical tanks is usually greater than that reached by fragments from spherical vessels. The following expressions have been suggested (Baum, 1988; Birk, 1995) for the prediction of the range of cylindrical tank projectiles (tube fragments):

$$\text{For tanks} < 5\ \text{m}^3 \text{ in capacity:} \qquad l = 90 \cdot M^{0.33} \tag{22.24a}$$

$$\text{For tanks} > 5\ \text{m}^3 \text{ in capacity:} \qquad l = 465 \cdot M^{0.1} \tag{22.24b}$$

where M is the mass of substance contained in the vessel (kg) and l is the range (m).

The difference between the two expressions is due to the reduced relative effect of drag (ratio of drag force to tank weight) as tank size increases (when tank capacities exceed 5 m^3, the tank size is increased by increasing the tank length, not the tank diameter) (Birk, 1995). These expressions were obtained assuming the tank is 80% full of liquid LPG at the time of failure, and for fragments launched at an optimum angle (45° to the horizontal). Because most fragments will not be launched at this angle, the real ranges will typically be less than those predicted by Eqs. (22.24*a*) and (22.24*b*).

Recently, Baum (1999) has proposed the following expressions, which agreed very well with his experimental data, to estimate the upper bond of velocities for fragments from horizontal pressure vessels containing a high-temperature liquid:

$$\text{End-cap missiles:} \qquad U = \frac{1.25 \cdot K^{0.375}}{\left(\dfrac{P_{sat}}{P_a}\right)^{0.085}} \cdot a_o \tag{22.25}$$

$$\text{“Rocket” missiles:} \qquad U = \frac{I}{W} \tag{22.26}$$

where

$$K = \frac{P_o \cdot \pi \cdot R_f^3}{W \cdot a_o^2} \tag{22.27}$$

and I is the impulse (integrated pressure history of the closed end), which can be calculated as follows:

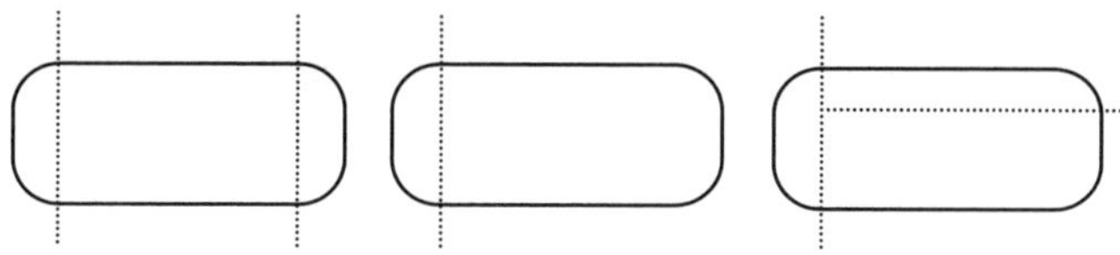

FIGURE 22.4 Common failure trends in cylindrical vessels.

$$\text{if } P_o > P_{sat} \qquad I = A \cdot (P_o \cdot t_w + P_{sat} \cdot (t_o - t_w) + P_{sat} \cdot t_e) \tag{22.28}$$

$$\text{if } P_o = P_{sat} \qquad I = P_{sat} \cdot A \cdot (t_o + t_e) \tag{22.29}$$

with

$$t_w = \frac{L}{a_w} \tag{22.30}$$

$$t_o = \left(\frac{W}{P_{sat} \cdot \pi \cdot R}\right)^{1/2} \tag{22.31}$$

$$t_e = \frac{L}{a_m} \tag{22.32}$$

Some years ago it was considered that projectiles could reach 500 m as a maximum; however, in the accident in Mexico City, a projectile from a large cylindrical vessel traveled 1,100 m and another one from a sphere went 600 m. The evacuation distance should therefore be 1,100 m for large cylindrical tanks.

In the case of spherical vessels, it is much more difficult to predict the number of fragments. The analysis of a reduced number of cases (Pietersen and Cendejas, 1985) gave an average of eight fragments per accident. This author obtained the following correlation:

$$N = -3.77 + 0.0096\ V_r \tag{22.33}$$

where N is the number of fragments and V_r is the volume of the spherical vessel, ranging from 700 to 2,500 m^3. However, taking into account the scattering of the data used by Pietersen, this correlation should not be considered very reliable.

The direction followed by the projectiles from a spherical vessel is practically impossible to predict. The analysis of several cases shows that the distribution is not symmetrical; this must be attributed probably to the special position of the contact flame/vessel in each case, although other aspects (construction details, for example) can also have an influence.

Finally, for projectile range, the analysis of 58 fragments from seven accidents has shown that 70% of the fragments reached distances less than 200 m. However, fragments from

TABLE 22.8 Probability of Projectile Launching in Cylindrical Vessels

Sector	Probability
1	0.62
2	0.38

Source: Holden and Reeves, 1985.

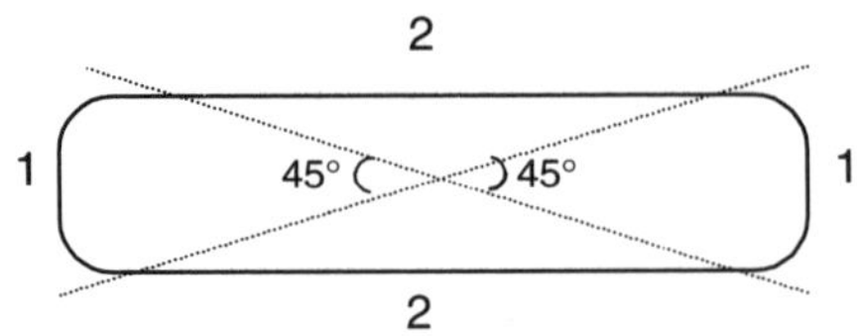

FIGURE 22.5 Distribution of projectiles from a cylindrical vessel.

spherical vessels have reached 600 m (Mexico City) and even 700 m. The distance reached is usually smaller in the case of fragments from spherical vessels because they are less aerodynamical than those from cylindrical tanks. Different theoretical models have been suggested for the prediction of these maximum distances, but they are not very practical, since to apply them the mass and shape of the fragment should be known.

22.5 PREVENTION MEASURES

In the case of an emergency that can lead to an accident of the BLEVE fireball type, it is very difficult to improvise adequate actions to control the situation. Any plan requiring the presence of people will be very dangerous because it is impossible to foresee when the explosion will occur. The actions should therefore be preventive and taken beforehand (Bestratén and Turmo, 1991b). The risk of a BLEVE can be reduced to tolerable levels if several of these measures can be taken at the same time. These are briefly discussed here.

22.5.1 Sloping Ground

The installation must be designed in such a way that any leak of a liquid (for example, liquefied petroleum gas) could be immediately removed from the area in which there is the tank that must be protected. The ground should be smooth and with a slope of 2.5% (1.5% minimum); a draining system must lead to a trench or a tank far enough away to avoid contact between the flames and the tank. It must be taken into account that in case of wind, the flames can have an inclination of 45° as well as a significant drag and that they can reach approximately twice the diameter of the trench (Kletz, 1977).

22.5.2 Thermal Insulation

If the walls of the tank are blanketed with a fireproof material (with a low thermal conductivity), the heating of the vessel, and therefore its pressure increase, by an eventual fire is significantly delayed. Furthermore, in long emergencies, thermal insulation reduces the heat flow to the system and makes it possible for the safety valve to prevent the explosion (fireproofing relies significantly on the correct operation of the safety valve). It must be taken into account that these valves are not designed to solve these types of emergencies on their own, as their cross section should be excessive. Fireproofing ensures protection for a limited time (usually four to five hours). It is the most suitable device for road or railway tanks (Londiche and Guillemet, 1991).

In any case, thermal insulation should be a complement, and other protective systems (for example, cooling of the vessel) should be installed. Another interesting point is that the structural elements (vessel legs) should also be insulated, to avoid the falling of the vessel under excessive heating (this is what happened with two of the spherical tanks in Mexico City; even after falling, however, surprisingly they did not explode). The thermal insulation should be installed in such a way that it could be effective in the event of a fire and also allow the tank surface and structural elements to be inspected periodically.

22.5.3 Cooling with Water

The usefulness of water sprinklers in protecting vessels exposed to the direct action of fire has been proven over many years. It is important to use the water from the first moments, with a layer of a certain thickness totally covering the wall to be cooled, especially those areas directly in contact with the flame. The required flowrate of water should be kept constant—in some cases, the action of firefighters and the consequent increase in water consumption have considerably decreased the pressure in the network and, thus, the water flowrate to the vessels—with a minimum value that will depend on the circumstances.

To protect a fire-engulfed tank, the water flowrate will depend on the circumstances. If the safety valve is correctly designed and works normally, the water rate (Londiche and Guillemet, 1991) should not be less than $8\ \mathrm{l \cdot m^{-2} \cdot min^{-1}}$; however, diverse authors (Maddison, 1989) consider that reducing the water flowrate below $10\ \mathrm{l \cdot m^{-2} \cdot min^{-1}}$ is dangerous if there is direct contact with the flame; a flowrate of $15\ \mathrm{l \cdot m^{-2} \cdot min^{-1}}$ has been recommended (Londiche and Guillemet, 1991; Nazario, 1988; Vílchez et al., 1993) as a general criterion. To have an efficient cooling effect, water should be applied before the temperature of the wall reaches 80°C. If there is no flame impingement, only thermal radiation, smaller flowrates can be used.

If there is flame impingement on the wall, the thermal flux will depend on the type of flame (for a pool fire it can be approximately $100\ \mathrm{kW \cdot m^{-2}}$, while for a highly turbulent flame it can reach $350\ \mathrm{kW \cdot m^{-2}}$). In this case, for the zone of the wall located above the liquid surface, flowrates even larger than $25\ \mathrm{l \cdot min^{-1} \cdot m^{-2}}$ may be required.

Another aspect to be taken into account is that all safety elements—valves, pipes, etc.—should be designed to resist the action of fire and the high temperatures that will be reached during the emergency; otherwise, they will collapse in the first moments, especially if there is direct contact with the flames.

22.5.4 Pressure Reduction

If pressure is reduced, the walls of the vessel will be exposed to less force and the risk of explosion if the temperature increases will be lower. As a general criterion, API recommends the installation of devices able to reduce the pressure up to approximately 7 bar (relative) or up to half of the design pressure in 15 minutes. If the ground is sloped and the vessel is thermally insulated, this time can be longer. The depressurization can require a remote control valve besides the safety valve. The released material should be eliminated in safe conditions (Shebeko et al., 1996), e.g., with a torch. It should also be taken into account that in some cases a strong depressurization can cause extremely low temperatures, leading to fragile conditions in the steel.

22.5.5 Mounding or Burying

The possibility of either totally or partially burying the vessel has been suggested. This provides good protection against thermal radiation for a very long time period, as well as against missiles impacts. However, this measure has many disadvantages, primarily the eventual corrosion in the tank walls.

22.5.6 Water Barriers

This is a relatively new system in which a set of sprayers generates curtains of fine water spray. The barriers retain the vapor released from the leak, thus reducing the possibility of ignition, and disperse them into the atmosphere.

22.5.7 Protection from Mechanical Impacts

Tanks containing materials stored at temperatures higher than their boiling temperatures at atmospheric pressure must be protected from impacts from cranes or other equipment or moving vehicles. A special case, not treated here, is the protection of tank cars.

22.5.8 Overflow

This is an incident that has caused a number of BLEVEs. Nowadays it is much less common, however, and adequate devices are installed to avoid it (level controls, safety valves).

22.5.9 Minimum Separation Distances

The minimum distances between vessels are usually established by regulations and will not be discussed here. They are important from the point of view of thermal radiation, and particularly to avoid direct contact between the flames from the fire in one piece of equipment and the wall of another vessel. They do not guarantee protection, however, in the case of an explosion (blast, projectiles).

22.5.10 Actuation on the Initiating Mechanisms

Diverse systems have been proposed to avoid homogeneous nucleation. These include installing aluminum mesh inside the tank and adding nuclei that initiate boiling. However, these systems are still being investigated, except for very specific applications.

22.6 EXAMPLE CALCULATION OF BLEVE FIREBALL EFFECTS

A tank with a volume of 250 m^3, 80% filled with propane (stored as a pressurized liquid at room temperature), is heated by a fire up to 55°C (~19 bar) and bursts. The thermal radiation and the pressure wave must be estimated at a distance of 180 m.

Data:
Room temperature = 20°C; H_R = 50% (partial pressure of water vapor, 1155 Pa); γ = 1.14; $H_v = 4.3 \cdot 10^5$ J · kg^{-1}; H_c = 46,000 kJ · kg^{-1}; T_{cr} = 369.8 K; $T_{\text{boil. atm. p.}}$ = 231.1 K; $\rho_{\text{liquid, 20°C}}$ = 500 kg · m^{-3}, $\rho_{\text{liquid, 55°C}}$ = 444 kg · m^{-3}; $\rho_{\text{vapor, 55°C}}$ = 37 kg · m^{-3}; Cp_{liquid} = 2,4 · 10^3 J · kg^{-1} · K^{-1}.

Solution:
First of all, the mass of propane involved is calculated:

$$M = V_l \cdot \rho_{\text{liq.,20°C}} = (0.8 \cdot 250 \text{ m}^3) \cdot 500 \frac{\text{kg}}{\text{m}^3} = 100{,}000 \text{ kg}$$

Schematic diagram:

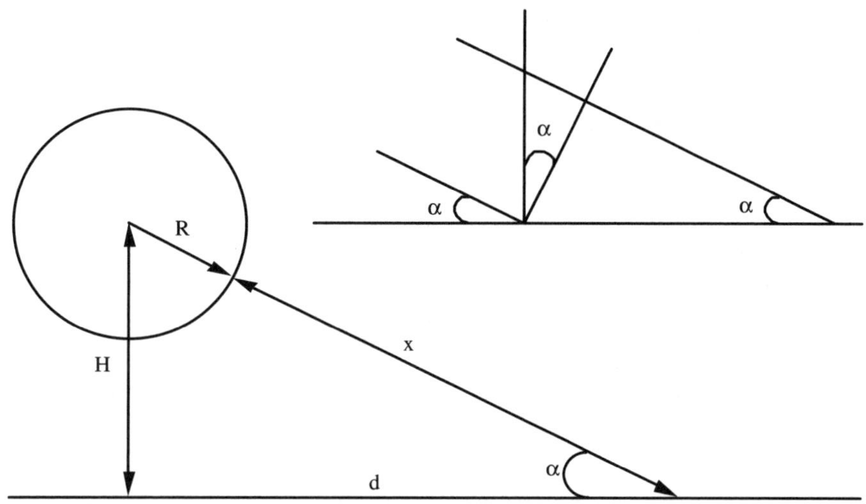

FIGURE 22.6 Position of fireball and target.

1. *Estimation of thermal radiation*
By using Eq. (22.8), the fireball diameter is estimated:

$$D = 6.14 \cdot M^{0.325} = 6.14 \cdot 100000^{0.325} = 259 \text{ m}$$

Its duration is estimated with Eq. (22.9)

$$t = 0.41 \cdot M^{0.340} = 0.41 \cdot 100000^{0.340} = 20.5 \text{ s}$$

And the height reached by the fireball is estimated by Eq. (22.11):

$$H = 0.75 \cdot D = 0.75 \cdot 259 = 194 \text{ m}$$

The distance between the flame and the target, according to Fig. 22.6, can be calculated as follows:

$$x = \sqrt{H^2 + d^2} - R = \sqrt{194^2 + 180^2} - 129.5 = 135 \text{ m}$$

The atmospheric transmissivity will be, according to Eq. (22.16):

$$\tau = 2.02 \cdot (P_w \cdot x')^{-0.09} = 2.02 \cdot (1155 \cdot 135)^{-0.09} = 0.69$$

The view factor is calculated with Eq. (22.15):

$$F = \frac{D^2}{4\,r^2} = \frac{259^2}{4 \cdot 264.5^2} = 0.24$$

Taking a value of $\eta \approx 0, 25$, the emissive power is [Eq. (22.14)]:

$$E_p = \frac{\eta\, M\, H_c}{\pi\, D^2 t} = \frac{0.25 \cdot 100000 \cdot 46000}{\pi \cdot 259^2 \cdot 20.5} = 266 \text{ kW} \cdot \text{m}^{-2}$$

The radiation intensity on a surface perpendicular to the radiation will be:

$$I = \tau \cdot F \cdot E_p = 0.69 \cdot 0.24 \cdot 266 = 44 \text{ kW} \cdot \text{m}^{-2}$$

On a vertical surface,

$$I_v = I \cdot \cos \alpha = 44 \cdot 0.68 = 30 \text{ kW} \cdot \text{m}^{-2}$$

And on a horizontal surface,

$$I_h = I \cdot \sin \alpha = 44 \cdot 0.73 = 32.3 \text{ kW} \cdot \text{m}^{-2}.$$

2. *Estimation of pressure wave*
The overpressure is estimated from Eqs. (22.20), (22.21), and (22.22):

$$f = 1 - e^{(-2.63 \cdot C_p / H_v \cdot (T_c - T_b) \cdot (1 - (T_c - T_o / T_c - T_b)^{0.38}))}$$

$$= 1 - e^{(-2.63 \cdot (2.4 \cdot 10^3 / 4.3 \cdot 10^5) \cdot (369.8 - 231.1) \cdot (1 - ((369.8 - 328)/(369.8 - 231.1))^{0.38}))} = 0.525$$

$$V^* = V + V_l \cdot f \cdot \left(\frac{\rho_l}{\rho_v}\right) = 50 + 200 \cdot 0.525 \cdot \frac{444}{37} = 1310 \text{ m}^3$$

$$W_{\text{TNT}} = \left(\frac{0.021\, P\, V^*}{\gamma - 1}\right) \cdot \left(1 - \left(\frac{P_a}{P}\right)^{(\gamma - 1)/\gamma}\right) = \left(\frac{0.021 \cdot 19 \cdot 1310}{1.14 - 1}\right)$$

$$\cdot \left(1 - \left(\frac{1}{19}\right)^{(1.14 - 1)/1.14}\right) = 1133 \text{ kg}$$

admitting that 40% of the released mechanical energy is transformed in pressure wave (ductile breaking of the vessel):

$$(W_{\text{TNT}})_{\text{overpressure}} = \beta \cdot W_{\text{TNT}} = 0,4 \cdot 1133 = 453 \text{ kg}$$

d_n is calculated with Eq. (22.23):

$$d_n = \frac{d}{(\beta \cdot W_{\text{TNT}})^{1/3}} = \frac{180}{453^{1/3}} = 23.4 \text{ m} \cdot \text{kg}^{-1/3}$$

With the TNT equivalence diagram (overpressure versus d_n, Fig. 22.3), an overpressure of 0.05 bar is found.

22.7 NOMENCLATURE

- A Constant in the Antoine equation (–); in Eqs. (22.28) and (22.29) bore cross-section area (m^2)
- a Constant (–)
- a_o Sound velocity in the vapor (m · s^{-1})
- a_m Velocity of flow at exit plane (m · s^{-1})
- a_w propagation velocity of the initial rarefaction wave (m · s^{-1})
- B Constant in the Antoine equation (–)
- b Constant (–)
- c Constant (–)
- C_p Specific heat at constant pressure (J · kg^{-1} · K^{-1})
- D Diameter of fireball (m)
- d Distance from the center of the vessel to the point at which the overpressure must be calculated (m)

d_n Normalized or scaled distance ($m \cdot kg^{-1/3}$)
E_p Emissive power ($kW \cdot m^{-2}$)
E_v Energy released in the vapor expansion (kJ)
e Constant (−)
F View factor (−)
f Vaporization factor (−)
H Height at which the center of fireball is located (m)
h Height at which the top of fireball is located (m)
H_c Heat of combustion ($kJ \cdot kg^{-1}$)
H_R Relative humidity (%)
H_v Enthalpy of vaporization ($kJ \cdot kg^{-1}$)
I Radiation intensity ($kW \cdot m^{-2}$). In Eqs. (22.26), (22.28), and (22.29) impulse applied to closed end ($bar \cdot m^2 \cdot s$)
l range of cylindrical tank projectiles (m)
L Length of "rocket" (m)
M Mass of fuel (kg)
m Mass of vapor existing initially (kg)
N Number of fragments (−)
P Vapor pressure (bar)
P_a Atmospheric pressure (bar)
P_c Critical pressure (atm)
P_o Relative pressure (bar). In Eqs. (22.28) and (22.29) rupture pressure (bar)
P_{sat} Liquid saturation pressure (bar)
P_v Partial pressure of water (Pa)
R Vessel bore radius (m)
R_f fragment radius (m)
r Distance from the center of fireball to the target (m)
T Temperature (K)
T_c Critical temperature (K)
T_b Boiling temperature (K)
T_g Superheat temperature limit at atmospheric pressure Eq. (22.5*a*) (K)
T'_g Superheat temperature limit at atmospheric pressure Eq. (22.5*b*) (K)
T_o Temperature of the substance at the moment of the explosion (K)
T_R Superheat temperature limit according to the tangent line to the vapor pressure-temperature curve at the critical point (K)
T_s Saturation temperature corresponding to atmospheric pressure (K)
t Time (s)
t_e Time required to expel the vessel content from the open end of the "rocket" (s)
t_o Time required to achieve a fully open breach (s)
t_w Time taken for the rarefaction wave to propagate from the break to the closed end of the "rocket" (s)
U Missile velocity ($m \cdot s^{-1}$)
u Internal energy of the vapor ($kJ \cdot kg^{-1}$)
V Volume of vapor in the vessel (m^3)
V_i Initial volume of vapor in the fireball (m^3)
V_r Volume of the spherical vessel (m^3)
V_l Volume of liquid in the vessel just before the explosion (m^3)
x Distance between the flame surface and the target (m)
W missile mass (kg)
W_{TNT} Equivalent mass of TNT (kg)
α Angle formed by the abscissa axis and the tangent to the saturation curve at the critical point (°); also, angle formed by the radius of the fireball and the horizontal (°)
β Fraction of the energy released converted in pressure wave (−)
γ Ratio of specific heats (−)

η Radiation coefficient (−)
ρ Density ($kg \cdot m^{-3}$)
ρ_l Liquid density ($kg \cdot m^{-3}$)
ρ_v Vapor density ($kg \cdot m^{-3}$)

22.8 REFERENCES

Appleyard, R. D. 1980. *Testing and Evaluation of the Explosafe System as a Method of Controlling the BLEVE,* Report TP2740, Transportation Development Centre, Montreal.

Armet, L. 1997. Private communication.

American Society of Testing and Materials (ASTM). 1983. "A Guide to the Safe Handling of Hazardous Materials Accidents," ASTM STP 825, ASTM, Philadelphia.

Bagster, D. F., and R. M. Pitblado. 1989. "Thermal Hazards in the Process Industry," *Chemical Engineering Progress,* vol. 85, pp. 69–75.

Baum, M. R. 1992. "Development of the Breach Generated by Axial Rupture of a Gas-Pressurized Steel Pipe," *Journal of Pressure Vessel Technology, Transactions of the ASME,* vol. 104, pp. 253–261.

Baum, M. R. 1988. "Disruptive Failure of Pressure Vessels: Preliminary Design Guidelines for Fragment Velocity and the Extent of the Hazard Zone," *Journal of Pressure Vessel Technology, Transactions of the ASME,* vol. 110, pp. 168–176.

Baum, M. R. 1999. "Failure of a Horizontal Pressure Vessel Containing a High Temperature Liquid: the Velocity of End-Cap and Rocket Missiles," *Journal of Loss Prevention in the Process Industries,* vol. 12, pp. 137–145.

Bestratén, M., and E. Turmo. 1991a. "Explosiones BLEVE (I): Evaluación de la Radiacin̄ Térmica," *Notas técnica de prevención NTP-293, Inst. Nac. Seg. Hig. Trab.,* Barcelona.

Bestratén, M., and E. Turmo. 1991b. "Explosiones BLEVE (II): Medidas Preventivas," *Notas técnica de prevención NTP-294, Inst. Nac. Seg. Hig. Trab.,* Barcelona.

Birk, A. M. 1995. "Scale Effects with Fire Exposure of Pressure-Liquefied Gas Tanks," *Journal of Loss Prevention in the Process Industries,* vol. 8, pp. 275–290.

Birk, A. M. 1996. "Hazards from Propane BLEVEs: An Update and Proposal for Emergency Responders," *Journal of Loss Prevention in the Process Industries,* vol. 9, pp. 173–181.

Birk, A. M., and M. H. Cunningham. 1996. "Liquid Temperature Stratification and Its Effects on BLEVEs and their Hazards," *Journal of Hazardous Materials,* vol. 48, pp. 219–237.

Calpe, J., and J. Casal. 1989. "BLEVE-Bola de Foc. Estudi Comparatiu de Models de Predicció d'Efectes," in *Proc. Conferència sobre seguretat ambiental,* Reus, Spain, pp. 389–404.

Capdevila, J. 1994. *Programa de càlcul per a l'estimació d'efectes i conseqüències de l'accident BLEVE/bola de foc,* Internal Report, CERTEC, Barcelona.

Casal, J., H. Montiel, E. Planas, and J. A. Vilchez. 1999. *Análisis del Riesgo en Instalaciones Industriales,* Edicions UPC, Barcelona.

Center for Chemical Process Safety (CCPS). 1994. *Guidelines for Evaluating the Characteristics of Vapor Cloud Explosions, Flash Fires and BLEVEs,* AIChE, New York.

Fay, J. A., and D. H. Lewis. 1977. "Unsteady Burning of Unconfined Fuel Vapor Clouds," in *16th International Symposium on Combustion,* pp. 1397–1405.

Hasegawa, K., and K. Kato. 1978. "Study of the Fireball Following Steam Explosion of n-Pentane," in *2nd. International Symposium on Loss Prevention,* Heidelberg, pp. 297–305.

Holden, P. L., and A. B. Reeves. 1985. "Fragment Hazards from Failures of Pressurized Liquefied Gas Vessels," *Chemical Engineering Symposium Series,* no. 93, pp. 205–217.

Kayes, P. J., ed. 1985. *Manual of Industrial Hazard Assessment Techniques,* Office of Environmental and Scientific Affairs, World Bank, Washington, DC.

Kletz, T. 1977. "Protect Pressure Vessels from Fire," *Hydrocarbon Processing,* vol. 56, August, pp. 98–102.

Lees, F. P. 1980. *Loss Prevention in the Process Industries,* vol. 1, Butterworth-Heinemann, London, pp. 519–528.

Lefin, Y., G. Mavrothalassitis, and J. P. Pineau. 1993. "Knowledge Gained from Hazard Studies on Accident Investigations," *Chemical Industry and Environment,* Girona, Spain, ed. J. Casal, vol. 1, pp. 93–106.

Lewis, D. 1991. "Crescent City, Illinois: 21 June 1970," *Loss Prevention Bulletin,* vol. 101, pp. 23–32.

Lihou, D. A., and J. K. Maunde. 1982. *Institute of Chemical Engineering Symposium Series,* no. 71, pp. 191–224.

Londiche, H., and R. Guillemet. 1991. "Comparison of Three Protective Devices for BLEVE Prevention," *Loss Prevention and Safety Promotion in the Process Industries,* vol. 1, pp. 551–564.

Maddison, T. E. 1989. "The Fire Protection of LPG Storage Vessels. The Design of Water Spray Systems," *LPGITA Seminar*, U.K. October.

Mañas, J. L. 1984. "BLEVEs, Their Nature and Prevention," *Fire,* June, pp. 27–39.

Mcdevitt, C. A., C. K. Chan, F. R. Steward, and K. N. Tennankore. 1990. "Initiation Step of Boiling Liquid Expanding Vapor Explosions," *Journal of Hazardous Materials,* vol. 25, pp. 169–180.

Nazario, F. N. 1988. "Preventing or Surviving Explosions," *Chemical Engineering,* August, pp. 102–109.

Pietersen, C. M., and S. Cendejas. 1985. *Analysis of the LPG Accident in San Juan Ixhuatepec, Mexico City,* TNO, Report 85-0222, The Hague.

Planas-Cuchi, E., J. Casal, A. Lancia, and L. Bordignon. 1996. "Protection of Equipment Engulfed in a Pool Fire," *Journal of Loss Prevention in the Process Industries,* vol. 9, pp. 231–240.

Prugh, R. W. 1991. "Quantify BLEVE Hazards," *Chemical Engineering Progress,* vol. 87, February, pp. 66–72.

Reid, R. C. 1976. "Superheated Liquids," *American Scientist,* vol. 64, pp. 146–156.

Reid, R. C. 1979. "Possible Mechanism for Pressurized-Liquid Tank Explosions or BLEVEs," *Science,* vol. 203, pp. 1263–1265.

Roberts, A. F. 1982. "Thermal Radiation Hazards from Releases of LPG from Pressurised Storage," *Fire Safety Journal,* vol. 4, pp. 197–212.

Satyanarayana, K., M. Borah, and P. G. Rao. 1991. "Prediction of Thermal Hazards from Fireballs," *Journal of Loss Prevention in the Process Industries,* vol. 4, pp. 344–347.

Shebeko, Y. N., A. P. Shevchuck, and I. M. Smolin. 1996. "BLEVE Prevention Using Vent Devices," *Journal of Hazardous Materials,* vol. 50, pp. 227–238.

Sigalés, B., and A. Trujillo. 1990. "Modelado de Estallido de Recipientes," *Ingeniería Química,* October, pp. 465–473.

Towsend, W., C. Anderson, J. Zook, and G. Cowgill. 1974. *Comparison of Thermally Coated and Uninsulated Rail Tank Cars Filled with LPG Subjected to a Fire Environment,* U.S. Department of Transport, Report no. FRA-OR8D, 75-32, Washington, DC.

Van den Berg, A. C., and A. Lannoy. 1993. "Methods for Vapor Cloud Explosion Blast Modelling," *Journal of Hazardous Materials,* vol. 34, pp. 151–171.

Vílchez, J. A., E. Planas-Cuchi, and J. Casal. 1993. "Safety Measures in LPG Storage Design," *Proceedings of the 6th Mediterranean Congress on Chemical Engineering,* Fira de Barcelona, vol. 1, pp. 353–354.

CHAPTER 23
U.S. DOE HAZMAT SPILL CENTER DATABASE

S. Bruce King, John S. Nordin, David Sheesley, and Thayne Routh
AristaTek, Inc. and Western Research Institute (WRI), Laramie, Wyoming

23.1 INTRODUCTION

During the late 1970s, when energy prices were escalating, a primary energy resource was the importation of liquefied natural gas (LNG) from the Middle East using cryogenic transport vessels. A concern was the consequence of an accidental release of LNG during the off-loading from a transport to onshore storage facilities. Minimal information was available regarding the behavior of such a release, and efforts were undertaken by the U.S. Department of Energy (DOE) to develop mathematical models based on large-scale controlled releases of LNG under different conditions to understand the dynamics of such an event. To conduct large-scale experiments of a potentially lethal substance such as LNG required access to a relatively isolated location with adequate control to exclude nonessential personnel.

The first two LNG experiments series performed by Lawrence Livermore National Laboratory (LLNL) under contract to the DOE were conducted at the Naval Weapons Center (NWC), China Lake, California. After completion of these two field studies, it was decided to build a permanent facility with capabilities of larger releases with better control. The U.S. Congress funded the construction of such a facility, called the Liquefied Gaseous Fuels Spill Test Facility (LGFSTF). The site selected was located on Frenchman Flat (a dry lakebed) at the Nevada Test Site (NTS) because of the persistent wind direction and its isolation from population centers. By the time the facility was commissioned and ready for use by research groups, the U.S. Congress also decided that more than just liquefied gaseous fuels needed to be investigated. There was a need for a research facility where various hazardous chemicals could be released under controlled conditions without endangering the public. To meet this need, it was decided to combine testing for both LNG and other hazardous chemicals at the new facility; its name was changed to reflect this new task and is now called the Hazmat Spill Center (HSC).

Being located at the Nevada Test Site has some unique advantages for the type of research performed at the HSC. Since most of the U.S. nuclear weapons tests were conducted at the NTS, the site meets many of the requirements for hazardous materials release studies. These are summarized below.

- The access to the NTS is restricted because of the type of research projects performed on-site and their relationship to national security. This precludes unauthorized personnel from approaching the release locations or accessing the downwind areas where concentrations may be dangerous to personnel.

- It is remote and isolated from any nearby population centers. The nearest metropolitan area is Las Vegas, Nevada, located about 75 miles southeast of the NTS. The nearest neighbor downwind from the NTS is the Nellis Air Force Range, which is also restricted to public access. This allows release of large volumes of toxic materials with adequate downwind areas for dispersion of these materials long before contact with publicly accessible areas.
- Because there is minimal vegetation or wildlife, the potential for environmental harm is reduced. While there is some vegetation and wildlife, the DOE maintains strict regulations on what can be released and has filed environmental impact statements (EIS) assessing the consequences of the release of those materials for which the site has currently been approved.
- Prior to the previous nuclear testing at the NTS, analysis was made of the meteorology for the area. This was critical in the original selection of the site, and continued collection of atmospheric data has allowed prediction of specific meteorological conditions that are hard to capture on any other site on a reliable basis. This is particular important when designing instrument arrays to capture dispersion clouds or specific atmospheric conditions.

When the HSC was completed in 1987, there were three large storage tanks, each 98 + m^3 (26,000 + gallons) in capacity. Two of these were insulated to allow storage of cryogenic materials, i.e., LNG, and were connected to other high-pressure tanks where N_2 was stored to allow pressurized discharge of the cryogenic liquid at high rates. The discharge was usually via one of two release lines in 30.5 cm (12 in) diameter that extended from the tank locations to the spill location about 152 m (500 ft) away. All of the control and measurement of this system was remotely monitored by the Command and Control Data Acquisition System (CCDAS), located in a control room about 1 km upwind from the tank farm. The CCDAS also allowed for additional instruments and sensors to record data as required for the specific experiment being conducted.

The following are the types of studies performed at the HSC over the years:

- Dispersion studies using large volume releases of chemicals
- Source characterization, or understanding how a material behaves when released from its normal storage container or a pressurized reactor
- Evaluation of mitigation techniques, i.e., mechanical vapor barriers, spray nozzles to knock down plumes, foam vapor barriers
- Evaluation of protective clothing when exposed to toxic and/or corrosive materials in a wind tunnel constructed on-site

23.2 HAZMAT SPILL CENTER CAPABILITIES AND FACILITIES

The HAZMAT Spill Center, located at the Department of Energy's Nevada Test Site near Mercury, Nevada, is a unique facility built to conduct hazardous materials testing and training under controlled conditions. The HSC, 75 miles northwest of Las Vegas, can accommodate both large- and small-scale testing. The HSC environmental impact statement allows live releases of hazardous materials for training purposes, field-test detection, plume dispersion experimentation, and equipment and materials testing. Tests are conducted from April through September, weather permitting.

This unique facility is available for private and public sector sponsors on a user fee basis to cover actual test costs plus clean-up costs to prepare the facility for other customers. With the HSC's unique release permits and spill rates up to 28,000 gallons per minute and spill volumes expandable to 75,000 gallons, the HAZMAT Spill Center is capable of the following:

"Live" HAZMAT testing

Applications:

- Test protective gear
- Test HAZMAT instrumentation
- Test equipment and procedures

Advanced HAZMAT Training

Applications:

- Train the trainers or leaders
- HAZMAT specialists
- Create confidence in equipment and procedures by allowing the use of live material
- First responder (state, municipal, and industrial)

Mitigation Technology

Applications:

- Test mitigation techniques using live material
- Test and validate computer mitigation models
- Conduct research for new techniques

Remote Sensing

Applications:

- Effluent analysis
- Stand-off HAZMAT identification

Source Term Definition—Dispersion Modeling

Applications:

- Computer model testing and validation
- Development of new computer models for HAZMAT dispersion

The HAZMAT Spill Center's facilities include:

- *Wind tunnel* (8 ft × 16 ft × 96 ft): an environmental test chamber, allows mixing and mitigation technology research as well as plume releases
- *Elevated stacks and spill pads:* available for realistic industrial release scenarios
- *Test cell areas:* allow materials and instrumentation to be exposed to known, high concentrations of test materials in a confined space
- *Tank farm:* contains systems for the release of large volumes of liquids, gases, and cryogenic liquids
- *Command, control, and data acquisition systems:* controls and records data from tests using monitoring and sensor stations
- *Other support facilities:*
    - Electronics shops
    - Maintenance shops
    - Conference facilities
    - Storage buildings
    - Chemical storage areas

23.3 THE U.S. DOE'S HAZMAT SPILL CENTER DATABASE

In 1986 Congress passed the Superfund Amendment and Reauthorization Act (SARA). Section 118(n) specified that funding should be made available to conduct technology transfer of the DOE LGFSTF Research Program. Western Research Institute (WRI) of Laramie, Wyoming, became the repository for the database developed from these experiments conducted at the HSC. The database is available via the Internet on WRI's website located at http://www.westernresearch.org/Projects/STFDB/DataPage.htm.

The data collected from these different experiments were primarily developed with public funds or provided by private companies or consortiums representing more than one industrial concern.

The data in these files are a combination of raw data and others that have been processed through reduction, calibration, and verification procedures from which individuals can begin their own analysis. To understand the organization of the data, its origin or location within a specific experiment, or its reliability, the user is directed to the References and Additional Reading at the end of the chapter.

23.4 DATABASE ORGANIZATION

The website is organized with a test series per page. Each test data page contains one executable file for each test in the test series as well as a file named DOC.EXE that contains information documenting the test series. These executable files are self-extracting Zip files that, when executed, create one or more comma-separated value (CSV) text files containing the actual test data or other test information.

Each TESTxx.EXE file contains one to three CSV text files, TESTxx.CSV, TESTxxW.CSV, and TESTxxG.CSV, where xx is the test number. The first file, that with no W or G suffix, is the data collected by the data acquisition system in the control room at the HSC; all of the tests on the site include this file. The second file, that with a W ("W" is for wind) suffix, is wind speed and direction data collected by 2-m meteorological towers during the tests. The final file, that with a G ("G" is for gas) suffix, is applicable only to the Silicone Health Council (SHC) test series and contains gas concentration information. Each of these files is structured in a similar fashion. The values in the first row are the tag names for each data channel. After the first row, each row in a file is a record of the recorded values at the time denoted by the timestamp located in the first column. A "????????" indicates that the data acquisition system did not record a valid value for that sensor at that time.

Each DOC.EXE file contains one or two text files, test.DBB, where "test" is the test series abbreviation, and TAGNAMES.TXT. The test.DBB file is a copy of the database used to control the data acquisition system during a test and has some information useful in understanding the sensors associated with the tag names. Where possible, a TAGNAMES.TXT file has been included that more fully describes the sensors associated with tag names, as well as other relevant test information.

Currently, the DOE HSC database contains only those experiments conducted after 1988, so most of the dispersion experiments conducted early in the research program are not available at the time that this article is being written. An effort is underway to make the earlier dispersion experiments available via the WRI website and hopefully that will occur in late 1999 or early 2000. Discussion of all database files in the WRI archive is provided, even though they may not currently be available on the WRI website. Currently 13 different test series are included in the DOE HSC database, although other experiments have been conducted at the HSC because of its unique location and capabilities. These experiments were not funded or conducted as investigations of hazardous materials behavior.

Because of the number of different experiments conducted and the resulting databases available, only an overview or quick summary of the different experiments can be provided in this article. The following information is provided for each of the experiments included in the DOE HSC database.

- Test name
- Sponsor
- Dates of experiments
- Chemicals released
- Brief description of experiments
- List of public domain references for data reports or analyses

It should be noted that, beginning with the earliest experiments, a tradition (which has not always been followed) was established by Lawrence Livermore National Laboratory of naming a test series with an animal name related to the NTS or the desert environment. This was done alphabetically beginning with the name Burro for the first LNG series conducted at China Lake, CA.

23.5 BURRO SERIES

Sponsors: U.S. Department of Energy and Gas Research Institute (GRI)

Dates of Experiments: June 6–September 17, 1980

Chemical Released: Liquefied natural gas (LNG)

Description of Experiments: During the summer of 1980, a series of LNG spill experiments was performed at the Naval Weapons Center (NWC), China Lake, California. The experiments involved eight LNG spills and one liquid nitrogen release (which is not included in the database files) onto a water surface. The release point was in the center of a small pond ~58 m in diameter, 1 m deep, with its surface ~1.5 m below the surrounding ground level. This was to simulate the accidental release of LNG during an offloading scenario in an aquatic environment where LNG would be spilled onto a water surface. Personnel from NWC and LLNL performed these experiments jointly.

The release point was instrumented to measure the instantaneous rate of LNG discharge from the 25-cm diameter insulated release line. Release rates varied from ~10 m^3/min to ~16 m^3/min. Predominate wind direction at the China Lake site is from the southwest or 225°. The upwind and downwind sensor arrays were located around a test grid centerline corresponding to the 225° azimuth. Downwind instrumentation was organized into arrays at different distances up to 800 m downwind from the spill point and located on towers to accommodate multiple levels of sensor measurements. Data acquisition provided for sensor capture before, during, and after cloud passage. Sensors were provided for determining meteorological measurements, including wind speed, wind direction, humidity, and temperature. These were positioned both upwind and downwind from the release point. At five stations (towers), bivane anemometers were installed to measure atmospheric turbulence within the test grid. These data were not provided by LLNL to the Western Research Institute for the DOE HSC database, but plots of some of this data are provided in the Burro Data Report (Koopman et al., 1982a).

TABLE 23.1 Summary of Burro Series

	Test 2	Test 3	Test 4	Test 5	Test 6	Test 7	Test 8	Test 9
Released material	LNG	LNG	LNG	LNG	LNG	LNG	LNG	LNG
Date	6/18/80	7/2/80	7/9/80	7/16/80	8/5/80	8/27/80	9/3/80	9/17/80
Time of day for start of spill (PDT)	15:59:28	15:00:06	14:07:27	16:19:36	16:05:06	18:12:21	19:09:22	18:37:08
Duration (sec)	173	166	175	190	128.5	174	107	79
Spill rate (m^3/min)	11.9	12.2	12.1	11.3	12.8	13.6	16.0	18.4
Spill volume (m^3)	34.3	34.0	35.3	35.8	27.5	39.4	28.4	24.2
Average wind direction (° from true north)	221	224	217.5	218	220	208.4	234	232
Average wind Speed (m/s)	5.4	5.4	9.0	7.4	9.1	8.4	1.8	5.7
Surface roughness—z_0 (m)	0.0002	0.0002	0.0002	0.0002	0.0002	0.0002	0.0002	0.0002
Monin-Obukhov length (m)	−11.3	−9.06	−37.1	−25.5	−45.0	−114.0	16.5	−140.0

Heat flux sensors to measure heat flow between the ground and the dispersing cloud were located on the desert surface at locations in the downwind arrays. Type K thermocouples were positioned on the downwind array towers to measure cloud temperature as it moved downwind. Open-path IR sensors for the measurement of LNG concentration were positioned at multiple levels on the downwind arrays. High-speed motion picture cameras and infrared imagers were also positioned to make visual recordings of the experiments.

23.6 COYOTE SERIES

Sponsors: U.S. Department of Energy and Gas Research Institute (GRI)

Dates of Experiments: June 30–November 24, 1981

Chemical Released: Liquefied natural gas (LNG)

Description of Experiments: The Coyote Series of LNG spill experiments was performed at the Naval Weapons Center, China Lake, California, during the summer and fall of 1981. Personnel from NWC and LLNL performed these experiments jointly. There were 10 Coyote experiments, five primarily for the study of vapor dispersion and burning vapor clouds and 5 for investigating the occurrence of rapid-phase-transition (RPT) explosions. When data were transferred from LLNL to WRI, data only for three Coyote experiments were provided, tests 3, 5 and 6. Most of the experiments released LNG, two used liquefied methane and one used liquid nitrogen. The three tests where data are available are described as vapor burn experiments where the cloud was ignited after the release. The releases were conducted on a water surface using the same small pond and release configuration as in the Burro Series. All of the LNG releases were conducted during neutral stabilities (Pasquill D) and one of the liquid nitrogen releases was close to stable stability (Pasquill E).

As in the Burro Series, the release point was instrumented to measure the instantaneous rate of LNG discharge from the 25-cm diameter insulated release line. Release rates varied from ~6 m^3/min to ~19 m^3/min, with total volumes spilled from ~3 m^3 to 28 m^3. Downwind instrumentation was organized into arrays at different distances of 110 up to 500 m downwind from the spill point and located on towers to accommodate multiple levels of sensor measurements. Sensor measurements were acquired before, during, and after cloud passage. Wind field measurements were made using two-axis cup-and-vane anemometers located at 20 locations at a 2-m height. These were positioned both upwind and downwind from the release point. Humidity sensors were located upwind and at multiple downwind locations from the spill point.

Bivane anemometers were installed to measure atmospheric turbulence at heights of 1.36, 3, and 8 m at two locations (62 m upwind of the spill point and 300 m downwind of the spill point). These data were not provided by LLNL to WRI for the DOE HSC database.

Heat flux sensors to measure heat flow between the ground and the dispersing cloud were located on the desert surface at locations in the downwind arrays. Type K thermocouples were positioned on the downwind array towers to measure cloud temperature as it moved downwind. Multiple levels of thermocouples were located just upwind of the spill point, as well as four resistive temperature devices (RTD), one defined as the absolute temperature (0.5-m height) and three to provide delta temperature measurements relative to the absolute reading (1, 2, and 4 m).

Because of the experience gained in the Burro Series experiments, the downwind gas concentration sensor arrays were modified. This was based on the facts that beyond 400 m from the spill point the very low concentrations made the sensors of limited value and that close to the spill point, water and mud thrown up by RPT explosions had adverse effects on the sensors. Most of the open-path IR sensors for the measurement of LNG concentration were positioned at multiple levels on the downwind arrays between 140 and 400 m. In addition, some alternative sensors from International Sensor Technology and Mine Safety Appliance for measuring concentration were installed.

Because of the objective to observe vapor cloud fires during some of the releases, sensors were installed to measure flame velocity during these vapor cloud fires. Four calorimeters manufactured by Hy-Cal Engineering were installed at four downwind stations at a height of 1 m. Twenty-seven LLNL-developed flame-velocity sensors were deployed at nine towers downwind of the spill point at three different levels.

Five high-speed 16 mm motion picture color cameras and side-on IR imaging video recorders were also positioned to make visual recordings of the experiments. Nine radiometers were deployed by the NWC personnel to measure both the radiative heat flux at several distances from the vapor cloud fires and the emissive power of the fire.

23.7 DESERT TORTOISE SERIES

Sponsors: U.S. Coast Guard and The Fertilizer Institute

Dates of Experiments: August 24–September 6, 1983

Chemical Released: Anhydrous ammonia (NH_3)

Description of Experiments: The Desert Tortoise Series of ammonia spill experiments was performed by LLNL at the Frenchman Flat area of the Nevada Test Site during the summer of 1983. There were four experiments for the study of the size, shape, and temperature of the ammonia clouds, as well as the extent of the hazardous concentrations downwind. In order to simulate simple cases, which later could be modeled, the tests were performed at constant pressure and on the dry lakebed, which provided a relatively flat and smooth surface.

The release point was instrumented to measure the instantaneous rate of ammonia discharge from the 6-in. diameter release line. The spill region was instrumented with three thermocouples placed at the soil surface at distances downwind along the centerline at 3, 5,

TABLE 23.2 Summary of Coyote Series

	Test 3	Test 5	Test 6
Released material	LNG	LNG	LNG
Date	9/3/81	10/7/81	10/27/81
Time of day for start of spill	15:38:26 PDT	12:08:50 PDT	16:43:23 PST
Duration (sec)	65	98	82
Spill rate (m^3/min)	13.5	17.1	16.6
Spill volume (m^3)	14.6	28.0	22.8
Average wind direction (° from true north)	205	229	220
Average Wind Speed (m/s)	6.0	9.7	4.6
Surface roughness—z_0 (m)	0.0002	0.0002	0.0002
Monin-Obukhov length (m)	−6.32	−26.5	73.6

and 9 m from the spill point. These were to provide a measure of soil temperature prior to a test and information on ammonia pooling. A Hy-Cal Engineering heat flux gauge was placed just below the soil surface at 3 m downwind of the spill point.

Wind field measurements were made using two-axis cup-and-vane anemometers located at 11 stations and positioned 2 m above the ground. These were positioned both upwind and downwind from the release point. Fifty m upwind of the spill point were four levels of 1,000 Ω platinum RTD mounted in aspirated solar shields. Bivane anemometers were installed to measure atmospheric turbulence at three heights at two locations (50 m upwind of the spill point and 100 m downwind of the spill point). The bivane data were not provided by LLNL to WRI for the DOE HSC database but are provided as plots in the Desert Tortoise Data Report (Goldwire et al., 1985).

The downwind instrumentation grid of multiple levels of sensors was defined by two arrays, a mass flux array at 100 m and dispersion array at 800 m downwind from the spill point. Up to eight additional portable ground-level stations were fielded in two distant arcs at distances of 1,400 or 2,800 m and, on occasion, at 5,500 m downwind.

Heat flux sensors to measure heat flow between the ground and the dispersing cloud were located on the desert surface at multiple locations in the downwind arrays. Type K thermocouples were positioned on the downwind array towers at multiple heights to measure cloud temperature as it moved downwind.

The primary sensors used to determine ammonia concentration on the 100-m mass flux array were MSA nondispersive IR (NDIR) gas sensors. International Sensor Technology

TABLE 23.3 Summary of Desert Tortoise Series

	Test 1	Test 2	Test 3	Test 4
Released material	Ammonia	Ammonia	Ammonia	Ammonia
Date	8/24/83	8/29/83	9/1/83	9/6/83
Time of day for start of spill (PDT)	16:37:47	11:20:56	15:37:47	18:15:07
Duration (sec)	126	255	166	381
Spill rate (m^3/min)	7.0	10.3	11.7	9.5
Spill volume (m^3)	14.9	43.8	32.4	60.3
Average wind direction (° from true north)	223.66	226.21	219.07	229.30
Average wind speed at 2 m (m/s)	7.42	5.76	7.38	4.51
Surface roughness—z_0 (m)	0.003	0.003	0.003	0.003
Monin-Obukhov length (m)	92.7	94.7	570.7	45.2

solid-state gas sensors were the primary means to determine gas concentration on the 800-m dispersion array. The open-path IR sensors developed by LLNL for the measurement of LNG concentration in the Burro and Coyote Series were sensitive to ammonia absorption when no water fog was present and were positioned at several locations on the downwind arrays at 100 and 800 m.

Photographic and video recordings of the experiments were provided with 16-mm movie cameras, programmable framing cameras, and black-and-white video coverage.

23.8 EAGLE SERIES

Sponsor: U.S. Air Force

Dates of Experiments: September17–October 30, 1983

Chemical Released: Nitrogen tetroxide (N_2O_4)

Description of Experiments: The Eagle Series of nitrogen tetroxide (N_2O_4) spill experiments was performed by LLNL during the fall of 1983. There were six experiments to determine the heavy gas dispersion aspects and source strength characteristics of large N_2O_4 spills. In addition, they were to provide N_2O_4 spills for the evaluation of a portable foam vapor suppression system (PFVSS). The dispersion and source strength experiments were under the sponsorship and direction of the Engineering and Services Center, Tyndall AFB. The Ogden Air Logistics Center, Hill AFB, with support from the USAF Space Division and the Strategic Air Command, directed the PFVSS efforts. Tests 1, 2, 3, and 6 were for the purpose of dispersion and source strength studies, and these are the source of the data provided by LLNL to WRI for the DOE HSC database. Tests 4 and 5 were for the evaluation of the PFVSS and are not included in the DOE HSC database.

Two different spill configurations were used during the experiments, a single-exit release and a multi-exit release. The single-exit, confined spill configuration was for studying evaporation rates as a function of liquid pool depth and wind speed. The multiexit unconfined configuration was designed to distribute N_2O_4 uniformly over a large area to allow it to evaporate rapidly. The single point release was used for test 1 (which was primarily a check-out and to determine source strength) and the PFVSS evaluation was conducted during tests 4 and 5. Real-time measurement of release rate was provided with a flow meter during test 1, but though the unit was supposed to be acid-proof it, was destroyed by the N_2O_4 and wasn't available for subsequent tests. Thereafter, the amount of N_2O_4 released was determined by measuring the pressure difference between the headspace and the bottom of the tanker vessel. Based on release duration and total volume released, an average flow rate was reported.

The temperature of the N_2O_4 was recorded just prior to its exit from the spill pipe. Three Hy-Cal Engineering heat flux gauges were placed just below the surface of the soil near the spill point. Adjacent to the spill point, three Type K thermocouples placed at three different heights (ground level and 2 and 4 cm) were used to measure vapor temperature during the dispersion and source characterization tests (tests 1, 2, 3, and 6). During these tests, the N_2O_4 was spilled directly onto the desert surface. During the PFVSS evaluation tests the N_2O_4 was spilled into a polyethylene plastic liner to help confine the liquid. The spills directly onto the lakebed playa surface exhibited considerable outgassing from the surface for several hours after the spill was terminated.

Wind field measurements were made with nine stations using two-axis cup-and-vane anemometers located 2 m above the ground, both upwind and downwind from the release point. Meteorological boundary layer measurements were made on a 20-m tower located 50

TABLE 23.4 Summary of Eagle Series

	Test 1	Test 2	Test 3	Test 6
Released material	Nitrogen tetroxide	Nitrogen tetroxide	Nitrogen tetroxide	Nitrogen tetroxide
Date	9/17/83	9/23/83	10/7/83	10/30/83
Time of day for start of spill	14:07 PDT	17:02 PDT	16:48 PDT	14:37 PST
Spill configuration	Multiexit	Multiexit	Multiexit	Multiexit
Duration (sec)	45	65	188	296
Average spill rate (m^3/min)	1.75	1.4	1.4	0.7
Spill volume (m^3)	1.3	1.5	4.2	3.4
Average wind direction (° from true north)	233	223	229	223
Average wind speed at 2 m (m/s)	6.15	5.80	3.13	4.96
Surface roughness—z_0 (m)	0.0003	0.0003	0.0003	0.0003
Monin-Obukhov length (m)	−20.2	−198.0	16.8	293.0

m upwind of the spill point. This tower was installed with four levels of 1,000 Ω platinum RTD mounted in aspirated solar shields at heights of 0.82, 2.46, 6.13, and 16.2 m. In addition, there were three Gill bivane anemometers installed at 3.36, 5.83, and 12.1 m. to measure atmospheric turbulence. A Hy-Cal Engineering heat flux gauge was installed at ground level. Additional meteorological measurements were made on a 10-m tower located 25 m downwind on the centerline of the test grid. This tower had three Gill bivane anemometers at 1.3, 3.0, and 6.0 m plus three Type K thermocouples for temperature measurements. The bivane data were not provided by LLNL to WRI for the DOE HSC database, but some of the data are provided as plots in the Eagle Series Data Report (McRae et al., 1987).

The downwind instrumentation grid of multiple levels of sensors was defined by two arrays, a mass flux array at 25 m and dispersion array at 785 m downwind from the spill point. Two additional portable NO_2 ground level stations were fielded at distances 2,800 m downwind, but the data from these sensors were not provided to WRI by LLNL.

In addition to the heat flux sensors located on the meteorological tower up wind of the spill point and near the spill point, two heat flux sensors were located on the mass flux array 25 m downwind from the spill point to measure heat flow between the ground and the dispersing cloud. Type K thermocouples were positioned on the downwind mass flux and dispersion array towers at multiple heights to measure cloud temperature as it moved downwind.

The primary sensors used to determine N_2O_4 and NO_2 concentration on the 25 m mass flux array were the open-path IR sensors developed by LLNL for the measurement of LNG concentration in the Burro and Coyote Series. They were capable of measuring both species without any modifications. The 785 m dispersion array was instrumented with NO_2 sensors manufactured by Energetic Sciences, Inc. and provided by the Shuttle Activation Task Force, Vandenburg AFB. Two additional NO_2 sensors manufactured by International Sensor Technology were also used on this dispersion array but were greatly affected by small variations in relative humidity and were not reported in the Eagle Data Report or in the data set provide by LLNL to WRI. The two portable NO_2 sensors were Interscan Model 140D electrochemical transducers providing an analog signal recorded on a strip chart recorder. These data were not provided to WRI by LLNL.

Photographic and video recordings of the experiments were provided with 16-mm movie cameras, programmable framing cameras, and black and white video coverage. There was only video coverage of the Eagle 6 test.

23.9 FALCON SERIES

Sponsors: U.S. Department of Transportation and Gas Research Institute (GRI)

Dates of Experiments: June 12–August 29, 1987

Chemical Released: Liquefied natural gas (LNG)

Description of Experiments: The Falcon Series of liquefied natural gas (LNG) spill experiments was performed by LLNL during the summer of 1987. These were the first tests performed at DOE's permanent Spill Test Facility. A series of five spills was performed on water within a vapor barrier structure as part of a joint government/industry study. These experiments were performed to evaluate the effectiveness of vapor fences as a mitigation technique for accidental releases of LNG. They also provided a database for the validation of wind tunnel and computer modeling simulations of vapor fence effects on LNG dispersion. Spills were made onto a water pond equipped with a circulation system to maximize evaporation to make the source evaporation rate as nearly equal to the spill rate as possible.

These experiments used the new tank farm, which contained two 26,000-gallon cryogenic storage tanks that were connected to the spill point with 500-ft-long spill pipes which were comprised of two 12-in. insulated lines and one 6-in. insulated line (which was not used in these experiments). Each spill pipe was equipped with a control valve at each end.

Prior to a release of LNG, the spill line(s) were cooled with liquid nitrogen, after which the LNG was delivered to the spill line by operating valves controlled remotely from the control room located approximately a mile from the release area. The LNG was pressure driven out of the storage tanks and through the spill pipes by means of nitrogen drive gas at 35 to 140 psig. The drive gas was supplied from a 2,000-psig, 2,400-ft^3 pressure vessel.

To provide uniform LNG distribution on the pond, a multiexit release configuration was built. Each of the four arms of the multiexit was approximately 11.6 m in length, oriented 90° from adjacent arms, and fitted with a restrictive orifice near the end to prevent flashing in the release pipe. The pond was 40 × 60 m and was filled to a depth of approximately 30 in (or 76 cm). Located upwind from the pond was a large "billboard" structure 17.1 m wide and 13.3 m high to simulate the turbulence created by a typical storage tank. The "billboard" was made of a proprietary fiberglass cloth impregnated with a mixture of silicon, Teflon,

TABLE 23.5 Summary of Falcon Series

	Test 1	Test 2	Test 3	Test 4	Test 5
Released material	LNG	LNG	LNG	LNG	LNG
Date	6/12/87	6/18/87	6/29/87	8/21/87	8/29/87
Time of day for start of spill (PDT)	19:47:56	18:09:09	18:52:02	19:27:04	18:58:00
Duration (sec)	131	78	154	301	78
Spill Rate (m^3/min)	28.7	15.9	18.9	8.7	30.3
Spill Volume (m^3)	66.4	20.6	50.7	44.9	43.9
Average wind direction (° from true north)	234.3	227.0	221.7	230.6	218.0
Average wind speed at 2 m (m/s)	1.7	4.7	4.1	5.2	2.8
Surface roughness—z_0 (m)	0.008	0.008	0.008	0.008	0.008
Monin-Obukhov length (m)	4.963	−103.4	−442.2	69.38	13.69

and graphite and reinforced with aluminum battens and suspended from 13.7-m aluminum pillars. A vapor fence was installed around the pond and simulated storage tank. This structure was fabricated from the same material used in the "billboard" and suspended from 9.1-m aluminum pillars with stainless steel cables. The vapor fence was 44 m wide, 88 m long, and 8.7 m high.

The spill area was instrumented with numerous Type K thermocouples to record temperatures at various locations within the vapor barrier fence. These were within the pond and upwind of the "billboard." Heat flux sensors were also placed within the vapor barrier fence area as well as at four locations downwind of the spill area outside of the vapor barrier fence. Humidity sensors were located both within and outside the vapor barrier fence. The spill line from the tank farm to the spill area was instrumented to provide instantaneous measurement of the flow rate of LNG to the spill area.

Wind field measurements were made with 19 stations using two-axis cup-and-vane anemometers (Met-One) located 2 m above the ground. These were positioned both upwind and downwind from the release point. Meteorological boundary layer measurements were made on a 20-m tower located 130 m upwind of the spill point and 25 m off the centerline of the test grid. This tower was installed with five levels of 1000 Ω platinum RTD mounted in aspirated solar shields at heights of 1, 2, 4, 8, and 16 m. In addition there were three Gill bi-vane anemometers installed at 1, 4, and 16 m to measure atmospheric turbulence. An additional 15 bi-vane anemometers were installed at heights of 1, 5, and 11 m on five towers located downwind of the spill point. The bivane data were not provided by LLNL to WRI, but some of the data are provided as plots in the Falcon Data Report (Brown et al., 1990).

The downwind instrumentation grid of multiple levels of sensors was defined by three arrays, at 50, 150, and 250 m from the downwind end of the vapor barrier fence. On these downwind array towers Type K thermocouples were positioned at four heights on the 50- and 150-m arrays and three heights on the 250-m array, to measure cloud temperature as it moved downwind. The exception to this was the tower located on the centerline of the 150-m array, which had four levels of RTD sensors rather than thermocouples. Between tests 3 and 4, two towers and their associated instrumentation were moved from the 50-m array and added to the 150-m array to increase the width of the 150-m array. This was done based on the wider than anticipated clouds created during the first three experiments.

A total of 77 gas concentration sensors were deployed during the experiments. There were 4 IR sensors developed by the Jet Propulsion Laboratory for use during the Burro and Coyote tests that measured concentration within the vapor barrier fence area. Thirty-five of the LLNL-IR sensors developed for the Burro Series were deployed at the lowest heights on the 50- and 150-m downwind arrays. Thirty-eight MSA catalytic sensors were located at higher heights at the 50- and 150-m arrays and at the 250-m arrays where gas concentrations were not anticipated to exceed 5%.

Photographic and video recordings of the experiments were provided with 16-mm movie 24 frame/second cameras, programmable framing cameras, and color video coverage.

23.10 GOLDFISH SERIES

Sponsor: Amoco Corporation and Allied Signal Corporation

Dates of Experiments: August 1986

Chemical Released: Anhydrous hydrofluoric acid (HF)

Description of Experiments: During the summer of 1986, Amoco Oil Company and LLNL conducted the Goldfish Series of anhydrous hydrofluoric acid (HF) spill experiments at the Frenchman Flat area of the Nevada Test Site. The objectives of the six experiments were threefold. First, to obtain basic information regarding the source characteristics during an atmospheric release of HF stored at an elevated pressure and temperature. Secondly, to provide downwind concentration measurements of HF in both the dense gas and toxic gas regions for comparison with the performance of dense gas dispersion models. Lastly, the final three tests were designed to provide information on the effectiveness of water spray systems for reducing the downwind concentration of HF. The data for the last three tests (4, 5, and 6) were not transferred to WRI from LLNL.

The spill release configuration was designed to release HF as a horizontal jet. The HF spill equipment consisted of the following elements: (1) spill tank connected to the discharge line, (2) liquid HF collection pad, (3) liquid HF collection pipe, and (4) vapor vent pipe. The spill tank was a 5,000-gal trailer connected to a 4-in. diameter spill line. The tank was fitted with electric heating so the liquid HF could be maintained at approximately 40°C. A load cell was positioned under one end of the tank to provide a continuous record of the weight of the trailer and allow calculation of the rate at which HF was released. A high-pressure gaseous nitrogen tube trailer provided dry nitrogen to pressurize the HF spill tank. The end of the release line was equipped with a remote-controlled spill valve that was used to start and stop the spill. An orifice plate located at the end of the release line controlled the flow rate.

A collection pad to collect and drain into the liquid collection system any HF that was not entrained into the atmosphere was constructed at the end of the release line. The pad was approximately 9 m wide and 61 m long and was constructed of 80-mil UV-stabilized polyethylene. During each release, the amount of HF released was measured by recording the HF tank weight, the orifice temperature and pressure, the HF temperature, and the drive gas pressure.

During Tests 3, 4, 5, and 6, additional atmospheric water vapor was provided to study the effects of increased relative humidity on the resulting cloud. This was attempted using two different methods. First, a steam boiler and water injection system was located upwind of the spill point. The injection system consisted of an array of spray nozzles, half for steam and half for warm water located approximately 25 m upwind of the release point. The spray system was 22 m long and was located approximately 3 m above the ground. The nozzles were oriented horizontally and pointed upwind from the spill point. The second technique involved a shallow pond, 250 m × 600 m, which was constructed upwind of the spill point.

Vertical wind profile and turbulence were measured using three Gill bivane anemometers installed at a tower 600 m upwind of the release point. The data from these sensors were not transferred from LLNL to WRI. On this same tower, five levels of platinum RTD sensors at heights of 1, 2, 4, 8, and 16 m were installed to measure the ambient temperature and temperature lapse rate. Eighteen (18) stations using two-axis cup-and-vane anemometers (Met-One) mounted 2 m above the ground were used to determine the characteristics of the wind field before, during and after each release.

Concentration measurements of HF downwind of the spill point were made at three arrays located 300, 1,000, and 3,000 m downwind from the spill point. The centerline of the array

TABLE 23.6 Summary of Goldfish Series

	Test 1	Test 2	Test 3
Released material	HF	HF	HF
Date	8/1/86	8/14/86	8/20/86
Duration (sec)	125	360	360
Spill rate (kg/s)	27.67	10.46	10.27
Spill mass (kg)	3459	3766	3697
Average wind speed at 2 m (m/s)	5.79	4.04	4.80
Pasquill stability class	D	D	D

was aligned with the 225° azimuth, which corresponded with the predominate wind direction at the Frenchman Flat site. The sensors on the 300-m array were located at heights of 1, 3, and 8 m on the centerline of the array and at ±15 and ±30 m from the centerline. Additional sensors were located at 1-m heights at ±45 and ±60 m from the centerline. After test 1, two additional sensors were located at ±75 m from the centerline. At the 1,000-m array the sensors were located at heights of 1, 3, and 8 m at the centerline, ±50, ±100, and ±150 m from the centerline. For test 1, sensors were located at 1 m for the ±200 and ±250 m, with additional sensors at 1 m for ±300 m for tests 2 through 6. For the 3,000-m array sensors were located at heights of 1, 3, and 8 m at the centerline of the array and at ±150 m. Additional sensors were located at the 1-m height at ±300, ±450, ±600, and ±750m for test 1. After test 1, the ±300 m location was augmented with sensors at 3-, and 8-m heights.

Two types of sensors were utilized in the downwind arrays. The first was an integrated filter sampler (IF) developed by Amoco and the second was the GMD HF analyzer. The IF data are reported in the HSC database, but the GMD HF analyzer data were not reported either in the references (Blewitt et al., 1987a, 1987b, and 1987c) nor provided in the data transferred from LLNL to WRI. The IF sampler consisted of 10 Gelman filter cassettes, which were treated with a sodium formate solution and loaded into a sampling manifold for each release. A sampling manifold had 10 solenoid valves, which opened in sequence to expose each cassette to an ambient air sample. The sampling time for each filter varied from 66.6 seconds to 100 seconds depending on the sampler location and the test conditions. After each test, the fluoride was chemically extracted from the filter in each cassette and analyzed to determine fluoride content. HF concentrations could be determined over a range from 0.3 to 42,000 ppm.

Photographic documentation was made with videotape recording from three different locations during the six tests. In addition, two framing cameras were used that took 35-mm pictures at 10- to 30-second intervals. During tests 2 to 6, aerial photographs were taken from a helicopter approximately 500 m upwind of the release point at an altitude of 500 m. Different types of recordings were utilized for different tests, including video, still photos, and an infrared thermal scanner.

23.11 DETERMINATION OF EFFECTIVENESS OF SPILL CONTROL AGENTS IN MITIGATING VAPORS FROM HAZARDOUS AND TOXIC CHEMICALS

Sponsor: Ansul Company

Dates of Experiments: April 3–7, 1990

Chemicals Released: Chlorosulfonic acid, 65% oleum, and sulfur trioxide

Description of Experiments: During the first week of April 1990, a study of the effectiveness of vapor mitigation for three different chemicals was performed at the HSC. There are no known reports presented in the open literature and the author is not aware of any internal reports generated by the participants. From reviewing the limited amount of documentation available, plus the data files, it appears that the objective was to investigate the evaporation rate for each of the three chemicals (chlorosulfonic acid, 65% oleum, and sulfur trioxide) under three different scenarios: (1) with no mitigation procedure applied, (2) with a foam blanket applied, and (3) with a combination of foam blanket and water applied.

The chemicals were placed in a square pan mounted on load cells that allowed determination of total amount of chemical released into the pan and monitoring weight change with time. Downwind concentration instrumentation was installed at three tower locations to measure the species released. The locations and heights of these sensors were not reported.

23.12 DETERMINATION OF EFFECTIVENESS OF AQUEOUS FOAMS IN EXTINGUISHING CHLOROSILANES FIRES AND VAPOR SUPPRESSION

Sponsor: Silicones Health Council, Inc.

Dates of Experiments: May 7–May 14, 1990

Chemicals Released: Trichlorosilane and silicon tetrachloride

Description of Experiments: During the second week of May 1990, a series of experiments was conducted at the HSC to evaluate the effectiveness of aqueous foams for suppressing vapors and mitigating chlorosilane fires. There are no known reports in the open literature. From reviewing the limited amount of information available, it appears that either trichlorosilane or silicon tetrachloride was placed in a square pan. After the pan was filled to a specified level, the system was allowed to stabilize and record downwind vapor concentrations. In some of the tests of trichlorosilane the chemical was ignited and then foam was applied for some duration that was probably sufficient to extinguish the fire. Since silicon tetrachloride is nonflammable, the foam application was used to determine effectiveness of vapor suppression. There appears to have been more than a single type of foam applied, but foam specifications are not provided. In some experiments the foam barrier is mechanically broken or swept back to allow exposure of the chemical to the atmosphere and the chemical is reignited with additional foam application. There also appears to have use of the Ansul neutralizer Spill-X and Spill-X-A, but whether this was a test of effectiveness or the standard method of neutralization is not clear.

The dispersal pan was mounted on load cells that allowed determination of total amount of chemical released into the pan. Heat flux sensors were placed at locations downwind and crosswind from the pan containing the released material. These were evidently to measure the amount of radiant heat. These locations had two heights at which measurements were recorded, 3 and 6 ft above ground level. In addition, temperatures were recorded at three locations: (1) the air temperature approximately 3 ft above the pan bottom, (2) the reaction temperature approximately 3 in above the chemical level, and (3) the chemical temperature approximately 2 in above the pan bottom.

Downwind concentration instrumentation was apparently installed at various distances and heights to measure a species, which is not identified but the author assumes to be HCl, since the chemicals react with moisture in the air to form HCl. The locations and heights of these sensors were not reported.

23.13 CHLORINE VAPOR MITIGATION TESTS

Sponsor: Dow Chemical U.S.A.

Dates of Experiments: June 18–25, 1990

Chemical Released: Liquid chlorine

Description of Experiments: In June 1990, Dow Chemical U.S.A. performed a series of experiments to collect data on the efficacy of aqueous foams and water sprays on suppression of chlorine vapors. The tests were run over a period of five days and involved releasing up to 150 lb/min of liquid chlorine into a 50 ft^2 insulated pan mounted on load cells. The supply cylinders from which the liquid chlorine was transferred were also mounted on load cells to measure material released. Temperature measurements were made at the pan and at downwind locations. Fifty-five chlorine sensors were mounted on 17 towers downwind of the release pan at heights of 5, 15, and 25 ft with some additional sensors at the 1-ft height. The towers were organized in three downwind arrays at downwind distances of 250, 500, and 750 ft from the release pan. Tower separation within an array was 150 ft. Two additional towers were located 0.5 miles downwind from the release pan, each with three sensors at heights of 5, 17, and 35 ft. The wind field measurements were made with five stations located across the site, each had a cup-and-vane anemometer mounted at the 6.6-ft height to provide wind speed and direction. The experiments were recorded with still cameras and a portable video camera.

A total of six tests was performed. Each involved transferring liquid chlorine from one or more 1-ton chlorine containers to the open pan. After the pool of liquid chlorine had stabilized, measurements were made of downwind concentration with no mitigation procedure applied. Then one of several different mitigation procedures was performed on the liquid pool. In the first and last tests, the only mitigation technique used was water spray nozzles to "knock down" the chlorine vapor cloud. In tests 2, 3, and 5, different vapor suppression foams were applied to the liquid pool surface in combination with water spray from the nozzle system and measurements were made of the downwind concentration to determine the effectiveness of the procedure. Test 4 used no foam but did use the water spray nozzles and portable water nozzles.

23.14 HYDROGEN FLUORIDE SUIT TESTS

Sponsors: U.S. Department of Energy, U.S. Department of Labor (OSHA), and U.S. Department of Transportation

Dates of Experiments: May 1, 1991–July 25, 1991

Chemical Released: Hydrogen fluoride

Description of Experiments: No documentation was available other than the data files, which don't describe the experimental objectives or the equipment, test conditions, suits tested, or sequence of operations for these experiments. A report from earlier work conducted by LLNL at the HSC (Johnson and Swearingen, 1989) described a series of experiments that were performed to test totally encapsulating chemical protective (TECP) suits under realistic worst-case exposures to hydrogen fluoride (HF). The authors assume that the data files from 1991 where HF was the chemical tested and the 1992 data files where chlorine and ammonia were tested were similar.

The tests described earlier were conducted using a static mannequin on which the suits were mounted. Other tests utilized another mannequin that articulated at the waist to provide more realistic test conditions. The mannequins were fitted with internal chemical sensors to determine internal HF concentration and a compressed air system to allow simulation of breathing. Suits were exposed to various concentrations of the released chemical for periods of about 10 minutes.

23.15 CHLORINE AND AMMONIA SUIT TESTS

Sponsors: U.S. Department of Energy, U.S. Department of Labor (OSHA), and U.S. Department of Transportation

Dates of Experiments: June 29, 1992–July 24, 1992

Chemical(s) Released: Chlorine and ammonia

Description of Experiments: See HF suit tests.

23.16 DUPONT MITIGATION WORKSHOP—DETERMINATION OF EFFECTIVENESS OF VARIOUS TECHNIQUES IN MITIGATING HAZARDOUS MATERIAL SPILLS AND TRAINING OF PERSONNEL TO MITIGATE HAZARDOUS MATERIAL SPILLS

Sponsor: E. I. du Pont de Nemours and Company

Dates of Experiments: April 7, 1992–May 14, 1992

Chemical(s) Released: 65% Oleum and chlorosulfonic acid

Description of Experiments: During the spring of 1992, DuPont sponsored a mitigation workshop to evaluate the effectiveness of certain mitigation techniques on spilled 65% oleum and chlorosulfonic acid. Different techniques and materials were evaluated while mitigating the released acid. The tests also provided an opportunity for industrial workers from facilities where these acids were used to have hands-on experience mitigating these released materials. There are no known publications or internal reports developed from these tests. The authors had access to some of the operations documents generated during the conduct of the tests, and the following description is based on those documents and discussions with Brad Kulesza of DuPont. From this series of tests, DuPont has developed a workshop that is conducted at the HSC approximately every two years for their customers to participate in this type of mitigation training.

The acids were transferred into an open pan mounted on load cells to monitor the mass of acid released. The pan was instrumented with Type K thermocouples to determine the air temperature above the pan (height not known), the temperature of the reaction was measured approximately 5 in. above the bottom of the pan bottom, and the temperature of the liquid (acid) which was measured at 1.5 in. above the bottom of the pan. Ambient air temperature, humidity, and wind speed and direction were also recorded during each test. Visual records were made using video cameras.

A number of different types of mitigation techniques or materials were tested and are listed in Table 23.7.

TABLE 23.7 Materials/Techniques Used During Mitigation Tests

Water	Foam—Universal Gold 3%
Water fog	Spill-X
Foam (not specified)	Spill-X-A
Foam—Universal Gold	Spill-X-A and foam (not specified)
Foam—Ansul 3 + 3	Spill-X-A and foam (not specified) and water
Foam—National Vapor Shield-AC	

23.17 DRI/WRI/EPA CO_2-I EXPERIMENTS

Sponsor: Environmental Protection Agency

Dates of Experiments: July 22–28, 1993

Chemical Released: Carbon dioxide

Description of Experiments: A series of four CO_2 vapor releases was conducted at the HSC during July 1993 by DRI and WRI personnel with support from the DOE site contractor, EG&G/EM, and EPA technical staff. This series was the first fieldwork in an EPA-sponsored research program to better understand dense gas dispersion under stable atmospheric conditions. These experiments documented two facts: (1) a demonstrated methodology for collecting high-resolution concentration data and (2) that dense gas behavior could be exhibited by releasing as little as 1–2 kg/s of the heavier-than-air surrogate gas, CO_2.

In 1990 the U.S. Congress passed the Clean Air Act Amendments (CAAA), of which section 112(r) mandated the EPA to develop regulations requiring industrial facilities that stored, produced, or used certain threshold quantities of hazardous materials to create a risk management plan (RMP). The RMP was to address issues related to accidental releases of these hazardous materials, one of which was the prediction of the consequences of a release of these materials under a worst-case scenario. The meteorological condition for the worst case scenario was defined as a Pasquill F stability with a wind speed of 1.5 m/s. Congress realized the available models to predict the consequences were unproven or tested for this worst-case scenario and therefore also decreed that research should be conducted at the HSC to address this fact along with evaluating the effectiveness of emergency response to accidental releases, CAAA sections 103(f) and 901(h). This research program was to be carried out by the Desert Research Institute (DRI), Reno, Nevada, and Western Research Institute, Laramie, Wyoming, and supported by the EPA and DOE.

In 1992, meetings at Research Triangle Park (RTP), North Carolina, between EPA, DRI, WRI, DOE, DOT, FEMA, and industry developed a consensus plan for addressing this research program. The first task was to demonstrate what was achievable at the HSC:

TABLE 23.8 Summary of DRI/WRI CO_2-I Experiments

	Test 1	Test 2	Test 3	Test 4
Date	7/22/93	7/26/93	7/27/93	7/28/93
Time	18:55:01	19:41:01	19:30:00	19:46:00
Duration (sec)	179	70	265	265
Mass of CO_2 released (kg)	81.9	112.9	171.9	165.0
Mass of SF_6 released (mg)	163.4	63.6	238.2	240.0
CO_2 release rate (kg/s)	0.458	1.613	0.649	0.623
Wind speed at 1 m height on 24-m tower (m/s)	N.R.[a]	N.R.	3.19	5.49
Pasquill stability	N.R.	N.R.	E	D
CO_2 mass balance (%) based on continuous sensors	114	53	94	99
CO_2 mass balance (%) based on gas sampling bags	95	65	91	97
SF_6 mass balance (%)	105	15	53	117

[a] N.R.—not reported

- That high-resolution (1 Hz or greater) concentration data from numerous sensors could be fielded at the HSC
- That dense gas effects could be produced using CO_2 as a dense gas surrogate without using massive release rates or volumes
- That it was possible to capture the required worst-case conditions specified in the CAAA section 112(r)

Based on prior meteorological data, a transition from neutral to stable conditions could be anticipated during the period from just before until just after sunset at the Frenchman Flat site. This series of four CO_2 vapor and tracer amounts of SF_6 were released during the late afternoon just prior to sunset. The releases lasted in duration from 70 to 265 seconds with CO_2 vapor release rates from ~0.5 to 1.6 kg/s. The experiments had been designed to use a horizontal jet source but were changed very late in the program to simulate a ground level low-momentum source.

Liquid CO_2 was supplied by BOC Gases in a portable 6-ton liquid storage tank that was connected to a 30-ton storage tank via a vaporizer to provide a large source of CO_2 vapor. The CO_2 vapor storage tank was connected to a 4-in. release line that extended to the original horizontal jet source, which had a 6-in. control valve attached. The late modification to the source allowed an 18-in. flexible hose to be connected to a 1-m × 1-m release box that was buried below ground level so that the top of the box was at grade level.

The mass of CO_2 released was computed by measuring the temperature and pressure of the 30-ton storage tank before, during, and after the release. With the fixed volume of the tank known, a mass calculation could be made to determine mass of CO_2 at any point during the tests. Temperatures within the release system were measured with Type K thermocouples to determine whether the release of CO_2 was isothermal.

In addition to the CO_2 released, a small measured and recorded amount of SF_6 was injected simultaneously into the release line during each release. The flow system was calibrated to release approximately 0.9 mg/s of SF_6.

Meteorological measurements were made on a 24-m permanent tower installed at the HSC. It was instrumented by DRI during the early part of July 1993 to record data for this test and during the year to characterize the meteorology of the site. The 24-m tower was instrumented with a dedicated data acquisition system (DAS) that also provided analog outputs to the HSC's CCDAS for acquisition during testing. These data were recorded every 10 sec during the testing periods. During the remainder of the year, when the data were being archived for later analysis, the data were acquired every 10 sec and stored as engineering units as a 15-min average. This data was automatically downloaded from the DAS via a modem connection to DRI's Data Processing Center in Reno (Coulombe et al., 1994).

The tower was instrumented with eight levels of R. M. Young Model QA-5305 anemometers to measure wind speed, direction (except during the CO_2 releases, when the R. M. Young Model RE-5701 anemometers were used), and RTD with aspirated solar-shielded air temperature sensors located at heights of 0.25, 0.5, 1, 2, 4, 8, 16, and 24 m. In addition, relative humidity measurements were made at three levels, 1, 4, and 16 m. Soil temperature was measured with a Fenwal Termistor at a depth of 7 cm near the tower, and barometric pressure was measured at the base of the tower. Solar radiation and net solar radiation were measured at the 2-m height on a small tower located about 20 m from the tower. This tower was located approximately 95 m upwind of the source and about 55 m to one side of the centerline of the test grid.

In addition to the 24-m meteorological tower, a R. M. Young Model RE-5701 anemometer was positioned 5 m in front of the first gas concentration sensor array on the centerline at a height of 0.5 m. Seven cup-and-vane anemometers mounted at a 2-m height measured wind speed and direction and were placed around the test grid to provide measurements of the wind field.

The dispersion of CO_2 was captured by a gas concentration sensor array containing 37 real-time gas sensors at 40 m downwind from the source. The sensors were positioned on towers across a total distance of 30 m to either side of the centerline of the array. The locations and range of the sensors are shown in Fig. 23.1. The position of the ground level source in front of the original position of the horizontal jetting source caused the distance from the source to the first instrument array to change from 40 m to approximately 38 m. A second gas concentration sensor array containing three sensors was located 100 m downwind from the source. The sensors were located at a height of 0.2 m with one on the centerline and the other two at 6.5 m on either side of the centerline.

The sensors were solid-state IR sensors that had a fast response time for changes in CO_2 concentration (<1 second). They could be ranged for different full-scale concentrations of CO_2. For these experiments two ranges were used, 0 to 1% and 0 to 10%. They provided a 0–5 VDC nonlinear output. Along with the fast response CO_2 sensors, gas-sampling bags were installed on the 40-m gas concentration array. The gas sampling bags were 20-L 2 mil Tedlar sample bags installed at 43 locations on this array. Two bags were installed upwind of the source to measure background of CO_2 and SF_6. The sample in each bag was assumed to be representative of the average ambient air passing a specific sampling point during the sampling period. The implementation of the gas sampling bags allowed for filling to be controlled by a remote signal to solenoid valves from the CCDAS. After each release, the bags were subjected to CO_2 and SF_6 analysis. This was done to provide a crosscheck against the real-time solid-state CO_2 sensors and for calculating a mass balance for each release.

23.18 KIT FOX SERIES

Sponsors: U.S. Environmental Protection Agency, Western Research Institute, U.S. Department of Energy, and Petroleum Environmental Research Forum Project 93-16

Dates of Experiments: August 22–September 15, 1995

Chemical Released: Carbon dioxide

Description of Experiments: In August and September 1995, a series of dense gas dispersion experiments was conducted by WRI and DRI as a continuation of the July 1993 DRI/WRI CO_2-I experiments. Funding for the Kit Fox collaborative research was provided by WRI with DOE Jointly Sponsored Research funds, together with the EPA's Office of Research and Development and Chemical Emergency Preparedness and Prevention Office, and the Petroleum Environmental Research Forum (PERF) Project 93-16.* The project, using CO_2 as a dense gas surrogate, was designed to develop a database of measurements during heavier-than-air releases that would simulate the worst-case scenario requirement of the CAAA 112(r). The gas releases were conducted under three different surface roughness configurations during neutral and stable atmospheric conditions. The Kit Fox field experiments required data on wind speed, ambient turbulence, wind direction, wind profiles as a function of height (from which a surface roughness can be calculated), temperature profiles as a function of height (from which a Monin-Obukhov length can be calculated), and humidity. A PERF 93-16 concern was the along-wind dispersion coefficient (σ_x) at the leading and tailing edge of the cloud and model validation of short duration releases, and therefore

*The 10 companies making up the Petroleum Environmental Research Forum 93-16 Project are Allied Signal Corporation; Amoco Corporation; Chevron Research and Technology Co.; CITGO Petroleum Corporation; Clark Oil and Refining Co.; Exxon Research and Engineering Co.; Marathon Corporation; Mobil Research and Development Co.; Phillips Petroleum Co.; and Shell Research and Development Co.

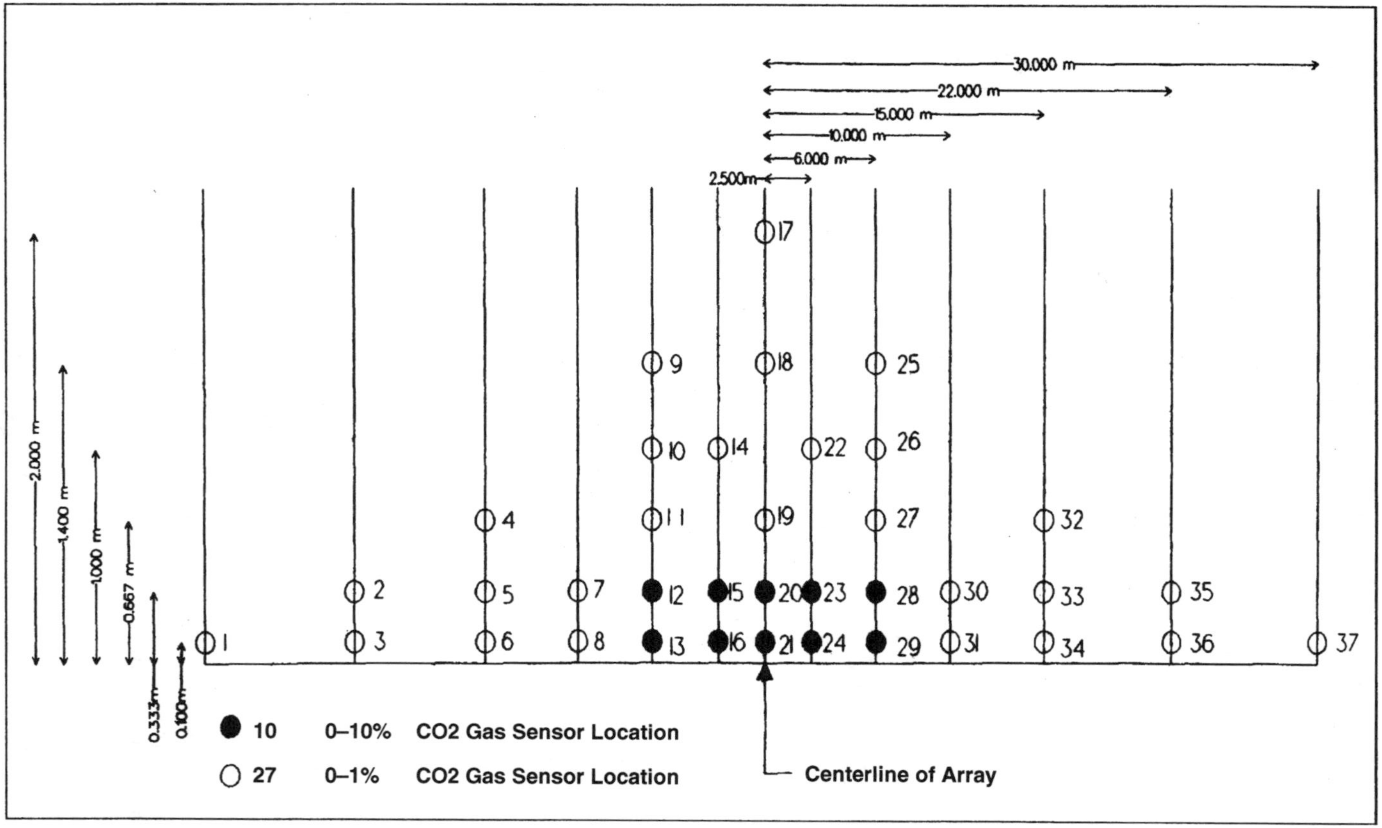

FIGURE 23.1 Real-time sensor locations for DRI/WRI CO_2-I experiments. (*Source: Egami et al., 1996*)

finite-duration releases (puffs) were incorporated into the test matrix. The DRI/WRI CO_2-II Experiments (Coulombe et al., 1999) were the smooth desert surface configuration of the Kit Fox Series conducted in September 1995 after the enhanced roughness portions of the experiments were completed in August.

The modeling, laboratory, and wind tunnel studies leading to the final design of the field project were documented by WRI (Sheesley et al., 1995) and are summarized below.

- The roughness obstacles, called the equivalent roughness pattern (ERP), simulate a refinery tank farm at about 1:10 scale and were developed from the Cermak Peterka Petersen, Inc. (CPP) wind tunnel work (Petersen and Cochran, 1995). The roughness pattern, called the uniform roughness array (URA), that simulated the surface surrounding a typical tank farm/refinery at about 1:10 scale was developed from the EPA wind tunnel work at Research Triangle Park (Snyder, 1995).
- The participants recognized that a full-scale mockup representing a typical refinery or chemical plant covering an area of about 100 acres (0.5 km × 1.0 km) would be impractical. The design used the techniques demonstrated at wind tunnel scale to demonstrate relationships to reproduce and/or predict full-scale phenomena at a scale that could be constructed in the field. By selecting appropriate scaling factors, a reduced-scale simulation of a dense gas release could be made to represent a release over a full-scale, real-world surface roughness that was outlined by Dr. Gary Briggs (EPA/NOAA Technical Advisor). The principal investigators of WRI, PERF, and EPA determined that a 1:10 physical scale and a 1:6 time scale simulation could be carried out in the field.
- The CPP and WRI 1:10 scale model development for the ERP was sized to simulate an artificial boundary layer or mixing depth that would be generated by a refinery with process towers, buildings, pipe racks, and tanks. To achieve this 1:10 scale model, each ERP baffle was about 2.4 m high with a roughness length, z_0, of about 0.2 m set up in a rectangular grid with a crosswind dimension of 39 m and an along-wind dimension of 85 m (see Figs. 23.2 and 23.3).
- A rectangular grid, measuring 314 m by 120 m, surrounded the central ERP grid. This URA was constructed to simulate a typical grass and low brush area surrounding an industrial facility. Again, the field scale was about 1:10. The URA baffle elements are about 0.2 m high with a calculated roughness length, z_0, of about 0.02 m (i.e. the URA z_0 was about 10% of the ERP z_0). The cross-wind size was designed to be large enough to contain the lateral spread of dense gas clouds based on tests conducted in the CPP wind tunnel. The along-wind dimension extended upwind of the ERP far enough (89 m) that the boundary/mixing layer would develop in the ERP and URA roughness up to heights of about 5 to 10 m at the source position.
- To reduce the number of URA elements required in the upwind fetch to develop atmospheric boundary layer depth in the simulated refinery grid, a wind tunnel technique was used. Trip spires were incorporated into the test grid upwind of the release point to decrease the distance required for boundary layer development.

The sensors for the Kit Fox tests included high-concentration and low-concentration CO_2 sensors. A low-concentration sensor could be spanned from 0 to 2,000 ppm (0 to 0.2%) up to 0 to 20,000 ppm (0 to 2%) full-scale, and a high-concentration sensor could be spanned from 0 to 50,000 ppm (0 to 5%) up to 0 to 150,000 ppm (0 to 15%) full-scale. The response time was less than one second for the high-range sensors and about three to four seconds for the low-range sensors.

The sensors were laid out in four arrays located downwind from the source and oriented perpendicular to the centerline of the predicted transport course of the cloud. The roughness conditions were: (1) ERP + URA, (2) URA-only, and (3) no roughness (or smooth desert surface). The array distances downwind and sensor concentration ranges were chosen for a 4-kg/s CO_2 release for the roughness condition with the ERP + URA, based on 1993

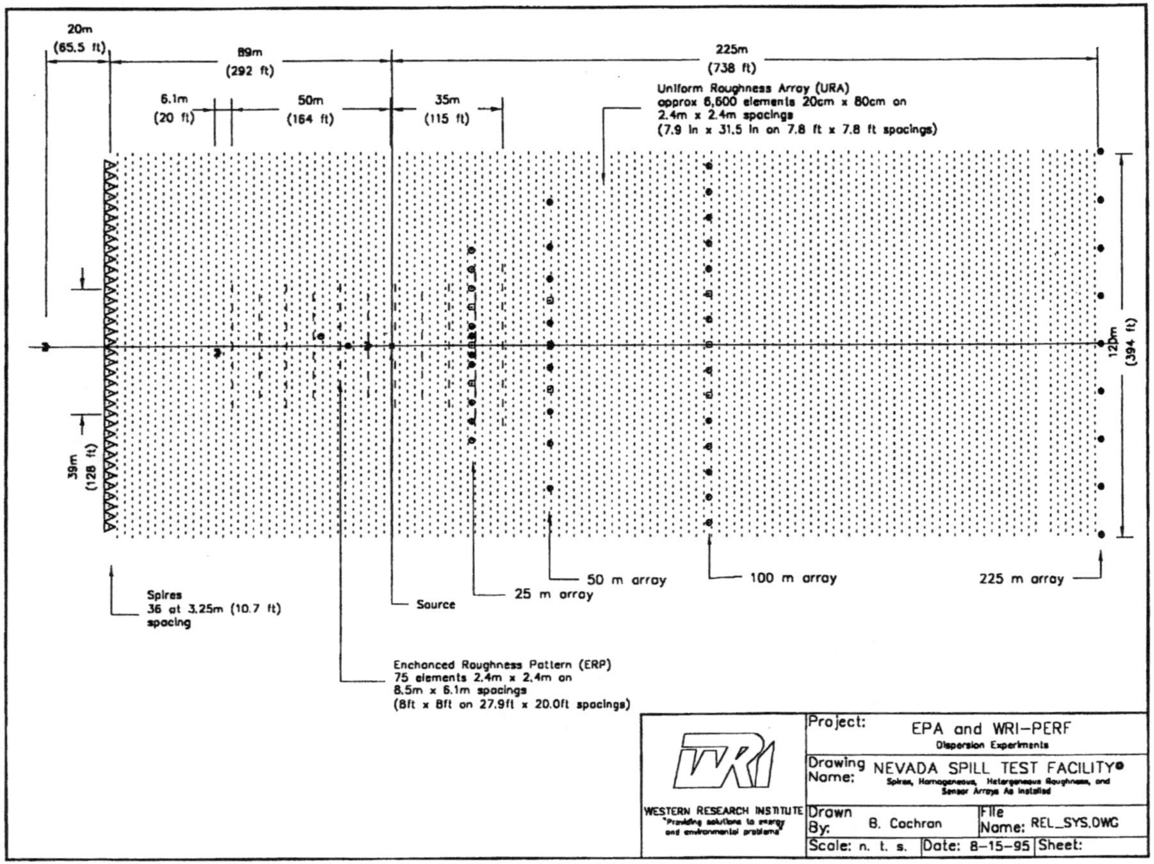

FIGURE 23.2 Source, monitoring array, ERP, and URA locations. (*Source: Sheesley et al., 1998*)

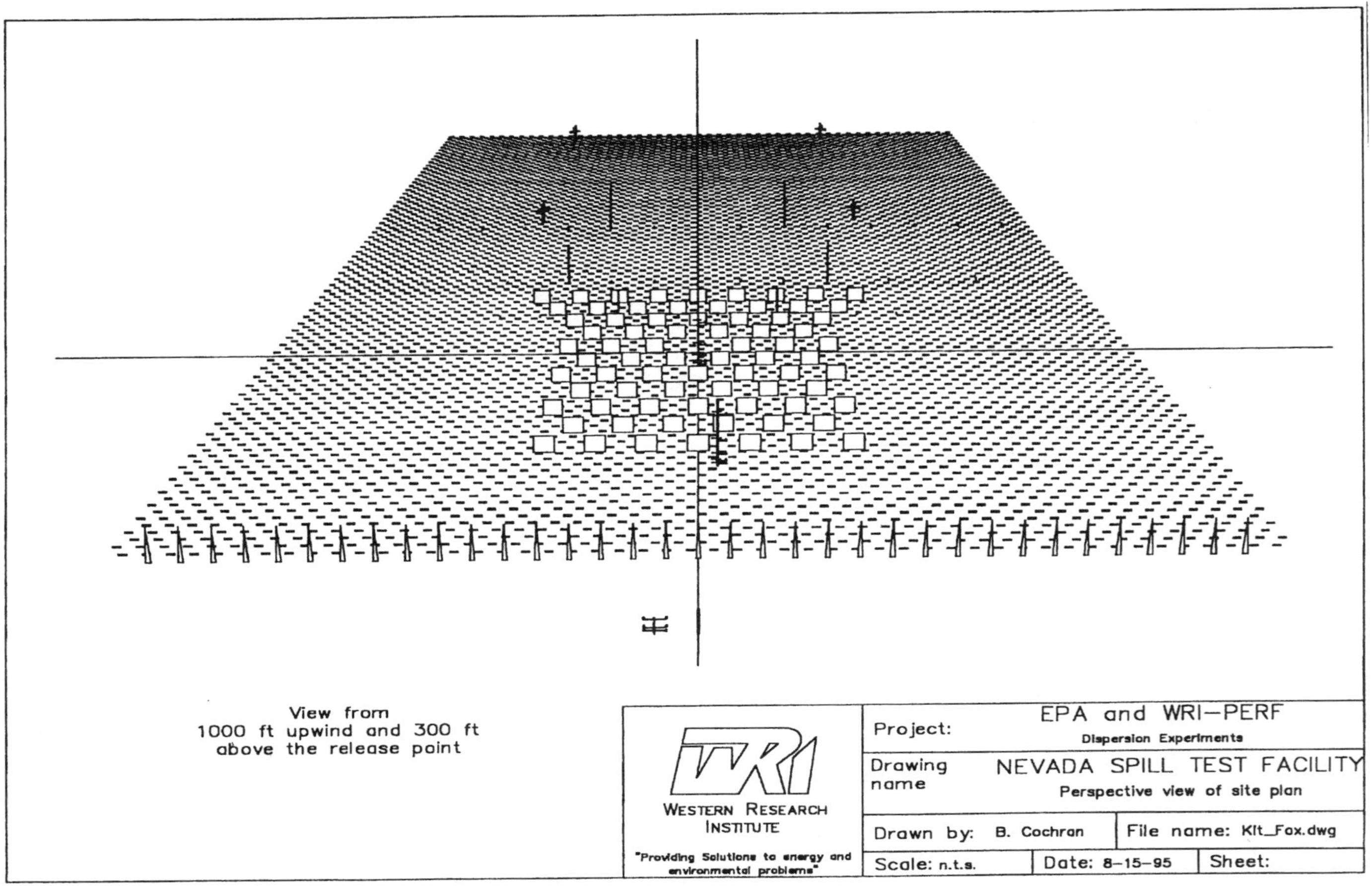

FIGURE 23.3 NTS experimental arrays viewed from upwind location. (*Source: Sheesley et al., 1998*)

WRI/DRI CO_2-I experiments data and reported modeling calculations (both SLAB and HEGADAS v3.0+ models were used). A 1.5-kg/s release rate was used for the URA-only and smooth (desert surface) experiments.

To provide an independent crosscheck of continuous sensor performance, during one release on a test day, integrated bag samples were collected from positions colocated at selected sensor locations. A high correlation between the integrated bag samples collected over a four- to nine-minute time period and the integrated response from the corresponding sensor for the same time interval provided a high confidence in the real-time sensor results.

A Kit Fox test objective was to provide adequate instrumentation to measure meteorological conditions both over the flat desert surface and within the dispersion grid containing the ERP and URA roughness. The meteorology measurements made on the permanent 24-m tower located southwest of the test grid were wind speed, wind direction, and temperature measured at 0.25, 0.5, 1, 2, 4, 8, 16, and 24 m above the ground. Solar and net radiation sensors were installed 2 m above the ground, 20 m south of the tower. Barometric pressure and soil temperature (1 cm below the surface) were measured near or at the base of the tower. Relative humidity was measured at 1, 4, and 16 m. Additional meteorology data were recorded on an 8-m tower with five levels of wind speed, direction, and air temperature (tower 2) located on the nominal 230° centerline in the URA; an 8-m tower also recorded five levels of wind speed, direction, and air temperature (tower 4) in the URA on the centerline downwind of the last obstacle (ERP); and four additional wind direction and speed measurements were recorded on 2-m high masts near the 100-m and 225-m arcs. Another short meteorology tower (tower 1) with three anemometers and temperature sensors was located upwind of the spire row or 110 m upwind of the source to be in a location in the flat desert upwind from the influence of the Irwin spires. The locations of the meteorology towers are shown in Figure 23.4.

One 3-axis sonic anemometer was provided from the 1993 field research program sponsored by EPA and conducted by DRI/WRI, and the remaining 3-axis and 2-axis sonic anemometers by provided by U.S. Army Dugway Proving Grounds (DPG) and funding from WRI. The EPA-DRI/WRI instrument and three of the DPG instruments were 3-axis instruments capable of measuring velocity fluctuations in the x, y, and z directions (u', v', and w'). Six other DPG instruments were 2-axis instruments that measured velocity fluctuations in two directions (u' and v').

The CO_2 release system was designed, constructed, and tested to provide a single source characteristic: a gas release velocity of less than or about 1 m/s to minimize vertical momentum. The system also provided gas temperature near ambient temperature, release rates of 1.5 kg/s or 4.0 kg/s, rapid on/off (less than one second), sufficient volume for a series of finite-duration (puff) and continuous releases in one evening of tests, and volume recharge logistics to provide for consecutive test days. The vapor storage tanks were recharged from a liquid carbon dioxide container and vaporizer supplied by the vendor, BOC Gases. Recharging took about eight hours and was done the night before or the morning of a test day.

The major components of the release system were:

- The existing HSC facility large-volume tanks, with a total capacity of 7,210 ft^3 and a 500-ft long 12-in. diameter release line
- A 580-ft extension to the existing release line
- The 1.5 m × 1.5 m × 1 m deep release box connected to control valves and flow measurement system all below ground level
- The release box doors at the ground level opening to obtain a rapid on/off

Three independent methods of measuring released CO_2 vapor were used. The first was the measurement of carbon dioxide flow rate by a vortex shedding flow meter, located just upstream of the 4-in. and 6-in. butterfly control valves. The second method was the measurement and recording of pressures and temperature at a 4.5-in. orifice installed downstream

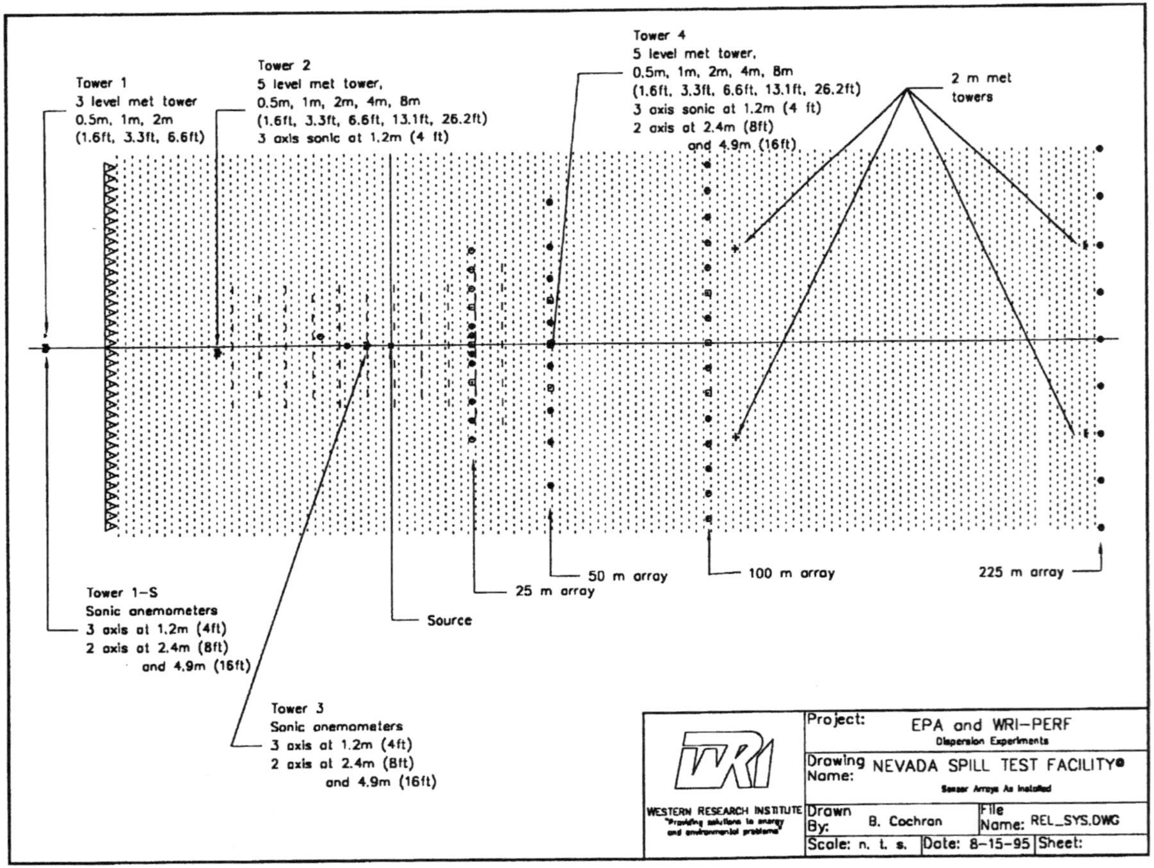

FIGURE 23.4 Locations of meteorological towers within the ERP and URA. (*Source: Sheesley et al., 1998*)

TABLE 23.9 Kit Fox Enhanced Roughness Test Matrix

Pasquill stability	ERP + URA		URA only	
	Continuous	Puffs	Continuous	Puffs
D	T5r3[a]	T2r1, T3r1, T3r2, T3r3, T5r1, T5r2	T6r4, T6r5, T6r6, T7r2, T7r3, T7r5, T8r5, T8r8, T8r11	T6r1, T6r2, T6r3, T7r4, T7r6, T7r8, T8r1, T8r2, T8r3, T844, T8r6, T8r7, T8r9, T8r10, T8r12
E	T2r5, T5r4	T3r4, T5r5	T6r9, T7r9, T7r12	T6r7, T7r10, T7r11
F	T3r5, T4r4, T5r8	T3r6, T3r7, T4r3, T5r6, T5r7		T6r8, T7r13, T7r14

[a]The naming convention for the individual releases was TXrY, where X is the test day and Y is the sequential release for that test day.

from the butterfly valves. The third was the continuous recording of supply tank pressure and temperature by the HSC's CCDAS and calculation of the total mass change during a test day.

From August 22 to August 31, WRI conducted 70 releases. Tests were made during the late afternoon and early evening as the strong and predominately southwest winds diminished and the atmospheric stability transitioned from neutral to very stable. Neutral conditions (Pasquill D) were achieved with wind speeds between 3 to 5 m/s, stable conditions (Pasquill E) in the 2 to 3 m/s range and very stable (Pasquill F) in the 1 to 2 m/s range. Release duration varied between two and five 5 minutes for the continuous releases and 20 seconds for the finite-duration (puff) releases. On a test day when favorable meteorological conditions were present, as many as 14 gas releases were conducted between the late afternoon before sunset until early evening after sunset.

These tests were conducted under three separate surface roughness configurations. The first series, from August 22 to 28, had the ERP + URA roughness elements installed. This was to simulate the refinery with its surrounding low vegetation. During the morning of August 29, the large obstacles (ERP) were removed, and from August 29 to 31 tests were conducted with an intermediate roughness configuration represented by the URA-only elements. During the week of September 4 to 8, the URA elements were removed and a week of releases with a smooth desert surface configuration was conducted from September 11 to 15. These smooth desert surface experiments are described in a separate report by DRI and WRI (Coulombe et al., 1999).

The August enhanced roughness test operations resulted in 74% (52 releases) of all releases being suitable for further data reduction, processing, and validation. Table 23.9 provides a matrix of the releases in the Kit Fox enhanced-roughness database with respect to surface roughness configuration, stability, and duration of release (continuous versus puff).

23.19 REFERENCES

Blewitt, D. N., J. F. Yohn, and D. L. Ermak. 1987b. "An Evaluation of SLAB and DEGADIS Heavy-Gas Dispersion Models Using the HF Spill Test Data," in *American Institute of Chemical Engineers, Proceedings, International Conference on Vapor Cloud Modeling,* Boston, November 2–4, pp. 56–80.

Blewitt, D. N., J. F. Yohn, R. P. Koopman, and T. C. Brown. 1987a. "Conduct of Anhydrous Hydrofluoric Acid Spill Experiments," in *American Institute of Chemical Engineers, Proceedings, International Conference on Vapor Cloud Modeling,* Boston, November 2–4.

Blewitt, D. N., J. F. Yohn, R. P. Koopman, T. C. Brown, and W. J. Hague. 1987c. "Effectiveness of Water Sprays on Mitigating Anhydrous Hydrofluoric Acid Releases," in *American Institute of Chemical Engineers, Proceedings, International Conference on Vapor Cloud Modeling,* November 2–4, Boston, pp. 155–180.

Brown, T. C., R. T. Cederwall, D. L. Ermak, R. P. Koopman, J. W. McClure, and L. K. Morris. 1990. *Falcon Series Data Report, 1987 LNG Vapor Barrier Verification Field Trials,* GRI-89/0138, Gas Research Institute, Chicago.

Coulombe, W., J. Bowen, R. Egami, D. Freeman, D. Sheesley, J. Nordin, T. Routh, and B. King. 1999. *Characterization of Carbon Dioxide Releases–Experiment Two,* DRI Doc. No. 97-7240.F, Desert Research Institute, Reno, NV.

Coulombe, W., J. Bowen, D. Freeman, R. Egami, B. Cristani, S. Schmidt, D. Sheesley, B. King, and T. Routh. 1994. *24M Meteorological Tower Data Report Period: August through December, 1993,* DRI Doc. No. 94-305.D3, Desert Research Institute, Reno, NV.

Egami, R., W. Coulombe, J. Bowen, D. Freeman, J. Watson, D. Koracin, D. Schorran, B. Zielinska, D. Sheesley, J. Nordin, T. Routh, and B. King. 1996. *Characterization of Carbon Dioxide Releases: Experiment One,* DRI Doc. No. 93-3305.3F, Desert Research Institute, Reno, NV.

Goldwire, H. C., Jr., T. G. McRae, G. W. Johnson, D. L. Hipple, R. P. Koopman, J. W. McClure, L. K. Morris, and R. T. Cederwall. 1985. *Desert Tortoise Series Data Report, 1983 Pressurized Ammonia Spills,* Lawrence Livermore National Laboratory, CA, UCID-20562.

Johnson, J. S., and P. M Swearingen. 1989. *Exposures of Totally-encapsulating Chemical Protective Suits to High and Very High Concentrations of Hydrogen Fluoride along with Future Experimental Plans,* Lawrence Livermore National Laboratory, UCRL-101717.

Koopman, R. P., J. Baker, R. T. Cederwall, H. C. Goldwire, Jr., W. J. Hogan, L. M. Kamppinen, R. D. Kiefer, J. W. McClure, T. G. McRae, D. L. Morgan, L. K. Morris, M. W. Spann, and C. D. Lind. *Burro Series Data Report, LLNL/NWC 1980 LNG Spill Tests,* Lawrence Livermore National Laboratory, UCID-19075, vols. 1, 2.

McRae, T. G., R. T. Cederwall, D. L. Ermak, H. C. Goldwire, D. L. Hipple, G. W. Johnson, R. P. Koopman, J. W. McClure, and L. K. Morris. 1987. *Eagle Series Data Report: 1983 Nitrogen Tetroxide Spills,* Lawrence Livermore National Laboratory, CA, UCID-20063, Rev. 1.

Petersen, R. L., and B. C. Cochran. 1995. *Wind Tunnel Determination of Equivalent Refinery Roughness Patterns,* CPP Report 94-1152, CPP, Fort Collins, CO.

Sheesley, D. C., J. S. Nordin, S. B. King, T. K. Routh, V. E. Smith, and B. C. Cochran. 1998. *Chemical Dispersion Modeling Project, 1995 Kit Fox Field Demonstration, DOE Nevada Hazardous Spill Center, Final Report,* vol. 1, *Experiment Description and Data Processing,* WRI-98-R019a, Western Research Institute, Laramie, WY.

Sheesley, D. C., S. B. King, T. K. Routh, J. S. Nordin, V. E. Smith, and B. C. Cochran. 1995. *Design Plan, 1995 Kit Fox: WRI-PERF-EPA Field-Scale Experiments at the Nevada Test Site,* WRI-95-R032, Western Research Institute, Laramie, WY.

Snyder, W. H. 1995. *Wind-Tunnel Roughness Array Tests,* Data Report, Code: RAT, Atmospheric Research and Exposure Assessment Laboratory, U.S. Environmental Protection Agency, Research Triangle Park, NC.

23.20 ADDITIONAL REFERENCES

Burro Series

Ermak, D. L., S. T. Chan, D. L. Morgan, and L. K. Morris. "Comparison of Dense Gas Dispersion Model Simulations with Burro Series LNG Spill Test Results," *Journal of Hazardous Materials,* vol. 6, pp. 129–160, 1982.

Koopman, R. P., R. T. Cederwall, D. L. Ermak, H. C. Goldwire, Jr., W. J. Hogan, J. W. McClure, T. G. McRae, D. L. Morgan, H. C. Rodean, and J. H. Shinn. "Analysis of Burro Series $40m^3$ LNG Spill Experiments," *Journal of Hazardous Materials,* vol. 6, pp. 43–83, 1982.

Morgan, D. L., Jr. 1988. "Dispersion Phenomenology of LNG Vapor in the Burro and Coyote LNG Spill Experiments," *Journal of Heat Transfer,* vol. 109, pp. 952–860. 1988.

Coyote Series

Goldwire, H. C., Jr., H. C. Rodean, R. T. Cederwall, E. J. Kansa, R. P. Koopman, J. W. McClure, T. G. McRae, L. K. Morris, L. Kamppinen, R. D. Kiefer, P. A. Urtiew, and C. D. Lind, *Coyote Series Data Report, LLNL/NWC 1981 LNG Spill Tests, Dispersion, Vapor Burn, and Rapid-Phase-Transition,* Lawrence Livermore National Laboratory, CA, UCID-19953, vol.1, 2, 1983.

Koopman, R. P., *Coyote Series for 40 m^3* Liquefied Natural Gas (*LNG*) *Dispersion, RPT, and Vapor Burn Tests,* Lawrence Livermore National Laboratory, CA, UCID-19211-Rev. 1, 1982.

Morgan, D. L., Jr. 1988. "Dispersion Phenomenology of LNG Vapor in the Burro and Coyote LNG Spill Experiments," *Journal of Heat Transfer,* vol. 109, pp. 952–960, 1988.

Rodean, H. C., W. J. Hogan, P. A. Urtiew, H. C. Goldwire, Jr., T. G. McRae, and D. L. Morgan, Jr., *Vapor Burn Analysis for the Coyote Series LNG Spill Experiments,* Lawrence Livermore National Laboratory, CA, DOE Report UCRL-53530, 1984.

Desert Tortoise Series

Goldwire H. C., Jr., "Large-Scale Ammonia Spill Tests," *Chemical Engineering Progress,* vol. 82, no. 4, pp. 35–41, 1986.

Kaiser, G. D., "A Review of Models for Predicting the Dispersion of Ammonia in the Atmosphere," *Plant/Operations Progress,* vol. 8, pp. 58–64, 1989.

Kansa, E. J., H. C. Rodean, S. T. Chan, and D. L. Ermak, *Atmospheric Dispersion of Ammonia: An Ammonia Fog Model,* Lawrence Livermore National Laboratory, CA, UCRL-88649, 1983.

Koopman, R. P., D. L. Ermak, and S. T. Chan, *Review of Recent Work in Atmospheric Dispersion of Large Spills,* Lawrence Livermore National Laboratory, CA, UCRL-97377, 1988.

Leitner, P., G. Miller, and J. H. Shinn, *Environmental Assessment for Spill Tests of Ammonia and Nitrogen Tetroxide at Frenchman Flat, Nevada Test Site,* Lawrence Livermore National Laboratory, CA, UCID-19822, 1983.

Spicer, T. O., J. A. Havens, and L. E. Key, "Evaluation of the DEGADIS Dispersion Model Using Data From Field Releases of Pressurized Ammonia," in *Proceedings, Air Pollution Control Association 80th Annual Meeting,* vol. 7, 1987.

Eagle Series

Koopman, R. P., D. L. Ermak, and S. T. Chan, *Review of Recent Work in Atmospheric Dispersion of Large Spills,* Lawrence Livermore National Laboratory, CA, UCRL-97377, 1988.

Koopman, R. P., T. G. McRae, H. C. Goldwire Jr., D. L. Ermak, and S. T. Chan, "Results of Recent 1983 NH_3 and N_2O_4 Spill Tests," in *Annual Technical Meeting*—Institute Environmental Sciences 1985, 1985.

McRae, T. G., *Analysis and Model/Data Comparisons of Large-Scale Releases of Nitrogen Tetroxide. Final Report June 1983–September 1984,* Lawrence Livermore National Laboratory, CA, UCID-20388, 1985.

McRae, T. G., *Evaluation of Source Strength and Dispersion Model Predictions with Data from Large Nitrogen Tetroxide Field Experiments,* Lawrence Livermore National Laboratory, CA, UCRL-91402, 1985.

Chlorine Releases

Thomerson, J., *Chlorine Vapor Suppression Tests,* Dow Chemical U.S.A., Chlor-Alkali Technology Center Internal Report, A-1230, Freeport, TX, 1990. (Copies of this report and a short video may still be available from the Chlorine Institute, Inc., Washington, DC.)

CO_2-I Experiments

Briggs, G. A., "Field-Measured Dense Gas Plume Characteristics and Some Parameterizations," in *International Conference and Workshop on Modeling and Mitigating the Consequences of Accidental Releases of Hazardous Materials,* New Orleans, September 26–29, 1995, American Institute of Chemical Engineers, New York, pp. 539–555.

Kit Fox Series

King., S. B., D. Sheesley, T. Routh, and J. Nordin. "The Kit Fox Field Demonstration Project and Data Set," in *AIChE's CCPS—International Conference and Workshop on Modeling the Consequences of Accidental Releases of Hazardous Materials,* San Francisco, September 29–October 1, 1999.

Sheesley, D. C., J. S. Nordin, S. B. King, T. K. Routh, V. E. Smith, and B. C. Cochran. *Chemical Dispersion Modeling Project, 1995 Kit Fox Field Demonstration, DOE Nevada Hazardous Spill Center, Final Report,* vol. 2, *Data Analysis for Enhanced Roughness Tests,* WRI-98-R019b, Western Research Institute, Laramie, WY, 1998.

CHAPTER 24
FUGACITY MODELING TO PREDICT LONG-TERM ENVIRONMENTAL FATE OF CHEMICALS FROM HAZARDOUS SPILLS

Rajesh Seth and Don Mackay

Canadian Environmental Modelling Centre, Trent University, Peterborough, Ontario, Canada

24.1 INTRODUCTION

As other sections of this Handbook indicate, there have been considerable efforts to model and predict the short-term fate of spilled hazardous materials over hours and days. These include estimates of spreading, evaporation, atmospheric dispersion, and flow in surface waters and groundwaters. A major incentive for such models is the protection of the public and remedial action personnel. Less attention is paid to the longer-term fate of the spilled material over months and years. In this section, we review the use of mass balance models to predict the long-term fate of spilled materials. Ultimately, the residual spilled material combines with the existing contaminant burden in the environment to raise general concentration levels and increase overall human exposure. The focus is thus on chronic exposure as distinct from short-term acute exposure, which is the primary initial concern of response agencies.

An incentive for developing the capability of predicting the long-term fate of spilled materials is that such information can be used to guide and prioritize emergency response measures. Persistent chemicals should merit more vigorous cleanup measures. A potential groundwater contaminant may justify strenuous efforts to clean up a soil and prevent entry to an aquifer. A chemical that is known to hydrolyze rapidly in water can possibly be allowed to enter water after due consideration to reaction products. On the contrary, one that is nonreactive and bioaccumulative can cause contamination problems and contact with water should be avoided.

It is suggested that multimedia mass balance models be used for this purpose. The specific class of models considered here are the fugacity models as described by Mackay (2001). Fugacity is used as a surrogate for concentration in these models because it simplifies the calculations. In the following sections we describe the fugacity concept and then introduce the approach of considering the environment as a set of connected well-mixed or homogeneous media or phases and the required input data of chemical properties. A series of progressively more complex models is then introduced, namely Level I, II, III, and IV calculations, it being suggested that the Level I–III models are adequate for general fate assessment of a chemical spill. For a more detailed site-specific evaluation, a Level IV or dynamic model is more appropriate, but implementation is more demanding. Such a model is detailed in a report by Mackay and Paterson (1985).

24.2 FUGACITY, Z-, AND D-VALUES

Fugacity can be regarded as the "escaping tendency" of a chemical substance from a phase and has units of pressure, preferably Pa. It is identical to partial pressure in ideal gases and is logarithmically related to the equilibrium criterion of chemical potential. Fugacity is to mass diffusion as temperature is to heat diffusion. The concept of fugacity has been shown to be useful in identifying the static and dynamic behavior of toxic substances in the environment.

Fugacity (f, Pa) is linearly related to concentration (C, mol/m^3) at low environmental concentrations and can be related by the following expression:

$$C = f \times Z \tag{24.1}$$

where Z is a proportionality constant, termed the "fugacity capacity" with units of mol/m^3 · Pa. This equation does not necessarily imply that C and f are always linearly related. As a rule of thumb, the linearity assumption may be considered to be valid for concentrations less than 10% of saturation. Nonlinearity at higher concentrations can be accommodated by allowing Z to vary as a function of C or f.

Z depends on temperature, pressure, the nature of the substance, and the medium in which it is present. It quantifies the capacity of a phase to dissolve or sorb a chemical. At a given concentration, if Z is low, then f is high and the escaping tendency is high, whereas if Z is high, then f is low and the escaping tendency is low. Substances thus tend to accumulate in phases where Z is high, or high concentrations can be reached without creating high fugacities and a correspondingly high tendency to escape. The expressions for Z-values of pure air and water phases are shown in Table 24.1. The Z-values for any other medium can be calculated using these Z-values (air or water) and the dimensionless partition coefficient between that medium and air or water. For example, the Z-value for organic carbon (OC) can be calculated as follows:

$$Z_{OC} = Z_W \times K_{OC} \tag{24.2}$$

where Z_W is the Z-value for pure water phase and K_{OC} is the dimensionless organic carbon/water partition coefficient. Extension of Z_{OC} to calculate the Z-value for bulk soil is shown in Table 24.1.

The third group of parameters used in fugacity calculations is D-values, which are transport and transformation rate parameters with units of mol/Pa · h. When multiplied by a fugacity, they give rates of transport or transformation. Fast processes have large D-values. D-values can be added when multiplied by a common fugacity. Thus, it becomes obvious which D-value, and hence which process, is most important. Essentially, a D-value is a fugacity rate constant. For fate and transport of a chemical in the environment, various D-values defining loss of chemical by advection, reaction, or intermedia transport can be obtained for each environmental medium. An example of each is included in Table 24.1.

24.3 ENVIRONMENT AS COMPARTMENTS

Mass balance models are based on Lavoisier's fundamental axiomatic law of the conservation of mass. The environment is very complex and changes in time and space. Since it is too

TABLE 24.1 Expression and Estimates for some Z and D-values at 25°C Calculated and Used by the Models

Parameter	Expression	Value	
		Styrene	Dinitrotoluene
Z-values			
Air (all levels)	$1/RT$	4.03×10^{-4}	4.03×10^{-4}
Water (all levels)	$1/H$	3.6×10^{-3}	11.15
Other, e.g., Soil			
Level I and II	$v_{OC}Z_{OC}$	7.95×10^{-2}	22.45
Level III	$v_A Z_A + v_W Z_W + v_{OC} Z_{OC}$	4.09×10^{-2}	14.57
D-values			
Advection (Level II and III) e.g., in air	$G_A Z_A$	4.03×10^{8}	4.03×10^{8}
Reaction (Level II and III) e.g., in soil (Level III)	$k_S V_S Z_S$	9.28×10^{5}	1.07×10^{8}
Intermedia transport (Level III) e.g., water to air	$K_v A Z_W$	1.65×10^{6}	2.01×10^{7}

R = gas constant ($Pa \cdot m^3 \cdot mol^{-1} \cdot K^{-1}$)
T = temperature (K)
H = Henry's law constant ($Pa \cdot m^3 \cdot mol^{-1} \cdot K^{-1}$)
v = volume fraction
G = flow rate ($m^3 \cdot h^{-1}$)
A = surface area (m^2)
K_v = overall water mass transfer coefficient ($m \cdot h^{-1}$)
k = reaction rate constant (h^{-1})
Subscripts: A = air; W = water; S = soil; OC = organic carbon

complex to describe in accurate detail, the art of modeling lies in selecting only the key media and processes. The fugacity models described here attempt to describe chemical fate in the entire relevant environment consisting of the atmosphere, terrestrial soils, water bodies, and bottom sediments. It proves useful to assemble evaluative environments that can later be used in calculations.

It is convenient to define two evaluative environments and undertake calculations of chemical entry, transport, and transformations in these environments. The simplest approach is to define a four-compartment (or medium) system of air, water, soil, and sediment. Second is a more complex eight-compartment system (including aerosols, suspended sediment, and terrestrial and aquatic biota), which is more representative of real environments and is correspondingly more data-intensive. For a detailed discussion on these environments or "unit worlds," the reader can consult Neely and Mackay (1982). The values and properties assigned to various environmental compartments or media can be modified if chemical fate in a specific region is required.

24.4 MODELS OF INCREASING COMPLEXITY

To develop a full understanding of how chemical and environmental properties affect the fate and transport of a chemical in the environment, a series of models of increasing complexity has been developed designated Levels I, II, and III. Salient features of Level I, II, and III models described by Mackay (1991) are presented below. These models are available for free download from the website of the Canadian Environmental Modelling Centre at

Trent University (http://www.trentu.ca/envmodel). Extension of Level III model to simulate unsteady-state conditions (Level IV model) is also discussed. Three types of chemicals are treated in these models, viz. chemicals that partition into all media (type 1), involatile chemicals (type II), and chemicals with zero or near-zero water solubility (type III). The models cannot treat ionizing or speciating chemicals. Chemicals that can potentially be discharged as spills are predominantly of type I, and illustrative application of Levels I–III models to two such industrial chemicals, styrene and dinitrotoluene, using EQC-standard environment (Mackay et al., 1996b) is presented. The models are also capable of treating type 2 and 3 chemicals, which differ slightly in terms of data requirement (Mackay et al., 1996b), and are not discussed here.

These models are totally transparent, easy to use and understand, and can be modified by the user to suit individual needs. Relevant physical-chemical properties of styrene and 2,4-dinitrotoluene (henceforth referred to as dinitrotoluene) are shown in Table 24.2. The data are from Mackay et al. (1992, vols. 3, 4).

The area selected for the evaluative environment of these models is 100,000 km^2, which is about the size of Ohio or England. Terrestrial land (or soil) is assumed to cover 90% of the area, the remaining 10% being water. This region is similar to that used in the EQC models described by Mackay et al. (1996a, b, c). The volumes and depths of the four primary media are included in Table 24.2. The calculated air, water and soil Z-values for styrene and dinitrotoluene are presented in Table 24.1. Values for one each of advection, reaction, and intermedia transport D-values are also included in Table 24.1 for the two chemicals.

24.4.1 Level I

Level I models describe the equilibrium distribution of a fixed quantity of conserved (non-reacting) chemical between different media. There is no input, output, degradation, or inter-media transport. The environmental medium in which the chemical is discharged is not important because the chemical is assumed to become instantaneously distributed to an equilibrium condition. This is useful for gaining an initial impression of a chemical's partitioning tendencies. Level I also gives an order-of-magnitude relative concentrations in different environmental media.

TABLE 24.2 Selected Properties for Styrene and Dinitrotoluene

Property		Styrene	Dinitrotoluene
Molecular mass (g/mol)		104.1	182.1
Water solubility (g/m^3)		300	270
Vapor pressure (Pa)		800	0.133
Log K_{OW}		3.05	2.01
Melting point (°C)		−30.6	70
Assumed reaction half-lives (h)	Air	5	17
	Water	170	55
	Soil	550	1700
	Sediment	1700	5500
Environmental Volumes (m^3)			
Air (depth = 1,000 m)		1×10^{14}	
Water (depth = 20 m)		2×10^{11}	
Soil solids[a] (depth = 0.1 m)		9×10^{9}	
Sediment solids[a] (depth = 0.01 m)		1×10^{8}	

[a] Same as bulk volume in Level I and II models

Mathematically, if there are four environmental media of volume V_i ($i = 1$ to 4), with respective Z-values Z_i, and the total amount of chemical present is M mol, then:

$$M = \sum (V_i C_i') = f \sum (V_i Z_i) \tag{24.3}$$

where C_i is concentration and f is the common fugacity, which can be readily calculated as $M/\Sigma(V_i Z_i)$. Each concentration C_i can then be calculated as $f\ Z_i$ and the amounts in each medium M_i as $V_i Z_i f$ or $V_i C_i$. Obviously ΣM_i will equal M.

In Level I calculations, soil and sediment are considered as solid phases, i.e., their porosity and the presence of interstitial air or water are ignored. Interstitial air and water are, however, included in Level III calculations. Apart from the four primary media, three other media, aerosols in air and suspended matter in water and fish, are included for interest. The volume of these media is small, but they often have the highest concentrations. In addition, the mass of chemical partitioning to aerosols can be quite significant.

Mass balance diagrams for 100,000 kg of styrene and dinitrotoluene in an evaluative environment are shown in Fig. 24.1. Some similarities and differences in the behavior of two chemicals due to differences in their properties immediately becomes obvious. Since neither of the chemicals is hydrophobic (low K_{OW}), only small amounts partition to phases containing organic matter (soil, suspended, and bottom sediment). Due to its relatively high vapor pressure, styrene tends to migrate to the air phase with more than 90% partitioning there at equilibrium. However, due to its low vapor pressure and moderate solubility in water, more than 90% of dinitrotoluene tends to be in the water phase.

These simple calculations readily show where the chemical is likely to partition and where concentrations are likely to be highest. This information can be useful in guiding remediation and monitoring efforts.

24.4.2 Level II

Level II includes the effects of advection and degradation reactions (represented as half-lives) in various media on the fate of chemicals. It describes a situation in which a chemical is discharged into the environment at a constant rate and achieves steady state (input equals output). The Level II model also assumes intermedia equilibrium, and thus rates of intermedia transport are again not considered. The environmental medium of discharge is therefore not important and the relative distribution of chemical among the various media is similar to Level I.

Mathematically, if the emission rate is E mol/h, then at steady state this must equal the total loss rate of $f\Sigma D_i$ where f is the common fugacity and each D-value represents a loss, either by degradation or advection. Concentrations and amounts can be calculated as for Level I. The ratio of the calculated total amount M in the system to E (or $f\Sigma D_i$) is the overall average residence time of the chemical. When M is divided by the degradation rate, the result is a residence time (longer than the overall residence time) attributable to degradation only, which is useful in assessing the chemical's persistence in the global environment.

Level II provides a measure of the relative rates of chemical loss by advection and reaction and a first estimate of overall chemical residence time or persistence, which is an important chemical property governing environmental exposures. The potential for global persistence is indicated by reaction persistence (controlled by reaction half-lives), whereas local persistence is indicated by overall persistence. The magnitude of chemical loss by advection in air gives an impression of the chemical's potential for long-range transport.

Level II diagrams for discharge of 1,000 kg/h of styrene and dinitrotoluene into an evaluative environment are shown in Fig. 24.2. Again there are similarities and differences in the behavior of the two chemicals due to differences in their properties. Although more styrene than dinitrotoluene is removed by advection in air, only less than 10% is lost by this process. Hence neither of the two chemicals is a candidate for long-range transport. The fate

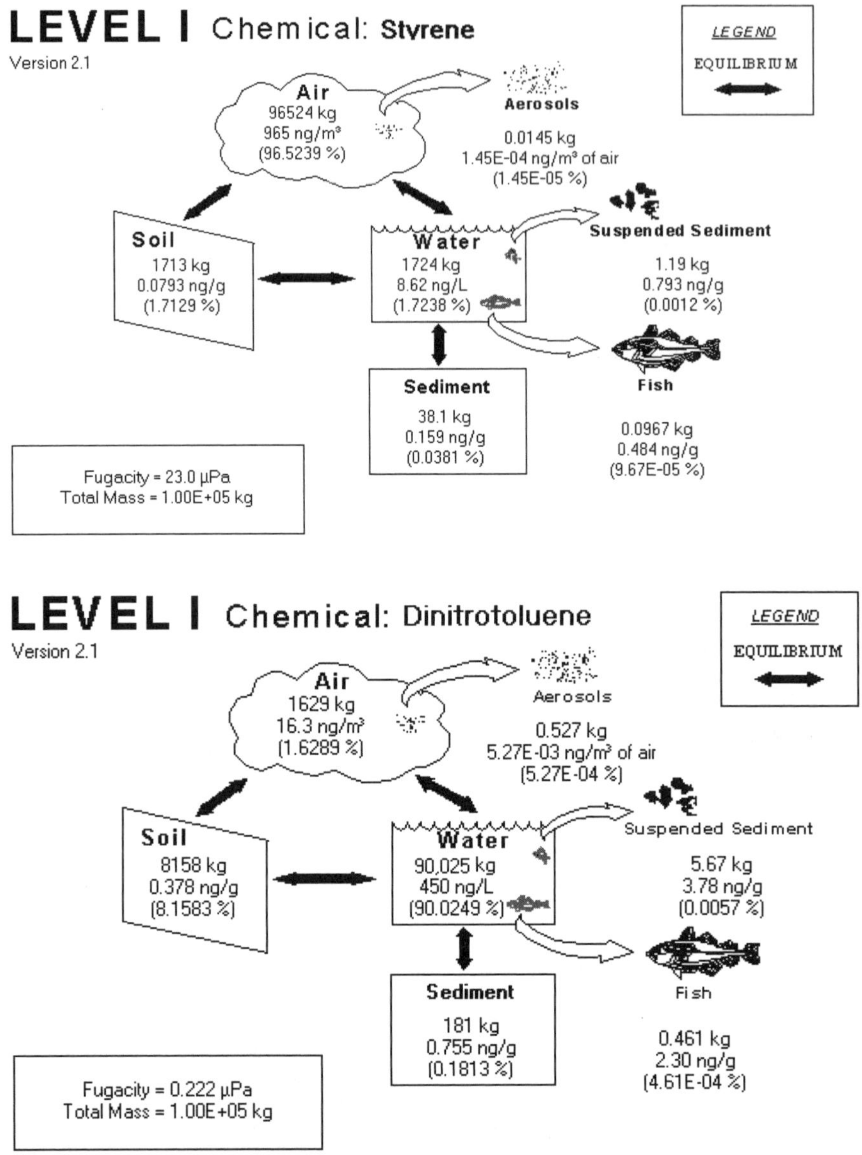

FIGURE 24.1 Level I diagram for partitioning of 100,000 kg of styrene and dinitrotoluene.

of both chemicals in the evaluative environment is primarily controlled by reaction (or degradation). However, reaction in air for styrene and reaction in water for dinitrotoluene are the dominant removal mechanisms and control their overall persistence in the environment. It is therefore important that accurate estimates of these half-lives, rate constants, and rates should be sought. Large uncertainties in values would translate into corresponding uncertainties in the estimated persistence. Due to differences in reaction rates, dinitrotoluene is about 10 times more persistent than styrene and their total masses in the environment at

Level II V 2.1 Chemical: Styrene

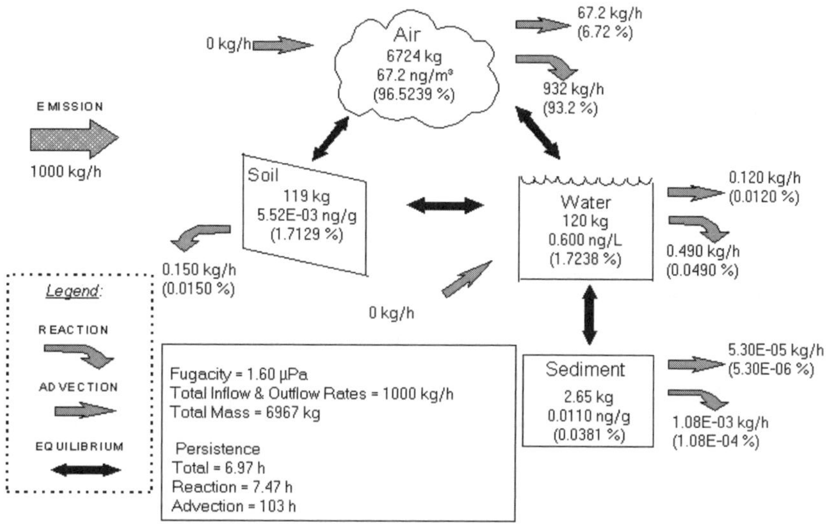

Level II V 2.1 Chemical: Dinitrotoluene

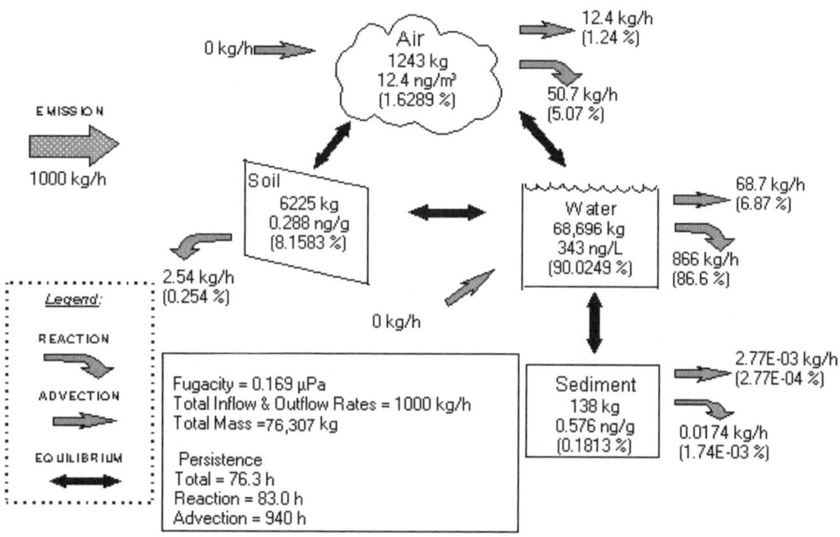

FIGURE 24.2 Level II diagrams for discharge of 1,000 kg/h of styrene and dinitrotoluene.

steady state differ by this ratio. Figure 24.2 shows that the environmental persistences of styrene and dinitrotoluene in the evaluative environment are low at 7.5 and 76 hours respectively. Since this persistence is dominated by reaction, the two chemicals are not very persistent in the global context.

Although spill situations are inherently dynamic in nature with continuously changing concentrations and it is usually desired to estimate recovery times, the Level II results provide indirect insights into these times. The persistence or residence time under steady-state conditions (calculated as mass in the system divided by total rate of input or output) is closely related to the recovery time. Indeed the dynamic half-life can be shown to be 69% of the steady-state residence time. The relative importance of various loss processes also becomes apparent.

24.4.3 Level III and IV

The Level III model includes all the important fate and transport processes in a real environment and is one step more complex than Level II. As in the Level II model, the chemical is discharged at a constant rate into the environment to reach a steady state (at which input equals output). Unlike Level II, equilibrium between different media is not assumed and rates of chemical transfer by intermedia transport processes are defined. The individual discharges to all environmental media must be specified because the distribution of the chemical between media now depends on how the chemical enters the system. Depending on the properties of a chemical, the mode of entry can also significantly alter chemical persistence or residence time in the environment to values that are quite different from Level II results. A series of 12 transport velocities control chemical transfer between the four primary environmental media (air, water, soil, and sediment). Equilibrium is assumed, however, within each medium. For example, suspended matter and fish are assumed to be at the same fugacity as water.

Mathematically, a steady-state mass balance equation is set up for each of the four compartments. These take the form for compartment i as follows

$$E_i + \sum (f_j D_{ji}) - \sum (f_i D_{ij}) = f_i \sum D_i \tag{24.4}$$

where E_i is emission into i, the second term is the sum of all intermedia transport process rates into i from other j compartments, the third term is the sum of all intermedia transport process rates from i to other j compartments, and the fourth term is the sum of all loss processes from compartment i by transport and transformation. There are thus four equations and four unknown fugacities thus a solution is possible. All concentrations and amounts can be calculated as before, leading to a mass balance diagram as illustrated in Figs. 24.3 and 24.4. The important new information is the intermedia fluxes.

The extension to a Level IV or unsteady-state model is obvious, the mass balance equations taking the form of differential equations, i.e.:

$$\frac{d(V_i Z_i f_i)}{dt} = E_i + \sum (f_j D_{ji}) - \sum (f_i D_{ij}) - f_i \sum D_i \tag{24.5}$$

These can be solved numerically or in some simple cases analytically by defining an initial condition and the time course of emission. This is essentially the approach described by Mackay and Paterson (1985).

Chemical fate and exposure following a spill is dynamic in nature, and thus a Level IV simulation is the preferred option. Much of the information that results from such a calculation can, however, be gleaned from inspection of Level III results. A Level III model is easier to use without site- and spill-specific input data and can provide important information for directing a spill response. It can identify dominant fate and exposure processes and the

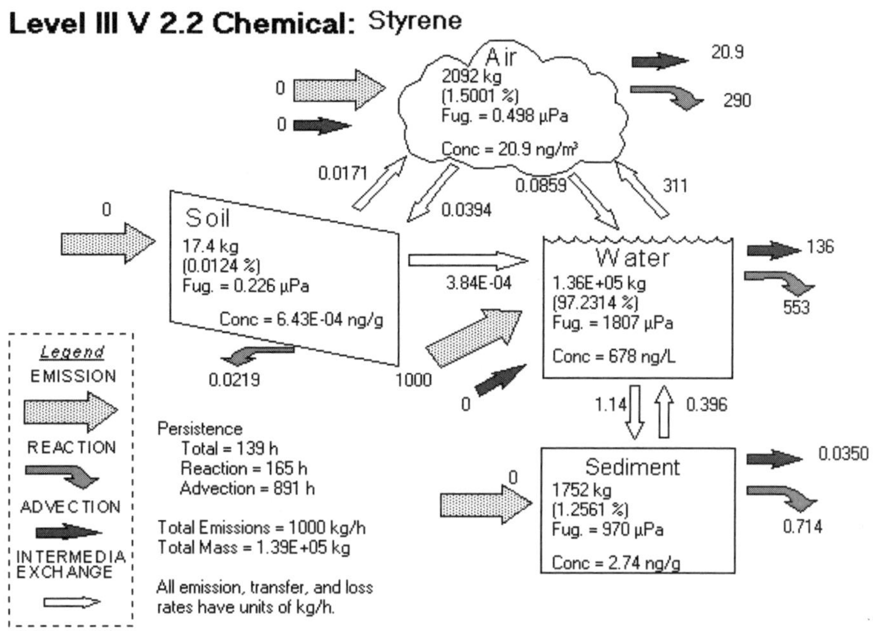

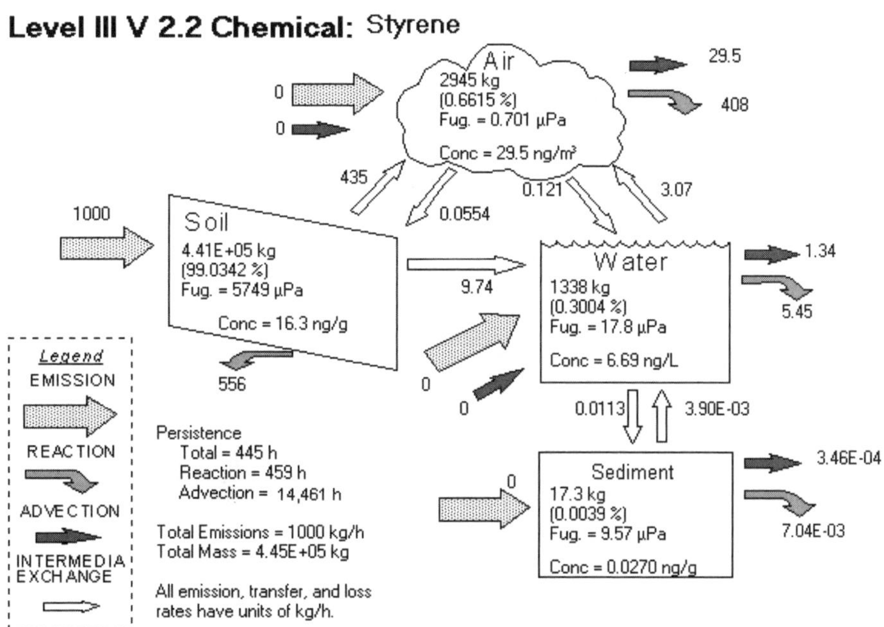

FIGURE 24.3 Level III diagrams for discharge of 1,000 kg/h of styrene into water and soil.

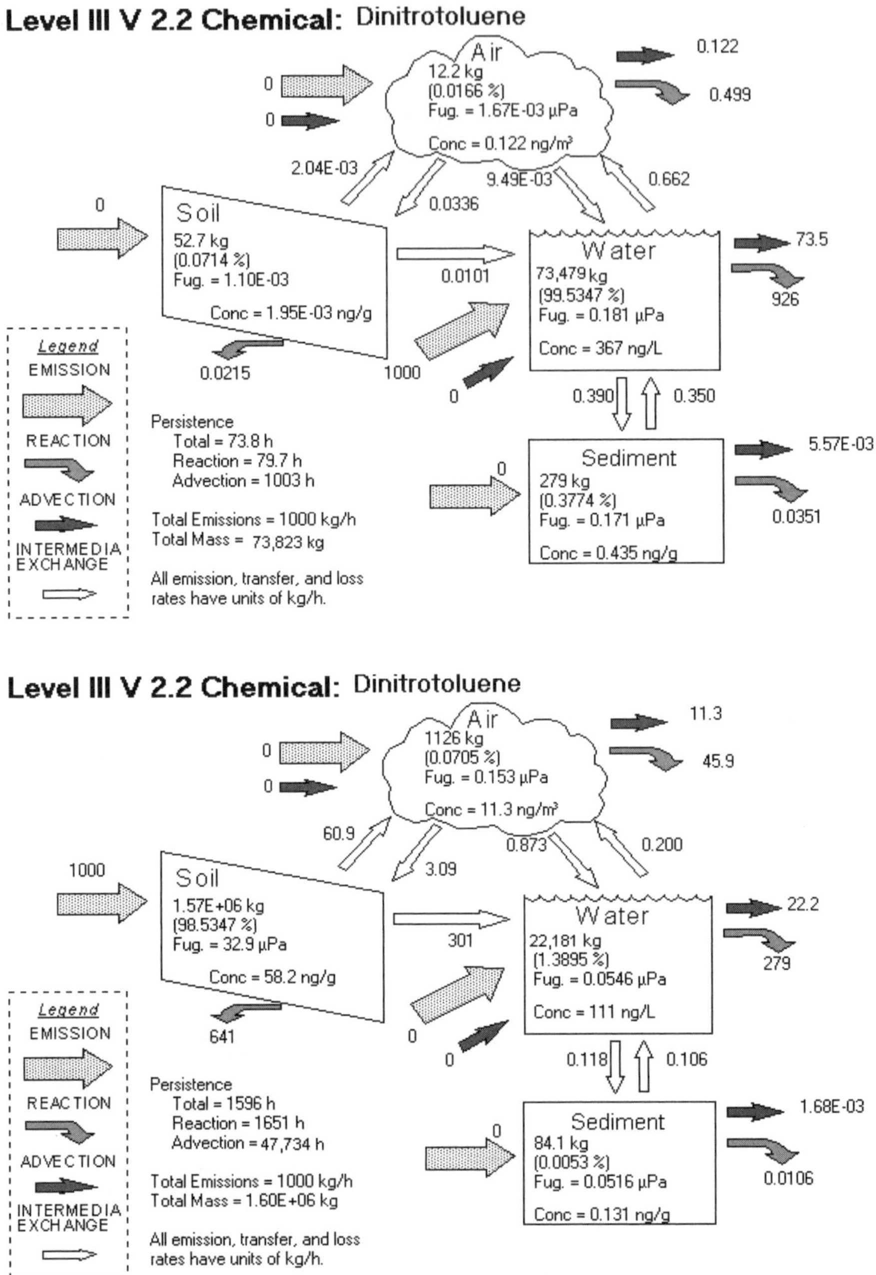

FIGURE 24.4 Level III diagrams for discharge of 1,000 kg/h of dinitrotoluene into water and soil.

associated environmental media, which can then be targeted for monitoring and/or remedial action. For example, it can show how fast a chemical will evaporate from soil or water into the atmosphere. Hence, a detailed site- and spill-specific dynamic simulation (Level IV model) is not always necessary. A well-tested version of a Level IV model is not yet available.

Level III diagrams for the discharge of 1000 kg/h of styrene in water and soil respectively are shown in Fig. 24.3 and those for dinitrotoluene are shown in Fig. 24.4. Since spill situations are of interest here, only discharges to soil and water are illustrated. It may be desired to estimate how long it will take spilled chemical to evaporate from soil. A Level III calculation gives a steady-state rate of evaporation, not a time. In Fig. 24.3 with a constant discharge of 1000 kg/h styrene in soil, the mass of styrene in soil at steady-state is 423,000 kg and the evaporation rate is 458 kg/h. The rate constant for evaporation is thus 458/423,000 or 1.08×10^{-3} h^{-1} and the corresponding half-life is 640 hours or about four weeks. We thus expect that half the initial styrene will evaporate every four weeks, assuming there is no degradation. After 20 weeks or five half-lives, only $1/2^5$ or 3% will remain.

The Level III model used here and available for download (http://www.trentu.ca/envmodel) does not include a separate groundwater compartment. However, concern for groundwater contamination can still be evaluated using this model as follows. Soil water leachate to groundwater can be considered equivalent in characteristics to soil water runoff included in the model, which is assumed to be in equilibrium (i.e., equifugacity) with soil pore water and in turn with soil. The soil runoff rate in the model can therefore be modified to include both leaching and runoff. For the illustrative examples presented here, the leaching rate was considered equal to that of runoff, and the default runoff rate of 0.00005 m/h (or $m^3/m^2 \cdot h$) in Level III model was modified to 0.0001 m/h to include both soil water leaching to groundwater and runoff. A groundwater depth of 1 m is considered.

Figure 24.3 shows that although styrene tends to evaporate, it is constrained to stay in the medium of discharge if emitted to soil or water. Slower degradation rates in these media increase the overall persistence of styrene in the environment to five days or more, which is more than an order of magnitude higher than that for the equilibrium situation in Level II. Overall, Fig. 24.3 shows that although discharge of styrene into water or soil tends to increase its persistence in the environment, it is still removed quite rapidly from the environment. If discharged as a spill, therefore, styrene is not likely to be of concern for long-term exposure. Calculations including the groundwater reveal that only about 1% or less of styrene is likely to be transferred to groundwater if discharged on soil and it may be of little concern as a potential groundwater contaminant. The key emergency response procedures for a styrene spill are to limit short-term exposure. The Level III model can be useful in directing such a response.

Figure 24.4 shows that, if discharged to water, dinitrotoluene would stay in water, with reaction in water being the dominant removal mechanism. The Level III diagram for discharge of dinitrotoluene to water is very similar to the Level II diagram (which assumes equilibrium between different media), which is expected due to the chemical's preference for the water phase at equilibrium. The overall persistence in the environment is expected to be a relatively short 74 hours, which is controlled by its reaction persistence in water. If discharged to soil, dinitrotoluene would tend to build up, increasing its overall environmental persistence to over two months. Inclusion of groundwater compartment in Level III calculations suggests that, if discharged to soil, dinitrotoluene could contaminate groundwater with persistence similar to that in soil. It must be noted that the estimated reaction half-lives of dinitrotoluene in soil (672 to 4,320 hours) and in groundwater (48 to 8,640 hours), and therefore its persistence, are quite uncertain and will be a function of local conditions such as temperature.

Overall Level III calculations suggest that the mode of entry of a dinitrotoluene spill can lead to very different environmental impacts. If spilled in surface water, it is expected to degrade quite rapidly. However, if spilled onto soil, dinitrotoluene is expected to persist in soil and contaminate groundwater for a period of several months unless preventive measures are taken. Hence, in such an event, the contamination of these compartments and the resulting potential exposure would have to be monitored for a period well beyond the initial emergency response. A useful response measure suggested by Level III results is to prevent soil leachate from contaminating the groundwater by, for example, directing it in a controlled manner to a surface water body where it will degrade rapidly.

24.5 SUMMARY

In summary, it has been shown that mass balance models can be used for assessing the long-term behavior of spilled materials and can guide and prioritize emergency response measures. Fugacity-based Level I, II, and III models are transparent, easy to use and understand, and can be used for general fate assessment of a chemical spill. There is a compelling incentive to use standard, well-tested, widely available models for such evaluations. One such source of Level I, II, and III models on the World Wide Web is the Canadian Environmental Modelling Centre (http://www.trentu.ca/envmodel). The models have been widely used and can be easily adapted to suit specific needs.

24.6 ACKNOWLEDGEMENTS

The authors are grateful to the Natural Science and Engineering Research Council of Canada (NSERC) and the consortium of chemical companies that support the Canadian Environmental Modelling Centre at Trent University.

24.7 REFERENCES

Mackay, D. 2001. *Multimedia Environmental Fate Models: The Fugacity Approach*, 2nd ed., Publ./CRC Press, Boca Raton, FL.

Mackay, D., and S. Paterson. 1985. *A Model of the Long-Term Fate of Chemical Spills*, Report EE-64, Environment Canada, Ottawa.

Mackay, D., A. Di Guardo, S. Paterson, G. Kicsi, and C. E. Cowan. 1996a. "Assessing the Fate of New and Existing Chemicals: A Five-Stage Process," *Environmental Toxicology and Chemistry,* vol. 15, no. 9, pp. 1618–1626.

Mackay, D., A. Di Guardo, S. Paterson, and C. E. Cowan. 1996b. "Evaluating the Fate of a Variety of Types of Chemicals Using the EQC Model," *Environmental Toxicology and Chemistry,* vol. 15, no. 9, pp. 1627–1637.

Mackay, D., A. Di Guardo, S. Paterson, G. Kicsi, C. E. Cowan, and D. Kane. 1996c. "Assessment of Chemical Fate in the Environment Using Evaluative, Regional and Local-Scale Models: Illustrative Application to Chlorobenzene and Linear Alkylbenzene Sulfonates," *Environmental Toxicology and Chemistry,* vol. 15, no. 9, pp. 1638–1648.

Mackay, D., W. Y. Shiu, and K. C. Ma. 1992–1997. *Illustrated Handbook of Physical–Chemical Properties and Environmental Fate for Organic Chemicals,* CRC Press, Boca Raton, FL, 5 volumes also available as a CD ROM, Chapman and Hall/CRCnetBASE 2000.

Neely, W. B., and D. Mackay. 1992. "Evaluative Model for Estimating Environmental Fate," in *Modeling the Fate of Chemicals in the Aquatic Environment,* ed. K. L. Dickson, A. W. Maki, and J. Cairns, Jr., Ann Arbor Science, Ann Arbor, MI, pp. 127–143.

CHAPTER 25
CONTAMINANT CONCENTRATION FLUCTUATIONS

Paul J. Sullivan
Department of Applied Mathematics,
The University of Western Ontario, London, Ontario, Canada

Philip Chatwin
Department of Applied Mathematics, University of Sheffield,
Sheffield, U.K.

25.1 INTRODUCTION

When a fluid, which will be called a *contaminant* fluid, is released into another miscible *host* fluid, one would like to describe the concentration $\Gamma(\mathbf{x}, t)$, in units of mass per unit volume at the position located by vector $\mathbf{x}$ at time t, of contaminant fluid as it evolves in time and space. The release may be sudden, such as a large rupture in a storage vessel resulting in a contaminant cloud, or continuous, such as a slow leak or smokestack emission providing a contaminant plume, jet, or wake. The fluids may be liquids or vapors, and differences in density between the two fluids may influence the flow structure. Chemical reactions may take place, which both affect density differences and generate new contaminants. Mixing between the host and contaminant fluids that results in a change of concentration values can only take place through the action of molecular diffusivity, κ. Although this presentation is restricted to miscible fluids, small solid particles, for example in smoke or small aerosols, can often be treated in the same framework with the effects of molecular diffusivity replaced by the effects of Brownian motion.

Almost all fluid motion is in a turbulent state. That is, all variables such as velocity, pressure, and contaminant concentration are random variables and the only reproducible entities are ensemble- or probability-averaged quantities. For example, the ensemble average mean concentration is

$$C(\mathbf{x}, t) = \langle \Gamma(\mathbf{x}, t) \rangle = \int_0^\infty \theta p(\theta; \mathbf{x}, t) d\theta \tag{25.1}$$

where $\langle , \rangle$ denotes an ensemble average and $p(\theta; \mathbf{x}, t)$ is the one-point, probability density function such that

$$p(\theta; \mathbf{x}, t) d\theta = \text{prob } \{\theta \leq \Gamma(x, t) \leq \theta + d\theta\} \tag{25.2}$$

The mean-square value—or variance—is

$$c^2(\mathbf{x}, t) = \langle(\Gamma(\mathbf{x}, t) - C(\mathbf{x}, t))^2\rangle = \int_0^\infty (\theta - C(\mathbf{x}, t))^2 p(\theta; \mathbf{x}, t) d\theta. \tag{25.3}$$

A very important observation is that the root-mean-square value of concentration or the *fluctuation* in concentration is typically as large as the mean concentration in turbulent flows. That is, the fluctuations are large and hence nonnegligible (see Fig. 25.1). The fact that contaminant concentration is a random variable has enormous consequences; one of these is that the equations that govern the evolution of the concentration field are intractable. These consequences will be explored in the following sections in order to expose the rationale for the conservative approach to describing the contaminant concentration field that will be developed here.

There will always be a considerable degree of uncertainty about the release and general flow conditions when a chemical spill occurs. Further, as will be explained at greater length, the scientific ability to predict the contaminant concentration field even under well-controlled laboratory conditions is severely limited. The objective of this chapter is to provide a general strategy to determine a conservative estimate of the magnitude of the fluctuations that is theoretically sound and consistent with the likely quality of available knowledge about the spill conditions. For a *conservative* estimate, combustible and toxic substances, including

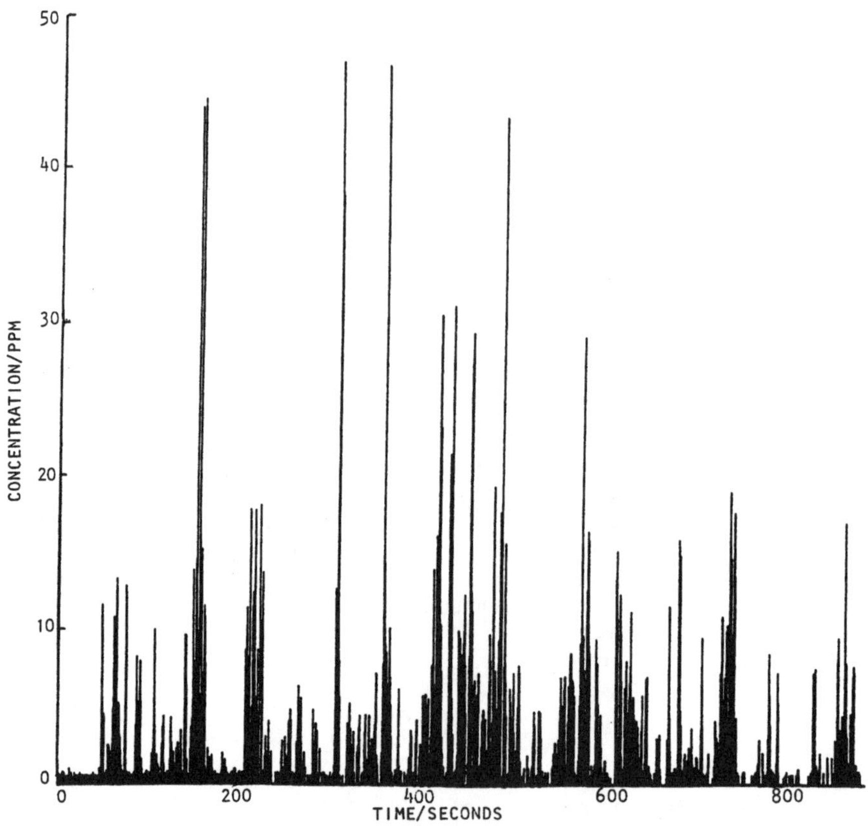

FIGURE 25.1 Sample concentration time series digitized at 200 Hz from a UVIC detector positioned approximately 30 m downwind of a continuously emitting source, during the November 1992 Dugway experiment under neutral conditions. (*Source:* Lewis and Chatwin, 1995a)

products of a chemical reaction, will be treated as though all of the release mass is at the most dangerous concentration at release and remains at that concentration thereafter. That is, it is assumed that no mixing (the $\kappa = 0$ situation) or chemical reaction takes place following release.

It will be shown, in general and without approximation, that the fluctuations $c(\mathbf{x}, t)$ and indeed all of the moments of the probability density function are given in terms of the mean concentration $C(\mathbf{x}, t)$ when $\kappa = 0$. It will also be shown that the effect of κ is to reduce fluctuations so that the $\kappa = 0$ case provides a conservative envelope. That is, all uncertainty in describing fluctuations for this conservative estimate will be confined to the uncertainty in describing the mean concentration. The mean concentration is the easiest to predict theoretically and to measure for experimental validation. Throughout the chapter an emphasis will be placed on the contrast between the difficulties in measuring or theoretically predicting the value of $C(\mathbf{x}, t)$ and the value of $c(\mathbf{x}, t)$. Although this chapter advocates the use of a safe, conservative approach to estimating the magnitude of concentration fluctuations at this time, there will be some discussion of a promising approach to derive a less conservative, though reliable, estimate of fluctuations in the future.

25.2 THE PROBABILITY DENSITY FUNCTION AND MOMENTS

In considering issues related to combustion, malodor, or toxicity or chemical reactions, one generally requires (at least) the one-point probability density function. For example, the probability of ignition (PI) of, say, a methane gas cloud is given by the probability that, at a position located by vector $\mathbf{x}$ at time t, concentrations between the lower θ_L and upper θ_u flammability limits are encountered. That is,

$$PI(\mathbf{x}, t) = \int_{\theta_L}^{\theta_u} p(\theta; \mathbf{x}, t) d\theta \tag{25.4}$$

and this information could be displayed using the probability contours of $PI(\mathbf{x}, t)$ (Birch et al., 1980). The probability density function is generally derivable from a knowledge of all the integral moments using for example a maximum entropy formalism or orthogonal polynomial expansion (Derkson and Sullivan, 1990). The moments are defined by

$$\mu_n(\mathbf{x}, t) = \langle(\Gamma(\mathbf{x}, t) - m_1(\mathbf{x}, t))^n\rangle = \int_0^\infty (\theta - m_1(\mathbf{x}, t))^n p(\theta; \mathbf{x}, t) d\theta \tag{25.5}$$

where

$$m_n(\mathbf{x}, t) = \langle(\Gamma(\mathbf{x}, t))^n\rangle = \int_0^\infty \theta^n p(\theta; \mathbf{x}, t) d\theta \tag{25.6}$$

The mean $m_1(\mathbf{x}, t) = C(\mathbf{x}, t)$ and the mean-square fluctuation $\mu_2(\mathbf{x}, t) = c^2(\mathbf{x}, t)$—see Eq. (25.3)—are simply the lowest two of an infinite number of moments of the probability density function. It is expected that a reasonable approximation to the overall shape of the probability density function can be found from as few as the first four lower ordered moments (Derkson and Sullivan, 1990). The equation governing the evolution of the moments (Chatwin and Sullivan, 1990a; Sullivan and Ye, 1993), excluding chemistry, is

$$\frac{\partial m_{m+1}}{\partial t} + \nabla \cdot (\langle \mathbf{u}\Gamma^{n+1}\rangle) = \kappa \nabla^2 m_{m+1} - \kappa n(n+1)\langle \Gamma^{n-1}(\nabla\Gamma)^2\rangle \tag{25.7}$$

where $\mathbf{u}(\mathbf{x}, t)$ is the fluid velocity. In particular, when $n = 0$,

$$\frac{\partial C}{\partial t} + \nabla \cdot (\langle \mathbf{u}\Gamma \rangle) = \kappa \nabla^2 C \tag{25.8}$$

and it is clear that an additional term, $\langle \mathbf{u}\Gamma \rangle$, is present due to taking an ensemble average of the convective-diffusion equation

$$\frac{\partial \Gamma}{\partial t} + \nabla \cdot (\mathbf{u}\Gamma) = \kappa \nabla^2 \Gamma \tag{25.9}$$

Further equations for the additional term lead to yet more additional terms. This is an example of the *nonclosure* problem that typifies turbulent phenomena. The equation for the probability density function (Chatwin, 1990; Mole et al., 1993) presents an even more challenging closure problem. There is no a priori justification for any scheme to close the equations so that they can be solved. The use of any such closure scheme or simulation must be thoroughly validated by experiment for each and every flow and contaminant release configuration.

The probability density function, or even many of the lower-ordered moments, are not usually available, so that an estimate is made from the mean and variance, which provide a measure of the location and width of the probability density function in the concentration range.

The term on the right-hand side of Eq. (25.8) is generally small enough that it can be neglected. The negligible effect of κ in determining $C(\mathbf{x}, t)$ can be illustrated using the classical G. I. Taylor Lagrangian autocorrelation formulation (Ye, 1995). The insensitivity of $C(\mathbf{x}, t)$ to κ is verified in flows with and without large velocity gradients and in flows where density effects are important. The fact that the mean concentration field is almost exclusively determined by the turbulent convective motions and not by κ renders this lowest order moment a more simple statistic than the higher-order moments. Conversely, a knowledge of the mean field $C(\mathbf{x}, t)$ reveals nothing about the reduction of contaminant concentration values, which can only take place through molecular diffusion, at an intensity determined by the molecular diffusivity κ.

The length scales of turbulent motion that govern the evolution of the mean field $C(\mathbf{x}, t)$ are invariably the large scales. In jets, wakes, and boundary layers, with and without density effects, the enlargement of the average contaminant region depends on the entrainment of surrounding fluid by the large-scale motions in these flows (Townsend, 1976; Turner, 1973). For flows where mean velocity gradients are not directly important, such as, a contaminant cloud far removed from the surface in the neutral atmospheric boundary layer, the mean field is determined by a one- fluid-particle analysis (Batchelor, 1949), and it is shown here that the largest scales of turbulent motion dominate the cloud spatial growth rate. The picture is different in center-of-mass coordinates (relative diffusion), where the dominant scales of motion for the lateral growth rate, such as in a continuous plume in the well-mixed layer near the surface of a lake or the ocean, are those scales comparable with the local plume width (Batchelor, 1952). These, however, soon become large scales of turbulent motion with respect to the scales that will be seen to govern concentration reduction through κ. In a confined flow such as a canal or river, or in a valley or street in the atmosphere, the most effective cloud enlargement is in the flow direction and is a result of the interaction between the mean velocity gradients and cross-stream transport due to the turbulent motions (Chatwin and Sullivan, 1982; Dewey and Sullivan, 1979). In this situation, the tendency of the velocity gradient to spread material in the flow direction is mitigated by the cross-stream mixing, which is dominated by the large-scale turbulent motion.

Fluctuations in concentration are generated by the same random, large-scale turbulent motions that cause the contaminant field to expand. Typically, in plumes, jets, wakes, and boundary layers, turbulent velocities with length scales comparable with the local flow width entrain external uncontaminated fluid from the periphery and transport this to the central flow region. The turbulent motion is generated mechanically through shear forces and/or through buoyancy forces where density differences are significant. In the case of a steady contaminant plume or cloud, one can distinguish between fluctuations, as observed at a fixed measuring station, caused by large-scale meandering of the entire plume and fluctuations that exist within the plume or cloud itself. Normally, anything that interferes with the action of the large-scale entraining motions will reduce fluctuations—for example, a continuous release of contaminants into a smooth boundary layer compared with the same flow and release when a homogeneous array of blocks is mounted on the boundary layer floor. In the latter case large-scale lateral motions are suppressed as well as fluctuations. For a phenomenological description of fluctuations see Wilson (1995).

It is useful to consider the release of a conserved scalar contaminant of mass Q and uniform release concentration θ_0 into a turbulent flow. We define a length scale $L(t)$ of the cloud by

$$QL^2(t) = \int_{\text{a.s.}} |\mathbf{x}|^2 C(\mathbf{x}, t) dV(\mathbf{x}) \tag{25.10}$$

where

$$Q = \int_{\text{a.s.}} \Gamma(\mathbf{x}, t) dV(\mathbf{x}) = \theta_0 L_0^3 \tag{25.11}$$

and the integral is taken over all space, i.e., the whole region available for dispersion. Consider the integral of Eq. (25.7);

$$\frac{d}{dt}\int_{\text{a.s.}} m_{n+1} dV(\mathbf{x}) = -\kappa n(n + 1) \int_{\text{a.s.}} \langle \Gamma^{n-1} (\nabla\Gamma)^2 \rangle dV(\mathbf{x}) \tag{25.12}$$

Setting $\kappa = 0$ for the purposes of this illustration and using the uniform release concentration θ_0 at $t = 0$ results in

$$\int_{\text{a.s.}} m_{n+1}(\mathbf{x}, t) dV)(\mathbf{x}) = \int_{\text{a.s.}} \theta_0^{n+1} \, dV(\mathbf{x}) = Q^{n+1} \, L_0^{-3n} \tag{25.13}$$

Specifically for $n = 1$ in Eq. (25.13) and recognizing that $m_2(\mathbf{x}, t) = C^2(\mathbf{x}, t) + c^2(\mathbf{x}, t)$,

$$\int_{\text{a.s.}} C^2(\mathbf{x}, t) dV(\mathbf{x}) + \int_{\text{a.s.}} c^2(\mathbf{x}, t) dV(\mathbf{x}) = Q^2 L_0^{-3} \tag{25.14}$$

From Eq. (25.10) it is clear that $C(\mathbf{x}, t) = O(QL^{-3})$ and the integral over $C(\mathbf{x}, t)^2$ in Eq. (25.14) goes to zero as $L(t) \rightarrow \infty$, leaving

$$c^2(\mathbf{x}, t) = 0(Q^2 L^{-3} L_0^{-3}) \tag{25.15}$$

The result presented in Eq. (25.15) illustrates an important difference between the mean and fluctuating concentration fields in that the latter depends on the initial conditions [the presence of L_0 in Eq. (25.14)] while the former does not. This fact, predicted in Chatwin and Sullivan (1979a), was observed in the experiments of Fachrell and Robbins (1982). An additional, simple, probabilistic argument for the above result is found in Chatwin and Sullivan (1980).

25.3 THE ROLE OF MOLECULAR DIFFUSIVITY

When a blob of contaminant is released in a turbulent flow, it is stretched into ever- thinning sheets and strands by the turbulent convective motion until this thinning is balanced by thickening due to molecular diffusion. The balance is reached when the sheets or strands have a thickness comparable with the conduction cutoff length $\lambda = (\nu\kappa^2/\epsilon)^{1/4}$, where ν is the kinematic viscosity and ϵ is the rate of turbulent energy dissipation per unit mass. λ has the value of $10^{-3} - 10^{-5}$ m in most flows. A contaminant cloud may extend over kilometers in an environmental flow, but the only mixing between host and contaminant fluid, resulting in a reduction of concentration values, takes place over the very small λ length scales by molecular diffusion. Extraordinarily well resolved experiments provide direct observation of this fine-scale texture of a diffusing scalar contaminant (Dahm et al., 1991; Dahm and Dimotakis, 1990; Corriveau and Baines, 1994).

It is instructive to consider the integrals given in Eq. (25.12). For a contaminant cloud, when $\kappa = 0$, it is observed from Eq. (25.13) that all of the spatially integrated moments m_n are conserved. Since the integral that appears on the right-hand side of Eq. (25.12) is intrinsically positive, it is clear that the only agency to take *moments out of the system* is molecular diffusion. For example, when $n = 1$ in Eq. (25.13), the entire process consists of converting

$$\int_{\text{a.s.}} c^2 dV(\mathbf{x})$$

from an initial value of 0 to a final values of $\theta_0^2 L_0^{-3}$.

Neglecting the small term $\kappa\nabla^2 m_{n+1}$ on the right-hand side of Eq. (25.7), and observing that the convective term $\nabla \cdot (\langle \mathbf{u}\Gamma^{n+1} \rangle)$ is not sensitive to κ as is the case for the mean concentration C in Eq. (25.8), we have, when $n = 1$,

$$\frac{\partial C^2}{\partial t} + \frac{\partial c^2}{\partial t} + \nabla \cdot (\langle \mathbf{u}\Gamma^2 \rangle) = -2\kappa\langle (\nabla\Gamma)^2 \rangle \tag{25.16}$$

Since the term on the right-hand side is intrinsically negative, we see that the $\kappa = 0$ result provides everywhere a conservative envelope or upper limit for the value of $c^2(\mathbf{x}, t)$. It should be noted that the gradient in Γ that appears on the right-hand-side of Eq. (25.16) can be large, due to the fine-scale texture of the concentration field discussed above, and cannot be neglected if accurate quantitative predictions are sought.

The probability density function, for a uniform source concentration θ_0, when $\kappa = 0$, is

$$p(\theta; \mathbf{x}, t) = \pi(\mathbf{x}, t)\delta(\theta - \theta_0) + (1 - \pi(\mathbf{x}, t))\delta(\theta) \tag{25.17}$$

where $\pi(\mathbf{x}, t) = C(\mathbf{x}, t)\theta_0^{-1}$ is the probability of the position located by vector $\mathbf{x}$ at time t being in contaminant fluid. $\delta(\cdot)$ is the Dirac delta function. Without mixing, an observer at any location and at any time is either in contaminant fluid at the release concentration θ_0 or in fluid of concentration zero. The simple expression given in Eq. (25.17) is completely general (provided $\kappa = 0$, a uniform source concentration, and there is no chemistry) and applies to all flows and release configurations for a conserved scalar including those mentioned in Section 25.1.

All of the moments, when $\kappa = 0$, are determined from Eqs. (25.17) and (25.5) in terms of the mean $C(\mathbf{x}, t)$. Specifically (Chatwin and Sullivan, 1990b), the first four moments, kurtosis $K = \mu_4\mu_2^{-2}$ and skewness $S = \mu_3\mu_2^{-3/2}$, are:

$$c^2 = C(\theta_0 - C) \tag{25.18}$$

$$c^3 = C(\theta_0 - C)(\theta_0 - 2C) \tag{25.19}$$

$$c^4 = C(\theta_0 - C)(\theta_0^2 - 3\theta_0 C + 3C^2) \tag{25.20}$$

$$S = (\theta_0 - 2C)(C(\theta_0 - C))^{-1/2} \tag{25.21}$$

$$K = (\theta_0^2 - 3\theta_0 C + 3c^2)(C(\theta_0 - C))^{-1} \tag{25.22}$$

The normalized moments skewness and kurtosis describe the shape of the probability density function. The skewness represents asymmetry about the mean value of concentration, and the kurtosis represent the "flatness" where, for example, the skewness for a symmetrical Gaussian p.d.f. is zero and the kurtosis is 3. It is also to be noted that Eq. (25.17) provides the lower bound (equality) for the general relationship between skewness and kurtosis (see, e.g., Wilkins, 1944):

$$K \geq S^2 + 1 \tag{25.23}$$

Modifications to Eqs. (25.18) to (25.22) to account for a nonuniform release concentration are given in Sawford and Sullivan (1995). There it is shown that the same expressions that appear in Eqs. (25.18) to (25.20) apply with the θ_0s replaced with constants that are determined from the release concentration distribution—one new constant for each higher moment.

It requires a finite time for the effects of κ to be appreciable so that the moment Eqs. (25.18) to (25.22) provide a small-time approximation, in the case of a cloud (or near-source in the case of a steady release), to the true $\kappa \neq 0$ result. The time period required for molecular diffusive effects to reduce concentration is generally large with respect to the time periods over which large-scale turbulent convective motions provide significant displacements of contaminant fluid. Thus, many qualitative features of the moment Eqs. (25.18) to (25.22) are observed in experimental flows.

In Fig. 25.2, a comparison is made between the measured centerline fluctuations for a plume in grid turbulence with Eq. (25.18). It is to be noted that Eq. (25.18) exhibits a maximum value for c^2 when $C = \theta_0/2$. In Fig. 25.2, the $\kappa = 0$ approximation appears to retain this feature. Another salient feature of Eq. (25.18) is that the distribution of c^2 across a steady flow at a given downstream location starts as a near-source bimodal distribution and changes to a unimodal distribution after the centerline mean value reaches and falls

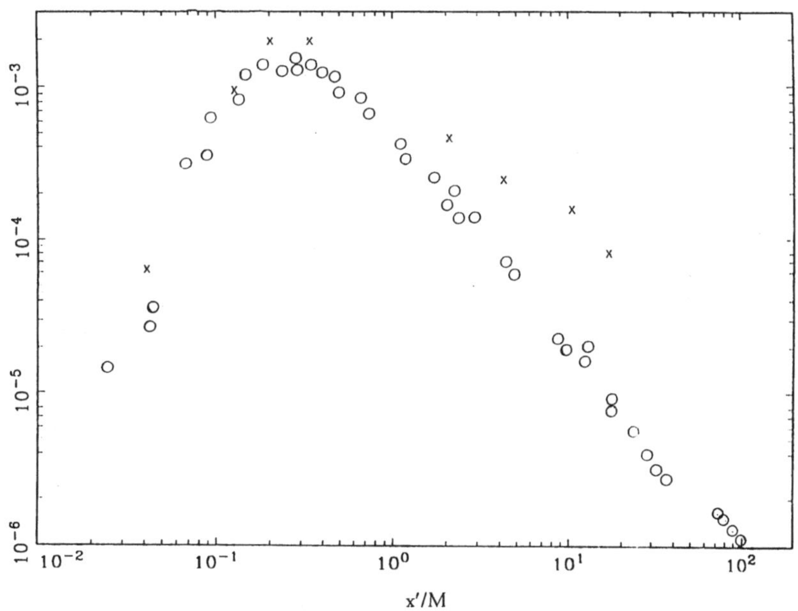

FIGURE 25.2 A comparison of centerline mean-square concentrations measured in a heated plume in grid turbulence, 0, with values calculated, **x**, for $\kappa = 0$ [see Eq. (25.18)]. (*Source:* Moseley, 1991)

below $\theta_0/2$. A sketch of the typical locus of the maximum value of c^2 in a steady configuration is shown in Fig. 25.3. It is to be noted in Fig. 25.3 that when $\kappa \neq 0$, c^2 returns to a bimodal distribution far downstream (Sawford and Sullivan, 1995; Moseley, 1991; Mole, 2001). Other qualitative features of the moment Eqs. (25.17) to (25.20) are discussed in Chatwin and Sullivan (1990b).

One further qualitative comparison between the $\kappa = 0$ result of equality in Eq. (25.23) and experiment deserves mentioning because of the remarkably widespread conditions over which it is observed. Experimental observations show the kurtosis values appear to fall on a quadratic curve of skewness that is slightly above the $K = S^2 + 1$ curve (Mole and Clarke, 1995; Lewis and Chatwin, 1995; Chatwin and Robinson, 1997). These experiments even include configurations where there are obstacles in the flow path.

The key equation in this presentation is a modification to Eq. (25.18) as

$$c^2(\mathbf{x}, t) = C(\mathbf{x}, t)(\hat{\theta}_0 - C(\mathbf{x}, t)) \tag{25.24}$$

where $\hat{\theta}_0$ is an estimated, equivalent, uniform release concentration. It is unlikely that suf-

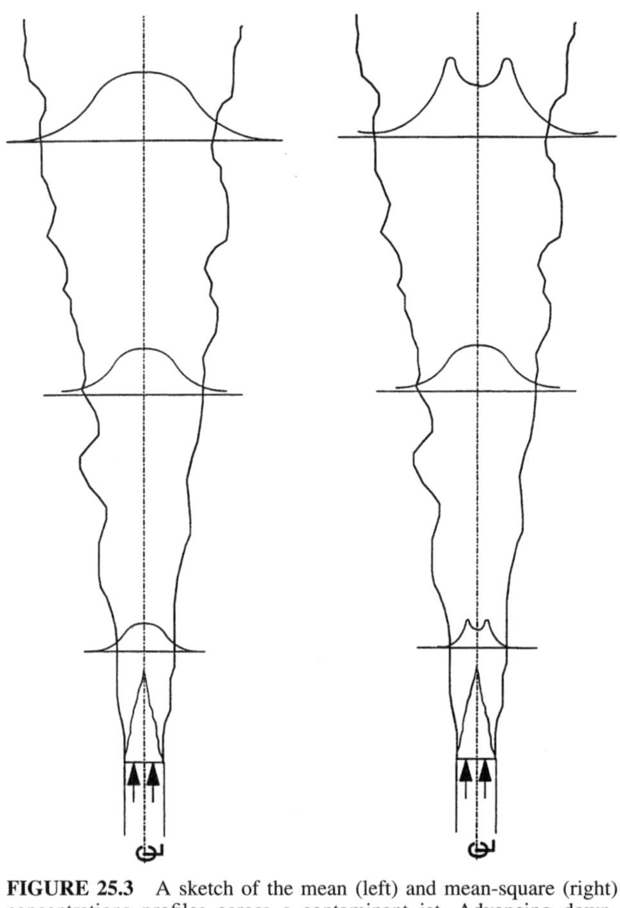

FIGURE 25.3 A sketch of the mean (left) and mean-square (right) concentrations profiles across a contaminant jet. Advancing downstream, profiles correspond to positions where $\alpha < 2$, $\alpha > 2$ and $\alpha < 2$, respectively. (See 25.28.)

ficient information would be available on the release conditions in a spill to enable the calculation of the constant $\hat{\theta}_0$ with the procedure outlined in Sawford and Sullivan (1995) so that a conservatively high value of θ_0 will be chosen for $\hat{\theta}_0$. The remarkable generality and simplicity of Eq. (25.24) leads to a scientifically sound, conservative estimate of concentration fluctuations.

25.4 MODELING AND EXPERIMENTAL VALIDATION

In principle, one could numerically solve the differential equations (so-called direct numerical simulation, DNS) governing the velocity and concentration, repeatedly, and form an ensemble average. This task is well beyond the computational power that will be available in the near future (see, e.g., discussion in Mole et al., 1993). Normally (see, e.g., Sykes et al., 1984), equations that describe the average quantities are solved using semiempirical closure hypotheses and these solutions must be carefully validated by experiment. Two questions immediately arise. First, how is molecular diffusion explicitly taken into account in the numerical solution? This is the only agency to take moments out of the system, and without this all of the moments are simply and exactly given in terms of the mean as in Eqs. (25.18) to (25.20). The second question relates to the fine-scale texture of the contaminant field. Has the model and numerical scheme achieved spatial and temporal continuum-scale resolution? It is also interesting to note the difference between the equation that governs the concentration field and the equation that governs a time- and/or space-averaged concentration field (Sullivan, 1984).

The above difficulties are compounded by the fact that in order to make progress in modeling, one must have access to reliable experimental data for validation. Measurements of concentration in a turbulent flow are difficult because ensemble averages must be approximated, because of the challenge to achieve adequate experimental temporal and spatial resolution, and because of the problem of distinguishing low values of concentration from instrument noise.

The problem of approximating an ensemble average can be difficult, particularly in environmental flows which are generally inhomogeneous and unsteady. One needs to observe the phenomena selected, that is, the prescribed ensemble, over a sufficient period of time (or space, if homogeneous) that the statistic in question will converge. For example, one may wish to measure the mean concentration at a fixed point downstream in the path of a steady plume or cloud in the atmospheric boundary layer. In both cases the time record of concentration taken at a fixed point is governed by the largest scales of turbulent motion. For example, the large excursions of a meandering plume, whose significance is greatly affected by atmospheric stability, is visibly the result of large-scale turbulent motion. In the case of a steady plume, one would require the flow to remain reasonably steady while a record length of many multiples of this turbulent time scale was compiled in order to calculate the estimate of the mean concentration $C(\mathbf{x}, t)$. In the case of a cloud, one must compile the record from many repeated releases during the course of a similarly long epoch. One generally requires an order of magnitude more record length (or number of realizations in the case of a cloud) for each higher-order moment. In the well controlled laboratory experiments of Hall et al. (1991), which were designed to replicate field experiments, a tent full of heavy gas was released into a logarithmic boundary layer and sampled at fixed locations along the centerline downstream. It was found in these experiments, consisting of 50 or 100 repeat contaminant cloud releases, that only the mean concentration could be reliably estimated (Heagy and Sullivan, 1995a).

One can change reference frame to the center-of-mass of a cloud (or cross section of a plume at a given downstream location) in each realization so that the important scales of turbulent motion are comparable to the local cloud (or plume cross-section) size. Thus the interval of time over which experimental realizations are taken can remain relatively small while the cloud or plume width remains small. In experiments on a continuous plume in the well mixed surface layer of Lake Huron, it was found that during the period in which current magnitude and direction remained reasonably constant, up to 25 plume crossings could be made. This was adequate to give a very good representation of the mean concentration profile but completely inadequate to represent the fluctuations (mean-square concentration) as shown in the discussion in Chatwin and Sullivan (1979b).

It is certainly conceivable that the statistical properties of a turbulent flow can change at a rate that is comparable with the time scale of the turbulent motion so that proper ensemble averages are required, i.e., repetitions under identical conditions, which of course cannot normally be done in environmental flows. Indeed, often the time-dependent nature of environmental flows, such as on-shore, offshore diurnal wind cycles or tidal flows, can be the most important feature. The main point is that the mean concentration requires fewer realizations (or record length for steady circumstances) than higher moments for an adequate approximation of an ensemble average and that the $\kappa = 0$ result of Eq. (25.18) is also correct for time-dependent flows over complex geometry.

The effects of instrument smoothing due to time and/or spatial averaging of a concentration signal by instrumentation can result in a significant apparent reduction in the value of the fluctuations and higher moments. The mean concentration is reasonably insensitive to temporal and spatial averaging (Schopflocher and Sullivan, 1998). However, as discussed in Chatwin and Sullivan (1993), measured values of the mean-square concentration on the centerline of a benchmark, laboratory, contaminant jet were observed to double as the probe sampling volume went from about $(0.54 \text{ mm})^3$ to $(0.10 \text{ mm})^3$. In Mylne and Mason (1991) it was demonstrated that significant changes occurred when their environmental concentration measurements were deconvoluted to account for time integration effects due to their instrumentation. The temporal and spatial resolution that is required to capture the fine-scale texture with significant concentration spikes (see Fig. 25.1) of the very small conduction cutoff length dimension is very demanding, particularly with the robust probes used in field experiments.

The retrieval of useful concentration information at low values of concentration when the signal is corrupted by instrument noise is often a real problem. In some measurements (e.g., Schopflocher, 1998; Lewis and Chatwin, 1995b) upwards of 70% of the concentration is shown at negative values of concentration. This is a clear indication of instrument noise corruption. If the measured signal Γ_m is simply the addition of the true signal Γ and independent noise η

$$\Gamma_m = \Gamma + \eta \tag{25.25}$$

then,

$$\langle \Gamma \rangle = \langle \Gamma_m \rangle - \langle \eta \rangle \tag{25.26}$$

and

$$\langle \Gamma^2 \rangle = \langle \Gamma_m^2 \rangle + \langle \eta^2 \rangle - 2\langle \eta \rangle \langle \Gamma_m \rangle \tag{25.27}$$

Thus, provided $\langle \eta \rangle$ and $\langle \eta^2 \rangle$ are small with respect to $\langle \Gamma \rangle$ and $\langle \Gamma^2 \rangle$ respectively, the measured signal provides a reasonable representation of the true value. This is clearly not the case in many data sets, including the aforementioned example. It is not an acceptable practice in general simply to ignore the negative concentration values or, worse, to ignore all measured values below an arbitrary positive threshold (see discussion in Robinson et al., 1985). A procedure using a maximum entropy inversion technique to deal with this problem is offered in Lewis and Chatwin (1995b).

Return to the point that a semiempirical scheme that purports to model or simulate the mean or fluctuating contaminant concentration field must be thoroughly validated by experiment. The mean field is more easily modeled and measured. $C(\mathbf{x}, t)$ requires fewer realizations to obtain an acceptable ensemble-average representation and is relatively insensitive to the effects of molecular diffusion and instrument smoothing. The fluctuations, or root-mean-square value of concentration, are very sensitive to the effects of molecular diffusion and require extraordinary spatial and temporal experimental resolution to have representative measurements. The conservative representation of fluctuations given by Eq. (25.18) has the decided advantage that it is expressed in terms of the mean, which is more easily modeled and validated.

25.5 TOWARDS A LESS CONSERVATIVE ESTIMATE

To achieve a more accurate and lower estimate of the size of fluctuations than by Eq. (25.24), one must incorporate the effects of molecular diffusivity κ. Based on the premise, which will be true in almost all flows, that turbulent convective motion will disperse marked fluid through space much more rapidly than molecular diffusion will reduce the concentration of that marked fluid, the $\alpha - \beta$ prescription for moments of the probability density function was provided in Chatwin and Sullivan (1990b). Specifically the second moment was given as

$$c^2(\mathbf{x}, t) = \beta^2 C(\mathbf{x}, t)(\alpha C_0 - C(\mathbf{x}, t)) \tag{25.28}$$

where C_0 is the largest value of mean concentration of a cloud at time t or the largest value on the cross-section of a continuous release at distance x downstream of the source. Equation (25.28) has the interpretation, by comparison with the $\kappa = 0$ result of Eq. (25.18), of a moving source with a local representative concentration αC_0 and a factor β used to account for the nonzero background concentration due to diffused contaminant. α and β are functions only of time for a cloud or distance downstream for a steady configuration. This prescription was shown to have a considerable amount of experimental validation in the original paper and since that time (Chatwin and Sullivan, 1990b). In Sawford and Sullivan (1995) the distribution of the first four moments over a plume in grid turbulence, investigated at 17 sampling distances from a heated line source, was shown to be in good agreement with this prescription.

As $x \rightarrow 0$ for a steady source and flow (or $t \rightarrow 0$ for a cloud), the $\kappa = 0$ result will be recovered from Eq. (25.28) and hence $\beta(0) = 1$ and $\alpha(0) = 1$ and also $\alpha(x) \sim \theta_0 C_0^{-1}$. The values of β and α are restricted to be $0 \leq \beta \leq 1$ and $\alpha \geq 1$. It should be noted that now, by contrast with the $\kappa = 0$ result, the locus of the maximum value of c^2 is found from $C = \alpha C_0/2$.

When $n = 1$ in Eq. (25.12),

$$\frac{d}{dt}\int_{\text{a.s.}} (C^2(\mathbf{x}, t) + c^2(\mathbf{x}, t)dV(\mathbf{x}) = -2\kappa \int_{\text{a.s.}} \langle(\nabla\Gamma)^2\rangle dV(\mathbf{x}) \tag{25.29}$$

Taking into account the fine-scale texture of the concentration field, it is assumed that all significant gradients occur over the conduction cutoff length λ such that

$$(\nabla\Gamma)^2 = A\,\frac{(\Gamma - \Gamma_t)^2}{\lambda^2} \tag{25.30}$$

where Γ_t is a background threshold concentration such that $\Gamma_t \rightarrow 0$ as $t \rightarrow 0$ and $\Gamma_t \rightarrow C$ as $t \rightarrow \infty$ for a cloud and A is a proportionality constant. Some validation for Eq. (25.30), using atmospheric data, is provided in Mole (1995). The use of Eq. (25.30) in Eq. (25.29) and also using the constraints imposed on α and β provide the analytic solution, for a uniform source concentration (Labropulu and Sullivan, 1995),

$$\alpha(\tau) = \frac{-e^{-\tau}}{QC_0\beta^2} \int_0^{\tau} e^{\xi} \frac{d}{d\xi} \int_{\text{a.s.}} C^2 dV d\xi + \frac{1}{QC_0} \int_{\text{a.s.}} C^2 dV \tag{25.31}$$

$$\beta^2(\tau) = \frac{\theta_0 C_0'}{C_0'(0)C} \tag{25.32}$$

where the prime in Eq. (25.32) denotes differentiation with respect to $\tau = 2A\kappa t/\lambda^2$. Figures 25.4 and 25.5 from Labropulu and Sullivan (1995), show $\alpha(\tau)$ and the center intensity c^2/C^2 for three typical cases of a contaminant cloud released in homogeneous (grid) turbulence, the inertial subrange, and in the atmospheric boundary layer. $\beta^2(\tau)$ is simply a monotonically decreasing function and is not shown. Other numerical solutions using Eqs. (25.29) and (25.30), which provide a similar result to Eqs. (25.31) and (25.32), are found in Moseley (1991), Mole (2001), and Clarke and Mole (1995). It is to be noted in Fig. 25.4 that as the values of $\alpha(\tau)$ are less than 2, initially the distribution of $c^2(\mathbf{x}, \tau)$ over the cloud will be bimodal. This behavior is followed by a period where $\alpha(\tau) \geq 2$ when the distribution is unimodal and followed thereafter by a return to a bimodal distribution corresponding to the value of $\alpha(\tau)$ falling below 2. This experimentally observed behavior was mentioned earlier and shown on the sketch given in Fig. 25.3. The solutions obtained by Eqs. (25.31) and (25.32) have been shown to compare favorably with some grid turbulence measurements. The fluctuations that would be experienced by a fixed observer for a cloud in the neutral atmospheric boundary layer, using the above result, are shown in Heagy and Sullivan (1995b). These results are promising; however, more thorough validation is required using data from specifically targeted experiments.

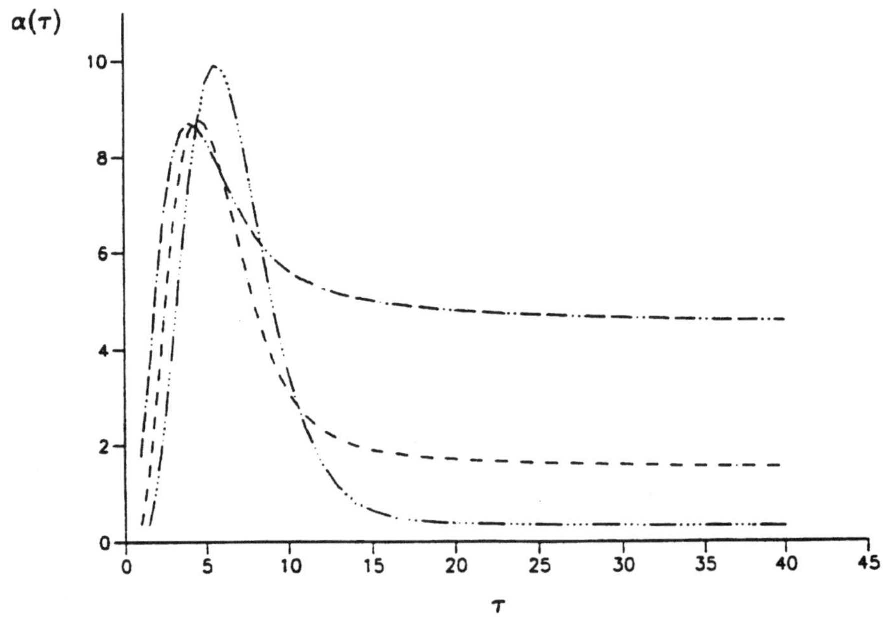

FIGURE 25.4 Values of $\alpha(\tau)$ for a cloud released: in the inertial subrange (divided by 7) —····—····—; in the neutral atmospheric boundary layer — — — —; in homogeneous grid turbulence (multiplied by 5) — —··— —··— —. (*Source:* Labropulu and Sullivan, 1995.)

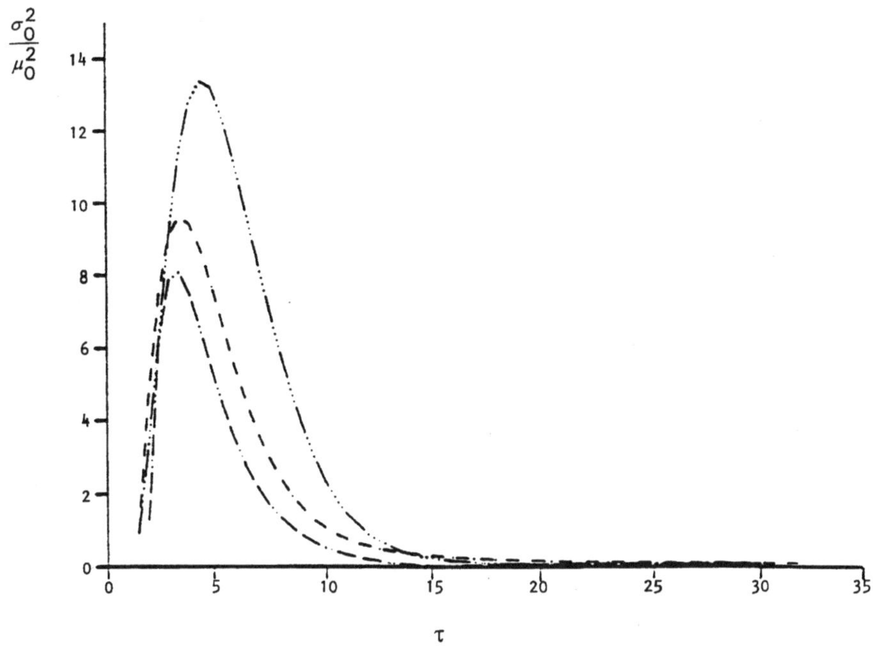

FIGURE 25.5 Values of center intensity for a cloud released: in the inertial subrange —····—····—; in the neutral atmospheric boundary layer (multiplied by 5) — — — —; in homogeneous grid turbulence (multiplied by 40) — —··— —··— —. (*Source:* in Labropulu and Sullivan, 1995)

25.6 CONCLUDING REMARKS

The use of the mean $C(\mathbf{x}, t)$ and fluctuations $c(\mathbf{x}, t)$ to gauge the range of concentration values that would be encountered by a fixed observer presupposes some sense of the shape of the probability density function. For example, about 98% of the concentration values will be less than $C + 2c$ with the familiar—although unrealistic in this application, as discussed in Chatwin (2000) and evident in Chatwin and Sullivan (1989)—Gaussian probability density function. In traversing the cross section of a plume, jet, or wake, one observes probability density functions that change significantly in shape. These change from more bell-shaped forms at the centerline to the more exponential-like shapes at the periphery. The very large values of concentration that are found in the high-concentration tails of the probability density function, although rare events, can of course have an inordinately large impact. Thus, one would like to have a better picture of the probability density function shape, and especially in the tails (see Munro et al., 2001).

All of the theoretical and experimental difficulties attached to the higher-order moments are more pronounced when considering the probability density function. The evolution equation is more complicated and the number of realizations and sampling duration required to experimentally approximate an ensemble average are large because one is dealing with rare events in the tails. The probability-density function can be approximated by using a finite number of moments, and here the convergence to an ensemble estimate will depend on the highest moment included. The quality of the representation of the tails will, in general, depend on how large a moment is included in the finite set. Spatial and temporal experimental resolution is very important in that the shape of the probability density function can change significantly with improved resolution (Schopflocher and Sullivan, 1998). Of course, for low concentrations, contamination with instrument noise has a direct effect on the observed probability density function. For independent noise, the measured probability density function, p_m, is simply the convolution of the true probability density function with that of the noise signal

$$p_m(\theta; \mathbf{x}, t) = \int_{-\infty}^{\infty} p(\theta; \mathbf{x}, t) q(\theta - \zeta; \mathbf{x}, t) d\zeta \tag{25.33}$$

where $q(\theta; \mathbf{x}, t)$ is the probability density function of the noise signal $\eta(\mathbf{x}, t)$. The objective is to find a parametric probability density function that represents the required features adequately, using as few parameters as possible, and to determine the parameters from the low-order measured or predicted moments.

Since concentration values are restricted to the finite range $0 \le \Gamma(\mathbf{x}, t) \le \theta_1$, where θ_1 is the largest concentration value at release, the probability density function is restricted to this range and

$$\int_0^{\theta_1} p(\theta; \mathbf{x}, t) d\theta = 1 \tag{25.34}$$

Because of the versatility to represent a large variety of shapes, including bimodal shapes, and to correspond to the fine-scale texture of the concentration field, the probability density function is usefully expressed as the mixture

$$p(\theta; \mathbf{x}, t) = \gamma(\mathbf{x}, t) f(\theta; \mathbf{x}, t) + (1 - \gamma(\mathbf{x}, t)) g(\theta; \mathbf{x}, t) \tag{25.35}$$

The f and g probability density functions in Eq. (25.35) represent high concentration values in the sheets and strands and the low background concentration in between that were discussed in Section 25.3. $0 \le \gamma(\mathbf{x}, t) \le 1$ is the mixture ratio. It was shown in Chatwin and Sullivan (1989) that all concentration probability density functions have this representation when f and g correspond to source and nonsource fluid respectively and Eq. (25.17) is recovered from Eq. (25.35) when $\kappa = 0$.

Evidence is mounting, both from atmospheric experiments over a wide range of stability classes (see Mole et al., 1995; Lewis and Chatwin, 1995b; Munro et al., 2000) and in a controlled plume experiment in grid turbulence (see Schopflocher, 1998), that a rather general result from extreme value analysis in statistics describes the high-concentration tails of the probability density function. Mixture models using this result show promise as good, generic models for the probability density function (see Lewis and Chatwin, 1995a, b; Schopflocher and Sullivan, 1999).

25.7 ACKNOWLEDGMENTS

Paul Sullivan received financial support from the National Science and Engineering Research Council of Canada. Philip Chatwin's contribution was partly supported by the Commission of the European Communities (CEC): COFIN project (contract number ENV4- CT97-0629).

25.8 REFERENCES

Batchelor, G. K. 1949. "Diffusion in a Field of Homogeneous Turbulence I. Eulerian Analysis," *Australian Journal of Scientific Research,* vol. 2, pp. 437–450.

Batchelor, G. K. 1952. "Diffusion in a field of Homogeneous Turbulence II. The Relative Motion of Particles," *Proceedings of Cambridge Philosophical Society,* vol. 48, pp. 345–362.

Birch, A. D., D. R. Brown, and M. G. Dodson. 1980. *Ignition Probabilities in Turbulent Mixing Flows,* Report MRS E 374, British Gas Corporation, Midlands Research Station, Solihull, West Midlands, U.K.

Chatwin, P. C. 1990. "Hazards Due to Dispersing Gases," *Environmetrics,* vol. 1, pp. 143–162.

Chatwin, P. C. 2000. "Some Remarks on Modelling the PDF of the Concentration of a Dispersing Scalar in Turbulence," submitted to *European Journal of Applied Mathematics.*

Chatwin, P. C., and C. Robinson. 1997. "The Moments of the PDF of Concentration for Gas Clouds in the Presence of Fences," *Il Nuovo Cimento,* vol. 29, no. 3, pp. 361–383.

Chatwin, P. C., and P. J. Sullivan. 1979a. "The Relative Diffusion of a Cloud of Passive Contaminant in Incompressible Turbulent Flow," *Journal of Fluid Mechanics,* vol. 91, no. 2, pp. 337–356.

Chatwin, P. C., and P. J. Sullivan. 1979b. "Measurements of Concentration Fluctuations in Relative Turbulent Diffusion," *Journal of Fluid Mechanics,* vol. 94, no. 1, pp. 83–102.

Chatwin, P. C., and P. J. Sullivan. 1980. "The Core-Bulk Structure Associated with Diffusing Clouds," in *Turbulent Shear Flows II,* ed. L. J. S. Bradbury, F. Durst, B. E. Launder, F. W. Schmidt, and J. H. Whitelaw, Springer-Verlag, Berlin.

Chatwin, P. C., and P. J. Sullivan. 1982. "The Effect of Aspect Ratio on Longitudinal Diffusivity," *Journal of Fluid Mechanics,* vol. 120, pp. 347–358.

Chatwin, P. C., and P. J. Sullivan. 1989. "The Intermittency Factor of Scalars in Turbulence," *Physics Fluids A,* vol. 4, pp. 761–763.

Chatwin, P. C., and P. J. Sullivan. 1990a. "Cloud-Average Concentration Statistics," *Mathematics and Computers in Simulation,* vol. 32, pp. 49–57.

Chatwin, P. C., and P. J. Sullivan. 1990b. "A Simple and Unifying Physical Interpretation of Scalar Fluctuation Measurements from Many Turbulent Shear Flows," *Journal of Fluid Mechanics,* vol. 212, pp. 533–556.

Chatwin, P. C., and P. J. Sullivan. 1993. "The Structure and Magnitude of Concentration—Fluctuations," *Boundary Layer Meteorology,* vol. 62, pp. 269–280.

Clarke, L., and N. Mole. 1995. "Modelling the Evolution of Moments of Contaminant Concentration in Turbulent Flows," *Environmetrics,* vol. 6, pp. 607–617.

Corriveau, F., and W. D. Baines. 1994. "Diffusive Mixing in Turbulent Jets as Revealed by a pH Indicator," *Experiments in Fluids,* vol. 16, pp. 129–136.

Dahm, W. J. A., and P. E. Dimotakis. 1990. "Mixing at Large Schmidt Number in the Self-Similar Far Field of Turbulent Jets," *Journal of Fluid Mechanics,* vol. 217, pp. 299–330.

Dahm, W. J. A., K. Southerland, and K. A. Buch. 1991. "Direct, High Resolution, Four Dimensional Measurements of the Fine Scale Structure of $sc << 1$ Molecular Mixing in Turbulent Flows," *Physics of Fluids,* vol. A3, pp. 1115–1127.

Derkson, R. W., and P. J. Sullivan. 1990. "Moment Approximations for Probability Density Functions," *Combustion and Flame,* vol. 81, pp. 378–391.

Dewey, R. J., and P. J. Sullivan. 1979. "Longitudinal Dispersion in Flows That are Homogeneous in the Streamwise Direction," *Journal of Applied Mathematics and Physics (ZAMP),* vol. 30, no. 4, pp. 601–613.

Fackrell, J. E., and A. G. Robbins. 1982. "Concentration Fluctuations and Fluxes in Plumes from Point Sources in a Turbulent Boundary Layer," *Journal of Fluid,* vol. 117, pp. 1–26.

Hall, D. J., R. A. Waters, G. W. Marsland, S. L. Upton, and M. A. Emmott. 1991. *Repeat Variability in Instantaneously Released Heavy Gas Clouds—Some Wind Tunnel Experiments,* Tech. Rep. LR 804(PA), National Energy Technology Centre, AEA Technology, Abingdon, Oxfordshire, U.K.

Heagy, W. K., and P. J. Sullivan. 1995a. "The Expected Mass Fraction," *Atmospheric Environment,* vol. 30, no. 1, pp. 35–47.

Heagy, W. K., and P. J. Sullivan. 1995b. "Fixed-Point Values of Contaminant Cloud Dilution," *Environmetrics,* vol. 6, pp. 637–641.

Labropulu, F., and P. J. Sullivan. 1995. "Mean-Square Values of Concentration in a Contaminant Cloud," *Environmetrics,* vol. 6, pp. 619–625.

Lewis, D. M., and P. C. Chatwin. 1995a. "A New Model PDF for Contaminants Dispersing in the Atmosphere," *Environmetrics,* vol. 6, pp. 583–593.

Lewis, D. M., and P. C. Chatwin. 1995b. "The Treatment of Atmospheric Dispersion Data in the Presence of Noise and Baseline Drift," *Boundary-Layer Meteorology,* vol. 72, pp. 53–85.

Lewis, D. M., and P. C. Chatwin. 1996. "A Three Parameter PDF for the Concentration of an Atmospheric Pollutant," *Journal of Applied Meteorology,* vol. 36, pp. 1064–1075.

Mole, N. 1995. "The α–β Model for Concentration Moments in Turbulent Flows," *Environmetrics,* vol. 6, pp. 559–569.

Mole, N. 2001. "The Large Time Behaviour in a Model for Concentration Fluctuations in Turbulent Dispersion," *Atmospheric Environment,* vol. 35, pp. 833–844.

Mole, N., and L. Clarke. 1995. "Relationships between Higher Moments of Concentration and of Dose in Turbulent Dispersion," *Boundary-Layer Meteorology,* vol. 73, pp. 35–52.

Mole, N., P. C. Chatwin, and P. J. Sullivan. 1993. "Modelling Concentration Fluctuations in Air Pollution," in *Modelling Change in Environmental Systems,* ed. B. Beck, M. McAleer, and A. J. Jakeman, John Wiley & Sons, Chichester, pp. 317–340.

Mole, N., C. W. Anderson, S. Nadarajah, and C. Wright. 1995. "A Generalized Pareto Distribution Model for High Concentrations in Short-Range Atmospheric Dispersion," *Environmetrics,* vol. 6, pp. 595–606.

Moseley, D. J. 1991. "A Closure Hypothesis for Contaminant Fluctuations in Turbulent Flow," Master's thesis, The University of Western Ontario, London, ON.

Munro, R. J., P. C. Chatwin, and N. Mole. 2001. "The High Concentration Tails of the PDF of a Dispersing Scalar in the Atmosphere," *Boundary-Layer Meteorology,* vol. 98, pp. 315–339.

Mylne, K. R., and P. J. Mason. 1991. "Concentration Fluctuation Measurements in a Dispersing Plume at a Range of up to 1000 m," *Quarterly Journal of the Royal Meteorological Society,* vol. 117, pp. 177–206.

Robinson, C., D. M. Lewis, and P. C. Chatwin. 1995. "The Pitfalls of Thresholding Atmospheric Dispersion Data," *Boundary Layer Meteorology,* vol. 73, pp. 183–188.

Sawford, B. L., and P. J. Sullivan. 1995. "A Simple Representation of a Developing Contaminant Concentration Field," *Journal of Fluid Mechanics,* vol. 289, pp. 141–157.

Schopflocher, T. P. 1998. "The Representation of the Scalar Concentration PDF in Turbulet flows as a Mixture," Ph.D. thesis, The University of Western Ontario, London, ON.

Schopflocher, T. P., and P. J. Sullivan. 1998. "Spatial Resolution in the Measurement of Concentration Fluctuations," *Boundary-Layer Meteorology,* vol. 87, pp. 27–40.

Schopflocher, T. P., and P. J. Sullivan. 1999. "Nonparametric Inference of the Mixture Components in the PDF for a Diffusing Scalar in a Turbulent Flow," in *Proceedings of International Conference and Work-shop on Modelling the Consequences of Accidental Releases of Hazardous Materials,* September 28–October 1, pp. 733–749.

Sullivan, P. J. 1984. "Whence the Fluctuations in Measured Values of Mean-Square Fluctuations?," in *Proceedings of 4th Joint Conference on Applications of Air Pollution Meteorology,* ed. G. A. Beals and N. E. Bowne, pp. 115–121.

Sullivan, P. J., and H. Ye. 1993. "Further Comments on 'Cloud Averaged' Concentration Statistics," *Mathematics and Computers in Simulation,* vol. 35, pp. 263–269.

Sykes, R. I., W. S. Lewellen, and S. F. Parker. 1984. "A Turbulent-Transport Model for Concentration Fluctuations and Fluxes," *Journal of Fluid Mechanics,* vol. 139, pp. 193–218.

Townsend, A. A. 1976. *The Structure of Turbulent Shear Flow,* Cambridge University Press.

Turner, J. S. 1973. *Buoyancy Effects in Fluids,* Cambridge University Press.

Wilkins, E. 1944. "A Note on Skewness and Kurtosis," *Annals of Mathematical Statistics,* vol. 15, pp. 133–135.

Wilson, D. J. 1995. *Concentration Fluctuations and Averaging Time in Vapor Clouds,* Center for Chemical Process Safety of American Institute of Chemical Engineers, New York.

Ye, H. 1995. "A New Statistic For the Contaminant Dilution Process in Turbulent Flows," PhD thesis, The University of Western Ontario, London, ON, March 1995.

CHAPTER 26

DEBRIS HAZARD FROM ACCIDENTAL EXPLOSIONS IN UNDERGROUND STORAGE FACILITIES: A Case Study on Modeling of Debris Throw

Peter O. Kummer
Bienz, Kummer & Partner Ltd., CH-8125 Zollikerberg, Switzerland

26.1 INTRODUCTION

Major accidental explosions of high explosives (HE) or products containing substantial amounts of HE, such as ammunition, stored in bulk in underground magazines in rock fortunately occur only rarely. However, the effects of such explosions, like air-blast and debris throw, have the potential to cause disastrous damage to property and seriously endanger people in the surrounding area. Therefore, such installations need to be situated with utmost care. In many countries it is common practice to situate such installations according to the Hazardous Distance (HD) or Quantity Distance (QD) concept. This concept stipulates that certain distances must be observed between the source of a possible event and inhabited buildings as well as roads, depending on the amount of hazardous substances to be handled or stored, and regardless of the probability of the event. The necessary distances stipulated in the regulations are usually based on general safety criteria such as that the air-blast overpressure shall not exceed 50 mbar or the density of hazardous debris shall not exceed one piece per 56 m^2. The hazardous distance concept proved effective in the past. With increasing population density and utilization of land, however, this concept is now proving to be conservative, unbalanced, and inflexible.

In Switzerland, with its high population density, a safety concept based on Quantitative Risk Analyses (QRA) was developed nearly 30 years ago for the storage of ammunition and explosives within the Swiss Department of Defence [1, 2]. This QRA concept takes into account not only the effects of an accidental explosion but also its probability. It has proven to be a very useful instrument, allowing the actual hazards of such installations to be calculated in terms of fatalities, taking into account site-specific conditions, such as the number of people living in the vicinity or passing by, leisure activities (place and duration), and the layout of the underground installation itself. Furthermore, it is an excellent tool for planning and evaluating safety measures. Because benefits and safety costs can easily be shown, it allows decisions about safety measures to be made on a consistent basis.

However, the risk-based safety concept also has some difficulties. One is that more accurate data and models for predicting the effects of an explosion over a whole range are needed than with the QD concept, which considers effects only at specific boundaries.

This chapter deals with this problem. Because for the QD approach only boundary values have been of interest, in many areas there is little reliable data for the development of suitable prediction models over a whole area. Information is particularly sparse about debris throw, especially from underground installations in rock with insufficient rock overburden leading to a crater in case of an explosion.

Although today many basic computer codes exist (hydrocodes like Autodyn and others) that allow the modeling of many physical effects, it has proved difficult to get reliable results from such calculations without being able to calibrate the calculated values against data from full-scale tests. Full-scale tests, however, are very expensive, and few countries have test sites where tens or even hundreds of tons of explosives can be detonated. In addition, in many areas model-scale tests are a useful, practicable, and comparatively cheap way to expand a database. In the field of rock-debris throw, however, small-scale tests are difficult to perform and evaluate because the scaling of gravity and geological conditions—being complicated in reality with joints, faults, fissures, and changing properties of the rock material within short distances—causes tremendous problems.

Therefore, it is important to take every other opportunity to enlarge the understanding of the effects of explosions. Although undesirable, accidental explosions are ideal opportunities for such a purpose. Despite the tragic consequences of such events, with often many victims, the physical evidence produced should be used as much as possible to help us understand the consequences of explosions and enhance the safety of such installations in the future.

On November 2, 1992, a detonation with an approximate energy of 225 tons of TNT occurred in a Swiss underground installation for the storage of old ammunition and explosives prior to their final destruction. Six people were killed and the installation was completely destroyed. The rock cover above the underground chamber broke off and rock as well as concrete pieces were thrown over a wide range of the surrounding area.

In Section 26.2, this accidental explosion is described in detail, followed by a description of the comprehensive evaluation of the rock-debris throw originating from the crater produced. Section 26.3 shows how these data, together with other sources, were used to enhance the current knowledge in this field and to develop a new prediction model for crater debris [3] to be used in the Manual of NATO Safety Principles and the Swiss regulation TLM 75 [4].

26.2 THE STEINGLETSCHER EXPLOSION AND ITS EVALUATION

26.2.1 The Installation

The storage magazine, called "Steingletscher" (Stone glacier) was located in the center of the Swiss Alps in an uninhabited area. It belonged to a state-owned ammunition factory and was used to store old delaborated ammunition, outdated explosives, and waste from the production of ammunition and explosives before their final destruction by open burning or detonation on the plain in front of the magazine. Figure 26.1 shows the location of the installation as well as the burning and detonation ground.

The general layout of the magazine is shown in Figs. 26.2 and 26.3. The magazine consisted of three major parts: the two storing chambers, the unloading area (accessible for trucks), and the building at the entrance containing all the technical installations. Figures

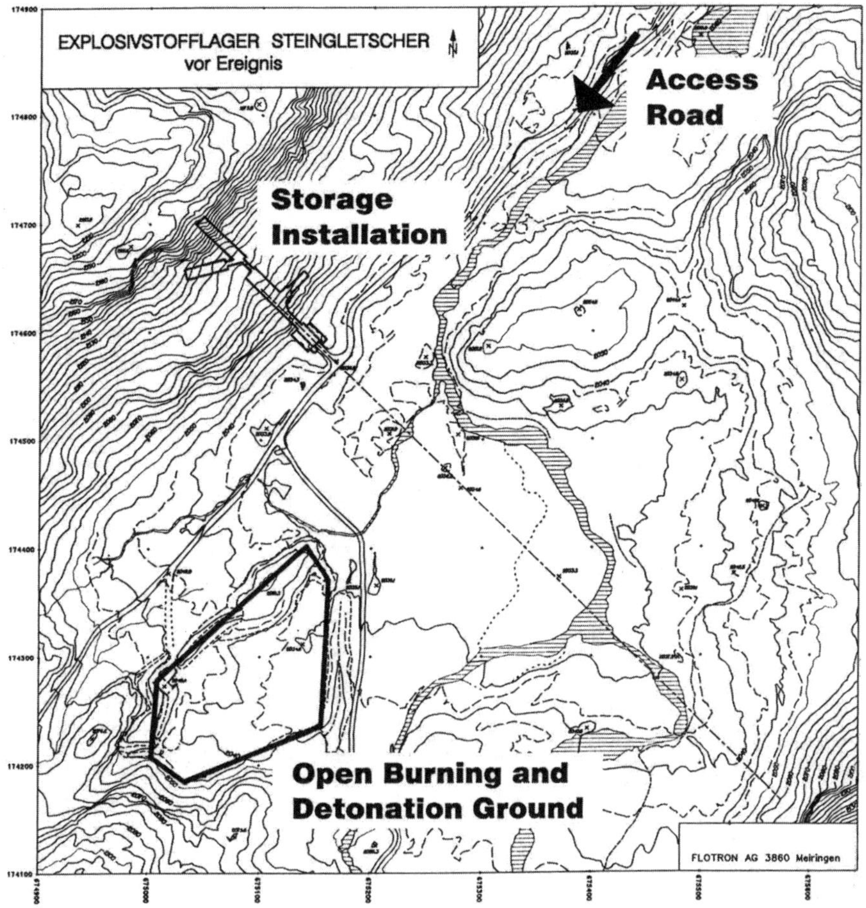

FIGURE 26.1 Installation before the event.

26.4 and 26.5 document the volume and cross-section of the different tunnel sections. The rock overburden above the storage chambers was at least 52 m thick and consisted generally of very good rock, mostly granite. A more detailed description of the installation can be found in Ref. 5.

On the day of the accident, the storage chambers were loaded with about 225 tons of explosives (TNT-equivalent). Most of this—about 190 tons—consisted of flaked TNT in cardboard drums. The rest consisted of a large variety of ammunition items, including pyrotechnical materials. The average loading density in the two chambers was around 45 kg/m^3.

26.2.2 Summary of the Accident

The usual operations had been underway on November 2, 1992. At the moment of the explosion, six persons were working inside the installation, at least one of them in one of the storing chambers, preparing material for destruction that day. Fifteen people were in

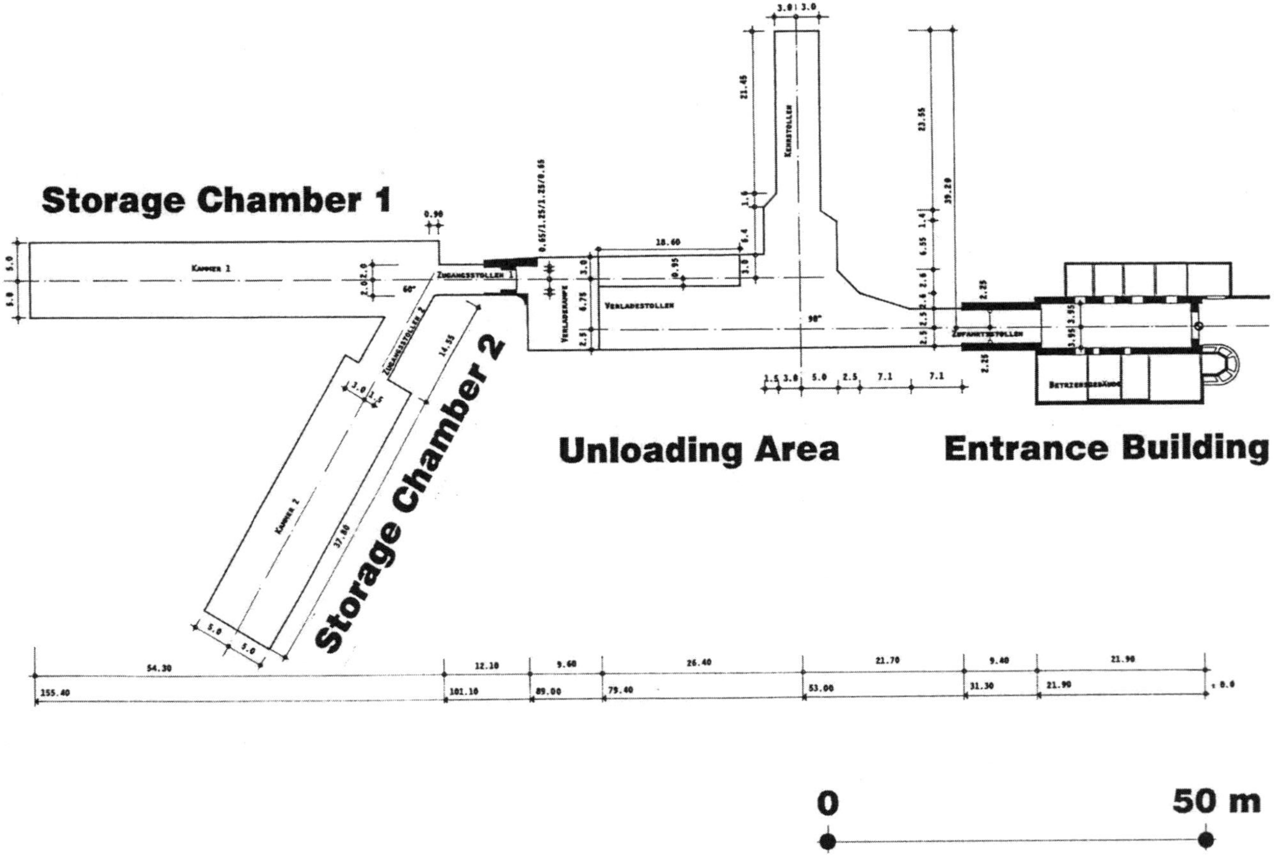

FIGURE 26.2 Magazine layout.

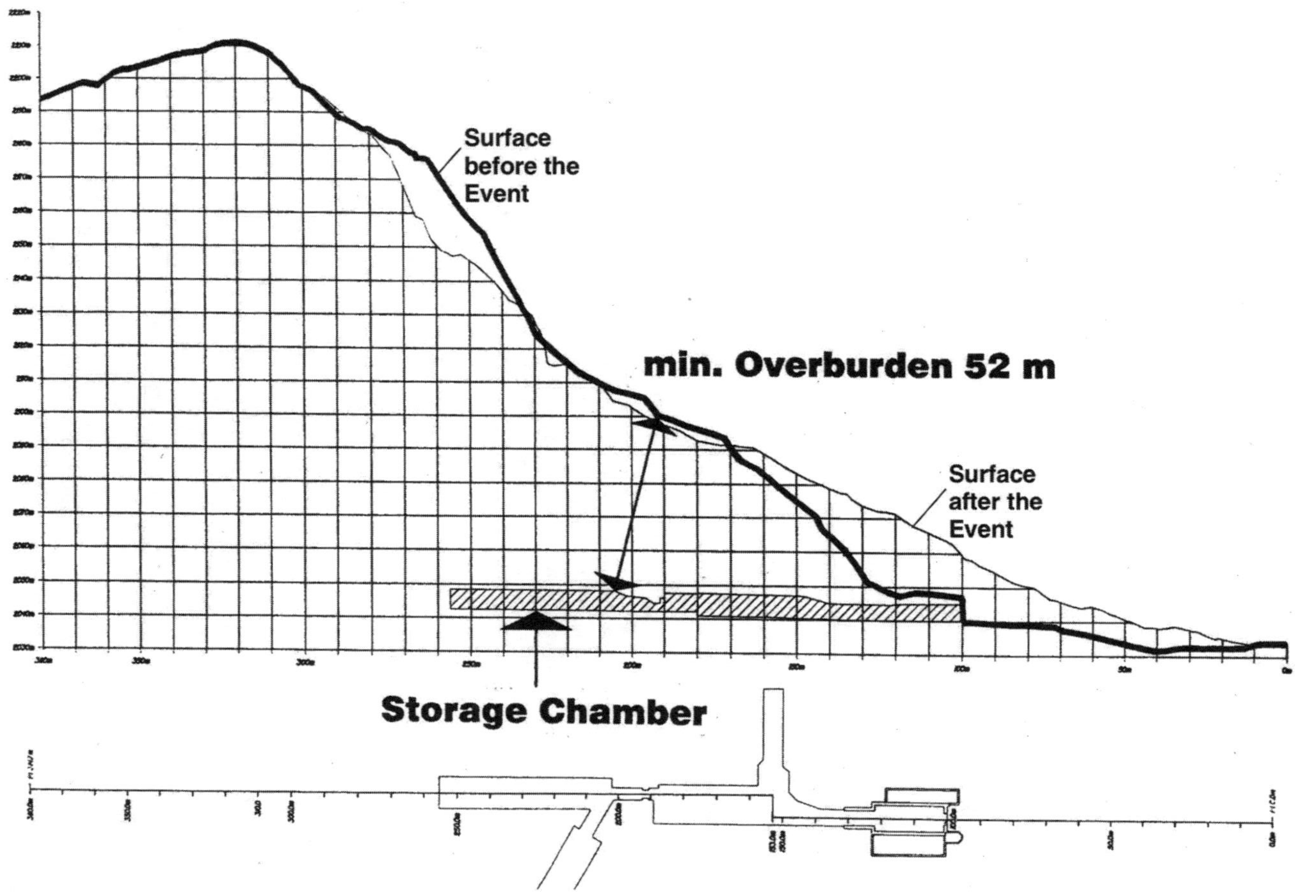

FIGURE 26.3 Longitudinal section before and after the event. (Grid size 10 m)

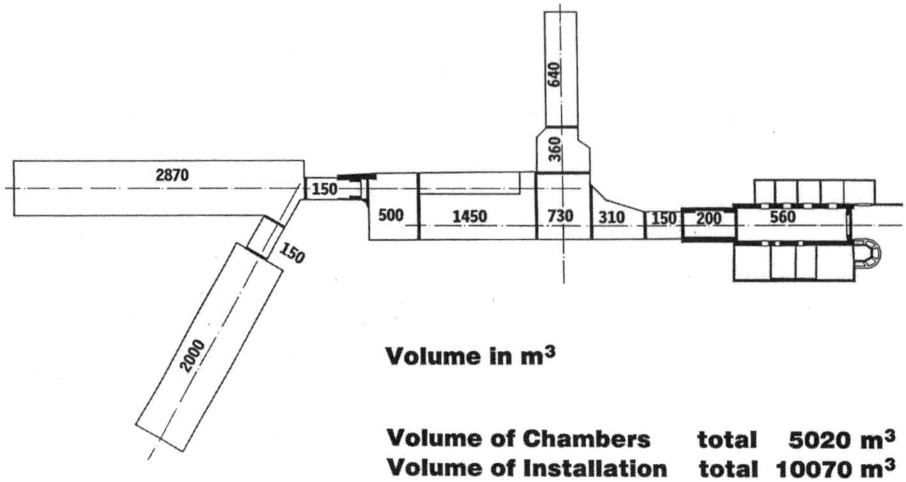

FIGURE 26.4 Volume of chambers and tunnel sections.

other areas outside the underground part of the installation. At about 4 p.m., a fire was reported by a worker inside the installation, and seconds later a huge explosion followed. The six people inside the installation were killed instantly; the workers in the surrounding area miraculously survived without any injuries.

The installation was completely destroyed and the force of the explosion uncovered part of the rock above the chamber. Afterwards, probably due to the ground shock and the dislocation of rock material due to the crater forming, a large quantity of rock material—about 100,000 m^3—broke loose from the top of the mountain and covered the area where the installation had been located. Figure 26.6 shows an overview of the area after the explosion. Rock debris from the crater were thrown into the surrounding area in all directions, up to distances exceeding 500 m. Along the axis of the access tunnel, the debris throw, consisting of rock material and concrete parts from the installation, especially from the entrance building, was even more dense up to a distance of about 700 m. Figure 26.7 shows a 15-ton block of concrete from the entrance building that was found 370 m from its original place. There was no damage due to air blast, however, as there were no above-ground structures such as houses in the immediate area. A more detailed description of the accident can be found in Refs. 6, 7, and 8.

26.2.3 The Evaluation—Course of Action

Despite the serious consequences and the lives lost in this accident, the Swiss Department of Defence decided to learn as much as possible from it. A private contractor (Bienz, Kummer & Partner Ltd.) was hired to evaluate the effects of the explosion and to study the impacts of this event on the existing regulations for the storage of ammunition by the military forces and the military administration in Switzerland.

Because the installation was relatively new (built in 1983), detailed information was available about the construction of the underground portions. Quite detailed lists of the contents of the chambers immediately before the explosion were also available. Because the basic conditions prior to the accident were quite clear, it was a unique situation for the investigation of an accidental event.

Because the installation had been built in a mountainous region, the surrounding area was not inhabited and no buildings—the nearest one being 1,500 m away—were affected by the

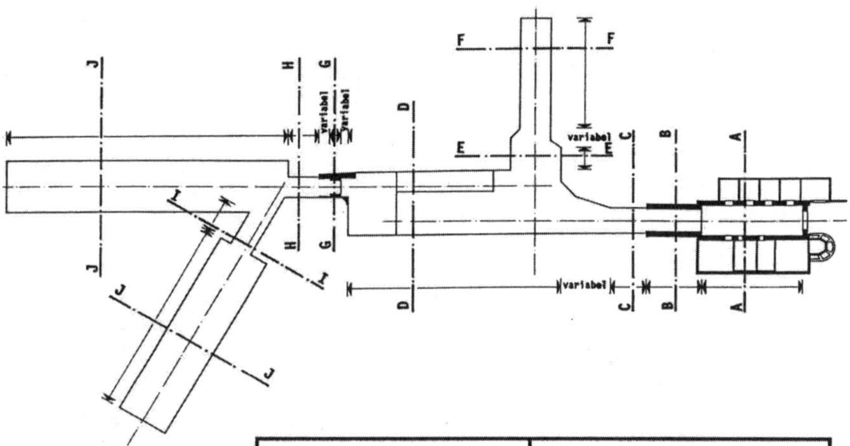

Section	Relevant Cross-Section Areas [m^2]
A - A	27.9
B - B	19.4
C - C	20.6
D - D	68.3
E - E	49.9
F - F	29.9
G - G	6.0
H - H	15.9
I - I	21.7
J - J	52.8

FIGURE 26.5 Relevant cross-section areas.

effects of the explosion. It was therefore decided that the main effort of the evaluation should be dedicated to analyzing the debris throw originating from the crater and the access tunnel. The debris throw from the crater will be described in detail below.

The collection of data began two days after the explosion. Because the ground was already covered with snow, however, it proved to be difficult to obtain reliable data (see Fig. 26.8). Furthermore, only three days later the real winter started in the Swiss Alps and everything was covered with masses of snow, making the place inaccessible for more than six months.

Most of the basic data were collected and documented during the summer of 1993, when the site was accessible again after the annual thaw. The detailed analysis of the basic data was performed in the following years.

Unfortunately, as often happens under such circumstances, the legal investigation, which was mainly concerned with the cause of and responsibilities for the event, and not in the effects produced by the explosion, took priority and caused many delays in the technical work. Thus, it was a long time before all the necessary data for the evaluation were made available to the technical people and results could finally be published, making the information also available to a broader audience [9, 10].

FIGURE 26.6 Overview of the scene seven months after the explosion.

FIGURE 26.7 Concrete block from entrance building, weight 15 tons, 370 m from original place.

FIGURE 26.8 Site of installation two days after the explosion.

26.2.4 Recovery and Documentation of Basic Field Data

As a basis for evaluating the physical effects of the explosion, the damage pattern was recorded by topographical maps, terrain sections, and aerial as well as terrestrial photos. Detailed documentation was elaborated for 53 pieces of large single debris and 40 debris collection fields [8]. The data from these debris fields were used for evaluating the debris throw from the crater, which is described in detail here.

The following were the main steps necessary for the recovery of the debris field data:

1. Suitable debris areas were selected that showed a characteristic debris distribution. This was quite a difficult task because new debris had to be found in a desert of stones. In the end, however, it was easier than expected because the shape of the crater debris, the vegetation under it, and the clean debris surface, without any lichens on it, made a distinction possible. Figure 26.9 shows two of the chosen fields.
2. The selected fields were marked, photographed, and surveyed. An area map showing the installation and all the fields where debris had been collected (Fig. 26.10: nos. 81 to 120 indicate the debris collection fields, nos. 1 to 53 show the location of the recorded large pieces of single debris) was produced.
3. Finally, all pieces of debris were collected and sorted out according to different materials (rock, concrete, metal parts, etc.), and size. Figure 26.11 shows an example. The debris were counted and weighed. A data sheet was prepared for each debris field, showing all details, and, as a first step in the evaluation, the debris mass density in kg/m^2 was calculated (Fig. 26.12). Figure 26.13 gives an overview of the data of all debris fields.

Twenty-five man-days were invested overall in the data recovery at the site. This is not as much as would have been desirable, but financial and time constraints made a more extensive site investigation impossible. Two things proved true again: First, after an accident there is always an urgent need to clean up the damage as fast as possible, not only at the scene but also to make all traces of the accident disappear and get the event out of the public mind. And secondly, it is much easier to get a million dollars for a sophisticated new test than 10,000 to evaluate an accident.

26.2.5 Analysis of Basic Data

Work Performed. Using the field data, the following evaluations were made:

1. Development of a debris mass density contour map
2. Determination of the number of hazardous debris per unit area (areal density)
3. First estimation of lethality based on the number of hazardous debris

These steps are described in the following subsections.

Development of a Debris Mass Density Contour Map. The distribution of debris from the crater above the storage chamber was shown by means of a debris mass density contour map. For the development of this map, only those debris fields could be used that were not influenced by debris from the access tunnel. Because the area was already covered with snow on the day of the explosion, and because crater debris and adit debris leave different traces in the snow at the places where they hit the ground, they could easily be distinguished. Most of the adit debris was lying within an angle of around 22° to either side of the axis of the tunnel entrance and were therefore not used for further evaluation of the crater debris throw. It can be mentioned that a 45° angle is also the area in which most of the debris coming from an access tunnel would be expected according to safety regulation TLM 75 [4, 11].

FIGURE 26.9 Debris collection areas and debris number 21 (see Fig. 26.7).

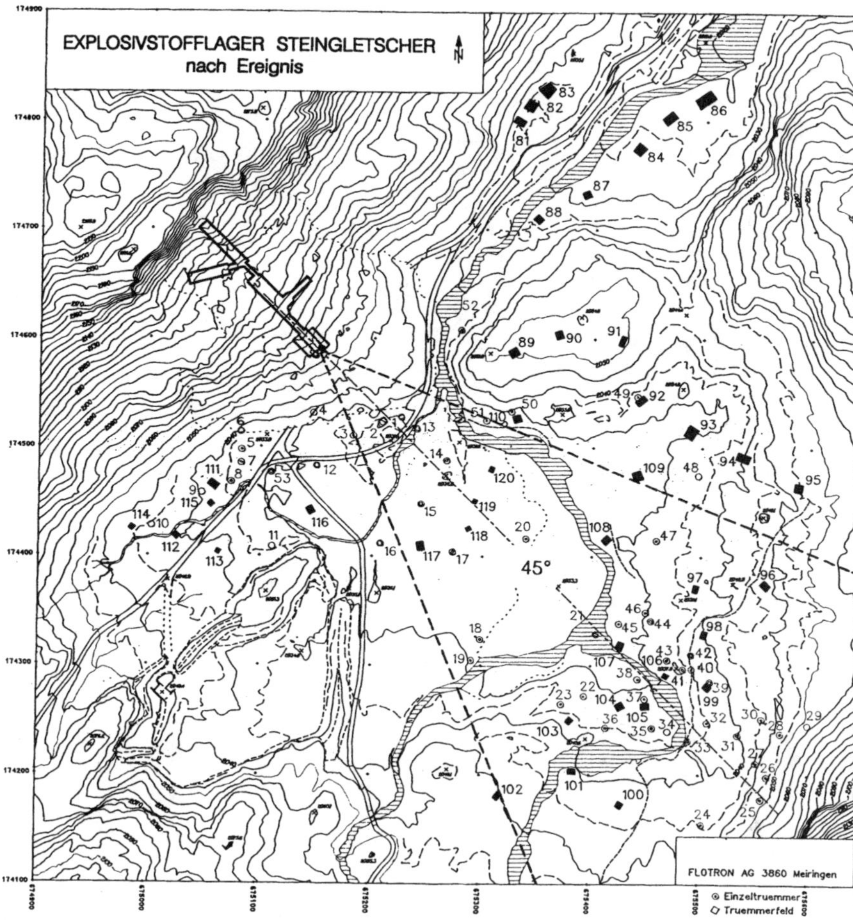

FIGURE 26.10 New area map with debris fields.

For the evaluation of the debris throw from the crater, only debris fields nos. 81 to 95 and 109 to 116, (a total of 23) were used. Together with the fact that the maximum crater debris throw distance was on the order of 600 to 700 m, the graph in Fig. 26.14 was developed. This shows the debris mass density in relation to the distance from the center of the crater. Although the data scattering was not as small as one would have liked it, this debris mass density versus distance curve represented the physical facts reasonably well. Based on this curve, the debris mass density contour lines in Fig. 26.15 could be drawn.

This curve does not yet show an angular dependency of the crater debris throw. Although some of the debris fields sideways to the axis showed somewhat smaller debris mass densities, it was decided at that time, because the number of data points was comparatively small, to draw a single curve through the data points as in Fig. 26.14. A more detailed investigation of the data points during the development of a new crater debris throw model showed, however, that there is a distinct dependency of the density of crater debris on the angle of the slope of the overburden in the area where the crater is formed. This effect is discussed in Section 26.3.

FIGURE 26.11 Debris from debris field no. 88.

Debris Field					**No. 88**
Description	: Green grassy field, 1/4 under water				
Location	: West of the installation, near glacial stream				
Dimension	: L = 7.45 m W = 6.35 m				
Area	: A = 47.30 m^2				
Type and Mass (kg) of Debris:					
Number of pieces	Weight Single	Weight Total	Weight Type Total	Type	Remarks
1	32.5	32.5		R	
1	20	20		R	
1	6	6		R	
2	4	8		R	
1	3	3		R	
3	2	6		R	
2	1.5	3		R	
2		2		R	
4		4		R	
12		8		R	
10		4		R	
10		3		R	
10		2		R	
Rest		10	111.5	R	
		0.05		M	
1	1	1	1.05	M	
		0.3	0.3	W	
Total		112.85			
Debris density (kg/m^2) Total: 2.38					
Rock: 2.36	Concrete: -	Wood: 0.0063	Metal-/ammunition parts: 0.022		
Remarks	: - 1 igniter of a 15,5 cm round, ca. 1 kg				
Date	: 28.7.93				

FIGURE 26.12 Data sheet of debris field no. 88.

Defining the Number of Hazardous Debris per Unit Area. Debris mass density contour lines are only one step in determining a safety distance or a lethality rate for a person exposed to this physical effect. In fact, it is always one or a couple of pieces of real debris that cause casualties, not an abstract value like "debris mass density."

Thus, the next step in the evaluation was to establish the relationship between the number of hazardous debris pieces per unit area (the areal density) and the debris mass density. Based on the data sheets of the debris fields (Fig. 26.12), a debris size summation curve was developed for each field. A summary of the curves of all 23 fields is shown in Fig. 26.16. A regression with these data points was made (Fig. 26.17), and a final average distribution of the debris size (mass) versus number of debris pieces—standardized for an area of one m^2 and a debris density of 1 kg per m^2—was the result. The data were evaluated to see if the distribution of the debris size depended on the distance from the crater or the angle from the tunnel axis, but neither were determined to be of significant influence within the range of interest of this study.

OVERVIEW DEBRIS FIELDS								
Debris Field no.	Location	Coordinates Y [m]	X [m]	Distance to Portal [m]	Distance to Center of Crater [m]	Altitude Above Sea Level [m]	Debris Density [kg/m2]
81	a	5328.0	4799.7	276.3	279.0	2033.5	2.45
82	a	5337.6	4814.1	293.5	294.5	2032.6	1.00
83	a	5352.9	4829.3	315.0	315.5	2032.1	0.38
84	a	5437.7	4774.5	342.2	370.3	2022.9	1.03
85	a	5464.8	4803.1	380.7	405.1	2021.1	0.23
86	a	5497.5	4820.0	417.2	441.7	2021.1	0.06
87	a	5390.3	4732.8	279.9	313.6	2024.1	0.91
88	a	5346.2	4709.9	230.3	265.8	2026.7	2.39
89	a	5325.8	4587.6	174.7	253.1	2050.5	2.17
90	a	5366.4	4604.1	215.9	288.2	2050.5	3.69
91	a	5425.0	4598.4	274.1	346.8	2050.3	1.32
92	a	5442.8	4543.5	295.0	378.0	2036.3	1.23
93	a	5488.8	4514.3	345.5	431.1	2035.2	1.15
94	a	5536.1	4491.5	396.7	483.4	2040.5	0.76
95	a	5587.9	4464.4	453.8	541.5	2047.1	0.77
96	i	5557.2	4374.6	458.5	554.2	2044.2	1.38
97	i	5495.1	4371.0	406.4	504.3	2038.9	1.33
98	i	5503.3	4328.7	437.0	536.4	2039.8	3.38
99	i	5506.8	4280.7	469.7	570.2	2037.9	4.80
100	i	5428.2	4172.3	499.1	599.7	2041.3	3.68
101	i	5383.5	4203.4	448.9	548.9	2040.7	3.09
102	(a)	5316.4	4180.5	439.2	536.0	2045.4	6.72
103	i	5380.6	4249.7	408.3	509.0	2038.7	5.86
104	i	5427.3	4263.2	425.9	527.2	2036.7	8.46
105	i	5450.2	4263.0	441.2	542.5	2035.6	4.63
106	i	5468.2	4290.9	434.1	535.0	2035.5	8.34
107	i	5426.6	4318.4	385.0	486.1	2034.3	9.29
108	i	5412.9	4415.2	313.3	411.9	2033.5	4.66
109	a	5440.6	4474.4	310.8	403.3	2034.0	1.38
110	a	5329.8	4526.5	188.8	281.1	2034.7	3.38
111	a	5057.8	4464.9	154.0	200.6	2039.7	4.26
112	a	5024.9	4417.2	211.9	253.6	2040.9	3.66
113	a	5062.8	4403.2	204.3	261.4	2041.2	3.54
114	a	4984.9	4424.8	232.5	258.8	2043.2	2.06
115	a	5055.8	4447.1	169.6	218.5	2039.7	8.30
116	a	5144.4	4441.5	146.1	230.1	2034.5	3.66
117	i	5243.4	4409.5	200.4	299.8	2033.6	5.39
118	i	5286.4	4425.5	211.0	312.3	2033.4	15.34
119	i	5292.1	4450.5	196.5	297.6	2033.1	11.22
120	i	5307.2	4480.1	189.4	288.7	2032.8	6.89
			Coordinates Portal:	5'151.13	Coordinates Cratercentre:		5084.42
				4'587.42			4663.68

a = debris fields outside 45° angle
i = debris fields inside 45° angle (area where debris from access tunnel dominates)

FIGURE 26.13 Data of all debris fields.

But what is hazardous debris? According to regulations in many countries, and especially in the NATO regulations [12], a piece of material with a kinetic energy of more than 79 joules qualifiies as hazardous debris. Taking into account a final ballistic velocity in the range of 35–50 m/s of such debris [13], it can be concluded that all debris pieces weighing more than 100–150 g are lethal. From Fig. 26.17, it can be concluded that for a debris density of 1 kg/m^2, the average would be 1 lethal debris/m^2 (on horizontal ground). Of course, this value is not a universal constant and is valid only for this explosion accident,

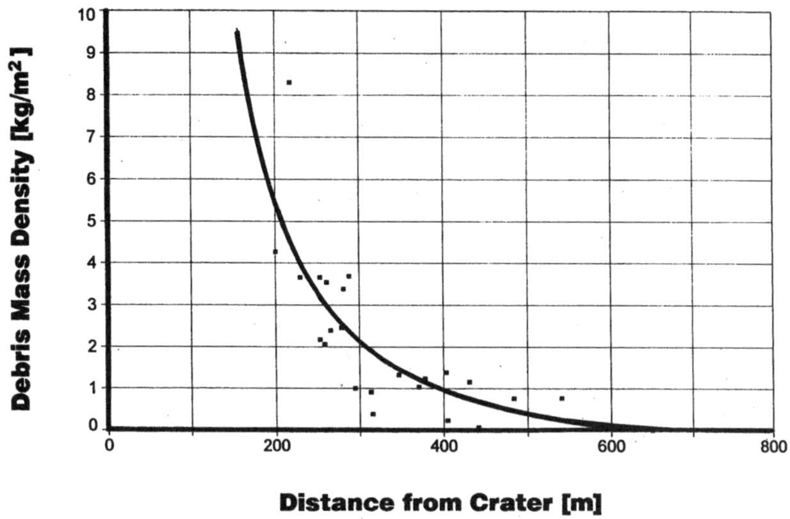

FIGURE 26.14 Debris mass density versus distance from center of crater.

but it is representative for locations with similar rock types, as will be shown in Section 26.3.

First Estimation of Lethality and Comparison with Existing Regulations. With the values from the initial evaluation (debris mass density and debris size distribution), it is easy to calculate safety distances. According to the NATO safety principles [12], the criterion for the required safety distance (Inhabited Building Distance, IBD) is a debris number density not exceeding one piece of hazardous debris/600 ft^2 or 55.7 m^2. According to Fig. 26.14 one would come up with a safety distance based on the actual debris number density measured of around 640 m (Fig. 26.18) (calculated values according to the NATO safety principles would lead to distances between 300 and 800 m, depending on the calculation model used [3]). Taking into consideration a lethal area for a person of 0.58 m^2 according to the NATO regulations (a relatively large area for a standing man facing the explosion or assuming that a person would immediately lie down flat on the ground in case of such an event), the lethality of persons standing in the open at that safety distance would be around 1%. For the contour line indicating a debris density of 1.0 kg/m^2, the respective lethality value would then be around 60%.

How do these lethality figures compare with reality? At the time of the explosion, 15 workers were standing in the open at point A in Fig. 26.18 and none of them were hurt! The probability of this happening is not zero, but it is very low. It was therefore concluded that the NATO safety criteria would have been overconservative in this case.

How would this situation be judged according to the current Swiss Safety Regulations, which do not apply safety distances but calculate the actual site-specific risk as outlined already in the introduction to this chapter (see also Refs. 1 and 2)? This risk-based approach gives a much better picture of what really happens. Thus, the technical models, such as lethality as a function of physical effects, have to be more detailed and cover a broader range than in a quantity–distance approach. Therefore, several years ago the lethality of persons due to debris throw was studied in depth [14, 15]. The impact angle of debris and the differing susceptibility of different parts of the body are examples of what was taken into account. Fig. 26.19 shows, as an example, that the impact of a piece of debris with an energy of 79 joules results in a considerable lethality rate only if the head of a person is hit.

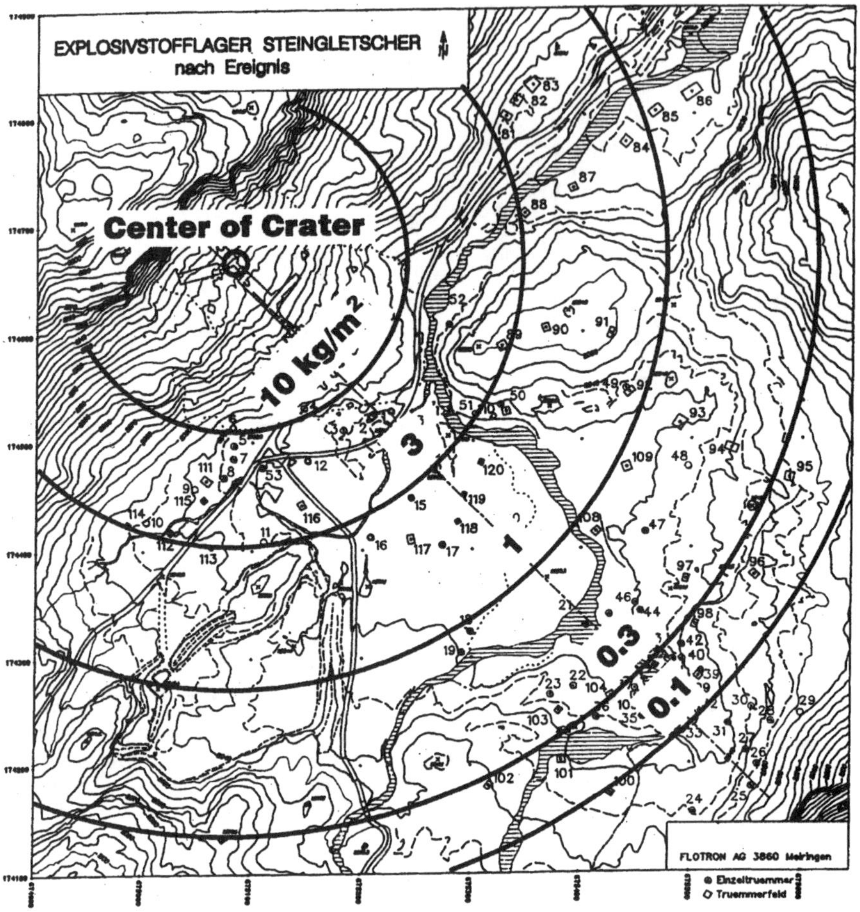

FIGURE 26.15 Debris mass density contour lines.

Other parts of the body are less sensitive to debris impact with respect to lethality. Based on that model, the lethality was calculated for the debris density measured at the Steingletscher site, taking into account the distribution of the debris size according to Fig. 26.17. The result was a lethality of slightly less than 10% at point A in Fig. 26.20, and it was concluded that this model is more realistic than the current NATO criteria.

26.2.6 Some Lessons Learned

On the technical level, it could be shown, based on a realistic case, that the widely used NATO safety criteria for debris throw might be too conservative, as already suspected by many experts. Furthermore, it is shown that the approach presented above gives more plausible results, which, together with a risk-based approach and respective safety criteria, would allow a more economic use of such installations.

In addition to the technical findings, this accident investigation again showed the following:

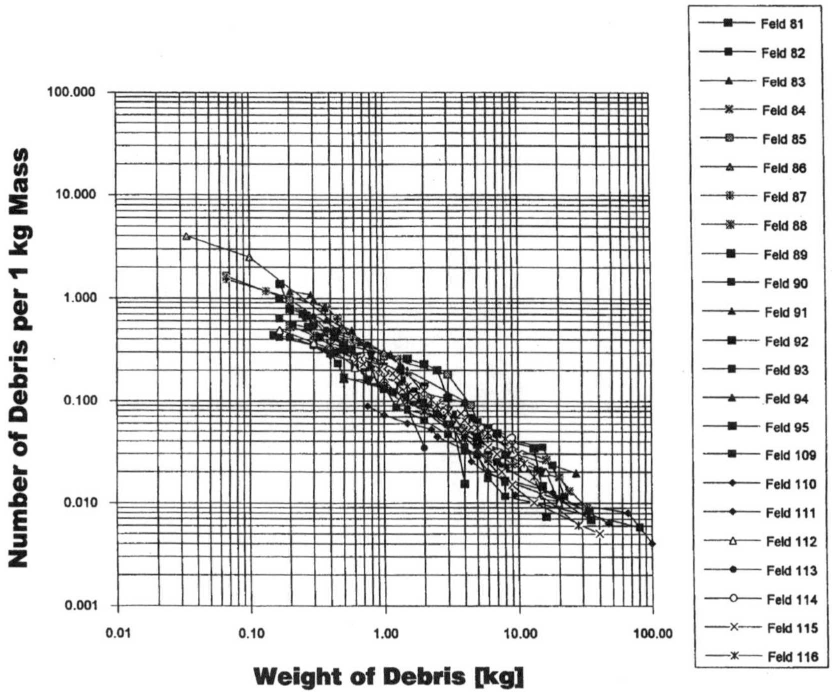

FIGURE 26.16 Summary of distribution of debris size for all debris fields.

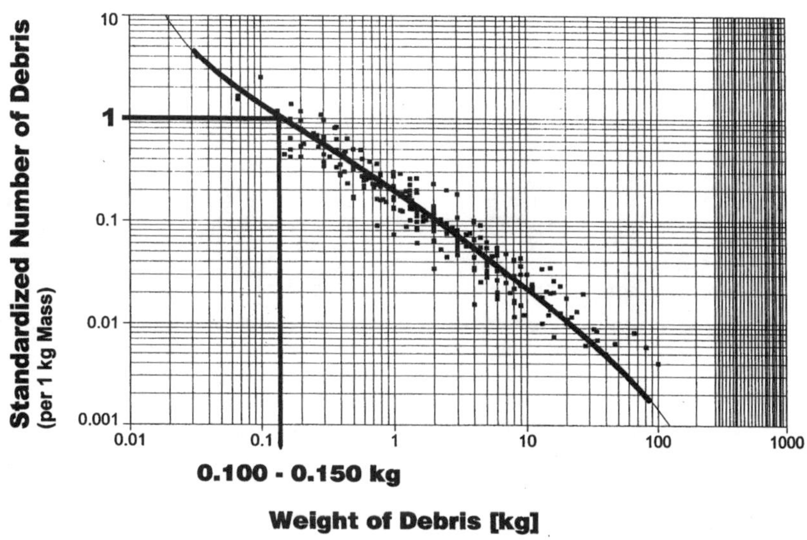

FIGURE 26.17 Mean debris size distribution.

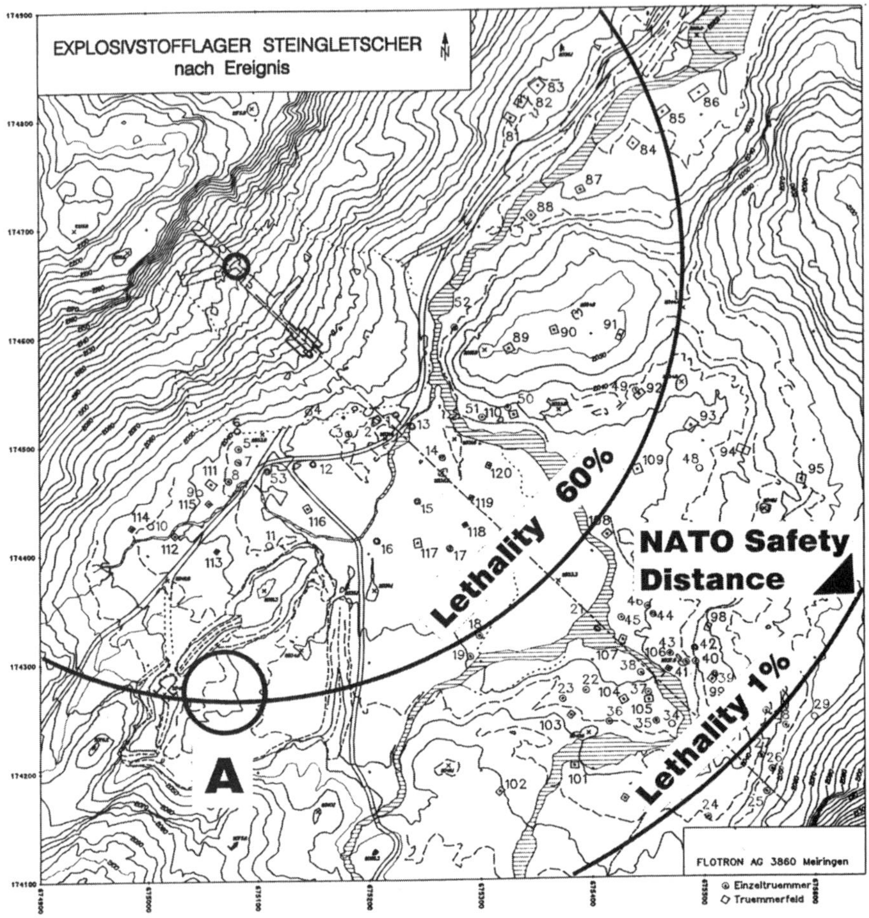

FIGURE 26.18 Lethality according to NATO safety criteria.

1. While an accident is a tragedy for the victims, it is a unique opportunity for safety specialists to check and improve their methodical and technical instruments for safety assessment, which improve, in the end, the overall safety of such installations.
2. At times it is not easy for a technical expert to get to the facts. Usually there is great social pressure to clean up the site immediately, and judges and lawyers tend to lock away important facts for a very long time.
3. Even with a limited set of data, valuable scientific findings can be made for which prohibitively expensive tests would otherwise be necessary.

26.3 THE DEVELOPMENT OF A NEW DEBRIS THROW MODEL

26.3.1 Historical Update

The Allied Ammunition Storage and Transport Publication, AASTP-1 "Manual of NATO Safety Principles for the Storage of Military Ammunition and Explosives" [12] contains the

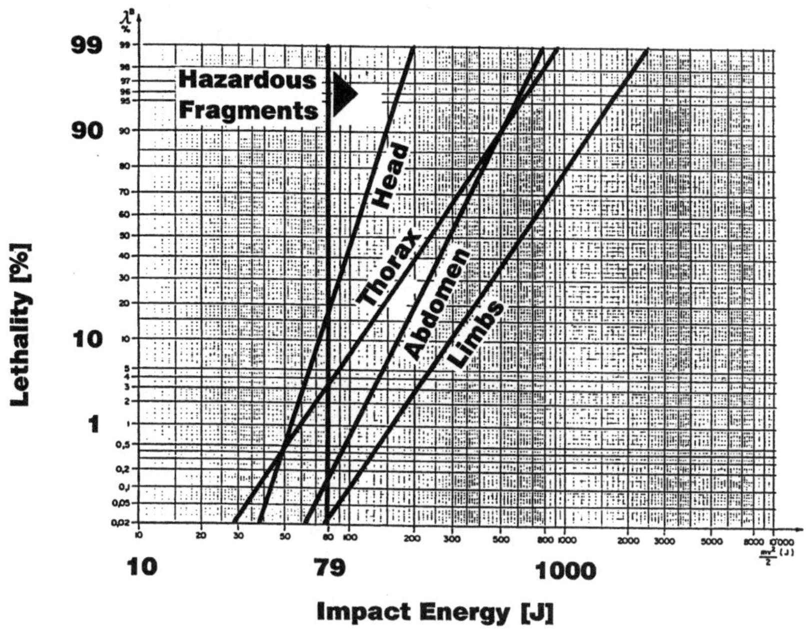

FIGURE 26.19 Lethality versus impact energy.

necessary information for the safe storage of ammunition and explosives based on a quantity-distance approach. This manual, although recently updated, still contains models dating back to the late 1950s. One of these older models is the one predicting IBDs (inhabited building distances) for debris throw from craters produced by accidental explosions in underground storage facilities in rock. This debris throw model is based on a paper written by D. E. Jarrett in 1957 [16].

Since that time, the model has not undergone any major improvements, as noted by Odello [17, 18]. Based on the data available, the formula developed by Jarrett was probably the most suitable at that time. From today's point of view, however, some shortcomings emerge. These include:

1. Most of his work was based on tests performed in the early 1950s with fully coupled explosions.
2. Most of these tests were performed in sand, dry and wet clay, or sandstone.
3. In the majority of the tests only small quantities of explosives, in the range of several hundred pounds were used.
4. Due to the type of ground at the test sites, in many of the tests only dustfall, and no real debris, was produced and recorded.

In contrast, today's underground installations are generally built in competent rock, the loading density would be low compared to a fully coupled explosion, and the amount of explosives to be stored would be in the range of tens to a few hundred tons per chamber. Therefore, questions have arisen concerning the scaling of the data from these tests and experiments with small quantities of explosives.

During the last decade, different parties have tried to enhance the tools available for predicting explosions. Substantial amounts of money have been spent to obtain additional data from full- and small-scale tests (such as the China Lake test [19]) or to perform theo-

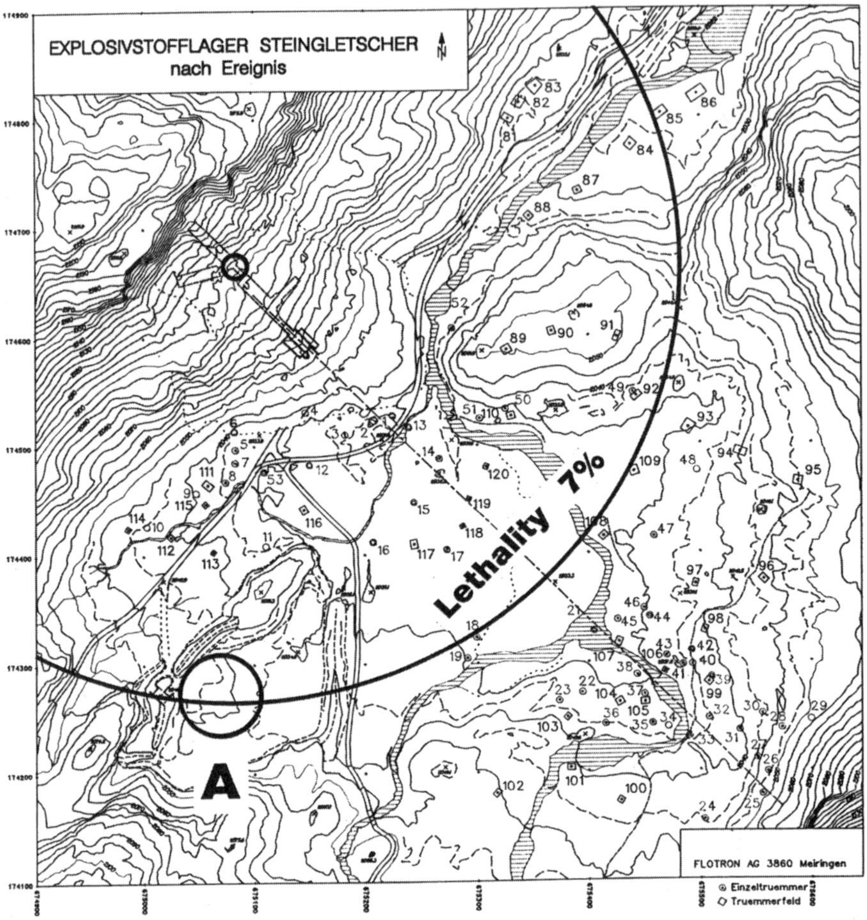

FIGURE 26.20 Lethality according to Swiss criteria.

retical investigations. Furthermore, accidental explosions in real installations, like the Steingletscher explosion, were investigated and the explosion effects were documented in detail to serve as a basis for enhancing the current knowledge.

This is why NATO AC/258 (Group of Experts on the Safety Aspects of Transportation and Storage of Military Ammunition and Explosives) decided to take advantage of this new knowledge and reinvestigate some of the explosion effect models for future inclusion in their safety manual [12] as well as in a proposed NATO risk analysis manual.

The development of the model, the basic data used to create it, and the model itself and its limitations are discussed in the following sections.

26.3.2 General Procedure

Modeling of debris throw from craters produced by explosions in underground storage installations is one of the most complex problems to be dealt with when analyzing the effects

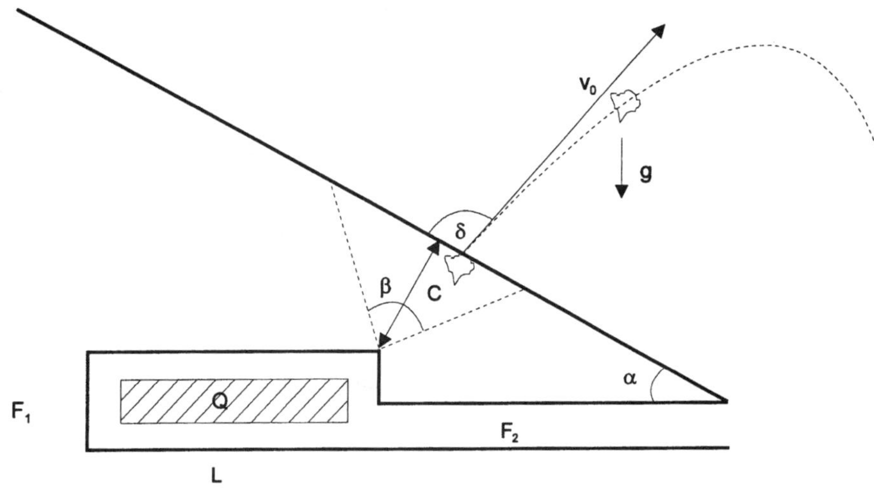

$V_c = F_1 * L$
$\gamma = Q / V_c$

Q = weight of explosives
L = length of chamber
F_1 = cross-section area of chamber
V_C = volume of chamber
γ = loading density
C = overburden, cover thickness
F_2 = cross-section area of adit tunnel
α = slope angle of overburden
ρ = density of overburden
c = seismic velocity
σ = tensile strength (dynamic) of overburden
S = shear strength (dynamic) of overburden
- water content of overburden
- cracks/joints in overburden

A = airdrag parameter
D = debris size parameter
v_0 = launch velocity of debris
g = gravity

(Listing not complete)

FIGURE 26.21 Parameters influencing debris throw from craters.

of such events. One main reason for this is that many parameters, such as the thickness of the overburden above the storage chamber, the quality of the rock surrounding the chamber, the weight of the explosives reacting, the slope angle of the overburden, and other factors may strongly influence the debris throw-out process (a more extensive list of parameters is given in Fig. 26.21). Another reason is that despite the new data mentioned above, the database for creating a prediction model taking into account all these parameters is still quite thin compared to the available data on other explosion effects, such as air blast, which means the problem cannot be treated on a statistical basis only.

Initially it was intended to tackle the problem along an analytical approach, taking into account as many parameters as possible. Due to the facts mentioned above, however, the need to limit the number of parameters to be used in a new prediction formula to the most important ones soon became clear and inevitable. Based on expert discussions within NATO AC/258 and other organizations, and taking into account real facts as well as engineering judgment, it was decided to include the following parameters in the new prediction model:

Q = effective weight of explosives involved in the explosion

γ = chamber loading density (kg of explosives per m^3 of chamber volume)

C = overburden/cover thickness (distance, representing a mass)

α = angle of overburden surface slope

Discussions and data showed that in addition to these factors, the rock quality (which influences the size of the crater) and the venting characteristics of the storage chamber (which influence the total energy available for the throw-out of the crater material) play some role. For several reasons, however, it was decided to develop a model valid only for strong (good) rock and unvented chambers as a first step. Since venting of the chamber reduces the pressure driving the debris throw, this is a conservative assumption.

For the analytical approach, the following procedure was proposed:

1. Define the launch velocity of the debris.
2. Calculate the trajectory of the debris.
3. Relate the calculated ejecta range to a debris density.

The launch velocity of the debris was established as a function of the scaled cover depth and the chamber loading density, based on reports about well documented experiments [19–23]. For the calculation of the trajectory of the debris, based on the launch velocity and the ejection angle, the code TRAJ [24–28], a two-dimensional trajectory program for personal computers, originally developed by the U.S. Naval Surface Warfare Center, was used. However, the results from the combination of these two steps did not correspond satisfactorily with observed values.

One explanation for this might be that the observed maximum debris launch velocities documented in many reports are not really the maximum velocities but only the maximum velocities of the debris throw front, and that part of the debris behind the front have a higher velocity due to a longer acceleration by the escaping gases of the explosion. A second weak point of the trajectory calculation approach is that it does not take into account the total amount of material that is thrown out of a crater. A simple example calculation using this two-part procedure shows this important fact.

Installation 1
$Q = 1$ t
$C = 3$ m
$\gamma = 50$ kg/m^3

Installation 2
$Q = 100$ t
$C = 14$ m
$\gamma = 50$ kg/m^3

Both installations have the same scaled cover depth of around 0.3 m/kg$^{1/3}$. This leads to the same maximum debris launch velocity of approximately 60 m/s and therefore, at the end, to the same ejecta range of approximately 150 m [19].

It can easily be understood that despite the debris throw distance being the same for both explosions, the debris density is not the same, as the total amount of debris displaced to the surroundings is much greater for the 100 t explosion.

This led to the conclusion that for establishing a relationship between a debris range and a debris density, an additional parameter would have to be introduced, taking into account the total mass of debris displaced. Furthermore, it was recognized that extensive calculations based on simulation techniques such as, Monte-Carlo simulations, would have been necessary to establish a proper debris density distribution.

Although this approach would have generally been preferred, it seemed impossible to come up with a sound model within the given time and financial limits. Therefore, an experiment-based relationship was finally developed directly from the available data. Simply stated, observed debris mass density distributions from accidents and full-scale tests were taken and correlated with the four main parameters: explosives weight, scaled cover depth, chamber loading density, and slope angle of the overburden.

26.3.3 Relevant Basic Data

Tests. Until now, only a few tests have been performed concerning debris throw from crater-forming mechanisms, some of them being only small-scale tests. Small-scale tests for debris throw are subject to problems of scaling laws (e.g., gravity effects) and the difficulty of simulating real rock material with its joints, cracks, and fissures.

For developing the new model, the data from the following trials were used:

- *China Lake test (1988):* So far, this has been the only large-scale scale test (actual scale 1:2) specially designed to study debris throw as one of the explosion effects of an underground magazine explosion. The data of this test are well documented in Refs. 19 and 29. Slope angle correction factors (see below, The Overburden Slope Angle Parameter) were applied to adjust the actual ranges to equivalent values for flat ground. The resulting data used can be found in Fig. 26.22.
- *Raufoss trials (1968):* A series of tests (scale approximately 1:3 to 1:4), the largest with 5400 kg of explosives producing a crater, was performed in Norway in 1968. Only a summary report with very sparse information concerning crater debris throw was available for this study. The few data reported are documented in Ref. 30. The crater area lay in a wooded area where it would have been difficult to collect all the debris, especially as the ground was covered with snow at that time. Furthermore, a substantial amount of debris probably would have been stopped by the trees.
- *Buckboard series, Underground Explosion Test Program (UETP) (1960, 1948):* These tests were primarily designed as cratering experiments. However, during some of these tests the debris density distribution was also measured, and some of the results have already been used for the existing crater debris formula in AASTP-1. For the current study, the debris density distributions documented in Ref. 31 of the Buckboard Tests No. 11/12/13 and UETP Tests No. 814/815/817 were used as reference values for fully coupled explosions.

Data from small-scale tests, as described in, for example, Refs. 22 and 32 to 37, were not used in this study mainly due to scaling problems. Furthermore in some of the tests it was difficult to differentiate between the debris coming from the adit and the debris coming from the crater, and sometimes only the debris launch velocity was measured.

Accidents. In general, data from accidents are more difficult to use for these types of analysis because often not all relevant parameters are known after the event, whether the total amount of explosives involved, the exact location of the stored explosives, or other important data. Nevertheless, some cases were fairly well documented and could be used as additional data points for developing the prediction model. Other, less documented accidents can at least be used for testing the new model.

Pad Nr.	Range [m]	Azimuth [deg]	Number [p.]	Density [p./m²]	Weight [kg]	Density [kg/m²]	Factor f_α	Range corr.
1	344	316	1	0.06	0.86	0.05	1.1	313
2	301	320	3	0.19	3.76	0.24	1.1	274
3	261	327	6	0.38	2.14	0.13	1.2	218
4	225	335	24	1.50	18.00	1.13	1.2	188
5	195	346	43	2.69	68.15	4.26	1.3	150
7	167	17	33	2.06	16.40	1.03	1.3	128
8	174	33	20	1.25	8.09	0.51	1.2	145
9	194	48	59	3.69	96.41	6.03	1.1	176
10	224	59	19	1.19	13.15	0.82	1	224
11	260	67	1	0.06	0.86	0.05	1	260

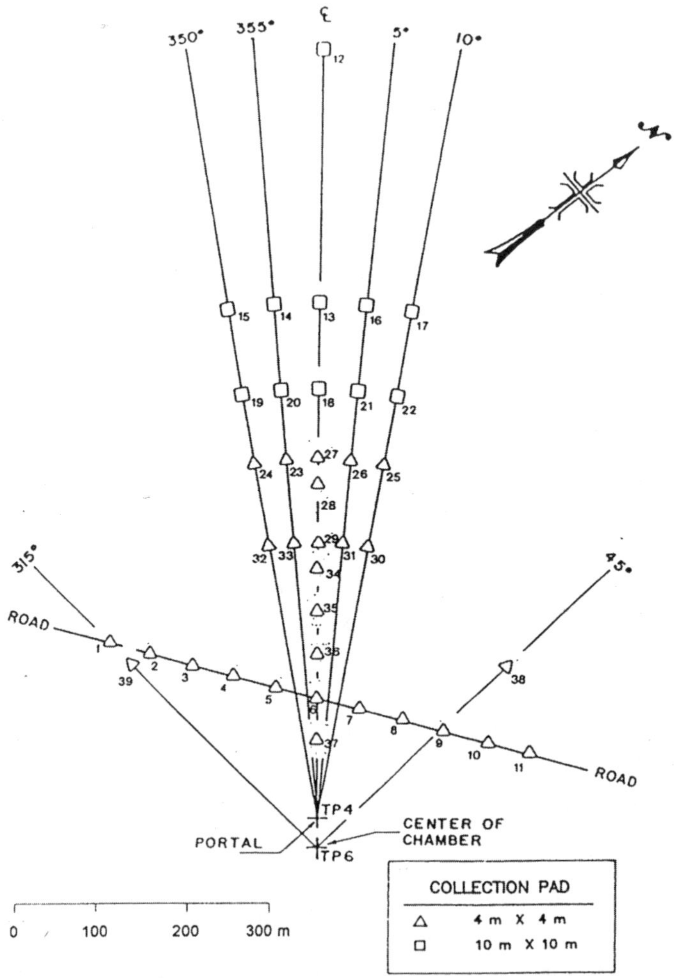

FIGURE 26.22 Debris data from China Lake test.

- *Susten/Steingletscher (1992):* One of the most recent explosions in an underground explosives storage installation in rock, discussed in detail in Section 2, took place in the Swiss Alps in November 1992. The stored types and amounts of explosives and ammunition on the day of the explosion and the installation itself are fairly well known, and a quite extensive investigation of the debris thrown to the surroundings was performed after the event (see Section 2). The results of this work are documented in detail in Refs. 5, 7 to 10, and 37.

  Because it was recognized that the slope angle of the overburden plays an important role for throw distances, the results [10, 37] concerning this aspect were reevaluated. The model for slope angle correction factors [38] was used. The result can be found in Table 26.1.

  One of the questions that could not be answered until today is the distribution of the explosives inside the storage installation. This is why there is some uncertainty about the effective overhead cover and the effective loading density needed for comparisons. For this study it was therefore decided to use upper and lower bound values for both parameters.
- *Fauld (1944):* This is one of the biggest accidents that has ever occurred in an underground explosives storage installation, involving up to 2,000 tons of explosives (mostly in bombs) in a single explosion. The event, including the damage to the surroundings, is also quite well documented [39–41]. The basic debris data used for this study can be found in Fig. 26.23. However, the storage installation itself (an old gypsum mine) and the rock material

TABLE 26.1 Debris Data from Susten/Steingletscher Accident

Pad number	Range (real) (m)	Azimuth (deg)	Debris density (kg/m^2)	Debris density (p./m^2)	Factor f_α	Range (corr.) (m)
81	279	287	2.45	2.5	1	279
82	294	285	1	1.0	1	294
83	315	284	0.38	0.4	1	315
84	370	297	1.03	1.0	1.1	336
85	405	295	0.23	0.2	1.1	368
86	442	294	0.06	0.1	1.1	402
87	314	302	0.91	0.9	1.1	285
88	266	305	2.39	2.4	1.1	242
89	253	331	2.17	2.2	1.4	181
90	288	325	3.69	3.7	1.3	222
91	347	325	1.32	1.3	1.3	267
92	378	333	1.23	1.2	1.4	270
93	431	334	1.15	1.2	1.4	308
94	483	335	0.76	0.8	1.4	345
95	541	336	0.77	0.8	1.4	386
102	536	19	6.72	6.7	1.4	383
109	403	342	1.38	1.4	1.4	288
110	281	343	3.38	3.4	1.4	201
111	201	50	4.26	4.3	1.2	168
112	254	57	3.66	3.7	1.1	231
113	261	49	3.54	3.5	1.2	218
114	259	67	2.06	2.1	1.1	235
115	218	50	8.3	8.3	1.2	182
116	230	27	3.66	3.7	1.4	164

For pad location see Fig. 26.10.

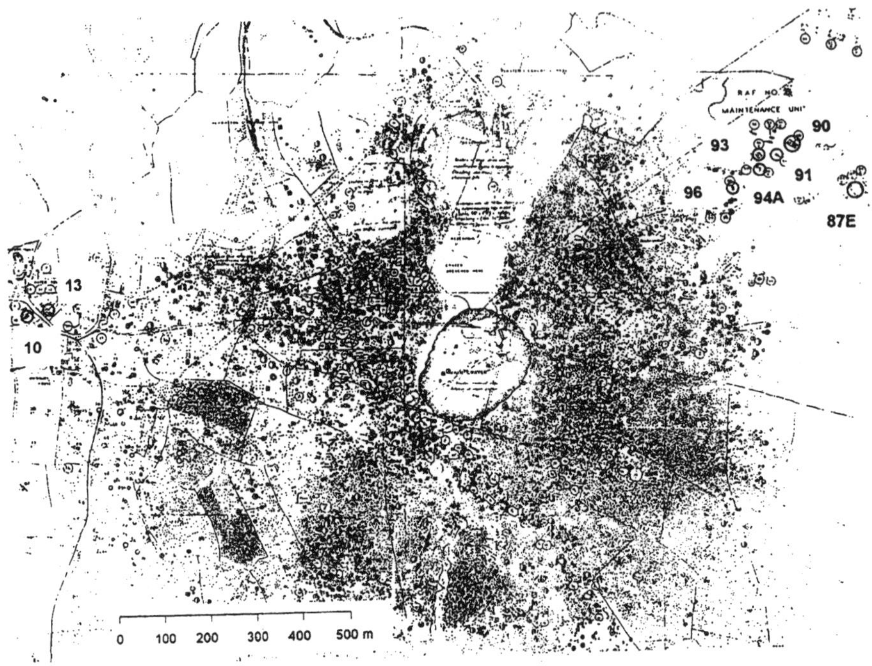

House number	Range (m)	Number (p.)	Corrected number (p.)	Area of house (m^2)	Debris density ($p./m^2$) Northeast	West
10	1050	10	15	97.5		0.15
13	960	8	12	72		0.17
87E	945	6	9	300	0.03	
90	870	6	9	225	0.04	
91	810	4	6	24	0.25	
93	780	12	18	150	0.12	
94A	750	6	9	27	0.33	
96	690	4	6	40	0.15	

FIGURE 26.23 Debris data from Fauld accident.

of the overburden were not very typical for an ammunition storage. Especially concerning the effective loading density several different estimations exist, ranging from 15 to 173 kg/m^3. For this study, the effective loading density was estimated to be around 50 kg/m^3, taking into account that less than the total volume of the widely branched tunnel system was effective as an expansion volume for the explosion gases within the appropriate time.

- *Uusikylä (1965):* This accident, also involving up to several hundred tons of munitions, is not documented very well, at least not regarding the debris throw to the surroundings. For the available information, see Refs. 42 and 43.

- *Other accidents:* For other accidents in underground storage installations, such as Prüm, Germany (1949) [44], Mitholz, Switzerland (1947) [45], and Waikalua, Hawaii (1946), either no crater or no typical crater was produced or the available information about crater debris throw was not sufficient to be used for developing a debris throw model.

Data Used for Calibration. In general, the most emphasis was laid on the China Lake test data because this test was specially designed for studying the effects of debris throw and the geometry of the installation was typical for an underground ammunition storage.

For the development of debris density versus range distributions, the data from the China Lake test and the Steingletscher accident were primarily used.

For establishing the decoupling factor for loading densities less than 1,600 kg of explosives per m^3, the data from the fully coupled Buckboard and UETP tests were compared with the data from the decoupled China Lake test.

26.3.4 The Model and Its Parameters

The General Model. Based on the main factors influencing debris throw from craters, the following empirical formula was proposed for calculating the distance for a given debris density D:

$$R = k \cdot f_q \cdot f_c \cdot f_\gamma \cdot f_\alpha$$

More generally, a debris density at a given location can be calculated using the following parameters:

$$D = f(R, f_q, f_c, f_\gamma, f_\alpha)$$

where D = debris density (pieces/m^2)
k = constant = $f(D)$ (—)
R = distance from center of crater (m)
f_q = explosives weight parameter (—)
f_c = cover depth parameter (—)
f_γ = loading density parameter (—)
f_α = overburden slope angle parameter (—)

The Explosives Weight Parameter. Based on common knowledge about scaling of explosion effects, it seemed obvious that the nondimensional explosives weight parameter f_q is proportional to the cube root of the explosives weight. The explosives weight in this formula is defined as the total amount of explosives reacting within a short time, expressed in an equivalent mass of the explosive TNT. This quantity is often also described as NEQ (net explosives quantity).

$$f_q = f(Q)$$

$$f_q = Q^{1/3} \qquad (\text{—})$$

where Q = weight of explosives (kg).

The Cover Depth Parameter. The total mass of rock material thrown into the surroundings is unquestionably a direct function of the crater volume produced by the explosion. The crater dimensions (depth, radius, volume) depend primarily on the depth and to some extent on the rock quality.

In order to take into account the crater volume for different scaled cover depths, a nondimensional parameter f_c as a function of the cover depth (C) was elaborated based on the information in Fig. 67 of Ref. 19, Ref. 31, and 46 to 51. This parameter is shown in Fig. 26.24.

$$f_c = f(C, Q)$$

$$f_c = 0.45 + 2.15 \cdot \left[\frac{C}{Q^{1/3}}\right] - 2.11 \cdot \left[\frac{C}{Q^{1/3}}\right]^2 \qquad (\text{—})$$

where C = cover depth (shortest distance between chamber and rock surface) (m)
Q = weight of explosives (kg)

The function f_c has its maximum (1.0) at the optimum depth of burst, i.e., where the crater volume reaches its maximum. In case of hard and moderately strong rock, this maximum occurs at a scaled cover depth of around 0.5 to 0.6 (m/kg$^{1/3}$).

For hard and moderately strong rock, it is assumed, based on the above-mentioned references, that for a scaled cover depth above 1.2 no cratering occurs that produces relevant debris throw into the surroundings. This does not imply, however, that there will be no explosion effects at all at the surface. Loose rock may be displaced, and spalling and a

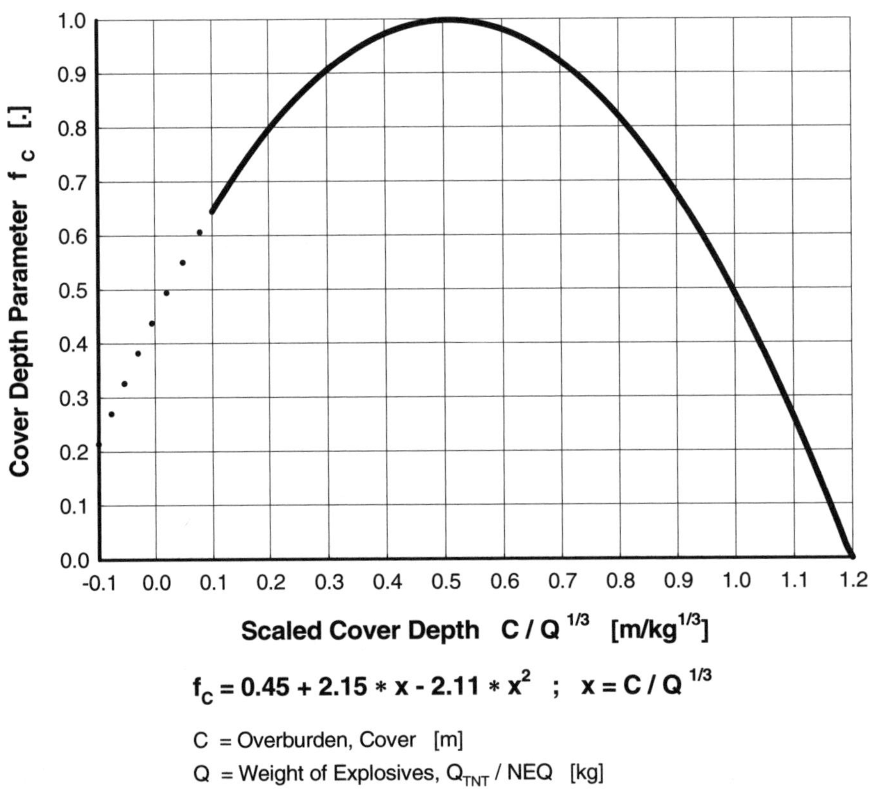

FIGURE 26.24 Cover depth parameter f_c.

certain mounding of rock rubble may still occur beyond this value, in addition to the ground shock effect.

The value for above-ground explosions, where the scaled cover depth is 0 and the cover depth parameter f_c is around 0.45, was derived from Refs. 46 to 48.

The Loading Density Parameter. The influence of the chamber loading density γ on the debris throw distance was taken into account by means of a nondimensional decoupling factor called loading density parameter f_γ (Fig. 26.25).

$$f_\gamma = f\,(Q, V_c)$$

$$f_\gamma = \left[\frac{(Q/V_c)}{1{,}600}\right]^{0.35} \qquad (—)$$

where Q = weight of explosives (kg)
V_c = chamber volume (m^3)

As mentioned above under Data Used for Calibration, this function was mainly derived by

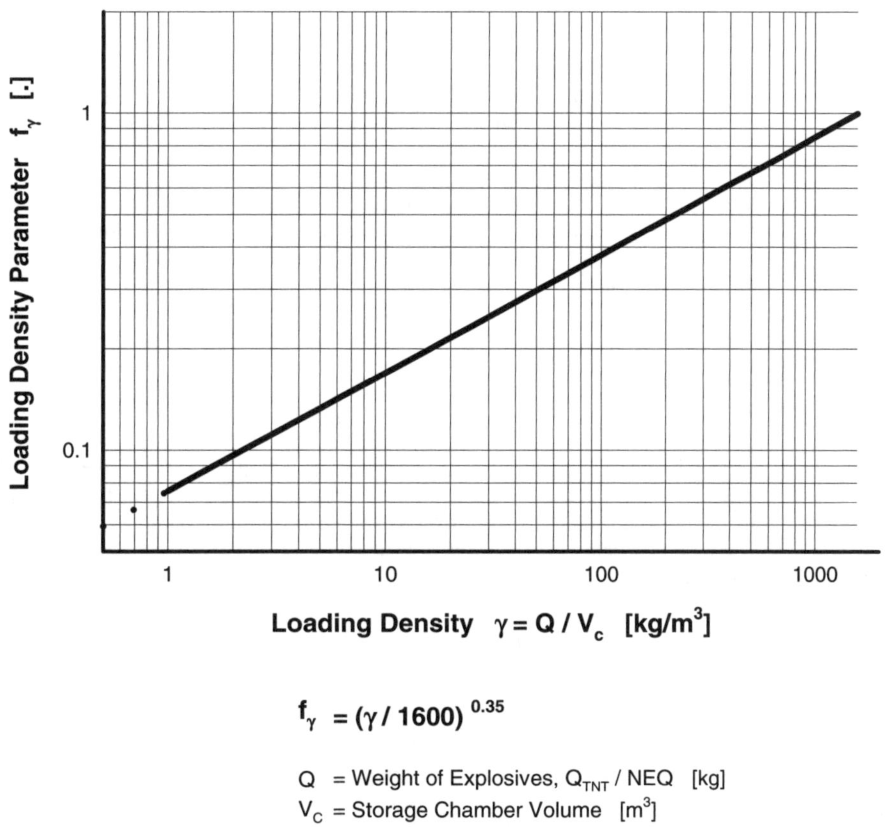

FIGURE 26.25 Loading density parameter f_γ.

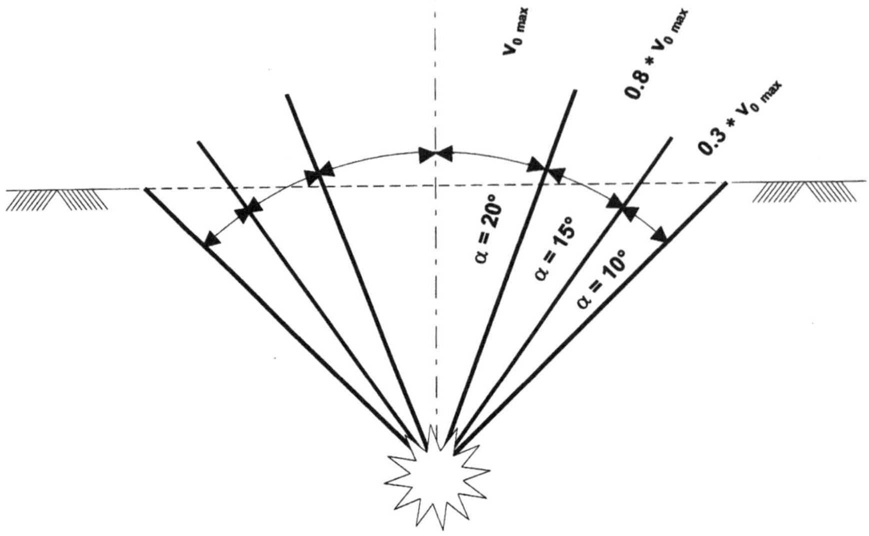

FIGURE 26.26 Debris launch velocity model for TRAJ calculations.

comparing the data from fully coupled explosions [31] with the data from the China Lake test [19, 29] (see also below, Debris Distribution versus Range).

The Overburden Slope Angle Parameter. In the current regulations, the influence of the slope angle of the overburden, which is a major factor, is treated in a binary way, which means, for slope angles below 45°, no increase of the hazardous distance is necessary. For slope angles above 45°, however, an increase factor of 1.5 has to be applied. It is thought that such a function in reality produces inconsistencies that can hardly be explained [38, 52].

Based on a set of calculations with the TRAJ computer model [24] for different overburden slope angles, the influence of a varying angle was studied in detail [38]. For the calculation of the throw distances, it was assumed that debris coming from the center of the crater has the highest launch velocity while the velocity decreases towards the edge of the crater. The model used for the calculations is shown in Fig. 26.26. The result of all calculations for different launch velocities and debris with a different mass is shown in Fig. 26.27, indicating an increase in the throw distance (compared to flat terrain) in the direction of the slope and a decrease of the throw distance backwards.

Based on these calculations, a simplified model for the non-dimensional overburden slope angle parameter f_α was developed (Fig. 26.28).

$$f_\alpha = f(\alpha)$$

Slope angle increase factor: $f_{\alpha I} = 1 + 0.02 \cdot \alpha, \qquad f_{\alpha I\max} = 1.5\ (—)$

Slope angle decrease factor: $f_{\alpha D} = 1 - 0.025 \cdot \alpha, \qquad f_{\alpha D\min} = 0.25\ (—)$

where α = slope angle of overburden (°)

This model shows constant values for slope angles above 25° respectively 30°. Although the calculation showed that for steeper slope angles than 45° the increase factor $f_{\alpha I}$ could be reduced slightly from its peak value of 1.5, it was decided to leave it on this level in order

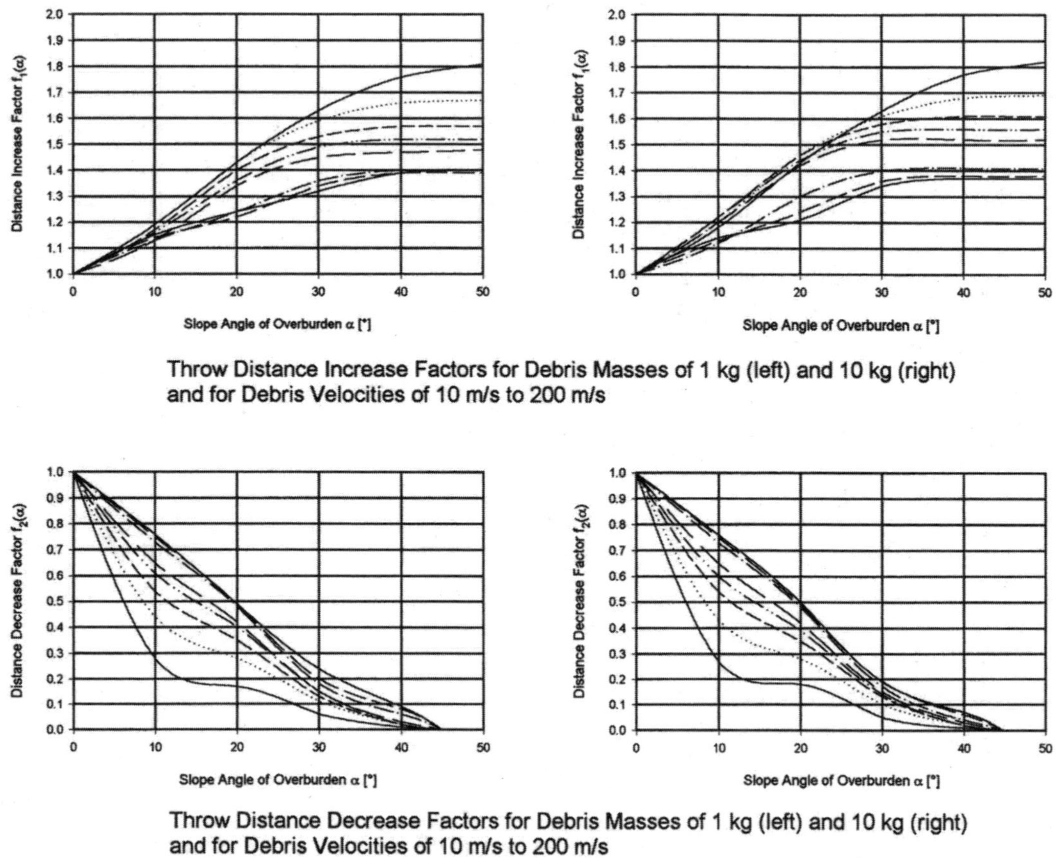

FIGURE 26.27 Debris throw increase and decrease factors. (Top line = 10 m/s; bottom line = 200 m/s)

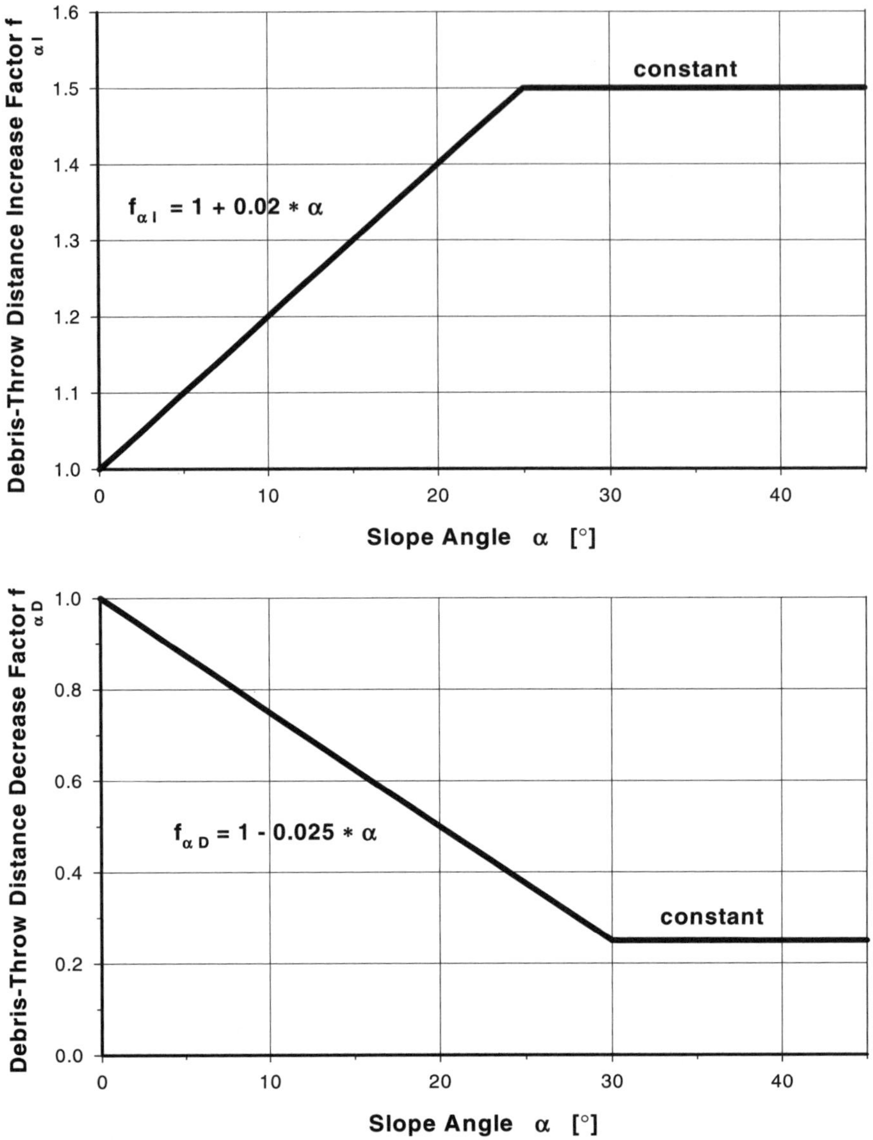

FIGURE 26.28 Overburden slope angle parameter f_α.

to take into account other effects, such as landslides from the overburden that might occur and influence the debris throw process in case of steep slopes. The same reason applies for not further reducing the decrease factor $f_{\alpha D}$ for slope angles above 30°.

Debris Distribution versus Range. As mentioned in Section 26.1, it was the aim to develop not a model that calculates only quantity distances, but a formula that predicts densities of hazardous debris over a wider range. Therefore, the debris to be considered as hazardous had to be defined and the relationship between debris areal density (number of debris pieces/m^2) and debris mass density (mass of debris/m^2) had to be established.

As explained in Section 26.2, hazardous debris includes all debris pieces that impact with an energy >79 J. The final (impact) velocity of debris with different masses was calculated with the computer code TRAJ. Taking these velocities into account, it can be concluded that all pieces of crater debris from explosions in such underground installations with a mass heavier than 150 g are hazardous (see also Fig. 26.19).

For most of the tests and accidents, raw data on debris density are presented as debris mass density in kg per m^2. For the China Lake test and the Steingletscher accident, an in-depth investigation of this relationship was performed. The surveyed data showed that for 1 kg of rock mass, 0.8 to 1.4 pieces of hazardous debris result (see also Fig. 26.17). For the development of the new model, an average value of one piece of hazardous debris per one kg of rock mass was finally assumed.

As already mentioned under Data Used for Calibration, the debris density distribution versus range from the center of the crater was mainly derived from the China Lake and Steingletscher data. To a lesser degree, data from the Buckboard and UETP tests and the Fauld accident were used to calibrate the data.

The following procedure was generally used to derive the debris density distribution:

1. The China Lake and Steingletscher raw data were taken and reduced to a flat terrain with the help of the overburden slope angle parameter f_α. For data points off-axis, reduced values for f_α were used. For the data from the Fauld accident (slope angle ~0°) this step was not necessary. When the raw data were available as debris mass densities, they first had to be transformed to debris areal densities (number of debris pieces per m^2).
2. The reduced data were then plotted against the range in a log-log diagram (Fig. 26.29).
3. In a third step, an initial debris density versus range relationship for these data was developed, based on assumptions concerning the general shape of such a curve. As can easily be seen, there were not enough data points from the Fauld accident to develop a specific curve. Thus, a curve with the same shape as that generated for the China Lake and Steingletscher data was used for the Fauld data points.
4. The cover depth parameter f_c, the loading density parameter f_γ, and the slope angle parameter f_α were then applied and the curve was scaled down to an explosives weight of 1 kg. The final result, presented in Fig. 26.30, shows the derived standardized debris density versus range curve.

From Fig. 26.30, the debris areal densities (pieces per m^2) can be calculated for any range (R), according to the following formula:

$$D = f(R^*)$$

$$D = \left[-1.31 + \frac{56.1}{R^*}\right]^2, \quad R^* < 42 \qquad (\text{pieces}/\text{m}^2)$$

where R^* = scaled range (m)

$$R^* = \frac{R}{f_q \cdot f_c \cdot f_\gamma \cdot f_\alpha} = \frac{R}{Q^{1/3} \cdot f_c \cdot f_\gamma \cdot f_\alpha} \quad (\text{m})$$

Figure 26.30 also shows a comparison of the new debris density versus range function with the data derived from the fully coupled explosion from the Buckboard and UETP test series [31]. There the curves are presented as straight lines in the log-log diagram; however, no upper or lower limits of validity are given.

As can be seen, the different models correspond quite well over a wide range, especially for the tests in hard rock. Unfortunately, this is not true for very low debris densities. Due

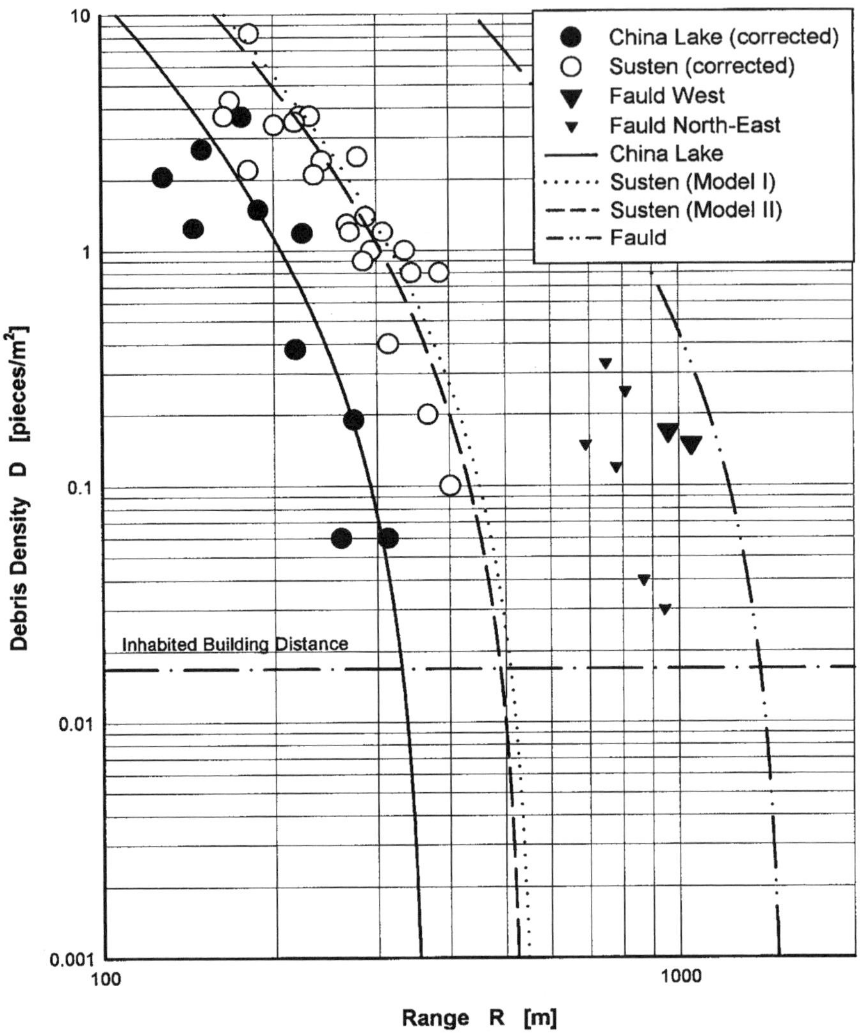

General form of curves: $D = \left(a + \frac{b}{R}\right)^2$

FIGURE 26.29 Debris density versus range relationship.

to ballistic throw limits, the debris density decreases more rapidly in this region. This fact can be clearly shown for the China Lake test and the Steingletscher event. But debris density versus range functions of other tests like ESKIMO 1 and Hastings [38, 53] also show a similar behavior. For the Buckboard and UETP test series, it is suspected that no real data were available for the low-density region.

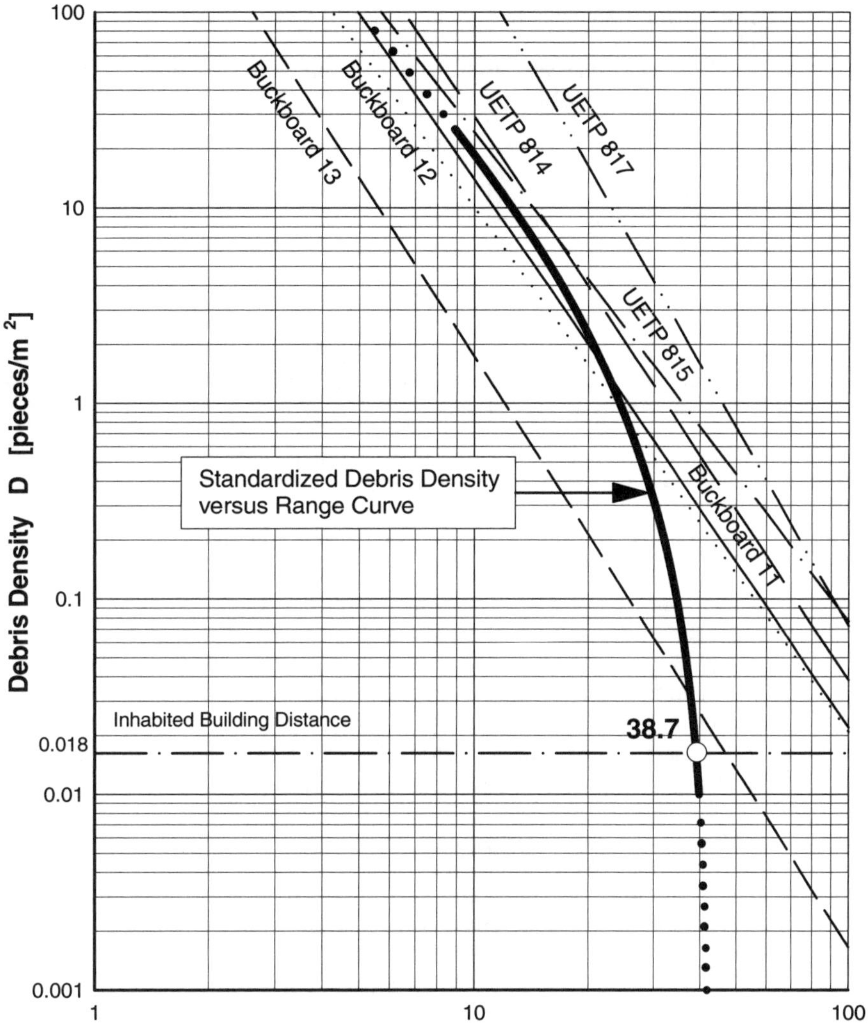

$$\text{Scaled Range } R^* = \frac{R}{Q^{1/3} * f_C * f_\gamma * f_\alpha} \quad [\text{m}]$$

$$D = \left(-1.31 + \frac{56.1}{R^*}\right)^2;\ (R^*<42)$$

FIGURE 26.30 Standardized debris density versus range curve.

26.3.5 The New Prediction Formula, Applications and Limitations

Inhabited Building Distances. For the time being, the NATO safety criteria concerning debris is still one piece of hazardous debris per 56 m^2 or 0.018 per m^2. Based on this value, the respective scaled distance can be derived from Fig. 26.30. The graph and the formula respectively show that the Inhabited Building Distance (IBD) corresponds with a scaled range of 38.7.

With this value, the IBD can be calculated for installations with a wide range of loading densities, cover depth, and explosives weight applying the following formula:

$$\text{IBD} = 38.7 \cdot f_q \cdot f_\gamma \cdot f_c \cdot f_\alpha \qquad \text{(m)}$$

Of course, with the function in Fig. 26.30, it is easily possible also to establish debris areal densities for other ranges. This is particularly helpful for risk analyses for such installations because they are already performed on a regular basis in Switzerland and some other countries.

As discussed earlier, the NATO also intends to introduce a risk analysis concept in the near future. Therefore, this model also serves as an advance investment in this direction.

Applying the Slope Angle Parameter. In applying the slope angle parameter, the following rules have to be observed:

- The inhabited building distance (IBD), or any other distance has to be measured as a horizontal distance from the crater center at the bottom of the crater (at the level of the installation).
- The slope angle α shall be established in the area where the crater center at the surface has to be expected.
- An average value for the slope angle α over the whole crater area shall be taken into account in case the surface is not plain in this area.
- The increase ($f_{\alpha I}$) as well as the decrease ($f_{\alpha D}$) factor have to be applied to the IBD (or any other distance) in the direction of the line with the largest gradient intersecting the center of the crater. This line does not necessarily coincide with the axis of the adit tunnel.
- No increase or decrease factor has to be applied to the sides of the crater.
- The IBD contour shall be elliptical in shape.

Comparison with Accident and Test Data. To show the accuracy and applicability of the new approach, data from accidents and tests were compared with the new and existing NATO debris throw model concerning hazardous distances in Table 26.3. Table 26.2 shows the main data of the installations used for the comparison. The new model seems to be quite accurate and fits the real data much better than the current one in AASTP-1.

Figure 26.31 shows a final comparison with the data and the preliminary model of the Steingletscher installation. As can be seen, the new model also fits the observed debris

TABLE 26.2 Main Data for Installations

Installation	Charge Q_{TNT}/NEQ (kg)	Scaled charge $Q^{1/3}$ (kg$^{1/3}$)	Loading density γ (kg/m^3)	Rock overburden C (m)	Scaled rock overburden (m/kg$^{1/3}$)	Slope of overburden α (°)
China Lake	22,000	28.0	68.3	8.7	0.31	20
Susten (Model I)	225,000	60.8	44.8	52.0	0.85	30
Susten (Model II)	225,000	60.8	25.0	17.0	0.28	45
Uusikylä	200,000	58.5	65.0	15.0	0.26	20
Fauld	2,000,000	126.0	50.0	27.5	0.22	0
Raufoss	5,400	17.5	90.0	13.0	0.74	~0

TABLE 26.3 Comparison of Observed Ranges from Tests and Accidents with the Current and the New Model

Installation	Observed range corrected to flat terrain (ORF)	Slope angle increase factor $f_{\alpha I}$	Weak rock correction factor	Observed range front OR = ORF * $f_{\alpha I}$	Observed range side OR = ORF	Current NATO model AASTP-1 (IBD)	Percentage of observed range IBD/OR	Proposed new NATO model (IBD_{new})	Percentage of observed range IBD_{new}/OR
China Lake	310	1.40	—	435	—	244	0.56	458	1.05
China Lake	310	1.00	—	—	310	244	0.79	327	1.05
Susten (Model I)	470	1.50	—	705	—	258	0.37	757	1.07
Susten (Model I)	470	1.00	—	—	470	258	0.55	505	1.07
Susten (Model II)	470	1.50	—	705	—	785	1.11	733	1.04
Susten (Model II)	470	1.00	—	—	470	523	1.11	488	1.04
Uusikylä	560	1.40	—	785	—	571	0.73	888	1.13
Uusikylä	560	1.00	—	—	560	571	1.02	635	1.13
Fauld	1,130	1.00	1.15	1,300	1,300	1,407	1.08	1,367	1.05
Raufoss	80	1.00	—	80	80	85	1.07	215	2.69[a]

[a] See comments above under Tests.

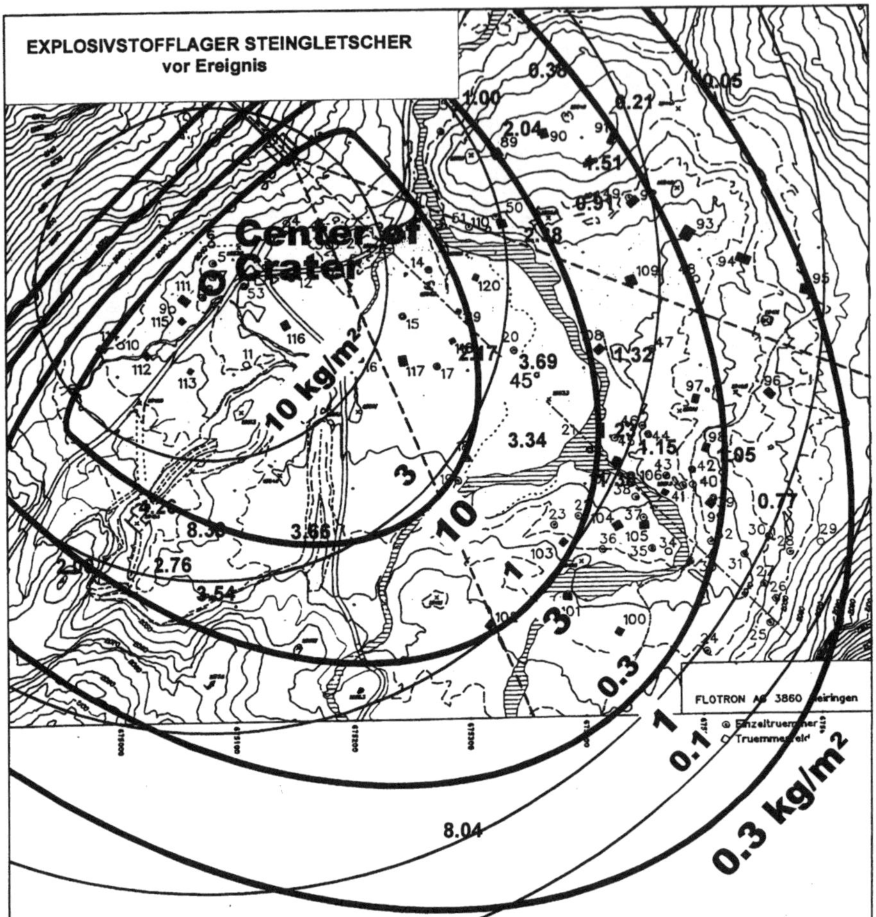

FIGURE 26.31 Comparison of debris mass density contour lines from the new model (bold) with the current model (thin, see Fig. 26.15).

densities much better than the preliminary model, which did not take into account the slope angle parameter.

Limitations. The model developed is based on data from a comparatively small number of tests and accidents. Furthermore, it is mainly based on an integral/empirical evaluation of the available data. The overall accuracy is therefore limited to the range of the investigated cases.

Thus, the crater debris throw model presented here may be used within the following boundaries only:

- Weight of explosives $Q = 1$ t–2000 t
- Chamber loading density $\gamma = 1$ kg/m³–300 kg/m³
- Scaled cover depth $C/Q^{1/3} \geq 0.1$

If parameters exceed these ranges, special care should be taken when applying the model.

Furthermore, the model is valid only for hard and moderately strong rock. For installations built into weak rock, distances should be increased by approximately 15%. As a conservative assumption, chambers/installations are assumed to be unvented.

The debris throw model presented here shall not be used to calculate maximum missile or debris ranges.

26.3.6 Conclusions

The newly developed crater debris throw model presented here was reviewed by the NATO AC/258 experts as well as by U.S. Waterways Experiment Station (WES), Vicksburg, Mississippi, and Swiss experts in the explosives safety field. It will be included in the revised NATO AASTP-1 safety manual as well as in the Swiss regulation for the storage of ammunition, TLM 75 [4, 11].

The development of the model illustrated the following points:

- It is important to learn as much as possible from accidents because they are often the only way of obtaining data that can be used to develop more reliable models for predicting the effects of explosions. This is especially true for effects for which simulation tests cannot be performed with reasonable cost and for which computer simulation tools either do not exist or are not reliable. Whenever possible, technical experts should insist on an extensive documentation of the real hard facts in any case where debris is produced by an explosion accident.
- Even when only limited data are available, it is often possible to develop suitable models for most common cases by combining the available data with sound engineering judgment and expert experience.
- When only limited data and knowledge are available, sound results can often only be achieved by international cooperation, such as within NATO AC/258.
- This article once more shows the importance of a free data exchange to enhance technical knowledge in fields important for safety.

26.4 REFERENCES

1. **How the Safety of the Ammunition and Explosives Storage and Handling is Managed in Switzerland**
Part I
Safety Concept, Regulations and Organisation
Bienz, Kummer & Partner AG
Bienz, Andreas F.; Kummer, Peter O.
Presented at the DDESB Explosives Safety Seminar 1992, 18.08.1992
2. **How the Safety of the Ammunition and Explosives Storage and Handling is Managed in Switzerland**
Part II
Risk Analysis of Ammunition Magazines
Bienz, Kummer & Partner AG
Kummer, Peter O.; Bienz, Andreas F.
Presented at the DDESB Explosives Safety Seminar 1992, 18.08.1992
3. **Debris Throw from Craters**
Proposed Changes to the NATO Safety Manual AASTP-1, Part III
Technical Background
Contribution to NATO–AC/258 Underground Storage Ad hoc Working Party

Defence Procurement Agency/CH
Bienz, Kummer & Partner AG
Kummer, P.; Nussbaumer, P.
TM 158-7/CH(ST)UGS/AHWP IWP 005, 30.10.1998

4. **Technische Vorschriften für die Lagerung von Munition (TLM 75)**
Teil 2 (Rev 90): Sicherheitsbeurteilung von Munitionslagern
Schweizerische Armee
01.07.1992, (CLASSIFIED)
Technical Regulation for the Storage of Ammunition (TLM 75)
Part 2 (Rev 90): Safety Assessment for Ammunition Magazines

5. **Scientific-technical Evaluation of the Explosion Event Susten/Steingletscher of November 2, 1992**
Description and Geometry of the Installation
Document Nr. 1
Swiss Committee for the Safety of the Handling of Ammunition and Explosives by Forces and Administration
Defence Technology and Procurement Agency
Bienz, Kummer & Partner
Kummer, Peter
TM 111-12, 01.09.1995

6. **Accident Investigation at the Swiss Steingletscher Installation**
Defence Technology and Procurement Agency, Switzerland
Bienz, Kummer & Partner AG, Switzerland
Air Force Operability Systems Management Office, Eglin Air Force Base/FL
Bakhtar Associates, Newport Beach/CA
Rytz, Hansjörg E.; Kummer, Peter O.; Jenus, Joseph Jr.; Bakhtar, Khosrow
Presented at the DDESB Explosives Safety Seminar 1994, 01.08.1994

7. **The Explosion in the Underground Explosives Storage Magazine in the Swiss Alps on November 2, 1992**
Swiss Committee for the Safety of the Handling of Ammunition and Explosives by Forces and Administration
Defence Technology and Procurement Agency
Bienz, Kummer & Partner
Kummer, P.
TM 111-7, 20.10.1993

8. **Scientific-technical Evaluation of the Explosion Event Susten/Steingletscher of November 2, 1992**
Documentation of the Observed Damage
Document Nr. 2
Swiss Committee for the Safety of the Handling of Ammunition and Explosives by Forces and Administration
Defence Technology and Procurement Agency
Bienz, Kummer & Partner
Kummer, P.
TM 111-11, 01.09.1995, (CLASSIFIED)

9. **Evaluation of the Debris Throw from the 1992 Explosion in the Steingletscher Installation in Switzerland**
Journal of Hazardous Materials, Elsevier Science B.V.
Kummer, P.O.
Journal of Hazardous Materials 56 (1997) 149-167/19.02.1997
0304-3894/97

10. **Evaluation of the Debris Throw from the 1992 Explosion in the Steingletscher Installation in Switzerland**
Bienz, Kummer & Partner CH-8125 Zollikerberg
Kummer, P.
Presented at the DDESB Explosives Safety Seminar 1996, 20.08.1996

11. **Technische Vorschriften für die Lagerung von Munition (TLM 75)**
Teil 1 (Rev 86): Allgemeine Grundsätze
Teil 2 (Rev 90): Sicherheitsbeurteilung von Munitionslagern
Teil 3 (Rev 86): Planung und Projektierung von Munitionslagern
Teil 4 (Rev 93): Friedensmässige Belegung von Munitionslagern
Teil 5: Sicherheitstechnische Vorschriften für den Umgang mit Munition bei der Truppe
Eidgenössisches Militärdepartement
01.01.1987, (CLASSIFIED)

12. **Allied Ammunition Storage and Transport Publication 1 (AASTP-1)**
Manual of NATO Safety Principles for the Storage of Military Ammunition and Explosives
NATO Headquarters, Brussels/Belgium
AASTP-1, 01.05.1992, NATO unclassified

13. **Letalität von Personen infolge Trümmerwurf aus Kratern von Felsanlagen**
Personen im Freien
TLM 75/Teil 2–Revision technische Anhänge
Generalstab, Sektion SUR
Bienz, Kummer & Partner AG
Willi, Walter
TM 150-5, 01.11.1999

14. **Letalität von Personen infolge Kraterauswurf**
Gruppe für Rüstungsdienste/Fachstellen und Forschung, Fachabteilung 26
Ernst Basler & Partner AG
B 3113-2, Überarbeitete Fassung, 01.10.1984

15. **Lethality of Unprotected Persons Due to Debris and Fragments**
Ernst Basler & Partner AG, Zürich
Janser, P.W.
24.08.1982

16. **Underground Storage Sites: Note on Effect of Overburden on Outside Safety Distance**
Part 2: Damage by Debris
Ministry of Defence/UK
Jarrett, D.E.
AR 115/67 21/7/Explos./41, 10.04.1957, (CLASSIFIED)

17. **Origins and Implications of Underground Explosives Storage Regulations**
Waterfront Structures Division of Naval Facilities Engineering Service Center, Port Hueneme/CA
Odello, Robert J.
01.01.1980

18. **Ursprünge und Bedeutung von Vorschriften über die unterirdische Lagerung von Explosivstoffen**
(USA)
Wehrbereichsverwaltung VI/II B 5–Zentrale Sprachmittlergruppe
Civil Engineering Laboratory, Port Hueneme, CA/USA
Odello, Robert J.
Auftragsnummer: 131/81 (Freigabe-Nr. 59125)

19. **Shallow Underground Tunnel/Chamber Explosion Test Program**
Summary Report
(China Lake)
US Department of Defense Explosives Safety Board (DDESB)/Klotz-Club
U.S. Army Engineer Waterways Experiment Station (WES), Vicksburg MS/USA
Joachim, Charles E.
Technical Report SL-90-10, 01.08.1990

20. **Debris Throw from Underground Ammunition Magazine in Granite**
Model Test
Fortifikationsförvaltningen Sweden
Hultgren, S.
C-Rapport nr 204 (119), 01.03.1982

21. **Above-Surface Earth Motion from Underground Explosions**
Norwegian Defence Construction Service, Oslo 1/Norway

Ingram Scientific Services, Vicksburg, MS/USA
Ingram, Leo F.
Final Report, 01.11.1983

22. **Development of Safety Criteria for Explosive Storage Structures**
Report for Period May 1995–February 1997
Aeronautical Systems Center/Range and Air Base Systems Program Office ASC/LIW (EHR)/Eglin AFB/FL
Bakhtar Associates, Newport Beach/CA
Bakhtar, Khosrow
ASC-TR-1997-1005, 01.04.1997, (CLASSIFIED)

23. **Shallow Underground Ammunition Magazine Trials at Naval Weapons Center, China Lake**
Department of Defense Explosives Saftety Board DDESB
Norwegian Defence Construction Service
Jenssen, Arnfinn; Krest, Ottar
Presented at the DDESB Explosives Safety Seminar 1988, 09.08.1988

24. **TRAJ—A Two Dimensional Trajectory Program For Personal Computers**
Department of Defense Explosives Safety Board (DDESB) USA
Naval Surface Warfare Center, Silver Spring MD/USA
Montanaro, Paul E.
Presented at the DDESB Explosives Safety Seminar 1990, 01.08.1990

25. **A Consequence Minimization Approach to Explosive Siting**
Indian Head Division/Naval Surface Warfare Center, Silver Spring/MD
Swisdak, Michael M. Jr.
Presented at the DDESB Explosives Safety Seminar 1994, 01.08.1994

26. **Technology Base of the NAVY Explosives Safety Improvement Program (NESIP)**
Naval Surface Weapons Center
Porzel, Francis B.
Presented at the DDESB Explosives Safety Seminar 1980, 09.11.1980

27. **Simulated Tomahawk Missile Handling Arc Test Results**
Naval Surface Weapons Center, Silver Spring/MD
Ward, J.M.
Presented at the DDESB Explosives Safety Seminar 1978, 01.08.1978

28. **Fragment Hazard Investigation Program**
DoD Explosives Safety Board/USA
Naval Surface Weapons Center, Dahlgren/VA
Ramsey, Richard T.; Powell, Joseph G.; Smith, William D.
NSWC/DL TR-3664, 01.10.1978

29. **Klotz Underground Magazine Trial Data Report**
(China Lake)
Naval Weapons Center, China Lake, California/USA
Halsey, Carl C.; Durbin, William F.; Berry, Sharon L.
NWC TM 6562, 01.07.1989

30. **Detonation Trials in a Tunnel System**
Forsvarets Forskningsinstitutt/Norwegian Defence Research Establishment
Stromsoe, E.
Intern rapport X-122, Reference: Job 203/738, 17.02.1969

31. **Graphic Portrayal of Discrete Explosion-Produced Crater-Ejecta Characteristics**
Chief of Engineers, U.S. Army, Washington, D.C.
Weapons Effects Laboratory U.S. Army Engineer, Waterways Experiment Station (WES), Vicksburg/MS/USA
Rooke, Allen D.
N-76-7/4A061101A91D, 01.06.1976, (CLASSIFIED)

32. **Brick Model Tests of Shallow Underground Magazines**
Final Report
US Department of Defense Explosives Safety Board/Directorate of Health and Safety, Ministry of Defence United

Kingdom/Norwegian Defence Construction Service
U.S. Army Engineer Waterways Experiment Station (WES), Vicksburg MS/USA
Joachim, Charles E.; Rubin de la Borbolla, George S.
Miscellaneous Paper SL-92-2, 01.03.1992

33. **Shallow Underground Ammunition Magazine Trials at Naval Weapons Center (NWC) China Lake**
Ad Hoc Model Tests
Approximate Scale 1:24.8/Draft
Norwegian Defence Construction Service/Office of Test and Development
Jenssen, Arnfinn; Krest, Ottar
01.02.1988

34. **Above-Surface Earth Motion from Underground Explosions**
Final Report
Norwegian Defence Construction Service, Oslo 1, Norway
Ingram Scientific Services, Vicksburg/MS
Ingram, Leo F.
01.11.1983

35. **Debris Throw from Underground Ammunition Magazines in Granite**
Model Test
Fortifikationsförvaltningen/Schweden
Hultgren, Staffan
C-rapport nr 204 (119), 01.03.1982

36. **Model Tests for Underground Ammunition Storage Facilities**
Results from Joint Swedish-Norwegian Tests
Royal Swedisch Fortifications Administration, Eskilstuna/Sweden
Vretblad, Bengt E.
Presented at the DDESB Explosives Safety Seminar 1982, 01.08.1982

37. **Wissenschaftlich-technische Auswertung des Explosionsereignisses Susten/Steingletscher vom 2.**
November 1992
Dokumentation Schadenerhebung vom Juli 1993
Dokument Nr. 2
Anhang
Ausschuss und Projektgruppe für die Sicherheit beim Umgang mit Munition und Explosivstoffen in Armee und Militärverwaltung (ASUME/PGUME)
Arbeitsgruppe 1.1 "Grundlagen"
Bienz, Kummer & Partner
Kummer, P.
TM 110-27A/101-21A, 01.08.1995

38. **Debris Throw from Craters**
AC/258 Underground Storage AHWP
CH Status Report as at March 6, 1998
Defence Procurement Agency–CH
Bienz, Kummer & Partner AG
Kummer, P.
TM 158-6/CH(ST)UGS/AHWP IWP 004, 06.03.1998

39. **Incident at Fauld, near Hanbury Village**
Explosion of R.A.F No. 21 Bomb Storage Dump
on 27th November 1944, England
Ministry of Home Security, Birmingham/UK
Brief No. 1748 R/4 & R/5. File No. Re. 5/5/7, Region IX., 01.01.1945, (CLASSIFIED)

40. **A Literature Review on the Effects of an Explosion in Royal Navy Underground Magazines**
Ministry of Defence, Ensleigh, Bath/GB
Royal Military College of Science, School of Engineering and Applied Science, Cranefield University
Barnes, R.A.; Kennedy, L.J.
SEAS/CMSE/RAB/08/96, 18.03.1996, (CLASSIFIED)

41. **Die Explosion im R.A.F. Munitionslager Fauld/Staffordshire, UK am 27. November 1944**
Gruppe Rüstung, Stab Wissenschaft und Technologie/Sicherheit/Umweltschutz
Bienz, Kummer & Partner AG
Nussbaumer, Peter; Kummer, Peter
TM 101-54
42. **Some Remarks on a 100 Ton Accidental Explosion in a Tunnel System in 1965 Uusikylä/Finland**
Ministry of Defence/FIN
Tuokko, Seppo
Presented at the DDESB Explosives Safety Seminar 1992, 01.08.1992
43. **Besichtigung einer durch Explosion teilweise zerstörten Munitionsanlage bei Uusikylä/Finnland sowie Ausspache über Fragen der Munitionslagerung–Reisebericht**
Gruppe für Rüstungsdienste, TA 6/CH
Aebi, F.; Baumann, D.
Ae/Wä-22.10/471, 10.02.1972
44. **Die Explosions-Katastrophe von Prüm**
Der Krater des Kalvarienberges bei Prüm
Technische Hochschule Aachen
Rode, Karl
45. **Zusammenfassende Darstellung der Explosionskatastrophe vom 19. und 20. Dezember 1947 in der unterirdischen Munitionsanlage Mitholz**
Gruppe für Rüstungsdienste, TUA 6, Thun/CH
Cruchand, M.
C 37509-00, 30.05.1969, (CLASSIFIED)
46. **Explosion Effects and Properties**
Part I–Explosion Effects in Air
Naval Surface Weapons Center, White Oak, Silver Spring/MD
Swisdak, Michael M.
NSWC/WOL/TR-75-116, 06.10.1975
47. **Explosive Ditching with TNT**
Office, Chief of Engineers, U.S. Army, Washington DC
US Army Engineers, Waterways Experiment Station (WES), Vicksburg/MS
Müller, Arno M.; Carleton, Hendrik D.
Miscellaneous Paper N-77-7, 01.07.1977
48. **Crater Dimensions in Soil from High-Explosive Surface Detonations**
Norwegian Defence Construction Service
Basler & Hofmann AG, Zürich
LUE/TS/HM/ms
B 997-1, 01.03.1979
49. **Project Sperre**
Report 2, Phase I: Literature Survey–Analysis of Literature Data for Explosive Cratering Tests in Rock
Bundesministerium der Verteidigung, Rue-III-8, Bonn/BRD
U.S. Army Corps of Engineers, Waterways Experiment Station (WES), Vicksburg/MS/USA
Davis, Landon K.; McAneny, Colin, C.
Technical Report SL-81-5, 01.07.1981
50. **Project Sperre**
Report 3, Phase II: Field Tests—Explosive Cratering Experiments in Granite and Sandstone
Bundesministerium der Verteidigung, Rue-III-8, Bonn/BRD
U.S. Army Engineer Waterways Experiment Station (WES), Vicksburg MS/USA
Davis, Landon K.; McAneny, Colin C.; Joachim, Charles E.
Technical Report SL-81-5/3, 01.07.1981, (CLASSIFIED)
51. **Project Sperre**
Report 5: Summary Report: Road Cratering in Rock with Preemplaced Charges
Bundesministerium der Verteidigung, Rue-III-8, Bonn/BRD
U.S. Army Corps of Engineers, Waterways Experiment Station (WES), Vicksburg/MS/USA

Davis, Landon, K.; McAneny, Colin C.
Technical Report SL-81-5/5, 01.07.1981, (CLASSIFIED)

52. **AC/258 Underground Storage AHWP**
Debris Throw from Craters
CH Status Report as at October 1, 1997
Defence Technology and Procurement Agency/CH
Bienz, Kummer & Partner AG
Kummer, P.
TM 158-4/CH(ST)UG/AHWP IWP 002, 24.09.1997

53. **A Reexamination of the Airblast and Debris Produced by Explosions Inside Earth-Covered Igloos**
DoD Explosives Safety Board/USA
Naval Surface Warfare Center, Silver Spring/MD
Swisdak, Michael M.
NAVSWC TR 91-102, 28.01.1991

CHAPTER 27

MODELING ATMOSPHERIC DISPERSION OF HEAVIER-THAN-AIR CLOUDS CONTAINING AEROSOL

Jaakko Kukkonen and Juha Nikmo
Finnish Meteorological Institute, Air Quality Research, Helsinki, Finland

27.1 INTRODUCTION

Since the early 1970s, numerous mathematical models have been presented for the dispersion of heavier-than-air clouds, and considerable progress has been made both theoretically and experimentally. Reviews of these developments have been written, e.g., by Blackmore et al. (1982), Wheatley and Webber (1984), Hanna and Drivas (1987), Koopman et al. (1989), Bricard and Friedel (1998), and Britter (1998). Clearly, practical modeling applications also require modeling of release rates, source terms, and effects for various conceivable accident scenarios, as shown in Fig. 27.1.

Modeling of heavier-than-air clouds involve several processes: e.g., gravity spreading, interaction with the surrounding atmosphere and the ground, cloud transport, the influence

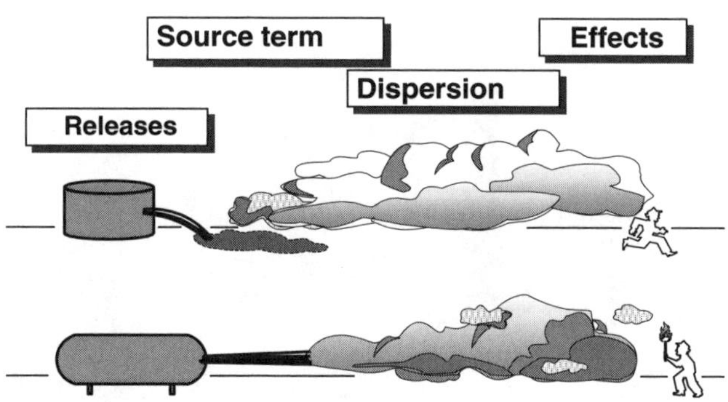

FIGURE 27.1 Schematic presentation showing modeling of releases, source terms, atmospheric dispersion and effects for various conceivable accident scenarios. (*Source:* David Webber, Integral Science and Software, U.K.)

of terrain and obstacles, deposition of material to the ground, and thermodynamic processes within the cloud. These processes are shown in Fig. 27.2. Several of these processes have been addressed in previous chapters of this Handbook. This article focuses on the influence of aerosol content of the cloud.

Releases of liquefied toxic or flammable gases often take place in aerosol form, consisting of vapor and liquid droplets of the released species, together with entrained humid air. This has been demonstrated in several laboratory and field-scale experiments (e.g., Koopman et al., 1986, Moodie and Ewan, 1990; Nolan et al., 1990; Schmidli et al., 1990; Nielsen et al., 1997). Aerosol phenomena may have a significant influence on the temperature and density evolution of the source term and on the subsequent heavy gas dispersion. In particular, the deposition of substance liquid droplets may, under certain conditions, cause a substantial decrease of concentration.

However, two-phase dynamics are commonly treated in a very simple way (or even ignored by some modeling systems) by both fairly simple integral models and more complex three-dimensional fluid models. Practically all of these models rely on the homogeneous equilibrium assumptions, implying that the liquid is uniformly distributed in the cloud and that the liquid and the gas are at a uniform temperature and in thermodynamical equilibrium.

This chapter presents the mathematical structure of a more general nonequilibrium model AERCLOUD (Kukkonen et al., 1989; Vesala and Kukkonen, 1992). The homogeneous equilibrium model is the limiting case of the aerosol model presented, when the cloud is well mixed and the droplet-gas equilibration processes are rapid compared to the rate at which other processes occur.

We then proceed to review results on the applicability of the homogeneous equilibrium assumptions (Kukkonen et al., 1993, 1994; Nikmo et al., 1994). We have compared thermodynamic model predictions with those of the heavier-than-air cloud dispersion model DRIFT (Webber et al., 1992). These investigations have shown that the simpler model does indeed provide a good description for many release situations and guidance is given on where the homogeneous equilibrium model is not likely to be adequate.

Mathematical models for droplet vaporization and condensation have also been developed by Hewitt and Pattison (1992), with the aim of incorporating the aerosol model into a heavy gas dispersion model. Woodward and Papadourakis (1991) have presented a two-phase jet model, including a description of droplet vaporization. Woodward et al. (1995) have devel-

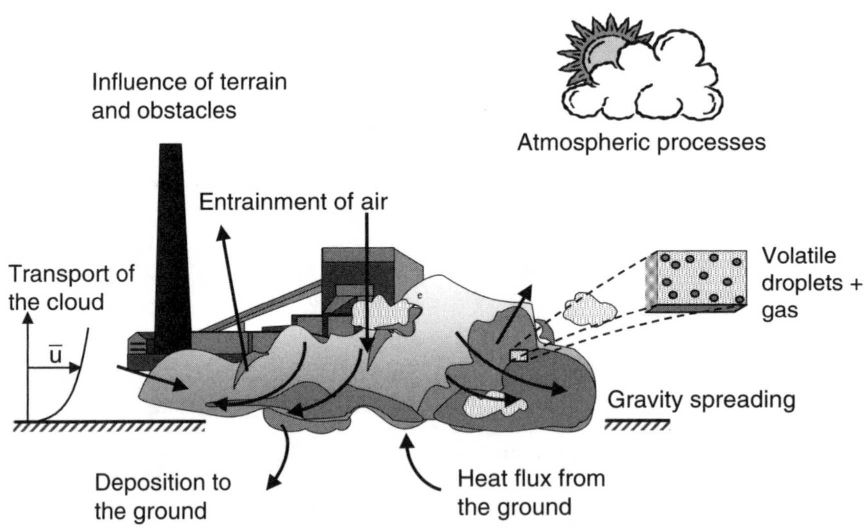

FIGURE 27.2 Processes important for the atmospheric dispersion of heavier-than-air clouds.

oped a model to simulate the dispersion and trajectories of a high-momentum aerosol jet, including a droplet evaporation model. Muralidhar et al. (1995) describes a thermodynamical nonequilibrium two-phase jet model for predicting the HF releases. Witlox and Holt (1999) have presented a dispersion model including a nonequilibrium thermodynamic model for droplet vaporization.

Most of the heavier-than-air cloud field experiments have focused on the dispersion of gaseous clouds. However, some experiments have been conducted with two-phase effects, e.g., the Desert Tortoise, Eagle, Goldfish series (Koopman et al., 1989) and FLADIS field experiments (Nielsen et al., 1997). In particular, the Desert Tortoise series was designed to study the dispersion of two-phase ammonia clouds. It was reported that two-phase effects had a dominating influence on the temperature and spreading behavior of the cloud and the extent of the heavy cloud dispersion regime (Koopman et al., 1986). Recently, available field data sets have been reviewed within the SMEDIS (Scientific Model Evaluation of Dense Gas Dispersion Models) project (Daish et al., 1998; Carissimo et al., 1999).

Nolan et al. (1990), Moodie and Ewan (1990), and Johnson (1991) have studied two-phase source terms on the laboratory scale. Johnson (1991) has presented results on the deposition of contaminant liquid versus liquid superheat for six substances. Nolan et al. (1990) and Moodie and Ewan (1990) have also presented measurements on droplet characteristics, including size and number concentration.

27.2 THEORY

27.2.1 Statement of the Problem

Consider a system consisting of binary (two-component) droplets together with the surrounding gas (such as air) and the vapors of species forming the droplet. In the following we refer to the gaseous phases of the two condensing or evaporating species simply as "the vapors" and to the mixture of air and the two vapors as "the surrounding gas," or "the gas." The total mass of humid air within the system is time dependent, while the total mass of contaminant is assumed to be constant, i.e., the deposition of droplets and the contaminant vapor are neglected. We refer to the system of droplets and gas as "the cloud."

The droplets are assumed to contain no impurities, to be spherical, and to be falling freely. They are also assumed to be well mixed; the validity of this assumption has been studied in detail by Vesala (1993). The gas is assumed to behave as an ideal gas. The total number of droplets is constant, i.e., the coagulation of droplets is neglected. The total pressure is assumed to be equal to atmospheric pressure at all times. Finally, the droplet population is assumed to be monodisperse and uniformly distributed in the gas.

In the following, we derive equations for the rate of change of quantities related to (1) the droplet population (the total mass of the droplets, the droplet composition, and the droplet temperature) and (2) the gas (the gas temperature, the droplet number concentration, and the gas composition, i.e., the partial vapor pressures in the gas). For a more detailed model description, the reader is referred to Nikmo et al. (1994).

Physically, the composition and temperature of the gas around the droplets are constantly changing due to the condensation and evaporation processes and to entrainment. For instance, the evaporation of contaminant tends to cool the gas and increase the concentration of contaminant vapor in the gas. The changes in the composition and temperature of the gas in turn have an influence on the behavior of the droplets.

27.2.2 Change of Quantities Related to the Droplet Population

The Change of the Total Mass of a Droplet. The droplet evaporation or condensation rate is described by means of the mass and heat fluxes from the droplet into the surrounding gas. In the following, the two condensing or evaporating species are denoted with the subscripts

1 and 2. In a steady state, the mass fluxes are constant according to the equation of continuity, and the quasisteady change of the droplet mass m_d is by definition equal to the sum of the mass fluxes of species 1 and 2 from the droplet surface into the gas:

$$\frac{dm_d}{dt} = I_1 + I_2 \tag{27.1}$$

The positive radial direction has been chosen to be into the droplet (i.e., condensation corresponds to a positive mass flux). Generally, the mass fluxes depend on the droplet radius, the total gas pressure, the droplet and gas temperatures, the mole fractions of the vapors just above the droplet surface and far from it, and the diffusivities of vapors in the gas mixture.

The Change of the Droplet Composition. In order to determine the mass fluxes, the composition of the droplet must be known. A direct consequence of the definition of the mass fluxes ($I_i = dm_i/dt$, $i = 1,2$) is that the change of the mass fraction X_i of species i in the droplet can be written as:

$$\frac{dX_i}{dt} = \frac{d}{dt}\left[\frac{m_i}{m_1 + m_2}\right] = \frac{I_i - X_i(I_1 + I_2)}{m_d} \tag{27.2}$$

where m_i is the mass of species i within the droplet.

The Change of the Droplet Temperature. In order to determine the mass fluxes, the droplet temperature must also be known. An equation for the quasisteady droplet temperature can be derived using energy conservation, yielding:

$$\frac{dT_a}{dt} = \frac{L_1I_1 + L_2I_2 + 4\pi ak(T_g - T_a)}{m_d\bar{c}_d} \tag{27.3}$$

where L_i = the heat of vaporization for species i
a = the droplet radius
k = the thermal conductivity of the gas
T_g = the temperature of the gas
$\bar{c}_d$ = the average specific heat capacity of the droplet

Kulmala and Vesala (1991) have discussed alternative methods for modeling the droplet temperature, also taking into account the temperature dependencies of the vapor phase enthalpies.

27.2.3 Change of Quantities Related to the Gas

Equations (27.1) to (27.3) form a complete set of equations for describing the evolution of a single droplet in an infinite volume of gas with no entrainment. In the following, the gas temperature and the partial vapor pressures in the gas are allowed to change due to the mutual influence of evaporating or growing droplets and the entraining humid air. It is assumed that the entrained air will be distributed sufficiently rapidly and uniformly into the cloud.

The Change of the Gas Temperature. The temperature of the gas changes due to the thermal conduction between the droplets and the gas and due to the entrainment of air, i.e.,

$$\frac{dT_g}{dt} = \frac{N\left(L_1I_1 + L_2I_2 - m_d\bar{c}_d\dfrac{dT_a}{dt}\right) + (T_{\text{amb}} - T_g)\bar{c}_{\text{amb}}G_{\text{ent}}}{\bar{c}_g m_g} \tag{27.4}$$

where N = the number of droplets
T_{amb} = the temperature of the ambient gas
$\bar{c}_{amb}$ = the average specific heat capacity of the ambient gas
G_{ent} = the mass flux of air into the gas cloud
$\bar{c}_g$ = the average specific heat capacity of the gas
m_g = is the mass of gas

The Change of the Droplet Number Concentration. The droplet number concentration is defined as the number of droplets in a unit volume of the cloud. The droplet number concentration changes due to (1) the change in the cloud volume caused by the temperature change of the gas, (2) the change in the number of gas molecules caused by evaporation and condensation, and (3) the change in the cloud volume caused by entrainment:

$$\frac{dC_N}{dt} = -\frac{C_N}{T_g}\frac{dT_g}{dt} + \frac{C_N N}{n}\left(\frac{I_1}{M_1} + \frac{I_2}{M_2}\right) - \frac{C_N G_{ent}}{n\bar{M}_{amb}} \tag{27.5}$$

where n = the number of moles in the gas
M_i = the molar mass of species i
$\bar{M}_{amb}$ = the average molar mass of the ambient air

The Change of the Partial Vapor Pressures. The partial vapor pressure changes due to evaporation and condensation processes and entrainment, i.e.,

$$\frac{dp_1}{dt} = \frac{N}{n}\left(\frac{I_1}{M_1}(p_1 - p) + \frac{I_2}{M_2}p_2\right) + \frac{(p_{amb,1} - p_1)G_{ent}}{\bar{M}_{amb}n}, \tag{27.6}$$

where p_i = the partial vapor pressure of species i
p = the total pressure
$p_{amb,1}$ = the partial vapor pressure of species 1 in ambient air

The equivalent equation for species 2 is found by interchanging subscripts 1 and 2 in Eq. (27.6).

27.2.4 Modeling Mass and Heat Transfer from Droplets to Surrounding Gas

Here we do not consider in detail the modeling of the mass and heat transfer processes from the droplet surface into the surrounding gas. The evaporation or condensation of droplets has been addressed in detail by Kukkonen et al. (1989), Kulmala and Vesala (1991), Vesala and Kukkonen (1992), and Vesala and Kulmala (1993); see also Kulmala et al. (1993). Vesala (1991) has discussed the validation of various mass flux and droplet temperature equations against laboratory-scale experimental data.

27.3 NUMERICAL RESULTS

27.3.1 Flash Fractions

The flash fraction is a useful concept in interpreting results on two-phase flow. Flashing can be defined as an isentropic evaporation process due to depressurization, in which the final pressure is equal to the atmospheric pressure. The flash-off vapor fraction (flash fraction) is the corresponding quality in the final state of the process. The term *partial flashing* refers to the same process when the final pressure is higher than the atmospheric pressure.

During flashing, the total entropy of the fluid is conserved:

$$x_{v,0}S_{v,0} + (1 - x_{v,0})S_{l,0} = x_{v,f}S_{v,f} + (1 - x_{v,f})S_{l,f}, \tag{27.7}$$

where x = the mass fraction
S = the specific entropy
Subscripts 0 and f = the initial and final states of the process
Subscripts v and l = the vapor and liquid phases

We assume that the fluid is in a saturated state initially and finally and that the fluid is initially pure liquid ($x_{v,0} = 0$). Then the flash fraction ($x_{v,f}$) can be computed numerically from Eq. (27.7). The final temperature of the fluid is the boiling point temperature of the species.

Figure 27.3 shows the flash fractions for some commonly occurring toxic and flammable substances as a function of temperature of storage. The curves for propane, chlorine, ammonia, sulfur dioxide, and hydrogen fluoride were computed numerically from Eq. (27.7) and the curves for ethylene and butane have been taken from Marshall (1987). The flash fraction increases with increasing storage temperature and vanishes at the boiling point of species. The flash fractions for the considered substances, except for ethylene and propane, are less than 0.20 for commonly occurring ambient temperatures.

The significance of flash fractions is that they give the maximum possible vapor fraction formed in an isentropic process for discharges into the atmosphere. This statement could be derived mathematically starting from Eq. (27.7) and equations for the temperature dependency of the vapor and liquid entropies in saturated state. For instance, in choked isentropic pipe flow the vapor fraction of the fluid at the aperture is always smaller than the flash fraction (because the exit pressure is higher than the atmospheric pressure).

27.3.2 Evaporation and Deposition of Ammonia–Water Droplets

We have written a computer program, AERCLOUD (for "aerosol cloud") for solving Eqs. (27.1) to (27.6), and the corresponding mass and heat transfer equations between the liquid

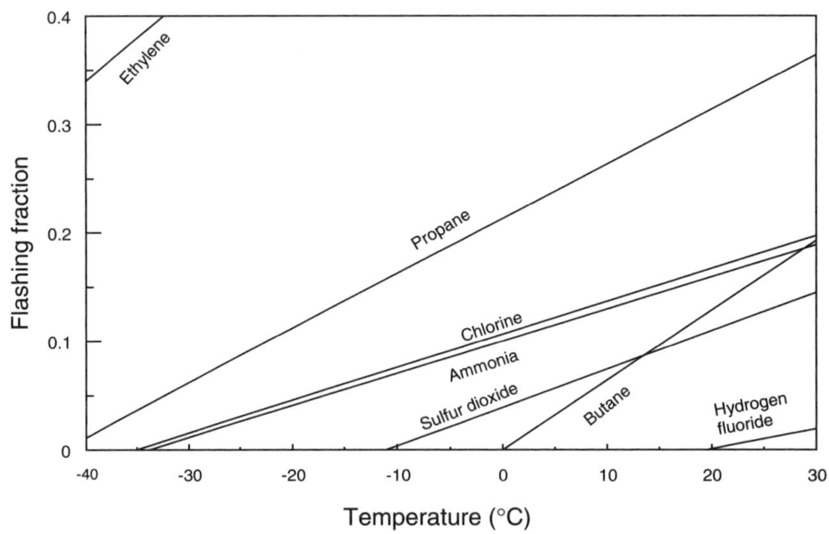

FIGURE 27.3 Flash fraction versus temperature of storage for some toxic and flammable substances. (*Source: Kukkonen, 1990*)

and gaseous components (Kukkonen et al., 1989; Vesala et al., 1989; Kukkonen 1990; Vesala, 1991; Vesala and Kukkonen, 1992; Nikmo et al., 1994).

Consider a droplet that consists initially of pure ammonia. Then ammonia evaporates and water vapor condenses into the droplet surface. The evaporation rate of ammonia depends on its vapor pressure at the droplet surface, which is a function of the droplet temperature, the ammonia mole fraction in the droplet, and the activity coefficient.

The condensation of water vapor has two simultaneous effects on the evaporation of ammonia, a thermal effect and a dilution effect. In the thermal effect, latent heat is released by the condensing water vapor, which tends to raise the droplet temperature and therefore enhance the evaporation of ammonia. In the dilution effect, the concentration of ammonia in the droplet decreases with increasing water vapor condensation. For a large dilution, the mole fraction of ammonia and the activity coefficient are small, reducing the evaporation rate of ammonia.

Figure 27.4 shows the radius of an ammonia droplet as a function of time for different relative humidities. Initially, the thermal effect of water vapor condensation is dominant and the rate of evaporation of ammonia is increased with increasing relative humidity. Later, however, the rate of evaporation of ammonia is slowed down due to the dilution effect.

Figures 27.5*a* and 27.5*b* show the drying and settling times of water and ammonia droplets versus the droplet radius. The curves have been shown for three combinations of gas temperatures and vapor pressures and for three initial heights. The simultaneous evaporation and settling of droplets have been taken into account in the computation of the curves. Each settling time curve corresponds to one drying time curve, and each intersection point of these curves corresponds to the maximum radius r_M of a totally evaporating droplet. The vapor pressures in Figs. 27.5*a* and 27.5*b* correspond to saturation ratios 0.0012 and 0.058 at a gas temperature of 20°C and to a saturation ratio of 0.12 at a gas temperature of 0°C, for both species.

The radii r_M range from approximately 40 to 110 μm for water droplets for the selected ambient conditions and initial heights. The corresponding radii for ammonia droplets range from 90 to 380 μm. The larger radii for ammonia droplets are mainly due to the shorter drying times. Evidently, the radius r_M increases with increasing release height because the gravitational settling time increases with height. For a given release height, r_M decreases with decreasing gas temperature and increasing vapor partial pressure because the drying

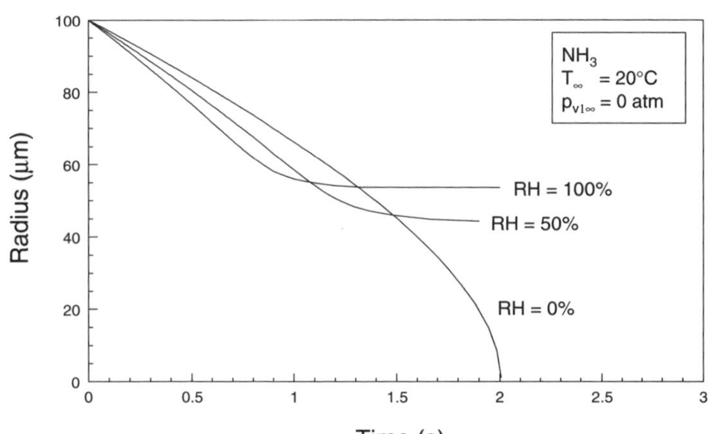

FIGURE 27.4 The influence of relative humidity on the evaporation of an ammonia droplet. The initial droplet radius is 100 μm, the gas temperature is 20°C, and the vapor pressure of ammonia in the gas is negligible. (*Source: Vesala and Kukkonen, 1992*)

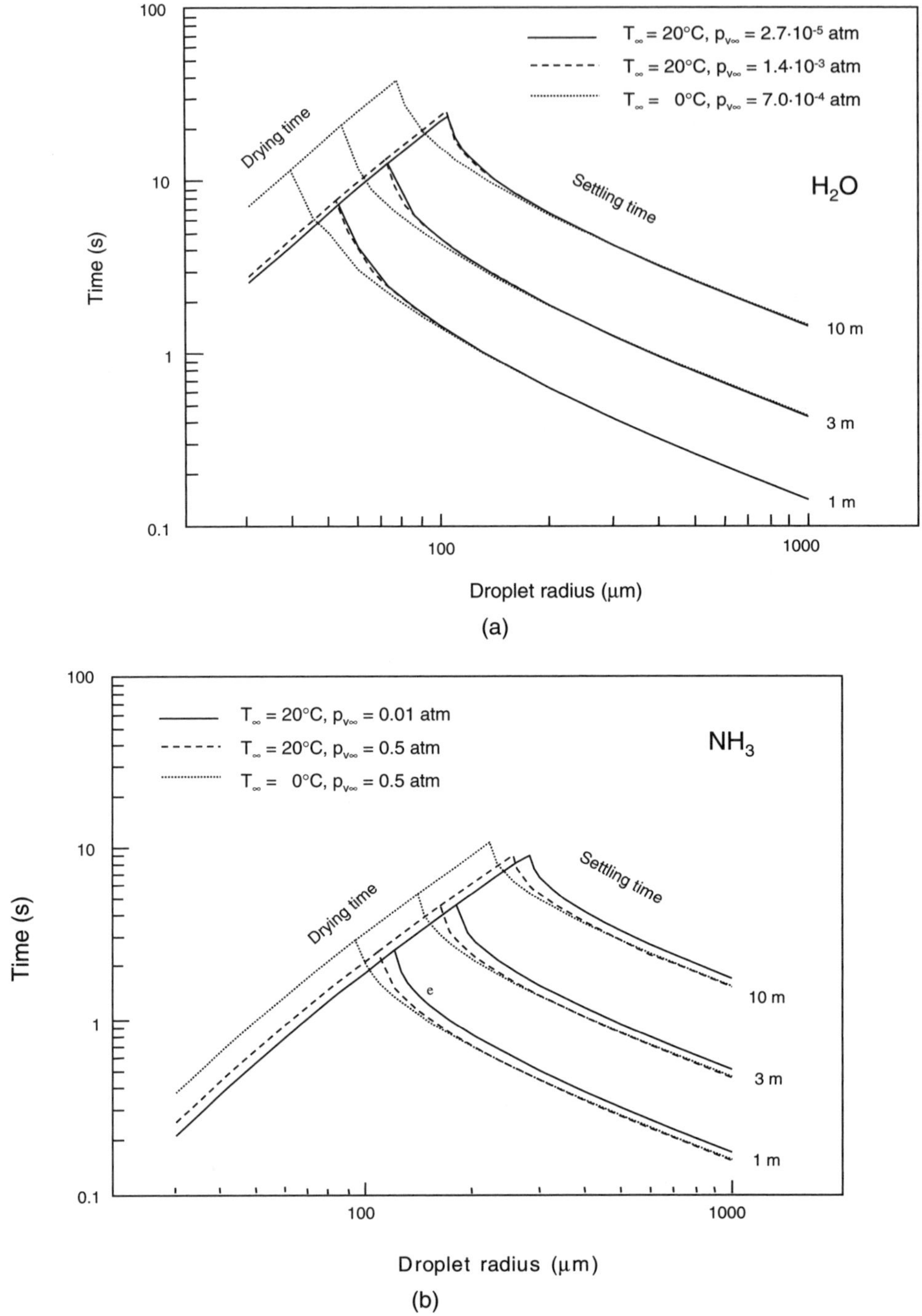

FIGURE 27.5 Drying time and gravitational settling time of water and ammonia droplets versus droplet radius. The curves are shown for three combinations of gas temperature and vapor pressure, and the initial height of droplets ranges from 1 m to 10 m. (*Source: Kukkonen et al., 1989*)

time is longer in these conditions. The curves in Figs. 27.5*a* and 27.5*b* have been calculated assuming a negligible droplet concentration. The effect of the simultaneous evaporation of a population of droplets is to increase the drying times, resulting in smaller r_M.

The droplet size regimes in which deposition or total evaporation is likely to take place in various conditions can be estimated from Fig. 27.5*a*. For instance, for a 100-μm droplet released from heights of from 0 to 3 m in the given conditions, the settling time is less than about 5 seconds and the drying time is about 20 seconds. The deposited mass fraction is therefore expected to be fairly large (the model computations give $X_d = 0.82$). For a 300-μm droplet, the settling time is much smaller than the drying time, and the deposition fraction is therefore expected to be very close to unity (the model computations give $X_d = 0.99$).

27.3.3 Evolution of a Two-Phase Ammonia Cloud

We have illustrated the model predictions by evaluating two-phase ammonia clouds released in dry and moist air. The numerical test cases are identical to those in Kukkonen et al. (1993), which presents a comparison of the model AERCLOUD and the thermodynamical submodel of the heavy cloud dispersion program DRIFT (Webber et al., 1992). DRIFT embodies the homogeneous equilibrium model, while AERCLOUD allows also for thermodynamic nonequilibrium effects. Both models will cope with ammonia interactions with moist air as well as with the simpler dry air problem.

We emphasize at the outset that our objective is a comparison of concepts—homogeneous equilibrium versus aerosol dynamics. It is done in the context of two models but is not intended to apply only to those models. By keeping everything in the two models the same, apart from the aerosol model, we shall focus sharply on the difference in predictions arising solely from the nature of the aerosol model.

We shall use predictions on the mass of entrained air from the code DRIFT. An aerosol cloud is modeled with the same initial content using the code AERCLOUD, importing air into it at the time-dependent rate given by DRIFT. In this way we need not couple the model AERCLOUD explicitly to a heavy gas dispersion model.

It is important to distinguish between vaporization and condensation, on the one hand, and deposition phenomena, on the other. The prime objective here is to examine the vaporization and condensation processes, and so we shall ignore deposition in the following, however large the droplets. The gravitational settling velocities of 10, 100, and 1,000 μm ammonia droplets in dry air are about 0.01, 0.5, and 5 m/s, respectively.

The mathematical model makes assumptions about homogeneity of the cloud; we are assuming homogeneity, or at least an approximately self-similar global inhomogeneity. However, equilibrium is regarded as the most questionable part of the homogeneous equilibrium assumption; in the instantaneously released cloud flow considered here, there is evidence (see, e.g., Brighton, 1985, for a discussion of the Thorney Island Trial data) that the strong frontal gravity current vortices mix the cloud reasonably well, at least in the early stages, when aerosol effects are likely to be most important.

The convective fluxes of mass and heat substantially increase the purely diffusive mass transfer and the purely conductive heat transfer for large droplets. The expressions for the convective fluxes are not valid for very high droplet number concentrations. We therefore believe it prudent to make estimates using two model options: including and excluding convective mass and heat transfer.

Selection of Cases. We have made predictions for instantaneous releases of a pure ammonia cloud into both dry and humid air. A summary of the initial and ambient conditions for the cases chosen is shown in Table 27.1. The ambient temperature was taken to be 15°C, the Pasquill class was D, the average wind velocity at a height of 10 m was 2 m/s, and the roughness length was 6 mm.

The total mass of contaminant was taken to be 10,000 kg, which is roughly two or three times the mass of the initial gas cloud in the Thorney Island field experiments (McQuaid

TABLE 27.1 A Summary of the Cases Selected for the Numerical Results

Case	Ambient air relative humidity (%)	Solution in liquid phase	Contaminant liquid fraction (%)
01	0.00	–	85
02	0.00	–	60
03	0.00	–	30
11	99.99	Ideal	85
12	99.99	Ideal	60
13	99.99	Ideal	30
21	99.99	Interactions	85
22	99.99	Interactions	60
23	99.99	Interactions	30

and Roebuck, 1985). The flashing has been assumed to have taken place, and therefore the pressure is atmospheric and the initial contaminant temperature is equal to the boiling point of ammonia (−33°C). The initial contaminant liquid fraction by mass was assumed to be 30, 60, or 85%. The value of 85% corresponds approximately to the largest possible liquid fraction after flashing has taken place, and the values of 30 and 60% correspond to cases where part of the released ammonia has been stored in vapor form.

In order to illustrate the effects of atmospheric moisture, we have included the case of dry air and that of 99.99% relative humidity. Two options were used for the interactions of ammonia and water in liquid phase: assuming an ideal solution and allowing for the actual interactions. Ammonia and water behave attractively in liquid phase, and the activity coefficients are therefore smaller than unity; respectively, the partial mixing enthalpies are negative (Vesala and Kukkonen, 1992; Wheatley, 1987). For an ideal solution, the activity coefficients are equal to unity and the partial mixing enthalpies vanish.

Numerical Results. We have evaluated case 21 in more detail. The droplet size regime considered ranges from 100 μm to 1,000 μm. The curves including and excluding the convective mass and heat transfer have been marked in the figures with "ventilation" and "no ventilation", respectively. The droplets consist initially of pure ammonia; ammonia then vaporizes and water vapor condenses onto the droplet surface.

Figures 27.6*a*, *b*, and *c* show the liquid, vapor, and total concentration of ammonia versus time for case 21. The curves are shown for three values of the initial droplet radius, including and excluding droplet ventilation. The total concentration of contaminant is almost identical for these three figures because the numerical tests were designed so that the entrainment of air versus time is the same by definition.

These figures show that the ammonia liquid is almost completely vaporized during 200 seconds in all three cases. The influence of droplet ventilation decreases with decreasing initial droplet radius, and for a droplet radius of 100 μm (or smaller), the influence of ventilation on the contaminant vaporization is negligible.

These figures also show that the influence of the initial droplet size on the vaporization rate of ammonia is fairly small, unless the droplet radius is on the order of a few hundred micrometers or larger. One reason for this phenomenon is that the evaporation/condensation process within a cloud is self-controlling, with negative feedback. For instance, consider the vaporization of fairly small droplets. The vaporization of contaminant is more efficient (compared to large droplets), which causes a lower cloud temperature and a larger contaminant

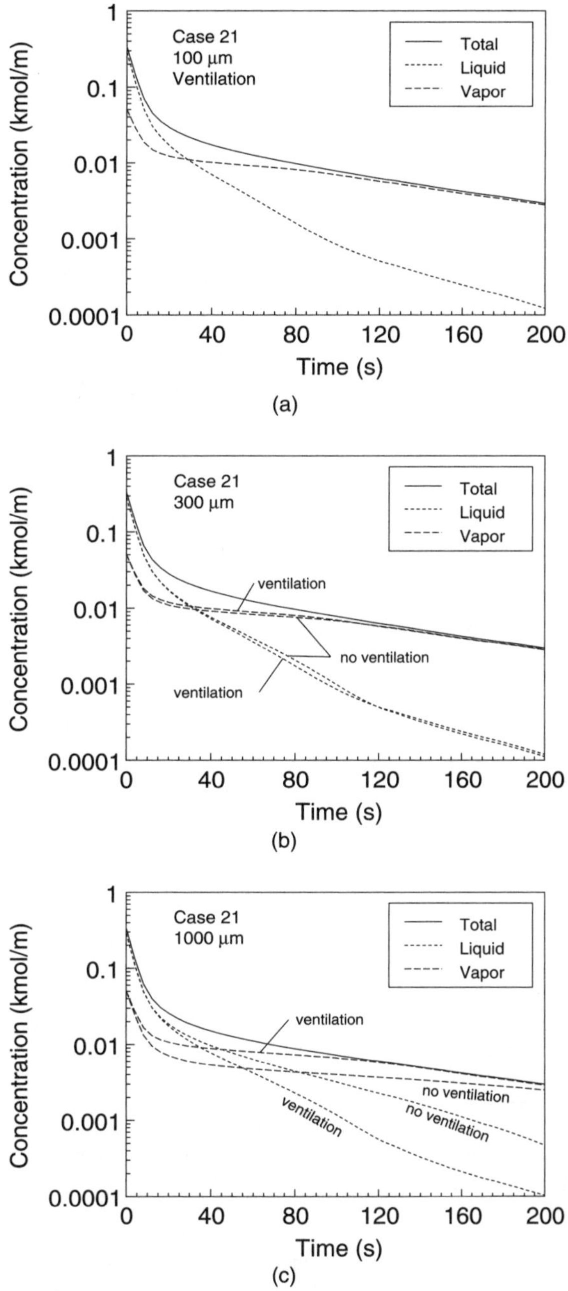

FIGURE 27.6 Contaminant molar concentration versus time for case 21 for three values of the initial droplet radius (100, 300, and 1000 μm). The curves are shown for liquid, vapor, and total concentrations. For an initial droplet radius of 100 μm, the curves including and excluding droplet ventilation are indistinguishable.

vapor pressure in the cloud. This in turn reduces the rate of vaporization. For large droplets, the changes in the gas properties due to vaporization are smaller, which tends to allow a correspondingly more efficient vaporization.

27.4 HOMOGENEOUS APPROXIMATION

27.4.1 Design of the Test

Our strategy for testing the homogeneous equilibrium model will be as follows. First we shall look at some predictions of the code DRIFT for instantaneous releases of ammonia into moist air. Following this, we shall model a gas cloud with the same initial content using the code AERCLOUD, importing air into it at the time-dependent rate given by DRIFT. In this way we can study the differences of model predictions for the thermodynamic behavior, excluding possible differences in modeling, such as entrainment, spreading, or transport processes. We also need not couple the aerosol model explicitly to a heavy gas dispersion model.

The HE model is the limiting case of the more complex aerosol behavior when the cloud is well mixed and the droplet-gas equilibration processes are rapid compared to the rate at which other processes occur. An important pretest condition, therefore, is that the two approaches be shown to be in agreement in the limit where the heat and mass transfer coefficients of the aerosol model are set to be sufficiently large to make these processes very rapid, i.e., the values predicted by AERCLOUD at the HE limit are expected to duplicate the respective results computed with DRIFT.

This is also a very useful verification of the accuracy of the derivation and coding of each of the models. Both models are fairly complex, and the model equations are solved using different numerical techniques. The models also use different, independently validated parametrizations for the physicochemical properties of substances, in particular for the properties of the mixtures of ammonia and water in liquid phase.

27.4.2 Predictions of the Model AERCLOUD

We are considering here only monodisperse aerosol populations. The initial droplet size is required by the aerosol model as input information; we have chosen the radii of 1,000 μm, 100 μm, and 10 μm. The droplet radius values in the laboratory-scale measurements in this context range typically from a few to a few hundred micrometers (e.g., Nolan et al., 1990; Moodie and Ewan, 1990; Schmidli et al., 1990). The droplets are assumed to consist initially of pure ammonia, and they are at the assumed source term temperature (−33°C).

We have selected as base cases those of dry air and of humid air with ammonia water interactions, both with the liquid fraction of $q = 0.85$. These were evaluated in more detail using the model AERCLOUD. Figures 27.7*a* to *d* show the computed numbers of moles of ammonia (vapor and liquid phases). The headlines of figures show also the selected initial droplet sizes (1,000 μm and 100 μm). The curves marked with "HE-limit" show the model predictions in the homogeneous equilibrium limit.

First, we compare the HE limit predictions of the model AERCLOUD (in Figs. 27.7*a* to *d*) with the corresponding results computed with the model DRIFT. The maximum differences for the numbers of moles of contaminant vapor and liquid, and for the temperature, are smaller than 4%; and for water vapor and liquid water the maximum differences are smaller than 9%.

The differences of various modeling options can be seen clearly in Figs. 27.7*a* to *d*. The rates of mass and heat transfer are slowest for the nonequilibrium computation, where droplet ventilation has been excluded ("no ventilation"). Including droplet ventilation in the none-

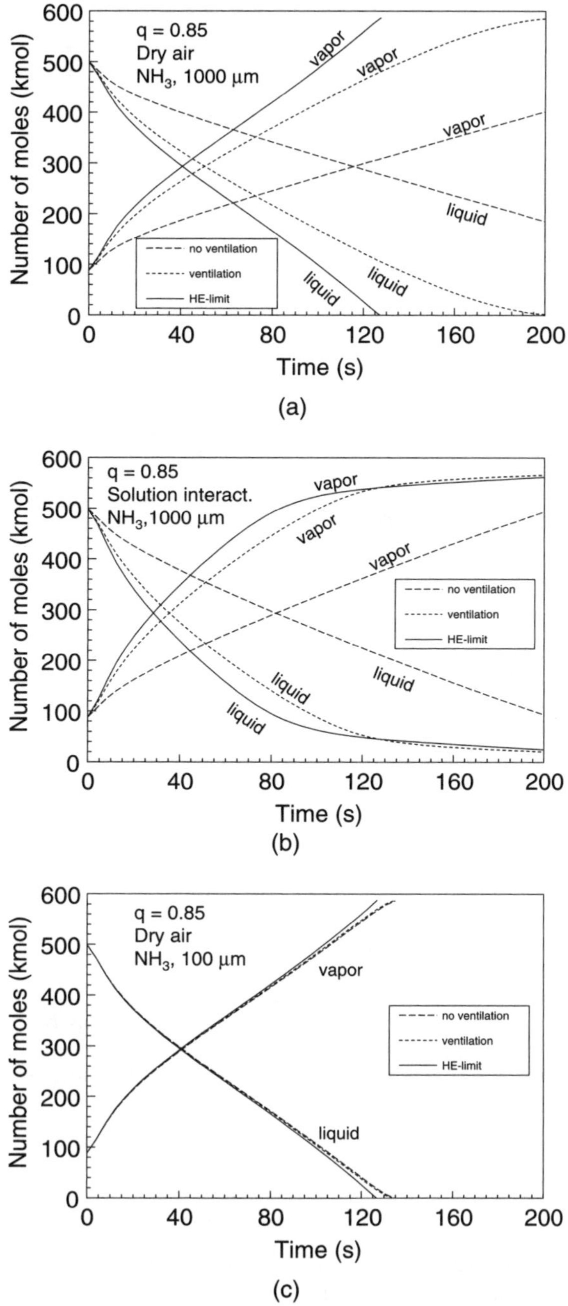

FIGURE 27.7 Number of moles of ammonia vapor and liquid versus time predicted by AERCLOUD, for 85% liquid releases in dry air and in 99.99% humid air allowing for ammonia–water interactions. The initial droplet radius ranges from 1000 μm to 100 μm. The curves have been computed using three model options: including and excluding droplet ventilation and in the homogeneous equilibrium limit.

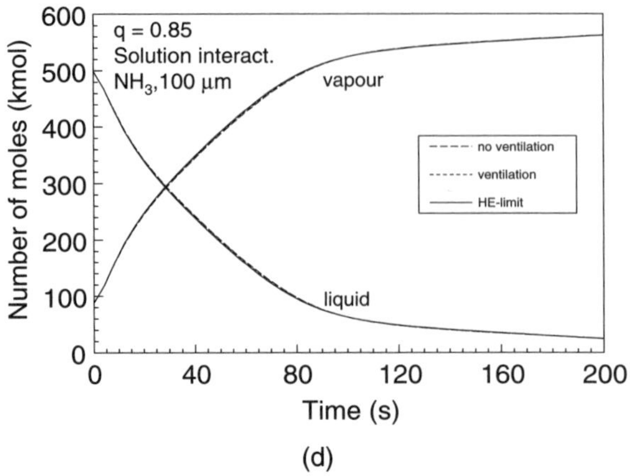

FIGURE 27.7 Number of moles of ammonia vapor and liquid versus time predicted by AERCLOUD, for 85% liquid releases in dry air and in 99.99% humid air allowing for ammonia–water interactions. The initial droplet radius ranges from 1000 μm to 100 μm. The curves have been computed using three model options: including and excluding droplet ventilation and in the homogeneous equilibrium limit. (*Continued*)

quilibrium computations enhances the mass and heat transfer. In the HE limit these transfer rates are the largest possible and ammonia is therefore vaporizing more quickly. The effect of ambient air moisture is again to increase the rate of vaporization at small times and suppress it at larger times.

For the initial droplet radii of 100 μm, the differences of model predictions of the three model options are less than 1% for both the cases considered. For a droplet radius of 10 μm, the curves are practically identical, implying that thermodynamic equilibrium prevails at all times. We can conclude that the HE model gives good results concerning contaminant vaporization for the selected cases if the droplet radius is not larger than about 100 μm. It can also be shown by simple qualitative arguments that the HE model is better applicable for small droplets compared to larger ones in the continuum regime (Kukkonen et al., 1994).

Figure 27.8 shows the relative difference of droplet vaporization times predicted by the aerosol model, compared to the HE limit, for ammonia–water interaction cases with 85% and 30% liquid initially. The vaporization time has been defined here as the time by which 80% of the liquid ammonia within the droplet has been vaporized. In accordance with the earlier results, the differences of the model predictions and the influence of droplet ventilation increase with the droplet size.

27.5 CONCLUSIONS

The homogeneous equilibrium model is often adopted in dispersion models of two-phase jets and clouds, but it has not previously undergone a critical test. We have therefore tested the homogeneous equilibrium model, in the context of the dispersion of a dense gas cloud, against a more sophisticated aerosol model.

The thermodynamical nonequilibrium effects in a two-phase cloud have two essential consequences: (1) the thermodynamical behavior of the mixture is different, in particular the temperature and density evolution, and (2) the deposition of contaminant liquid may cause

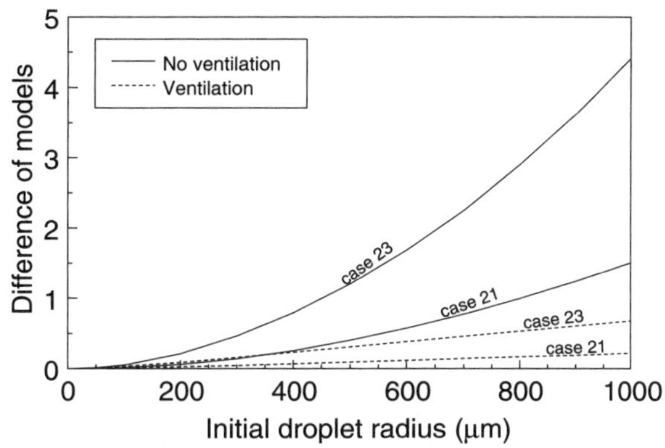

FIGURE 27.8 Relative difference of AERCLOUD predictions compared to the homogeneous equilibrium limit, i.e., $(t_{am} - t_{he})/t_{he}$, where t is the droplet vaporization time. The subscript "am" refers to the aerosol model, with or without the convective mass and heat fluxes, and "he" refers to the homogeneous equilibrium limit.

a decrease of concentration. Our prime objective here was to examine the vaporization and condensation processes, and we have therefore ignored deposition. However, the deposition of droplets is certainly an important process in some conditions; clearly, deposition is highly sensitive to droplet size.

We find that if the droplets are small enough, then the heat and mass transfer processes to the droplet are fast enough (compared to the overall dispersion rate) to make homogeneous equilibrium a good approximation. In the heavy gas dispersion tests done here, 100 μm droplets are seen to be small enough. The question of the adequacy of the homogeneous equilibrium model is thus reduced to that of estimating droplet sizes. It is clear that in many processes involving violent flashing of pressure-liquefied gases, the drop sizes will be expected to be small enough to admit homogeneous equilibrium as a good approximation in the subsequent advection and dispersion of a heavy cloud.

27.6 REFERENCES

Blackmore, D., M. Herman, and J. Woodward. 1982. "Heavy Gas Dispersion Models," *Journal of Hazardous Materials,* vol. 6, pp. 107–128.

Bricard, P., and L. Friedel. 1998. "Two-phase Jet Dispersion," *Journal of Hazardous Materials,* vol. 59, pp. 287–310.

Brighton P. W. M. 1985. "Area Averaged Concentrations, Height Scales, and Mass Balances," *Journal of Hazardous Materials,* vol. 11, pp. 189–208.

Britter, R. E. *Recent Research on the Dispersion of Hazardous Materials,* European Communities Report EUR 18198 EN, Luxembourg.

Carissimo B., S. F. Jagger, N. C. Daish, A. Halford, S. Selmer-Olsen, K. Riikonen, J. M. Perroux, J. Würtz, J. Bartzis, N. J. Duijm, K. Ham, M. Schatzmann, and R. Hall. 1999. "The SMEDIS Database and Validation Exercises," presented at SMEDIS Workshop on the 6th International Conference on Harmonisation within Atmospheric Dispersion Modeling for Regulatory Purposes, Rouen, October 11–14.

Daish, N. C., R. E. Britter, P. F. Linden, S. F. Jagger, and B. Carissimo. 1998. "SMEDIS: Scientific Model Evaluation of Dense Gas Dispersion Models," in *Proceedings of the 5th International Conference on Harmonisation within Atmospheric Dispersion Modelling for Regulatory Purposes,* May 18–21, Rhodes, Greece, National Centre for Scientific Research ''Demokritos,'' Athens, pp. 54–61.

Hanna, S. R., and P. J. Drivas. 1987. *Vapor Cloud Dispersion Models,* Center for the Chemical Process Safety of the American Institute of Chemical Engineers, New York.

Hewitt, G. F., and M. J. Pattison. 1992. Modeling of Release and Flow of Two-phase Jets. The Safe Handling of Pressure Liquefied Gases (IBC), London, November 26–27.

Johnson, D. W. 1991. ''Prediction of Aerosol Formation from the Release of Pressurized, Superheated Liquids to the Atmosphere,'' in *Proceedings of International Conference and Workshop on Modeling and Mitigating the Consequences of Accidental Releases of Hazardous Materials,* New Orleans, May 20–24, American Institute of Chemical Engineers, New York, pp. 1–34.

Koopman, R. P., T. G. McRae, H. C. Goldwire, Jr., D. L. Ermak, and E. J. Kansa. 1986. ''Results of Recent Large-Scale NH3 and N2O4 Dispersion Experiments,'' in *Heavy Gas and Risk Assessment III: Proceedings of Third Symposium on Heavy Gas and Risk Assessment,* Bonn, November 12–13, ed. S. Hartwig, D. Reidel, Dordrecht, pp. 137–156.

Koopman, R. P., D. L. Ermak, and S. T. Chan. 1989. ''A Review of Recent Field Tests and Mathematical Modelling of Atmospheric Dispersion of Large Spills of Denser-than-Air Gases,'' *Atmospheric Environment,* vol. 23, pp. 731–745.

Kukkonen, J. 1990. *Modelling Source Terms for the Atmospheric Dispersion of Hazardous Substances,* Commentationes Physico-Mathematicae 115, The Finnish Society of Sciences and Letters, Helsinki.

Kukkonen, J., T. Vesala, and M. Kulmala. 1989. ''The Interdependence of Evaporation and Settling for Airborne Freely Falling Droplets,'' '*Journal of Aerosol Science,* vol. 20, no. 7, pp. 749–763.

Kukkonen, J., M. Kulmala, J. Nikmo, T. Vesala, D. M. Webber, and T. Wren. 1993. *Aerosol Cloud Dispersion and the Suitability of the Homogeneous Equilibrium Approximation,* AEA Report AEA/CS/HSE R 1003/R, Warrington, Cheshire.

Kukkonen, J., M. Kulmala, J. Nikmo, T. Vesala, D. M. Webber, and T. Wren. 1994. ''The Homogeneous Equilibrium Approximation in Models of Aerosol Cloud Dispersion,'' *Atmospheric Environment,* vol. 28, pp. 2763–2776.

Kulmala, M., and T. Vesala. 1991. ''Condensation in the Continuum Regime,'' *Journal of Aerosol Science,* vol. 20, pp. 337–346.

Kulmala, M., T. Vesala, and P. E. Wagner. 1993. ''An Analytical Expression for the Rate of Binary Condensational Particle Growth,'' *Proceedings of the Royal Society of London A,* vol. 441, pp. 589–605.

Marshall, V. C. 1987. *Major Chemical Hazards,* Ellis Horwood, New York.

McQuaid, J., and B. Roebuck. 1985. *Large Scale Field Trials on Dense Vapor Dispersion, Final Report on the Heavy Gas Dispersion Trials at Thorney Island 1982–1984,* Report EUR 10029, Commission of the European Communities, Brussels, pp. 417.

Moodie, K., and B. C. R. Ewan. 1990. ''Jets Discharging to the Atmosphere,'' *Journal of Loss Prevention in the Process Industries,* vol. 3, pp. 68–76.

Muralidhar, R., G. R. Jersey, F. J. Krambeck, and S. Sundaresan. 1995. ''A Two-Phase Release Model for Quantifying Risk Reduction for Modified HF Alkylation Catalysts,'' *Journal of Hazardous Materials,* vol. 44, pp. 141–183.

Nielsen, M., S. Ott, H. E. Jørgensen, R. Bengtsson, K. Nyrén, S. Winter, D. Ride, and C. Jones. 1997. ''Field Experiments with Dispersion of Pressure Liquified Ammonia,'' *Journal of Hazardous Materials,* vol. 56, pp. 59–105.

Nikmo, J., J. Kukkonen, T. Vesala, and M. Kulmala. 1994. ''A Model for Mass and Heat Transfer in an Aerosol Cloud,'' *Journal of Hazardous Materials,* vol. 38, pp. 293–311.

Nolan, P. F., G. N. Pettitt, N. R. Hardy, and R. J. Bettis. 1990. ''Release Conditions Following Loss of Containment,'' *Journal of Loss Prevention in the Process Industries,* vol. 3, pp. 97–103.

Schmidli, J., S. Banerjee, and G. Yadigaroglu. 1990. ''Effects of Vapour/Aerosol and Pool Formation on Rupture of Vessels Containing Superheated Liquid,'' *Journal of Loss Prevention in the Process Industries,* vol. 3, pp. 104–111.

Vesala, T. 1991. ''Binary Droplet Evaporation and Condensation as Phenomenological Processes,'' Ph.D. thesis, University of Helsinki; Commentationes Physico-Mathematicae 127, The Finnish Society of Sciences and Letters, Helsinki.

Vesala, T. 1993. "On Droplet Evaporation in the Presence of a Condensing Substance: The Effect of Internal Diffusion," *International Journal of Heat Mass Transfer,* vol. 36, no. 3, pp. 695–703.

Vesala, T., J. Kukkonen, and M. Kulmala. 1989. *A Model for Evaporation of Freely Falling Droplets,* Publications on Air Quality 6, Finnish Meteorological Institute, Helsinki.

Vesala, T., and J. Kukkonen. 1992. "A Model for Binary Droplet Evaporation and Condensation, and its Application for Ammonia Droplets in Humid Air," *Atmospheric Environment,* vol. 26A, pp. 1573–1581.

Vesala, T., and M. Kulmala. 1993. "Comparisons of Uncoupled, Film Theoretical and Exact Solutions for Binary Droplet Evaportion and Condensation," *Physica A,* vol. 192, pp. 107–123.

Webber, D. M, S. J. Jones, G. A. Tickle, and T. Wren. 1992. *A Model of a Dispersing Dense Gas Cloud and the Computer Implementation DRIFT: I. Near-instantaneous Releases,* AEA Report SRD/HSE R586.

Wheatley, C. J. 1987. *Discharge of Liquid Ammonia to Moist Atmospheres: Survey of Experimental Data and Model for Estimating Initial Conditions for Dispersion Calculations,* UKAEA Report SRD/HSE/R410, London.

Wheatley, C. J., and D. M. Webber. 1984. Aspects of the Dispersion of Denser-than-Air Vapours Relevant to Gas Cloud Explosions, UKAEA, Safety and Reliability Directorate, Commission of the European Communities Report EUR592.

Witlox, H. W. M., and A. Holt. 1999. "A Unified Model for Jet, Heavy, and Passive Dispersion Including Droplet Rainout and Re-evaporation," in *International Conference and Workshop on Modeling the Consequences of Accidental Releases of Hazardous Materials,* San Francisco, September 28–October 1, American Institute of Chemical Engineers, New York, pp. 315–344

Woodward, J. L., J. Cook, and A. Papadourakis. 1995. "Modeling and Validation of a Dispersing Aerosol Jet," *Journal of Hazardous Materials,* vol. 44, pp. 185–207.

Woodward, J., and A. Papadourakis. 1991. "Modeling of Droplet Entrainment and Evaporation in a Dispersing Jet," in *Proceedings of International Conference and Workshop on Modeling and Mitigating the Consequences of Accidental Releases of Hazardous Materials,* New Orleans, May 20–24, American Institute of Chemical Engineers, New York, pp. 147–167.

P · A · R · T · 6

DISPOSAL

CHAPTER 28
DISPOSAL OF WASTE

K. E. Sellers
AMEC Earth & Environmental, Inc., Westford, Massachusetts

28.1 INTRODUCTION

Spill response often begins with actions intended to remove or stabilize a spill. The objective of these actions may be to prevent acute human health and environmental effects and/or to prevent further migration of the spilled material. After the situation stabilizes, long-term remediation may be necessary to restore the soil and groundwater to acceptable conditions.

The options for waste disposal typically depend on the characteristics of the waste, the governing laws and regulations, and the cost of disposal options. Time constraints, the availability of materials, labor, or equipment, and limits on the work area can also affect the choice of waste treatment or disposal.

This chapter describes options for the treatment and disposal of waste generated from the initial response actions. It also briefly describes some of the treatment or disposal options for long-term response.

28.2 WASTE CHARACTERISTICS

The physical characteristics of a substance can determine how it moves through the environment after a spill (Chapter 11). The physical and chemical characteristics also determine the health and safety precautions that must be taken during spill cleanup (Chapters 8 and 10). Finally, those characteristics determine how a waste resulting from spill cleanup can be treated or disposed. The material safety data sheet (MSDS) provides fundamental information about the composition and characteristics of commercial products.

28.3 WASTE TRANSPORTATION[1]

Depending on the waste classification, certain limitations apply to storing and transporting waste material off-site. The limitations of environmental laws in the United States are dis-

[1]Adapted and reprinted with permission from K. Sellers, *Fundamentals of Hazardous Waste Site Remediation*, CRC Press/Lewis Publishers, Boca Raton, FL, 1999, p. 217. Copyright CRC Press, Boca Raton, Florida.

cussed in this section. For information on the applicable laws and regulations of other countries, see Chapters 1 through 3.

In the United States, the Department of Transportation (DOT) regulates the shipment of hazardous waste and hazardous materials (49 CFR Chapter I, Subchapter C, Hazardous Materials Regulations). The DOT regulations identify hazardous materials for the purpose of transportation and dictate requirements for labeling, packaging, and quantity limits aboard aircraft and stowage of hazardous materials aboard vessels (49 CFR 172). DOT regulations also include requirements for vehicle marking and manifests (49 CFR 171).

Bulk soils and solid waste may be shipped off-site for disposal or treatment by truck or, when available, by rail. Depending on their size, dump trailers can carry 15 to 28 m^3 (20 to 36 yd^3 (CY)) of soil; 15 m^3 (20 CY) trailers are commonly used. Roll-off boxes can carry 15 to 23 m^3 (20 to 30 CY). Loading can be volume-limited or weight-limited (by road weight limits). For example, a 20-CY dump trailer can carry only 18 tons of soil per load on certain roadways.

Shipping costs are commonly billed per loaded kilometer (mile) (i.e., distance between the site and the treatment or disposal facility). Charges can also include demurrage, or an hourly cost for time spent idling at the site waiting for the truck to be filled or paperwork to be filed. Ancillary charges may include a fee for waste acceptance at the treatment or disposal facility, a charge for a truck liner, or a surcharge to manage the presence of free liquid in the shipment once it reaches the facility.

Depending upon the waste, it may be regulated in the United States as a hazardous waste under the Resource Conservation and Recovery Act (RCRA) or analogous state regulations. RCRA regulations require that the waste be shipped under a manifest which provides information about the physical and chemical nature of the waste, its origin, and destination. The generator must fill out a manifest form for each truckload of waste.

28.4 OFF-SITE DISPOSAL IN LANDFILLS[2]

Off-site disposal in a landfill can be the most cost-effective option for certain wastes. It is still the easiest, cheapest option in most cases for handling small amounts of soil or waste. The cost of landfilling soils or solid wastes depends on the waste classification, treatment required to meet land disposal restrictions, if any, and proximity to a disposal facility. The decision to dispose of soil must balance these cost factors against the potential liabilities from transportation and disposal of contaminated material.

28.4.1 Waste Classification

Not all of the waste from a spill is hazardous waste. "Waste" soil or sludge can be classified as hazardous waste or nonhazardous waste under applicable regulations. In the United States, a waste is a hazardous waste under federal RCRA regulations when it is (1) *listed* in the regulations (40 CFR 261.30–261.33); or (2) a *characteristic* hazardous waste, determined by chemical testing (40 CFR 261.20–261.24). State regulations can be more stringent than federal regulations. (See Chapters 1 through 4 for additional information on pertinent laws and regulations.)

A disposal facility typically requires a waste profile before it will accept waste for disposal. The profile includes information on the source of the waste, its physical state and chemical composition, the volume, and the type of shipping. The generator must also certify

[2]Adapted and reprinted with permission from K. Sellers, *Fundamentals of Hazardous Waste Site Remediation*, CRC Press/Lewis Publishers, Boca Raton, FL, 1999, p. 218. Copyright CRC Press, Boca Raton, Florida.

that the waste meets or does not meet regulatory criteria for waste classification under applicable regulations, such as RCRA, DOT regulations, and/or regulations promulgated under the Toxic Substances Control Act (TSCA). The disposal facility may also request a sample of the waste.

28.4.2 Disposal Requirements

Not all hazardous wastes can be put in a landfill. In the United States, Land Disposal Restrictions promulgated under RCRA (40 CFR 268) require that most hazardous wastes meet certain concentration limits or be treated using a specified technology before disposal. Certain hazardous wastes cannot be placed in a landfill under any circumstances.

In the United States, RCRA regulations govern the construction, operation, and closure of hazardous waste landfills (40 CFR 264, Subpart N). The permit requirements generally include the following conditions:

- The landfill must have a double liner system that can prevent any migration of wastes out of the landfill to the adjacent soil, groundwater, or surface water. The liner system must include a leachate collection and removal system between the two liners.
- Operators must monitor and inspect the landfill during operation to ensure, for example, that the leachate collection and removal system operates correctly.
- Upon closure, the owner or operator must cover the landfill with a cap. The cap must be designed and constructed to achieve specified performance requirements.
- The owner or operator must monitor and maintain the landfill for a postclosure period, typically 30 years.

28.5 OFF-SITE TREATMENT

This section describes relatively common fixed-base treatment options. Table 28.1 lists those methods and the contaminants that they can treat.

28.5.1 Incineration[3]

Incineration destroys organic materials by thermal oxidation. Organic compounds oxidize to form carbon dioxide, water, sulfur dioxide, nitric oxide, ash, and/or products of incomplete

TABLE 28.1 Common Fixed-Base Treatment Options

Type of treatment	Compounds commonly treated
Incineration	Organic compounds
Thermal desorbtion	Petroleum hydrocarbons; organic solvents
Solidification/stabilization	Metals

[3]Adapted and reprinted with permission from K. Sellers, *Fundamentals of Hazardous Waste Site Remediation*, CRC Press/Lewis Publishers, Boca Raton, FL, 1999, p. 273–279. Copyright CRC Press, Boca Raton, Florida.

combustion (PICs). Incineration does not destroy metals. Metals concentrate in the ash or enter the off-gas treatment system.

Virtually any organic contaminant can be incinerated. Because of the relatively high cost of handling and heating soil to incineration temperatures, other technologies, such as soil vapor extraction and thermal desorption, are sometimes used to separate organic contaminants from soil so that the contaminants can be incinerated in vapor or liquid phase. However, concentrated liquid or solid hazardous wastes are often incinerated directly. Incineration is the Best Developed Available Technology for certain organic wastes under the U.S. RCRA Land Disposal Restrictions (40 CFR 268). In the United States, commercial incinerators must be permitted to handle RCRA hazardous wastes. TSCA regulations require that concentrated polychlorinated biphenyl (PCB) wastes (>500 mg/kg) must be incinerated (40 CFR 761). Incineration of certain PCB-contaminated materials is regulated under TSCA (40 CFR 761.70). The Federal Insecticide, Fungicide, and Rodenticide Act (FIFRA) regulates the incineration of certain pesticides (40 CFR 165).

Process Description. An incineration system consists of three main units: feed preparation and handling, primary combustion unit, and off-gas treatment. Feed preparation usually includes screening to remove oversized particles that could clog or jam the combustion unit. Oversized waste may be shredded so that it can be incinerated, or treated or landfilled separately. Sludge is usually dewatered to make the material easier to feed to the incinerator and to limit the fuel cost associated with heating, volatilizing, and treating the water in the incinerator.

The combustion unit is designed to provide the temperature, turbulence, and time necessary to oxidize contaminants. Several types of combustion units are used to treat soils or sludge containing hazardous waste. Rotary kilns are the most common units. Fluidized bed incinerators and other designs, such as infrared incineration systems, are also used.

The combustion unit in a rotary kiln incinerator is a refractory-lined cylinder mounted on a slight incline. Wastes and auxiliary fuel injected into the high end of the kiln move downward through the combustion zone as the kiln slowly rotates. The retention time in the kiln is typically 30 to 60 minutes. At the operating temperature of 650 to 980°C (1,200 to 1,800°F), organic materials oxidize to gaseous compounds and ash. The ash is removed from the lower end of the kiln. It may be sprayed with water to cool the hot ash and control dust. The off-gases contain water vapor, some desorbed contaminants, volatile metals, particulates, and the products of combustion. The off-gases pass through a secondary combustion chamber, then through air pollution control units for particulate and acid gas removal.

Ash leaving the combustion unit must be landfilled. If the ash contains high levels of metals, it may require treatment before disposal. The quantity of ash (as a percentage of the feed) varies but depends on the level of combustible organic material in the original waste.

Limitations. Incinerators are commonly and successfully used to treat a variety of wastes. However, incineration is subject to the following limitations:

- Wastes with a high water content can be expensive and difficult to burn.
- High levels of halogens such as chloride cause acid emissions. While the emissions control system is designed to remove some acids, removal of high concentrations can be difficult and expensive.
- Volatile metals (such as lead, mercury, and arsenic) or their oxides will volatilize. These metals cannot be easily removed from off-gases using conventional air pollution control equipment.
- Trivalent chromium can be oxidized to the more toxic and mobile hexavalent form in an incinerator.

- High metal concentrations in the ash can render the ash a hazardous waste, making treatment of the ash before disposal necessary.
- A high proportion of fine particles such as clays and silts in the soil fed to an incinerator can result in high particulate loading in off-gases and thus clogging.

Waste can be incinerated on a hazardous waste site or at an off-site commercial incinerator permitted to handle such wastes. The choice between on-site and off-site incineration depends on logistics, safety issues, and cost. Mobilizing an incinerator to a site and obtaining a permit for its use require months of planning and preliminary work. Therefore, it is not practical as a quick response to a spill. Public reaction to on-site incineration in a populated area can also make that option infeasible. For relatively small quantities of waste, off-site incineration is far more cost-effective than on-site incineration.

Waste Characterization. The physical and chemical parameters used to characterize waste for incineration depend on the treatment facility. They commonly include:

- Contaminants and levels, in particular of PCBs, metals, cyanide, and pentachlorophenol
- Waste classification under RCRA and TSCA
- Proximate analysis, which includes the moisture content, ash content, volatile matter, fixed carbon, and heating value (BTU/LB) of the waste
- Ultimate analysis, which provides the levels of the elements C, H, N, S, Cl as a percentage of the waste material
- Viscosity and density, for sludge, tars, or similar materials

Commercial facilities may also require other information.

28.5.2 Thermal Desorption[4]

Thermal desorption is the process of heating soil or sludge to volatilize and remove contaminants from the soil. The off-gases are collected and treated. Soil can be excavated and treated or treated in situ. The treatment of excavated waste is described in this section.

Thermal desorption is the Best Developed Available Treatment (BDAT) for certain wastes under the RCRA Land Disposal Restrictions (40 CFR 268). Depending on the operating temperature of the unit, thermal desorption can remove volatile organic compounds, such as benzene; petroleum hydrocarbons; creosote or coal tar; polycyclic aromatic hydrocarbons (PAHs); certain herbicides and pesticides; PCBs; and dioxins.

Excavated soil can be treated using a mobile unit brought to the site or taken off-site to a permitted fixed-base unit. As for incineration, the decision to treat waste on- or off-site depends on the logistics, volume of soil, and cost.

Process Description. The basic components of a thermal desorption system include pretreatment and feed, thermal processor and discharge, and air emissions control. Each of these components is described briefly here. The critical parameters that affect treatment include (1) the type of soil and moisture content and (2) temperature, residence time, mixing, and the sweep gas flow rate in the thermal processor.

Thermal desorption systems typically have particle size limitations of 2 to 10 cm (0.75 to 4 in.), depending on the components in the system. Soil containing oversized particles must be screened to remove large particles before treatment. The oversized particles can be

[4]Adapted and reprinted with permission from K. Sellers, *Fundamentals of Hazardous Waste Site Remediation*, CRC Press/Lewis Publishers, Boca Raton, FL, 1999, p. 245–250. Copyright CRC Press, Boca Raton, Florida.

crushed and treated by thermal desorption or, depending on the level of contamination, backfilled on site or landfilled. Soil or sludge may also be pretreated to remove or absorb water before treatment. Several vendors suggest that soil or sludge should contain no more than 20 to 50% water, depending on the system, if it is to be treated by thermal desorption.

Soils are loaded into the heating chamber using several types of feed equipment. In many systems the soil conveyor passes through an air lock to minimize the air entering the thermal processor. Soils are commonly dumped into a hopper using a front-end loader, then fed to the thermal processor using an auger or screw conveyor. Some systems use belt conveyors.

Thermal desorption systems are classified by the temperature at which they operate and the type of equipment used to heat the soils. In general, low-temperature thermal desorption units operate at temperatures between 150 and 430°C (300 and 800°F). Low-temperature units treat soil that contains contaminants with relatively low boiling points, such as volatile organic compounds and petroleum hydrocarbons. High-temperature thermal desorption units generally operate between 320 and 650°C (600 and 1,200°F). Such units are used to treat soils containing contaminants with higher boiling points, such as PAHs and PCBs. The distinction between high-temperature and low-temperature units is not precise, and it refers to different operating conditions rather than different types of thermal processors. High- and low-temperature thermal desorption systems may be constructed of different materials, as appropriate for the operating temperatures, and include different emission-control technologies, as appropriate for the target contaminants that are typically treated at those operating temperatures.

Several different types of equipment are used to heat soils, a rotary drier or drum being the most common. A rotary drier is a steel cylinder mounted on a slight incline. The soil is fed to the high end of the drier or drum. As the drier rotates, the soil moves downward through the unit. The design can incorporate internal mixers or lifters that lift the soil as the drier rotates, carry the soil to the top of the cylinder, then drop it through the hot gases. The retention time in the drier is generally 20 minutes or less.

Other thermal desorption systems use a hollow-flight auger to heat the soils. One or more hollow-flight augers or screws rotate within a jacketed trough to move the waste through the unit. A heat-transfer fluid such as oil or molten salt flows through the center of the hollow flights of the screw augers and through the trough jacket to provide the thermal energy needed to desorb contaminants. The heat transfer fluid circulates through a heater fired by propane, natural gas, or fuel oil. Finally, a few vendors use infrared heating or fluidized bed reactors.

Regardless of the type of thermal processor, the goal of thermal desorption is to separate, not to incinerate, the contaminants. Technicians achieve this goal by operating a thermal desorption unit at temperatures lower than an incinerator, and often by using a sweep gas in the thermal processor that contains little or no oxygen. This gas stream, which may be pure nitrogen or an air stream with some oxygen removed (commonly combustion gas), sweeps the desorbed contaminants from the thermal processor to the off-gas treatment system. Once treated, an inert gas such as nitrogen is recycled back through the thermal processor.

The rate at which soils can be treated depends primarily on the size and type of the thermal processor, the moisture content of the soils, the type of soils, and the level of contamination. Wetter soils must be processed more slowly than dry soils. All else being equal, sandy soils can be treated at roughly twice the rate of clay soils. Many mobile systems can treat soils at a rate between 20 and 40 tons per hour (tph), with processing rates for some full-scale systems as low as 3 tph or as high as 100 tph. (For efficiency, thermal desorption systems can operate 24 hours per day rather than cooling and reheating the system each work day.) A mobile system occupies an area roughly 21 m by 24 m (70 ft by 80 ft) to 60 m by 90 m (200 ft by 300 ft), depending on the system and not including soil stockpile areas or ancillary facilities.

Treated soil typically exits the thermal processor by a screw conveyor. The clean, hot soils are misted or mixed with water—sometimes obtained from the air pollution control system—to cool the soil and suppress dust. Finally, the soils are stockpiled for testing and eventual backfilling on site or landfilling.

The type of emissions control varies with the type of system and the waste stream being treated. Off-gas treatment may include:

- A baghouse or cyclone, to remove particulates
- A condenser, to cool the gas stream and remove water and/or liquid hydrocarbons from the gas stream. Water must be treated, and condensed organics may be recycled or incinerated
- An afterburner, to destroy organic contaminants
- Activated carbon, to adsorb organic contaminants

Limitations. Operating concerns include the type of soil, the type of contamination, and regulatory and public concerns. The process has the following limitations:

- Oversized soil particles or debris must be removed before treatment.
- Soils with a high proportion of fines (clays) can cake and can generate high levels of particulates in off-gases. Some vendors suggest a limit of 30% fines. Above that level, soil must be fed more slowly or should be mixed with coarser soils to decrease the level of fines.
- A high water content increases the cost of treatment and can cause materials-handling problems.
- In systems that contain some oxygen, operators may limit the acceptable concentration of petroleum hydrocarbons in the feed (to approximately 1 to 3%) in order to keep the contaminant concentrations in the vapors below the lower explosive limit. This limit can be met by blending soils.
- The vapors from treating wastes that contain relatively high levels of sulfur or chlorine can corrode system components and may require treatment to remove acid gases before discharge to the atmosphere. Many thermal desorption systems do not include that type of pollution control equipment.
- Thermal desorption does not remove most heavy metals. Portland cement or another stabilizing agent can be added to treated soil when it is mixed with water in order to limit the mobility of metals in the treated soil.
- Thermal desorption can result in partial breakdown and reformation of contaminant molecules. As a result, the residuals can (but do not always) contain new contaminants of concern.
- Some systems are noisy, and since they run 24 hours per day, the level of noise can cause problems if operated near residential areas or in other sensitive areas.
- Regulatory requirements vary. Thermal desorption systems that vaporize and then burn organic contaminants in hazardous waste are considered incinerators for the purpose of RCRA regulations in the United States. Thermal desorption systems that condense the organics so they can be recycled rather than incinerated can be considered recycling units that are exempt from RCRA regulations on treatment systems.

28.6 RECYCLE/REUSE OPTIONS

Some waste materials can be reused, recycled, or used for energy recovery. The ability to reuse or recycle a material depends upon several factors:

TABLE 28.2 Options for Reuse/Recycle

Option	Spill/waste type
Distillation/rerefining	Petroleum products; solvents
Energy recovery	Organic solvents; petroleum products
Asphalt batching	Petroleum products
Mercury recovery	Elemental mercury

- The type of waste
- The level of impurities resulting from spilling the material into the environment and then recovering the material
- The availability of recovery facilities, market for the material, and economics of reusing or recycling the material

Table 28.2 lists common reuse/recycle options and the applicable wastes. Each technique is briefly described in this section.

28.6.1 Distillation/Rerefining

Some relatively pure liquids recovered from the environment can be distilled or rerefined and reused. For example, petroleum products recovered from some waste sludge by thermal desorption or waste oils collected from automobiles can be rerefined to make a usable product. The potential to recover usable product from material spilled into the environment can be limited, however. Once spilled, a solvent or petroleum product begins to change physically and chemically. The liquid can become "contaminated" with water, dirt, or other substances. After some time, chemical and biochemical reactions will alter the chemical composition of the material. Further, the volume recovered from spill cleanup may be relatively small and a rerefining facility may not be located nearby. As a result, it may be more feasible simply to ship the waste for energy recovery.

28.6.2 Energy Recovery

Petroleum wastes or other flammable liquid wastes are often burned for energy recovery. Relatively small amounts recovered from a spill can be shipped to a fuel blender, who mixes flammable waste liquids to achieve acceptable physical and chemical properties and markets the resulting material as fuel. Flammable wastes are sometimes burned at cement kilns for energy recovery because cement manufacture is highly energy-intensive. In the United States, facilities that recover energy from hazardous wastes are regulated under the RCRA Boiler and Industrial Furnace (BIF) Rules.

28.6.3 Asphalt Batching[5]

Soil contaminated with petroleum hydrocarbons, typically from a leaking underground fuel storage tank, can be incorporated into asphalt. The soil is substituted for a portion of the

[5]Adapted and reprinted with permission from K. Sellers, *Fundamentals of Hazardous Waste Site Remediation*, CRC Press/Lewis Publishers, Boca Raton, FL, 1999, p. 224–226. Copyright CRC Press, Boca Raton, Florida.

stone aggregate ordinarily used to make asphalt concrete. Asphalt-batching processes are characterized as either hot mix or cold mix. Each type of process is briefly described below.

Hot-Mix Process. In the hot-mix process, the aggregate (soil) is dried at temperatures ranging from 150 to 315°C (300 to 600°F) to remove moisture. The drier is akin to the rotary driers used in thermal desorption and typically includes air pollution control systems to remove particulates and organics from the off-gases. After drying, the soil is then mixed with hot liquid asphalt. The product cures, or hardens, as it cools.

When petroleum-contaminated soils are mixed with the clean aggregate and dried, volatile compounds volatilize and, depending on the drier temperature and retention time, oxidize. Most asphalt-batching plants have extensive facilities to control air emissions. Volatilization of volatile organic compounds (VOCs) during the drying step is critical. Large quantities of the lighter petroleum hydrocarbons act as solvents and soften the final asphalt product, resulting in an inferior product.

Volatilization and thermal destruction are the primary methods of remediation of lighter hydrocarbons in the hot-mix asphalt process. Solidification/stabilization is the secondary method of remediation. Heavier hydrocarbons remaining in the soil after the drying step are encapsulated within the asphalt/aggregate matrix.

In many states in the United States, commercial hot-mix asphalt plants are permitted to accept contaminated soils containing petroleum hydrocarbons as a recycling operation. A commercial plant is typically responsible for the ultimate use of the asphalt.

Cold-Mix Process. In the cold-mix asphalt-batching process, aggregate is mixed with an asphalt emulsion in a pug mill. (A pug mill is a large self-contained unit with a horizontal-blade mixing apparatus. Pug mills are used in common construction operations, for example, to mix cement.) The asphalt emulsion contains particles of asphalt suspended in an aqueous solution. Emulsifying agents (surfactants) are used to keep the asphalt in suspension. The proportions are approximately, by weight, 50 to 65% heated liquefied asphalt, 35 to 50% water, and 1 to 2% emulsifying agent. Mixing the emulsion with soil causes the emulsion to break, or separate into water and asphalt phases. The asphalt droplets coat the aggregate particles with a thin film.

When the asphalt emulsion/aggregate mixture is applied to the area to be paved and compacted, the water is expelled under pressure. The asphalt and aggregate mixture cures, developing into bituminous concrete, as the remaining water evaporates. High humidity, low temperature, and rainfall hinder curing. The final volume of paving material is roughly the same as the in-place volume of contaminated soil.

Cold-mix asphalt batching occurs at ambient temperature. The ambient temperature has two effects. First, the product cannot be made or applied when the ambient temperature is below 0°C (32°F). Second, volatile compounds remain in the mixture, in contrast to the hot-mix process. Large proportions of VOCs can soften the final product. Leaching of these VOCs from the final product may also be a concern.

In many states in the United States, mobile cold mix asphalt plants can be brought on-site to make asphalt concrete from contaminated soils. The product must then be used-on site. Cold-mix asphalt can be used as base materials for private parking lots and roads. It does not have the strength of a hot-mix product.

Cold-mix asphalt batching may be subject to state regulations. For example, those regulations may limit the concentrations of contaminants in the feed, the materials used for the asphalt, the properties of the product, and the conditions under which the asphalt can be placed.

Limitations. In addition to the regulatory restrictions already described, the following factors may limit the application of asphalt batching:

- Only nonhazardous waste containing total petroleum hydrocarbons (TPHs) may be treated.
- Hot-mix plants may have upper limits on the level of TPH based on a state-issued recycling permit. Higher prices may be charged for highly contaminated soils. The plant's permit limits may preclude acceptance of highly contaminated soils. Such limits vary but can be on the order of 30,000 to 60,000 mg/kg TPH.
- In cold climates, hot-mix plants may not operate in the winter, since the use of the product in construction slows during the winter.
- A high proportion of fines in soil can result in unacceptable product. Fines cannot be readily and uniformly coated by the asphalt binder and tend to separate into uncoated pockets in the bituminous concrete. Soils with a high proportion of clays are unacceptable.
- Stones or debris larger than 5 to 8 cm (2 to 3 in.) in diameter must be screened out of the soil before asphalt-batching.
- Long-term liability may remain.
- The volume of soil that can be treated may be limited by the use for the product.

28.6.4 Mercury Recovery

Mercury can be recovered for reuse from soil or other wastes contaminated from a spill of mercury. In the United States, thermal treatment to recover mercury is the Best Developed Treatment Technology for various RCRA nonwastewater wastes containing more than 260 mg/kg mercury (40 CFR 268).

A variation of thermal desorption called retorting is used to recover elemental mercury from waste. In a batch process, operators place the waste material in a retort (or furnace).

TABLE 28.3 Options for On-site Treatment

Option	Description	Spill/waste type
Air sparging/soil vapor extraction	Air is injected through wells screened below the water table to strip volatile compounds from groundwater and/or stimulate biodegradation. Contaminated soil vapors are extracted from the vadose zone for treatment above ground.	Organic solvents; petroleum products
Bioremediation	In a land farm, lined treatment unit, bioreactor, or in situ (biosparging), oxygen or other supplements are added to optimize biodegradation.	Petroleum products; certain solvents
Multiphase extraction	A vacuum applied to light nonaqueous phase liquid (LNAPL) through a "slurp tube" recovers LNAPL, vapors, and some groundwater for treatment at the surface.	LNAPL solvents; petroleum products
Phytoremediation	A variation of bioremediation that relies on plants, rather than microorganisms, to stabilize or degrade contaminants.	Certain metals; certain organic compounds
Pump and treat	Groundwater and/or LNAPL are pumped from the ground for treatment.	LNAPL; groundwater contamination
Solidification/stabilization	Additives such as portland cement are mixed with soil in situ or in an above-ground treatment unit to limit the mobility of contaminants.	Metals

The furnace is sealed and heated under vacuum to a temperature over 700°C (1,300°F). Mercury that vaporizes from the waste is collected in the off-gas and condensed to a liquid. This liquid mercury is distilled to further purify the material. Off-gases are further treated to remove water vapor and traces of mercury (U.S. EPA, 1994; Bethlehem Resource Recovery Division). As an alternative to the batch process, mercury can be recovered in a continuous retort process utilizing a rotary kiln (AERC/MTI).

Mercury-bearing wastes are typically transported to permitted, fixed-base facilities for treatment. However, mobile units that can be brought to the site of a spill are available. One vendor of a transportable unit, Mercury Recovery Services, Inc., has indicated that treatment can achieve concentrations as low as less than 1 mg/kg mercury in the treated solids (Mercury Recovery Services, Inc., 1997).

28.7 OTHER TREATMENT OPTIONS

Table 28.3 lists some other common on-site treatment options. Many of these techniques are commonly used for long-term response to a spill rather than short-term cleanup. These techniques may be used with the disposal options described in this chapter, or instead of those disposal options. Chapter 8 provides additional information regarding on-site treatment techniques.

28.8 REFERENCES

Advanced Environmental Recycling Group/Mercury Technologies International (AERC/MTI). 1999. "Continuous Retort Processing," http://www.aercmti.com/retort/.

Bethlehem Resource Recovery Division, Bethlehem Apparatus Company, Inc., "Mercury Recovery/Recycling Service," http://www.bethapp.thomasregister.com/olc/bethapp/merc01.htm.

Mercury Recovery Services, Inc. 1997. Profile in VISITT 6.0 (database), Technology Innovation Office, Office of Solid Waste and Emergency Response, December.

Sellers, K. 1999. *Fundamentals of Hazardous Waste Site Remediation*, CRC Press/Lewis, Boca Raton, FL, pp. 217–279.

U.S. EPA. 1994. *Handbook: Recycling and Reuse of Material Found on Superfund Sites*, EPA/625/R-94/004, Office of Research and Development, Washington, DC, September, pp. 51–53.

CHAPTER 29
DISPOSAL OF SMALL-SCALE SPILLS

Margaret-Ann Armour
Department of Chemistry, University of Alberta, Edmonton

29.1 INTRODUCTION

29.1.1 Applicability

Small-scale spills of hazardous chemicals can occur in laboratories in educational institutions, quality control and testing laboratories, hospitals, greenhouses, and wherever small quantities of hazardous chemicals are handled. Such spills have less potential to cause widespread problems than spills on an industrial scale. However, they pose a risk to the health of workers in the laboratory or at the site of the spill from the inhalation of fumes, from the potential hazards of reactive chemicals, and, in the case of liquids, from slipping on wet floors. It is recognized that small spills should be cleaned up as quickly, responsibly, and efficiently as possible. Workplaces must have a protocol for handling spills and provide spill kits in appropriate locations. It is important that such kits be readily available in locations where they are easily accessed when needed. Furthermore, a plan needs to be in place to guide what is to be done with the residues from cleanup of the spills (National Research Council, 1995).

Since 1981, we have been developing and testing in the laboratory, methods for the on-site disposal of small quantities of a variety of hazardous chemicals (Armour, 1996a). We have been able to apply many of the methods to disposing of the residues from the cleanup of spills of these chemicals. The application of several of the methods to spills of commonly used hazardous chemicals is described in subsequent paragraphs. The quantity of spilled material to which the disposal procedures are applicable varies with the hazard of the material spilled. In general, these procedures can be applied to the quantities of chemicals normally handled and transported in laboratories. As a guide, the disposal procedures are applicable to spills of up to about 4 L of liquid.

29.1.2 Requirements for Effective Handling of Small-Scale Spills

One of the most important requirements in handling small-scale spills is speed of reaction. The site of the spill should be isolated and workers not directly involved in the spill cleanup kept at a safe distance. For identifiable spills, the hazards are considered and appropriate personal protection worn with breathing apparatus if necessary. For spills of unknown substances it is best to wear full personal protection, including breathing apparatus. Solids that do not present a health or reactive hazard can be scooped up and either disposed of on-site using an appropriate method or packaged and labeled for disposal off-site. Reactive solids or those posing a health hazard should be covered with a powder form of absorbent and the mixture should be scooped into a container and disposed of according to an acceptable method. Liquid spills are absorbed using absorbent pillows or powdered absorbent, and the mixture is scooped into a container and disposed of appropriately. The area of the spill is washed with soap and water or with a suitable reagent solution.

29.1.3 Factors in the Disposal of Residues from Spill Clean-up

Cleanup of a spill results in residues of the chemical itself, or of absorbent material containing the spilled chemical. These require disposal. In deciding how to dispose of the absorbed material, a critical question is related to the reactivity of the spilled chemical. Workers who are knowledgeable about the properties of the spilled chemical are often the best people to perform the cleanup and disposal of the spill. If the spill was of a single known chemical or its solution, it may be possible for the laboratory worker to denature the chemical for on-site disposal, thus avoiding the risks of storage and transportation for off-site disposal. Specific procedures that can be used to convert the hazardous material at the laboratory bench to nonhazardous products are provided later in the chapter. Alternatively, the residue can be packaged and labeled for off-site disposal. If the spill was of a mixture or the identity of the spilled chemical is not known, the disposal may present a more difficult problem. The issue of spilled unknowns is discussed in more detail in Section 29.3.9.

29.2 SPILL MIX FOR THE ABSORPTION AND COVERAGE OF SMALL SPILLS

Many chemical supply companies sell spill kits that include absorbents. These may be in the form of loose powders or pillows. Frequently there are different absorbents for different types of spill. For example, one pillow is used for aqueous acid spills, another for aqueous alkaline spills, and a third for organic liquids. We have developed an absorbent that we have found to be practical and effective for all types of spills. It is easily prepared and is inexpensive so that containers of it can be available wherever hazardous chemicals are being used. The spill mix is particularly useful in schools and in the teaching laboratories of academic institutions since it can be made available at many locations.

29.2.1 Contents of Spill Mix

The spill mix contains sodium carbonate (soda ash, often available from swimming pools), bentonite (clay cat litter), and dry sand (Armour et al., 1999). The rationale for choosing these components for the spill mix was as follows. The sodium carbonate at least partially neutralizes any acid in the spill, thus rapidly reducing the corrosive impact of the acid on the surface where the spill occurred. The bentonite absorbs liquid and vapors effectively, and the sand moderates and helps to smother any reaction that may occur (for example, in a spill of a solution of lithium aluminum hydride in ether, there is a risk of fire).

29.2.2 Preparation of Spill Mix

The preparation of the spill mix is very simple. Equal volumes of sodium carbonate, bentonite, and dry sand are placed in a plastic container with a lid. The container is shaken until the components are mixed. It is also good to shake the container before use.

29.2.3 Method for Testing Spill Mix

We have tested the spill mix in the laboratory on simulated spills of over 250 hazardous chemicals and their solutions. Three characteristics were measured. First, we looked at the ability of the spill mix to absorb liquids and eliminate toxic and/or flammable fumes. Secondly, we assessed its effectiveness in reducing the danger of highly reactive solids catching fire. Thirdly, we determined methods by which the spill mix containing the absorbed spill could be converted to environmentally safe products for disposal.

Samples of the hazardous chemicals were spilled in a large evaporating basin. Spill mix was added to the basin until the liquid was completely absorbed or the solid was covered. With a plastic scoop, the mixture was transferred to a plastic or glass container and the container placed in the fume hood. Depending on the nature of the chemical that had been spilled, a method was selected to convert the hazardous chemical into nontoxic and environmentally acceptable products. The residues after the conversion reaction were identified. The concentration of the metal ion remaining after precipitation by sodium silicate was measured by ICP (Leco Plasmarray ICP Spectrometer, argon gas plasma, or, for lead ions, ICP/MS Elan Model 250, argon gas plasma). From the results of testing in the laboratory, appropriate treatment methods for specific spills have been documented.

29.2.4 Use of Spill Mix

The general procedure for using spill mix is to pour the spill mix onto the spill of liquid or reactive solid so that all of the liquid is absorbed or all of the reactive solid is completely covered. The residue is scooped up using a plastic scoop and placed in a plastic pail. For off-site disposal, the residue is packaged, labeled appropriately, and sent for disposal. In many cases the residue can be converted at the laboratory bench to nontoxic and nonhazardous residues for easier disposal. Some of the methods developed and tested in the laboratory to facilitate on-site disposal are described in Section 29.3. To ensure a fast cleanup, spill kits including gloves, goggles, plastic pail, plastic scoop, and step-by-step cleanup procedures should be readily available.

29.3 DISPOSAL OF RESIDUES FROM SPILL CLEANUP

29.3.1 Liquid Spills of Mineral Acids

To treat a spill of a mineral acid, such as concentrated hydrochloric, sulfuric, or nitric acids, breathing apparatus must be worn in addition to appropriate protective equipment. If the spill is of a dilute aqueous solution of a mineral acid, breathing apparatus may not be necessary. The spill is covered with spill mix until all of the liquid is absorbed. A plastic pail is three-quarters filled with water and the residue is transferred into the pail of water using a plastic scoop. The pail is transported to a fume hood. The pH of the solution is measured and, if necessary, the solution is neutralized by the addition of sodium carbonate or dilute (5%) aqueous sodium hydroxide solution.

Table 29.1 lists the approximate quantities of spill mix required for spills of 100 mL of the given liquids and the approximate volumes of 5% aqueous sodium hydroxide needed to neutralize the resultant solution when the spill mix is added to 5 L of water. After standing in the fume hood until the solids have settled, the liquid containing sodium chloride, sodium sulfate or sodium nitrate in aqueous solution is poured into the drain. The solids, bentonite and sand, are treated as normal garbage. When the aqueous acidic solution is neutralized with sodium carbonate or aqueous 5% sodium hydroxide solution, the acids are converted to their sodium salts as shown in Eqs. (29.1) and (29.2) respectively

TABLE 29.1 Quantities of Spill Mix Required to Absorb Some Acids and Bases and Volumes Required to Neutralize the Resulting Solutions

Liquid "spilled" (100 mL)	Weight of spill mix used	Volume of 5% aqueous NaOH to give pH 7	Volume of 5% aqueous HCl to give pH 7
Sulfuric acid	255 g	1.68 L	
Nitric acid	165 g	360 mL	
Phosphoric acid	220 g	700 mL[a]	
Hydrochloric acid	185 g	72 mL	
20% Sodium hydroxide	125 g		250 mL
Glacial acetic acid	150 g	600 mL[a]	

[a] Phosphoric acid and acetic acid form buffer solutions with aqueous sodium hydroxide. For these acids, enough 5% aqueous sodium hydroxide solution should be added to bring the solution to pH 5 rather than pH 7. This is the quantity indicated in the table.

$$2HCl + Na_2CO_3 \rightarrow 2NaCl + H_2O + CO_2 \quad (29.1)$$

$$H_2SO_4 + 2NaOH \rightarrow Na_2SO_4 + 2H_2O \quad (29.2)$$

29.3.2 Liquid Spills of Bases

Spills of aqueous solutions of sodium and potassium hydroxide and of ammonia are discussed in this section. Aqueous solutions of bases such as sodium or potassium hydroxide are covered with sufficient spill mix so that all of the liquid is absorbed. With a plastic scoop, the mixture is slowly added to a plastic pail about three-quarters full of cold water. The pail is transported to the fume hood. The pH of the solution is measured and it is brought to a pH of between 5 and 8 by the addition with stirring of solid sodium bisulfate or 5% aqueous hydrochloric acid.

Table 29.1 lists the approximate quantity of spill mix required for a spill of 100 mL of 20% sodium hydroxide solution and the volume of 5% aqueous hydrochloric acid required to bring the solution formed by the addition of the absorbent to 5 L of water to pH 7. After the solids have settled, the neutral aqueous liquid is decanted to the drain and the solid residue, consisting of sand and bentonite, can be discarded as normal garbage or washed with water, dried, and reused in spill mix. Neutralization of the spilled base with sodium bisulfate or 5% aqueous hydrochloric acid solution occurs according to Eqs. (29.3) and (29.4) respectively.

$$NaOH + NaHSO_4 \rightarrow Na_2SO_4 + H_2O \quad (29.3)$$

$$KOH + HCl \rightarrow KCl + H_2O \quad (29.4)$$

Spills of ammonia present two hazards. Laboratory workers may slip in the spill or be overcome by the vapors, especially if the ammonia is a concentrated solution. Thus, the worker handling the spill should wear breathing apparatus and other people should remain a safe distance away. The spill is covered with sufficient spill mix to absorb all of the liquid and maintain a dry layer of spill mix on top. The mixture is scooped into a container and transported to the fume hood. In the fume hood it is slowly added to a plastic pail of cold water. The liquid is neutralized by the addition, with stirring, of either solid sodium bisulfate or 5% aqueous hydrochloric acid. The aqueous solution is decanted into the drain and the solid, sand/bentonite, disposed as normal garbage or washed with water, dried, and reused to prepare spill mix. The addition of sodium bisulfate to the ammonia solution yields ammonium sulfate [Eq. 29(5)], while the addition of hydrochloric acid yields ammonium chloride [Eq. (29.6)].

$$2NH_4OH + NaHSO_4 \rightarrow (NH_4)_2SO_4 + 2H_2O \tag{29.5}$$

$$NH_4OH + HCl \rightarrow NH_4Cl + H_2O \tag{29.6}$$

29.3.3 Spills of Compounds with Highly Toxic Fumes

Breathing apparatus must be worn to manage spills of chemicals that have highly toxic fumes. Bromine is an example of a liquid with this characteristic, and iodine is an example of a solid. Everyone except those managing the spill must be instructed to stay at a safe distance from the spill.

A spill of bromine is covered as quickly as possible with spill mix to eliminate the fumes. When all of the bromine has been absorbed, the mixture is transferred using a plastic scoop to an empty plastic or glass container. The container is placed in the fume hood and the bromine/spill mix mixture is slowly added to a pail of cold water. During stirring, a 10% aqueous solution of sodium bisulfite is added to the liquid in the pail until it is colourless. The pH is measured and, if necessary, the liquid is neutralized by the addition, with stirring, of solid sodium carbonate or 5% aqueous sodium hydroxide solution. The liquid is decanted to the drain and the solid residue, consisting of a mixture of sand and bentonite, can be treated as normal garbage. The area of the spill can be washed with a 10% aqueous solution of sodium bisulfite to remove any remaining traces of bromine. When sodium bisulfite is added to the bromine/spill mix/water mixture, the bromine is converted to sodium bromide according to Eq. (29.7).

$$Br_2 + 4NaHSO_3 \rightarrow 2NaBr + Na_2SO_4 + 3SO_2 + 2H_2O \tag{29.7}$$

Iodine is a solid at room temperature, but it sublimes readily and the fumes are highly toxic. Therefore, covering the spilled crystals of iodine with spill mix avoids the danger of inhalation of the fumes. Spills of solutions of iodine can be treated similarly. The mixture of iodine and spill mix is scooped into a container and, in the fume hood, added to a pail of cold water. During stirring, a 10% aqueous solution of sodium bisulfite is added to the liquid in the pail until it is colorless and no crystals of iodine remain. The mixture is allowed to stand in the fume hood for at least 12 hours, stirred, and if the color of iodine has returned, more sodium bisulfite solution is added until the liquid is again colorless. If necessary, the aqueous solution is neutralized by the addition, with stirring, of solid sodium carbonate or 5% aqueous sodium hydroxide solution. The liquid is decanted to the drain and the solid residue, consisting of a mixture of sand and bentonite, is treated as normal garbage. As in the case of the bromine spill, the area of the iodine spill can be washed with a 10% aqueous solution of sodium bisulfite to remove remaining traces of iodine. When sodium bisulfite is added to the iodine/spill mix/water mixture, the iodine is converted to sodium iodide according to Eq. (29.8)

$$I_2 + 4NaHSO_3 \rightarrow 2NaI + Na_2SO_4 + 3SO_2 + 2H_2O \tag{29.8}$$

29.3.4 Spills of Flammable Solvents

Examples of flammable solvents are methanol, ethanol, acetone, diethyl ether, and hexane. Several of these liquids have toxic fumes but the greatest hazard of spills of these highly flammable liquids is the risk of fire. When a spill occurs, all ignition sources should be immediately shut off. Covering the spilled liquid with spill mix rapidly eliminates further evaporation and so reduces the risk of ignition. Sufficient spill mix should be used to completely absorb the liquid and provide a top layer of dry solid. Using a plastic scoop to avoid the chance of sparks, the mixture is transferred to a container and transported to the fume hood, where it is packaged and labeled for off-site disposal. In some jurisdictions the spill mix with the absorbed liquid may be allowed to stand in an open container in the fume hood until all of the liquid has evaporated. The remaining solid is added to water to dissolve the sodium carbonate and the residue is treated as normal garbage.

29.3.5 Spills of Organic Bases

Many organic bases have highly toxic vapors; pyridine has an especially nauseating odor. The following is an example of the oxidative destruction of an absorbed spill of pyridine or aniline using acidic aqueous potassium permanganate. The spill is covered with sufficient spill mix to absorb the liquid. The mixture is scooped into a container with a plastic scoop and transported to the fume hood. Enough water is added, with stirring, to dissolve the sodium carbonate present in the spill mix. After the solids have settled, the liquid is decanted into a glass or plastic container. Enough 6 M sulfuric acid is added to the liquid (frothing will occur) to give a pH of 2. (To prepare 6 M sulfuric acid, slowly add 10 mL of concentrated acid, with cooling if necessary, to 20 mL of cold water. Always add the acid to water, never the other way around.) Sufficient solid potassium permanganate is added to the stirred solution that the liquid remains purple. (A drop of the liquid placed on a piece of filter paper will show a purple ring.) The mixture is allowed to stand at room temperature for 48 hours and more permanganate is stirred in if the color is no longer purple. Solid sodium bisulfite is stirred into the solution until it is colorless, and then the solution is neutralized by adding solid sodium carbonate (frothing will occur) or 10% aqueous sodium hydroxide solution. The clear solution is decanted into the drain and any brown solid (manganese dioxide) discarded with normal garbage.

29.3.6 Spills of Aqueous Solutions Containing Heavy Metal Ions

Spills of aqueous solutions containing heavy metal salts present a disposal problem since many of these salts are highly toxic and should not be disposed in such a way that they will contaminate groundwater. The spill is covered with sufficient spill mix to absorb all of the liquid. The mixture is scooped into a container and transported to the fume hood. Sufficient water is added to dissolve the sodium carbonate in the spill mix. The solids are allowed to settle and the supernatant liquid decanted into another container. The solids are washed with water, allowed to settle, and the washings are added to the decanted liquid. An aqueous solution of sodium metasilicate ($Na_2SiO_3 \cdot 5H_2O$, 12.5 g per 100 mL of water) is added, with stirring, until there is no further precipitation of solid.

The pH of the solution containing the metal silicate should be adjusted, using 2 M sulfuric acid, to the pH shown in Table 29.2 for the metal ion that was present in the spilled aqueous solution. The pH listed in Table 29.2 is that at which maximum precipitation of the metal ion occurs. (To prepare 2 M sulfuric acid, slowly add 10 mL of concentrated acid, with cooling if necessary, to 50 mL of cold water. Always add the acid to water, never the other way around.) The mixture is allowed to stand at room temperature for 12 hours to ensure complete precipitation has occurred. The solid can be collected by filtration or the liquid allowed to evaporate in a large evaporation dish in the fume hood. The dry solid is packaged and labeled for disposal in accordance with local regulations. The reaction is shown in Eq. (29.9).

TABLE 29.2 The Optimum pH for the Precipitation of Metal Ions as Silicates

Metal ion	pH	Metal remaining in solution (ppm)
Pb(II)	10.5	0.8
Fe(II)	9.5 to 10	5
Fe(III)	10 to 10.5	2
Zn(II)	8.5	<0.5
Al(III)	8.5	<2
Cu(II)	10.5 to 11	0.03
Co(II)	9.5 to 10	0.08
Mn(II)	9.5 to 10	0.2
Ni(II)	9.5 to 10	0.3

$$\underset{\text{divalent metal ion}}{M^{2+}} + \underset{\text{sodium silicate}}{Na_2SiO_3} \rightarrow \underset{\text{metal silicate (insoluble)}}{MSiO_3} + 2Na^+ \tag{29.9}$$

Dichromate salts are particularly toxic and can be present in dichromate cleaning solutions or in plating baths. From our laboratory testing, we have developed a method for the handling of spills of aqueous solutions containing dichromate ions that precipitates the dichromate as a flocculent (rather than a gelatinous) precipitate of highly insoluble chromium hydroxide. The spilled aqueous solution containing dichromate ions is covered with sufficient spill mix to absorb all of the liquid. The mixture is scooped into a container and transported to the fume hood. Enough water is added to the mixture to dissolve the sodium carbonate in the spill mix. After the solids have settled, the liquid is decanted into a second container. The solids are washed with water, the solid is allowed to settle, and the washings are added to the decanted liquid. The acidity of the liquid is adjusted to pH 1 with 6 M sulfuric acid. (To prepare 6 M sulfuric acid, slowly add 10 mL of concentrated acid, with cooling if necessary, to 20 mL of cold water. Always add the acid to water, never the other way around.) During stirring, solid sodium thiosulfate ($Na_2S_2O_3 \cdot 5H_2O$) is added until the solution changes from orange to blue and cloudy. The solution is neutralized by the addition of sodium carbonate, and after a few minutes a blue-grey flocculent precipitate forms. The precipitate can be allowed to settle and the liquid decanted into the drain. The solid is allowed to dry in a large evaporating dish in the fume hood. The decanted liquid contains less than 0.5 ppm of dichromate. The dry solid is packaged and labeled for disposal in accordance with local regulations. The reaction of dichromate with thiosulfate proceeds according to Eq. (29.10).

$$\underset{\text{dichromate}}{Cr_2O_7^{2-}} + \underset{\text{sodium thiosulfate}}{3Na_2S_2O_3} + 2H_3O^+ \rightarrow \underset{\text{chromium hydroxide (insoluble)}}{2Cr(OH)_3} + 3Na_2SO_4 + 3S \tag{29.10}$$

29.3.7 Spills of Some Potentially Carcinogenic Materials

When a chemical is especially hazardous to the worker handling the spill, it is advisable to try to neutralize the spill before absorption onto spill mix. Such treatment is especially recommended when the spilled material is a known human carcinogen. Many of the chemicals used in chemotherapy are known to be mutagenic and therefore have the potential to be carcinogens. For example, daunorubicin is deactivated with acidic potassium permanganate solution (Castegnaro et al., 1985). If a solution containing daunorubicin is spilled, a solution of freshly prepared acidic potassium permanganate is poured over the spill. (This solution is prepared by adding 17 mL of concentrated sulfuric acid to 83 mL of water, cooling, and adding 4.7 g of potassium permanganate.) The appropriate quantity of potassium permanganate solution to use can be found by estimating the weight of daunorubicin in the spilled liquid and allowing 10 mL of solution per 50 mg of daunorubicin. The area of the spill must be isolated since the reaction should be allowed to proceed for at least two hours. A saturated solution of sodium bisulfite (10 g of sodium bisulfite, $NaHSO_3$, in 35 mL of water) is poured onto the spill to decolorize the potassium permanganate solution. The liquid is absorbed on spill mix and the mixture scooped into a container with a plastic scoop and transported to the fume hood. The clear liquid can be decanted to the drain and the solid residue treated as normal garbage.

29.3.8 Spills of Pesticides

Commercial concentrated pesticide formulations may be spilled when workers are preparing the diluted solution for treatment of plants. In some instances the surface on which the spill occurred may be somewhat porous. Then, in addition to the spill being cleaned up, it is important that the site of the spill be decontaminated so that the pesticide is not carried on the footwear of workers walking through the area. Spill management methods for Diazinon, Chlorpyrifos, and 2,4-D are described here (Armour et al., 1996).

Diazinon is an insecticide used to control sucking and chewing insects on a wide variety of crops. Chlorpyrifos is used to control insects and mites in stored grain and various pests on fruit, vegetables, and other crops. When a spill of the concentrated formulation of Diazinon or Chlorpyrifos occurs, it is suggested that the liquid be absorbed on spill mix and the solid scooped into a plastic pail. If the liquid has been spilled on a porous surface, the site of the spill should be decontaminated by isolating the area and pouring household chlorine bleach onto the site. The excess liquid is absorbed on spill mix and the solid residue combined with the solid from absorption of the pesticide in the plastic pail. The pail is transported to a fume hood or to a very well ventilated area, preferably out of doors. The volume of concentrate spilled is estimated, and for each 1 mL, 100 mL of household chlorine bleach is added to the solids. The mixture is stirred and allowed to stand in a secure and very well ventilated area for at least 18 hours. The liquid is poured into the drain and the solids washed with water and treated as normal garbage.

The herbicide 2,4-D is widely used for weeding cereal and other crops. The most commonly used form is the iso-octyl ester. Spills of the ester are treated by absorbing on spill mix and either packaging and labeling the residue for off-site disposal or scooping the residue into a plastic pail for on-site denaturation. To decompose the absorbed 2,4-D iso-octyl ester, enough water is added to the residue in the pail to dissolve the sodium carbonate in the spill mix. A solution of aqueous acidic potassium permanganate is prepared as follows. Concentrated sulfuric acid (17 mL) is added slowly to 70 mL of cold water. The solution is allowed to cool to room temperature and 4.7 g of potassium permanganate is dissolved in the solution. The volume of the solution is brought to 100 mL with cold water. The amount of 2,4-D iso-octyl ester in the spill is estimated, and 100 mL of acidic potassium permanganate solution is added for each 200 mg of 2,4-D iso-octyl ester. Foaming will occur due to the presence of sodium carbonate from the spill mix. The mixture is allowed to stand in a secure, well ventilated place (a fume hood is ideal) for 96 hours. Solid sodium bisulfite is added with stirring until the liquid is colorless. The pH is adjusted to between 6 and 8 by the addition with stirring of either solid sodium carbonate or 5% aqueous sodium hydroxide. The solids are allowed to settle, the liquid is decanted to the drain, and the solids are treated as normal garbage.

29.3.9 Spills of Unknown Substances

Spills of unknown substances produce a difficult situation. Fortunately, they are a rare occurrence, and even if the exact identity of the spilled chemical is unknown, it is usually not difficult to narrow the possibilities.

For spills of solid unknown substances, a judgment has to be made about the probable hazard of the spill. It is always wise to assume the worst-case scenario and wear complete personal protective equipment with breathing apparatus. If it is known that the solid is not dangerously reactive, then it can be scooped into a container, packaged, and labeled as accurately as possible for disposal. If there is a possibility that the spill is of a metal hydride or other solid that may spontaneously catch fire, it should be covered with spill mix, scooped into a plastic pail using a plastic scoop, and transported to the fume hood. Slowly and cautiously, butanol is added with stirring to the mixture. The weight of hydride spilled is estimated and about 40 mL of butanol is allowed for each 1 g of mixture

If the spill is of a liquid, it should be covered with spill mix as quickly as possible until all the liquid is absorbed and there is a dry layer of spill mix on top. If the spill is known to be of a flammable organic solvent, although the exact identity of the liquid is unknown, the adsorbed residue can be treated as described under Section 29.3.5. Similarly, if the spill is known to be of an aqueous solution containing heavy metal ions, the procedure described in Section 29.3.7 can be followed. If the nature of the liquid is completely unknown, either it can be analyzed to discover its identity and the residue can then be treated appropriately, or the residue from absorption on spill mix can be packaged, labeled as accurately as possible, and sent off-site for disposal.

29.4 CONCLUSIONS

For the over 250 hazardous chemicals that we have spilled and tested with spill mix, it has been found that liquids are rapidly absorbed and fuming is quickly eliminated. Because the spill mix is relatively inexpensive, containers of it can be kept in laboratories or other areas where hazardous chemicals may be spilled. This allows for very rapid covering of the spill with spill mix and the control of many of the risks associated with the spill. The spill mix we have developed is versatile and provides a practical way of reducing the danger of spills of hazardous chemicals. When the identity of the spilled chemical is known, it is often possible to convert the absorbed mixture to environmentally acceptable products using reactions that have been tested in the laboratory.

29.5 REFERENCES

Armour, M. A. 1996. *Hazardous Laboratory Chemicals Disposal Guide,* 2nd ed., CRC Press, Boca Raton, FL.

Armour, M. A., C. Nelson, P. Sather, Y. Briker, E. Crown, K. B. Rigakis, and N. Kerr. 1996. "Decontamination of Spills and Residues of Pesticides and Protective Clothing Worn during Their Handling," in *Proceedings of the Pacific Basin Conference on Hazardous Waste,* Pacific Basin Consortium for Hazardous Waste Research and Management, Kuala Lumpur, Malaysia, pp. 44–54.

Armour, M. A., D. Ashick, and J. Konrad. 1999. "Tested Methods for the Handling of Small-Scale Spills," *Chemical Health and Safety*, vol. 6, no. 1, pp. 24–27.

Castegnaro, M., J. Adams, M. A. Armour, J. Barek, J. Benvenuto, C. Confaloneri, C. Goff, S. Ludeman, D. Reed, E. B. Sansone, and G. Telling. 1985. *Laboratory Decontamination and Destruction of Carcinogens in Laboratory Wastes: Some Antineoplastic Agents,* International Agency for Research on Cancer, Lyons, France.

National Research Council. 1995. *Prudent Practices in the Laboratory. Handling and Disposal of Chemicals,* National Academy Press, Washington, DC.

P • A • R • T • 7

SAFETY

CHAPTER 30
REVIEW OF PERSONAL PROTECTIVE EQUIPMENT FOR SPILL SITUATIONS

Merv Fingas
Emergencies Science Division, Environment Canada, Environmental Technology Centre, River Road, Ottawa, Ontario

30.1 INTRODUCTION

The use of personal protective equipment is very important in spill situations because response personnel face exposure to chemicals that could cause serious injury or illness. The hazards are accentuated by the unknown physical environment of the spill site and the random nature of spills. The multiple hazards of spill sites distinguish such incidents from spills in the laboratory or exposure to chemicals in the workplace. Spill responders entering a site cannot always predict what chemicals or concentrations they will encounter and what other hazards may be present. There is also a lack of standards and guidelines for selecting and using personal protective equipment for chemical spill emergency situations. Most standards are written for use in the workplace and may not be fully applicable to emergency situations.

Preventing exposure to toxic chemicals is a primary concern at spill sites. Substances can enter the unprotected body by inhalation, ingestion, and skin absorption or permeation. Ingestion can occur by transferring absorbed contaminant during eating or smoking or by other forms of contact with the mouth. Some chemicals readily pass through the skin in a process known as permeation.

The most common types of protective equipment necessary for spill response are clothing and respirators. Totally encapsulated chemical protection suits (TECPS) or gas-tight suits are used when the contaminant is unknown or when a skin-permeating or skin-attacking chemical is present. The self-contained breathing apparatus (SCBA) provides the highest protection against the inhalation of chemical contaminants and is the most commonly used form of respiratory protection in the initial phases of a spill.

The response levels used in North America are commonly accepted among spill response organizations, including Environment Canada, the U.S. Environmental Protection Agency, and the U.S. Coast Guard. Level A is the first-response or entry level. The SCBA and totally encapsulated suit are used to protect against high or unknown levels of chemicals. Chemical substances that permeate or otherwise attack the skin may be present, and the totally encapsulated suit is necessary to protect against them.

Level B involves the use of the SCBA with liquid-tight chemical-protection clothing, or splash gear, as it is sometimes known. Level B gear is used to enter a site where it is known that no skin-penetrating or damaging chemicals are present or where high levels of contaminant may be present that a standard air-purifying device would not protect against. There is a variety of acceptable options, but the clothing usually consists of a special watertight suit or watertight rain gear.

Level C includes respirators and standard protective clothing, which typically consists of liquid-repellent coveralls made of polypropylene, or Tyvek. This level of response is very often used by cleanup crews when the situation has stabilized and concentrations are known and are not likely to rise above the capability of the respirator. No skin-penetrating materials are present. Gloves and boots are used throughout the first three protection levels.

Level D is applicable to spills where there are no airborne contaminants of concern and where there is a minimum likelihood of harm by contact with the spilled material. Many organizations provide treated cotton coveralls or similar garb for working in such frequently occurring situations.

30.2 THE SPILL SITUATION

The most important consideration when selecting protective equipment is the target chemical. While it is not possible to be prepared for all types of chemical spills, it is possible to be prepared for the most frequently spilled chemicals.

Canada has had a national database for a number of years (Beach, 1982; Fingas, 1996). Such a database is useful in assessing priorities. Table 30.1 is a list of some common chemicals spilled from 1984 to 1993 and their frequency and volume (Fingas et al., 1991; Fingas and Lau, 1996). Only about 40 materials have been spilled more than 10 times, and about 150 materials have been spilled more than 3 times. The data show that about 5% of the

TABLE 30.1 Frequently Spilled Substances (data from 1974 to 1993)

Chemical	Numbers of spills	Spill volume (t)
PCBs	334	89
Sulfuric acid	155	13,000
Hydrochloric acid	123	3,300
Ammonia	107	470
Sodium hydroxide	92	8,200
Sulphur	68	70,000
Ammonium nitrate	63	4,200
Fenitrothion	49	100
Nitric acid	48	140
2,4-dichlorophenoxyacetic acid (2,4-D)	37	130
Chlorine	36	120
Potassium chloride	31	12,000
Ethylene glycol	31	590
Vinyl chloride	31	180
Styrene	24	5,000
Sodium chlorate	23	7,700
Calcium chloride	20	3,700

spills are single or one-time incidents with a low probability of repeat. About 90% of the incidents are spills of common industrial chemicals. Fifty substances account for about 80% of the spills and 100 substances for about 90% of the spills. It can be seen that most spills are of high-volume production industrial chemicals, while less common chemicals account for only a small number of spills.

A priority list of spill substances has been prepared using spill and toxicity data. A summary of the protection requirements necessary for the highest-priority materials is provided in the Appendix. The assessments are given in terms of whether or not SCBAs or encapsulated suits are required. It is important to stress that the protection levels listed in the Appendix are for typical spill situations. When chemicals and/or their concentrations are unknown, SCBAs and totally encapsulated suits must be worn. The same is true for enclosed or confined spaces. On the other hand, once chemical concentrations are known and the situation is under control, lower levels of protection might be appropriate.

30.3 EQUIPMENT OVERVIEW

A wide variety of respirators is available on the market. Respirators are common and are of two basic types: air-supplied and air-purifying. Air-supplied respirators are used when oxygen levels are in question as well as when contaminant levels are unknown or expected to be high. Air-purifying respirators are used when contaminant levels are relatively known and are below the maximum capacity of the respirator.

Protective clothing is quite diverse, but the most common items used for spill response are boots, gloves, totally encapsulated suits, splash gear or firefighters' bunker suits, and coveralls. Hearing protection and hard hats are used where necessary.

30.4 SELECTION OF RESPIRATORS

Selecting respiratory protective equipment for use in the workplace has been the topic of several well-known references (NIOSH, 1985; OE, 1991; CSA, 1993). Some publications have dealt with selecting respirators for spill situations (EPA, 1992; Fingas, 1996). The differences between selecting a respirator for the workplace and selecting a respirator for use in spill situations hinge on the certainty with which both the substances present and their concentration are known. In spill situations, maximum protection must often be used because of the possible presence of high concentrations of chemicals.

Respirators provide protection from three dangers: chemical contaminants, particulate matter, and oxygen deficiency. Chemical contaminants include a wide variety of materials, as listed in the Appendix. Particulate matter consists of small matter that can be inhaled. Respirable particulate matter has particulate sizes less than 10 μm, below which size, particulates lodge in the lungs and can cause permanent damage. Particulates are also important in view of chemical spill response because many chemicals can be absorbed by or adsorbed to particulate matter. Special types of filter, known as HEPA, or high efficiency particulate filters, remove the small particulates and are part of many air-purifying devices. Oxygen deficiency is a common danger associated with chemical contamination, particularly in confined spaces where the air is displaced by the chemical. Oxygen deficiency can only be protected against by using air-supplied respirators.

Respiratory protective devices consist of a facepiece connected to either an air-source or an air-purifying device. Facepieces for respirators are available in two basic configurations: half facepieces and full facepieces. Half facepieces fit around the nose and do not have an integrated eye-shield. A wide variety of devices are available on the market, some of which are listed in Table 30.2. This table also lists a protection factor, which is the ratio of the concentration of the contaminant outside the facepiece to the concentration of the contaminant inside the facepiece. The protection factors presented in the table represent an average value for a large number of individuals. Such values can be much lower, however, if the facepiece does not fit properly. For example, beards can cause leakage around a facepiece, reducing the protection factor by as much as 10 times.

Protection factors are important criteria for selecting respiratory protective equipment. The protection factor must be sufficiently high to reduce the contaminant inside the facepiece to an acceptable level, usually taken as the TLV, or threshold limit value. The Appendix lists the TLV values for commonly spilled materials. These data are used in the following manner (ACGIH, 1999; NIOSH, 1994). A spill of a substance with a TLV of 5 ppm, which according to calculations could rise as high as 5,000 ppm, would require a respirator with a protection factor of at least 1,000. For a safety factor of 2, a protection factor of 2,000 would be required. Pressure-demand SCBAs, which have a protection factor of about 10,000, represent the ultimate in safety and are generally used at spill scenes because the exact substance and the level of the contaminants are not known for certain until measurements have been made.

Air-purifying respirators are limited in the concentrations that they can handle or absorb. The top level at which an air-purifying respirator is useful is referred to as the "immediately dangerous to life and health" (IDLH) level. This is also the level at which a chemical can cause severe damage. The IDLH level represents the point at which one must either switch from an air-purifying respirator to an air-supplying respirator or escape from the environment. The IDLH values for commonly spilled chemicals are listed in the Appendix. Another requirement for air-purifying respirators is that they be used at contaminant concentrations less than the specified maximum use concentration (MUC). The MUC is based on the capability of the sorbent in air-purifying respirators to deal with high concentrations of materials.

TABLE 30.2 Respiratory Protection Equipment and Associated Protection Factors

Respirator	Protection factor[a]
Air-purifying respirators—dust masks	
Single-use dust mask	5
Quarter mask	5
Half mask	10
Full-facepiece dust mask	50
Powered dust mask	1,000
Air-purifying respirators—chemical cartridge	
Half-facepiece mask	10
Full-facepiece mask	50
Supplied-air respirators	
Demand half facepiece mask	10
Demand full facepiece mask	50
Pressure-demand half-facepiece mask	1,000
Pressure-demand full-facepiece mask	2,000
Continuous-flow helmet or suit	2,000
SCBAs (*self-contained breathing apparatus*)	
Open-circuit demand	50
Open-circuit pressure demand	10,000
Closed-circuit, oxygen cylinder type	50
(*all are full facepiece*)	

[a] Protection factor is the ratio of the contaminant concentration outside the facepiece to the concentration inside.

The selection process for a respirator is shown in Fig. 30.1. The following guidelines can simplify the selection process for respiratory protective devices for use at a spill scene.

- The SCBA should be used for entry into an unknown situation, if unknown or high levels of a toxic chemical are present, or if there is any possibility of an oxygen shortage.
- The air-purifying respirator can be used when the situation is stable and when the levels of chemicals are below the IDLH, with very little possibility of rising.

In both cases, the selection should be verified by taking measurements and calculating concentrations inside the facepiece.

The protection factor can now be measured for a given mask and a given person. This is called fit testing. In the past, a number of qualitative measures using indicators such as banana oil and smoke were used that provided a rough indication of mask fit. In the past 10 years, many quantitative fit-testing apparatuses have come onto the market. These directly measure the protection factor for a given mask by measuring the concentration of ambient particulate inside and outside a mask or by measuring the ingress of air using sensitive pressure transducers. The measurement devices now include computers or built-in computing devices to directly calculate protection factors and the percentage of facepiece leakage. Fit testing should be done for new users as well as on an ongoing basis to ensure adequate protection. Sometimes certain facial features, such as a scar, facial hair, or a very thin face, result in a lack of proper protection. Individuals with those features should not be put in a position that requires respiratory protection.

Problems can arise when respirators are used in extreme cold or hot weather. In cold weather, the icing of the regulator and fogging of the facepiece window are concerns. Icing of the regulator becomes serious because the decompression of the air can cause the moisture to precipitate and then freeze. The amount of water in the starting air is usually measured by the dew point, which is the temperature at which moisture begins to form droplets of water. Table 30.3 shows the dew point and water content of air. The CSA standard requires extensive filtration to remove the moisture (CSA, 1985). When supplied-air respirators are

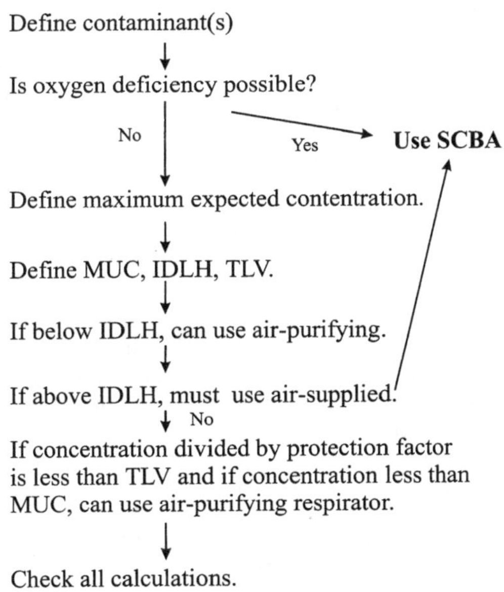

FIGURE 30.1 Respirator selection.

TABLE 30.3 Approximate Moisture Content in Compressed Breathing Air[a]

Atmospheric dew point (°C)		Pressure dew-point (°C)	
30- or 60-minute apparatus	Water content at atmospheric pressure (ppm)	30-minute apparatus at 2,216 psig	60-minute apparatus at 4,500 psig
−45.5	68	7	18
−48.5	48	2	12
−51	34	−2	8
−53 (CSA standard)	27	−5	4
−54	24	−6	3
−56.5	17	−10	−2
−59.5	11.5	−14	−6
−62	8	−18	−10
−65	5.5	−22	−15
−68	3.5	−27	−20
−70.5	2.3	−31	−24
−73	1.5	−34	−27

[a] Data adapted from CSA, 1985.

used in cold weather, a number of precautions are taken, including leaving tanks in heated vehicles before use. The fogging of lenses by exhaled moisture is readily prevented by using a device known as a nose cup, which directs moist air to the exhalation valve and away from the lens. The use of respirators in very hot conditions does not cause life-threatening events, but the user can tire quickly and excessive sweating can cause face-seal problems with the mask.

Another issue when using respirators is the need for prescription glasses in the facepiece. Regular glasses cannot be worn because anything that interferes with the face piece seal is dangerous and illegal. Special lens holders are available that fit into the face piece. Contact lenses, once not recommended for use in a respirator, can now be used because new contact lenses allow for gas exchange and thus do not dry out and stick to the eyeball.

For air-supplied respirators, the quality of the air is also an issue. Standards exist for the minimum air purity (CSA, 1985). The moisture content in the air is also an issue, especially when operating in cold climates, as this can cause the respirator to freeze.

30.5 PROTECTIVE CLOTHING

Clothing, gloves, goggles, boots, and other such items are required to prevent the chemical from contacting the skin or eyes. In the case of vapors that can permeate the skin, gas-tight protection is required. In the case of chemicals that are corrosive or absorbed as liquids through the skin, protection is required to prevent contact with the substance itself. Some chemicals pose both dangers.

Chemicals gain access to the wearer or affect clothing material through three processes.

1. *Permeation* is the process by which liquid or gaseous chemicals move through clothing material on a molecular basis. It is the most important indicator of the usefulness of a particular clothing material. Some chemicals can permeate through clothing material in

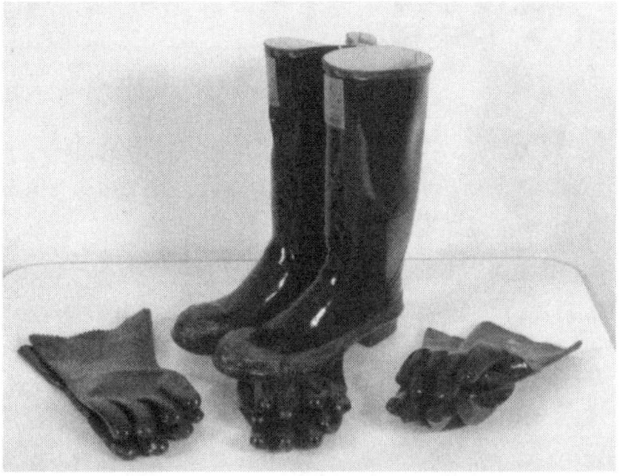

Boots and gloves are the most important pieces of protective clothing because of frequent contact with the contaminant.

Protective clothing, especially boots, often requires decontamination after use.

A team working in level A gear. It is important to work in "buddy" groups of two or three.

FIGURE 30.2 Examples of personal protective equipment.

Trainees in level B practice using a simulated spill.

An oil cleanup crew goes through decontamination.

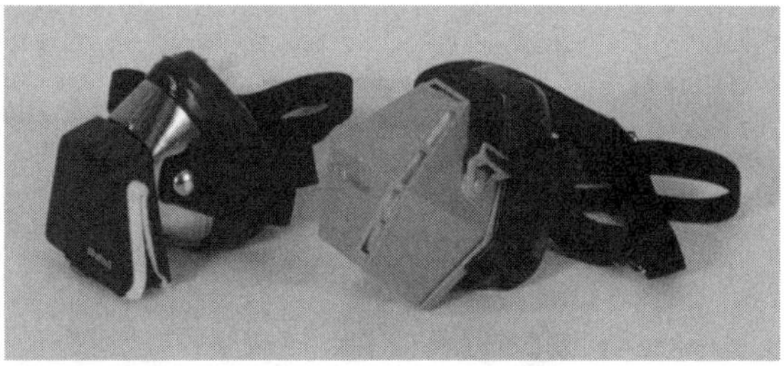

Dust masks used for exposure to nonrespirable particles.

FIGURE 30.2 Examples of personal protective equipment. (*Continued*)

only a few seconds. If these chemicals are toxic, then the clothing material is not useful for chemical protection.

2. *Penetration* occurs when liquid or gaseous chemicals flow through closures, seams, pinholes, or other similar openings in the clothing. Penetration can occur regardless of the type of material selected, although certain types of materials are more or less resistant to puncture mechanisms, such as abrasion or pinholing, depending on the conditions the clothing is subjected to.

A full-facepiece mask with chemical cartridges.

Gas masks are similar to air-purifying respirators but have larger chemical-absorbing canisters.

This cleanup team equipped in level A uses a variety of analytical equipment to assess a burned chemical warehouse.

FIGURE 30.2 Examples of personal protective equipment. (*Continued*)

Response team assesses a burned-out chemical warehouse.

A half-facepiece mask with chemical cartridges.

Worker in level C has retrieved a label from a chemical drum.

FIGURE 30.2 Examples of personal protective equipment. (*Continued*)

A "class photo" of a group in training.

Spill response worker in level C takes water sample near an incident.

Spill response worker in level C prepares an air sampler downwind of an incident.

FIGURE 30.2 Examples of personal protective equipment. (*Continued*)

3. *Degradation* is the deterioration of clothing material caused by the action of the chemical. Degradation may change bulk properties such as tensile strength or cause small areas of the material to dissolve. In the past, most data on a material's resistance to a chemical related to degradation because there were no standards for measuring other types of chemical intrusion. Many different measurements were lumped together and termed *chemical compatibility.* As will be shown later, degradation data, although important, are not usually as crucial as permeation data.

Permeation is the most important of the entry mechanisms in terms of spill response. It is chemical dependent and changes with material thickness, temperature, and the presence of other solvents. It has been found that mixtures of different chemicals can sometimes permeate the clothing material much faster than any of the chemicals alone.

Existing data on permeation are primarily measured using the ASTM (American Society for Testing and Materials) procedure or the ISO (International Standards Organization) procedure. These prescribe the use of a standard test cell that consists of two spherical halves. The clothing material to be tested is placed as a divider between these two halves. The challenge liquid is placed on one side, with air on the other side. The air is then monitored and breakthrough is said to occur when the chemical can be measured in the air space. Because the clothing material is completely immersed in the challenge liquid, the test provides a conservative measure.

The time it takes for some chemicals to permeate some typical clothing materials is presented in Table 30.4. Data are compiled from Forsberg and Keith (1995.) The clothing materials presented here are those commonly used in totally encapsulated suits or protective gloves. In both cases, permeation data is very important for selection. Permeation times of less than 30 minutes imply that the material has little application to spills because this is the time usually spent in an encapsulated suit. In some cases, however, there is no material with a long permeation time and the material with the best permeation time would be used.

Specific permeation data on the clothing material actually used in the manufacture of the clothing item should be obtained whenever possible. As already noted, permeation data vary greatly, even with similar materials. Some of the variance may be due to the thickness of the material. The thicker the material, the longer the permeation time. In fact, thickness is so important that materials, such as those used in light clothing, that show significant permeation in thin sheets of typically 0.05 cm can have no or little permeability at thicknesses of about 0.5 cm. Thus, permeation through very thick synthetic materials, such as those in SCBA facepieces, may not be a serious concern. Permeation times also vary with material fabrication, and other differences may be due to erroneous data. It is important to verify the permeation data for a given material with more than one source.

Boots and gloves are the two most important pieces of protective clothing because they most often come into contact with contaminated materials. Glove technology has progressed immensely in the past few years, and many companies have good selection charts based on permeation data for their specific materials. The manufacture of boots has not progressed much and very little specific permeation data is available for these clothing items.

Another important piece of clothing for chemical spills is the totally encapsulated chemical protection suit (TECPS), which is used if the materials encountered are unknown or known to be skin-permeating. A few cautions should be noted in selecting totally encapsulated suits.

1. There are few standards governing the design and manufacture of these suits and those documents that do exist are applicable primarily to suits for firefighters. The buyer must therefore ensure that the suit purchased is appropriate for use in a chemical spill response.
2. Any permeation data provided by the manufacturer must have been measured for the actual suit material and be generated by a standard method, preferably the ASTM method.

TABLE 30.4 Permeation Times through Clothing Materials (in minutes)

Chemical	BETEX	Butyl	Rubber	Neoprene	Nitrile	PVC	Teflon	Viton
Acetic acid	>360	180	120	360	360	180	>480	120
Acetic anhydride	>360	>240	60	210	180	90	>180	60
Ammonia, anhydrous	>360	>480	2V	>180	250V	15V	>300	
Ammonium hydroxide	>360	>480	120	360	360	180		>60
Benzene	15	30V	3V	12V	15V	1V	>200	9V
Chlorine	>360	>480	>480	>480	>480	30	>300	>480
Ethylene glycol	>360	>480	360	360	360	360	>480	
Formaldehyde	>360	>480	60	120	>360	70V	>180	>480
Hexane	5	15V	5	50V	360	30	>300	>480
Hydrogen peroxide	>360		>360	6	>360	>360		
Hydrochloric acid	300	>480	360	>360	360	360	>480	>480
Hydrofluoric acid	>480	>480	150V	360	120V	360	>480	>480
Methanol	100	>480	15	10V	180V	2V	>480	60
Nitric acid	>360	>480	360	150	100V	240		60
Pentachlorophenol			150	6-360V	>360	180		>480
Perchloroethylene	20	4V	0	15V	40V	15V	>180	>480
Phenol	>480	>480	60V	180	60	20V	>180	>480
PCBs		>480	60	>480	150V		>480	>480
Sodium hydroxide	>360	>480	360	360	360	>360	>480	>480
Sodium hypochlorite	>360	>360	360	360	360	360		
Styrene	10	30V	1V	12	30	30	>240	>180
Sulfuric acid	>360	>480	80	>360	10V	10V		>240
Toluene	<10	10V	5V	10V	20V	10V	>180	>180
Toluene diisocyanate		>480	7V	0-240V	240	480	>480	>480
Vinyl chloride					300			260
Xylenes	10	30V	2V	4V	60V	1V	>180	>60

- All permeation times are in minutes
- BETEX = Butyl on neoprene
- V indicates highly variable data
- 0 indicates that the material should not be used, usually because material degrades
- Blank indicates no testing performed
- Rubber = natural rubber
- PVC = polyvinyl chloride

3. Suits that interfere with the face-seal of the self-contained breathing apparatus (SCBA) should not be purchased. Such practice is dangerous and contravenes most occupational health laws.
4. Gas-tight suits are sometimes made of several materials. Permeation of the weakest material is the limiting factor. The permeation of each material should be measured, as well as the joint between each type of material. It may be wise to avoid suits made of more than one type of material.
5. The suit should allow access to the controls of the SCBA, whether the SCBA is worn inside or outside the suit. Many responders prefer that the SCBA be worn outside the suit so that they have unrestricted access to the controls.
6. Caution should be observed when dealing with sales staff for totally encapsulated suits. Many sales staff are not aware of the intricacies of spill response, the respiratory protection required, safety requirements at spill scenes, and permeation data for the clothing needed.
7. The question of whether a SCBA should (or can be) worn inside or outside the suit is still controversial. Sales staff may indicate that suits with built-in face masks are not safe or legal, but this is not correct.

8. Any prospective suits should be tested for application in different weather conditions. Many suits cannot be used in cold weather because they are too stiff and some suits do not allow enough movement to perform many tasks.
9. Other users of the suit should be surveyed to ensure that the suit has performed well in actual spill situations.

The use of totally encapsulated chemical protection suits is an extensive topic. Wearing TECPS is often physically and mentally stressful. In hot weather, a person wearing such a suit can lose as much as 2 kg of water through perspiration. Organizations such as the EPA have suggested that TECPS not be worn more than for more than one hour per day and that this be done in two half-hour sessions.

The selection of other clothing material is less critical. Coveralls and such clothing are not worn when there is a skin-penetrating material spilled. Disposable coveralls are now frequently used at spill scenes and are very useful for minimizing contact with chemicals. Treated cellulose fabrics are now more popular than Tyvek because of their greater comfort. Goggles are used occasionally at the spill scene if there is a danger of material getting into eyes and if respirators with full facepiece masks are not worn. Splash guards are also occasionally used, but their use is not recommended, as they were originally designed to protect construction workers from sparks and projectiles projected directly at the guard. In the case of liquid spills, the materials are often at ground level and the open area at the bottom of the device can actually direct liquids to the face. Hard hats and ear protectors should be used as required in any hazardous situation. Some types of protective clothing, such as totally encapsulated suits, do not readily allow for the use of hard hats, but most responders would not require both forms of protection at once.

Decontamination of clothing and breathing apparatus has been a controversial issue. There has been too much fixation on this task, and far too complicated procedures have been proposed in the past. Totally encapsulated suits rarely need decontamination unless contact with a material has been made. Procedures for specific chemicals should then be followed. Generally, if the contaminant is a gas, the suit should be hung in a well ventilated area, away from the sun and away from people, for about two weeks. If the material is a liquid that permeates, the suit may have to be destroyed or special procedures followed in consultation with the manufacturer. Nonpermeating liquids or solids can be washed from the suit by placing the person in the suit under a specially designed shower or other water spray. If the material is oily or can be removed more easily with a detergent, this can be used. Only dishwashing detergents are mild enough not to affect the suit material. More damage can be caused to the suit by a harsh decontaminating mixture than if the suit is simply washed with water. Special detergents are available for decontaminating for bacteria and viruses, but mild chlorine and peroxide solutions have also been used.

Boots must almost always be decontaminated after use at a chemical spill incident. A child's wading pool with a mild detergent is suitable for many situations. The water must often be treated as contaminated waste, as is the water from the suit decontaminations noted above. Gloves can be washed separately in a bucket in a manner similar to boots.

If it is suspected that equipment is contaminated, it is far better to dispose of it or put it into secure containers for further assessment than to try to reuse it without assurance that it is in good condition. Management will certainly be sympathetic and replace any equipment. In many organizations, this occurs so infrequently that the occasional loss is not a serious economic problem.

30.6 CONFINED SPACES

Confined spaces are those areas with poor air circulation where gas concentrations can rise far beyond danger levels and there is a high potential for oxygen deficiency. There is also the danger of entrapment and fire. Confined spaces include sewers, closed rooms, and silos. Special occupational health and safety rules govern entry and work in confined spaces because they pose a high hazard. In many jurisdictions, the entry into confined spaces and details of activities and safety procedures followed must be recorded.

30.7 THE PERSONAL PROTECTION PROGRAM

Response to chemical spills requires a complete program that includes medical testing, training, retraining, and practice. Purchasing, maintaining, upgrading, and replacing equipment are also part of the program. One or more persons in the organization should be responsible for supervising, coordinating, and developing the program. A body of literature is available to help with establishing a recognized and systematic program (EPA, 1992; OE, 1991). The program should be based on a carefully developed policy concerning spill site entry procedures and minimum requirements for training and equipment. Organizations such as Environment Canada and the U.S. Environmental Protection Agency have had such programs and policies for many years. For example, anyone entering a spill site must have a minimum of one week of training in using the assigned equipment and must take a refresher course of at least two days every year.

Environment Canada has issued its spill responders with an SCBA and a totally encapsulated suit with all the accessories. This equipment is signed out by the employees and kept until they leave the program. The equipment is repaired and replaced on a regular basis.

Training at regular intervals is very important. In the United States, 40 hours of training are required before an individual can enter a site. Annual refresher courses are also required. Similar requirements should be put into place by any response organization. It is also important to recognize that first-responders are generally like first-aiders: they perform limited emergency duties at the site but refer to specialists for further advice and follow-up action. Just as the first-aider does not perform surgery at the site, those who respond to chemical spills should not be required to perform tasks beyond their training and capability. Professionals in the fields of chemistry and site remediation should be consulted and a network of information sources should be built up by any response organization.

In conclusion, there are many requirements for a thorough chemical spill response program. These include record-keeping, development of checklists for equipment testing, donning and doffing of equipment, and site entry procedures. Detailed site safety plans and standard operating procedures must be developed. A specification should be in place to ensure that persons required to respond are fit to do so, and this should include regular medical checkups. Regular practice programs and technical refresher programs are essential.

30.8 REFERENCES

American Conference of Governmental Industrial Hygienists (ACGIH). 1999. *1999 TLVs and BEIs,* ACGIH, Cincinnati, OH.

Beach, R. A. 1982. "Information Systems and Reporting—Canadian Systems," in *Hazardous Materials Spills Handbook,* ed. G. F. Bennett, F. S. Feates, and I. Wilder, McGraw-Hill, New York, NY, pp. 3–45–3–54.

Canadian Standards Association (CSA). 1985. *Compressed Breathing Air and Systems,* CAN3-Z180.1-M85, CSA, Toronto, ON.

Canadian Standards Association (CSA). 1993. *Selection, Care and Use of Respirators,* Z94.4-93, Toronto, ON.

Fingas, M. F., and C. Lau. 1996. "Recent Chemical Spill Data," in *Proceedings of the Thirteenth Technical Seminar on Chemical Spills,* Environment Canada, Ottawa, ON, pp. 327–369.

Fingas, M. F. 1996. "Personal Protective Equipment for Spill Situations," *Spill Technology Newsletter,* vol. 21, Nos. 3–4, pp. 1–12.

Fingas, M. F., N. Laroche, G. Sergy, B. Mansfield, G. Cloutier, and P. Mazerolle. 1991. "A New Chemical Spill Priority List," in *Proceedings of the Eighth Technical Seminar on Chemical Spills,* Environment Canada, Ottawa, ON, pp. 223–332.

Forsberg, K., and L. H. Keith. 1995. *Chemical Protective Clothing,* Lewis, Boca Raton, FL.

National Institute for Occupational Safety and Health (NIOSH). 1985. *Occupational Safety and Health Guidance Manual for Hazardous Waste Site Activities,* National Institute for Occupational Safety and Health (NIOSH), Occupational Safety and Health Administration (OSHA), U.S. Coast Guard (USCG) and U.S. Environmental Protection Agency (EPA), U.S. Government Printing Office, Washington, DC.

National Institute for Occupational Safety and Health (NIOSH). 1994. *NIOSH Pocket Guide to Chemical Hazards,* NIOSH, U.S. Government Printing Office, Washington, DC.

Operating Engineers (OE). 1991. *Occupational Health and Safety Guidance Manual for Contaminated Site Activities,* OE Local 115, Joint Apprenticeship and Training Program, Burnaby, BC.

U.S. Environmental Protection Agency (EPA). 1992. *Standard Operating Safety Guides,* EPA, Washington, DC.

APPENDIX

Name	Significant vapor at normal temperature	Health exposure limits						Concern with skin contact	Maximum use capacity for respirators[a] (PPM)				Need TECPS under normal conditions?[b]
		TLV (ppm)	TLV (mg/m^3)	STEL (ppm)	STEL (mg/m^3)	IDLH (ppm)	IDLH (mg/m^3)		Canister half-face	Regular canister	Gas mask	SCBA	
Acetic acid	√	10		15		50		corrosive	10	25	50	unlimited	[c]
Acetic anhydride	√	5				200		corrosive	50	125	200	unlimited	[c]
Acetaldehyde	√	100		150		2000			400	1000	2000	unlimited	
Acetone	√	500		750		2500			500	1000	2500	unlimited	
Acrylonitrile	√	2		10		85			20	50	85	unlimited	
Aldrin		0.02	0.25	0.05	0.75	2	25	absorbed	0.5	1	2	unlimited	
Aluminum sulfate													
Aminocarb													
Ammonia	√	25		35		300			50	150	300	unlimited	
Ammonium chloride	as fume	5	10	9	20				10	25	50	unlimited	
Ammonium hydroxide	as ammonia								50	150	300	unlimited	
Ammonium nitrate													
Ammonium phosphates													
Ammonium sulfate													
Aniline		2				100		permeation	20	50	100	unlimited	
Atrazine		0.6	5						2	5	10	unlimited	
Azinphosmethyl		0.02	0.2			0.8	10	absorbed	0.2	0.4	0.8	unlimited	
Benzene	√	0.5		2.5		500		permeation	10	50	200	unlimited	[c]
Benzene hexachloride (lindane)		0.04	0.5	0.1	1.5	4	50	absorbed	0.4	2	4	unlimited	
Benzoic acid													
Biphenyl		0.2		0.6		16			2	8	16	unlimited	
2-butanol		100		150		2000			200	1000	2000	unlimited	
Butyl acetate		150		200		1700			200	1000	1700	unlimited	
Butyl alcohol				25		1400			150	750	1400	unlimited	
Calcium chloride													
Calcium cyanide													
Calcium hydroxide		1.6	5						dust only	dust only	11	unlimited	
Calcium hypochlorite													
Calcium oxide		0.9	2	2	5	11	25		dust only	dust only	11	unlimited	
Calcium phosphate													
Carbon dioxide	√	5000		30000		40000			asphyxiant			unlimited	
Carbon monoxide	√	25		200		1200			100	500	1200	unlimited	
Carbaryl		0.6	5	1.2	10	12	100		1	6	12	unlimited	
Carbon tetrachloride	√	5		10		200		permeation	20	100	200	unlimited	
Chlorpyrifos		0.01		0.03		10		permeation	1	5	10	unlimited	
Chlordane		0.03	0.5	0.1	2	6	100	absorbed	1	3	6	unlimited	
Chlorine	√	0.5		1		10		corrosive	5	5	10	unlimited	√
Chlorodifluoromethane	√	1000		1250		2500			500	1500	2500	unlimited	
Chromic acid								corrosive					[c]
Cobaltous nitrate													

APPENDIX (Continued)

Name	Significant vapor at normal temperature	Health exposure limits TLV (ppm)	TLV (mg/m³)	STEL (ppm)	STEL (mg/m³)	IDLH (ppm)	IDLH (mg/m³)	Concern with skin contact	Maximum use capacity for respirators[a] (PPM) Canister half-face	Regular canister	Gas mask	SCBA	Need TECPS under normal conditions?[b]
Copper sulfate													
Cresol		5		10		250			25	125	250	unlimited	
Cyclohexane	√	200		400		1300			250	700	1300	unlimited	
Demeton		0.1		0.3		0.9		absorbed	0.5	0.5	0.9	unlimited	
Diazinon		0.01	0.1	0.02	0.3			absorbed	0.2	0.5	1	unlimited	
Dicamba													
1,2-dichlorobenzene		25		50		200			50	100	200	unlimited	
1,2-dichlorotetrafluoroethane	√	1000		1250		15000			asphyxiant			unlimited	
Dichlorvos		0.10	0.9	0.3		11	100	absorbed	2	5	10	unlimited	
Dieldrin		0.016	0.25	0.05	0.75	3.16	50	absorbed	1	2	3	unlimited	
Diethylamine	√	5		15		200		irritation	50	100	200	unlimited	[c]
Dimethylamine	√	5		15		500		irritation	100	250	500	unlimited	[c]
Dinoseb													
Diquat		0.03	0.5					absorbed					
Ethyl alcohol	√	1000		5000		3300			400	2000	3300	unlimited	
Ethyl acetate	√	400		1000		2000			1000	1500	2000	unlimited	
Ethyl acrylate	√	5		15		300			50	150	300	unlimited	
Ethylbenzene	√	100		125		800			150	400	800	unlimited	
Ethylene	√	10000							asphyxiant only			unlimited	
Ethylene dichloride	√	10		20		50			5	25	50	unlimited	
Ethylene glycol		10							5	25	50	unlimited	
Ethylhexanol													
Fenitrothion													
Ferric chloride													
Formaldehyde	√	0.75		2		20			5	10	20	unlimited	
Formic acid	√	5		10		30			5	15	30	unlimited	
Hexane	√	50		100		1100			not effective			unlimited	
Hydrazine	√	0.02		0.1		50			limited usefulness		2.5	unlimited	
Hydrochloric acid	√	5		10		50		corrosive	10	25	50	unlimited	[c]
Hydrofluoric acid	√	3		6		30		corrosive	5	15	30	unlimited	[c]
Hydrogen peroxide	√	1		2		75		corrosive	limited usefulness		75	unlimited	[c]
Isopropyl alcohol		200		400		2000			500	1000	2000	unlimited	
Malathion		0.7	10	1.5	20	18	250	absorbed	5	10	18	unlimited	
Maleic anhydride		0.25		0.75		2.5	10		limited usefulness		2.5	unlimited	
Mercury		0.006	0.05	0.06	0.5	1,2	10	absorbed	0.2	0.5	1.2	unlimited	
Methoxychlor		0.70	10	1.4	20	350	5000		70	200	350	unlimited	
Methyl alcohol	√	200		250		6000		permeation	not effective			unlimited	
Methyl chloride	√	50		100		2000		permeation	not effective			unlimited	
Methyl ethyl ketone	√	200		300		3000		defatting	600	1500	3000	unlimited	
Methyl methacrylate		50		100		1000			200	500	1000	unlimited	

Methylene chloride	√	100		300		2300		defatting	limited usefulness			unlimited	
Naphthalene		10		15		250			50	150	250	unlimited	
Nitric acid	√	2		4		25		corrosive	limited	15	25	unlimited	c
Nitrogen (liquefied)	√							freezing	asphyxiant only			unlimited	
Nonylphenol													
Oxygen (liquefied)	√							freezing				unlimited	
Paraquat		0.05	0.5			0.09	1	absorbed	0.09	0.09	0.09	unlimited	
Parathion		0.01	0.1	0.02	0.3	0.8	10	absorbed	0.2	0.4	0.8	unlimited	
PCBs		0.01	0.1					absorbed	2	5	10	unlimited	
Pentachlorophenol		0.05	0.5	0.14	1.5	0.2	2.5	absorbed	0.2	0.2	0.2	unlimited	
Perchloroethylene	√	25		100		150		defatting	limited usefulness		150	unlimited	
Phenol	√	5		15.6		250		absorbed	50	125	250	unlimited	
Phosphamidon												unlimited	
Phosphoric acid		0.25	1	0.7	3	245	1000	corrosive	50	125	245	unlimited	c
Phosphorus		0.02				1	5		not effective			unlimited	
Phthalic anhydride		1		4		10	60	sensitizer	2	5	10	unlimited	
Picloram		1	10	2	20				20	50	100	unlimited	
Potassium chloride													
Potassium cyanide													
Potassium hydroxide		0.9	2					corrosive	20	40	80	unlimited	
Potassium permanganate													
Propylene glycol													
Propylene oxide	√	5		10		400		irritation	not effective		400	unlimited	
Pyridine	√	5		10		1000		irritation	50	250	1000	unlimited	
Sodium carbonate													
Sodium chlorate													
Sodium chloride													
Sodium cyanide (as CN^-)						11	25	moisture produces toxic CN^-			11	unlimited	
Sodium hydroxide		1.2	2	2.4	4	6	10	corrosive	6	6	6	unlimited	
Sodium hypochlorite								corrosive					
Sodium silicate													
Styrene	√	20		40		700		irritation	100	200	700	unlimited	
Sulfur													
Sulfur dioxide	√	2		5		100			20	50	100	unlimited	
Sulfuric acid	√	0.2	1	0.7	3	4	15	corrosive	2	2	4	unlimited	
Tetraethyl lead		0.007	0.1	0.02		3	40	absorbed			3	unlimited	
Thallium sulfate													
Toluene	√	50		150		500		permeation	100	250	500	unlimited	
Toluene-2,4-diisocyanate	√	0.005		0.02		2.5		sensitization	escape only		2.5	unlimited	
Trichloroethane	√	10		20		100		permeation	escape only		100	unlimited	
Trichloroethylene	√	50		100		1000		de-fatting	escape only		1000	unlimited	
Trichlorofon								absorbed					
Trifluralin								absorbed					
Trinitrotoluene		0.01	0.1	0.02		53	500	permeation	escape only		50	unlimited	
Vinyl acetate		10		15					20	50	100	unlimited	
Vinyl chloride	√	1		2.3				hazardous	escape only		25	unlimited	
Xylene	√	100		150		900			200	450	900	unlimited	
Zinc chloride (fumes)		0.2	1	0.4	2	9	50		2	5	9	unlimited	

APPENDIX (Continued)

Name	Significant vapor at normal temperature	Health exposure limits						Concern with skin contact	Maximum use capacity for respirators[a] (PPM)				Need TECPS under normal conditions?[b]
		TLV (ppm)	TLV (mg/m^3)	STEL (ppm)	STEL (mg/m^3)	IDLH (ppm)	IDLH (mg/m^3)		Canister half-face	Regular canister	Gas mask	SCBA	
Zinc cyanide													
Zinc oxide (dust/fumes)		1.5	5	2.955	10	150	500		50	75	150	unlimited	
Zinc sulfate													

[a] Maximum use concentration is taken from commercial literature or it is generally taken that MUC for a gas mask is equal to IDLH, for a full-facepiece respirator is half of IDLH, and for a half-facepiece, air-purifying respirator is 1/5 IDLH. It is important to note that MUC may vary with manufacturer and model. It is important to consult with the manufacturer for specific data.

[b] Entry into spill situations where the substance and its concentrations are unknown should always be done with a totally encapsulated chemical protective suit (TECPS).

[c] Conditional on potentially high vapor concentrations.

CHAPTER 31
RISK COMMUNICATION OF CHEMICAL ACCIDENTS

H. Montiel, J. A. Vílchez, E. Planas, J. Arnaldos, and J. Casal
Centre d'Estudis del Risc Tecnològic (CERTEC),
Universitat Politècnica de Catalunya—Institut d'Estudis Catalans,
Barcelona, Catalonia, Spain

31.1 INTRODUCTION

The increasing number and size of industrial installations, the consequent larger stockpiles of toxic and flammable substances, and the increase in their transportation have given rise to a growing increase in the number of accidents involving hazardous substances (Vílchez et al., 1995).

Current legislation in the chemical industry includes measures designed to increase the safety of chemical installations and the security of the civil population, as well as making it obligatory to inform local populations of the risks involved in living near a chemical plant. This should include informing the local population and training them to act in the event of an accident.

Informing a civil population of such risks is an extremely delicate process (Renn, 1992). Providing the information is meant to prepare the population to act accordingly in an emergency. The security that this implies should, in turn, reassure the population. Nevertheless, an information campaign carried out incorrectly may simply spread alarm and fear among the population.

Risk communication has been defined as (Covello et al., 1986):

> *any purposeful exchange of information about health or environmental risks between interested parties. More specifically, risk communication is the act of conveying or transmitting information between parties about:*
> *a) Levels of health or environmental risks*
> *b) The significance or meaning of health or environmental risks*
> *c) Decisions, actions, or policies aimed at managing or controlling health or environmental risks*
> *Interested parties include government agencies, corporation and industry groups, unions, the media, scientists, professional organizations, public interest groups and individual citizens.*

The manner in which a group of people perceives risk depends on a number of factors. One of the primary factors is whether the risks are known or unknown, or assumed volun-

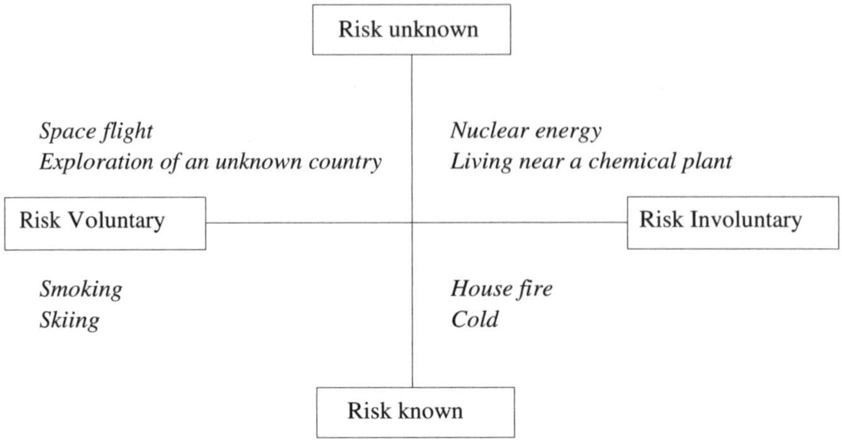

FIGURE 31.1 Perception of risk.

tarily or involuntarily (Fig. 31.1). The fear of the unknown is common, while we tend to feel in control of situations which we partake of voluntarily, i.e., we can escape when we want. Slovic describes these four probabilities that affect people's perception of risk (Slovic, 1987).

31.2 COMMUNICATION OF RISK

The general objectives of risk communication should be to change behavioral responses and provide information about possible dangers and remedies. Three general-purpose categories can be described (Renn, 1992):

1. *Understanding risk messages:* to make sure that all receivers of the message are able and capable of understanding and decoding the meaning of the messages sent to them
2. *Persuasion:* to persuade the receivers of the message to change their attitudes or their behavior with respect to a specific cause or class of risk
3. *Conditions for a rational risk discourse:* to provide the conditions for a rational discourse on risk issues so that all affected parties can take part in an effective and democratic conflict-resolution process

Communication of risk can be described, in a simple manner, by the classic sender–receiver model. In this, a message composed by a source is sent to the receiver by means of a transmitter. The receiver may respond to the message by initiating feedback communication or behavioral reactions. Table 31.1 shows different groups which can act as sources, transmitters, and receivers.

Information must be provided that familiarizes people with the risk and gives each individual some sense of control. Other important factors to be clarified are the real dimension of the danger, including the normally low probability of an accident, and the possibility that individuals, by means of a few simple actions, can notably increase their safety and that of their family in the event of an emergency.

TABLE 31.1 Sender–Receiver Communication Model*

Sources	Transmitters	Receivers
• Scientists	• Media	• General public
• Scientific institutions	• Other public institutions	• Specific audiences
• Public agencies	• Interest groups	• Professional groups
• Interest groups	• Opinion leaders	• Avocation activities groups
• Eyewitnesses		• Value groups

*Renn, 1992.

31.2.1 Educational Principles

When a campaign of risk communication is prepared, some basic principles of communication must be taken into account, including (Handmer and Penning-Rowsell, 1990):

- *Credibility:* The source of information must have credibility; this may mean building a climate of cooperation and belief between the public and those communicating the risk.
- *Normalcy:* The education program must form part of the normal environment of the public.
- *Consistency:* Any information must be *consistent* and *repeated* and the context should not contradict the message.
- *Relevance:* The message must be *relevant* to the needs of the public.
- *The communication channel:* The communication channel used must be *respected* and *used* by the audience. This may *not* mean the *usual* or most *obvious*.

Also, the message must take into consideration other aspects, such as audience habits, previous experience, literacy, current knowledge, level of outrage and the time that has passed since the last message. Table 31.2 shows the major concepts used in communicating risks (Handmer and Penning-Rowsell, 1990)

31.2.2 Suggestions for Successful Risk Communication

Several authors have developed guidelines for designing and evaluating risk communication on the basis of risk perception and risk communication studies (Casal et al., 1997; Covello et al., 1986). Some of them have proposed essential criteria for developing information programs. The most important of these guidelines are outlined in Table 31.3 (Renn, 1992).

TABLE 31.2 Major Concepts Used in Communication of Risks

• Use a *multimodal* presentation including verbal and pictorial information.
• Use a *specific* message.
• Make the message *positive* rather than negative.
• Encourage *activity* rather than passivity.
• Encourage *sociability* rather than isolation.

TABLE 31.3 Guidelines for Successful Communication

- Be clear about your intentions and make them the central message of your communication effort.
- Simplify your message as much as possible without being inaccurate.
- Place your simple messages at the beginning of a text and gradually add the complex issues.
- Anticipate the interests and concerns of your target audiences and design your communication program to match their needs. You can convince them only if your message addresses their concerns and interests.
- Devise different communication programs for different target audiences but do not change the message.
- Distribute your message through different channels and stimulate and encourage feedback as much as possible.
- Be honest, complete, and responsive in the composition of your message.
- Use a personal approach and match the communication to the personal experience of your target audience.
- Allocate enough time for packaging your message, but do not change your message in order to make the package more attractive.
- Be careful in selecting the right cues for appealing to the peripheral audience without offending your target audience.
- Explain the risk rationale to your audience and demonstrate the logic and adequacy of this rationale in a straitforward manner.
- Place risk in a social context and report numerical probabilities only in conjunction with actual examples.
- Remember that institutional performance is the major key to trust and credibility.
- Remember that risk managers have to learn from the public as much as the public can learn from them.

31.3 DESIGN AND IMPLEMENTATION OF RISK COMMUNICATION

Industrial development has brought both positive and negative aspects to the population. Industries have contributed to the increase in the quality of life, but, unfortunately, serious accidents occur causing deaths, injuries, and damage. Therefore, industries and governments must work together to reduce the risk to an acceptable level. As an important aspect of this task, authorities must prepare civil population for emergency response in case of accidents.

Precedents for addressing risk communication come from all around the world. Although these campaigns were initially created for natural risks, they have been implemented in other situations, such as harbors, natural emergencies, and chemical emergencies. Some studies in the 1980s analyzed such aspects as evacuation behavior problems and decision making in disasters and stated the need to clarify the roles of different institutions and the planning for the emergency (Quarantelli, 1980, Perry et al., 1981; Gray, 1981; Ikeda, 1982; Quarantelli, 1984). Fortunately, subsequent studies and campaigns have demonstrated a positive evolution.

One example is the Seveso Directive, created in 1988 for European Union (EU) countries. The Seveso Directive requires that information be provided to populations affected by a major accident on the site. This information should include:

- Activities on-site
- Dangerous substances involved
- Major accident hazards
- Potential effects on the population and the environment
- How a warning of an accident will be given
- Actions people should take in the event of an accident

In EU countries, certain improvements to the legislative underpinnings of these arrangements have been made in the context of the implementation of the European Union's Seveso II Directive. This involves greater consultation with the public during the preparation of off-site emergency plans and requirements for local authorities and industry to test their plans at regular intervals.

Another example is the Awareness and Preparedness for Emergencies at Local Level (APELL) program, also launched in 1988. APELL was originally based on the Community Awareness and Emergency Response (CAER) program, which later became part of the chemical industry's Responsible Care initiative. It includes elements of the EU Seveso Directive and the U.S. Right-to-Know initiative (UNEP, 1997).

Within the context of activities of international organizations on accident prevention and preparedness, APELL has become known as a practical program that assists industry, public authorities, and communities at the local level to increase community awareness of technological hazards and to prepare integrated plans for effective response should accidents occur (UNEP, 1997). Of the countries that have implemented APELL programs, some have implemented emergency planning and communication to the population and others have implemented only the technical part and the risk communication forms part of the second stage of the program.

The preparation and execution period of a campaign falls into the following categories:

- Before the campaign
- During the campaign
- After the campaign

31.3.1 Before the Campaign

Before the campaign begins, a certain period of time is needed to prepare the information and instruct the representatives about the campaign's strategy. The campaign is usually promoted by the authorities (autonomous governments, national governments) or by the industries (local industries, multinationals) according to the laws.

A campaign of risk communication to a civil population starts when the authorities, according to the laws, identify a certain industrialized area that must be subjected to a risk communication plan. The first step will be carried out by the authorities and will consist of establishing which zones are at risk in the event of an accident. This can be done by using the information presented by the various industrial plants to the administration, corresponding to the fulfillment of the laws from the Seveso Directive or other laws. This information allows the intervention and alarm distances to be determined. From this information, the number of industries, municipalities, and inhabitants affected will be defined.

Once the study of people and areas affected is made, the next step will be the design of the campaign's strategy, which will include:

- Designing and installing a warning system
- Organizing the information sessions
- Preparing the information for the sessions
- Arranging and coordinating the people involved, such as communicators and the public

Some previous examples of communication of chemical risk can be useful in designing the strategy. The steps in the campaign strategy are set out below.

Designing and Installing a Warning System. The authorities must design and install a warning system for the local population in the event of an emergency. The system must be audible in the affected areas but not in unaffected areas, to prevent unnecessary concern.

Various systems are available, such as acoustic systems (sirens installed in industrial plants and public buildings), loudspeakers, and radio alarms. Because in some areas within the affected zones the population may be more dispersed and out of the range of the sirens, a routine can be established by which the police notify individuals personally. Once the system has been selected, the emergency, end of emergency, and test signals must be defined and demonstrated to the population to familiarize them with the system.

Preparing the Sessions. A sociological study of people affected by the industry can be useful for identifying their perception of risk and to detect possible opposition groups. The way in which these people will respond to the campaign will depend on how the message is delivered and on any previous incidents that might have occurred with local industry. In almost all cases, however, one advantage is that some of the local population usually work in the industry in question, which means that the local economy depends to some extent on that industry.

Preparing the Information for Sessions. As mentioned before, in the preparation of the campaign, some educational principles must be taken into account. One is that the source of information must have credibility. Because an institution's performance is the major key to trust and credibility, the message has to be transmitted by technical staff from a university or other public and impartial institution.

These institutions must design and impart the oral informative sessions. To do this, a series of meetings must be held with the groups at which the message is to be directed. Then the content of the information sessions for each different group must be designed individually.

The methods that can be used to disseminate information among the local population include public meetings, audiovisual material, distribution of documentation (leaflets), and door-to-door visits (de las Alas, 1991). The best method probably combines all these possibilities, as shown in Fig. 31.2.

Below are the most common methods used for risk communication.

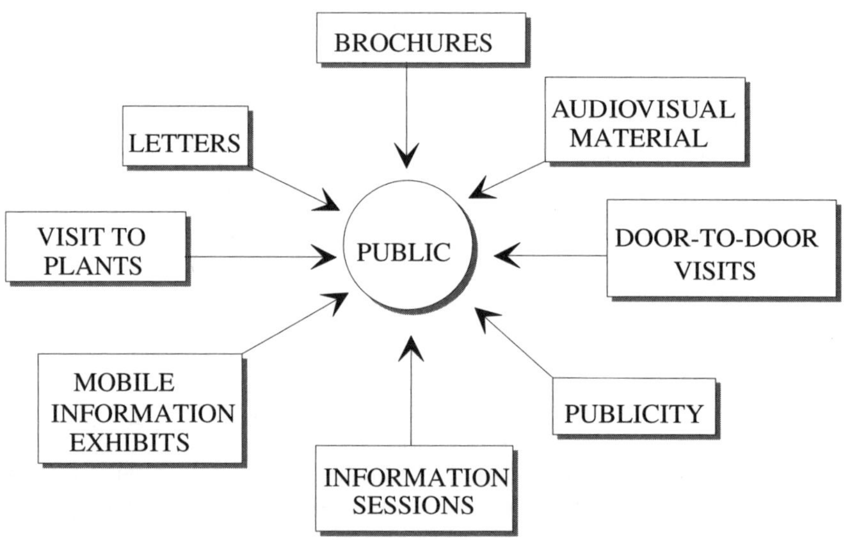

FIGURE 31.2 Methods for disseminating information.

Information Sessions. The information sessions are the central point of this campaign. In these sessions, individuals must be educated about the potential risks. The way in which the message is transmitted must follow the suggestions for successful risk communication outlined in Section 31.2.2 and must include the following aspects:

- *Perception of risk:* a description of the individual risk of each industry and municipality, using consistent and credible language and not creating fear.
- *Explanation of the risk communication campaign:* an introduction of the campaign, its objectives, the source of information (public authorities), and the action in case of risk.
- *Educational principles:* practice and recommendations to conclude the campaign

These sessions can take the form of conferences, round table sessions, or speech presentations. They must be held in various locations within the affected area so that all the population is reached. This includes the general public, industry, police, schools, and other interest groups.

Audiovisual Material. Some audiovisual material can be used during oral sessions to reinforce the message. These can be slides, videos, or computer demonstrations.

Documentation for Civil Population. The documentation given out can consist of booklets, charts, etc. This material should include a summary of the oral presentation. It must be written in simple language and aimed especially at those people who are unaware of the situation. The chart can be a summary of the brochure. These documents should contain all the necessary information presented by the campaign, but it must be designed as a backup rather than as a complete source of information.

Door-to-door visits, mobile exhibits, and visits to the industrial plants are also used in risk communication.

31.3.2 During the Campaign

The information sessions must be carefully planned and organized. After consultation with the local administration (town councils of the affected municipalities), the social groups at which the sessions are to be aimed must be identified. This includes the administration's technical staff, neighborhood associations, teachers, cultural and sports associations, as well as the workers in the industries affected.

The information must be delivered to all people affected. Usually when there are a large number of people, a first session is organized exclusively for members of the local administration, the people responsible for authorizing the campaign in their municipality. A number of press conferences can also be organized to maximize the campaign's impact.

The sessions should be presided over by a platform of representative people, including the speaker (from an impartial institution), a representative from the local administration and one from the local authorities, and a representative from the affected industries. People attending must have the opportunity to ask questions.

31.3.3 After the Campaign

After each information session a report should be produced in which any incidents or questions must be noted. It is also recommended that after a certain number of sessions representatives from all the groups involved in the campaign (authorities, institutions) meet to

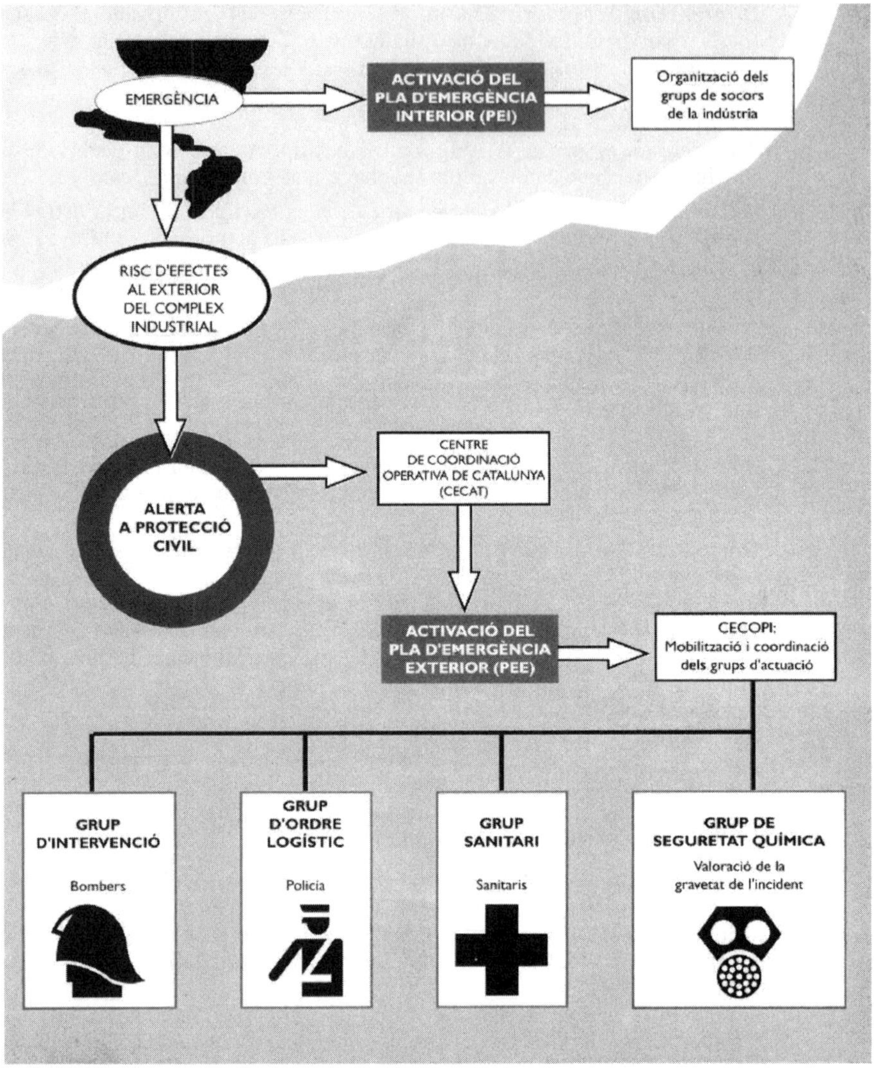

FIGURE 31.3 Emergency planning in case of accident. (*From the risk communication campaign in Baix Llobregat, Spain.*)

study the progress of the project, the running of the sessions, the public's positions, and the questions asked.

31.4 CASE STUDY: RISK COMMUNICATION OF CHEMICAL ACCIDENTS IN BAIX LLOBREGAT, SPAIN

As a result of the application of the Seveso Directive in Catalonia, Spain, in 1995, the Autonomous Government of Catalonia, by means of the Civil Protection Service, launched

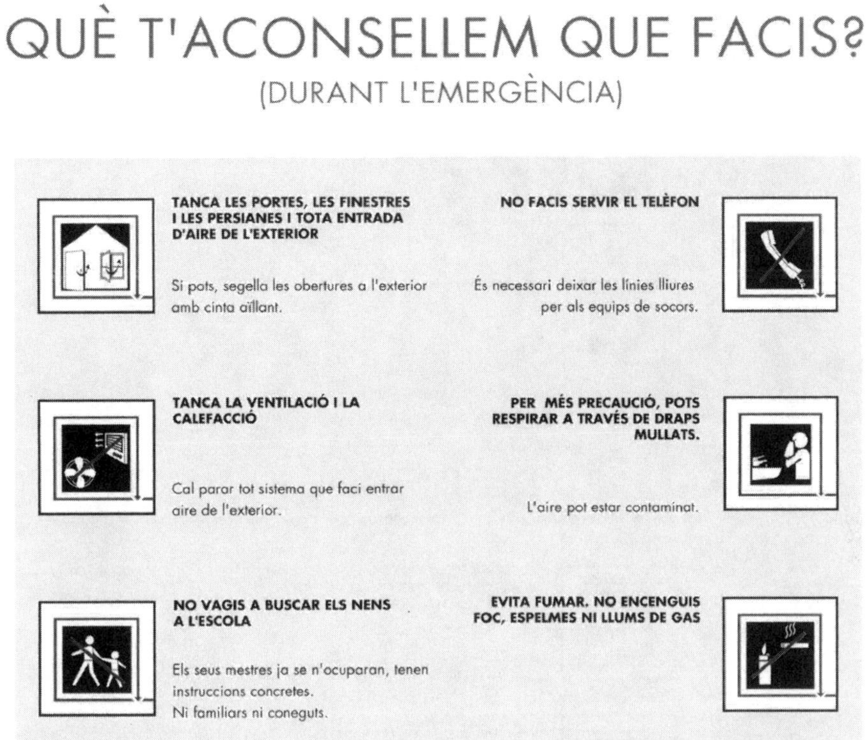

FIGURE 31.4 What must you do in case of accident? (*From the risk communication campaign in Baix Llobregat, Spain.*)

a public risk communication information campaign. This was the first of its kind in Spain and was therefore extremely informative. The first phase of this campaign was undertaken in Baix Llobregat, a heavily industrialized area covering nine different municipalities. The Center for Studies of Technological Risk (CERTEC) of the Universitat Politècnica de Catalunya (UPC) took part in this campaign. As a result, CERTEC was involved in other campaigns, namely communication of chemical risk in Galicia, Spain, covering a large number of villages (1998).

Campaigns were also launched in other parts of Europe following the Seveso Directive. The preparation of these campaigns was essentially the same. The following describes the campaign in Baix Llobregat, Spain. Some information from other campaigns has also been included.

31.4.1 Before the Campaign

The campaign was promoted by the Civil Protection Service of the Autonomous Government of Catalonia and was coordinated by a public relations company, which arranged the information sessions and coordinated all people involved.

The first step taken by the Autonomous Government was to establish which zones were at risk in the event of an accident in the industrialized area of Baix Llobregat. This was done by using the information presented by the various industrial plants in the area to the Administration in the declarations corresponding to the fulfillment of the Catalan laws from

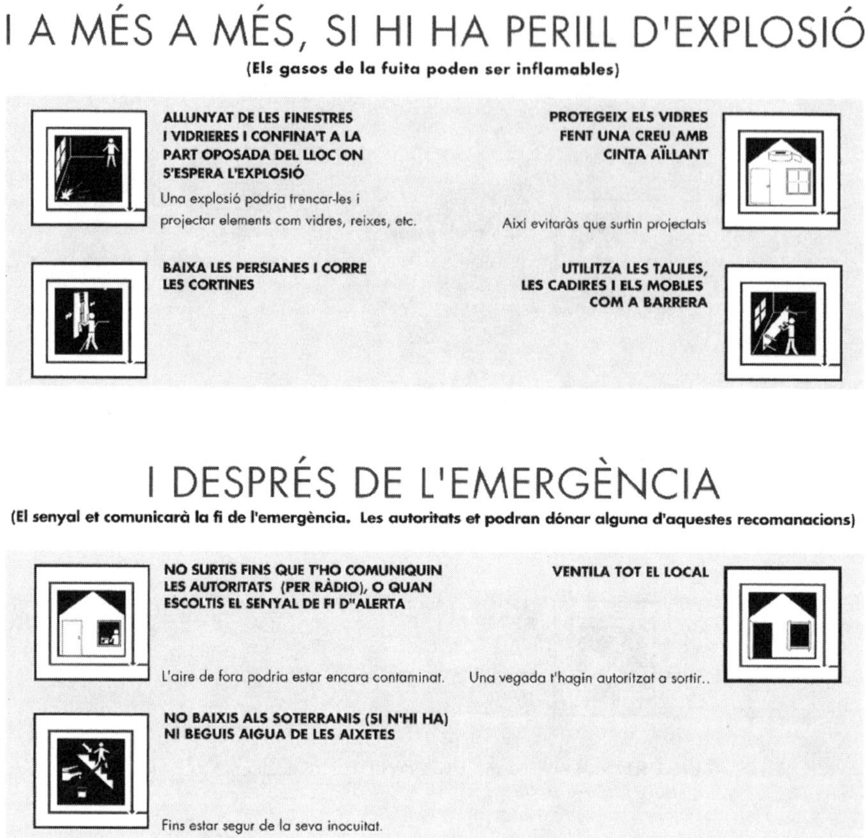

FIGURE 31.5 What must you do if there is danger of explosion after the emergency? *(From the risk communication campaign in Baix Llobregat, Spain.)*

the Seveso Directive. This information was used to determine the intervention and alarm distances. As a result, nine different municipalities were involved, with five villages affected by industries (126,000 inhabitants) and four municipalities located in areas that could be affected (with 110,000 inhabitants).

Once the study of people and areas affected was done, the next step was the design of the campaign's strategy (Casal et al., 1997). The previous experience of communication of chemical risk in Mantua, Italy, was very useful. The following is a description of the tasks involved.

Designing and Installing a Warning System. The Civil Protection Service designs and installs a warning system for the local population in the event of an emergency. In this case an acoustic system was chosen. The location and strength of the sirens were calculated for each municipality. The sirens were installed in the industrial plants and in public buildings. The objective of the system was for all the affected areas to be able to hear the acoustic alarm and for areas not affected not to be able to hear it, to prevent unnecessary concern. Because some areas within the affected zones were out of range of the sirens, a routine by which the police give personal alarms was established.

FIGURE 31.6 Do's and don't's in case of accident. (*From the APPELL program in Bahia Blanca, Argentina.*)

The sirens are activated by microwaves (IRT2000; frequency 2 GHz), and the instructions to sound them can come from the industry involved in the emergency, the municipal authority, or from the Civil Protection Service.

Three acoustic signals were defined: emergency (an oscillating alarm, lasting one minute, repeated three times with 5-second pauses between each burst), end of emergency (continuous for 30 seconds), and test (continuous for 10 seconds).

Preparing the Sessions. The public relations company conducted a sociological study of people affected by the chemical industry in the Baix Llobregat area. Two initial positions could be found: people who are unaware of the risk or who have never taken it into consideration, and people with a certain degree of ''risk culture.'' The first group consists of people in urban areas affected by industrial installations that are not close to any large-scale industrial complexes. In general, it has not been taken into account in these areas that certain types of accidents might affect the local population.

The second group consists of people who perceive more or less correctly the risk and potential effects of a chemical emergency. Generally, they live in urban areas near relatively large industrial plants and have experience of emergency alarms.

The way in which these people will respond to the campaign will depend on how the message is communicated and on any previous incidents that might have occurred with local industry. However, in almost all cases there was one factor that had a favorable effect: part

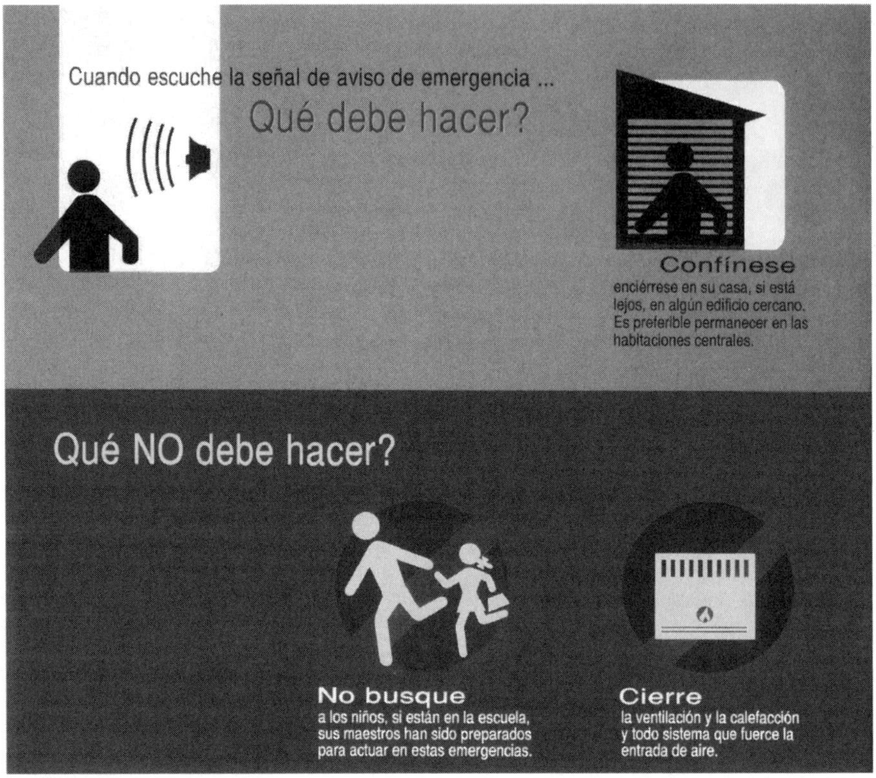

FIGURE 31.7 Do's and don't in case of accident? (*From the APPELL program in Bahia Blanca, Argentina.*)

of the local population worked in the industry in question, which meant the local economy depended more or less on that industry.

Preparing the Information for the Sessions. From the beginning of the campaign, the Civil Protection Service thought that the message to the civil population should be transmitted by an impartial body. For this purpose, the CERTEC was chosen. CERTEC designed and delivered the message at the information sessions. In a series of previous meetings with the administration, the public relations company and CERTEC analyzed the population to which the message was to be directed and designed the information contents.

Oral Sessions. In the oral and video sessions the outline was essentially the same. It can be divided into three phases.

First, a general description of the positive aspects of the chemical industry was given by using examples of essential products that this sector provides, such as detergents, paint, gasoline, and medicine. Examples of negative aspects were environmental damage, and noise. The main characteristics of these risks were then explained: fire, explosion, and toxic clouds.

Second, the mechanisms and operations that are to be put into effect were explained: immediate notification of the Administration, the activation of internal and external emergency plans, the participation of the emergency services (police, fire brigade, medical teams, and special chemical risk teams), and their coordination by the Civil Protection Service (Figure 31.3).

Third, the actions to be taken by the population in the event of an emergency were described. The message essentially consisted of the recommendation that in the event of a chemical emergency people remain inside because buildings are a good refuge. These actions can be summarized in the following basic points:

- Take refuge at home or in the nearest building.
- Do not go to pick children up at school.
- Prevent outside air from entering (close windows and all openings, shut off any air-circulation systems).
- Take cover from possible explosions (seek protection from fragments of broken glass).
- Do not use telephones (to avoid overloading phone lines).
- Listen to the radio (information was given on which radio programs should be listened to).
- Do not smoke; turn off all lights.
- Do not leave cover until the emergency is over. Once it is over, air the house, but do not enter basements.

The presentation was shown with 15 slides that had been designed by CERTEC and the public relations company. Figures 31.4 and 31.5 reproduce slides from the campaign in Baix Llobregat, Spain, and Figs. 31.6 and 31.7 reproduce similar material from the campaign in Bahia Blanca, Argentina.

Video. The aim of the video was to reinforce the message conveyed during the oral sessions. Filmed by a group of experts with all necessary means at their disposal, the video made a general and brief presentation of the different types of risk, such as forest fires and floods, but concentrated above all on chemical risks. The introduction showed the different types of major actions that could occur, and then a simulation of an emergency situation presented the role of the emergency services and the centralized coordination of the emergency. Finally, the video illustrated various safety measures to be followed, such as confinement and listening to the radio.

Documentation for the Public. The documentation given out consisted of a brochure and a chart. The brochure was a graphic summary of the oral presentation: the types of accidents and instructions for the population. It was written in simple language and aimed especially at people who were unaware of the problem. It also listed which radio stations should be listened to in the event of an emergency and the telephone numbers to call in each town for more information.

The chart was a graphic backup to the information presented in the brochure. Although these documents contained all the necessary information presented by the campaign, they were designed as complementary rather than a complete source of information.

31.4.2 During the Campaign

The public relations company organized the information sessions. This company consulted the local administrations (town councils of the affected municipalities) and identified the social groups at which the sessions were to be aimed. This included people from the administration, neighborhood associations, teachers, cultural and sports associations, and the workers in the industries affected. The same company convened the meetings and took charge of proceedings.

In all the affected towns, the first session was exclusively for members of the local administration, the people responsible for authorizing the campaign in their municipality. A number of press conferences were also organized to maximize the campaign's impact.

The sessions were presided over by a platform of four persons, consisting of a representative from the local administration, a representative from the Catalan Autonomous Government (globally responsible for the campaign), a member of CERTEC, and a representative from one of the affected local industries.

The sessions began with a 1-minute introduction by the local administration person, followed by a short message from the representative of the Civil Protection Service. Next, a member of the university gave a talk with slides for about 17 minutes, followed by the video (about 20 minutes). Finally, people attending were given the opportunity to ask questions. In general, the sessions were kept short to avoid getting into other problems such as the environment.

A total of 55 information sessions were held. There was a great deal of variety, as much in the number of participants as in their attitudes towards the campaign, the chemical industry, and the administration.

31.4.3 After the Campaign

After each session a report was produced in which any incidents or questions were noted. After a certain number of sessions, representatives of the Civil Protection Service, the public relations company, and CERTEC met to study the progress of the project, the running of the sessions, the public's positions, and the questions asked. These questions are reviewed here.

Questions. The most often repeated questions concerned the installations of the industry, the control of emergencies by the industry and Administration, and the duration of the possible emergencies. The most frequent questions have been summarized in Table 31.4.

Representative People. The presence of representatives from the various industries was beneficial. They generally gave information on the chemical products they dealt with and the dangers involved. Then the CERTEC representatives often intervened and explained in a simple fashion the possible effects of these products in the event of an escape.

Time. The amount of time dedicated to each section was kept under control and attempts were made to keep the sessions relatively short. However, this was not always possible. Neither was it always possible to prevent some representatives from the local administrations from trying to take over some of the sessions; it should be noted that the information campaign was interrupted by a local election campaign.

TABLE 31.4 Common Questions about Chemical Risk

- Which products would escape in the event of an accident?
- What dangers would be implied?
- Will we hear the alarm sirens if we are indoors with the windows closed? Are the sirens reliable?
- External emergency plan: How is it set in motion? How does it work?
- Emergency control: Who coordinates it? What resources are available?
- How will the accident affect water supplies, plants, and crops?
- What security systems does the company have?
- Are simulations carried out? When? How?
- How long will the emergency last?
- How will the rest of the population be informed?
- Why haven't we been warned before?

31.4.4 Conclusions

The campaign, which was the first risk information campaign on chemical accidents carried out in Spain, had positive results. Overall, it was an interesting experience and provided very useful information for future campaigns. The conclusions of the campaign can be summarized as follows:

1. The campaign's message was understood by the population without alarming them. There were no surprises and the situation was as expected.
2. In general, the public reacted positively and was receptive, although there were signs of distrust of the administration and its ability to handle an emergency effectively.
3. The collaboration between industry and the administration, essential in such a campaign, was very satisfactory.
4. Some dialogue between people and industry representatives has resulted and encouraged open-door policies that are useful in improving the relationship between the public and industry.
5. The number of people attending the information sessions was the only disappointing aspect of the campaign. For future campaigns these sessions must be better publicized, especially in those areas in which the population is not aware of the risks of large accidents.

31.5 REFERENCES

Casal, J., H. Montiel, E. Planas-Cuchi, J. A. Vílchez, J. Guamis, and J. Sans. 1997. "Information on the Risks of Chemical Accidents to the Civil Population: The experience of Baix Llobregat," *Journal of Loss Prevention in Process Industries,* vol. 10, no. 3, pp. 169–178.

Covello, V., D. von Winterfeldt, and P. Slovic. 1986. "Risk Communication: A Review of the Literature," *Risk Abstracts,* vol. 3, pp. 171–182.

de las Alas, E. 1991. "La comunicación al público de ciertos riesgos industriales," *Seguritecnia,* pp. 48–52.

Gray, J. 1981. "Characteristic Patterns of and Variations in Community Response to Acute Chemical Emergencies," *Journal of Hazardous Materials,* vol. 4, pp. 357–365.

Handmer, J., and E. Penning-Rowsell. 1990. *Hazards and the Communication of Risk,* Gower Technical, Aldershot, Hants., England, and Brookfield, VT.

Ikeda, K. 1982. "Warning of Disaster and Evacuation Behavior in a Japanese Chemical Fire," *Journal of Hazardous Materials,* vol. 7, pp. 51–62.

Perry, R. W., M. K. Lindell, and M. R. Greene. 1981. "Evacuation Planning in Emergency Management," D.C. Health, Lexington, KY.

Quarantelli, E. L. 1980. *Evacuation Behavior and Problems: Findings and Implications from the Research Literature,* Disaster Research Center, Ohio State University, Columbus, OH.

Quarantelli, E. L. 1984. "Chemical Disaster Preparedness at the Local Community Level," *Journal of Hazardous Materials,* vol. 8, pp. 239–249.

Renn, O. 1992. "Risk Communication: Towards a Rational Discourse with the Public," *Journal of Hazardous Materials,* vol. 29, pp. 465–519.

Slovic, P. 1987. "Perception of Risk," *Science,* vol. 236, pp. 280–285.

United Nations Environment Program (UNEP). 1997. Industry and Environment, vol. 2, no. 3.

Vílchez, J. A., S. Sevilla, H. Montiel, and J. Casal. 1995. "Historical Analysis of Accidents in Chemical Plants and in the Transportation of Hazardous Materials". *Journal of Loss Prevention in the Process Industries,* vol. 8, no. 2, pp. 87–96.

P • A • R • T • 8

PERSPECTIVES ON SPECIFIC CHEMICALS

CHAPTER 32
PERSPECTIVES ON SPECIFIC SUBSTANCES: AMMONIA

Richard Lawuyi and Merv Fingas
Emergencies Science Division, Environment Canada, Environmental Technology Centre, River Road, Ottawa, Ontario

32.1 OVERVIEW OF PRODUCT AND INDUSTRIAL USES

More than ever, the government and the chemical industry are committed to the prevention of accidental spills and the management of hazards and risks associated with these substances throughout their life cycles. Increasing awareness of the deleterious effects of hazardous chemicals has reduced the environmental load of pollutants, and this trend will no doubt continue. Reduction of chemical releases into the environment is a shared responsibility that we should all participate in and that will be of great benefit to us all.

More than a decade ago, the Emergencies Branch of Environment Canada prepared a chemical spill priority list of over 500 chemicals, each chemical being ranked according to its supply volume, reported spill frequency, historical spill volume, and toxicity. Our main objective then was to identify the minimum number of chemicals that would account for the maximum number of spills. The use of the list would be to act as a focus for the development of countermeasures, analytical methods, and spill manuals. This function has been well served. The list has since been reviewed, in 1990 (Fingas et al., 1991).

One other approach that was considered was ranking chemicals by groups according to their physical and chemical properties. A number of properties were selected, for example, LD_{50}, bioaccumulation, and persistence as the main criteria for toxicity to the environment and humans. Substances were then classified according to the range they fell within (Ministers' Advisory Panel, 1995). This approach was used for the Canadian Environmental Protection Act (CEPA) assessments. For example, polychlorinated dibenzodioxins, polycyclic aromatic hydrocarbons, inorganic cadmium compounds, benzidine, trichloroethylene, and a host of others were concluded to be toxic. Others, such as chlorobenzene, toluenes, xylenes, and dibutyl pthalate, were concluded to be nontoxic. Others, such as aniline, styrene, crankcase oils, and pentachlorobenzene, do not have sufficient information for classification.

Ammonia is a rather ubiquitous substance that occurs naturally but can also be manufactured quite readily. Demand for ammonia has peaked since the late 1980s, and total supply both in Canada and the United States has leveled throughout the 1990s simply because no new uses have been found for this product (CIS, 1998).

Ammonia has also been found to be a widespread contaminant in sewers and aquatic environments and has often been a nuisance around wastewater outfalls and in agricultural

TABLE 32.1 Priority List Ranking of the Top Hazardous Chemicals

Chemical	Ranking	Spill number	Spill volume	Supply volume
Ammonia	1	107	470	3,700
Chlorine	2	36	120	1,700
Tetraethyllead	3	4	72	26
Styrene	4	24	5,000	630
PCBs	5	334	89	–
Sulfuric acid	6	155	13,000	3,700
Sodium cyanide	7	3	83	12
Hydrochloric acid	8	123	3,300	170
Potassium chloride	9	31	12,000	–
Pentachlorophenol	10	19	110	1.5
Phenol	11	10	14	68
Zinc sulfate	12	3	68	1,500
Phosphorus	13	16	46	68
Toluene	14	13	110	430

wastes and industrial effluents. The ammonia produced in most of these cases is mostly nonanthropogenic. The main natural source is the decomposition of nitrogenous substances by bacteria. In Canada and the United States, some criteria and guidelines have been determined for ammonia in all kinds of water uses. Ammonia concentrations in air and some water columns have been shown to be well above the guideline values (Lusis and Phillips, 1990; NAQUADAT, 1991; Saskatchewan Environment Data, 1991; Fenske, 1993; Williamson, 1990). Large-scale ammonia releases can result in gross water, soil, and atmospheric contamination and hence constitute risk to public health and safety. Enormous fish kills and horrific human injuries and fatalities have also been reported during accidental spills (Environment Canada, 1999; Markham, 1987).

While there have been many accidents involving ammonia at industrial facilities all over the world, very few have occurred during transportation, such as by rail, road, or water (Environment Canada, 1999; Markham, 1987). The reason for this is quite clear—transporters and distributors are better trained in handling dangerous goods than in the past. Handlers are now aware of the dangerous consequences of chemical spills and hence more respectful. Another reason is that there are more guidelines and regulations laid down by governments and corporations on handling and transportation of chemicals apart from emergency response planning activities that are often required. Significant progress has also been made in the safe storage of these chemicals. The phrase "responsible care" has been reverberating across North America for the last decade in meetings, lecture halls, conferences, and symposiums.

Research on ammonia toxicity has been slowing down considerably since the early 1990s; only about 458 relevant papers could be found in the last decade. The most significant areas of immediate concern are: (1) human inhalation effects; (2) effects on aquatic biota; (3) effects on land fauna and microflora; and (4) effects and extent of water pollution. Of lesser importance is human ingestion since this is an improbable route of exposure. The concentration/exposure effects of ammonia in many areas such as odor recognition, respiratory and eye irritation, and death have not been clearly defined and tend to vary with the researcher. Markham (1987), in his review, discussed the absence in the literature of basic data on exposure times/concentrations of ammonia that affect aquatic biota, terrestrial animals, and humans as well as occasions where little or no damage has occurred or those that show a threshold below which any effects are unlikely to occur. There has also been very little published on potential effects of low-level, long-term exposure and the sensitivity of certain high-risk sectors of the population, especially the sick, the very old, and infants. These sectors of the population are often ignored for most studies. How best to determine hazardous and lethal exposures for humans of hazardous chemicals is still a formidable problem.

32.1.1 Modern Industrial Uses

Since the late 1960s, many technological advances and easy access to abundant natural gas have made the production of ammonia cheaper and the chemical itself more readily available. The resulting rapid consumption of ammonia fertilizers has resulted in a green revolution. For example, the annual consumption of ammonia-based fertilizer grew by 11.5% in 1960–1970, 6.7% in 1970–1980, 3.8% in the 1980s, and less than 3.0% in the 1990s. A slump in the world market in the 1980s led to downsizing and reduction in production in the late 1980s. Over 85% of ammonia produced worldwide is used as nitrogen-based fertilizers, with the remaining 15% going into industrial products such as pharmaceuticals, various fibers, animal feeds, and explosives. Aside from its extensive use as a fertilizer, other main uses of ammonia include:

1. Starting materials for the production of chemicals such as nitrates, ammonium salts, urea, amines, and nitrocompounds
2. As an industrial refrigerant
3. As a reactant in many industrial processes, e.g., nylons, plastics, isocyanates, pesticides, detergents, and neutralization of acids
4. In water disinfectant to supplement the chlorination process
5. In bleaching, removing stains, extraction, printing, and photocopying
6. In various metal-treating processes where ammonia atmosphere is required, such as metal-nitriding, carbo-nitriding, bright annealing, sintering, furnace-brazing, and atomic hydrogen welding
7. As a source of hydrogen for the hydrogenation of fats and oils
8. As a source of pure nitrogen
9. As a common household cleaning agent
10. For equipment protection such as in condensers, storage tanks, heat exchangers, and bubble plate towers
11. In the rubber industry as a stabilizer of raw latex to prevent coagulation
12. In mining to extract metals such as copper, nickel, and molybdenum from their ores
13. In the pulp industry as a bisulphite to improve the quality of pulp
14. As a catalyst in the manufacture of phenol-formaldehyde and urea-formaldehyde resins
15. In fruits for preservation during warehousing

About 20% of total ammonia produced in Canada, the United States, and Western Europe is used for industrial purposes, but the figure represents only about 10% in Russia, 5% in Latin America, and 1 to 10% in Asia. While the largest exporter of ammonia is Russia (34%), the largest importer is the United States (34%). Ammonia is an ideal fertilizer and has many benefits over others in that:

1. It is easy to apply.
2. It has the highest nitrogen content (82%).

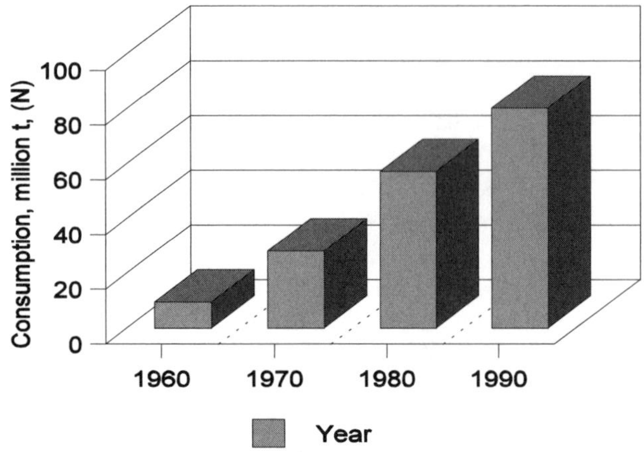

FIGURE 32.1 Fertilizer nitrogen consumption, $\times 10^6$ t.

3. Because of its anhydrous nature, it can increase the soil capacity to hold water and also the soil tilth.
4. It reacts with soil, organic, inorganic, and crop constituents, thus enhancing its fertilizing power, and retards leaching.
5. When applied below the soil, it spreads under its own pressure over a wide area, thus reaching more plant roots.

32.2 INTRODUCTION

The main differences in the fate and behavior of various chemicals, both organic and inorganic, are due to their physical and chemical properties, such as solubility in various media, octanol/water partition coefficient, vapor pressure, biodegradation, chemical reactivity, and persistence. The need for physical properties data on behavior and environmental fate of chemicals in scientific research, emergency response, including firefighting, and government regulations and guidelines cannot be overemphasized. While the data in this chapter came from different sources, as shown in the references, the most reliable authors have been included. It is also recommended that the reader consult the original reference in case of updates. Some of these data have also been generated or verified in our laboratory. Some estimates have been made where necessary as well as comparisons between different chemicals when such exist.

32.3 PHYSICAL AND CHEMICAL PROPERTIES AND GUIDELINES

At room temperature, ammonia is one of the most stable compounds in commerce (Czuppon et al., 1992), but it decomposes only at very high temperatures (above 450°C). One of its main hazards arises from the fact that it is transported as a liquefied gas under pressure (cryogenic) and on release quickly evaporates and dissipates. It has also been known to explode or catch fire. Many models have treated ammonia as a heavy gas because the vapor cloud often absorbs moisture and eventually moves close to the ground. When released as a vapor from a gaseous source, it behaves as a lighter-than-air vapor, with its density a function of its vapor temperature only. When released as a liquid from a pressurized source, it will atomize and droplets will autorefrigerate, resulting in a cold mixture containing droplets of ammonia, water vapor, and water vapor with dissolved ammonia. This mixture is generally heavier than air. Ammonia will also undergo a number of reactions with other chemicals, including oxidation and neutralization of acidic compartments on release. Its main properties are as shown below (Compressed Gas Association, 1990; RTECS On-Line, 2000; Transport Canada, 1985; Braker and Mossman, 1980; AIHA, 1988).

Common name: Ammonia, anhydrous or aqueous ammonia

Molecular formula: NH_3

Molecular weight: 17.031

CAS number: 7664-41-7

UN number: 1005 (>50%)
2672 (>10% ≤ 50%)

RTECS number:

Anhydrous: BO0875000
10–35%: BO0876000
35–50%: BO0877000

Labels: Compressed gases, corrosive gases

Specific hazards:

Alkali—ALK
Corrosive—COR
Use no water with liquefied gas

Synonyms and trade names:

Ammonia gas:
- Ammonia anhydrous
- Ammoniac (French)
- Ammonia gas
- Ammoniak (German)
- Spirit of hartshorn
- Anhydrous ammonia (DOT)
- Nitrosil

Ammonia solutions:
- Ammonium hydroxide (<28% aqueous ammonia)
- Ammonia water
- Aqueous ammonia

Grades:

Grade	Weight % minimum
Commercial	99.5
Agricultural	99.7
Refrigeration	99.95
Anhydrous	99.99
Metallurgical	99.995
Electronic	99.998

Impurities are usually water, oil, and noncondensable gases or inerts such as hydrogen, argon, nitrogen, and methane carried over usually from the different steps.

Physical data:

Anhydrous ammonia (NH_3):

State (room temp., 1 atm)	Gas
Boiling point (101.325 kPa)	−33.4°C
Melting point (1 atm.)	−77.72°C
($NH_3 \cdot H_2O$)	−78.85°C
($2NH_3 \cdot H_2O$)	−78.68°C
Density, liquid (−33.7°C)	0.6828 kg/L
Vapor pressure (21.1°C)	888.0 kPa
Autoignition temperature	651°C
Vapor density (−33.4°C, 1 atm)	0.88983 kg/m^3
Flammability limits in air	15 to 28% by volume
Gas density (gas, 0°C, 1 atm)	0.771 kg/m^3
Odor threshold range	2–10 mg/m^3
Ignition temperature	850.0°C

Ammonium hydroxide (ammonia water, aqua ammonia, about 25% ammonia):

State (room temp., 1 atm)	Gas
Boiling point	36°C
Melting point/freezing point	−77°C
Relative density (water = 1)	0.9
Relative vapor density (air = 1)	0.6
Vapor pressure (20°C)	153 mbar
Solubility in water	Infinity
Relative molecular mass	35.1
Heat of solution (28%/w)	4.999×10^5 J/kg
Heat of solution (0%/w)	8.081×10^5 J/kg

Hazards and toxicity: Mixtures of ammonia and air in certain proportions (15 to 28% ammonia) can ignite or even explode at high temperatures (above 648.9°C) if a source of ignition is present. Ammonia can react to form explosives with some chemicals.

Human health: Anhydrous ammonia is very corrosive, with high affinity for water. In high concentration, inhalation can cause paralysis of the respiratory system and death.

Environment: Ammonia can adversely affect many microorganisms, plants, and aquatic life. Ammonia toxicity to the biota is often due to its severe systemic and metabolic disruptions.

Behavior in air: Anhydrous ammonia will react with water vapor in the air to form dense fog. Ammonia will burn with a yellow flame when exposed to an open flame in air or oxygen. Ammonia–air mixtures can also explode.

Behavior in water: Anhydrous ammonia dissolves readily and reversibly in water to produce aqueous ammonia, ionized and un-ionized ammonia. The ratio of ionized and un-ionized ammonia depends on the pH. High concentrations in water will adversely affect aquatic organisms.

32.3.1 Emergency Response

Victims should be quickly removed from spill site to an uncontaminated area where there is an abundant supply of fresh air. A doctor should be called.

Inhalation: Give artificial respiration if breathing is difficult or has ceased.

Skin: Wash and flush with water immediately. Remove clothing. Keep on flushing for at least another 15 minutes.

Eyes: Eyes should be well rinsed with large quantities of water or 5% boric acid.

Nose and mouth: Should be flushed with large quantities of water continuously.

Ingestion: This is very rare. Victims should be given copious amounts of water to drink. Seek medical attention immediately.

Spill control: Self-contained breathing apparatus should be worn for all emergency measures involving anhydrous ammonia and full protective clothing for liquefied ammonia.

Leaks: Most leaks can be detected by smell, wet red litmus paper, or an open bottle of hydrochloric acid. Try to seal leak if possible.

Fires: Anhydrous ammonia can burn but is very difficult to ignite. In the presence of air it can also explode when exposed to a flame or a source of ignition. Use dry chemicals or carbon dioxide to put out small fires. Use cold spray to cool container.

Spills: Evacuate spill site. Eliminate ignition sources. Provide ventilation and stay upwind if possible. Neutralize ammonia spills with carbon dioxide. Special foams are also available.

Soils: Flood site with plenty of cold water. Build barriers (dikes and lagoons) to contain large spills. Allow small spills to vaporize and absorb with sand or vermiculite. Use carbon dioxide foams to neutralize spills.

Water: Contain by damming or water diversion. Use carbon dioxide to neutralize spills.

Air: Let spill evaporate. Follow the emergency response planning guidelines (ERPG) of the American Industrial Hygiene Association (AIHA, 1988):

ERPG-3. 1,000 ppm. This is the maximum airborne concentration below which it is believed that nearly all individuals could be exposed for up to one hour without experiencing or developing life-threatening health effects.

ERPG-2. 200 ppm. This is the maximum airborne concentration below which it is believed that nearly all individuals could be exposed for up to one hour without experiencing or developing irreversible or other serious health effects or symptoms which could impair an individual's ability to take protective action.

ERPG-1. 25 ppm. This is the maximum airborne concentration below which it is believed that nearly all individuals could be exposed for up to one hour without experiencing other than mild, transient adverse health effects or without perceiving a clearly defined objectionable odor.

32.4 INDUSTRIAL ASPECTS AND PRODUCTION IN THE UNITED STATES, CANADA, AND WORLDWIDE

The name "ammonia" derives ultimately, through the Latin *sal ammoniacum* [salt of ammonia], from the Egyptian god Amon, or Ammon. It was near an oasis beside a temple to the god that ammonia is said to have first been used, several thousand years ago, by Arab farmers spreading burned camel dung on their fields. Ammonium salts were observed in Persia about 900 A.D.

Ammonia gas was first produced in pure form by J. B. Priestley in 1774, followed by C. L. Berthollet in 1784, who identified its elements. In 1809, W. B. Henry determined its composition and the formula to be NH_3. It wasn't until the end of the last century that fertilization of the soil with ammonia became a widespread farming practice. Ammonia gas was also produced in 1828 by heating animal refuse with lime. In the same year, Wohler observed that the evaporation of a solution of ammonium cyanate when heated produced urea. Ammonia is the largest-volume basic, inorganic substance in use today and is generally considered the most important agricultural chemical. Anhydrous ammonia is usually stored and transported as a liquefied gas under pressure.

As the use of ammonia grows and gains general acceptance, so grows the number of accidents. An examination of the origin and trend of these accidents will be useful in understanding and preventing further accidents.

32.4.1 Manufacture of Ammonia

The basic synthetic process is the Haber-Bosch method. As presented in Fig. 32.2, ammonia is produced by the reversible reaction of hydrogen with nitrogen. The reactants mixture, often called the synthesis gas, consists of hydrogen and nitrogen in roughly 3:1 molar ratio and impurities such as argon, methane, carbon dioxide, carbon monoxide, and water vapor. Nitrogen is separated from the air.

Hydrogen, on the other hand, can be obtained from a variety of sources, such as hydrocarbons, coal, water, natural gas, coke oven gas, and mixtures of these. However, the most common and cheapest source is by steam reforming of natural gas, in which case the overall reaction is:

$$7CH_4 + 10H_2O + 8N_2 + 2O_2 \Rightarrow 16NH_3 + 7CO_2 \qquad (32.1)$$

Table 32.2 shows the different types of feedstocks and their contributions to the world manufacture of ammonia (Czuppon et al., 1992).

The exothermic reversible reaction of hydrogen and nitrogen to form ammonia occurs according to the following equation:

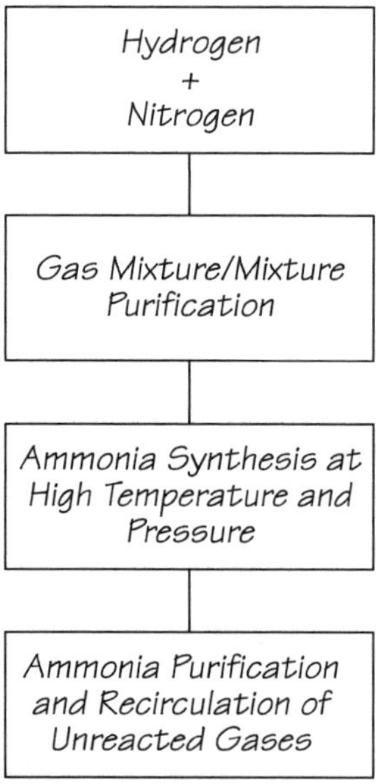

FIGURE 32.2 Manufacturing of ammonia.

TABLE 32.2 Raw Materials for Hydrogen Production

Feedstock	1987 as %
Natural gas	69.8
Naphtha and fuel oil	5.8
Naphtha and fuel oil condensate	10.5
Coke oven, refinery gas, hydrogen	3.9
Coal	10
Total	100

$$N_2 + 3H_2 \rightleftharpoons 2NH_3 \qquad \Delta H = -91.44 \text{ kJ/mol} \tag{32.2}$$

The product yield depends on the pressure, temperature, reactant ratios, and type of catalyst used. Canada has an abundant supply of the most important raw material for the manufacture of ammonia-natural gas. In Canada, ammonia is manufactured by the high-pressure catalytic steam reforming of natural gas. The main source of nitrogen is air, while hydrogen is also obtainable from a variety of other sources, such as crude oil, coal, and other hydrocarbons. The crude gas is processed in a series of operations consisting of steam regeneration, washing, compression, sulfur removal, shift conversion, carbon dioxide removal, nitrogen scrubbing, compression, and synthesis. The most important stages are as follows:

1. Purification of feedstock
2. Primary and secondary reforming
3. Shift conversion
4. Carbon dioxide removal
5. Synthesis gas purification
6. Ammonia synthesis and recovery

The reaction proceeds at very low temperatures or high pressures, but the use of catalysts produces satisfactory yields at approximately 350°C. The rate of conversion per pass is about 25 to 35% and much greater in some cases.

Coke was used as the main source of hydrogen in the original Haber-Bosch process. Today coal is still being used in some processes, such as the Lurgi process and the Koppers-Totzek process. However, with the discovery of large gas reserves, coal use is rapidly declining. Over 80% of ammonia produced today is obtained by steam-reforming, and about 70% uses natural gas feedstock.

Purification of Feedstock. The feedstock is first desulfurized or purified. This is a very crucial step in ammonia manufacture. It is required in order to maintain the life of the catalyst.

Impurities such as chlorine, arsenic, and nickel used in the manufacturing process tend to shorten the life of the catalyst. The usual methods for desulfurization are by activated carbon adsorption at about 15 to 50°C or oxidation with zinc oxide or both. The use of zinc oxide is often preferred when large quantities of mercaptans and highly condensable hydrocarbons are present, which may quickly saturate the catalysts. The main function of the zinc oxide is to remove the hydrogen sulfide, mercaptans and chlorine. Combining activated carbon and zinc oxide is very effective in removing the different types of sulfide compounds.

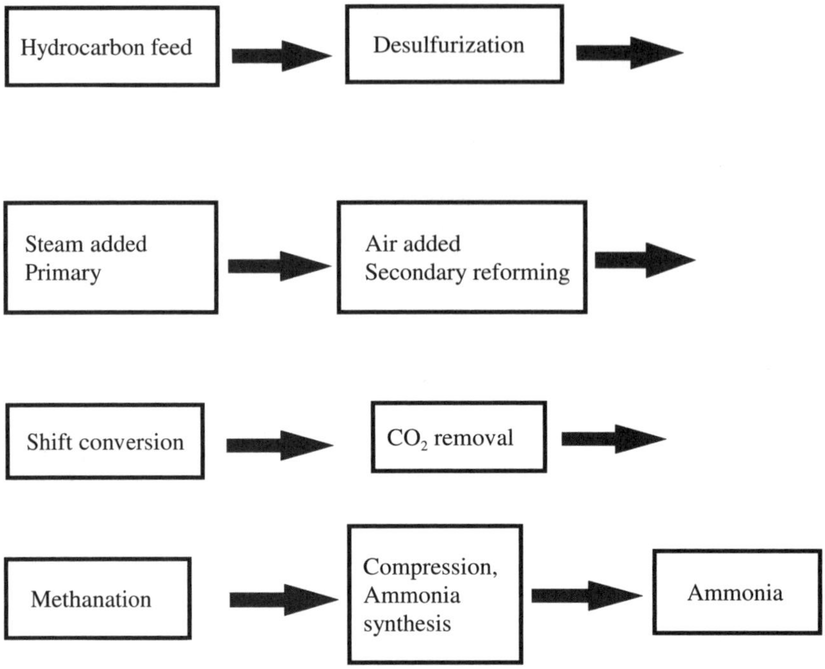

FIGURE 32.3 A typical scheme for the production of ammonia.

Other methods include hydrodesulfurization and absorption-stripping followed by zinc oxide treatment.

$$ZnO + H_2S \rightarrow ZnS + H_2O \tag{32.3}$$

Natural gas containing large amounts of hydrogen sulfide is often subjected to an absorption-stripping operation followed by zinc oxide treatment. Organic sulfur compounds are often removed by hydrodesulfurization, in which the organic sulfur is converted to hydrogen sulfide in the presence of hydrogen nickel molybdate and cobalt catalysts at 340 to 400°C. The process also includes some zinc oxide as a supporting catalyst.

Reforming. Reforming of the natural gas feedstock takes place in two distinct catalytic reaction steps. The first step is the primary reforming and is carried out in a furnace in the presence of steam to produce a partially reformed gas. The second step constitutes the secondary reforming, in which the reaction is carried out in a refractory-lined pressure vessel to produce a low-methane product.

Primary reformer

$$CH_4 + H_2O \rightarrow CO + 3H_2 \qquad \Delta H = 206.08 \text{ kJ/mol at } 25^\circ\text{C} \tag{32.4}$$

Secondary reformer

$$CO + H_2O \rightarrow CO_2 + H_2 \qquad \Delta H = 206.08 \text{ kJ/mol at } 25^\circ\text{C} \tag{32.5}$$

The primary reformer reaction is largely endothermic, and considerable heat must be supplied to this process. Nickel supported on alumina is generally the primary reformer catalyst, the reaction temperature being in the 750 to 850°C range at a pressure of 2,860 to 3,550 kPa. Calcium aluminate is often the catalyst used at higher operating temperatures and is often supported by sodium or potassium oxides. The following side reactions will lead to the deposition of carbon. The presence of these oxides leads to reduced carbon deposition and increased life of the catalysts.

$$CH_4 \rightarrow C + 2H_2 \tag{32.6}$$

$$2CO \rightarrow C + CO_2 \tag{32.7}$$

$$CO + H_2 \rightarrow C + H_2O \tag{32.8}$$

$$CO_2 + 2H_2 \rightarrow C + 2H_2O \tag{32.9}$$

The catalyst often used in the secondary reformer is nickel supported on alumina. Air is added in adequate quantities to supply the required amount of nitrogen.

Shift Conversion. Byproducts such as carbon oxides that deactivate the catalysts used in the synthesis have to be removed or converted to the inert species before the actual synthesis occurs. In the shift conversion process, carbon monoxide is converted to carbon dioxide. The reaction is exothermic and reversible.

$$CO + H_2O \rightarrow CO_2 + H_2 \qquad \Delta H = -41.17 \text{ kJ/mol at } 25°C \tag{32.10}$$

This reaction equilibrium is favored at low temperatures, and most of the carbon monoxide is converted to carbon dioxide in a high-temperature shift furnace (HTS) operating at 350 to 450°C. This step is followed by low-temperature conversion of the remaining carbon monoxide to carbon dioxide in a low-temperature shift converter (LTS) after cooling. The usual catalyst for the low-temperature shift converter is copper oxide supported on zinc oxide and alumina.

Removal of Carbon Dioxide and Methanation. About 17 to 19% of the effluent gases is carbon dioxide. This undesirable byproduct is usually removed by a number of methods, such as physical absorption, chemical reaction, or a combination of both. Removal methods include monoethanolamine process for removing acid gases, activated carbonate process based on absorption of carbon dioxide by potassium carbonate, and physical absorption systems in which solvents and agents that remove carbon dioxide are used. Generally, the choice of carbon dioxide removal depends on the overall plant design and the integration procedures needed.

The final purification step employs the methanation process to convert carbon oxides to methane. In the very final cleaning up, the cryogenic purifier system is often used. In the cryogenic process, the gas from methanation is dried to a low dew point, cooled, and then expanded in a turbine to liquefy some portion of the gas followed by further cooling. The vapor emerging after this step is scrubbed in a rectifying column to remove most of the argon and almost all of the methane. Methanation reactions are as follows:

$$CO + 3H_2 \rightleftharpoons CH_4 + H_2O \qquad \Delta H = -206 \text{ kJ/mol} \tag{32.11}$$

and

$$CO_2 + 4H_2 \rightleftharpoons CH_4 + 2H_2O \qquad \Delta H = -165 \text{ kJ/mol} \tag{32.12}$$

Synthesis of Ammonia and Recovery. After the various purification steps, the synthesis gas, consisting of hydrogen and nitrogen, is ready for compression and ammonia manufacture usually in a synthesis loop. The operating pressure is normally in the range of 13,785 to 34,475 kPa. Most modern large-scale ammonia plants use centrifugal compression for the synthesis. The synthesis loops fall into two main groups: those that recover ammonia after make-up gas-recycle compression and those that recover ammonia before the recycle compression.

The synthesis gas is then delivered to the converter. There are two types of converters: tubular and multiple bed. The feed gas is fed at a 3:1 ratio, hydrogen to nitrogen. The synthesis is carried out on a catalyst surface based on metallic iron (magnetite) that has been promoted with iron oxides. An increase in pressure increases the equilibrium percentage of ammonia as well as the reaction rate. An increase in temperature accelerates the reaction, decreases the equilibrium amount of ammonia, and also degrades the catalyst. The ammonia product is usually recovered by condensation, and any unreacted feed gas is recycled through the synthesizer. Modern ammonia plants are designed and built to reflect some degree of integration between the process and energy systems. Any waste heat in the process is used to provide energy for the boiler-feed water heating and steam generation. Thus, in modern plants it is not difficult to run all the processes without employing external power.

Environmental Concerns. Natural gas contains low quantities of sulfur that are easily removed by stripping in a conventional adsorption/absorption system. Estimates of effluent emissions from ammonia plants have revealed the results shown in Table 32.3. (Committee on Medical and Biological Effects, 1979).

Other feedstocks, such as coal and naphthas, may present some environmental problems. For example, coal contains significant quantities of sulfur and metals requiring extra steps for removal and safe disposal. Byproducts such as tars, phenols, and naphthas must also be safely disposed of.

Apart from feedlots and other natural sources, which account for 95.5% of total fugitive emissions of ammonia, anthropogenic sources of ammonia emitted into the environment during the production of fertilizers and industrial chemicals are shown in Table 32.4.

32.4.2 Production in Canada, the United States, and Worldwide

Because of the cheap and abundant natural gas, most of the ammonia produced in Canada comes from western Canada and some from Ontario (CIS, 1998). Plants are located in Calgary, Carseland, Joffre, Redwater, Sombra, Trail, Niagara Falls, Maitland, Belle Plaine, Hamilton, Nanticoke, Medicine Hat, Fort Saskatchewan, Redwater, Belle Plaine, Brandon, Kitimat, and Courtright. The major market for anhydrous ammonia is in western Canada. Ammonia is shipped from different manufacturing sites by rail to other manufacturing facilities, atmospheric storage tanks or rail transload locations; by truck or road transport to atmospheric storage tanks and retail storage locations; and by retail truck to farm wagons.

About 5,000 kilotons of anhydrous ammonia were produced in Canada in 1997, of which 3,500 were utilized as fertilizer products and 2,000 applied directly to the soil as urea. About 14 kilotons were imported mostly from the United States (CIS, 1998). Ammonia prices and

TABLE 32.3 Pollutants from Ammonia Production

Pollutant	Emission factor (kg/t)	Total emissions (t/year)[a]
Sulfur dioxide	0.4	5,900
Nitrogen dioxide	0.6	8,900
Carbon monoxide	6.0	89,200
Ammonia	1.3	19,300

[a] The abbreviation "t" is used to denote the metric ton of 1,000 kg (2204.0 lb).

TABLE 32.4 Ammonia Emissions from Production of Fertilizers and Industrial Chemicals

Source	Emission rate (tons/year)
Direct application	168,000
Ammonium nitrate	59,000
Petroleum refineries	32,000
Sodium carbonate (Solvay)	14,000
Diammonium phosphate	10,000
Ammoniator-granulators	10,000
Urea	4,000
Miscellaneous	2,000
Beehive ovens	1,000
Total	300,000

sales reached a record high in 1996. Ammonia plants continue to expand as a result of a cheap supply of natural gas. The price range per ton in western Canada was $350 to 420 in 1997. The major Canadian producers are Agrium, Beker Industries, Canadian Fertilizers, Cominco, Cyanamid, Nitrochem, Pacific Ammonia, Saskferco Products, Sherritt International, Simplot Chemical, Stelco, Terra International, and Western Cooperative.

The total production of anhydrous ammonia in the United States in 1997 was about 17,000 kilotons, in addition to around 3,000 kilotons that was imported. Prices at the U.S. Gulf Coast ranged from $225 to 230 in 1997. Ammonia plants are located in several cities including Borger, Pace Junction, Hopewell, Fortier, Geismar, Walla Walla, Donaldsonville, Girard, Cheyenne, Freeport, St. Helens, Beulah, Beaumont, Beatrice, Coffeyville, Dodge City, Enid, Fort Dodge, Lawrence, Pollock, Port Comfort, Creston, Faustina, East Dubuque, Pocatello, Wells, Sterlington, Cherokee, Yazoo City, Dumas, Augusta, Clinton, Geismar, Laplatte, Lima, Memphis, Gordon, Luling, Beaumont, Blytheville, Port Neal, Verdigris, Woodward, Finley, Kenai, and Pryor.

The major U.S. producers are Agrium, Air Products, Allied Signal, Avondale Ammonia, Borden Chemicals, BTU Energy, CF Industries, COGA Industries, Coastal Chem, Coastal Refining, Coastal St. Helens Chemical, Dakota Gasification, Dupont, Farmland Industries, Formica Plastics, Green Valley Chemical, IMC-Agrico, IMC AgriBusiness, J R Simplot, Koch Industries, LaRoche Industries, Mississippi Chemical, Nitromite Fertilizer, PCS Nitrogen, Shoreline Chemical, Solutia, Terra Nitrogen, Triad Nitrogen, Unocal, and Wil-Grow Fertilizer.

Other countries actively producing anhydrous ammonia in North America are Mexico, which produces 26,995 kilotons, and Trinidad, which produces about 2,370 kilotons. On a global scale, ammonia consumption as a fertilizer has been growing steadily since 1960, as shown in Table 32.5 (Lancaster, 1989; IFDC, 1989). The use of ammonia fertilizers is steadily growing in developing countries as the soil quality deteriorates over time. Urea is the most common form of ammonia fertilizer that is exported by the Commonwealth of Independent States (CIS) and Romania (43%).

While the United States is the largest importer of ammonia (34%), CIS is the world's largest producer and the largest exporter of ammonia (34%), followed by China, India, Mexico, Indonesia, and the Middle East. There are still large variations in ammonia supply, use, and distribution in different regions of the world. This probably depends on affordability, agricultural practice, and the industrial base in various parts of the globe. The demand for ammonia in different parts of the world is shown in Table 32.6 (Lancaster, 1989).

TABLE 32.5 Global Ammonia Fertilizer Use

Year	Ammonia fertilizer use (kilotons)
1960	9.54
1970	28.17
1980	57.19
1990	80.3

32.4.3 Production and Transportation Costs

The total production of ammonia worldwide in 1990–1991 was 98.84 megatons, even though the nominal capacity was 119.18 megatons. Industrial activities utilized 10.93 (11%) of this amount. Production costs and profitability largely depend on the type of feedstock being used, labor, capital investment costs, and energy prices (Sheldrick, 1987; *Green Markets,* 1990). Profitability also depends on the selling price of ammonia, which in turn is a function of supply and demand dynamics. Estimation of the total production costs in a 1000 metric tons per day plant at the United States Gulf Coast was $106.38/ton. This rises to $150/ton when taxes, storage, and investment costs are included. Transportation costs are separate from the costs of manufacturing. The transportation and use of ammonia are at their peak during the spring fertilizer season. In North America, ammonia is usually transported for long distances by rail, barge, and pipeline or by truck for very short distances from local dealers or distributors to farm locations. Long-distance transport is generally cheaper than for shorter distances. For example, costs estimates for long-distance hauling (1,600 km) by barge, pipeline, and rail are $0.0153, $0.0161, and $0.0215 per ton per kilometer, respectively, while for short distances (160 km), it is about $0.0365 per ton per kilometer. Shipping liquefied ammonia across the Atlantic costs about $35 a ton.

32.5 CHEMISTRY

Due to its reactivity, ammonia will take part in a variety of reactions. Only those of some environmental significance or relevance to spill situations will be discussed here.

The main driving force for the reactions of ammonia is the lone pair of electrons on the nitrogen atom, which is so oriented that it adds to the nitrogen-hydrogen dipole moment. Hence, ammonia is an extremely polar molecule.

TABLE 32.6 Global Demand for Ammonia in Different Regions in 1990

Region	Percentage of global distribution
Africa	3
Asia	35.4
Latin America	5.3
North America	13.8
Eastern Europe	9.7
Western Europe	11.5
CIS	21.5

32.5.1 Reactions

Water. The reaction between ammonia and water is a reversible one in which very small amounts of the hydroxide are formed. This is clearly evident from the equilibrium constant (K_{eq}) of the reaction (Jolly, 1964):

$$NH_3 + H_2O \rightleftharpoons NH_4^+ + OH \qquad (32.13)$$

$$K_{eq} = \frac{(NH_4^+)(OH^-)}{(NH_3)(H_2O)} = 1.75 \times 10^{-5} \text{ at } 18°C \qquad (32.14)$$

Spectroscopic data provide evidence for the presence of the dihydrate $NH_3 \cdot 2H_2O$, the monohydrate $NH_3 \cdot H_2O$, the hemihydrate $2NH_3 \cdot H_2O$, and unhydrated ammonia in aqueous ammonia solutions. This is also shown in the water–ammonia (Eq. 32.15). Ammonia and water will mix in all proportions above 0°C. The maxima shown in the diagrams indicate the existence of these hydrates. The monohydrate melts at −79.0°C, the hemihydrate at −78.8°C, and the dihydrate at −98°C.

$$NH_3 + H_2O \rightleftharpoons NH_3 \cdot H_2O \rightleftharpoons 2NH_3 \cdot H_2O \rightleftharpoons NH_3 \cdot 2H_2O \rightleftharpoons NH_4^+OH^- \qquad (32.15)$$

Oxidation. When ammonia is ignited, it burns to produce water and nitrogen. Powerful oxidizing agents such as potassium permanganate will also oxidize ammonia to elemental nitrogen. In the presence of catalysts, ammonia is used in the production of nitric acid.

$$2NH_3 + O_2 \rightarrow N_2 + 3H_2O \qquad (32.16)$$

Decomposition. Even though quite stable at high temperatures, ammonia will decompose to hydrogen and nitrogen. The decomposition range is about 300 to 600°C.

$$2NH_3 \rightleftharpoons N_2 + 3H_2 \qquad (32.17)$$

Neutralization. The acid-base properties of ammonia are very useful in spill situations. This is an important reaction often used in spill response and disposal. Ammonia will react with acids to give salts and water. The product, an ammonium salt, often makes a good fertilizer.

$$NH_3 \cdot H_2O + HNO_3 \rightarrow NH_4NO_3 + H_2O \qquad (32.18)$$

Another reaction of great importance in spill response is of ammonia with carbon dioxide to yield ammonium carbamate, which then decomposes to urea.

$$2NH_3 + CO_2 \rightarrow NH_2CO_2NH4 \qquad (32.19)$$

$$NH_2CO_2NH_4 \rightarrow NH_2CONH_2 + H_2O \qquad (32.20)$$

Reactions with Metals. Liquid ammonia possesses a remarkable ability to dissolve some electropositive metals, mostly the alkali and the alkali earths metals such as lithium, sodium, potassium, cesium, and copper. It is therefore advisable not to store these chemicals together. For example, sodium dissolves in excess ammonia to produce sodamide, a white solid, while magnesium, when heated in ammonia, produces magnesium nitride.

$$2Na + 2NH_3 \rightarrow 2NaNH_2 + H_2 \qquad (32.21)$$

$$3Mg + 2NH_3 \rightarrow Mg_3N_2 + 3H_2 \qquad (32.22)$$

Reactions with Nonmetals

Halogens. Halogens react with ammonia to yield haloamines. The eventual trihalides formed can react with excess ammonia to produce unstable ammoniates, which can decompose to give ammonium salt and nitrogen. This reaction has some implication for water purification using chlorine as disinfectant. Aqueous solutions of chloramine, NH_2Cl, are formed by the reaction of aqueous ammonia and chlorine or hypochlorite solutions:

$$2NH_3 + Cl_2 \rightarrow NH_2Cl + NH_4Cl \quad (32.23)$$

$$NH_3 + OCl^- \rightarrow NH_2Cl + OH^- \quad (32.24)$$

$$NH_3 + Cl \rightarrow NH_2Cl \rightarrow NHCl_2 \rightarrow NCl_3 \quad (32.25)$$

$$NCl_3 + NH_3 \rightarrow NCl_3 \cdot NH_3 \quad (32.26)$$

$$NCl \cdot NH_3 + 3NH_3 \rightarrow N_2 + 3NH_4Cl \quad (32.27)$$

Reaction with Phosphorus. Ammonia will react with phosphorus vapor at red heat to produce nitrogen and phosphine.

$$2NH_3 + 2P \rightarrow 2PH_3 + N_2 \quad (32.28)$$

Reaction with Sulfur. Sulfur reacts with anhydrous ammonia to form nitrogen sulfide.

$$10S + 4NH_3 \rightarrow 6H_2S + N_4S_4 \quad (32.29)$$

Reaction with Metallic Salts. Ammonia solutions will react with some metallic salts to form complexes.

$$CuSO_4 + 2NH_3 \cdot H_2O \rightarrow Cu(OH)_2 + (NH_4)_2SO_4 \quad (32.30)$$

$$Cu(OH)_2 \rightarrow Cu^{2+} + 2OH^- \quad (32.31)$$

$$4NH_3 + Cu^{2+} \rightleftharpoons [Cu(NH_3)_4]^{2+} \quad (32.32)$$

32.6 ENVIRONMENTAL FATE

Over 99.5% of atmospheric and aquatic ammonia is from natural sources, namely from biodegradation or decomposition of organic matter. Ammonia also enters the atmosphere from anthropogenic sources such as during production and use. Other sources include waste incineration, domestic heating, internal combustion engines, and other industry-related sources.

32.6.1 Terrestrial Fate

Anhydrous ammonia is an extremely volatile compound, and when released on land it does not stay in the same place for very long periods of time. It will tend to volatilize very rapidly, albeit some of it will be adsorbed onto the soil. The portion that is sorbed to the soil will undergo further reactions such as nitrification and eventual plant uptake. Because it is poorly adsorbed to the soil, leaching into underground water may also occur.

32.6.2 Atmospheric Fate

Comparatively, ammonia has a residence time of 5 to 10 days in the atmosphere, depending on the location and weather conditions. The main fraction that evaporates into the air can undergo a variety of reactions, such as (1) decomposition, thermal, and photochemical reactions, (2) aqueous phase reactions that produce aerosols, and (3) heterogeneous reactions with soot particles.

32.6.3 Photodecomposition

Ammonia is known to decompose when exposed to radiation in the far ultraviolet to produce the reactive amino and imino radicals.

$$NH_3 + h\mu \rightarrow NH_2 + H \quad (32.33)$$

$$NH_3 + h\mu \rightarrow NH + 2H \quad (32.34)$$

There is a theory that ammonia decomposes at high temperatures (above 450°C) to produce nitrogen and hydrogen, which often causes explosions during ammonia fires.

With photochemically excited species:

$$NH_3 + OH \rightarrow NH_2 + H_2O \quad (32.35)$$

$$NH_3 + O \rightarrow NH_2 + OH \quad (32.36)$$

$$2NH_3 + 4O_3 \rightarrow 4O_2 + H_2O + NH_4NO_3 \quad (32.37)$$

32.6.4 Aqueous Phase Reactions

With acid aerosols, ammonia forms salt clusters:

$$8NH_3 + 8HNO_3 \rightarrow 8NH_4NO_3 \quad (32.38)$$

With sulfur dioxide, it initially forms amido sulfurous acid.

$$nNH_{3\ (g)} + SO_{2\ (g)} \rightleftharpoons (NH_3)_nSO_{2\ (s)} \quad (32.39)$$

$$NH_3SO_{2\ (g)} + H_2O_{(g)} \rightarrow NH_3SO_{2(aq)} \quad (32.40)$$

$$\rightarrow \rightarrow \rightarrow (NH_4)_2SO_4 \quad (32.41)$$

32.6.5 Rainfall and Precipitation

Washout of ammonia by rainfall, precipitation, water vapor, and reactions with other reactive species reduces the amounts of ammonia in the air and also ensures quick return to the soil. Levels of ammonia in precipitation have been shown to be much higher in rural areas than in urban centers and have shown seasonal variation (Lusis and Phillips, 1990).

32.6.6 Aquatic Fate

If anhydrous ammonia is discharged into water, while most of it will volatilize into the air (35%) depending on the wind and mixing conditions, some will stay in the water. Ammonia is extremely soluble in water. It will dissolve to produce un-ionized ammonia, such as ammonia (NH_3), ammonia hydrates $NH_3 \cdot H_2O$, $NH_3 \cdot 2H_2O$, $2NH_3 \cdot 2H_2O$, NH_4OH, and the ionized ammonia $NH_4^+ \; OH^-$. It is known that the un-ionized ammonia is toxic to fishes and aquatic organisms.

$$NH_3 + H_2O \rightleftharpoons NH_3 \cdot H_2O; NH_3 \cdot 2H_2O; 2NH_3 \cdot H_2O \rightleftharpoons NH_4^+ + OH^- \quad (32.42)$$

The relative proportion of ionized and un-ionized ammonia formed depends on pH, temperature, the ionic strength of the receiving waters, and the prevailing weather conditions. One would therefore expect different fractions of these ammonia species when the same quantity of ammonia is released in sea water as in fresh water.

Any of these solution products can react with chemical species already present in the water, such as humic substances, metals, and metal complexes. As for the rest of the dissolved ammonia, while some will bind to sediments, suspended particles, and organic matter available in the water, most will undergo nitrification by the *Nitrosomonas* and *Nitrobacter* bacteria species to yield nitrates that can be uptaken by aquatic plants and organisms. Many algae and phytoplankton have been found to utilize ammonia directly as their source of nitrogen. Because of its low octanol/water partition coefficient, ammonia is not expected to adsorb strongly to sediments.

32.7 BEHAVIOR

Because of many variations in conditions such as cold products versus hot products, weather and wind conditions, various leak geometries, instantaneous releases, massive tank failures, continuous versus intermittent releases, and pipeline leaks versus railcar punctures, there is no universal model to make accurate quantitative predictions of dispersion of ammonia (Pederson and Selig, 1989; Kaiser and Walker, 1978). All models have described the behavior of ammonia on sudden release from pressurized tank as follows:

1. Liquefied ammonia will definitely flash off with simultaneous entrainment of liquid droplets depending on the pressure drop in the storage tank or size of leak.
2. There is a possibility of gravity-driven slumping, becoming very much broader than would be expected for a cloud that, e.g., is not denser than air.
3. There is likely to be a period of low-level ground hugging during which the vertical rate of growth of the cloud is less than that expected for a neutral cloud.
4. The cloud is not expected to become buoyant except when the wind speed is small.

32.8 HUMAN AND ENVIRONMENTAL TOXICITY

It has been known for more than a century that ammonia is toxic. In 1893 the main effects of ammonium chloride salt on humans were described as convulsions, twitches, tremors, salivation, irregular respiration, somnolence, and lassitude. Toxic effects of other ammonia derivatives, such as urea, have been described as being characterized by restlessness, ataxia, dyspnea, collapse, edema, heart abnormalities, muscle spasms, tetany, and death.

Because the odor threshold of ammonia gas is very low (approximately 5 ppm), it provides ample warning properties. Ammonia attacks wet tissue such as the eyes and mucous membranes of the respiratory system. Two types of injuries can result from ammonia spills. One is direct contact, in which the victim is sprayed or splashed resulting in skin burns or loss of vision. Such injuries are often restricted to the immediate surroundings of the release point. Ammonia can also be released as a gas or cloud, which can spread over a wide area and affect many people adversely as it is inhaled. In this case, injuries are often confined to the respiratory system (Markham, 1987; Czuppon et al., 1992; Braker and Mossman, 1980; Pedersen and Selig, 1989). At very high concentrations, bronchial spasm or edema of the lungs can result in permanent lung damage or death in a few minutes.

The range of effects of ammonia exposure to humans can be divided into three groups:

1. *Mild effects:* Exposure to concentrations lower than 5,000 ppm for a few minutes.
   - *Symptoms:* Restricted to the eyes and upper respiratory tract, smarting sensation in the eyes and mouth, pain when swallowing, marked hoarseness and tightness of the throat, slight cough.
   - *Signs:* Reddening of the conjunctivae, lips, mouth, and tongue with swelling of the eyelids and edema of the throat. No clinical signs of lung involvement.
   - *Consequences:* There may be spontaneous recovery without pulmonary complications ensuing.
2. *Moderate effects:* Exposure to concentrations from 5,000 to 10,000 ppm for a few minutes. There is evidence of the progression of effects deeper into the respiratory system with involvement of the bronchi and bronchioles.
   - *Symptoms:* Exaggeration of the symptoms described under mild effects. Feeling of tightness in the chest, difficulty in swallowing, sometimes complete loss of voice. Cough with copious sputum, sometimes bloodstained.
   - *Signs:* Distress, increase in pulse and respiration rates. Marked swelling of the eyelids with spasm and lacrimation. Moderate edema of the oropharynx with burning of the mucous membranes and resultant stripping of the epithelium to reveal dark red glazed patches. Examination of the chest reveals diminished air entry with the presence of moist sounds.
   - *Consequences:* Fatalities due to obstruction of airways or complications such as lung infection cannot be ruled out.
3. *Severe effects:* Exposure to concentrations in excess of 10,000 ppm for a few minutes.
   - *Symptoms:* Symptoms of oropharynx and eyes similar to those described under moderate effects. Persistent cough with copious frothy sputum.
   - *Signs:* Shock, restlessness, and obvious distress. Rapid pulse of poor volume. Cyanosis and great difficulty in breathing. Generalized moist sounds in the chest. Death as a result of asphyxiation may be expected. Survivors may die later as a result of complications, such as lung infections.

32.8.1 Environmental Toxicity

In dilute or moderate concentrations, ammonia is good for the soil and microorganisms. During a spill on the soil, all acid constituents are neutralized. Liquefied ammonia, on the

TABLE 32.7 Typical Effects of Gaseous Ammonia on Humans

Concentration (ppm)	Exposure time (min)	Effects
72	5	Some irritation.
330	30	Concentration tolerated.
600	1 to 3	Eyes streaming within 30 seconds.
1,000	1 to 3	Eyes streaming instantly, vision impaired but not lost. Breathing intolerable to most participants after exposure.
1,500	1 to 3	Instant reaction is to get out.

other hand, is very corrosive, and excessive discoloration of plants has been observed during large spills of liquefied ammonia. Fatalities in ponies and hogs have been reported at concentrations of 5,000 to 20,000 ppm (assumed) and 700 to 10,000 ppm, respectively. Birds and small wildlife have also been reported dead at 2,300 ppm; trees and ground vegetation have withered.

Ammonia is also produced from burning of coal and bacterial decomposition of proteinaceous organic matter. Conversion of ammonia to ammonium salts will occur rapidly both in the atmosphere and in the soil but will result in less acidity of the soil. It is not known to what extent ammonia reduces excess greenhouse gases such as carbon dioxide in the atmosphere to form ammonium carbamate. Microbial assimilation of ammonium compounds consumes alkalinity of the soil as well as converting ammonium ions to nitrate. Nitrification is carried out by many aerobic bacteria. Nitrification is pH sensitive. Denitrification, which is an anaerobic process, is also pH sensitive.

32.9 PAST SPILLS, LESSONS LEARNED, AND COUNTERMEASURES APPLIED

Three well-researched accidents involving instantaneous releases of massive quantities of anhydrous ammonia from pressurized containers are discussed in this section.

32.9.1 Houston, Texas, May 11, 1976

A road tanker carrying approximately 19 tons of anhydrous ammonia crashed through a barrier at an elevated section of motorway near Houston, Texas (Pedersen and Selig, 1989; Kaiser and Walker, 1978). The pressurized tank burst on hitting the roadway below. Five people were killed and 178 people injured. High concentrations of ammonia were confined to within a few hundred meters of the source of the accident. All the fatalities and permanent injuries were within 70 m of the release source. The concentration contours predicted from calculations and models to account for the fatalities and injuries are shown in Table 32.8.

Prevention may help in reducing road accidents of this nature. Although the cause of the crash was not determined, it was due either to mechanical failure or human error. Sound maintenance programs should be in place for these chemical tankers in accordance with the industry guidelines. Drivers should be given medical examinations and regularly tested for fitness and drug use. Long routes that could result in fatigue should also be prevented. Weather is also an important factor. Chemical tanker drivers should limit their activities on the road during inclement weather conditions.

TABLE 32.8 Predicted Concentration Contours for the Houston Accident

Isopleth (ppm)	Approximate length (m)	Approximate width (m)
1,200	1,130	400
2,500	875	420
5,000	835	430
10,000	600	350

Source: Pedersen and Selig, 1989.

32.9.2 Potchefstroom, South Africa, July 13, 1973

In the Potchefstroom accident, 18 people died and hundreds were injured as a result of the release of 38 tons of anhydrous ammonia following a sudden failure of a pressurized ammonia storage tank (Pedersen and Selig, 1989). The cloud was visible about 450 m downwind and was about 300 m wide. Most of the fatalities were within 200 m of the release point. Model-predicted concentration contours and calculations for this spill are shown in Table 32.9.

Lessons Learned. After an inquiry and follow-up, it was determined that the accident happened as a result of equipment failure. It was apparent that proper guidelines and codes of conduct were not followed. For example, pressure vessels containing ammonia should have been subjected to periodic examination for strains and cracks. The following are the causes of the spill:

1. The vessel was not stress-relieved after manufacture.
2. The vessel material had been weakened by strain and aging.
3. The weld repairs as well as hydraulic pressure testing had induced more stresses on the vessel wall and were not followed by stress relief.
4. Failure of the tank could also have been caused by temperature fluctuations during decanting operations.

The following recommendations were made.

1. Employees must have emergency response training and be taught to recognize emergency alarms and obey them.
2. It may be a lot safer to stay in a closed room if the route to a safe area is blocked or falls within a hot spot. Strategically located rooms should be set aside and equipped and clearly identified as gas-proof rooms.
3. Air supply and self-contained-breathing apparatus (SCBA) must be made available to every employee.
4. Driving through a dense ammonia cloud is very risky and should not be attempted.
5. Emergency plans must be in place for worst-case scenarios, not just for leaks.

Communications between different departments and the outside world must be made possible during catastrophic accidents such as this, and plant control rooms must be habitable.

TABLE 32.9 Predicted Concentration Contours of the Potchefstroom Accident

Isopleth (ppm)	Approximate length (m)	Approximate width (m)
1,200	2,670	680
2,500	1,400	650
5,000	910	640
10,000	730	660

Source: Pedersen and Selig, 1989

32.9.3 Lithuanian Ammonia Accident, March 20, 1989

This accident started when a 10,000-tons unpressurized liquid ammonia tank failed. (Andersson, 1991). The ammonia gas released caught fire and ignited a fertilizer store containing 15,000 tons of fertilizer, NPK. The ammonia vapor and fertilizer decomposition products formed a 400-km^2 cloud. After 12 hours, all of the ammonia had evaporated, but the fertilizer continued to burn (decompose) for three days, producing large quantities of nitrogen oxide fumes. There were 7 fatalities, 57 injured, and 32,000 people evacuated. The following are the causes of the failure according to the official Commission of the Commonwealth of Independent States (CIS):

- The tank experienced an overpressure due to delivery of 14 tons of warm ammonia (10°C) at the bottom of the tank because of an operating mistake in the plant.
- The warm ammonia accumulated at the bottom of the tank in the form of lenses or a layer of warm unstable ammonia.
- It did not evaporate at once because of the hydrostatic pressure.
- This layer or lenses tilted over the surface of the liquid.
- The warm liquid ammonia evaporated to lower its temperature to −33°C, and this caused the overpressure.
- All refrigeration compressors were out of operation at the time and ammonia did catch fire.

The following recommendations were made:

1. Human error seems to have been the cause of the accident. Safe and standard operating procedures and practices should be established. In this case, the ammonia could have been added at the top of the tank and not at the bottom.
2. The fertilizer plant should be built at some distance from other facilities to avoid fire spread.
3. The tank should be surrounded by steel retainer walls.
4. Pressure tanks with higher relief valve specifications could have helped.
5. Frangible roof designs and continuous circulation of ammonia would reduce risks.
6. Emergency response planning must be in place.

32.10 ENVIRONMENTAL CONCENTRATIONS AND STANDARDS

32.10.1 Ammonia Levels in Different Media

The highest levels of ammonia in the environment result from natural sources. The levels of ammonia in various media in Canada have been measured by Atmospheric Services, Environment Canada. These are shown in Table 32.10. Values obtained are generally very low for most environmental compartments except for spills, sewages, outfalls, and animal-rearing houses.

32.10.2 Threshold Limit Values

The American Conference Governmental Industrial Hygienists have determined the threshold limit values for TWA (time-weighted average) to be 25 ppm (17 mg/m^3) and STEL (short-term exposure limit) to be 35 ppm (24 mg/m^3).

TABLE 32.10 Environmental Levels in Different Media in Canada

Compartment	Levels (mean in ppm)
Surface water	0.76
Drinking water	0.1
Precipitation	0.59
Aerosols	1.8×10^{-3}
Food	Varies extensively
Soil	Varies extensively 478 μg/g (0–18 cm) 413 μg/g (75–112 cm)

When the threshold limit levels are compared with the levels found in the Canadian environment, one can only conclude that the Canadian public is not significantly exposed to dangerous levels of ammonia and is therefore not at great risk. High exposures are often occupational or purely accidental.

32.11 CONCLUSIONS

Ammonia is one of the most beneficial industrial chemicals to mankind in its use as a fertilizer, or precursor, reactant, and intermediate in many syntheses. The manufacture of ammonia is a relatively clean process except for greenhouse gases and ammonia emissions.

The driving force for most of its reactions is the lone pair of electrons on the nitrogen atom. The toxic component in water is the un-ionized ammonia.

A person caught in a large spill should stay indoors with doors and windows closed and ventilation shut down. Only in circumstances where there is little alternative or where success is sure should one run away from the cloud. In any case, respiratory breathing apparatus is recommended. The most serious injuries often occur within a couple of hundred meters from the spill source. Most risks from ammonia exposure are accidental or occupational.

Epidemiological studies with respect to ammonia exposures are still inadequate. Since the aquatic ecosystems are the most sensitive to high ammonia concentrations, long-term monitoring of ammonia is needed in underground sewage systems and outfalls. Effects of other chemical species in the presence of ammonia should also be studied.

32.12 REFERENCES

American Industrial Hygiene Association (AIHA). 1988. *Emergency Response Planning Guidelines: Ammonia,* AIHA, Akron, OH.

Andersson, B. O. 1991. "Lithuanian Ammonia Accident, March 20, 1989," *Icheme Symposium Series,* no. 124, pp. 15–17.

Braker, W. and A. L. Mossman. 1980. *Matheson Gas Data Book,* 6th ed., Matheson Gas Products, East Rutherford, NJ, pp. 23–33.

Camford Information Services, Inc. (CIS). 1998. "Ammonia," in *CPI Product Profiles,* CIS, Scarborough, ON.

Committee on Medical and Biological Effects of Environmental Pollutants. 1970. "Chemical and Biological Effects of Pollutants," in *Ammonia,* University Park Press, Baltimore, MD, pp. 115–176.

Compressed Gas Association, Inc. 1990. *Handbook of Compressed Gases,* 3rd ed., Van Nostrand Reinhold, New York, NY, pp. 231–232.

Czuppon, T. A., S. A. Knez, J. M. Rovner, and M. W. Kellog Company. 1992. "Ammonia," in *Encyclopedia of Chemical Technology,* ed. M. Howe-Grant. 4th ed., John Wiley & Sons, New York, pp. 638–688.

Environment Canada. 1999. National Analysis of Trends in Emergency Situations (NATES): *Chemical Accidents Reports Database,* Hull, QC.

Fingas, M., N. Laroche, G. Sergy, B. Mansfield, G. Clouthier, and P. Mazerolle. 1991. "A New Chemical Spill Priority List," in *Proceedings of the 8th Technical Seminar on Chemical Spills,* Environment Canada, Ottawa, ON, pp. 223–237.

Fenske, B. 1993. *Manitoba Ambient Air Quality Annual Report—1989,* Air Quality Report No. 93-02.

Green Markets. 1990. McGraw-Hill, New York, NY.

International Fertilizer Development Center (IFDC). 1989. *Global Fertilizer Perspective, 1960–1995,* IFDC, Muscle Shoals, AL.

Jolly, W. L. 1964. *The Inorganic Chemistry of Nitrogen,* W. A. Benjamin, New York, NY, pp. 1–42.

Kaiser, G. D., and B. C. Walker. 1978. "Releases of Anhydrous Ammonia from Pressurized Containers—The Importance of Denser than Air Mixture," *Atmospheric Environ.,* vol. 12, pp. 2289–2300.

Lancaster, J. M. 1989. " New Developments and Growth in World Fertilizer, Demand, Supply and Trade," in *Fertilizer Latin American International Conference,* British Sulphur Corporation Ltd., Caracas, Venezuela.

Lusis, M. A. and M. L. Phillips, eds. 1990. *The 1990 Canadian Long-Range Transport of Air Pollutants and Acid Deposition Assessment Report, Part 3: Atmospheric Sciences,* Federal/Provincial Research and Monitoring Coordinating Committee (RMCC), Environment Canada, Downsview, ON.

Markham, R. S. 1987. "Review of Damages from Ammonia Spills," *Ammonia Plant Safety,* vol. 27, pp. 137–149.

Ministers' Expert Advisory Panel on the PSL2. 1995. *Report of the Ministers' Expert Advisory Panel on the Second Priority Substances List,* Environment Canada, Hull, QC.

National Water Quality Data Bank (NAQUADAT). 1991. Water Quality Branch, Inland Waters Directorate, Environment Canada, Ottawa, ON.

Pedersen, F. and R. S. Selig. 1989. "Predicting the Consequences of Short-term Exposure to High Concentrations of Gaseous Ammonia," *Journal of Hazardous Materials,* vol. 21, pp. 143–159.

RTECS On-Line. 2000. *Registry of Toxic Effects of Chemical Substances,* Department of Health and Human Services, Centers for Disease Control, National Institute for Occupational Safety and Health, Washington, DC.

Saskatchewan Environment Data (WQB). 1991. *Esquadat Summary Report,* pp. 1–16.

Sheldrick, W. L. 1987. *World Nitrogen Survey,* Technical Paper No. 59, World Bank, Washington, DC.

Transport Canada. 1985. *Transport of Dangerous Goods Act and Regulations,* Ottawa, ON.

Williamson, D. A. 1990. *The Development and Use of Water Quality Objectives in Manitoba,* Water Standards Section Report 90-2.

CHAPTER 33
PERSPECTIVES ON SPECIFIC SUBSTANCES: CHLORINE

Richard Lawuyi and Merv Fingas
Emergencies Science Division, Environment Canada,
Environmental Technology Centre, River Road,
Ottawa, Ontario

33.1 OVERVIEW OF PRODUCT AND INDUSTRIAL USES

Chlorine is a greenish-yellow gas that is 2.5 times heavier than air. It does not occur naturally as a gas. It makes up about 0.0314% of the earth's upper crust, where it exists mainly in the form of sodium, potassium, and magnesium chlorides. The major proportion is present in the oceans, constituting about 2%.

For many decades, the growth in chlorine use has often been equated with the strength of a nation's economy. This is largely due to the various chemical industries and chemical products that are often generated. The most important sources of man-made emissions and releases of chlorine into the environment are those involving its production, transportation, handling, and use.

Chlorine was placed second in a priority list ranking of hazardous chemicals determined by the Emergencies Branch of Environment Canada (Fingas et al., 1991). Some of the criteria used in ranking chemicals are historical spill volume, supply volume, reported spill frequencies, and toxicity, including persistence and bioaccumulation. This substance should therefore be of great concern.

33.1.1 Modern Industrial Uses

The largest use of chlorine is in plastics. As shown in Fig. 33.1, chlorine is used extensively in the manufacture of several chlorinated aliphatic solvents, organic and inorganic intermediates, plastics, and in the pulp and paper industry (CIS, 1999). Chlorine and chlorine dioxide are still used in the bleaching of pulp, although to a lesser degree than in the past in order to reduce the potential for the formation of polychlorinated dioxins. The disinfection of drinking and wastewater, sewage, and biofouling in cooling and power plants represent other important uses of chlorine. The complex chemistry and implications for the environment will be discussed later.

33.2 INTRODUCTION

Chlorine does not occur naturally except during volcano eruptions, but it is found combined with metals in the form of sodium, potassium, and magnesium chlorides. Chlorides make

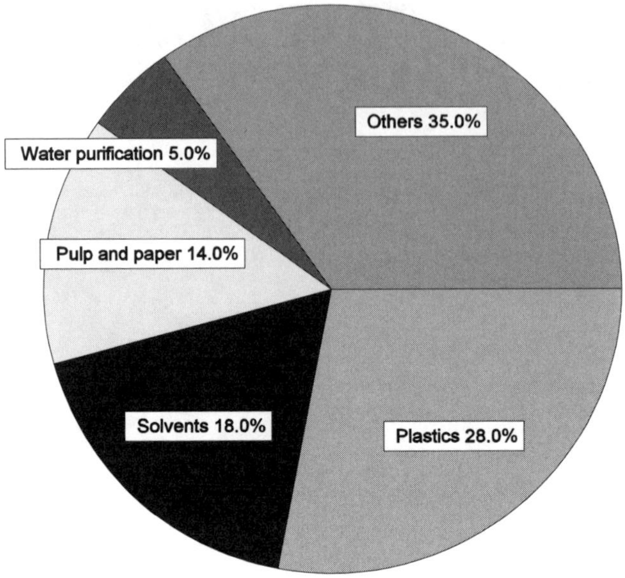

FIGURE 33.1 Major uses of chlorine.

up approximately 0.0314% of the earth's upper crust. A substantial portion is found in the oceans, which contain about 2% chloride. Some chlorides are also present in the atmosphere in the form of aerosols.

Chlorine gas finds its way into the environment during production and use, as well as from accidental releases during transportation and handling. Chemical spills occur frequently in Canada and most other countries in the industrialized world. These spills often involve very large volumes of chemicals. Many spills happen in stationary facilities, and only a small proportion occur during transfers or while in transit. In Canada between 1974 and 1984, 36 major spills of chlorine were recorded in the NATES (National Analysis of Trends in Emergencies System), involving about 121 tons of liquefied chlorine (Environment Canada, 1999). According to the DGAIS (Dangerous Goods Analytical Information System) database, there were about 24 major chlorine spills in 1991. This is shown in Table 33.1. According to the 1993 National Pollutant Release Inventory, 4,859 tons of chlorine were released in 1993.

TABLE 33.1 Spill Profile of Chlorine in Canada

Year	Number of spills	Quantity (tons)
1974–84	36	121
1985	9	2,146
1986	18	414,000
1987	19	49,802
1988	4	9
1989	5	0
1990	8	11
1991	24	–

33.3 PHYSICAL AND CHEMICAL PROPERTIES AND SUMMARY OF GUIDELINES

33.3.1 Physical State Properties

At room temperature, chlorine is a pungent, greenish-yellow, nonflammable gas with a distinctive irritating odor. It is about 2.5 times as heavy as air. On the other hand, chlorine liquid is a clear amber color and about 1.5 times as heavy as water (CGA, 1990).

Common name: Chlorine

Molecular formula: Cl_2

Atomic weight: 35.453

Molecular weight: 70.906

CAS number: 7782-50-5

UN number: 1017

RTECS number: F02100000

Labels: Corrosive gas

Specific hazards: Corrosive

Synonyms and Trade Names (RTECS On-Line, 2000):

Bertholite
Chloor
Chlor
Chlore
Cloro
Chlorine mol
Molecular chlorine

Grades and purities (Braker and Mossman, 1980):

Grade	Minimum purity, mole % (liquid phase)
Research	99.96
Ultrahigh purity	99.9
High purity	99.5

33.3.2 Physical Data (all information from Braker and Mossman, 1980, unless noted otherwise)

Usual shipping state: Liquefied gas under its own vapor pressure, 590 kPa, (21.1°C)

Physical state at room temperature and pressure: Gas

Melting point (1 atm): −100.98°C (CGA, 1990)

Boiling point (101.325 kPa): −34.05°C
−34.6° (Lide, 1999)

Vapor pressure (21.1°C): 689.0 kPa

Densities:

Absolute density, gas at 101.325 kPa and 20°C: 2.9800 kg/m^3
Relative density, gas at 101.325 kPa and 25°C, (air = 1): 2.473

Density, liquid at saturation pressure and −40°C: 1,574 kg/m^3
Specific gravity of gas at 0°C and 1 atm (air = 1): 2.485 (CGA, 1990)
Density, 0°C: 3.214 kg/m^3 (Lide, 1999)

Fire properties (from CGA, 1990):

Flammability: Nonflammable classification by U.S. DOT
Behavior in fire: Will support combustion of some materials

Other properties:

Molecular weight: 70.90540 (Lide, 1999)
Grades (minimum purity % chlorine):

Research	99.96%
Ultrahigh purity	99.9%
High purity	99.5%

Refractive index:
Gas, 1.000713 at 25°C and 101.325 kPa
1.000768, gas (Lide, 1999)
1.367, liquid (Lide, 1999)
Viscosity:
Gas, 0.01327 mPa · s at 20°C and 101.325 kPa
Liquefied, 0.385 mPa · s at 0°C
Surface tension, 0°C: 21.90 mN/m
Latent heat of fusion at −100.98°C: 90.341 kJ/kg
Latent heat of vaporization at −34.05°C: 287.84 kJ/kg
Specific heat dry gas (−1.1 to 27°C at or below 100 psia):
Cp 0.473 kJ/kg · °C (CGA, 1990)
Cv 0.348 kJ/kg · °C
Specific heat ratio (gas 101.325 kPa, 25°C) Cp/Cv: 1.308
Critical temperature: 144°C
Critical pressure: 7,710 kPa
Critical volume: 1.745 dm^3/kg
Critical density: 0.573 kg/dm^3
Critical compressibility factor: 0.276
Triple point temperature: −34.05°C
Specific volume (21.1°C, 101.325 kPa): 337.1 dm^3/kg
Thermal conductivity, gas (26.7°C, 101.325 kPa): 0.0088 W/m · °K
Solubility in water (0°C, 101.325): 4.610 cm^3/cm^3 of water
0°C: 1.46 g/100 cc of water (liquid) (Lide, 1999)
10°C: 310 cc/100 cc of water (gas) (Lide, 1999)
30°C: 177 cc/100 cc of water (gas) (Lide, 1999)
30°C: 0.57 g/100 cc of water (liquid) (Lide, 1999)
Solubility in other solvents: Alkali (Lide, 1999)
Diffusivity: 1.44 × 10^{-5} cm^2/s (water, 25°C) (Perry et al., 1984)

Some physical constants of the chlorine atom (Cl) (Bailar et al., 1975):

Property: Values
Mass number of naturally occurring isotopes:
35 (75.77%)
37 (24.23%)
35.453
Atomic number: 17
Electronic configuration and ground state: [Ne]3s^2 3p^5
First ionization potential (kcal): 299
Electron affinity at 298°K (kcal): 84.8

Electronegativity: 3.0
Dissociation energy of Cl_2 molecule at 298°K, $D(X_2)$(kcal): 58.0

Thermodynamic properties:

$\Delta H_f°$ [X(g)] at 298°K (kcal): 28.989
Single-bond covalent radius (Å): 0.99
Ionic radius of X^- ion (NaCl structure) (Å): 1.82
$\Delta H_f°$ [X^- (g)] at 298°K (kcal): −55.7
$\Delta H°_{hydration}$ [X^- (g)] at 298°K (kcal g ion^{-1}): −88
E° ($0.5X_2/X^-$), aqueous solution (volts): +1.356

Hazards and toxicity: For emergency response to a spill of chlorine, the immediate concerns are shown in the following box.

Immediate Concerns

Hazard: Compressed, liquefied, greenish-yellow gas with an irritating odor. May react with other chemicals, causing fires and explosions. Fumes are very toxic.

Humans: Severe irritant to skin, eyes, and respiratory system. May cause burns. Toxic by all routes, death may result.

Environment: Toxic to aquatic life and harmful to the environment.

Protection: Complete self-contained breathing apparatus recommended with face mask, suit, boots, and gloves.

Human health: On inhalation, chlorine can cause irritation of the mucous membranes, watering of eyes, nasal discharge, sneezing, coughing, breathing difficulties, headaches, nausea, muscular weakness, and pulmonary edema. Liquid will burn skin and eyes on contact.

Environment: Chlorine is toxic to aquatic life even in low concentrations and harmful to animals and vegetation. When moist, chlorine will react with many metals, e.g., brass, copper, aluminum, water, and turpentine.

Behavior in air: Because chlorine gas is much heavier than air, it tends to settle in low-lying areas during spills. It may react with moisture to produce toxic and corrosive hydrogen chloride fumes, which often cause coughing in humans.

Behavior in water: Slightly soluble in water. May react to produce hypochlorous and hydrochloric acid, raising the pH. Toxic to aquatic organisms.

33.3.3 Emergency Response

Move victim to fresh air and call for a doctor.

Inhalation: If breathing is difficult or has ceased, give artificial respiration.

Skin: Remove and isolate contaminated clothing and shoes at the site. In case of contact with material, immediately flush skin or eyes with running water for at least 15 minutes. Keep victim calm and monitor the temperature.

Eyes: Should be well rinsed with large amounts of water for at least 15 minutes.

Nose and mouth: Should be continuously flushed with large amounts of water.

Ingestion: Ingestion of chlorine is very rare. If it does occur, victims should be given large amounts of water to drink. Seek medical attention immediately.

Spill Control: Self-contained breathing apparatus with full face mask should be worn for all emergency measures and total encapsulated suits for liquefied gas spills.

Leaks: Call for emergency assistance. If leak is around valve stems, try to tighten the packing nut or packing gland. If leak does not stop, close container valve. If possible, reduce pressure in the container by removing the chlorine as a gas (not liquid). Move container to an isolated area.

Fires: Chlorine itself does not burn or burns with difficulty. Try to extinguish fire with a drying agent well suited to the surrounding materials. Cool containers with large quantities of water. Use water spray to reduce vapor.

Spills: Evacuate site. Avoid breathing vapors. Avoid bodily contact. Stay upwind. Provide ventilation.

Soils: Flood site with copious amounts of water and neutralize with dilute sodium hydroxide, special foams, or soda ash. Keep away from water sources and sewers. Build barriers such as dikes, walls, and lagoons to contain spills.

Water: Contain by damming or water diversion. Use dilute soda ash or sodium hydroxide to neutralize the water.

Air: Knock down vapor with water spray or fine mist. Follow the emergency response guidelines (ERPG) of the American Industrial Hygiene Association as shown here (AIHA, 1988). Evacuate if concentrations are above guidelines.

ERPG-3 = 20 ppm: The maximum airborne concentration below which it is believed that nearly all individuals could be exposed for up to one hour without experiencing or developing life-threatening health effects

ERPG-2 = 3 ppm: The maximum airborne concentration below which it is believed that nearly all individuals could be exposed for up to one hour without experiencing or developing irreversible or other serious health effects or symptoms that could impair an individual's ability to take protective action

ERPG-1 = 1 ppm: The maximum airborne concentration below which it is believed that nearly all individuals could be exposed for up to one hour without experiencing other than mild, transient adverse health effects or without perceiving a clearly defined objectionable odor.

33.4 INDUSTRIAL ASPECTS AND PRODUCTION

Chlorine is shipped as a liquefied compressed gas. Many grades of purity are available, such as high-purity grade with a minimum purity of 99.5%; ultrahigh-purity grade with a minimum purity of 99.9%; and research grade with a minimum purity of 99.96%. Since chlorine and caustic soda are produced by the same process, sales of caustic soda are dependent on chlorine production. However, caustic soda is not as great an environmental concern as chlorine.

A large portion of the chlorine produced in Canada and the United States is for domestic use. Chlorine production reached its peak in 1982 and then started a downward trend. Some producers stopped production, and in 1989 chlorine capacity had to be expanded again. To date there are about 50 chloralkali plants in Canada and the United States with a total capacity of about 15 million tons/year of chlorine and 17 million tons/year of caustic soda. About 95% of total chlorine production is achieved by the electrolysis of brine, which also yields an almost equal amount of caustic soda. There are over 500 chloralkali producers worldwide, with a nameplate capacity of 45 million tons of chlorine/year and 50 million tons of caustic soda. Demand for chlorine is quite stable, with only little or modest growth expected.

33.4.1 Manufacture of Chlorine

Over 95% of the world's chloralkali production is achieved by the electrolysis of sodium chloride solution. The products are chlorine, caustic soda, and some hydrogen. The main types of electrolytic cells currently available are mercury, membrane, and diaphragm. The most commonly used in North America is the diaphragm cell (CIS, 1999). In Western Europe, the mercury cell is most commonly used, and in Japan the membrane cell is used because it requires much less energy.

Industrially, chlorine is manufactured by the electrolysis of chloride salts (Curlin and Bommaraju, 1991). Three basic types of processes are used to produce chlorine, each process representing a different method of isolating the chlorine produced at the anode from the caustic soda and hydrogen produced at the cathode (Curlin and Bommaraju, 1991). Electrolytic processes currently in use are:

1. The diaphragm cell process (Griesheim cell, 1885)
2. The mercury cell process (Castner-Kellner cell, 1892)
3. The membrane cell process (1970)

Chlorine is manufactured in Canada by the electrolysis of brine, which produces chlorine and caustic soda in about a 1:1.1 ratio, as well as some hydrogen. There are about 14 chloralkali plants in Canada, with a total nameplate capacity of about 1,293 kt per year. The ICI plant in Cornwall, Ontario produces caustic potash as well as chlorine, while the PGG's plant at Beauharnois, Quebec, also produces sodium hypochlorite. Saskatoon Chemicals in Saskatchewan produces chloralkali and calcium hypochlorite.

In the chloralkali electrolysis process, an aqueous solution of sodium chloride is decomposed electrolytically by direct current to produce chlorine, hydrogen, and sodium hydroxide according to the following equations:

$$2NaCl + 2H_2O \rightarrow Cl_2 + H_2 + 2NaOH \tag{33.1}$$

Evolution of chlorine takes place at the anode and is formed by the process:

$$2Cl^- \rightarrow 2Cl + 2e^- \rightarrow Cl_2 + 2e^- \tag{33.2}$$

While the basic electrolytic reaction in all three processes is the same, the methods for collecting chlorine generated at the anode are different. Each process represents a variation on keeping the chlorine produced at the anode separated from the caustic soda and hydrogen by-products. In all three processes, nearly saturated purified brine is introduced into the electrolytic cell at the beginning.

Mercury Cell Process. The mercury cell process is not energy-efficient and is falling out of favor as a result of government regulations restricting the use of mercury. Less than 16% of Canadian chloralkali production currently uses mercury cells, and the process is gradually being phased out.

In this process, sodium amalgam is produced at the cathode. The amalgam is reacted with water to produce hydrogen gas and caustic soda in a separate reactor, the decomposer. The brine is recirculated and charged with fresh solid salt. Very pure products are produced by this method. The reactions at the cathode are as follows:

$$Na^+ + e^- + Hg_x \rightarrow NaHg_x \quad (33.3)$$

$$2\ NaHg_x + 2\ H_2O \rightarrow 2\ NaOH + H_{2(g)} + 2\ Hg_x \quad (33.4)$$

The mercury cell actually consists of two cells. The brine cell or electrolyzer is a long, slightly sloped trough containing purified brine. At the bottom of this trough is a thin sheet of metallic mercury that flows under the brine and constitutes the cathode. Projecting above the cathode is the anode assembly, which consists of horizontal blocks of graphite or titanium-based, dimensionally stable anode (DSA).

During electrolysis, the brine is decomposed by the current passing through the electrodes to produce chlorine gas at the anode. The chlorine gas then leaves the anode assembly for purification. At the cathode, however, sodium ions are discharged to form a dilute sodium amalgam. The amalgam then leaves the cells and flows into the decomposer, which consists of vertical steel columns packed with graphite, where it reacts with water to form 50% caustic soda, hydrogen, and mercury. The sodium-free mercury is recirculated for reuse. These reactions can be illustrated by the following equations.

$$\text{Decomposer:} \quad Na\text{—}Hg + H_2O \rightarrow NaOH + 1/2\ H_2 + Hg \quad (33.5)$$

$$\text{Anode and cathode:} \quad 2NaCl + 2H_2O \rightarrow \underset{\text{Cathode}}{2NaOH_{(aq)}} + \underset{\text{Cathode}}{H_{2(g)}} + \underset{\text{Anode}}{Cl_2\ (g)} \quad (33.6)$$

Because of mercury losses, the cathode has to be constantly topped up. The mercury cells pose an environmental hazard because mercury is constantly released into the air and effluents are released into water systems. Excessive formation of chloroorganic compounds is another concern with this process.

Diaphragm Cell Process. The use of diaphragm cells to produce chloralkali has increased dramatically in recent years. The process now accounts for over 80% of chloralkali produced in Canada. In this process, the brine is fed into the anode compartment where chlorine is produced. The anode is separated from the cathode by a permeable asbestos-based diaphragm. Sodium ions pass through the diaphragm to the cathode, where sodium hydroxide is formed. The use of a diaphragm permits the free flow of ions and prevents the diffusion of electrolytic products. However, the sodium chloride solution can migrate through the pores of the diaphragm and the caustic soda can migrate back through the diaphragm to the anode. The end result is a mixture of caustic soda and sodium chloride. Some oxygen is also produced, which dilutes the chlorine and attacks the anode.

Purified, saturated brine is fed in at the anode compartment. High voltage and hydrostatic pressure are required on the brine feed to allow a controlled rate of entry into the cathode. As the brine flows through the diaphragm into the cathode assembly, it undergoes electrolysis to produce chlorine, caustic soda, and hydrogen according to the following equation:

$$2NaCl + 2H_2O \rightarrow \underset{\text{Anode}}{Cl_2} + \underset{\text{Cathode}}{2NaOH} + \underset{\text{Cathode}}{H_2} \quad (33.7)$$

A secondary reaction at the anode is the formation of oxygen as a result of decomposition of water and hypochlorous acid. Chlorine produced in this process is therefore often contaminated with oxygen. The chlorine is discharged at the dimensionally stable anodes and is collected with some oxygen and small amounts of hydrogen. Sometimes carbon dioxide is also produced as a result of oxidation of the graphite. Sodium ions migrate to the cathode, where hydrogen and hydroxide ions are also produced from the electrolysis of water. The hydrogen is removed from the cell and cooled to remove water. The cell liquor in the cathode contains about 10% caustic soda and 14% sodium chloride. The liquor is then concentrated in multiple-effect evaporators to produce a 50% caustic soda, with small amounts of salt.

In the anode compartment, formation of hypochlorous acid is often suppressed by the addition of hydrochloric acid, which also neutralizes any hydroxyl ions that have back-migrated. Many chlorinated and unchlorinated pollutants have been identified in the effluents of diaphragm cell plants.

To overcome the disadvantages of the diaphragm cell, new ion-selective membranes have been developed to separate the anode and cathode. These new membranes allow the production of higher strength caustic and reduce the need for investment in downstream cleanup situations. The modified diaphragm is a mixture of asbestos and a fibrous fluorocarbon polymer. It is the most common diaphragm cell in use today.

Membrane Cell Process. Less than 5% of chloralkali production in Canada is done using the membrane cell process. In this process, a cation-permeable ion-exchange membrane separates the anode from the cathode and only sodium ions and a little water can pass through the membrane. The brine is dechlorinated and recirculated, thus requiring solid salt for resaturation as in the mercury process. The chloride content in the caustic soda is similar to that in the mercury process. The chlorine gas is purified either by liquefaction or evaporation because it contains some oxygen.

This is the most energy-efficient process, and there are no known environmental problems. The membrane cell represents a great improvement over the other processes from an environmental point of view. Mercury, asbestos, graphite particles, sodium sulfate, and sodium chloride often produced in the other processes are all eliminated. Production of chlorinated hydrocarbons is also reduced.

In this process, a cation-exchange membrane such as perfluorocarboxylic acid membrane is used to partition the anode from the cathode assembly. The cation-exchange membrane allows only the hydrated positive ions, the sodium and hydronium ions, into the cathode compartment. The membrane also prevents the hydroxide ions generated in the cathode or chlorine ions in the anode from migrating across.

Saturated brine is fed into the anode compartment but cannot pass through the membrane into the water-filled cathode assembly. Hydrogen is liberated at the cathode from the electrolysis of water. More water is often added to the cathode compartment to maintain a constant concentration. Some of the catholyte stream, which constitutes about 21% sodium hydroxide, is removed and concentrated. Chlorine gas generated at the anode is collected. In terms of environmental pollution, capital investment, and operating costs, the membrane process is preferred over the other processes.

Environmental Concerns from Chlorine Use. The major environmental problem with chlorine is the chlorinated organic pollutants that are created in many processes in which it is used. Some of these byproducts are very toxic and persistent. For example, even though chlorine dioxide is now favored, several studies have reported that most pulp mills that use chlorine for bleaching discharge effluents that are acutely lethal to fish especially at point of discharge. On many occasions, effluents discharged into rivers do not dilute rapidly enough and kills of aquatic animals have been observed at locations far away from point of discharge.

In 1989, Canadian pulp mills using chlorine bleached 678 tons of pulp per mill a day. It was estimated that about 86,000 tons of organic chlorine compounds were discharged into the receiving waters. Extremely high levels of dichlorophenols, trichlorophenols, trichloroguaiacol, and tetrachloroguaiacol have been detected in fish in the vicinity of effluent discharges.

Chlorinated organic compounds are formed when chlorine used in bleaching the pulp reacts with phenols of the wood and humic substances during the pulping step. These pollutants are discharged with the effluents into the aquatic environment. It has been found that some of these effluents are lethal even at concentrations as low as 3.2% of the effluent. Chronic effects, including gross deformities, embryo and larval mortalities, behavioral modifications, and reproductive abnormalities, have been documented even at very low concentrations of effluent.

Environment Canada estimates that 100 to 150 g per year of 2,3,7,8-tetrachlorodibenzodioxin and 2,000 to 3,000 g per year of 2,3,7,8-tetrachlorodibenzofuran are discharged in bleached pulp mill effluents (Phenicie, 1993). It is estimated that only 10 to 40% of small chlorinated organic compounds in bleached pulp mill effluents have been identified. Some of the other substances detected are: chlorinated phenols, chlorinated acids, alcohols, aldehydes, ketones, sugars, aliphatic and aromatic hydrocarbons, trihalomethanes, chlorobenzenes, and aromatic thio ethers. Concentrations of individual chlorinated compound such as dichlorophenol, trichlorophenol, trichloroguaiacol, and tetrachloroguaiacol have been measured at 23%, 29%, 50%, and 95% of their respective LC_{50}s.

Toxicological effects of these effluent chloroorganics on fish and crustaceans include reduced gonad size, disorientation, high level of embryo and skeletal deformities, high parental and larvae mortality, steroid hormone imbalance, many physiological dysfunctions, disruptions in enzyme levels, liver enlargement, and fin and gill erosion.

Another process of great concern involving chlorine is the production of chlorinated pesticides, plastics, and pharmaceuticals. Such pesticides include atrazine, oxychlordane, pentachlorophenol, and 1,1,1-trichloro-2,2-bis[4-chlorophenyl ethane (DDT)], all of which have now been banned. There is overwhelming evidence that DDT is responsible for reproductive failure in many birds and wildlife. Many of these substances are also extremely persistent and toxic and could disrupt the endocrine system of both wildlife and humans, resulting in dysfunctional development and behavior. Other similar chlorinated substances are polychlorinated biphenyls (PCBs) and their byproducts, such as polychlorinated dioxins and furans. In Germany and Austria, the use of polyvinyl chloride (PVC), which releases dioxins on burning, has been stopped. Another worrisome group of chlorinated substances is the chloroflurocarbons (CFCs), which are responsible for the destruction of the earth's protective ozone layer. CFCs are being phased out gradually under the Montreal Protocol.

Another area of concern is the use of chlorine in disinfecting water for drinking purposes. Known or suspected carcinogens that have been detected in samples of finished (chlorinated) drinking water are listed in the left column of Table 33.3 (Feige et al., 1980). Those compounds listed in the right column of the table are being investigated. Most of these substances are also found in chlorinated waste, sewage, cooling water, and pulp mill effluents.

Health Canada has recommended a maximum acceptable level of 350 $\mu g/L$ for trihalomethanes in drinking water. The U.S. EPA recommendation is 100 $\mu g/L$. Risk estimates for exposure to trihalomethanes at this concentration based on 2 L of water per day for a 70-kg man is 1 in 2.5 million per year. Chloroform poses the highest cancer risk.

TABLE 33.2 Concentration Range and LC_{50}s of Bleached Pulp Mill Compounds found in Biologically Treated Effluents from Kraft Pulp Mills

Compound	96 h LC_{50} (ppb)	Effluent concentration (ppb)
2,4-Dichlorophenol	2,800	9 to 15
Dichloroguaiacols	2,300	22 to 100
Dichlorocatechol	500 to 1,000	12 to 90
2,4,6-Trichlophenol	450 to 2,600	1 to 51
Trichloroguaiacols	700 to 1,500	10 to 340
Trichlorocatechols	1,000 to 1,500	120 to 270
Tetrachlorophenol	500	–
Tetrachloroguaiacol	200 to 1,700	10 to 620
Tetrachlorocatechol	400 to 1,500	22 to 420
Pentachlorophenol	200	–
Dehydroabietic acid	500 to 2,000	–
Chlorodehydroabietic acid	600 to 900	10 to 750

TABLE 33.3 Compounds Occurring in Drinking Water as a Result of Chlorination

Chloroform	Trichloroacetic acid
Bromodichloromethane	Trichloroacetaldehyde
Chlorodibromomethane	Chlorophenols
Bromoform	Chlorobenzenes
Dichloromethane	α-Chloroketones
Bis(2-chloroethyl) ether	Chlorinated aromatic acids
Trichloroethylene	Chlorinated purines
Tetrachloroethylene	Chlorinated pyrimidines
Chlorophenols	Nonhalogenated compounds

Food chlorination is another source by which chlorinated and nonchlorinated organics enter the human body. Chlorine is used to treat cake flour to prevent crumbling on removal from the oven. It has been estimated that over 1.5% of the total flour production in the United Kingdom is chlorinated. Trihalomethanes have also been detected in human milk and serum. Levels detected are 0.30 ± 0.032 μg/mL for human milk and 0.88 ± 0.012 μg/mL for the serum, thus showing the extensive burdens of these trihalomethanes in the human body.

33.4.2 Production in Canada, United States, and Worldwide

There are about 14 chloralkali plants in Canada and 36 in the United States, for a total of 50 (CIS, 1999). The total chlorine capacity in North America is about 15 million tons per year and 17 million tons per year of caustic soda. There are over 500 chloralkali producers worldwide, at more than 500 locations, and a total nameplate capacity of 45 million tons of chlorine per year and 50 million tons per year of caustic soda.

In global terms, Dow Chemical is currently one of the largest chlorine producers, accounting for 13% of global production. More than 90% of this is used to make other products within the same company and most is used for domestic purposes. Dow has six manufacturing sites, with one in Canada, two in the United States, two in Germany, and one in Brazil.

Chlorine prices often fluctuate in a relatively mature market. In 1998, prices declined by 30 to 40% as a result of a slump in the vinyl chloride industry and market prices have so far not improved. In the long-term, a modest growth of 1.5 to 2.5% per year has been predicted. New demands for polyvinyl chloride have been projected to grow at an above average rate of 5%. On the other hand, the pulp and paper industry will continue to use less and less chlorine for environmental reasons, and this may adversely affect the chlorine market. For example, the Canadian output for elemental chlorine-free pulp grew by 4% to 9.1 million tons in 1998, which is about 76% of the market and this trend is expected to continue. In 1998, in the United States, another 2.3 million tons of elemental chlorine-free pulp entered the market. These negative impacts will no doubt continue to grow except in the developing countries.

The total nameplate capacity for chlorine in 1998 was 1,293 kilotons per year in Canada and 13,837 in the United States. Total imports to Canada in 1998 were 29 kilotons/year and 388 kilotons to the United States from all countries. The total supply of chlorine in Canada and the United States in 1998 was 1,049 and 12,073 kilotons/year respectively.

The Canadian and American chlorine producers are listed here.

Canadian producer	Plant location
Avenor	Dryden, ON
CXY Chemicals	Vancouver, BC
Canadian Occidental	Squamish, BC and Nanaimo, BC
Canso Chemicals	Abercrombie Point, NS
Dow Chemical Canada	Sarnia, ON, and Fort Saskatchewan, AB
ICI Canada	Cornwall, ON
Norsk Hydro	Becancour, QC
PCI Chemicals Canada	Becancour, QC, and Dalhousie, NB
PPG Canada	Beauharnois, QC
St. Anne Chemical	Nackawic, NB
Sterling Pulp Chemicals	Saskatoon, SK

U.S. producer	Plant location
ASHTA Chemicals	Ashtabula, OH
Bayer	Baytown, TX
Dow Chemical	Plaquemine, LA, and Freeport, TX
DuPont Chemicals	Niagara Falls, NY
Elf Atochem NA	Portland, OR
Formosa Plastics	Two locations
Fort Howard	Three locations
General Electric	Mt. Vernon, IN, and Burkville, AL
Georgia Gulf	Plaquemine, LA
Georgia Pacific	Bellingham, WA
Holtra Chem	Orrington, ME, and Acme, NC
La Roche Chemicals	Gramercy, LA
Magnesium Corp.	Rowley, UT
Occidental Chemical	Nine locations
Olin	Four locations
Olin/Geon	McIntosh, GA
Oregon Metallurgical	Albany, OR
Pioneer	Three locations
PPG Industries	Two locations
Titanium Metals	Henderson, NV
Vicksburg Chemical	Vicksburg, MS
Vulcan Chemicals	Three locations
Westlake Polymers	Calvert City, KY
Weyerhaeuser	Longview, WA

The world consumes over 30 million tons of chlorine every year, with North America using the most (34.0%), followed by Western Europe (28.7), Eastern Europe (14.8%), Japan (8.5%), Asia (6.8%), and Latin America (4.3%). While some countries, such as the United States, the Western European countries, Brazil, Saudi Arabia, and Canada, export chlorine derivatives, others like Venezuela, Taiwan, Indonesia, Korea, Japan, Thailand, and South Africa do import some. The world's chlorine consumption and capacity by region is shown in Table 33.4.

33.4.3 Transportation

Within the premises of a manufacturing plant, chlorine can usually be transported as a liquefied gas or a gas or transported by rail, road, or pipeline for several kilometers. In

TABLE 33.4 Chlorine Consumption and Capacity Worldwide

Region or country	Consumption × 10^3 t	Total capacity × 10^3 t
North America	12,050	11,710
United States	10,240	10,340
Canada	1,810	1,370
Western Europe	11,540	9,890
Eastern Europe	6,840	5,110
Japan	3,930	2,940
Asia and Pacific	3,390	2,350
Latin America	1,790	1,500
Middle East	680	590
Africa	460	420

commerce, chlorine is usually transported as a liquid either in small quantities in cylinders and drums or in bulk in railway tank cars, ton-containers, tank motor vehicles, and tank barges. While the capacity of railway tank cars ranges from 15 to over 90 tons, the road motor vehicle tank cars can carry only 15 to 20 tons. Barges are usually of the open-hopper type, with cylindrical, uninsulated pressure vessels with a total capacity of 600 to 1,200 tonnes. Some ships with a capacity of up to 1,000 tons are also specially designed to transport chlorine in Europe.

Transport Canada and the Department of Transport (DOT) in the United States have developed regulations governing the design, construction, labeling, handling, loading and unloading, testing, and other specifications for containers carrying chlorine. The Chlorine Institute in the United States provides advice and information on handling and transportation of chlorine.

33.5 CHEMISTRY

Chlorine is a very reactive element and its chemistry is extremely complex. It is a strong oxidizing agent and belongs to the halogen group in the periodic table. Its reactivity is determined by its seven electrons in the outer shell. A negatively charged chloride ion is formed when it gains one electron to complete its octet. Its various oxidation states are manifested by sharing one or more electrons with other elements. During chlorination of water, several complex reactions occur and the actual products of chlorination in raw water can never be predicted. The products that are formed, as well as their concentrations, depend not only on the chemical constituents of the water, which vary with the origin and time of the year, but also on the amount of chlorine added, temperature, pH, and reaction time. The reactions of dry chlorine are very different from those of wet chlorine.

Chlorine can undergo the following four main types of reactions.

33.5.1 Hydrolysis

When chlorine is added to water, it rapidly hydrolyzes to produce hypochlorous and hydrochloric acids. The hypochlorous acid is rather unstable and generally regarded as the major reactive species, often assuming the role of an electrophilic reagent which can react with several naturally occurring compounds to form a wide variety of organochlorine compounds. Adding chlorine to seawater, however, does not result in the formation of hypochlorous acid but rather of hypobromous acid due to high concentrations of bromide ions in seawater. Hypobromous acid is far more reactive than the sister hypochlorous acid. Hypochlorous acid and the hypochlorite anions are often referred to as free residual chlorine and are very toxic. The reactions of chlorine with water and of bromide with hypochlorous acid are illustrated in the following equations:

$$Cl_2 + H_2O \rightarrow HOCl + HCl \tag{33.8}$$

$$Br^- + HOCl \rightarrow HOBr + Cl^- \quad \text{(seawater)} \tag{33.9}$$

33.5.2 Addition Reactions

Dry chlorine and wet chlorine (hypochlorous acid) will react with organic substances containing double or triple bonds to yield vicinal dichlorides and chlorohydrins respectively. For example, hypochlorous acid will react with fatty acids in lipids such as oleic acid at pH 2 to 10, to produce 9-chloro-10-hydroxystearic acid, while dry chlorine will produce a mixture of chlorinated products.

With fatty acids in plants and animals:

$$\underset{\text{Oleic acid}}{CH_3(CH_2)_7\ CH{=}CH\ (CH_2)_7\ COOH} + HOCl \rightarrow \tag{33.10}$$

$$\underset{\text{9-Chloro-10-hydroxystearic acid}}{CH_3(CH_2)_7\ CHCl{-}CHOH\ (CH_2)_7\ COO}$$

$$\underset{\text{Oleic acid}}{CH_3(CH_2)_7\ CH{=}CH\ (CH_2)_7\ COOH} + Cl_2$$

$$\downarrow \tag{33.11}$$

$$\underset{\text{9,10-dichlorostearic acid}}{CH_3(CH_2)_7\quad CHCl{-}CHCl\ (CH_2)_7\ COOH}$$

With sulfur dioxide and carbon monoxide in the air, dry chlorine also reacts to give sulfuryl chloride and carbonyl chloride or phosgene respectively.

$$SO_2 + Cl_2 \rightarrow SO_2Cl_2 \tag{33.12}$$

$$CO + Cl_2 \rightarrow COCl_2 \tag{33.13}$$

33.5.3 Oxidation Reactions

As has been pointed out, chlorine is a strong oxidizing agent. Oxidation is the most predominant type of reaction that occurs between hypochlorous acid and all the naturally occurring organic constituents of water and wastewater (Ghanbari et al., 1982; Helz et al., 1978; Jolley and Carpenter, 1983; NRC, 1979; Rook, 1980). The most common example of this group is the oxidation of alcohols, phenols, and aldehydes to their corresponding ketones, aldehydes, or acids. The oxidation products of phenols by chlorine in pulp and paper can further condense to produce substances like dioxins.

$$RCHO + HOCl \rightarrow RCOOH + H^+ + Cl^- \tag{33.14}$$

$$RCHOHR^1 + HOCl \rightarrow RCOR^1 + H_2O + H^+ + Cl^- \tag{33.15}$$

33.5.4 Substitution Reactions

Chlorine reacts vigorously with amines and ammonia. Both wet and dry chlorine react with many aromatics, aliphatics, and amines.

$$Cl_2 + 4\ NH_3 \rightarrow NCl_3 + 3\ NH_4Cl \qquad (33.16)$$

Chlorine reacts with amine or ammonia constituents in water or in air, often replacing the hydrogens. Formation of chloramines, trihalomethanes, and other chlorinated organic compounds during natural water chlorination and wastewater treatment are the result of substitution reactions, as illustrated by the following equations:

$$RNH_2 + HOCl \rightarrow RNHCl + H_2O \qquad (33.17)$$

Further substitution can occur to give the di- and trichloramines.

$$RCOCH_3 + 3HOCl \rightarrow RCOOH + CHCl_3 + 2H_2O \qquad (33.18)$$

33.5.5 Reactions with Specific Chemicals

Chlorine will also react with the following chemicals: ammonia, acetylene, butadiene, benzene, hydrogen, sodium carbides, turpentine, and finely divided powdered metals. In some cases, these reactions may be violent and explosive mixtures may be formed.

33.6 ENVIRONMENTAL FATE

When chlorine is released into the air from a container, liquefied chlorine expands 470 times to the gaseous phase. It is quickly dispersed and at a faster rate by air motion if a strong wind is blowing. Chlorine is heavier than air and can accumulate in low-lying regions. Under wet conditions it is quickly hydrolyzed, or, in bright sunlight, it is photolyzed to form chlorine radicals. Some heavy vapor dispersion models for chlorine have been developed and found deficient. Some models are deficient because methods used to estimate the dispersion of heavy gas clouds apply to the case of dispersion over flat, uninterrupted terrains. It has always been assumed that the concentration-time profile follows the ''top-hat'' behavior.

Once released into the environment, chlorine can undergo some of the reactions discussed in Section 33.5. The most important reaction is the formation of chlorides and hypochlorites with water which will eventually end up in the soil or water bodies.

Many chlorine derivatives, including chlorofluorocarbons, can be transformed into a variety of compounds, mostly through oxidation and photolysis by visible solar radiation to produce oxidized species and chlorine radicals respectively. Chlorine radicals are of great interest because they have been specially implicated in the depletion of the ozone layer, as shown with the following equations:

$$RCl + h\nu \rightarrow R\cdot + Cl\cdot \qquad (33.19)$$

$$Cl_2 + h\nu \rightarrow 2Cl\cdot \qquad (33.20)$$

$$HCl + OH. \rightarrow Cl\cdot + H_2O \qquad (33.21)$$

$$1/2Cl_2 + O_2 + M \leftrightarrow ClOO + \qquad (33.22)$$

$$Cl\cdot + O_3 \rightarrow ClO\cdot + O_2 \text{ ozone removal step} \qquad (33.23)$$

$$Cl_2F_2C + h\nu \rightarrow Cl\cdot + \cdot ClF_2C \text{ etc.} \qquad (33.24)$$

Increased ambient concentrations of organochlorine compounds, including chlorofluoromethanes (freons), carbon tetrachloride, and chloroorganic compounds, constitute a great hazard to the ozone layer. The chlorofluorocarbon radicals are very stable and can persist for a very long time.

33.7 BEHAVIOR

Chlorine is about 2.5 times heavier than air and on release will generate vapor clouds that remain close to the ground. According to the Chlorine Institute in the United States, a failure of a 1-in. chlorine gas line with an infinite supply of chlorine generating a chlorine cloud containing in excess of 25 ppm will travel from 304.8 to 1,219.2 m (1,000 to 4,000 ft), depending on atmospheric conditions. If the valve is struck and sheared off a one-ton container, a cloud will be produced containing in excess of 25 ppm and will travel from 914 to 3,352.8 m (3,000 to 11,000 ft), depending on atmospheric conditions. A 90-ton chlorine tank car derailment with a puncture that produces a chlorine cloud containing in excess of 25 ppm will travel from 1,828.8 to 5,486.4 m (6,000 to 8,000 ft), depending on atmospheric conditions.

The behavior and impact of an accidental release of chlorine can be estimated by three groups of models:

1. Emission rate models, which focus on the chemical as it is being released from its container into the air
2. Dispersion models, which predict the downwind concentrations and distances as a function of time
3. Risk assessment models, which analyze the toxicity and exposure of a chemical to a specific receptor

33.8 HUMAN AND ENVIRONMENTAL TOXICITY

During accidental spills, exposure to chlorine is often moderate and short-term in nature due to chlorine's pungent odor. Full recovery often occurs. Also, chronic to low-level exposures, as often occur in occupational settings, do not seem to produce any long-term toxic effects. Some symptoms of irritation and coughing have been reported, however, after hours or days of exposure. No serious risks such as mutation, cancer, reproductive, and development impairment have been documented.

Contact with liquefied chlorine will definitely cause frostbite, burns, ulcerations, or necrosis. A concentration of 1,000 ppm by volume of chlorine in air is rapidly fatal after a few deep breaths. The least amounts that will result in throat irritation and coughing are 15 and 30 ppm respectively. The maximum level that can be inhaled for one hour without serious effects is 5 ppm. Concentrations of 3 to 5 ppm by volume in air are readily detectable. The effects of different concentrations of chlorine on humans are shown in Table 33.5. The maximum exposure levels of chlorine according to various guidelines are shown in Table 33.6.

33.9 SURVEY OF PAST SPILLS, LESSONS LEARNED, AND COUNTERMEASURES APPLIED

Most chlorine spills reported in Canada involve small volumes. For example, 25% involve volumes of less than 10 L and about 50% have some environmental or human impact. Spills

TABLE 33.5 Effects of Chlorine on Humans

Concentration (ppm)	Effects
1	Minimum concentration causing slight symptoms after several hours
3.5	Minimum concentration detectable by odor
4	Maximum concentration that can be breathed for one hour without damage
15	Minimum concentration causing throat irritation
30	Minimum concentration causing coughing
40 to 60	Concentration dangerous in 30 minutes
1,000	Concentration likely to be fatal after a few deep breaths

from manufacturing and processing facilities account for over 50% of reported spills. Because equipment failure and operator error are the major reasons for spills, spills could probably be prevented by proper equipment maintenance and staff training.

33.9.1 The Mississauga Train Derailment, 1979

The largest chlorine spill in Canada occurred in November 1979. The following is a summary of the accident report (Grange, 1980). Twenty-four cars from a CP Rail train derailed in Mississauga, Ontario. The cars were carrying chemicals, including toluene, propane, and chlorine. Fire spread through most if not all of the derailed cars. Those loaded with propane exploded and considerably damaged neighbouring property. The accident was caused by a fault in a lubricator pad, which caused the bearing and journal to come in direct contact, the bearing to subsequently overheat, and a tank car to collapse, causing the derailment.

The emergency response team arrived at the scene six and a half hours after the accident occurred. A smell of chlorine was detected immediately on arrival at site, but the team was

TABLE 33.6 Maximum Exposure Level Guidelines

Guideline	Concentration and time
Threshold limit value (TLV)	1 ppm. Time-weighted average for 8-hour working day
Short time public exposure limit (STPL)	1 ppm for 10 minutes 0.5 ppm for 30 minutes 0.5 ppm for 60 minutes
Public exposure limit (PEL)	3 ppm for 10 minutes 2 ppm for 30 minutes 2 ppm for 60 minutes
Emergency exposure limit (EEL)	7 ppm for 5 minutes 5 ppm for 15 minutes 4 ppm for 30 minutes 3 ppm for 60 minutes 0.5 ppm for 24 hours
Continuous exposure limit (CEL)	0.1 ppm for 90 days
Emergency response planning guidelines (ERPGs)	1 ppm (ERPG-1) 3 ppm (ERPG-2) 20 ppm (ERPG-3)

unable to locate the exact car, even with a helicopter. The Ontario Ministry of Energy and Environment analyzed chlorine levels in the area and found very low concentrations. Due to fear of the chlorine escaping, however, almost a quarter of a million people were evacuated from their homes and businesses for periods of up to five days. When the tank car carrying the chlorine was finally located, it was found to have a 2.5-ft hole in its outside covering and was covered with a tarpaulin. A command post was set up and manned by staff of the Attorney General.

When the fire died down, the response team attempted to patch the hole with a Proctor patch and to empty the car of its chlorine contents. When pressure was applied, the patch did not hold. The tank car was finally repaired three and a half days after the accident. Technical personnel from Canadian Industries Limited vacuumed out the chlorine as opposed to pressurizing the tank car. The liquid was removed to a tank truck and then to a tank car and the chlorine vapor was neutralized with caustic soda.

Lessons to Be Learned. The Commission of Inquiry concluded that the response team sent by the chemical company were not qualified or adequate. They did not know how much chlorine remained in the tank car after the initial escape and did not know how to calculate it. They failed to take measures to vent the car at the beginning to prevent chlorine from escaping. They did not know how to empty the chlorine from the tank cars. Furthermore, the response team incorrectly instructed the firefighters at the scene to water the chlorine car, and in fact team members watered the tank car themselves. And finally, the response team gave poor advice to the command post team, particularly with regard to the rate of vapor and liquid being emptied, the method of emptying the chlorine from the tank car, and the dangers of release of gas after ice was broken.

33.9.2 Chlorine Leak at Pioneer Plant, 1991

Officials at Pioneer Chlor Alkali Co., Inc. near Henderson, Nevada, investigated an accidental chlorine leak early one morning. The chlorine gas leak was detected by the air monitoring system about 1:10 a.m. and the 10 employees working at the time were evacuated by 2:30 a.m. As a precaution, many residents were evacuated until the chlorine gas dissipated later in the morning.

The accident apparently began when a narrow tube in the heat exchanger, which kept the chlorine's temperature low as it was converted from gas to liquid, broke and water coolant leaked into the chlorine. Much of the water–chlorine mixture travelled into one of the eight 150-ton storage tanks. Chlorine and water, which mix to make hydrochloric acid, corroded a hole in a discharge pipe coming out of the storage tank. Employees closed the valve on the tank to stop the flow of chlorine through the corroded hole. But the water and chlorine in the storage tank corroded at the tank valve and the valve did not hold. The chlorine and water mixture continued to leak until a piece of metal was blinded into the pipe at about 7 a.m. No serious injuries were reported in this incident.

Lessons to Be Learned

1. This incident demonstrated the need for fail-safe equipment systems and new technologies using stainless steel equipment.
2. Preventive maintenance must be carried out on critical process equipment and vessels.

3. Early warning signs are necessary when such failures occur.
4. Emergency response needs to be improved, i.e., it should not have taken more than 30 minutes to evacuate 10 employees.

33.10 REFERENCES

American Industrial Hygiene Association (AIHA). 1988. *Emergency Response Planning Guidelines—Chlorine,* AIHA, Akron, OH.

Bailar, J. C., H. J. Emeleus, R. Nyholm, and A. F. Trotman-Dickenson. eds. 1975. "Halides," in *Comprehensive Inorganic Chemistry,* Pergamon Press, New York, NY, pp. 1124–1280.

Braker, W. and A. L. Mossman. 1980. "Chlorine," in *Matheson Gas Data Book,* Matheson, Gas Products, East Rutherford, NJ, pp. 155–159.

Camford Information Services, Inc. (CIS). 1999. "Chlorine," in *CPI Product Profiles,* CIS, Scarborough, ON.

Compressed Gas Association, Inc. (CGA). 1990. *Handbook of Compressed Gases,* 3rd ed., Van Nostrand Reinhold, New York, NY, pp. 231–232.

Curlin, L. C. and T. V. Bommaraju. 1991. "Alkali and Chlorine Products," in *Encyclopedia of Chemical Technology,* ed. M. Howe-Grant, 4th ed., vol. 1, John Wiley & Sons, New York, NY, pp. 938–1025.

Environment Canada. 1999. National Analysis of Trends in Emergency Situations (NATES): *Chemical Accidents Reports Database,* Hull, QC.

Feige, M. A., E. M. Glick, J. W. Munch, D. J. Munch, R. L. Noschang, and H. J. Brass. 1980. "Potential Contaminants Introduced into Water Supplies by the Use of Coagulant Aids," in *Water Chlorination Environmental Impact and Health Effects,* ed. R. L. Jolley, W. A. Brungs, R. B. Cumming, and V. A. Jacobs, Ann Arbor Science Publishers, Ann Arbor, MI, pp. 789–799.

Fingas, M., N. Laroche, G. Sergy, B. Mansfield, G. Clouthier, and P. Mazerolle. 1991. "A New Chemical Spill Priority List," in *Proceedings of the Eighth Technical Seminar on Chemical Spills,* Environment Canada, Ottawa, ON, pp. 223–237.

Ghanbari, H. A., W. B. Wheeler, and J. R. Kirk. 1982. "Reactions of Aqueous Chlorine and Chlorine Dioxide with Lipids: Chlorine Incorporation," *Journal of Food Science,* vol. 47, pp. 482–485.

Grange, S. G. M. 1980. *Report of the Mississauga Railway Accident Inquiry,* Supply and Services, Hull, QC.

Helz, G. R., R. Sugam, and R. Y. Hsu. 1978. "Chlorine Degradation and Halocarbon Production in Estuarine Water," in *Water Chlorination: Environmental Impact and Health Effects,* ed. R. L. Jolley, H. Gorchev, and D. H. Hamiltion, Jr., vol. 2, Ann Arbor Science Publishers, Ann Arbor, MI, pp. 209–222.

Jolley, R. L., and J. H. Carpenter. 1983. "A Review of the Chemistry and Environmental Fate of Reactive Oxidant Species in Chlorinated Water," in *Water Chlorination: Environmental Impact and Health Effects,* ed. R. L. Jolley, W. A. Brungs, J. A. Cotruvo, R. B. Cumming, J. S. Mattice, and V. A. Jacobs, vol. 4, bk 1, Ann Arbor Science Publishers, Ann Arbor, MI, pp. 3–47.

Lide, D. R., ed. 1999. *CRC Handbook of Chemistry and Physics,* 80th ed. CRC Press, New York, NY.

Nuclear Regulatory Commission (NRC). 1979. *The Chemistry of Disinfectants in Water, Reactions and Products,* PB-292776, U.S. Department of Commerce, Washington, DC.

Perry, R. H., D. W. Green, and J. O. Maloney, eds. 1984. *Perry's Chemical Engineer's Handbook,* 6th ed. McGraw-Hill, New York, NY.

Phenicie, D. K. 1993. *Virtual Elimination in the Pulp and Paper Industry,* Report to the Virtual Elimination Task Force.

Rook, J. J. 1980. "Possible Pathways for the Formation of Chlorinated Degradation Products during Chlorination of Humic Acids and Resorcinol," in *Water Chlorination: Environmental Impact and Health Effects,* ed. R. L. Jolley, W. A. Brungs, and R. B. Cumming, vol. 3, Ann Arbor Science Publishers, Ann Arbor, MI, pp. 85–98.

RTECS On-Line. 2000. *Registry of Toxic Effects of Chemical Substances,* Department of Health and Human Services, Centers for Disease Control, National Institute for Occupational Safety and Health, Washington, DC.

CHAPTER 34
PERSPECTIVES ON SPECIFIC SUBSTANCES: METHYL tert-BUTYL ETHER (MTBE)

Richard Lawuyi and Merv Fingas
Emergencies Science Division, Environment Canada, Environmental Technology Centre, River Road, Ottawa, Ontario

34.1 OVERVIEW OF PRODUCT AND INDUSTRIAL USES

The primary function of methyl tert-butyl ether (MTBE) as an additive in gasoline is to enhance the octane level of unleaded gasoline. By virtue of the oxygen atom it contains, it increases the oxygen-to-fuel ratio in gasoline. It has also been shown that adding MTBE results in lower emissions of carbon monoxide and hydrocarbon as well as polycyclic aromatic hydrocarbons (PAHs). MTBE oxidation, initiated by hydroxyl radicals, can yield a number of products such as tert-butyl formate, formaldehyde, methyl acetate, and acetone, depending on the pathway.

The environmental and health risks associated with the use of MTBE are not clear. To date, overall toxicity data show that MTBE has low toxicity in animals even though it is an irritant to the skin, eyes, lungs, and mucous membranes. Since it has only recently been introduced into commerce, the long-term effects are not known. Some toxicologists have reported that the short-term, high-level exposure that often occurs during accidental spills can cause euphoria, headache, dizziness, drowsiness, blurred vision, tremors, respiratory arrest, and death. In view of its high production level, widespread use and exposure, and inadequate and conflicting data on toxic effects, more research is needed.

Even though MTBE is thought to be quite efficient, there are environmental problems associated with its use. In the United States, MTBE has been banned as a gasoline additive in many states due to its widespread contamination of groundwater.

34.1.1 Modern Industrial Uses

MTBE came into prominence as a substitute for tetraethyl lead in gasoline. It was first used as an oxygenate in gasoline in the late 1980s. Other than that, MTBE is relatively unknown and used only in organic synthesis as a rather unreactive solvent with a high boiling point. It is also used as a degreasing agent and in solvent extraction.

34.2 INTRODUCTION

Before the early 1980s, the oil industry had very little interest in oxygenates as constituents of gasoline. However, mandatory phasing out of lead in gasoline in the mid-1970s led to a flurry of research to improve the octane rating and antiknock properties of gasoline. MTBE has since been fulfilling this role. Studies have shown that commercial gasoline is the most commonly released hazardous chemical (Industrial Economics, Inc., 1985). Most gasoline contains up to 15% MTBE. The studies also indicated that 74.8% of the gasoline releases were in fixed facilities, while 25.2% occurred in transit. The U.S. Environmental Protection Agency (U.S. EPA) has updated these data, and although the newer statistics vary slightly from these figures, the general picture remains basically the same.

The main push for the use of MTBE in gasoline in Canada has been coming from the United States with the passage of the amended Clean Air Act in 1990 providing a permanent use for oxygenates in gasoline. As a gasoline additive, it has been demonstrated that MTBE reduces automobile hydrocarbon emissions. However, some toxic byproducts are also produced, which is often not mentioned (Fadope, 1995; Environment Canada, 1995).

The mandate to keep the environment free of toxic chemicals falls under the Canadian Environmental Protection Act (CEPA), which is administered by both Environment Canada and Health Canada. Even though MTBE has been assessed as nontoxic under the CEPA Priority Substances List, its effects on humans are still controversial (Environment Canada and Health Canada, 1992). Even though many investigators have found that MTBE does not pose a substantial health risk to humans, the effects of MTBE and other oxygenates are still relatively unknown (Duffy et al., 1992; Silvak and Murphy, 1991; ATSDR, 1996). No real studies have been done on the effects of this substance on human health.

In 1992, there were reports of people in Fairbanks, Alaska, experiencing headaches, nausea, breathing difficulties, and rashes after driving or refueling cars with gasoline containing up to 15% MTBE (ATSDR, 1996). The United States Centers for Disease Control and Prevention found elevated levels of MTBE in the blood of these motorists and gasoline service attendants at the end of their shifts (CDC, 1993a). There could be a link between these elevated MTBE levels and health complaints. In addition, gasoline workers surveyed in Alaska and New Jersey indicated that exposure to MTBE from stationary sources caused illness (ATSDR, 1996).

MTBE is currently used as a gasoline blendstock based mainly on its exceptional octane-enhancing characteristics. Advantages of MTBE include:

- While the blending Reid vapor pressure (RVP) of an oxygenate depends on the oxygenate itself, the amount present, and the RVP or nature of the base blend, the RVP of MTBE (115 to 123 psi) is much lower than that of methanol (127 to 136 psi) and ethanol (120 to135 psi). Lower-blending RVP is required to reduce the evaporative emissions from motor vehicles. Low-blending RVP also allows more volatile butane and olefines to be added.
- MTBE can readily be produced within refineries.
- It blends easily without phase separation in gasoline.
- Reformulated mix can be transferred through existing pipelines.
- Since MTBE does not contain any benzene rings, it does not contribute to the undesirable emission of aromatics.
- In general, hydrocarbon emissions from oxygenated gasolines have reduced photochemical reactivity and thereby reduced the potential for ozone formation.
- Because of its low boiling point (55°C), MTBE increases the octane number of gasoline.
- MTBE-blended fuels are compatible with all materials used in automobile manufacture, e.g., pumps, gaskets, lacquers, elastomers, and components of carburetors. Therefore, existing vehicles do not require modification.

- It lowers the gasoline 50% ignition point, thus enhancing cold engine warmup performance.
- Its relatively low taste and odor thresholds provide ample warning of its presence.
- It has very low toxicity and does not bioaccumulate.

Some disadvantages of MTBE are that is does not biodegrade readily, and when spilled, it is extremely mobile in the soil and often reaches the water table very quickly.

34.2.1 Spill Profile

There is only one MTBE maufacturing plant in Canada, but more are being planned. In North America, large quantities of MTBE enter the environment, mostly as gasoline spills, often polluting groundwater and surface water. Gasoline contains oxygenates such as MTBE and is the most commonly spilled chemical. For example, of about 12,444 release incidents of petroleum products, over 70% were due to gasoline alone (USEPA, 1986a). Furthermore, over 74.8% of the releases were fixed-facility incidents while 25.2% were in transit. Other releases of MTBE occur during blending at refineries and refueling at gas stations, in storage, at distribution centers, manufacturing plants, leaks from gasoline underground tanks and pipes, and transport accidents including barges and tankers. The relative contribution of these sources has not been quantitated in Canada. There is potential for extremely high concentrations of MTBE buildup during spills, especially in confined areas. People who work around garages, gasoline stations, manufacturing plants, and refineries are constantly exposed to low-level background concentrations of MTBE.

Only one large MTBE spill has been reported so far in Canada. It occurred on a ship in Courtney Bay, New Brunswick, on February 19, 1993. The 1,365-L spill was contained on the deck and cleanup was completed within hours (Environment Canada, 2000). As previously mentioned, gasoline spills are common, however, and sometimes occur in high volumes.

The annual spill frequency of gasoline with up to 15% MTBE in Canada from 1985–1991 is shown in Fig. 34.1. The amounts seem to be decreasing slowly each year. It should also be noted that the data for Ontario region cease to be available as of May 1989.

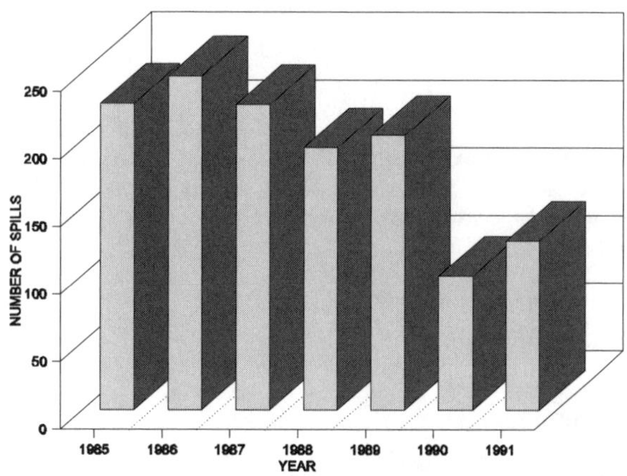

FIGURE 34.1 Annual spill frequency of gasoline with MTBE (7–15%) 1985–1991.

TABLE 34.1 Priority List Ranking of Hazardous Chemicals

Chemical	Ranking	Spill number	Spill quantity (t)	Supply quantity (kt)
Ammonia	1	107	470	3,700
Chlorine	2	36	120	1,700
Tetraethyl lead	3	4	72	26
Styrene	4	24	5,000	630
PCBs	5	334	89	–
Sulfuric acid	6	155	13,000	3,700
Sodium cyanide	7	3	83	12
Hydrochloric acid	8	123	3,300	170
Potassium chloride	9	31	12,000	–
Pentachlorphenol	10	19	110	1.5
Phenol	11	10	14	68
Zinc sulfate	12	3	68	1,500
Phosphorus	13	16	46	68
Toluene	14	13	110	430
MTBE	15	1	1	500

34.2.2 Priority List Ranking

MTBE falls in the same group of gasoline additives as methanol, ethanol, benzene, toluene, ethylbenzene, and xylenes. However, it is less toxic and more persistent in the soil than these compounds. The spill number, spill volume, and supply volume of MTBE are also much less than those of these compounds.

As shown in Table 34.1, MTBE is tentatively ranked as 15th on Environment Canada's priority list of hazardous chemicals (Fingas et al., 1991). The primary objective of the list was to determine the minimum number of hazardous substances that were most frequently spilled. The list was developed by a simple ranking of reported spill frequency; supply volumes; historical spill volumes; and toxicity data, stability, accumulation, and persistence.

34.3 PHYSICAL AND CHEMICAL PROPERTIES AND GUIDELINES

34.3.1 Physical Data

MTBE is a colorless, low-viscosity liquid with characteristic terpene-like odor. It is flammable and nonautoxidizable. It is miscible in all proportions with gasoline, scarcely soluble in water, and presents no phase-separation problems in existing distribution systems. It is stable in storage and has no extreme effect on automotive fuel elastomers and plastics at the commercial levels used.

Molecular formula: $(CH_3)_3COCH_3$

Molecular weight: 88.15

CAS number: 1634-04-4

UN number: 2398

STCC number: 4908224

Labels: Flammable liquid

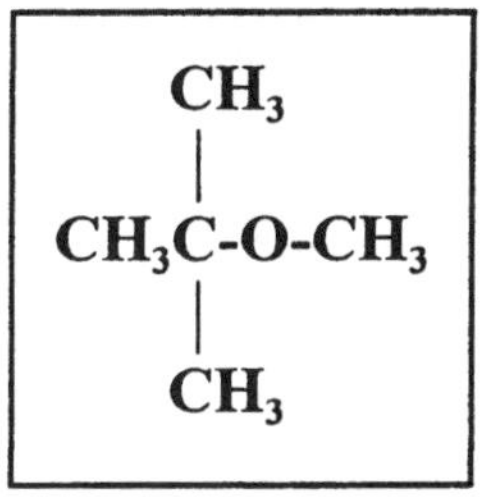

FIGURE 34.2 Structure of MTBE.

RTECS number: KN5250000

Synonyms and trade names (RTECS On-Line, 2000)

tert-butyl methyl ether
Methyl tert-butyl ether
Methyl 1,1-dimethylethyl ether
Propane, 2-methoxy-2-methyl-
2-Methyl-2-methoxypropane
2-Methoxy-2-ethylpropane
(2-Methyl-2-propyl)methyl ether
Ether, tert-butyl methyl
tert-butoxymethane
t-butyl methyl ether
2,2-MMOP
2,2-methylmethoxypropane
MTBE
MBE

Grades and purities: There are two main grades of MTBE: (1) the usual commercial purity, which is 98 to 99 wt %, and (2) a special quality with 99.95% MTBE, marketed under the trade name Driveron S. Other grades are HPLC and anhydrous. Impurities are usually water, hydrocarbons (C_5, C_6, and diisobutenes), and alcohols (methanol and tert-butanol) (Elvers et al., 1990).

Grade	Purity %
Special	99.95
HPLC	99.8
Anhydrous	99.7
Commercial	98

Density: 0.7405 at 20°C

Vapor pressure: 245 mm Hg (25°C)

Appearance: Clear, colorless liquid with terpene-like odor

Usual shipping state: Liquid

Physical state at room temperature and pressure: Liquid

Boiling point: 55.3°C

Melting point: −108.6°C

Relative vapor density (air = 1): 3.1

Fire properties:

Flammability:	Vapors are highly flammable
Behavior in fire:	Carbon dioxide and carbon monoxide are produced
Decomposition temperature:	>200°C
Decomposition products:	Oxides of carbon
Flash point:	−28°C
Flammable limits:	1.6 to 8.4 vol %
Autoignition temperature:	460°C
Extinguishing:	Dry chemical, foam, or carbon dioxide. Do not use water jet. Keep containers cool with water spray.
Reactivity:	Forms flammable vapor mixture with air.

Other properties:

Solubility in water.	23.2 to 54.4 g/L (at 25°C)
Solubility in organic solvents.	Very soluble in alcohol and ether, 4.8 g/100 g in water
Log octanol/water partition coefficient.	1.30
Refractive index (20°C):	1.3689

Binary azeotropes with MTBE:

Azeotrope	Boiling point °C	MTBE content, weight %
MTBE-water	52.6	96
MTBE-methanol	51.6	86
MTBE-methanol (1.0 MPa)	130	68
MTBE-methanol (2.5 Mpa)	175	54

34.3.2 Summary of Chemical Properties and Behavior

Even though very stable, MTBE decomposes in contact with strong acids and alkalis above 200°C. Unlike other ethers, it does not form peroxides during storage. In spite of its high water solubility (40 to 48 g/L) compared with the relative solubility of BTX (benzene 1.8, toluene 0.5, and xylene 0.2), MTBE is expected to partition largely to air. Photooxidation (Japer et al., 1990, 1991), volatilization (Thomas, 1982), and slow biodegradation (Fujiwara et al., 1984) are the possible mechanisms by which it is removed or transported in the environment. Any MTBE present in surface water will have a half-life of about nine hours before volatilizing. Furthermore, most of the MTBE released to the environment will be emitted directly to the air during transfer operations. The low octanol/water partition coefficient and relatively high water solubility of MTBE indicate little tendency for significant partitioning to soils, sediments, or biota. Degradation by hydroxy radicals has been shown to be the main pathway for the disappearance of MTBE in the air. Atmospheric half-life of MTBE is calculated to be about 3.5 days. The products of the oxidation are likely to include t-butyl formate (major product), acetone, and methyl radical. The half-life of photooxidation reactions with hydroxy radicals has been estimated to be 20.7 to 265 hours (Atkinson, 1985; Wallington et al., 1988).

MTBE is hydrophobic in a ternary system of fuel, water, and MTBE itself, and under aquifer conditions it will concentrate approximately 80%. As an additive in gasoline, it is more soluble than benzene and has been shown to travel faster than gasoline in groundwater. By all accounts, its presence in groundwater indicates the presence of gasoline. One would expect MTBE to be the first to be detected due to its higher mobility. It is still not clear whether it shows any cosolubility effects with BTX (benzene, toluene, and xylene) compounds. Speculations are that the greater solubility of MTBE in water, coupled with about 100% solubility of gasoline constituents in MTBE, may lead to an increase in the total sum of all dissolved organic gasoline components in groundwater, thus increasing the total dissolved hydrocarbons in groundwater.

MTBE will reside mainly in the air and in water because of its high vapor pressure, high water solubility, and low octanol/water partition coefficient. Since the sorption of organic compounds is inversely proportional to their solubility in water, MTBE should have a low sorption to soil particles. Another consequence of MTBE solubility is that a plume of MTBE in groundwater would be more extensive than a plume of other gasoline components. The outer fringes of the total plume are expected to have MTBE as the only contaminant that is detectable. The MTBE plume will show up as a halo around the dissolved gasoline plume, which in turn will appear as a halo around the free product plume. Because of its assumed low toxicity and low taste and odor detection thresholds, it is a good indicator of gasoline spills.

There are scanty and conflicting reports on the biodegradability of MTBE. The formation of a halo of MTBE is good evidence of its rapid mobility and its persistence in the environment. All indications are that it is very persistent and does not biodegrade readily under aquifer conditions. It has also been suggested that biodegradation is possible in water. The half-life has been estimated to be between 28 and 180 days for aerobic biodegradation in surface waters and 112 to 720 days for anaerobic biodegradation in deep water or groundwater (Fujiwara et al., 1984). However, it has also been found that the presence of MTBE had little effect on the biodegradability of blended gasoline.

34.3.3 Main Hazards

MTBE forms a flammable vapor mixture with air. High concentrations of vapor are irritating to the eyes and respiratory tract and may cause dizziness and headaches. Inhalation of high concentrations causes sleepiness and finally narcosis.

Immediate Concerns

Hazard: Flammable, colorless liquid. Forms flammable vapor mixture with air. May be ignited by heat, sparks, or flames. Vapor may travel to source of ignition and flash back. Container may explode in heat or fire. Vapor explosion hazard indoors, outdoors, or in sewers. Runoff to sewer may create fire or explosion hazard. Decomposes in contact with acids and alkalis. Decomposes in temperatures above 200°C.

Humans: High concentrations of vapor irritate eyes and respiratory tract and may cause dizziness and headaches. Inhalation of very high concentrations causes sleepiness and finally narcosis. Splashes in the eye cause irritation. May cause anesthesia.

Environment: Dangers of explosion when material enters sewers. Nontoxic except in large spill situations.

Protection: Wear self-contained breathing apparatus. Use PVC or rubber gloves. Wear impervious clothing and goggles.

Human health:

TLV: none established
TWA: 1 to 3 ppm (8-hour) reported
Drinking water advisory: 20 to 40 μg/L

Exposure effects:

Contact and Inhalation: Contact may irritate or burn skin and eyes. High concentrations of vapor are irritating to the eyes and respiratory tract. May cause dizziness and headaches. Inhalation of high concentrations causes sleepiness and finally narcosis. Contact of the liquid with the eye causes irritation. The foul odor of MTBE serves as a warning for water contamination by gasoline. MTBE is considered to be of relatively low toxicity.

Environment: Not toxic to biota except at accidental spill situations. Rapidly degraded in the atmosphere. May be fairly persistent in groundwater. According to one study, MTBE may be more mobile in the soil than gasoline.

Behavior in air: Forms flammable vapor mixture with air.

Behavior in water: Will exist briefly as a discrete floating layer but will dissipate quickly into the air, leaving no hydrocarbon residue. Soluble in water (40 to 43 g/L at 25°C).

Emergency response:

Symptoms:

Local: Not a primary irritant. Slight erythema and irritation have been noted in animals when applied to abraded skin or under occlusive dressing.

Respiratory: Respiratory depression may be noted. Labored and irregular breathing were associated with anesthetic effects in rats. Aspiration pneumonitis is possible following ingestion of gasoline–MTBE mixtures.

Gastrointestinal: Nausea and vomiting have been observed in animals and humans.

Treatment:

Inhalation:

Move patient to fresh air.

Monitor for respiratory distress.

If cough or difficulty in breathing develops, evaluate for respiratory tract irritation, bronchitis, or pneumonitis.

Administer 100% humidified supplemental oxygen with assisted ventilation as required.

Skin:

Wash exposed areas extremely thoroughly with soap and water.

Remove contaminated clothing.

Shower thoroughly.

A physician may need to examine the area if irritation or pain persist.

Eyes:

Flush eyes with water for 15 minutes.

Hold eyelids open while washing and continue to irrigate.

If irritation, pain, swelling, lacrimation, or photophobia persist, determine nature and degree of corneal damage. Notify ophthalmologist.

Ingestion:

Wash out mouth with water.

Do not induce vomiting.

If any material enters the lungs, e.g., during vomiting, seek medical help immediately.

Give large quantities of water.

Observe vital signs.

Spill control:

Spill or leaks: Remove all ignition sources. Keep all unprotected personnel clear of area. Contain spill with earth and sand. Do not allow spilled material to enter confined spaces, drains, sewers, waterways, and soil. There is a danger of explosion if material enters sewers. Inform appropriate authorities if a major spill occurs. Ventilate area. Use water spray to protect workers trying to stop a leak. Absorb small quantities on paper towels.

Fire: Use dry chemical or carbon dioxide for small fires; alcohol foam or polymer foam for large fires. Use water spray to keep fire-exposed containers cool, disperse vapors, and protect workers trying to stop the leak. Water spray may be used to flush spills away from exposures. Wear goggles and self-contained breathing apparatus.

Countermeasures—Emergency control procedures:

Soil: Contain spill with earth and sand or other noncombustible absorbent material and place into containers for later disposal. Construct barriers to contain spill or divert to impermeable holding area. Remove material manually or by mechanical means.

Water: Contain spill by damming, water diversions, sandbag barriers, or natural barriers. MTBE will evaporate.

Reaction with acids: MTBE reacts with sulfuric acid when heated to produce an olefin. The mechanism involves formation of a carbonium ion followed by elimination.

$$CH_3\ OC(CH_3)_3 \ + H_2SO_4 \rightarrow (CH_3\ )_2C{=}CH_2 \ + \ CH_3OH \qquad (34.1)$$

Physiological Effects: High concentrations of vapor irritate eyes and respiratory tract and may cause dizziness and headaches. Inhalation of very high concentrations causes sleepiness and finally narcosis. Other effects are nausea, vomiting, and mild inflammatory changes in the gall bladder after repeated exposure. Other temporary effects include loss of appetite, weakness, and loss of coordination. Prolonged or repeated skin contact may cause the skin to become dry or cracked from defattening with some reddening. Splashes in the eye cause irritation, but no adverse long-term effects are known to exist.

Estimates of potential exposure at process units and terminals range from about 3 ppm to below 0.6 ppm, while those at service stations are much below these values. To date, health effects clearly indicate a low toxicity except at high exposure levels, which often exceed levels at refueling stations or even in the occupational environment.

34.4 INDUSTRIAL ASPECTS AND PRODUCTION IN THE UNITED STATES, CANADA, AND WORLDWIDE

34.4.1 Manufacture of MTBE

The only MTBE producer in Canada is based in Edmonton. It has a total capacity of 1,590 tons. There are 43 MTBE producers in the United States, 67 locations with a total capacity of 16,710 tons in 1998. The MTBE produced in Canada is mainly for export to the Western United States, and only a small amount is used for domestic purposes in British Columbia.

MTBE is commercially produced by the reaction of isobutylene with methanol in the presence of an acidic ion-exchange resin as catalyst, usually in the liquid phase and at temperatures below 100°C. A typical catalyst is sulfonated styrene/divinylbenzene resin catalyst. Other solid acid catalysts such as bentonites are also effective and other novel catalysts have recently been discovered. Isobutylene is obtained from field butane by initial isomerization of n-butane to isobutane, followed by dehydrogenation to isobutylene. Commercial preparations of MTBE are 95.03 to 98.93% pure. Impurities are methanol ($<$0.43%), *t*-butyl alcohol ($<$0.80%), and diisobutylene ($<$0.25%).

34.4.2 Manufacturing Process

Raw butane is first fractionated to separate the isobutane fraction from the normal butane in the deisobutanizer (DIB). Isomerization of the normal butane to isobutane takes place in the isomerization unit. Any unreacted n-butane is removed in the deisobutanizer and reisomerized. The isobutane produced is then fed to the catalytic dehydrogenation unit to yield a mixture of isobutylene and unreacted isobutane, which flows into the synthesis unit, where the reaction with methanol occurs to yield MTBE. Any unreacted products and residues containing isobutane are retreated in the oxygenate removal unit, enter a saturation unit to remove any olefines, and are then fed to the isobutanizer.

Isobutene, the main feedstock, is obtained in the form of raffinate from steam crackers, which make up an estimated 40% of MTBE feedstock throughout the world. Isobutene in the form of butene-butane fractions from fluid catalytic crackers represents 28% of MTBE feedstocks; isobutene from dehydrogenation of isobutane represents 12% of MTBE feedstocks; and isobutene by dehydration of tert-butanol represents 36% of MTBE feedstocks. The Butamer process is often used for the primary butane isomerization, while the Catofin and Olefex processes are commonly used for the isobutane dehydrogenation.

Methanol, the other reagent in the synthesis, is produced from desulfurized natural gas feedstock at a purity of over 99.9% and is used directly for the synthesis without further treatment. An overall conversion of around 97% of the isobutylene is often obtained.

$$(CH_3)_2C{=}CH_2 + CH_3OH \underset{\text{catalyst}}{\rightarrow} (CH_3)_3CCOCH_3 \qquad (34.2)$$

The synthesis process for manufacturing MTBE is shown in Fig. 34.3.

34.4.3 Manufacturers in Canada and the United States (CIS, 1995)

The major buyers in Canada are Chevron, Husky Oil, Imperial Oil, PetroCanada, Shell Canada, Suncor, and Turbo Resources.

The major buyers in the United States are BP America, Chevron USA, Cibro Petroleum Products, Crysen Corporation, Giant Industries, Holly Corporation, Howell Hydrocarbons,

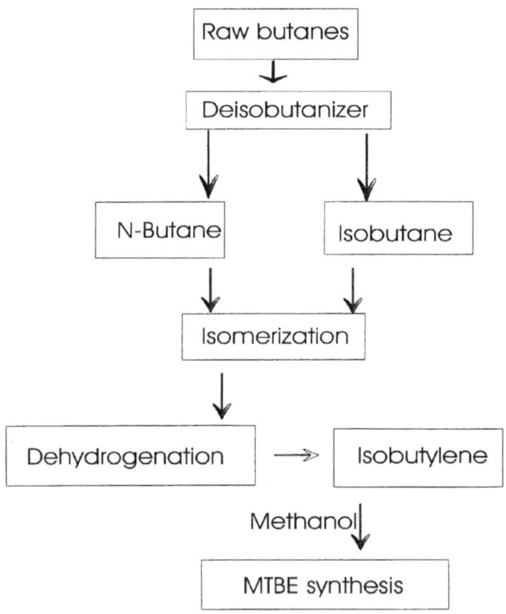

FIGURE 34.3 MTBE synthesis process.

Indiana Farm Bureau Coop, Koch Refining, Mapco Petroleum, Murphy Oil USA, Tosco Corporation, and Total Petroleum.

The following is a list of manufacturers of MTBE in Canada and the United States.

Canadian manufacturers	Location
Alberta Envirofuels	Edmonton, AB

U.S. manufacturers	Location
Amerada Hess	Port Reading, NJ
American Petrofina	Big Spring, TX
Amoco Petroleum	Whiting, IN, and Yorktown, VA
Arco Chemical	Channelview and Corpus Christi, TX
Arco Petroleum	Carson/Watson, CA
Ashland Chemical	Catlettsburg, KY, and St. Paul Park, MN
Belvieu Environmental Fuels	Mont Belvieu, TX
BP America	Belle Chasse, LA, Lima, OH, and Marcus Hook, PA
Chevron	El Segundo and Richmond, CA, and Pascagoula, MS
Citgo Petroleum	Corpus Christi, TX and Lake Charles, LA
Clean Air Fuels	Channelview, TX
Coastal Chemicals	Cheyenne, WY
Conoco	Lake Charles, LA, and Ponca City, OK
Crown Technology	Houston, TX
Diamond Shamrock	Dumas, TX
Enron (EPG)	Morgan's Point, TX
Exxon	Baton Rouge, LA, Baytown, TX, and Benicia, CA
Global Octanes	Deer Park, TX
Huntsman	Port Neches, TX
Koch Refining	Corpus Christi, TX, and Rosemont, MN
Lyondell Petrochemical	Channelview, TX
Marathon	Detroit, MI, and Robinson, IL
Mark West Hydrocarbon	South Shore, KY
Mobil	Beaumont, TX, and Chalmette, LA
Oxychem	Chocolate Bayou, TX
Oxyfuels	Beaumont, TX
Phibro (Hill)	Houston and Texas City, TX, and Krotz Springs, LA
Phillips	Borger and Sweeny, TX
Shell Oil	Deer Park, TX, Martinez, CA, Norco, LA, and Wood River, IL
Southwestern (Kerr-McGee)	Corpus Christi, TX
Star Enterprise	Delaware City, DE, and Convent, LA
Sun Oil	Marcus Hook, PA, and Toledo, OH
Texaco	El Dorado, KS, Los Angeles, CA, and Puget Sound, WA
Texaco Chemical	Port Neches, TX
Texas Partners	Mont Belvieu, TX
Texas Petrochemical	Houston, TX
Tosco	Martinez, CA
Total Petroleum	Alma, MI, and Ardmore, OK
TransAmerican	Norco, LA
Unocal	Wilmington, CA
Valero	Corpus Christi, TX

34.4.4 Production and Transportation

In 1998, 975 kilotons of MTBE were produced in Canada, of which 670 kilotons were exported. In the United States, the total production of MTBE was 12,310 kilotons, with only 235 kilotons exported. In Europe, the use of MTBE is not mandatory and no data are available.

In transport, MTBE is classified as a flammable liquid. It is transported in Canada by rail from Alberta to British Columbia and then to the western United States. In the United States, it is transported by rail, road motor vehicle tanks, and barges.

34.5 CHEMISTRY

34.5.1 Combustion

In general, oxygenates require less oxygen to burn than their hydrocarbon counterparts. For example, as shown in Eqs. (34.3) to (34.9), 176 g of MTBE will require 480 g of oxygen for complete combustion or 100 g of MTBE will require 273 g of oxygen, whereas 100 g of heptane, which is very similar to gasoline in terms of stoichiometric air–fuel ratio, would require 352 g of oxygen. The excess oxygen is used to reduce the formation of carbon monoxide and any unburnt hydrocarbons as long as the mixture is within the explosive limit. When complete combustion occurs, carbon dioxide and water are produced.

In a fire, MTBE will produce mostly carbon dioxide and water in the presence of an abundant supply of oxygen. Carbon monoxide can also be produced with insufficient oxygen present.

$$\underset{176\text{ g}}{2CH_3OC(CH_3)_3} + \underset{480\text{ g}}{15O_2} \rightarrow 10CO_2 + 12H_2O \tag{34.3}$$

$$\underset{200\text{ g}}{2CH_3(CH_2)_5CH_3} + \underset{704\text{ g}}{22O_2} \rightarrow 14CO_2 + 16H_2O \tag{34.4}$$

34.5.2 Reaction with Acids

MTBE reacts with concentrated sulfuric acid to produce an olefine. The mechanism involves the formation of a carbonium ion followed by proton elimination. Methanol is also formed. This is the exact reverse of the synthesis of MTBE.

MTBE is a very stable substance and does not undergo common reactions like other hazardous substances. Formation of peroxides, unlike that of other ethers, has not been substantiated.

The ether linkage is broken to give alcohol and a halide:

$$(CH_3)_3COCH_3 + HI \rightarrow (CH_3)_3COH + CH_3I \tag{34.5}$$

The tert-butyl alcohol can further dehydrate to produce isobutene:

$$CH_3OC(CH_3)_3 + H_2SO_4 \rightarrow (CH_3)_2C{=}CH_2 + CH_3OH \tag{34.6}$$

Atmospheric reactions of MTBE are often initiated by hydroxyl radicals that abstract hydrogen from the methyl groups to form water and organic radicals. There are various kinds of models and mechanisms postulated. The following are two possible initial reactions. For hydrogen radical abstraction at the methoxy end:

$$CH_3\text{–}O\text{—}C(CH_3)_3 + OH^{\bullet} \rightarrow {}^{\bullet}CH_2\text{—}O\text{—}C(CH_3)_3 + H_2O \tag{34.7}$$

and at the butoxy end:

$$\rightarrow CH_3\text{–}O\text{—}C(CH_3)_2\text{—}CH_2^{\bullet} + H_2O \tag{34.8}$$

One theory is that peroxy radicals are formed in the presence of oxygen:

$${}^{\bullet}CH_2\text{–}O\text{—}C(CH_3)_3 + O_2 \rightarrow {}^{\bullet}OOCH_2\text{—}O\text{—}C(CH_3)_3 \tag{34.9}$$

$$CH_3\text{–}O\text{—}C(CH_3)_2\text{—}CH_2^{\bullet} + O_2 \rightarrow CH_3\text{—}O\text{—}C(CH_3)_2\text{—}CH_2OO \tag{34.10}$$

After a series of steps, the final products formed are HCHO; $CH_3\text{–}CO\text{—}CH_3$; and $(CH_3)_3C\text{—}O\text{–}CHO$. $CH_3\text{—}COOCH_3$ has also been suggested.

34.6 ENVIRONMENTAL FATE AND EFFECTS

The major environmental concern with MTBE is the potential contamination of drinking water supplies if it leaks from underground gasoline pipelines and storage tanks. The presence of MTBE in water gives the water a strong and unpleasant odor and taste. As little as 15 ppb can be smelled and tasted in drinking water, and people in many American states are calling for a ban on its use.

The chief pathways to account for the disappearance of MTBE in the environment are atmospheric reactions and biodegradation. By virtue of its sterically hindered structure, MTBE has lower reactivity than other hydrocarbons in gasoline. The atmospheric half-life of MTBE has been estimated at 4 to11 days (Carter et al., 1991). MTBE will react with hydroxyl (OH) radicals in the atmosphere. The rate of atmospheric reactivity of hydroxyl radicals with MTBE has been determined to be close to 2.8×10^{-12} cm^3 $molecule^{-1}sec^{-1}$. While the ultimate product of degradation of MTBE is carbon dioxide and water, in laboratory experiments, observed products are tertiarybutyl formate, formaldehyde, methyl acetate, and acetone. Organic nitrate has also been noted when nitrogen oxides are present.

There are very few reports on biodegradation of MTBE. MTBE is a very persistent chemical (Yeh and Novak, 1994). All indications are that it will biodegrade slowly in suitably acclimated conditions. Suflita and Mormile (1993) have found that oxygenates containing a tertiary or quaternary carbon atom were much more recalcitrant than their unbranched or not so hindered counterparts. The strong influence of chemical structure on biodegradation has been known for decades. For example, *n*-butyl methyl ether has been shown to degrade at a faster rate than MTBE under similar suitably acclimatized systems.

The legislated or suggested concentration limits in the environment and those relating to human health are given in Tables 34.1 and 34.2. The results of aquatic and environmental toxicity studies conducted on MTBE are summarized in Table 34.3.

34.7 BEHAVIOR

MTBE is a volatile (vapor pressure 26.8 kPa at 20°C; boiling point 55.3°C) flammable, and colorless liquid at room temperature with a terpene-like odor. Reported Henry's law constant varies from 59 to 305 Pa m^3 mol^{-1} but is usually less than 100. It is miscible with gasoline and is soluble in water, alcohol, esters, and ethers. It forms azeotropic mixtures (constant temperature-constant composition boiling mixture) with some solvents such as methanol. It

TABLE 34.2 Concentration Limits in the Environment

Criteria	Air	Water	Soil	Biota	Surface water
Legislated limits	SUGGESTED in EXPERIMENTS Rats: NOEL 300; 797; 800 ppm Rabbits: NOEL 1,000 ppm Mice: NOEL 1,000 ppm	RECOMMENDED USEPA draft drinking water lifetime health advisory: 20 to 200 μg/L Drinking: 0.2 mg/L and 50 μg/L Groundwater: 0.2 mg/L Odor threshold: 680 μg/L NOEL: 100 mg/L EPA maximum contaminant level (MCL): 1,260 to 1,610 μg/L	No data available	No data available	No data available
Typical ambient levels	Refineries: >30 μg/m³ 1.5 ng/m³ Vehicle exhaust emissions: 0.9 to 81 mg/km Urban: 0.025 to 8.4 ppb Gas station median: 3 to 14 ppb. Max:140 ppb	Groundwater: 0.2 to 23,000 μg/L Median: <0.2 μg/L(urban) 690 ppb; 1.96 μg to 236 mg/L Deeper groundwater: 1.3 μg/L Stormwater: 1.5 μg/L	100 mg/L and less depending on depth and type of soil.	No data available	Lakes: 6 ng/L; 0.12 ng/L (approx.) Sediments and estuaries: 5 ng/L
LC_{50}/LD_{50} levels to biota	LC_{50}: rats, 23,576 ppm; 120 mg/L; 85 mg/L, 4 h (ihl). LC_{50}: mice, 650 mg/m³ (10 min). LD_{50}: rats 3.0 to 3.8 g/kg (oral) LD_{50}: rabbits 10 g/kg (dermal)	No data available	No data available	LD_{50}—rats: 4,000 mg/kg (oral) 3.0 to 3.8 g/kg LD_{50}—rabbits: 6,800 mg/kg and >10,200 mg/kg dermal)	LC_{50}—*Daphnia:* 513 mg/L, 96 h. LC_{50}—fathead minnow: 672; 691; 706 mg/L, 96 h LC_{50}—rainbow trout: 714 mg/L, 96 h. LC_{50}—bleak 1,700 to 1,800 mg/L, 24 h (marine)
Half-life and Bioconcentration factor (BCF)	15 days 5.6 days 3.5 days BCF 1.5; 1.1; 1.08	No data available	No data available	No data available	Surface water, river: 4.1 h (volatilization); 9 h; Pond: 2.0 d Surface water Aerobic biodegradation: 28 to 180 d Anaerobic: 112 to 720 d in deep water and groundwater
Action cleanup levels	No data available	Groundwater: 50 ppb	No data available	No data available	No data available

TABLE 34.3 Concentration Limits Relevant to Human Health

Criteria	Air	Water
Legislated limits	TLV-TWA: 40 ppm (144 mg/m^3) MCLG (maximum contaminant level goal): 0.03 mg/L/d	RECOMMENDED Drinking water: 0.2 mg/L Interim drinking water standards in U.S.: 5 to 100 μg/L NOEL: 100 mg/L
Normal levels	<1 ppm; 0.6 to 3 ppm Refineries: <30 μg/m^3	106 μg/L Groundwater: 1.96 μg/L to 236 mg//L (spills)
Short-term or emergency exposure limits	0.6 to 3 ppm	No data available

also has a high water solubility (23.2 to 54.4g/L at 25°C) and a low octanol/water partition coefficient (log K_{ow} = 0.94 to1.24) (Funasaki et al., 1985; Fujiwara et al., 1984).

Its strong terpene-like odor and low taste and odor threshold levels (taste detection threshold in water—700 ppb; odor detection threshold in water—680 ppb) often provide ample warning during spills, leaks, and especially when groundwater is contaminated. A number of studies have indicated that MTBE does not increase the solubility of other organic components in gasoline, e.g., the aromatics and BTEX compounds, as once thought (cosolvency) (Piel, 1989; API, 1991). Other studies say the opposite (Garrett, 1987; Garrett et al., 1986; Mihelcic, 1990; Poulsen et al., 1992). Furthermore, MTBE does not affect the movement of other constituents of gasoline in groundwater but usually advances in the solvent front or at the same speed as the water. It has also been shown that it does not alter the biodegradation rate of other gasoline components (Fujiwara et al., 1984).

The fate of all organic chemicals entering the environment depends largely on a number of physicochemical and biological processes. The most important ones are:

- Solubility in aqueous and organic solvents
- Sorption
- Volatilization
- Biodegradation and persistence
- Photodegradation

MTBE is highly volatile and does not sorb to soil particles as other organic chemicals do. Because of its volatility, one of the most successful and cheapest remediation technologies to clean MTBE-contaminated water is air stripping or air stripping coupled with aqueous phase carbon adsorption. Removal rates for MTBE have been shown to be between 56 and 99.9%. While the vapor pressure curve of MTBE resembles the classical vapor pressure plot of organic compounds, the more relevant plots of evaporative exposure versus temperature (weathering) during spills are very different.

All reports indicate that MTBE will be very slow to form peroxide, if at all. This is probably due to absence of acidic hydrogens that are readily abstracted by radicals (OGJ, 1995). Such highly encumbered compounds are usually unreactive and hard to biodegrade by microbes (Suflita and Mormile, 1993).

34.7.1 MTBE Spills on Land

The primary pathways for the dissipation of an MTBE spill on land have been described as follows: after the initial settling on the soil, evaporation starts to occur. The rate of evaporation is directly dependent on the wind speed and temperature. Similar fugitive emissions are the chief source of MTBE releases from manufacturing facilities, storage of MTBE, and vehicle exhaust.

While the major bulk of the spill will evaporate into the air because of its poor adsorption characteristics and partial solubility in water, MTBE is extremely mobile in the soil and will often reach groundwater aquifer very quickly. Any remaining traces of the spill on the soil surface will either be dissolved and washed away by precipitation and runoffs to rivers and lakes or pushed farther downwards into the ground.

TABLE 34.4 Environmental Toxicity of MTBE

Freshwater species	Concentration (mg/L)	Effects	Reference
Goldfish	894	LC_{50}, 96 h	QSAR, 1989[a]
Bluegill	770	LC_{50}, 96 h	QSAR, 1989[a]
Leuciscus idus melanotus	1,000	LC_{50}, 48 h	
(golden orfe)	2,000	LC_{100}, 48 h	
Mosquito fish	746	LC_{50}, 96 h	QSAR, 1989[a]
Pimephales promelas (fathead minnow)	706	LC_{50}, 96 h	
Pimephales promelas (fathead minnow)	691	LC_{50}, 96 h	QSAR, 1989[a]
Pimephales promelas (fathead minnow)	672	LC_{50}, 96 h	Geiger et al., 1988
Rainbow trout	714	LC_{50}, 96 h	QSAR, 1989[a]
Marine species	**Concentration (mg/L)**	**Effects**	**Reference**
Alburnus alburnus (bleak)	1,700 to 1,800	LC_{50}, 24 h	
Alburnus alburnus (bleak)	>1,000	LC_{50}, 96 h	Bengtsson and Tarkpea, 1983[b]
Amphibian species	**Concentration (mg/L)**	**Effects**	**Reference**
Rana temporaria (frog tadpoles)	2,500	LC_{50}, 48 h	Paulov, 1987
Copepod species	**Concentration (mg/L)**	**Effects**	**Reference**
Nitocra spinipes (harpacticoid copepod)	>10,000	LC_{50}, 96 h	
Nitocra spinipes (harpacticoid copepod)	>1,000	LC_{50}, 96 h	Bengtsson and Tarkpea, 1983[b]
Daphnia (daphnia)	513	LC_{50}, 96 h	QSAR, 1989[a]

[a] QSAR values are estimations from models based on chemical structure.
[b] Insufficient method and result documentation (USEPA database, AQUIRE).

34.7.2 MTBE Spills on Water

MTBE is less dense than water (0.7407 g/cc at 20°C) and will therefore tend to float on the surface initially, where evaporation will take place as with gasoline and hydrocarbons. In addition, because of its partial solubility in water, some of it will dissolve in water, depending on turbulence and air and water temperatures. In small-scale laboratory experiments, MTBE was found to evaporate preferentially when placed over enough water to dissolve it before it actually completely dissolves, but dissolution will occur when large quantities of water are present. But just as MTBE can dissolve in water, so water can also dissolve in MTBE. The weight of MTBE in water decreases with temperature while that of water in MTBE increases with temperature (Elvers et al., 1990). Small quantities of water will also dissolve in large amounts of MTBE. Reuter et al. (1998) have determined the half-life of MTBE spilled on water to be 193 days (1.2 kg/day decline) during the boating season and 14 days half-life (8.1 kg/day loss) at the end of the boating season.

In underground water, MTBE usually moves with the solvent front or at the same speed as the water, but generally in front of other gasoline components thus providing advance warning of a polluted aquifer. In a monitoring study in which the occurrence of MTBE in groundwater was investigated along with other organochlorine chemicals, of the 210 urban wells and springs sampled, 28% contained chloroform, 27% contained MTBE, 18% contained tetrachloroethane, 10% contained trichloroethane, 7% contained cis-1,2-dichloroethane, 5% contained 1,1-dichloroethane, and 5% contained benzene. The maximum concentrations of MTBE detected in shallow groundwater in urban areas were over 100 μg/L, while in shallow groundwater in agricultural areas concentrations were 1.3 μg/L (Squillace et al., 1996).

34.8 HUMAN AND ENVIRONMENTAL TOXICITY

34.8.1 Exposure to Humans

The toxicity characteristics associated with MTBE seem to be mild or low relative to other gasoline components. The most common toxicological findings are the general anesthetic effect often induced by this chemical. The low organoleptic property threshold provides a large margin of safety for people in exposure situations. No human deaths have been reported as a result of this chemical.

While no real studies have been done in humans on the effects of MTBE on human health, acute symptoms from exposure to low levels of MTBE in gasoline are widely reported (Anderson, 1993). People exposed to MTBE while operating or fueling their motor vehicles or those occupationally exposed, such as mechanics, service station workers, manufacturers, and technicians at refineries, have complained of irritation of the ear, nose, throat, and sinus, vomiting and nausea, severe headaches, anxiety, skin rash, dizziness, disorientation, and breathing difficulties (White et al., 1995; Moolenaar et al., 1994; Chandler and Middaugh, 1992; CDC 1993a,b; Beller and Middaugh, 1992). This has been the experience of some motorists in Alaska and the southern United States. At present the additive is not used in Alaska.

MTBE is a probable human carcinogen. One theory is that the exhausts of automobiles using gasoline treated with MTBE emit large quantities of formaldehyde, which is known to be carcinogenic and leukomogenic. The Centers for Disease Control and Prevention have also documented elevated levels of MTBE (about 10 times higher than normal) in the blood of some Alaskan motorists and gasoline service station workers at the end of their shifts. The U.S. EPA has classified MTBE as a possible human carcinogen as a result of several studies on experimental animals.

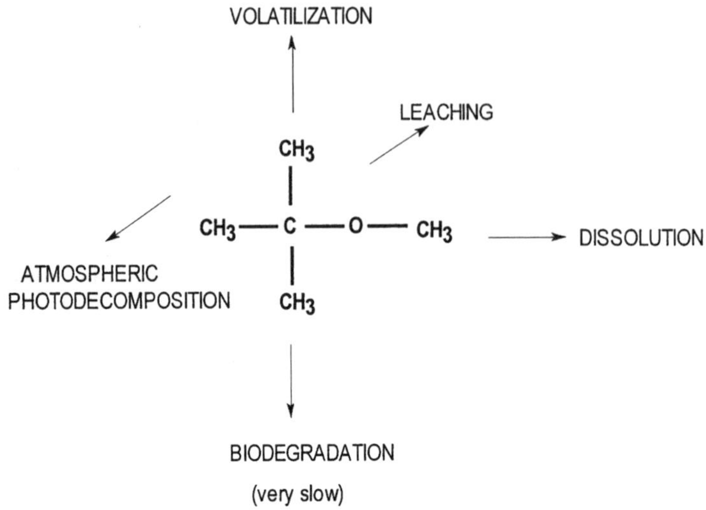

FIGURE 34.4 Behavior and fate of MTBE.

34.8.2 Exposure in Experimental Animals

MTBE has been found to cause kidney cancers, leukemias, liver cancers, lymphomas, and testicular cancers in animals. Many laboratory experiments have been conducted on short-term acute exposure, but very few chronic studies have been done on animals. Animals tested include rats, rabbits, and mice. Most of these experiments identified the central nervous system as the most sensitive target organ. The following toxic effects occurred in these animals at various doses and times.

Rats. The following toxic effects occurred in rats:

1. Increased renal tubule cell epithelial proliferation and decreased body weight gains (Chun and Kintigh, 1993)
2. Irregular respiration, eye irritation, lack of coordination, prostration, and hyperemia of lung tissue; LC_{50} 85 to 120 mg/L (U.S. EPA, 1986b)
3. Depression of central nervous system, tremors, labored breathing, lack of coordination, hypoactivity, muscular weakness, hyperpnea, lacrimation, prostration, inflammation of the stomach and small intestine, and hyperemia of the lung tissue, LD_{50} 3800 mg/kg (Mastri et al., 1969);
4. Increased liver weight; other target organs are nasal mucosa and trachea (Biodynamics Inc., 1981)
5. Decreased body weight gain, no effects on lungs (Dodd and Kintigh, 1989)
6. Labored respiration, CNS depression, loss of muscle control, and lacrimation (ARCO, 1980)
7. Neurological effects, NOAEL, minimal risk level (MRL) 0.4 mg/kg/day during exposure
8. Diarrhea (Bioresearch Laboratories, 1990a,b)
9. Chronic progressive nephropathy and increased mortality in male rats (Chun et al., 1992)
10. Increased mortality and dysplastic proliferation of lymphoreticular tissues observed in female rats at the lowest dose tested (Belpoggi et al., 1995)

11. The main target organs identified: liver, gastrointestinal tract, blood, serum cholesterol, immune system, endocrine system, CNS, etc. (ITT Research Institute, 1992; Robinson et al., 1990; Biles et al., 1987; Chun and Kintigh, 1993; Greenough et al., 1980)

Mice. Toxic effects in mice are similar to those in rats and include the following.

1. Mortality was increased in male mice due to obstructive uropathy. Kidney weight increased, but body weight gains decreased and there was an increased incidence of hepatocellular adenoma (Burleigh-Flayer et al., 1992).
2. There were no effects on the liver or body weight (Conaway et al., 1985).
3. Breathing was labored and maternal weight gain was reduced (Tyl and Neeper-Bradley, 1989).
4. The MRL for acute inhalation neurotoxicity was 2 ppm (Gill, 1989).

In rats and mice exposed to MTBE, unchanged MTBE and tert-butyl alcohol were shown to be the main respiratory excretion products. MTBE has also been found to be metabolized to formaldehyde, methanol, formic acid, carbon dioxide, 2-methyl-1,2-propanediol, and α-hydroxy-isobutyric acid. MTBE has not been shown to bioaccumulate in the tissues (ATSDR, 1996).

Rabbits. Toxicological investigations of the effects of MTBE on rabbits are not very common, but in one study it was found that liver weight increased and maternal weight gain was reduced (Tyl, 1989).

The final products of metabolism in animals are carbon dioxide, methanol, and α-hydroxy-isobutyric acid. The toxicities of these products should also be considered.

34.8.3 Environmental Guidelines (Reuter et al., 1998)

Canada has no guidelines on safe levels of MTBE in drinking water.

U.S. EPA drinking water advisory: 20 to 40 μg/L
California drinking water action level: 35 μg/L
Primary drinking water standard: 5 μg/L
Based on odor and taste concerns: 14 μg/L

34.9 SURVEY OF PAST SPILLS, LESSONS LEARNED, AND COUNTERMEASURES APPLIED

Many reported spills of MTBE in Canada involve small volumes. In the United States, leaks from underground pipes and tanks often become large before they are discovered. The major reasons for spills are failure of equipment such as underground tanks and pipes and operator error.

34.9.1 Tank Barge Leaks 1,200 Barrels of MTBE (*Golob's Oil Pollution Bulletin,* 1995)

On April 6, 1995, about 1,200 barrels of methyl tert butyl ether (MTBE) leaked from a hole in a cargo tank of the 280-ft tank barge *STCO 510* into the Gulf Intracoastal Waterway (ICW) near Cypremort, Louisiana. The crew of the tugboat *Aurora,* which was pushing the barge, discovered the leak and notified the U.S. Coast Guard. The owner of the barge, Sabine Transportation Co. (Grove, Texas), responded to the spill and determined that the MTBE was leaking from an 8-in. by 4-in. hole in the cargo tank but was unable to determine how or when the tank was damaged. American Pollution Control (Lafayette, Louisiana), a contractor hired by Sabine Transportation, deployed 900 ft of containment boom and sorbent material around the barge. According to John Grez of the U.S. Coast Guard's Marine Safety Office in Morgan City, Louisiana, the boom was deployed primarily as a precautionary measure in the event of a much larger MTBE release from the barge and most of the spilled cargo dispersed rapidly. The MTBE remaining in the damaged tank was transferred to other tanks aboard the barge and Sabine Transportation offloaded the entire cargo to another barge on April 7. The spill occurred in a man-made section of the Gulf Waterway and the MTBE did not pose a threat to natural resources along the waterway. Vessel traffic in the area was suspended until April 13, 1995.

34.9.2 Case Studies of MTBE/Gasoline Plumes in Maine (Garrett et al., 1986)

In 1984, it was found that a farmer's tank in rural Maine was leaking product directly into fractured schistose bedrock. Due to the configuration of the bedrock surface, the product plume was located on the other side of the road, about 150 ft away from the tank. Product recovered from the pump hose and plume was chromatographically identical and found to be a leaded gasoline containing about 3% MTBE by volume.

The 80-ft-deep, drilled well of the farmer's neighbor rather suddenly began producing water contaminated with 126,000 ppb total gasoline including MTBE. Because the spill occurred near the crest of a ridge, the dissolved plume spread in two directions downslope. Within months, other homeowners only a few hundred yards from the spill site were complaining of water smelling not of gasoline, but of "a funny chemical smell." Two years later and with the plume either stable or enlarging only slowly, most of the plume was an MTBE-only plume occurring as a halo around the dissolved gasoline plume. Water from some wells was analyzed as containing as much as 690 ppb MTBE with no detectable gasoline. About 30 other sites in Maine were identified at which MTBE was a component of the spilled gasoline. About 90% of these analyses were routed through the Public Health Lab.

Attempts were also made to treat the groundwater contaminated with MTBE. Filtration through activated carbon beds was found to be too expensive. Air stripping provided an effective means of removing this chemical when high air-to-water ratios were applied. The investigators made the following observations:

1. Concentrations of gasoline and MTBE in groundwater at the center of the plume can be extremely high. The record high concentration so far is over 600,000 ppb in one household well with the intake pipe beneath floating product in a sand and gravel aquifer. This contrasts with the usual maximum concentration for similar situations without MTBE of about 10 to 30,000 ppb.
2. MTBE can occur as the only contaminant above detection limit over large areas of the plume. In one plume, believed to have originated from a small driveway spill, MTBE was the only detected contaminant of the spill.
3. The MTBE plume seems to occur as a halo around the gasoline-plus-MTBE plume. Where the plume is expanding, MTBE is detected before gasoline in contaminated wells.

Lessons Learned

1. MTBE is very mobile in the ground. It is more soluble and spreads faster than other gasoline components. Monitoring of underground water is required during both land and water spills.
2. When spilled, MTBE evaporates very quickly.

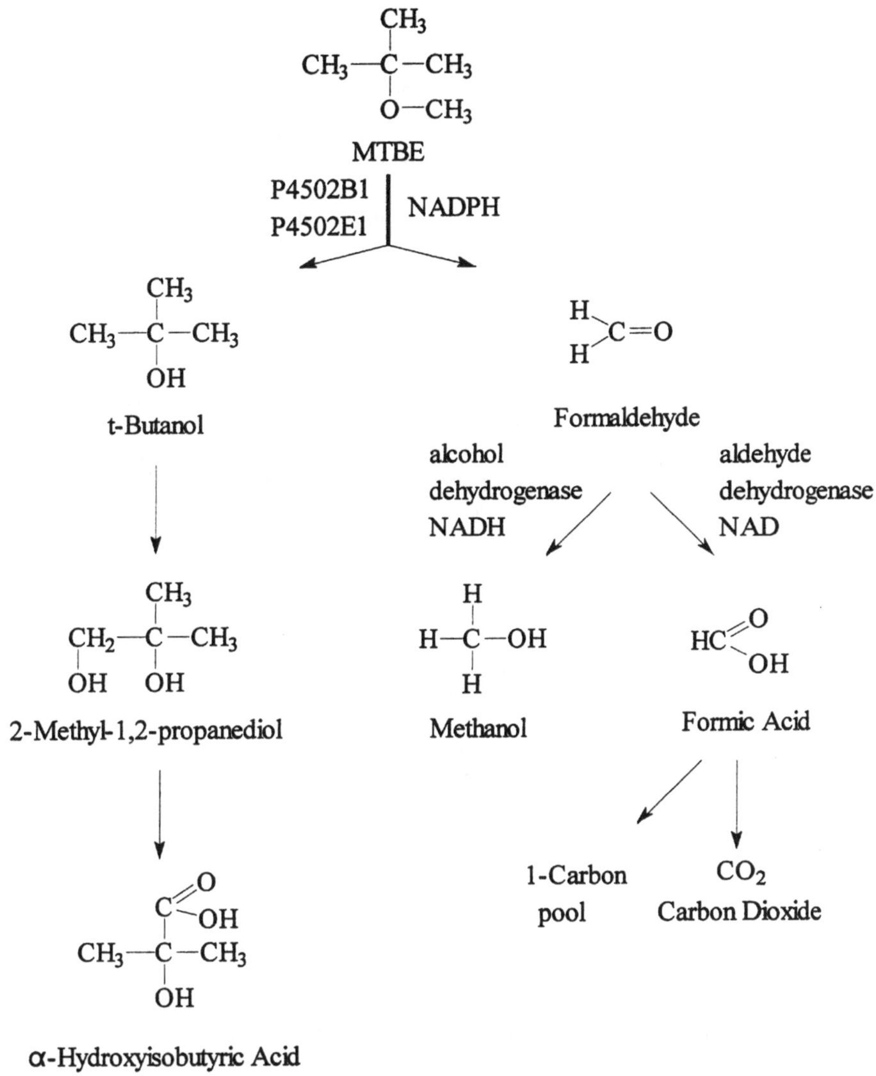

FIGURE 34.5 MTBE metabolic pathway in rats. (*Source: Biodynamics Inc., 1984; Bioresearch Labs, 1990a, b, c; Brady et al., 1990; Snyder et al., 1979*).

3. The presence of MTBE in spilled gasoline increases dissolved concentrations of gasoline in groundwater in the immediate vicinity of the spill to about an order of magnitude above typical values for spills in which there is no MTBE.
4. It is more difficult to remove MTBE from contaminated water than it is to remove the other components of gasoline during remediation.
5. MTBE is very easy to detect and recognize because of its odor.
6. Appropriate personal protection clothing and breathing apparatus must be worn during spill cleanup.
7. High concentration buildup can occur in restricted areas and should be avoided.

34.10 CONCLUSIONS

It is unfortunate that MTBE has been banned as an oxygenate, not because it is ineffective but because it is being spilled along with other components in gasoline into drinking water supplies. No strong evidence could be located that links exposure to MTBE with an acute or chronic illness. More research should be done to determine the toxic properties, both acute and chronic, of this substance. In addition, precautions should be investigated to prevent spills and leaks from underground tanks and pipes.

34.11 REFERENCES

Anderson, E. V. 1993. "Health Studies Indicate MTBE Is Safe Gasoline Additive," *Chemical and Engineering News,* vol. 71, no. 38, pp. 9–18.

American Petroleum Institute (API). 1991. *Chemical Fate and Impact of Oxygenate in Ground Water: Solubility of BTEX from Gasoline Oxygenate Mixtures,* API Publication 4531.

ARCO Chemical Company. 1980. "MTBE: Acute Toxicological Studies," ARCO Ressearch and Development, Glenolden, PA.

Atkinson, R. 1985. "Kinetics and Mechanism of the Gas-Phase Reactions of the Hydroxyl Radical with Organic Compounds under Atmospheric Conditions," *Chemical Reviews,* vol. 86, no. 1, pp. 69–201.

Agency for Toxic Substances and Disease Registry (ATSDR). 1996. *Toxicological Profile for Methyl t-Butyl Ether,* U.S. Department of Health and Human Services.

Beller, M., and J. Middaugh. 1992. *Potential Illness Due to Exposure to Oxygenated Fuels,* State of Alaska Department of Health and Social Services, Anchorage, AK.

Belpoggi, F., M. Soffritti, and C. Maltoni. 1995. "MTBE, A Gasoline Additive, Causes Testicular and Lymphohaematopoietic Cancers in Rats," *Toxicology and Industrial Health,* vol. 11, pp. 119–149.

Bengtsson, B. E., and M. Tarkpea. 1983. "The Acute Aquatic Toxicity of Some Substances Carried by Ships," *Maritime Pollution Bulletin,* vol. 14, no. 6, pp. 213–214.

Biles, R. W., R. E. Schroeder, and C. E. Holdsworth. 1987. "MTBE Inhalation in Rats: A Single Generation Reproduction Study," *Toxicology and Industrial Health,* vol. 3, no. 4, pp. 519–534.

Biodynamics Inc. 1981. *Nine Day Inhalation Toxicity Study of MTBE in the Rat,* Project No. 80-1452, East Millstone, NJ, report submitted to American Petroleum Institute, Washington, DC.

Biodynamics Inc. 1984. *The Metabolic Fate of MTBE following an Acute Intraperitoneal Injection,* Project No. 80089, East Millstone, NJ, report submitted to American Petroleum Institute, Washington, DC.

Bioresearch Laboratories. 1990a. *Mass Balance of Radioactivity and Metabolism in Male and Female Fischer-344 Rats after Intravenous, Oral and Dermal Administration of ^{14}C-methyl tertiary-butyl ether,* Report No. 38843, Senneville, QC.

Bioresearch Laboratories. 1990b. *Pharmacokinetics of MTBE and tert-butyl alcohol (TBA) in Male and Female Fischer-344 Rats after Single and Repeated Inhalation Nose-Only Exposures to MTBE,* Report No. 38844, Senneville, QC.

Bioresearch Laboratories. 1990c. *Disposition of Radioactivity and Metabolism of MTBE in Male and Female Fischer-344 Rats after Nose-Only Inhalation to ^{14}C-MTBE,* Report No. 38845, Senneville, QC.

Brady, J. F., F. Xiao, and S. M. Ning. 1990. "Metabolism of Methyl tertiary-Butyl Ether by Rat Hepatic Microsomes," *Archives of Toxicology,* vol. 64, no. 2, pp. 157–160.

Burleigh-Flayer, H. D., J. S. Chun, and W. J. Kintigh. 1992. *MTBE: Vapor Inhalation Oncogenicity Study in CD-10 Mice,* Project No. 91N0013A, Bushy Run Research Center, Export, PA.

Camford Information Services Inc. (CIS). 1995. *Methyl tert-Butyl Ether (MTBE),* in CPI Product Profiles, Scarborough, ON.

Carter, W. P. L., E. C. Tuazon, and S. A. Aschmann. 1991. *Investigation of the Atmospheric Chemistry of MTBE,* Auto/Oil Air Quality Improvement Research Program, Atlanta, GA.

Centers for Disease Control (CDC). 1993a. *MTBE in Human Blood after Exposure to Oxygenated Fuel in Fairbanks, Alaska,* CDC, National Center for Environmental Health, Atlanta, GA.

Centers for Disease Control (CDC). 1993b. *An Investigation of Exposure to MTBE among Motorists and Exposed Workers in Stamford, Connecticut,* CDC, National Center for Environmental Health, National Institute for Occupational Safety and Health, Atlanta, GA.

Chandler, B., and J. Middaugh. 1992. *Potential Illness Due to Exposure to Oxygenated Fuels, Anchorage, Alaska,* State of Alaska Department of Health and Social Services, Anchorage, AK.

Chun, J. S. and W. J. Kintigh. 1993. *MTBE: Twenty-eight Day Vapor Inhalation Study in Rats and Mice,* Laboratory Project ID 93N1241, Bushy Run Research Center, Export, PA.

Chun, J. S., H. D. Burleigh-Flayer, and W. J. Kintigh. 1992. *MTBE: Vapor Inhalation Oncogenicity Study in Fischer-344 Rats,* Project No. 91N0013B, Bushy Run Research Center, Export, PA.

Conaway, C. C., R. E. Schroeder, and N. K. Snyder. 1985. "Teratology Evaluation of MTBE in Rats and Mice," *Journal of Toxicology and Environmental Health,* vol. 16, pp. 797–807.

Dodd, D. E., and W. J. Kintigh. 1989. *Methyl tertiary Butyl Ether (MTBE): Repeated (13 week) Vapor Inhalation Study in Rats with Neurotoxicity Evaluation,* Project Report 52-507, Bushy Run Research Center, Export, PA.

Duffy, J. S., J. A. Delpup, and J. J. Kneiss. 1992. "Toxicological Evaluation of Methyl tertiary Butyl Ether (MTBE): Testing Performed under TSCA Consent Agreement," *Journal of Soil Contamination,* vol. 1, no. 1, pp. 29–37.

Elvers, B., S. Hawkins, and G. Schulz, eds. 1990. *Ullmann's Encyclopedia of Industrial Chemistry,* vol. A16, VCH, New York, NY, p. 543.

Environment Canada. 1995. *Phase 1 Final Report: Auto/Oil,* Air Quality Improvement Research Program, Ottawa, ON.

Environment Canada. 2000. "NATES Data Base, National Analysis of Trends in Emergencies Systems," Environment Canada, Ottawa, ON, 2000.

Environment Canada and Health Canada. 1992."Priority Substances List—Assessment Report No. 5, Methyl tertiary-Butyl Ether," Ottawa, ON.

Fadope, C. M. 1995. "Washington Meeting Examines the Safety of MTBE," *Chemical Engineering News,* April.

Fingas, M., N. Laroche, G. Sergy, B. Mansfield, G. Clouthier, and P. Mazerolle. 1991. "A New Chemical Spill Priority List," in *Proceedings of the 8th Technical Seminar on Chemical Spills,* Environment Canada, Ottawa, ON, pp. 223–237.

Fujiwara, Y., T. Kinosita, H. Sato, and I. Kojima. 1984. "Biodegradation and Bioconcentration of Alkyl Ethers," *Yukagaku,* vol. 33, pp. 111–114.

Funasaki, N., S. Had, and S. Neya. 1985. "Partition Coefficients of Aliphatic Ethers—Molecular Surface Area Approach," *Journal of Physical Chemistry,* vol. 89, pp. 3406–3409.

Garrett, P. 1987. "Oxygenates as Ground Water Contaminants," paper presented at 1987 Conference on Alcohols and Octane, San Antonio, TX.

Garrett, P., M. M. Moreau, and J. D. Lowry. 1986. "MTBE as a Ground Water Contaminant," in *Proceedings of the NWWA/API Conference on Petroleum Hydrocarbons and Organic Chemicals in Ground Water-Prevention, Detection, and Restoration,* NWWA-API, National Well Water Association-American Petroleum Institute, Houston, TX.

Geiger, D. L., D. J. Call, and L. T. Brooke, eds. 1988. *Acute Toxicities of Organic Chemicals to Fathead Minnows* (Pimephales promelas), vol. 4, Center for Lake Superior Environmental Studies, University of Wisconsin-Superior.

Gill, M. W. 1989. *MTBE: Single Exposure Vapor Inhalation Neurotoxicity Study in Rats,* Project Report 52-533, Bushy Run Research Center, Export, PA.

Golob's Oil Pollution Bulletin. 1995. "Holed Tank Barge Leaks 1,200 Barrels of MTBE in Waterway near Cypremort, Louisiana," vol. 7, no. 8, pp. 5–6.

Greenough, R. J. P. McDonald, and P. Robinson. 1980. *MTBE (Driveron) Three Month Inhalation Toxicity in Rats,* IRI Project No. 413038, Inveresk Research International, Edinburgh, Scotland.

Industrial Economics, Inc. 1985. Acute Hazardous Events Data Base. Industrial Economics Inc., Cambridge, MA.

ITT Research Institute. 1992. *28-day Oral Toxicity Study of Methyl tert-Butyl Ether in Rats,* Project No. L08100, ITT Research Institute, Chicago.

Japar, S. M., T. J. Wallington, J. F. O. Richert, and J. C. Ball. 1990. "The Atmospheric Chemistry of Oxygenated Fuel Additives: t-Butyl Alcohol and t-Butyl Ether," in *Proceedings of 83rd Annual Meeting of the Air and Waste Management Association,* Pittsburgh, PA, June 24–29.

Japar, S. M., T. J. Wallington, S. J. Rudy, and T. Y. Chong. 1991. "Ozone-Forming Potential of a Series of Oxygenated Organic Compounds," *Environmental Science and Technology,* vol. 25, pp. 415–420.

Mastri, C., M. Keplinger, and O. Fancher. 1969. *Acute Toxicity Studies on X-801-25 IBT No. A6809,* Report to Sun Oil Co., study conducted by Industrial Bio-Test Laboratories, Inc.

Mihelcic, J. R. 1990. "Modeling the Potential Effect of Additives on Enhancing the Solubility of Aromatic Solutes Contained in Gasoline," *Ground Water Monitoring Review,* vol. 10, no. 3, pp. 132–137.

Moolenaar, R. L., B. J. Hefflin, and D. L. Ashley. 1994. "Methyl tert-Butyl Ether in Human Blood after Exposure to Oxygenated Fuel in Fairbanks, Alaska," *Archives of Environmental. Health,* vol. 49, no. 5, pp. 402–409.

Oil and Gas Journal (OGJ). 1995. "Rumorbusters," vol. 93, March 20.

Paulov, S. 1987. "The Effect of the Antiknock Compound tertiary-Butyl Ether on the Model Species *Rana temporaria,* L.," *Biologia (Bratislava),* vol. 42, pp. 185–189 (in Czechoslovakian, English abstract).

Piel, W. J. 1989. "Fuel Oxygenates Affects on Aromatics Solubility in Water," presented before Division of Environmental Chemistry, American Chemical Society, Miami Beach, FL.

Poulsen, M., L. Lemon, and J. F. Barker. 1992. "Dissolution of Monoaromatic Hydrocarbons into Ground Water from Gasoline-oxygenated Mixture," *Environmental Science and Technology,* vol. 26, no. 12, pp. 2483–2489.

Quantitative Structure Activity Relationships (QSAR). 1989. Montana State University, Bozeman, MT.

Reuter, J. E., B. C. Allen, R. C. Richards, J. F. Pankow, C. R. Goldman, R. L. Scholl, and J. S. Seyfried. 1998. "Concentrations, Sources, and Fate of the Gasoline Oxygenate Methyl tert-Butyl Ether (MTBE) in a Multiple-Use Lake."

Robinson, M., R. H. Bruner, and G. R. Olson. 1990. "Fourteen- and Ninety-Day Oral Toxicity Studies of MTBE in Sprague-Dawley Rats," *Journal of the American College of Toxicology,* vol. 9, pp. 525–540.

RTECS On-Line. 2000. *Registry of Toxic Effects of Chemical Substances,* Department of Health and Human Services, Centers for Disease Control, National Institute for Occupational Safety and Health, Washington, DC.

Silvak, A., and J. Murphy. 1991. "A Review of Health Effects Data on the Oxygenated Fuel Additive MTBE," in *International Alcohol Fuels Symposium,* Florence, Italy, November.

Snyder, R. 1979. "Studies on the Metabolism of t-Butyl Methyl Ether (TBME)," Thomas Jefferson University, Jefferson Medical College, Philadelphia, PA.

Squillace, P. J., J. S. Zogorski, W. G. Wilber, and C. V. Price. 1996."Preliminary Assessment of the Occurrence and Possible Sources of MTBE in Ground water in the United States 1993–94," *Environmental Science and Technology,* vol. 30, pp. 1721–1730.

Suflita, J. M., and R. Mormile. 1993. "Anaerobic Biodegradation of Known and Potential Gasoline Oxygenates in the Terrestrial Subsurface," *Environmental Science and Technology, vol.* 27, pp. 976–978.

Thomas, R. G. 1982. "Volatilization," in *Handbook of Chemical Properties Estimation Methods,* ed. W. J. Lyman, W. J. Reehl, and D. H. Rosenblatt, McGraw-Hill, New York, NY.

Tyl, R. W., and T. L. Neeper-Bradley. 1989. "Developmental Toxicity Study of Inhaled MTBE in CD-1 Mice," Project Report 51-266, Bushy Run Research Center, Export, PA.

Tyl, R. W. 1989. *Developmental Toxicity Study of Inhaled MTBE in New Zealand White Rabbits,* Project Report 51-268, Bushy Run Research Center, Export, PA.

U.S. Environmental Protection Agency (USEPA). 1986a. *Nineteenth Report on the Interagency Testing Committee to the Administrator; Receipt and Request for Comments Regarding List of Chemicals, 51 F.R. pp. 41417–41432, November 14.*

U.S. Environmental Protection Agency (USEPA). 1986b. Method 8015, in *Test Methods for Evaluating Solid Waste, Laboratory Manual Physical/Chemical Methods SW-846,* 3rd ed., vol. 1B, U.S. EPA, Office of Solid Waste and Emergency Response, Washington, DC.

Wallington, T. J., P. Dagaut, R. Liu, and M. J. Kurylo. 1988. "Gas-phase Reactions of Hydroxyl Radicals with the Fuel Additives Methyl tert-Butyl Ether and tert-Butyl Alcohol over the Temperature Range 240–440°K," *Environ. Sci. Technol.,* vol. 22, no. 7, pp. 842–844.

White, M. C., C. A. Johnson, and D. L. Ashely. 1995. "Exposure to MTBE from Oxygenated Gasoline in Stamford Connecticut," *Archives of Environmental Health,* vol. 50, no. 3, pp. 183–189.

Yeh, C. K., and J. T. Novak. 1994. "Anaerobic Biodegradation of Gasoline Oxygenates in Soils," *Water Environment Research,* 66, pp. 744–752.

CHAPTER 35
PERSPECTIVES ON SPECIFIC SUBSTANCES: SULFURIC ACID

Richard Lawuyi and Merv Fingas
Emergencies Science Division, Environment Canada,
Environmental Technology Centre, River Road,
Ottawa, Ontario

35.1 OVERVIEW OF PRODUCT AND INDUSTRIAL USES

Sulfuric acid is a corrosive, oily, and colorless liquid well known since the Middle Ages. It is a good dehydrating and drying agent and has been an important commercial commodity for more than three centuries. Until the late 1960s, sulfuric acid production and consumption were generally accepted throughout the industrialized world as an accurate barometer of a nation's commercial activity and wealth. Nowadays, because of its extensive use in the agriculture industry, sulfuric acid no longer measures a country's wealth. Nevertheless, more sulfuric acid is produced and spilled in the chemical industry today than any other chemical. It is indispensable in many chemical and industrial processes. The market for sulfuric acid is still very strong and will continue to grow.

Canada has an abundant supply of sulfur deposits, and sulfur occurs naturally in natural gas and metal ores. Sulfuric acid can therefore be produced cheaply for both domestic use and export. Sulfuric acid is used primarily to produce phosphate fertilizers by the phosphoric acid route. It also has many uses in the processing industry, including kraft bleaching; in copper, petroleum, and uranium refineries; to manufacture various salts such as magnesium sulfate; to manufacture soaps and detergents; in electroplating and metallurgy; and to produce various chemical reagents such as phenols, alums, alkylation catalysts, explosives, propellants, fibers, electronic chips, pharmaceuticals, and acid batteries.

35.1.1 Modern Industrial Uses

The estimated consumption of sulfuric acid in 1996 by the industrial sector in the United States is shown in Figure 35.1. In Canada, while fertilizer consumption is not as prominent, consumption of sulfuric acid follows a similar pattern to that in the United States: fertilizers—68%; mining—5.8%; miscellaneous—10.6%; inorganics—5.1%; others, including petroleum refining and products, synthetic rubber and plastics, pulp mills and other paper products, and industrial organic chemicals—10.5% (CIS, 1997). Sulfuric acid consumption is very stable and should continue to be so.

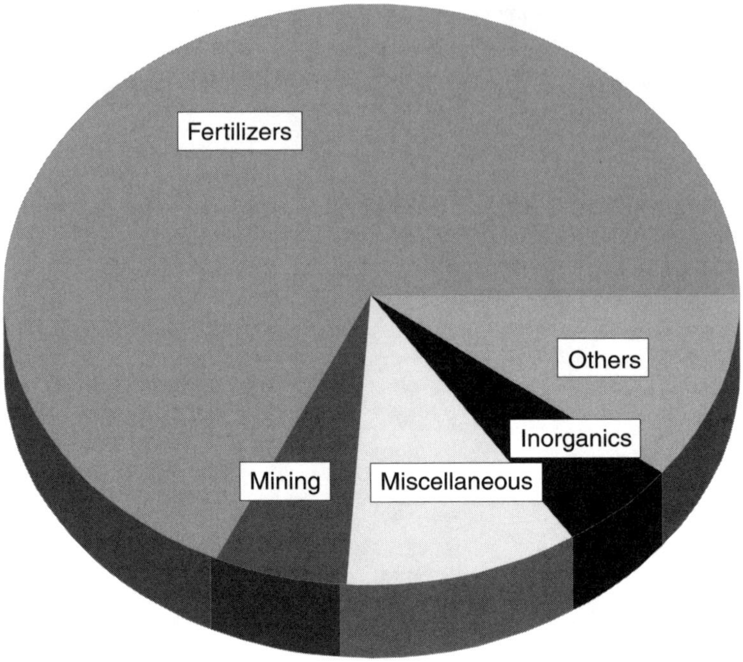

FIGURE 35.1 Uses of sulfuric acid in the United States in 1996.

35.2 INTRODUCTION

Sulfuric acid was apparently discovered in the 8th century by a Persian alchemist who distilled niter (potassium nitrate) with green vitriol (ferrous sulfate crystals) obtained from weathered iron pyrites. Several alchemists, including Jabir ibn Hayyan, Vincentius de Beauvais (1240), Albertus Magnus (1193–1280), Paracelsus, Gerhard Dornaeus (1570), Andreas Libavius (1595), Angelus Sala (1613), Nicholas le Fevre (1666), Nicholas Lemery, and Cornelius Drebbel, tried to improve on his methodology.

By the middle of the 12th century, Occidental alchemists were also producing sulfuric acid from sulfur and pyrites. Following Lavoisier's discovery in the 18th century that sulfur is a chemical element and not a mixture, production of sulfuric acid from sulfur and pyrites was commercialized in many parts of the world (Duecker and West, 1959).

In approximately 1740 in Great Britain, Ward began to produce the acid on a large-scale by burning sulfur with potassium nitrate. In 1746, Dr. Roebuck of Birmingham introduced the lead chamber process and built a factory in Scotland to manufacture the acid. This practice quickly spread throughout Europe and North America

In North America, commercial production of sulfuric acid began in 1797 when John Harrison built a sulfuric acid plant in Philadelphia. Much research has since been done on the nature of the catalyst and feedstock. In the late 1800s, the lead chamber process was gradually replaced by the contact process, patented by Phillips in 1831. In 1875, Emil Jacob successfully demonstrated the new process and the modern contact acid manufacture of the acid began with a pyrite-burning gas as the source of sulfur dioxide.

Today, a large proportion of the sulfuric acid produced in the world is what is termed "fatal" acid, which is manufactured to prevent substantial amounts of waste sulfur dioxide formed in metallurgical and smelting processes, such as nonferrous metal smelting and iron production from pyrites, from entering the environment (Inco Limited, 1985; Muller, 1997).

Many of the environmental and technical problems associated with the large-scale production, handling, and shipment of sulfuric acid are now fairly well understood. Today, emissions, effluent discharge, handling, and shipping are governed by national and international regulations and codes.

35.2.1 Spill Profile

For the last few years, sulfuric acid has been one of the most commonly spilled substances (Environment Canada, 2000). The annual frequency of spills of sulfuric acid from 1990 to 1996 is shown in Figure 35.2. It can be seen that the number of spills is slowly decreasing. Chemical burns are the most prevalent hazard from spills of sulfuric acid. Its main properties, including physical, chemical, toxicological, behavioral, and environmental fate, are outlined in Section 35.3.

35.2.2 Priority List Ranking

According to a study to determine the minimum number of hazardous chemicals most frequently spilled, sulfuric acid is sixth on the list, with the highest supply volume (Fingas et al., 1991). It also has one of the highest numbers of spills. The priority list was developed by a simple ranking of: (1) reported spill frequency; (2) supply volumes; (3) historical spill volumes; and (4) toxicities. Table 35.1 shows how sulfuric acid ranks in Environment Canada's priority listing of hazardous chemicals.

35.3 PHYSICAL AND CHEMICAL PROPERTIES AND GUIDELINES SUMMARY

O
‖
HO–S–OH
‖
O

Sulfuric acid: Sulfuric acid is a colorless (when pure) to dark brown, oily liquid with a sharp, penetrating odor.

Molecular formula: H_2SO_4

Molecular weight: 98.08

CAS number: 7664-93-9

UN number: 1830

Fuming: 1831
Spent: 1832

STCC number: 4930040

Spent: 4930042

OHM-TADS number: 7216915

Labels:

Corrosive
Corrosive and poison (fuming)–IMO

Synonyms and trade names (RTECS On-Line, 1999):

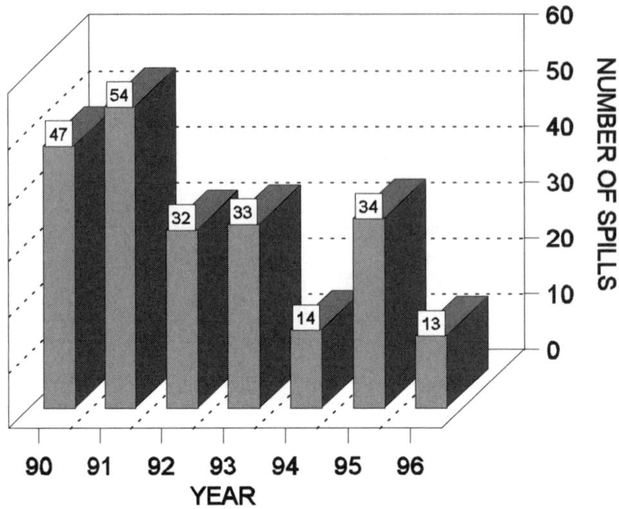

FIGURE 35.2 Spill profile of sulfuric acid (1990–1996).

Sulfuric Acid
- Acide sulfurique
- Battery acid
- Electrolyte acid
- Fertilizer acid
- Hydrogen sulfate
- Oil of vitriol
- Spirit of sulfur
- Sulphuric acid
- Acido solforico
- Bov

TABLE 35.1 Priority List Ranking of Sulfuric Acid

Chemical	Ranking	Spill numbers	Spill volume (thousand tons)	Supply volume (million tons)
Ammonia	1	107	470	3,700
Chlorine	2	36	120	1,700
Tetraethyllead	3	4	72	26
Styrene	4	24	5,000	630
PCBs	5	334	89	–
Sulfuric acid	**6**	**155**	**13,000**	**3,700**
Sodium cyanide	7	3	83	12
Hydrochloric acid	8	123	3,300	170
Potassium chloride	9	31	12,000	–
Pentachlorophenol	10	19	110	1.5
Phenol	11	10	14	68
Zinc sulfate	12	3	68	1,500
Phosphorus	13	16	46	68
Toluene	14	13	110	430

Dipping acid
Matting acid
Nordhaausen acid
Spent sulfuric acid
Schwefelsäurelöesungen
Vitriol brown oil
Zwavelzuuroplossingen

Fuming sulfuric acid (sulfuric acid with dissolved SO_3):
Disulfuric acid
Dithionic acid
Fuming sulfuric acid
Fuming sulfuric acid
Oleum
Pyrosulfuric acid

Grades and Purities (Kirk-Othmer, 1983; Budavari, 1989): Pure sulfuric acid is a colorless, oily liquid. When impure, it is brownish. The pure acid decomposes into sulfur trioxide and water at 340°C. It is soluble in water and alcohol with evolution of heat. Spent sulfuric acid is a black oily liquid. It is also soluble in water with release of heat. Impurities are iron, arsenic, sulfur dioxide, nitrogen compounds, chloride, and fluoride.

Grade	Minimum purity mole %[1] (liquid phase)
Commercial (technical)	75; 78; 88; 93; 96; 98 to 99; and 100
Electrolyte-battery acid	>93.19
Chemically pure (CP)	95.5 to 96.5
USP	–
Sulfuric acid (fuming)	Sulfuric acid and SO_3 20% oleum (104.5%); 40% oleum (109.0%); 65% oleum (114.63%)

[1]Impurity is mostly water and some metals.

Strengths:

Sulfuric acid:

Degrees Baumé	% H_2SO_4	Specific gravity at 17.78°C	Freezing point (°C)
52	65.13	1.5591	−40.0
58	74.36	1.6667	−44.0
60	77.67	1.7059	−8.0
66	93.19	1.8354	−32.0
–	98.00	1.8438	3.0
–	100.00	1.8392	10.0

Fuming sulfuric acid:

Oleum, % free SO_3	% Equivalent H_2SO_4	Specific gravity (37.78)	Freezing point (°C)
20.0	104.50	1.8820	−9.0
30.0	106.75	1.9156	15.5
40.0	109.00	1.9473	33.0
65.0	114.63	1.9820	3.6
100.0	122.50	1.8342	16.8

Physical data:

Concentrated sulfuric acid:
- Physical state: Liquid (Budavari, 1989)
- Boiling point: 330 ± 0.5°C (100%)
- Melting point: 10.36°C (100%)
- Solubility (water): Infinity (heat evolved)
- Density, liquid: 1.841 (96 to 98%)
- Vapor pressure (mm Hg): 1 at 146°C
- Specific gravity (water = 1): 1.841
- Flammability: Not flammable
- Vapor density (air = 1): 3.4
- Behavior (in water): Sinks, reacts with heat evolution
- Odor threshold and range: Low concentration is odorless; in fire, high concentration emits SO_3 with an odor threshold >1 mg/m^3

Sulfuric acid solution in water 93 to 98%
- State (15°C, 1 atm): Liquid
- Melting point: −32°C (93%)
- Relative density: 1.8 (water = 1)
- Solubility in water: Infinity

Sulfuric acid is a dense, colorless liquid at room temperature. In the past, its concentration was described as a function of its specific gravity in degrees Baumé (°Bé). In the United States, the Baumé scale is calculated using the following formula:

$$°\text{Bé} = 145 - (145/\text{specific gravity})$$

In Germany and France, the Baumé scale is calculated using 144.3 as the constant. The Baumé scale includes only concentrations in the range of 0 to 93.19% H_2SO_4. The scale does not extend to higher concentrations ranging from 93 to 100% H_2SO_4.

Sulfuric acid:

Appearance: Clear to cloudy fuming liquid
Usual shipping state: Liquid
Physical state at room temperature and pressure: Liquid

Melting point:

% H_2SO_4	°C	Bé
65	−64	
65.13	−40.0	52°
74.36	−44.0	58°
75	−12	
77.67	−8.0	60°
78	−38	
88	−20	
93	−35	
93.19	−32.0	66°
96	−14	
98.00	3.0	
100.00	10.0	

% oleum	°C
20.0	−9.0
30.0	15.5
40.0	33.0
65.0	3.6
100.0	17.2

Boiling point:

% H_2SO_4	°C
98	340
96	~310
93	279
75	~193

% oleum	°C
20	142

Decomposition temperature: 340°C (Budavari, 1989)

Vapor pressure (mm Hg):

% H_2SO_4	mm Hg
96	1 at 146°C
93	1 at 145.8°C

Densities:

Specific gravity:

%H_2SO_4	Density
96	1.84
93	1.835
88	1.50
75	~1.674 at 15.5°C
+20% SO_3	1.92

Vapor density (air = 1):

% H_2SO_4	V.D.
96	<0.3 at 25°C
93	3.4

Fire properties:

Flammability: Nonflammable. Highly reactive and capable of igniting finely divided, combustible materials on contact.
Behavior in fire: Emits toxic fumes.
Decomposition temperature: 340°C (Budavari, 1989).
Decomposition products: Toxic fumes of sulfur oxides.

Other properties:

Molecular weight: 98.08 (Budavari, 1989)
Grades (minimum purity %):

Commercial (technical)	75%
	78%
	93%
	96%
	98 to 99%
	100%

Electrolyte battery acid	>93.19
Chemically pure (CP)	95.5 to 96.5
USP	–
Fuming sulfuric acid (oleum)	80% H_2SO_4 + 20% SO_3
	60% H_2SO_4 + 40% SO_3
	35% H_2SO_4 + 65% SO_3

Refractive index: 1.4297 (for 100% H_2SO_4)
Viscosity:

Viscosity (cp) (for 100% H_2SO_4)	Temp. °C
48.4	0
32.8	15
25.4	20
15.7	30
11.5	40
8.82	50
7.22	60
6.09	70
5.19	80

Surface tension, 25°C (compared to water):

% wt H_2SO_4	mN/m
4.11	72.21
8.26	72.55
12.18	72.80
17.66	73.36
21.88	73.91
29.07	74.80
33.63	75.29

Surface tension, 20°C (with vapor or air):

% wt H_2SO_4	mN/m
98.5	55.1

Hygroscopicity: Hygroscopic
Latent heat of fusion: 2360 cal/g mole (9.8 kJ/mole at melting point)
Latent heat of vaporization: 56 kJ/mole (at boiling point)
Heat of formation: −813.9 kJ/mole (25°C)
Heat of solution: −971.5 kJ/kg
Heat capacity (Cp) constant pressure: 138.9 J/mole°C (25°C)
Coefficient of thermal expansion (20°C): 0.5758×10^{-3} (100% solution); 0.2835×10^{-3} (10.9% solution)

Thermal conductivity (30°C):	
90%	0.21
60%	0.25
30%	0.30

Diffusivity: 1.97×10^{-5} cm^2/s (water, 25°C)
pH of aqueous solution: 0.3, 1 N solution at 25°C
Eutectic compositions: 36% aqueous solution (freezing point, −64°C)

Solubility:

Water: Soluble in water. Reacts violently with evolution of heat. Spattering occurs when water is added to sulfuric acid.

Properties of sulfur trioxide (*Kirk-Othmer*, 1983):

Critical temperature: 217.8°C
Critical pressure: 8,208 kPa
Critical density: 0.630 g/cm^3
Triple point temperature (γ phase): 16.8°C
Triple point pressure (γ phase): 21.13 kPa
Normal boiling point: 44.8°C
Melting point (γ phase): 16.8°C
Transition temperature: −183.0°C
Liquid density (γ phase at 20°C): 1.9224 g/cm^3
Solid density (γ phase at −10°C): 2.29 g/cm^3
Liquid coefficient of thermal expansion, 18°C: 0.002005 per °C
Liquid heat capacity at 30°C: 3.222 kJ/(kg°C)
Heat of formation of gas at 25°C: −395.76 (MJ · kg)/mol
Free energy of formation of gas at 25°C: −371.07 (MJ · kg)/mol
Entropy of gas at 25°C: 0.25666 (MJ · kg)/(mol · °C)
Heat of dilution: 2.109 MJ/kg
Heat of fusion, α: 324.0 kJ/kg
β: 151.6 kJ/kg
γ: 94.07 kJ/kg
Heat of sublimation, α: 0.8518 MJ/kg
β: 0.7269 MJ/kg
γ: 0.7029 MJ/kg
Heat of vaporization (γ liquid): 0.5843 MJ/kg
Diffusion in air at 80°C: 0.000013 m/s
Liquid dielectric constant at 18°C: 3.11
Electric conductivity: Negligible

35.3.1 Summary of Chemical Properties and Behavior (Bailar et al., 1975)

Sulfuric acid is a strong mineral acid, one of the most important compounds of sulfur and one of the best known nonaqueous protonic solvents. The sulfur atoms are surrounded symmetrically by four oxygen atoms and are therefore highly associated because of the strong intermolecular hydrogen bonding. At the boiling point, sulfuric acid solutions containing less than 85% H_2SO_4 evaporate water exclusively, while those containing more than 35% free SO_3 (oleum) evaporate sulfur trioxide exclusively. The vapor of sulfuric acid solutions in between these two concentrations (85% H_2SO_4 to >35% oleum) is a mixture of sulfuric acid, water, and sulfur trioxide.

Sulfuric acid is miscible with water in all concentrations. The heat of hydration is very high, about 210 kcal/mole at infinite dilution. Sulfuric acid forms hydrates, such as $H_2SO_4 \cdot H_2O$, $H_2SO_4 \cdot 2H_2O$, and $H_2SO_4 \cdot 3H_2O$, in the presence of water.

35.3.2 Physiological Effects of Sulfuric Acid

The physiological effects of exposure to sulfuric acid fumes depend on the particle size of the aerosol. Thus, for a constant sulfuric acid aerosol concentration, the irritant action of the aerosol increases with aerosol particle size. Other factors are humidity, temperature, and previous exposure. Several studies have shown that prolonged exposure to sulfuric acid causes tooth erosion, while overexposure to sulfuric acid aerosols leads to pulmonary edema, chronic pulmonary fibrosis, residual bronchiectasis, and pulmonary emphysema. The threshold limit value (TLV) for sulfuric acid mist for humans is 1 mg/m^3. This threshold is recommended to prevent injury to the teeth and pulmonary irritation at particle sizes likely to occur in industrial settings. Eye contact may result in loss of vision, while skin contact may produce severe necrosis. If ingested, gastric perforation and peritonitis may occur, possibly followed by circulatory collapse. A few drops may be fatal if the acid gains access to the trachea.

35.3.3 Main Hazards

General: Sulfuric acid is a very corrosive, colorless liquid that may exhibit violent chemical changes at elevated temperatures and pressures and may react violently with water. Inhalation of vapor may cause serious lung damage, and contact with eyes may result in total loss of vision. Skin contact may cause burning and severe necrosis. Sulfuric acid is highly reactive and capable of igniting finely divided combustible materials on contact. It will char wood and many other organic materials on contact and emits toxic fumes when heated.

Human health: Sulfuric acid is corrosive to all body tissues. Inhalation can paralyze the respiratory system, contact with eyes may result in loss of vision, and skin contact may result in severe burns and necrosis. Swallowing may cause severe injury or death. Between one teaspoonful and half an ounce of the concentrated acid may be fatal if swallowed, and an even smaller quantity may be fatal if inhaled. Chronic exposure may cause tracheobronchitis, stomatitis, conjunctivitis, and gastritis. Gastric perforation and peritonitis may occur and may be followed by collapse of the circulatory system. Pulmonary fibrosis, bronchiectasis, and emphysema have been reported from acute exposure to fuming sulfuric acid and sulfuric acid mist. Chronic exposure usually results in erosion of the teeth, particularly the incisors.

Environment: Sulfuric acid is a powerful acidic oxidizer and a strong dehydrating agent. It is harmful to many microorganisms, plants, and aquatic life in concentrations as low as 6 mg/L.

Behavior in Air: Sulfuric acid is not combustible, but many reactions may cause fire and explosion and produce fumes in the air.

Behavior in Water: Concentrated sulfuric acid reacts violently with water and produces heat.

35.3.4 Emergency Response–Human Health

Symptoms:

Local: Conjunctivitis, corneal necrosis, dermatitis, skin burns, ulceration, fainting. May lead to pulmonary edema, nausea, and vomiting.

Respiratory: Irritation of the nose and throat, coughing, sneezing, laryngeal edema, bronchitis, pneumonitis, and pulmonary edema.

Gastrointestinal: Dental erosion, shock, anuria; burning in the mouth, throat, and abdomen; nausea, vomiting of blood and eroded tissue; perforation of gastrointestinal tract; and albumin, blood, and casts in urine.

Treatment:

Inhalation:

Move victim to fresh air.
If not breathing, give artificial respiration.

If breathing is difficult, give oxygen.
During transportation:
Continue giving oxygen by 40% venturi mask.
Observe vital signs.
In emergency room:
Check arterial blood gas.
Perform a chest X ray.
Observe for pulmonary edema and bronchospasms.

Skin:
Wash and flush with water immediately.
Remove clothing.
Shower thoroughly.
During transportation:
Cover affected areas with sterile wet dressings immediately.
In emergency room:
Treat as burn.

Eyes:
Rinse well with large quantities of water for 15 minutes.
Hold eyelids open while washing.
During transportation:
Continue eye rinsing.
In emergency room:
Continue eye rinsing (use Mediflow lens if available).
Determine nature and extent of corneal damage.
Notify an ophthalmologist.

Ingestion:
Do not induce vomiting.
During transportation:
Keep victim comfortable in an upright position.
Support circulation.
In emergency room:
Do not lavage.
Treat as acid ingestion.

35.3.5 Spill Control

Wear self-contained breathing apparatus and full protective clothing for all emergency measures involving sulfuric acid.

Leaks: Keep the acid away from water sources and sewers. Build dikes to contain flow as necessary. Neutralize spilled material with agricultural lime (CaO), crushed limestone ($CaCO_3$), soda ash, or sodium bicarbonate. Keep material out of water sources and showers and spray large liquid spills with vapor-smothering foams.

Fires: Sulfuric acid does not burn by itself. Use a dry chemical to smother small fires involving combustibles. Use water in flooding quantities as fog to cope with large fires. Cool affected containers with flooding quantities of water. Apply water from as far away as possible.

Spills: Wear appropriate chemical protective gloves, boots, and goggles. Avoid contact with contaminated material. Evacuate spill site and restrict access. Issue warning: "Corrosive, Poison." Notify manufacturer and relevant authorities. Contain spill if it is safe to do so. Avoid contact with liquid or vapor. Stay upwind. Vapor cloud collects in low-lying areas. Keep contaminated water from entering sewers or watercourses. For small liquid spills, neutralize with lime. For large liquid spills, contain spill as much as possible. Neutralize spilled material with crushed limestone, soda ash, or lime.

Land spill: Dig a pit, pond, lagoon, or holding area to contain the acid. Dike surface flow using soil, sand bags, foamed polyurethane, or foamed concrete. Absorb bulk liquid with fly ash or cement powder. Neutralize with agricultural lime (CaO), crushed limestone ($CaCO_3$), or sodium bicarbonate ($NaHCO_3$).

Air spill: Apply water spray or mist to knock down vapors. Liquid produced is corrosive and toxic; it should be diked and neutralized.

Water spill: Neutralize with agricultural lime (CaO), crushed limestone ($CaCO_3$), or sodium bicarbonate ($NaHCO_3$).

Immediate Concerns

Hazard: Extremely corrosive. Rapidly destroys human tissues on contact. Causes severe burns. Sharp, penetrating odor. May react with other chemicals, causing fires and explosions. Fumes are very toxic. Powerful oxidizer. May react violently with water. May exhibit violent chemical changes at elevated temperatures and pressures.

Humans: Corrosive to skin, eyes, and respiratory system. Will cause burns. Toxic by all routes, causing severe damage to tissues. Death may result.

Environment: Harmful to aquatic life and the environment.

Protection: Complete self-contained breathing apparatus recommended with face mask, suit, boots, and gloves (rubber, neoprene, or plasticized PVC).

35.3.6 Guidelines

The following are some emergency response guidelines provided by the American Industrial Hygiene Association (AIHA) for use during large releases of sulfuric acid (AIHA, 1988).

ERPG-3: 30 mg/m^3 (as sulfuric acid mist): The maximum airborne concentration below which it is believed that nearly all individuals could be exposed for up to one hour without experiencing or developing life-threatening health effects.

ERPG-2: 10 mg/m^3 (as sulfuric acid mist): The maximum airborne concentration below which it is believed that nearly all individuals could be exposed for up to one hour without experiencing or developing irreversible or other serious health effects or symptoms that could impair an individual's ability to take protective action.

ERPG-1: 2 mg/m^3 (as sulfuric acid mist): The maximum airborne concentration below which it is believed that nearly all individuals could be exposed for up to one hour without experiencing other than mild, transient adverse health effects or without perceiving a clearly defined objectionable odor.

Threshold Limit Values

The American Conference of Governmental Industrial Hygienists (ACGIH) has determined the threshold limit values for TWA(time-weighted average) to be 1 mg/m^3 and STEL (short-term exposure limit) to be 3 mg/m^3. The ACGIH also recommends a TLV of 1 mg/m^3 for sulfuric acid to prevent pulmonary irritation and injury to the teeth.

The Occupational Safety and Health Administration (OSHA) permissible exposure limit (PEL) is 1 mg/m^3. The National Institute for Occupational Safety and Health (NIOSH) recommended exposure limit (REL) is 1 mg/m^3 (10-hr TWA).

35.4 INDUSTRIAL ASPECTS AND PRODUCTION IN THE UNITED STATES, CANADA, AND WORLDWIDE

35.4.1 Manufacture of Sulfuric Acid

Until the late 1800s, sulfuric acid was produced mainly by the lead chamber process, which uses nitrogen oxides as homogenous catalysts for the sulfur dioxide oxidation step. The feedstocks were sulfur, iron pyrites, nonferrous pyrites, hydrogen sulfide, and spent sulfuric acid. Nowadays, approximately 99% of all production is by the contact process. Sulfuric acid is also manufactured to prevent the substantial quantities of waste sulfur dioxide produced in metallurgical processes, such as the smelting of non-ferrous metals like nickel and copper and iron production from pyrites, from entering the environment. Sulfur dioxide produced during roasting of sulfide ores is sometimes referred to as roaster gas. Hydrogen sulfide gas is another useful raw material for the manufacture of sulfuric acid. The following are the steps involved in the production of SO_2 for the contact process.

1. As the sulfide ores (pyrites) are roasted, the roaster gas and the metal oxides produced are separated from each other in cyclones.

$$3\ Fe_7S_8 + 38\ O_2\ 2\ Fe_3O_4 + 24\ SO_2 \quad (35.1)$$

$$Ni_3S_2 + 2\ O_2 \rightarrow 3\ Ni + 2\ SO_2 \quad (35.2)$$

2. The roaster gas is mixed with additional air to complete the combustion of all volatilized products.
3. Any remaining metal oxides in the roaster gas are removed by cooling and contacting with sulfuric acid solution.
4. The roaster gas is then passed through a further stage of washing and drying, followed by a wet precipitation stage.
5. Following the primary converter, the process gas passes through an interpass absorption tower. This tower removes SO_3 from the process gas to provide improved equilibrium conditions for further oxidation of SO_2 to SO_3 in a second converter. SO_3 produced in the second converter is absorbed in the final absorber.
6. The roaster gas is catalytically oxidized to sulfur trioxide in a fixed bed converter which operates adiabatically with each catalyst pass.

$$2SO_2 + O_2 \rightleftharpoons 2SO_3 \quad (35.3)$$

7. The gas is now cooled and allowed to flow into the packed towers, where it is absorbed. The production of fuming sulfuric acid (oleum), however, requires sulfur trioxide absorption in special absorption towers irrigated with oleum. The reaction is exothermic.

$$SO_3 + H_2O \rightarrow H_2SO_4 \quad (35.4)$$

35.4.2 Manufacturers in Canada and the United States

Canadian and American manufacturers of sulfuric acid are listed here. It should be added that some of these plants have been mothballed and others may have changed hands or closed.

Canadian manufacturers	Plant location
Sherritt	Fort Saskatchewan Redwater, AB
Alcan Smelters and Chemicals	Jonquière, QC
Allied Chemical Canada	Valleyfield, QC
Border Chemical	Transcona, MB
Brunswick Mining and Smelting	Belledune, NB
Cameco	Key Lake and Rabbit Lake, SK
Canadian Electrolytic Zinc	Valleyfield, QC
Cogema Resources	McClean Lake, SK
Cominco	Trail and Kimberly, BC
Falconbridge	Sudbury, ON
Gaspé Copper Mines	Murdochville, QC
Hudson Bay Mining	Flin Flon, MB
ICI Canada	Beloeil, QC
Inco	Coppercliff, ON
International Minerals and Chemicals	Port Maitland, ON
Kidd Creek Mines	Kidd Creek, ON
Kronos Canada	Varennes, QC
Lever	Toronto, ON
Marsulex	Fort Saskatchewan, AB
Noranda	Rouyn-Noranda, QC
Solv-Ex	Fort McMurray, AB
Sulco Chemicals	Elmira, ON
Sulconam	Montreal East, QC
Westcoast Energy	Prince George, BC
Western Co-operative Fertilizers	Calgary and Medicine Hat, AB

American manufacturers	Plant location
Akzo Chemicals	Le Moyne, AL
Amoco Chemical	Texas City, TX
Arcadian	Geismar, LA
Asarco	Three locations
Big River Zinc	Sauget, IL
Boliden Intertrade	Copperhill, TX
Cargill Fertilizer	Bartow and Riverview, FL
CF Industries	Plant City, FL
Chevron Chemical	El Segundo, CA, and Honolulu, HI
Citgo Petroleum	Lake Charles, LA
Climax Molybdenum	Fort Madison, IA
Coulton Chemical	Cairo and Oregon, OH
Cyprus Minerals	Claypool, AZ
Cytec	Fortier, LA
Doe Run	Herculaneum, MO
Dupont Chemicals	Six locations
El Dorado Chemical	El Dorado, AR
Farmland	Pierce, FL
General Chemical	Three locations

American manufacturers	Plant location
IMC-Agro	Five locations
Jersey Miniere Zinc	Clarkesville, TN
J R Simplot	Lathrop, CA, and Pocatello, ID
Kemira	Savannah, GA
Kennecott	Salt Lake City, UT
Koch Sulfur Products	Five locations
Magma Copper	San Manuel, AZ
Marsulex	Sayreville, NJ
Mobil Mining and Minerals	Pasadena, TX
Nu-West	Conda, ID
Olin Hunt Speciality Products	Beaumont, TX, and Shreveport, LA
PCS Phosphates	Aurora, NC, and White Springs, FL
Peridot Chemicals	Augusta, GA, and Newark, NJ
Phelps Dodge Refining	Three locations
Public Service Co of New Mexico	Waterflow, NM
PVS Chemicals	Chicago, IL and Buffalo, NY
Rhone-Poulenc Basic Chemicals	Six locations
Rohm and Haas	Deer Park, TX
SFI Phosphates	Rock Springs, WY
Southern States	Savannah, GA
Tampa Electric	Tampa, FL
Tosco Refining	Martinez, CA
Unocal Chemicals	Wilmington, CA
U.S. Agri-Chemicals	Fort Meade, FL
U.S. Army	Radford, VA
Zinc Corp. of America	Bartlesville, OK and Monaca, PA

The major buyers in Canada are Abitibi-Consolidated, Alcan Smelters and Chemicals, Canadian Copper Refiners, Canadian Forest Products, Domtar, E.B. Eddy Forest Products, Explosive Technologies, Imperial Oil, Kimberly-Clark of Canada, MacMillan Bloedel, Procter and Gamble, Shell Canada, Scott Maritimes, and Weyerhaeuser Canada.

The major buyers in the United States are AGP Refineries, Agrium, Air Products and Chemicals, Alabama River Pulp, Boise Cascade, Champion International, Consolidated Papers, Federal Paper Board, Georgia Pacific, Glatfelter, International Paper, Procter and Gamble, and Warren and Weyerhaeuser Paper.

35.4.3 Production and Transportation

In 1996, the total nameplate capacity for sulfuric acid in Canada was 5,681 kilotons and in the United States it was 36,306 kilotons. In 1993, the total global production of sulfuric acid was close to 135.3 megatons. Some growth is expected in the Middle East and North Africa. In Canada and the United States, sulfuric acid is transported primarily by rail and truck. Long-distance transport of sulfuric acid from smelters in remote locations makes freight costs extremely high.

35.5 CHEMISTRY

Some of the chemical reactions often encountered during sulfuric acid spills are discussed here. One common element in all these reactions is the large quantity of heat that is given off. As these reactions are typically exothermic, some caution is required during emergency response operations and spill mitigation.

35.5.1 Neutralization

Like most strong mineral acids, sulfuric acid will react with bases to form salt and water. Therefore, substances such as slaked lime or sodium hydroxide are often added to sulfuric acid spills.

$$H_2SO_4 + 2\ NaOH \rightarrow Na_2SO_4 + 2\ H_2O \tag{35.5}$$

$$H_2SO_4 + Ca(OH)_2 \rightarrow CaSO_4 + 2H_2O \tag{35.6}$$

35.5.2 Reaction with Water and Hygroscopicity

Concentrated sulfuric acid is a strong dehydrating agent with an enormous affinity for water. It will extract water and elements of water from most materials, e.g., organic and inorganic with evolution of heat. Sometimes enough heat is generated to ignite surrounding combustible materials or vapor.

$$H_2SO_4 + H_2O \rightarrow H_2SO_4 \cdot H_2O \tag{35.7}$$

$$(C_6H_{10}O_5)_X + H_2SO_4 \rightarrow 6C + H_2SO_4 \cdot 5H_2O \tag{35.8}$$

35.5.3 Fire and Intense Heat

Sulfuric acid itself is not combustible, but under certain conditions, it can produce enough heat when reacting with other substances to ignite or decompose.

$$4\ H_2SO_4 \rightarrow 4\ H_2O + 2\ SO_2 + 2\ SO_3 + O_2 \tag{35.9}$$

35.5.4 Reaction with Metals

Under certain conditions, sulfuric acid can react with many metals to produce salt and flammable, explosive hydrogen gas. It will also corrode many materials to form innocuous substances. It will react with many sulfides, oxides, and carbonates.

$$2\ Al + 3\ H_2SO_4 \rightarrow Al_2(SO_4)_3 + 3\ H_2 \tag{35.10}$$

$$2\ Fe + 3\ H_2SO_4 \rightarrow Fe_2(SO_4)_3 + 3\ H_2 \tag{35.11}$$

$$FeS + H_2SO_4 \rightarrow FeSO_4 + H_2S \tag{35.12}$$

35.5.5 Oxidation

Concentrated sulfuric acid is a strong oxidizing agent and can oxidize carbon, nonmetallic elements, and many metals. These reactions generally occur when the acid is in high concentrations and at high temperatures. The potential hazard of these reactions is due to the formation of sulfur dioxide, a toxic gas.

$$C + 2\ H_2SO_4 \rightarrow CO_2 + 2\ SO_2 + 2\ H_2O \tag{35.13}$$

$$Pb + 3\ H_2SO_4 \rightarrow Pb(HSO_4)_2 + SO_2 + 2\ H_2O \tag{35.14}$$

$$Cu + 2\ H_2SO_4 \rightarrow CuSO_4 + SO_2 + 2\ H_2O \tag{35.15}$$

35.5.6 Hazardous Reactions

Sulfuric acid will undergo many double-displacement reactions to produce new substances that may be hazardous. These substances should therefore be stored at distant locations.

$$2\ NaBr + 2\ H_2SO_4 \rightarrow Br_2 + SO_2 + NaSO_4 + 2\ H_2O \tag{35.16}$$

$$3\ NaClO_3 + 3\ H_2SO_4 \rightarrow 3\ NaHSO_4 + HClO_4 + 2\ ClO_2 + H_2O \tag{35.17}$$

$$2\ HClO_4 \rightarrow Cl_2O_7 + H_2O \quad \text{violent explosion} \tag{35.18}$$

$$8\ HI + 3\ H_2SO_4 \rightarrow H_2S + 4\ I_2 + 4\ H_2O \tag{35.19}$$

35.6 BEHAVIOR

Sulfuric acid is a heavy, viscous, water-soluble, and very corrosive liquid. Although sulfuric acid is not a volatile substance, fuming sulfuric acid (referred to as oleum) is. Since it does not evaporate extensively, it is hard to detect by smell. Sulfate aerosols and mist may form in the atmosphere.

35.6.1 Terrestrial Fate

When spilled on land, sulfuric acid will settle for a few minutes before flowing away as it finds the lowest levels. It will sorb to the soil and char any vegetation with which it comes in contact. Because of its high viscosity, concentrated amounts will not rapidly leach into the soil unless it rains or other precipitation occurs. The process of soil acidification involves replacing exchangeable base cations such as calcium and magnesium with protons and aluminum ions. In other words, removing bases and mobilizing aluminum are the key processes. As sulfuric acid penetrates the soil, some of it will be neutralized by bases and carbonates, some will react with silicates and organic materials, some will exchange with metal cations, and the rest may leach into groundwater.

35.6.2 Aquatic Fate

If spilled into water, sulfuric acid will mix convectively and dissolve in the water column as it sinks. Vigorous turbulent mixing will occur as the sulfuric acid front advances, resulting in entrainment. As the entrainment grows, the density differences will become smaller. Two kinds of acid-spill scenarios can occur: (1) those in which no appreciable amount of acid aerosols are produced, and (2) those in which acid aerosols form and provide the major hazard.

In both cases, large amounts of heat are liberated at the early stages of the spill. Examples of the first kind of spill are a capsized barge with all the hatches wide open and a gradual leak from a ship in a collision underwater. The second kind involves a spill of fuming sulfuric acid or oleum over water surfaces and shallow surface water. The heat released as a result of dilution provides latent heat, and the heat of vaporization produces acid mist of sulfur trioxide and water to form sulfuric acid. Sulfuric acid will also react with any bases and organic matter in the water. It has been shown that very low pH is extremely toxic to fish and aquatic organisms.

35.6.3 Study of a Sulfuric Acid Spill near Springhill, Nova Scotia

On December 13, 1978, a railway tank car of concentrated sulfuric acid (93%) was spilled as a result of a train derailment about 10 km northwest of the town of Springhill, Nova Scotia. The spill caused visible damage to vegetation over a limited area. In the fall of 1985, seven growing seasons after the spill, Environment Canada contracted MacLaren Plansearch Limited, in association with P. Lane and Associates Limited, to undertake a study to document the effects of the spill and make recommendations on appropriate response measures for future spills of sulfuric acid (Environment Canada, 1986). Some Environment Canada scientists were also involved in the research project. This spill and the results of the study are discussed in Section 35.8.2.

35.7 HUMAN AND ENVIRONMENTAL TOXICITY

35.7.1 Effects on Humans

The most common types of injuries to people during spills of concentrated sulfuric acid are burns as a result of direct contact with this acid. These burns often cause marked scarring of the skin. The concentrated form of sulfuric acid destroys organic matter as a result of its severe dehydrating action. It is also a severe irritant to the eyes, respiratory tract, and skin. The respiratory system and teeth are usually damaged as a result of chronic exposure to the acid aerosols and mists.

Accidental exposure to liquid fuming sulfuric acid can result in skin burns as well as pulmonary edema from inhalation. Pulmonary fibrosis, residual bronchitis, and pulmonary emphysema have also been reported. A single overexposure to sulfuric acid may lead to acute laryngeal, tracheobronchial, and pulmonary edema. Concentrations of around 5 mg/m^3 have been found quite objectionable, often causing coughs and respiratory dysfunctions. The data in Table 35.2 summarize human responses to various levels of concentration of sulfuric acid mist.

Investigations of respiratory effects of sulfuric acid during acute exposures, as in some spill cases, have had mixed results because there are many factors that influence its toxicity. These factors include particle size of the mist, humidity, presence of particulate, synergistic and protective agents, and preexisting conditions of victims (Amdur, 1989; Amdur and Chen, 1989; Linn et al., 1989; Lippmann, 1989; and Schlesinger et al., 1984). Bronchospasm in asthmatics has been shown to be a major concern. These effects are caused mainly by

TABLE 35.2 Human Responses to Sulfuric Acid Mist

Concentrations (mg/m^3)	Response
0.5 to 2.0	Barely noticeable irritation
3.0 to 4.0	Coughing, easily noticeable
6.0 to 8.0	Decidedly unpleasant, marked alterations in respiration

inhalation of the acid aerosols deposited on the surface of the respiratory tracts. The smallest aerosols (less than 10 μm) often cause the greatest alteration in pulmonary function because they can penetrate farther into the alveoli, whereas the larger particles are deposited in the upper respiratory tract. Increase in airway resistance at high acid concentrations has been demonstrated by several authors. Presence of ammonia in expired air has been reported to afford some protection in humans. Synergism has been demonstrated when sulfuric acid is deposited on zinc oxide dusts in the presence of sulfur dioxide, and also when combined with nitrogen dioxide or sulfur dioxide, ozone, metallic aerosols, and sulfuric acid.

Chronic exposure of the teeth to the corrosive action of sulfuric acid as in battery acid workers has resulted in etching or total loss of teeth substance. A number of studies have also shown some association between chronic exposure to sulfuric acid and laryngeal cancer (U.S. Department of Health, Education and Welfare, 1974). For example, a 13-fold excess risk of laryngeal cancer was found among chemical refinery workers with the highest exposure, and a 4-fold risk was found for those moderately exposed, as opposed to very low exposure. Repeated exposures to sulfuric acid mists have been reported to cause dermatitis, stomatitis, conjunctivitis, and tracheobronchitis.

35.7.2 Animals

A considerable body of evidence exists on the sensitivity of laboratory animals to sulfuric acid. As discussed in Section 35.7.1, inhalation of sulfuric acid causes changes in pulmonary flow resistance, which are sometimes irreversible. These changes could be seen as the first stage of bronchitis. The smaller the particles (<2 μm), the more damage is done. The results of some animal studies are shown in Table 35.3.

35.7.3 Aquatic Species

The main cause of death of species in acid lakes is the excessive loss of sodium ions, which cannot be rapidly replaced by active transport. The disruption of sodium/potassium pump mechanism has been attributed to the presence of high concentrations of hydrogen ions.

TABLE 35.3 Effects of Sulfuric Acid on Animals

Concentration ($\mu g/m^3$), time	Species	Effects
100, 1 hour	Guinea pig	Pulmonary resistance increased 47%, pulmonary compliance decreased 27%.
500, 1 hour	Dog	Slight increases in tracheal mucociliary transport velocities immediately and one day after exposure. One week later clearance was significantly decreased.
510, 1 hour	Guinea pig	Pulmonary resistance increased 60%. Pulmonary compliance decreased 33%.
1,000, 1 hour	Guinea pig	Pulmonary resistance increased 78%. Pulmonary compliance decreased 40%.
190 and 1,400, no time	Donkey	Bronchial mucociliary clearance was slowed.
1,000, 1 hour	Dog	Depression in tracheal mucociliary transport rate persisted one week after exposure.

TABLE 35.4 Effects of Acidity on Aquatic Organisms

pH	Effects
6	Crustaceans, molluscs, etc. disappear; white moss increases.
5.8	Salmon, char, trout, and roach die. Sensitive insects, phytoplankton, and zooplankton die.
5.5	Whitefish and grayling die.
5	Perch and pike die.
4.5	Eels and brook trout die.

In 1975, sulfuric acid was added to an experimental lake in Canada, causing the pH in the lake to drop from 6.8 to 5, which killed many fish species. Shrimp and minnows died at about pH 5.8, followed by the young trout. At pH 5.6, crayfish began to die as their exoskeletons lost their calcium and became infested with parasites. The sensitivities of aquatic organisms to the lowering of pH based on studies in Scandinavian lakes are summarized in Table 35.4.

35.7.4 Plants

Direct contact of concentrated sulfuric acid with plants will result in perforation of the plant tissue and the plant may subsequently die. The most common response of plants to acidic precipitation is low growth and the formation of foliar lesions or areas of dead tissue on the upper surface of the leaves. Necrotic spotting of the epidermis of the leaves after exposure to sulfuric acid mist has been reported in previous investigations. The effects of acid precipitation on plants are shown in Table 35.5.

35.8 SURVEY OF PAST SPILLS, LESSONS LEARNED, AND COUNTERMEASURES APPLIED

Sulfuric acid is the most frequently spilled high-volume hazardous chemical in industry. Unlike spills of other chemicals, some spills are large, especially those that involve rail and ship transport. Spills from pipes, bottles, and cylinders are often small. While very few deaths have been reported as a result of these spills, respiratory distress is common.

TABLE 35.5 Effects of Acid Precipitation on Plants

pH	Plant	Effects
2.5	Bean	Foliar aberrations, decrease in growth
3.1	Yellow birch	Foliar lesions, decrease in growth
3.1	Bean, sunflower	Foliar lesions
3.4	Hybrid poplar	Foliar lesions
3.4	Sunflower	Foliar lesions
4.04	Bean	Reduction in dry weight

35.8.1 Sulfuric Acid Spill at a CPR Train Derailment Near MacTier, Parry Sound, Ontario

On February 12, 1984, 14 railcars were fully derailed and another jumped the tracks, spilling sulfuric acid into Beers Lake, 29 km south of Parry Sound in Ontario. Each car contained 93 tons of sulfuric acid. Three cars went through the ice, 9 cars were on the bank leading to the lake, and 1 straddled the tracks. Four cars were leaking from the dome and 1 lost half its contents

On the night of the spill, a pH reading of 6.5 was measured at the outlet of the lake. The next day, tests were taken across the lake and down the center and pH levels ranged from 5.0 to 6.0, with one location at 4.5. Officials at the scene believed that most of the acid remained on the shore and in the water nearest the accident. Emergency crews worked through the night to contain the spill. One outlet was blocked with sandbags to prevent any flow of acidic water. Three cars were offloaded during the night. The Canadian Transport Commission investigated a broken wheel as a possible cause of the derailment. Sampling and pH measurements continued as officials from the Ontario Ministry of the Environment monitored the spill. Work crews from the Canadian Pacific Railways (CPR) were dispatched and 15 cars of limestone were sent from Toronto and two truckloads from Sudbury to neutralize the acid spill.

35.8.2 Sulfuric Acid Spill Near Springhill, Nova Scotia

On December 13, 1978, an eastbound CN freight train travelling at approximately 40 mph (64 km/h) derailed close to Springhill, Nova Scotia. A total of 51 cars were derailed. One railway tank car containing concentrated sulfuric acid (93%) ruptured and almost its entire contents spilled. The acid pooled in a swale parallel to the south side of the tracks and flowed downslope along three distinct paths (one major, two minor). Most of the spilled acid flowed into an underground pit or cavity of unknown origin. Because of the highly fractured nature of the bedrock in the area, most of the acid that reached the pit quickly probably found its way into the groundwater, although no adverse effects on groundwater were reported.

Overland flow of the acid was fairly restricted. Although the spill occurred at the top of a slope, eyewitnesses from Environment Canada (R. Simmons, L. Tripp) indicated that the three rivulets of acid flowed only partway down the slope. No acid reached the ditch on the north side of the highway at the bottom of the slope. The following factors may have contributed to this:

1. The amount of acid flowing overland was greatly reduced because most of it flowed into the underground cavity.
2. The surface materials on the slope were highly permeable.
3. Although it was early winter and the ground was lightly snow-covered, the ground was not yet frozen. This was confirmed by members of the spill response team, who indicated there was no frost in the ground when an emergency road was bulldozed to the site.

The Environment Canada report on the spill indicated that a CN employee and a local resident were injured as a result of stepping in a pool of acid (Environment Canada, 1986).

After the derailment, the priority was to restore service on the mainline. No action was taken to neutralize the spilled acid until a truckload of sodium hydroxide solution arrived on the site on December 14, 25 hours after the spill occurred. By the time the truckload of sodium hydroxide was pulled up to the trackside the following day, almost two days had passed since the spill. Neutralization was done selectively where pockets of acid could be reached. Sodium hydroxide was allowed to percolate through the new roadbed material to neutralize the acid underneath. Sgnificant amounts of sodium hydroxide were pumped into the cavity where much of the acid had flowed. (A solid block of sodium sulfate, the product of neutralizing sulfuric acid with sodium hydroxide, remains in the pit to this day.)

The top of the slope where the spill occurred was severely impacted by physical disturbance associated with response to the spill. All of the natural vegetation was bulldozed and the site covered with up to 90 cm of sandy fill. Much of this fill remained almost barren of vegetation, which is the major long-term disturbance associated with the spill.

The direct effects of the spill itself are relatively subdued after seven growing seasons. The ground vegetation in acidic areas is superficially indistinguishable from unaffected areas. Shrubs are reinvading areas where acid-induced shrub mortality was high. The main signs of the spill are the standing remains of dead shrubs. It is surprising that even after seven years all the spilled acid has not leached through. This clearly shows that the hydrogen ions are held in place by more than simple adsorption forces. Some effects on exchangeable soil minerals were also studied.

35.8.3 Sulfuric Acid Leak

In October, 1993, a 2.5-cm transfer line suddenly failed as workers were transferring sulfuric acid from a 3,785-L storage tank to a 378-L day tank (*Loss Prevention*, 1993). The leak caused an 18-m spray from where the leak occurred. A worker was sprayed by the acid mist, which penetrated his clothing and caused second-degree burns on his back. The victim was washed down in a shower and later taken to a medical facility for treatment.

The bulk storage tank, which contained a 93% sulfuric acid solution, was connected to two tanks, the day tank and an acid regeneration tank (also called the dilute tank), by 2.5-cm carbon steel lines. The valve in the line connecting the bulk storage tank and the dilute tank was located near the dilute tank. The acid spray was caused by a failure in the line where it was connected to the valve. The procedure for transferring acid from the bulk tank to the day tank required that the valve at the dilute tank be closed and a transfer pump be used to facilitate the transfer of acid from the bulk tank to the day tank. When the accident occurred, the valve at the dilute tank was closed and the transfer pump had been started. The pump built up pressure in the pipe, causing the acid to spray out.

The failed line was constructed of carbon steel and appeared to be a Schedule 40 pipe, although the engineering drawings specified use of Schedule 160 pipe, which is about twice as thick. In addition, it was known that the flow of acid through the line normally reduces the thickness of the pipe wall by about 5 μm per year. The section of the line that failed was approximately 10 years old. As soon as the leak was discovered and the transfer pump shut down, the area was barricaded and thoroughly washed. All piping was subsequently inspected using nondestructive evaluation (NDE) techniques, and pipes of insufficient thickness were replaced.

Lessons Learned

The following lessons were learned from this incident:

1. When system components are replaced or repaired, all engineering documents must be checked to ensure that the correct materials are used.
2. A preventative maintenance program, including a replacement schedule, must be in place. Management must ensure that all hazardous materials and processes are identified and that procedures are developed and implemented to ensure safety.
3. Implementation of relevant standards related to mechanical integrity, procedures, and training would have prevented the use of incorrect schedule piping.

35.8.4 Sulfuric Acid Spill in Richmond, California

In July 1993, 20 to 50 tons of fuming sulfuric acid spilled at the General Chemical Corp. plant in Richmond, California, a major industrial center near San Francisco. The release occurred when oleum was being loaded into a nonfuming acid railroad tank car that contained only a rupture disk. The tank car was overheated and this rupture disk blew. The resulting cloud of sulfuric acid drifted northeast with prevailing winds over a number of populated areas. More than 3,000 people subsequently sought medical attention for burning eyes, coughing, headaches, and nausea. Almost all were treated and released on the day of the spill. By the day after the release, another 5,000 people had sought medical attention. The spill forced the closure of five freeways in the region as well as some Bay Area Rapid Transit System stations. Health officials stated that most of the people affected would suffer no long-term effects from the exposure and that environmental effects would be minimal.

Lesson Learned

Again, a sound preventative maintenance program is crucial to eliminate such accidental spills.

35.9 CONCLUSIONS

Sulfuric acid is an important substance that is widely used in industry. The effects of sulfuric acid spills are very localized, and evidence suggests that sulfuric acid and oleum do not constitute a serious threat to the Canadian public at large.

35.10 REFERENCES

Amdur, M. O. 1989. "Sulfuric Acid: The Animals Tried to Tell Us," *Applied Industrial Hygiene,* vol. 4, pp. 189–197.

Amdur, M. O., and L. C. Chen. 1989. "Furnace Generated Acid Aerosols: Speciation and Pulmonary Effects," *Environmental Health Perspectives,* vol. 79: pp. 147–150.

American Industrial Hygiene Association (AIHA). 1988. *Emergency Response Planning Guidelines,* AIHA, Fairfax, VA.

Bailar, J. C., H. J. Emeleus, R. Nyholm, and A. F. Trotman-Dickenson, eds. 1975. "Halides," in *Comprehensive lnorganic Chemistry*, Pergamon Press, New York, NY, pp. 1124–1280.

Budavari, S. 1989. *The Merck Index: An Encyclopedia of Chemicals, Drugs, and Biologicals,* 11th ed. Merck & Co., Rahway, NJ.

Camford Information Services, Inc. (CIS). 1997. "Sulfuric Acid," in *CPI Product Profiles*, CPI, Scarborough, ON.

Duecker, W. W., and J. R. West, eds. 1959. *The Manufacture of Sulfuric Acid,* Robert E. Krieger, Huntington, NY, pp. 1–134.

Environment Canada. 1986. *Follow-up Study of Sulfuric Acid Spill Near Springhill,* Nova Scotia.

Environment Canada. 2000. *National Air Pollution Surveillance (NAPS) Network—Annual Summary for 1998,* Report EPS 7/AP/31, Environmental Protection Service, Ottawa, ON.

Fingas, M., N. Laroche, G. Sergy, B. Mansfield, G. Cloutier, and P. Mazerolle. 1991. "A New Chemical Spill Priority List," in *Proceedings of the 8th Technical Seminar on Chemical Spills,* Environment Canada, Ottawa, ON.

Inco Limited. 1985. *Sulfur Dioxide Abatement Project, Final Report-December 1988,* Sudbury Smelter Complex, SO_2 Emission Control Regulation 660/85, December 12.

Kirk-Othmer Encyclopedia of Chemical Technology. 1983. 3rd ed., vol. 22, Wiley-Interscience, New York, NY.

Linn, W. S., E. L. Avol, K. R. Anderson, D. A. Shamoo, R. C. Peng, and J. D. Hackney. 1989. "Effect of Droplet Size on Respiratory Responses to Inhaled Sulfuric Acid in Normal and Asthmatic Volunteers," *American Review of Respiratory Disease,* vol. 140, pp. 161–166.

Lippmann, M. 1989. "Background on Health Effects of Acid Aerosols," *Environmental Health Perspectives,* vol. 79, pp. 3–6.

Loss Prevention. 1993. "Acid Line Breaks—Worker Burns," p. 4.

Muller, T. L. 1997. "Sulfuric Acid and Sulfur Trioxide," in *Encyclopedia of Chemical Technology,* 4th ed., ed. M. Howe-Grant, vol. 23, John Wiley & Sons, New York, NY, pp. 363-407.

RRETCS On-Line. 1999. *Registry of Toxic Effects of Chemical Substances*, Department of Health and Human Services, Centers for Disease Control, National Institute for Occupational Safety and Health, Washington, DC.

Schlesinger, R. B., L.-C. Chen, and K. E. Driscoll. 1984. "Exposure Response Relationship of Bronchial Mucociliary Clearance in Rabbits, following Acute Inhalations of Sulfuric Acid Mist," *Toxicology Letters,* vol. 22, pp. 249–254.

U.S. Department of Health, Education and Welfare. 1974. *Occupational Exposure to Sulfuric Acid, Criteria for a Recommended Standard.*

CHAPTER 36
PERSPECTIVES ON SPECIFIC SUBSTANCES: POLYCHLORINATED BIPHENYLS (PCBs)

Richard Lawuyi and Merv Fingas
Emergencies Science Division, Environment Canada, Environmental Technology Centre, River Road, Ottawa, Ontario

36.1 OVERVIEW OF PRODUCT AND INDUSTRIAL USES

Even though polychlorinated biphenyls (PCBs) are no longer imported into Canada or manufactured in many parts of the globe, large quantities are still being used in transformers or kept in storage facilities for future disposal. First developed and synthesized in 1881 and introduced commercially in 1929, PCBs were manufactured in response to the North American electrical industry's urgent need for a more stable and effective transformer and capacitor fluid. Unfortunately, their chemical and physical stability have also led to widespread and persistent environmental contamination. As a result, the further import and use of PCBs were banned in Canada in 1977.

Because efficient and cheap destruction technologies were lacking in the late 1970s and early 1980s, containment was the obvious effective means of control. PCBs and PCB-contaminated materials were kept in temporary warehouses and special storage facilities. In Canada, as of December 1993, there were 11,505 tons (net weight) of askarel (commercial PCB mixtures used for transformers) still in use in transformers and capacitors, 15,247 tons (gross weight) of waste askarel and askarel equipment, 2,161 tons (net weight) of in-use PCB-contaminated mineral oil, 3,787 tons (net weight) of waste PCB-contaminated mineral oil, and 107,991 tons (gross weight) of other PCB waste, including soil, fluorescent lamp ballasts, and drained equipment (Environment Canada, 1995).

There were more than 150 spills and fires involving PCBs in 1999, some of which could have been prevented. Of these, 41 could be considered major. PCBs enter the environment through leaks in piping systems, tanks, and storage containers and during routine disposals, manufacturing, and fires. The danger with PCB fires is the formation of more toxic substances such as polychlorinated dibenzodioxins and polychlorinated dibenzofurans. Two case histories of incidents involving PCBs will be outlined.

36.2 INTRODUCTION

The introduction of PCBs for use in electric transformers and capacitors in 1929 represented a major breakthrough in the history and technology of dielectric fluids (Durfee et al., 1976). Essentially, PCBs were used as a fire safety measure by virtue of their good dielectric characteristics, chemical and physical stability, and inertness. Unfortunately, these qualities have also contributed to their persistence and bioaccumulation in the environment (Jensen, 1986; Holmes et al., 1967; Risebrough et al., 1969).

As a group, PCBs may contain from 1 to 10 chlorine atoms attached to the biphenyl rings in 209 different ways or 209 congeners. As the number of chlorine atoms increases, so does the degree of bioaccumulation and persistence.

Pure PCBs are white crystalline solids, but PCBs are often liquid because of impurities. Because of the need for this dielectric fluid to transfer heat, PCBs were often diluted with chlorinated benzenes to yield askarels, which are less viscous (Durfee et al., 1976; Miller, 1982). While PCBs are no longer manufactured, they continue to be released into the environment.

36.2.1 Spill Profile

PCBs have become identified in the public mind as one of the most hazardous compounds on earth. One possible reason for this is that they are ubiquitous and man-made, and aside from two incidents in Japan (Yusho and Yucheng), not much is known about the long-term effects of low-level exposure on humans. Of all the various toxic effects demonstrated in animals, only one or two have been clearly identified in humans.

Because of several government and industry initiatives and increased public awareness, the frequency of PCB spills has decreased dramatically. While the Canadian government banned further production and import of PCBs into Canada in 1977, there are still large quantities of PCBs in Canada. As of December 1993, there were 11,505 tons (net weight) of askarel still in use in transformers and capacitors; 15,247 tons (gross weight) of waste askarel and askarel equipment, 2,161 tons (net weight) of in-use PCB-contaminated mineral oil; 3,787 tons (net weight) of waste PCB-contaminated mineral oil; and 107,991 tons (gross weight) of other PCB waste, including soil, fluorescent lamp ballasts, and drained equipment (Environment Canada, 1995).

There were about 41 major PCB spills in 1999, some of which could have been prevented (Environment Canada, 2000). The annual spill frequency of PCBs from 1992 to 1995 is shown in Fig. 36.1. Minor spills are not included in the figure.

Most spills occur at stationary situations, from leaking pipes, transformers, and capacitors, and during retrofitting. Such electrical equipment may occasionally fail, resulting in fluid release. Spills can also occur as a result of accidental damage from vehicle accidents and human error (Environment Canada, 2000). Some spills have also been reported during normal maintenance of large tranformers and other equipment involving dielectric fluid such as retrofilling, bushing repair, and topping up.Yet others occur through unlawful disposal.

As shown in Table 36.1, PCBs are ranked as fifth on Environment Canada's priority list of hazardous chemicals (Fingas et al., 1991). The primary objective of the list was to determine the minimum number of hazardous chemicals that were most frequently spilled. The list was developed by a simple ranking of reported spill frequency; supply volumes; historical spill volumes; and toxicity data, stability, accumulation, and persistence.

36.3 PHYSICAL AND CHEMICAL PROPERTIES AND GUIDELINES

Polychlorinated biphenyls (PCBs) are light, straw-colored liquids with the usual chlorinated aromatic odors. The trichlorinated biphenyls are mobile liquids, while the pentachlorinated

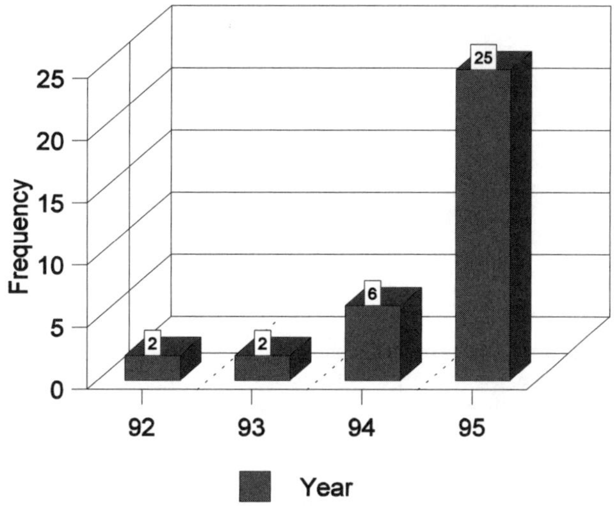

FIGURE 36.1 Annual spill frequency of PCBs (1992–1995).

biphenyls are viscous. The most commonly used are the trichlorobiphenyls, containing 42% chlorine, and the pentachlorobiphenyls, containing 54% chlorine

Two hundred and nine possible isomers can be obtained by substituting chlorine for the hydrogen on the aromatic rings. From 40 to 70 different chlorinated derivatives can be present in the higher-chlorinated commercial mixtures. For instance, arochlor 1254 contains 69 different PCB molecules differing in number and position of chlorine atoms, as well as

TABLE 36.1 Priority List Ranking of PCBs

Chemical	Ranking	Spill number	Spill volume	Supply volume
Ammonia	1	107	470	3,700
Chlorine	2	36	120	1,700
Tetraethyl lead	3	4	72	26
Styrene	4	24	5,000	630
PCBs	**5**	**334**	**89**	—
Sulfuric acid	6	155	13,000	3,700
Sodium cyanide	7	3	83	12
Hydrochloric acid	8	123	3300	170
Potassium chloride	9	31	12,000	—
Pentachlorophenol	10	19	110	1.5
Phenol	11	10	14	68
Zinc sulfate	12	3	68	1,500
Phosphorus	13	16	46	68
Toluene	14	13	110	430

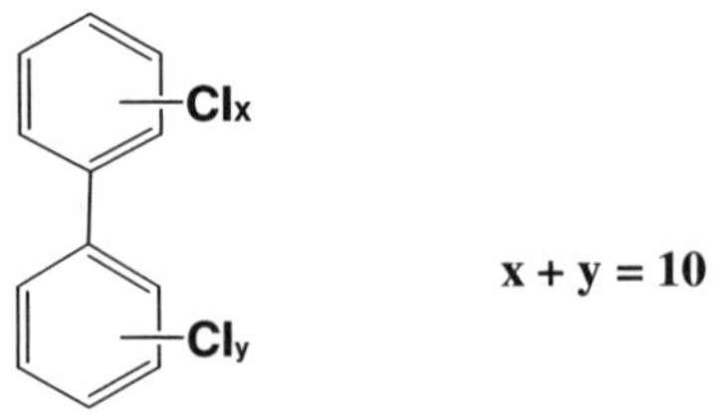

FIGURE 36.2 Chemical structure of PCB.

other compounds such as chlorinated naphthalenes, dioxins, and dibenzofurans. PCBs are very stable, persistent, thermoplastic, and nonflammable. The chemical structure of PCBs is shown in Figure 36.2.

Molecular formula: $C_{12}H_{10\text{-}X}Cl_X$
Molecular weight: Varies with number of chlorine atoms
UN number: Each congener has a number

Aroclor	CAS no.	RTECS no.
1242 (42% Cl)	53469-21-9	1356000
1248 (48% Cl)	12672-29-6	1358000
1254 (54% Cl)	11097-69-1	1360000
1260 (60% Cl)	11096-82-5	1362000
1262 (62% Cl)	37324-23-5	1364000
1268 (68% Cl)	11100-14-4	1366000

Synonyms and trade names (RTECS On-Line, 2000):

Askarels
PCBs
Chlorodiphenyls
Aroclors
Kanechlors
Pyranol
Pyroclor
Phenochlor
Pyralene
Clophen
Elaol
Santotherm
Fenchlor
Apirolio
Sovol

Grades and purities: PCBs are usually manufactured as mixtures of aroclors, chlorobenzenes, and oxygen-scavenging additives. Different aroclors are characterized by the composition of the isomer groups they contain. Other impurities are chlorinated naphthalenes, dibenzofurans, and dioxins.

36.3.1 Physical Data (NRC, 1979; WHO, 1976):

PCB isomer group	Physical state	Vapor pressure
Monochlorobiphenyl	Solid/liquid	1.1 Pa at 25°C
Dichlorobiphenyl	Solid/liquid	0.24
Trichlorobiphenyl	Solid/liquid	0.054
Tetrachlorobiphenyl	Solid/liquid	0.012
Pentachlorobiphenyl	Solid/liquid	2.6×10^{-3}
Hexachlorobiphenyl	Solid/liquid	5.8×10^{-4}
Heptachlorobiphenyl	Solid/liquid	1.3×10^{-4}
Octachlorobiphenyl	Solid/liquid	2.8×10^{-5}
Nonachlorobiphenyl	Solid/liquid	6.3×10^{-6}
Decachlorobiphenyl	Solid	1.4×10^{-6}

PCB isomer group	Melting point (°C)	Boiling point (°C)
Monochlorobiphenyl	25 to 77.9	285
Dichlorobiphenyl	24.4 to 149	312
Trichlorobiphenyl	28 to 87	337
Tetrachlorobiphenyl	47 to 180	360
Pentachlorobiphenyl	76.5 to 124	381
Hexachlorobiphenyl	77 to 150	400
Heptachlorobiphenyl	122.4 to 149	417
Octachlorobiphenyl	159 to 162	432
Nonachlorobiphenyl	182.8 to 206	445
Decachlorobiphenyl	305.9	456

PCB isomer group	Specific gravity	Viscosity (mPa.s)
Monochlorobiphenyl	1.1	20
Dichlorobiphenyl	1.3	28
Trichlorobiphenyl	1.4	56
Tetrachlorobiphenyl	1.5	200
Pentachlorobiphenyl	1.5	1.5×10^{3}
Hexachlorobiphenyl	1.6	2.9×10^{4}
Heptachlorobiphenyl	1.7	$>10^{6}$
Octachlorobiphenyl	1.7	$>10^{6}$
Nonachlorobiphenyl	1.8	$>10^{6}$
Decachlorobiphenyl	1.8	$>10^{6}$

PCB isomer group	Molecular weight	Water solubility
Monochlorobiphenyl	188.7 (g/mol)	4.0 (g/m³ @ 25°C)
Dichlorobiphenyl	223.1	1.6
Trichlorobiphenyl	257.5	0.65
Tetrachlorobiphenyl	292.0	0.26
Pentachlorobiphenyl	326.4	0.099
Hexachlorobiphenyl	360.9	0.038
Heptachlorobiphenyl	395.3	0.014
Octachlorobiphenyl	429.8	5.5×10^{-3}
Nonachlorobiphenyl	464.2	2.0×10^{-3}
Decachlorobiphenyl	498.7	7.6×10^{-4}

PCB isomer group	LOG K_{ow}	Bioconcentration factor (fish)
Monochlorobiphenyl	4.7	2,500
Dichlorobiphenyl	5.1	6,300
Trichlorobiphenyl	5.5	1.6×10^4
Tetrachlorobiphenyl	5.9	4.0×10^4
Pentachlorobiphenyl	6.3	1.0×10^5
Hexachlorobiphenyl	6.7	2.5×10^5
Heptachlorobiphenyl	7.1	6.3×10^5
Octachlorobiphenyl	7.5	1.6×10^6
Nonachlorobiphenyl	7.9	4.0×10^6
Decachlorobiphenyl	8.3	1.0×10^7

PCB isomer group	Evaporation rate ($g/m^2 \cdot h$ @ 25°C) (approximately)
Monochlorobiphenyl	0.25
Dichlorobiphenyl	0.065
Trichlorobiphenyl	0.017
Tetrachlorobiphenyl	4.2×10^{-3}
Pentachlorobiphenyl	1.0×10^{-3}
Hexachlorobiphenyl	2.5×10^{-4}
Heptachlorobiphenyl	6.2×10^{-5}
Octachlorobiphenyl	1.5×10^{-5}
Nonachlorobiphenyl	3.5×10^{-6}
Decachlorobiphenyl	8.5×10^{-7}

Average Composition of Some Aroclors. Aroclors are essentially mixtures of different isomer groups. Some composition data are shown in Tables 36.2 to 36.5.

Askarels. Askarels are mixtures of many isomer groups of PCBs and chlorobenzenes and may also contain phenoxypropene oxide or diepoxide. They are mainly used as transformer oils and are of many types, namely, A, B, C, D, E, F, and G. Physical properties of some askarels are shown in Tables 36.6 and 36.7.

TABLE 36.2 Composition of Aroclors

PCB isomer group	1221	1232	1016	1242	1248	1254	1260
Biphenyl	10						
Monochlorobiphenyl	50	26	2	3			
Dichlorobiphenyl	35	29	19	13	2		
Trichlorobiphenyl	4	24	57	28	18		
Tetrachlorobiphenyl	1	15	22	30	40	11	
Pentachlorobiphenyl				22	36	49	12
Hexachlorobiphenyl				4	4	34	38
Heptachlorobiphenyl						6	41
Octachlorobiphenyl							8
Nonachlorobiphenyl							1
Decachlorobiphenyl							

TABLE 36.3 Physical Properties of Some Aroclors

Property	1016	1242	1248	1254	1260
Appearance	Clear, mobile oil	Clear, mobile oil	Clear, mobile oil	Light yellow, viscous liquid	Light yellow, soft sticky resin
Physical state at 25°C, 1 atm	Liquid	Liquid	Liquid	Liquid	Solid
Distillation range (°C)	323–356	325–366	340–375	365–390	385–420
Pour point (°C)	−14	−19	−7	10	31
Vapor pressure at 25°C (Pa)	0.10	0.091	0.023	6.7×10^{-3}	6.4×10^{-4}
Viscosity at 25°C (mPa.s) (Saybolt Universal s)	45 200	69 270	280 1.0×10^{3}	2.0×10^{3} 6.7×10^{3}	1.9×10^{5} 6.0×10^{5}
Specific gravity at 25°C (approx.)	1.4	1.4	1.4	1.5	1.6
Fluid density at 25°C (kg/m^3)	1,370 at 20°C	1,383	1,445	1,539	1,621

Immediate Concerns

Hazard: Readily absorbed into the body by all routes of exposure. May cause irritation to nose, throat, and lungs.

Humans: May produce irritation to nose, throat, and lungs. Inhalation may result in all symptoms of long-term exposure. Skin absorption may produce a rash after a few days of exposure.

Environment: Environmentally hazardous substance. Very persistent substance. Will bioaccumulate and biomagnify in the food chain.

Protection: Wear self-contained breathing apparatus (SCBA) and appropriate clothing impermeable to PCBs.

TABLE 36.4 Physical Properties of Some Aroclors

Property	1242	1248	1254	1260
Average molecular weight (g/mol)	261	297	327	375
Solubility in water at 25°C (g/m^3)	0.75	0.32	0.14	0.035
Log K_{ow}	4.5 to 5.8	5.8 to 6.3	6.1 to 6.8	6.3 to 7.5
Bioconcentration factor	1.6×10^{3}–3.2×10^{4}	3.2×10^{4}–1.0×10^{5}	6.3×10^{4}–3.2×10^{5}	1.0×10^{5}–1.6×10^{6}
Evaporation rate at 25°C (g/m^2/h)	0.029	8.3×10^{-3}	2.7×10^{-3}	2.9×10^{-4}
Dielectric constant at 25°C	5.8	5.6	5.0	4.3
Dielectric strength (kV)	>35	>35	>35	>35
Volume resistivity at 100°C, 500 V D.C. (ohm-cm)	$>500 \times 10^{9}$	$>500 \times 10^{9}$	$>500 \times 10^{9}$	$>500 \times 10^{9}$
Power factor at 100°C (%)	<0.1	<0.1	<0.1	<0.1

TABLE 36.5 Other Physical Properties of Some Aroclors

Property	1242	1248	1254	1260
Flash point (COC) (°C)	176 to 180	193 to 196	>boiling point	>boiling point
Fire point (COC) (°C)	>boiling point	>boiling point	>boiling point	>boiling point
Refractive index at 20°C	1.628	1.631	1.640	1.648
Coefficient of thermal expansion ($cm^3/cm^3/°C$)	0.00068 (25 to 65°C)	0.00070 (25 to 65°C)	0.00066 (25 to 65°C)	0.00067 (20 to 100°C)
Maximum moisture content (ppm)	50	50	50	50

36.3.2 Countermeasures

Spills on land:

Clean up PCB spills immediately to prevent further contamination.
Restrict persons not wearing protective equipment from spill area.
Ventilate the area of spill or leak.
Absorb liquids in vermiculate, dry sand, earth, or a similar material and deposit in sealed containers. A temporary holding area would consist of a truck box, a dumpster with impermeable lining, or an area on the ground with soil berms or dykes with appropriate impermeable lining.
Store in a cool, well-ventilated area away from strong oxidizers.

Spills on water:

Spills on water require immediate action to minimize widespread contamination.
PCBs less dense than water, such as PCB-contaminated oils, will float on the water surface and require usual oil-recovery techniques while those that are heavier will sink rapidly in water and will require dredging of sediments.

36.4 INDUSTRIAL ASPECTS AND PRODUCTION IN THE UNITED STATES, CANADA, AND WORLDWIDE

36.4.1 Manufacture of Polychlorinated Biphenyls

PCBs are no longer manufactured anywhere in the world, but large quantities still remain in transformers and in storage in contaminated form. In Canada, the import of PCBs was prohibited in 1977. The following are highlights of the national inventory as of December 1993 (Environment Canada, 1995).

TABLE 36.6 Physical Properties of Some Askarels

Property	Type A	Type B	Type C	Type D
Average molecular weight	258.6			263.4
Water solubility at 25°C (g/m^3)	0.016			0.056
Evaporation rate at 25°C ($g/m^2/h$)	7.8×10^{-5}			4.8×10^{-4}
Dielectric constant at 100°C, 1 kHz	3.7–4.0	3.8–4.3	4.4–4.9	4.3–4.6
Dielectric breakdown voltage, min. (kV)	35	35	35	35
Resistivity at 100°C (ohm-cm)	100×10^{-9}	100×10^{-9}	100×10^{-9}	100×10^{-9}

TABLE 36.7 Physical Properties of Some Askarels

Property	A	B	C	D
Appearance	Clear, mobile oil	Clear, mobile oil	Clear, mobile oil	Clear, mobile oil
Physical state at 25°C, 1 atm	Liquid	Liquid	Liquid	Liquid
Pour point, max. (°C)	−32	−44	−30	−30
Vapor pressure at 25°C (Pa)	2.5×10^{-4}			1.5×10^{-3}
Kinematic viscosity at 37.8°C (Saybolt universal s)	54	43	53	59
Absolute viscocity at 25°C (mPa.s)	31			33
Specific gravity at 15.5°C	1.564	1.566	1.419	1.523
Density at 25°C (kg/m³)	1563			1524
Fire point (COC) °C	>boiling point	>boiling point	>boiling point	>boiling point

- There were 11,505 tons of askarel in use, 9,649 tons of which were in transformers, 1,697 tons in capacitors, and 159 tons in other equipment.
- There were 15,247 tons of waste askarel and askarel equipment, 6,131 tons of which were in transformers, 2,576 tons in bulk storage, and 125 tons in other equipment.
- There were 2,161 tons of in-use PCB-contaminated mineral oil, 2,083 tons of which were in transformers and 77 tons in other equipment.
- There were 3,787 tons of waste PCB-contaminated mineral oil, 418 tons of which were in transformers, 3,364 tons in bulk storage, and 5 tons in other equipment.
- There were 107,991 tons of other PCB wastes, consisting of 95,718 tons of soil, 6,328 tons of fluorescent lamp ballasts, 1,581 tons of drained equipment, and 4,364 tons of other wastes.
- In the United States, more than 340,000 tons of PCBs were in service in over 900 million pieces of equipment in the mid-1970s.

36.5 CHEMISTRY

There are two main reactions of PCBs that significantly affect the environment. These are the stripping off of the chlorine atom/atoms by sodium metal to produce biphenyl and sodium chloride (Fig. 36.3) and the formation of polychlorinated dibenzofurans and polychlorinated dibenzodiozins at elevated temperatures, such as during fires (Fig. 36.4).

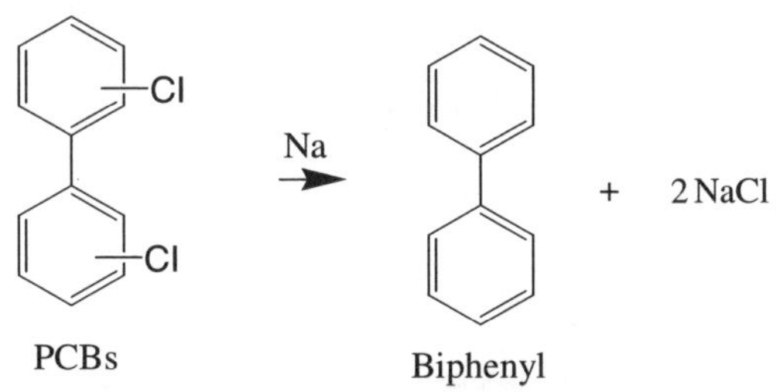

FIGURE 36.3 Formation of biphenyl and sodium chloride.

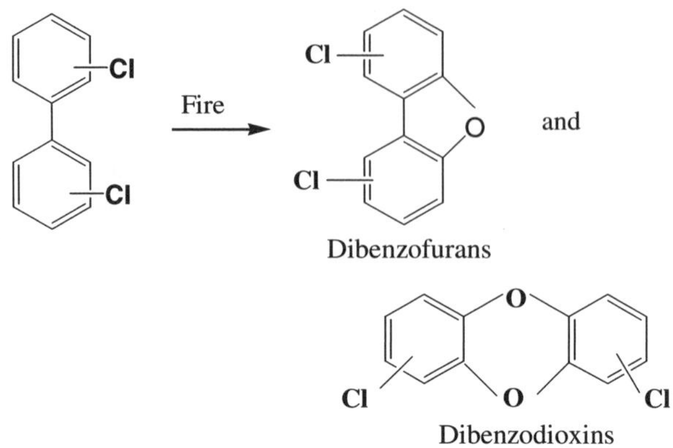

FIGURE 36.4 Formation of dibenzofurans and dibenzodioxins.

36.6 ENVIRONMENTAL FATE AND EFFECTS

36.6.1 Environmental Fate and Transport

The predominating factors in terms of the environmental fate of a chemical when spilled are its physicochemical properties as well as the location of the spill and weather conditions in the area. PCBs are transported by air, water, insects, fish, birds, and humans and are deposited by rain, snow, and dry fallout. Polychlorinated biphenyls have been detected in all corners of the globe and in almost every medium, including soil, air, sediment, water, human milk, fish, seagulls, meat, and even polar bears.

While very low concentrations of PCBs have been found in the air and water, extremely high levels have been reported in soils, sediments, and aquatic biota. This is not surprising given the persistence of these chemicals and the various forms of transport involved. The ultimate sink for PCB spills is the sediments. Various types of models have been used to predict the environmental long-term fate of spilled chemicals. In fate and transport studies, the following types of chemical releases are possible: volatilization, run-off, leaching to groundwater, and fugitive dust emissions.

Volatilization. Even though the vapor pressure of PCBs is very low, small amounts of spilled PCBs will slowly evaporate into the overlying air (evaporation rate is about 4.8×10^{-4} to 7.8×10^{-5} g/m^2/h) (NRC, 1979). This can happen for spills on both water and soils. Calculations of average mass losses through evaporation could be computed for many spills. As for PCB mixtures that contain chlorobenzenes and mineral oils, the lighter components consisting mainly of chlorobenzenes and mineral oils will evaporate first, leaving behind the thicker PCBs. The evaporation rate is dependent on the wind speed and temperature. However, the net result of volatilization is the removal or transference of PCBs from one environmental compartment to another without any net loss from the environment.

Runoffs. In the case of spills that occur near coastlines or water bodies, PCBs can be transported to receiving surface waters by runoff of surface soil from the spill site. The amount of runoff will depend on the properties of the surface soil, the location, the slope or proximity to the shoreline, and the nature of vegetation. The mass of PCBs that is lost from a site has been roughly estimated from the average concentration in surface soil, the amount of runoff, and soil erosion.

Leaching to Groundwater. Because PCBs are highly adsorptive and relatively immiscible in soil, small spills (about 1 to 2 L) are not likely to leach into groundwater. As for larger spills, it is possible that the PCBs will migrate through the soil to shallow groundwater. Soil penetration is enhanced by rainfall that infiltrates through the soil downward to the water table. Some leaching can also occur through water movement as a result of partitioning of PCBs, but this is not considered to be a major mechanism, as the partitioning coefficient of PCBs to water is very low.

Fugitive Dust Emissions. Fine dust contaminated with PCBs can be stirred up at spill and station sites during heavy construction and high winds. Because particles of smaller sizes travel much farther than those of larger sizes, it is not surprising to find PCBs in the Arctic as well as the Antarctic. PCB dust emissions are transported by air, water, human, animals, fish, birds, and microorganisms. Precipitation from air can also occur by rain, snow, vapor condensation, and dry fallout.

36.6.2 Destruction and Degradation

Abiotic Degradation. Incineration is the most common method of destroying PCBs. This method is carefully regulated to ensure that PCBs are completely destroyed since incomplete destruction can lead to more toxic products such as polychlorinated dibenzofurans and dioxins, which are often detected during fires, use, or aging of askarels. The destruction efficiency for incinerators has been specified as 99.9999% for emissions of all PCBs. This efficiency can be achieved with high heat, long residence time, and various pollution control systems. High efficiency boilers have also been approved for combustion of PCBs (Hunt et al., 1984).

Abiotic nonthermal methods include chemical processes such as reaction with molten sodium, sodium naphthalide, sodium salt in amine, catalytic dechlorination, wet air oxidation, ozonation, and physical methods including adsorption, microwave plasma, and photolysis.

Photolysis is another possible route of environmental breakdown of PCBs. The process appears to occur by C–Cl bond cleavage to produce biphenyl-free radical species, which then abstract hydrogens. Photolytic degradation depends on the degree of chlorination, position of chlorine substitution in the ring, and the solvent used for the PCB dissolution. For example, while the half-life of monochlorobiphenyl ranges from 0.62 to 1.4 days, that of pentachlorobiphenyl is over 67 days. It has been shown that the photolysis of the o-chlorine is probably due to steric hindrance to the preferred excited state geometry. Therefore, the more highly chlorinated commercial PCBs often photolyze preferentially in environmental samples. Bunce et al. (1978) reported that in shallow waters, at least one chlorine atom from highly chlorinated PCBs is photolyzed annually.

Biodegradation. The only form of biodegradation is microbial degradation and this includes different bacteria methods, activated sludge, and trickling filters. Microbial degradation has been shown to depend on the degree and position of chlorination (Moolenaar, 1983). For example, the half-life of monochloro biphenyls has been shown to be about 2 to 3 days and over 3,000 days for the pentachloro biphenyls. The smaller the number of chlorine atoms, the higher the efficiency of degradation (Moolenaar, 1983). The substitution of chlorine also affects the extent of dechlorination; it has been reported that ortho substitution markedly decreases the rate of degradation. Typical end products found were hydroxychlorinated biphenyls and chlorobenzoic acid.

36.6.3 Bioaccumulation, Bioconcentration, and Biomagnification

Bioaccumulation. The hydrophobic and lipophilic nature of PCBs is a strong indication of their tendency to partition into the fatty tissues of animals. Their stability or persistence is an additional factor that contributes to their accumulation since they cannot be readily degraded by enzymes. It has also been shown that species with high lipid contents tend to accumulate relatively large concentrations of PCBs. PCBs exhibit bioaccumulation tendencies very similar to DDT and some chlorinated pesticides, and the higher the degree of chlorination, the more readily it will be accumulated (Hansen et al., 1983). Other factors that affect the bioaccumulation of PCBs are:

1. Duration of exposure
2. Temperature
3. Solubility
4. Species age
5. Concentration of PCBs
6. Adsorption
7. Lipid content

Bioconcentration. Uptake of PCBs from surrounding water and sediments has been demonstrated for many aquatic animals, such as young oysters, clams, and worms.

Biomagnification. Several authors have reported that PCB concentration in the components of the food web shows a stepwise increase at each trophic level (NRC, 1979). PCBs are accumulated by lipid-water partitioning via the food chain. It is possible that there are different PCB-uptake mechanisms at different trophic levels.

36.7 BEHAVIOR

36.7.1 Behavior and Environmental Fate during a Spill

There are no known natural sources of PCBs. Sources are mainly anthropogenic. PCBs were prepared commercially by the chlorination of biphenyl with anhydrous chlorine in the presence of catalysts, followed by some purification. Usually, the longer the purification step, the more chlorine substitution takes place on the rings.

"Behavior" includes all the visible characteristics displayed by a chemical. PCB is a viscous to mobile, colorless to brown liquid that slowly spreads when spilled, depending on the ambient temperature and exact formulation. Indoor spills usually occur in small quantities on concrete, and the vapor will adsorb onto walls, carpeting, drapery, and metal objects. All forms of exposure should be avoided and the airborne concentration carefully monitored. On the other hand, outdoor spills can range from a few drops from transformers to full trailer tanks and can take place on either land or water.

From an environmental impact standpoint, the four most important physical properties of PCBs are very low water solubility and vapor pressure, high octanol/water partition coefficients, and stability or persistence (NRC, 1979). The five mechanisms by which PCBs can be transported in the soil are as a dissolved material in the water; by sorption; as an emulsion with water; as an immiscible oily liquid phase; and as a discrete fluid.

PCB fluids are generally insoluble in water and, when spilled on the soil, usually settle first as a pool on the ground surface before infiltration and downward movement through multilayered soil, water, and soil gases. The liquid fills the pores at the soil surface and begins to penetrate downward. Some of the lighter components will evaporate. The downward transport of oils and immiscible liquids has been extensively reviewed (Greenkorn, 1983).

The following are the different stages of PCB fluid penetration through the soil:

1. Spill occurs.
2. Soil penetration begins.
3. Spill pool is dissipated.
4. Saturated plug continues to move downward, leaving behind a constant residual saturation in the soil.
5. Saturated slug is exhausted as residual saturation and downward movement stop.
6. Light PCB oils form a pancake and float on the water table.
7. Heavy PCB oils sink into the water table.

Extensive soil and water sampling is necessary to determine the nature and extent of contamination when a spill occurs (NRC, 1979). Soil sampling should always be directed first at determining the presence of PCBs. Several quick tests are now available. Once the qualitative results are known, the quantitative measurements can be made. Water table contamination after a PCB spill can also be determined based on knowledge of the elapsed time, stratigraphy of the soil, permeability coefficient, and penetration depth. The area of the spill pool will increase with time and depth of soil penetration for a given location. Soil penetration also depends on how long the spill pool has been at the location. The maximum soil penetration generally occurs where the spill originates.

When spilled on water, PCB fluids are less dense than water (density of water = 1,000 kg/m^3; average density of PCBs = 1,500 kg/m^3), such as contaminated mineral oils, will float and assume a round or pancake shape (NRC, 1979). This will subsequently be broken down into globules due to turbulence or the action of waves. On the other hand, PCB fluids heavier than water will sink and adhere to the sediments, where they will be taken up by aquatic organisms. Some of the PCB fluids that float, even though not very volatile, will evaporate into the atmosphere, depending on the wind speed, temperature, atmospheric stability, and type of fluid.

When human exposures are likely, protective respiratory equipment should be used. Table 36.8 shows the distribution and fate of an Aroclor 1248 spill after 30 days (NCR, 1979). It is interesting to note that most askarels spilled on land end up in the soil, while the bulk of askarels spilled on water will eventually be adsorbed onto the sediments.

36.7.2 Fires

Because PCB molecules have two benzene rings or 12 carbon atoms as well as some chlorine atoms, they will burn at high temperatures when ignited. The main danger from a PCB fire is not the PCB itself, but the formation of extremely toxic byproducts of combustion, such as polychlorinated dibenzodioxins, polychlorinated dibenzofurans, and hydrogen chloride gas, which often produce immediate respiratory effect. The formation of dioxins and furans

TABLE 36.8 Distribution of Spilled Aroclor 1248, 30 days after Spill

Medium	% PCB remaining (land spill)	% PCB remaining (water spill)
Air	0.005	0.4
Soil	96.3	0.04
Water	0.001	0.4
Bottom sediment	0.001	90.1
Groundwater	0.003	0

from thermal degradation of PCBs has been reviewed (Choudhry and Hutzinger, 1983; Erickson et al., 1984).

Dioxins are frequently formed during thermal breakdown as a result of molecular rearrangement of chlorinated organic compounds during incineration. The actual formation of dioxins takes place after flue gases, fly ash, and other byproducts have begun to cool down and condense at about 300°C. Incomplete combustion of PCBs usually produces furans. Its synthesis often takes place during the postcombustion phase. Generally, PCDDs in soot from fires involving PCBs range from nondetectable to 19 ppm. There are three theories about the presence of dioxins and furans after a PCB fire (Choudry and Hutzinger, 1983).

1. Dioxins and furans are present all along and are able to survive the thermal stress.
2. Dioxins and furans are formed as a result of thermal breakdown and molecular rearrangement of the starting materials to form precursors very similar to dioxins and furans, such as chlorinated phenols, polychlorinated biphenyls, and chlorinated benzenes.
3. Dioxins and furans are formed from precursors that bear no resemblance to these products. For example, burning different nonprecursor substances such as cellulose, lignin, chlorinated plastics, hydrogen chloride gas, and other petroleum products has produced dioxins.

36.8 HUMAN AND ENVIRONMENTAL TOXICITY

36.8.1 Chronic Toxic Effects on Humans

PCBs are suspected human carcinogens (IARC, 1994). They are moderately toxic for short-term exposure by ingestion and skin exposure. There are no reported deaths of humans as a result of a single ingestion. However, experiments in animals suggest that an ingestion of 6 to 10 fluid oz would cause death to a healthy 150-lb adult. Much lower dose levels for chronic exposures have been found to cause various toxic effects that are very similar to those of polychlorinated dibenzofurans and other halogenated aryl hydrocarbons and include the following symptoms:

- Chloracne, blackheads, and skin lesions in many laboratory animals and humans
- Immunotoxic effects
- A wasting syndrome
- Reproductive and fetal toxicity
- Carcinogenesis and fetal toxicity
- Porphyria and hepatotoxicity
- The induction of various drug-metabolizing enzymes

Other effects include unusual eye discharge by all routes, liver damage and digestive disturbance, burning feeling in the eyes, nose, and face, dry throat, lung and throat irritation, nausea, and dizziness (Kimbrough, 1974).

The toxicity of PCBs depends not only on the degree of chlorination but also on the types of isomer. For example, PCBs devoid of orthosubstitution but heavily substituted at the meta and para positions can assume a planar conformation that can interact with the same receptor as TCDD. Examples of these isomers are: 3,3′,4,4′-tetrachlorobiphenyl; 3,3′,4,4′,5-pentachlorobiphenyl; and 3,3′,4,4′,5,5′-hexachlorobiphenyl (NIOSH, 1977).

The Yusho and Yucheng incidents in Japan have provided much of the information on the effects of PCBs on humans (WHO, 1976; NIOSH, 1977; Gaffey, 1983a, b). Some health effects on workers exposed to PCBs are outlined in Table 36.9. Environmental standards and criteria for PCBs are given in Table 36.10.

TABLE 36.9 Health Effects on Workers Exposed to PCBs

Duration (years)	Environmental level (mg/m^3)	Blood level (ppb)	Effects
Not known	10	–	Unbearable irritation
0.33–0.66	5–7	–	Chloracne
1–20	0.2–1.6	370	Chloracne, hyperpigmentation, liver disease
2.5	–	820	Chloracne
2.5–18	0.013–0.27	36–286	Irritation, liver disease
1.17	0.1	—	Chloracne, liver disease
2–23	0.32–1.44	>200	Chloracne, liver disease
Up to 15	–	7–300	Chloracne, elevated triglycerides

36.8.2 Chronic Toxic Effects on Animals

Animals respond in a variety of different ways to the same chemical. Toxicity studies on PCBs are very difficult for the following reasons:

- PCBs are extremely heterogeneous, each formulation consisting of several isomers.
- No two formulations are identical.
- Samples of the same formulation may be different.
- The composition of the same sample often changes when exposed to the environment (NRC, 1979).

Some effects on animals are given in Table 36.11.

36.8.3 Estimated Dietary Intake of PCBs in Canada

Most meat and dairy products contain some amount of PCBs. Major exposure routes are contaminated water, vegetation, and feed. Dietary intake of PCBs from various food sources is given in Table 36.12.

TABLE 36.10 Environmental Standards and Criteria for PCBs

Environmental medium	Region	Concentration	Notes/comments
Air–ambient	Ontario	35 ng/m^3 (1 Y average)	Air quality criteria
Water–ambient	Great Lakes–IJC	1 ng/L	Aquatic life
Water–ambient	U.S. EPA	0.79 to 0.79×10^{-3} ng/L	Human health
Water–recreation	Quebec	0.1 $\mu g/L$	Standard recommendation
Water–drinking	Nova Scotia	3 $\mu g/L$	Maximum concentration
Sediments	Canada	1 mg/kg	Disposal (marine)
Soil	Saskatchewan	<5 mg/kg	Cleanup
Solid waste	Canada	≥50 mg/kg	PCB wastes
Fish	Canada	2 mg/kg	Health Canada
Food (TDI)[a]	Canada	1 $\mu g/kg/d$	Estimated tolerance

[a] = tolerable daily intake.

TABLE 36.11 Chronic Toxic Effects on Animals

Test (chronic feeding)	Effects
Aquatic species	Threshold effects in egg hatchability of vertebrates and invertebrates at levels of 2–5 μg/L. Embryo toxicity evident at 50 μg/L.
Terrestrial species	Mice: Some liver changes with exposure to high chlorine-containing products, 300–500 μg/g. Rats: Some liver changes, minimal reproductive effects, 100–500 μg/g. Monkeys: Yusho symptoms, altered reproductive cycles, hyperplastic gastritis and ulceration, 2.5–5 μg/g. Chickens: Some morphologic deformity, reproduction decline, subcutaneous edema, 20–50 μg/g. Mink: Dose response relationship in growth and reproduction, 10 μg/g. Pelicans: Some hepatocellular changes, 100 μg. Dogs: Reduced growth, some liver changes, 100 μg. Wildfowl: Some reproduction changes, varies with species, 50–200 μg/g.
Teratogenicity	Effects seen in avian species, 50–200 μg/g.
Mutagenicity	Chromosomal abnormalities: negative Dominant lethal mutations: negative Ames test: 1221, tetrachloro biphenyl significantly mutagenic.
Oncogenicity	Highly chlorinated compounds produced tumors in rats and mice, relationship with PCBs not always clear.

Source: NIOSH, 1977; NRC, 1979.

36.9 SURVEY OF PAST SPILLS, LESSONS LEARNED, AND COUNTERMEASURES APPLIED

The two worst-case histories of a PCB spill and a fire are presented here. These exemplify some aspects of the behavior and environmental fate of PCBs as described in the previous sections. A number of small spills of PCBs and askarels have been described in Garrett (1983).

TABLE 36.12 Dietary Intake of PCBs

Foodsource	Food intake (g/person/day)	PCBs intake (μg/person/day)
Dairy	32.8	6.6
Meat	48	9.6
Poultry	3.6	1.8
Eggs	34	3.4
Fish	20	40
Total	**138.4**	**61.4**

36.9.1 PCB Spill in Regina, Saskatchewan

In the summer of 1976, an estimated 6,800 to 21,000 L of a transformer oil formulation (Inerteen 70-30) containing PCBs (Aroclor 1254), trichloro, and tetrachlorobenzenes were spilled when an underground pipe broke at the Federal Pioneer Limited manufacturing plant in Regina, Saskatchewan (Haug and Atwater, 1992; NRCC, 1980). The spill site was located on a major aquifer that supplies drinking water to the city. There was also the potential for contamination of future water supply wells and another aquifer close to the site.

Methods were sought to determine the extent of the spill and its surface and subsurface distribution. Detailed test drilling and sampling identified the following major hydrostratigraphic units: a thin surface fill, Regina clay, Condie silt, the till zone, the interglacial silt, and the Regina aquifer system. The permanent water table is situated below the Condie silt and the sandy glacial till. It was reported that an active perched groundwater had developed within the thin fill layer beneath the floor of the plant. The spread and extensive contamination of the PCBs were the result of the lateral flow of the groundwater. It was also found that PCBs had moved downwards through fractures in the Regina clay and the upper part of the Condie silt. It has been suggested that the potential for transport as a dissolved phase in the groundwater may be the mechanism for the migration of the PCBs downward to the aquifer or laterally to the surrounding sites.

Shallow sumps and deep wells were dug to dewater the site. The effectiveness was measured by a network of piezometers. The extent of groundwater contamination was also determined. The concentration of PCBs in the water collected from sumps on the site ranged from 1 to 500 ppm. High concentrations of approximately 5,000 ppm were detected in some cases. Soil containing over 50 ppm of PCBs was excavated and transported to a nearby storage facility.

The remedial measures to contain and mitigate the effects of the spill consisted of constructing a cutoff wall extending down through the clay to the silt on the spill site to contain the spill, installing a thick seal over the entire surface, and practicing active dewatering to prevent further downward migration of PCBs since one of the driving forces is the downward movement of moisture through the clay.

It has been suggested that the contamination of the deeper wells may have been due to the fact that the PCBs were dissolved in mineral oil. The PCBs moved downward through fractures in the Regina clay and in the upper part of the Condie silt. It was also mentioned that the possible existence of fracture connections to the Regina aquifer and the transport in the dissolved phase may have allowed not only downward migration but also lateral off-site flow. A large amount of the PCBs found at great depths could have moved as discrete fluid.

36.9.2 PCB Storage Depot Fire at St-Basile-le-Grand, Quebec

On August 23, 1988, a warehouse containing about 3,800 barrels of PCBs and PCB-contaminated oils (22,400 gallons of askarels in 45-gallon drums) caught fire (Journault et al., 1989; Phaneuf et al., 1995; Emergency Preparedness Canada, 1988). The warehouse is located approximately 50 km east of Montreal on the south shore of the St. Lawrence River at St-Basile-le-Grand, which has a population of 12,000. It took four fire departments and two police forces seven hours to put out the fire. More than 3,800 people were evacuated in the incident. The heat generated by the fire was so great at one point that the roof and walls collapsed and the drums of PCB were exploding.

When the fire subsided, a team of scientists from Environment Canada's Environmental Technology Centre in Ottawa went to the site to determine the extent of contamination. The Quebec Environmental Department, the TAGA 3,000 team of the Ontario Ministry of the Environment, and the TAGA 6,000 research group of the New York State Department of Environmental Conservation were also involved. Contaminant levels in air, water, soil, vegetation, and livestock were measured. The following are some of their findings.

1. All the air samples from the schools and residences were below the detection limits (100 mg/m^3).
2. The maximum concentration of PCBs detected in the air samples was 11,512 ng/m^3.
3. Mostly T_4CDF, P_5CDF, and H_6CDF were found and small quantities of dioxins. The dioxins were generally H_6CDD, H_7CDD, and O_8CDD.
4. Most of the determinations were below the detection limits. One or two samples of PCBs that exceeded the guidelines toxicity equivalents (15 pg/L for surface water; 1 μg/kg for soil) were found in the soil and surface waters.
5. None of the positive values for the PCBs and dioxins detected off-site presented any health hazards, since they were well below the guidelines (0.25 ng/100 cm^2).
6. The levels of both dibenzodioxins and dibenzofurans close to the site were as high as several hundreds of ppm.
7. Many oil droplets that vaporized in the fire and particulates were found to contain PCBs and furans two times below the exposure limit.

Complete combustion of PCBs produces hydrogen chloride gas and carbon dioxide, while incomplete combustion yields plenty of soot and smoke and far more toxic products such as polychlorinated dibenzofurans (mostly) and polychlorinated dibenzodioxins. This is why incineration of PCBs is strictly regulated. The guideline is 99.9999% combustion.

Lessons to Be Learned

1. These incidents demonstrate the cause, mechanisms, remedial measures, and ways to minimize the impact and prepare for similar occurrences in the future.
2. While no lives were lost in these incidents, disasters involving hazardous materials can be very expensive. The St-Basile-le-Grand fire cost over $6.6 million, while costs of the Regina spill are over $5 million. In both cases, regular maintenance and vigilance are all that is needed to prevent such accidents.
3. A sprinkler system could have helped put out the fire at the initial stages, or at least cool down the warehouse structures. A heat- or smoke-activated chemical foam and fire alarm system could also have been installed in storage places. Materials for cleanup, self-contained breathing apparatus, protective clothing, and portable fire extinguishers should be available on-site.
4. The broken pipe at Federal Pioneer in Regina could have been prevented with an accurate leak detection system. Because the pipe had been reported leaking for some time, the leak could have been detected much earlier.
5. Steps should also be taken to incinerate or otherwise dispose of all PCBs and PCB-contaminated oils that are in storage to avoid fires such as this.

36.10 CONCLUSIONS

The human health risks from environmental exposure through the skin and by inhalation have been assessed with occupationally exposed workers as shown in Table 36.9. It has been found that these chronically exposed workers exhibit few adverse effects. One can only conclude that the potential adverse human health effects that can result from accidental exposure and low-level environmental impact are very low. The chronic low-level, long-term impact on the environment is still unclear.

36.11 REFERENCES

Bunce, N. J., Y. Kumar, and B. G. Brownlee. 1978. "An Assessment of the Impact of Solar Degradation of Polychlorinated Biphenyl in Aquatic Environment," *Chemosphere,* vol. 2, pp. 155–164.

Choudhry, G. G., and O. Hutzinger, eds. 1983. *Mechanistic Aspects of the Thermal Formation of Halogenated Organic Compounds Including Polychlorinated Dibenzo-p-dioxins,* Gordon & Breach, New York, NY.

Durfee, R. L., G. Contos, F. C. Whitmore, J. D. Barden, E. E. Hackman, III, and R. A. Westin. 1976. *PCBs in the United States Industrial Use and Environmental Distributions,* Report No. EPA 560/6-76-005 (NTIS No. PB-252 012), prepared for the U.S. E.P.A. Office of Toxic Substances, Washington, DC.

Emergency Preparedness Canada. 1988. *Fire in a Warehouse Containing PCBs at St-Basile-le-Grand August 23, 1988, Federal Involvement,* Minister of Supply and Services, Ottawa, ON.

Environment Canada. 1995. *National Inventory of PCBs in Use and PCB Wastes in Storage in Canada,* Ottawa, ON.

Environment Canada. 2000. *National Analysis of Trends in Emergencies Systems,* Ottawa, ON.

Erickson, M. D., C. J. Cole, J. D. Flora, Jr., P. G. Gorman, C. L. Haile, G. D. Hinshaw, F. C. Hopkins, and S. E. Swanson. 1984. *Thermal Degradation Products from Dielectric Fluids,* EPA 560/5-84-009; NTIS No. PB 85/38535, U.S. EPA, Washington, DC.

Fingas, M., N. Laroche, G. Sergy, B. Mansfield, G. Cloutier, and P. Mazerolle. 1991. "A New Chemical Spill Priority List," in *Proceedings of the Eighth Technical Seminar on Chemical Spills,* Environment Canada, Ottawa, ON.

Gaffey, W. R. 1983a. "The Epidemiology of PCBs," in *PCBs: Human and Environmental Hazards,* ed. F. M. D'Itri and M. A. Kamrin, Butterworth, Boston, MA.

Gaffey, W. R. 1983b. "Recent Epidemiological Studies of PCBs," in *Advances in Exposure, Health and Environmental Effects Studies of PCBs, Symposium Proceedings,* ed. R. J. Davenport and B. K. Bernard, Report No. LSI-TR-507-1373, NTIS No. PB 84-135771 189-201, Office of Toxic Substances, U.S. EPA, Washington, DC.

Garrett, C. L. 1983. *An Overview of PCBs and Their Current Status in British Columbia,* Report 83-16, Environmental Protection Service, Pacific and Yukon Region, Environment Canada.

Greenkorn, R. A. 1983. "Single-Fluid Flow through Porous Media," in *Handbook of Fluids in Motion,* ed. N. P. Cheremisinoff and R. Gupta, Butterworth, Boston, MA, pp. 279–309.

Hansen, L. G., M. T. Tuinstra, C. A. Kan, J. J. Stirk, and J. H. Koeman. 1983. "Accumulation of Chlorobiphenyl in Chicken Fat and Liver after Feeding Aroclor 1254 Directly or Fat from Swine Feed Aroclor 1254," *Journal of Agriculture Food Chemistry,* vol. 31, p. 254.

Haug, M. D., and J. Atwater. 1992. "Reflections on Polychlorinated Biphenyl and Polychlorinated Benzene Movement at Regina, Saskatchewan," *Subsurface Contamination by Immiscible Fluids,* ed. K. U. Weyer, Balkema, Rotterdam, pp. 477–487.

Holmes, D. C., J. M. Simmens, and J. O. G. Tatton. 1967. *Nature,* vol. 216, p. 227.

Hunt, G. T., P. Wolf, and P. F. Fennely. 1984. "Incineration of Polychlorinated Biphenyls in High-Efficiency Boilers: A Viable Disposal Option," *Environmental Science and Technology,* vol. 18, pp. 171–179.

International Agency for Research on Cancer (IARC). 1994. *IARC Monographs on the Evaluation of Carcinogenic Risk of Chemicals to Man,* World Health Organization, IARC, Lyon, France.

Jensen, S. 1986. "Report of a New Chemical Hazard," *New Scientist,* vol. 32, p. 612.

Journault, C., R. Perrier, and F. Laperriere. 1989. "The PCB Storage Depot Fire at Saint-Basile-le-Grand, Quebec," *Canadian Electricity Forum, Industry and PCBs, Toward A Final Solution,* ed. R. W. Hurst, vol. 2, pp. 16–25.

Kimbrough, R. D. 1974. "The Toxicity of Polychlorinated Polycyclic Compounds and Related Chemicals," *CRC Critical Review of Toxicology,* vol. 2, pp. 445–498.

Miller, S. 1982. "The Persistent PCB Problem," *Environmental Science and Technology,* vol. 16, no. 2, pp. 98A–99A.

Moolenaar, R. J. 1983. "Distribution and Fate of Chlorobiphenyls in the Environment," in *Advances in Exposure, Health and Environmental Effects Studies of PCBs, Symposium Proceedings,* ed. R. J. Davenport and B. K. Bernard, Report No. LSI-TR-507-1373, NTIS No. PB84-135771 67-96, Office of Toxic Substances, U.S. EPA, Washington, DC.

National Institute for Occupational Safety and Health (NIOSH). 1977. *Criteria for a Recommended Standard: Occupational Exposure to Polychlorinated Biphenyls (PCBs),* U.S. Department of Health, Education, and Welfare, Public Health Service, Centers for Disease Control and National Institute for Occupational Safety and Health DHEW (NIOSH), Publication No. 77-225.

National Research Council (NRC). 1979. *Polychlorinated Biphenyls,* National Academy of Sciences, Washington, DC.

National Research Council of Canada (NRCC). 1980. *A Case Study of a Spill of Industrial Chemicals—Polychlorinated Biphenyls and Chlorinated Benzenes,* Publication No. 17586.

Phaneuf, D., J. L. DesGranges, N. Plante, and J. Rodrigue. 1995. "Contamination of Local Wildlife Following a Fire at a Polychlorinated Biphenyl Warehouse in St-Basile-le-Grand, Quebec, Canada," *Archive of Environmental Contamination and Toxicology,* vol. 28, pp. 145–153.

Risebrough, R. W., P. Reichle, and H. S. Olcott. 1969. *Bulletin of Environmental Contamination and Toxicology,* vol. 4, p. 192.

RTECS On-Line. 2000. *Registry of Toxic Effects of Chemical Substances,* Department of Health and Human Services, Centers for Disease Control, National Institute for Occupational Safety and Health, Washington, DC.

World Health Organization (WHO). 1976. *Polychlorinated Biphenyls and Terphenyls, Environmental Health Criteria 2,* WHO, Geneva.

CHAPTER 37
SULFUR TRIOXIDE AND OLEUM

T. Kapias and R. F. Griffiths
Environmental Technology Centre,
Department of Chemical Engineering, UMIST,
Manchester, United Kingdom

37.1 INTRODUCTION

Sulfur trioxide (SO_3) and oleums are generally aggressive materials with complicated properties. They are widely used in the process industries. Their main feature is the violent reaction of SO_3 and water that occurs in both the liquid and the vapor phase. Oleum (or fuming sulfuric acid) is a solution of sulfur trioxide in 100% sulfuric acid. Its strength is expressed as weight percentage free SO_3; thus 30% oleum contains 30% SO_3 and 70% H_2SO_4 by weight. The most commonly industrially used strengths are 65%, 30%, and 20% oleum; their hazardous potential increases with increasing strength.

37.1.1 Uses—Main Properties

Sulfur trioxide is a clear, oily, colorless liquid. As shipped, it may contain a haze and be off-white to light brown in appearance. Oleum is normally turbid and off-white in appearance. Their main uses are in the production of plastics, detergents, and dyestuffs (*Kirk-Othmer,* 1983; Duecker and West, 1971).

The properties of these substances are complicated and are dependent on the percentage free SO_3 and the temperature. Some of these properties are summarised in Table 37.1. Their

TABLE 37.1 Properties of SO_3 and Oleums

Properties	SO_3	65% oleum	30% oleum	20% oleum
Melting point (°C)	17	2	19	−10
Boiling point (°C)	45	58	121	142
Density at 30°C (kg m^{-3})	1,880	1,975	1,920	1,880
Heat capacity at 20°C (kJ kg^{-1} K^{-1})	2.70	1.85	1.43	1.41
Vapor pressure at 20°C	0.228	0.141	<0.001	<0.001
Vapor pressure at 50°C (atm)	1.316	0.789	0.049	0.021
Viscosity at 20°C (cP)	2.2	55	48	39

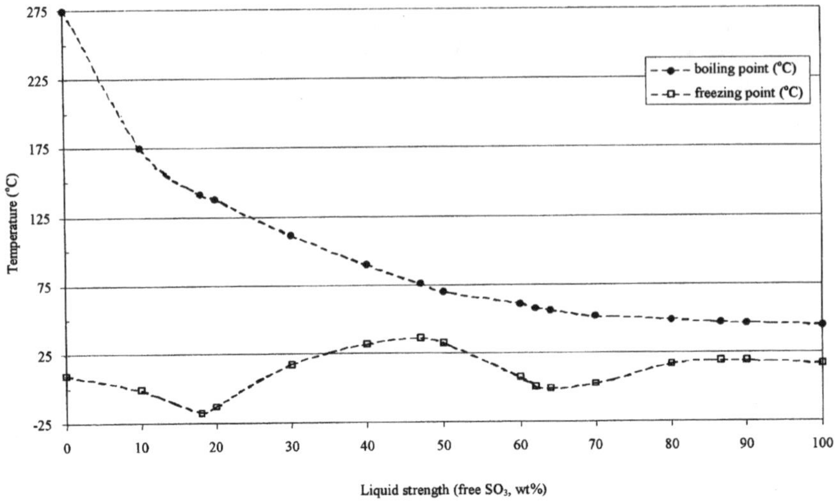

FIGURE 37.1 Boiling and freezing curves of oleums. (*Source: Kirk-Othmer, 1983; Duecker and West, 1971*)

boiling and freezing curves are shown in Fig. 37.1. More detailed descriptions can be found elsewhere (*Kirk-Othmer,* 1983; Duecker and West, 1971; Du Pont Chemicals, 1988).

37.1.2 Legislation

The hazardous nature of SO_3 and oleum is recognized in various items of legislation relating to industrial safety. In the United Kingdom, SO_3 is included in the Notification of Installations Handling Hazardous Substances Regulations (NIHHS, 1982) and in the list of substances covered by the Control of Industrial Major Accident Hazards Regulations (CIMAH Regulations, 1984). Under the new Seveso II European Union (EU) Directive, all substances that react violently with water or liberate toxic gases when in contact with water, are described as major hazards and should be included in the list of substances covered by this directive (European Union, 1996). Therefore, each member state should include not only SO_3 but all strengths of oleums in its new regulations on major hazards. This EU Directive has been implemented in the United Kingdom as the COMAH Regulations, which came into force on April 1, 1999 (COMAH Regulations, 1999).

37.2 EFFECTS ON HUMANS AND THE ENVIRONMENT

On escape to the atmosphere, SO_3 reacts violently and highly exothermically with water, producing fumes of sulfuric acid mist. Information on the acute effects of exposure to sulfuric acid mist is available from animal studies and from studies on human volunteers involving both healthy and asthmatic subjects. A detailed review of the health effects of sulfuric acid aerosols can be found in an IChemE monograph (Griffiths, 1996). This review is mainly concerned with acute effects, as is appropriate for the accidental exposures of interest in major hazards. With regard to long-term exposures, it is noteworthy that a report of the International Agency for Research on Cancer concludes that long-term occupational exposure to strong inorganic mists containing H_2SO_4 is carcinogenic to humans (IARC, 1992).

37.2.1 Human Toxicity Data

From volunteer studies, it was shown that irritant responses to H_2SO_4 mist generally occur at exposure concentrations above 1 mg m^{-3}. It was also shown that this response is strongly dependent on droplet size, with droplets less than 1 μm in diameter producing no respiratory effects. Studies using aerosols with droplet diameters in the range of 1 to 1.54 μm have produced a variety of concentration-dependent respiratory changes and symptoms of irritancy. It should be noted, though, that the studies with submicron-sized mists of sulfuric acid were conducted for exposures of short duration (usually 15 minutes) and at very low concentrations (0.1 to 1 mg m^{-3}). Thus, no clear conclusion can be drawn concerning the possible effects of mists of submicron particle size at higher exposure concentrations and longer times. Human responses observed during these tests are summarized in Table 37.2.

37.2.2 Animal Toxicity Data

Most of the work on toxicity data of H_2SO_4 mist has been done using the guinea pig because it appears to be a better model for man than other rodents. Consistent with evidence from human studies, the toxicity of H_2SO_4 mist in animals is also dependent on the ambient humidity and the aerosol particle size. The relative humidity affects particle size due to the hygroscopic nature of the acid droplets (Wolff et al., 1979).

There is little doubt in the literature that the guinea pig is considerably more sensitive to the effects of inhalation of H_2SO_4 mist than other laboratory species and other larger animals.

TABLE 37.2 Human Toxicity Data

Exposure conditions	Human responses
0.1 mg m^{-3} for 4 hours	No changes in pulmonary function in normal subjects (Kerr et al., 1981).
0.35 mg m^{-3} for 5 to 15 minutes	Decreases in tidal volume and increases in respiratory rate (Amdur et al., 1952a).
0.4 mg m^{-3} for 1 hour	No changes in pulmonary function attributable to H_2SO_4 mist were seen; subjects undertook alternate 10-minute periods of exercise and rest during exposures (Linn et al., 1986).
0.45 mg m^{-3} for 16 minutes	Increases in airway resistance in asthmatics (Utell et al., 1983).
1 mg m^{-3} for 10 minutes	No changes in pulmonary function were seen in normal and asthmatic subjects (Sackner et al., 1978).
1 mg m^{-3} for 5 to15 minutes	Odor threshold (Amdur et al., 1952a).
1 and 2 mg m^{-3} for 1 hour	Normal and asthmatic subjects undertook three 10-minute exercise periods during exposure; no changes in pulmonary function were seen at 1 mg m^{-3}; slight effects and symptoms of irritancy at 2 mg m^{-3} (Avol et al., 1988).
1 to 5 mg m^{-3} for 5 to 15 minutes	Irritation, coughing, rapid shallow breathing (Amdur et al., 1952a; Sim and Pattle, 1957).
1.4 mg m^{-3} for 2 hours	No pulmonary changes were seen, but throat irritation and coughing were experienced; subjects undertook intermittent exercise during exposure (Horvath et al., 1981).
20.8 mg m^{-3} for 30 minutes	Initially intolerable with intense coughing, lacrimation and rhinorrhoea which did not cease entirely throughout the exposure period (Sim and Pattle, 1957).
39 mg m^{-3} for 1 hour	Bronchoconstriction, rales, coughing; increases in airway resistance up to 100% above normal (Sim and Pattle, 1957).

The mouse, rabbit, monkey, cat, sheep and donkey have all been subjected to a variety of exposure regimes and are all less responsive to H_2SO_4 mist than the guinea pig (Cameron, 1954; Treon et al., 1950). Overall, the 8-hour LC50 for the guinea pig lies in the range 18 to 30 mg m^{-3}. At a concentration of 8 mg m^{-3}, there was no lethality to the guinea pig, even with an exposure duration of 72 hours (Amdur et al., 1952b). A more detailed overview of animal toxicity data can be found in the IChemE monograph (Griffiths, 1996).

37.2.3 Relationship between Animal and Human Toxicity

It should be noted that there are no quantitative data concerning human exposures to acutely toxic or lethal concentrations of H_2SO_4 mist. Animal studies indicate that only the guinea pig shows a degree of respiratory tract sensitivity to H_2SO_4 mist similar to human responses. However, the guinea pig appears to be particularly susceptible to the lethal effects of H_2SO_4 mist due to an intense bronchoconstriction and possible laryngospasm mechanism that occurs in the species. After detailed investigation of the toxicity data, the working party of the IChemE monograph concluded that the combinations of airborne concentrations and exposure times leading to 50% lethality in the guinea pig correspond to conditions which would cause severe and possibly disabling respiratory tract irritancy in humans and a possible low rate of deaths in the more vulnerable (asthmatic or bronchitic) members of the population (Griffiths, 1996).

37.2.4 Toxicity in the Context of Major Hazards

There is continuing debate on how to establish a description of a set of exposure conditions that properly represents those likely to be encountered in the event of accidental releases and reflects an appropriate level of concern for human populations in relation to such hazards. This issue has been examined in detail in a report by the European Chemical Ecology and Toxicity Centre (ECETOC, 1991).

The U.K. Health and Safety Executive (HSE) has published a document on the toxicology of H_2SO_4 mist (Turner and Fairhurst, 1992). This document is one of a series of reports on the toxicology of substances, which are primarily directed at establishing toxicity values for use in land-use planning decisions involving major hazards and are based on an approach set out in related publications by the same authors (Turner and Fairhurst, 1993; Fairhurst and Turner, 1993). In this scheme, toxicological data are reviewed and used to derive a measure referred to as the dangerous toxic load (DTL). The DTL is generally a combination of concentration, c, and duration of exposure, t, in the form $c^n t$ = a constant, where the index n takes values typically in the range 1 to 4. The DTL is associated with a level of effect, which is designated the specified level of toxicity (SLOT). The effects associated with the SLOT are:

- Severe distress to almost everyone
- A substantial fraction requires medical attention
- Some people are seriously injured, requiring prolonged treatment
- Any highly susceptible people might be killed

For H_2SO_4 mist, the DTL is based principally on an examination of LC50 data from the guinea pig, and is given a value:

$$\text{DTL} = c^2 t = 2.16 \times 10^5 \ (\text{mg m}^{-3})^2 \text{ min}$$

where c is in mg m^{-3}, t in minutes and the index $n = 2$. The concentration corresponding to the above DTL for an exposure duration of 60 minutes is 60 mg m^{-3}.

The American Industrial Hygiene Association (AIHA) has derived Emergency Response Planning Guidelines (ERPGs) for oleum, SO_3, and H_2SO_4 (AIHA, 1989). Three levels of ERPG are defined, the ERPG-3 (the highest of the three) being the "maximum airborne concentration below which it is believed that nearly all individuals could be exposed for up to one hour without experiencing or developing life-threatening effects." The concentrations are expressed as H_2SO_4 mist, the ERPG-3 being set at 30 mg m^{-3}, based on animal data as well as human experience; it is recognized that unusually sensitive individuals in the population may be affected more severely.

The AIHA ERPG-3 and the HSE DTL differ in their basis; the former asserts that for nearly all individuals the specified one-hour exposure would not be life-threatening, whereas the DTL is possibly fatal to very susceptible individuals and is intended to apply over a range of durations. It is noteworthy that the concentrations corresponding to these two reference levels for a 60-minute exposure (30 mg m^{-3} and 60 mg m^{-3}, respectively) differ by a factor of only 2.

37.3 BEHAVIOR ON RELEASE

When SO_3 or oleum is spilled onto the ground, a highly exothermic and violent reaction between SO_3 and water takes place. The products of the reaction are H_2SO_4 and SO_3. The volume of SO_3 that reacts per unit volume of water depends on the percentage of free SO_3 in the pool (Hermann, 1941). The energy produced by the reaction increases the temperature of the pool and thus increases its vapor evolution rate. Depending on the pool temperature and composition (expressed as % free SO_3 in the pool), the liquid may boil (usually when there is high availability of water), and under different conditions (usually when there is low availability of water) it may solidify. The pool composition and temperature are not constant; they change continuously, and so do all the properties of the liquid, such as the boiling and freezing points. When the spill is onto a concrete surface, another exothermic reaction occurs between H_2SO_4 and calcium hydroxide (Plum and Hammersley, 1984). The pool contains SO_3, H_2SO_4, and water (when there is an excess of water). SO_3 and possibly H_2SO_4 vapor will evolve from the pool, the amount of which depends on the composition, the temperature, and other characteristics of the liquid.

There are three sources of water available for reaction: as the pool spreads it meets free groundwater; it also absorbs atmospheric moisture and it extracts substrate water. Apart from SO_3, which has a high vapor pressure, H_2SO_4 may also evolve under certain conditions (pool composition <35% free SO_3 and pool temperature >333 K) and always in smaller amounts than the SO_3 vapor (due to its much lower vapor pressure) (*Kirk-Othmer,* 1983). The main interactions that occur in the liquid phase inside the pool are depicted in Fig. 37.2.

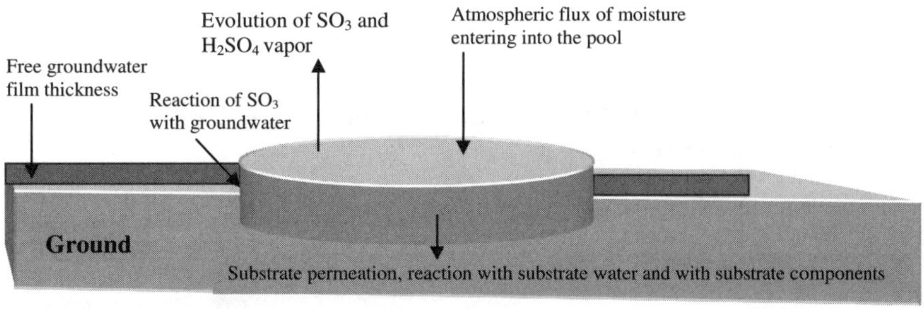

FIGURE 37.2 Main interactions occurring in the pool (some dimensions are exaggerated).

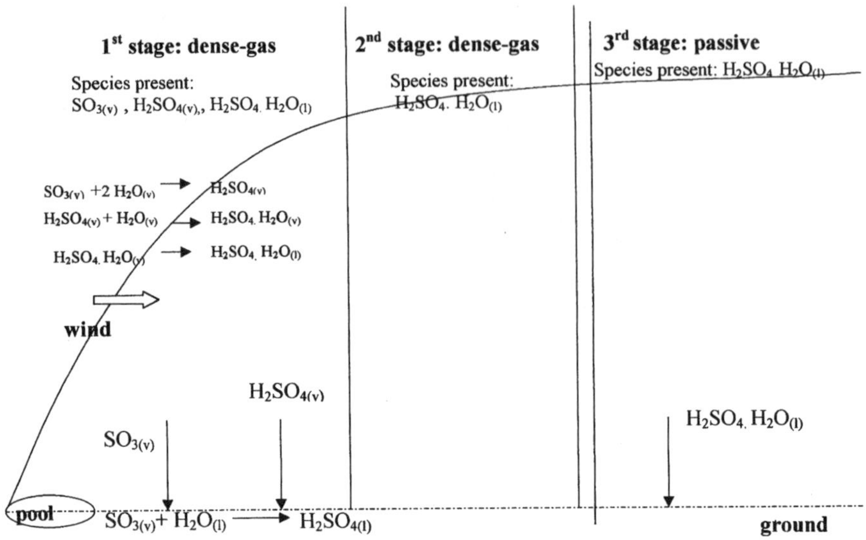

FIGURE 37.3 Behavior and processes that occur in a cloud generated from pools of SO_3 and oleum.

The evolved SO_3 and H_2SO_4 vapors will react with the atmospheric moisture in a complicated way. The final product of these reactions is H_2SO_4 aerosol. In most cases the cloud will initially be denser than air, with numerous processes occurring in it. Only after some distance of travel downwind, allowing adequate dilution with air, will transition to passive behavior occur. By this point, all the SO_3 and H_2SO_4 vapor is consumed and only H_2SO_4 aerosol is present in the cloud. The processes that occur in the cloud can be considered in three stages, as shown in Fig. 37.3.

SO_3 or oleums are usually stored and transported in their liquid form. Therefore, almost all of the accidents that have occurred involved the generation of a liquid pool (with the exception of the Richmond accident [Basket et al., 1994]). Although there are numerous pool evolution models in the literature, most of them deal with nonreactive liquids, with boiling points either much lower or much higher than typical ambient temperatures. The regime of behavior is then clearly either that of a boiling pool or the evaporation of a liquid of low volatility.

The behavior of pools of liquids with boiling points close to ambient temperatures presents a more complicated modeling problem, especially if the liquid properties are critical, as is the case here, where the temperature range between boiling and freezing points is quite narrow. The chemical reactions that are involved add further to this complexity.

37.4 SPILL TESTS, MITIGATION TECHNIQUES, ACCIDENTS, LESSONS LEARNED

Almost all the experiments that have been conducted with SO_3 or oleum have been designed to test the effectiveness of substances that are used to control such spills. The mitigation techniques that were mainly tested are the application of water spray, foam, and Spill-X-A agent.

37.4.1 Spill Tests and Mitigation Techniques

In 1978 and 1980, the Mond Division of ICI, in collaboration with several other companies, carried out a series of spillage trials at Porton Down in Wiltshire, U.K. (Braley, 1980). The purpose of the trials was to examine the effectiveness of different treatments and remedies for spills of corrosive acid materials, to study the effectiveness of fume knockdown using water spray systems, and to find the most effective way of dealing with the residues of the treatment. Although SO_3 was not tested, up to 750 kg of 65% oleum were spilt onto both asphalt and concrete trays.

A variety of treatment methods was checked. The most effective was the rapid addition of Kemsol (anhydrous sodium sulfate) to the pool. It was observed that the initial vapor cloud traveled 6 to 7 km downwind and dispersed very slowly. The conditions of the tests were very humid (relative humidity $\approx$ 84%). It took 13 minutes of application of Kemsol to suppress the fumes (the rate of application was not specified). Treatment of the cloud with water spray seemed to be ineffective. Addition of water to the pool produced very large quantities of fume, and this method appeared to be suitable only for handling small spills. After the trial, the concrete tray was eroded by 2 mm and the asphalt tray by 25 mm. It seemed that the upper surface of the trays was rapidly attacked, but at depth erosion proceeded very slowly.

In 1990, the Ansul Company tested its product Spill-X-A (magnesium oxide blended with other chemical additives) on mitigating spills of SO_3, 65% oleum, and chlorosulfonic acid (Behnke, 1999). Six tests that involved spillages of 65% oleum and SO_3 have been carried out. In each test, 55 gallons of SO_3 or 65% oleum were spilled into a 50 ft^2 (4.6 m^2) test pan. The generated pool was 1.75 in (4.1 cm) deep. The choice of a confined spill (rather than onto flat ground) was made because this would be a more severe test of the effectiveness of Spill-X-A agent. In the test report summary, it is stated that some previous testing on small (one to five gallons) flat spills showed that application of Spill-X-A is an effective way of mitigating and neutralizing acid vapor releases in a one-step process. It was found that application of Spill-X-A agent along with an intermittent water fog spray proved to be a fast and effective means of dealing with spills of these acids. As shown in Table 37.3, it was most effective on chlorosulfonic acid, followed by 65% oleum, and then sulfur trioxide.

The reaction efficiency was taken as the percentage ratio of the mass of reacted acid to the mass of acid spilled. During the sulfur trioxide test series, a test of an existing control technique (direct application of water spray) was run to examine acid cloud generation during application. This profile was compared with Spill-X-A agent application. This comparison showed that both techniques increase the mist concentration during initial application. However, the concentration spike from the Spill-X-A agent application was less intense and shorter in duration than the water fog application. Visual examination during several tests showed a large reduction in the fuming five minutes after the Spill-X-A agent application began. Complete control, with no potential for fuming, was accomplished in 10 to 20 minutes.

In the summary of this test report it is stated that reaction coefficients and control times will improve for flat spills normally encountered during an accidental release. The advantage

TABLE 37.3 Ansul Tests Results

Acid tested	Reaction efficiency range
Chlorosulfonic	78 to 86%
65% oleum	44 to 63%
Sulfur trioxide	38 to 54%

TABLE 37.4 DuPont Spill Tests Results

	Average temperature before control (°C)	Maximum temperature before control (°C)	Control time (s)
Water spray	192.2	260	141
Foam	111.7	195	70
Spill-X-A	183.9	276.7	326

of this technique is that it mitigates and neutralizes in one step, leaving a nonhazardous waste to dispose of after cleanup (in compliance with the Resource Conservation and Recovery Act).

The Du Pont Specialty Chemicals company carried out field tests on mitigation methods for small spills of chlorosulfonic acid and oleum at the Liquefied Gaseous Fuels Spill Test Facility in Nevada during April and May 1992.[1] Sixteen different tests were conducted by spilling about 25 U.S. gallons of 65% oleum. Three different control agents were tested: water spray, two new types of foam, and Spill-X-A agent. The pan was 1.5 in. deep. The results of these tests are summarized in Table 37.4.

It should be noted that the control time was taken as the time from the start of the mitigation to the point at which fuming has essentially stopped. The following conclusions were drawn from these mitigation tests:

- High temperatures (over 260°C) may result from the mitigation/neutralization of 65% oleum.
- Both types of foam were very effective on 65% oleum (control time was 50% of the water spray mitigation).
- The temperatures on foam mitigation tests were lower than on water mitigation tests.
- Mitigation with Spill-X-A agent resulted in higher temperatures and longer control times. This technique has been reported as probably best suited for small spills spread out over a larger area.

The generated plume had the appearance of an opaque white cloud even though the prevailing relative humidity was quite low (about 15%). The heat of reaction of SO_3 with the applied mitigating agent resulted in large temperature rises in the pan, as shown in Table 37.4. These field tests further demonstrated that the reaction between SO_3 and water is extremely violent and that application of the control agent should be made with all the necessary precautions by an experienced person wearing protective equipment.

The application of water spray and foam produced a hot nonfuming solution of H_2SO_4 at the end of the mitigation procedure. This solution should be neutralized before disposal. Different substances, such as lime, soda ash, pot-ash, lime stone, and Spill-X-A, could be applied for this purpose. Spill-X-A neutralization was selected in these tests because of the ease of application into the solution and also because it has been formulated to minimize the heat of reaction and result in a nonhazardous waste.

The U.K. Health and Safety Executive (HSE) carried out some small-scale (a few kg) spillage trials at their Buxton site, in Derbyshire (Carter, 1996). The trials were conducted by spilling 20%, 30%, and 65% oleum in concrete trays. Although the scale of the tests was

[1]CSA/oleum mitigation tests, 4, 5/92 (file no. 179-08-92), obtainable in VHS format at U.S. $15 per copy, PAL or C-CAM format at U.S. $30 per copy, from W. Lazor, Du Pont Specialty Chemicals, Chestnut Run Plaza, P.O. Box 80709, Wilmington, DE 19880–0709, U.S.A.

very small, it was observed that the pool was gaining weight rather than losing it because the SO_3 in the pool was absorbing atmospheric water. The conclusion was that there must be a flux of moisture entering the pool from the windfield. Particle size measurements were made, and they showed that the diameter of the H_2SO_4 mist droplets was around 0.65 μm. This small value of the droplet diameter is the cause of the great opacity observed. The vapor release from the spillages of 20% and 30% oleum ceased very quickly. The visible plume from the spillage of 65% oleum extended 200 m downwind of the release point.

37.4.2 Accidents

Accidental releases also give an indication of the behavior of spillages of SO_3 or oleum. There have been several incidents, and a few of the most important are described in this subsection. It should be noted that the level of detail in the accident reports is highly variable, limiting the usefulness of the information. More detailed descriptions of accidents involving these substances can be found elsewhere (Griffiths, 1996; Kapias and Griffiths, 1998a–e; Kapias, 1999).

On December 4, 1985, in Delhi, India, a tank containing 65% oleum collapsed due to corrosion of the metal supports (Singh et al., 1990). This resulted in a leakage from a 3-in. diameter pipe. According to eyewitness estimates, a liquid pool about 4 m in diameter was formed. The leak continued for about 40 minutes, during which 55 tons of oleum were released. In an attempt to minimize the toxic effects of the leakage, large quantities of water, lime, and other alkalis were added to the spillage. The process generated sufficient heat to produce a buoyant plume. As a result, people in the factory complex and immediate vicinity remained relatively unaffected. However, the toxic fumes dispersed into highly populated areas of the city due to the prevailing wind direction. Medical reports show that 297 persons reported for treatment to various hospitals in the city and 15 of them were detained. Singh and coworkers presented a mathematically modeled simulation of this spillage (Singh et al., 1990).

A leak of 35% oleum occurred from a 100-ton rail tanker in Richmond, California, in July 1993 (Basket et al., 1994). The cause of the leak was the application of heat to reduce the viscosity of oleum for ease of pumping. As a result, a safety valve ruptured on a tank car. SO_3 was released to the air under high pressure and quickly condensed to a whitish H_2SO_4 mist due to the prevailing values of relative humidity (78 to 92%). The release formed a visible cloud 300 m high stretching for about 15 to 20 km. The leak occurred in a densely populated industrial area and more than 20,000 people sought hospital treatment. The responsible company received claims for compensation from more than 60,000 people and it paid out about $180 million. A dispersion modeling simulation of this accident was published by the Atmospheric Release Advisory Capability (ARAC) (Basket et al., 1994).

In August 1995, in Kentucky, a cast iron pipe on an oleum storage tank sheared off from a connecting valve, releasing more than 23,000 U.S. gallons of oleum over four hours. Between 1,000 and 1,400 people in two counties were evacuated and 50 to 70 sought hospital treatment. The company was prosecuted on the grounds that it failed to design and maintain a safe operation for handling hazardous substances. It also failed to contain the leak once it occurred and to notify the authorities promptly. Nearly 250 people filed federal suits. The company spent nearly $600,000 on penalties and improvements at the plant (Washington dateline general news, 1997).

37.4.3 Lessons Learned

The following are the main lessons learned from field tests and accidental releases of SO_3 and oleum:

- The direct application of water to mitigate vapors from accidental spills of SO_3 and oleum is shown to be ineffective except on very small spills (less than 20 kg).
- The application of foam or Spill-X-A agent is a generally effective technique in handling such spills.
- The reaction between SO_3 and water is extremely violent and therefore mitigation and neutralisation procedures should be carried out by experienced personnel, wearing the right protective equipment.

37.5 REVIEW OF PREVIOUS APPROACHES TO THE MODELING OF SPILLS

Early work on hazard assessment of spillages of SO_3 and oleum was prepared by the Technology Division of the HSE with assistance from the Safety and Reliability Directorate of the UKAEA and was published in 1990 (Grint and Purdy, 1990). This article contains descriptions of both the pool and the cloud behavior. The pool model consists of two different models, RASP for spills of SO_3 and 65% oleum and TOEM for spills of 20% oleum. For spills of SO_3 and 65% oleum, it has been assumed that there are two different evaporation mechanisms, one due to the reaction between SO_3 and water on the ground and the other due to the wind flow. The pool-spreading behavior is not taken into account, and the pool radius is calculated simply from the volume of the liquid present in the pool. It is assumed that the pool temperature remains constant and is arbitrarily set equal to an extremely low value (no energy balance is incorporated). It is also assumed that the pool composition remains constant throughout the release duration. The heat generated by the reaction with water on the ground is a function of the percentage-free SO_3, and it is assumed to be entirely used in the vaporization process. Boiling, solidification, interaction with the substrate, and H_2SO_4 evolution are not included. The wind-driven evaporation rate is estimated by Sutton's correlation. For spillages of 20% oleum, the heat generated by the reaction is assumed to be used entirely in warming up the bulk liquid. For those spillages, evolution of vapor occurs only due to the wind flow. For spills of different strengths of oleum (other than 20%, 65% oleum, and pure SO_3), interpolation of results is necessary. Grint and Purdy assumed that there is always enough atmospheric moisture above the pool for complete and rapid reaction of the SO_3 vapor to H_2SO_4 aerosol. They also assumed that the cloud will behave as passive for purposes of dispersion modeling, that it contains only H_2SO_4 aerosol, and that it has reached its chemical end point instantaneously. The conventional Gaussian plume model is used to describe the dispersion behavior. Gravitational settling and deposition effects are ignored. Although this work was a useful first step, it described the behavior in a very oversimplified manner.

In 1990, Singh and coworkers published a mathematical model developed to describe the 65% oleum leakage that occurred in Delhi in 1985 (Singh et al., 1990). Their modeling approach is composed of two stages. The first stage deals with a possible mechanism that led in this particular accident to a rapid initial plume rise. In this stage, they estimated a temperature rise through thermodynamic heat-exchange relationships. The plume rise is calculated by a process that includes some very approximate assumptions. In the second stage, a time-dependent advection–diffusion equation is solved analytically in order to estimate the downwind aerosol concentration. The authors adopted the basic assumption that the dispersing vapor is H_2SO_4 mist, ignoring interactions and processes that occur in the vapor phase. It was also assumed that the cloud will behave passively. The model incorporated the particular local topographical and meteorological conditions of the place where the spill occurred.

The Atmospheric Release Advisory Capability (ARAC) team has used a dispersion model to simulate the behavior of the accidental release of 35% oleum that occurred in 1993, in Richmond, California (Basket et al., 1994). In their report, ARAC provides some information on the topographical and meteorological conditions of the site and of the conditions and reasons for the accident. It should be noted that the authors revised the initial source estimate of 97.5 tons of 35% oleum over 90 minutes to a final estimate of 8 tons of SO_3 over 225 minutes. The SO_3 vapor was assumed to condense immediately into a fine sulfuric acid mist with a median particle diameter of 1 μm (initial estimate). In their final assessment, the value of the median particle diameter of the mist was revised to 0.4 μm. A three-dimensional, diagnostic, finite-difference dispersion modeling system for estimating consequences of accidental atmospheric releases was employed. This system is used by ARAC for any accidental atmospheric release. However, this general approach ignores the complexity of the behavior of releases of SO_3 and oleum. A number of the release characteristics were revised in order to approximate the measurements of mist concentration from a monitoring station located at about 2.3 km downwind.

In more recent work, Kapias improved the initial RASP and TOEM pool models (Kapias, 1996). The basic improvements presented, compared to RASP and TOEM, were the incorporation of a more recent and reliable evaporation model (Brighton, 1987), the introduction of a flux of atmospheric moisture entering the pool, and the use of more reliable values for the properties of the liquids. It was also shown that there is not usually enough atmospheric moisture above the pool for complete and rapid reaction of the SO_3 vapor to H_2SO_4 aerosol. Therefore, the cloud will initially be denser than air with gradual reactions taking place, and it will become passive only after some distance downwind. This work identified the particular requirements for the new modeling procedure described in Section 37.6.

The IChemE monograph (Griffiths, 1996) reviews the hazards of accidental releases of SO_3 and oleum. The monograph gives a detailed review of the relevant literature on toxicity and challenges the assumption that there is enough atmospheric moisture in the wind flow above the pool for complete and rapid conversion of the SO_3 vapor to H_2SO_4 aerosol.

37.6 ASTO: A NEW MODEL FOR ACCIDENTAL SPILLS OF SULFUR TRIOXIDE AND OLEUM—BRIEF DESCRIPTION

Kapias and Griffiths have developed ASTO, a new model that describes accidental spills of SO_3 and oleum. Detailed description of the model and its results can be found elsewhere (Kapias and Griffiths, 1998a–e; 1999a; Kapias, 1999a). This model describes both the pool and the cloud behaviour. It can be used for releases of different strengths of oleums, either continuous or instantaneous, and under varying values of input parameters. This model incorporates and describes realistically the phenomena that occur in the liquid and vapor phase.

37.6.1 Model Description

The pool model consists of different submodels that describe different features of the behavior of the liquid:

- *Pool spreading:* The spreading behavior of the pool is described by differential equations that are based on the principles of mass, momentum, and energy conservation (TNO, 1992; Leonelli et al., 1994).
- *Availability of water:* The amount of water available for reaction is estimated by calculating the available amount from each source. The free water lying on the ground is modeled according to previous work (Grint and Purdy, 1990), while the atmospheric and substrate water fluxes are estimated by a new procedure (Kapias and Griffiths, 1998a; Kapias, 1999).
- *Reaction of SO_3 and water:* The volume of SO_3 liquid that reacts per unit volume of water depends on the percentage free SO_3 in the pool (Hermann, 1941).

- *Reaction of SO_3 and H_2SO_4 with the substrate:* Both these liquids are highly aggressive and could react with some of the components of any type of substrate (i.e., in the case of concrete, H_2SO_4 reacts exothermically with calcium hydroxide: Plum and Hammersley, 1984).
- *Energy balance:* An energy balance that takes into account all the possible energy sources is implemented and the pool temperature is calculated (Kawamura and MacKay, 1987).
- *Evaporation:* The evaporation rate is calculated by Brighton's model, which is an analytical model that accounts for high vapor pressures (Brighton, 1987).
- *Boiling and solidification:* The boiling and solidification scenarios are described by new methods (Kapias and Griffiths, 1998a; Kapias, 1999).

The cloud modeling procedure includes the following submodels (Kapias and Griffiths, 1999a):

- *Thermodynamics:* A new thermodynamic model that describes all the significant processes and interactions that occur in the vapor phase has been developed. It includes an energy balance in order to calculate cloud temperature.
- *Dense-gas dispersion:* The dense-gas regime is described by the thermodynamic model and a box-type dispersion model (Raj et al., 1987; Mohan et al., 1994).
- *Transition:* Different criteria are used to check whether transition to passive behavior occurs. When transition takes place, a transition model is used to match the final dense-gas cloud characteristics to those of the initial passive cloud (Jagger, 1981).
- *Passive dispersion:* The passive regime is described by the partial reflection model, which is a Gaussian-based model that accounts for gravitational settling and deposition effects (Overcamp, 1976).

A summarized algorithm of the modeling procedure is depicted in Fig. 37.4.

37.6.2 Model Inputs and Outputs

The inputs of the model are:

- The spill rate of SO_3 or oleum (M' in kg s^{-1}) [or the quantity spilled (M in kg)]
- The time step (δt in s) (used by the pool model)
- The distance increment (dx in m) (used by the dispersion model)
- The spill duration (t_{s1} in s)
- The type of release (rel) (SO_3 or different strengths of oleum)
- The maximum duration of the release to the atmosphere to be modeled (t_{s2} in s)
- The maximum radius of the pool (r_{max} in m)
- The free water film thickness on the ground (w_g in m)
- The air temperature (T_a in K)
- The location defined by the longitude (LG in degrees) and the latitude (LA in degrees)
- The time and day (τ and d)
- The wind speed at a reference height of 10 m [$U(10)$ in m s^{-1}]
- The type of surface (s) (whether it is concrete or asphalt, etc.)
- The roughness length of the surface (z_o in m)
- The atmospheric radiation and cloud cover factors (r and c)
- The atmospheric relative humidity (% RH)

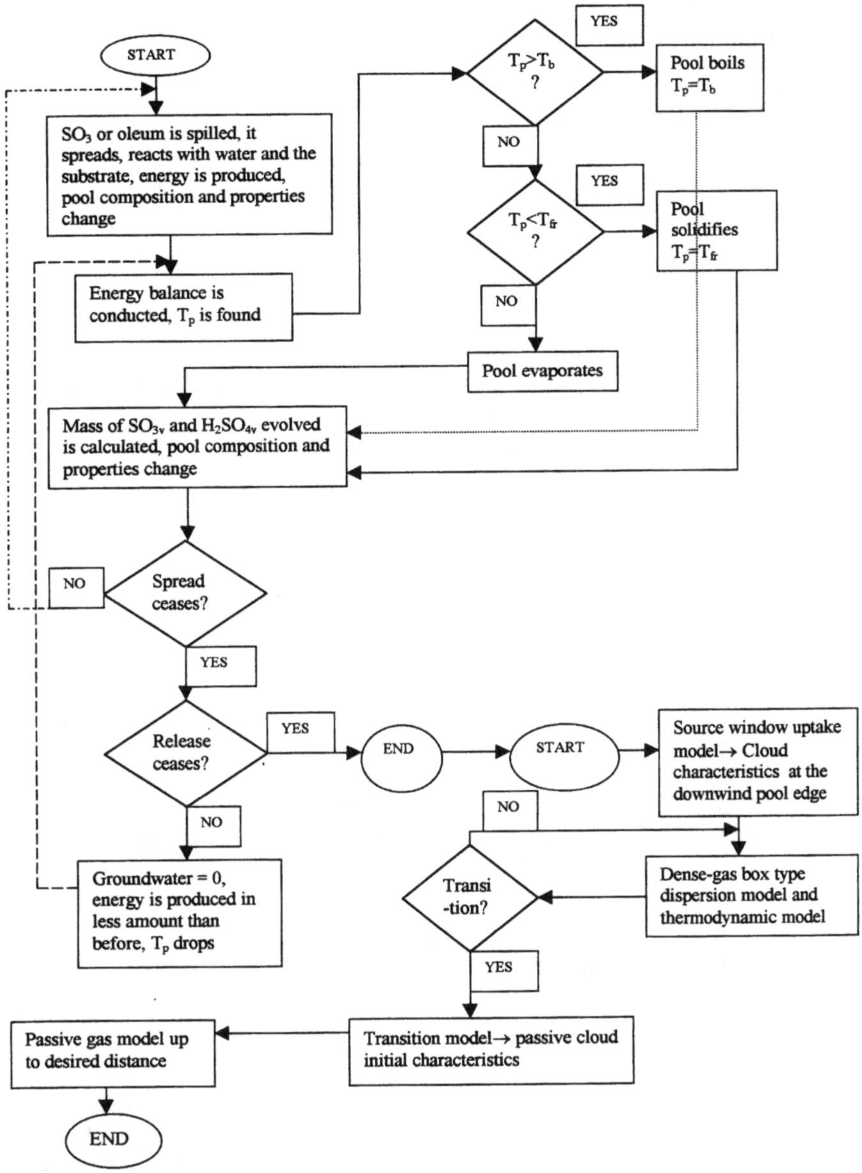

FIGURE 37.4 Logical representation of the algorithm of the modeling procedure.

- The standard deviation of the horizontal wind direction (σ_θ in degrees)
- The average ground surface temperature (T_{grd} in K)

The main outputs of the pool model are:

- The total mass of SO_3 vapor evolved in the three regimes (spreading duration, duration after spreading ceases, and total release duration)

- The total mass of H_2SO_4 vapor evolved in the three regimes
- The average pool temperature, radius, and vapor evolution rates in the three regimes
- The spreading duration (which can be different to the spill duration t_{s1})
- The total amount of liquid left in the pool
- The % free SO_3 of the liquid left in the pool
- The pool temperature versus time, boiling point versus time, and freezing point versus time (indicating at which stages the pool boils or evaporates or solidifies)
- The pool radius versus time
- The vapor evolution rates versus time

The main outputs from the cloud model are:

- The cloud width and height in the dense-gas regime and the vertical and lateral dispersion coefficients in the passive regime
- The total mass rate of the toxic substances (SO_{3v}, H_2SO_{4v}, and H_2SO_4 aerosol) versus distance in both regimes
- The cloud density and the Richardson number in the dense-gas regime
- The cloud composition in the dense-gas regime

For spills of liquid SO_3 or oleum, the pool model should be run first. The results of the pool model are automatically set as inputs to the dispersion model. The dispersion model is subsequently run. For releases of SO_3 vapor, the dispersion model should be run directly. For varying input parameters, the distance- or time-varying functions should be defined.

37.6.3 Model Results

The new model has been implemented in a computer code in Microsoft Visual Basic 5.0. The code has a user-friendly interface providing all the necessary help and information on the release behavior. Some representative results are given in Figs. 37.5 to 37.9. The pool behavior is governed by the amount of water available for reaction as shown in Fig. 37.5. This figure shows that the total sulfuric acid evolution rate usually increases with increasing free groundwater film thickness. The total sulfuric acid evolution rate is calculated by the following equation:

$$\text{Total } H_2SO_4 \text{ evolution rate: } \frac{H_2SO_4 \text{ evolved}}{\text{release duration}} + \frac{(98/80) \times (SO_3 \text{ evolved})}{\text{release duration}}$$

The most important water source is the free water lying on the ground. Atmospheric water is usually a less significant source, and its contribution to the total water used increases with decreasing values of film thickness of the free groundwater. Concrete water is usually the least significant source, and its contribution increases with decreasing film thickness of the free groundwater (Kapias, 1999; Kapias and Griffiths, 1998b). The wind speed also has a relatively strong effect, while the effect of the rest of the pool model input parameters is usually less significant.

The pool behavior during the pool spreading phase is quite different from that after spreading ceases. The vapor evolution rates are much higher during spreading due to the greater amount of water available for reaction, thereby generating higher pool temperatures, resulting in higher vapor pressures and vapor evolution rates. After spreading ceases, there are only two water sources available for reaction: atmospheric and substrate water. Moreover, in this phase, it is more likely that the pool will solidify, terminating the reaction between SO_3 and water (Kapias, 1999; Kapias and Griffiths, 1998b).

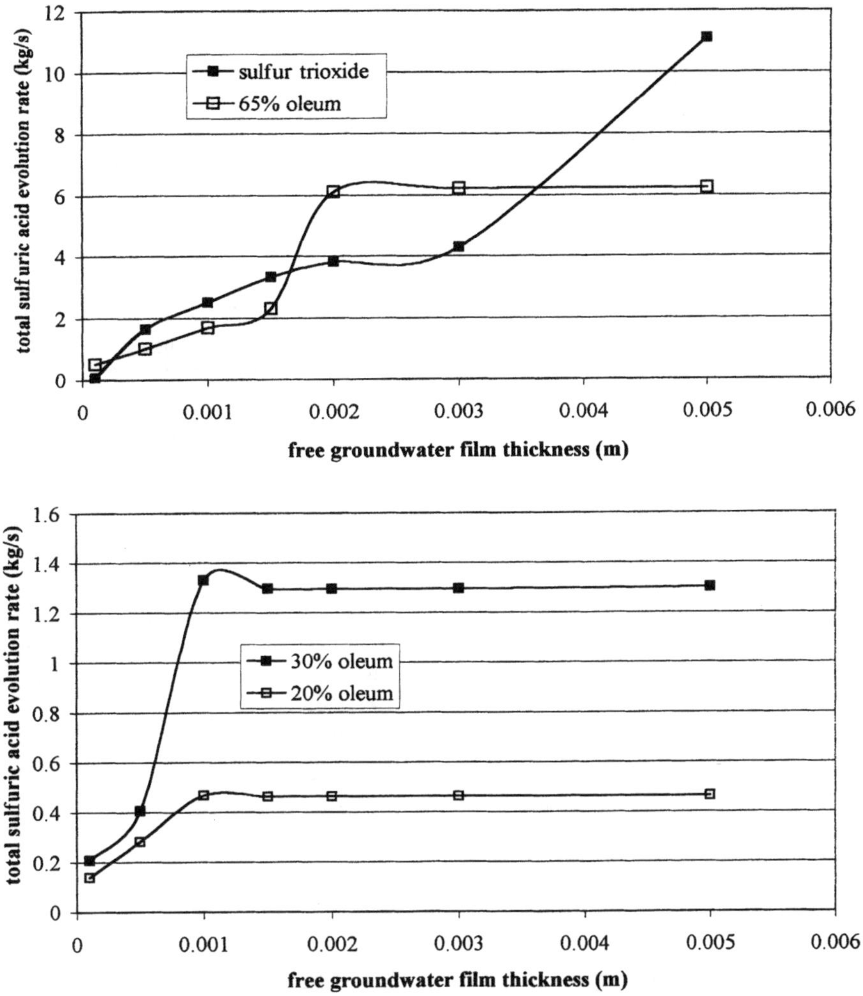

FIGURE 37.5 Effect of free groundwater film thickness on the total H_2SO_4 evolution rate for the total release duration (1,800 s) for spills of 16 kg s^{-1} of SO_3, 65% oleum, and 30%, 20% oleum for a spill duration of 600 s [$U(10)$ = 5 m s^{-1}].

As mentioned, the pool composition (expressed as percentage-free SO_3 in the liquid) changes continuously, and so do all the pool properties. Different composition profiles for continuous spills of SO_3 under different values of free water film thickness lying on the ground are given in Fig. 37.6.

Whether the pool is predicted to boil or evaporate or solidify is a function of many parameters. One case that involves all three phenomena is depicted in Fig. 37.7. In this scenario, the pool boils in the early stage of the spill, then it evaporates during the rest of the spreading period, and just after spreading ceases, it solidifies. The pool temperature in this scenario varies from 265 K to 330 K.

The cloud behavior is affected mostly by the wind speed, the vapor evolution rates from the pool, the atmospheric stability class, and the atmospheric relative humidity (Kapias, 1999; Kapias and Griffiths, 1999a). In the majority of the cases, the cloud will initially be denser

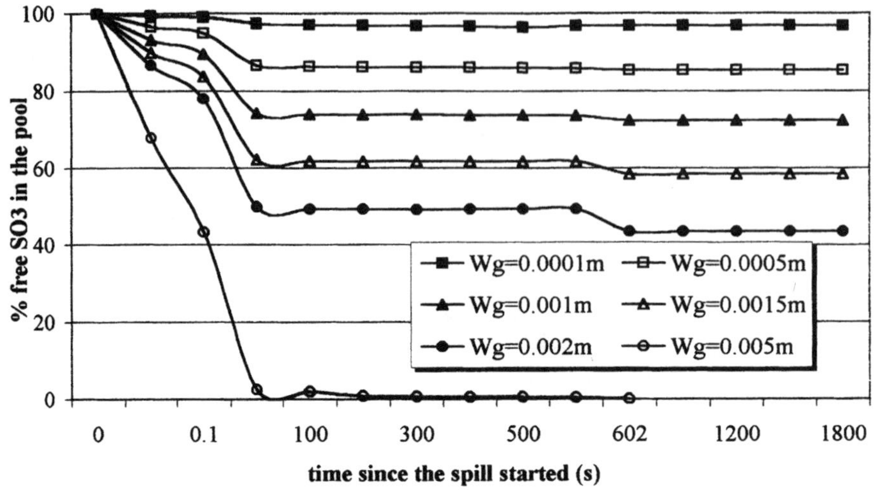

FIGURE 37.6 Pool composition profiles for spills of 16 kg s^{-1} of SO_3 for 600 s [$U(10)$ = 5 m s^{-1}].

than air and transition to passive dispersion behavior will occur only after some distance downwind. Only when the vapor evolution rates from the pool are very low and the wind speed quite high will transition to passive behavior occur rapidly. The duration of the dense-gas regime increases with decreasing wind speed and with increasing stability. It also increases with increasing values of free groundwater film thickness (or increasing strength of spilled liquid). In the dense-gas regime, SO_3 vapor, H_2SO_4 vapor, and H_2SO_4 aerosol are present in the cloud. Both the SO_3 and H_2SO_4 vapors interact with the atmospheric moisture, yielding H_2SO_4 vapor and H_2SO_4 aerosol respectively. The interactions that occur in the

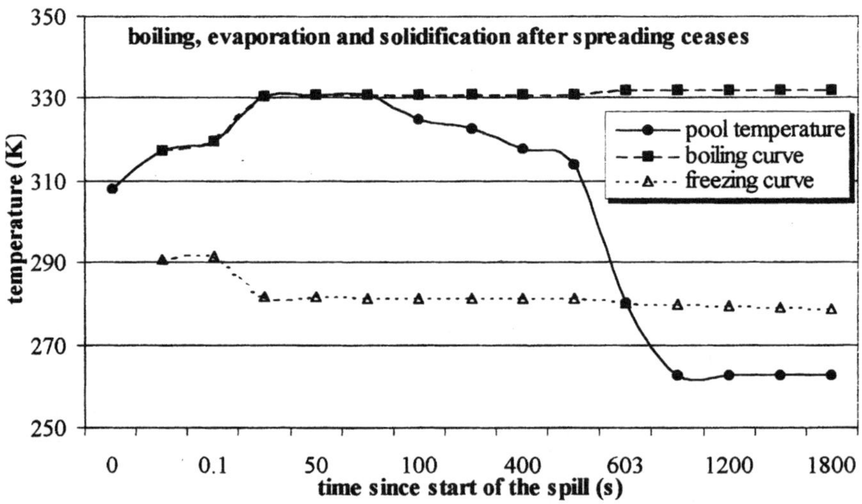

FIGURE 37.7 Temperature profile for a spill of 16 kg s^{-1} of SO_3 for 600 s [w_g = 0.001 m, $U(10)$ = 5 m s^{-1}].

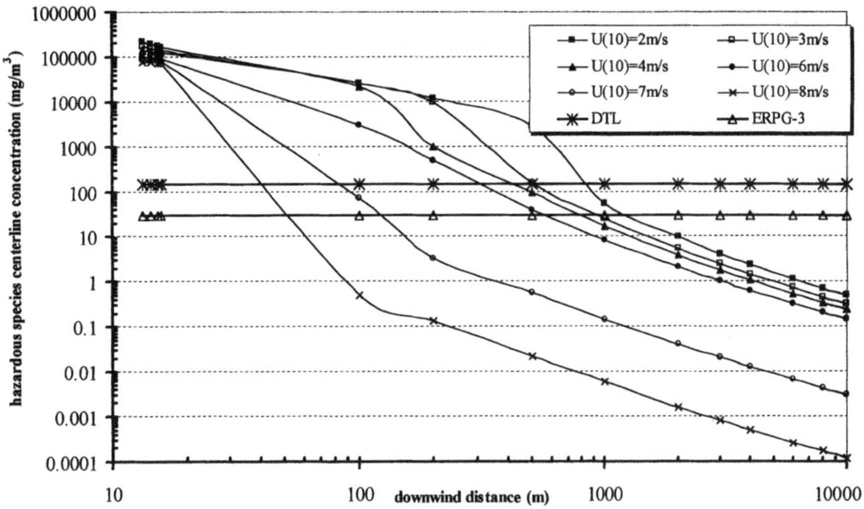

FIGURE 37.8 Wind speed effect on the hazardous species centerline concentration for spills of 16 kg s^{-1} of SO_3 for 600 s (stability class D).

cloud usually finish before transition to passive dispersion behavior occurs. Generally, the cloud composition in the dense-gas regime is affected mainly by the wind speed, the source strength, and the relative humidity (Kapias, 1999; Kapias and Griffiths, 1999a).

When transition to passive dispersion behavior occurs, only H_2SO_4 aerosol is present in the cloud and some of it is lost on the ground due to deposition and gravitational settling effects (Kapias, 1999; Kapias and Griffiths, 1999a).

The effects of wind speed and atmospheric stability class on the hazardous species centreline concentration in the cloud are depicted in Figs. 37.8 and 37.9. It should be noted that

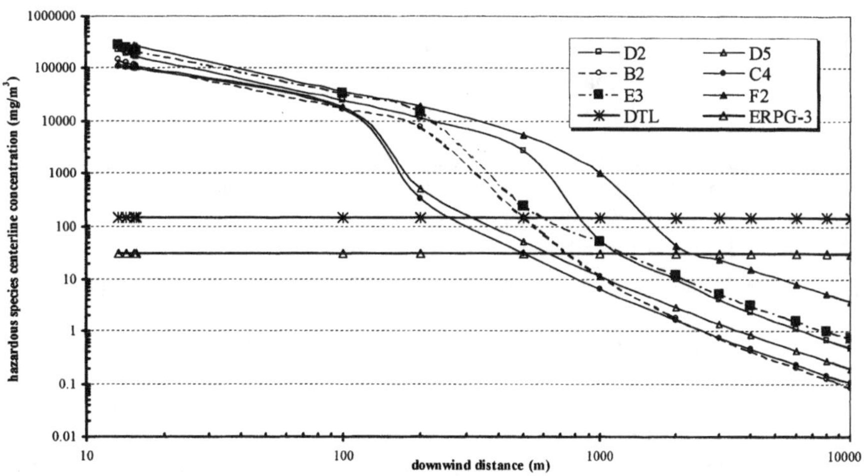

FIGURE 37.9 Atmospheric stability class and wind speed effect on the hazardous species centerline concentration for spills of 16 kg s^{-1} of SO_3 for 600 s.

the results depicted in these figures correspond to the period before spreading ceases (a little over 10 minutes in this case), where the vapor evolution rates from the pool are much higher. For the period after spreading ceases, the vapor evolution rates are much lower, transition to passive behavior occurs much faster, and the hazardous species centerline concentration is much lower (Kapias, 1999; Kapias and Griffiths, 1999a).

In Figs. 37.8 and 37.9, the dangerous toxic load (DTL) concentration, equal to around 147 mg m^{-3} for a period of 10 minutes, and the ERPG-3 (for exposures up to 60 minutes), equal to 30 mg m^{-3}, are also shown for comparison. The exposure duration for calculating the DTL concentration was chosen equal to the spreading period of about 10 minutes because it is the period of main exposure (higher vapor evolution rates).

Overall, the hazardous species centerline concentration increases with decreasing wind speed, increasing stability, and increasing source strength. Figure 37.8 shows a very big effect of the wind speed on the centerline concentration as compared with simple Gaussian plumes, where for a decrease of 1 m s^{-1}, the centerline concentration decreases by a factor of about 4. This big change shows the significance of the physical and chemical processes that occur in the initially dense-gas plume.

In the scenarios depicted in Figs. 37.5 to 37.9, some of the input parameters were constant and set equal to: $M' = 16$ kg s^{-1}, $\delta t = 0.01$ s, $dx = 0.01$ m, $t_{s1} = 600$ s, $t_{s2} = 1800$ s, $r_{max} = 50$ m, $T_a = 288$ K, location = Manchester, UK, τ = 09:00 hrs, day = October 25, s = concrete, rough, $z_0 = 0.1$ m, RH = 70%, $T_{grd} = 285$ K.

37.7 DISCUSSION, CONCLUSIONS, AND RECOMMENDATIONS FOR FURTHER INVESTIGATION

Sulfur trioxide and oleums of all strengths are important substances in the field of major hazards. Their main feature is the violent and exothermic reaction between SO_3 and water that occurs in both the liquid and vapor phase.

With regard to available mitigation techniques, the application of water is appropriate only when handling small spills (less than 20 kg). The other available techniques (application of foam or other agents) are reported to be quite effective in mitigating spills of SO_3 and oleum.

Previous models of spillage and dispersion behavior described the releases of these substances in an unrealistic way. The new model ASTO described in Section 37.6 is much more realistic and is suitable for use by industry and enforcing authorities in fulfilling the requirements of international legislation on major hazards risk assessment. The following are the main advantages of the new model.

- It describes the pool and the cloud behavior realistically, taking into account significant processes that occur in both the liquid and the vapor phases.
- Although the overall release behavior is very complicated, calculation times are very satisfactory (usually less than 15 minutes on a Pentium II).
- It can incorporate a wide variety of input parameters and can model both instantaneous and continuous releases (steady or varying in the latter case).
- It can be used for releases of any strength of oleum as a liquid spill or as a vapor released directly to the atmosphere.
- It can readily be adapted for use with other reactive chemicals, especially those with high affinity for water.
- ASTO is very easy to use. It has a user-friendly interface providing extensive help and information on the release behavior.

The pool model has been modified to include other water reactive chemicals such as phosphorus trichloride and oxychloride, silicon tetrachloride, chlorosulfonic acid, and acetyl chloride. REACTPOOL is a new model that can be used for both instantaneous and continuous liquid releases under a wide range of input parameters (steady or varying). It can be used for all liquids irrespective of their volatility or reactivity and it also describes pools consisting of more than one liquid that can have changing composition and properties (Kapias and Griffiths, 1999b; 2001a and b; Kapias et al., 2001a and b).

Unfortunately, although a number of accidents involving spillages of SO_3 and oleum have occurred, almost no experimental data on release behavior are available. Therefore, it is strongly recommended that experiments be carried out, especially to determine the following aspects, which are the main areas of uncertainty in order of importance:

1. Aerosol nucleation and growth under conditions of high concentrations of the aerosol components, including particle deposition behavior
2. More precise kinetics on the gas phase reaction of SO_3 and H_2SO_4 with water
3. Atmospheric water entering into the pool

37.8 REFERENCES

American Industrial Hygiene Association (AIHA). 1989. *Emergency Response Planning Guidelines: Oleum, Sulfur Trioxide and Sulfuric Acid,* AIHA, Fairfax, VA.

Amdur, M. O., L. Silverman, and P. Drinker. 1952a. "Inhalation of Sulfuric Acid Mist by Human Subjects," *Archives of Industrial Hygiene and Occupational Medicine,* vol. 6, pp. 305–313.

Amdur, M. O., L. Silverman, and P. Drinker. 1952b. "Toxicity of Sulfuric Acid Mist to Guinea Pigs," *Archives of Industrial Hygiene and Occupational Medicine,* vol. 5, 716–734.

Avol, E. L., W. S. Linn, L. H. Wightman, J. D. Whynot, K. R. Anderson, and J. D. Hackney. 1988. "Short-Term Respiratory Effects of Sulfuric Acid in Fog: A Laboratory Study of Healthy and Asthmatic Volunteers," *Journal of the Air Pollution Control Association,* vol. 38, pp. 258–263.

Basket, R. L., P. L. Vogt, W. W. Schalk, and B. M. Pobanz. 1994. *ARAC Dispersion Modeling of the July 26, 1993 Oleum Tank Car Spill in Richmond, CA,* EG&G Energy Measurements, Pleasanton, CA.

Behnke, J. 1999. Senior Technical Service Engineer, Ansul Company, private communication.

Braley, G. K. 1980. In *Proceedings of the 1980 National Conference on Control of Hazardous Material Spills,* p. 104.

Brighton, P. W. M. 1987. *Evaporation from a Plane Liquid Surface into a Turbulent Boundary Layer,* Report No: SRD R375, UK Atomic Energy Authority, Safety and Reliability Directorate.

Cameron, G. R. 1954. "Toxicity of Chlorosulfonic Acid-Sulfur Trioxide Mixture Smoke Clouds," *Journal of Pathology and Bacteriology,* vol. 68, pp. 197–204.

Carter, D. 1996. Health and Safety Executive (HSE), private communication, 1996.

Control of Industrial Major Accident Hazards (CIMAH) Regulations. 1999. SI 1984, No. 1902, HMSO.

Control of Major Accident Hazards (COMAH) Regulations. 1999. SI 1999, No. 743, HMSO.

Duecker, W. W., and J. R. West. 1971. *The Manufacture of Sulfuric Acid,* R. E. Krieger, New York.

Du Pont Chemicals. 1988. *Sulfur Trioxide and Oleum: Storage and Handling.*

European Chemical Ecology and Toxicity Centre (ECETOC). 1991. *Emergency Exposure Indices for Industrial Chemicals,* ECETOC Technical Report No. 43, Brussels.

European Union. 1996. *Council Directive 96/82 of 9 December 1996 on the control of major-accident hazards involving dangerous substances* (Seveso II Directive).

Fairhurst, S., and R. M. Turner. 1993. "Toxicological Assessments in Relation to Major Hazards," *Journal of Hazardous Materials,* vol. 33, pp. 215–227.

Griffiths, R. F., ed. 1996. *Sulfur Trioxide, Oleum and Sulfuric Acid Mist,* Major Hazards Monograph, Institution of Chemical Engineers.

Grint, G., and G. Purdy. 1990. "Sulfur Trioxide and Oleum Hazard Assessment," *Journal of Loss Prevention in the Process Industries,* vol. 3, pp. 177–184.

Hermann, C. V. 1941. "Heat of Formation of Oleums from Sulfur Trioxide and Water," *Industrial Engineering Chemistry,* vol. 33, pp. 898–899.

Horvath, S. M, L. J. Folinsbee, and J. F. Bedi. 1981. *Effects of Sulfuric Acid Mist Exposure on Pulmonary Function,* EPA 600/1-81-044, Office of Research and Development, U.S. Environmental Protection Agency, Research Triangle Park, NC.

International Agency for Research on Cancer (IARC). 1992. *Occupational Exposure to Mists and Vapors from Strong Inorganic Acids and Other Industrial Chemicals,* IARC Monographs on the Evaluation of Carcinogenic Risks to Humans, vol. 54, World Health Organization, Geneva.

Jagger, S. F. 1981. *Development of CRUNCH: A Dispersion Model for Continuous Releases of a Denser-than-Air Vapor into the Atmosphere,* UKAEA, SRD R 229.

Kapias, T. 1996. "Modelling the Behaviour of Pools of Sulfur Trioxide and Oleum," M.Sc. dissertation, Department of Chemical Engineering, UMIST, U.K.

Kapias, T. 1999. "Modelling Accidental Releases of Sulfur Trioxide and Oleum," PhD thesis, UMIST, Manchester, U.K.

Kapias, T., and R. F. Griffiths. 1998a. "A Model for Spills of SO_3 and Oleum, Part I: Model Description," *Journal of Hazardous Materials,* vol. 62, pp. 101–129 (see also Erratum, *Journal of Hazardous Materials,* vol. 66, pp. 307–308).

Kapias, T., and R. F. Griffiths. 1998b. "A Model for Spills of SO_3 and Oleum, Part II: Results, Conclusions and Discussion," *Journal of Hazardous Materials,* vol. 62, pp. 131–142.

Kapias, T., and R. F. Griffiths. 1998c. "Modelling Accidental Releases of SO_3 and Oleum," in *Proceedings of the 9th International Symposium on Loss Prevention and Safety Promotion in the Process Industries,* Barcelona, vol. 3, pp. 987–996.

Kapias, T., and R. F. Griffiths. 1998d. "SO_3 and H_2SO_4 Releases from Accidental Spills of SO_3 and Oleum," in *Proceedings of the International Conference on Protection and Restoration of the Environment IV,* Halkidiki, Greece, vol. 1, pp. 428–435.

Kapias, T., and R. F. Griffiths. 1998e. "A New Model for Accidental Releases of SO_3 and Oleum (Cloud Behaviour)," in *Proceedings of the 23rd NATO/CCMS International Technical Meeting on Air Pollution Mitigation and Its Application,* Varna, Bulgaria, vol. 2, pp. 361–369.

Kapias, T., and R. F. Griffiths. 1999a. "Dispersion and Thermodynamics of Clouds Generated from Spills of SO_3 and Oleum," *Journal of Hazardous Materials,* vol. A67, pp. 9–40.

Kapias, T., and R. F. Griffiths. 1999b. "A New Model for Accidental Releases of Water Reactive Chemicals," in *Proceedings of CCPS International Conference and Workshop on Modeling the Consequences of Accidental Releases of Hazardous Materials,* San Francisco, pp. 133–145.

Kapias, T., and R. F. Griffiths. 2001a. "Spill behaviour using REACTPOOL Part I. Results for accidental releases of chlorosulfonic acid (HSO_3Cl)," *Journal of Hazardous Materials,* vol. A81, pp. 19–30.

Kapias, T., and R. F. Griffiths. 2001b. "Spill behaviour using REACTPOOL Part III. Results for accidental releases of phosphorus trichloride (PCl_3) and oxychloride ($POCl_3$) and general discussion," *Journal of Hazardous Materials,* vol. A81, pp. 223–249.

Kapias, T., R. F. Griffiths, and C. Stefanidis. 2001a. "REACTPOOL: A code implementing a new multi-compound pool model that accounts for chemical reactions and changing composition for spills of water reactive chemicals," *Journal of Hazardous Materials,* vol. A81, pp. 1–18.

Kapias, T., R. F. Griffiths, and C. Stefanidis. 2001b. "Spill behaviour using REACTPOOL Part II. Results for accidental releases of silicon tetrachloride ($SiCl_4$)," *Journal of Hazardous Materials,* vol. A81, pp. 209–222.

Kawamura, P. I., and D. MacKay. 1987. "The Evaporation of Volatile Liquids," *Journal of Hazardous Materials,* vol. 15, pp. 343–364.

Kerr, D. H., T. J. Kulle, B. P. Farrell, L. R. Sauder, J. L. Young, D. L. Swift, and R. M. Boroshuk. 1981. "Effects of Sulfuric Acid Aerosol on Pulmonary Function in Human Subjects: An Environmental Chamber Study," *Environmental Research,* vol. 26, pp. 42–50.

Kirk-Othmer Encyclopedia of Chemical Technology. 1983. "Sulfuric Acid and Sulfur Trioxide," in 3d ed., vol. 22, John Wiley & Sons.

Leonelli, P., C. Stamigioli, and G. Spandoni. 1994. "The Modelling of Pool Vaporization," *Journal of Loss Prevention in the Process Industries,* vol. 7 no. 6, pp. 443–449.

Linn, W. S., E. L. Avol, D. A. Shamoo, J. D. Whynot, K. R. Anderson, and J. D. Hackney. 1986. "Respiratory Responses of Exercising Asthmatic Volunteers Exposed to Sulfuric Acid Aerosol," *Journal of the Air Pollution Control Association,* vol. 36, pp. 1323–1328.

Mohan, M., T. S. Panwar, and M. P. Singh. 1994. "Development of Dense Gas Dispersion Model for Emergency Preparedness," *Atmospheric Environment,* vol. 29, pp. 2075–2087.

Notification of Installations Handling Hazardous Substances (NIHHS) Regulations. 1982. SI 1982, No. 1357, HMSO.

Overcamp, T. J. 1976. "A General Gaussian Diffusion-Deposition Model for Elevated Point Sources," *Journal of Applied Meteorology,* vol. 15, pp. 1167–1171.

Plum, D. R., and G. P. Hammersley. 1984. "Concrete Attack in an Industrial Environment," *Concrete,* vol. 18, no. 5, pp. 8–11.

Raj, P. K., and J. A. Morris. 1987. *Source Characterization and Heavy Gas Dispersion Models for Reactive Chemicals,* AFGL-TR-88-3 (I), Technology and Management Systems Inc., Burlington, MA.

Sackner, M. A., D. Ford, R. Fernandez, J. Cipley, D. Perez, M. Kwoka, M. Reinhart, E. D. Michaelson, R. Schreck, and A. Wanner. 1978. "Effects of Sulfuric Acid Aerosol on Cardiopulmonary Function of Dogs, Sheep and Humans," *American Review of Respiratory Disease,* vol. 118, pp. 497–510.

Sim, V. M., and R. E. Pattle. 1957. "Effect of Possible Smog Irritants on Human Subjects," *JAMA,* vol. 165, pp. 1908–1913.

Singh, M. P., M. Kumari, and S. Ghosh. 1990. "A Mathematical Model for the Recent Oleum Leakage in Delhi," Atmospheric Environment, vol. 24A, no. 4, pp. 735–741.

TNO Bureau of Industrial Safety. 1992. *Methods for the Calculation of Physical Effects Resulting from Releases of Hazardous Materials (Liquids and Gases),* CPR 14E, chap. 5, TNO, The Netherlands.

Treon, J. F., F. R. Dutra, B. S. Cappel, H. Sigmon, and W. Younker. 1950. "Toxicity of Sulfuric Acid Mist," *Archives of Industrial Hygiene and Occupational Medicine,* vol. 2, pp. 716–734.

Turner, R. M., and S. Fairhurst. 1992. *Toxicology of Substances in Relation to Major Hazards: Sulfuric Acid Mist,* Health and Safety Executive, HMSO, London.

Turner, R. M., and S. Fairhurst. 1993. *Assessment of Toxicity of Major Hazard Substances,* Health and Safety Executive Specialist Inspector Report No. 21.

Utell, M. J., P. E. Morrow, D. M. Speers, J. Darling, and R. W. Hyde. 1983. "Air-way Responses to Sulfate and Sulfuric Acid Aerosols in Asthmatics," *American Review of Respiratory Disease,* vol. 128, pp. 444–450.

Washington dateline general news. February 9, 1997.

Wolff, R. K., S. A. Silbaugh, D. G. Brownstein, R. L. Carpenter, and J. L. Mauderley. 1979. "Toxicity of 0.4 and 0.8 μm Sulfuric Acid Aerosols in the Guinea Pig," *Journal of Toxicology and Environmental Health,* vol. 5, pp. 1037–1047.

CHAPTER 38
PERSPECTIVES ON SPECIFIC SUBSTANCES: PENTACHLOROPHENOL

Richard Lawuyi and Merv Fingas
Emergencies Science Division, Environment Canada, Environmental Technology Centre, River Road, Ottawa, Ontario

38.1 OVERVIEW OF PRODUCT AND INDUSTRIAL USES

Pentachlorophenol (PCP) and its salts are purely anthropogenic substances that are widely used for wood and leather preservation due to their broad spectrum of antimicrobial characteristics. Industrially, their application as fungicides, bacteriacides, herbicides, insecticides, ovicides, algicides, and molluscicides is well known.

Pentachlorophenol enters the environment and works its way into the soil, surface water, groundwater, sediments, food, and air primarily from treated utility poles and fences, during production, from treatment facilities, hazardous waste sites, and accidental spills, and from its use as a pesticide. The major contaminants in commercial pentachlorophenol can be very toxic and include polychlorodibenzodioxins, polychlorodibenzofurans, polychlorodiphenyl ethers, and chlorophenols.

Many toxicologists have reported that short-term, high-level exposure, which often occurs during accidental spills, can damage the liver, kidneys, nervous system, and blood and cause death in humans. Other researchers have shown that these exposures cause a decrease in the number of offspring in animals. The use of PCP is now restricted in both Canada and the United States.

38.1.1 Modern Industrial Uses

While pentachlorophenol is no longer manufactured in Canada, it is still imported and used as a preservative for utility poles, paints, adhesives, textiles, leather, and furs. Other uses such as pest control and preserving fence posts and crossties were stopped in 1990. The environmental impact of pentachlorophenol and its uses are being reviewed by regulatory agencies.

38.2 INTRODUCTION

PCP and its salts were introduced into commerce in 1936 and were one of the most heavily used substances in North America for many years. They were used primarily as a timber preservative to control molds, wood-boring insects, termites, and a variety of fungal rots. Their widespread use was due to their solubility in both aqueous and organic solvents. They were so successful that their use was extended to other applications such as in petrochemical drilling fluids, paints, oils, leather, masonry, paper mill systems, agricultural seeds, rope, and in cooling tower water.

Unfortunately, PCP often contains impurities that are toxic not only to fungi and bacteria but also to other living organisms. Its environmental impact includes effects on human health as well as on plants and other environmental organisms, such as aquatic species and wildlife. Its impurities include the less chlorinated phenols, polychlorinated phenoxy phenols, polychlorinated dibenzo-p-dioxins, and polychlorinated furans. By the late 1980s, pentachlorophenol and its impurities had become so ubiquitous in the environment that its use has now been restricted.

38.2.1 Spill Profile

PCP and its salts enter the environment through a variety of sources, primarily during wood treatment, spills, discharge at dump sites, other disinfecting uses, as well as during manufacture. Most spills are due to one or more of the following causes (Environment Canada, 2000):

1. Overflow
2. Leaks from containers, pipes, underground storage tanks, and valves and fittings
3. Intentional discharges
4. Process upsets
5. Unknown causes

The annual spill frequency of PCP from 1982 to 1992 is shown in Fig. 38.1. As can be seen, there were very few spills—a total of 31 in 11 years. The frequency of spills has been decreasing since 1987, when there were major concerns in Canada about the byproducts of PCP.

Once it is introduced into the environment, PCP transport is very fast. The sodium salt is very soluble in water, while PCP itself is soluble in organic solvents. Transport often occurs through runoff from treatment sites, as well as from leaching and volatilization.

38.2.2 Priority List Ranking

Pentachlorophenol is very toxic and more persistent in the soil than most organic compounds and the spill number, spill volume, and supply volume are much less. It is ranked tenth, just below potassium chloride, on Environment Canada's priority substance list of hazardous chemicals (Fingas et al., 1991). This is shown in Table 38.1.

The primary objective of the list was to determine the minimum number of hazardous substances that were most frequently spilled. The list was developed by a simple ranking of: (1) reported spill frequency, (2) supply volumes, (3) historical spill volumes, and (4) toxicity data, stability, accumulation, and persistence.

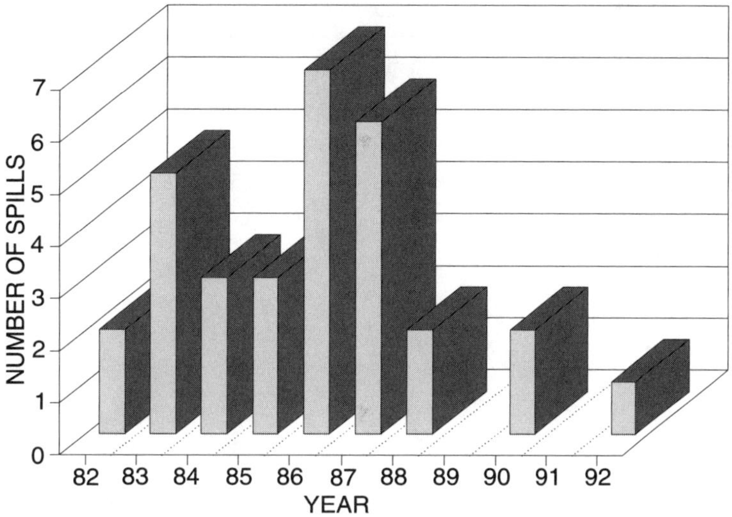

FIGURE 38.1 PCP—Annual spill frequency (1982–1992).

38.3 PHYSICAL AND CHEMICAL PROPERTIES AND GUIDELINES SUMMARY

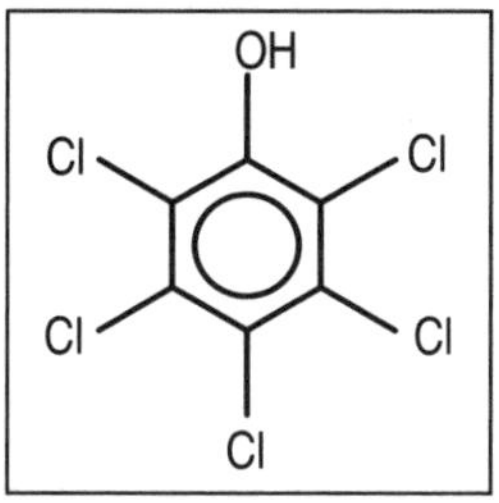

38.3.1 Physical State Properties

Pentachlorophenol is a colorless to light or dark brown crystalline solid or powder with a weak or phenolic odor, but with a strong pungent odor when heated.

Common name: Pentacholorphenol (PCP)
Molecular formula: Cl_5C_6OH
Molecular weight: 266.35
CAS number: 87-86-5
UN number: 2020
STCC number: 4961380
Labels: Poisonous and infectious substances

TABLE 38.1 Priority List Ranking of Hazardous Chemicals

Chemical	Ranking	Number of spills	Spill quantity (t)	Supply quantity (kt)
Ammonia	1	107	470	3,700
Chlorine	2	36	120	1,700
Tetraethyl lead	3	4	72	26
Styrene	4	24	5,000	630
PCBs	5	334	89	—
Sulfuric acid	6	155	13,000	3,700
Sodium cyanide	7	3	83	12
Hydrochloric acid	8	123	3,300	170
Potassium chloride	9	31	12,000	—
Pentachlorophenol	10	19	110	1.5
Phenol	11	10	14	68
Zinc sulfate	12	3	68	1,500
Phosphorus	13	16	46	68
Toluene	14	13	110	430
MTBE	15	1	1	500

RTECS number: SM6300000

Synonyms and trade names (RTECS On-Line, 2000)

Antimicrobial*
Block Penta
Chem-Tol
Chion
Chlon
Chlorophen
Cryptogil
Dow Pentachlorophenol DP-2* Antimicrobial
Dowicide 7
Dowicide EC-7*
Dowicide-G
Dura Treet II
Durotox
EP 30
Fungifen
GLAZD*
Glazed Penta
Grundier arbezol
Lauxtol
Lauxtol A
Laustrol
Liroprem
Monsanto Penta
PCP
Penchlorol*
Penta
Penta EC 30*
Penta Ready*
Penta WR*
Pentacon*
Penta Pres 1-10*

Penwar*
Permasan
Permatox
PKHF
POL NU
Preventol
Santobrite*
Santophen 20*
Sinituho*
Weedone*
Woodtreat A

*Discontinued names.

Grades and Purities: Pentachlorophenol (PCP) is available in two main grades: (1) technical grade which is about 86% pure, being contaminated by a mixture of mostly polychlorodibenzodioxins and phenol ethers; and (2) a pure quality grade consisting of 99% PCP. The main impurities are chlorinated phenols and polychlorodibenzodioxins. PCP is available in solution or as a dry powder. For example, technical grade pentachlorophenol contains traces of hexa, hepta, and octachlorodibenzo-p-dioxins; hexa-, hepta-, and octachlorodibenzofurans; and hexachlorobenzene.

Grade	Purity
Technical	86%
Special	99%

38.3.2 Physical Data

Pentachlorophenol is a colorless to light or dark brown crystalline solid or powder with a weak or phenolic odor, but a strong, pungent odor when hot. It melts at between 187 and 189°C and boils with decomposition at 310°C. The odor threshold in water is 1.6 mg/L and the taste threshold in water is 30 μg/L. It is soluble in alcohol, ether, benzene, and slightly soluble in water. It is stable at room temperature, in closed containers, and under normal conditions. Photodegradation can occur in sunlight. It is also available as sodium pentachlorophenate.

Appearance: White, monoclinic, crystalline solid; technical grade dark grey to brown (Verschueren, 1996)

Usual shipping state: Solid (Transport Canada, 1994)

Physical state at room temperature and pressure: Solid (Verschueren, 1996)

Boiling point: 309 to 310°C (decomposes)

Melting point: 188 to 191°C

Vapor pressure: 0.00011 mm Hg at 25°C or 0.100 Pa

Specific gravity (kg/m^3): 1.98

Relative vapor density (air = 1): 9.2

Fire properties:

Flammability: Depends on the solvent; substance itself is not flammable
Behavior in fire: Produces toxic gases and vapor
Decomposition temperature (°C): >191
Decomposition products: Hydrogen chloride, chlorinated phenols, and carbon monoxide
Flash point: –

Flammable limits: Not flammable
Autoignition temperature: –
Extinguishing: Dry chemical, foam or carbon dioxide, water spray
Keep containers cool with water spray
Reactivity: Stable

Other properties:

Molecular weight: 266.35
Solubility in water:
Slightly soluble, 0.5 mg/100 mL (0°C)
1.4 mg/100 mL (20°C)
3.5 mg/100 mL (50°C)
8.5 mg/100 mL (70°C)
Solubility in organic solvents: Alcohol, ether, and benzene
Log octanol/water partition coefficient: 5.01
Refractive index (20°C): 1.3689
Henry's law constant (Pa · m^3/mol): 0.284; 0.00248; 0.0127; 0.277; 0.079
Molar volume (cm^3/mol): 207.9

Summary of Chemical Properties and Behavior. As a group, the phenols are the aromatic counterparts of the alcohols although they are far more acidic. The phenoxide anion is stabilized by the delocalization of electrons into the aromatic ring. Any substitution by electron withdrawing groups such as chlorine increases the acidity. The pK_a of PCP at 25°C is 4.7, while that of ethanol is 16.00. Some substituted phenols such as the nitrophenols are even more acidic than the carboxylic acids. At pH 4, PCP is fully protonated and highly lipophilic, while at pH 9, it is ionized and least toxic.

38.3.3 Main Hazards

Pentachlorophenol is a poison that is highly toxic by skin absorption, inhalation, and ingestion and irritating to the eyes, skin, throat, and lungs. Repeated exposure can cause skin rashes. Ingestion can result in severe systematic poisoning. Acute exposures may cause rapid heartbeat, rapid breathing, hypertension, fevers, muscular weakness, anorexia, sweating, dizziness, and nausea. High doses can cause unconsciousness, convulsions, and death due to cardiac arrest. Excessive skin exposure has caused human death. Chronic exposure can cause chloracne and bronchitis. Liver and kidney damage have been found in animals.

When heated or during a fire (above 200°C), PCP emits toxic fumes of hydrogen chloride, chlorine, chlorinated hydrocarbons, and chlorodibenzodioxins.

38.3.4 Human Health

TLV: 0.5 mg/m^3 (skin)

TWA: 0.5 mg/m^3 (8-hour) (skin)

IDLH: 150 mg/m^3

Probable human carcinogen (U.S. EPA)

Possibly carcinogenic to humans (IARC, 1994)

Exposure Effects

Contact and inhalation: The main routes of exposure are inhalation, skin absorption, ingestion, and eye contact. Exposure will cause irritation of the skin, throat, eyes, and lungs. Acute exposure may cause rapid heartbeat, rapid breathing, hypertension, fevers, sweating, dizziness, nausea, muscular weakness, and anorexia. High doses may cause seizures, unconsciousness, or death. Chronic effects include chloracne and bronchitis. Liver and kidney damage have been found in animals.

Environment: PCP is toxic to plants, fungi, bacteria, algae, fish, and aquatic life.

Behavior in air: PCP tends to associate with particulate matter in air and eventually settles on the ground. PCP vapor is lost by photolysis, or by reaction with hydroxyl radicals.

Behavior in water: PCP will dissociate in water at ambient pH levels. The dissociated species will photodegrade (half-life of hours to days). Adsorption to sediments can occur. Photolysis can occur in water.

38.3.5 Emergency Response

Symptoms: Irritating to the eyes, throat, skin, and nose. Exposure can cause weakness, nausea, vomiting, shortness of breath, headaches, dizziness, fever, and high temperature.

Local: Skin contact is irritating and could provide efficient way for the chemical to enter the body and cause systemic poisoning. Skin rashes may occur.

Respiratory: Acute exposure is irritating. Respiratory tract irritation. It may cause rapid breathing, rapid heartbeat, high blood pressure, fevers, muscular weakness, sweating, dizziness, and nausea. Large doses may cause unconsciousness, seizures, and death.

Gastrointestinal: Ingestion may cause severe systemic poisoning.

Treatment:

Inhalation:

At the spill site, move patient to fresh air and call for medical assistance. Monitor for respiratory distress. If breathing has stopped, give artificial respiration (not mouth to mouth) or CPR if indicated. If breathing is labored, give oxygen. If cough or difficulty in breathing develops, evaluate for respiratory tract irritation, bronchitis, or pneumonitis. During transportation, continue to administer oxygen by 40% venturi mask.

In the emergency room, continue giving oxygen. Monitor arterial blood gas. Evaluate for pulmonary involvement. Observe for liver or kidney damage.

Skin:

At the spill site, wash exposed areas extremely thoroughly with soap and water and remove contaminated clothing. Shower thoroughly.

In the emergency room, treat as skin irritant. Prolonged exposure may cause acneiform dermatitis.

Eyes:

At the spill site, flush eyes with water for 15 minutes. Hold eyelids open while washing and continue to irrigate. Continue eye irrigation during transportation.

In the emergency room, continue giving oxygen (Mediflow lens if available); determine nature and degree of corneal damage. Notify ophthalmologist.

Ingestion:

At the spill site, give large quantities of water. Induce vomiting.

During transportation, observe vital signs.

In the emergency room, induce vomiting or lavage. Observe for liver or kidney damage. Observe urine for pentachlorophenol.

Hazards and toxicity: For emergency response to a spill of pentachlorophenol, the immediate concerns are shown in the following box.

Immediate Concerns

Hazard: Material itself is not flammable, but may be mixed with flammable solvents in commercial formulations. Toxic gases and vapors, e.g., hydrogen chloride gas, chlorinated phenols, chlorodibenzodioxins, furans, and carbon dioxide are released when product is heated.

Humans: Poison. Irritating to eyes, nose, and throat. May cause weakness, nausea, vomiting, shortness of breath, chest pain, headaches, dizziness, and increased temperature and heartbeat. May cause death or permanent injury after exposure to small quantities. Possible human carcinogen.

Environment: Toxic to fish and plant life. Could disrupt stable ecological relationships among localized populations. Adverse effects on various animals and plant species.

Protection: Wear self-contained breathing apparatus and totally encapsulated suits, rubber gloves, boots, and goggles.

38.3.6 Spill Control

Spills or leaks: Keep all unprotected personnel clear of area. Wear protective clothing. Evacuate area. Ventilate area. Contain spill with earth and sand. Prevent spilled material from entering drains, sewers, waterways, and soil. Inform appropriate authorities if a major spill occurs. Absorb small quantities on paper towels. Do not allow to enter confined space or sewer.

Fire: PCP is not flammable, even though it may be mixed with flammable solvents in commercial preparations. Use dry chemical or carbon dioxide, water spray, or foam to fight fires. Use water spray to keep fire-exposed containers cool, disperse vapors, and protect workers trying to fight the fires. Water spray may be used to flush spills away from exposures. Wear goggles and self-contained breathing apparatus. When heated to decomposition, toxic fumes of hydrogen chloride, carbon dioxide, dioxins, and furans are produced.

38.3.7 Countermeasures

Emergency control procedures:

Soil:
- Stop or reduce discharge if safe to do so.
- Contact manufacturer or supplier for help.
- Notify Canutec and provincial authorities.
- Contain spill with earth, sand, or other noncombustible absorbent material and place into containers for later disposal.
- Construct barriers to contain spill or divert spill to impermeable holding area. Remove material by manual or mechanical means.
- Recover undamaged containers.
- Remove contaminated soil for disposal.

Water:
- Stop or reduce discharge if safe to do so.
- Contact manufacturer or supplier for help.
- Notify Canutec and provincial authorities.

Contain discharge by damming or water diversion.
Dredge or vacuum pump to remove liquids, contaminants, and contaminated bottom sediments.

38.4 INDUSTRIAL ASPECTS AND PRODUCTION IN THE UNITED STATES, CANADA, AND WORLDWIDE

The pentachlorophenol market is quite small and has been declining for some time for environmental reasons. PCP prices have been very stable. The main producers worldwide are Vulcan (U.S.A.), Idacon (U.S.A.), Rhone-Poulenc, and Chapman (CIS, 1995). The global market for pentachlorophenol has been estimated at about 25,000 tons/year. In 1994, the imports of pentachlorophenol to Canada jumped by more than 110% probably due to extra demand for utility poles after many years of austerity. The main buyers of pentachlorophenols are listed here.

38.4.1 Main Buyers in Canada

Bell Pole	Carseland, AB
Bios KMS (GMI)	L'Annonciation, QC
Canada Cedar Pole Preservers	Galloway, BC
Goodfellow	Delson, QC
Newfoundland Hardwoods	Clarenville, NF
Northern Pressure Treated Wood	Kirkland Lake, ON
Northern Wood Preservers	Thunder Bay, ON
Stella-Jones	Prince George, BC, Truro, NS, Cochrane, AB, and Delson, QC
Trans Canada Pole	Thornton, ON
Wood Preservation Industries	Mascouche, QC

38.4.2 Manufacture of Pentachlorophenol

The most common method for manufacturing pentachlorophenol is the progressive chlorination of phenols in the presence of Lewis acid catalysts. Another method specific for 2,4,5-trichlorophenol and pentachlorophenol is by hydrolysis of chlorobenzenes such as penta- and hexachlorobenzene. In general, the hydrolysis method always produces more impurities. The final product is a mixture often containing a couple or more of the following compounds:

Dichlorophenol
Trichlorophenol
Tetrachlorophenol
Hexachlorobenzene
Tetrachlorodibenzodioxin
Hexachlorodibenzodioxin
Heptachlorodibenzodioxin

Octachlorodibenzodioxin
Pentachlorodibenzofuran
Hexachlorodibenzofuran
Heptachlorodibenzofuran
Octachlorodibenzofuran
Heptachlorohydroxydiphenyl ether
Octachlorohydroxydiphenyl ether

38.5 CHEMISTRY

Pentachlorophenol, while essentially an aromatic alcohol, is a rather acidic substance because the pentachlorophenate anion is resonance stabilized by the aromatic ring resulting in increased stability for this anion. The electron-withdrawing chlorines also delocalize the negative charge (electrons) on the oxygen, thus making the proton easily removable and hence much more acidic ($pK_a = 4.7$).

38.5.1 Formation of Dioxins and Furans

When heated, PCP will form chlorinated diphenyl ethers which can further react to produce chlorinated dioxins and furans. This is why wood treated with PCP should only be burned at controlled temperatures so that dioxins and furans are not produced. The reactions are illustrated in Figs. 38.2 to 38.5.

38.5.2 Acid–Base Reactions

Pentachlorophenols are extremely acidic and will react with bases to produce a salt, sodium pentachlorophenate, and water. PCP can also be converted to esters and several other aromatic and alcoholic derivatives. In a fire, formation of dioxins and furans is also possible, as shown in Fig. 38.4.

38.5.3 Formation of Other Ethers

Pentachlorophenol will also take part in nucleophilic displacement reactions to produce the nontoxic conjugates such as sulfates and anisoles. This is the principal process for the elimination of PCP in fish and many mammals (Kobayashi et al., 1977; Glickman et al., 1977).

38.6 ENVIRONMENTAL FATE AND EFFECTS

Pentachlorophenol is extremely toxic to the environment and is still the subject of much research and many government regulations. It has a broad spectrum of bactericidal activities and will kill many algae and fungi. It also has phytotoxic properties. In aquatic environments, the LC_{50} for fish is 0.03 to 0.6 mg/L, causing renal and hepatic lesions at lesser concentrations.

Pentachlorophenol has been found to be widespread in the soil and surface water in areas where utility poles are used due to washing of the poles by rain and snow and exposure to

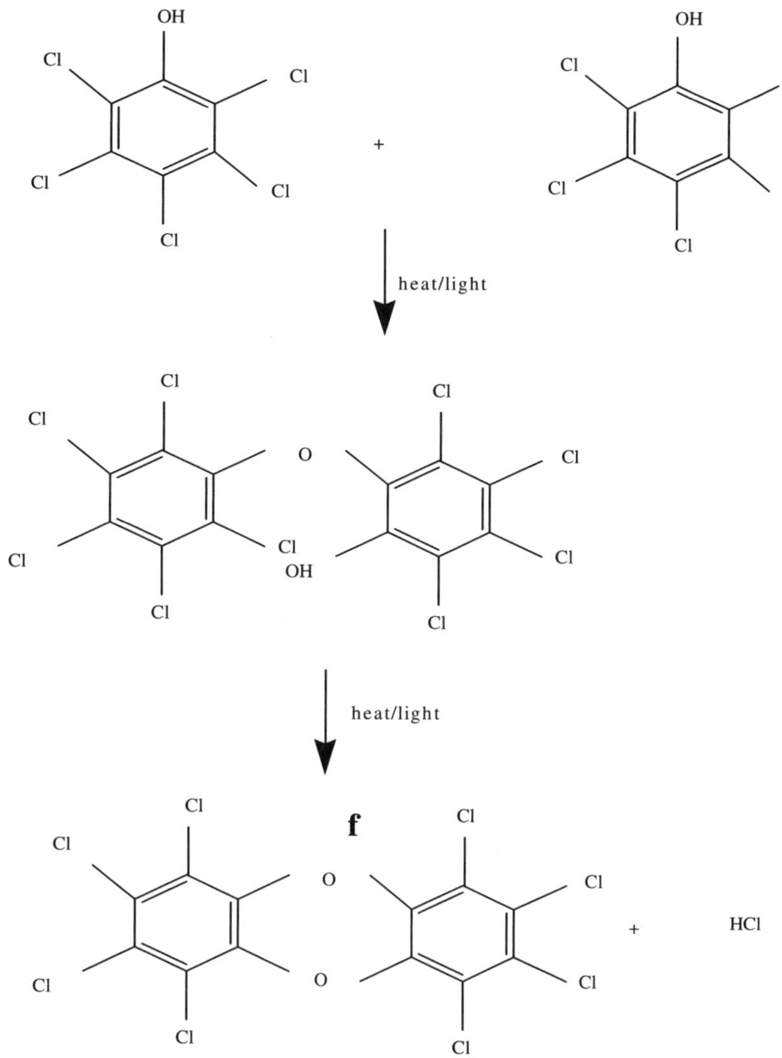

dibenzo-p-dioxin

FIGURE 38.2 Formation of polychlorodibenzo-p-dioxins.

wind. Pentachlorophenol has been detected in air, rainwater, and snow and at hazardous waste sites in Canada and the United States and continues to be an exposure risk chemical. Formation of polychlorodibenzo-p-dioxins and dibenzofurans could occur as a result of environmental transformation.

Levels of pentachlorophenol in the soil decrease through biodegradation and in the air, by photodegradation. PCP has a high adsorption coefficient ($K_{oc} \sim 3000$ to 4000) and it is therefore strongly adsorbed to the soil particles and sediments. It will also bioconcentrate to a small extent in fish and animals. It is also known to volatilize from water surfaces and soil. The environmental fate of pentachlorphenol is shown in Fig. 38.6.

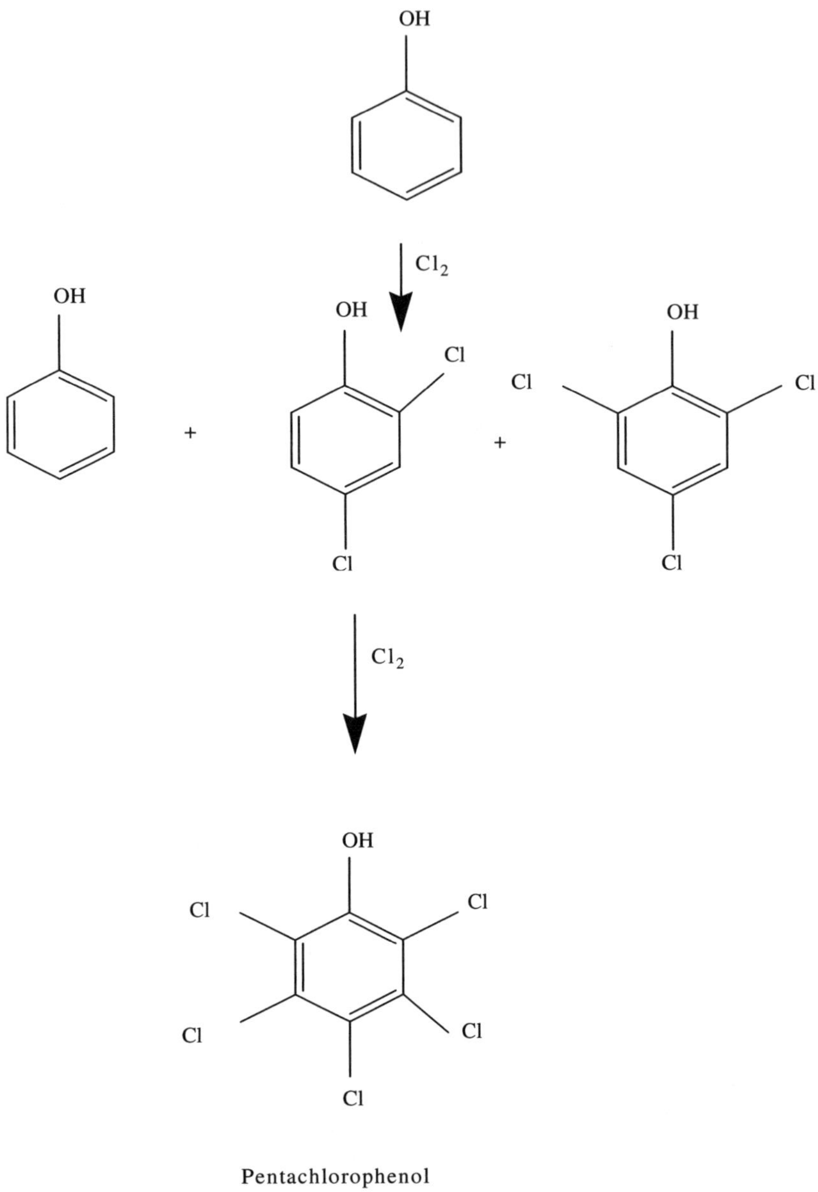

FIGURE 38.3 Manufacture of pentachlorophenol.

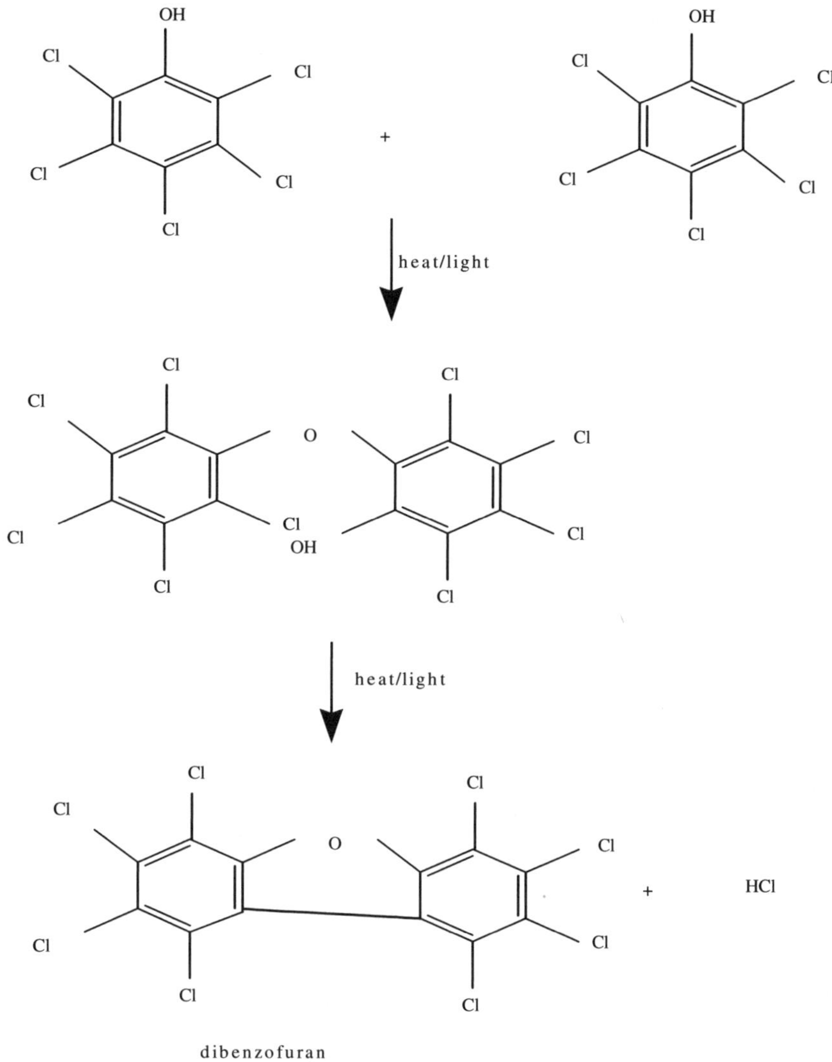

FIGURE 38.4 Formation of polychlorodibenzofuran.

38.7 BEHAVIOR

38.7.1 Solubility in Aqueous and Organic Solvents

The solubilities of PCP and its salts tend to complement each other. While PCP is readily soluble in most organic solvents such as oils, hydrocarbons, ethers, alcohols (ethanol 470 to 520 g/L), and esters, it is only slightly soluble in water (14 mg/L at 20°C). On the other hand, sodium pentachlorophenate, its salt, is very soluble in water (330 g/L at 25°C) and only slightly in organic solvents. The solubilities of PCP and its salts are pH-dependent. For

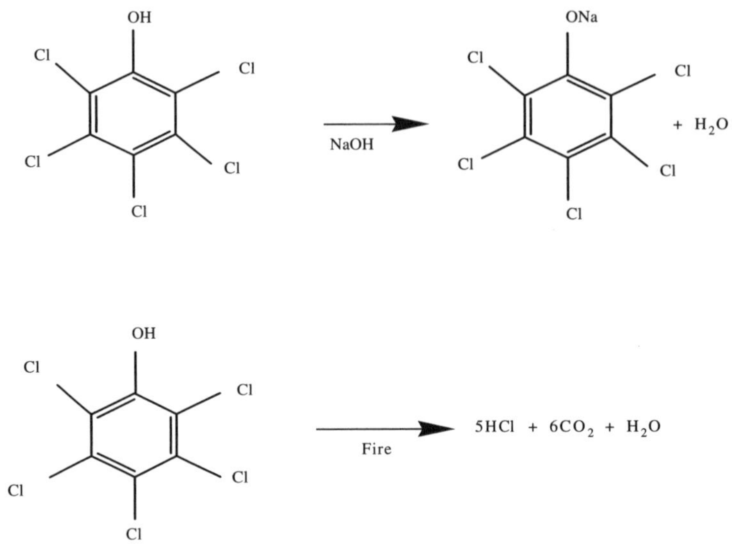

FIGURE 38.5 Reactions of pentachlorophenol with NaOH and in fires.

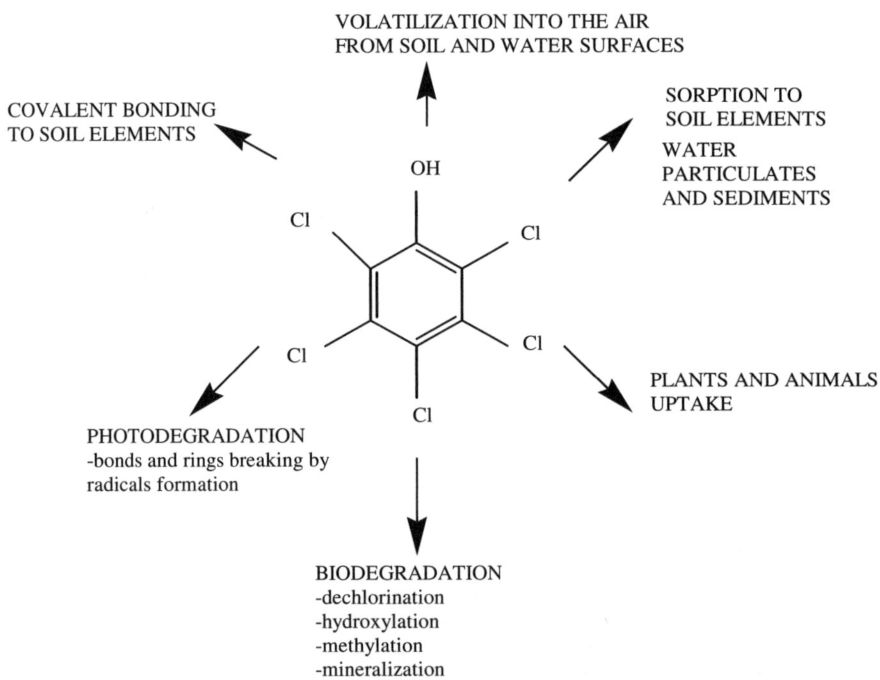

FIGURE 38.6 Environmental fate of pentachlorophenol.

example, the solubility of sodium or potassium pentachlorophenate in water increases from 79 mg/L at pH 5 to >4 g/L at pH 8.0. High concentrations of these salts would, of course, have profound and adverse effects during exposure at high pHs. Differences in lipid solubilities also reflect different bioaccumulation potentials. Clearly, the highly lipophilic pentachlorophenol will be more prone to bioaccumulate than the salts. The toxic contaminants, such as tri-, tetra-, and other chlorophenols isomers, dibenzofurans, dioxins, hexachlorobenzene, and phenoxyphenols, which are often present in the technical grades, are also lipophilic.

In terms of transport, during precipitation, the readily soluble pentachlorophenate salts will most likely be dissolved and washed away in runoff from spill and wood-treatment sites to rivers, lakes, and streams or leached through the soil to groundwater, while the lipophilic PCP will easily adsorb to soil material and sediments (Wan, 1992).

38.7.2 Sorption

The adsorption of pentachlorophenol by soils strongly depends on the chemical and physical properties of the soil, pH of the soil–water systems, the chemical species, and their affinity for the soil surface (Callahan, 1979). Soil properties that can affect adsorption are grain size, soil pH, clay, and organic carbon content. The organic matter includes mostly biodegradation products and humic and fulvic acids that usually cover the surface of soil particles. High organic matter content tends to increase adsorption. Clay soil consists of fine particles with large surface area and large solid–water interfaces on which PCP molecules can adhere. The measured adsorption coefficient (K_{oc}) is 3,000 to 4,000.

The type of chemical species present will also influence adsorption. Pentachlorophenol behaves like a weak organic acid ($pK_a = 4.74$, same as acetic acid). It ionizes in solution according to the following equilibrium:

$$C_6Cl_5OH \rightleftharpoons C_6Cl_5O^- + H^+ \tag{38.1}$$

The equilibrium constant for the reaction increases with temperature and pH. At low pH, the undissociated form of PCP is the predominant species, while at high pH, the reverse exists, the pentachlorophenate being the predominant species. At low pH, the adsorption coefficient is high because the undissociated PCP is strongly adsorbed onto the soil particles by virtue of its affinity. On the other hand, at high pH above the pK_a, the pentachlorophenate ions (PCP^-) that have little affinity for the soil surface predominate and adsorption is greatly reduced. Another reason is that pentachlorophenate is more soluble at high pH than at lower pH.

When PCP in solution is added to the soil, the PCP molecules adsorb on to the surface of the soil particles until an equilibrium is reached at the liquid and soil particle interface. The model that is often used for solid–liquid systems is the Freundlich equation:

$$q = K_f C^{1/n} \tag{38.2}$$

where: q = the amount of chemical adsorbed per unit weight of adsorbent
K_f and $1/n$ are constants
C = the equilibrium concentration in the liquid phase

These adsorption isotherms seem to follow the Freundlich behavior. The solute sorption at pH 7 is much greater than at pH 10. The adsorption isotherms vary for different types of soils. The adsorption capacity in medium sand seems to be the least, while that of loam is the highest.

38.7.3 Volatilization

PCP solution is not a volatile substance even though it is easy to detect in the air where ever it is being used (the vapor pressure is 0.00011 mm at 20°C; methanol is 92 mm at 20°C). One reason is that most aqueous solutions are at pH 7 when PCP is already in ionized form. Transport by evaporation is therefore not an important process (Callahan, 1979). However, it has been shown to volatilize in special pesticide formulations such as aerosols or mists. Significant quantities have been detected in air and precipitation (Boyd, 1989; Weiss et al., 1982a). PCP has been found to evaporate at high temperatures and humidity, and volatilization is a probable means of transference into the air (Kozak et al., 1979).

38.7.4 Biodegradation and Persistence

The microbial degradation of pentachlorophenol has been well studied and the pathways determined (Boyd, 1989; Haggblom and Valo, 1995). The consensus is that, although it is rather persistent in the soil and groundwater, it is degradable by photochemical, microbiological, or chemical processes. PCP has been biodegraded by different strains of bacteria and fungi to produce differing products, and a number of these bacteria have been isolated. The degradation pathways seem to differ with different microbial groups (Tabak et al., 1981). A number of these are aerobic, while others are obligately anaerobic. Polychlorinated phenols are degraded primarily by initial dechlorination, followed by hydroxylation and reductive dechlorinations. Ring cleavage often occurs after most of the chlorines have been removed.

Chlorinated catechols, quinones, and hydroquinones are the central intermediates in the biodegradation of PCP. Methylation of PCP produces anisoles, which are more hydrophobic, more resistant to degradation, and may bioaccumulate. In many organisms, PCP is rapidly bioaccumulated and excreted. Biomagnification does not occur in the food chain. Biodegradation is the predominant process for PCP removal from the environment. PCP has been known to persist in soils for 15 days to more than 60 days, depending on the type of soil. In warm, moist soils PCP was still detectable after 12 months. The half-life of PCP in water ranged from 0.15 to 15 days; in the presence of bacteria alone it was 5 to 12 hours.

38.7.5 Photodegradation

PCP readily absorbs sunlight in the UV region (wavelength = 320 nm). Irradiation of a dilute aqueous solution of PCP with sunlight or UV light has been found to be an effective means of degradation, yielding chlorinated phenols, trichlorobenzoquinones, and nonaromatic fragments such as dichloromaleic acid (Callahan, 1979; Weiss et al., 1982b; Ghoshal et al., 1992). Octachlorodibenzo-*p*-dioxin and furans have also been detected when PCP and sodium pentachlorophenate were irradiated. Photodegradation is definitely significant in PCP removal from the environment.

38.7.6 Spills on Land

When pentachlorophenol solution in oil or the aqueous solution of the sodium salt is spilled on land, after initial settling, some overland flow will occur depending on the quantity spilled. The pentachlorophenol oil solution will mostly be adsorbed to the soil while the aqueous solution will be much less adsorbed. Leaching to groundwater can occur in both cases. Little evaporation will occur, but some biodegradation and photodegradation will, depending on the moisture content, pH, organic matter content, and the bacteria strains. Some of the spill will, however, be washed off in runoffs during precipitation to rivers and streams.

One incident involving a PCP spill occurred when telephone poles were resprayed. In 1975, a company sprayed the base of a number of telephone poles to prevent rot. In one month, 15 fish kills were reported. The fish kills were all found to be caused by the wash-off from the sprayed poles. The poles were all close to the stream where the kill occurred, heavy rains having washed the solution some distance to the stream.

38.7.7 Spills on Water

When released into water bodies, PCP will readily dissociate at ordinary pHs (for PCP, $pK_a = 4.74$) to produce the pentachlorophenate, sodium, and hydrogen ions in the water (Paasivirta et al., 1990). Any undissolved PCP will eventually end up in the sediments. Some photolysis and biodegradation will occur. Evaporation will be negligible. An incident of PCP pollution occurred after a 4-in. rainfall at a lumber treatment site. The treatment tank, containing 10,000 ppm penta- and tetrachlorophenol, overflowed into a stream, killing 5,000 adult coho salmon.

38.8 HUMAN AND ENVIRONMENTAL TOXICITY

38.8.1 Toxicological Profile

The toxic effects of PCP are due not only to the undissociated PCP but also to other impurities, such as the polychlorinated dioxins, furans, and ethers (Ahlborg and Thunberg, 1980; Rao, 1978). PCP formulations in oil are therefore more toxic than in aqueous media.

Acute Effects during Spills. PCP causes severe irritation of the eyes, skin, and upper respiratory tract. Even at moderate concentrations, coughing, violent sneezing, pain in the nose and throat, dizziness, convulsions, and unconsciousness can develop, and in extreme cases PCP can be fatal. Acute toxicity is the result of uncoupling of mitochondrial oxidative phosphorylation (Eisler, 1989; ATSDR, 1992). The eventual stimulation of cell activities can produce heat stress.

Other acute effects include increase in alkaline phosphatase, blood urea nitrogen, and serum creatinine. Excessive exposure can cause damage to the liver, kidneys, skin, blood, gastrointestinal tract, nervous and immune system, and death. Many deaths have been reported as a result of unnecessary exposure and improper handling of treated lumber and sawdust. Some symptoms of poisoning are general weakness, vomiting, nausea, abdominal pain, headaches, anorexia, intense thirst, pain in the extremities, tachycardia (rapid heartbeat), tachypnea (rapid breathing), hypertension, and fevers (Wood, 1983).

Chronic Exposure. Acute effects are often pronounced during chronic exposure as in occupational settings. These include conjunctivitis, chronic sinusitis, bronchitis, dermatitis, and polyneuritis. The presence of impurities such as dioxins and furans can cause chloracne (ATSDR, 1992; Choudhry et al., 1986). Many animal studies have shown that the immune system is adversely affected on prolonged exposure (Bevenue and Beckman, 1967; Roszell and Anderson, 1994). Increased susceptibility to tumor growth was observed in mice fed diets containing 50 to 500 ppm PCP. Exposure of newborn guppies to the toxicant caused decreased growth, increased mortality, and delayed sexual maturity (Crandall and Goodnight, 1962). When administered to pregnant rats in doses of 5 to 50 mg/kg bw/d, symptoms of fetotoxicity were observed (Schwetz, 1974). These include dilated ureters, resorptions, subcutaneous edema, and anomalies in the bone structure.

Evidence of carcinogenicity for some PCP mixtures has also been found in mice. Mice fed diets with 100 to 200 ppm technical grade PCP showed increased incidences of hemangiosarcomas and hepatocellular neoplasms. Another group of male and female mice given up to 600 ppm PCP showed increased incidences of adrenal medullary and hepatocellular neoplasms. However, no consensus on its carcinogenicity has been reached, and so far the weight of evidence supports it not being carcinogenic.

38.9 SURVEY OF PAST SPILLS, LESSONS LEARNED, AND COUNTERMEASURES APPLIED

After the fish kill incident noted in Section 38.7.6, the company complied with the order to stop spraying.

As a result of the incident noted in Section 38.7.7 and similar incidents, Environment Canada encouraged the use of enclosed tanks or putting roofs over open tanks. It was also recommended that drainage areas be constructed to collect overflow in secondary tanks for holding purposes.

38.10 CONCLUSIONS

Pentachlorophenol has a variety of uses. It is an effective bacteriacide, algicide, fungicide, insecticide, and herbicide. It is used in adhesives, rubber, paints, oils, textiles, carpet shampoos, gaskets for food containers, beehives, and poultry products.

There is no doubt that the persistence of PCP and its impurities has been of concern since the late 1970s. As with many chlorinated organics, the long-term effects of this substance and its contaminants have not been clearly defined, even though some are known. It is now shown that although PCP is readily metabolized in animal tissues and in the environment, many of its metabolites, such as the anisoles and impurities such as dioxins, furans, and diphenyl ethers, are even more toxic or persistent. Not much is known about the toxicity of the tetra-, tri-, and dichlorophenol precursors. Use of PCP as a pesticide is now restricted. Monitoring of PCP in the environment should continue due to its widespread contamination of the marine and terrestrial ecosystems.

38.11 REFERENCES

Ahlborg, U. G., and T. M. Thunberg. 1980. "Chlorinated Phenols: Occurrence, Toxicity, Metabolism and Environmental Impact," *CRC Critical Review of Toxicology,* vol. 7, pp. 1–35.

Agency for Toxic Substances and Disease Registry (ATSDR). 1992. *Toxicological Profile for Pentachlorophenol:* U.S. Department of Health and Human Services.

Bevenue, A., and H. Beckman. 1967. "Pentachlorophenol: A Discussion of Its Properties and Its Occurrence as a Residue in Human and Animal Tissues," *Residue Reviews,* vol. 19, pp. 83–134.

Boyd, S. A. 1989. "Chlorophenols in Soils," in *Reactions and Movement of Organic Chemicals in Soils,* ed. B. L. Sawhney and K. Brown, Soil Science Society of America, Madison, WI, pp. 209–228.

Callahan, M. A. 1979. *Water-Related Environmental Fate of 129 Priority Pollutants,* 440/4-79-029b pp. 87–94.

Camford Information Services, Inc. (CIS). 1995. "Pentachlorophenol," *CPI Product Profiles,* CIS, Scarborough, ON.

Choudhry, H., J. Coleman, L. T. De Rosa, and J. F. Stara. 1986. "Pentachlorophenol: Health and Environmental Effects Profile," *Toxicology and Industrial Health,* vol. 2, pp. 483–571.

Crandall, C. A., and C. J. Goodnight. 1962. "Effects of Sublethal Concentrations of Several Toxicants on the Growth of the Common Guppy, *Lebistes reticulatus,*" *Limnology and Oceanography,* vol. 7, pp. 233–239.

Eisler, R. 1989. "Pentachlorophenol Hazards to Fish, Wildlife and Invertebrates: A Synoptic Review," *USDI/Contaminants Hazard Reviews,* vol. 85, pp. 1–17.

Environment Canada. 2000. *NATES—National Analysis of Trends in Emergencies Systems,* Hull, QC.

Fingas, M., N. Laroche, G. Sergy, B. Mansfield, G. Cloutier, and P. Mazerolle. 1991. "A New Chemical Spill Priority List," in *Proceedings of the Eighth Technical Seminar on Chemical Spills,* Environment Canada, Ottawa, ON.

Ghoshal, S., S. K. Banerji, and R. K. Bajpai. 1992. "Role of Photodegradation in Pentachlorophenol Decontamination in Soils," *Annals of the New York Academy of Sciences,* vol. 665, no. 7, pp. 412–422.

Glickman, A. H., C. N. Statham, A. Wu, and J. J. Lech. 1977. "Studies on the Uptake, Metabolism and Disposition of Pentachlorophenol and Pentachloroanisole in Rainbow Trout," *Toxicology and Applied Pharmacology,* vol. 41, p. 649.

Haggblom, M. M., and R. J. Valo. 1995. "Bioremediation of Chlorophenols Wastes," in *Microbial Transformation and Degradation of Toxic Organic Chemicals,* ed. L. Y. Young and C. E. Cerniglia, Wiley-Liss, New York, NY, pp. 389–434.

International Agency for Research on Cancer (IARC). 1994. *IARC Monographs on the Evaluation of Carcinogenic Risk of Chemicals to Man,* IARC, World Health Organization, Lyon, France.

Kobayashi, K., S. Kimura, and E. Shimizu. 1977. "Studies on the Metabolism of Chlorophenols in Fish—IX. Isolation and Identification of Pentachlorophenols-β-glucuronide Accumulated in Bile of Goldfish," *Bulletin of Japanese Society of Scientific Fisheries,* vol. 43, pp. 601–607.

Kozak, V. P., G. V. Simsiman, G. Chesters, D. Stensby, and J. Harkin. 1979. *Reviews of the Environmental Effects of Pollutants: XI Chlorophenols,* EPA-600/1-79-012, U.S. EPA Health Effects Research Laboratory, Cincinnati, OH.

Paasivirta, J., H. Hakala, J. Knuutinen, T. Otollinen, J. Sarkka, L. Welling, R. Paukku, and R. Lammi. 1990. "Organic Chlorine Compounds in Lake Sediments. III. Chlorohydrocarbons, Free and Chemically Bound Chlorophenols," *Chemosphere,* vol. 21, pp. 1355–1370.

Rao, K. R., ed. 1978. *Pentachlorophenol Chemistry, Pharmacology, and Environmental Toxicology,* Plenum Press, New York, NY.

Roszell, L. E., and R. S. Anderson. 1994. "Inhibition of Phagocytosis and Superoxide Production by Pentachlorophenol in Two Leukocyte Subpopulations from *Fundulus heteroclitus,*" *Marine Environmental Research,* vol. 38, pp. 195–206.

RTECS On-Line. 2000. *Registry of Toxic Effects of Chemical Substances,* Department of Health and Human Services, Centers for Disease Control and National Institute for Occupational Safety and Health, Washington, DC.

Schwetz, B. A. 1974. "The Effect of Purified and Commercial Grade Pentachlorophenol on Rat Embryonal and Fetal Development," *Toxicology and Applied Pharmacology,* vol. 28, pp. 151–161.

Tabak, H. H., S. A. Quave, C. A. Mashni, and E. F. Barth. 1981. "Biodegradability Studies with Organic Priority Pollution Compounds," *Journal of the Water Pollution Control Federation,* vol. 53, pp. 1503–1518.

Transport Canada. 1994. *Transport of Dangerous Goods Regulations (TDG),* Ottawa, ON.

Verschueren, K. 1996. *Handbook of Environmental Data on Organic Chemicals,* 3d ed., Van Nostrand Reinhold, New York, NY, pp. 1464–1480.

Wan, M. T. 1992. "Utility and Railway Right-of-Way Contaminants in British Columbia: Chlorophenols," *Journal of Environmental Quality,* vol. 21, pp. 225–231.

Weiss, U., P. Moza, I. Scheunert, A. Hague, and F. Korte. 1982a. "Fate of Pentachlorophenol-^{14}C in Rice Plants under Controlled Conditions," *Journal of Agricultural and Food Chemistry,* vol. 30, pp. 1186–1190.

Weiss, U., I. Scheunert, W. Klein, and F. Korte. 1982b. "Fate of Pentachlorophenol-^{14}C under Controlled Conditions," *Journal of Agricultural and Food Chemistry,* vol. 30, pp. 1191–1194.

Wood, S. 1983. "Pentachlorophenol Poisoning," *Journal of Occupational Medicine,* vol. 25, pp. 527–530.

CHAPTER 39
SODIUM CYANIDE: PROPERTIES, TOXICITY, USES AND ENVIRONMENTAL IMPACTS

Kerry Ketcheson and Merv Fingas
Emergencies Science Division, Environment Canada, Environmental Technology Centre, River Road, Ottawa, Ontario

39.1 PROPERTIES

Sodium cyanide (CAS No. 143-33-9) was first produced in 1834 by F. and E. Rogers, who heated Prussian blue (highly colored mixed oxidation state $Fe^{II/III}$ derivatives) and sodium carbonate together (King, 1994). The mixture was cooled and sodium cyanide was extracted with alcohol. The compound had no intended purpose until J. S. MacArthur and the Forrest brothers patented a process for extracting gold and silver from ores in 1887 (*Kirk-Othmer,* 1979a).

By 1899, half the cyanide manufactured in Europe was produced by the Beilby process. From 1900 to 1961, another process, called the Castner process, was used to make sodium cyanide. The Castner process is shown below:

$$2Na + 2C + 2NH_3 \rightarrow 2NaCN + 3H_2 \quad (39.1)$$

Today, a process based on the neutralization of sodium hydroxide (aqueous) and hydrogen cyanide (gas) is the preferred method, as shown below (*Kirk-Othmer,* 1993):

$$HCN + NaOH \rightarrow NaCN + H_2O \quad (39.2)$$

A few chemical reactions that occur with sodium cyanide should be mentioned. In moist air, carbon dioxide (CO_2) slowly decomposes sodium cyanide to hydrogen cyanide according to this reaction:

$$2NaCN + CO_2 + H_2O \rightarrow Na_2CO_3 + 2HCN \quad (39.3)$$

The carbonate salt produced becomes a brownish color due to the formation of polymerization products of HCN (*Ullmann's,* 1987).

At elevated temperatures, oxygen reacts with sodium cyanide to yield nitrogen, carbonate, and carbon dioxide according to the two-step reaction (*Ullmann's,* 1987):

$$2NaCN + O_2 \rightarrow 2NaOCN$$
$$2NaOCN + 3/2O_2 \rightarrow Na_2CO_3 + CO_2 + N_2 \quad (39.4)$$

When sodium cyanide is either stored for a long time or heated, there is a slow hydrolysis of the C≡N bond as per the equation (*Ullmann's,* 1987):

$$NaCN + 2H_2O \rightarrow HCOONa + NH_3 \quad (39.5)$$

Some basic physical properties of sodium cyanide are shown in Table 39.1. It should be noted that sodium cyanide has only a faint odor and is not detectable at concentrations that would provide a margin of safety. Sodium cyanide also sinks and is quite soluble in water—480,000 ppm at a temperature of 10°C. There are two different hydrates of sodium cynanide: the more common $NaCN \cdot 2H_2O$ (below 35°C) and $NaCN \cdot ½\ H_2O$ (>35°C) (Environment Canada, 1984; *Ullmann's,* 1987).

39.2 TOXICITY

It is estimated that the average lethal dose (LD_{50}) of sodium cyanide for humans is 2.86 mg/kg (Corn, 1993). Cyanide, originating by disassociation of sodium cyanide, complexes with the ferric iron ion (3+) in metalloenzymes, resulting in anoxia through inhibition of cytochrome oxidase, which inhibits cellular oxygenation. The enzyme becomes unable to catalyze the reactions in which electrons would be transferred from reduced cytochrome to oxygen. Thus, cellular oxygen utilization is impaired. The inhibition of oxygen causes oxygen hunger to rise in peripheral tissues. Thus, oxyhemoglobin is carried in the venous blood (TPF, 1997). This causes the pink or red discoloration of the skin that is seen in cyanide poisoning, which is the oxyhemoglobin entering the skin to provide the necessary oxygen.

Not all the cyanide that is absorbed by the body undergoes oxygen inhibition. It is known that 80% of the absorbed cyanide is detoxified in the liver by the enzyme mitochondrial rhodanase. The enzyme catalyzes the transfer of a sulfur atom to combine with the cyanide to form the less toxic thiocyanate, which is excreted in the urine. In high exposures to cyanide, the sulfur donors are rapidly depleted and cyanide metabolism is slowed (Pope and Rall, 1995). The half-life for the conversion of cyanide to thiocyanate from a nonlethal dose in humans is between 20 and 60 minutes (CHEM-BANK, 1999).

Other pathways to elimination of cyanide are possible. About 1 to 2% of cyanide is lost through the lungs by exhalation. Another 15% is released by conversion to 2-aminothiazaline-4-carboxylic acid, incorporation into the one-carbon metabolic pool, or the combination with hydroxycobalamin (vitamin B_{12a}) to form cyanocobalamin (Pope and Rall, 1995).

In normal, healthy human organs, cyanide is present at concentrations ranging up to 0.5 mg/kg (CHEM-BANK, 1999) and the relative proportion of thiocyanate to cyanide in body fluids is about 1,000:1. Normal plasma thiocyanate levels are 0.01 mg/mL for both smokers and non-smokers. Lethal values of thiocyanates range from 0.05 to 0.2 mg/mL (Pope and Rall, 1995). Smokers are included in the data because heavy smoking increases the urinary concentrations of thiocyanate levels and can be misleading for cyanide exposure (Corn, 1993).

Sodium cyanide is known to be very toxic to humans. Some human toxicity data are shown in Table 39.3. To put the data into perspective, for oral—human LD_{Lo} 2.857 mg/kg, a human of 70 kg (154 lbs) would only have to ingest 0.2 g (200 mg) in order to be in the range of a lethal dose. The data in Table 39.3 demonstrate that even small amounts of sodium cyanide can cause harm to the human body.

Other data from the testing of sodium cyanide on animal species also show that sodium cyanide is very toxic. For example, as shown in Table 39.4, a small dose of 0.0085 g of sodium cyanide per kilogram administered to quail has killed half of the test animals.

TABLE 39.1 Physical Properties of Sodium Cyanide

Properties	Sodium cyanide
Appearance	White, granular or crystalline solid (*NIOSH Pocket Guide to Chemical Hazards*)
Odor	Faint almond-like odor (*NIOSH Pocket Guide to Chemical Hazards*)
Odor threshold	0.2 to 5 ppm (as HCN), poor warning properties, not detectable at concentrations providing a margin of safety (MSDS Database A and B)
Reaction with water	Sinks and mixes with water (CHRIS Database)
Molecular weight	49.015
Melting point	561.7°C (98 wt%), 562°C (*Kirk-Othmer,* 1993), 560°C (CHEMtox, 1999)
Boiling point	1500 ± 10°C, 1530°C (*Kirk-Othmer,* 1993), 1499.9°C (CHEMtox Database, 1999)
Density of solid	1.620 g/cm^3 (6°C, rhombic)
Density of solid	1.595 g/cm^3 (20°C, cubic), 1.60 (*Kirk-Othmer,* 1993)
Density of solid	1.19 g/cm^3 (850°C, fused)
Density of 10% (w/v) aqueous solution	1.047 g/mL (25°C) (*Kirk-Othmer,* 1993)
Density of 20% (w/v) aqueous solution	1.098 g/mL (25°C) (*Kirk-Othmer,* 1993)
Density of 30% (w/v) aqueous solution	1.150 g/mL (25°C) (*Kirk-Othmer,* 1993)
Viscosity	4 cP (30°C) (26% aqueous solution)
Biological oxygen demand (BOD)	(theoretical) 6%, 7 days (CHRIS Database)
Bioaccumulation and tainting	No evidence to support any rating (CHRIS Database)
Specific heat capacity (Cp)	1.667 kJ/kg · K (0.1°C)
Specific heat capacity (Cp)	31.630 kJ/kg · K (15.5°C)
Specific heat capacity (Cp)	1.402 kJ/kg · K (25.6°C)
Enthalpy at 25°C (H)	−89.9 kJ/mol
Heat of fusion	314 kJ/kg
Heat of formation	−89.9 kJ/mol (*Kirk-Othmer,* 1993)
Heat of vaporization (ΔH_v)	3185 kJ/kg, 3041 kJ/kg (*Kirk-Othmer,* 1979b), 3190 kJ/kg (Hazardous Substances Data Bank)
Heat of solution (ΔH_{soln})	−1.548 kJ/mol (*Kirk-Othmer,* 1993), 1.51 kJ/mol (Hazardous Substances Data Bank)
Hydrolysis constant, K_h at 25°C	0.0000251 (*Kirk-Othmer,* 1993)
Vapor pressure (P_v) at 800°C	0.1 kPa, 0.1013 kPa (*Kirk-Othmer,* 1993)
Vapor pressure (P_v) at 1,000°C	1.65 kPa
Vapor pressure (P_v) at 1,200°C	11.98 kPa
Vapor pressure (P_v) at 1,350°C	39.10 kPa
Vapor pressure (P_v) at 1,360°C	41.8 kPa (*Kirk-Othmer,* 1993)
Index of refraction	1.452
Solubility (10°C)	480,000 ppm
Solubility (35°C)	820,000 ppm
Solubility of $NaCN \cdot 2H_2O$ per 100 g of saturated solution	26.01 (−15°C) (*Kirk-Othmer,* 1993)
Solubility of $NaCN \cdot 2H_2O$ per 100 g of saturated solution	32.8 (10°C) (*Kirk-Othmer,* 1993)
Solubility of $NaCN \cdot 2H_2O$ per 100 g of saturated solution	34.2 (15°C) (*Kirk-Othmer,* 1993)
Solubility of $NaCN \cdot 2H_2O$ per 100 g of saturated solution	45 (34.7°C) (*Kirk-Othmer,* 1993)
Solubility in 100 g of ethanol (100%)	1.235 g (25°C)
Solubility in 100 g of ethanol (95%)	2.445 g (25°C)
Solubility in 100 g of anhydrous methanol	6.44 g (15°C)
Solubility in 100 g of anhydrous methanol	4.58 g (25°C)
Solubility in 100 g of anhydrous methanol	4.10 g (67.4°C)
Solubility in 100 g of NH_3, liquid	58 g (−31°C)

Source: *Kirk-Othmer,* 1979a; *Ullmann's,* 1987; CHEM-BANK, 1999.

TABLE 39.2 Blood Cyanide Effects

Whole blood cyanide level concentrations (mg/mL)	Effect on body
Less than 0.0002	No symptoms of toxicity
0.0004 (in smokers)	No symptoms of toxicity
0.0005 to 0.001	Untreated patients may be conscious, flushed, and tachycardic
0.001 to 0.0025	Stupor and agitation
Greater than 0.0025	Coma and potentially fatal without treatment

Source: Pope and Rall, 1995.

When sodium cyanide mixes with water and comes in contact with aquatic species, the results are also detrimental to the health of the species. When mixed with water, sodium cyanide can undergo two possible reactions (*Kirk-Othmer,* 1993).

$$NaCN + H_2O \leftrightarrows NaOH + HCN \text{ (room temperature)} \tag{39.6}$$

$$NaCN + 2H_2O \leftrightarrows NH_3 + HCOONa \text{ (> 50°C)} \tag{39.7}$$

As in the case of most accidental spills, Eq. (39.6) would be the reaction since the water is generally not above 50°C. Thus, a spill of sodium cyanide in water would form hydrocyanic acid and sodium hydroxide.

In water, the pH is affected by how much sodium cyanide is added. The concentration of sodium cyanide added correlated to the pH of the water is shown in Table 39.5. The aqueous solutions are strongly alkaline. At pH 9.4, in an aqueous solution of sodium cyanide, half of the total cyanide is present as hydrogen cyanide as in Eq. (39.6) (*Ullmann's,* 1987).

The toxicity of sodium cyanide on freshwater aquatic life is shown in Table 39.6. Sodium cyanide is detrimental to aquatic life because of the dual toxicity effect of an acid and a base impacting the environment. A range of tolerance is shown in the table. For example, bluegill at 96 hours TLm is 0.15 mg/kg, whereas oligochaete at 96 hours LC_{50} is 11 mg/L. Thus, it takes 73 times more sodium cyanide to kill half the oligochaete than it takes to kill half of the bluegill.

It should also be noted that cyanide toxicity to fish increases with any reduction in dissolved oxygen below 100%. The presence of zinc and ammonia results in a greater than additive increase in toxicity (Moran, 1998). Fish are about 1000 times more sensitive to cyanide than are humans. Even if the levels are less than lethal, there are still toxic effects, including physiological and pathological responses. For example, cyanide can reduce swimming ability, which will leave fish more vulnerable to predators, or interfere with reproductive capacity, which can lead to deformed offspring. It should be noted that cyanide toxicity in fish increases three- fold with a 12°C decrease in temperature. As well, 17 parts per thousand (ppt) of chloride ion or 8.8 ppt (Moran, 1998) is known to decrease the survival time (UNEP/OCHA, 2000).

TABLE 39.3 Human Toxicity Data

Oral—man LD_{Lo} 6.557 mg/kg (gastrointestinal, gastritis)
Oral—human LD_{Lo} 2.857 mg/kg
Unreported—human LD_{Lo} 2.206 mg/kg
Oral—man TD_{Lo} 0.714 mg/kg (hallucinations, distorted perceptions, muscle weakness)

Source: CHEMtox Database, 1999; RTECS Database.

TABLE 39.4 Animal Toxicity

Oral—rat LD_{50} 6.44 mg/kg
Intraperitoneal—rat LD_{50} 4.3 mg/kg
Oral—quail LD_{50} 8.5 mg/kg
Oral—wild bird species LD_{50} 4 mg/kg

Source: CHEMtox Database, 1999; RTECS Database.

There is some saltwater toxicity data as well, although it is not as extensive as freshwater data. This is shown in Table 39.7.

39.3 CATEGORIZATIONS AND GUIDELINES

A compilation of classifications, guidelines, and legislative limits is given in Table 39.8. The information from Transport of Dangerous Goods (TDG) states that sodium cyanide is a poison. Other references, such as the National Fire Protection Association (NFPA) and the Hazardous Materials Identification System (HMIS), identify sodium cyanide as a serious health hazard. The classifications suggest that it is not a flammability hazard and is not very reactive (stable). Also note that the drinking water standards have been set at 0.2 mg/L for safe drinking water. At a sodium cyanide level of 0.2 mg/L, not all of the fish are safe. Note that at this concentration, from Table 39.6, half of the bluegill will die after 48 hours and half of the rainbow trout after 144 hours.

39.4 INDUSTRIAL USES

During the 1950s, almost all sodium cyanide was used in case hardening and electroplating. Today, case hardening is a minor use, while electroplating or metal plating, gold and silver extraction, and chemical synthesis are major applications of sodium cyanide (*Kirk-Othmer,* 1979a).

Figure 39.1 shows a comparison of how much sodium cyanide is consumed in these processes relative to the other major uses from 1976 to 1997. It can be seen that gold cyanidation is the most prominent process. A description of the different processes will follow.

TABLE 39.5 pH of Sodium Cyanide

Concentration of sodium cyanide (mol/L)	Concentration of sodium cyanide (ppm)	pH
1	49,015	11.64
0.1	4,902	11.15
0.01	490	10.67
0.001	49	10.15
0.0001	4.9	9.6
0.00001	0.5	8.9

Source: *Ullmann's,* 1987.

TABLE 39.6 Freshwater Toxicity of Sodium Cyanide

Species	Concentration (mg/kg)	Exposure (hours)	Effect
Apple snail	1.6–2.9 (mg/L)	96	LC_{50}
Fathead minnow	0.25	24	TLm
Fathead minnow	0.24	48	TLm
Fathead minnow	0.23	96	TLm
Yellow perch	0.0758–0.337 (mg/L)	96	LC_{50}
Bluegill	5	2	Killed
Bluegill	0.1959 (mg/L)	3.36	LT_{50}
Bluegill	0.16	48	LC_{50}
Bluegill	0.16 (mg/L)	72	LC_{50}
Bluegill	0.15	96	TLm
Eurasian water milfoil	22.4 (as CN)		IL_{50}, root weight
Eurasian water milfoil	28.6 (as CN)		IL_{50}, root length
Eurasian water milfoil	20 (as CN)		IL_{50}, stem weight
Eurasian water milfoil	27.3 (as CN)		IL_{50}, stem length
Archiannelid	8.9–11 (mg/L)	24	LC_{50}
Archiannelid	7.1–9.5 (mg/L)	48	LC_{50}
Archiannelid	6.7–8.5 (mg/L)	72	LC_{50}
Archiannelid	5.9–7.6 (mg/L)	96	LC_{50}
River snail	940 (mg/L)	24	LC_{50}
River snail	760 (mg/L)	48	LC_{50}
Water flea	<3.4		Immobilized
Water flea	0.09–0.3 (mg/L)	96	LC_{50}
Water flea	0.02–0.05 (mg/L)	120	LOEC REP
Polycelis nigra	30		Threshold concentration
Trout	0.08		Nontoxic
Trout	2	1	100% mortality
Trout	0.05	124	100% mortality
Oligochaete	11 (mg/L)	96	LC_{50}
Pond snail	3.3 (mg/L)	24	LC_{50}
Pond snail	3.3 (mg/L)	48	LC_{50}
Pond snail	2.4–2.5 (mg/L)	96	LC_{50}
Fish (*Labeo bata*)	0.4–1.9 (mg/L)	96	LC_{50}
Rohu	0.3–1.9 (mg/L)	96	LC_{50}
Catla	0.5–1.7 (mg/L)	96	LC_{50}
Cyclopoid copepod	0.1–0.3 (mg/L)	96	LC_{50}
Calanoid copepod	0.1–0.3 (mg/L)	96	LC_{50}
Scud	0.9 (mg/L)	96	LC_{50}
Rainbow trout	0.05–0.09 (mg/L)	96	LC_{50}
Rainbow trout	0.098 (mg/L)	144	LC_{50}
Minnows	0.4	1	Stopped eating
Minnows	0.33–0.35	2.5	Lethal
Minnows	0.5 to 0.7	24	25% mortality
Minnows	0.75	24	50% mortality
Minnows	0.8	24	100% mortality

TABLE 39.6 Freshwater Toxicity of Sodium Cyanide (*Continued*)

Species	Concentration (mg/kg)	Exposure (hours)	Effect
Pouch snail	2.4 (mg/L)	24	LC_{50}
Pouch snail	135 (mg/L)	48	LC_{50}
Stickleback	2		Lethal
Hardy carp	4.3		Paralyzed
Shiner	10	0.06	Killed
Shiner	2	0.2	Killed
Snail	2.9 (mg/L)	96	LC_{50}
Carp, hawk fish	0.37 (mg/L)	96	LC_{50}
Trout fingerlings	0.02		No effect
Gold and green sunfish	0.5	4 to 6	Lethal
Sea lamprey	5.0	2	Ill, but survived

Source: Sax, 1989; Aquire Database, 1994.

Cyanidation is a chemical process used to recover gold and other metals that are very finely distributed in the ore, or gold that is complexed with sulfides. The ore or tailings are added to a cyanide solution (0.2 to 0.05%), which is aerated to provide oxygen to dissolve the gold as $Na[Au(CN)_2]$. A typical concentration of cyanide usage is 1 to 2.5 kg NaCN per metric ton of ore (*Kirk-Othmer,* 1995b), or 0.25 to 1.0 kg (Ingles and Scott, 1987). This is known as the MacArthur process and is shown in the following equation:

$$4NaCN + 2Au + \tfrac{1}{2} O_2 + H_2O \leftrightarrows 2NaAu(CN)_2 + 2NaOH \qquad (39.8)$$

The solution is then treated and clarified and the gold is precipitated by the addition of zinc dust according to the following reaction (Clayton and Clayton, 1991; CIS, 1998; *Kirk-Othmer,* 1980):

$$2Au(CN)_2^{2-} + Zn \leftrightarrows 2Au + Zn(CN)_4 \qquad (39.9)$$

Sodium cyanide is used in the extraction of gold and silver from ores because of the very stable linear cyano complexes formed with these metals. The molecular bonding for silver cyano complexes is drawn as $[:N{\equiv}C{-}Ag{-}C{\equiv}N:]^-$ and similarly for gold complexes $[:N{\equiv}C{-}Au{-}C{\equiv}N:]^-$ (King, 1994).

Sodium cyanide is also used in the electroplating industry. Cadmium is usually plated from cyanide solutions from a still plating process (90 to 120 g/L NaCN), or barrel plating

TABLE 39.7 Saltwater Toxicity of Sodium Cyanide

Species	Concentration (mg/kg)	Exposure (hours)	Effect
Aquatic sowbug	1.7 (mg/L)	96	LC_{50}
Prawn	0.25	48	LC_{50}
Cockle	>25	48	LC_{50}
Three-spine stickleback	0.2254 (mg/L)	3.36	LC_{50}
Three-spine stickleback	0.1312 (mg/L)	13.68	LC_{50}
Crab	75	48	LC_{50}

Source: Sax, 1989; Aquire Database, 1994.

TABLE 39.8 Legislative Limits and Classifications

Topic	Limit
TDG classification	6.1, 9.2 (9.2 applies if regulated limit is exceeded) Regulated Limit (Schedule XII) S4, UN1689 Regulated Limit (Schedule XIII) 0.5 kg UN1689 (CCOHS, 2000e)
Packing group	I
Emergency response assistance planning requirements of Sections 7.16 to 7.19 of TDG	For quantities exceeding 1,000 kg or L net per consignment
Consumer commodity	Prohibited
Limited quantity	Prohibited
Passenger road/railway vehicles	Maximum net quantity per package is 5 kg
Standard industry trade classification	52381 (CHRIS Database)
Stability during transport	Stable (CHRIS Database)
RCRA hazardous waste	No. P106 (*Genium's,* 1999)
Fire diamond (NFPA hazard rating)	3 health hazard (extreme danger) 0 Flammability (will not burn) 0 Reactivity (stable and not reactive with water) (*Genium's,* 1999)
The Hazardous Materials Identification System (HMIS)	H 3 (serious health hazard) F 1 (slight flammability hazard) R 1 (slight reactivity hazard) (*Genium's,* 1999)
Wilson's Risk Scale of Material Hazards (RISK)	R 1 (reactivity: stable at room temperature; may be unstable at elevated temperatures) I 2 [inhalation: TLV 101 to 500 ppm (vapor) or 1.1 to 10 mg/m^3 (solid)] S 3 (skin contact: severe irritation; tissue corrosion within short time period) K 0 (kindling: will not burn) (*Genium's,* 1999)
Workplace Hazardous Materials Information System (WHMIS)	D1A (very toxic material) D2B (other toxic effects that are not immediate and serious) E (corrosive) (*Chemical Advisor*)
IDLH	25 mg/m^3 (as CN)
TLV-TWA (ceiling exposure limit)	4.7 ppm to 10 minutes (as CN) skin designation
NIOSH recommendations	4.7 ppm (hydrogen cyanide and cyanide salts)
Federal drinking water guidelines	0.2 mg/L (cyanide ion)

Source: CHRIS Database.

(70 to 90 g/L NaCN). Because cadmium is very harmful to the environment, the use of plating with cadmium is declining (*Kirk-Othmer,* 1979b).

Reactions that are of commercial importance in electroplating require sodium cyanide. In the reaction shown in the following equations, silver is dissolved by sodium cyanide solutions as the anode,

$$Ag + 2NaCN \rightarrow NaAg(CN)_2 + Na^+ + e \qquad (39.10)$$

or in the presence of oxygen to form $NaAg(CN)_2$ (*Kirk-Othmer,* 1984):

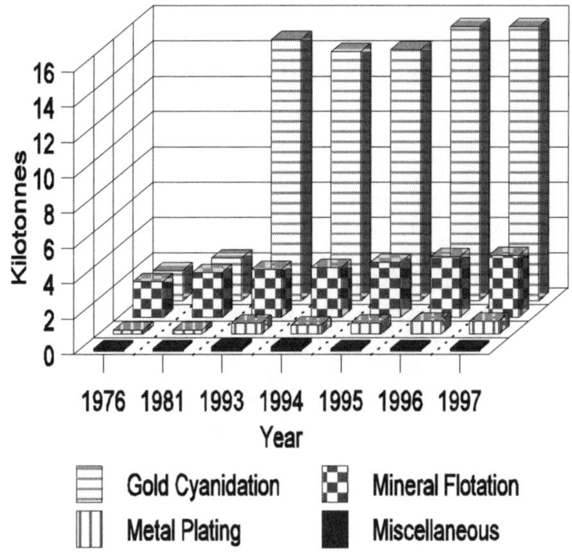

FIGURE 39.1 Uses of sodium cyanide—1976–1997. (*Source: CIS, 1998.*)

$$2Ag + 4NaCN + H_2O + 0.5O_2 \rightarrow 2NaAg(CN)_2 + 2NaOH \tag{39.11}$$

Brass is plated not only for decorative purposes but also to ensure a good adhesion of rubber to steel wire, which is used in tires. An example of such a plating bath uses a concentration of sodium cyanide at 90 g/L for a 70/30 brass plating. For white brass plating, the amount of sodium cyanide is 52 to 60 g/L (*Kirk-Othmer,* 1979b). In the late 1980s, cyanides were largely removed from plating baths due to the environmental hazards of using them. Since 1995, cyanide-containing cleaners are rare and are used only for special purposes.

Sodium cyanide is also used as a depressant in the mining industry. A depressant is a reagent that selectively prevents the reaction between a collector and a mineral. In other words, a depressant prevents the mineral from flotation. The use of sodium cyanide, in this case, is to selectively depress sphalerite (zinc sulfide) and pyrite (iron sulfide) but not galena. Sodium cyanide therefore enhances flotation of galena (*Kirk-Othmer,* 1995b).

In the process known as cyaniding, carbon and alloy steels are combined at high temperatures in a liquid bath from which they absorb both carbon and nitrogen simultaneously. The purpose of the nascent nitrogen is to increase the metal's surface hardness. The typical composition of the liquid bath is 30% sodium chloride, 40% sodium carbonate, and 30% sodium cyanide. The sodium cyanide concentration may range from 97 to 45%. The following are equations for producing nitrogen and carbon.

$$2NaCN + O_2 \rightarrow 2NaCNO \tag{39.12}$$

The sodium cyanate created breaks down to form nascent nitrogen:

$$4NaCNO \rightarrow Na_2CO_3 + 2NaCN + CO + 2N \tag{39.13}$$

The carbon monoxide (CO) produces carbon dioxide:

$$2CO \rightarrow CO_2 + C \tag{39.14}$$

The carbon dioxide reacts with sodium cyanide:

$$NaCN + CO_2 \rightarrow NaCNO + CO \tag{39.15}$$

The carbon monoxide (CO) further reacts to produce more carbon (C).

The following three reactions lead to the decomposition of the sodium cyanate (NaCNO) (*Kirk-Othmer,* 1981):

1. $NaCNO + C \rightarrow NaCN + CO$
2. $4NaCNO + 4CO_2 \rightarrow 2Na_2CO_3 + 2CO + 4N + 2C$
3. $4NaCNO + 4CO_2 \rightarrow 2Na_2CO_3 + 6CO + 4N$

Sodium cyanide is used for case hardening of steels. Trace amounts of iron or nickel oxide are used to cause a rapid oxidation of the cyanide in air, first to the cyanate, then to carbonate, by the following reactions:

$$2NaCN + O_2 \rightarrow 2NaCNO \tag{39.16}$$

$$2NaCNO + 3/2\ O_2 \rightarrow Na_2CO_3 + N_2 + CO_2 \tag{39.17}$$

Case hardening of steel using a sodium cyanide molten bath depends on the above reactions where the active carbon and nitrogen are absorbed into the steel surface. Sodium cyanide is a good reducing agent and the oxides of tin, lead, copper, or manganese are readily reduced (*Kirk-Othmer,* 1993).

39.4.1 Chemical Synthesis Using Sodium Cyanide

For chemical synthesis of a phosphorus–carbon bond, sodium cyanide can be used as shown below (*Kirk-Othmer,* 1982a):

$$PCl_3 + 3NaCN \rightarrow P(CN)_3 + 3NaCl \tag{39.18}$$

Ethylene cyanohydrin is obtained from ethylene chlorohydrin and sodium cyanide in this reaction (*Kirk-Othmer,* 1978a):

$$HOCH_2CH_2Cl + NaCN \rightarrow HOCH_2CH_2CN + NaCl \tag{39.19}$$

Sodium cyanide reacts with 1,2-dichloroethane to produce 2-chloropropionitrile in this reaction (*Kirk-Othmer,* 1979a):

$$ClCH_2CH_2Cl + NaCN \xrightarrow{\text{In methanol (50°C)}} HOCH_2CH_2CN + NaCl \tag{39.20}$$

Thiocyanocarbons are prepared using sodium cyanide shown below (*Kirk-Othmer,* 1979a):

$$CS_2 + NaCN \rightarrow NCC({\equiv}S)SNa \rightarrow (SNa)_2{-}C{=}C{-}(CN)_2 \tag{39.21}$$

A chemical process to synthesize malonates uses sodium cyanide. In this procedure, a 25% solution of sodium cyanide in water is heated to 65 to 70°C. The reaction is shown below (*Kirk-Othmer,* 1995a):

$$ClCH_2COONa + NaCN \rightarrow NCCH_2COONa + NaCl \quad (39.22)$$

$$NCCH_2COONa \xrightarrow[H_2SO_4, H_2O]{+ROH} ROOCCH_2COOR \quad (39.23)$$

Sulfuric acid reacts with sodium cyanide according to the reaction (*Ullmann's,* 1987):

$$2NaCN + H_2SO_4 \rightarrow Na_2SO_4 + 2HCN \quad (39.24)$$

39.4.2 Miscellaneous Uses of Sodium Cyanide

Sodium cyanide is also used as a radioactive tracer, sodium [^{14}C]-cyanide (*Kirk-Othmer,* 1996). In the United States, it is used as an active ingredient in a predator collar around a sheep's neck to poison coyotes (*Kirk-Othmer,* 1982b).

For the purpose of sterilization, sodium cyanide is used to remove fish, nuisance vegetation, and predators of fry by applying a concentration of 5 ppm to the water. Since sodium cyanide has a rapid rate of deterioration, fry can be restocked soon after the chemical has sterilized the pond (*Kirk-Othmer,* 1978b).

39.5 IMPORTS AND COSTS

Countries that imported sodium cyanide from 1976 to 1997 are shown in Fig. 39.2. It can be seen that the United States is a major user of sodium cyanide, followed by the United Kingdom.

As shown in Table 39.9, the cost ranges within each year depend on supply and demand. Also from 1993 to 1997, costs have risen 77 cents per kilogram from the extreme low to the extreme high end.

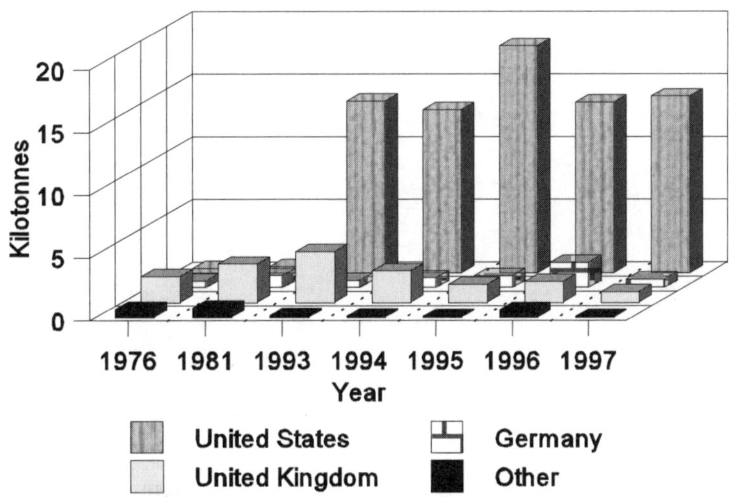

FIGURE 39.2 Imports of sodium cyanide by country. (*Source: CIS, 1998.*)

TABLE 39.9 Price History

Year	Bulk, tons/literdelivered to mine site: cents/kilogram	
	High	Low
1993	153	148
1994	153	153
1995	185	153
1996	225	185
1997	225	225

Source: CIS, 1998.

39.6 SPILL DATA

Cyanide spills in general were tracked using data from National Analysis of Trends in Emergencies Systems (NATES) for 1974 to 1995 (Environment Canada, 2000). This is shown in Fig. 39.3 and includes spills of cyanide leachates, sodium cyanide, cyanide not otherwise specified (nos), and cyanide/copper tailings. As can be seen in Fig. 39.4, cyanide nos is the largest of the spills and may contain any cations with cyanide or the cyanide ion itself.

The number of reported spills of sodium cyanide from 1974 to 1995 was extracted from the same NATES databank. It can be seen in Fig. 39.3 that no more than 2 spills of sodium

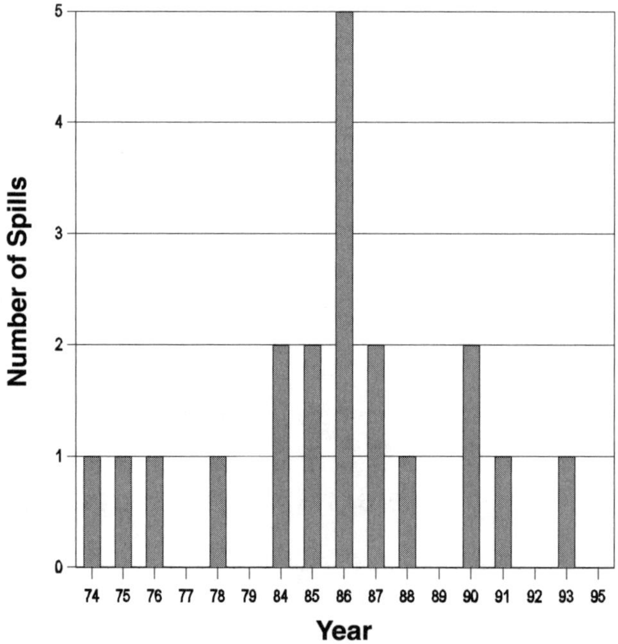

FIGURE 39.3 Number of sodium cyanide spills from 1974–1995. (*Source: Environment Canada, 2000.*)

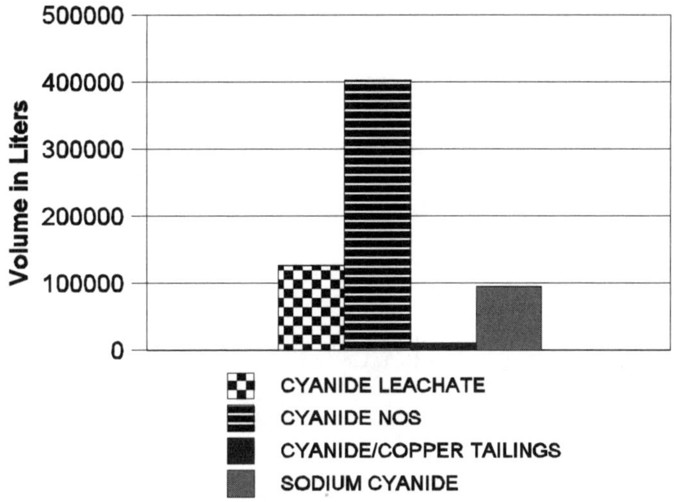

FIGURE 39.4 Cyanide spills from 1974–1995. (*Source: Environment Canada, 2000.*)

cyanide typically occur each year. However, Fig. 39.4 shows that sodium cyanide spills released just under 100,000 L into the environment from a total of 20 spills recorded in Fig. 39.3.

39.7 POTENTIAL COMPLEXES OF SODIUM CYANIDE IN A SPILL

Cyanide will form complexes of very stable, insoluble, or slightly soluble simple cyanides with heavy metals. Unidentate cyanide always bonds to the metal via its carbon atom, but the M—C≡N—M links can be seen in many of the solid cyanides (King, 1994). There may be some settling of heavy metal cyanide complexes to the sediment. Some of the more stable complexes and their respective dissociation properties are shown in Table 39.10.

When metal cyanide complexes are formed and released into the environment, they begin to decompose at varying rates. The weak cyanide complexes, such as zinc and cadmium, will decompose faster than strong complexes such as copper, nickel, iron, cobalt, gold, and silver cyanides. Strong cyanide complexes do not break down in the presence of strong acids, such as hydrochloric acid or sulfuric acid, but will decompose when exposed to various wavelengths of light. Cyanide ions will be released when the decomposition process begins (Moran, 1998).

39.8 SPILLS OF SODIUM CYANIDE

39.8.1 Cyanide Spill at Gold Mine, Romania

On January 30, 2000, the Baia Mare gold mine in Romania reported a massive cyanide spill when the tailings dam collapsed due to pressure from snow buildup. Baia Mare is owned and operated by the Romanian company Aurul and Esmeralda Explorations Limited of Perth,

TABLE 39.10 Cyanide Dissociation Equilibria

Dissociation equilibria	$pK_{dissociation}$	Toxicity range for freshwater fish (mg/L)
$[Pb(CN)_4]^{2-} \leftrightarrows Pb^{2+} + 4CN^-$	10.3	
$[Cd(CN)_4]^{2-} \leftrightarrows [Cd(CN)_3]^- + CN^-$	2.5	0.02–0.3
$[Cd(CN)_3]^- \leftrightarrows Cd^{2+} + 3CN^-$	14.7	0.02–0.3
$[Zn(CN)_4]^{2-} \leftrightarrows [Zn(CN)_3]^- + CN^-$	1	0.02–0.3
$[Zn(CN)_3]^- \leftrightarrows Zn^{2+} + 3CN^-$	17.9	0.02–0.3
$[Ag(CN)_2]^- \leftrightarrows Ag^+ + 2CN^-$	20.9	
$[Ni(CN)_4]^{2-} \leftrightarrows Ni^{2+} + 4CN^-$	22	0.4 (pH 6.5), 730 (pH 8)
$[Cu(CN)_4]^{3-} \leftrightarrows [Cu(CN)_3]^{2-} + CN^-$	1.5	0.4–4.0
$[Cu(CN)_3]^{2-} \leftrightarrows [Cu(CN)_2]^- + CN^-$	5.3	0.4–4.0
$[Cu(CN)_2]^- \leftrightarrows Cu^{2+} + 2CN^-$	23.9	0.4–4.0
$[Fe(CN)_6]^{3-} \leftrightarrows Fe^{2+} + 6CN^-$	36	300 (in darkness), <0.2 (in light)
$[Au(CN)_2]^- \leftrightarrows Au^{2+} + 2CN^-$	37	
$[Hg(CN)_4]^{2-} \leftrightarrows Hg^{2+} + 4CN^-$	40.5	
$[Fe(CN)_6]^{4-} \leftrightarrows Fe^{2+} + 6CN^-$	42	
$[Co(CN)_6]^{4-} \leftrightarrows Co^{2+} + 6CN^-$	64	
$[Pt(CN)_4]^{2-} \leftrightarrows Pt^{2+} + 4CN^-$	40	
$[Pd(CN)_4]^{2-} \leftrightarrows Pd^{2+} + 4CN^-$	42	

Source: *Ullmann's,* 1987; Moran, 1998.

Australia. The spill was estimated to be 26 million gallons (98,410,000 L) (OSHS, 2000a) of wastewater, highly contaminated with cyanides and heavy metals (WWF, 2000a, b). The tailings flowed into the Szamos River in Hungary, then proceeded to flow down the Tisza River and into the Serbian Republic to the larger Danube upstream of Belgrade in Yugoslavia.

The wave of cyanide-contaminated tailings took several days to travel the rivers to Belgrade. As can be seen in Fig. 39.5, samples taken from the Szamos River at Csenger on February 1, 2000 were 32.6 ppm cyanide. On February 11, samples from the Tisza River at Tiszasziget showed concentrations of 1.49 ppm cyanide. These concentrations could kill some fish species. Along the way, the authorities closed water intakes and warned the public against any contact with the river water or dead fish. The city of Szolnok, which depends entirely on the Tisza for drinking water, was forced to use additional chlorine and ozone as a neutralization method. (OSHS, 2000b).

The company treated the spillage with sodium hypochlorite to neutralize the cyanide. The desired reaction is shown below (McGill and Comba, 1990):

$$NaCN + NaClO \rightarrow NaCNO + NaCl \qquad (39.25)$$

The wave of cyanide had a negative impact on life in the Tisza River and killed an estimated 80% of the fish in the river (see Fig. 39.6).

On February 8, it was found that eight drinking wells were contaminated with cyanide (UNEP/OCHA, 2000). A week later, the water appeared to be safe to drink. On March 6, 17 scientists from seven countries finished sampling the water of the Danube River, 800 km downstream from the mine. A total of 180 tons of dead fish were removed from Hungary's Szamos and Tisza Rivers. According to Hungary's Minister of the Environment, the recovered fish represented only 15% of the total fish killed, while an estimated 15% drifted down the Tisza into Yugoslavia and another 15% were still trapped under ice.

The Australian mining company maintained that there was no scientific evidence linking extensive fish kills with the spill of tailing pond water (OSHS, 2000a, b). This comment is in stark contrast to that of György Gado, Conservation Director of the World Wildlife Fed-

FIGURE 39.5 Tracking the sodium cyanide spill from Baia Mare. (*Source: MTI, 2000.*)

eration in Hungary, who stated, "This spill has, in practical terms, eradicated all life from a stretch of up to 400 km of the Tisza River" (WWF, 2000c).

The chairman of the Hungarian government's Environment Committee was very concerned that the heavy metals in the environment could pose a far greater health threat in the long term (Ottawa Citizen Online, 2000). Given the complexes that cyanide can form with heavy metals, the sediment could remain highly toxic for some time. It was found that the contamination in the sediments decreased rapidly with increased distance from the source of the spill. The concentrations of heavy metals in the Lapus River were high or extremely high (UNEP/OCHA, 2000).

FIGURE 39.6 Fishermen from the Tisza region collecting dead fish. (*Source: WWF, 2000a, b, c.*)

The breached tailings pond at the gold mine, which was 93 ha in area and about 20 m deep, was lined with a plastic membrane and held approximately 4.43 million tons of flotation solid wastes. Due to an imbalance between water burden, unexpected quantities of water from the addition of snow and rain, and safe storage volume, the capacity of the pond was insufficient to safely store the slimes. It was estimated that 50 to 100 tons of cyanide were released into the river (UNEP/OCHA, 2000).

Although the company was operating with a government permit, no contingency plan was in place. The emergency preparedness and response procedures were rudimentary considering the quantities of chemicals involved and the mine's proximity to human population and the river systems (UNEP/OCHA, 2000).

It was recommended that a risk assessment be carried out on the entire system, that the process water ponds be reduced to sizes that could be handled in emergencies, and that emergency water supplies be available (UNEP/OCHA, 2000).

The following are the impacts of the spill:

- 180 tons (ETK, 2000) or 1,000 tons (UNEP/OCHA, 2000) of dead fish
- Cyanide and cyanide-heavy metal contamination in riverbeds of Tisza and Szamos Rivers
- Otters are now extinct in the Tisza and Szamos Rivers
- Birds and animals died from eating contaminated fish
- Groundwater resources contaminated

It is estimated that it will take 1 to 2 years to rehabilitate the water and 10 years for the riverbed to recover (ETK, 2000).

39.8.2 Cyanide Spill at Omai Gold Mines, Guyana

An uncontrolled release of cyanide-contaminated wastewater occurred on August 19, 1995, at a gold mine in central Guyana (see Fig. 39.7). The mine was operated by Omai Gold Mines Limited (OGML), a Canadian- (Cambior Inc.) and U.S.-owned company. The failure of the dam was caused by massive loss of core integrity from internal erosion of the dam fill, a process called piping (Vick et al., 1995). On August 24, the spill was contained and diverted to an open pit mine. It was estimated that 2.6 to 2.7 million m^3 of contaminated water flowed into the Essequibo River. This is the country's major river, used for drinking, bathing, and fishing (NEEC Alert, 1995).

The spill was colored red by the clay saprolite core material of the dam. The release flowed into the Omai River, which is 3 to 5 m wide, before entering the Essequibo River. It was estimated that the dilution factor would be 150:1 in the Essequibo River and that this

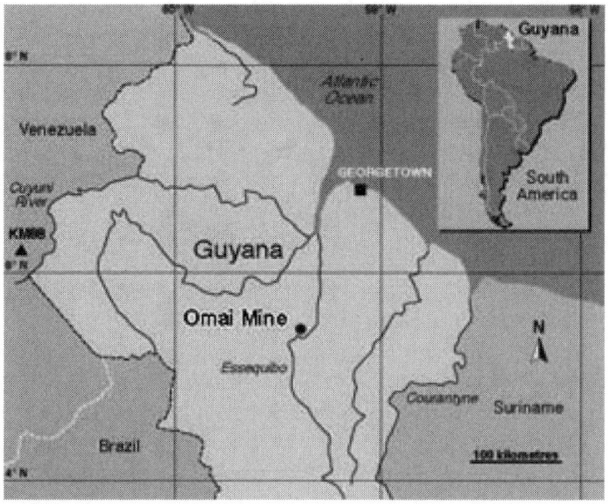

FIGURE 39.7 Omai Mine in Guyana. (*Source: Geocities.com, 1995.*)

would increase downstream as the river expanded. The initial concentration of total cyanide was 25 mg/L from the water discharging from the tailings pond. In the Essequibo River the average cyanide level was 0.12 mg/L. After the spill, 200 dead fish were sighted in the Omai River (Feasby, 1995), but there were no human casualties.

There was concern that 5 ppm of cyanide in water could be lethal to a 7-kg infant. A lethal dose for an adult weighing 60 kg would be a concentration of 34 mg/L of cyanide in the water (Feasby, 1995). On August 21, the Pan American Health Organization (PAHO) reported cyanide levels in the Omai River to be 5.42 ppm and rising. Officials informed the people not to drink the water and began distributing drinking water to affected communities. Helicopters were mobilized to inform people of the potential dangers (NEEC Alert, 1995).

According to Environment Canada, the long-term impact of the spill will be negligible since cyanide degrades rapidly. Also, the high volume of the Essequibo River diluted the cyanide levels so that there were no visible environmental impacts downstream from the entry point of the spill, where the Essequibo and Omai Rivers intersect (Trip and Blakeman, 1995).

39.8.3 Cyanide Spill in Central Asia

On May 20, 1998, in Kyrgyzstan in central Asia, a truck carrying 20 tons of sodium cyanide crashed through the railing of a narrow bridge into the Barskaun River (Fig. 39.8). This river empties into Lake Issyk-Kul. The truck was en route to the Kumtor gold mine.

Because the police were not immediately notified of the spill, five hours passed before the public was notified. Company employees were preoccupied with attempts to rescue the driver from the truck and getting the container out of the water. It was found that 1,762 kg of cyanide briquets were missing. The Ministry of Environmental Protection found cyanide levels in the river to be 1,600 times the maximum acceptable level during the spill, but a day later the concentrations fell to acceptable levels.

A water sample taken hours after the accident 20 m from the spill site showed free cyanide concentrations of 79.5 mg/L. This sample did not include cyanates, thiocyanates, cyanogen, cyanogen chloride, ammonia, chloramine, or metal cyanide complexes that are forms of cyanide-related compounds (Moran, 1998).

FIGURE 39.8 The bridge where the truck broke through and the river as seen from the bridge. (*Source: CBC, 1998.*)

Before the spill, the pH level of the river was probably less than 9. Most of the dissolved cyanide would therefore have formed hydrocyanic acid (HCN), a toxic gas that would have dissipated into the air. After a few hours, sodium hypochlorite was added, which would have resulted in the formation of cyanate and cyanogen chloride. Cyanogen chloride can cause throat and eye irritations and, being heavier than air, could have traveled significant distances from the spill (Moran, 1998).

Two days after the spill, villagers started showing up at hospitals, which evidently confused the responders. A representative of Cameco (owner of the mine) stated that it was "highly unlikely, if not impossible" that the villagers' symptoms were due to the cyanide spill. The argument was that "cyanide poisoning would manifest itself almost immediately" (OSHS, 1998). A pediatrician at the children's hospital in Kyrgyzstan reported that one patient was in serious condition and others had enlarged livers (CBC, 1998). Another report suggested that the spill led to a fish kill, but there were no confirmed reports of sick villagers and livestock (OSHS, 1998).

The comments in Table 39.11 demonstrate the confusion and conflicting reports on the spill.

39.9 NEUTRALIZATION OF CYANIDE SPILLS

A typical cyanide neutralization is shown below.

For alkaline chlorination:

1. $NaCN + 2NaOH + Cl_2 \rightarrow NaCNO + 2NaCl + H_2O$
2. $2NaCNO + 4NaOH + 3Cl_2 \rightarrow 6NaCl + N_2 + 2H_2O + 2CO_2$

This procedure uses sodium hydroxide and pH has to be maintained at above 9, preferably pH 9 to pH 11, to avoid formation of toxic nitrogen trichloride (Kirk-Othmer, 1993).

Hypochlorite neutralization is shown here (McGill and Comba, 1990):

$$NaCN + NaClO \rightarrow NaCNO + NaCl \tag{39.26}$$

The cyanide solutions are oxidized to the less toxic cyanate (Corn, 1993). Cyanates are toxic to trout at concentrations ranging from 13 to 82 mg/L cyanate (Moran, 1998).

Cyanide spills can also be neutralized with peroxide by the Degussa method, in which the pH is maintained above 9 (between 9.0 and 10.5) to avoid formation of toxic NCl_3, nitrogen trichloride. The Degussa method is shown below:

TABLE 39.11 Conflicting Reports on Cyanide Spill

Number of dead	Number effected
Two died.	" . . . 2,533 poisoned, most from local village of Barskaun." (Ling, 1998)
"No one died as a result of the spill of sodium cyanide."	"Although there was potential for up to 16 people to become exposed as a result of cyanide . . . no medical evidence has been supplied to the commission to support these cases as being affected by cyanide." (Cameco, 1998)
"A second person has died from cyanide poisoning."	"444 people remain in hospital since the May 20 accident, including six who are in grave condition." (Perreaux, 1998)
"A woman has died after a river has been contaminated with cyanides."	"Kyrgyz officials . . . say a total of 506 local residents had been hospitalized by June 3." (*Financial Post,* 1998)
"Doctor blames cyanide for a woman's death."	"Hundreds flooded to hospitals complaining of being sick." (CTV, 1998)
"She died of sodium cyanide poisoning that aggravated a previous illness."	"Some 600 people have sought medical treatment for poisoning. About 100 of those have been hospitalized." (*Globe and Mail,* 1998)
"No cyanide-related deaths."	"Maximum of 16 potential cases of direct cyanide exposure. . . . " (CNW, 1998)

1. Oxidation:

$$CN^- + H_2O_2 \rightarrow CNO^- + H_2O$$

$$2Cu(CN)_3{}^{2-} + 7H_2O_2 + 2OH^- \rightarrow 6CNO^- + 2Cu(OH)_2 + 6H_2O$$

2. Precipitation:

$$Fe(CN)_6{}^{4-} + 4Na^+ + 2CuSO_4 \rightarrow Cu_2Fe(CN)_6\downarrow + 2Na_2SO_4$$

3. Hydrolysis of cyanate:

$$CNO^- + 2H_2O \rightarrow CO_3{}^{2-} + NH_4{}^+$$

4. Side reactions (McGill and Comba, 1990):

$$S^{2-} + H_2O_2 \rightarrow S^\circ\downarrow + 2OH^-$$

$$SCN^- + H_2O_2 \rightarrow S^\circ\downarrow + CNO^- + H_2O$$

When 200 L of cyanide were spilled into the River Wey in Surrey, England in May 1999, peroxide was added to the river to save the fish. The peroxide breaks down into water and oxygen and raises the pH of the polluted water, causing the cyanide to break down more rapidly. The fish were saved by the additional oxygen added. The Environment Agency's Thames Region now includes peroxide crystals, or sodium carbonate peroxyhydrate, in their response kits for cyanide spills (OSHS, 1999).

Peroxide was used to increase the oxygen content of a small pond of fish about 0.1 ha in size. The oxygen content was at 13%, but within two hours of about 200 kg of liquid sodium peroxide being added, the oxygen content rose to 35% (OSHS, 1999).

39.10 CONCLUSIONS

Sodium cyanide is highly toxic to humans and the environment. Its usage in the mining industry has called into question the responsibility of the owners and operators to create a safe environment. As can be seen in the case studies, cyanide spillages into waterways are catastrophic. A recent comment on March 22, 2000 regarding a helicopter spillage of sodium cyanide shows the changing tide for tolerating cyanide spills. According to the Mineral Policy Institute in Sydney, Australia, "There is mounting evidence that using cyanide to extract gold is inherently dangerous. The high risk of cyanide accidents outweigh the dubious benefits of mining gold. We are being forced to question whether cyanide usage in gold mining should be limited or even banned" (Evans, 2000). These are questions that could change the future of the usage of sodium cyanide.

39.11 ACKNOWLEDGMENTS

The authors acknowledge Carol Lau from the National Environmental Emergency Centre (NEEC) of Environment Canada for providing information from the NATES database.

39.12 REFERENCES

Aquire Database. 1994. vols. 1–6, June.

Cameco. 1998. "Information on the May 20, 1998 Truck Accident and Sodium Cyanide Spill," Saskatoon, SK, September 29, http://www.cameco.com/investor/news_releases/index.html.

Camford Information Services, Inc. (CIS). 1998. "Sodium Cyanide," in *CPI Product Profiles,* April, pp. 1–3.

Canada NewWire (CNW). "The International Scientific Commission Presents Its Final Report on the Cyanide Spill," September 25, http://www.newswire.ca/releases/September1998/25/c6477.html.

Canadian Broadcasting Corporation (CBC). 1998. "Poisoned Partnership," *The National Online,* June 24, http://www.tv.cbc.ca/national/pgminfo/ppartner/index.html.

CHEM-BANK. 1999. "Sodium Cyanide," SilverPlatter Health & Safety Publishing, Norwood, MA (CD-ROM).

Chemical Advisor, The. "Sodium Cyanide," ChemADVISOR, Inc., Pittsburgh, PA (on The Chemical Advisor CD-ROM, Canadian Centre for Occupational Health and Safety [CCOHS]).

Chemical Hazards Response Information System (CHRIS) Database. "Sodium Cyanide," Canadian Centre for Occupational Health and Safety (CCOHS) (on CHEMpendium CD-ROM, CCOHS).

CHEMtox Database. 1999. "Sodium Cyanide," CHEMTOX, Brentwood, TN (CD-ROM).

Clayton, G. D., and F. E. Clayton. 1991. *Patty's Industrial Hygiene and Toxicology,* vol. 1, pt A, John Wiley & Sons, p. 2021.

Corn, M. 1993. *Handbook of Hazardous Materials,* Academic Press, San Diego, CA, pp. 167–169.

CTV News and Current Affairs. 1998. "A Canadian Company is at the Centre of an Environmental Mishap Overseas," Thursday, June 4.

Environment Canada. 1984. "Sodium Cyanide," in *Manual for Spills of Hazardous Materials,* Environmental Protection Service, Environment Canada, Ottawa, ON.

Environment Canada. 2000. *NATES—National Analysis of Trends in Emergencies Systems,* Environment Canada, Ottawa, ON.

ETK. 2000. *Impacts of the Romanian Mine Accidents,* http://www.zpok.hu/~jfeiler/baiamare/impacts.htm, March 11.

Evans, G. 2000. "PNG Cyanide Accident: Water Testing Results," Mineral Policy Institute, Australia, Media Release, http://www.zpok.hu/jfeiler/baiamare/News/mpi2203.htm, Wednesday, March 22.

Feasby, G. 1995. "The OMAI Tailings Water Spill, An Uncontrolled Release of Cyanide-Contaminated Water," Natural Resources Canada, Health Canada, September 1, pp. 1–8.

Financial Post. 1998. "Death Reported after Cameco Spill in Kyrgyzstan," Reuters News Service, Friday, June 5.

Genium's Handbook of Safety, Health, and Environmental Data for Common Substances. 1999. "Sodium Cyanide," MSDS update 52 (CD-ROM), McGraw-Hill, New York, NY.

Geocities.com. 1995. wyxiwyg://17/http://www.geocities.com/RainForest/2651/.

Globe and Mail [*Toronto*]. 1998. "Cameco Denies Cyanide Spill Killed Woman. Doctors in Kyrgyzstan Say Previous Illness was Aggravated by Poison in Nearby River," June 5.

Hazardous Substances Data Bank. "Sodium Cyanide," U.S. National LIbrary of Medicine, Bethesda, MD (on CHEMpendium CD-ROM, Canadian Centre for Occupational Health and Safety [CCOHS]).

Ingles, J., and J. S. Scott. 1987. *State-of-the-Art of Processes for the Treatment of Gold Mill Effluents,* Environmental Protection Service, Environment Canada, Ottawa, ON, March.

King, B. R. ed. 1994. *Encyclopedia of Inorganic Chemistry,* vol. 2, John Wiley & Sons, New York, NY, p. 558.

Kirk-Othmer Encyclopedia of Chemical Technology. 1978a. 3d ed., vol. 1, John Wiley & Sons, New York, NY, p. 345.

Kirk-Othmer Encyclopedia of Chemical Technology. 1978b. 3d ed., vol. 3, John Wiley & Sons, New York, NY, p. 201.

Kirk-Othmer Encyclopedia of Chemical Technology. 1979a. 3d ed., vol. 5, John Wiley & Sons, New York, NY, pp. 725–726.

Kirk-Othmer Encyclopedia of Chemical Technology. 1979b. 3d ed., vol. 8, John Wiley & Sons, New York, NY, p. 841.

Kirk-Othmer Encyclopedia of Chemical Technology. 1980. 3d ed., vol. 11, John Wiley & Sons, New York, NY, p. 973.

Kirk-Othmer Encyclopedia of Chemical Technology. 1981. 3d ed., vol. 15, John Wiley & Sons, New York, NY, p. 316.

Kirk-Othmer Encyclopedia of Chemical Technology. 1982a. 3d ed., vol. 17, John Wiley & Sons, New York, NY, p. 497

Kirk-Othmer Encyclopedia of Chemical Technology. 1982b. 3d ed., vol. 19, John Wiley & Sons, New York, NY, p. 634.

Kirk-Othmer Encyclopedia of Chemical Technology. 1984. 3d ed., vol. 21, John Wiley & Sons, New York, NY, p. 197.

Kirk-Othmer Encyclopedia of Chemical Technology. 1993. 4th ed., vol. 7, John Wiley & Sons, New York, NY, p. 768.

Kirk-Othmer Encyclopedia of Chemical Technology. 1995a. 4th ed., vol. 15, John Wiley & Sons, New York, NY, p. 75.

Kirk-Othmer Encyclopedia of Chemical Technology. 1995b. 4th ed., vol. 16, John Wiley & Sons, New York, NY, p. 741.

Kirk-Othmer Encyclopedia of Chemical Technology. 1996. 4th ed., vol. 20, John Wiley & Sons, New York, NY, p. 861.

Ling, C. 1998. *Famed Kyrgyz Lake Fights Aftermath of Chemical Spill,* http://www.nando.com/newsroom/ntn/health/070798/health4_5314_noframes.html, Reuters News Service, 1998.

McGill, S. L., and P. G. Comba. 1990. *A Review of Existing Cyanide Destruction Practices,* U.S. Bureau of Mines, Reno Research Center, Nevada Mining Association, Nevada Department of Wildlife, Wildlife/Mining Workshop, Reno, NV, March, pp. 11, 16, 29.

Moran, R. 1998. *Cyanide Uncertainties, Observations on the Chemisty, Toxicity, and Analysis of Cyanide in Mining-Related Waters,* Mineral Policy Center, MPC Issue Paper No. 1, http://www.zpok.hu/~jfeiler/baiamare/docs/cyanideuncertain.htm, pp. 7–12.

MSDS Database A and B. "Sodium Cyanide," Canadian Centre for Occupational Health and Safety (CCOHS) (on MSDS CD-ROM, CCOHS).

MTI. 2000. *Spread of the Cyanide Spill from Baia Mare,* Ministry for Environmental Protection, Hungary, Environmental Inspectorates, MTI Co. Media Data Bank, http://www.zpok.hu/~jfeiler/images/mapen.jpg, February 20.

National Environmental Emergency Centre (NEEC) of Environment Canada. 1995. NEEC Alert, Urgent fax from United Nations, Environment Desk of Department of Humanitarian Affairs, "Information Report No. 1—Guyana Cyanide Mine Waste Spill," NEEC # 1503, August 22.

NIOSH Pocket Guide to Chemical Hazards. "Sodium Cyanide," National Institute for Occupational Safety and Health (NIOSH), Washington, DC (on CHEMpendium CD-ROM, Canadian Centre for Occupational Health and Safety [CCOHS]).

Oil Spill Hazardous Substances (OSHS). 1998. "Panic Reported after Cyanide Spill in Kyrgyzstan," Global News on the Marine Chemical Response Industry by Cutter Information Corp., vol. 1, no. 4, June 4.

Oil Spill Hazardous Substances (OSHS). 1999. "UK Using Peroxide to Stem Fish Kills from Sewage, Cyanide," Global News on the Marine Chemical Response Industry by Cutter Information Corp., vol. 2, no. 12, June 17.

Oil Spill Hazardous Substances (OSHS). 2000a. "UN Team Calculates Cyanide Damage," Global News on the Marine Chemical Response Industry by Cutter Information Corp., vol. 3, no. 5, March 9.

Oil Spill Hazardous Substances (OSHS). 2000b. "Cyanide Spill Purges Hungarian River en Route to Danube," Global News on the Marine Chemical Response Industry by Cutter Information Corp., vol. 3, no. 3, February 10.

Ottawa Citizen Online. 2000. "River Killed 'for Many Years to Come' after Cyanide Spill, Official Says," February 15, http://www.southam.com/ottawacitizen/newsnow/cpfs/world/000215/w21546.html.

Perreaux, L. 1998. "Kumtor Boss Quits after Second Death Blamed on Spill," *The Saskatoon Star-Phoenix,* June 9.

Pope, A. M., and D. P. Rall. 1995. *Environmental Medicine: Integrating a Missing Element into Medical Education,* National Academy Press, Washington, DC, pp. 317–323.

RTECS Database. "Sodium Cyanide," National Institute for Occupational Safety and Health (NIOSH), Washington, DC (on RTECS CD-ROM, Canadian Centre for Occupational Health and Safety [CCOHS]).

Sax, R. J. 1989. *Dangerous Properties of Industrial Materials,* 7th ed., vol. 9, no. 5, Van Nostrand Reinhold, New York, NY, pp. 76–77.

TPF. 1997. "Toxicological Profile for Cyanide," U.S. Department of Health and Human Services, Public Health Service, Agency for Toxic Substances and Disease Registry, Division of Toxicology/Toxicology Information Branch, Atlanta, GA, p. 70.

Trip, L., and B. Blakeman. 1995. Advice to Minister on Subject: *Cambior Inc. Cyanide Spill in Guyana,* Letter from Environmental Protection Service, Environment Canada, February 14, pp. 1–2.

Ullmann's Encyclopedia of Industrial Chemistry. 1987. 5th ed., vol. A8, VCH, New York, NY, pp. 165–171.

United Nations Environment Program/Office for the Co-ordination of Humanitarian Affairs (UNEP/OCHA). 2000. *Spill of Liquid and Suspended Waste at the Aurul S.A. Retreatment Plant in Baia Mare,* Assessment Mission, Romania, Hungary, Federal Republic of Yugoslavia, February 23–March 6, pp. 7–55.

Vick, S., R. Squires, R. Dundee, R. Pedroni, and N. Watson. 1995. *Preliminary Report on Technical Causation OMAI Tailings Dam Failure,* Submitted to Guyana Geology and Mines Commission, http://www.antenna.nl/wise-database/uranium/mdgr.html, November 16.

World Wildlife Federation (WWF). 2000a. "The Biggest Ever Freshwater Disaster in Central and Eastern Europe," World-Wide Global Network, press release, http://www.panda.org/news/press/news.cfm?id=1850, February 9.

World Wildlife Federation (WWF). 2000b. "Dam Must Be Secured to Prevent Further Damage, Warns WWF," World-Wide Global Network, press release, http://www.panda.org/news/press/news.cfm?id=1852, February 18.

World Wildlife Federation (WWF). 2000c. "WWF Demands Immediate Government Action on Toxic Spill," World-Wide Global Network, press release, http://www.panda.org/news/press/news.cfm?id=1848, February 14.

P • A • R • T • 9

CASE HISTORIES

CHAPTER 40
LONG-RANGE CONSEQUENCES OF A MAJOR AMMONIA ACCIDENT AT IONAVA, LITHUANIA, IN 1989

Jaakko Kukkonen
Finnish Meteorological Institute, Air Quality Research, Helsinki, Finland

40.1 INTRODUCTION

We evaluated the atmospheric dispersion of ammonia released in a major chemical accident near the town of Ionava in central Lithuania. The course of the accident has been briefly described by Kletz (1991). More detailed reports have been prepared by the Lithuanian State Committee for Environmental Protection (1989) and by Andersson (1991). For a more complete description of this study, the reader is referred to Kukkonen et al. (1993).

The accident occurred at the chemical plant Azot, which produces mainly fertilizers. This plant is situated about 5 km northeast of the town of Ionava. A 10,000-ton-capacity tank containing 7,000 tons of refrigerated ammonia at its boiling point (−33°C) split without warning. The rupture was caused by an erroneous filling of the tank with relatively warm (+10°C) liquid ammonia, which formed a layer at the base of the tank. The warm ammonia then suddenly rose to the surface and evaporated, and the pressure rise overwhelmed the relief valves.

The liquid ammonia released formed a pool that was 70 cm deep in places. According to the chemical plant itself, about 1,400 tons of the spilled liquid ammonia evaporated. The pool caught fire, and the fire spread to a fertilizer store containing 15,000 tons of NPK 11-11-11. It was theoretically estimated that a further 700 tons of ammonia was released into the environment due to the self-sustaining thermal decomposition of this material (Lithuanian State Committee for Environmental Protection, 1989). Substantial amounts of nitrogen oxides were also released from the fire. Seven people were killed, 57 injured, and about 32,000 evacuated in the course of the accident.

The objective of this study was to analyze whether a release of this magnitude might cause long-range effects, on a scale of tens or hundreds of kilometers. The long-range consequences of major accidents should also be known when formulating international conventions on transboundary transport of chemically toxic substances.

Our computations are based on the release estimates provided by the authorities (Lithuanian State Committee for Environmental Protection, 1989). We have applied a long-range dispersion model that uses a Gaussian plume model up to a specified transition distance, and subsequently the gradient-transfer diffusion theory. Our analyses address mainly downwind distances greater than about 10 km. Because the release estimates contain fairly large uncertainties, computations were made for two release scenarios. Our objective has been to obtain order-of-magnitude estimates of the actual concentrations.

40.2 RELEASE SCENARIOS

The exact time of occurrence of the accident was 8:15 a.m. UTC (Universal Time Coordinated), March 20, 1989. All times given in this chapter are expressed as UTC. Intensive evaporation of the liquid pool lasted for up to about eight hours, and the fire was totally extinguished by 5:00 p.m. on March 22 (Lithuanian State Committee for Environmental Protection, 1989). The fire also resulted in releases of nitrogen oxides, fluoride, and chlorine into the air and escape of liquid pollutants into a nearby river. However, because their released amounts have not been quantitatively estimated in the available references, we address only the dispersion of ammonia in the air in this chapter. We also do not address the ammonia releases from the fire.

In scenario 1, we assume that the total amount of ammonia vapor released into the air was 1,400 tons and that the duration of the release was eight hours. Scenario (1) is a best-guess estimate according to available release data (Lithuanian State Committee for Environmental Protection, 1989). However, because the rate of ammonia evaporation from the pool most likely decreased with time, most of the ammonia may have evaporated within a much shorter time. We therefore assume in scenario 2 a release of 1,000 tons of ammonia within one hour. Scenario 2 is a worst-case estimate for the release of ammonia into the air from the liquid pool. We have assumed a constant release rate at the ground level in both cases.

40.3 COMPUTATIONAL METHODS

Unfortunately, the available data were not sufficiently accurate for the application of mathematical models governing liquid pool evaporation and spreading. An evaporating liquid pool of ammonia does not produce a heavier-than-air gas cloud, as ammonia vapor at its boiling point is lighter than air at commonly occurring ambient temperatures (0 to 20°C). Therefore, a heavy gas cloud could only be formed if there was significant aerosol formation, which is unlikely in the reported conditions.

We have therefore applied a long-range dispersion model, which uses gradient-transfer diffusion theory for vertical dispersion. In particular, this approach more readily takes into account the change of the vertical concentration profile brought about by the dry deposition, compared to the Gaussian approach. The model has been described in detail by Nordlund et al. (1985), Nordlund and Savolainen (1983), and Nordlund and Tuovinen (1986). A brief overview of the model and its application will be given in this section. A more refined version of the model has recently been addressed by Nikmo et al. (1999).

The model applies a Gaussian plume model up to a specified transition distance, which is dependent on atmospheric stability: 200 m for unstable, 500 m for neutral, and 1000 m for stable stratification, respectively. Outside this nearby zone, the gradient-transfer approach is used for computing the diffusion in the vertical direction.

We assume steady-state release and atmospheric conditions. Further, neglecting diffusion in the wind direction, the atmospheric diffusion equation reduces in terms of the crosswind integrated concentration C^y to (e.g., Nikmo et al., 1999)

$$u \frac{\partial C^y}{\partial x} = \frac{\partial}{\partial z} K_z \frac{\partial C^y}{\partial z} + R(C^y) + S \qquad (40.1)$$

where u = mean wind speed
K_z = vertical eddy diffusion coefficient
R = rate of generation by chemical reactions
S = sink or source term
x, y, and z = downwind distance, crosswind distance, and height, respectively

The crosswind integrated concentration (unit kg/m^2) is defined as:

$$C^y(x, z) = \int_{-\infty}^{\infty} C(x, y, z)\, dy \tag{40.2}$$

The profiles of the vertical eddy diffusion coefficients for different stability categories were estimated using the correlations presented in Pasquill (1974).

The boundary condition for differential Eq. (40.1) in the horizontal direction is:

$$C^y(x, z) = 0, \; x \rightarrow \pm\infty \tag{40.3}$$

In the vertical direction, the boundary conditions are defined as:

$$\begin{aligned} K_z \frac{\partial C^y}{\partial z} &= v_d C^y, \qquad z = z_s \quad \text{and} \\ K_z \frac{\partial C^y}{\partial z} &= -v_e C^y, \qquad z = h \end{aligned} \tag{40.4}$$

where v_d = dry deposition velocity of gas or particles
v_e = corresponding velocity for the flux through the top of the mixing layer
z_s = reference height just above the ground
h = height of atmospheric boundary layer

The vertical boundary conditions Eq. (40.4) define the rate of loss of contaminant to the ground surface and through the top of the mixing layer.

Horizontal diffusion outside the nearby zone is also computed using a Gaussian approach with correlations of the dispersion coefficients presented by Hanna et al. (1982) up to the distance of 100 km. This distance is clearly larger than the experimentally validated regime of these correlations; the results are therefore applicable only for order-of-magnitude estimates at larger distances. The above-mentioned correlations were modified to allow for a reduced diffusion rate for longer distances (Nordlund et al., 1985). However, this procedure cannot explicitly allow for the influence of synoptic-scale atmospheric turbulence on the horizontal diffusion. For the longer release scenario (with a release duration of eight hours), this might be important. However, in the actual meteorological conditions of this study, the divergence between successive trajectories was fairly small, and consequently, the effect of synoptic scale fluctuations is not large.

The atmospheric transformation and deposition processes of ammonia have been addressed by Asman and van Janssen (1987), Iversen et al. (1991), Derwent et al. (1989), and Duyzer et al. (1987). In European conditions, most of the gaseous ammonia, NH_3, will react with H_2SO_4 to form ammonium, NH_4^+. The conversion rate of this reaction depends on the concentrations of acidic components and on the atmospheric mixing characteristics; most of the NH_3 is transformed into NH_4^+ on a time scale of a few hours.

In the computations, we have used a constant deposition velocity appropriate for ammonia vapor, 1.0 cm/s. At long distances this choice overpredicts dry deposition, as part of the NH_x is in the form of particulate ammonium (for which the dry deposition velocity is smaller, approximately 0.1 m/s, Asman and Janssen, 1987). However, the resulting error is moderate because dry deposition is dominated by ammonia vapor deposition. It has been assumed that the vertical mass flux vanishes at the upper boundary, i.e., $v_e = 0$.

The occurrence of precipitation could cause a substantial decrease in concentration. However, in the meteorological conditions of this study, there was no precipitation during the transport of pollutants to a downwind distance of up to about 500 km.

Air parcel trajectories were computed on a two-dimensional horizontal grid using numerically analyzed meteorological data. The trajectory model has been described in detail by Nordlund et al. (1985). The meteorological data analyses are based on simultaneous synoptical surface observations and sounding and satellite data from the whole of the northern hemisphere. These data are numerically analyzed onto constant pressure levels on a rectangular mesh system. The mesh is about 150 km wide, and the values within each grid square are computed by linear interpolation procedures.

By combining trajectories at different levels, we can study the transport of air masses in the whole of the atmospheric surface layer. The variation of transport conditions in time and space is also allowed for in these computations. Because the pollutants considered here were released near ground level, we have made most of the computations for the 1,000 hPa pressure level, corresponding approximately to the lowest 200 to 300 m layer of the atmosphere. Some computations were also made for the 950 hPa pressure level, corresponding approximately to a height of 500 m. Two-dimensional (instead of three-dimensional) trajectories were considered to be sufficient due to the relatively low transport altitudes.

40.4 RESULTS

40.4.1 Meteorological Conditions during Transport

Dispersion conditions change with time and location during transport. In particular, the time of day and the nature of the underlying surface affect the meteorological situation. We address first the dispersion conditions from the accident location to the southern coast of Finland. The corresponding straight-line distance is about 500 km (Fig. 40.1). No ice cover existed over the Gulf of Finland (situated between Finland and Estonia) at the time, except for sea areas very near the coastline.

On March 20, 1989, light southerly winds (wind speed at 10 m height $\leq$ 2 m/s) prevailed in the accident area. In general, during the transport of pollutants, neutral and stable atmospheric stability conditions were dominant and mainly moderate southerly winds (2 to 7 m/s at 10 m height) prevailed. No significant rainfall took place during the transport. Consequently, atmospheric conditions were unfavorable for the rapid mixing of contaminants.

We have selected two characteristic dispersion conditions for our analyses: (1) stable atmospheric stratification (Pasquill-class F) with a wind speed of 2 m/s and (2) neutral atmospheric stratification (Pasquill-class D) with a wind speed of 5 m/s. The mixing layer was estimated to be 200 m for stable and 500 m for neutral stratification. Although the actual meteorological conditions vary continuously during the transport of pollutants, most of the time the actual dispersion conditions varied between the two selected cases.

40.4.2 Trajectories of Air Masses

The source-oriented trajectories of air masses immediately after the estimated time of the accident (8:15 a.m. UTC on March 20, 1989) are shown in Fig. 40.2. The trajectories shown cover the period of intensive evaporation of the liquid ammonia pool.

Most of the releases from the accident were transported directly to the southern coast of Finland. The scatter of the different trajectories is quite small, reflecting the slowly changing synoptical meteorological situation at the time. The computed transport times along the various trajectories from the accident location to the southern coast of Finland were about 27 to 36 hours. For instance, air parcels released from ground level on March 20 at times

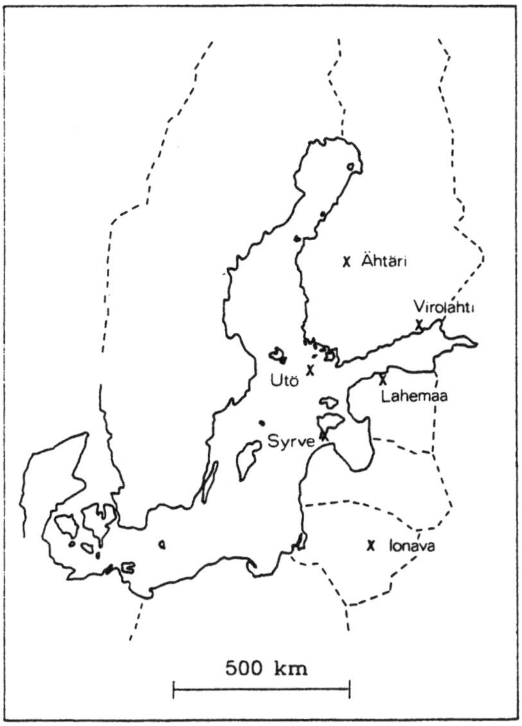

FIGURE 40.1 The accident location and the relevant European Monitoring and Evaluation Program (EMEP) stations.

from 6:00 a.m. to 9:00 p.m. UTC would reach the southern coast of Finland between approximately 3:00 p.m. on March 21 and 3:00 a.m. on March 22.

Trajectories computed for later times were directed northeast, east, and southeast, into the area of the former Soviet Union. These trajectories were computed for the lowest atmospheric layer, with a height less than about 200 to 300 m. Because some pollutants may have risen to higher altitudes, we also computed the trajectories of air masses in other atmospheric layers. The transport directions at higher altitudes turn in a clockwise direction. The contribution of these pollutants to observational data in Finland is therefore small.

Clearly, the transport times at higher levels are smaller than those at lower levels. Computed transport times from the accident location to the southern coast of Finland along various trajectories at the 925 hPa level were about 13 to 30 hours. For instance, air parcels released from ground level on March 20 at times between 6:00 a.m. and 9:00 p.m. would reach the southern coast of Finland at approximately 7:00 a.m. to 12:00 p.m. on March 21.

40.4.3 Predicted Concentrations

Figure 40.3 shows the maximum concentration on the plume centerline at the ground level of NH_x components in the air (NH_3 + NH_4^+) for the selected release scenarios. The order of magnitude of the concentrations at a distance of about 500 km was 0.1 to 1.0 mg/m^3 for release scenario 1 and 1 to 10 mg/m^3 for release scenario 2.

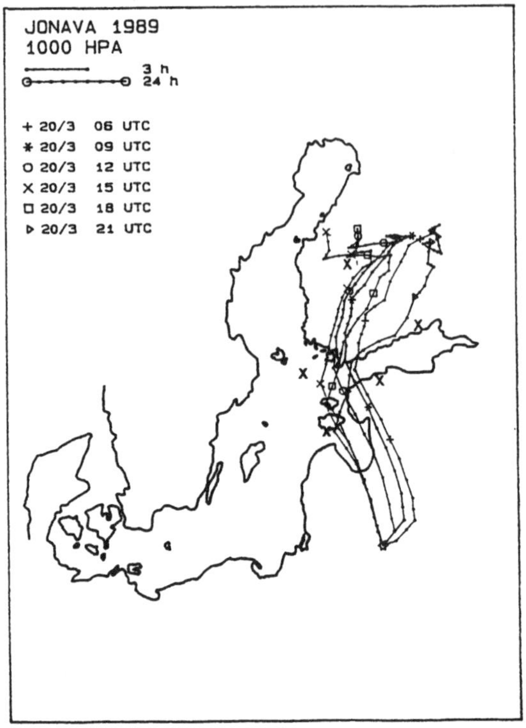

FIGURE 40.2 Trajectories in the lowest atmospheric layer (below a height of 200–300 m) originating from the accident site on March 20, 1989. The legend shows the three-hourly and daily intervals and the starting times of the trajectories.

In these computations we have assumed a constant dry deposition velocity, although it is actually dependent, among other factors, on wind speed and atmospheric stability. In particular, the dry deposition velocity is typically smaller in stable conditions. However, the loss of pollutant from the plume depends also on the vertical concentration distribution. For stable conditions, the plume is substantially more shallow; dry deposition therefore causes an increased decrease of concentration at larger distances, compared with the corresponding case in neutral stability. The apparent convergence of the concentration curves for the two stability classes is fortuitous.

40.5 COMPARISON OF NUMERICAL RESULTS AND OBSERVATIONS

The Lithuanian authorities reported some ammonia measurements on March 20, 1989 (Lithuanian State Committee for Environmental Protection, 1989). The largest measured concentration was 200 mg/m^3 about 5 km downwind of the accident site at 3:00 to 4:30 p.m. UTC, and concentrations of about 20 to 25 mg/m^3 were measured at distances of from 5 to 12 km from the source. These values are not inconsistent with the concentration curves of release scenario 1 in Fig. 40.3.

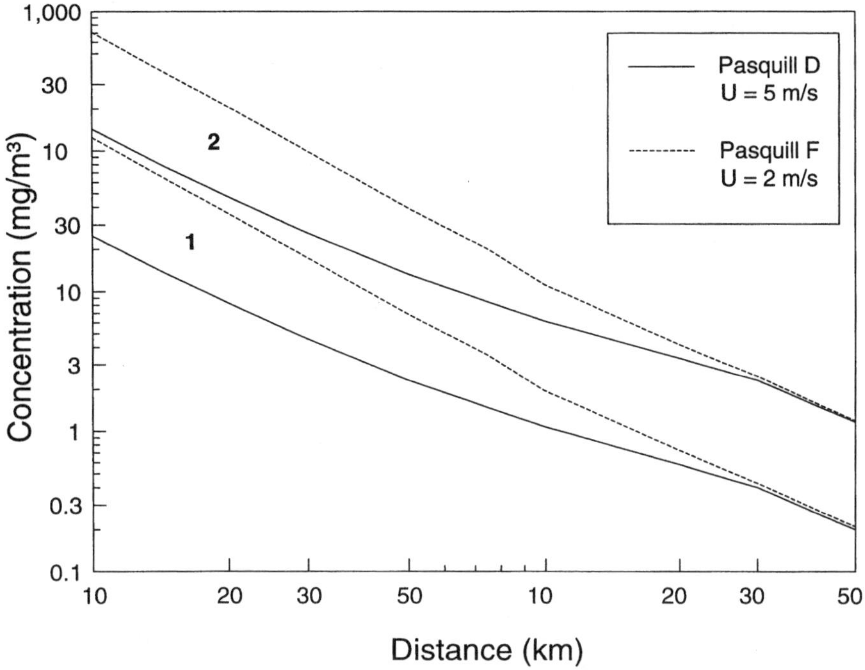

FIGURE 40.3 The contaminant concentration at the ground level as a function of downwind distance for release scenarios 1 and 2 for two combinations of atmospheric stability and wind velocity.

Ammonia measurements are made in the EMEP stations (Leinonen and Juntto, 1990); these include the concentration of NH_x components in the air and in bulk precipitation samples. Listed from north to south, the Finnish stations are Ähtäri, Virolahti, and Utö. These are shown in Fig. 40.1. However, NH_x concentrations in the air were not available from the Utö station at the time, and daily measurements were not available from the Estonian stations (Syrve and Lahemaa). Both the concentration and precipitation samples are collected for 24 hours starting at 6:00 a.m.

The measured ammonia and ammonium concentrations at the Finnish EMEP measurement stations of Virolahti and Ähtäri show a maximum on March 21, 1989; this time corresponds to computed arrival times. However, such maximum values are not uncommon in springtime. A detailed analysis of backward trajectories, computed for these measurement stations, and the measured concentrations do not show conclusively that the measured maximum values would have been caused by the accident (Kukkonen et al., 1993). Most of the NH_x due to the accident may have escaped the available measurement stations.

We also received a number of phone calls from individual citizens in Finland after the accident, mainly reporting eye irritation. These observations were made during the evening of March 21 in a limited area, on or near the southern coastline of Finland, about 10 to 50 km west of Helsinki. The location and time of these observations are consistent with the results of the trajectory computations in the lowest atmospheric layer (Fig. 40.2). We received the phone calls during March 22 to 25, before any information about the accident had been disseminated to the public in Finland.

The least detectable odor level of ammonia vapor varies from 1 to 50 ppm (for NH_3 at 20°C: 1 ppm = 0.711 mg/m^3), and 20 ppm has caused complaints and discomfort in uninjured workers (American Conference, 1991). It has also been reported that maximum acceptable concentration at the working place without severe complaints is 20 to 25 ppm (American Conference, 1991). Ammonia vapor causes irritation of eye, skin, and mucous membranes. According to the computations, the order of magnitude of the NH_x concentrations on the southern coastline of Finland was 0.1 to 1.0 mg/m^3 for release scenario 1 and 1 to 10 mg/m^3 for release scenario 2. We conclude that these observations of irritation may have been caused by the Lithuanian accident.

40.6 CONCLUSIONS

Toxic substances released accidentally from industrial processes may be transported substantial distances in the atmosphere. Information that is as accurate and reliable as possible should be disseminated to the public, and over a much wider area than the zone of immediate health effects.

We have analyzed the possible long-range effects of an accident at a chemical plant near the town of Ionava, Lithuania, on March, 20 1989. The accident was unusual in several respects.

1. The releases into the atmosphere were exceptionally large—the estimated ammonia releases were 1,400 tons from pool evaporation and about 700 tons from the resulting fire. Intensive evaporation of the liquid pool lasted about eight hours, and the fire continued for three days.
2. Second, atmospheric conditions were unfavorable for the rapid mixing of contaminants. Neutral or stable atmospheric stability conditions and mainly moderate or low wind speeds prevailed, and no significant rainfall took place during the transport of pollutants to a distance of up to about 500 km.

The gas cloud was transported directly northwards. Transport times along various trajectories in the lowest atmospheric layer from the accident location to the southern coast of Finland were about 27 to 36 hours. According to the numerical computations, the maximum NH_x concentrations in Finland may have been 0.1 to 10.0 mg/m^3. The corresponding commonly occurring background concentrations are in the μg/m^3 range.

Clearly, several uncertainties affect these estimates. The masses of the released pollutants and the release duration are uncertain. The transformation and dry deposition processes were not modeled in detail. The dispersion computations do not include the changes of meteorological conditions with time and location along each trajectory.

The results indicate that most of the NH_x due to the accident escaped the EMEP monitoring stations in Finland. The EMEP network was designed for monitoring background air quality and is not well suited for detecting chemical emergencies. The daily sampling time is too long for such rapidly occurring releases. Only a few of the hundreds of toxic gases commonly used by industry are measured in the EMEP network. A more closely spaced measurement network, equipped with instruments sampling with a sufficient time resolution, would be needed for reliable detection of accidentally released chemical compounds.

40.7 ACKNOWLEDGMENTS

The author wishes to thank his coauthors on the paper on which this compilation is based: Ms. A. L. Savolainen, Mr. I. Valkama, Ms. S. Juntto, and Dr. T. Vesala; and Mr. Juha Nikmo for his comments.

40.8 REFERENCES

American Conference of Governmental Industrial Hygienists, Inc. 1991. *Documentation of the Threshold Limit Values and Biological Exposure Indices,* 6th ed. vol. 1, Cincinnati, Ohio.

Andersson, B. O. 1991. "The Lithuanian Ammonia Accident, March 20, 1989."

Asman, W. A. H., and A. J. Janssen. 1987. "A Long-Range Transport Model for Ammonia and Ammonium for Europe," *Atmospheric Environment,* vol. 21, pp. 2099–2119.

Derwent, R. G., O. Hov, W. A. H. Asman, J. A. van Jaarsveld, and F. A. A. M. de Leeuw. 1989. "An Intercomparison of Long-Term Atmospheric Transport Models: The Budgets of Acidifying Species for the Netherlands," *Atmospheric Environment,* vol. 23, pp. 1893–1909.

Duyzer, J. H., A. M. H. Bouman, H. S. M. A. Diederen, and R. M. van Aalst. 1987. *Measurements of Dry Deposition Velocities of NH_3 and NH_4^+ over Natural Terrain,* Report R87/273, TNO Division of Technology for Society, Delft, The Netherlands.

Leinonen, L., and S. Juntto. 1990. *Results of Air Quality at Background Stations, January–June 1989,* Finnish Meteorological Institute, Air Quality Department, Helsinki.

Hanna, S. R., G. A. Briggs, and R. P. Hosker, Jr. 1982. *Handbook on Atmospheric Diffusion,* DOE-TIC-11223, Technical Information Center, U.S. Department of Energy.

Iversen, T., N. E. Halvorsen, S. Mylona, and H. Sandnes. 1991. *Calculated Budgets for Airborne Acidifying Components in Europe, 1985, 1987, 1988, 1989 and 1990,* Det Norske Meteorologiske Institutt, Technical Report no. 91.

Kletz, T. 1991. "Ammonia Incidents," *Journal of Loss Prevention in the Process Industries,* vol. 4, p. 207.

Kukkonen, J. A., A. L. Savolainen, I. Valkama, S. Juntto, and T. Vesala. 1993. "Long-Range Transport of Ammonia Released in a Major Chemical Accident at Ionava, Lithuania," *Journal of Hazardous Materials,* vol. 35, pp. 1–16.

Lithuanian State Committee for Environmental Protection. 1989. *A Report on the Ammonia Accident at Ionava,* Vilna (in Russian).

Nikmo, J., J.-P. Tuovinen, J. Kukkonen, and I. Valkama. 1999. "A Hybrid Plume Model for Local-scale Atmospheric Dispersion," *Atmospheric Environment,* vol. 33, pp. 4389–4399.

Nordlund, G., and A. L. Savolainen. 1983. "Application of non-Gaussian Calculation Schemes for Evaluating Atmospheric Transfer of Radionuclides," in IAEA-SR-85/20, vol. 1, International Atomic Energy Agency, pp. 57–65.

Nordlund, G., and J.-P. Tuovinen. 1986. "Modelling Long-Term Averages of Sulphur Deposition on a Regional Scale," in *Proceedings of the WMO Conference on Air Pollution Modelling and Its Application,* vol. 44, WMO/TD 187, pp. 53–61.

Nordlund, G., J. P. Partanen, J. Rossi, I. Savolainen, and I. Valkama. 1985. "Radiation Doses Due to Long-Range Transport of Airborne Radionuclides Released by a Reactor Accident—Effects of Changing Dispersion Conditions during Transport," *Health Phys.,* vol. 49. pp. 1239–1249.

Pasquill, F. 1974. *Atmospheric Diffusion,* John Wiley & Sons, New York.

CHAPTER 41
HISTORICAL ANALYSIS OF NATURAL GAS ACCIDENTS

H. Montiel, E. Planas-Cuchi, J. Casal, J. Arnaldos, and J. A. Vílchez
Centre d'Estudis del Risc Tecnològic (CERTEC), Universitat Politècnica de Catalunya—Institut d'Estudis Catalans, Barcelona, Catalonia, Spain

41.1 INTRODUCTION

Natural gas is currently one of the most widely used sources of energy, and its use is growing. As a fuel it has several advantages that justify its wide use. Among these are its ease of distribution at low pressures and, most important, its clean combustion characteristics. This can be seen in Table 41.1.

Natural gas networks have been installed and maintained in complex piping systems commonly located in highly populated areas to transport and distribute gas. Due to these circumstances, accidents caused by the loss of containment of natural gas tend to involve substantial economic losses and a considerable number of victims among the population.

It is therefore very important to study the main features and consequences of these accidents. An interesting approach for carrying out this study is historical analysis. This analysis should lead to a better definition of accidents and to the design of the organizational steps needed to reduce the probability of occurrence of accidents.

Several reports on large accidents in industrial activities have been published (Garrison, 1988; Davenport, 1977, 1983). Recently, diverse reports have been devoted to accidents in chemical plants and in the transportation of hazardous materials. Some of these reports analyze a relatively large amount of data concerning all the types of accidents (Haastrup and Brockhoff, 1990; Vílchez et al., 1995), and others focus on accidents in specific fields (Rømer et al., 1993; Planas-Cuchi et al., 1997). Other studies have also been published on accidents involving natural gas (Lees, 1986; Knowles et al., 1977; Montiel et al., 1996).

This report describes the results of a historical analysis carried out on a sample of 413 accidents involving natural gas. The accident cases were taken from the MHIDAS (Major Hazard Incident Data Service) database, developed and managed by the Safety and Reliability Directorate (SRD) on behalf of the Major Hazard Assessment Unit of the U.K. Health and Safety Executive. The MHIDAS database includes accidents from a number of countries, and, although it was created in 1986, it includes accidents occurring at the beginning of this century and is periodically updated.

The survey for this study was carried out in April 1999. At that time the database included 10,089 entries, and of these 413 accidents (4.1% of the total) involved natural gas.

TABLE 41.1 World Consumption of Natural Gas

Year	World consumption (billion m^3)
1988	1846.7
1990	1967.5
1992	2005.5
1994	2059.0
1996	2227.2
1998	2240.5

Source: BP-Amoco, 1999.

41.2 MATERIAL INVOLVED

Among the accidents involving natural gas, the type of material involved was known in 411 cases (97%). The distribution can be seen in Table 41.2.

As can be seen, a large percentage of the total (80%) is natural gas in a gaseous state, which is due to the fact that natural gas is mainly transported and distributed in pipes in its gaseous state. Liquified gas, both pressurized and refrigerated, constitutes much lower percentages (7.5% and 6% of total, respectively).

41.3 ORIGIN OF THE ACCIDENTS

The origin of accidents is described in the MHIDAS database by two codes: the first describes the general origin of the accident (transport, processing plant, etc.), and the second describes in greater detail the circumstances surrounding the accident. The origin of accidents is shown in Table 41.3.

Most accidents (66.8%) occurred during transportation and distribution, 15% occurred in process plants, and the remaining origins have a much lower frequency.

TABLE 41.2 Material Type Involved in Accidents

Material type	Total records	%
Gas	331	80.0
Liquefied refrigerated gas	31	7.5
Liquefied pressurized gas	25	6.0
Liquid	8	2.0
Liquefied gas (not specified)	6	1.5
Unknown	12	3.0
Total	413	100.0

TABLE 41.3 Origin of Accidents

Origin	Number of accidents	% of accidents
Transport	276	66.8
Process plant	62	15.0
Storage plant	31	7.5
Domestic/commercial	17	4.0
Loading/unloading	8	2.0
Warehouse	1	0.2
Unknown	18	4.5
Total	413	100.0

41.4 DISTRIBUTION OF ACCIDENTS ACCORDING TO DATE

The decades of the 20th century were used to study the distribution. Table 41.4 shows the distribution of accidents according to date for both the overall number of entries in the database and the entries related to natural gas.

There is a significant increase in accidents from 1970 onwards, as is shown in Fig. 41.1. This increase can be attributed to two factors: (1) increase in consumption of natural gas at this time in many countries, with a consequent increase in piping used for transportation and distribution; and (2) improvements in the reporting of accidents and the difficulties of data collection in the past.

This can be seen in both parts of the study: total entries and natural gas entries. However, some distinctions have to be made with natural gas entries. Because natural gas is mainly used from the 1960s, there are no accidents involving natural gas before the 1940s. From

TABLE 41.4 Distribution of Accidents According to Date

Decades	Total registers	Natural gas registers
Before 1910	14	0
1910–1919	11	0
1920–1929	12	0
1930–1939	30	0
1940–1949	89	1
1950–1959	159	1
1960–1969	450	14
1970–1979	1,391	108
1980–1989	2,657	76
1990–1999	5,277	213
Total	10,090	413

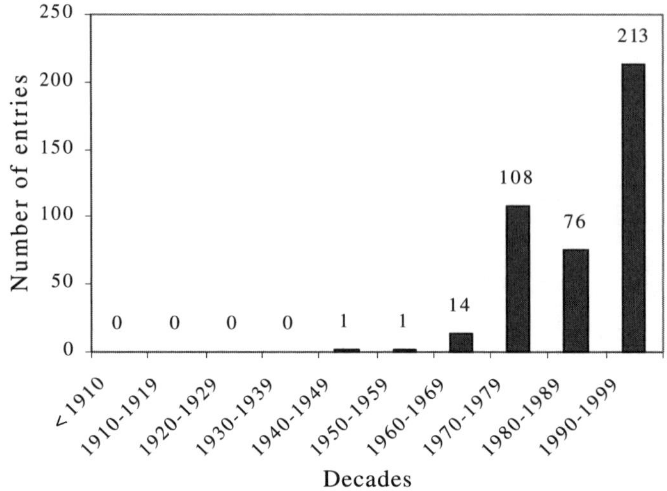

FIGURE 41.1 Distribution of accidents according to time.

the 1960s, accidents with natural gas increase due to the two aforementioned factors. However, this trend is somewhat different in the 1980s.

41.5 ACCIDENT TYPE

The database classifies accidents based on four different types: loss of containment, fire, explosion, and gas cloud. The results of the survey are summarized in Table 41.5. As can be seen, the accident type was known in 383 cases (93%). In the remaining 30 cases, the accident type is not specified.

The detailed analysis of the unknown accidents (going back to the description included in the database) shows different results for them. First of all, 10 accidents in the database were accidents with no consequences (i.e., without release, fire, explosion, or gas cloud). In 12 entries the accident was not specified, but in the accident abstract the type was implicit. In most of these cases the loss of containment is implicit but not specified. The last 8 entries have no abstract to describe the accident, which probably means these are not accurate.

The first part of Table 41.5 shows how many events involved the different types of accidents; because two of these types could exist in each event, the percentage can be greater than 100. Three of these types—explosion, loss of containment, and fire—have high percentages; in almost half of the accidents there was an explosion (43% of total), while accidents with fire and loss of containment have similar percentages (27.4% and 27.1% of total, respectively). There were 14 (2.5%) accidents involving gas cloud, which constituted the less frequent accident type.

The second part of the table shows the specific events for the three main types of accidents (explosion, fire, and loss of containment). The type of explosion was not specified in most cases (39%); for those in which it was specified, unconfined explosions and confined explosions have the same frequency, much lower than for physical explosions.

A similar situation was found for accidents involving fire, where the type of fire was specified in only 13 entries. Of these, there were the same number of gas cloud fires and jet fires (6 entries, 1.1% of total), and a pool fire was formed in only one entry.

TABLE 41.5 Types of accidents

Accident type	Number of accidents	% of total
Explosion	241	43.0
Fire	154	27.4
Loss of containment	152	27.1
Gas cloud	14	2.5
Accident type	**Number of accidents**	**% of total**
Accidents with at least one of the following types of explosion		
Unconfined explosion	12	2.1
Confined explosion	11	2.0
Physical explosion	218	38.9
Unknown	0	0.0
Accidents with at least one of the following types of fire		
Gas cloud fire	6	1.1
Jet fire	6	1.1
Pool fire	1	0.2
Unknown	141	25.0
Accidents with at least one of the following types of loss of containment		
Continuous release	84	15.0
Instantaneous release	5	0.9
Unknown	63	11.2

Finally, in accidents involving a loss of containment, this was a continuous release in most cases (84 entries, 15% of total). The release was instantaneous only in 5 cases (1% of total).

In order to obtain a distribution in which the sum of the percentages is 100, a classification with specific categories was done. Table 41.6 summarizes the results obtained.

Accidents in which an explosion has occurred are the most frequent (106 entries, 25.7%). Accidents in which an explosion and a fire have occurred are almost as frequent (97 entries, 23.5%), followed by "loss of containment" (72 entries, 17.4%). The other categories have lower percentages. Accidents in which a gas cloud was formed and those with loss of containment and pool formation are the least frequent (1 entry, 0.2%, respectively), which demonstrates that accidents involving natural gas without fire or explosion are very rare.

41.6 HUMAN DAMAGE

Because distribution networks are usually located in highly populated areas, the probability of human injury and death in case of accidents is very high. This has been examined in this

TABLE 41.6 Accident Type According to Categories

Accident type	Total registers	% of accidents
Explosion	106	25.7
Explosion + fire	97	23.5
Loss of containment	72	17.4
Loss of containment + explosion	36	8.7
Loss of containment + fire	32	7.7
Fire	25	6.1
Loss of containment + gas cloud	11	2.7
Gas cloud + explosion	2	0.5
Gas cloud	1	0.2
Loss of containment + pool	1	0.2
Unknown	30	7.3
Total	413	100.0

study. This classification includes those accidents in which the number of deaths was greater than zero. Of the overall number of accidents (413), the number of deaths was known in 291 cases (70.5%). Table 41.7 shows this distribution according to arbitrary categories (0 deaths, 1–10 deaths, etc.). In 69 cases (23.7% of known) there were no deaths, while in 201 accidents (69.1% of known) the number of deaths ranged from 1 to 10.

The maximum number of deaths from an accident involving natural gas occurred in Cleveland, Ohio, in 1944. There were 131 deaths and more than 200 people injured.

41.6.1 Human Damage According to Date

The ratio between the number of deaths caused by accidents and the number of accidents according to date is shown in Table 41.8.

If the evolution of this ratio as a function of time is analyzed, the general trend shows a decrease in deaths for each accident. This implies that the number of deaths becomes progressively less, which means that safety measures, control, prevention, and protection measures lead to safer installations and accidents become less grave.

TABLE 41.7 Distribution of Accidents Involving Injuries, According to the Data

Number of deaths	Number of entries	% of known
0	69	23.7
1–10	201	69.1
11–20	12	4.1
21–30	1	0.4
31–40	3	1.0
41–50	2	0.7
>50	3	1.0
Total	291	100

TABLE 41.8 Number of Deaths to Number of Accidents over Time

	Total registers		Natural gas registers	
Decades	Number of deaths	Ratio	Number of deaths	Ratio
Before 1910	9	0.64	0	—
1910–1919	10	0.91	0	—
1920–1929	8	0.67	0	—
1930–1939	20	0.67	0	—
1940–1949	45	0.51	1	1
1950–1959	54	0.34	0	0
1960–1969	163	0.36	4	0.29
1970–1979	411	0.30	46	0.43
1980–1989	373	0.14	27	0.36
1990–1999	763	0.14	33	0.15

Some conclusions can be drawn from the values of the ratio (number of deaths/number of accidents). On one hand, this ratio decreases from 1910 onwards. This is continuous in total entries, while in natural gas it is continuous from 1970. It is observed that values for natural gas entries are greater than values of total entries. This implies that while most accidents involving natural gas caused a number of deaths ranging between 1 and 10, they caused comparatively more deaths than the average value corresponding to the whole of entries of the database.

41.6.2 Human Damage According to Population Density

As mentioned, gas distribution networks are usually located in highly populated areas. The influence of this situation can be again studied from the database. This includes an information field on the population density of the areas where the accident took place. A qualitative value is used, with four possibilities: high population (town), low population (village), dispersed population (rural), and unknown. The results are shown in Table 41.9.

Approximately 70% of the entries do not include this information; it is known in only 119 cases. About half of the accidents (48%) occurred in highly populated areas, 35% in rural areas, and only 17% in low-populated areas.

Although the larger percentage in highly populated areas is due to the location of most industrial plants near large towns, the percentage obtained here should be considered with

TABLE 41.9 Population Density

	Number of accidents	% of total
Town	57	14
Village	20	5
Rural	42	10
Unknown	294	71
Total	413	100

TABLE 41.10 Number of Deaths According to the Accident Type

	Accident type							
	Explosion		Fire		Release		Gas cloud	
Number of deaths	Number of entries	% of type	Number of entries	% of type	Number of entries	% of type	Number of entries	% of type
0	42	17.4	20	13.0	33	21.7	4	28.6
1–10	72	30.0	43	28.0	24	15.8	2	14.3
11–20	11	4.6	5	3.2	2	1.3	0	0.0
21–30	1	0.4	0	0.0	0	0.0	0	0.0
31–40	3	1.2	2	1.3	0	0.0	0	0.0
41–50	2	0.8	1	0.6	1	0.6	0	0.0
>50	3	1.2	1	0.6	0	0.0	1	7.1

caution due to the low number of known entries. Furthermore, an accident occurring in a highly populated area is much more evident than a similar one in a rural zone, and therefore the probability of being included in a database is higher for the first case than for the second.

41.6.3 Human Damage According to Accident Type

The number of victims in an accident can be related to the accident type in order to show which type of accident is more dangerous. Table 41.10 shows the results obtained in this case.

As can be observed, most of the cases with deaths correspond to accidents where an explosion took place. This is partly due to the fact that explosion is the most frequent accident type. Although this trend can be observed in all categories, including that of no victims (0 deaths), explosion is the accident type with the highest number of victims.

41.7 GENERAL CAUSE OF ACCIDENTS

Two kinds of causes are taken into account in the database: the general cause and the specific cause. Of the 413 accidents studied in this survey, the cause was known only in 252 cases (61% of total). This can be explained by the fact that it can be very difficult to establish the cause after the accident. The various general causes are summarized in Table 41.11. Because one accident can be due to more than one cause, the sum of the percentages is again greater than 100.

The four most general causes show a similar contribution: impact failure (37.3%) is the first one, followed by mechanical failure (33.3%), human factor (27%), and external events (22.2%). There are far fewer accidents from the other causes.

Because accidents may have more than one cause, an interesting study is that of classifying accidents as due to one single cause or to more than one cause. This is shown in Fig. 41.2. In this case, more than one cause was given for 54 accidents, and in 198 there was only one cause. Impact failure was recorded as the only cause for 68 cases (27% of known), and in 26 cases (10.3% of known) it was associated with other causes. The corresponding figures for mechanical failure are 53 entries (21% of known) and 31 entries (12.3% of

TABLE 41.11 Distribution of Accidents According to General Cause

Cause	Number of accidents	% of known
Impact failure	94	37.3
Mechanical failure	84	33.3
Human factor	68	27.0
External events	56	22.2
Violent reaction	6	2.4
Process conditions changes	4	1.6
Instrument failure	3	1.2
Services failure	3	1.2

known), respectively. A different situation is found in the accidents caused by human error; it was the only cause in 25 cases (10% of known) while in 43 cases (17.1% of known), human error was directly associated with another cause (e.g., impact due to excavating machinery).

The summary of one of these accidents with more than one cause supposes that percentages will be greater than 100.

41.7.1 Specific Cause of Accidents

Some accidents have more than one specific cause for the same general cause. Therefore, in the corresponding analysis, all percentages are greater than 100. The results can be seen in Table 41.12.

Studying the most frequent general cause (impact failure) yields some interesting information. Almost half of these accidents (43%) were caused by excavating machinery; fewer accidents were caused by collisions: 17% (vehicle) and 14% (ship–coast impact). The same

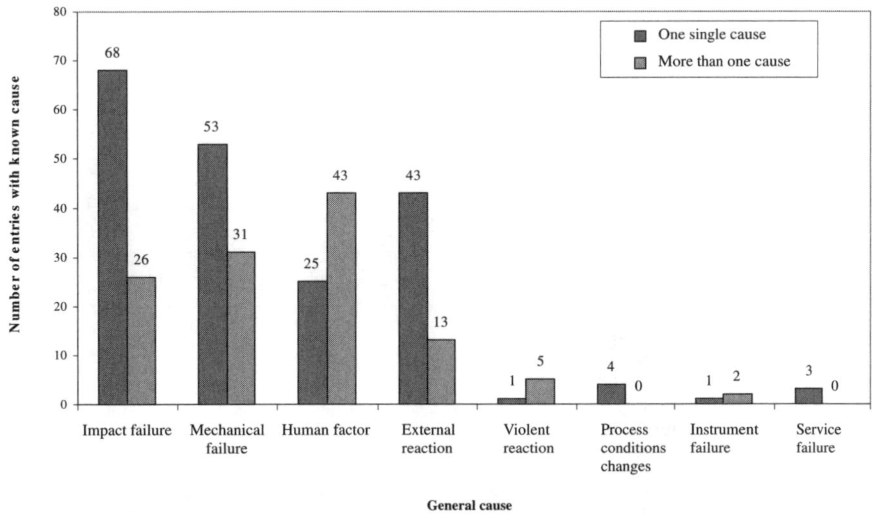

FIGURE 41.2 Distribution of general causes of accidents.

TABLE 41.12 Distribution of Accidents According to Specific Causes

Cause	Number of entries	% of general cause	% of known
Specific cause of accidents due to impact failure			
Excavating machinery	40	42.6	51.9
Vehicle	16	17.0	20.8
Impact ship–coast	13	13.8	16.9
Heavy object	9	9.6	11.7
Road accident	9	9.6	11.7
Crane	1	1.1	1.3
Not specified	17	18.1	22.1
Specific cause of accidents due to mechanical failure			
Corrosion	13	15.5	26.5
Overpressure	12	14.3	24.5
Leaking valve	11	13.1	22.4
Leaking flange	7	8.3	14.3
Other metallurgical failure	6	7.1	12.2
Weld failure	5	6.0	10.2
Fatigue	4	4.8	8.2
Ground subsidence	3	3.6	6.1
Excavating machinery	3	3.6	6.1
Overloading	1	1.2	2.0
Brittle failure	1	1.2	2.0
Relief valve failure	1	1.2	2.0
Leaking gland or seal	1	1.2	2.0
Not specified	23	27.4	46.9
Specific cause of accidents due to human error			
General operations	31	45.6	63.3
Maintenance	11	16.2	22.4
Procedures	5	7.4	10.2
Communications	4	5.9	8.2
Construction failure	3	4.4	6.1
Installation failure	3	4.4	6.1
Coupling failure	3	4.4	6.1
Management failure	3	4.4	6.1
Insulation failure	1	1.5	2.0
Sabotage/vandalism	1	1.5	2.0
Design	1	1.5	2.0
Not specified	19	27.9	38.8
Specific cause of accidents due to external events			
Sabotage/vandalism	25	44.6	48.1
Ground subsidence	7	12.5	13.5
Extreme temperatures	5	8.9	9.6
Flooding	4	7.1	7.7
External fire	4	7.1	7.7
High winds	3	5.4	5.8
Lightning	2	3.6	3.8
Earthquake	2	3.6	3.8
Not specified	4	7.1	7.7

number of accidents were caused by heavy objects and road accidents (9 cases, 10% of general cause).

Of the accidents caused by mechanical failure, the most frequent specific cause was corrosion (13 cases, 15.5% of general cause), followed by overpressure (12 entries, 14.3%). Specific causes, including metallurgical failures, constitute significant percentages as a whole. Among them is a leak in a valve (11 entries, 13.1%), a leak in a flange (7 cases, 8.3%), weld failure (5 cases, 6%), brittle failure, relief valve failure, and a leaking gland or seal (one entry, 1.2% respectively) and other, unspecified failures (6 cases, 7.1%).

Four accidents were caused by fatigue (4.8% of general cause), and only one was caused by overloading (1.2%).

Of the accidents caused by human error, in 28% of cases there was no information available on the type of error. For the rest, the most frequent source of error was general operations (31 cases, 45.6% of general cause), followed by maintenance (11 entries, 16.2%), and communications (4 cases, 5.9%).

Finally, for cases caused by external events, the most important contributions were those of sabotage/vandalism (25 cases, 44.6%), followed by ground subsidence (7 cases, 12.5%). Fewer accidents were caused by natural disasters, which include flooding, extreme temperatures, external fires, highwinds, lightning, and earthquakes.

41.8 ECONOMIC LOSSES

Information about economic losses is not known at the time of the accident, and after the investigations it is often not easy to obtain. Here the economic losses are known in 38 cases (9.2% of total); the data are shown in Table 41.13.

The small size of the sample does not allow a significant statistical treatment. Most of these accidents have economic losses between $1 million and $10 million, and in 9 cases they are greater than $10 million. This shows that the economic losses in an accident of this type can be considerable. The accident with the highest economic losses occurred in Hanau, Germany, in 1991, when a natural gas explosion in the glassworks of a metal-processing firm caused $264 million in damages.

However, there is an inherent difficulty in comparing economic data from different periods. Inflation implies a continuous change in the economic cost of any given damage, and the MHIDAS database does not update the economic losses for each accident; the values are those obtained directly from the information source.

TABLE 41.13 Economic Losses

Economic losses (U.S.$)	Number of entries	% of known
0	1	2.6
1–1,000	0	0
1,001–10,000	0	0
10,001–100,000	3	8.0
100,001–1,000,000	8	21.0
1M–10M	17	44.7
>10M	9	23.7

TABLE 41.14 Ignition Source

Ignition source	Number of accidents	% of accidents
No ignition	71	17.2
Electric	17	4.1
Flames	10	2.4
Friction sparks	7	1.7
Hot surfaces	6	1.5
Unknown	302	73.1
Total	413	100.0

41.9 IGNITION SOURCE

When the ignition source of an accident is sought, a surprisingly high number (73%) of "unknown" cases is found. This is because the information often is not evident, and different sources may be possible. Because installations used to be destroyed after an accident, the investigation, if done, can hardly clear up this aspect. The results obtained in this study are summarized in Table 41.14.

For those accidents where the information was available, there was no ignition in 17.2%. For those in which there was actually ignition, the most frequent source was electrical sparks (4.1%), followed by naked flames (2.4% of known), friction sparks (1.7%), and hot surfaces (1.5%). This trend is also observed in similar studies. One of these corresponds to the data published by Planas-Cuchi in a historical analysis for pool fires (Planas-Cuchi et al., 1997). In this case, the information was available in a low percentage of the accidents as well (24%). The percentages are maintained: electrical sparks are the most frequent ignition source (8%), followed by naked flames (7%) and hot surfaces (4%). Friction sparks are the less frequent ignition source, with only 2% of cases.

41.10 REFERENCES

British Petroleum-Amoco. 1999. BP Statistical Review of World Energy 1999, BP-Amoco home page: http://www.bpamoco.com/worldenergy/naturalgas/index.htm.

Davenport, J. A. 1977. "Survey of Vapor Cloud Incidents," *Chemical Engineering Progress,* vol. 73, no. 9, pp. 54–63.

Davenport, J. A. 1983. In *4th International Symposium on Loss Prevention and Safety in the Process Industries,* European Federation of Chemical Engineering, Harrogate, UK, September.

Garrison, W. G. 1988. *One Hundred Largest Losses—A Thirty Year Review of Property Damage Loss in the Hydrocarbon-Chemical Industries,* 11th ed., M&M Protection Consultants, New York.

Haastrup, P., and L. Brockhoff. 1990. "Severity of Accidents with Hazardous Materials," *Journal of Loss Prevention in the Process Industries,* vol. 3, pp. 395–405.

Knowles, A. E., F. Tweedle, and J. L. Van der Post. 1977. *The Background and Implications of IGE/TD,* 2nd ed., Comm. 1044, Institution of Gas Engineers, London.

Lees, F. P. 1986. *Loss Prevention in the Process Industries,* vol. 2, Butterworths, London, pp. 790–793.

Montiel, H., J. A. Vílchez, J. Arnaldos, and J. Casal. 1996. "Historical Analysis of Accidents in the Transportation of Natural Gas," *Journal of Hazardous Materials,* vol. 51, pp. 77–92.

Planas-Cuchi, E., H. Montiel, and J. Casal. 1997. "Survey of the Origin, Type and Consequences of Fire Accidents in Process Plants in the Transportation of Dangerous Materials," *Transactions of the Institution of Chemical Engineers,* Part B, vol. 75, pp. 3–8.

Rømer, H., L. Brockhoff, P. Haastrup, and H. J. Styhr-Petersen. 1993. In *Journal of Loss Prevention in Process Industries,* vol. 6, pp. 219–225

Vílchez, J. A., S. Sevilla, H. Montiel, and J. Casal. 1995. "Historical Analysis of Accidents in Chemical Plants and in the Transportation of Hazardous Materials," *Journal of Loss Prevention in Process Industries,* vol. 8, pp. 87–96.

CHAPTER 42
FLIXBOROUGH: THE DISASTER AND ITS AFTERMATH

J. E. S. Venart
Department of Mechanical Engineering,
University of New Brunswick, Fredericton,
New Brunswick, Canada

If we do not understand the past we cannot make proper sense of the present or the future.

42.1 BACKGROUND

On June 1, 1974, a fire and massive explosion occurred at the Nypro (UK) Ltd. works near Flixborough, North Lincolnshire (Fig. 42.1). An accidental release of cyclohexane at 0.96 MPa and between 150 to 155°C resulted in an unconfined vapor cloud explosion (UVCE) that caused 28 fatalities, the destruction of the plant, severe injury to inhabitants, and sig-

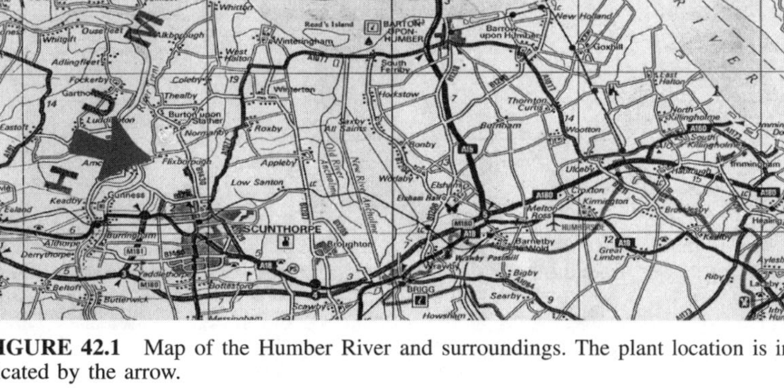

FIGURE 42.1 Map of the Humber River and surroundings. The plant location is indicated by the arrow.

nificant damage to many buildings in the surrounding countryside (Secretary of State for Employment, 1975).

The plant was located on the east bank of the River Trent, a tributary of the Humber, about 50 km west of the fishing port of Grimsby. On the west bank of the river, opposite the NYPRO site, was the village of Amcotts; the village of Flixborough was on a small rise to the East; the town of Scunthorpe was southeast of the location. All suffered major distress though no loss of life.

The catastrophe was initiated in section 25A of the process stream (Fig. 42.2) during start-up while the cyclohexane feedstock, inerted with nitrogen, was under hot recycle through the reactor train, R1 to R6. Only about one-third of the 70 kg/s full production flow was recirculating at the time.

A portion of the plant is shown in Fig. 42.2. Section 25A was south and midway between the laboratory and the works and main office buildings. Adjacent was section 7 (the caprolactam plant) and section 27. The reactor train (R1 to R6), which was considered the source of the release, was aligned primarily north-south. A temporary 20-in. (0.5-m) pipe assembly bridged reactors R4 and R6, and the removed reactor, R5, had been placed in waste ground in front of the works office and alongside of the section. At the time the wind was blowing to the northeast at about 4 to 6 m/s. Shown also is an estimated outline of the flammable cloud. The perceived center of the explosion, as determined by several investigators, was identified as being at the roadway just in front of the main office.

In the oxidation sequence there had been six reactors (Fig. 42.3). However, the fifth reactor, R5, had been removed because of damage and a temporary 304L stainless steel (SS) 0.5-m diameter pipe bridge, which used the two original 0.71-m diameter bellows (B) and 316L SS stub pipe (P) assemblies, had been installed between R4 and R6.

The reactors were made of 13-mm steel with a 3-mm thick SS steel liner. Each was about 5 m high by 3.5 m in diameter and was fitted with an interior overflow weir and baffle plate to ensure level control and bottom feed, proper oxidation, and mixing. All but one, R4, had an agitator (Artingstall, 1974).

The original 0.71-m interconnecting pipes and their bellows had permitted the hot pressurized feedstock to communicate by open channel flow over the outlet weir in the upstream reactors, through the bellows connection into the downstream reactors with a fixed difference in elevation of 0.178 m and at a velocity of less than 0.3 m/s through the nearly full pipes under full production flow. The adaptation (Fig. 42.4), however, doubled the head between R4 and R6 to 0.356 m. There was also a significant weir "waterfall" (velocities from 1.8 to

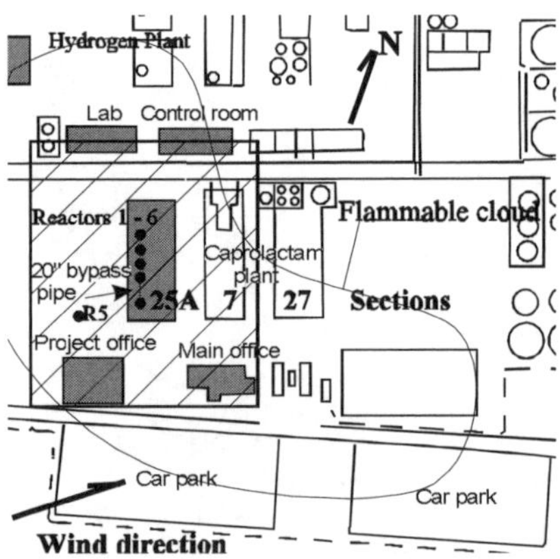

FIGURE 42.2 Portion of the Nypro site plan showing section 25A and the reactor train.

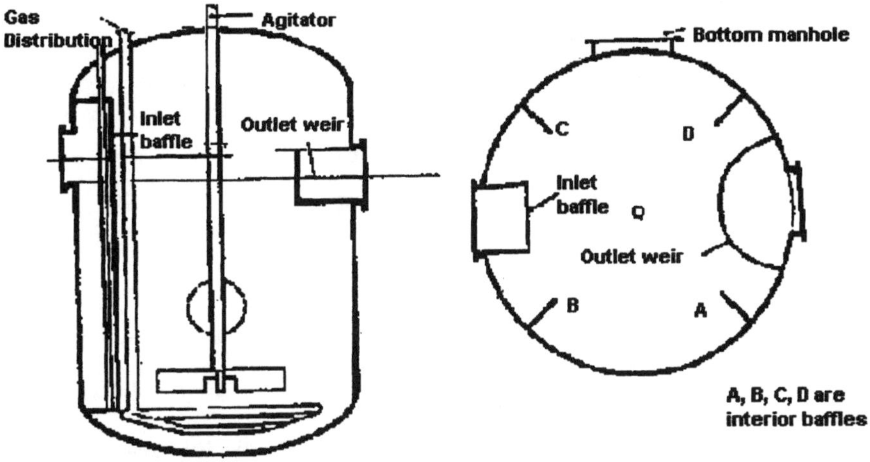

FIGURE 42.3 Reactor vessel details; elevation and plan views. (*Source: Artingstall, 1974.*)

2.6 m/s) in addition to two hydraulic jumps in the very complex flow structure (Fig. 42.5). In the R4 to B4 section, the 0.71-m pipe was now nearly half full and under full production the velocity would have been about 0.5 m/s. Immediately after the transition into the 0.5-m section, the flow accelerated and became supercritical, ending in a significant hydraulic jump near the lower miter bend, with flow slugging into the lower B6 bellows and thence into R6. Here the flow was quiescent and filled the rest of the 0.5-m pipe (velocity at full production about 0.3 m/s) (Teng-yang et al., 2000).

The modification made was thus more than just a geometric alteration. Significant changes were made to the statics, the dynamics, and the flow in the assembly. First, due to the differences in diameter and the dogleg shape, we have a couple which loads the pipe latterly in the vertical direction. Second, we have a mass spring system with very little support

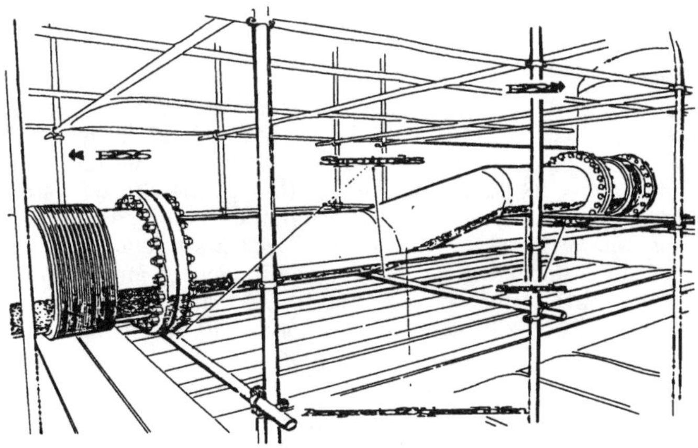

FIGURE 42.4 Reconstruction of the R5 bypass assembly showing the two 0.71-m diameter bellows and stub pipes, the 0.5-m dogleg pipe, and the support scaffolding used. (*Source: Secretary of State for Employment, 1975.*)

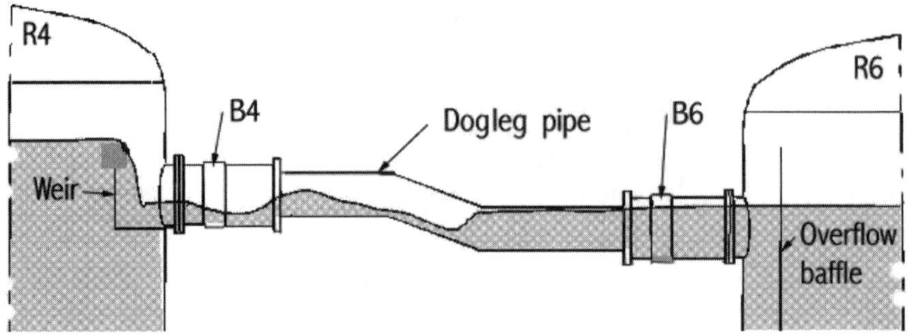

FIGURE 42.5 Reactor train vessels R4 to R6 with the temporary bypass pipe with the liquid levels as for full flow. (*Source: Teng-yang et al., 2000.*)

provided by the scaffolding, especially upon vertical thermal expansion of the reactors. And finally, there are the significant discharge and flow alterations noted earlier.

Each bellows had a first natural frequency of about 65 to 71 Hz axially and between 325 to 416 Hz in the transverse direction; the system frequency, as determined for a straight pipe, was 3.9 Hz axial, 15.7 Hz lateral (ends of the spool pipe in phase), and 27.2 Hz for the rocking mode.

The excitation of this very complex spring-mass system was now such that resonances with the flow were possible (Teng-yang et al., 2000). Most of the lower frequencies could have been excitable by the very broad range of flow velocities present, particularly near the two hydraulic jumps (EJMA, 1993). Additionally, the jetting at the base of the weir outfall could possibly result in axial resonance of the bellows. Coherent turbulence may also have been able to excite the higher modes.

With flow-induced vibration, no significant response is observed up to some critical velocity, and then, as the velocity is increased further, the bellows can become unstable. For uniform mean flows the critical Strouhal number[1] is St = 0.45 (Weaver and Ainsworth, 1989); for nonuniform flows, as at the exits of elbows leading to bellows and perhaps the weir outfall region, St = 0.57 (Jakubauskas and Weaver, 1998a, b).

The flexibility of expansion joints makes them susceptible to self-excited vibration, resulting in bending stresses that may produce fatigue failure (Weaver and Ainsworth, 1989). Gigacycle[2] fatigue damage to the thin wall (~1-mm thick: Foley, 1974a) 12-convolution bellows could well have occurred since between 64 to over 300 million cycles could have been possible within the range of potential excitations present during the two-month operating period. Indeed, it had already been noted that the assembly was unstable as it lifted off its supports during operation (Secretary of State for Employment, 1975).

The process fluid, Cyclohexane (C_6H_{12}, MW 84) is a colorless liquid similar in some of its properties to gasoline (flash point 20°C, auto-ignition temperature (AIT) 245°C). The fluid has a normal boiling point of 80°C and a saturation pressure, p_{sat}, at 150°C of 0.55 MPa. When the liquid was saturated with nitrogen, or air as when oxidation was occurring in the reactor stream, the pressure was 0.96 MPa. Under these conditions the liquid was 99.8% C_6H_{12} and 0.2% N_2 or air; the vapor would have consisted of 82.6% C_6H_{12} and 17.4% N_2

[1] St $(fp)/V$, where f is the bellows natural frequency in Hz, p is the bellows pitch (m), and V is the mean flow velocity (m/s).

[2] The fatigue testing of steels usually stops before 100 million cycles since a fatigue limit is registered between 1 and 10 million cycles. However, tests conducted with high-frequency random vibration (similar to those associated with the natural resonance of a structure) show that the fatigue limit of steels disappears and that fatigue failures occur (Miller, 1999; Au-Yang, 1999).

FIGURE 42.6 Looking north over the rubble of the main office building into section 27. (*Photograph courtesy the Scunthorpe Evening Telegraph.*)

or air. The bubble point pressure of this mixture, $p_{sat}(150°C)$, was 0.956 MPa. The lower flammable (LFL) and upper flammable limits (UFL) of a vapor–air mixture by weight under atmospheric conditions are 3.46 and 19.8 wt % (1.3 vol. %; 7.8 vol. %), respectively.

The scale of the disaster and damage can be appreciated from photographs taken soon after the explosion and shown in Figs. 42.6 and 42.7.

In Fig. 42.6 we are looking north over the rubble of the main office building at section 27. Notice the crushed columns to the right of the section and the (amazingly) still near-vertical lamppost, all indicators that this location was near the center of the explosion. Note also that the columns and vessels on the west (left) side of the section and to the north are uncrushed.

In Fig. 42.7 the parking lot to the southeast of the main office building is shown. This spot is near where the main entrance and gate house would earlier have been. In the background, notice the overturned tanker truck (estimates of the pressures necessary to crush its tank range up to 1 MPa: Gugan, 1978). Note that the vehicles in the parking lot have been shredded, also indicative of extremely large overpressures.

42.2 THE COURT OF INQUIRY

An Official Court of Inquiry was immediately appointed by the Secretary of State for Employment, Michael Foot, "to establish the causes and circumstances of the disaster and to point out any lessons . . . to be learned therefrom." The essential problem faced by the Court was "to determine what had caused the rupture of the by-pass." The possibilities investigated were: "i) rupture of the by-pass assembly through internal pressure; ii) rupture of the assembly in two stages; a *small tear in the bellows* (bellows B4) leading to an escape and a minor explosion causing final rupture; and iii) rupture of the '8 inch' line at the 50-inch split leading also to a minor explosion causing rupture of the by-pass assembly."

FIGURE 42.7 The Nypro parking lot to the southeast of the main office building. (*Photograph courtesy of the Scunthorpe Evening Telegraph.*)

Processes ii) and iii) were interpreted to require *external* minor explosions to trigger events; no physical evidence for these could be determined. Despite the physical evidence of the internal tear in bellows B4 (Foley, 1974a) and compelling eyewitness evidence, only three paragraphs in the final report and a few pages in the Court's record and *no* definitive studies were devoted to the consideration of the two-stage rupture possibility that would have included the likelihood of a fatigue-induced initiation to the sequence of events. Only processes i) and iii) were examined in depth, with nearly 50% of the Report of the Inquiry, the Record of the Court, and the supporting studies being devoted to the consideration of one or the other of these cases. Other possibilities considered for an *internal* initiating event (e.g., peroxide decomposition and the operation of a nitrogen purge) were excluded. The likelihood of a water/cyclohexane superheated explosive boiling interaction (King, 1975, 1977) was also not considered, and recent experiments undertaken by the U.K. Health and Safety Executive (HSE) on behalf of the Minister of the Environment (King, 1999), could suggest what role, if any, water trapped in the bottom of one of the reactors may have played in the event.

On the basis of expert evidence by Professor D. E. Newland (1976) based on a dynamic analysis and assessment of energy requirements for both bellows to squirm and the bypass pipe to buckle as one event, the Court concluded, albeit with *low probability*, "that the disaster resulted from a one stage failure of the 20-inch assembly" as a result "of conditions of pressure and temperature more severe than any which had previously prevailed but no higher than careful and conscientious plant operators could be expected to permit." The Court did, however, state that this conclusion "would be readily displaced if some [other event of] greater probability to account for the rupture could be found" (Secretary of State for Employment, 1975).

Extensive full-scale experiments were undertaken in attempts to reproduce the failure (Games and Waterhouse, 1974; Batstone, 1975). These employed a similar pipe assembly (as shown in Fig. 42.4) as well as other experiments and bellows tests evaluated at the same and more severe operating conditions. All these experiments were unable to reproduce the failure process accepted.

In view of the Court of Inquiry's qualified conclusion, the cause of the accident has been the subject of considerable controversy, especially as to the actual failure process (e.g., Ball, 1975, 1976; Butler, 1975; Cox, 1976; Gugan, 1976; King, 1977; Warner, 1975; Warner and Newland, 1975); the amount of cyclohexane released, and whether the unconfined vapor cloud formed in the release detonated (e.g., Gugan, 1978, 1980; Ale and Bruning, 1980a, b; Fu and Eyre, 1980; Phillips, 1981). The debate and argument continue to this day (e.g., Gugan, 2000; Hoiset et al., 2000; King, 2000; Kletz, 2000; Swan, 2000).

In the present reanalysis, the bellows, the bridging pipe, and its assembly into the bypass are modeled and a release scenario that considers the likely sequence of events initiated by the crack noted by the Court in possibility ii) above is advanced. These considerations utilize finite-element analyses (FEA) of the bellows and pipe assembly as well as computational fluid mechanics (CFD) models of the two potential release secenarios: that accepted by the Court and the one considered here. The results of both examinations demonstrate that the proposed two-step failure of the pipe bridge is more probable based on the physical and eyewitness evidence available. The localized explosion damage additionally supports the conclusion that the cloud was detonated.

42.3 INTRODUCTION

The initiating mechanism proposed first requires an initiating tear near the top of the upstream, B4, bellows. This bellows was the only one that was found with an internal tear (Secretary of State for Employment, 1975; Foley, 1974a) in the areas of the bellows most highly stressed; i.e., near the topmost convolutions of the bellows closest to R4 (Venart and Tan, 2000).

The various steps in the failure process are detailed in Fig. 42.8 with a tabulation of the sights and sounds that would have been experienced in Table 42.1. The very complex process can be summarized as follows:

- The instability of the assembly and the near resonance of the upstream, B4, bellows with the forcing fluid flow initiates fatigue damage, which accumulates in the most highly stressed, and least damped, regions of the unit.
- Sufficient damage accumulates to cause a tear, which further opens, due to pressure, and local vapor-gas depressurization occurs (Fig. 42.8*a*). There is a bang and a sound like a pressure relief valve going off (Table 42.1).
- B6 next rapidly expands and B4 compresses due to this unequal pressure distribution (Fig. 42.8*b*), and the now superheated and supersaturated cyclohexane rapidly (~50 ms) boils up. The impact of this two-phase swell results in a head-space impact and dynamic re/overpressurization (Gromles, 1984; Venart and Ramier, 2000). This immediately causes both the compression of the downstream, B6, bellows *and* the initiation of upwards 'squirm' of the B4 bellows (Fig. 42.8*c–e*). First there is a thud as B4 is compressed, and then a minor rumbling of the boiling process in the B4 region of the pipe bridge, and then a further thud as B6 is compressed by the repressurization occurring near B4 (Table 42.1).
- The compression of B6 by this local overpressure constrains the dogleg pipe at its R6 connection (Snedden, 1985) and causes the process of B4 squirm to act over the full 356-mm eccentricity (Fig. 42.8*d*) with much less force than that required for the 178-mm offset required in Newland's single-stage analysis (Newland, 1976).
- This process results in the commencement of collapse by buckling of the 0.5-m pipe as the B4 bellows continues to squirm, then balloon, burst and rip off *at its normal operating pressure* due to the p-V work of the expanding two-phase fluid (Fig. 42.8*f*).
- The ripped off topmost section of the bellows by this process is projected upwards and sails into the air and appears to some (Harry, 1974), as a blown-off "tank top"; it is found

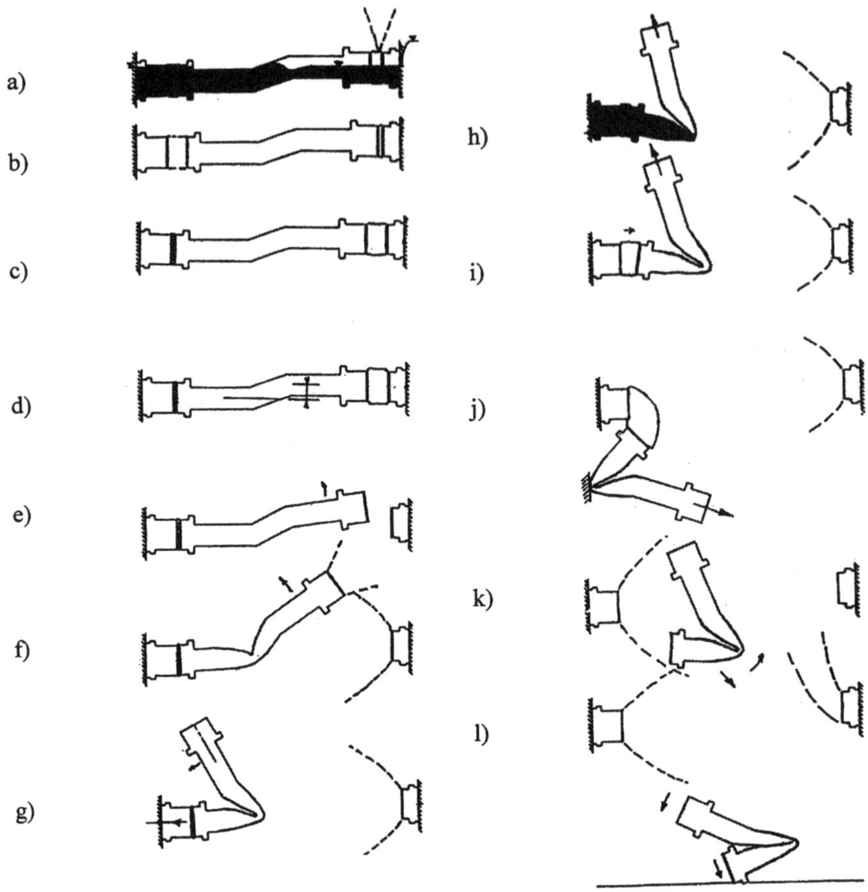

FIGURE 42.8 Successive stages of a crack-initiated two-stage bellows and pipe bridge failure.

only two months later *under* extensive debris between the line of reactors and the after-reactors (Foley and Nicholson, 1974); *all other portions of both the B4 and B6 bellows had been immediately located by investigators* (Foley, 1974a).

- The p-V work of the expanding two-phase supersaturated cyclohexane in the squirming and expanding B4 bellows, with the now greater constraint offered by the now compressed B6 bellows, buckles and completely collapses the 0.5-m pipe at its lower miter bend (Fig. 42.8*g*).
- The shape of the buckled pipe is now momentarily like the "swan neck" (Fig. 42.8*g*,*h*) noted by some (Marshall, 1974; Gugan, 1974b) with a significant vertical jet discharge. The B6 bellows through all of this is, however, still intact, though possibly severely damaged.
- In the process of suddenly buckling (Fig. 42.8*g*,*h*), the 250-kg liquid contents in the lower end of the 0.5-m diameter pipe are forcibly ejected into R6, and this impulse tears off its inlet baffle and causes this to impact and bend the internal 154-mm diameter SS agitator shaft (Artingstall, 1974; Gugan, 1978).

TABLE 42.1 Sights and Sounds in the Flixborough Bypass Pipe Bridge Failure (Fig. 42.8)

Time (s)	Event (Fig. 42.8)	Sound	Sight	Comment
1–2	(a)	PRV vent	Whisp to jet of "steam"	Many observers
3	(b)	"Thud"	Jets of "steam"	
3–4	(c)–(d)	Bang and rumble	Two-phase jet	
4+	(e)	Greater rumble, larger jet		Bellows squirms
5−	(f)–(g)	More severe rumbling	Upward debris, large kettle spout	Bellows off, pipe buckles
5	(h)–(i)	Roar "like a jet engine"	Vertical venting "swan-neck" pipe	Jet discharge, R4
5+	(j)	Pipe bangs R6	Vertical flames	Dent in R6
6–25	(j)	Continuing roar, first explosion	"Thrashing" jet flame	Jet discharge, R4, diminishes
26	(k)	Detonation	B6 tears off	R6 jet discharge
27+	(l)	Crash and thud	Buckled bypass impacts R5 plinth	R6 jet fire on B4 remnants and the 8-in. lines

- This process would have been very noisy because the bellows burst and the dog-leg pipe buckles sending the overhead scaffolding upwards. There would have additionally been the development of a roar due to the massive two-phase discharges from both R4 and the B4 end of the pinched pipe bridge and the rumbling of the boiling of the superheated Cyclohexane in R4 (Table 42.1).
- The choked two-phase discharge, which occurs through the B4 end of the pipe bridge via the 0.5-m pipe pinch—as the R6 internal pressure now expands B6 and frees its constraint (Fig. 42.8*h*) allows the reaction of the escaping two phase jet to bend and distort the bellows, forcibly expanding and stretching its topmost elements to their full extent (~1 m) so that the swinging pipe can sweep the scaffolding below (Fig. 42.3) toward the base of the R6 reactor (Plates 8 and 9 of Secretary of State for Employment, 1975). The position of the crimple points in B6 differs significantly from those in B4 (Foley and Nicholson, 1974) indicating a different distortion process for this bellows.
- In this operation the pipe deflects and the lower surface of the bend impacts the side of R6, denting it (Artingstall, 1974), flattening the bottom of the pipe after the lower miter bend farther (Foley, 1974a) (Fig. 42.8*j*), and effectively cutting off flow through the pinch. Additionally, the topmost portion of B6 is stretched out flat, without any convolutions being left, unlike the companion piece of B4, where some convolution creases were still intact (Foley, 1974b).
- In this dynamic process, the B6 bellows are now torn off and the rebound from the R6 impact *and* the impulse of the two-phase release from the now open R4 end of the pipe spins the assembly downwards (Fig. 42.8*k*) such that the R6 end flanges, impact the top of the 0.46-m-thick reinforced concrete R5 plinth, puncturing it (Secretary for State for Employment, 1975). The R6 stub end pipe and flanges are severely distorted at the point of impact by this process (Foley, 1974a).
- These later sequences (Fig. 42.8*h–k*) occur *after* the major explosion and explain the somewhat vertical but moving jet flame film taken by a witness (Goodchild, 1975) and observed earlier by others (Carter, 1974).

- Once the B6 bellows has blown off the resulting horizontal, northward developing torch flame envelopes the now blast damaged 8-in. pipes and causes the observed creep swelling and zinc contamination (Foley and Nicholson, 1974) of these as well as the high temperature carburization found *only* on the attached remnants of the B4 bellows of the R4 stub pipe (Foley, 1974a).

The major consequence of this sequence of events is that a discharge through reactor R4 *alone* occurs *prior to* the main explosion. This is sufficient only to provide a release of some 370 kg/s of cyclohexane for probably less than the 30 seconds determined for the initial events (Cox, 1976). This vent, of some 10–15 tons (Leclude 1996; Leclude and Venart, 1996), is very much smaller than the 30–60-ton estimates previously attributed to the explosion (Sadee et al., 1977; Roberts and Pritchard, 1982; Bjerketvedt et al., 1997).

Thus far, the postulated complex movements of the bellows-pipe assembly have been described such as might have occurred with a fatigue-initiated tear in the B4 bellows. The detailed finite-element analysis of the bellows and pipe bridge behavior supporting this interpretation is described next, and then the remainder of the section examines the discharge processes for the case postulated by the Court and that which would have resulted under the circumstances outlined above. We will thus compare the developments of the vapor cloud noted by witnesses and the consequences of its ignition.

42.4 FINITE-ELEMENT ANALYSES

The elastic lateral stiffness of a bellows decreases linearly with increasing pressure (EJMA, 1993; Newland, 1964; Snedden, 1985) and, upon sufficient increase, the bellows can become unstable, usually resulting in very rapid (10 to 50 ms) gross lateral deflection or "squirm." The lateral stiffness and response of a universal expansion joint, i.e., two bellows interconnected by a straight length of pipe or spool piece, has been considered by Newland (1964). The lateral stiffness for two bellows connected by an offset spool piece, such as here, was undertaken by Newland (1976). No determination appears to exist for the determination of natural frequency with an offset spool piece.

ANSYS 5.3 (Swanson Analysis, 1994) nonlinear elastic-plastic large deformation finite-element (FE) models were made of the bellows and assembly illustrated in Fig. 42.8. The bellows and 0.71-m OD stub pipes were fabricated from 316L SS; the 0.508-m OD dogleg pipe bridge was made from four sections of 4.8-mm-thick 304L SS pipe with two 80° miter bends (Foley, 1974a).

42.4.1 The Bellows

The bellows (Fig. 42.9*a*) was originally formed from 1.19-mm-thick sheet. In the forming process, the thickness was reduced to 1 mm, (t) and the material cold worked. Each bellows had 12 convolutions (N) and was about 230 mm long overall with a 750-mm mean diameter (R_m = 375 mm, t = 1 mm, p = 33.9 mm, h = 37 mm).

The mechanical properties used in the models were determined from samples cut (in both longitudinal and transverse rolling directions) from a 1.2 by 2.4-m 316L sheet 1.19 mm thick. Mechanical properties were also determined after a single-pass cold-roll reduction to 1 mm. Twenty samples (five for each direction, both as received and single pass cold rolled) were utilized. The mean results are given in Table 42.2. Directional variation was between 2.6 and 3.4 and 4.8 and 6.7% on strength for the as-received and cold-rolled specimens respectively.[3]

[3] ASTM subsize sheet type E 8M 89b specimens; Instron servohydraulic testing machine, load cell in series with the specimen.

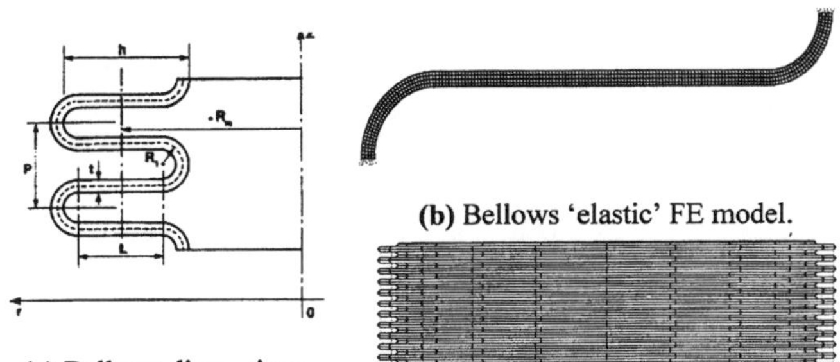

(a) Bellows dimensions

(b) Bellows 'elastic' FE model.

(c) Bellows full-size FE model.

FIGURE 42.9 Bellows dimensions and models. (*Source: Venart and Tan, 2000.*)

For small deflections (i.e., elastic behavior), a bellows can be simplified by only half a convolution (Fig. 42.9*b*). The axial and transverse stiffness from such an analysis can be regarded as a reference base for the verification of any later more detailed, and therefore necessarily less refined, FE model. An analysis using 2688 eight-node isoparametric 2D symmetric elements, PLANE82, with 8753 nodes, resulted in a determined axial stiffness, K_{axial} of 403.8 kN/m (2,306 lbf/in) with a transverse stiffness, $K_{\text{transverse}}$ of 5,558 kN/m (31,738 lb_f/in.). Experimental values obtained from the manufacturer (Batstone, 1975) were stated to be between 485 and 578 kN/m and 7,567 and 9,015 kN/m for the axial and transverse directions respectively. The differences could be due to nonisotropy due to the forming process and the choice of a uniform wall thickness in the model.

With large transverse deflections, contacts between adjacent convolutions occur and the whole bellows must be modeled. The shape of the bellows is asymmetric, i.e., the overall shape of the deformed structure possesses 180° rotational symmetry about the midpoint of the neutral axis. Three-dimensional shell elements (SHELL43) are thus necessary (Fig. 42.9*c*). An element size of 12 × 12 mm yields 11,760 elements and 11,979 nodes to mesh only one half a bellows. The computational requirements for this are large. Increases in element size result in an increase in axial stiffness.

Figure 42.10 illustrates the relative transverse deflection of the present model as element numbers are increased. When the sum of elements exceeds between 500–650 a near asymptote is reached. A maximum element size of 125 mm was therefore chosen using 720 elements with 803 nodes to model the bellows (Fig. 42.9*c*).

Figure 42.11*a* shows the deformed shape of the bellows joint with an axial extension of 5 mm at an internal pressure of 0.96 MPa and a transverse displacement of 75 mm. Figure 42.11*b* shows the transverse loads required to effect the displacement as a function of internal pressure (0.1 and 0.96 MPa). Four main stages can be observed. The first stage, stage I, is the linear elastic range; here the load-deflection curve is linear and the stress in the bellows is below the yield strength. The second stage, II, is a primary yielding range; localized

TABLE 42.2 Mean Mechanical Properties for As-Received and Cold-Rolled 316L SS Sheet

	Yield strength s_y MPa (kpsi)	Ultimate strength s_u MPa (kpsi)	Elongation (%)
Bulk	275 (39.9)	607 (88.0)	55
Cold rolled	648 (93.9)	766 (111.0)	37.2

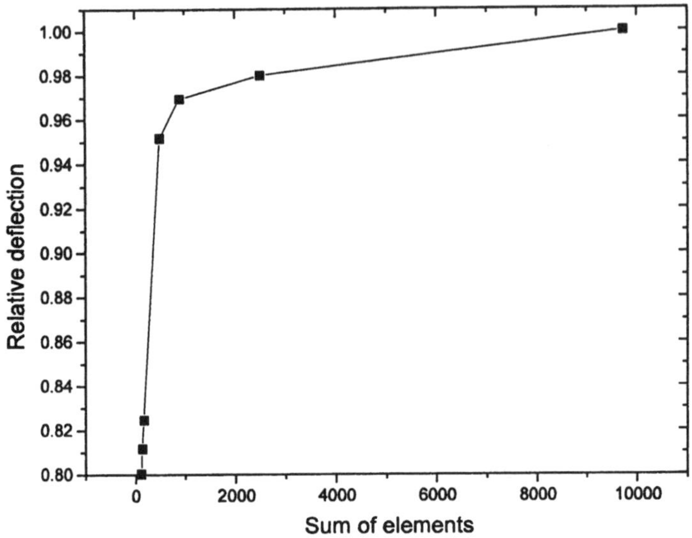

FIGURE 42.10 Total element numbers versus relative transverse deflection (stiffness).

yielding occurs at the tops or bottoms of the convolutions where stress levels are high. With increased loading, the bellows go into a third stage, III, where contact between convolutions occurs. This stiffens the bellows in the lateral direction and so the slope of load-deflection curve increases. The final stage, IV, is of general plastic deformation where the whole bellows yields and the slope of the load-deflection curve decreases until final rupture.

Snedden (1985) experimentally obtained similar load-deflection curves and developed simplified analyses to determine the collapse pressure and evaluate the first three ranges noted above. He considered the lateral stiffness once convolution contact occurred to be infinite and thus any constraint to be rigid, though as can be seen from his results, this is far from the case (Fig. 7 of Snedden, 1985) and a more gradual approach to failure develops.

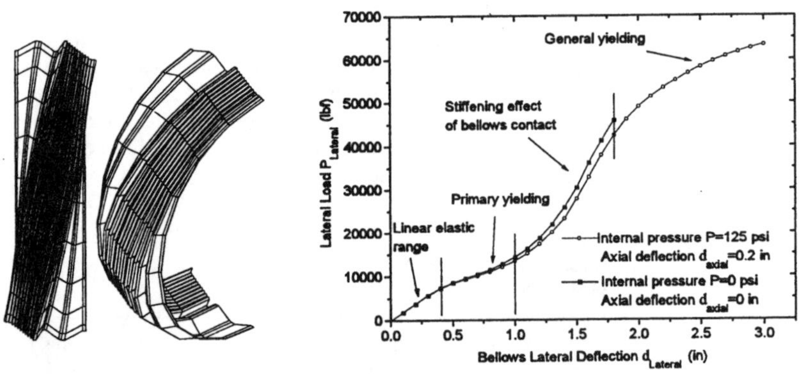

(a) (b)

FIGURE 42.11 Large-scale deformation of bellows: (*a*) with imposed axial deflection under internal pressure and lateral loading; (*b*) load in lb_f, pressures in psig, deflections in inches.

The limiting internal design pressures, based on column instability (both ends rigidly supported) and no axial extension, are estimated by EJMA (1993) methods to be 2.1 MPa (290 psig) and 1.8 MPa (246 psig) for the cold and hot bellows respectively.

42.4.2 The Bellows and Stub Pipes

Batstone (1975) conducted four individual water-filled bellows tests with a facility designed to simulate half the dogleg pipe/double bellows assembly (Fig. 42.12). Because the distance from the midpoint between the two miter bends to the first convolution of each bellows differed by only 17.5 mm, only the lower, R6, bellows was evaluated. Axial and shear forces on the bellows were measured as functions of internal pressure along with vertical deflection measurements. The pivot point of the pipe assembly was fixed 0.178 m above the centerline of the bellows at zero internal pressure at a distance of 2.4 m to the first convolution. The ends of the bellows were welded to stub pipes and these bolted to blank flanges, the fixed end being bolted to the support frame and the free end being welded to an I-beam that supported the pivot point of the assembly.

The bellows all failed, i.e., squirmed, at an internal pressure of about 1.2 MPa. In one of the tests, pressures were cycled over selected ranges of up to 3,456 times. In this test, cracks did not occur until further pressurization to 1.5 MPa. These cracks were sealed and the test cycles continued for 900 cycles (1.0 to 1.1 MPa) and a further 400 cycles (1.07 to 1.2 MPa); there was only a slight increase (2 mm) in the three cracks that had formed at the crimple points due to squirm. In a few experiments a 48-mm-diameter 1.82-m-long scaffold support was provided with a 6.4-mm space between stub pipe flange in order to evaluate the influence these had on the behavior.

The large deformation FE model for these experiments was set according to Fig. 42.13. The bellows and stub pipe were simulated with 654 3-D 4-node elastic-plastic shell elements (bellows 514, pipe 140 elements). The stub pipe flanges were simulated by different thicknesses in order to maintain appropriate gravity loading with the fluid contents. The linkage between the flange and pivot was simulated by a group of very stiff beam elements.

Internal pressure in the model was increased gradually and both the axial and vertical loads at the pivot as well as the vertical deflections of the bellows were computed. A beam element and a contact element, with a set gap of 6.4 mm, could be used to simulate the scaffold support.

A comparison between measured and simulated loads in both the axial and vertical directions shows that these were similar. Figure 42.14 indicates the vertical deflections as a function of internal pressure. Although there are some differences, agreement is reasonable,

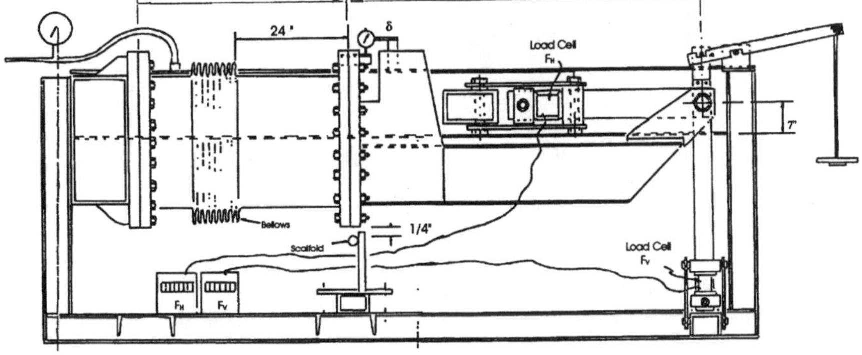

FIGURE 42.12 Bellows test facility, dimensions in inches. (*Source: Batstone, 1975.*)

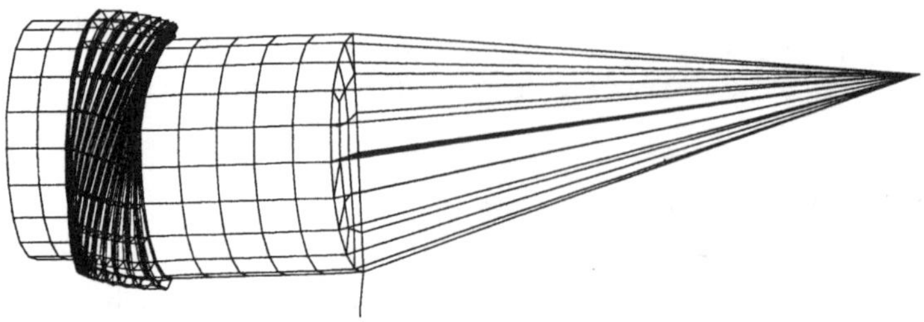

FIGURE 42.13 Finite-element model of the bellows test.

with the FE model generally more compliant. One possible reason for this is that the bellows used in the experiment had been cycled extensively with pressure, which could cause a certain degree of strain-hardening before the final cycle loading when the bellows squirmed. The final measured and simulated squirming pressures are, however, very nearly the same (1.24 MPa).

42.4.3 The Bridging Pipe

Experiments on the 4.8-mm-thick wall dogleg pipe were also carried out (Batstone, 1975) (Fig. 42.15). Two 75-mm pin joints were located at the center of the bellows at each end of the original assembly, where the bending moment was zero. An axial load was applied horizontally at the lower end through two precalibrated hydraulic rams, each of 100 ton capacity with 152-mm stroke. In order to support the vertical force on the ram, two double-row, self-aligning ball bearings were located on the extended pin of the joints. These were

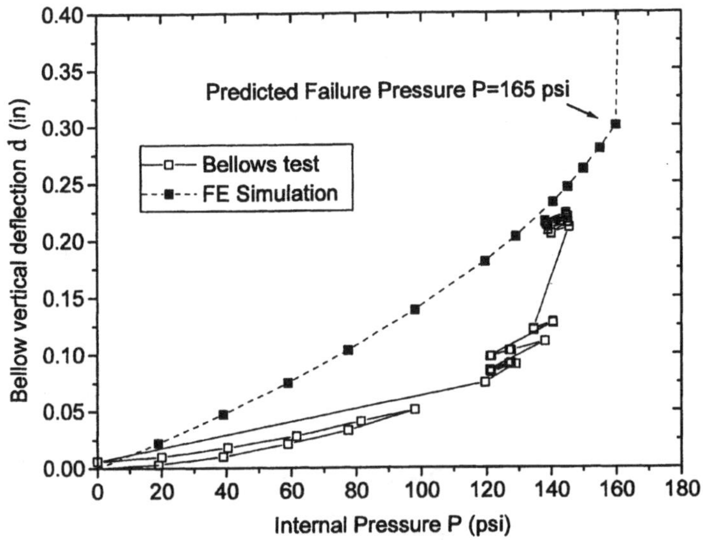

FIGURE 42.14 Bellows vertical deflection as a function of pressure; test, Batstone (1975) and FE simulation, pressure in psig, deflection in inches.

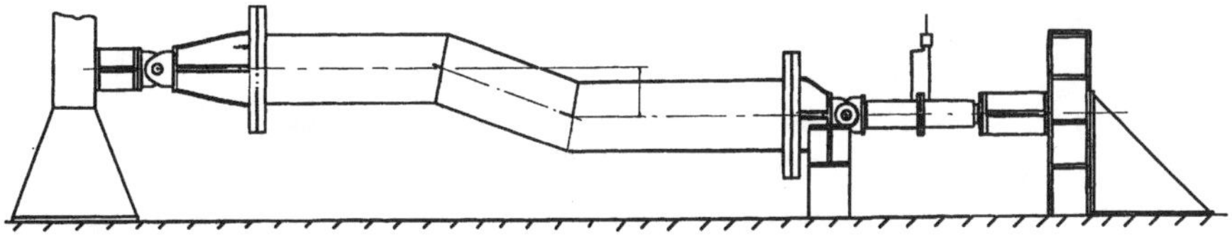

FIGURE 42.15 Bypass pipe experimental facility. (*Source: Batstone, 1975.*)

free to move on two horizontal supports mounted on the floor as the rams extended but were prevented from twisting. Two strengthened adapter pieces were welded to blank flanges bolted to both ends of the pipe and pivoted on pin joints fixed to vertical support columns. The pipe itself was filled with water and internal pressure applied with a hand pump. The axial load applied via the ram maintained the internal pressure according to the relationship:

$$F = PA \tag{42.1}$$

where F was the horizontally applied axial load, P the internal pressure, and A the effective cross-sectional area of bellows ($d_{\text{eff}} = 747$ mm). The internal pressure and deflections at several positions were measured and recorded.

The FE simulation of this test was simpler than that for the earlier bellows model. The pipe was represented with shell elements (SHELL43) and the strengthened adapter pieces, which connected the blank flanges and the pin joints, were represented by beam elements as shown in Fig. 42.16. The weights of the two heavy flanges, each with 28 bolts, nuts, and washers, the adapters, and the water-filled pipe were converted into equivalent thickness of shell and beam elements similar to that in the bellows simulation. Internal pressure and axial forces were applied to the pipe instantaneously.

A comparison of the model results to those of the experiments shows that the test data exhibit a greater yield load in addition to a prestrain (Fig. 42.17). This might have occurred because a previous load, beyond yield, was experienced in earlier testing of the pipe used (Games and Waterhouse, 1974). The measured load-deflection curve, shifted by 7.6-mm, post yield, appears to correspond well to the FE simulation; both indicate a final failure by buckling at an internal pressure of about 1.93 MPa (265 psig).

In order to evaluate the influence of end constraint on pipe behavior, a further FE study was conducted using this model. This employed a "hard" or fixed constraint at the B6 end that restricted rotation of the pipe and thus simulated the compression of the B6 bellows corresponding to the stage III bellows model of Snedden (1985). This showed that an internal pressure of only 0.76 MPa (95 psig) caused buckling failure of the pipe at its lower miter bend. Relaxing the B6 constraint would obviously increase the buckling pressure. Such a result indicates that, had a differential pressure occurred across the two reactors, such as could have happened by only one bellows suddenly tearing and depressurizing, squirm of that bellows *and* buckling of the pipe assembly was possible at operating pressure (Venart, 1999).

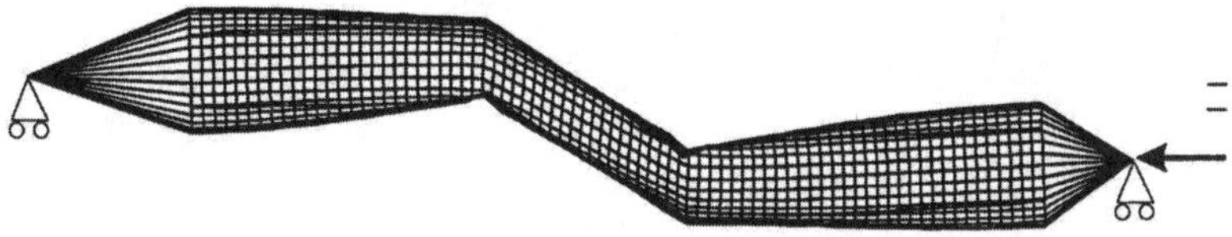

FIGURE 42.16 FE model of bridging pipe. (*Source: Venart and Tan, 2000.*)

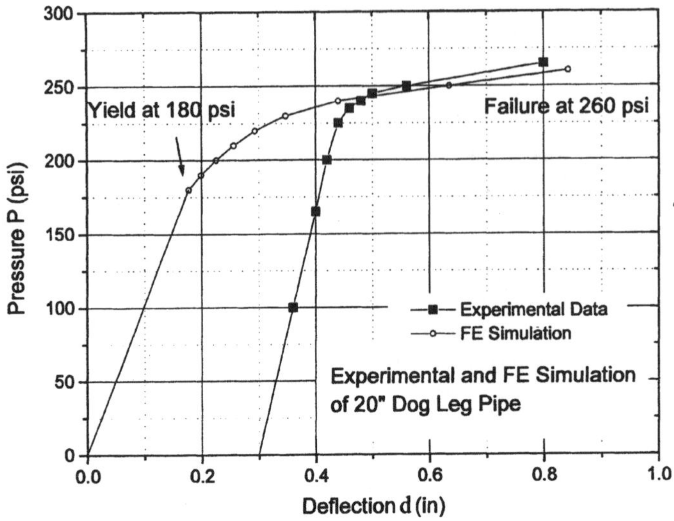

FIGURE 42.17 Load deflection of the bypass pipe; experiment (Batstone, 1975) and FE model, pressure psig, deflection in inches. (*Source: Venart and Tan, 2000.*)

42.4.4 The Bridging Pipe–Bellows Assembly

Both static and dynamic FE analyses for the assembly were made based on the previous models. The model setup, loads, and weight conversions were similar to those used earlier. The main difference was that the pipe was not fully filled with cyclohexane (Teng-yang et al., 2000; Venart, 1999) and that a scaffold system could be set to support the pipe assembly (Figs. 42.4 and 42.18).

The Experiments. A test of the assembly was conducted at Flixborough using a new pipe built according to the original design and inserted in place of reactor 3 (R3) (Secretary of State for Employment, 1975; Games and Waterhouse, 1974). The bellows utilized were new, with axial stiffness of 498 (2841) and 482 kN/m (2753 lbf/in) for the B4 and B6 bellows respectively. Scaffolding similar to that in the original installation could also be used. Axial and lateral deflections were measured, both cold and hot (155 and 160°C), with a chain used to simulate the dead load of the fluid inventory. Pressure was applied by nitrogen gas. Strains in the pipe were also measured. There were eight test runs, three at ambient temperature, three at 155°C, and two at 160°C. The maximum pressure utilized was 0.964 MPa in all but

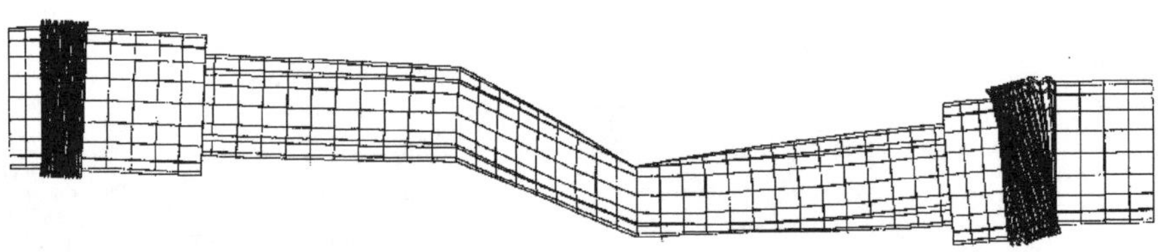

FIGURE 42.18 FE model of the Flixborough bypass assembly at an internal pressure of 0.964 MPa (125 psig); deformation scale exaggeration is 100. (*Source: Venart and Tan, 2000.*)

the last two cases; for these the bellows were squirmed at 1.06 MPa and then burst at 1.53 MPa.

In the experiments at ambient temperature, internal pressure causes the B4 end of pipe to move upward and the scaffolds there are only lightly loaded. With increase in pressure, the B4 end rises off the scaffold support as the B6 end moves downwards and the scaffolds there carry greater load. However, under operating conditions, 150 to 155°C, the assembly was mostly unsupported since the reactors expand upwards about 6 mm. The scaffold supports are thus neglected in the FE analysis under operating conditions, though they were needed to commence the analysis.

The FE Model (Fig. 42.18). Four load steps were used in the FE simulation of the complete assembly:

1. The by-pass pipe rested on the scaffolding without internal pressure.
2. The system was next pressurized.
3. The assembly was set free due to the upward thermal expansion of the reactors (6.4 mm).
4. The pipe was allowed to thermally expand axially (approximately 10 mm) along with the radial expansion of each reactor and its stub pipe (about 2.5 mm). At the lower or R6 end contact and deformation of the scaffolding was permitted if the downward assembly deflection exceeded 6.4 mm.

In the model it was assumed that the upper pipe was half full and the lower pipe full of liquid cyclohexane. It was also assumed that the scaffolds were made of normal structural steel; these were simulated by beam elements with contact elements to prevent any loading when the pipe contact was lost. Elastic-plastic load-deflection relationships were used to simulate the pipe, the bellows, and the scaffolding.

The vertical deflections at the two ends of the pipe, obtained for an internal pressure of 0.964 MPa, are listed in Table 42.3. Compared to the experimental measurements, the trends are similar. However, there are some differences in the values, particularly for the upper bellows where the experimental deflections exceed the FE results by some 35%.

TABLE 42.3 FE Analysis of the Assembly and Comparison with Test Data[a] Vertical Deflections[b]

Case	Condition	Deflection at upper, B4, end of pipe, mm (in)	Deflection at lower, B6, end of pipe, mm (in)
FE initial (a)	Full scaffold support no internal pressure	−0.72 (−0.028)[c]	−0.77 (−0.030)
FE at 20°C (b)	Partial scaffold	3.41 (0.134)	−3.77 (−0.148)
FE at 150°C (c)	Partial scaffold support, 0.964 MPa	6.93 (0.273)	−4.49 (−0.177)
(a)–(b) FE Cold	Scaffold support, 0.964 MPa, Cold	4.13 (0.163) [10.66 (0.4195)][d]	−3.00 (−0.118) [−3.26 (−0.1285)]
(a)–(c) FE Hot	Partial scaffold support, 0.964 MPa	7.65 (0.301) [10.27 (0.4042)]	−3.72 (−0.147) [−3.26 (−0.1285)]

[a] Games and Waterhouse, 1974.
[b] Venart and Tan, 2000.
[c] + upwards, − downwards displacement.
[d] Average values from experiment, Games and Waterhouse, 1974.

The maximum von Mises stresses, 231 MPa (33.5 kpsi), occur along the tops and bottoms of the convolutions, particularly at the second-to-last convolutions of both ends at normal operating conditions. One of the regions is apparently coincident with parts of the internal tear noted by the Court (Secretary of State for Employment, 1975; Foley, 1974a) (Fig. 42.19). Stress levels in the lower bellows are similar to those in the upper bellows. The stresses in the experimental bellows at B4 would, however, have exceeded these values since deflections were larger.

At temperature and pressure, the upper (R4) end of the pipe moves towards R2524 by 0.67 mm; the absolute axial displacement of the other end of the bellows is away from R2524 by 2.89 mm, so that the axial compression of the upper bellows is 3.55 mm. At the lower (R6) end of the pipe the assembly moves away from R2526 by 0.037 mm with the absolute axial displacement of other end of the bellows moving away from R2526 by 3.08 mm; the axial compression of this bellows (B6) is thus 3.05 mm. The pipe, between bellows, shortens under these conditions by 3.51 mm compared to a measured value of between 5 and 7 mm (Games and Waterhouse, 1974).

FE Modal Analysis of the Bypass Pipe Assembly. In the pipe model previously described, the mass of fluid was converted into a density increase as a function of location. The total weight of the pipe, flanges, bolts, bellows, and insulation was about 1.13 tons and its liquid inventory some 600 kg, of which at least 375 kg were distributed in the lower half of the pipe under operating conditions.

The first three natural frequencies and vibration modes were determined using modal analysis. The first natural axial frequency was found to be $f_1 = 3.98$ Hz. The second natural, and transverse (vertical) frequency (coupled with bending of the pipe) was $f_2 = 12.35$ Hz. The third natural frequency was $f_3 = 30.16$ Hz; this mode of vibration was the rocking mode of the assembly. The equivalent determinations based on EJMA (1993) are 3.93, 15.7, and 27.2 Hz respectively. The axial mode agrees well, but the inclusion of the offset pipe appears to reduce the transverse and increase the rocking modes.

Three main harmonic forces are induced with the flow system (Teng-yang et al., 2000). One is the axial load that results from the flow fluctuations at the diameter change in the

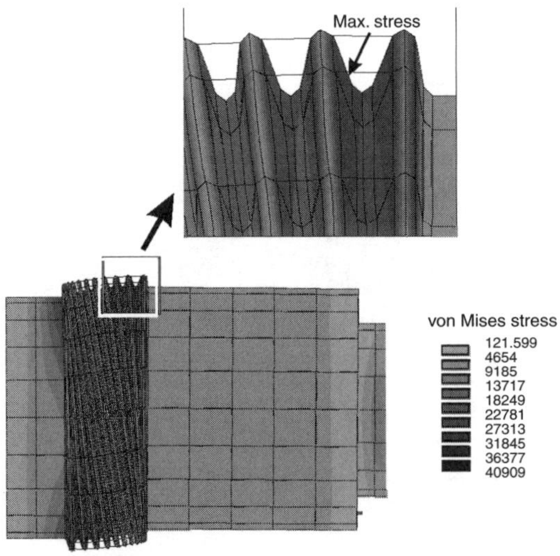

FIGURE 42.19 Static equivalent (von Mises) stress upper B4 bellows, stress in psi.

TABLE 42.4 The Harmonic Response of Bypass Assembly as a Function of Frequency

Frequency f (Hz)	1.15	1.52	1.89	2.26	2.44
Response Amplitude: s_{eq}; MPa (psi)	31.1(4514)	31.5(4570)	31.9(4626)	32.3(4682)	32.5(4710)

upstream pipe. Another is the transverse loading due to the change in fluid mass at the lower hydraulic jump. Both these forces are in phase and at the same frequency as the hydraulic head change. Due to the hydraulic jump after the upper miter bend the mass, in the lower half of the pipe would have oscillated by about 60 kg (~3.5% of the total system mass) at a frequency of between 1 and 2.5 Hz. A third force is the level fluctuation in the weir waterfall pools; levels in similar-sized outfalls vary with a frequency of about 7 Hz (Rajaratnam and Chamani, 1995). Further, the plunge pool jet velocity, even for the lowest flows, is sufficient to excite axial resonance of the bellows as noted earlier.

The lowest natural frequency of the system was earlier determined to be about 4 Hz. Since the frequency of driving excitation is 50% of the lowest natural frequency, a simple sensitivity analysis was performed in order to determine its influence. It was found that as the driving frequency increased from 1.15 Hz to 2.44 Hz, the harmonic response increased only 5%. The results are listed in Table 42.4.

Damping is important in any harmonic analysis. Due to the complexity of this fluid-structure interaction problem, the damping of the system was difficult to determine and an approximation was made based on a sensitivity analysis. First damping is very low for stainless steel; values of 10^{-3} to 10^{-5} are normally assumed; damping in gas or vapor spaces would likewise be small, i.e., the top of B4. For the liquid-structure coupled portion of the system, damping could be large (e.g. B6 and the bottom half of B4) and was therefore assumed to be 0.5. An analysis within this range is illustrated in Table 42.5, which indicates that the amplitude of the equivalent stress is quite stable for damping coefficients in the range of 0.001 to 0.3.

The comparison between the finite-element models and the experiments undertaken above indicate that the models reasonably represent the behavior of the bellows, bellows and stub pipes, as well as the bridging pipe. The interpretation of how the assembly could have behaved with a failure in the B4 bellows follows closely that shown in Fig. 42.8, discussed earlier.

42.4.5 Summary of the Finite-Element Analyses

Finite-element models of the Flixborough bellows, bellows and stub pipe, bypass pipe, and the complete bypass assembly have been made. The analyses permit the following conclusions:

TABLE 42.5 Harmonic Response of Bypass Assembly with Damping Coefficient ζ

Damping coefficient ζ	0.001	0.01	0.1	0.3	0.5
Response amplitude: s_{eq}; MPa (psi)	31.1(4509)	31.1(4509)	31.0(4486)	29.7(4315)	25.9(3761)

- The FE model of bellows under large lateral deflection reasonably replicates the experimental behavior observed by Batstone, Snedden, and others.
- The model of the bellows and stub pipe assembly behaves in a similar manner to that in the experiments; failure, that is, squirm, occurs at similar internal pressures.
- The model of the bridging pipe corresponds well to the experimental behavior of the pipe with the imposed end constraints; with increased end constraint at the lower, B6, end of the pipe, the structure can buckle at a considerably lower pressure.
- The deflections of the complete bypass assembly model replicate the trends of the experimental data obtained by Games and Waterhouse for both the ambient and hot experiments. The maximum von Mises stresses occur along the topmost bellows elements. Some of these points appear to coincide with the locations of portions of the internal tear found in one of the recovered B4 bellows pieces.
- Modal analysis of the complete bypass assembly indicates that excitation by the flows was possible, and a harmonic response analysis, considering the possibilities of damping, indicates that some greater concern for fatigue failure should have been considered.
- A possible failure mechanism for the bypass pipe assembly that only involved the release from the upstream, R4, reactor has been indicated.

42.4.6 The CFD Analyses

A transient three-dimensional CFD model was developed for a 130 (x) by 130 (z) by 30 (y, elevation) m volume shown in Fig. 42.2 (Venart, 1999). The region considered encompassed the reactors and after-reactor vessels and their plumbing (especially the 8-in. lines) in Sections 25A and 7 (and a portion of Section 27), the laboratory, and the control room, as well as the main and works office buildings.

The time marching computations were carried out using the commercial code ***FloSYS*** (Flomerics, 1994), which is a CFD solver operating in Cartesian coordinates.

The solution was performed throughout the rectangular cuboid domain for the velocities u, v, w in the x, y, and z directions respectively, fluid static pressure, p, temperature, T in the gas and solid regions, and concentration of the secondary medium (Cyclohexane), c. Several tests were evaluated for solution convergence, and as a compromise among accuracy, computational requirements, and resources, a grid system of $98 \times 41 \times 68$ nodal points in the x, y, and z directions respectively was selected. The smallest cell size, commensurate with the smallest scale of detail, the 8-in. pipe, was 0.1 m in the x and z directions and 0.05 m in the y direction; the largest cell sizes were on the upper domain corners; these were typically $3 \times 2.3 \times 3.3$ m in the x, y, and z directions, respectively. The grid, where required, was set automatically to reconcile with internal solid surfaces particularly in the reactor near field.

The equations solved formulate the principles of conservation of mass, momentum in the x, y, and z directions, thermal energy, and mass of secondary medium (cyclohexane). Turbulence modeling is performed by using a standard two-equation (k-e) turbulence model with near-wall grid node wall functions used to evaluate surface friction and surface heat transfer. The effects of turbulence are represented via a turbulent viscosity and thermal diffusivity where these quantities are linked by the turbulent Prandtl number, which takes a constant value of 0.9.

Buoyancy effects due to thermal expansion and the heavy vapor dispersion of the evaporated cyclohexane are included in the y direction as a gravitational source term. The variation of properties, in this particular case, density, is assumed to be governed by the ideal gas laws. The molecular viscosity, thermal conductivity, and diffusivity were taken to be those for air and constant.

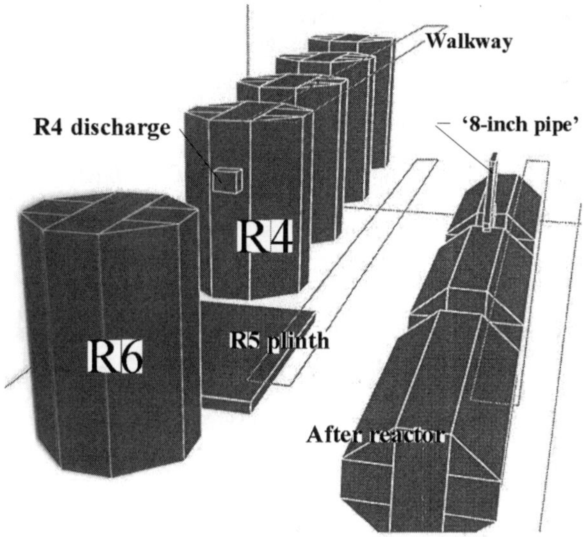

FIGURE 42.20 A near-field northwest view of the reactor train and the after-reactors as might have been seen from the roof of the main office building. (*Source: Venart, 1999.*)

Plant equipment, pipe racks, and supporting structures, other than the five reactors and the associated after vessels and plumbing, were treated as flow obstacles with a uniform resistance and porosity of 0.5. Buildings and other large structures (the main office, the works office, the laboratory, and the control room) were assumed to be impervious to the flow and adiabatic. The prevailing wind was assigned by steady vector components as boundary conditions on the free boundaries of the domain. The surface of the ground was set to a uniform roughness of 0.01 m.

The mass source terms, representing the flashing two-phase discharges from the reactors, were evaluated from prior choked flow determinations and the assumption of sonic, vapor-only flow at the reactor, R4 and R6, outlet nozzles (Leclude and Venart, 1996). The properties of cyclohexane and its dissolved nitrogen were obtained from the NIST (1990) mixture property program, STRAPP.

A near-field isometric view of the reactor train is shown in Fig. 42.20 as represented in the Flomerics code. In the figure, the walkways along the reactor train are represented by porous resistance areas, and the 8-in. pipes connecting the two after-vessels together are indicated. The cuboid representations of the vessels are also shown.

The FloSYS ASCI output files were used as input files to be further processed by Tecplot (Amtec Engineering Inc., 1996) to extract the isocontours for the lower (LFL) and upper (UFL) flammable limit (0.037 and 0.198 wt. % respectively) and determine interactive visualization through time-sequenced plots of the isosurfaces.

42.4.7 CFD Results

Simulations were preformed for two cases: (1) the simultaneous discharge from both R4 and R6, as the base case representing that process accepted by the Court, and (2) a discharge from only R4 prior to the explosion, as the situation explored here. In both instances the discharges were taken as choked sonic cyclohexane vapor at 370 kg/s and 150°C. The results are available in the form of movies[4] or animations both for the far-field and close-in regions of the computational domain.

The transient development of the Cyclohexane vapor cloud at its lower and upper flammable limits (LFL and UFL respectively) over a period of 30 seconds will be shown as a Tecplot movie. The UFL is represented as the paneled interior surface, and that for the LFL, the wire-frame exterior surface in the figures which follow.

Discharge from Both R4 and R6 (Fig. 42.21, Movie 1). The first frame illustrates the computational domain, as indicated in Fig. 42.2, to a height of 30 m, showing the placement of portions of the laboratory, the main control room, the main office and the works office. The reactor train is indicated along with the after-reactor and treatment vessels in addition to the 8-in. lines. Also included are the access walkways along the reactor train and the rest of plant, approximated as porous areas and volumes respectively, both with porosities and resistance coefficients set to 0.5 as estimated from the site photographs in the Report of the Inquiry and other sources. These areas are indicated as transparent boxes or areas outlined in color.

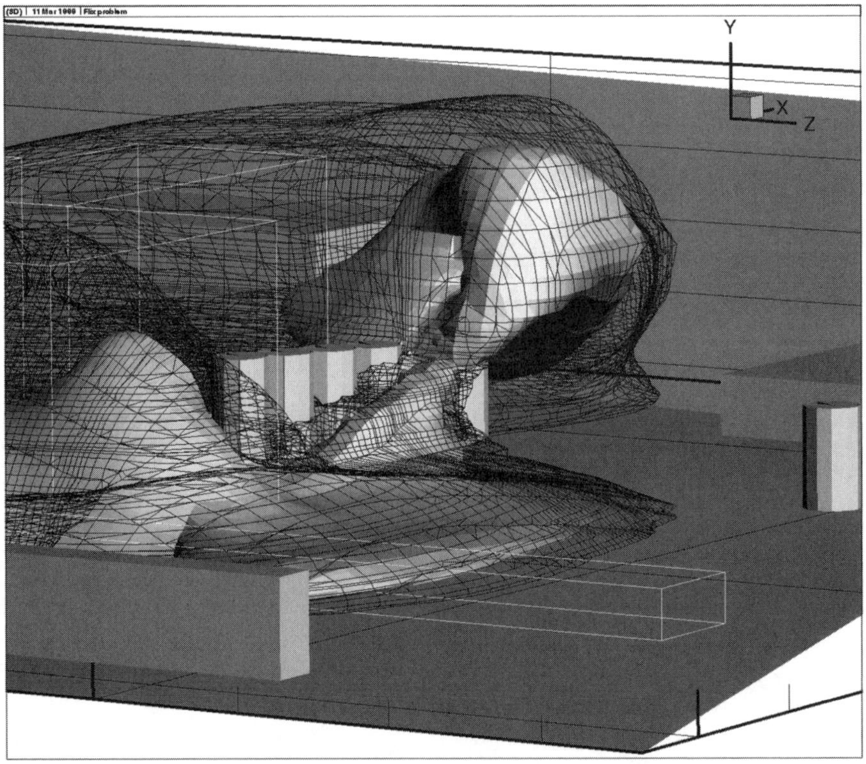

FIGURE 42.21 Southeast view from the top of the hydrogen plant looking towards the main office building: R6 and R4 discharges at 12 seconds, LFL and UFL envelopes. (*Source: Venart, 1999.*)

[4] The movies are available from the author and may be played using a supplied utility.

The sequence of all frames is at one-second intervals.

It can be seen that the simultaneous discharges, even at one second into the event, result in an aerial cloud from the discharge of R4, riding up and over both the jet discharge from R6 and the top of this vessel and reaching to a height of about 20 m. This cloud proceeds southerly into the prevailing wind and towards the opening between the two office blocks.

At the same time, the R6 discharge is split by R4 into two jets and channeled by the R4 jet above, the reactor train walkways, and the after-reactor and reactor vessels to form a dense gas cloud moving predominately at ground level along the reactor train in a NNW direction.

After only two seconds, the LFL of this cloud has reached the northern end of the reactor train, i.e., R1; there is relatively little westward development of this cloud due to the intense momentum of the R6 discharge split by the R4 reactor.

At 8 seconds, the R4 cloud is still airborne (~24 m elevation at the LFL) and much larger and starting to slump towards the ground in a southerly direction with the LFL, making a footprint around the northwest corner of the main office building and commencing to exit from the domain in a southerly direction between the two buildings; the cloud does not, however, contact the works office. The R6 cloud, though now shown to have developed its LFL envelope up to some 16 m elevation, is, at its north-northwest ground boundary, approaching the laboratory with little westerly development. The height of this cloud would no doubt have been restricted by the floors in the rest of the open plant structure of section 25A and so the 16 m vertical extent of the LFL is perhaps an artifact induced by the use of uniform porosity and resistance.

By 12 seconds, part of the R4 UFL envelope has touched down between the office buildings and the LFL cloud has proceeded out into the car park at a mean velocity of some 4 m/s. The other portion of the R4 LFL cloud has exited to the east from the space between the north face of the three-story main office building and the southern end of sections 25 and 7. The R6 LFL envelope by this time has started to surmount the two-story laboratory building and is commencing contact with the control room; the westerly spread of the LFL has reached the westernmost end of the laboratory.

After 18 seconds, the R6 UFL envelope has reached the control room and has topped the two-story laboratory building. Both the LFL and UFL envelopes have exited from the western boundary of the domain towards the hydrogen plant. The R4 discharge LFL cloud continues to exit between the main office block and the southerly ends of sections 25 and 7 as well as southerly between the space with the works office; the UFL of this envelope has now mounted the northwest corner of the office block.

By 22 seconds, the R6 LFL and UFL envelopes have both surmounted the laboratory and the control room and are proceeding across the western domain boundary towards the hydrogen plant. The center of gravity of the main body of this cloud is situated just north of R1. Both the LFL and UFL R4 envelopes are now exiting from the southern domain boundary between the two offices and into the car park; the eastward LFL cloud still exits from the eastern boundary in the space between the main office and the southern end of section 7.

At 30 s, the R6 discharge clouds have proceeded northerly past the control room and laboratory. The R4 discharge cloud is still split by the main office block into two sections; one exiting into the car park, the other to the east between the office and the southern ends of sections 25A and 7.

Single Discharge from R4 (Fig. 42.22, Movie 2). The transient development of this vapor cloud differs significantly from that above. At the commencement of the event, a series of choked vapor discharges from a slit, set to be approximately similar to that noted in a piece of the B4 bellows by Foley (1974a), are shown for 0.2-second intervals up to 1 second followed by 1-second snapshots thereafter.

The first four frames portray the discharge from the slit in the R4 bellows as a westward-proceeding jet developing over a time span of 1 second. The next sequences show the

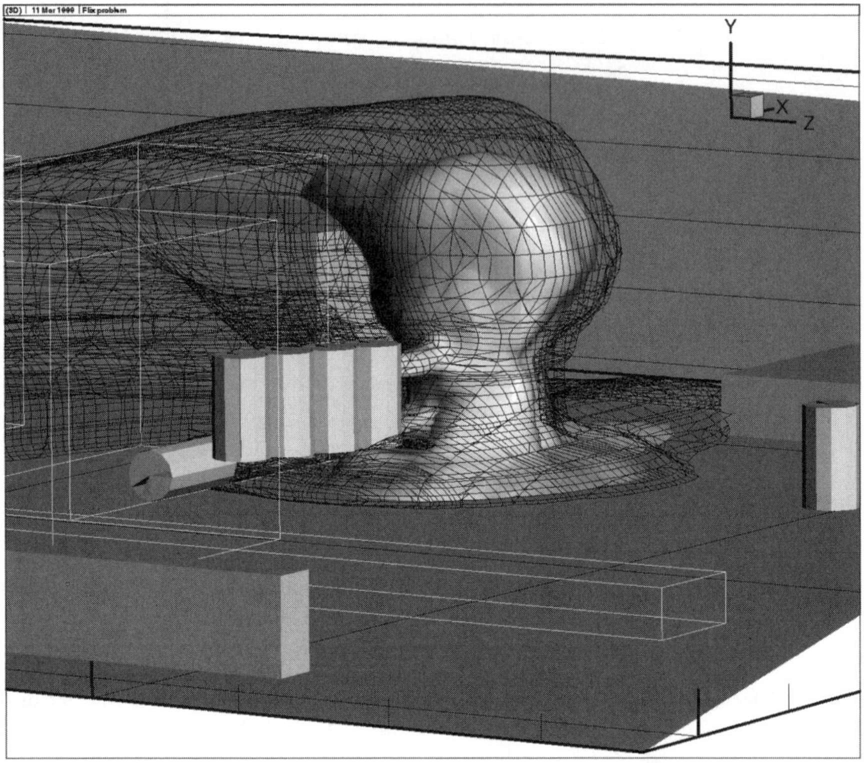

FIGURE 42.22 Southeast view from the top of the hydrogen plant looking towards the main office building: R4 discharge only at 12 seconds, LFL and UFL envelopes. (*Source: Venart, 1999.*)

interaction of the jet impact from R4 as it hits R6, when the B4 bellows is blown off and the jet formed is deflected upwards. By 4 seconds, the shape of both the LFL and UFL envelopes are roughly cylindrical; the height of the LFL cloud is about 25 m high with a cylindrical ground footprint of some 30 m diameter for the LFL and 15 m for the UFL.

At 6 seconds, the prevailing wind is distorting the LFL envelope; both the UFL and LFL footprints commence a westward progression across the open space in front of the works office and northward to the base of the R2 reactor (LFL). The LFL southern envelope approaches the NW corner of the Main Office building.

In 8 seconds, the LFL envelope's SE extremity has impacted the northwest corner of the Main Office building and has commenced exiting from the eastern domain boundary between the southern ends of section 25A and 7 and the northern wall of the office building. The base of the LFL footprint has progressed northward to the base of R1 with the northern boundary of the UFL just past the base of R4.

At 12 seconds, the momentum of the jet has pushed the UFL envelope to the northwest corner of the office building. The LFL envelope in its eastern branch is now greater in elevation than the office building, with a southern branch pushing past the west wall of the office building and commencing to exit from the southern domain boundary into the parking lot. The northern extent of the LFL footprint has developed just midway of the last after-reactor. Westward development of the cloud has just reached the northeast corner of the works building with the edges of both the LFL and UFL in line with the east-west axis of R6 pushing some 20 m past R6 at somewhat less than 1 m/s.

Within 16 seconds, due to the continuing momentum of the R4 jet, the UFL envelope detaches from the eastern wall of the office block to progress also in the direction of the parking lot. The LFL southern branch fills the space between the two offices and further pushes out into the parking lot at over 8 m/s; the eastern LFL development has topped the main office block and is exiting eastward at some 4 to 5 m/s. Westward development of both envelopes now has hit the NE corner of the works office and progressed westward some 8 m along its northern wall in the case of the LFL.

By 19 seconds, the LFL envelope at ground level has progressed westward some 20 m along the works office northern wall with the UFL cloud only 8 m behind. The northern development of the LFL envelope has only just reached within 22 m of the laboratory as a rather flattened pancake-shaped cloud. Both the LFL and UFL envelopes have pushed southward out into the parking lot; the UFL cloud has become completely detached from the main office block and slides along the eastern wall of the works office. The LFL envelope continues its eastern progression over the top of the main office block and around behind sections 25A and 7, exiting eastward at some 4 to 6 m/s.

In the sequence of frames following 19 seconds, pseudo-steady state has developed for the southern and easternmost exiting domain discharges; there is a strong UFL and LFL ground jet into the parking lot, along with a less strong, but volumetrically greater, one at the LFL to the east channeled by the north wall of the main office building. Aside from these two discharges to the west, both the LFL and UFL continue to enlarge with little northern development.

42.4.8 Discussion of CFD Results

Of the permanent staff of 550 employed by the company, approximately 70 of these were on-site June 1 (Westgate, 1975) . There were 28 fatalities in the explosion; of these, 18 individuals were in the control room at the time as well as 3 each in the gate house and warehouse/flaking plant (Humberside Police, 1974). In the laboratory (separated by less than 12 m from the control room, Fig. 42.2) there were eight survivors who directly witnessed the events prior to the explosion; from what they saw and heard they were able to appreciate the threat of the observed escaping cyclohexane and flee. Other witnesses, farther away, likewise directly observed the developing vapor cloud, the resultant fires, and the explosion.

In this section the eyewitness accounts of developments are briefly compared to the results of the CFD simulations of events. Additionally, site examinations of the explosion damage and fire evidence (Secretary of State for Employment, 1975; Sadee et al., 1977; Gugan, 1978; Roberts and Pritchard, 1982) and other witness statements strengthen the case for an initial single discharge from R4, followed by further follow-on occurrences that indicated that the pipe bridge failure was by a two-step process and not the single-stage event accepted by the Court.

Eyewitness Evidence. Of the eight survivors from the laboratory, seven gave evidence. Of these, five were specific in the description of a *slowly* developing vapor cloud moving to the west. This cloud was described as being about 1.5 to 2 m high and moving at 5 to 6 m/s along the ground across the clear space between the laboratory and the works office. The origin of the vapor was described as either being near the R4 reactor or its base or at least between the two (R4 and R6) reactors. *None* of the witnesses described a northward developing cloud. One described debris being thrown *up* into the air very early into the event.

Others, one about 90 m east and some 60 m north of the main office building, described a cloud of dust and smoke coming round the southern end of Section 27 (Colquhoun, 1974). Another, some 500 m to the west, observed the development of the cloud from its initiation as a "whisper of 'steam' " to its enlargement into a "roaring" jet going up river (i.e., south) and ending up "like a fog" across the car park (Ayre, 1974).

Various off-site witnesses, e.g., King (1974), about 500 m away, reported seeing "dirty steam" going up into the air, followed by "what appeared to be an explosion that blew vapor sideways *into* the works" and then the major explosion; Dickinson (1974), about 200 m away, saw a flash like a large "creamy colored torch" coming towards him (i.e., eastward) before the explosion.

Physical Evidence. The enormity of the disaster can only be appreciated by reference to the photographs included in references (Secretary of State for Employment, 1975; Sadee et al., 1977; Gugan, 1978; Roberts and Pritchard, 1982) and others available in reports prepared in the investigation. Photographs from the press, particularly those taken by the Scunthorpe *Evening Telegraph,* reveal the extent of the destruction especially out into the car park south of the offices where most automobiles were completely crushed (Fig. 42.7).

The main and works office buildings were totally destroyed, as were the laboratory and main control room. The 0.4-m-thick concrete second floor of the southeast corner of Section 27 was displaced downwards (Roberts and Pritchard, 1982), indicating an aerial flammable cloud at this location. A tanker truck (cf. Fig. 42.7) located about 150 m east of the southern end of Section 27 was completely crushed by an overpressure estimated to be about 1 MPa (Gugan, 1976). A manhole cover located on the roadway nearly in front of the main office building and midway between the gate house and section 25A was broken by a similar overpressure (Gugan, 1976).

Physical damage to steel chemical plant components, such as reactors and strong pressure vessels located in sections 25A, 7, and 27, and closest to the main office building, exhibited crushing of their skirts or bases, which indicated overpressures on the order of 1 MPa. R5, located in the waste ground north of the works office, only exhibited a crushed skirt; the upper and weaker cylindrical section of which was uncrushed (cf. Plate 7 in Secretary of State for Employment, 1975). Similar crushing was also evident on the base of R6 but not the upstream reactors (cf. Plate 8 in Secretary of State for Employment, 1975). Process columns to the southeast in Sections 25A, 7 and 27 exhibited considerable crushing at a variety of elevations (e.g., Fig. 42.6; Plate 8 in Secretary of State for Employment, 1975, sections 25A and 27; and Fig. 2 in Roberts and Pritchard, 1982). This damage is consistent with the explosion of both ground-level (around R5 and R6) and elevated flammable clouds. Pressures sufficient to crush the bases of the reactors can only be obtained with the detonations of fuel drop-air vapor clouds (Fishburn et al., 1981; CCPS, 1994; Hoiset et al., 2000). Pressures in *closed* vessel deflagrations do not normally exceed 0.8 MPa.

That only the southern and southeast outer perimeter of plant appeared to have been subjected to such great pressures indicates that the initial extent of the fuel air cloud had not penetrated extensively into the northern sections of the plant such as would have been the result of the simultaneous dual-bellows failure scenario. Furthermore, the determined center of the explosion, just immediately north of the main office building, and stated to have been at an elevation of about 20 to 30 m (Sadee et al., 1977), coincides closely with the cloud development and its center of mass for the R4-only discharge.

Detonations of unconfined fuel-air clouds require significant ignition energy (Fishburn et al., 1981; CCPS, 1994), thus requiring, perhaps, the prior gas explosion (Harris, 1983) of the main office and/or the works office building to act as a precursor or trigger for the detonation.

Ionospheric measurements made in the close range (Jones and Spracklen, 1974) indicated that there might have been a precursor, or smaller disturbance, to the main explosion such as could have occurred with one or both of these buildings exploding. Photographs of the rubble and debris of both (Secretary of State for Employment, 1975; Sadee et al., 1977) suggest that an interior explosion occurred here first followed by the major blast. Far-range ionospheric and seismic measurements were less detailed (Grover, 1974).

The fact that all eight of the laboratory personnel saw *only* the development of a westward-moving cloud (along with the noise and other activity, some of which is noted in Table 42.1) and appreciated the threat to themselves, whereas the 18 control room staff apparently did not see anything to warn them of their great danger, indicated also that there could not have been an extensive fuel cloud within the northern portions of sections 25A and 7 prior to the explosion.

42.4.9 Summary of the Results of the CFD Simulations

There are clear indications from the previous material that if there had been the simultaneous discharge from two reactors, as required by the Court of Inquiry, *both* the laboratory and control room staff should have appreciated the threat and been able to escape. In addition, blast damage to the northern portions of sections 25A and 7 should then have been more extreme since this northerly developing cloud should also have exploded.

This CFD analysis, coupled with the eyewitness and physical evidence, unquestionably supports the case for a two-stage failure of the pipe bridge. Such a failure only allows for an initial discharge from reactor R4 prior to the main explosion and thus a much smaller release than the 40 to 60 tons that had been previously attributed to the event (Secretary of State for Employment, 1975; Sadee et al., 1977; Gugan, 1978; Roberts and Pritchard, 1982; Bjerketvedt et al., 1997).

The facts support the case that the major explosion may have been caused by a blast in the main office building, the works office building, or both and that this may have been the trigger to the detonation of the cloud as evidenced by the physical damage and according to some observers. If true, this is one of a very few genuine UVCE detonations.

Since, as outlined earlier, flow-induced fatigue failure of the upstream bellows is here thought to be the initiating event, it is worthwhile to review the concerns of the Court on these matters. The only serious questioning was raised by Dr. Pope, Deputy Chair of the Court, in his questions of Dr. Foley (Foley, 1974d). Drs. Gill and Kitching (Gill, 1974; Kitching, 1974) as well as Dr. Ryder (Ryder, 1974) also raised concerns regarding flow-induced and other vibrations and fatigue. Dr. Foley had earlier stated that "detecting fatigue failure in thin sheets is difficult" (Foley, 1974e). To determine evidence for fatigue, over 26 m of failure surface would have had to have been carefully examined; by Dr. Foley's own statements this inspection was only made by "eye and a small lens" and thus the possibility of fatigue, especially at the interior tear in the upstream B4 bellows, cannot be ruled out. Indeed, the earlier unexplained consumption of nitrogen (Secretary of State for Employment, 1975) may have been as a result of several small existing cracks that caused the losses and then suddenly, due to fatigue damage accumulation, joined up to form the tear and thus commence the disaster.

The two CFD simulations of the possible accidental cyclohexane releases in the Flixborough disaster permit the following conclusions:

- In both cases, the R4 discharge results in ground-level vapor clouds that exit to the east, from around the southern end of sections 25A and 7, and to the south as a jet channeled between the works and main office buildings out into the parking lot.
- As well, both instances exhibit conspicuous vertical cloudlike accumulations near the southern ends of sections 25A, 7, and 27 and located on the determined focus of the blast.
- Only in the case of the single discharge from R4 is a significant westward development of the ground-level cloud displayed that corresponds to witness recollections.
- In the situation with the two simultaneous discharges, there is an extremely strong split jet (from R6) that develops into a rapidly northward moving ground-level cloud, with little westward progression. This cloud reaches the laboratory and control room areas in less than 12 seconds.
- On the basis of eyewitness reports from the laboratory technicians, who were able to escape the building before the explosion, as well as physical examinations made after the accident, it is concluded that the cause of the accidental release must have been a two-stage event,

initiated by a release from only the upstream, or R4, reactor. The initiator for this could have been a release through fatigue-generated cracks in the R4 bellows, leading to the collapse of the bridging pipe as described here.

- Further, it is possible, based upon the sizes and distributions of the vapor cloud produced, that the precursor event triggering the detonation of the cloud was a confined gas explosion that took place in either the main office or the works office or both together.

Finally, it is concluded that only about 10 to 16 tons of cyclohexane were involved in the explosion and not the 30 to 60 tons previously attributed to the event.

42.5 CONCLUSION

Finite-element and computational fluid dynamics analyses of the accident at the NYPRO Works at Flixborough have been conducted. The results suggest that the cause of the catastrophe was flow-induced fatigue of one of the bellows forming the bypass assembly that resulted in the initiation of a complex sequence of events that released only 10 to 16 tons of cyclohexane to form an unconfined vapor cloud that was detonated.

42.6 DEDICATION

It is hoped that this work may provide closure for the families of the individuals who lost their lives, and in particular remove the stigma of inappropriate action from the process control technicians killed in the event.

42.7 ACKNOWLEDGMENTS

Many individuals have contributed to the preparation of this work. Specifically, the staff at the Health and Safety Laboratory Library in Sheffield, and in particular Linda Heritage, are thanked for their help in making the Minutes of Proceedings of the Court of Inquiry and many other documents and reports dealing with the disaster available over an extended period of time; C. E. Nicholson gave permission to reference many of the SMRE and other reports available. The Scunthorpe *Evening Telegraph,* and in particular Ms. Jackie Cole, provided access to many photographs and material from their archives dealing with their staff's coverage of the accident and its aftermath. Professors T. B. Jones and Terry Robinson of the University of Leicester gave freely of their time to explain and elaborate on the ionospheric effects of the Flixborough explosion. Dr. John I. Cox provided details of his precis of the statements and reenactments of eyewitness made for the Inquiry. Ralph King and his colleague, Ronald Hirst, have provided further details regarding what influence water in the reactors may have had on the accident; the HSE are currently reporting on experiments to evaluate this. The personnel at the BBC Film Archive Unit in London assisted in the review of the background video and cine film records for the event. BBC North, and especially Christine Hamill and *Close-up North* series producer Ian Cundall made arrangements for an interview with one of the eyewitnesses, Brian Pettitt, and prepared a program historically reviewing the event and its consequences. My colleagues at the University of New Brunswick and many graduate students have all substantially contributed through discussion, critiques, and other studies. D. M. Tan conducted the ANSYS finite element (FEM) analyses and assisted in the preparation of the Tecplot movies of the Flomerics CFD and the FEM results. Any errors and omissions are this author's alone.

42.8 REFERENCES

Ale, B. J. M., and F. Bruning. 1980a. "Unconfined Vapour Cloud Explosions," Letter, *The Chemical Engineer*, vol. 352, pp. 47–49.

Ale, B. J. M., and F. Bruning. 1980b. "Unconfined Vapour Cloud Explosions," Letter of reply, *The Chemical Engineer*, vol. 358, p. 494.

Amtec Engineering Inc. 1996. *Tecplot User's Manual*, version 7, Bellevue, WA.

ANSYS. 1994. Swanson Analysis Systems Inc., Houston, PA.

Artingstall, G. 1974. *Appraisal of the Damage in the Reactor Vessels in Section 25A,* Report No. 4, Part I, Safety in Mines Research Establishment (SMRE), September 11.

Au-Yang, M. K. 1999. "Common Mistakes and Misconceptions in Flow-Induced Vibration and Coupled Fluid-Structural Dynamics Analysis," *ASME PVPD-29*, pp. 6.1–6.17.

Ayre, E. 1974. Minutes of Proceedings at the Court of Inquiry; into the disaster which occurred at Nypro (UK) Ltd., Flixborough, Thirty-fourth Day, November 15.

Ball, J. G. 1975. "After the Flixborough Report: Do We Know the Real Truth?" *Process Engineering*, vol. 56, no. 12, pp. 39–46.

Ball, J. G. 1976. "Was This What Caused the Flixborough Explosion?" *Process Engineering*, vol. 57, no. 1, pp. 53–60.

Batstone, R. J. 1975. "Description and Results of Bellows Test Program and 20″ Dog Leg Pipe Tests," Cremer and Warner Report No. 7, January 7.

Bjerketvedt, D., J. R. Bakke, and K. van Wingerden. 1997. "Gas Explosion Handbook," *Journal of Hazardous Materials,* vol. 52, pp. 1–150.

Butler, P. 1975. "The Lessons to Be Learned from the Flixborough Enquiry," *The Engineer* (*GB*), vol. 241, no. 6247, pp. 20–24.

Carter, B. 1974. Minutes of Proceedings at the Court of Inquiry; into the disaster which occurred at Nypro (UK) Ltd., Flixborough, Fourth Day, September 12.

Center for Chemical Process Safety (CPPS). 1994. *Guidelines for Evaluating the Characteristics of Vapor-Cloud Explosions, Flash Fires, and BLEVEs,* American Institute of Chemical Engineers, New York, App. F, pp. 363–382.

Colquhoun, D. C. 1974. Minutes of Proceedings at the Court of Inquiry; into the disaster which occurred at Nypro (UK) Ltd., Flixborough, Third Day, September 11.

Cottrell, A. H., and P. R. Swan. 1974. "A Metallurgical Examination of the Eight-Inch Line," *The Chemical Engineer*, April, pp. 266–274.

Cox, J. I. 1976b. "Flixborough—Some Additional Lessons," *The Chemical Engineer*, vol. 387, pp. 353–358.

Dickinson, J. 1974. Minutes of Proceedings at the Court of Inquiry; into the disaster which occurred at Nypro (UK) Ltd., Flixborough, Second Day, September 10.

Expansion Joint Manufacturers Association (EJMC). 1993. *Standards for the Expansion Joint Manufacturers Association, Inc.*, 6th ed., EJMA, Tarrytown, NY.

Fishburn, B., N. Slagg, and P. Lu. 1981. "Blast Effect from a Pancake Shaped Fuel-Drop-Air Cloud Detonation (Theory and Experiment)," *Journal of Hazardous Materials,* vol. 5, pp. 65–75.

Flomerics Ltd. 1994. *FloVent Reference Manual,* Document No. FloVent/RM/0994/1/0, Flomerics Ltd., Surrey, U.K.

Foley, J. H. 1974a. *Metallurgical Examination of the Damaged Pipes from Section 25A,* Report no. 1, Part I, Safety in Mines Research Establishment (SMRE), September 10.

Foley, J. H. 1974b. Minutes of Proceedings at the Court of Inquiry; into the disaster which occurred at Nypro (UK) Ltd., Flixborough, Thirty-third Day, November 14, pp. 34–35.

Foley, J. H. 1974c. Minutes of Proceedings at the Court of Inquiry; into the disaster which occurred at Nypro (UK) Ltd., Flixborough, Thirty-fourth Day, November 15.

Foley, J. H. 1974d. Minutes of Proceedings at the Court of Inquiry; into the disaster which occurred at Nypro (UK) Ltd., Flixborough, Thirty-seventh Day, pp. 26–27.

Foley, J. H. 1974e. Minutes of Proceedings at the Court of Inquiry; into the disaster which occurred at Nypro (UK) Ltd., Flixborough, Thirty-seventh Day, p. 21, para 5.

Foley, J. H., and C. E. Nicholson. 1974. *Metallurgical Examination of the Damaged Pipes from Section 25A,* Report no. 1, Part III, Safety in Mines Research Establishment (SMRE), November 8.

Fu, W. B., and J. A. Eyre. 1974. "Unconfined Vapor Cloud Explosions," Letter, *The Chemical Engineer*, vol. 359/360, p. 368.

Games, G. A. C., and D. Waterhouse. 1974. *Construction of and Tests on a Reconstructed Bridging Pipe Assembly,* Report no. 2, Part II, Safety in Mines Research Establishment (SMRE), November 10.

Gill, S. S. 1974. Minutes of Proceedings at the Court of Inquiry; into the disaster which occurred at Nypro (UK) Ltd., Thirty-seventh Day, pp. 51–52.

Goodchild, M. B. 1975. Minutes of Proceedings at the Court of Inquiry; into the disaster which occurred at Nypro (UK) Ltd., Flixborough, Fifty-fourth Day, January 24, pp. 24–26.

Gromles, M. A. 1984. "A Simple Approach to Transient Two-Phase Level Swell," in Multi-phase Flow and Heat Transfer III, Miami Beach, FL, vol. 1, pp. 527–538.

Grover, F. H. 1974. *Infrasonic and Seismic Wave Records from the Flixborough and St. Bridget Explosions,* AWRE Report 046/74.

Gugan, K. 1974. Minutes of Proceedings at the Court of Inquiry; into the disaster which occurred at Nypro (UK) Ltd., Flixborough, Thirty-ninth Day, p. 17, November 22.

Gugan, K. 1976. "Flixborough—A Combustion Specialist's View," *The Chemical Engineer*, pp. 341–352.

Gugan, K. *Unconfined Vapor Cloud Explosions,* IChemE, Rugby.

Gugan, K. 1980a. "Unconfined Vapor Cloud Explosions," Letter of reply, *The Chemical Engineer*, vol. 358, pp. 491–493.

Gugan, K. 1980b. "Unconfined Vapor Cloud Explosions," Letter of reply, *The Chemical Engineer*, vol. 359/360, p. 567.

Gugan, K. 2000. "Flixborough Kicking," Letter, *The Chemical Engineer*, May 25, p. 30.

Harris, R. J. 1983. *The Investigation and Control of Gas Explosions in Buildings and Heating Plant,* E & FN Spon, London (in association with British Gas).

Harry, L. 1974. Minutes of Proceedings at the Court of Inquiry; into the disaster which occurred at Nypro (UK) Ltd., Flixborough, Third Day, September 11.

Hoiset, S., B. H. Hjertager, T. Solberg, and K. A. Malo. 2000. "Flixborough Revisited—An Explosion Simulation Approach," *Journal of Hazardous Materials*, vol. 77, nos. 1–3, pp. 1–9.

Humberside Police. 1974. Explosion, Nypro (UK) Ltd., June 1.

Jakubauskas, V. F., and D. S. Weaver. 1998a. "Transverse Vibrations of Bellows Expansion Joints. Part I: Fluid Added Mass," *Journal of Fluids and Structures*, vol. 12, pp. 445–456.

Jakubauskas, V. F., and D. S. Weaver. 1998b. "Transverse Vibrations of Bellows Expansion Joints. Part II: Beam Model Development and Experimental Verification," *Journal of Fluids and Structures*, vol. 12, pp. 457–473.

Jones, T. B., and C. T. Spracklen. 1974. "Ionospheric Effects of the Flixborough Explosion," *Nature*, vol. 250, pp. 719–721.

King, E. 1974. Minutes of Proceedings at the Court of Inquiry; into the disaster which occurred at Nypro (UK) Ltd., Flixborough, Second Day, September 10.

King, R. 1975. "Flixborough-Role of Water Re-examined," *Process Engineering*, September, pp. 69–72.

King, R. 1977. "Latent Superheat—A Hazard of Two-Phase Liquid Systems," *Chemical Process Hazards,* IChemE Symposium Series no. 49, pp. 139–145.

King, R. 1999. "Flixborough, 25 Years on," in *Process Engineering*, June, pp. 27–28.

King, R. 2000. "Flixborough: Re-Open Public Inquiry," Letter, *The Chemical Engineer*, May 25, p. 30.

Kitching, R. 1974. Minutes of Proceedings at the Court of Inquiry; into the disaster which occurred at Nypro (UK) Ltd., Flixborough, Thirty-seventh Day, pp. 51–52.

Kletz, T. 2000. "Flixborough, No Whitewash," Letter, *The Chemical Engineer*, April 20, p. 18.

Leclude, J., and J. E. S. Venart. 1996. "Single and Two-Phase Discharge from a Pressurized Vessel," *Revue Generale du Thermique,* vol. 35, pp. 503–516.

Leclude, J., and J. E. S. Venart. 1997. "Determination of the Amounts Discharged in the Flixborough Disaster," unpublished work, University of New Brunswick.

Marshall, H. 1974. Official Statement to Humberside Police, H Division, Scunthorpe, June 9.

Miller, K. J. 1999. Private communication, March 30.

National Institute for Standards and Technology (NIST). 1990. Standard Reference Database 4; Thermophysical Properties of Hydrocarbon Mixtures, Program Supertrapp, version 1.07, NIST, Gaithersburg, MD.

Newland, D. E. 1964. "Buckling of Double Bellows Expansion Joints under Internal Pressure," *Journal of Mechanical Engineering Science,* vol. 6, pp. 270–277.

Newland, D. E. 1976. "Buckling and Rupture of the Double Bellows Expansion Joint Assembly at Flixborough," *Proceedings of Royal Society of London,* series A, vol. 351, pp. 525–549.

Phillips, H. 1981. "Unconfined Vapour Cloud Explosions: A New Look at Gugan's Book," *The Chemical Engineer*, vol. 369, pp. 286–287.

Rajaratnam, N., and M. R. Chamani. 1995. "Energy Loss at Drops," *Journal of Hydraulic Research*, vol. 33, pp. 373–384.

Ramier, S., and J. E. S. Venart. 2000. "Boiling Liquid Expanding Vapour Explosions: Dynamic Repressurization and Two-Phase Discharge," in *Hazards XV,* IChemE Symposium Series no. 147, Rugby, pp. 527–537.

Roberts, A. F., and D. K. Pritchard. 1982. "Blast Effect from Unconfined Vapour Cloud Explosions," *Journal of Occupational Accidents,* vol. 3, pp. 231–247.

Ryder, D. A. 1974. Minutes of Proceedings at the Court of Inquiry; into the disaster which occurred at Nypro (UK) Ltd., Flixborough, Thirty-seventh Day, 1974, pp. 57–58.

Sadee, C., D. E. Samuels, and T. P. O'Brien. 1977. "The Characteristics of the Explosion at the Nypro (UK) Flixborough Plant on 1 June 1974," *Journal of Occupational Accidents,* vol. 1, pp. 203–235.

Secretary of State for Employment. 1975. *The Flixborough Disaster*, HMSO, London.

Snedden, N. W. 1985. "Analysis and Design Guidance for the Lateral Stiffness of Bellows Expansion Joints," *Thin-Walled Structures,* vol. 3, pp. 145–162.

Swan, R. 2000. "This Is Beginning to Look Like a Cover-up," Editorial, *The Chemical Engineer,* April 6, p. 2.

Teng-yang, R., K. F. Sollows, and J. E. S. Venart. 2000. "Froude Modeling of the Flixborough 'By-pass' Pipe," *Hazards XV,* IChemE Symposium Series no. 147, Rugby, pp. 139–154.

Venart, J. E. S. 1999. "Flixborough: A Final Resolution?" in INTERFLAM'99, ed. S. Grayson, vol. 1, pp. 257–272.

Venart, J. E. S., and D. M. Tan. 2000. Flixborough; Twenty-five Years after; the Final Resolution?" in *Proceedings of ICPVT-9,* Sydney, April 9–14, vol. 2, pp. 525–539.

Warner, F. 1975. "The Flixborough," *Chemical Engineering Progress*, vol. 71, no. 9, pp. 77–84.

Warner, F., and D. E. Newland. 1975. "Flixborough Explosion—Mechanical Engineering Provides the Key," *Chartered Mechanical Engineer* (*GB*), June, pp. 76–81.

Weaver, D. S., and P. Ainsworth. 1989. "Flow Induced Vibration in Bellows," *ASME Journal of Pressure Vessel Technology,* vol. 111, pp. 402–406.

Westgate, K. 1975. "Flixborough—the Human Response," University of Bradford, Disaster Research Unit, Occasional Paper no. 7, January (reprinted September 1976).

CHAPTER 43
CHEMICAL SPILLS AT SEA: CASE STUDIES

Michel Marchand
Rue Alain Colas, BP 20 413, 29604 BREST (France)

43.1 MARITIME TRANSPORT OF CHEMICAL PRODUCTS

Over the last 20 years there has been considerable development in the transport and handling of hazardous chemical products. Ships that transport chemical products carry a whole range of products that often pose a number of problems and risks in the case of accidents. Hazardous substances can be transported by sea either in bulk or packaged form. Products in bulk are transported either by chemical carriers, as in the case of liquid substances at an ambient, or in some cases elevated, temperature, or by gas carriers if gaseous substances are involved. The capacity of tankers for chemical products varies from 400 m^3 to 40,000 m^3, and tanks vary from 70 m^3 to 2,000 m^3. The capacity of ships carrying liquefied gases also varies and can reach 100,000 m^3. International construction norms for ships carrying chemical products are defined by the International Maritime Organization: the IBC code for chemical products in bulk (IMO, 1998a), the IGC code for liquefied gases in bulk (IMO, 1993), and the BC code for solids in bulk (IMO, 1998b).

The most common chemical products to be transported in bulk can be classified into the following categories:

- Chemical products made in large quantities, the most common being sulfuric acid, phosphoric acid, nitric acid, chlorhydric acid, caustic soda, and ammonia
- Molasses and alcohols
- Vegetable oils (e.g., soya, palm nut, sunflower) and animal oils (e.g., lard, fish oils)
- Petrochemical products, e.g., benzene, xylene, phenol, styrene
- Coal tar products (e.g., phenol, naphthalene)

Annex II of the MARPOL 73/78 Convention deals with pollution caused by hazardous liquid substances transported in bulk. These substances are classified into four categories (A, B, C, D) according to hazards (decreasing) to the marine environment.

Regulations imposed on the maritime transport of chemical substances carried in packaged form are described in detail in the IMDG Code, integrated into Annex III of the MARPOL 73/78 Convention. The IMDG Code classifies hazardous substances transported by sea into nine classes of products:

1. Explosives
2. Gases
3. Flammable liquids
4. Flammable solids
5. Oxidizing substances and organic peroxides
6. Toxic and infectious materials
7. Radioactive materials
8. Corrosive materials
9. Miscellaneous dangerous substances and articles

With regard to the chemical risk involved in transport by sea, many operational guides have been published by various international, regional, and national authorities. Examples of these are the IMO manuals (IMO, 1991, 1999), the REMPEC (Regional Marine Pollution Emergency Response Centre for the Mediterranean Sea) Manuals (REMPEC, 1996, 1999), the Helsinki Baltic Sea Convention (HELCOM, 1991), the North Sea Bonn Agreement Manual (Bonn Agreement, 1994), and other national documents, such as the Dutch-developed ELSA (Emergency Level Scale Procedure) software (Bonn Agreement, 1990).

43.2 METHODOLOGY

In the first stage, before studying chemical incidents or accidents, we have taken into consideration statistics on the most frequently accidentally spilled chemical products and facilities available in ports. This was made possible by using the statistical data of the U.S. Coast Guard (U.S. Coast Guard, 1999).

In the second stage, we collected incidents that were brought to our attention and had occurred in the Bonn Agreement zone. We extended the analysis to other recorded incidents in European waters and elsewhere so as to be able to have enough representative information concerning chemical risks. The framework used, according to the information available, is as follows:

- Scene of the accident: place and causes
- Chemical products involved: product(s), type of transport, properties, and hazards
- Development and organization of response actions: personnel, risk evaluation, plan adopted, and means developed
- Environmental impact: chemical monitoring and impact on the environment
- Report: comparison of comments on what should have been done and what was done, the organization of the response actions, the evaluation of the impact, and a general report

An almost complete analysis of 23 incidents, according to information available, has been possible. For convenience we have classified them into the following groups according to the type of maritime transport (packaged form or in bulk) and, if products were transported in bulk, according to behavior after being spilled at sea:

1. Products transported in packaged form
2. Dissolvers in bulk
3. Floaters in bulk

4. Sinkers in bulk
5. Gases or evaporators in bulk

These incidents are listed in Table 43.1. More details about the incidents are provided in the Appendix.

43.3 STATISTICAL APPROACH

The statistical study made by the U.S. Coast Guard (U.S. Coast Guard, 1999) in the United States from 1992 to 1996 lists 423 spills of hazardous substances from ships or port installations, for an average of 85 spills each year. The total of these spills is 7,500 tons, half in sulfuric acid. The nine most frequently spilled products are those that dissolve in water (sulfuric acid, phosphoric acid, caustic soda), products that evaporate and dissolve in water (acrylonitrile, vinyl acetate), and petrochemical-based products that float and/or evaporate (benzene, toluene, xylene, styrene). These are shown in Table 43.2.

Fifty-four percent of the spills are from ships (mainly carrier barges) and 66% are from facilities (when the spill comes from the facility itself or from a ship in dock). A complementary study made over a period of 13 years (1981 to 1994) on the 10 most important port zones reports 288 spills of hazardous substances, representing 22 incidents each year.

As far as we know, such statistics are not available for Europe or, more specifically, for the area covered by the Bonn Agreement.

43.4 EVENTS AND RISKS

43.4.1 Events

Most events reported happened at sea while the vessel was moving. Of all the chemical incidents reported (Table 43.1), only four happened in port or in nearby zones; the entrance to the port of Zhanjiang (the *No. 1 Chung Mu* accident in China), in Tokyo Bay (the *Yuyo Maru No. 10* accident in Japan), in Rio Grande do Sul (the *Bahamas* accident in Brazil), and in the Elba River (the *Oostzee* accident in Germany). The last event was the only one that happened in the Bonn Agreement zone. The conditions of these incidents have been classed into two broad groups, one linked to navigation risks, usually following bad weather conditions causing the loss of part of the cargo, a shipwreck, or a collision, the second linked to an initial internal event on board ship, such as fire, a faulty structure onboard, or a false maneuver (the ballast of the ship, the stowage of the cargo, even the deliberate dumping of substances).

The main elements of these conditions (21 cases reported out of the 23 treated) are summarized in Table 43.3. The difference between the two types of maritime transport, transport in packaged form (38% of incidents) and transport in bulk (62% of incidents), can be seen in the table. Bad weather conditions and the resulting consequences (13 cases out of 21, 62%) are the main causes of incidents including, in order of importance, ship collisions (5 cases), shipwrecks (3 cases), grounding (3 cases), and loss of cargo during navigation (2 cases).

43.4.2 Risks

Broadly speaking, there are two types of risks involved in chemical incidents, which, to some extent, are independent of the type of maritime transport of the products, whether

TABLE 43.1 Summary of Chemical Incidents at Sea

Name of ship	Year	Chemical products	Country	Maritime zone
1. Products transported in package				
Sinbad	1979	Chlorine	Holland	North Sea
Cason	1987	Sodium, aniline, creosol, diphenyl methane di-isocyanate	Spain	Atlantic
Perintis	1989	Lindane	France	Channel
Oostzee	1989	Epichlorhydrin	Germany	North Sea Elba River
Ariel	1992	White spirit	Holland	North Sea
Sherbro	1993	Pesticides	France	Channel/North Sea
Rosa M	1997	Hazardous materials	France	Channel
Apus	1998	Flammable solids (fire-lighters)	Holland	North Sea
Ban-Ann	1998	Sulfur-phosphine	Holland	North Sea
Ever Decent	1999	Hazardous materials	Great Britain	North Sea
2. Dissolvers in bulk				
Anna Broere	1988	Acrylonitrile (DE) Dodecylbenzene (F)	Holland	North Sea
Alessandro Primo	1991	Acrylonitrile (DE) Dichloroethane (SD)	Italy	Mediterranean
Panam Perla	1998	Sulfuric acid (D)	—	Atlantic
Bahamas	1998	Sulfuric acid (D)	Brazil	Atlantic
3. Floaters in bulk				
Lindenbank	1975	Coconut oil (F)	Hawaii	Pacific
Kimya	1991	Sunflower oil (F)	Great Britain	North Sea
Grape One	1993	Xylene (FE)	Great Britain	Channel
No. 1 Chung Mu	1995	Styrene (FE)	China	China Sea
Allegra	1997	Palm nut oil (F)	France	Channel
4. Sinkers in bulk				
Norafrakt	1992	Lead sulfur (S)	Holland	North Sea
Fenes	1996	Wheat (S)	France	Mediterranean
5. Gases or evaporators in bulk				
Yuyo Maru No. 10	1974	Propane (G) Butane Naphtha	Japan	Tokyo Bay
Ascania	1999	Vinyl acetate (ED)	Great Britain	North Sea

(D): dissolver; (DE): dissolver that evaporates; (ED): evaporator that dissolves; (F): floater; (FE): floater that evaporates; (G): gas; (S): sinker; (SD): sinker that dissolves.

TABLE 43.2 Accidental Spills of Hazardous Substances (U.S.A., 1992–1995). Total: 423 spills

Frequently spilled products	Number of spills	Behavior classification
Sulfuric acid	86	D
Toluene	42	FE
Caustic soda	35	D
Benzene	23	E
Styrene	20	FE
Acrylonitrile	18	DE
Xylene	18	FE
Vinyl acetate	17	FD
Phosphoric acid	12	D

(D): dissolver; (DE): dissolver that evaporates; (E): evaporator; (FD): floater that dissolves; (FE): floater that evaporates.

carried in bulk or in packaged form. These are risks involving the personnel (crew, response personnel, the local population) and the environment.

Risks that affect human health come from reactive substances (reactivity with air or water or between the products themselves) and toxic substances. Maximum hazard was clearly shown in the grounding of the *Cason* (Spain, 1987), following a fire on board the ship, which was carrying a number of toxic and strongly reactive flammable substances (reactivity of sodium, which in contact with water forms highly flammable hydrogen gas) and whose product identification and classification (IMDG Code) were unknown during the first few hours after the accident. The evaluation of the chemical risks of ships in difficulty when they are carrying diverse hazardous materials is also a priority in certain incidents, such as the *Rosa M* (France, 1997), and the *Ever Decent* (Great Britain, 1999).

Certain substances that are transported in large quantities can pose very serious risks to human health. Cylinders (1 ton each) of chlorine, a highly reactive, toxic, and corrosive gas lost by the *Sinbad* (Holland, 1979) are an example of the problems involved in transport in

TABLE 43.3 Origins of Chemical Incidents

Causes of incidents	Transport in packaged form	Transport in bulk	Total
Bad weather conditions	4	9	13
Navigation and loss of cargo	2	–	2
Grounding	–	3	3
Shipwreck	1	2	3
Collision	1	4	5
Incident on board ship	4	4	8
Fire	1	1	2
Faulty construction	–	2	2
False maneuvers[a]	2	1	3
Deliberate dumping	1	0	1

[a] Arrival of the cargo, pumping operation, ballast balancing.

packaged form. Fumes of epichlorhydrin leaking from the damaged drums on the *Oostzee* (Germany, 1989) seriously damaged the health of the ship's crew; as several years later cases of cancer, probably linked to the incident, were diagnosed in several people, some of whom died soon after. As far as transport in bulk is concerned, four types of products must be noted as being particularly reactive and hazardous. Acrylonitrile (*Alessandro Primo,* Italy, 1991; *Anna Broere,* Holland, 1988) is a toxic product, both flammable and explosive (it may polymerize spontaneously). Styrene (*No. 1 Chung Mu,* China, 1995) can polymerize in the form of a violent exothermic reaction. Leakages of sulfuric acid on board ship (*Panam Perla,* Atlantic, 1998; *Bahamas,* Brazil, 1998) cause risks to the ships themselves, diluted acid being much more corrosive to steel than concentrated acid. A mixture of acid with water also gives off explosive hydrogen. Vinyl acetate is a flammable and polymerizable plasticizing product; in the case of the *Multitank Ascania* incident (Great Britain, 1999), the exclusion zone was evaluated as 2 km long and 1 km wide.

Hazards to the environment are varied, especially when there is a loss of pesticide products, as in the case of almost 200,000 packages of thiocarbamate (*Sherbro,* France, 1993), which drifted from the coast of France to the coasts of Belgium, Holland, and Germany; the loss of a container of 5 tons of lindane (*Perintis,* France, 1989); and a spill of 1,600 tons of lead sulfur (*Nordfrakt,* Germany, 1992) resulting in an input of lead equal to the global amount of the metal over the whole of the North Sea. Substances considered as nonpollutants such as vegetable oils (*Lindenbank,* Hawaii, 1975; *Kimya,* Great Britain, 1991; *Allegra,* France, 1997) can also lead to the mortality of certain marine species in the environment with coating of feathers of birds, or skin of mammals, or to disturbances affecting the use of local amenities.

Even a substance as inoffensive as wheat, a food product (*Fenes,* France, 1997), can cause risks. Wheat fermentation in the marine environment, in an anoxic environment, results in the release of hydrogen sulfide (H_2S), a highly toxic gas which makes it necessary for intervening personnel to wear protective masks on the site.

43.5 RESPONSE ACTIONS AND CONCLUSIONS

Response actions taken during reported events obviously differ according to the conditions of the incidents or accidents, the offending chemical products, and the risks involved. It is possible, however, to show a certain number of significant or specific elements in all chemical incidents at sea.

Information concerning the ship's cargo and an evaluation of chemical risks are of primary importance before any operational decision is made, especially when the ship is carrying a wide variety of chemical products in packaged form. Information concerning the cargo is not always immediately available, as shown in the case of the *Cason* (Spain) and, to a lesser extent, in the case of the *Ever Decent* (Great Britain). The method of loading the containers is also to be taken into account although the rules are not always respected (*Rosa M,* France). The evaluation of the chemical risk involved is an essential element which can rely on national chemical emergency centres (*Multitank Ascania,* Great Britain) and on direct cooperation with the chemical industry (*Sherbro,* France; *Sindbad,* Holland). The establishment of a specialized intervention team to deal with chemical risks was underlined by the incident of the *Multitank Ascania* (Great Britain) and has been recommended since the incident of the *Rosa M* (France).

Initial response action can be carried out by the ship's crew, whose effectiveness depends on their professional competence and that of the captain. It may be excellent, as seen during the *Multitank Ascania* (Great Britain) incident before they abandoned ship, or a disaster, as seen in the incident during the unloading of the sulfuric acid from the *Bahamas* (Brazil).

Outside response action varies. A number of incidents (*Multitank Ascania,* Great Britain; *Rosa M,* France) show the importance of action taken during emergency towing. Delays in the transporting of means of intervention and bad weather conditions can seriously affect the efficiency of previously planned operations (*Cason,* Spain). In a port, the facilities required to lighten the weight of a chemical carrier in difficulty, whether onshore or onboard, are not always foreseen, as shown in the transfer of a cargo of sulfuric acid from the *Bahamas* (Brazil).

Specific or nonspecific tools used in chemical incidents are also mentioned. Different ways of dispersing a chemical pollutant in water and in air are used to evaluate risks to human health (air) and impact on the environment (water) (*Anna Broere,* Holland; *Alessandro Primo,* Italy) as well as the different ways in which floating objects or products drift (*Sherbro,* France; *Allegra,* France). Remote sensing with sensors normally used for detecting oil slicks is also effective when following drifts of vegetable oil (*Allegra,* France). An evaluation of the state of the shipwreck (*Alessandro Primo,* Italy) or a search for spilled products, such as cylinders of chlorine (*Sindbad,* Holland) can be undertaken by underwater remote control vehicles or by sound. Fires on board chemical carriers necessitate certain safety measures, taking into account the risk of explosion. An evaluation of hot points shown by an infrared (IR) camera was used during the incident of the *Multitank Ascania* (Great Britain).

Intervention onboard a ship in difficulty involves a large quantity of equipment, often very specialized, and including protective clothing, the monitoring of any contamination, information concerning emergency procedures, and the organization of means of evacuation (*Anna Broere,* Holland; *Alessandro Primo,* Italy).

The efficiency of the response action depends on the competence of each member of the personnel involved and the coordination of the actions taken. The experience gained during the intervention operation on the shipwrecked *Anna Broere* (Holland) to recover acrylonitrile proved to be useful some three years later during a similar operation, after the *Alessandro Primo* (Italy) shipwreck. The organization of the crisis of the *Multitank Ascania* (Great Britain) incident was made much easier by an antipollution exercise on chemical risks that had been practiced two weeks earlier.

Time involved in the operations can vary from 2 to 3 months (*Anna Broere,* Holland; *Cason,* Spain; *Alessandro Primo,* Italy), 8 months (*Fenes,* France), 10 months (*Bahamas,* Brazil), or several years, as in the case of research carried out on sunken cargo (*Sinbad,* Holland).

Communication is extremely important during chemical incidents, first on an operational level, between different personnel involved (the refusal of dialogue between the captain of the *Bahamas* (Brazil) and the port authorities had a disastrous effect compared with the initial incident), and secondly with the public, who are always very anxious about chemical risks. The *Cason* (Spain) and to a lesser extent the *Rosa M* (France) are examples of what happens when communication is poor.

Toxic effects of a chemical incident can be extremely harmful to intervening personnel during the intervention phase (wheat fermentation and the production of hydrogen sulfide (H_2S) during the recovery operation of the cargo of the *Fenes,* France), in the long term it may involve medical monitoring over several years (exposure of the crew of the *Oostzee,* Germany, to toxic fumes of epichlorhydrin).

Impact on the environment necessitates a monitoring of the quality of the environment and a study of the eventual effects on the flora and fauna, as in the long-term monitoring of a toxic substance spill (*Cason,* Spain; *Perintis,* France). This is equally necessary for substances considered nonpollutants, such as vegetable oils (*Kimya,* Great Britain). Indirect effects on the environment must also be considered as in the remobilization of toxic metals absorbed in sediments because of a decrease in pH caused by an acid spill (*Bahamas,* Brazil) or the production of hydrogen sulfide as a consequence of wheat fermentation by sulfate-reducing bacteria (*Fenes,* France).

APPENDIX—Conditions of Reported Incidents, Risks Involved, and Operational Action

Ship	Events	Chemical products	Risks	Action	Reports
① Transport in packaged form					
*Sinbad** (1979—Holland)	Loss of part of the cargo at sea during bad weather conditions.	Loss of 51 cylinders of chlorine (1 ton each).	Risks involved in the recovery of the cylinders in fishermen's trawler nets. Product toxic, harmful to human health (inhalation) and reactive, producing a corrosive acid with water.	Locating by sound and rapid recovery of 7 cylinders of chlorine. Recovery of 5 other cylinders by fishermen (lack of safety measures). Search five years later for other cylinders: 27 found and destroyed in the zone (unacceptable risk involved in a recovery exercise due to corrosion). Thirteen cylinders still missing.	Search for lost cargo in the zone. Operation carried out over several years (five years). Destruction of chlorine cylinders in the zone.
Cason (1987—Spain)	Fire on board ship. Grounding on the coast.	A number of hazardous substances including diphenyl methane diisocyanate (MDI) orthocresol, aniline, sodium.	Cargo initially unknown, products toxic to the environment and harmful to human health, reactive products leading to explosive reactions.	European cooperation (task force) and IMO expertise for identification of the cargo. Plans to unload hazardous materials from the ship hampered by bad weather conditions and fire on board. The operation necessitated three months' work. Monitoring of water and air contamination. Evacuation of the local population.	Difficulty in evaluating risks involved without having any immediate information concerning the cargo. Delay in transporting intervention equipment to the site. Intervention operation hampered by bad weather conditions. Communication with the public.
*Perintis** (1989—France)	Shipwreck during a storm.	Transport of pesticides, lindane: (5.8 tons), permethrine (1.0 tons), cypermethrine (0.6 tons). Loss of lindane containers during towing.	Contamination of the environment.	Search (unsuccessful) for a lindane container. Immediate recovery, by the British MPCU, of drums of permethrine and cypermethrine found near the wreck. Surveillance of the contamination of the marine environment in the zone where the container was presumed lost.	Franco-British cooperation. Cooperation on a national level between operational personnel (Navy, Cedre, Ifremer, etc.)

*Oostzee** (1989—Germany)	Bad stowing and deterioration of drums because of bad weather conditions. Incident on the Elba River.	Epichlorhydrin leakage from damaged drums.	Toxic substance (fumes) of a carcinogenic nature. A flammable product.	Inspection of the ship, the crew taken to hospital for medical checks. Towing and unloading of the ship in safe conditions. Cleaning operation on board.	Long-term effects (toxic fumes) on the crew (several years).
*Ariel** (1992—Holland)	Loss of drums at sea (cause unknown)Loss of 45 drums of white spirit when ship grounded on the coastline.	Contamination of the environment from drum leakages.	Recovery of drums grounded on the coastline.		
*Sherbro** (1993—France)	Loss of 88 containers after bad weather conditions at sea.	10 containers of hazardous materials, mainly pesticides, including thiocarbamate (Apron Plus) (188,000 packets).	Grounding of packets along the French, Dutch, and German coasts. A danger to the environment and a product which reacts with water, forming a toxic gas (phosphine).	Unloading of badly stowed containers in the port of Brest. Special storage (under tarpaulin) of containers of hazardous materials classified 3, 4, 5, 6, and 8, while respecting the compatibility of the products. Calculation of the drifting of 15 containers lost at sea. Three containers of pesticides were spilled at sea, 91% recovered on beaches in France, Germany, and Holland.	Cooperation between the countries concerned. Technical advice from the chemical industry. Cooperation of personnel during the operation to ensure the safety of unloading the ship.
*Rosa M** (1997—France)	Ship in trouble (listing more than 20°), after a fault in the balancing of the ballast.	70 tons of hazardous materials (liquid gases, flammable solids, corrosive substances, and oxidized substances).	Intervention work on board ship during a potential chemical risk.	Information concerning the type of cargo on board. Towing and voluntary grounding of the ship outside the port zone (Cherbourg). Water pumped from the ballast to rebalance the ship correctly.	Rapid towing of the ship in difficulty. Need for a specialized team to evaluate chemical risks involved.
*Apus** (1998—Holland)	Loss of a trailer on board a ferry.	Loss of a trailer's container of 2,100 boxes of fire lighters, grounded on the coastline.	Product harmful to human health (ureum-formaldehyde). Pollution on the environment (kerosene).	Recovery of fire lighters over several m^3 of sand on the beaches. Costs reimbursed by the ship's owner.	
*Ban-Ann** (1998—Holland)	Deliberate dumping of packets of chemical products at sea.	Packets of an active product containing sulfurphosphine (Detia-Ex-B) used against vermin.	Reactive product, reacting to humidity and causing a toxic gas.	Recovery and destruction of packets grounded on the coastline. Proceedings taken against the ship's owner and captain.	

APPENDIX—Conditions of Reported Incidents, Risks Involved, and Operational Action (*Continued*)

Ship	Events	Chemical products	Risks	Action	Reports
*Ever Decent** (1999—Great Britain)	Collision of a container ship with a liner. Fire on-board the container ship.	A number of containers of hazardous materials, especially cyanide, organic lead, and pesticides.	Chemical risk from fire on board ship.	Towing. Fire fighting. Control of air contamination (fear of cyanide).	Difficulty in knowing the exact type of cargo.
② Transport in bulk—dissolvers					
*Anna Broere** (1988—Holland)	Collision with another ship.	Acrylonitrile (547 tons) (DE) Dodecylbenzene (500 tons) (F)	Risks linked to acrylonitrile (cf. above).	Establishing of a safety perimeter (10 miles for navigation, 300 m for air transport). Unsuccessful attempts to lift the ship. Ship cut into two and cargo lightened. Environmental monitoring during the whole operation. Work done is under safe conditions (protective clothing, chemical control).	Time involved in the complex operation: two months. Effective cooperation between various teams of intervention personnel.
Alessandro Primo (1991—Italy)	Shipwreck as the result of a storm.	Acrylonitrile (549 barrels) (DE). Dichloroethane (3,013 tons) (SD).	Acrylonitrile toxic product, flammable and explosive. Gives off toxic fumes (HCN) in the event of fire. Dichloroethane: product harmful to human health.	Localization of the wreck (−108 m) by a remote control underwater vehicle. Beginning of a cargo recovery operation two months after the accident (priority given to the acrylonitrile). Total time of the operation: three months. Recovery of 900 tons of acrylonitrile, 2,750 tons of dichloroethane, and 50 tons of water for cleaning.	Advantages from a past experience (the *Anna Broere* shipwreck). Maximum safety recovery work (fire alarms, emergency training of intervention personnel, protective clothing, and medical and evacuation services on site).
Panam Perla (1998—Atlantic)	Incident on board ship, tank no longer watertight.	Sulfuric acid (100 tons) (D).	Corrosive product, reactive with water with a risk of ignition and explosion (formation of hydrogen). Harmful to human health, adverse effect on the environment	Recovery operation involving pumping the acid (100 tons) from the double hull. Operation completed one week after being told of the leakage. Neutralization of acid lost (3.4 tons) by bicarbonate.	Speed of operations carried out.

Bahamas (1998—Brazil)	Incident on board ship (error in the manipulation of the pumping system during unloading) culminating in abandoning ship.	Sulfuric acid (1,700 tons) (D).	(Cf. above).	An internal crisis on board ship was kept secret, resulting in a catastrophe. No means of stocking the diluted acid (very corrosive) on land or on another ship, neutralization impossible because no basic neutralizing agent was available. Cargo slowly pumped and dumped in the port as the tide was going out with a chemical monitoring of pH. Court order to dump the cargo out at sea (operation to lighten the weight using another chemical carrier): 12,000 tons of acid. Towing and scuttling in international waters. Total time of the incident: 10 months.	Initial errors leading to a crisis linked to the ship's dilapidated condition (28 years old) and to the incompetence of the crew. Lack of communication with the port authorities. Need for a local port structure and an emergency plan to treat chemical incidents. Impact on the environment: direct and indirect impact by the reactivation of toxic metals absorbed in the sediment.
③ Transport in bulk—floaters					
Lindenbank (1975—Hawaii)	Grounding on an atoll.	Coconut oil (18,000 tons) (F) and other food products.	Impact on the environment.	Raising of the ship stranded on a coral reef. Unloading of 18,000 tons of cargo in the water.	Impact on the environment due to a layer of coconut oil.
*Kimya** (1991—Great Britain)	Stranding during a storm.	Sunflower oil (1,500 tons) (F).	Impact on the environment.	Raising of the ship.	Environmental monitoring confirms the impact of such a spill on an intertidal environment.
*Grape One** (1993—Great Britain)	Fault on board ship (ballast) and shipwreck.	Xylene (3,041 tons) (FE).	Weak, but very flammable pollutant.	Crew evacuated and winched to safety. Ship stranded and shipwrecked with the cargo in the Channel.	
No. 1 Chung Mu (1995—China)	Collision with another ship at the entrance of the port of Zhanjang.	Spill of 230 tons of styrene (FE).	Reactive product (exothermic polymerization), flammable, and irritant. Impact on the environment (fishing and marine culture produce).	Attempt to limit the leakage and to stop the spill by using dams. Organoleptic tests carried out on fishing produce. Environmental monitoring of species affected by the pollution.	Intervention to seal a nonprotected hole (diving equipment adapted).

APPENDIX—Conditions of Reported Incidents, Risks Involved, and Operational Action (*Continued*)

Ship	Events	Chemical products	Risks	Action	Reports
*Allegra** (1997—France)	Collision with another ship in the Channel due to foggy weather.	Palm nut oil (700 tons) (F).	Impact on the environment.	Monitoring of the stranding of the product at sea (observation by air and by sea). Recovery of oil residues on the coast (12 tons).	Franco-British cooperation. Positive use of remote sensing.
④ Transport in bulk—sinkers					
*Nordfrakt** (1992—Germany)	Shipwreck due to the displacement of the cargo.	Lead sulfur (1,600 tons) (S).	Impact on the environment.	Raising of the wreck and its cargo.	
Fenes (1996—France)	Grounding.	Loss of 2,600 tons of wheat (S).	Wheat fermentation. Intoxication risk of gaseous fermentation products (H_2S) on intervention personnel.	Pumping of the ship's fuel and oil. Pumping of the cargo (wheat), dumped over 20 m on the sea bed; transferred onto a barge and reimmersed into the sea to very low depths. Displacing the shipwreck. Total time of the operation: eight months.	Toxic risk (hydrogen sulfide) linked to the fermentation of wheat in the sea, necessitating protective clothing (masks) during personnel intervention.
⑤ Transport in bulk—gases or evaporators					
Yuyo Maru No. 10 (1974—Japan)	Collision in Tokyo Bay.	Propane, butane, naphtha (G).	Flammable gaseous products.	Towing and anchoring out at sea in Tokyo Bay. Ship caught on fire.	
*Multitank Ascania** (1999—Great Britain)	Fire on board ship due to bad weather conditions.	Vinyl acetate (1,750 tons) (ED).	Flammable and polymerizable product, natural product possibly carcinogenic.	Evacuation of the crew because of a fire on board ship. Ship stranded at least half a mile from the coast. Establishment of a temporary 5 km exclusion zone, necessitating the evacuation of 200 inhabitants. Reconnaissance by helicopter of the hot spots on board ship (risk of explosion) using the ship's IR camera. Towed by a tug.	Advantages of a "chemical" exercise two weeks earlier. Positive action taken by the crew before abandoning ship. Importance of an intervention team in times of chemical risks. Importance of the disposal of emergency towing equipment.

(D): dissolver; (DE): dissolver that evaporates; (ED): evaporator that dissolves; (F): floater; (FE): floater that dissolves; (G): gas; (S): sinker; (SD): sinker that dissolves.

43.6 REFERENCES

Bonn Agreement. 1990. "Introduction of the Emergency Level Scale Procedure for Accidental Spills," in 20th Meeting of BAWG OTSOPA, Brussels, January 16–19 Bonn Agreement Working Group OTSOPA, London.

Bonn Agreement. 1994. "Bonn Agreement: Counter-Pollution Manual," ed. Bonn Agreement, London.

Helsinki Baltic Sea Convention (HELCOM). 1991. *Manual on Co-operation in Combating Marine Pollution within the Framework of the Convention on the Protection of the Marine Environment of the Baltic Sea Area, 1974 (Helsinki Convention)*, vol. 3, *Response to Incidents Involving Chemicals.*

International Maritime Organization (IMO). 1991. *Manual on Chemical Pollution,* Section 2, "Search and Recovery of Packaged Goods Lost at Sea," IMO, London.

International Maritime Organization (IMO). 1993. International Code for the Construction and Equipment of Ships Carrying Liquefied Gases in Bulk (IGC Code), IMO, London.

International Maritime Organization (IMO). 1998a. International Code for the Construction and Equipment of Ships Carrying Dangerous Chemicals in Bulk (IBC Code), IMO, London.

International Maritime Organization (IMO). 1998b. Code of Safe Practice for Solid Bulk Cargoes (BC Code), IMO, London.

International Maritime Organization (IMO). 1998. *Manual on Chemical Pollution,* Section 1, "Problem Assessment and Response Arrangements," IMO, London.

Regional Marine Pollution Emergency Response Centre for the Mediterranean Sea (REMPEC). 1996. *Compendium of Notes on Preparedness and Response to Maritime Pollution Emergencies Involving Hazardous Substances,* REMPEC, IMO, and UNEP, Malta.

Regional Marine Pollution Emergency Response Centre for the Mediterranean Sea (REMPEC). 1999. *Practical Guide for Marine Chemical Spills,* REMPEC, IMO, and UNEP, Malta.

U.S. Coast Guard. 1999. "Most Spills Near Gulf; Most Spillage Inland; Most Were Acid," *Hazardous Substances Spill Report,* vol. 2, no. 8, April.

P • A • R • T • 10

CHEMICAL SPILL DATA

CHAPTER 44
DEVELOPMENT OF A CHEMICAL SPILL PRIORITY LIST FOR ENVIRONMENT CANADA

Merv Fingas and Kerry Ketcheson
Emergencies Science Division, Environment Canada, Ottawa, Ontario

Carol Lau and Mary-Ann Spicer
Environmental Emergencies Branch, Environment Canada, Hull, Quebec

44.1 INTRODUCTION

In 1979, the Environmental Emergencies Branch embarked on an accelerated program to improve the response and countermeasures technology for spills of materials other than oil. As an essential part of this program, a methodology was devised to develop a priority list of chemicals on which to focus further research efforts. A list was compiled of 150 hazardous materials with significant potential to be spilled in Canada. This list was then used to develop specialized countermeasures, analytical techniques, and spill manuals for these chemicals. Individual manuals called EnviroTIPS were prepared for the first 50 priority substances. Tests of the utility (mathematical efficiency) of the priority list showed that a large portion of the spills were those of the higher priority substances.

Statistical spill data are extremely useful for setting priorities and identifying project needs for spill prevention and preparedness. This was recognized in 1972, and a database was developed to consolidate all Canadian spill data. This database is known as NATES, short for National Analysis of Trends in Emergencies System. The NATES database now contains data on over 150,000 spill events. About 60% of these spill events are related to oil and petroleum products. About 20% involve wastes and aggregate materials. Only about 10% involve the spillage of "pure" chemicals.

The first spill priority list was developed in 1980 (Fingas, 1984). Lists were developed of the top 10, 50, and 150 chemicals. Several mathematical approaches to the development of the priority list were tried. However, it was found that a simple ranking of supply volume, reported spill frequency, historical spill volumes, and toxicity could result in a satisfactory list. The main objective was to identify the minimum number of chemicals that would account for the maximum number of spills. The use of the list would be as noted above, to act as a focus for the development of countermeasures, analytical methods, and spill manuals.

The first priority list (1980) was successful in that a chemical could account for many spills. The first 10 chemicals accounted for 37% of the reported spill events by number, 83% of the spill volume, and 50% of the volume of chemical supply in Canada.

A new list was developed in 1990 (Fingas et al., 1990, 1991a, b). While this list used a similar procedure, more extensive toxicity data was used. The utility of this new list was analyzed as was the 1980 list. The first 10 priority substances account for 43% of spill numbers, 25% of spill volume, and 15% of supply volume. The first 50 substances account for 80% of spill numbers, 97% of spill volume, and 65% of supply volume. The 150 top priority chemicals account for 97% of spill numbers, 98% of spill volume, and 82% of supply volume. This utility was much higher than the previous priority list.

This article presents the third update of the spill priority list based on chemical spill data over a 10-year period from 1987 to 1997.

44.2 DEVELOPMENT OF A NEW PRIORITY LIST

The objective of developing lists is to provide a minimum number of chemicals that account for the most spill numbers and concerns. A large part of the concern relates to the environmental and human toxicity of the spilled material. Therefore, in addition to considering the spill numbers, it is very important to examine the toxicities of the spilled material.

The following objectives and principles were set for developing the new list.

1. The list should be developed in a systematic and mechanical fashion—that is, the placement of individual compounds will be based only on the algorithm used and not on human judgment.
2. The high-volume and frequently spilled chemicals should appear at the top of the list, but toxicity should affect their relative placement.
3. Toxicity will be divided into two considerations, mammalian and aquatic, encompassing both environmental and health concerns.
4. As in the first two lists, one of the prime objectives is to produce a list that contains as few chemicals as possible to deal with the largest number and volume of spills.
5. All possible candidates will be evaluated, but only materials with a spill history will be in the top lists.
6. Completely innocuous materials will not be considered to ensure that they do not appear in any final priority lists.
7. Several sizes of groupings (e.g., 10, 25, 50, etc.) will be prepared to allow use for a variety of purposes.

The procedure employed for developing the new list is different than previous methods. Candidate chemicals and hazardous substances other than oil were not taken from all known spill priority lists but only from materials known to have been spilled in the past 15 years. This was done to eliminate the possibility that chemicals that have never been spilled or have a low spill potential might be present on the new priority list, as occurred in the past list. The problem is that many lists contain materials that have no spill history. This is because these chemicals may not be transported or consumed in the form noted on the particular priority list, they may be a degradation or reaction product, or they may not be used commercially at the present time. Furthermore, there is an extensive spill history at this time, as evidenced by a sizeable database, and substances that have a potential for spillage are likely to be listed. Few new chemicals appear to be produced in high volumes.

After the candidate chemicals were accumulated, they were reviewed and synonyms and substances that were different forms of each other were removed. Materials that could not be classified were separated. They are presented in Table 44.1, which shows many materials that could not be specified as a pure chemical. Examples of this include pesticides and unknown substances. There are also a large number of other materials that are best listed on a table such as this, e.g., the metal concentrates, construction materials, and food. Because some of these materials do not pose a simple chemical hazard, they are best considered as aggregate materials, as noted in the table. Table 44.1 shows the diversity of materials that are typically dealt with by emergency crews as environmental concerns. Some materials that may appear to be innocuous to the public may pose a serious environmental threat. Examples of this include soil spilled into a salmon-spawning area. This apparently harmless incident could result in serious consequences to the salmon spawn in that area.

Data on aquatic and mammalian toxicity were collected from standard reference sources. Aquatic data included the results of acute lethal toxicity testing on rainbow trout, bluegill, fathead minnow, and *Daphnia magna*. Aquatic data on similar species were collected if none of the above species data was available. Aquatic data were found for over 50% of the named compounds. Oral and inhalation toxicity to the rat were collected for the mammalian toxicity. Toxicity data were available for over 70% of substances on the list. These data on the substances and the remaining priority substances are shown in Table 44.2. If the toxicity data were for a slightly different species or time range, data were still included.

Similarly, the spill numbers and spill amount of the chemicals were ranked by simple numbers. The sum of the ranking for each of the four categories (spill numbers, spill volume, aquatic toxicity, and mammalian toxicity) was simply added to yield an overall value. The value for spill frequency was added in twice (as a square) to ensure that this factor was rated higher than other factors. These overall values are then directly related to the order of priority, lowest number first.

The top priority chemicals are listed in Table 44.3.

The utility of this new list was analyzed in a similar manner to the previous list and a graph of this is shown in Fig. 44.1. The first 10 priority substances account for 45% of spill numbers and 24% of spill volume. The first 50 substances account for 80% of spill numbers and 79% of spill volume. The 150 top priority chemicals account for 92% of spill numbers and 91% of spill volume. This utility is about the same as the previous priority list and easily satisfies the starting criteria.

Table 44.4 lists the top 10, 50, and 150 priority substances.

The Appendix contains a list of priority substances from various organizations and shows how this new priority list relates to these substances and to previous priority lists. This list is abbreviated to remove substances that appeared on only one list and are not produced in large quantities. This reduction amounts to a list one-third shorter than it would otherwise be. An examination of the table in the Appendix shows that many chemicals that are on priority lists are also a concern for frequent spillage.

44.3 REFERENCES

Fingas, M. F. 1984. "The Development of Improved Countermeasures for Chemical Spills in Canada," *Proceedings of the 1984 Hazardous Materials Conference,* Hazardous Material Institutes, Rockville, MD, pp. 255–260.

Fingas, M. F., N. Stone, D. LeBlanc, G. Sergy, B. Mansfield, G. Cloutier, and P. Mazerolle. 1990. "Development of a New Chemical Spill Priority List," in *Proceedings of the Seventh Technical Seminar on Chemical Spills,* Environment Canada, Ottawa, ON, pp. 115–189.

Fingas, M. F., N. Laroche, G. Sergy, B. Mansfield, G. Cloutier, and P. Mazerolle. 1991a. "The Development of a New Spill Priority List," *Spill Technology Newsletter,* Vol. 16, no. 3, pp. 1–8.

Fingas, M. F., N. Laroche, G. Sergy, B. Mansfield, G. Cloutier, and P. Mazerolle. 1991b. "A New Chemical Spill Priority List," in *Proceedings of the Eighth Technical Seminar on Chemical Spills,*

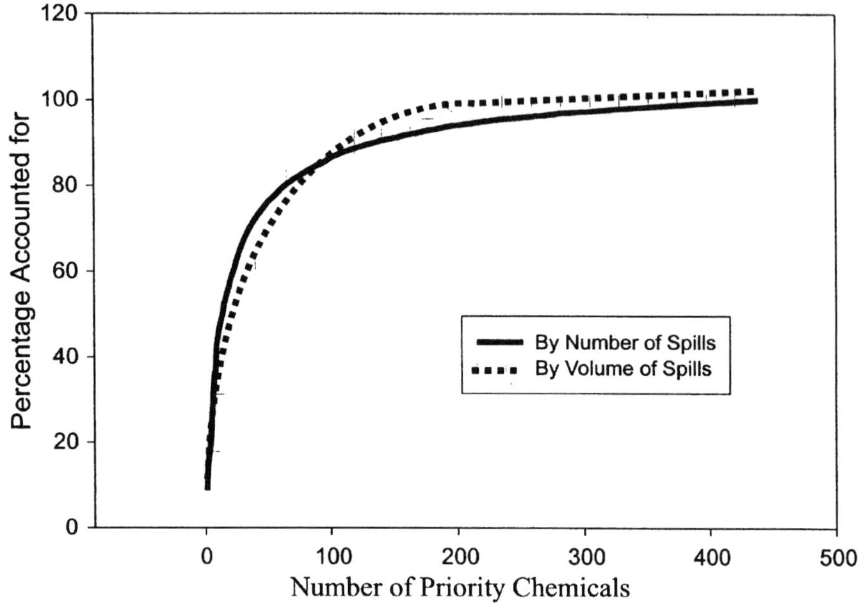

FIGURE 44.1 The priority chemicals.

TABLE 44.1 Aggregate Materials Spilled (1987–1997)

	Number	Mass (tons)
Acid, unspecified	52	89
Acrylic materials	22	0.3
Adhesives	44	127
Alcohols	10	1
Alkylamines	2	0
Aluminum	3	15
Bases, unspecified	29	982
Burn emissions	40	0
Calcium lignosulfonate	7	35
Chromium materials	71	65
Clay	20	4
Cleaners	69	455
Coal and clinker	4	0.02
Cobalt	3	4
Construction materials	58	813
Consumer products	11	0.03
Copper and concentrates of copper	17	4,803
Copper/nickel concentrates	8	45
Drilling muds	17	662
Dyes	118	94
Explosives	2	0
Fatty acids	4	6
Fertilizers, unspecified	151	1,675

TABLE 44.1 Aggregate Materials Spilled (1987–1997) (*Continued*)

	Number	Mass (tons)
Firefighting foam	12	12
Food	64	67
Fungicides, unspecified	5	0.003
Gas releases	51	45
Gasoline additives	20	0.50
Herbicides, unspecified	77	2
Industrial materials, various	158	149
Iron and iron oxide	11	32,909
Latex	43	34
Lead ore and concentrates	12	1,100
Lead/zinc concentrates	1	48
Manure	27	138
Mercaptan liquid	4	0
Metals, unspecified	19	306
Mine material	6	64
Molybdates	11	33
Natural materials	13	0
Nickel and nickel concentrates	21	4,739
Oil additives	9	9
Paints	401	312
Pesticides, unspecified	134	128
Phenolic resins	29	39
Phosphates	7	125
Phossy water	24	189
Plastics and plastic materials	37	96
Polymers, miscellaneous	18	9
Polyols	6	2
Radioactives	5	0.30
Resins	82	15
Sewage	171	4,425
Silicone materials	12	30
Silver concentrate	1	17
Soils and sediments	62	100
Solvents, unspecified	94	80
Steel mill liquors	6	0.17
Surfactants	74	315
Tin and tin concentrates	7	2
Tire fires	8	0
Unknown or not clearly specified	777	134,046
Waste materials	143	643
Waste water	62	11,482
Wood waste and products	37	61
Total	**3,523**	**201,650**

TABLE 44.2 The Spilled Substances and Their Toxicity

Name	Number of spills	Spill amount (tons)	Aquatic toxicity (in ppm)				Mammalian toxicity (in ppm)			
			Rainbow trout	Bluegill	Fathead minnow	*Daphnia magna*	Orl-rat LD_{50} (mg/kg)	Ihl-rat LC_{50} 1 hour	Ihl-rat LC_{50} 2 hour	Ihl-rat LC_{50} 4 hour
1,1-dichloroethane	1	0.00004696				1,000	725			13,000
1,2-propanediamine	1	0.03			1,010		2,230			
1,3-butadiene	4	0.22887825					5,480			128,250
1,4-dioxane	3	0.001			10,000		7,120		12,788	12,788
1-decene	1						10,000			
1-methyl-2-pyrrolidone	1	0.09766					3,914			
2,4-D	51	14.3636885	200	263	130	120	375			
2-chloro-4-hexanone	1	0.0224								
2-ethylhexanoic acid	1	0.00908					3,000			
2-ethylhexanol	4	7.68	35		28.2		3,730			
2-ethylpyridine	1	0.008918			414					
2-methyl-5-ethylpyridine	1	0.0018368								
2-thioethanol	1	0.000007					244	4,070		
3-methoxypropylamine	1									
8-hydroxyquinoline benzoate	1	0								
9,10-anthraquinone	1	0.35					6,000			150
Acephate	2	0.23	500	600			700			
Acetaldehyde	1			53	30.8	10,000	661			13,300
Acetic acid	9	0.15352542	200	11	82		3,310	5,620		
Acetic anhydride	10	29.1043246		75		45	1,780			1,000
Acetone	4	0.19834325	5,540	8,300	7,000	10	5,800			18,568
Acetonitrile	4	39.003148		1,850	1,300	100	2,460			5,655
Acetylene	3	0	0.4							
Acrylamide	5	0.245		12	14		124			
Acrylic acid	2	0.276275					33.5		1,802	
Acrylonitrile	4	6.7		12	16		78			425
Adipic acid	1	2.75		330	97		11,000			
Aldrin	1	0.0000288	0.036	0.013		0.028	39			
Aluminum alkyl compounds	2	2.004								
Aluminum chloride	8	29.4462016	0.56				3,450			
Aluminum fluoride	1	3					10,103			
Aluminum oxide	3	100.47146								
Aluminum phosphide	1	0								
Aluminum sulfate	52	223.2571452					6,207			
Aminocarb	3	0.41	13.57	3.1	8.5		30			
Amitrole	1	0.2				30	1,100			
Ammonia	225	178.7559487	1	1.5	4	80				2,000
Ammonium bisulfate	1									
Ammonium bisulfite	1	0.2								
Ammonium chloride	4	3.03879704	81	0.9	144	13	1,650			

Ammonium hydroxide	123	692.0088598	6.25				350			
Ammonium nitrate	98	1510.210443		800			2,217			
Ammonium persulfate	1					120	689			
Ammonium phosphate	1									
Ammonium sulfate	23	22.40344236	23		20	202	2,840			
Ammonium thiosulfate	3	6.700322		440	833	530	2,890			
Ammonium xylenesulfonate	1	0.00574								
Antimony trioxide	1	0.2			80		34,600			
Argon	3	4.006								
Arsenic	7	16.242	0.55		9.9	3.8	763			
Arsenic trioxide	2	1	12		109		14.6			
Asbestos	11	64.1762								
Atrazine	16	1.51708692	11	18,476	15	31	672			587.6
Avenge	5	1.86					206			
Azinphos-methyl	2	0.624	0.0062	0.0061	0.093	0.0059	7	5.3		
Bacillus thuringiensis	12	7.91304					20,000			
Barium	3	0.05	42.7			410				
Barium carbonate	1	1.188					418			
Barium chlorate	1	5.5								
Barium sulfate	6	124.2	10,464							
Bensulide	2	0.6014					271			
Bentazon	1	0.000504					1,100			510
Benzaldehyde	1	0.00169556	6	1.07	10		1,300			
Benzene	50	88.2	5	63	24	15	930			8,000
Benzene phosphorus dichloride	1	0.0056			62					
Benzenesulfonic acid	2	4.50182			18		0.89			
Benzoic acid	1	0.9				1,540	1,700	5		
Benzoyl chloride	3	1.345761			34		1,900	325		
Boric acid	1	0.028694	100			133	2,660			
Bromacil	1	0					641	432		
Bromadiolone	1	0.00098					0.49			
Bromine	1	0			16,500	1	2,600	405		
Bromoxynil	1	0.2					190			
Busan (2-bromo-4′-hydroxyacetophenone)	7	0.04676								
Butadiene dimer	1	0.014					3.08	6,095		
Butyl acetate	4	0.7843475		100	327	4,730	10,768			390
Butyl acrylate	4	0.001157					900			2,730
Butyl alcohol	8	7.9319975			1,910	2,802.5	790			8,000
Butyl cellosolve	2	0.000728					470			450
Butyl mercaptan	1			5.5			1,500			4,020
Butyl methacrylate	3	0.005					16,000			4,910
Butylate	4	0.905056					4,000		2,102	
Butylene	2	9.30922								
Cadmium oxide	1						72		47	
Calcium	3	4.90378								
Calcium carbide	5	1.26								
Calcium carbonate	27	268.251644								
Calcium chloride	52	498.6678977		10,650		649	1,000			

TABLE 44.2 The Spilled Substances and Their Toxicity (*Continued*)

Name	Number of spills	Spill amount (tons)	Aquatic toxicity (in ppm): Rainbow trout	Aquatic toxicity (in ppm): Bluegill	Aquatic toxicity (in ppm): Fathead minnow	Aquatic toxicity (in ppm): *Daphnia magna*	Mammalian toxicity (in ppm): Orl-rat LD_{50} (mg/kg)	Mammalian toxicity (in ppm): Ihl-rat LC_{50} 1 hour	Mammalian toxicity (in ppm): Ihl-rat LC_{50} 2 hour	Mammalian toxicity (in ppm): Ihl-rat LC_{50} 4 hour
Calcium hydroxide	18	45.219788					7,340			
Calcium hypochlorite	21	19.04265416	0.07	0.0623			850			
Calcium nitrate	2	0.7		2,400			3,900			
Calcium oxide	34	536.086238								
Calcium silicon	1									
Calcium sulfate	1	0.002368		2,980						
Caprolactam	4	0.41223					1,210		66	
Carbamate	2	0.20028								
Carbaryl	3	1.1012	4.3	6.8	9	0.005	230			
Carbofuran	10	0.477258	0.28	0.24	0.87		5	9		
Carbon 14	1									
Carbon black	11	21.31074					15,400			
Carbon dioxide	6	30.247725								
Carbon disulfide	8	80.320112				2.1	2,780		8,025	
Carbon monoxide	7	0.02							1,807	
Carbon tetrachloride	1		1.97	125	4	35	2,350			8,000
Carboxin	4	1.6					430	2,044		
Cesium	1	7.048E-07					1,700			
Cesium chloride	1	0.00126	181				2,600			
Chlordane	1		0.026	0.06	0.1	0.153	283			
Chlorine	124	30.61175392	0.02		0.09	0.12		293		
Chlorine dioxide	9	17.11925928					292			
Chlorine trifluoride	1	0.2409						299		
Chlorobenzene	1	0.02	350	15	28	21.36	1,110	2,965		
Chloroethylphosphonic acid		0.0111475					4,000			
Chloroform	4	0.00053424	2.09	100	58		695			9,779
Chlorophenol	2	0	3	6	10	10	200	2		
Chlorosulfonic acid	1						50			
Chlorothalonil	1	0.00028					10,000	28		
Chlorpyrifos	14	0.64855842	0.0104	0.015	0.15	0.0004	82			14
Chromic acid	45	38.6058304	24			0.01				
Copper chloride	2	0.0426636	0.042	1.1		0.1				
Copper chromium arsenate	2	0.015925								
Copper cyanide	2	0.013432	0.1	0.2	1.2	12.311	1,265			
Copper sulfate	8	36.2489256	0.45	1.1	0.075	0.05	300			
Creosote	18	0.005212					725			
Cresol	4	0.44	8.4	10.9	12.8		1,454			
Cresylic acid	5	0.003540236					1,500			
Cyanide	10	1.3050336								
Cyclohexylamine	5	1.679747					11	1,852		

Cygon	3	0.00008185					60			
Cymenes	1					6.5	2,800	1,846		
Dacthal	2	0.02					3,000			413
Dalapon	1	0.54		105						
Decanol	1	0.00014			2.4		4,720		618	
Demeton	1	0.01		0.11	3.2		1.7			
Diacetone	1	2.3					2,520			
Diammonium sulfate	1	0.931658	36.7		20	202	2,840			
Diazinon	17	0.555316635	1.35	0.12	10.3		66			277
Dibenzoyl peroxide	1	0.00122728					7,710			
Dibutyl-p-phenylene	1	0.000014								
Dibutyltin compounds	1	0.033								
Dicamba	5	0.2364182	28.4	50			1,039			
Dichloromethane	2	0.1824075		220	99		1,600	14,924		
Diclofop-methyl	6	0.1					512			585
Dicyanodiamide	1						500			
Diethanolamine	10	8.7585731		2,100	100	1.4	0.62			
Diethylamine	3	3.605025	104		855	56	540			4,000
Diethylene glycol	3	2.37459			100	0.6	12,565			
Diethylene triamine pentaacetic acid	1	0.0603288								
Diisobutyl dithiophosphinate	1	0.0979								
Diisobutylene	3	0.8007227								
Diisocyanate	3	2.24315464								
Diisononyl phthalate	1	0.004312								
Diisopropanolamine	1	0.4945					4,765			
Diisopropyl ether	1	0.000056								
Diisopropylamine	1	0.032265					770		1,152	
Dimethoate	7	4.96025	6.2	28			60			
Dimethyl disulfide	1	2384					200			805
Dimethyl disulfonate	1	0.01								
Dimethyl sulfide	3	0					3,300	40,250		
Dimethyl terephthalate	1	0.416					3,200			
Dimethylacetamide	1	1.1					4,300	2,475		
Dimethylamine	6	0.16806	68				698		4,725	
Dimethylcyclohexylamine	2	0.256								
Dimethylphosphine	1	0.00000056								
Dinitrophenol	1	0.00837	12	20	19.4					
Dinitrotoluene	1	540					750			
Dinoseb	1				0.36	0.24	25			
Dioctyl phthalate	9	6.604734					30,000			
Dioxin	3	0.02					0.022			
Diphenyl	1	0.24					2,140			
Diphenyl oxide	2	0.87					2,450			
Diphenylmethane	3	0.4183274					2,250			
Diquat	1	0.024	11.2	35	14		231			
Dithiocarbamate	5	1.8200168								
Diundecyl phthalate	1	0.0112								
Dodecylbenzene	1	0.00112								

TABLE 44.2 The Spilled Substances and Their Toxicity (*Continued*)

Name	Number of spills	Spill amount (tons)	Aquatic toxicity (in ppm) Rainbow trout	Bluegill	Fathead minnow	*Daphnia magna*	Mammalian toxicity (in ppm) Orl-rat LD_{50} (mg/kg)	Ihl-rat LC_{50} 1 hour	Ihl-rat LC_{50} 2 hour	Ihl-rat LC_{50} 4 hour
EPTC	4	0.41148					916			
Ethanethiol	1	0.002					682			4,420
Ethanol	45	352.2149308	13,000		14,000	1,000	7,060			20,748
Ethanolamine	9	4.714200848	150	329.16	2,070	140	1,720			
Ethyl acetate	6	0.1265968	452		260	716.8	5,620			19,600
Ethyl acrylate	9	0.53495792			2.5		800			1,414
Ethyl hexyl phthalate	2	0.91872								
Ethyl mercaptan	1						682			4,420
Ethyl nitrite	1	0.09								160
Ethylamine	7	0.05902582				110	400	5,540		
Ethylbenzene	58	149.7739	9	60	29		3,500			
Ethylene	1	7								
Ethylene dibromide	2	3.5000868		18			108			
Ethylene glycol	349	199.950697					4,700			
Ethylene glycol monobutyl ether	4	0.30968385		1,490			2,400			
Ethylene oxide	6	0.004193135	19	27	84	2,000	72			800
Ethylenediamine	6	0.27989			119	0.88	1,200	123		
Fenitrothion	7	0.2027032	100	2.6	4		250			33
Ferric chloride	47	23.32664057	3.4			15	450			
Ferric nitrate	4	225.4					3,250			
Ferric oxide	6	170.9723341								
Ferric phosphate	1	2.2								
Ferric sulfate	12	7.29677094								
Ferrous chloride	7	3.26948393					450			
Fluoboric acid	2	0.68472								
Fluorescein	6	126.1263184					300			
Fluorine	3	0.618775	10,002.3					185		
Formaldehyde	41	47.39029797	50	40	24.1	52	100	164		
Formic acid	6	9.046065		175			1,100			
Fosamine ammonia	2	0.007028					11,000	8,065		
Freon	73	14.41362437								
Fumaric acid	2	0.0390882		5,000						
Furfural	3	0.0541776			32		65			175
Glycidol	1	0.00056					420			580
Glyphosate	24	3.2947473	8.3		2.3		4,873			1,736
Guthion	1	0.00154					7	5		
Halon	26	0.656748494								
Heavy water	1	0								
Helium	2	0								

Heptene	2	0.022295				50			
Hexamethylenediamine	17	1.1973695					750		
Hexamethylenetetramine	1	0.0061468			49,800		569		
Hexanoic acid	1				200		3,000		433
Hexene	1	0.126							32,000
Hydrazine	10	24.81776616		1.3	5.98	0.81	60		570
Hydrobromic acid	1	0						2,858	
Hydrochloric acid	377	770.7733863		3.5				3,124	
Hydrocyanic acid	2	0.0033024	0.057		0.167		3.7		
Hydrofluoric acid	10	0.9343108						1,276	
Hydrofluorosilicic acid	20	101.3329621							
Hydrogen	12	8.551							
Hydrogen chloride	12	17.608425						3,124	
Hydrogen fluoride	2	10.07						1,276	
Hydrogen peroxide	40	96.41022267					2,000		1,440
Hydrogen sulfate	1	9							
Hydrogen sulfide	4	0.22295	0.4	0.3	0.6			634	
Inorganic chromate	3	3.2							
Iron sulfite	2	1.15							
Isobutyl alcohol	2	0.180675					2,460		
Isobutyl isocyanate	1								
Isobutyl xanthate	1	0.0504							
Isocyanates	7	0.751							
Isodecanol	1	0.00381672					6.5		
Isophoronediamine	1	0.0001472							
Isophthalic acid	19	1175.876189					10,400		
Isoprene	1	0.000644		42.54	80				63,554
Isopropane	1	15							
Isopropanol	26	11.98710775			11,130		5,045		16,000
Isothiazol	1	0.7							
Isovaleric acid	1	0.0005					2		
Lead acetate	2	0.4		250					
Lead chloride	1		0.200	200	200		1,947		
Lead chromate	2	0.05					12,000		
Lead oxide	3	0.035			1,000				
Lead sulfate	4	48.622032		3,800	1,300				
Linuron	1	0.01911					1,146		5
Lithium metal	3	0.01821612	9.28						
Magnesium	1	0.0056							
Magnesium bisulfate	2	2.22264							
Magnesium chloride	3	34.4914	1,355			888.4	2,800		
Magnesium hydroxide	1	1.7					8,500		
Magnesium oxide	3	2.0756							
Magnesium sulfate	10	21.4245875		3,800		459.3			
Malathion	14	0.684090036	0.17	0.11	9		290		3
Maleic anhydride	27	6.09313938		138			400		
Maneb	2	0.1	3			1	3,000		
Manganese	2	0.164					9,000		
MAPP	1								
MCPA	5	2.41584		100			700		164

TABLE 44.2 The Spilled Substances and Their Toxicity (*Continued*)

Name	Number of spills	Spill amount (tons)	Aquatic toxicity (in ppm)				Mammalian toxicity (in ppm)			
			Rainbow trout	Bluegill	Fathead minnow	*Daphnia magna*	Orl-rat LD_{50} (mg/kg)	Ihl-rat LC_{50} 1 hour	Ihl-rat LC_{50} 2 hour	Ihl-rat LC_{50} 4 hour
MDI	17	5.8416075					9,200	17		
Mecoprop	1	0.0014					650			
Mercuric chloride	1		0.11	0.118	0.16	0.025				
Mercuric iodide	1	0.0810264					18			
Mercury	65	103.2176028	0.005							
Metachlor	1	3.7					930			2,086
Methane	1									
Methanol	80	1290.688963	20,000	13,500	28,400	100	5,628			64,000
Methoxychlor	2	0.1128005	0.0447	0.032–0.075	0.0075–0.039	0.006	1,855			
Methoxypropylamine	1									
Methyl chloride	3	35.24		550			1,800			2,525
Methyl chloroform	1	0.27416025					9,600			18,000
Methyl iodide	1	0.00228					76			220
Methyl isobutyl ketone	2	0.00365911					2,080			24,000
Methyl methacrylate	7	13.0100768		240	260	1,760	7,872			18,738
Methyl parathion	1	0.5	2.75	5.72	8.9		6.01			3
Methylacrylamide	1	0.14								
Methylene bisthiocyanate	1	0.00315								
Methylnaphthalene	1	4.631708								
Methylphosphoric acid	2	0.2								
Metolachlor	11	0.057604	0.00335				2,200			148
MMT	2	0.6					8			8
m-nitro-*p*-toluidine	1						6,860			
Molybdenum trioxide	1	0.1					2,689			976
Monomethylamine	3	14.2					100			1,858
Morpholine	3	0.00429264	380	350		100	1,450			8,000
MTBE	3	0.02952885					4,000			23,576
Naphtha	47	33.3951287		3	4.9	1.5	5,000			
Naphthalene	5	0.8	4		6.14	4.6	490			64
Naphthalenesulfonic acid	1	0.2	420							
n-butylamine	1	0.148	180		268		366		263	
Neopentylglycol	1	0.22								
Nickel carbonate	1	3.12								
Nickel carbonyl	4	0.28721856						35		
Nickel sulfate	5	0.2743808	33.97	188						
Nitric acid	134	23.79410482								
Nitrilotriacetic acid	1	0.454	98	198	100	950	1,100			
Nitrobenzene	1	0.00006384	0.002	43	80	27				556

Nitrocellulose	6	0.023625					5,000			
Nitrogen	32	213.8722559								
Nitrogen dioxide	2	4.467698								88
Nitrohydrochloric acid	1	0.02								
NO_x	4	0.08753452								
Oleum	16	99.22728524		24				347		
Ortho-dichlorbenzene	3	0.023751	1.58		33		500			
Oxygen	7	194.0841								
Paraquat	4	5.02	15				100			
Pcbs	367	21.46503764	0.054			0.1	1,350			
Pentachlorophenol	17	40.10874	1	0.13	0.32	0.53	27			32
Pentane	1									121,325
Perchloroethylene	7	0.07360975			18.4		2,629			4,961
Permethrin	1		0.025				383			30
Phenol	47	287.7430848	8.3	20	45	4	317			81
Phenolsulfonic acid	7	21.20865					1,900			
Phenylisocyanate	1	0.00219								
Phosgene	1	0								
Phosphoric acid	66	64.3174429					1,530	209		
Phosphorus	11	0.0212			3					
Phosphorus pentoxide	2	71.4						206		
Phthalic anhydride	3	2.256906	44.2				1,530	34		
Picloram	8	1.34628052	11.6				8,200			
Picric acid	2	0					200			
Piperonyl butoxide	1	0.000126	5	0.0042		2.83	6,150			
Potassium acetate	2	0.258265					3,250			
Potassium amylxanthate	1						99			
Potassium carbonate	4	0.49774					1,870			
Potassium chloride	22	8191.685796		1,200		14.5	2,600			
Potassium chromate	3	0.0010206			45.6	0.18	180			
Potassium cyanide	1	0.0001672	0.064	0.14	0.34		5			
Potassium hydroxide	25	13.73254924					273			
Potassium nitrate	1	0.2		800		10.2	3,750			
Potassium permanganate	6	0.016504	0.45	1.2			1,090			
Potassium sulfide	1	0.000168								
Propanil	1	0.00042			8.6	4.2	367			
Propargyl alcohol	1	0.002			1.53		20		873	
Propyl acetate	3	2.3075305			60	511	9,370			
Propylene	1	0.002								
Propylene glycol	17	18.60701244	10,000	1,700						
Propylene oxide	3	3.605524		141			380			4,000
Pyridine	5	0.015	4.6		105	1,320.7	891	8,824		
Rhodamine B	2	0.000168	217				887			
Silane	1									9.600
Silica	4	9.3119232								
Silicic acid	1	4.5								
Simazine	1	0.000272					971			1,169
Sodium	4	12.00261								
Sodium aluminate	2	1.5								

TABLE 44.2 The Spilled Substances and Their Toxicity (*Continued*)

Name	Number of spills	Spill amount (tons)	Aquatic toxicity (in ppm)				Mammalian toxicity (in ppm)			
			Rainbow trout	Bluegill	Fathead minnow	*Daphnia magna*	Orl-rat LD_{50} (mg/kg)	Ihl-rat LC_{50} 1 hour	Ihl-rat LC_{50} 2 hour	Ihl-rat LC_{50} 4 hour
Sodium azide	1		0.8	1.6			27			
Sodium bisulfite	10	242.9140045				90	2,000			
Sodium borohydride	2	0.0642								
Sodium carbonate	9	1.95612088		320		524	2,050			
Sodium chlorate	52	134.2943894	2,750				1,200	6,328		
Sodium chloride	35	4058.703523		12,946	7,650	3,114	3,000	17,288		
Sodium chlorite	3	265.4494					165			61
Sodium chromate	6	0.05156645	600			0.05				
Sodium cyanide	11	49.38545	0.06	0.28	0.15	0.09	6.44			
Sodium dichromate	21	20.94946024	69	410		4	50			
Sodium dithionite	2	12.1								
Sodium ethyl sulfate	1	0.4								
Sodium ferrocyanide	1	0.0040824				540				
Sodium hydrosulfite	19	8.14583559								
Sodium hydroxide	339	3964.468778		9.9		100				
Sodium hypochlorite	78	50.009554					5,800			
Sodium hyposulfite	1	0.00674								
Sodium isopropylxanthate	2	0.165								
Sodium metabisulfite	1	0.056					2,000			
Sodium nitrite	13	47.74242								
Sodium pentachlorophenate	8	3.9328		0.16			126		20	
Sodium phosphate	4	0.079912				126				
Sodium silicate	9	1.1975834				49.6	1,153			
Sodium stearate	2	0.052652								
Sodium sulfate	17	93.851647		8,000		800	5,989			
Sodium sulfide	10	0.8838013		25	1.38	9				
Sodium sulfite	5	209.71452				52	820			
Sodium thiocyanate	3	45.5600351					764			
Sodium thiosulfate	1	0.64				805				
Sorbitol	1	0.1					15,900			
Stannic chloride	1	0.155								
Strychnine	1	0.00544		0.87			2.35			
Styrene monomer	36	74.60043097	2.5	25.05	30	40				
Sulfamic acid	2	2.8147032			42		3,160			
Sulfonic acid	2	0.00294								
Sulfur	114	6612.294188		10,000						1,246
Sulfur chloride	1	0.000676								150
Sulfur dioxide	24	6.40805512						2,520		

Sulfuric acid	506	5310.450003		42			2,140		125	
Sulfurous acid	2	0.10336272								
Tannic acid	1	0.0000198					2,260			
TCA	2	0.066			2,000	3,100	0.2			
Tebuthiuron	2	0.001					644			389
Terbufos	1	0.22					1.6			
Terephthalic acid	2	0.062514					6,400			
Tert-butyl peroxide	2	0.011					25,000			4,100
Tetrabromomethane	1									
Tetraethyl lead	3	3.07		0.23			12.3	63		
Tetrahydrofuran	3	0.1426956			2,160	10,000	1,650			21,000
Tetrahydronaphthalene	1	0.1					1.62			275
Thioglycolic acid	2	0.1			30		114			
Tin fluoborate	1	1								
Titanium dioxide	5	3.13744								
Ttitanium tetrachloride	2	0.000016								
Toluene	58	100.6274276	0.02		36.2		636			12,791
Toluene diisocyanate	11	2.4065432			164.5		5,800			14
Toluene sulfonic acid	1	0.00047775								
Treflan	1	0.014					1,930	200		
Triallate	3	0.07					800			
Trichlorfon	2	7.587	4.85	3.8	109		560			121
Trichloroethane	27	1.026632284		40	60	40	9,600			18,000
Trichloroethylene	6	1.4468872	42	45	52	58.1	5,650			8,450
Trichlorotrifluoroethane	2	0.014508								
Triclopyr	2	0.000056					630			
Tricresyl phosphate	2	0.06279448	0.26	1,000			3,000			
Triethanolamine	1	0.4			11,800	1,390	4.92			
Trifluralin	13	0.3736	0.12	0.12	0.105	0.193	1,930	200		
Trimethyl phosphate	2						840			
Trimethylamine anhydrous	1	0.0006356					500			7,729
Triphenyl phosphite	1	0.0035					444			
Urea	25	138.674167				10,000	8,471			
Urea formaldehyde	6	14.62965144								
Urea nitrate	12	1.6451712								
Vanadium pentoxide		0.09	62				10			17
Vinyl acetate	19	68.78090788		18	29	330	2,900			3,185
Vinyl chloride	42	84.38357011					500			
Viscose (cellulose xanthate)	8	60.32								
Xanthate	3	0.07182								
Xylenes	42	82.10627652	14	19	40	0.233				
Zinc chloride	4	0.241186475	4	7		0.06791	350			
Zinc oxide	1	0.088614	6				7,950			
Zinc phosphate	1	0.02726636	0.09				1,990			
Zinc stearate	1	0.0438					10,000			
Zinc sulfate	6	2039.34151	2.08	3.3	14		2,000			
Zirconium	1			7						

TABLE 44.3 The Top-Priority Chemicals

Name	Priority number	Name	Priority number
Sulfuric acid	1	Pentachlorophenol	50
PCBs	2	Halons	51
Ammonium hydroxide	3	Vinyl acetate	52
Chlorine	4	Diazinon	53
Hydrochloric acid	5	Isophthalic acid	54
Ammonia	6	Chlorpyrifos	55
Sodium hydroxide	7	Atrazine	56
Toluene	8	Malathion	57
Ethylene glycol	9	Oleum	58
Ammonium nitrate	10	Sodium hydrosulfite	59
Sulfur	11	Calcium hydroxide	60
Methanol	12	MDI	61
Phosphoric acid	13	Sodium sulfate	62
Sodium chlorate	14	Propylene glycol	63
Nitric acid	15	Hexamethylenediamine	64
Mercury	16	Trifluralin	65
Phenol	17	Sodium cyanide	66
Ethylbenzene	18	Creosote	67
Sodium hypochlorite	19	Carbofuran	68
Benzene	20	Hydrazine	69
2,4-D	21	Diethanolamine	70
Calcium chloride	22	Sodium nitrite	71
Carbon disulfide	23	Hydrogen chloride	72
Formaldehyde	24	Metolachlor	73
Ferric chloride	25	Acetic anhydride	74
Freons	26	Copper sulfate	75
Aluminum sulfate	27	Toluene diisocyanate	76
Ethanol	28	Sodium pentachlorophenate	77
Naphtha	29	Bacillus thuringiensis	78
Vinyl chloride	30	Hydrogen	79
Chromic acid	31	Sodium bisulfite	80
Xylenes	32	Ferric sulfate	81
Hydrogen peroxide	33	Asbestos	82
Sodium chloride	34	Urea nitrate	83
Styrene monomer	35	Chlorine dioxide	84
Maleic anhydride	36	Carbon black	85
Calcium oxide	37	Ethanolamine	86
Nitrogen	38	Dimethoate	87
Potassium hydroxide	39	Magnesium sulfate	88
Calcium carbonate	40	Aluminum chloride	89
Trichloroethane	41	Arsenic	90
Isopropanol	42	Ethyl acrylate	91
Glyphosate	43	Hydrofluoric acid	92
Urea	44	Acetic acid	93
Ammonium sulfate	45	Sodium carbonate	94
Sodium dichromate	46	Butyl alcohol	95
Calcium hypochlorite	47	Sodium sulfide	96
Sulfur dioxide	48	Fenitrothion	97
Potassium chloride	49	Sodium silicate	98

TABLE 44.3 The Top-Priority Chemicals (*Continued*)

Name	Priority number	Name	Priority number
Hydrofluorosilicic acid	99	Hydrogen sulfide	149
Zinc sulfate	100	Monomethylamine	150
Phosphorus	101	Cresol	151
Cyanide	102	Propylene oxide	152
Picloram	103	Aminocarb	153
Dioctyl phthalate	104	Sodium chromate	154
Fluorescein	105	Ferric nitrate	155
Ferrous chloride	106	Methyl chloride	156
Viscose (cellulose xanthate)	107	Busan (2-bromo-4′-hydroxyacetophenone)	157
Dimethylamine	108		
Ethylamine	109	Caprolactam	158
Naphthalene	110	Nickel sulfate	159
Phenolsulfonic acid	111	Magnesium chloride	160
Methyl methacrylate	112	Chloroform	161
Cyclohexylamine	113	Benzoyl chloride	162
Formic acid	114	Lead sulfate	163
Ethylene oxide	115	Arsenic trioxide	164
Ethylenediamine	116	Fluorine	165
Oxygen	117	Sodium thiocyanate	166
Barium sulfate	118	Methyl parathion	167
MCPA	119	Titanium dioxide	168
Acrylonitrile	120	Benzenesulfonic acid	169
Paraquat	121	Nitrocellulose	170
Trichloroethylene	122	Butylate	171
Tetraethyl lead	123	Butyl acetate	172
Sodium sulfite	124	Furfural	173
Potassium permanganate	125	Ortho-dichlorbenzene	174
Pyridine	126	Ethylene dibromide	175
Perchloroethylene	127	Dithiocarbamate	176
Diclofop-methyl	128	Calcium carbide	177
Acrylamide	129	EPTC	178
Azinphos-methyl	130	Ethylene glycol monobutyl ether	179
Ferric oxide	131		
Sodium chlorite	132	Chlorophenol	180
Avenge	133	Dimethyl disulfide	181
Dicamba	134	MMT	182
Carboxin	135	Nickel carbonyl	183
Carbon dioxide	136	Acetone	184
Zinc chloride	137	Ammonium thiosulfate	185
Phthalic anhydride	138	Potassium carbonate	186
Isocyanates	139	Hydrocyanic acid	187
Ethyl acetate	140	Cresylic acid	188
Urea formaldehyde	141	Sodium	189
Carbon monoxide	142	Silica	190
Acetonitrile	143	Acrylic acid	191
Ammonium chloride	144	Dioxin	192
Diethylamine	145	Vanadium pentoxide	193
Carbaryl	146	Phosphorus pentoxide	194
2-ethylhexanol	147	Aluminum oxide	195
Trichlorfon	148	Butyl acrylate	196

TABLE 44.3 The Top-Priority Chemicals (*Continued*)

Name	Priority number	Name	Priority number
Methoxychlor	197	Diisobutylene	246
Dichloromethane	198	Terbufos	247
n-butylamine	199	Calcium nitrate	248
Morpholine	200	Copper chloride	249
Acephate	201	Diammonium sulfate	250
1,3-butadiene	202	Chlordane	251
TCA	203	Acetylene	252
Dinitrotoluene	204	Barium carbonate	253
Hydrogen fluoride	205	Diphenyl oxide	254
Potassium chromate	206	Sodium dithionite	255
Propyl acetate	207	1,4-dioxane	256
Tetrahydrofuran	208	Rhodamine B	257
Bensulide	209	Butylene	258
Copper cyanide	210	Butyl cellosolve	259
Thioglycolic acid	211	Dimethyl sulfide	260
Diethylene glycol	212	Tebuthiuron	261
Triallate	213	Zinc phosphate	262
Diphenylmethane	214	MTBE	263
Calcium	215	Dimethylacetamide	264
Tricresyl phosphate	216	Mercuric iodide	265
Cygon	217	Bromoxynil	266
Argon	218	Lead oxide	267
Permethrin	219	Diisopropylamine	268
Maneb	220	Methyl iodide	269
Potassium cyanide	221	Lead acetate	270
Inorganic chromate	222	Xanthate	271
Aluminum fluoride	223	Magnesium bisulfate	272
Tetrahydronaphthalene	224	Isobutyl alcohol	273
Benzoic acid	225	Aluminum alkyl compounds	274
Sodium phosphate	226	Cadmium oxide	275
Barium	227	Potassium acetate	276
NO_X	228	Picric acid	277
Diisocyanate	229	Sodium aluminate	278
Sulfamic acid	230	Dacthal	279
Lithium metal	231	2-thioethanol	280
Butadiene dimer	232	Benzaldehyde	281
Magnesium oxide	233	Iron sulfite	282
Diquat	234	Butyl methacrylate	283
Triethanolamine	235	Diacetone	284
Aldrin	236	Treflan	285
Chlorobenzene	237	Isodecanol	286
Guthion	238	Adipic acid	287
Nitrogen dioxide	239	Bromadiolone	288
Demeton	240	Linuron	289
Metachlor	241	Triclopyr	290
Nitrilotriacetic acid	242	Fluoboric acid	291
Propargyl alcohol	243	Bromacil	292
Sodium azide	244	Zinc oxide	293
Strychnine	245	Dinoseb	294

TABLE 44.3 The Top-Priority Chemicals (*Continued*)

Name	Priority number	Name	Priority number
Fosamine ammonia	295	Bentazon	345
Boric acid	296	Sodium stearate	346
Isovaleric acid	297	1,2-propanediamine	347
Isopropane	298	Mecoprop	348
Lead chloride	299	Cymenes	349
Carbon tetrachloride	300	Simazine	350
Chlorine trifluoride	301	Heptene	351
Amitrole	302	Sodium metabisulfite	352
Hydrogen sulfate	303	Dinitrophenol	353
Glycidol	304	Decanol	354
Methyl isobutyl ketone	305	Piperonyl butoxide	355
Nitrobenzene	306	Tin fluoborate	356
Manganese	307	Ethyl mercaptan	357
9,10-anthraquinone	308	Ethyl hexyl phthalate	358
Acetaldehyde	309	1,1-dichloroethane	359
Ethylene	310	Copper chromium arsenate	360
Trimethyl phosphate	311	Ethyl nitrite	361
Fumaric acid	312	Trichlorotrifluoroethane	362
Propanil	313	Isothiazol	363
Potassium nitrate	314	Methyl chloroform	364
Barium chlorate	315	Sodium thiosulfate	365
Hexamethylenetetramine	316	Mercuric chloride	366
Dimethylcyclohexylamine	317	Dicyanodiamide	367
Terephthalic acid	318	1-methyl-2-pyrrolidone	368
Carbamate	319	Chlorothalonil	369
Molybdenum trioxide	320	Antimony trioxide	370
Methylphosphoric acid	321	Ammonium persulfate	371
Methylnaphthalene	322	Sodium ethyl sulfate	372
Silicic acid	323	Sulfonic acid	373
Naphthalenesulfonic acid	324	Hexanoic acid	374
Dalapon	325	Neopentylglycol	375
Sodium isopropylxanthate	326	Ammonium bisulfite	376
Diphenyl	327	2-ethylhexanoic acid	377
Butyl mercaptan	328	Hexene	378
Ethanethiol	329	Stannic chloride	379
Chlorosulfonic acid	330	Sorbitol	380
Magnesium hydroxide	331	Methylacrylamide	381
Triphenyl phosphite	332	Chloroethylphosphonic acid	382
Nickel carbonate	333	Diisobutyl dithiophosphinate	383
Sulfurous acid	334	Cesium	384
Cesium chloride	335	Titanium tetrachloride	385
Trimethylamine anhydrous	336	Zinc stearate	386
Dimethyl terephthalate	337	Hydrobromic acid	387
Potassium amylxanthate	338	Helium	388
Sodium borohydride	339	Isoprene	389
Bromine	340	Diethylene triamine pentaacetic acid	390
Ferric phosphate	341		
Lead chromate	342	Isobutyl xanthate	391
Diisopropanolamine	343	Calcium sulfate	392
Tert-butyl peroxide	344	Sulfur chloride	393

TABLE 44.3 The Top-Priority Chemicals (*Continued*)

Name	Priority number	Name	Priority number
Tannic acid	394	Silane	416
Dibutyltin compounds	395	Potassium sulfide	417
Benzene phosphorus dichloride	396	Isophoronediamine	418
		1-decene	419
2-ethylpyridine	397	Diisopropyl ether	420
Zirconium	398	Dibutyl-p-phenylene	421
1-chloro-4-hexannone	399	Pentane	422
Nitrohydrochloric acid	400	Dimethylphosphine	423
Diundecyl phthalate	401	3-methoxypropylamine	424
Dimethyl disulfonate	402	8-hydroxyquinoline benzoate	425
Sodium hyposulfite	403	Aluminum phosphide	426
Dibenzoyl peroxide	404	Ammonium bisulfate	427
Ammonium xylenesulfoanate	405	Ammonium phosphate	428
Magnesium	406	Calcium silicon	429
Sodium ferrocyanide	407	Carbon 14	430
Diisononyl phthalate	408	Heavy water	431
Methylene bisthiocyanate	409	Isobutyl isocyanate	432
Phenyl isocyanate	410	MAPP	433
Propylene	411	Methane	434
2-methyl-5-ethylpyridine	412	Methoxypropylamine	435
Dodecylbenzene	413	Phosgene	436
m-nitro-p-toluidine	414	Tetrabromomethane	437
Toluene sulfonic acid	415		

TABLE 44.4 The Top-Priority Materials

The top 10 chemicals	
Ammonia	Hydrochloric acid
Ammonium hydroxide	PCBs
Ammonium nitrate	Sodium hydroxide
Chlorine	Sulfuric acid
Ethylene glycol	Toluene
The top 50 chemicals (next 40 chemicals)	
2,4-D	Methanol
Aluminum sulfate	Naphtha
Ammonium sulfate	Nitric acid
Benzene	Nitrogen
Calcium carbonate	Pentachlorophenol
Calcium chloride	Phenol
Calcium hypochlorite	Phosphoric acid
Calcium oxide	Potassium chloride
Carbon disulfide	Potassium hydroxide
Chromic acid	Sodium chlorate
Ethanol	Sodium chloride
Ethylbenzene	Sodium dichromate
Ferric chloride	Sodium hypochlorite
Formaldehyde	Styrene monomer
Freons	Sulfur
Glyphosate	Sulfur dioxide
Hydrogen peroxide	Trichloroethane
Isopropanol	Urea
Maleic anhydride	Vinyl chloride
Mercury	Xylenes
The top 150 Chemicals (next 100 chemicals)	
2-ethylhexanol	Carboxin
Acetic acid	Chlorine dioxide
Acetic anhydride	Chlorpyrifos
Acetonitrile	Copper sulfate
Acrylamide	Creosote
Acrylonitrile	Cyanide
Aluminum chloride	Cyclohexylamine
Ammonium chloride	Diazinon
Arsenic	Dicamba
Asbestos	Diclofop-methyl
Atrazine	Diethanolamine
Avenge	Diethylamine
Azinphos-methyl	Dimethoate
Bacillus thuringiensis	Dimethylamine
Barium sulfate	Dioctyl phthalate
Butyl alcohol	Ethanolamine
Calcium hydroxide	Ethyl acetate
Carbaryl	Ethyl acrylate
Carbofuran	Ethylamine
Carbon black	Ethylene oxide
Carbon dioxide	Ethylenediamine
Carbon monoxide	Fenitrothion

TABLE 44.4 The Top-Priority Materials (*Continued*)

The top 150 Chemicals (next 100 chemicals)	
Ferric oxide	Phthalic anhydride
Ferric sulfate	Picloram
Ferrous chloride	Potassium permanganate
Fluorescein	Propylene glycol
Formic acid	Pyridine
Halons	Sodium bisulfite
Hexamethylenediamine	Sodium carbonate
Hydrazine	Sodium chlorite
Hydrofluoric acid	Sodium cyanide
Hydrofluorosilicic acid	Sodium hydrosulfite
Hydrogen	Sodium nitrite
Hydrogen chloride	Sodium pentachlorophenate
Hydrogen sulfide	Sodium silicate
Isocyanates	Sodium sulfate
Isophthalic acid	Sodium sulfide
Magnesium sulfate	Sodium sulfite
Malathion	Tetraethyl lead
MCPA	Toluene diisocyanate
MDI	Trichlorfon
Methyl methacrylate	Trichloroethylene
Metolachlor	Trifluralin
Monomethylamine	Urea formaldehyde
Naphthalene	Urea nitrate
Oleum	Vinyl acetate
Oxygen	Viscose (cellulose xanthate)
Paraquat	Zinc chloride
Perchloroethylene	Zinc sulfate
Phenolsulfonic acid	
Phosphorus	

Chemical		2000 ESD	1997 NPRI	1990 EPS	1981 EPS	MIAC 1996	CEPA 2000	CEPA List II	TDG Sch. 2	TDG Sch. 12	Spilled in Canada	ARET Lists	EPA 1985	EPA 1980	U.S. Railroad	U.S. Reportable	USCG Spill	EPA Ex Danger	U.S. top 100	AAR Hazard.	SARA ex. Dang.	CERCLA Haz.	SARA Ex. Toxic	RCRA Haz.	Tox & Dis Reg	OSHA 29CFR
(2-Chlorophenyl)thiourea																x					x	x		x		
(bis(2-chloroethyl)amino)uracil, 5-			Z													x										
(p-di(chloroethyl)amino) phenyl-L-alanine, 3-																x						x		x		
(Trifluoromethyl)benzenamine, 3-																		x			x					
Acenaphthene													x	x		x						x			x	
Acenaphthylene													x	x		x						x				
Acetaldehyde		x	Y		x		x	x	x			B3			x	x	x		x	x		x	x	x		
Acetamide												B3				x							x			
Acetic acid	Z		x	x			x	x		x					x	x	x		x	x		x				
Acetic anhydride	Z		x	x				x		x					x	x	x		x	x		x				
Acetone		x	x	x			x	x		x					x	x	x		x	x		x	x	x	x	
Acetone cyanohydrin			Y					x	x							x		x	x		x	x		x		
Acetonitrile	Z	x						x							x	x	x					x	x	x		
Acetophenone			Z													x						x		x		
Acetyl bromide								x								x						x				
Acetyl chloride			Z					x	x							x	x					x		x		
Acetylene			x	x	x			x	x	x																
Acrolein			Y				x	x	x			B3	x	x		x	x	x			x	x	x	x	x	
Acrylamide	Z	x	Y				x	x				B3				x		x			x	x	x	x		
Acrylic acid		x						x							x	x	x		x			x	x			
Acrylonitrile	Z	x	x	x			x	x	x	x		B3	x	x	x	x	x	x	x	x	x	x	x	x		
Adipic acid			Z	x				x								x			x			x				
Adiponitrile			Z					x									x	x			x					
Aldrin			x							x			x	x		x		x			x	x	x	x	x	
Allyl alcohol		x	Y					x								x	x	x			x	x		x		
Allyl chloride		x	Z					x								x						x	x			
Allylamine								x										x			x					
Aluminum		x					x*																x		x	
Aluminum alkyl halides								x	x																	
Aluminum chloride	Z		Y	x				x																		
Aluminum phosphate			Z							x													x			
Aluminum phosphide								x	x							x		x			x	x		x		
Aluminum sulfate	Y		x	x				x		x						x	x		x			x				

Chemical		2000 ESD	1997 NPRI	1990 EPS	1981 EPS	MIAC 1996	CEPA 2000	CEPA List II	TDG Sch. 2	TDG Sch. 12	Spilled in Canada	ARET Lists	EPA 1985	EPA 1980	U.S. Railroad	U.S. Reportable	USCG Spill	EPA Ex Danger	U.S. top 100	AAR Hazard.	SARA ex. Dang.	CERCLA Haz.	SARA Ex. Toxic	RCRA Haz.	Tox & Dis Reg	OSHA 29CFR
Aminobiphenyl, 4-							x					B3				x							x		x	
Aminocarb			x	x						x																
Aminopyridine, 4-			Z																		x	x		x		
Amiton			Z															x			x					
Amitrole																x						x	x	x		
Ammonia	x	x	x	x	x		x	x	x	x					x	x	x	x	x	x	x	x	x		x	
Ammonia solutions					x			x	x						x											
Ammonium acetate								x								x						x				
Ammonium benzoate								x								x						x				
Ammonium bicarbonate								x								x						x				
Ammonium bisulphite																x						x				
Ammonium carbamate								x								x						x				
Ammonium carbonate								x								x						x				
Ammonium chlorate								x																		
Ammonium chloride	Z		x	x			x	x		x						x						x				
Ammonium chromate								x								x						x				
Ammonium citrate								x								x						x				
Ammonium dichromate								x								x						x				
Ammonium fluoride								x								x						x				
Ammonium hydrogen fluoride								x								x						x				
Ammonium hydrogen sulfate								x																		
Ammonium hydroxide	x		x							x						x	x					x				
Ammonium nitrate	x		x	x				x	x	x									x	x			x			
Ammonium nitrate fertilizers								x	x																	
Ammonium oxalate								x							x	x						x				
Ammonium phosphates			x	x						x																
Ammonium picrate			Z					x	x							x						x		x		
Ammonium sulfamate			Z					x								x	x					x				
Ammonium sulfate	Y		x	x						x									x				x			
Ammonium sulfide								x							x	x										
Ammonium sulfite								x								x										
Ammonium tartrate								x								x						x				
Amyl acetate								x								x						x				

Aniline		x	x					x		x	B3			x	x	x	x	x	x	x	x	x	x	
Aniline hydrochloride								x																
Anisidine								x			B2				x							x		
Anthracene		x	Z								B1	x	x		x						x	x		x
Antimony		x*					x*					x	x*		x						x	x		x
Antimony pentachloride								x							x						x			
Antimony pentafluoride								x	x								x			x				
Antimony potassium tartrate								x							x						x			
Antimony trichloride								x							x						x			
Antimony trifluoride															x						x			
Antimony trioxide			Y					x							x						x			
Antimycin A																	x			x			x	
Antu																	x			x	x		x	
Arsenic	Z	x*	Z					x				x	x*		x						x	x	x	x
Arsenic acid			Y					x						x	x						x		x	x
Arsenic pentoxide			Z					x							x		x			x	x		x	
Arsenic trichloride								x							x					x	x			
Arsenic trioxide			Z					x						x	x	x								x
Arsenous oxide																	x			x	x		x	
Arsine			Y	x	x			x	x	x							x			x				x
Asbestos	Z	x	Y			x		x		x	B2	x	x		x						x	x		x
Atrazine	Z		x							x														
Auramine															x						x	x	x	
Avenge	Z																							
Azaserine															x						x		x	
Azinphos ethyl			Y														x			x				
Azinphos methyl	Z		x					x	x	x							x			x	x			
Bacillus thuringiensis	Z																							
Barban			Z							x					x									
Barium								x														x	x	x
Barium cyanide								x							x						x		x	
Barium sulfate	Z		Y	x						x														
Benz[c]acridine															x						x		x	
Benzal chloride															x		x			x	x	x	x	
Benzene	Y	x	x	x	x	x		x		x	B3	x	x	x	x	x		x	x		x	x	x	x
Benzenedicarboxylic acid anhydride, 1,2-																				x	x	x		
Benzenedicarboxylic acid, [bis(2-ethylhexyl)]ester, 1,2-																			x	x	x			
Benzenedicarboxylic acid, diethyl ester, 1,2-																			x	x	x			
Benzenesulfonyl chloride								x									x			x	x		x	
Benzidine			Z			x		x			B3	x	x								x	x	x	x
Benzo[a]anthracene											A1	x	x		x						x		x	x
Benzo[a]pyrene											A1	x	x		x									x
Benzo[b]fluoranthene											A1				x									x
Benzo[ghi]perylene											A1	x	x		x						x			
Benzo[k]fluoranthene											A1	x	x		x									

Chemical		2000 ESD	1997 NPRI	1990 EPS	1981 EPS	MIAC 1996	CEPA 2000	CEPA List II	TDG Sch. 2	TDG Sch. 12	Spilled in Canada	ARET Lists	EPA 1985	EPA 1980	U.S. Railroad	U.S. Reportable	USCG Spill	EPA Ex Danger	U.S. top 100	AAR Hazard.	SARA ex. Dang.	CERCLA Haz.	SARA Ex. Toxic	RCRA Haz.	Tox & Dis Reg	OSHA 29CFR
Benzoic acid			x	x				x		x						x						x				
Benzonitrile			Z					x								x						x				
Benzophenone			Z																							
Benzotrichloride								x								x		x			x	x	x	x		
Benzotrifluoride			x					x		x																
Benzoyl chloride		x	Z					x							x	x						x	x			
Benzyl chloride		x	Z				x	x				b2			x	x	x	x			x	x	x	x		
Benzyl butyl phthalate		x	Z				x						x	x		x						x	x		x	
Beryllium			Z					x	x			B2	x	x*		x						x	x	x	x	
Beryllium nitrate								x								x						x				
Biphenyl		x	Y				x			x						x										
bis(2-chloroethoxy) methane													x	x		x										
bis(2-ethylhexyl) phthalate	Z	x				x						B1	x	x		x		x			x	x	x	x	x	
bis(chloromethyl) ether						x						B3	x	x		x										
Boron trichloride								x	x									x			x					
Boron trifluoride								x	x									x			x					
Bromine			Z		x			x	x									x			x					
Bromine solutions					x			x	x																	
Bromo-4-phenoxybenzene, 1-													x	x		x										
Bromoacetone								x								x						x		x		
Bromodichloromethane													x	x		x						x	x		x	
Bromoform			x					x		x			x	x		x						x	x	x	x	
Brucine			Y					x								x						x		x		
Butadienes		x	Z	x			x	x	x			B3				x		x	x	x	x		x			
Butane					x			x	x																	
Butyl acetate							x	x							x	x	x					x				
Butyl acrylate		x						x									x		x				x			
Butyl alcohol	Z	x	x	x				x		x						x	x		x	x		x	x	x		
Butyl formate			Y					x		x																
Butyl vinyl ether								x										x			x					
Butyl-2,4,6-trinitro-m-xylene, 5-tert-							x	x																		
Butylamine			Z					x								x	x					x				
Butylene			Z	x				x	x										x							

Butylene oxide		x						x							x							x		
Butyl-N-nitroso-1-butanamine, N-															x						x	x	x	
Butyraldehyde		x	Z	x				x								x		x	x			x		
Butyric acid			Z					x							x	x					x			
C.I. acid green 3		x															x					x		
Cacodylic acid								x							x						x		x	x
Cadmium		x*	Y									x	x*		x						x	x	x	x
Cadmium chloride			Z												x		x							
Cadmium compounds: respirable & soluble inorganic forms				x							A2													
Calcium arsenate			Y					x		x					x		x			x	x			
Calcium carbide			Y	x				x	x	x				x	x				x		x			
Calcium carbonate	Y		Z	x																				
Calcium chloride	Y		x	x			x			x								x						
Calcium cyanamide		x					x	x																
Calcium cyanide			x				x	x	x	x					x						x		x	
Calcium hydroxide	Z		x	x						x						x								
Calcium hypochlorite	Y		x	x				x		x					x						x			x
Calcium manganese silicon								x																
Calcium nitrate			Y					x		x														
Calcium oxide	Y		x	x				x		x								x						
Calcium phosphate			x	x						x														
Camphechlor																	x			x	x	x	x	
Camphor oil			Z					x																
Cantharidin			Z														x			x				
Caprolactam			Z	x			x								x			x						
Captan															x						x	x		
Carbamic acid, ethyl ester															x						x	x	x	
Carbamidoselenic acid															x						x		x	
Carbaryl	Z		x	x						x					x						x	x		
Carbendazim															x									
Carbofuran	Z		x	x						x					x		x			x	x			
Carbon black	Z																	x						
Carbon dioxide	Z		Y	x				x		x								x	x					
Carbon disulfide	Y	x	Y	x			x	x	x					x	x		x		x	x	x	x	x	x
Carbon monoxide								x	x															
Carbon oxyfluoride															x						x		x	
Carbon tetrachloride		x	x	x		x		x		x	B2	x	x		x	x		x			x	x	x	x
Carbon-14							x*																	x
Carbonyl fluoride								x	x															
Carbonyl sulfide			Z					x	x						x									
Carboxin	Z																							
Catechol		x													x							x		
Cesium-137			Y				x*			x														x
Chlorambucil															x						x		x	

Chemical		2000 ESD	1997 NPRI	1990 EPS	1981 EPS	MIAC 1996	CEPA 2000	CEPA List II	TDG Sch. 2	TDG Sch. 12	Spilled in Canada	ARET Lists	EPA 1985	EPA 1980	U.S. Railroad	U.S. Reportable	USCG Spill	EPA Ex Danger	U.S. top 100	AAR Hazard.	SARA ex. Dang.	CERCLA Haz.	SARA Ex. Toxic	RCRA Haz.	Tox & Dis Reg	OSHA 29CFR
Chlordane			x	x				x	x	x			x	x		x	x	x			x	x	x	x	x	
Chlorine	x	x	x	x	x		x	x	x	x					x	x	x	x	x	x					x	
Chlorine dioxide	Z	x	Y				x	x		x		B3														
Chlornaphazine																x						x		x		
Chloro-3-methylphenol, 4-							x	x					x	x		x						x		x		
Chloroacetaldehyde								x								x		x			x	x		x		
Chloroacetic acid		x						x								x	x	x								
Chloroacetophenone								x								x							x			
Chloroacetyl chloride								x													x	x		x		
Chloroaniline			Y					x		x						x										
Chlorobenzene		x	Z					x					x	x		x	x					x	x	x	x	
Chlorodibromomethane												B2	x	x		x						x			x	
Chlorodifluoroethane								x	x																	
Chlorodifluoromethane			Z					x		x															x	
Chloroethane		x	x					x		x			x	x		x				x		x	x		x	
Chloroethyl vinyl ether, 2-			Z										x	x		x						x		x		
Chloroform		x	Y	x			x	x				B2	x	x		x	x	x			x	x	x	x	x	
Chloromethyl ether																		x			x	x	x	x		
Chloromethyl methyl ether						x										x		x			x	x	x	x		
Chloromethylbenzenamine hydrochloride																						x		x		
Chloronaphthalene													x	x		x						x		x		
Chlorophacinone			Z							x								x			x					
Chlorophenol			x					x		x			x	x		x						x		x	x	
Chlorophenyl phenyl ether, 4-							x						x	x		x										
Chloroprene								x								x							x			
Chloropropionitrile, 3-																x		x			x	x		x		
Chloropropyl octyl sulfoxide, 3-																		x			x					
Chlorosulfonic acid								x	x							x						x				
Chlorpyrifos	Z		Y													x						x			x	
Choline chloride			Z																							
Chromic acid	Y		x					x		x						x						x			x	
Chromic sulfate			Z					x								x										
Chromium		x*	Z			x*						B2	x	x*		x						x	x	x	x	
Chrysene												A1	x	x		x						x		x	x	

Coal tar							x																	x
Cobalt		x*					x*										x			x		x		x
Cobaltous bromide			Z					x							x						x			
Cobaltous formate								x							x						x			
Copper		x*					x*					x	x*		x						x	x		x
Copper chloride			Y					x							x						x			
Copper cyanide			x					x	x	x					x						x		x	
Copper diacetate			Z					x							x						x			
Copper nitrate			Z												x						x			
Copper oxide										x														
Copper sulfate	Z		x	x				x		x					x	x					x			
Copper tartrate								x							x						x			
Coumaphos			Y					x	x						x		x			x	x			
Creosote	Z					x		x						x	x									x
Cresols	Z	x*	x	x						x					x	x				x	x	x	x	x
Cresylic acid								x		x					x		x			x	x	x	x	
Crimidine			Z														x			x				
Crotonaldehyde			Z					x							x		x			x	x		x	
Cumene		x	Y				x								x		x	x						
Cyanides	Z	x					x			x	B2	x	x		x						x	x	x	x
Cyanogen								x	x						x						x		x	
Cyanogen bromide			Y					x		x					x		x			x	x		x	
Cyanogen chloride								x	x						x						x		x	
Cyclohexane		x	x	x	x			x		x				x	x	x		x	x		x	x	x	
Cyclohexanone			Z				x	x							x						x	x	x	
Cycloheximide			Y														x			x				
Cyclohexylamine	Z		Y					x		x							x			x				
Cyclopentane								x									x			x				
Cyclophosphamide															x						x		x	
Cyclopropane								x	x															
Cyclotrimethylenetrinitramine								x	x															x
Dalapon			Z													x					x			
Daunomycin															x						x		x	
DDD												x	x		x						x		x	x
DDE												x	x		x					x				x
DDT			Y									x	x		x						x		x	x
Diallate			Y							x					x						x	x	x	
Diaminotoluene		x																			x	x	x	
Diazinon	Z		x							x					x						x			x
Dibenzo[a,h]anthracene											A1	x	x		x						x		x	x
Dibenzo[c,g]carbazole, 7,H-											A1													
Dibenzofurans						x		x			A1				x		x					x		x
Dibenzopyrene, 1,2:7,8-											A1				x						x		x	
Diborane								x	x								x			x				

Chemical		2000 ESD	1997 NPRI	1990 EPS	1981 EPS	MIAC 1996	CEPA 2000	CEPA List II	TDG Sch. 2	TDG Sch. 12	Spilled in Canada	ARET Lists	EPA 1985	EPA 1980	U.S. Railroad	U.S. Reportable	USCG Spill	EPA Ex Danger	U.S. top 100	AAR Hazard.	SARA ex. Dang.	CERCLA Haz.	SARA Ex. Toxic	RCRA Haz.	Tox & Dis Reg	OSHA 29CFR
Dibromo-1-propanol phosphate, 2,3-			Z																			x	x	x		
Dibromo-3-chloropropane, 1,2-												B3				x						x	x	x	x	
Dibromomethane			Z				x	x								x						x	x	x		
Dibutyl ether			Z					x									x									
Dibutyl phthalate		x											x	x		x		x			x	x	x	x	x	
Dicamba	Z		x	x						x						x						x				
Dichlobenil			Z													x						x				
Dichlone			Y													x						x				
Dichloromethyl ether																					x	x	x	x		
Dichloro-(1,1′-biphenyl)-4,4′-diamine, 3,3′-																					x	x	x			
Dichlorobenzene		x		x				x				A2	x	x		x						x	x	x	x	
Dichlorobenzidine						x						B1	x	x		x									x	
Dichlorodifluoromethane			Z					x					x	x		x						x		x		
Dichloroethane, 1,1-								x				B2	x	x		x						x		x	x	
Dichloroethane, 1,2-		x	x	x	x	x		x		x			x	x	x	x			x			x	x	x	x	
Dichloroethyl ether			Z									B2	x	x		x		x			x	x	x	x	x	
Dichloroethylene, 1,1-		x					x	x					x	x	x	x	x			x		x	x	x		
Dichloroethylene, 1,2-							x	x					x	x	x	x						x		x	x	
Dichloroisopropyl ether								x					x	x		x										
Dichlorophenol, 2,4-		x	Z				x					B3	x	x		x						x	x	x	x	
Dichlorophenol, 2,6-																x						x				
Dichlorophenoxyacetic acid, 2,4-	Y		x	x						x						x	x					x	x	x	x	
Dichlorophenyl isocyanate, 3,4-								x	x									x								
Dichloropropane, 1,1-			Z													x						x				
Dichloropropane, 1,2-		x					x						x	x		x	x					x	x	x	x	
Dichloropropene, 1,2-													x	x		x										
Dichloropropene, 1,3-							x					B3				x						x	x	x	x	
Dichlorvos			x							x						x		x			x	x	x		x	
Diclofop-methyl	Z																									
Dicrotophos			Y															x			x					
Dieldrin			x							x			x	x		x						x		x	x	
Diepoxybutane, 1,2:3,4-																x		x			x	x	x	x		
Diethanolamine		x	x							x						x	x						x			

Diethyl phthalate		x	Z								x	x		x				x	x	x	x
Diethyl sulfate		x					x							x			x				
Diethyl sulfide							x														
Diethyl-p-nitrophenyl phosphate			Z											x				x		x	
Diethylamine	Z		x				x		x				x	x	x			x			
Diethylhydrazine, N,N′-														x				x		x	
Diethylnitrosoamine														x				x	x	x	
Diethylstilbestrol														x				x		x	
Dimefox			Z											x		x	x				
Dimethoate	Z		Z											x		x	x	x		x	
Dimethoxy-(1,1′-biphenyl)-4,4′-diamine, 3,3′-																	x	x	x		
Dimethyl aminobenzene														x				x	x	x	
Dimethyl ether				x			x	x													
Dimethyl ethyl amine			Z						x												
Dimethyl hydrazine																x	x	x	x	x	
Dimethyl phthalate		x	Z								x	x		x		x	x	x	x	x	
Dimethyl sulfate		x	Z			x	x							x		x	x	x	x	x	
Dimethyl sulfide							x									x	x				
Dimethyl-(1,1′-biphenyl)-4,4′-diamine, 3,3′-																	x	x	x		
Dimethyl-1,2-benzanthracene, 7,12-										B1				x				x		x	
Dimethylamine	Z		x			x	x	x	x					x	x			x		x	
Dimethylaniline, N,N-		x	Z				x		x					x							
Dimethylbenzy-1-hydroperoxide, α,α-														x				x	x	x	
Dimethylcarbamoyl chloride							x							x				x	x	x	
Dimethyldichlorosilane							x	x								x		x			
Dimethylformamide						x	x							x							
Dimethylhydrazine, 1,1-							x	x						x							
Dimethylhydrazine, 1,2-							x	x						x				x		x	
Dimethylphenol, 2,4-											x	x		x							x
Dinitroaniline			Z				x														
Dinitrobenzene			Z				x	x						x				x			x
Dinitrocresols		x	Y			x	x			B2	x	x		x		x	x	x	x	x	x
Dinitrophenol							x	x						x				x			
Dinitrophenol, 2,4-			Y			x					x	x		x				x	x	x	x
Dinitrotoluene		x					x					x		x				x			x
Dinitrotoluene, 2,4-		x	Z							B3	x			x				x	x	x	x
Dinitrotoluene, 2,6-		x								B3	x			x				x	x	x	x
Dinoseb			x	x					x					x		x	x	x		x	
Dioxane		x			x	x	x			B2				x				x	x	x	x
Dioxolane							x									x	x				
Diphacinone			Z													x	x				
Diphenylamine			Z			x			x												
Diphenylhydrazine, 1,2-										B3	x	x		x				x	x	x	x
Diphenylmethane-4,4′-diisocyanate			Z	x		x								x					x		

Chemical	2000 ESD	1997 NPRI	1990 EPS	1981 EPS	MIAC 1996	CEPA 2000	CEPA List II	TDG Sch. 2	TDG Sch. 12	Spilled in Canada	ARET Lists	EPA 1985	EPA 1980	U.S. Railroad	U.S. Reportable	USCG Spill	EPA Ex Danger	U.S. top 100	AAR Hazard.	SARA ex. Dang.	CERCLA Haz.	SARA Ex. Toxic	RCRA Haz.	Tox & Dis Reg	OSHA 29CFR
Dipropylamine								x							x						x		x		
Dipropylaminobenzenediazonium zinc chloride, 4-						x																			
Diquat			x							x					x						x				
Disulfoton			Y					x	x						x		x			x	x		x	x	
Dithiobiuret			Z												x		x			x	x		x		
Diuron			Z												x						x				
Endosulfan			Y									x	x		x		x			x	x		x	x	
Endosulphan sulfate												x			x									x	
Endothall			Y												x						x		x		
Endrin			x									x	x		x		x			x	x		x	x	
Epichlorohydrin		x	Z				x	x			B3				x	x	x	x		x	x	x	x		
EPN			Z														x			x					
Ethanethioamide															x						x	x	x		
Ethanolamine	Z		Z	x				x										x							
Ethion			Y					x	x						x	x	x			x	x			x	
Ethoprophos																	x			x					
Ethoxyethanol, 2-		x					x	x							x							x			
Ethyl 4,4′-dichlorobenzilate																					x	x	x		
Ethyl acetate	Z		Z				x	x						x	x	x					x		x		
Ethyl acrylate	Z	x	Y				x	x		x				x	x	x			x		x	x	x		
Ethyl alcohol	Y		x	x			x	x		x	B3					x		x							
Ethyl chloroformate		x						x														x			
Ethyl mercaptan			Y				x	x		x															
Ethyl methacrylate								x							x						x		x		
Ethyl methanesulfonate															x						x		x		
Ethylamine	Z		Y					x	x	x					x	x					x				
Ethylbenzene	Y	x	x	x	x		x	x		x		x	x	x	x	x		x			x	x		x	
Ethylene		x	x	x	x			x	x	x								x	x			x			
Ethylene cyanohydrin																x				x	x				
Ethylene dibromide			Z	x			x	x			B3				x						x	x	x	x	
Ethylene glycol	x	x	x	x			x			x					x	x		x				x			
Ethylene oxide	Z	x	x	x	x		x	x	x		B2			x	x		x	x	x	x	x	x	x		
Ethylene thiourea		x	Z								B3				x										

Ethylenediamine	Z							x							x	x	x							
Ethylenediamine tetraacetic acid						x							x						x					
Ethyleneimine			Z					x	x						x		x			x	x	x	x	
Ethylhexanol	Z		x	x						x								x						
Ethyl-N-nitrosocarbamide, N-																					x	x	x	
Famphur															x						x		x	
Fenitrothion	Z		x	x						x							x			x				
Fensulphothion			Z														x			x				
Ferric ammonium citrate								x							x						x			
Ferric ammonium oxalate								x							x						x			
Ferric chloride	Y		x	x				x		x				x	x	x			x		x			
Ferric hydroxide			Z							x														
Ferric oxide	Z		Y							x														
Ferric sulfate	Z							x		x					x									
Ferrous chloride	Z														x						x			
Ferrous sulfate			Y					x		x					x	x					x			
Fluenetil			Z														x			x				
Fluoranthene			Z								A1	x	x		x									x
Fluoren-2-yl acetamine, N-9H-																					x	x	x	
Fluorene												x	x		x						x			x
Fluorine			Z		x			x	x						x		x			x	x		x	x
Fluoroacetamide			Z												x					x	x		x	
Fluoroacetic acid			Z					x												x				
Fluoroscein	Z																							
Fonophos			Y														x			x				
Formaldehyde	Y	x	x	x			x	x		x	B3			x	x	x	x	x	x	x	x	x	x	x
Formic acid	Z		x	x			x	x		x					x						x		x	
Formparanate															x		x			x				
Freon (as class)	Y																							
Fumaric acid			Z					x							x						x			
Furfural			Z				x	x							x	x								
Gasolines					x		x	x											x					
Glycerine			Z													x		x						
Glycidaldehyde								x							x						x		x	
Glyphosate	Y		Y							x														
Halons	Z					x																		
Heavy water			Y							x														
Heptachlor												x	x		x						x	x	x	x
Heptachlor epoxide												x			x						x			x
Hexachlorobenzene			Z			x		x			A1	x	x		x						x	x	x	x
Hexachlorobutadiene			Y				x	x				x	x		x							x		x
Hexachlorocyclohexane, alpha-											A1	x	x		x									x
Hexachlorocyclohexane, gamma-		x							x		A1	x	x		x		x			x	x	x	x	x
Hexachlorocyclohexane, mixed isomers										x	x		x									x		

Chemical		2000 ESD	1997 NPRI	1990 EPS	1981 EPS	MIAC 1996	CEPA 2000	CEPA List II	TDG Sch. 2	TDG Sch. 12	Spilled in Canada	ARET Lists	EPA 1985	EPA 1980	U.S. Railroad	U.S. Reportable	USCG Spill	EPA Ex Danger	U.S. top 100	AAR Hazard.	SARA ex. Dang.	CERCLA Haz.	SARA Ex. Toxic	RCRA Haz.	Tox & Dis Reg	OSHA 29CFR
Hexachlorocyclopentadiene		x	Z				x	x				B1	x	x		x		x			x	x	x	x	x	
Hexachloroethane		x	Z					x					x	x		x						x	x	x	x	
Hexachloronaphthalene							x											x			x		x			
Hexachlorophene			Y					x								x						x		x		
Hexaethyl tetraphosphate			Z					x								x						x		x		
Hexahydrobenzene																						x	x	x		
Hexamethylenediamine	Z							x											x	x						
Hexane			x	x				x		x						x	x		x						x	
Hexanoic acid								x	x																	
Hydrazine	Z	x	x	x				x	x	x		B3				x		x			x	x	x	x	x	
Hydrochloric acid	x	x	x	x	x		x	x	x	x					x	x	x	x	x	x	x	x	x			
Hydrocyanic acid		x	Y					x	x							x	x	x	x	x	x	x	x	x		
Hydrofluoric acid	Z	x	x	x	x			x	x	x					x	x	x	x		x	x	x	x	x		
Hydrofluorosilicic acid	Z																									
Hydrogen	Z		Z	x				x	x																	
Hydrogen chloride, anhydrous	Z				x			x	x											x						
Hydrogen fluoride, anhydrous					x			x	x																	
Hydrogen peroxide	Y		x	x			x	x	x	x					x		x	x		x	x					
Hydrogen selenide								x	x									x			x					
Hydrogen sulfide	Z		Y	x	x		x	x	x			B3				x		x			x	x		x	x	
Hydroquinone		x	Y					x								x		x			x		x			
Indeno(1,2,3-cd)pyrene												A1	x	x		x									x	
Isobutyl acetate								x							x							x				
Isobutyl alcohol			Z				x	x								x						x		x		
Isobutylamine								x							x							x				
Isobutylene								x	x										x							
Isobutyraldehyde		x						x															x			
Isobutyric acid								x								x						x				
Isobutyronitrile								x										x			x					
Isodrin			Y													x		x								
Isofluorophate																		x			x	x		x		
Isophorone			Z				x						x	x		x						x				
Isophorone diisocyanate								x										x			x					

Isophthalic acid	Z																					
Isoprene								x					x						x			
Isopropyl alcohol	Y	x	x	x				x		x					x	x	x			x		
Isopropylmethylpyrazoyl dimethylcarbamate			Z																x			
Isosafrole		x											x						x		x	
Kelthane													x						x	x		x
Kepone			Y										x						x		x	x
Lactonitrile			Z												x			x				
Lasiocarpine													x						x		x	
Lead		x*				x	x*				x	x*	x						x	x	x	x
Lead acetate								x					x						x		x	
Lead arsenate			Z					x					x						x			
Lead azide, wetted								x	x													
Lead chloride			Z					x					x						x			
Lead iodide								x					x						x			
Lead oxide			x	x						x												
Lead phosphate													x						x		x	
Lead stearate								x					x						x			
Leptophos							x*								x			x				
Lithium hydride								x							x			x				
Magnesium hydroxide			Y	x						x												
Magnesium oxide			Z							x												
Magnesium sulfate	Z																					
Malathion	Z		x	x						x			x	x					x			x
Maleic acid			Z					x					x	x					x			
Maleic anhydride	Y	x	x	x				x		x			x	x		x			x	x	x	
Maleic hydrazide													x						x		x	
Malononitrile			Z					x					x		x			x	x		x	
Maneb			x					x		x												
Manganese		x					x*						x									x
MCPA	Z		x	x						x												
Merchloroethamine															x			x		x		
Mercuric acetate			Y					x							x			x				
Mercuric arsenate								x														
Mercuric chloride			Y					x							x			x				x
Mercuric cyanide								x	x				x						x			
Mercuric sulfate								x					x						x			
Mercuric sulfide			Z							x												
Mercuric thiocyanate			Y					x					x									
Mercurous nitrate			Y					x					x						x			
Mercury	Y	x*	x	x	x	x	x*	x		x	x	x*	x						x	x	x	x
Mercury fulminate								x	x				x						x		x	
Methacrylonitrile								x					x		x			x	x		x	
Methane					x		x	x	x	x												x
Methapyrilene													x						x		x	

Chemical		2000 ESD	1997 NPRI	1990 EPS	1981 EPS	MIAC 1996	CEPA 2000	CEPA List II	TDG Sch. 2	TDG Sch. 12	Spilled in Canada	ARET Lists	EPA 1985	EPA 1980	U.S. Railroad	U.S. Reportable	USCG Spill	EPA Ex Danger	U.S. top 100	AAR Hazard.	SARA ex. Dang.	CERCLA Haz.	SARA Ex. Toxic	RCRA Haz.	Tox & Dis Reg	OSHA 29CFR
Methiocarb			Y															x			x	x				
Methomyl										x						x		x			x	x				
Methoxychlor			x							x						x						x	x	x	x	
Methoxyethanol, 2-		x					x																x			
Methoxyethylmercuric acetate			Y															x			x					
Methyl acrylate		x					x	x									x						x			
Methyl alcohol	Y	x	x	x			x	x		x					x	x	x	x	x	x		x	x	x		
Methyl bromide		x	Z			x		x	x				x	x		x		x			x	x	x	x		
Methyl butyl ether		x						x															x			
Methyl tert-butyl ether								x								x									x	
Methyl chloride		x	x				x	x	x	x			x	x	x	x	x		x	x		x	x	x	x	
Methyl chloroformate								x								x		x			x	x		x		
Methyl ethyl ketone		x	x	x						x					x	x	x		x	x		x	x	x	x	
Methyl ethyl ketone peroxide			Z						x	x						x						x		x		
Methyl hydrazine								x								x		x			x	x	x	x		
Methyl iodide		x	Z					x								x					x	x	x	x		
Methyl isobutyl ketone		x	Z	x				x			B3					x	x					x	x	x	x	
Methyl isocyanate								x								x		x			x	x	x	x		
Methyl isothiocyanate								x										x			x					
Methyl mercaptan								x	x							x		x			x	x		x		
Methyl methacrylate	Z	x	x	x		x		x		x					x	x	x		x	x		x	x	x	x	
Methyl parathion			Y													x	x				x	x		x	x	
Methyl phenkapton			Z														x				x					
Methyl phosphoric dichloride								x	x									x			x					
Methylamine	Z		x	x			x	x	x	x					x	x						x				
Methylbenzenamine hydrochloride																x						x	x	x		
Methylbenzene																						x	x	x		
Methyldinitrobenzene																						x	x	x		
Methylene chloride		x	x	x		x		x		x	B2		x	x		x	x		x			x	x	x	x	
Methylenebis(2-chloro)benzenamine, 4,4′-																					x	x	x			
Methylenebis(2-chloroaniline), 4,4′-		x									A1					x									x	
Methylenebis(phenyl isocyanate)	Z	x																					x			
Methylenedianiline, 4,4′-		x														x										

Chemical	1	2	3	4	5	6	7	8	9	10	11	12	13	14	15	16	17	18	19	20	21	22	23
Methylenedioxy-4-propylbenzene, 1,2-													x						x		x		
Methylethylbenzene																				x	x	x	
Metolcarb														x		x			x				
Mevinphos			Y				x	x						x		x			x	x		x	
Mexacarbate			Z											x		x			x	x		x	
Mitomycin C														x		x			x	x		x	
Monocrotophos			Z													x			x				
Morpholine			Z	x			x																
Motor fuel anti-knock compounds							x	x										x					
Muscimol														x		x			x	x		x	
Mustard gas																x			x		x		
Naled			Y											x	x					x			x
Naphtha	Y		Z	x	x		x						x										
Naphthalene	Z	x	x			x	x		x		x	x		x	x		x			x	x	x	x
Naphthalenedione, 1,4-														x						x		x	
Naphthenic acids							x							x						x			
Naphthylamine						x	x			B2				x						x	x	x	
Naphthylthiourea							x							x									
Nickel		x*									x	x*		x		x			x	x	x		x
Nickel carbonyl			Y				x	x	x					x		x							
Nickel chloride			Z				x							x						x			
Nickel cyanide							x													x		x	
Nicotine							x							x		x			x	x		x	
Nicotine sulfate			Z				x									x			x				
Nitrating acids							x	x					x										
Nitric acid	Y	x	x	x	x		x	x	x				x	x	x	x	x		x	x	x		
Nitric oxide							x	x						x		x							
Nitrilotriacetic acid		x	Z	x		x															x		
Nitroanilines			x				x		x					x									
Nitrobenzene		x	Z				x				x	x		x	x	x	x		x	x	x	x	
Nitrogen	Y		x				x		x								x						
Nitrogen dioxide			Y			x	x	x	x					x	x	x							x
Nitroglycerin		x	Z	x		x	x	x						x						x	x	x	
Nitrophenols		x	Z				x				x	x		x						x			x
Nitropropane		x	Z				x			B2				x						x	x	x	
Nitrosodimethylamine, N-			Z			x				B3	x	x		x		x				x	x		x
Nitrosodiphenylamine, N-		x	Z							B3	x	x		x						x	x		x
Nitrosodipropylamine, N-										B3	x	x		x						x	x	x	x
Nitrosomorpholine, N-										B3				x							x		
Nitrosopiperidine, N-														x						x	x	x	
Nitrosopyrrolidine, N-														x						x		x	
Nitrotoluenes			Z				x							x						x			
Nitrous oxide			Y				x		x														
Nitroxylenes							x																
Nonane			Z				x		x														

Chemical		2000 ESD	1997 NPRI	1990 EPS	1981 EPS	MIAC 1996	CEPA 2000	CEPA List II	TDG Sch. 2	TDG Sch. 12	Spilled in Canada	ARET Lists	EPA 1985	EPA 1980	U.S. Railroad	U.S. Reportable	USCG Spill	EPA Ex Danger	U.S. top 100	AAR Hazard.	SARA ex. Dang.	CERCLA Haz.	SARA Ex. Toxic	RCRA Haz.	Tox & Dis Reg	OSHA 29CFR
Nonylphenol			x	x			x			x																
Octamethylpyrophosphoramide																x					x	x		x		
Oleic acid			Z				x																			
Oleum	Z		Z	x																						
Osmium tetroxide								x								x		x			x	x	x	x		
Oxamyl																x		x			x					
Oxathiolane-2,2-dioxide, 1,2-																						x	x	x		
Oxygen	Z		x					x	x	x									x							
Ozone							x											x			x					
Paraldehyde								x								x						x				
Paraquat	Z		x							x								x			x					
Parathion			x					x	x	x						x	x	x			x	x	x	x	x	
Parathion methyl																		x			x	x				
PCBs	x		x	x		x		x		x		A1	x	x		x	x				x	x	x		x	
Pentachlorobenzene			Z													x						x		x	x	
Pentachloroethane								x								x		x			x	x		x		
Pentachloronitrobenzene																x						x	x	x		
Pentachlorophenol	Y		x	x			x	x		x		A1	x	x		x	x	x			x	x	x	x	x	
Peracetic acid		x																x			x		x			
Perchloromethyl mercaptan								x										x			x	x		x		
Phenanthrene			Z									A1	x	x		x						x			x	
Phenol	Y	x	x	x			x	x	x	x		B3	x	x	x	x	x	x	x	x	x	x	x	x	x	
Phenolsulfonic acid	Z		Y					x		x																
Phenyldichloroarsine																		x			x	x		x		
Phenylenediamine, p-		x	Z					x		x						x							x			
Phenylmercuric acetate			Y					x		x						x		x			x	x		x		
Phenylphenol, 2-		x	Z																				x			
Phenylsiltrane																		x			x					
Phenylthiourea			Z													x		x			x	x		x		
Phorate			Z													x		x			x	x		x	x	
Phosgene		x	Z				x	x	x	x					x	x		x	x		x	x	x	x		
Phosmet			Z															x			x					
Phosphamidon			x							x								x								

Phosphine			Z				x	x	x					x		x			x	x		x	
Phosphoric acid	Y	x	x	x			x		x				x	x	x		x	x		x	x		
Phosphorus	Z	x	x	x		x	x		x				x	x	x	x	x		x	x	x		x
Phosphorus oxychloride							x						x	x	x	x							
Phosphorus pentasulfide							x							x						x		x	
Phosphorus pentoxide			Y				x		x							x			x	x	x		
Phosphorus trichloride							x						x	x		x			x	x			
Phosphorus, white		x					x	x										x					
Phthalic anhydride	Z	x	x	x			x		x								x			x	x	x	
Picloram	Z		x						x														
Picoline, 2-							x							x						x		x	
Piperidine							x							x		x			x				
Potassium arsenate							x							x	x								
Potassium arsenite							x							x		x							
Potassium chloride	Y		x	x					x														
Potassium chromate							x							x	x					x			
Potassium cyanide			x				x	x						x	x	x			x	x		x	
Potassium dichromate			Z											x						x			
Potassium hydroxide	Y		x	x			x		x				x	x	x		x	x		x			
Potassium nitrate							x						x					x					
Potassium permanganate	Z		x				x		x				x	x	x					x			
Potassium silver cyanide														x		x			x	x		x	
Promecarb														x		x			x				
Propane					x		x	x															
Propargyl alcohol			Y											x						x		x	
Propiolactone														x		x			x		x		
Propionaldehyde		x	Z				x							x							x		
Propionic acid			x				x		x					x	x					x			
Propionic anhydride			Z				x							x						x			
Propionitrile							x							x		x			x	x		x	
Propyl alcohol						x	x								x								
Propylbenzene, n-						x	x																
Propyl chloroformate							x									x			x				
Propylamine			Z				x							x									
Propylene		x	x	x			x	x	x								x				x		
Propylene chlorohydrin			Z				x		x														
Propylene glycol	Z		x	x					x								x						
Propylene oxide		x	x	x	x	x	x	x	x				x	x	x	x	x	x	x	x	x		
Propyleneimine							x									x			x	x	x	x	
Pyrene										A1	x	x		x		x			x	x			x
Pyridine	Z	x	x				x		x				x	x									x
Quinoline		x	Z				x			B3				x						x	x		
Quinone		x	Z			x	x							x						x	x	x	
Reserpine														x						x		x	
Resorcinol			Z				x							x	x					x		x	

Chemical		2000 ESD	1997 NPRI	1990 EPS	1981 EPS	MIAC 1996	CEPA 2000	CEPA List II	TDG Sch. 2	TDG Sch. 12	Spilled in Canada	ARET Lists	EPA 1985	EPA 1980	U.S. Railroad	U.S. Reportable	USCG Spill	EPA Ex Danger	U.S. top 100	AAR Hazard.	SARA ex. Dang.	CERCLA Haz.	SARA Ex. Toxic	RCRA Haz.	Tox & Dis Reg	OSHA 29CFR
Selenious acid																x		x			x	x		x		
Selenium		x*					x*	x					x	x*		x									x	
Selenium dioxide																x						x		x		
Selenium disulfide								x														x		x		
Selenium oxychloride								x										x			x					
Silver		x*					x*						x	x*		x						x	x	x	x	
Silver cyanide								x							x	x						x		x		
Silver nitrate			Z					x								x						x				
Sodium			Z					x		x						x	x			x		x				
Sodium aluminate			Y	x				x		x																
Sodium anthraquinone sulfonate			Z															x			x					
Sodium arsenate			Z					x								x		x			x	x				
Sodium arsenite			Y	x				x		x						x	x	x			x	x			x	
Sodium azide			Y					x								x		x								
Sodium bisulfite	Z		Z	x			x								x	x						x				
Sodium borohydride			Y	x				x		x																
Sodium cacodylate								x										x			x					
Sodium carbonate	Z		x	x						x									x							
Sodium chlorate	Y		x	x	x		x	x		x									x	x						
Sodium chloride	Y		x	x						x																
Sodium chlorite	Z						x	x																		
Sodium chromate			Y					x		x						x						x				
Sodium cyanide	Z		x	x			x	x	x	x					x	x	x	x			x	x		x		
Sodium dichloroisocyanurate				x												x										
Sodium dichromate	Y		Y	x												x	x					x				
Sodium fluoride			x					x		x						x						x				
Sodium fluoroacetate			Z					x										x				x		x		
Sodium fluorosilicate			Z					x																		
Sodium hydrosulfide			Z					x	x	x						x	x									
Sodium hydrosulfite	Z		x												x											
Sodium hydroxide	x		x	x				x		x					x	x	x		x	x		x	x			
Sodium hypochlorite	Y		x	x						x						x						x				
Sodium methylate								x								x						x				

Sodium nitrate			Y					x		x				x		x			x					
Sodium nitrite	Z							x							x						x			
Sodium pentachlorophenate	Z							x									x			x				
Sodium phosphate, tribasic			Z					x							x						x			
Sodium phosphates			Y	x				x							x	x					x			
Sodium selenite			Z												x		x				x			
Sodium silicate	Z		x	x						x								x						
Sodium sulfate	Z		Z	x														x				x		
Sodium sulfide	Z							x								x								
Sodium sulfite	Z		x	x						x														
Stannic chloride			Z					x		x														
Strychnine			Y					x							x	x	x			x	x		x	
Strychnine sulfate			Z														x			x				
Styrene	Y	x	x	x				x		x				x	x	x		x	x		x	x		x
Styrene oxide		x													x							x		
Sulfotepp			Z														x			x	x		x	
Sulfur	Y		x	x				x		x						x								
Sulfur chloride			Z					x		x				x	x	x					x			
Sulfur dioxide	Y		x	x	x			x	x	x							x		x	x				
Sulfur tetrafluoride									x								x			x				
Sulfur trioxide								x	x								x			x				
Sulfuric acid	x	x	x	x	x		x	x	x	x				x	x	x	x	x	x	x	x	x		
Sulfuryl chloride			Z	x				x		x														
Terephthalic acid			x	x						x								x				x		
Terphenyls			Z	x		x				x														
Tetrachlorobenzene, 1,2,4,5-			Z												x						x		x	
Tetrachlorodibenzo-p-dioxin, 2,3,7,8-						x					A1	x	x		x						x			x
Tetrachloroethane		x					x	x				x	x		x						x	x	x	x
Tetrachloroethylene	Z	x	x	x		x		x		x		x	x		x	x		x						x
Tetrachlorophenol, 2,3,4,6-			Y								B2				x						x		x	x
Tetraethyl lead	Z		x	x	x					x	B2			x	x	x	x			x	x		x	
Tetraethyl pyrophosphate			Y												x					x	x		x	
Tetrahydrofuran								x						x							x		x	
Tetramethyl lead			Z														x			x				
Tetranitromethane								x	x						x					x	x		x	
Thallic oxide															x		x			x	x		x	
Thallium			Z				x*					x	x*		x						x	x		
Thallium acetate			Z												x						x		x	
Thallium carbonate															x		x			x	x		x	
Thallium chloride															x					x	x		x	
Thallium nitrate								x							x									
Thallium peroxide																				x	x		x	
Thallium sulfate			x							x					x		x							
Thioazin																	x			x	x		x	
Thiobis-(4,6-dichlorophenol), 2,2′-			Z																	x				

Chemical		2000 ESD	1997 NPRI	1990 EPS	1981 EPS	MIAC 1996	CEPA 2000	CEPA List II	TDG Sch. 2	TDG Sch. 12	Spilled in Canada	ARET Lists	EPA 1985	EPA 1980	U.S. Railroad	U.S. Reportable	USCG Spill	EPA Ex Danger	U.S. top 100	AAR Hazard.	SARA ex. Dang.	CERCLA Haz.	SARA Ex. Toxic	RCRA Haz.	Tox & Dis Reg	OSHA 29CFR
Thiofanox																x		x			x	x		x		
Thioglycolic acid			x					x		x								x								
Thionyl chloride			Z					x		x																
Thiophenol																x		x			x	x		x		
Thiosemicarbazide			Z													x		x			x	x		x		
Thiourea		x	Z				x					B2				x										
Thiram			Y							x															x	
Titanium dioxide			x	x						x									x							
Titanium tetrachloride		x						x								x					x		x			
Toluene	x	x	x	x	x			x		x			x	x	x	x	x		x	x		x	x	x	x	
Toluene-2,4-diamine			x							x						x						x	x	x		
Toluene-2,4-diisocyanate	Z	x	x	x			x	x		x		B3			x	x		x	x	x	x	x	x			
Toluene-2,6-diisocyanate		x										B3						x			x	x	x			
Toluidine			Z				x	x								x										
Toxaphene			Y				x						x	x		x	x								x	
TP acid esters, 2,4,5-																x								x		
TP acid, 2,4,5-			Z													x						x		x	x	
Tri(1-aziridinyl) phosphine oxide								x																		
Trichlorfon	Z		x	x						x						x		x			x	x	x			
Trichloro-2,2-bis(p-methoxyphenyl)ethane, 1,1,1-																						x	x	x		
Trichloroacetyl chloride								x										x			x					
Trichlorobenzene		x	Z					x					x	x		x						x	x		x	
Trichloroethane, 1,1,1-	Y		x	x		x		x		x			x	x		x	x		x			x	x	x	x	
Trichloroethane, 1,1,2-		x											x	x		x						x	x	x	x	
Trichloroethylene, 1,1,1-																						x	x	x		
Trichloroethylene, 1,1,2-	Z	x	x			x		x		x		B3	x	x		x	x								x	
Trichloroethylsilane								x								x		x			x					
Trichlorofluoromethane				x									x	x		x						x		x		
Trichlorophenol			Z													x						x				
Trichlorophenol, 2,4,5-							x									x						x	x	x	x	
Trichlorophenol, 2,4,6-												B1	x	x		x									x	
Trichlorophenoxyacetic acid, 2,4,5-			x							x						x						x		x		
Trichlorophon																					x	x	x			

Triethanolamine			Y							x					x									
Triethanolamine dodecylbenzenesulfonate								x							x						x			
Triethylamine			Z					x						x	x						x			
Trifluralin	Z		x	x						x					x							x		x
Trimethylamine			Z					x	x	x					x						x			
Trimethylchlorosilane								x									x			x				
Trinitrobenzene								x	x						x						x		x	x
Trinitrophenol								x	x													x		
Trinitrotoluene			x	x			x	x	x	x														x
Trypan blue															x						x		x	
Turpentine			Z	x				x								x								
Uranium							x*	x																x
Uranium (233, 234, 235)																								x
Uranium dioxide			Z	x						x														
Uranyl acetate								x							x						x			
Uranyl nitrate			x							x					x									
Urea	Y		Y	x														x						
Urea formaldehyde	Z																							
Urea nitrate	Z							x	x															
Vanadium		x					x*															x*		
Vanadium pentoxide			Y					x							x		x			x	x		x	
Vinyl acetate	Z	x	x	x				x		x				x	x	x	x	x	x	x	x	x		
Vinyl bromide								x	x		B3										x			
Vinyl chloride	Y	x	x	x	x	x		x	x	x		x	x	x	x			x	x		x	x	x	x
Warfarin			Z												x		x			x	x		x	
Xylene	Y	x	x	x	x			x		x				x	x	x		x	x					x
Xylenol			Z					x							x	x					x			x
Zinc		x*					x*					x	x*		x						x	x		x
Zinc acetate			Z					x							x						x			
Zinc borate								x							x						x			
Zinc bromide			Z					x		x					x						x			
Zinc carbonate								x							x						x			
Zinc chloride	Z		x	x				x		x					x	x					x			x
Zinc cyanide			x					x		x					x						x		x	
Zinc fluoride								x							x						x			
Zinc nitrate								x							x						x			
Zinc oxide			x	x						x														
Zinc phenolsulfonate								x							x						x			
Zinc phosphide								x	x						x		x			x	x	x		
Zinc sulfate	Z		x	x				x		x					x									

APPENDIX List of Lists—Abbreviated (*Continued*)

Legend: Code

2000 ESD—2000 Priority List: Top 10 (x), 11-50 (Y), 51-150 (Z) (Environment Canada—Emergencies Science Divison [ESD])

1997 NPRI—1997 National Pollutants Release Inventory (NPRI) (Environment Canada)

1990 EPS—1990 Priority List: Top 150 (x), 151-250 (Y), 251-500 (Z) (Environment Canada)

1981 EPS—1981 Priority List (Environment Canada)

MIAC 1996—MIAC 1996 Priority List

CEPA 2000—Canadian Environmental Protection Act (CEPA) Priority Substance List (2000)

CEPA List II—CEPA Priority Substance List PSLII (1995) (Environment Canada)

TDG Sch. 2—TDG Schedule 2 (1996)

TDG Sch. 12—TDG Schedule 12 (1994)

Spilled in Canada—Spill History in Canada (1974–1984)

ARET Lists—ARET Substance List (A1, etc) denotes list (1995)

EPA 1985—EPA list of top 129 contaminants

EPA 1980—EPA list of top 65 contaminants

U.S. Railroad—AAR spill priority list

U.S. Reportable—EPA Reportable Quantities List, 40 CFR parts 117 & 302 (1997)

USCG Spill—U.S. Coast Guard CHRIS List (1992)

EPA Ex Danger—EPA Acutely Toxic or Potentially Dangerous Chemicals List (1986)

U.S. top 100—1990 list of top 100 industrial materials in the USA

AAR Hazard.—AAR Hazardous Materials Data Base (1981)

SARA ex. Dang.—SARA Section 302 Extremely Hazardous Substances (1987)

CERCLA Haz.—CERCLA Hazardous Substances (1987)

SARA Ex. Toxic—SARA Section 313 Toxic Chemicals (1987)

RCRA Haz.—RCRA Hazardous Wastes (1987)

Tox & Dis Reg—Agency for Toxic Substances and Disease Registry Priority List (1995)

OSHA 29CFR—OSHA 29 CFR 1910 List of Highly Hazardous Chemicals, Toxics and Reactives

CHAPTER 45

SEARCHING FOR INFORMATION ON CHEMICAL PRODUCTS ON THE INTERNET

Michel Marchand

Rue Alain Colas, BP 20 413, 29604 BREST (France)

45.1 INTRODUCTION

In cases of accidental spills of chemical products, competent authorities must rapidly have the precise information to be able to evaluate the risks to humans and the environment and to plan emergency action, in maximum safety conditions, for the intervention team concerning the emergency measures to be taken against fire, accidental spreading, handling, storing and elimination of dumped products.

Two types of documents generally provide this information: safety guides (Environment Canada, 1984; CEFIC, 2000; GORSAP, 1992; Transport Canada, 1996) and material safety data sheets (MSDS) on chemical products, which now constitute a regulatory obligation in the European Union (European Directive, 1991). This information is accessible on paper and more and more on computer. Other sources of information include databases on chemical products, either general or specific, such as CHEMINFO (Chemical Information), HSDB (Hazardous Substances Database), CHRIS (Chemical Hazards Response Information System), CESARS (Chemical Evaluation Search and Retrieval System), RTECS (Registry of Toxic Effects of Chemical Substances), AQUIRE (Aquatic toxicity Information Retrieval), NIOSH (National Institute for Occupational Safety and Health), IUCLID (International Uniform Chemical Information Database), and ECDIN (Environmental Chemicals Data and Information Network). Some of these can be ordered individually (i.e., IUCLID, and ECDIN from the European Commission), others can be grouped together to form operational information systems, computerized without CD-ROM, i.e., the CCINFO system edited by the Canadian Centre for Occupational Health and Safety.

Today, another source of information is available to operators in the form of the Internet. A huge and almost unlimited quantity of information of all sorts is now available on the Web. Information on chemical products is therefore available, suitable for answering emergency calls in cases of accidents or spills involving hazardous materials. While the results of this research certainly do not claim to be complete, this section invites the reader to become aware of a certain number of interesting websites and demonstrates the operational aspect of the Internet in emergencies.

45.2 SEARCHING ON THE WEB

To begin searching for information on the Internet, a number of search engines can be used, i.e., AltaVista, Euroseek, Excite, Galaxy, Go, Google, HotBot, Lycos, Magellan, Nomade, Nothernlight, WebCrawler, Yahoo, Yelloweb, or PC seekers, i.e., All4one, Profusion, InferenceFind, Metacrawler, SavySearch, Ixquick, Dogpile, Copernic. To start searching, you enter one or more key words and click the command "search."

Until now, several key words, such as "database," "chemical products," "reactivity," "toxicity," and "MSDS," have been used. Certain key words give access to websites that offer university courses and scientific articles on a choice of subjects but are not precise enough for our purposes. In the same way, the key word "database" opens access to sites that introduce databases, but without access. The key word "MSDS" (material safety data sheet) gives access to interesting addresses that provide databases about chemical products. The user soon learns that a number of sites are grouped together on the same site. This is the case with a site set up by an American organization, ILPI (Interactive Learning Paradigms Incorporated) (http://www.ilpi.com/msds/index.chtml). This site offers files on safety data sheets; such as what they are for, why they are there, where to find them. It also offers a list of more than 60 Internet addresses classified according to different categories: general sites, government sites, manufacturers and suppliers of chemical products, pesticides, and miscellaneous. Some of these sites cannot be visited, either because entry requires a customer code or for internal reasons of accessibility to the sites themselves. Another interesting site, set up by the French Institut National de Recherche et de Sécurité (INRS), summarizes websites related to chemical products (http://www.inrs.fr/actualites/Siteschim.html).*

We classified the sites, including those grouped by ILPI and others, into the following categories:

- Emergency guides
- Safety and information data on general chemical products
- Specific information, either according to a classified subject or according to the products or the manufacturers

45.3 EMERGENCY GUIDES

A chemical spills emergency guide is a reference book to be used during an emergency. It details the immediate risks and action to be taken in case of accidents during the transport of hazardous materials or when toxic substances are spilled. Examples of the types of information provided in several such guides are provided in Table 45.1. The Internet addresses for five emergency guides for chemical spills are provided in Table 45.2.

45.4 SAFETY AND INFORMATION DATA CARDS

Material safety data sheets (MSDS) provide basic information on chemical products as shown in Table 45.3. The websites listed in Table 45.4 provide access to information or safety data

*At the international level, the IPCS iNCHEM Programme was developed in 1997 as a cooperative undertaking between the International Programme on Chemical Safety (IPCS) and the Canadian Centre for Occupational Health and Safety (CCOHS). This very important and valuable chemical safety information collection from international agencies has been publicly available through a paid subscription. From June 2001, the public access to the IPCS iNCHEM Web service is free-of-charge (http://www.inchem.org).

TABLE 45.1 Information Given in Four Emergency Guides

North America Emergency Response Guidebook (NAERG)	*Geneva Fireman's Guide* (GOR SAP)	Emergency Response Information Cards (ERIC)	*Manual for Spills of Hazardous Materials* (Environment Canada)
• Potential risks (fire or explosion, health) • Public safety • Protective clothing, evacuation • Emergency measures for fire, spills or leaks, and first aid	• Scale of dangers • Description • Risks • Fire • Protective equipment • Spillage • Poisoning	• Characteristics • Hazards • Personal protection • Intervention • First aid • Essential precautions for product recovery • Precautions after intervention	• Description (characteristics, transport and storage, physical and chemical properties • Risks (health, fire, reactivity, environment) • Emergency measures • Measures to protect the environment

TABLE 45.2 Emergency Guides

Guide	Information	Address
North American Emergency Response Guidebook (NAERG)	North-American Emergency Guide available in English, French, and Spanish. 3,714 chemical products. Plan: potential risks, public safety, emergency action.	(U.S.A.) http://hazmat.dot.gov/gydebook.htm (Canada) http://www.tc.gc.ca/canutec/
National Institute for Occupational Safety and Health (NIOSH)	Taken from the "NIOSH Pocket Guide to Chemical Hazards." English version. Spanish version (in progress). 677 chemical products.	(U.S.A.) http://www.cdc.gov/niosh/npg/pgdgst gov./art.html
International Chemical Safety Cards (ICSC) International Programme on Chemical Safety Project	Basic information cards, available in eight languages. 869 chemical products.	Site on NIOSH (U.S.A.) http://www.cdc.gov./niosh/ipcs/icstart.html
Emergency Response Information Card (ERIC)	Emergency Cards available in four languages: English, French, Spanish, German. More than 2,000 chemical products.	(Europe) http://www.cefic.org/activities/logistics/logistics.htm http://www.ericards.net

TABLE 45.3 Plan of Material Safety Data Sheet

1. Product identification
2. Composition/information or ingredients
3. Hazards identification
4. First aid measures
5. Firefighting measures
6. Accidental release measures
7. Handling and storage
8. Exposure controls/personal protection
9. Physical and chemical properties
10. Stability and reactivity
11. Toxicological information
12. Ecological information
13. Disposal considerations
14. Transport information
15. Regulatory information
16. Other information

TABLE 45.4 Safety and Information Data Sheets on Chemical Products

Addresses	Characteristics
http://hazard.com/msds/index.html	Safety data (MSDS) approximately 180,000 chemical products classified in Safety Information Resources, Inc. (SIRI).
http://www.fishersci.com/catalogs	Safety data sheets of 61,000 chemical products from Acros Organics and Fisher Chemical.
http://msds.pdc.cornell.edu/issearch/msdsrch.htm	Safety data sheets on 325,000 classified products from Cornell University.
http://www.sigma-aldrich.com/saws.nsf/sigma-aldrichmsds_help?OpenForm	Grouping of several companies' safety data sheets of: Sigma, Aldrich, Fluka, Supelco. List of 90,000 chemical products. Necessity to fill in a questionnaire to obtain a password. Sheets are sent only by fax.
http://ecdin.etomep.net/	European database: Environmental Chemicals Data and Information Network (ECDIN). Very varied information (characteristics, detection, pharmacology, toxicology, legislation, etc.).
http://www.msdsonline.com/	Safety data sheets on 250,000 products. Access charge.
http://www.chemfinder.com	Database of 60,000 products from Cambridge Soft Corp (Internet Chemistry Software Leader).
http://www.chemexper.be.	Database giving the list of suppliers and safety sheets for more than 4,000 chemical products.
http://physchem.ox.ac.uk/msds	Safety data sheets on 2,650 chemical products. Site of Oxford University.
http://www.olis.oecd.exichem.nsf	EXICHEM database of the Organization for Economic Cooperation and Development (OECD), which gives information on chemical products. A correspondent's e-mail address can provide information concerning a given product.

sheets which are adapted to an emergency situation, i.e., clear presentation, easy to read, etc. To retain the general character of this type of information, only sites with more than 2,000 products are listed in this table. Some other sites are also included but are classified in a specific way product/manufacturer.

45.5 SEARCHING BY SPECIFIC TOPIC

The following topics have been chosen to facilitate the search for precise information on chemical products:

- Reactivity/compatibility of chemical products
- Toxicology/ecotoxicology
- Carcinogenics
- Pathogenic agents
- Manufacturers and distributors of chemical products

The website addresses where information on the subjects can be found are provided in Table 45.5.

45.6 SPECIFIC SEARCHING BY PRODUCT/MANUFACTURER

There are many Internet sites covering relatively specific information by range of products, often referring to one or more manufacturers. This type of information allows access to safety data sheets concerning products that are sometimes only known by their commercial name or the way they are used. Eighteen categories have been selected from products used for treating and cleaning surfaces. Obviously, the list given in Table 45.6 is not to be considered as complete.

45.7 CONCLUSION

A huge amount of information is available on the Internet through an impressive number of sites. In order to use the Internet as an operational tool in searching for information during an emergency, a preliminary selection must be made of the most interesting (important information) or the most relevant (selective information) sites. These sites must then be classified in order to find information as quickly as possible. Using search engines or PC seekers may give immediate results but may not be very useful in identifying the most relevant site addresses. We recommend using the website of Interactive Learning Paradigms Incorporated (ILPI), which groups together about 60 websites. It is advised that each person should proceed according to his or her own interests, selecting the best sites and exploring new sites to take into account the constant changes in accessible information on the Internet.

TABLE 45.5 Information on Specific Subjects

Subject	Address	Comments
Reactivity/compatibility of chemical products	http://response.restoration.noaa.gov/chemaids/react/react.html	Information on reactivity of more than 4,000 chemical products, compatibility or incompatibility of chemical products (only two to two).
	http://hazmat.dot.gov/gydebook.htm or http://www.tc.gc.ca/canutec	*Safety Guide* (NAERG): substances reacting with water.
Toxicology/ecotoxicology	http://www.state.nj.us/health/eoh/rtkweb/stkhsfs.htm	703 products used in companies in New Jersey (U.S.A.).
	http://www.atsdr.cdc.gov/toxfaq.html	Agency for Toxic Substances and Disease Registry (ATSDR) 75 products.
	http://ntp-server.niehs.nih.gov/	1,000 classified products in toxicology data National Toxicology Program (NTP) U.S.A.
	http://epa.gov/ecotox/	Ecotoxicology in an aquatic environment. Database ECOTOX of the EPA (U.S.A.). 6,500 products. Bibliographic data.
Carcinogenics	http://ehis.niehs.nih.gov/roc/toc8.html	Report on suspected or known carcinogenic products.
	http://ntp-db.niehs.nih.gov/	National Toxicology Program (NTP) (U.S.A.). 200 substances.
	http://193.51.164.11/defaut.html http://www.iarc.fr/	International Agency on Cancer Research (IARC) Lyon (France). 800 substances.
	http://www.ozemail.com.au/paulr/cancer.html	List of carcinogenic substances established by the IARC and the NPT.
Pathogenic agents	http://www.hc-sc.gc.ca/hpb/lcdc/biosafty/index.html	Safety data sheets of approximately 150 pathogenic agents.
Manufacturers/distributors of chemical products	http://www.chem.com	Site linking 500 manufacturers (suppliers of chemical products).

TABLE 45.6 Specific Information per Product/Manufacturer

Subject	Manufacturer	Internet address	Comments
Biochemistry	Roche Molecular Biochemicals	http://www.biochem. boehringer-manheim.com/	3,600 products
Isolation construction	Schuller Corp	http//www.schuller.com/msds/index.html	419 products
Electronics	Electronic Space Products International	http://www.espi-metals.com/msds.htm	425 products (metals, alloys)
	Center for Advanced Microstructure and Devices	http://camd.Isu.edu/msds/MSDS-home1.html	450 sheets
Gases	Boc gases	http://www.boc.com/gases/msds_us/index.html	218 products
	Messer MG Industries	http://www.mgindustries.com/msds.asp	200 products
	Matheson Tri-Gas, Inc.	http;//www.mathesongas.com/msds	115 products
	Hopper Welding Supply	http://www.HopperSupply.com/msds/index.htm	6 products (welding gases)
Printing, printers	Hewlett-Packard	http://www.hp.com/abouthp/envrnment/contents/env facts/htpmsds.htm	77 products
	Apple Computer, Inc.	http://17.254.3.191/slg/msds.html	48 products
	Digital Equipment Corp.	http://www.digital.com/info/ehs/msds/home.htm	24 products
	Xerox Corporation	http://www.xeros.com/ehs/msds/index.htm	—
Laboratories	Howard Hughes Medical Institute	http://www.hhmi.org/science/labsafe/lcss/less.html	88 products + practical sheets
Microbiology	Difco Laboratory	http://www.difco.com/	274 products
		http://www.bdms.com/	70 products (culture environment)
Paints/varnishes	ProSciTech	http://www.proscitech.com.au/msds.htm	120 products
	Star Autobody Product	http://www.autobodypro.com/5s-msds.htm	103 products
	General Point		
	General Point	http://www.generalpaint.com/msds.html	50 products
	Cabot Stains	http://www.cabostain.com/top-htm/masada.htm	45 products
	The Sherwin-Williams Company	http://www.sherwin.com/PickPros/SWListAll.asp	300 products

TABLE 45.6 Specific Information per Product/Manufacturer (*Continued*)

Subject	Manufacturer	Internet address	Comments
Pesticides and fertilizers	Crop Data Management Systems	http://cdms.net/manuf/manuf.asp	1,180 products
	C & P Press Greenbook	http://www.greenbook.net/free.asp	1,300 products
	Dupont Agricultural Products	http://www.dupont.com/cgi-bin/ag/prodsearch/start.cgi	81 products
	Automatic Rain Company	http://autorain.com/MSDS_sheets1.htm	1,345 products
	Zeneca Agricultural Products Professionnal	http://www.zenecaagproducts.com/1blsmsds/	52 products
	Novartis Crop Protection, Inc.	http://www.cp.us.novartis.com/msds-general-frame.html	81 products
	Rohm and Haas	http://www.rohmhaas.com/businesses/AgChem/MSDS.html	43 products
Petroleum products	Environment Canada	http://www.etcentre.org/spills	Crude oil and refined products: 412
Photography	Eastman Kodak	http://kodak.com/vs/en/corp/hse/prodSearchMSDS.shtml	1,500 products
Plumbing	Hercules Plumbing Products	http://www.herchem.com/msds/msds.cfm	98 products
Pharmaceutical products	PharmInfoNet	http://pharminfo.com/drugdb/db_mnu.html	300 produits
Pyrotechnics	John Ray	http://www.jps.net/johnray/msds	144 products
Silicates	Silmcaco	www.silmaco.com/Main/safety.htm	5 products
Surface treatment and cleaning	S.C. Johnson Professional	http://www.MSDS/parsed/index.htm	344 products
	ABC Compounding	http://www.abccompounding.com	256 products
Miscellaneous	Alfa Aesar	http://alfa.com/cgi-bin/odc_webcat/pl?begin	14,000 products
	EM Sciences	http://www.emscience.com	7,000 products
	BDH Inc.	http://bdhinc.com	—
	JT Baker	http://www.jtbaker.com/cgi-bin/pasi.pl?begin	2,100 products
	Mallinckrodt Laboratories	http://www.mallechem.com/cgi-bin/pasi-pl?begin	2,100 products
	Alken-Murray Corp.	http://alken-murray.hypermart.net/msds.html	50 products
	Redox Chemicals Pty Ltd	http://www.redox.com.au/msds/	198 products
	Dojindo Laboratories	http://www.dojindo.com.msds.html	200 products
	VWR Scientific Products	http://www.vwtsp.com/catalog/	—

45.8 REFERENCES

CCINFO, established by the Canadian Centre for Occupational Health and Safety (CCOHS), 250 Main Street, East Hamilton, ON.

Environment Canada, *Manual for Spills of Hazardous Materials,* Environmental Protection Service, Ottawa, ON, 1984.

European Chemical Industry Council (CEFIC), Emergency Response Intervention Cards (ERIC), European Chemical Industry Council, Brussels, http://www.cefic.org, or http://ericards.net, 2000.

European Directive 91/155/CEE, *Journal Officiel des Communautés Européennes,* March 5, 1991.

Guide Orange des Sapeurs-Pompiers Genevois (GORSAP), *Répertoire des Produits Dangereux, Guide Orange des Sapeurs-Pompiers Genevois, ed. Service d'incendie et de secours de la ville de Genève,* 3d ed., Geneva, 1992.

IUCLID, ECDIN, CD-ROM available from the European Chemicals Bureau (ECB) Joint Research Centre ISPRA (va) (Italy) http://ecb.ei.jrc.it/.

Transport Canada, U.S. DOT, *1996 North American Emergency Response Guidebook* (NAERG) or *Guide Nord Américain des mesures d'urgence* (GNAMU), Transportation of Dangerous Goods, Ottawa, ON (Canada) or Department of Transportation, Office of Hazardous Materials Initiatives and Training, Washington, DC (U.S.A.), 1996.

INDEX

ABOUT THE EDITOR-IN-CHIEF

Merv Fingas, Ph.D., is Chief, Emergencies Sciences Division, Environment Canada. His research specialties are the properties, behavior, and modeling of hazardous materials, especially oil. A member of the editorial board of the *Journal of Hazardous Materials,* he also edits *Spill Technology Newsletter,* published by Environment Canada. He is a member of the executive committee of the ASTM F-20 on spill standards. He has over 450 publications.